Cell Structure & Function

AN INTEGRATED APPROACH

Third Edition

Ariel G. Loewy, *Haverford College*

Philip Siekevitz, *Rockefeller University*

John R. Menninger, *University of Iowa*

Jonathan A. N. Gallant, *University of Washington*

 SAUNDERS COLLEGE PUBLISHING
Philadelphia   Ft. Worth   Chicago   San Francisco
Montreal   Toronto   London   Sydney   Tokyo

Text Typeface: 10/12 Bembo
Compositor: The Clarinda Company
Acquisitions Editor: Julie Levin Alexander
Developmental Editor: Lloyd W. Black
Managing Editor: Carol Field
Project Editor: Maureen Iannuzzi
Copy Editor: David Watt, Charlotte Nelson
Manager of Art and Design: Carol Bleistine
Art Director: Christine Schueler
Art and Design Coordinator: Doris Bruey
Text Designer: Diane Pella
Cover Designer: Lawrence R. Didona
Text Artwork: J & R Technical Services
Layout Artist: NSG Design
Director of EDP: Tim Frelick
Production Manager: Tim Frelick
Marketing Manager: Marjorie Waldron

Cover Credit: Courtesy of S. J. Wright

Printed in the United States of America

Cell Structure and Function, 3/e

ISBN: 0-03-047439-6

Library of Congress Catalog Card Number: 90-053237

1234   061   987654321

THIS BOOK IS PRINTED ON **ACID-FREE**, **RECYCLED** PAPER

About the Cover

Recent technological advances have contributed to a renaissance in light microscopy. Improved optical design, sensitive imaging cameras, new fluorescent dyes, and advances in computers have revolutionized the way in which biological information can be obtained. As a result of these powerful new capabilities, a new field, *video microscopy,* has emerged.

This technique has many advantages. Living cells, which previously were difficult to study because of their inherent low contrast and faint fluorescence, can be observed for long periods with video microscopy. More sensitive detectors allow clear observation of the minute intracellular movements of fluorescently labeled cellular structures or ions that were not detectable with previous methods. Special computers are used to enhance the video images and highlight specific cellular features.

The cover shows three-dimensional DNA profiles of living oöcytes from the surf clam: *Spisula solidissima.* The oöcytes are in the prophase of meiosis, in which stage they are arrested when spawned. The chromosomes (white peaks) lie in the periphery of the nucleus (germinal vesicle). After staining the DNA with a fluorescent dye (Hoechst 33342) in vivo, the cells were viewed with a video camera through an ultraviolet microscope. Sixty-four frames of the video image were summed using special computer hardware, and a three-dimensional representation of the image was generated by transforming fluorescent brightness into height. Further computer processing was used to color this image, and it was then photographed from the computer's monitor. The fluorescently labeled chromosomes are easily identified by highlighting the brightest fluorescence signal. Both the color and height of the peaks reflect the brightness of each image point. (Courtesy of S. J. Wright.)

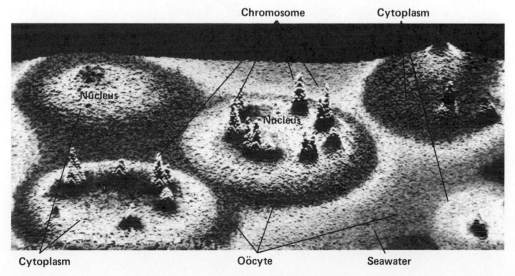

The harmonious co-operation of all beings arose, not from the orders of a superior authority external to themselves, but from the fact that they were all parts in a hierarchy of wholes forming a cosmic pattern, and what they obeyed were the internal dictates of their own natures.

Chuang Tzu
(Third Century B.C.)

Preface
To the Third Edition

Almost three decades have passed since publication of the first edition of this book. Momentous developments of the intervening years have fully justified the faith we expressed then in the significance of early discoveries regarding the "molecular machinery of the cell."

This third edition reflects recent practice of the "new biology" in a variety of ways, for instance in the increased number of pages necessary to encompass the major themes of the subject. Also, a contemporary explication of cell structure and function must incorporate important *qualitative* changes into an introductory text.

The first two editions attempted to integrate the results obtained by the then-new high-resolution optics with those that had emerged from years of biochemical investigation. The present edition recognizes that genetic manipulation and genetic analysis have now become inextricably interwoven with structural and biochemical approaches. By introducing genetics we have transformed this book into a basic self-sufficient text, suitable for beginning students provided they can use the elementary organic chemistry required for understanding the molecules of living things.

Another change we have introduced in this edition is the use of simple mathematical models—the formulae are on the whole less complicated than those found in prerequisite chemistry courses. Our reason is that numerical computations and mathematical models are essential tools for investigating important areas of cell biology. As we explain in greater detail in the Introduction, this new edition attempts to reflect the reality of mathematical analysis in contemporary biology.

Although cell biology has increased tremendously in scope and level of complexity, the last paragraph of the Preface to the First Edition still reflects our views:

> It was not the intention of the authors to make this an "elementary" book in the sense of limiting it necessarily to the simpler and more accessible aspects of cell biology. Yet we hope that it is a book for beginners because we believe it is precisely the beginner who deserves an initial statement that is fully representative of the contemporary quality and mode of the field.

Acknowledgements

Preparing this volume has been a long task, and a list that included all those who made contributions would tax even the permissive spirit of our present publisher. Nevertheless, we cannot miss the opportunity to recognize some who shared the labor. Ken Getman, the original editor of the first and second editions, began the process of creating this third edition—truly a man before his time. We are especially grateful to Don Schumacher and Dan Schiller, editors who read the manuscript and were true believers in the project. Dan Schiller needs additional thanks for his instruction in how, exactly, to tell a story that students will want to read. Eleanor Milspaugh helped keep a complicated project on track, and Cathleen Petree lowered many administrative barriers while urging us ever onward. Stacy Simpson organized the final "kilo-art" and David Watt the "kilo-page" manuscript. Lesley Menninger contributed to the production of this book in numerous and important ways, not least by producing the index, and Martha R. Hurt helped us deal with a number of legal complications.

We are also grateful to the many reviewers of the manuscript. Those who praised encouraged us to press forward. Those who criticized will, we hope, appreciate this final version. Since editors often followed the custom of preserving anonymity, we aren't sure of the identities of many reviewers. We can, however, acknowledge particularly helpful comments by Michael Bishop, Tony Cashmore, David Dawson, William Sistrom, David Shappirio, Michael Solursh, Rocky Tuan, and Alex Tzagaloff. We appreciate the long hours taken from busy schedules to produce such reviews.

Numerous colleagues responded generously and rapidly to our requests for help in the form of discussions, micrographs, or drawings. Many of these contributions are recognized in the figure legends. We describe the work of many others as a group effort and their names are not specifically mentioned. To all we are grateful. It is a source of wonder and pride to be reminded that we work in a sharing community. We hope this cooperative spirit endures, since it makes cell biology an even more attractive field for students to enter.

Our families, associates, and students have put up with frequent physical and psychological absences during the nurturing of this text (some of our children regard it, not always fondly, as a sibling). We thank them for their patience and for filling in often while we were lost in the embraces of the muse. We're ready now to return to the "normal" life of being a scientist, hopeful that this text will help students to better understand what that is all about.

A.G.L.
Haverford, Pennsylvania

P.S.
New York, New York

J.R.M.
Iowa City, Iowa

J.G.
Seattle, Washington

January 1991

Preface
To the First Edition

The past decade has witnessed an explosive accumulation of insights into the molecular machinery of the cell.

We are beginning to discern a molecular pattern that includes such phenomena as the self-duplication of the cellular hereditary material and the control it exerts in the formation of the catalysts of the cell. Biological specificity, a property so characteristic of the world of life, is in the process of being related to the structure and interaction of macromolecules. The electron microscope has suddenly created a new world, rich and intricate in detail, which is actively being interpreted in molecular terms. The regulation of this very complicated cellular machinery is now beginning to be examined in a manner that, at the very least, has operational rigor.

These developments have resulted in the realization that there is an inextricable relationship between cellular structure and cellular function. The more classical disciplines of cytology, cell physiology, biochemistry, and biophysics are becoming fused into one common structure that is often referred to as the "molecular biology of the cell."

The purpose of this book is to document these exciting developments and to make them accessible to the introductory student. It is our conviction that since the cell is the "common denominator" of living systems, it is extremely important that the beginner should become acquainted with the major facts and theories of cell biology. This introduction will serve a dual purpose: it will provide the student with a firm basis from which to examine the other manifold phenomena of the world of life, and it will also serve to convince him, at an early stage of his development, of the crucial importance of the physical sciences to the study of living systems.

Because of its limited size, this book both omits certain topics and treats others with unavoidable brevity. However, the material contained in the suggested readings at the end of chapters will considerably broaden the scope of the text proper.

It was not the intention of the authors to make this an "elementary" book in the sense of limiting it necessarily to the simpler and more accessible aspects of cell biology. Yet we hope that it is a book for beginners because we believe it is precisely the beginner who deserves an initial statement that is representative of the contemporary quality and mode of the field.

A.G.L.
Haverford, Pennsylvania

P.S.
New York City

June 1963

Contents

Introduction

Some forty years ago a change began in experimental biology, the results of which now shape the vision and control the practice of those of us who place investigation of the cell at the center of the quest to understand life. This new approach has enlarged the concept of the cell as the basic unit of biological structure and function. It has also broadened and blended the experimental methods—cytological or physical, biochemical or genetic, mathematical or physiological—that cell biologists use to understand the molecular basis of life. We intend this book to reflect that change in direction, to trace its course, and to communicate to students the excitement of the revolution in cell biology.

About Cell Biology

The earliest subjects for the study of living matter were, of course, organisms. Our ancestors focused their attention on plants and animals (for food, companionship, and other amenities) long before they became aware that cells are the basic units of biological function or that populations are the fundamental units of evolutionary change.

During the long period of preeminence of the organism as the subject of biological interest, a variety of experimental approaches developed, each with its own methods of investigation. The field of anatomy began with dissection and observation, and continued to prosper as higher resolution was achieved through increasingly powerful optical methods. The field of physiology began with measurements of phenomena such as the flow of blood and has progressed over the years with the development of ingenious procedures for the study of whole living systems down to the cellular level. The study of genetics began with the breeding of plants and animals, and the insights into the mechanism of heredity which this has yielded are as powerful as ever. The study of biochemistry began with the techniques of organic chemistry and enzymology to investigate metabolic processes in organisms, and the success of this approach has not abated.

Thus, each of the various subdisciplines of experimental biology began with the study of organisms. Each developed intellectual orientations, channels of communication, institutional connections, pedagogical structures, and even vested interests of its own. In consequence, the concept of the cell as the fundamental unit of biological structure and function was rediscovered independently in each of the disciplines.

In the last forty years, there has been a revolution in cell biology. It began with insights into the molecular basis of heredity and the fine structures within cells, and continues today to deepen our vision of many other aspects of cell structure and function. This revolution is significant not only for the new knowledge we have but also because its practice has brought about *a fusion of the traditional approaches of cytology, biochemistry, genetics, and cell physiology*. In contrast to what had been recorded in textbooks and codified in curricula, the reality of the material under investigation, and consequently the practice of the research laboratory, led the biochemist to recognize the usefulness of understanding cell structure and of employing genetic analysis, seduced the geneticist to consider the structure of macromolecules and of cell organelles, and made the cytologist aware that the ultimate aim of understanding cell structure is to relate it to the functioning of macromolecules.

Not surprisingly, integrated use of all the methods of experimental biology has become accepted routine in the laboratory, simply because this approach has constituted a recipe for success. Unfortunately, until recently, the content of curricula and of textbooks written to serve them have responded but slowly to the demands generated by the developments in research. At the introductory level, genetics, cytology, biochemistry, and cell physiology are often still taught as if each were an isolated, self-sustaining discipline. In this text, we propose to treat cell biology in the same integrated fashion that cells are in fact treated in the research laboratory.

About This Book

This is a basic textbook in cell biology; it is intended to be accessible to students who have had college-level introductory courses in chemistry and who are taking organic chemistry concurrently. Just as introductory textbooks on botany and zoology have traditionally presented a broad vision leading to a general understanding of plants and animals, so we have endeavored to write *a beginning cell biology text,* one that presents a balanced view of the entire range of phenomena lying between molecules and interacting cells.

This book has two aims. First, we want to acquaint the student with a broad range of cell biological phenomena and to show how insights have been obtained by integrating a variety of approaches chosen from the entire spectrum of experimental disciplines. Second, we want students to develop an understanding of the experimental evidence on which our knowledge is based. The devising of experiments and the reasoning one uses to make conclusions about the world around us are the basic stuff of science and possibly more emblematic of the human spirit than the conclusions themselves.

To achieve these aims we have not only unified the various fields of cell biology but also explained the experimental approaches used in the various traditional disciplines. Thus, Chapter 2 covers the optical approach; Chapter 3 reviews the necessary background in organic chemistry; Chapters 4, 5, 10, and 12 deal with the properties of biological macromolecules; Chapter 6 illustrates the subject and methods of traditional biochemistry; and Chapter 9 presents the content and experimental approach of formal genetics. We have interwoven the classical and the modern, placing related topics near each other, rather than presenting subjects in a temporal sequence. For example, the discussion of nucleic acids is grouped with other chapters on the flow of genetic information, rather than with the chapter on proteins.

Unfortunately, no grouping of chapters or ordering of presentation will fit neatly into all existing courses. A carefully prepared index and numerous cross-references to other parts of this book nevertheless make the appropriate information easily available to the student, whatever organization of lectures is chosen.

It is clear that unifying the various disciplines of cell biology and providing experimental evidence for results could easily give rise to an overly large text. We have used the following principles to keep from exceeding reasonable limits:

- emphasize general rather than comparative aspects of cell biology,
- focus primarily on processes that are susceptible to molecular analysis,
- reduce redundancy as much as possible.

Rather than offering, as more advanced textbooks do, a complete review of the broad range of findings in this rapidly expanding field, *our aim is to provide a foundation of information and of analytical procedure on which future study can build.*

Because we see this book as the first step in learning about experimental cell biology, we take seriously a thorough grounding in the physical sciences. Excluding certain useful areas because they might be called "physics" or "chemistry" makes little sense in the context of modern cell biology; it disguises the historical development of the field and raises unproductive barriers in the minds of students. This is why we present quantitative aspects of cell structure and function, including the molecular properties of proteins and nucleic acids. We are convinced that we should train students not just to understand what is known today, but also to be capable of incorporating the developments of the future.

Attempting to master a large set of scientific data can be a daunting prospect. We do not provide a comprehensive collection in this text. Rather, we present the basic principles and ideas that modern cell biologists use to think about and work on problems. Since science is, in essence, the application of a method, we give many examples of the experiments that led to our current beliefs about biology. Scientific knowledge is incomplete—likely a permanent state of affairs—so we tend to think in terms of *models* of how biology works at the cellular and molecular levels.

Models in Cell Biology

Models are useful to scientists. They not only summarize the current data set—valid models must at least do that—but they also suggest further experiments to test their own validity. As experimental results accumulate, models evolve: those contradicted by data are discarded or revised and the others receive more intense scrutiny through further experimentation.

Biological models can take many forms. Some models are structural: bent wires that sit on the laboratory bench or a rotating set of contiguous balls that appears on a computer's video display terminal are used to model a protein molecule. Some models resemble chemical reactions: the sequential transformation of precursors into products is used to mimic the movement of signals from the environment to the nucleus of a cell. Some models are expressed in mathematical form: equations can portray energy relations in a metabolic pathway. There are even examples from the realm of quantum mechanics: models can explain the interaction of light with pigments during photosynthesis. The different forms of models reflect their origins in different experimental contexts or the different states of our knowledge about the biological phenomena being modeled.

Because cellular and molecular biologists use many different models, the student needs to become comfortable with a variety of them. Biology is the study of living things; whatever techniques are effective are the ones that modern cellular and molecular biologists use. Therefore, we see no value in saying that *this* is biology (and thus relevant) and *that* is chemistry or physics or mathematics (and thus can be ignored). To understand cellular and molecular biology today means to study the methods that are effective. The former distinctions among various disciplines are no longer helpful.

Features of the Text

We include some features that should help make studying cell biology more comfortable. Most terms are defined in the text where they first appear. Some terms are used throughout and are so important that we have collected them and their definitions in a Glossary at the end of the book. This is where you should look first if you find a term whose meaning you don't remember, and especially if you read the book in an order different from the one in which it was printed. Glossary terms are printed in boldface when they first appear.

The next place to look for definitions is the early page references to the term that are found in the Index. We have tried to make the index extensive so students will be able to find what they need regardless of the order of topics presented by the instructor. After all, cell function is not a linear process and the order of presenting topics is largely personal.

Most chapters contain passages set off from the main text in boxes. These are discussions of special topics, examples that illustrate the text, or derivations of the equations that make up a mathematical model. You may skip over the boxed material and some instructors may not wish to emphasize it. But the boxed information *is* important to a knowledge of cell biology. In the case of mathematical derivations, for example, a boxed step-by-step approach allows a leisurely contemplation of the methods and should inspire confidence that the student can follow even the more complex material.

The advantages of putting mathematical models in a text, as opposed to a lecture, is that the student can go over them as slowly or as quickly as desired, and can return to them later for whatever further study is necessary. Equations in a text are stable and don't require verbatim stenographic skills. The mathematical models we have included require only a good knowledge of algebra and a very modest experience with calculus. Including derivations is done deliberately to reassure students that these models do not spring fully formed from the brow of Zeus, or from even more god-like biology instructors. It is not only possible to understand the origin of the models, but one can become comfortable working with them. To help with this, we have stressed the relations between models used in different places to describe biological processes in different contexts, relations often obscured by different conventions of symbols and/or signs.

Problems appear at the end of each chapter. These are of varying difficulty. Few require mere memorization of data presented in the immediately preceding chapter. Many will require a combination of facts and skills developed in previous chapters or courses. The student should try to work them out; the effort spent will be repaid many times over in terms of ability to use the knowledge gained by reading the text.

We frequently discuss the historical context in which discoveries were made, not because history as such is of central importance for our purpose, but because an understanding of the process of scientific discovery is crucial for the student's development as a mature and creative thinker. When an individual scientist's contribution has influenced the field in a significant way we usually include his or her name, acknowledging that the results of research in cell biology have been obtained by people. Perhaps in the future, important results will be obtained by those who are readers of this book. (We apologize for omitting, because of the constraints of page space, many names.)

Recurring Themes

One exciting thing about modern molecular and cellular biology is how much of physical science has become integrated into it. The material we cover includes not only biological concepts, but also those originating and expressed in the languages of chemistry, mathematics, and physics. We make these relations explicit as we go and refer to other parts of the text where related discussions are presented.

Another theme is that biological phenomena can be understood in terms of flows of mass, energy, and information. By "mass" we mean the materials of which biological structures are made. By "energy" we mean the forces that drive the flows. By "information" we mean the influences that control the flows. All these are characteristic of biological processes, including the assembly of biological structures. Mass and energy are studied by chemists and physicists. Biologists need also to study information, most of which is present in the form of genes that are passed on from one generation to the next over evolutionary time. That illustrates the major difference between biology and the other sciences: Living systems have a history; they have evolved. The current state of our knowledge of evolution is the main message of this text, but to understand the way in which we have arrived at that knowledge is also very important. It not only represents a glorious chapter in the history of ideas, but it also illuminates the activities of today's creators of tomorrow's biology.

Summary

This book reflects the biological revolution that has occurred in recent years in which a variety of approaches have combined to produce a powerful and unifying molecular biology of the cell. *An integrated treatment, therefore, should serve as the introduction to modern cell biology, before focusing on the special concerns of individual experimental disciplines.* And now, let us put these ideas into practice!

Living *Stylonychia,* a fresh-water ciliate protozoan, about 100 μm long, viewed with Nomarski microscopy. The cilia of this cell are bunched in groups on the ventral surface (some are visible at the bottom); the cell "walks" around on the substratum using these cilia. (Courtesy of J. Frankel and E. M. Nelsen.)

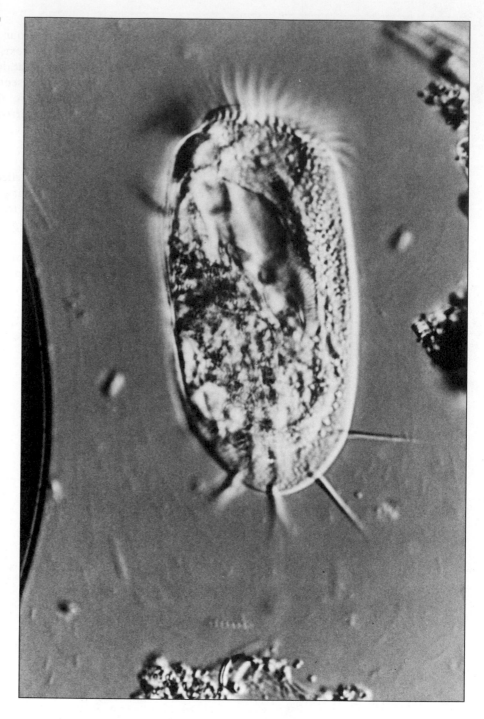

We discuss the cell in relation to the properties of living matter as a whole.

We introduce two basic cell types which are the "common denominators" of almost all living matter and familiarize the student with the idea that the cell is the fundamental unit of biological activity. A presentation of cell structure introduces the object of our study and provides a visual base for further discussion.

We also introduce the concept of continuity between nonliving and living things by arguing that living matter is a special (and interesting) form of physical matter, obeying the laws of physics and chemistry. In studying life in terms of molecules, however, we do not reduce the meaning of life; rather, we elevate the concept of the molecule.

The theme of this part is that biological structure and function are intimately related, being two sides of the same coin.

THE CELL— COMMON DENOMINATOR OF LIVING MATTER

Cell Function—The Behavioral Basis of Biological Organization

The Unity of Life

The observation and description of living objects, and especially their *diversity*, have probably been going on since humans first noticed their environment. The idea that there is a *unity* underlying the diversity, and encompassing most processes of life, is a relatively recent development. Systematic exploration of the *similarities* among organisms had to wait until the advent of the scientific method. And when biological chemists turned their attention to the molecules of life they realized almost at once that outward differences disguised chemical identities. Although our ancestors recognized easily the differences between a bat and a bird, we now are struck by the similarities between the molecules of a mushroom and those of a human.

Three major theories, each less than a century and a half old, illustrate different aspects of the unity of life:

- The Schleiden and Schwann **cell theory** (1838) generalized the view that *all living organisms are composed of cells and of cell products.*
- The Darwin-Wallace **theory of evolution** (1858) suggested that **natural selection** could explain how the diversity of living forms "descended by modification" from common ancestry.
- **Mendel's laws of heredity** (1866) proposed a general mechanism whereby **information** in genes is **segregated,** assorted—also, we now know, **recombined**—and transmitted by each generation.

Each of these theories provoked intense controversy when introduced and the repercussions haven't subsided yet. For example, the cell theory now seems to be self-evident, but we must not forget that we have lived with it for some time. To Schleiden and Schwann's predecessors the idea that the organism is the fundamental unit of biological activity was an easily observable, well-established, common-sense principle. That organisms are in turn composed of smaller subunits, each of which is endowed with an individual identity, was a counterintuitive insight. Even as late as 1858 the great biologist T. H. Huxley wrote in his review of cell theory that cells ". . . are not the instruments but indications, . . . they are no more the producers of vital phenomena than the shells scattered in orderly lines along the seabeach are the instruments by which the gravitational force of the moon acts on the oceans. Like these, the cells mark only where the vital tides have been and how they have acted."[1] It took both time and intensive investigation to convince the scientific community that understanding the cell was both interesting and important for understanding biology.

The remarks of Huxley are also interesting for another reason. He used language which evokes an older view of the essence of life. That the cell is merely a vessel for vital forces (as opposed to physical forces) which account for life is a notion that persists even to this day, though not often among biologists. The assumption that living systems are fundamentally different from nonliving systems is the basis for the set of beliefs called *vitalism*. Originally, vitalists believed in special forces that set living things apart from other physical objects. More recently the concerns of vitalism have shifted; now the "vital tides" are thought to influence the molecules within the cell rather than the cell as a whole. In Huxley's time vitalists argued that you could not "reduce" the properties of the organism to considerations of interactions between cells. Some vitalists today argue that you cannot "reduce" the properties of the cell to descriptions of interactions between molecules. As we shall see repeatedly in this book, the study of the molecules of life does not so much reduce life to "mere" molecular phenomena as it enlarges our concept of molecules. We now understand properties of cellular molecules that no one would have considered possible had studies been restricted to molecules of the nonliving world. *The study of modern cell biology has not diminished the concept of the cell; it has, rather, elevated the concept of the molecule!*

The extent to which cells, although part of an organism, have *individuality* and are capable of *independent existence*—originally central postulates of the cell theory—was by no means clear even to Schleiden and Schwann. It was only later that biologists began to recognize these properties derive from the fact that all cells possess in their own right the structures and processes one might call the "machinery of life." Whether from a protozoan, a plant, or a painter, they all show remarkably similar internal architecture and even greater similarity of function. All cells duplicate their genetic material, synthesize proteins, transfer energy, regulate the movement of materials, convert chemical

energy into mechanical motion, and so on, in essentially the same ways. In fact, it has been disconcerting to those interested in differentiation, or in the pathology of cancer cells, that so few fundamental differences can be detected among cells of various types. One purpose of this book is to emphasize the common features of the structure and function of all cells.

Cells also have specializations, ranging over the entire breadth of diversity in biology. A "generalized cell," containing only the common features, does not exist as such in nature. The concept of a generalized cell, however, does provide a convenient abstraction, and it is this abstraction that most of the later chapters describe. In this chapter we wish to emphasize certain themes that will recur and to place the cell in its proper context as part of the physical and chemical world. Even as there is a unity of cellular structures and processes, there is a unity of sciences. Biology and its phenomena are part of physics and chemistry. Equally important, knowledge of physics and chemistry is necessary to gain knowledge of biology.

Before turning to the study of the cell, which is the main subject of this text, it is appropriate to note that cell biology and the methods that are used to investigate cells by no means exhaust the interesting features of biology. That cell biology is so prominent a feature of the current scientific landscape testifies to the success of its practitioners. But many other problems of central importance to biology are peculiar to the operation of whole organisms or to the dynamics of populations of organisms. For these problems, the study of the environment of the organism, and its past behavior and experience, may be as important as the study of cells.

What Is a Cell?

Cells may be defined as objects that possess at least two major components of the machinery of life: *information transfer machinery*, which stores, distributes, reads out, and above all reproduces the information that controls the processes of life, and **energy transduction machinery**, which changes **energy** from one form to another, stores it, and distributes it to run the processes of life.

Admittedly, some entities might lack small portions of either of these major components but stay alive with the help of other cells or cell products to supply the missing functions. We might or might not think of these cases as cells. There are some self-reproductive forms—the chlamydiae—which have membranes but are generally incapable of growth outside living cells and lack major portions, probably all, of the energy transduction machinery. Even simpler are **viruses,** which have none of the energy transduction machinery and only a bit of the information transfer machinery. There is general

[1]*Life and Letters of Thomas Henry Huxley*, vol. I. (1979) New York: AMS Press. (Reprint of D. Appleton and Company, 1900 Edition.)

agreement that viruses are not cells. Chlamydiae are related to free-living bacterial cells and probably represent successors that lost the ability to derive energy from the nonliving environment. Chlamydiae are intracellular parasites which get their energy from the host and are certainly simpler than most cells.

Cells of Procaryotes and Eucaryotes

The development of electron microscopy has revealed two basic plans of cellular organization, the simpler one found in **procaryotes** and the more complex one found in **eucaryotes.** These plans have a number of major differences. The transition from procaryotic to eucaryotic organisms is so abrupt, in fact, that it represents one of the outstanding evolutionary steps in the history of life. The fundamental distinction between procaryote and eucaryote cells is the presence in the latter of a nucleus (pro-caryote = before nucleus; eu-caryote = good or true nucleus).

Procaryotes are believed to have appeared on this earth some three and one-half to four billion (four thousand million) years ago. They originated from nonliving matter which had undergone extensive *chemical evolution* since the formation of the earth from hot gases some four and one-half billion years ago. It has been shown that many of the molecules formerly regarded as resulting from life processes can in fact be produced through purely inorganic chemical reactions of the sort that might be expected to have occurred in the earliest conditions on earth. These most likely included abundant water in which was dissolved ammonia, carbon dioxide, hydrogen, methane, and nitrogen. A troubling feature of the chemical evolution scenario, however, is that it occurred over such a remarkably short interval (short, that is, for geological events); perhaps as little as three-quarters of a billion years. An alternative hypothesis, not easily tested by experiment, is that life was transported here from elsewhere, perhaps another planet where chemical evolution could have proceeded more leisurely.

Procaryotes are relatively simple cells; today, they comprise only two groups of organisms: the *bacteria* and the *blue-green algae (or Cyanobacteria)*. Biologists regard bacteria as being the descendents of the first cells capable of processing pre-existing reduced carbon compounds as a source for their own organic molecules. Cells—like our own—that use such compounds as a source of carbon are called **heterotrophs,** and they doubtless were the earliest cells on this planet, using at the outset compounds formed by purely nonbiological chemistry. (A "reduced" molecule can also be oxidized to release the energy necessary to maintain life.) The blue-green algae are probably descendents of the first cells that used light

energy to convert inorganic compounds, especially carbon dioxide, into living matter. Those organisms capable of making living matter by using carbon dioxide as their primary source of carbon are called **autotrophs.** The autotrophs most likely arose after the heterotrophs had become limited by the availability of sufficient reduced carbon compounds in their environment. *Photoautotrophs* —like green plant cells—use light as the source of energy necessary to convert CO_2 into biological macromolecules; *chemoautotrophs*—for example, the hydrogen bacteria—get this energy from the oxidation of inorganic compounds (in this example, hydrogen gas).

For a long time, the similarities in structural design among procaryotes seemed to argue for classifying them together as a single primary kingdom. More sophisticated analyses have challenged this view. It is now agreed there are two kingdoms among the present-day procaryotes, and it can be persuasively argued that this dichotomy was also present at the earliest times during the evolution of life. C. Woese and his collaborators have named the present-day descendents of one group of these ancient forms *archaebacteria*. Their biochemistry and the detailed structure of their large molecules are as different from those of other procaryotes (eubacteria) as from those of eucaryotes. The surviving species include some of the more exotic bacteria alive today: *methanogens* that synthesize methane gas from carbon dioxide; extreme *halophiles* that thrive in strong saline solutions (5.5 M NaCl); and *thermoacidophiles* that grow in acidic hot springs (pH 2, 80°C). More research will be needed to determine whether these species actually are, as their name implies, relatively less evolved survivors of the first and, therefore, oldest procaryotes.

The eucaryotes include all the forms of life other than the procaryotes: the unicellular *protists* such as *algae* and *protozoa,* and all the multicellular plants, animals, and fungi that have evolved from these. Eucaryotic cells can also be divided into autotrophs and heterotrophs. A membrane system bounds the distinctive nucleus and separates it from the rest of the cell. Indeed, the presence of numerous membrane-bound compartments in eucaryotic cells is another property that distinguishes them from procaryotic cells. The bulk of hereditary material of the procaryotic cell is stored in a single molecule of **deoxyribonucleic acid (DNA)**; the nuclei of eucaryotic cells store their larger amount of hereditary information in more than one DNA- and protein-containing **chromosome**. During cell division, each daughter cell must get a complete set of chromosomes. There is a **mitotic apparatus** which controls the complicated choreography of **mitosis** as it occurs in the eucaryotic cell.

A striking characteristic of eucaryotic cells is that they are as much as 10 to 20 times larger in diameter and thousands of times greater in mass than most procaryotic cells (Figure 1–1). This difference in size is not just the

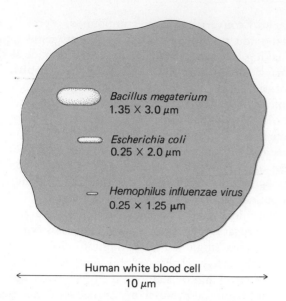

Bacillus megaterium
1.35 × 3.0 μm

Escherichia coli
0.25 × 2.0 μm

Hemophilus influenzae virus
0.25 × 1.25 μm

Human white blood cell
10 μm

FIGURE **1–1**

Size Differences Between Various Procaryotic Cells and a Eucaryotic Cell (Human White Blood Cell).   A ten-fold difference in diameter between two spherical cells is equivalent to a thousand-fold difference in mass.

result of a greater quantity of internal structure in the eucaryotic cell but represents entirely new principles of organization which probably took some one and one-half to two billion years to evolve (Figure 1–2). The cytoplasm of procaryotes, for example, does not appear to exhibit any systematic internal movement, so these organisms must depend on diffusion as a means of distributing materials and energy about the cell. But in eucaryotes the cytoplasm *is* capable of internal motion, freeing the eucaryotic cell from some of the limitations of diffusion and making possible an enormous increase in size and complexity.

Procaryotic cells as a group are generally characterized by wide distribution, rapid growth, short generation time, tremendous biochemical versatility, and genetic flexibility. Experimental biologists have exploited these convenient properties to great advantage—for example, the use of *Escherichia coli* bacteria in genetic engineering techniques. In their apparently modest way, procaryotes represent one kind of evolutionary success story, transcending their dependence on the environment by their fecundity and by their versatility in adapting to the extreme variations in environmental conditions on our earth.

Eucaryotic cells carry out most, if not all, of the functions of procaryotic cells. In addition, however, many of them have evolved the potential for cooperative existence as subunits of multicellular differentiated organisms. There are many other detailed differences in the structures of procaryotic and eucaryotic cells that will be discussed in the next chapter.

Eucaryotic cells are believed to have originated some one and one-half billion years ago. The earliest eucaryotes for which fossil evidence exists had cell walls. It is reasonable to assume that these cells contained **chloroplasts,** the specialized **organelles** for carrying out photosynthesis, and were ancestors of the green algae. An interesting question, then, is how such an organelle came to reside inside a cell.

Since paleontologists have not yet discovered surviving intermediate forms or fossil records of a transition, it is only possible to speculate how eucaryotic cells might have evolved. The *endosymbiont theory*—first suggested by C. Mereschkowsky in 1910—holds that present-day eucaryotic organelles such as chloroplasts and **mitochondria** represent the remains of procaryotic cells which fused with another (presumably procaryotic?) cell, with the combination adapting to a *symbiotic* mode of existence. Fusion events of this kind gave rise to the line of eucaryotic cells. The supporters of the endosymbiont theory cite evidence such as the presence in chloroplasts of **ribosomes** resembling those of procaryotes. The great similarities between the photosynthetic apparatus of *Cyanobacteria* and that of today's eucaryotic red algae argue for an evolutionary relationship between them. It is, however, not easy to see how a cell like present-day procaryotes, by fusing with another such cell, could lead to the radically different organization of the genetic material that characterizes the eucaryotic cell nucleus. It has therefore been hypothesized that a *protoeucaryotic cell* must have diverged from the procaryotes, both the archaebacterial and the eubacterial lineages, very early during evolution. Recent data obtained by analyzing the macromolecules of procaryotes and eucaryotes supports the idea of early divergence of the three lineages. Although not accepted by everyone, the endosymbiont theory is now the most popular theory to account for the origin of organelles in eucaryotic cells.

Fossil evidence shows that some 900 million years ago there were eucaryotes with sexual reproduction, but it was not until 600–800 million years ago that the first known fossils of multicellular organisms were formed. Then, in just a few million years—the Cambrian epoch— a remarkable evolutionary acceleration occurred: within this "short" period most of the major phyla of invertebrate animals evolved. The origins and causes of the *Cambrian explosion* had concerned Darwin and they remain mysterious today. A plausible suggestion by S. Stanley is the appearance of a cell with the ability to eat the other large eucaryotic plant cells that had lived unchallenged for hundreds of millions of years.

Certain evolutionary developments would have had to take place to produce a primitive plant-eating cell. To engulf another cell, this first animal cell would have had to give up its stiff outer wall, requiring some strategy for balancing water movement in and out of the cell (see

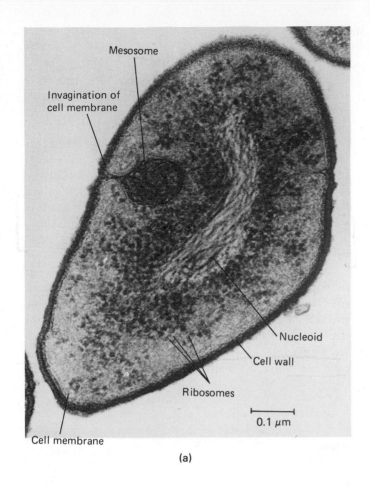

Mesosome

Invagination of
cell membrane

Nucleoid

Cell wall

Ribosomes

0.1 µm

Cell membrane

(a)

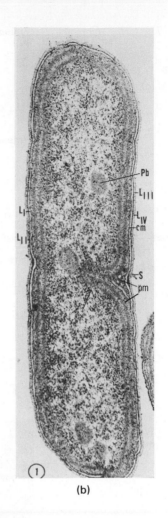

Pb

L_III

L_IV
cm

L_I

L_II

S
pm

①

(b)

FIGURE 1–2
**The Basic Types of Cellular
Organization.** (a) The bacterium *E. coli,* a
procaryotic cell. (Courtesy of J. Jamieson.)
(b) The cyanobacterium *Aphanocapsa 6308,* a
blue-green algal procaryotic cell. L_I–L_IV =
layers of cell wall; cm = cell membrane; pm
= photosynthetic membranes; Pb = polar
body. × 72,000. (Courtesy of M. M. Allen.)
(c) Germ cell from dragonfly testis. The
well-defined nucleus, surrounded by a
nuclear envelope, identifies this as a
eucaryotic cell. In the nucleus are the
nucleolus and chromatin; in the cytoplasm
are mitochondria, ribosomes, microtubules,
and internal membranes. × 28,000.
(Courtesy of R. Kessel.)

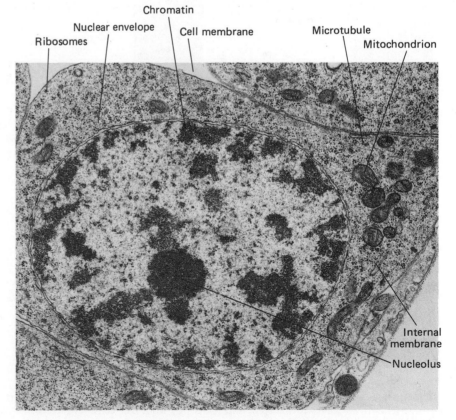

Chromatin

Nuclear envelope

Cell membrane

Microtubule

Mitochondrion

Ribosomes

Internal
membrane

Nucleolus

(c)

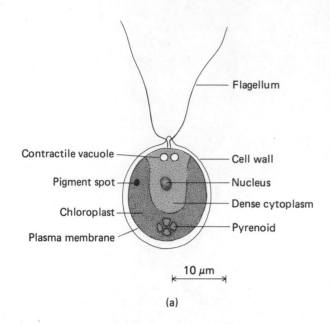

(a)

Chapter 7). It also would have had to develop a mechanism for engulfing large particles, in turn requiring contractile structures to allow for some form of movement of the cytoplasm (see Chapter 16).

Once such animal cells developed the ability to ingest plant cells, a vast source of materials and energy became available to them, which Stanley suggests formed the basis of rapid evolution not only in the animal kingdom but also in the plant kingdom. As the result of selection by the unfolding diversity of the predatory animal world, there was also an increase in plant diversity.

Although there is a general pattern of structure and function common to all of today's surviving eucaryotic cells, there are nevertheless two distinguishable types—the walled-in cells of green plants and fungi, and the naked cells of the animal world. Plant cells (Figure 1–3) are surrounded by a stiff envelope usually made of cellulose. This cell wall prevents the cell from bursting

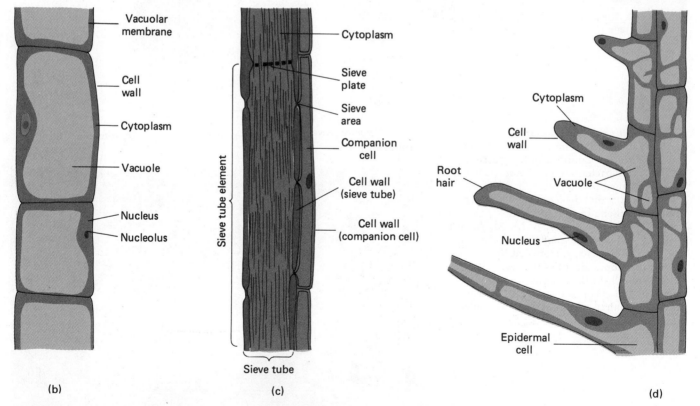

(b)

(c)

(d)

FIGURE **1–3**
A Variety of Plant Cell Types. The drawings illustrate some of the different structures that can be observed in the eucaryotic cells of plants. (a) *Chlamydomonas,* a unicellular motile green alga. The single chloroplast is cup-shaped and occupies a large fraction of the cell volume. The pigment spot is a light receptor and allows the organism to move towards light. (b) Parenchyma cells. These relatively unspecialized cells are found in the roots, stems, and leaves of higher plants. (c) Phloem sieve tubes and companion cells. This structure is specialized for transport of nutrients from leaves to the stem and roots. (d) Root hair cells. These cells absorb water and minerals for use by the plant.

when living in a medium which is lower in **solute** concentration than the cell's interior. In most mature plant cells, a thin layer of cytoplasm lies just inside the **plasma membrane** and surrounds a large membrane-bound body of liquid, or **vacuole.** The most characteristic feature of green plant cells is *chloroplasts,* the organelles which carry out the process of photosynthesis. Animal cells (Figure 1–4) are not usually surrounded by a stiff envelope, are lacking in chloroplasts, are frequently motile, and often change their shapes. When living in a

medium with solute concentration lower than that of their cytoplasm, they must have some way to prevent net water uptake, which would otherwise eventually cause swelling and bursting.

Specialization Derives from Unity

The *protozoa* are a distinct evolutionary line of unicellular animals. Instead of developing into multicellular organisms composed of different cell types, the

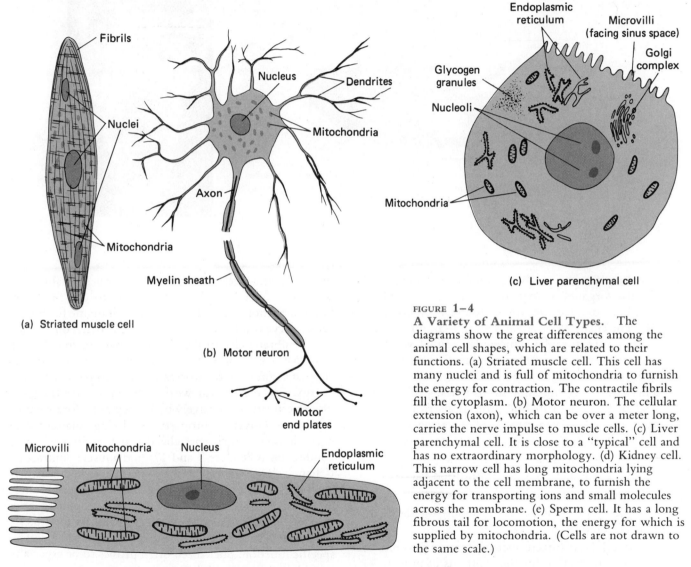

(a) Striated muscle cell

(b) Motor neuron

(c) Liver parenchymal cell

(d) Kidney cell (proximal convoluted tubule)

FIGURE **1–4**
A Variety of Animal Cell Types.   The diagrams show the great differences among the animal cell shapes, which are related to their functions. (a) Striated muscle cell. This cell has many nuclei and is full of mitochondria to furnish the energy for contraction. The contractile fibrils fill the cytoplasm. (b) Motor neuron. The cellular extension (axon), which can be over a meter long, carries the nerve impulse to muscle cells. (c) Liver parenchymal cell. It is close to a "typical" cell and has no extraordinary morphology. (d) Kidney cell. This narrow cell has long mitochondria lying adjacent to the cell membrane, to furnish the energy for transporting ions and small molecules across the membrane. (e) Sperm cell. It has a long fibrous tail for locomotion, the energy for which is supplied by mitochondria. (Cells are not drawn to the same scale.)

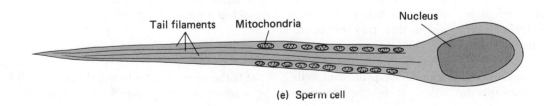

(e) Sperm cell

FIGURE **1–5**
The Unicellular Protozoan *Paramecium*.   (a) Drawing showing the complex structures that are crudely analogous to those of certain multicellular organs (indicated in parentheses) in the metazoa. (b) Scanning electron micrograph of a freeze-dried *Paramecium*. The specimen was fixed very rapidly in OsO_4-$HgCl_2$ while swimming forward. The small cells at the upper right are yeast that are present in the culture medium. A-P = anterior-posterior axis; D-V = dorsal-ventral axis. (Courtesy of S. L. Tamm.)

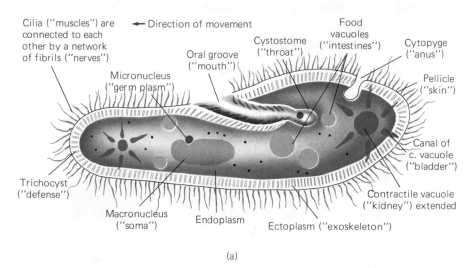

(a)

(b)

zoa have managed to evolve a number of complicated intracellular organelles that function, in a simplified manner, like some of the organs found in multicellular animals. The ciliate *Paramecium* (Figure 1–5), for instance, is an exquisitely complex unicellular animal which, in addition to its typical eucaryotic organelles, contains:

- a network of cytoplasmic fibrils that coordinates the beat of its cilia—hair-like projections from the cell surface—and is even capable of reversing the direction of their beat during avoidance reactions
- specialized structures for the ingestion of food (oral groove and gullet) and digestion (food vacuoles)
- a device for excretion of solids (cytopyge)
- two "kidneys" (contractile vacuoles) for the excretion of liquid
- somatic DNA (macronucleus) carrying copies of genetic information for the synthesis of proteins
- germ plasm DNA (one or more micronuclei) responsible for hereditary continuity
- an apparent "defensive" apparatus (trichocysts) composed of little spears which this tiny organism releases when stimulated

These unusual specializations allow protozoan cells to adapt to their specific roles. It is important to recognize, however, that many apparently specialized cell functions, including those of the protozoa, find their bases in phenomena that occur in less dramatic form in all cells. Nerve conduction, for instance, is based on varying the membrane electrical potentials that are found in all cells. Muscular contraction is based on generally occurring structures and processes that convert chemical energy into mechanical work. The molecular transport work performed intensively by kidney cells, for example, is also a universal property of living matter. Our excursions into such specialized cell functions as nerve conduction (Chapters 7 and 17) and muscle contraction (Chapter 16) are motivated by the fact that our understanding of the basic principles is most advanced in these highly specialized cells. Generally, it is found to be a valuable practice in experimentation to select for study a cell system that is very efficient in carrying out a certain specific function. Such a cell often proves to be especially suitable as experimental material for the analysis of that function. In the end our knowledge of these more specialized cells is merged into the single conceptual model of a generalized cell.

On Defining Life

If the properties of living things can all be described in molecular terms, one could argue that life is not unique, that it is indistinguishable from the nonliving physical world. Common sense, however, tells us that we can usually distinguish easily between the living and the nonliving. Even as children we were able to identify objects in our games as "animal, vegetable, or mineral." If living things are then so easily identifiable, it seems reasonable to ask upon what criteria such judgments are normally based.

Evolutionary History—A Unique Criterion of Life

Are there any properties which are uniquely characteristic of living matter? Many have suggested properties such as movement, reproduction, metabolism, sensitivity, and growth. When each of these is carefully defined, however, we find that there are nonliving systems that satisfy the formal definitions. This is not surprising in an age of automated systems. Moreover, although engineers have had a tendency to design specialized machines to carry out specific functions, it is conceivable that they could combine all these functions in a single machine which simultaneously moves, grows, metabolizes, is sensitive to external stimuli, and even reproduces itself. The space laboratory deposited on Mars in 1976 certainly went part way toward creating such a robot, and the properties of growth and reproduction have been mimicked by other even less sophisticated devices.

Is there then no essential physical difference between humans and robots? Probably not. The critical difference is not physical but historical: *the origins of humans and robots are different.* Humans built the robots and not vice versa.

Life is an evolutionary *process*. From relatively simple beginnings it has gained in complexity through a series of events, the sum of which comprises the process called evolution. This series has been going on for a long time and finds its present and apparently most complex expression in the form of *Homo sapiens,* the living being capable of building machines with the appearance of life.

If one defines life as that which evolved and robots as the products of human thought and action, what about the effects of human thought and action on biological organisms? Are rust-resistant strains of wheat, hybrid corn, or a genetically "engineered" bacterium not alive? And what about the effects on our own future? It is conceivable, after all, that someday we shall be able to apply our understanding of biological principles to the regulation of human evolution itself.

Despite these apparent contradictions, an evolutionary history is still the appropriate criterion for defining living systems. After all, intelligence is itself the result of natural selection. At some point this characteristic must feed back upon and influence the evolutionary process from which it emerged. The organization associated with human evolution is at present being extended to our physical surroundings. Different forms of organization, as concrete as machines and buildings and as abstract as knowledge and wisdom, are being accumulated at an increasing rate. Whether this new form of organization that we call civilization is stable, leading to developments yet undreamed of in the social evolution of mankind, or whether it is permeated with biologically rooted and historically transmitted contradictions, incapable of solution at a social level, only the future will determine.

Compactness and Energetic Economy as Properties of Life

Although their history defines their critical difference from the nonliving, there are two other properties that are characteristic of living things. They are the extraordinary *compactness* and *energetic economy* of living matter.

The almost infinitesimally minute scale in which biological organization has evolved is one of the most dazzling wonders of nature. Even present-day electronic computers are larger and use many times more energy than the vastly more sophisticated human brain. The tremendous compactness and efficiency of living systems is possible, of course, because their structures are built on a molecular scale. As will be discussed throughout this book, the cell is capable of **replication, transcription,** and **translation** of information stored in the structure of single molecules which, though larger than the molecules generally encountered in the nonliving world, are still extraordinarily tiny compared with the components of modern machines. Furthermore, the **enzymes** that are the cell's regulatory devices are also single molecules, many orders of magnitude smaller than the devices built by process engineers.

Among the astonishing consequences of this compactness are the tremendous amplification effects exhibited by living systems. The 5×10^{-15} kg of DNA present in the nucleus of a fertilized whale egg determines most of the hereditary qualities of an animal which will eventually have a mass of 5×10^4 kg. The albinism of a whale like the fictional Moby Dick was doubtlessly caused by a tiny change, perhaps as small as one base in five thousand million, in the whale's DNA. An effect of almost 29 orders of magnitude is a feat which electronic engineers might well envy! Amplifications like these are due, of course, to the growth of biological systems. Less staggering amplifications provide the basis for sensitive modern assays of molecular events. The compactness and energy economy of biological systems will be exploited more in the future as societies attempt to conserve scarce resources.

Life, Order, and the Laws of Thermodynamics

The world of life evolves, maintains, and extends an impressive degree of organization. Perhaps the process that increases organization is a unique property of living matter. After all, the **second law of thermodynamics** —a fundamental law of the physical universe—says that when **systems** in isolation and free of internal force fields change, they always move toward states of greater *disorganization*. At first glance it might seem that the second law does not apply to living matter; this is what G. N. Lewis, one of the creators of the thermodynamic theory, suspected. Cells do, in fact, obey the second law: even though their own disorganization *decreases*, that is outweighed by a greater *increase* in the disorganization of their environment. This statement, while true, is not self-evident and requires a thorough discussion to be meaningful. To examine this problem more carefully and to provide an appropriate framework for discussing cellular energy transductions we need more precise statements of the laws of thermodynamics and their meaning in biology.

The First Law: Conservation of Energy

The **first law of thermodynamics** says that various forms of energy—chemical, mechanical, electrical, or radiant—are interconvertible and that equal amounts of these forms of energy are totally extractable as equal amounts of heat. It also says that, during the process of conversion, energy is neither created nor destroyed. This implies that the total energy content (called the *internal energy*) of an isolated system—one that cannot exchange energy or mass with its **surroundings**—remains constant. For an open system like a cell, which is exchanging both energy and materials with its environment, a constant internal energy would imply that the input of energy must equal the output (Figure 1–6). It has been demonstrated repeatedly with very careful measurements that living systems behave in accordance with the first law. In fact, all living things, including cells, are very active in *transducing* energy. (To *transduce* is to change from one form into another.) They transduce chemical energy into kinetic energy, into mechanical potential energy, into electrical energy, and even (in special cases) into light energy. Many cells can also transduce light into chemical energy. Above all, they can take radiant or chemical energy from the environment and use it to increase cellular organization. How they are able to do this is the question that we shall attempt to answer—in principle here, and in concrete detail in later chapters.

During the historical development of thermodynamics, when physicists were concerned with heat engines,

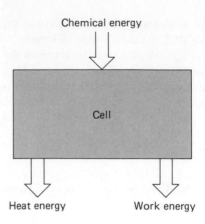

FIGURE 1–6

The Cell as an Energy Transducer. Cells act in accordance with the first law of thermodynamics: The energy flowing into a cell—either the chemical energy available by oxidizing reduced compounds (or light energy in the case of photoautotrophs)—is exactly equal to the energy flowing out of a cell, whether in the form of heat or various kinds of work. Since cells act in accordance with the second law of thermodynamics, there is always some heat in the energy output.

energy flows were divided into two types: *heat,* energy which flows between a system and its surroundings because of a *difference in temperature,* and *work,* all the other kinds of energy flow. A system like a heat engine has a certain amount of internal energy (U) that depends on the *state of the system.* The state of a system is defined by the values of certain parameters such as the temperature, pressure, and chemical concentrations. In a heat engine, the *change of internal energy* in going from one state to another can be represented as

$$\Delta U = Q - W$$

where Q is the heat flowing *into* the engine, W is the work flowing *out,* and ΔU is the change in internal energy of the engine. (We use the symbol Δ to signify "the difference in" a given quantity that exists between two states, destination minus origin. By convention, Δ indicates an *increase* in the specified quantity. Thus, a negative value for ΔU represents a *decrease* in the internal energy of the system in going from the origin state to the destination state.) The first law asserts that if the work output is greater than the heat input, there must be a decrease in the internal energy ($\Delta U < 0$) to account for the difference. Although the form of the first law is reminiscent of its origins in the study of heat engines, it is generally true for all physical systems, including cells. The change in internal energy is a *state function;* in a system changing between two states, ΔU depends only on the two states and not on the path taken between them. The heat Q and the work W, however, are not state functions; they depend on the path as well as on the properties of the two states.

The Second Law: Spontaneity and Equilibrium

The first law provides an overall balance sheet for the energy conversions that occur during a biological process. It does not, however, distinguish between processes that are *spontaneous*—those that can occur without the input of energy—and processes that are *nonspontaneous*—those that require an input of energy to occur. According to the first law, the molecules of an isolated cell could divide themselves into two regions, one hot and the other cold, so long as the total energy of the cell remained unchanged. It is the *second law of thermodynamics* that says such a thing does not occur spontaneously. The first law would be satisfied if sweetened coffee segregated out a lump of sugar at the bottom of the cup but the second law, and our common experience, says this never happens spontaneously. (In fact, the second law is nothing more than a precise statement of our experience.)

The second law of thermodynamics, then, is concerned with the *direction* of change. It is capable of predicting whether or not a given process can occur spontaneously. A process like concentrating dissolved sugar to a lump, which is the reverse of the spontaneous process of dissolving a lump of sugar and increasing its disorganization, is described as *nonspontaneous*. A spontaneous process can occur without input of energy; a nonspontaneous process will not occur unless energy— for example, the heat necessary to evaporate water to recover dissolved sugar—is supplied by a second (spontaneous) process. In the latter case the combination of the two processes will be spontaneous. Put another way, spontaneous processes can be used as sources of energy and thus can do work. Work must be done to force an otherwise nonspontaneous process to occur. The second law also covers the case when a process is neither spontaneous nor nonspontaneous. This occurs when the system is in **equilibrium** with its surroundings. In this case, the disorganization is at a maximum, no further spontaneous processes can occur, and no work can be obtained from such a system.

Entropy: The Measure of Disorganization

Spontaneous processes always lead to an increase in total disorganization, according to the second law. But how can one measure disorganization and how can one predict whether some imaginable process will be spontaneous or not?

The organization or disorganization of a system depends on how the parts of that system can be arranged. The more possible arrangements that give the same temperature, pressure, concentrations, and total energy

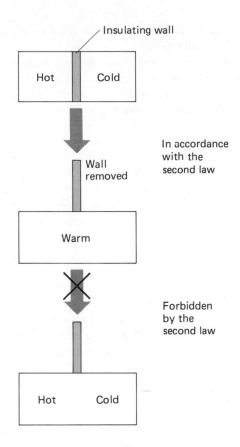

(i.e., the same state), the greater the disorganization. In the study of statistical mechanics one learns how to count these arrangements exactly. We can. get a qualitative feeling for such analyses by considering some special cases.

In a crystalline solid there is only one possible arrangement for the molecules, so the degree of disorganization is very low. Given the position of one molecule and the knowledge that it is part of a crystal, one can predict with high confidence the positions of the remaining molecules. When the solid melts, the molecules occupy a greater variety of the possible positions and disorganization increases. Knowing the position of one molecule in a liquid, one cannot predict so exactly where other molecules will be. When a liquid evaporates, the positions of the different molecules in the resultant gas are even less predictable, and disorganization is therefore even greater.

Such considerations can be applied to chemical reactions as well as changes of state. Other things being equal, there are more ways of arranging a larger number of molecules than of arranging a smaller number. A chemical reaction which yields products that are more numerous than the reactants will generally therefore lead to greater disorganization. (Of course, there are uncertainties other than the numbers and positions of the molecules—for instance, their vibrational and rotational states—and these can also influence the overall measurement of disorganization.)

This way of thinking about disorganization is *microscopic* in that it focuses attention on the smallest parts of a system. There is also a *macroscopic* measure of disorganization which is called the **entropy.** The relation between the two can be exemplified by considering the melting of a crystalline solid. In an isolated mixture of ice and water at melting temperature, the solid and liquid are in equilibrium (Figure 1–7). That is, there is no spontaneous change either toward further melting or toward further solidification. If one raises the temperature of the surroundings a very little and slowly transfers heat into an *un*isolated system of this type, there is further melting and an increase in disorganization. The heat absorbed, *Q,* divided by the (melting) temperature, *T,* is found to be a measure of the disorganization and is called the *change in entropy, ΔS.* In mathematical terms $\Delta S = Q/T$ for this process. Thus, for this special case of reversible melting, the entropy change can be evaluated by macroscopic parameters and the change in disorganization can be measured. The change in entropy is a state function, like the change in internal energy.

We described the melting as a *reversible process* because it was done slowly to avoid friction and sudden macroscopic changes in the system and the surroundings. In principle, during reversible processes the system is in *thermodynamic equilibrium,* in the sense that there is a single value of the temperature that applies to the whole system; likewise, the pressure and the concentrations of components are the same throughout the system. Since the temperature of the surroundings is only infinitesimally different from the temperature of the system, the system is practically in *thermal equilibrium* with the surroundings

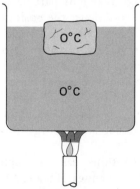

FIGURE 1–7
Water and Ice in Equilibrium at 0°C. If a small amount of heat is added slowly to the mixture, the temperature and pressure will remain the same throughout but some of the ice will melt. The increase in the entropy of this system, which is due to the greater disorder in the water from the newly melted ice, is the heat flow divided by the temperature, 273 K (0°C).

during the reversible melting. For reversible processes, the change in system entropy is the reverse of the change in the entropy of the surroundings. It turns out that for *all reversible* processes, the heat flow divided by the temperature is a measure of the system entropy change:

$$\Delta S = \frac{Q_{rev}}{T}.$$

For an irreversible process between the same two states, ΔS is the same but the heat flow divided by the temperature is less than the change in entropy of the system. Irreversible processes involve friction and other similar phenomena; they are, therefore, more realistic models of our experience. Although reversible processes are ideal, and not achievable in our real world, they are nevertheless useful models to guide our thinking because they are simpler.

Entropy and the Second Law

For open systems like the cell, a process is usually accompanied by a change in the entropy of both the system and its surroundings. The second law thus states: "During a spontaneous change, the sum of the change in entropy of the system and the change in entropy of the surroundings is always positive." Generalizing this statement leads to the ideas that the entropy of the universe always increases and that our ultimate fate would be an "entropic death" of maximum disorganization. Increase in entropy is the directional arrow which seems to dominate the historical development of our universe as we know it at the present time.

At this point the basic question becomes: are living things exempt from the second law? Organisms after all are constantly synthesizing large, complex molecules from simpler components—a process that appears to represent a decrease in entropy. More generally, if the entropy of a system plus that of its surroundings must always increase, how can one explain the maintenance of order in living systems, much less the increase of order during the development of the fertilized egg to the mature adult? Furthermore, how can one reconcile the second law with the evolution of life from the simplest organic molecules to man?

The way biological *systems* like the cell can maintain and even increase their organization is by decreasing even more the organization of their *surroundings.* The second law does not forbid a local decrease in entropy, provided it is accompanied by an equal or larger increase in entropy elsewhere.

To get a feeling for the way in which cells are able to purchase local order at the expense of greater disorder in their environment, let us consider a simplified example. Yeast cells can grow in an aerated broth containing only very simple chemicals: glucose, ammonia, phosphate, and sulfate ions, other ions (some in very small amounts), and water. This disorganized system gives rise to substantial quantities of complicated, highly organized yeast cells (Figure 1–8). Is this a violation of the second law?

The answer can be seen in part by considering the fate of carbon atoms during growth of the yeast. Well-aerated yeast cells growing in such a simple broth will utilize 1400 g of glucose carbon atoms to produce 1000 g of carbon atoms in new yeast cells. When one searches for the missing 400 g of carbon, one finds essentially all of it in the carbon dioxide gas given off during growth. This reveals part of the solution to the thermodynamic mystery: the entropy of some glucose carbons was diminished in forming the complex organization of carbon atoms in yeast cell molecules, but this was overbalanced by a greater entropic increase in forming many very simple molecules of CO_2. Moreover, the carbon dioxide is formed from glucose as part of an oxidation reaction which is accompanied by the release of large amounts of heat. The flow of heat from a system to its surroundings causes an increase in the entropy of the surroundings.

A complete quantitative account of this example, of course, must include all the other atoms involved, but the main point is clear with even a qualitative description. Many of the "simple" glucose molecules are broken down into carbon dioxide molecules of even greater simplicity and number, with the release of heat to the environment. The overall increase in entropy that this involves is more than enough to compensate for the local

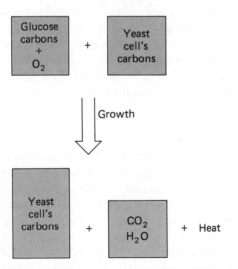

FIGURE 1–8
Flow of Energy Through Culture of Growing Yeast.
The entropy of the yeast cells is smaller than that of the glucose of which they are largely made, but the increase in the entropy of the surroundings more than compensates. This is because there is heat flow into the surroundings and some of the glucose is converted into less ordered CO_2.

decrease in entropy represented by changing glucose into yeast cell molecules. In fact, the force of the second law is such that one ought always to search, when confronted by a local apparent decrease in entropy, for a larger system, a system in which the total entropy always increases.

Free Energy Change (ΔG) and Spontaneity

The inevitable increase in entropy predicted by the second law is true in general for all spontaneous processes. In practice, however, using entropy changes to determine whether a specific process—such as a chemical reaction in a cell—is spontaneous has a number of drawbacks. For one thing, it is necessary to evaluate the entropy change of the system *and* its surroundings—in the worst case, of the entire universe. The change of entropy of the universe is, not surprisingly, difficult to measure directly. Moreover, we often want to know not only whether a given cellular reaction will take place spontaneously, but also *how far* it will go; that is, at what point it will reach equilibrium. Finally, since it is the work obtained from spontaneous chemical processes that powers all life, we would very much like to be able to predict *how much work* a given process can yield.

All these questions can be answered, under conditions of constant temperature and pressure, with the aid of another thermodynamic function, called **free energy** and symbolized by **G.** Since most cellular processes occur under just such conditions, the free energy change (ΔG) is a valuable measure both of the spontaneity of a process and of its potential value to a living system. Moreover, the free energy change has the great advantage of being able to predict the direction of a process solely in terms of the state of the *system* being studied; there is no need to attempt to measure the entropy of the universe or, indeed, any property of the surroundings.

Energy and Enthalpy

Although a rigorous account of the free energy function is beyond our scope, we can get an intuitive idea of why it has these very useful properties. Let us start by imagining a process (such as a chemical reaction) taking place at constant pressure, and use the first law to do a little energy bookkeeping. In general, there will be a change in the internal energy of the system, and a flow of heat into or out of the system. The *heat flow* associated with a reaction *at constant pressure* is called the change in **enthalpy** (**ΔH**). When ΔH is positive, heat flows into the system and the process is called *endothermic;* when ΔH is negative, heat is given off and the process is called *exothermic.* The change in enthalpy is a state function; it depends only on the initial and final states of the system and not on the path taken between the states. The enthalpy change associated with a process can be measured in a calorimeter, which is simply a container so well insulated from its surroundings that the heat produced or absorbed can be directly computed from the temperature change and the heat capacity of the container and the system (Figure 1–9). When used to measure ΔH, the calorimeter is constructed in such a way that the reaction takes place at constant pressure—generally atmospheric pressure, as in most living systems. In many chemical and biological processes of interest, the only way that work can be done by a system on its surroundings at constant pressure is by pushing against them during a volume increase. During most chemical and biological processes there is almost no change in volume, so the work done, W, is essentially zero and the heat flow alone is a good estimate of the internal energy change of the system: $\Delta U \cong \Delta H$.

Now we might expect that if a reaction is exothermic it will necessarily be spontaneous, since the flow of heat from the system, as we asserted earlier, will increase the entropy of the surroundings by Q/T. We might also expect that the enthalpy change represents energy that the cell could use to do work, if it could harness it by some appropriate means. In fact, neither of these expectations is completely correct. Some processes that involve the release of heat—exothermic processes—are not spontaneous, while some endothermic processes *are* spontaneous. When one dissolves urea in water, for instance, one finds that the temperature of the solution *decreases*, so that heat flows into the system from its surroundings ($\Delta H_{sys} > 0$). This seems puzzling, for one would not ordinarily expect to find a system gaining energy, any more than one would expect to find a ball falling up. Nor can we assume

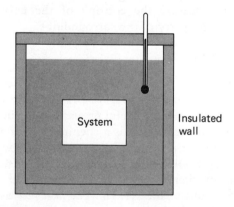

FIGURE 1–9

A Calorimeter for Measuring Heat Flow Between a System and Its Surroundings. Heat flows through the walls containing the system and causes a change in the temperature of the surrounding fluid. If the volume of the fluid is great the change in temperature is very small and the heat flow occurs practically at constant temperature.

that the amount of energy available for the performance of cellular work is always equal to the enthalpy change; sometimes it is greater, and sometimes less. Indeed, it may even be zero.

The reason the enthalpy change does not tell the whole story is that for a process to be spontaneous, the *total* entropy change—ΔS for the system *plus* ΔS for the surroundings—must be positive. It is true that any flow of heat from the system to its surroundings increases the entropy of the surroundings, so that ΔS for the surroundings is positive. But there may also be entropy changes within the system itself—changes stemming from differences between the product and reactant molecules in number, physical state, structure, and so on. Such system entropy changes may be positive or negative, depending on the particular chemical reactions involved, and they must be taken into account in any calculation of the total entropy change.

Suppose, for example, that a chemical reaction in a cell is accompanied by a negative system entropy change—that is, the disorder of the system *decreases* as a result of the reaction. (This would typically be the case, for example, if a large, complex, cellular molecule were synthesized from smaller, simpler precursors.) Such a reaction can still be spontaneous, but only if it releases enough heat to produce an even greater increase in the entropy of its surroundings. In other words, some of the energy change of the reaction must flow from the system as heat, and so is *not* available to perform useful work.

Using our earlier formulation of the relationship between entropy change and heat flow, we can easily calculate how much energy must be "lost" as heat in order for the entropy increase of the surroundings to offset the entropy decrease of the system. Let us designate the decrease of entropy within the system as $-\Delta S$. In order for the entropy of the surroundings to increase by ΔS, Equation 1–1 tells us that the heat flow from the system to the surroundings $(-Q)$ must be equal to $-T\Delta S$:

$$-Q = -T\Delta S. \qquad (1-1)$$

This equation defines the equilibrium condition, in which the net entropy change of the universe is zero. For a process to be spontaneous, the net entropy of the system plus its surroundings must increase; this requires that the heat flow from the system, $-Q$, be greater (more negative) than $-T\Delta S$.

We have referred to this energy as "lost" because it *must* flow from the system as heat, and so is not available for the performance of useful work. It also is energy that must be expended to pay for an increase of order within the system, by decreasing the orderliness of the rest of the universe. We can therefore think of it as representing the "cost," in energy, of creating local order.

Although we shall not attempt to give a rigorous derivation, it turns out that by extending and formalizing the kind of analysis presented above it is possible to define a state function that represents the balance of the energy change: the amount of energy that *can* be taken from the system in the form of work. Under conditions of *constant temperature and pressure,* this *free energy change* of a process, ΔG, is given by the expression

$$\Delta G = \Delta H - T\Delta S. \qquad (1–2)$$

Note that of the four quantities in this equation, only one—the temperature T—is measured directly on an absolute scale. The enthalpy, entropy, and free energy are all measured only as changes between two states of the system. The inverse of the enthalpy change, $-\Delta H$, represents the amount of energy that the reaction yields in the form of heat under conditions of constant pressure. The $-T\Delta S$ term, when it is negative, tells us how much of this energy *must* in fact flow from the system as heat, and so is not available to do useful work. The free energy change is then that portion of the enthalpy change which *need not* leave the system as heat. This energy is "free" in that it can be used to perform work for the cell (e.g., to synthesize a molecule or to transport an ion across a membrane) provided, of course, that the cell has some means of transducing this energy into a usable form. (How cells do this is the subject of Chapters 6, 7, and 8.)

These ideas can be emphasized by recasting Equation 1–2 to show how it is related to our statement of the second law. Dividing by $-T$:

$$-\frac{\Delta G}{T} = -\frac{\Delta H}{T} + \Delta S.$$

The last term on the right is the increase in the entropy of the system. The first term on the right measures the increase in the entropy of the surroundings, since it is the heat flow out of the system $(-\Delta H)$ under our conditions of constant pressure, divided by the constant temperature. The term on the left is thus the total entropy change, which must for a spontaneous process be positive, $-\Delta G/T > 0$, as stated by the second law. Thus, $\Delta G < 0$ also defines a spontaneous process.

Properties and Uses of ΔG

The free energy change that accompanies a chemical reaction can tell us many useful things about that reaction.

- If ΔG is negative, the process can occur spontaneously.
- If ΔG is positive, the process is nonspontaneous and

cannot occur without an input of energy; the *reverse* process, however, will then be spontaneous.
- If $\Delta G = 0$, the free energy of the reacting system has attained its minimum, and the system is at *equilibrium.*
- For a process with a negative value of ΔG, the *magnitude* of the free energy change indicates the theoretical *maximum amount of work* that can be obtained from the process. The farther the system is from equilibrium, the more work can be obtained from it. (This assumes the process is carried out reversibly. In the real world, the actual amount of work will always be somewhat less than this value.) Conversely, the magnitude of a positive ΔG indicates the amount of energy that must be invested to make this process take place.

These properties of the free energy change, together with Equation 1–2, enable us to understand some of the puzzling phenomena mentioned earlier. Consider, for example, an endothermic process—one that absorbs heat, so that ΔH is positive. You can see from Equation 1–2 that it is still possible for ΔG to be negative; that is, for free energy to be released and the process to be spontaneous. All that is needed is that the $T\Delta S$ term be large enough. In other words, the process must involve a sufficiently great increase in the entropy of the system. Such a process (melting ice at 1°C, for example, or dissolving urea in water) will proceed spontaneously even while absorbing heat from its environment.

Now suppose that a chemical reaction yields products with an entropy greater than that of the reactants (ΔS is positive) so that the disorder of the system increases. Equation 1–2 tells us that in such a case, the term $-T\Delta S$ is negative, so ΔG is more negative than ΔH. This means that the increase in the entropy of the system contributes to the energy available for the performance of work. This is actually the case when a cell oxidizes glucose for energy. Under certain conditions, the free energy change of -686 units/mole is composed of -670 units/mole (98%) contributed by the change in enthalpy and of -16 units/mole (2%) contributed by the $-T\Delta S$ term. This last can be attributed to the greater entropy of carbon dioxide and water compared with glucose and oxygen. (One can see that the heat released to the environment contributes much more to the free energy than does the increase in entropy of the system. This is typical of cellular reactions.)

In summary, then, Equation 1–2 tells us that

1. Energy must be used to purchase increased organization in the system.
2. A decrease in organization of a system can provide usable energy.

Life and the Steady State

A labile system like the cell, which is maintained at an apparently constant level of organization by a continuous supply of free energy, is often described as being in a **steady state.** It does not represent an equilibrium state in which the system has achieved the lowest possible internal energy and the highest possible disorganization. A nonliving example of a steady state is the surface of a calmly flowing river (Figure 1–A). The energy input is provided by the difference in hydrostatic pressure between an upstream location and one downstream. The characteristic and relatively unchanging appearance depends on a flux of energy and materials moving in the river. A pond could be imagined to have the same appearance but the pond in equilibrium can easily be distinguished from the river in a steady state; one has only to mount a waterwheel to tell the critical difference. No work can be extracted from a system in equilibrium, but work can be obtained from a system in a steady state by harnessing the source of energy and materials that flow through the system.

The equations that model the laws of thermodynamics are not, strictly speaking, applicable to nonequilibrium systems like the flowing river or like the cell. Instead, the mathematically more complex model of *nonequilibrium thermodynamics,* ap-

FIGURE 1–A
Equilibrium and the Steady State.   Various parameters are constant in both these states, as are the levels of water in the pond and the mill stream. Parameters remain constant in an equilibrium because the forces affecting the values of the parameters are balanced; in the pond, loss by evaporation is balanced with gain by condensation. Parameters remain constant in a steady state because there is a flow of mass and materials through the system; in the placid stream, there is an equal flow into and out of the region with constant water level. The difference can be seen by trying to extract work from systems in the two kinds of states. A waterwheel will not turn if it is immersed in the pond but will be driven by the flow of water in the mill stream.

plicable to steady-state phenomena, has been developed. Luckily, *equilibrium is often a useful approximate model of the actual steady state.* For some phenomena in cell biology—enzyme kinetics, for example—the steady state is traditionally used as a model description.

Where Does the Energy Come From?

We can now return to our original question—How do living things increase organization without violating the second law?—and answer it in a more concrete way. As we have seen, the second law permits a process to be spontaneous, even though it produces a local increase in organization, so long as the $-T\Delta S$ is "paid for" by a sufficiently large decrease in enthalpy, $-\Delta H$. This is indeed what must occur when large, highly organized **macromolecules** are synthesized from small **monomers.** It's the release of internal energy (as ΔH) that does the job. It has to be sufficiently negative (released) not only to outweigh $\Delta S < 0$, but also to force $\Delta G < 0$, so the process is spontaneous. The way in which the cell manages to perform this trick is to build monomers with high internal energy and then release a great deal of it in the course of the polymerization reaction. Such monomers can be spontaneously ($\Delta G < 0$) knitted together into a **polymer** even though this may entail an overall decrease in entropy, since the decrease in enthalpy is large enough to "pay" for it.

Now it is necessary to ask, where does the high internal energy of these monomers come from? One way that cells activate (increase the energy of) reactant molecules is by attaching phosphate groups to them (Chapter 6); when these groups are subsequently split off, the energy needed to drive the polymerization reaction is released. The synthesis of **high-energy phosphate** compounds, in turn, must be driven by some other energy-yielding process, usually an **oxidation** reaction, such as the oxidation of glucose to CO_2 and H_2O (Chapter 8). If one follows the trail of these energy transductions far enough, one eventually discovers a source of energy which is external to the cell. In the case of animals it is chemical free energy in the foods eaten. These energy-rich foods are all ultimately derived from green plants, which synthesize them by capturing radiant energy (Chapter 8) emanating from the sun, produced by the nuclear fusion reactions occurring in it.

Here we have the clue to the mystery of the apparent disregard of the second law, which G. N. Lewis and others thought to be characteristic of living matter. Because of the apparently autonomous behavior of living matter, they were led to think of it as being autonomous also from the point of view of energy when, in fact, nothing could be further from the truth. *Living matter is capable of preserving its apparently highly improbable individuality only at the expense of large amounts of free energy extracted from the environment and by releasing large amounts of heat to the environment.* As soon as the supply of free energy is cut off, living systems proceed spontaneously to a greater state of disorganization (equilibrium), which means death.

The laws of thermodynamics for equilibrium states concern themselves only with the *initial* and *final* states of a process, not with the particular path taken in traveling between these states. The second law predicts whether a certain reaction *could* occur spontaneously; that is, without input of energy from its surroundings. The second law does not, however, predict that a reaction involving a decrease in free energy *will* in fact occur at a measurable rate; there may be no reaction pathway available. Despite the celebrated high internal energy of the compound ATP and the large negative free energy change that would occur if ATP were simply hydrolyzed, this never happens in cells. There are no enzymes to catalyze the simple hydrolysis of ATP and the uncatalyzed rate is very low. There are, however, enzymes which catalyze the release of ATP's high internal energy in reactions other than

simple hydrolysis; we shall learn more about them in Chapter 6.

In summary, the second law says that the entropy of an isolated system never decreases. If one observes a decrease in the entropy of a system (as often is the case with biological systems), then the system is not isolated; instead, it is a portion of a larger system and the $T\Delta S$ term must be outweighed by a larger negative ΔH term. When all the energy sources are included in the system, the laws of thermodynamics are obeyed.

A constant supply of free energy is only one requirement for the maintenance of a steady state. There must also be a *structural organization* capable of absorbing and channeling the energy in a usable manner. During biological evolution many such organizations have arisen, each capable of extracting free energy from an extremely varied environment. As evolution has proceeded, the overall organization of the biosphere has become increasingly complex, and the amount of energy utilized has continually increased. This also should not be viewed as a violation of the second law. The machinery of natural selection, utilizing the phenomena of reproduction, sexuality, inherited change, and differential survival, provides a means to direct the plentiful supply of free energy from the sun into organizations of increasing complexity.

Cell Function — The Dynamic Expression of Cell Organization

The cell *utilizes energy to maintain and extend its organization.* And the cell also *has organized structure so that it can utilize energy.* These two properties of the cell are inextricably interrelated, as will be revealed in detail in the later chapters of this book. Treating them separately, however, allows us to stress their different conceptual bases. These concepts can then be merged into a more realistic view of the cell. In this section, we shall show how cell function can be described in terms of (1) the flow and transduction of **mass,** (2) the flow and transduction of *energy,* and (3) the flow and transduction of *information.* In the next chapter we shall describe the structure of the cell which acts as a container for these flows and transductions.

It is important to bear in mind that the flow of mass and energy on the one hand and of information on the other hand are also closely interrelated, as indeed they must be in any system capable of transducing energy in a controlled manner. Take as an example of all these interrelations a simple thermoregulated water bath designed to maintain its temperature several degrees above the temperature of its surroundings (Figure 1–10). This device is analogous to a living system in the sense that it maintains constant a given characteristic (its own temper-

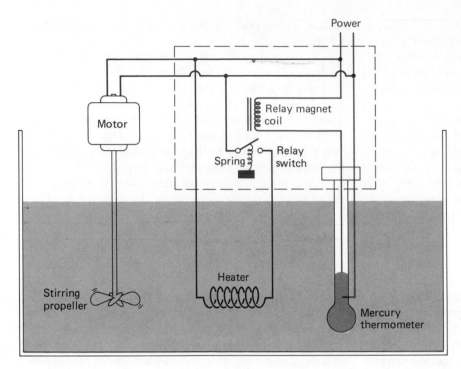

FIGURE **1–10**
A Thermoregulated Water Bath. The power supply furnishes energy for the heater, the relay, and for stirring the bath. All three energy-requiring processes are necessary to ensure that all parts of the bath are at the same temperature. As heat is lost to the environment, the level of mercury in the thermometer falls and the electrical current passing through the mercury is interrupted, which in turn interrupts the current through the relay coil. Without current the relay switch closes and allows current to flow through the heater. As the temperature rises again, the relay coil circuit is closed and the relay switch is opened, turning off the heater. This device will maintain a constant water bath temperature so long as heat is lost to a colder environment.

ature) in a variable environment (ambient temperature). Furthermore, like living systems, it maintains this steady-state temperature by utilizing an external source of energy. To do so, however, it must have a structural organization which can

1. transduce electrical energy into heat (the heater)
2. sense temperature (the mercury thermometer)
3. transmit a signal to the heater (the mercury contact switch, wires, and heater relay)
4. ensure that the temperature registered by the sensor is always the same as that which exists throughout the bath (the stirrer)

The flow of energy to the heater is accomplished by a flow of electrons along wires. The flow of information concerning temperature in the bath includes a flow of mercury and of electrons. As we shall see later, flows of energy and information in cells are also carried by flows of material substances, namely special molecules. It should be stressed that in this system energy is utilized not only to *establish* a temperature above ambient, but also to *regulate* the flow of energy so that the temperature can be maintained at a steady state. The thermoregulated water bath is an example of the use of energy to achieve *negative* **feedback** *control*. As energy flows from the heater, the temperature of the water rises. Information about the water temperature is fed back to the heater in the form of electrical energy by means of the thermometer contact, wires, and heater relay. Since continued

output of the heater leads to an inhibition of further heating, the feedback is said to be negative. There are numerous examples of negative feedback control in later chapters of this text, which discuss its use in controlling the flows of mass, energy, and information.

The Flow of Mass

A striking characteristic of many living systems is that they grow. Since living systems obey the *law of conservation of mass* and cannot achieve their growth in a vacuum, this requires an influx of material, which we call food. Even cells that are apparently not growing experience a flow of mass; food is taken in to replace cellular material that is constantly broken down and excreted. The food is not generally in the same form in which mass is found in cells. There must be *transductions* of the mass taken in, by processes we call *metabolic chemistry*, to maintain and increase cellular material.

All cell materials are composed of chemical elements, and the atoms of elements are not altered during metabolism by living systems. Thus, there is a law of conservation of elements that is applicable to cells. The most important element in the cell is carbon. It can enter a cell in various molecular forms to be used for growth. One form is organic molecules, produced by other living systems, in which the carbon is relatively reduced. Another form is carbon dioxide. This latter is completely adequate for the photoautotrophs because the energy necessary to convert the oxidized CO_2 into more reduced

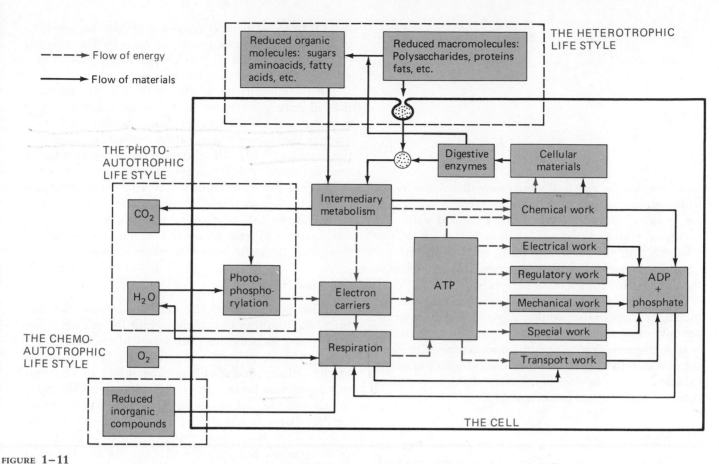

FIGURE **1–11**
The Flows of Mass and Energy in Cells. Only the most important pathways are
shown. Not illustrated, but quite important, are the outflow of heat in all energy
transduction steps, the production of ATP by fermentation during intermediary
metabolism, the material input for the synthesis of ATP (namely, ADP and inorganic
phosphate), and the intermediary metabolism and respiration of chemoautotrophs. The
flow of mass with energy is actually intermingled at every step except the absorption of
light energy by photoautotrophs. The color arrows illustrate the most obvious
transductions where chemical energy moves through the cell.

cellular compounds is supplied by light. The bacteria
called chemoautotrophs also use CO_2 as their sole carbon
source; the energy required to reduce the carbon comes
from the oxidation of inorganic molecules.

The **reduction, or fixation, of CO_2 not only
requires a source of energy but also a source of electrons.**
Green plant photoautotrophs obtain these electrons from
water, producing molecular oxygen as a by-product. The
electron source varies among the chemoautotrophs.
Purple sulfur bacteria, for example, use hydrogen sulfide
and excrete elemental sulfur.

Most of the other required elements (H, O, N, P,
etc.) are obtained by heterotrophic cells in the form of
organic molecules such as sugars, amino acids, nucleic
acids, etc. (Figure 1–11). Autotrophs can get their oxygen
from CO_2. There are specialized procaryotic cells—the
nitrogen-fixing bacteria that live free in the soil or in the

root nodules of legumes—that can convert molecular
nitrogen gas into ammonia, which can be used in the
synthesis of various biologically important substances.
Other bacteria and all plants can reduce sulfate ions to fix
sulfur into biological compounds. By such means a
continued supply of the major elements required for life
is assured.

Even though the law of conservation of mass
requires only that elements be supplied for growth, most
foods are not elements, but molecules. This is necessary
because the chemistry of cells, while impressive in its
versatility, cannot perform every imaginable transduction of mass. Those molecules which the cell cannot
synthesize must be supplied from outside. Indeed, cells
often have strict requirements for certain molecules, even
though they contain only elements already present in
excess. Among the most famous of these required

Thiamine (vitamin B₁)

molecules are the *vitamins,* which were discovered by the pathological consequences to living systems when they are omitted from the diet. Many animal cells, for example, require thiamine (vitamin B₁). Lack of this relatively simple substance leads in humans to the condition known as beriberi, a serious health problem in the Far East for over a thousand years.

The details of some of the requirements and the metabolic transformations that allow the synthesis of the panoply of biological molecular forms will be presented in later chapters. For the present, students should remember that *all* cells obey the law of conservation of mass and exhibit a continuing flow of materials as a consequence of their growth and maintenance activities.

The Flow of Energy

All cells obey the *law of conservation of energy* as well. As mentioned above, growth and maintenance of cell function requires a continuous input of energy to drive the material transformations which, in its absence, would not be spontaneous. While many different forms of energy emerge from cells, we know of *only two forms of energy that are acceptable as input: radiant energy and the chemical energy contained in reduced molecules.* Most of the energy for metabolism comes from the strongly spontaneous oxidations of reduced foods. In fact, one of the first consequences of the useful absorption of a photon of light by a phototroph is the production of a reduced, intracellular, organic molecule. It is the subsequent oxidation of this molecule that provides the energy for the metabolism of photoautotrophs. Luckily for the student of cell biology, the successive oxidations of reduced organic molecules by autotrophs and heterotrophs share many similarities.

All the heterotrophs have their energy input in the same molecular form as their mass input: reduced food molecules such as sugars, amino acids, etc. (Figure 1–11). The flows of mass and energy in those cases are apparently indistinguishable. But they are different; *the food molecule destined to be an energy source does not enter the material substance of the cell.* For example, a glucose molecule entering the cell's energy metabolism is oxidized to yield up its chemical energy, and the products (CO_2 and H_2O)

are excreted. The energy obtained can be used to incorporate other glucose molecules into the material structures of the cell (as part of a glycogen granule or a cellulose fiber, for example).

Although it is the oxidation of reduced molecules that ultimately energizes the biological system, cells funnel much of the energy from various oxidations through an intermediate carrier molecule, adenosine triphosphate (ATP). The energy in this ubiquitous fuel is used to "drive" many energy-requiring processes of the cell. These processes usually represent a *transduction* of one form of energy into another, or in many cases, a change of one kind of chemical energy into another. We can also think of these processes as occuring in specialized machines, the useful output of which is *work.*

Chemical Work

The *metabolism* of the cell, its set of chemical reactions, furnishes two things which the cell needs for maintaining its activities. The subset of reactions called *anabolism* builds up *materials* from which the components of the cell are manufactured. The subset of reactions called *catabolism* breaks down molecules to provide *energy* which is used in all the energy-transducing processes of the cell. These subsets overlap partially, since a metabolic reaction may both break down a particular molecule, releasing usable energy, and yield a product that is used as part of cellular material.

Figure 1–11 indicates one process that is capable of using the energy from ATP to perform chemical work. The cell actually contains many such processes, many of which are understood in detail. The cell performs chemical work whenever it utilizes energy (often derived from ATP) to synthesize compounds that contain more energy than the building blocks from which these compounds were constructed—for example, the synthesis of the complex cholesterol molecule from the simple precursor acetyl-Co A. The cell has biosynthetic mechanisms for nucleic acids, proteins, fats, polysaccharides, and so on, and each of these includes the chemical means of transducing some of the energy of ATP, or some other energy storage compound, into chemical work.

Transport Work

Living systems differ greatly from their environment in both the relative abundances and absolute concentrations of their component materials. To bring about and maintain this situation, the cell must perform **active transport** work against concentration gradients. That is, it must perform work both to *accumulate* materials (e.g., molecules and ions) found in low concentrations in its environment, as well as to *eliminate* other materials found in high concentrations in its surrounding medium. Not only do accumulation and elimination themselves require work, but also the *maintenance* of this thermodynamically improbable situation requires work. The amount of energy required to maintain this distribution of materials is reduced by the presence of the plasma membrane, which represents an extremely efficient (though not perfect) barrier to diffusion, thus restricting severely the movement of water-soluble molecules and ions in to and out of the cell.

Eucaryotic cells must perform additional transport work. They must establish and maintain different concentrations of many molecules and ions in the various compartments created by their interior membranes. Some of the energy for transport and electrical work arises directly from the respiratory metabolism of procaryotic cells and the mitochondria of eucaryotes. The other major energy source is, as indicated in Figure 1–11, the ubiquitous ATP.

Electrical Work

The plasma membrane is never equally permeable to all of the various cations and anions found in cells and their environments. Therefore, once differences of concentrations have been established by transport work, flows of positive and negative ions in to and out of cells are never exactly equal. For this reason all cells exhibit a slight separation of electrostatic charge across their membrane, usually leaving the inside negative with respect to the outside. Nerve cells are specialized to permit a rapid and transient change of this charge separation in response to certain stimuli. The result is an electrical impulse, known as the *action potential,* that can be utilized for rapid communication over distances as great as a meter or more. The separation of charges across the membrane requires energy (transport work) to be established and releases energy when abolished. In the case of the electric eel, very large amounts of energy are generated by a special organ, the *electroplax,* consisting of thousands of cells stacked in a column. This energy is released suddenly (up to 1/8 amperes at more than 500 volts) to kill or stun prey.

Mechanical Work

All of life is characterized by motion of some sort, whether it is the contraction of muscle cells, the beat of cilia, the flowing movement of the amoeba, the cyclosis (circling movement) of the cytoplasm in a plant cell, or the separation of chromosomes during mitosis (see Chapter 16). Motion requires energy funneled into processes that can convert chemical energy into mechanical work. Even certain viruses exhibit a form of motion, as does T4 bacteriophage, for example, when it injects its DNA into a bacterial cell (see Chapter 15).

Regulatory Work

The cell must incorporate energy during the biosynthesis of complex, highly organized macromolecules, but it must also regulate their interactions once they are synthesized. Regulation requires the input of energy because, in limiting the possible ways of arranging components of a cell, regulation reduces the cell's entropy. Furthermore, not only chemical work, but osmotic, electrical, and mechanical work must also be regulated. And all these forms of work must be regulated with respect to each other.

Special Forms of Work

There are a number of specialized forms of energy transduction which, although not present in all cells, may nevertheless be very important. Some microorganisms, and some macroorganisms (such as the firefly), carry out the process of converting chemical energy into light. This "cold light" (produced without high temperatures) represents a potentially very useful industrial application of a biological phenomenon. Cells also utilize energy while interacting with each other as parts of organisms. The activities of individual cells, when arranged in tissues and organs, are transduced into macroscopic motion, the generation of pressure, the flow of fluids, the production of biologically useful heat, etc.

From an evolutionary standpoint, perhaps the most significant recent development in the thermodynamics of life is that some of the organization of the cell is transferred outside its cellular limits. That is, living systems can utilize energy to organize parts of their surroundings. As an example, consider the establishment by a bird of an appropriate environment for incubating an egg: building a nest, providing the correct temperature, and warding off potential predators. This expression of biological evolution occurs extensively at the organismic and population levels and is most evident in human activities.

The Flow of Information

As discussed earlier, the growth of a cell or the development of an organism represents an increase in organization which is "paid for" by energy supplied to the cell from its surroundings. It is important to recognize, however, that not all of the organization is newly generated in the cell; some organization is maintained and transmitted from cell to cell. The substance that carries this organization, which we shall call the *hereditary material,* represents information that has slowly accumulated over billions of years during evolution. We shall define the hereditary material as those molecules or structures of the cell which store the information necessary for their own reproduction.

The hereditary material represents the starting point of the flow of information in the cell. There is by now little question that most of the hereditary information is stored by eucaryotic cells in their nuclei, and by procaryotic cells in their nucleoid regions, in the form of deoxyribonucleic acid (DNA). But other cellular structures, the chloroplasts and mitochondria, also contain DNA. Furthermore, certain viruses store their hereditary information in **ribonucleic acid** (**RNA**).

The question arises whether DNA and, in some cases, RNA represent the only structures that store hereditary information, or whether the cell possesses other levels of storage that pass biological information on from generation to generation. This important question is discussed in Chapter 15. In any case, the best understood system is the one that starts with DNA. The elucidation of this system of information flow represents a triumph of modern molecular biology. Its elegance and beauty is a tribute to the cell which over billions of years of evolution has become the magnificent thing that it is. The basic facts about DNA are:

- It has a *stable structure* that is ideally suited to the storage of information.

- It has a *linear structure* so that the information included in it is arranged in a linear sequence.

- It is capable, in the presence of the appropriate enzymes, of extremely accurate *self-duplication (replication)*.

- It lies at the beginning of a process which culminates in the synthesis of proteins, both catalytic and structural. The chemical properties of proteins result from their three-dimensional organization which in turn derives from the linear amino acid sequence that is specified by the DNA. Thus, the cell exhibits a flow of information in which there is a transformation from *linear to three-dimensional organization*.

Finally, it must be emphasized that DNA is also capable of *responding* to information. DNA is not an absolute beginning of information flow; the activity of DNA can be regulated by messages reaching it from other parts of the cell and from outside the cell. This complexity would not have been surprising to the Taoist philosopher Chuang Tzu, who saw the world as a harmonious interplay of its components rather than as a rigid, hierarchical array of subordinate sets (see epigraph to this text).

The process of information flow from DNA to protein occurs over several stages, and each stage is regulated by "feedback loops" and other regulatory messages from the cell. As Figure 1–12 shows, the information in DNA is transferred to messenger RNA by the mechanism of linear *transcription*, often followed by various modification processes, all of which can be regulated by the cell. This is followed by *translation*, a regulated process in which the information carried by messenger RNA is translated from the linear nucleic acid code, based on four letters (the nucleotide bases), into the linear protein code, based on 20 letters (the amino acids). Each protein in the cell can arrange itself spontaneously into a unique *three-dimensional structure*, determined primarily by the unique linear sequence of its amino acids. This latter process generates a new property not encountered in the linear topology of the DNA and RNA: the highly specific chemical reactivity of individual proteins. It is this property that endows the cell with its uncanny ability to promote and regulate its chemical processes, endows the millions of biological species with their distinctness, and endows the billions of biological organisms with their individuality.

The three-dimensional arrangement that determines the chemical specificity of proteins is the most important result of the information flow from the DNA molecule. Once the protein has been synthesized and its three-dimensional structure generated, it can become an enzyme, catalyzing a chemical reaction, a structural protein to be incorporated as part of cellular architecture, a "repressor" that combines with the DNA to prevent it from being transcribed, or some other kind of regulatory protein, the existence of which we may as yet not even suspect. These many different kinds of proteins—3000 in a bacterium and probably as many as 50,000 in a human cell—function together in an integrated way. One of the purposes of molecular cell biology is to study the properties of these component pieces and to understand the mechanism of their interactions.

Other regulated processes illustrated in Figure 1–12 are: the replication of DNA, the transcription of DNA to form ribosomal RNA and transfer RNA (both used in protein synthesis), the spontaneous assembly of ribosomes (also used in protein synthesis) from proteins and RNA, and the *directed assembly* of cellular structures. (Directed assembly, discussed in Chapter 15, requires the pre-existing presence of a structure to make more of that structure; for example, short lengths of microtubules which can grow by addition of the protein tubulin.) Not illustrated are the details of the regulatory processes that occur during transcription and translation (see Chapters 11–14).

The above description of the flows of information applies to procaryotic cells. Eucaryotic cells are not yet as well understood. It is likely that they contain most of the procaryotic processes of information transfer as well as some distinctive eucaryotic mechanisms. As of now, there are known to be at least four additional regulatory mechanisms:

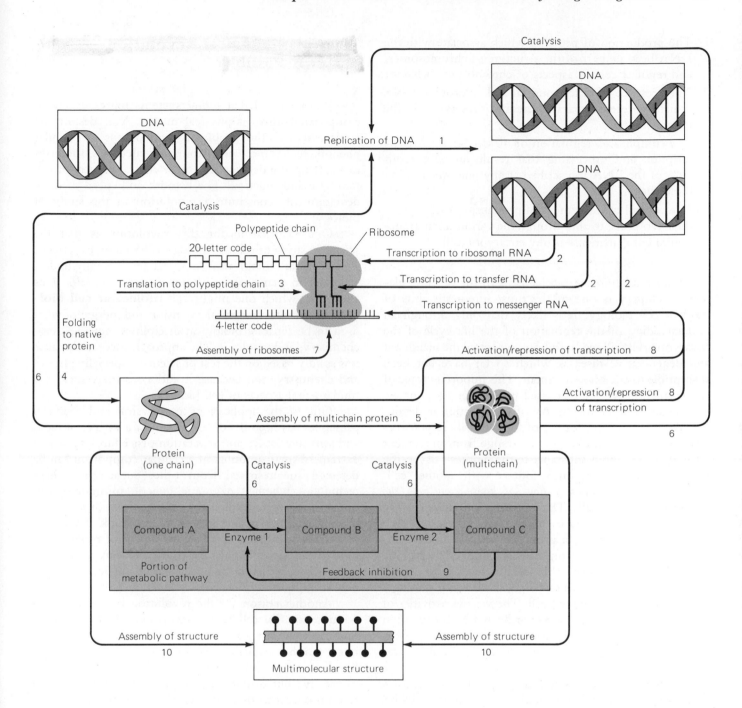

FIGURE **1–12**
The Flow of Information in the Cell. There are two major pathways of
information flow: (a) The replication of DNA, in which one molecule of DNA
provides the information for the synthesis of two molecules like the original (1),
and (b) the synthesis of proteins by transcription of the information from DNA to
RNA (2) and translation of the information in RNA into protein (3). Once the
three-dimensional structures of protein molecules have been formed (4), they can
participate in a number of regulatory processes: assembly of multichain proteins
(5), catalysis (6), assembly of ribosomes (7), activation and repression of
transcription (8), feedback inhibition of metabolic pathways (9), self-assembly and
directed assembly into multimolecular structures (10) and other processes discussed
elsewhere.

1. The production of proteins which associate with the DNA become part of the structure of chromosomes, and regulate certain aspects of chromosome behavior
2. The presence of mitochondrial and chloroplast DNA which represent separate hereditary reservoirs in the eucaryotic cell
3. The complicated regulation of messenger RNA transcription and processing that results in only about 10% of the RNA transcribed at any one time leaving the nucleus
4. The control of movement of proteins, during or after their synthesis, to and through the various membrane-bound compartments of the eucaryotic cell

Understanding the mechanisms of the cell's regulatory machinery is one of the most important aims of modern cell biology. It is likely that only through an understanding of the regulation of the life cycle of the eucaryotic cell will we learn how to control the malignant and degenerative diseases which so far have not been susceptible to effective treatment. This important topic of cell regulation will be discussed in Chapter 14, after we have elucidated the machinery of cell function in detail.

Another striking feature of the cell is its responsiveness to external stimuli. This can range from movement away from a noxious substance to the synthesis of specific proteins after coming into contact with another cell. Responsiveness implies a *flow of information from the environment to the cell*. This information flow usually crosses the cell surface membrane via a material substance—a hormone, a group of ions, or an environmental molecule, for example. Often, the cell membrane has on its outer surface a specific *receptor* to which environmental molecules can bind, the binding causing a further alteration inside the cell. The various activities of a multicellular organism are coordinated by information transfer between cells via hormones and neuronal conduction. These topics will be covered in detail in later chapters.

Although we have stressed independently the flows of mass, energy, and information, they almost always occur together. The information in DNA resides in the linear sequence of material substances, the nucleotide bases. The energy of a lump of sugar resides in the oxidation state of its component atoms. The biosynthesis of the amino acid alanine requires energy and information, as well as the carbon, hydrogen, nitrogen, and oxygen atoms that make up alanine's mass. The value of separating these flows for purposes of discussion is twofold. First, it reminds us that cells obey the laws of conservation of mass and of energy. Second, it provides a framework on which to build our knowledge about all biological processes, because they all require mass, energy, and information.

The Cell Biological Revolution—What Makes It Possible?

We have said that living systems represent a very complicated form of physical matter. Yet, despite the complexity of the problem, the analysis of cellular phenomena has met with astounding successes, successes in studying the deepest and most central aspects of cell structure and function. The breadth and speed of these developments constitute a revolution in the study of biology.

One explanation for this revolution is that the scientific study of cellular biology is based on experience showing that biological phenomena are generally understandable in terms of *physical and chemical principles*. This approach, which one might call **molecular cell biology,** represents the natural extension and merger of what used to be separate biological disciplines: genetics, biochemistry, and cytology. The approach succeeds because it is solidly based on the rest of science, especially physics and chemistry, and because it addresses the central and fundamental aspects of cell biology.

One of the applications of physical and chemical principles has been the development of a variety of rapid and sensitive assays and separation procedures by which extremely small amounts of a cellular component can be detected and resolved from other substances. These techniques, whether enzymatic, genetic, immunological, optical, chromatographic, electrophoretic, or radioisotopic, will be discussed throughout this book. They have played a critical role in the development of modern cell biology. In fact, the progress of the field may almost be measured in terms of the introduction and application of new techniques.

Another reason for the revolution has been a conscious effort by cell biologists to choose for intensive investigation only a few different kinds of cells from the huge number available. Focusing attention on simple living systems and ignoring, for the time being, the impressive but distracting diversity of biology allowed rapid progress to be made. Thus, the bacterium *Escherichia coli,* the bacterial viruses T4 and lambda, the budding yeast *Saccharomyces cerevisiae,* the human red blood cell, and the cells of rat liver became the objects of study of whole schools of scientists. Because they were working on the same cell or cell product, they could easily and productively exchange ideas, techniques, and data. The benefits of this style of investigation have become so well known that conferences on other areas of biology have sometimes discussed the desirability of choosing, for example, a "T4 of neurobiology." The risk in this approach was that structures and functions might have proven to be very different from cell to cell. But as it has turned out, fundamental features are shared largely in common and the knowledge gained from seemingly arbitrary choices of cells is broadly applicable. Of course, the "fundamental features" needed to be recognized as such. The ability to do that, to ask the right questions, also helped to ensure rapid advances. In retrospect, it seems almost obvious that elucidating the similarities among cells should have been the easiest place to begin an understanding. But the early attention of biologists was more drawn to the diversity than to the common features. The qualities of genius that we recognize in the leaders of modern cell biology include the intuition necessary to ask the right questions and the stamina necessary to study deeply a single cell.

There developed a synergism between the methodological advances and the questions asked. Investigating the common basic problems led to the necessity for new techniques which, when applied, revealed yet more interesting questions. Now we are in a phase of rapid expansion; spiraling out from a central core of knowledge, we can turn our attention to more specialized cell characteristics and more specialized cells as well as continue the detailed study of the "standard" cells and cell products.

A cell is a remarkable object. Not only is it composed of tens of thousands of different kinds of molecules, functioning in a volume of a few cubic micrometers, but some of these molecules occur in very few copies per cell. Predicting the behavior of molecules —in chemical reactions, for example—usually depends on there being so many of them that their average properties dominate and variation from the average is not great. This can not be expected when the number of molecules is small. But the bacterium *E. coli* carries all its hereditary information in a single giant molecule of DNA and the bacterium nevertheless has a very high probability of growing and dividing into two cells even when placed in a variety of quite different environments. An *E. coli* cell can grow and divide at 15°C and at 35°C, for example, even though temperature affects the rate of different chemical reactions differently. No wonder the physicist E. Schrödinger marveled, in his famous little book *What Is Life?*, that the cell behaves like a precise piece of Cartesian clockwork, even though its component pieces are frequently not present in sufficient numbers to expect from them statistically predictable behavior.

How the cell manages to perform this feat is becoming increasingly clear and is discussed in later chapters. What concerns us now, however, is not *how* this happens, but *that* it happens. What are the common features of these processes that account for their reproducibility? It is the *specificity* of cellular processes that gives rise to their precision. Cellular processes involve specific molecules interacting in specific ways with certain other specific molecules and no others. As we shall see, this specificity results from the evolution of complementary structures in molecules: in cellular conditions, only certain molecules interact because only those molecules can fit together sufficiently well to interact at all. The fortunate thing is that specificity also makes the phenomena particularly susceptible to analysis, by methods which utilize precisely the same specificity. In this sense, modern biology can use the cell to make it reveal its own secrets.

Historically, two aspects of biological specificity were exploited first in developing this powerful approach. They were the employment of gene mutation in the study of heredity and the employment of enzyme specificity in the study of metabolism. The specificity of the protein enzyme for its reactant **substrate,** for example, makes possible the detection of the catalytic activity of the enzyme when it is present in a complex mixture of other proteins. After all, it is just this property that in the cell allows a particular enzyme to catalyze the reactions of only certain molecules. The specificity of the catalytic activity allows one to follow the enzyme during its purification. In favorable cases, it is even possible to immobilize a reactant substrate and, by its strong specific interaction, also immobilize the enzyme; after washing away all other proteins, the enzyme can be released, thus purifying it in a single operation. The specificity of these gigantic enzyme molecules is in turn made possible by the cell's remarkably accurate synthesis of their structures, structures sufficiently precise to allow us to separate enzymes from one another, to crystallize and study them in great detail by **x-ray diffraction,** and to relate their structures to their functions.

SUMMARY

The impetus for the recent and dramatic progress in cell biology has resulted from the fruitful combination of three formerly separate approaches: genetic, biochemical, and cytological. But the success of modern cell biologists depends ultimately not on their breadth of perspective, or ingenuity, or technical virtuosity so much as on the immensely more impressive system of specific functional relationships in the cell itself! Understanding ourselves can only be complete when we understand in detail the operation of cells. Since organisms interact with one another and since all cells seem to share the central features of structure and function, we are led to study any and every cell from which more knowledge can readily be obtained. The successful study of cell biology proceeds from the solid base of chemistry and physics. The laws of conservation of mass and of energy imply that cell processes must require both mass and energy. The second law of thermodynamics implies that biological organiza-tion must also require information. Flows of mass, energy, and information characterize the structure and function of cells.

KEY WORDS

Active transport, autotroph, cell theory, chloroplast, chromosome, deoxyribonucleic acid (DNA), energy, energy transduction, enthalpy (H), entropy, enzyme, equilibrium, eucaryote, evolution, feedback, free energy (G) (Gibbs), heterotroph, high-energy phosphate, information, macromolecule, mass, Mendel's laws of heredity, mitochondrion, mitosis, mitotic apparatus, molecular cell biology, monomer, natural selection, organelle, oxidation, plasma membrane, polymer, procaryote, recombination, reduction, replication, ribonucleic acid (RNA), ribosome, segregation, solute, steady state, substrate, surroundings, system, first law of thermodynamics, second law of thermodynamics, transcription, translation, vacuole, virus, vitalism, x-ray diffraction

PROBLEMS

1. Your friend Pat has been asked to invest in a new genetic engineering company. They claim to have isolated different bacterial cell lines with unusual energy properties that make them particularly suit-able for animal food. Pat has consulted you for advice on the cells. After reading the prospectus you are to write Pat a brief note giving your opinion of the two cell lines described below.
    (a) Cell line X is supposed to have a very high food conversion ratio. For every 5000 units of energy in the food, there are 4000 units of energy in the resulting bacterial cells and 2000 units of energy given off as heat during the growth.
    (b) Cell line Y is supposed to have a better efficiency. For every 5000 units of energy in the food, there are 5000 units of energy in the resulting bacterial cells and no loss of energy as heat.

2. There have been recent reports of objects called "prions" which contain protein but no nucleic acid. Are these cells?

3. In 1976 Jeons and Jeons reported experiments in which they infected the large eucaryotic cells of *Amoeba proteus* with bacteria. After nine years of growth, the infected *Amoeba* cells could no longer survive without the bacteria. Is this a good model for the original formation of eucaryotic cells?

4. It is possible, at least in science fiction, to imagine the construction of an artificial cell that can metabolize and reproduce itself. Using the discussion of the definition of life in this chapter, discuss whether such an object satisfies the definition.

5. Consider the change of a system between two states. The origin state is methane (CH_4) plus oxygen at room temperature and pressure; the destination state is CO_2 plus water at room temperature and pressure: $CH_4 + 2 O_2 \rightleftarrows CO_2 + 2 H_2O$. Burning methane can occur in a sealed metal container so only heat can flow to the surroundings. Or the burning can occur in the cylinder of an internal combustion engine, in which case heat flows and the piston will move outward, doing work on the surroundings. Are the destination states the same in these two examples? Are the heat flows the same? Is your description consistent with the thermodynamic laws?

6. A mixture of ice (the system) and water (the sur-roundings) at atmospheric pressure and 10°C (283 K) provides an interesting example of a thermodynamic process. At a standard set of conditions the enthalpy change ΔH is +6.4 units/mole and the entropy change ΔS is + or − 0.023 units/mole·K. In which direction does the heat flow? Is the sign of the ΔS term positive or negative? What is the free energy change for this process? How can this be interpreted according to thermodynamic laws?

7. Stretch a rubber band and, after waiting a minute for its temperature to equilibrate with the surrounding air, let it contract while in gentle contact with your lips. What is the temperature of the rubber band during contraction, relative to your lips? If heat is flowing between your lips and the rubber band during the contraction, in which direction does the flow occur? What is ΔH for the contraction? What is ΔG for the contraction? What must be ΔS for the contraction?

8. Compare the earth orbiting around the sun at a steady rate and an automobile moving down the highway at constant speed. To which case would an equilibrium model be appropriate and to which a steady state?

9. What are the mass, energy, and information requirements for riding a bicycle?

10. A molecule of glucose synthesized by one cell in a multicellular organism may be utilized by another cell at a distant location. The movement of this glucose molecule could represent a flow of mass, or energy, or information. Give examples by which each kind of flow might be a reasonable description of this movement.

SELECTED READINGS

Chaisson, E. (1981) *Cosmic dawn— The origins of matter and life*. Boston: Little, Brown and Company.

Dickerson, R. E. (1978) Chemical evolution and the origin of life. *Scientific American* 239:62–78.

Dyer, B. D., and Obar, R., eds. (1985) *The origin of eukaryotic cells*. New York: Van Nostrand Reinhold Company.

Folsome, C. E., ed. (1979) *Life: Origin and evolution*. *Readings from* Scientific American. San Francisco: W. H. Freeman.

Fox, S. W., and Dose, K. (1977) *Molecular evolution and the origin of life*. New York: Marcel Dekker.

Gray, M. W. (1983) The bacterial ancestry of plastids and mitochondria. *Bioscience* 33:693–99.

Schopf, J. W. (1978) The evolution of the earliest cells. *Scientific American* 239:110–38.

Schrödinger, E. (1944) What is life? The physical aspect of the living cell. New York: Cambridge University Press.

Searcy, D. G., Stein, D. B., and Searcy, K. B. (1981) A mycoplasma-like archaebacterium possibly related to the nucleus and cytoplasm of eukaryotic cells. *Annals of the New York Academy of Science* 361:312–23.

Cell Structure—The Organizational Basis of Cell Function

2

The previous chapter emphasized that the cell uses energy to drive the processes that maintain its highly individual structure in the midst of an ever-changing environment. It is also possible, however, to look at this phenomenon from the other direction and say that, without an existing structure, energy could not be transduced to maintain, extend, and duplicate the living system. Thus, *energy is necessary to maintain and extend structure, but structure also is necessary to capture and utilize the energy.*

A dramatic demonstration of the importance of biological structure was provided by the experiments of A. I. Skoultchi and H. J. Morowitz, who cooled the eggs of the brine shrimp *Artemia* to temperatures below 2 K (−271°C) for varying periods of time. They showed that, upon rewarming, the hatch rate of the chilled eggs was the same as that of control eggs held at room temperature. At a temperature of 2 K biological structure would presumably remain intact, but all chemical processes would slow to a virtual halt. It therefore seems reasonable to conclude that appropriate structure is a sufficient condition for initiating biological function. In the case mentioned above, the brine shrimp structures necessary for the hatch rate were apparently not significantly altered by freezing and thawing. It would thus appear that living processes could be generated by putting together the proper structure. If so, the synthesis of life would then become "merely" a very complicated exercise in organic chemistry.

Just how complicated it would be is the subject of this chapter. Here we introduce the fine structure of the cell, structure that provides the organization necessary to capture and transform energy and that represents the locations of the processes associated with the living cell.

Cell Size and Shape

One of the early investigators of size and shape in biological systems was the British biologist D'Arcy Thompson, who pointed out in his 1941 book, *On Growth and Form,* the relations between biological structures. For example, if an elephant were twice as large in its linear dimensions, it would have eight times the original weight. To support that weight the leg bones would have to be increased in cross-sectional area by an equivalent amount which would require not twice but almost three times the original diameter. In spite of D'Arcy Thompson's widely read, interesting book, biologists on the whole have been slow in recognizing the significance of *absolute* dimensions. In the physical world, relations between objects are very much influenced by absolute dimensions, since structures ultimately depend for their stability on the forces holding molecules together. These forces are the same in a small object as in a larger one of the same molecular composition, which is why, as structures increase in size, new stabilizing principles are necessary to withstand the greater destabilizing forces. One can see this principle at work in man-made ropes: the single fiber of hemp cannot withstand the force of an ocean-going ship but a properly twisted cable of such strands can.

The world of life falls into a size range that from a general viewpoint is restricted to a narrow band of magnitudes. Figure 2–1 is a summary of the sizes of observed structures, from cosmic to atomic. It illustrates the fact that biological systems fall into a series of distinct but overlapping size ranges. For example, the largest cell is larger than the smallest mammal, the largest virus larger than the smallest cell, and the largest macromolecule larger than the smallest virus.

Since cells lie in a certain size range, it is possible that there are laws governing the extent of the range. Consider a cell as a unit of biological activity delimited by a differentially permeable membrane and capable of self-reproduction in a medium free of other living systems. The smallest free-living cells are found among the bacteria. The bacterium *Dialister pneumosintes,* for

instance, has dimensions of approximately $0.5 \times 0.5 \times 1.6 \ \mu m$ (see Table 2–1 for the units used in the measurement of cells and organelles). One can argue that this cell may be near the lower size limit for independent biological activity. Assuming a 75% water content, this amounts to 2.8×10^{-14} g of dry weight per cell. Table 2–2 shows the composition of this bacterium in terms of the major classes of compounds generally found in living systems.

As will be seen in Chapters 11 and 12, double-stranded DNA must have approximately 20 times the mass of the protein for which it provides information. Thus, if one does not consider other functions for DNA, the 6.5×10^8 **daltons (Da)** of *Dialister* DNA will carry the information for at the very most 0.33×10^8 Da of different protein species. If one assumes an average molecular weight for the proteins of 40,000, one obtains a maximum estimate of $(0.33 \times 10^8)/(4 \times 10^4)$ or approximately 800 different kinds of proteins in this cell. Ignoring for the moment the presence of structural proteins that do not perform enzymatic functions, there are likely to be in the cells of this bacterium no more than 800 separate enzymes that must catalyze an equal number of biochemical reactions. Over 1000 genes have been identified and close to 3000 different proteins detected in the larger and more complicated *Escherichia coli.* Anything very much smaller than the *Dialister* bacterium is likely to have insufficient DNA to make the enzymes required to catalyze the 500–1000 biochemical reactions that possibly constitute the minimum number necessary for the maintenance of independent life.

It is possible, of course, that somewhat smaller free-living organisms can exist. There are, for instance, a group of organisms called *mycoplasmas* or PPLO

TABLE 2–1
Definition of Units of Measure

Unit	Symbol	Conversion
Meter	m	1000 mm
Centimeter	cm	10 mm (10^{-2} m)
Millimeter	mm	1000 μm (10^{-3} m)
Micrometer	μm	1000 nm (10^{-6} m)
Nanometer	nm	1000 pm (10^{-9} m)
Angstrom units	Å	100 pm (10^{-10} m)
Picometer	pm	— (10^{-12} m)

Note: The shaded units are those used in the SI *(Système international d'unités).* Others are included because of historical importance and common usage.

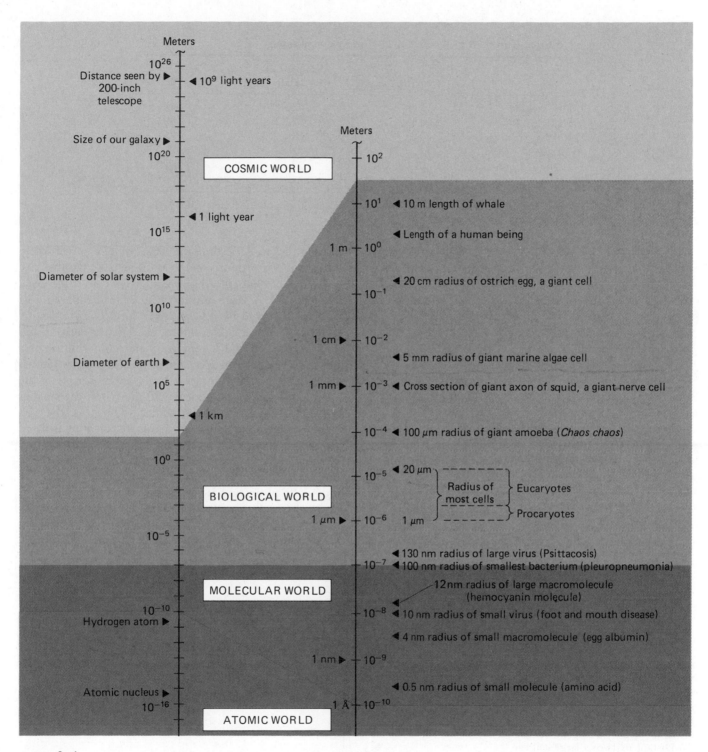

FIGURE **2–1**
Size Relationships at the Atomic, Molecular, Biological, and Cosmic Levels.

(pleuropneumonia-like organisms), some of which have a DNA content of only 5×10^8 Da and a dry weight of only 0.5×10^{-14} g. Perhaps even smaller independent organisms will eventually be discovered, but it is not likely that they will be *very* much smaller; a minimum would soon be reached below which there would be an insufficient diversity of enzymes to catalyze the reactions necessary for an independent cellular existence. Viruses, on the other hand, rely for their self-duplication processes largely on the biochemical versatility of their living hosts.

TABLE 2–2

Composition of the Bacterium *Dialister pneumosintes*

	$(\times 10^{-14}\ \text{g})$	Mass per Cell $(\times 10^{-4}\ \text{Daltons})$	Percent of Dry Weight
Dry weight	2.80	168	100
DNA	0.13	7.8	4.6
RNA	0.30	18	10.1
Protein	1.20	72	43.0
Carbohydrate	0.47	28	16.8
Lipid	0.61	37	21.8

Note: The dalton is a unit of mass nearly that of a hydrogen atom $(1.661 \times 10^{-24}\ \text{g})$. One gram is equal to 6.020×10^{23} daltons.

The amount of their DNA, therefore, tends to be much smaller. One of the largest viral DNAs (that of *Vaccinia*) is on the order of 1.6×10^8 Da, while the DNA of the bacterial virus λ is only 0.32×10^8 Da. Even smaller are the infectious agents known as plant viroids $(0.0013 \times 10^8$ Da), which are made up of RNA and may lack protein altogether.

The problems affecting the *upper* limit in cell size are very different (Figure 2–2). First, there is the relation between the nucleus and the rest of the cell. As we shall see later, the nucleus produces "messengers" that determine the synthesis of the cell's proteins. A given nucleus can control only a certain amount of cytoplasm. Some cells manage to transcend this limitation by having more than one nucleus. Examples are the multinucleate giant cells of the amoeba *Chaos chaos* (100 μm diameter) or the green alga *Nitella* (10,000 μm diameter). Second, there are the relationships between the various parts of the cell. The larger the cell, the greater the problem of communicating, because the time for a molecule to travel solely by diffusion is proportional to the square of the distance traveled: doubling the distance increases the time by a factor of four. In large multinucleate cells like *Nitella,* this problem is solved by a very active form of cytoplasmic streaming (Figure 2–2b). Third, there is the relationship between the cell and its environment. The volume of a sphere increases in proportion to the cube of its radius, while its surface area increases only as the square of the radius. The larger a spherical cell, therefore, the smaller its surface/volume ratio becomes. Thus, as the cell enlarges it has a tendency to become increasingly isolated from its environment, a handicap that the cell sometimes overcomes by a number of anatomical modifications. One solution is to deform the sheet-like surface by folding or projecting it locally inward *(invagination)* or outward *(evagination)* (see Figure 2–13). Extensive deformations of this kind increase significantly the surface area as compared to a smooth sheet. Another anatomical

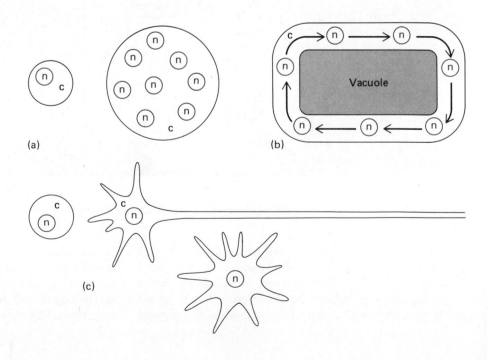

FIGURE 2–2

Solutions to the Problem of Large Cell Size. (a) A cell with many nuclei (n) can be large and still have the same volume of cytoplasm (c) controlled by a single nucleus. (b) Distant parts of a large cell can communicate by energy-driven cytoplasmic streaming (in this case the circular motion of the cytoplasm—cyclosis—carries along with it multiple nuclei). (c) The unfavorable surface-to-volume ratio of a large cell can be improved if the cell geometry is nonspherical. The surface area can be increased by evaginations or the general shape can be dominated by a cylindrical form, as in nerve cells.

feature, found in nerve cells (which in larger animals can be meters in length), is to adopt a thread-like geometry (Figure 2–2c). In a long thin thread (as opposed to a sphere) increase in surface area remains roughly proportional to an increase in volume, and thus the degree of contact between the nerve cell and its environment is not reduced. Last, although the vast majority of cells lie in the range of 0.5 to 20 μm in diameter, there are some truly giant cells. The existence of these cells, however, is always related to some very special biological circumstance. Yolks of the eggs of reptiles and birds, for example, are gigantic cells which store large amounts of food materials and water for the developing embryo. As soon as development begins, however, they subdivide into smaller cells.

There are also advantages to small size, advantages having to do with the problem of control of cellular function. The information necessary for controlling many of the cell's activities is manifest as the concentration and amount of small molecules. When the concentration of a small molecule is reduced by its use in some process, the change is quickly apparent throughout the cell volume as the new lower amount is equilibrated by diffusion. There is no need for wires or pipes to convey information from the consumers to the producers. Permanent one-to-one connections between functioning units of the cell, being unnecessary, are rare. The cellular machines that respond to such information, the catalytic enzymes, are therefore free to evolve independently to optimize the function of the whole cell or organism. This kind of control is very different, for example, from an automobile assembly factory, because the spread of information is made possible by diffusion in a small cell volume.

Even if one sets aside for a moment the special cells that attain unusually large dimensions through a series of special anatomical modifications, one finds that the average eucaryotic cell is about 1000 times larger in volume than the average procaryotic cell. This startling fact, still not satisfactorily explained, reinforces the idea that the evolution of the eucaryotic cell was an evolutionary step of major importance. The increase in size and complexity was accompanied by the appearance of an internal cellular structure, a mesh of filaments called the *cytoskeleton* (Figure 2–38). Also, the increase had to be accompanied by an increase in DNA content. This last development required an entirely new set of machinery for reapportioning the hereditary material, the mitotic apparatus. Advantageous specialization occurred by separating with membranes certain biochemical functions into special compartments—for example, the Golgi apparatus, mitochondria, chloroplasts, and lysosome organelles. It seems, therefore, that there are not one but two size ranges into which most cells fall, and although they overlap, they are governed by principles of structure

and function which are specific, respectively, to the procaryotic and eucaryotic worlds. See box entitled The Optical Approach to Structure.

Optical methods represent a major historical tradition of cell biology. Another such tradition is the isolation of cell structures for biochemical analysis. In the early days of developing this method, the only assay to prove one had isolated the correct structure was to look at the preparation under the microscope. That is why we introduce the basic techniques here; extensive discussion of particular structures will be given in later chapters. See box entitled The Biochemical Approach to Structure, page 50.

To begin the study of cell biology it is useful to have a broad understanding of the morphology of cells. This chapter will, therefore, give a general description of the structure of cells and their organelles. In later chapters, many of the structures mentioned only briefly here will be elaborated in greater detail when their function is also described.

The Fine Structure of Procaryotes

All bacteria, archaebacteria, and blue-green algae are procaryotes. Their main distinguishing characteristics are that the genetic material is not separated from the rest of the cell by a membrane and that they are small (diameter 0.2–5 μm; volume 0.004–65 $μm^3$). In addition, procaryotic cells (1) are almost always surrounded by a rigid cell wall, (2) lack membrane-bound organelles, (3) lack a mitotic apparatus, (4) have ribosomes of a distinctive size and structure, and (5) when motile, usually have simple flagella, whose bodies mainly consist of a single kind of protein.

Bacteria

The eubacteria are classified into two major and a number of minor groups. The two major groups are distinguished by their ability to retain Gram's stain (using the dye crystal violet), which is used to visualize bacteria under the light microscope. *Staphylococcus, Streptococcus,* and *Clostridium* stain deeply and are examples of *gram-positive bacteria; Escherichia, Azotobacter,* and *Pseudomonas* stain much less intensely and are examples of *gram-negative bacteria.* The minor groups include *Actinomycetes, Spirochaeta, Mycoplasma,* and *Rickettsia.*

The bacterium *Bacillus subtilis* (Figure 2–3, p. 48) is a gram-positive straight rod. Ultrastructurally, it consists of (1) a prominent wall, (2) an inner membrane, (3) from one to several nuclear areas *(nucleoids),* (4) many ribosomes, and (5) a mesosome (discussed below). At the inner side of the cell wall is the *plasma membrane,* the primary permeability barrier of the cell.

(text continued on page 48)

The Optical Approach to Structure

The classical approach to the optical analysis of biological structure, still productive today, has been description based on direct visual observation. This process began with the use of the unaided eye even before the dawn of recorded history and was given a tremendous boost by A. van Leeuwenhoek's invention of the single lens "microscope" (actually only a magnifying glass—a microscope as such has at least two lenses or, with the eye, three). Optical analysis finally came into its own, however, through the development of the compound microscope, a two-lens system capable of visualizing objects smaller than 100 μm. The first microscopes, of course, used visible light refracted by glass lenses. But in the last few decades smaller and smaller objects have been visualized by the use of shorter wavelengths of light and the even shorter wavelengths associated with particles (e.g., electrons). Furthermore, the definition of the object examined under the microscope has been improved by special manipulations of the material or modifications of the optical system.

The Compound Microscope

In spite of the tremendous increase in sophistication of techniques for optical analysis, the principle of the compound microscope has remained the basis of the optical approach to structure. The compound microscope has been frequently characterized as consisting of a two-lens system, but as Figure 2–B shows, it really may be thought of as a three-lens system since the lens of the eye or camera is required to focus the image on the retina or photographic film.

The microscope consists first of an *objective lens* which receives light from an object placed beyond the focal point (F_0) of the lens. This lens is used like the lens of a projector which can project an *inverted real image* on a screen (Figure 2–Ab).

In this instance, however, one does not use a screen. Instead, there is an *eyepiece lens* through which one looks at the real image. The eyepiece (ocular) lens is placed in such a way that the real image it examines is located between the lens and its focal point (F_E). This lens, analogous to a magnifying glass (Figure 2–Ac), causes the light rays to diverge less. In such a system, an additional lens like the *lens of the eye* or a *camera* is needed to focus the still diverging light and form an image on the retina or on the photographic plate. To the observer, the object then appears as a *virtual image* lying in a plane beyond the real object and having the same orientation as the *real image* (Figure 2–Ac). Since a compound microscope inverts an object only once, it will appear inverted to the observer. (To be sure, Figure 2–B shows that the eye lens inverts the object once more, but our brain compensates for this since it normally interprets right side up images on the retina as being formed from objects that are upside down.)

The magnification achieved by a compound microscope (Figure 2–C) is the product of the magnification produced by the objective lens (usually ×10 or ×40) and by the eyepiece lens (usually ×10). Thus, typical compound microscopes operate at a magnification of ×100 (low power) or ×400 (high power).

The object examined by a compound microscope is usually sandwiched between two pieces of glass, the slide and the cover slip. The air space between the cover slip and the objective lens allows some light to be lost by refraction because air has a *refractive index* different from that of glass. Replacing the air space by a drop of oil with a refractive index similar to that of glass eliminates this loss of light (Figure 2–D), and it is then possible to use a ×90 objective lens, thereby achieving a magnification of ×900.

Of course, magnification can be increased well beyond that by focusing the light coming out of the eyepiece and projecting it onto a screen or photographic plate. There is, however, a limit to how much can be gained from that approach since no amount of magnification will increase the *resolution* of the microscope. Resolution (R), which sets the limit to the amount of detail that can be recognized, is defined as the inverse of the shortest distance (d) between two adjacent points that can be seen as separate. The theoretical limit to the resolution of a microscope is described by the equation:

$$\frac{1}{R} = d = \frac{0.61\,\lambda}{NA}$$

where λ is the wavelength of the light employed and NA is the numerical aperture of the objective lens. (It is the recombination of light diffracted by the object that forms the image, and fine detail causes diffraction at wider angles. The numerical aperture measures the angular distribution of light emanating from the object that can be captured by the objective lens. The greater the angular distribution,

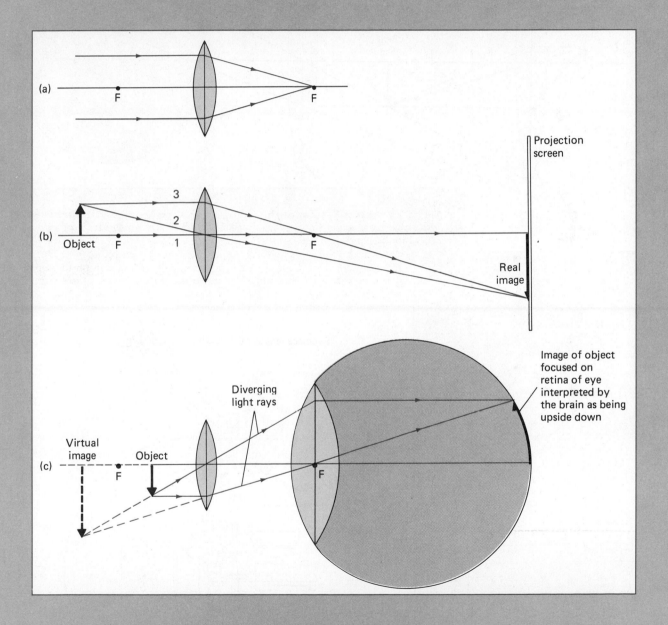

FIGURE 2–A
Optics of the Compound microscope.   The behavior of light passing through a convex lens is illustrated. (a) The focal point of the lens (F) is where a beam of parallel light rays come together and focus. (b) A real image is obtained when the object is located beyond the focus. This image can be seen on a screen just like the image produced by a slide or movie projector. Three rays define the real image: (1) the ray going through the optic axis, (2) the ray going through the center of the lens, which is therefore not bent, and (3) the ray which is parallel to the optic axis and therefore goes through the focal point. (c) A virtual image is obtained if the object is closer to the lens than the focal point. The virtual image cannot be projected on a screen. It can only be "seen" if diverging light rays are first focused by the lens of the eye or camera, and projected onto the retina or film, respectively.

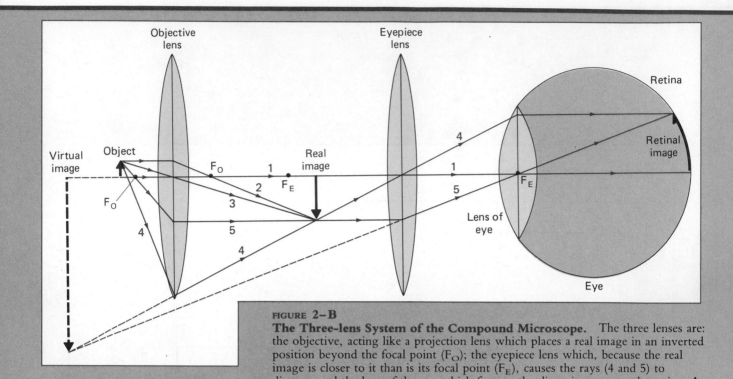

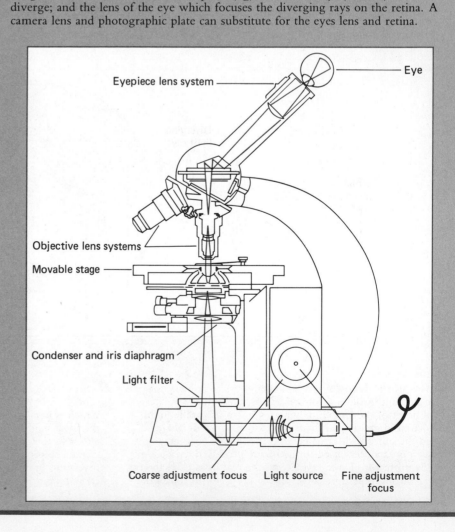

FIGURE **2–B**
The Three-lens System of the Compound Microscope.   The three lenses are: the objective, acting like a projection lens which places a real image in an inverted position beyond the focal point (F_O); the eyepiece lens which, because the real image is closer to it than is its focal point (F_E), causes the rays (4 and 5) to diverge; and the lens of the eye which focuses the diverging rays on the retina. A camera lens and photographic plate can substitute for the eyes lens and retina.

FIGURE **2–C**
Diagram of Compound Microscope.

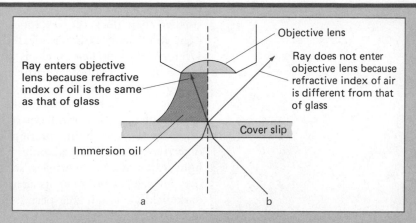

FIGURE **2–D**
Optics of Oil Immersion Lens.   The diagram shows the paths of two light rays, one of which (a) does not pass through immersion oil and therefore misses the objective lens because of refraction. The other ray (b) passes through immersion oil and enters the lens. The use of immersion oil with a refractive index similar to that of glass thus increases the light gathering power of the objective lens.

the greater the resolving power of a lens.) Under the very best circumstances of visible light microscopy—that is, by using an objective lens with an NA of 1.40 and a violet light filter that allows the passage of light with a wavelength of 400 nm—one can achieve a resolution of 175 nm.

There are two approaches to increasing the effectiveness of microscopy: (1) making objects stand out more clearly—that is, increasing contrast—and (2) making smaller objects susceptible to observation by using radiation of shorter wavelengths. Both of these can be achieved through the use of specialized types of microscopes and various methods of preparing the materials for optical study.

A Variety of Microscopes

The material found in cells is, with few exceptions, transparent and colorless. There are, however, marked differences in refractive index (velocity of light through a substance relative to the velocity in a vacuum) between various cell components. The *phase contrast microscope* is designed to detect differences in refractive index, especially where these

differences are abrupt, and to transform them into differences of light intensity. The differences in refractive index lead to differences in the phase of light waves passing through different parts of the object, relative to light waves that are not altered in phase by the object. The perturbed light waves are directed through a path different from that taken by the unperturbed waves, although both paths are through the same lens. When an image is formed by the objective lens, these two types of light *interfere* with one another and the image has differences in intensity corresponding to the differences in the phase of light perturbed by the object. Because it renders the chromosomes clearly visible, phase contrast microscopy—especially the variant known as *differential contrast interference microscopy,* but more often simply by the name of its inventor, *Nomarski*—has been particularly useful for the observation of living cells undergoing nuclear division (Figure 2–E).

Another microscope capable of creating visible contrast from differences in refractive index is the *interference microscope,* which is more versatile than the phase contrast mi-

croscope but also more complex and expensive. The principle of operation is the same as in the phase contrast system, but the unperturbed beam of light is routed to the image plane by an entirely separate path whose properties can be varied by the investigator. An additional virtue of the interference microscope is that it can be used quantitatively for determining the mass of the objects under observation. This is possible because the refractive index of an object is determined in part by the density of the object—that is, the denser an object, the greater the difference in the phase of light waves after passing through the object. Since the paths taken by perturbed and unperturbed light waves can be manipulated separately, the difference in phase can be measured and, knowing the thickness of the object, its density can be estimated. The density times the volume (thickness times cross-sectional area) gives the mass. A number of interference microscopes have been designed. One of these, designed by Nomarski and based on a principle called "interference contrast," has been particularly successful in enhancing small differences in refractive index. It produces images with an illusion of three-dimensionality.

Some biological materials contain molecules in ordered arrays. When illuminated with polarized light, such objects can not only be rendered visible, relative to their disordered surroundings, but information can also be obtained regarding their internal organization. *Polarizing microscopes* have, for instance, been used for studying the structure of the mitotic spindle (Figure 2–40a).

As we have said, the material in cells is usually colorless. By this we mean that it does not usually absorb visible light. Classical histology got around this difficulty by the development of dyes and stains that would bind specifically to some cellular structures and not to others. The

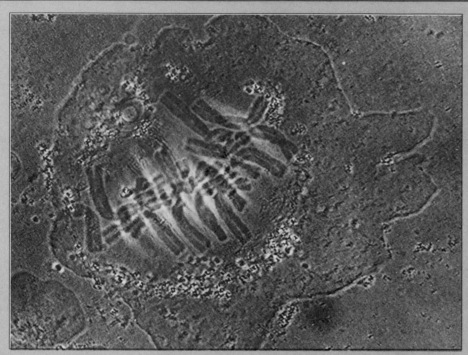

FIGURE 2–E
Nomarski (differential interference contrast microscopy) Image of Endosperm Cell from the Fruit of the African Blood Lily *Haemanthus katherinae.*   Notice that the giant lily chromosomes can be seen quite clearly. Chromosome pairs are just about to be separated along the mitotic spindle. × 800. (Courtesy of S. Inoue and *Chromosoma.*)

microscopist can easily see the location of the stain, provided the stained object is sufficiently large, because it strongly absorbs visible light. There are, however, certain cellular components that normally absorb ultraviolet light. Chief among these are proteins and, even more so, nucleic acids. *Ultraviolet microscopy* not only allows one to visualize objects such as chromosomes, which absorb ultraviolet light intensely, but also increases the resolution to about 120 nm because this light has a wavelength shorter than that of visible light (260 nm is used to visualize nucleic acids, for example). Since the usual glass lenses also absorb ultraviolet light, microscopes using this radiation must have such components made of quartz.

Fluorescent compounds are often introduced into the cell. The introduced compounds—for example, antibodies raised against one or another cell component—are chosen because

they bind specifically to cellular structures. Fluorescent compounds, when excited with an electromagnetic wave of one wavelength, emit light of a longer wavelength. *Fluorescence microscopes* use short wavelength visible or ultraviolet light for excitation, visible light being emitted by the object for observation. The use of fluorescently labeled antibodies, *immunofluorescence,* has been developed further with the use of laser light sources and appropriate computer processing. This allows the elimination of background fluorescence and improves the contrast of the structure of interest. With the *confocal microscope,* one can achieve high clarity of the object at different focal levels by using a complex scanning process that produces "optical sections" of approximately one micrometer thickness.

There is another form of microscopy which uses visible light and yet permits the detection of particles

smaller than does the resolution of light microscopy (that is, less than 175 nm). *Dark field microscopy* is capable of detecting colloidal particles of 50 nm or less by illuminating the particle with light, for example, at right angles to the optical axis of the microscope. The light entering the objective lens is derived only from light scattered by the particles, so that they are seen as bright spots against a dark field. Though this permits detection of the particles, it is not possible to observe their shape, content, or surface structure. The value of dark field microscopy is that it can be used for counting or detecting the motion of structures smaller than the resolving power of the microscope. Dark field microscopy is especially useful today for counting photographic grains in autoradiography (see discussion later in this chapter).

We have said that by using shorter wavelengths we can increase resolution; unfortunately, new problems arise when we reduce the wavelength below that of *near ultraviolet* (240 nm). The reason for this is that almost everything, including water, air, and the material of which the lenses are made, absorbs light in the *far ultraviolet*. It is therefore necessary to go to even shorter wavelengths where differential absorption of radiation again permits us to make observations (Figure 2–F). *X-rays* are of convenient wavelengths for detection of small objects and have the virtue of not being strongly absorbed by water, thus permitting the visualization of living material. Unfortunately, most useful materials have very similar refractive indices for x-rays so simple lenses cannot be used to create an image. X-rays, therefore, have been used for visualizing large objects (such as organs and tissues) where magnification is not required, or for ordered structures (such as crystals) where the diffraction and scattering of the radiation can be measured and the shape of the object calculated. The

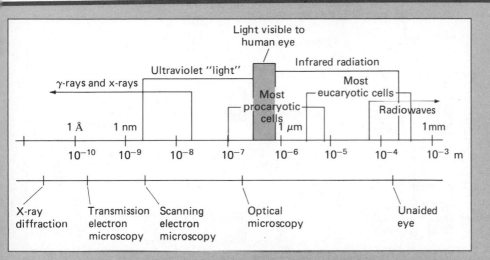

FIGURE 2–F
Electromagnetic Spectrum. The ranges of wavelengths of radiation used for biological studies are shown, including the minimum wavelength accessible to various optical instruments, using practical methods.

method of x-ray diffraction is most useful at the molecular level and will be discussed in Chapters 4 and 10.

There is, however, one kind of short wavelength radiation which has been extremely useful for adaptation to direct microscopy. This is the radiation associated with small particles. The most widespread use of this optical approach exploits the waves associated with electrons in an instrument known as the *transmission* **electron microscope** (Figure 2–G). When a tungsten filament is heated it emits electrons. Since these have a negative charge, they can be accelerated by an electric field, as well as focused with magnetic lenses, into a beam of electrons. Furthermore, because of their charge, electrons are readily absorbed by biological material. In practice the object to be visualized must be *dry* (water absorbs electrons) and *thin* (to allow some electrons to penetrate it). Instead of the eye, the recording system is a fluorescent screen or photographic plate.

Note that we have referred to electrons as being associated with short wavelength radiation. One of the key insights of quantum mechanics is that entities we are accustomed

to think of as particles, such as electrons and protons, also exhibit wave properties. (Conversely, electromagnetic radiation can also be thought of as streams of particles called photons, a fact that will take on special significance when we discuss photosynthesis in Chapter 8.) The wavelength of an electron beam is determined by the voltage of the electric field accelerating the electrons. A 50 kV instrument will produce radiation with a wavelength of 5 picometers (pm); that is, 0.005 nm. However, the present development of electromagnetic lenses is such that resolving powers of less than 0.5 nm have seldom been achieved. In recent years 1000 kV instruments have been built, increasing the penetrating power of electrons, and allowing the use of greater thicknesses of material (Figure 2–38) and stereoscopic techniques for three-dimensional analysis of cell structures. Because electron beams have very short wavelengths, there is a tremendous potential to improve the resolution of electron microscopes, and we can expect steady progress in this area for years to come. Such improvements, however, have limited value without comparable improvements in methods of staining

very small biological structures to achieve sufficient contrast. This is an area of continuing investigation.

The *scanning electron microscope* (Figure 2–G) scans a specimen with an electron beam and detects *secondary electrons*—the electrons dislodged by the impact of the primary electron beam—emitted by the specimen. The information thus obtained is displayed on a cathode ray tube. This method is useful for examining *surfaces* of biological specimens. The disadvantage of scanning electron microscopy is that the resolution is low, relative to transmission electron microscopes (less than 10 nm, 100 Å); the advantage is that areas of the surface at different distances from the beam source are still in focus, giving images with a startling degree of three-dimensionality (Figure 2–H).

In recent years, a powerful approach to the structural study of large molecules and molecular assemblies has been developed which combines electron microscopy and diffraction. Structures which appear repetitively can be photographed in the electron microscope and, using *optical diffraction* techniques, can be converted into a diffraction pattern. The advantage of this is that the information in the diffraction pattern derives from all of the large number of repeated structures. Using "optical reconstruction" and computer analysis techniques very similar to those used in x-ray diffraction studies, a high-resolution (1.5 nm) image of the single structure can be obtained which, in a sense, averages in all the common (i.e., relevant) information derived from the repeated structures and eliminates the "noise" that degrades the image of any one of them (Figure 2–I). This method has developed into an important tool for macromolecular structure analysis.

This brief review of microscopes should make the student aware that techniques of optical analysis, reaching back to biology's beginnings, are

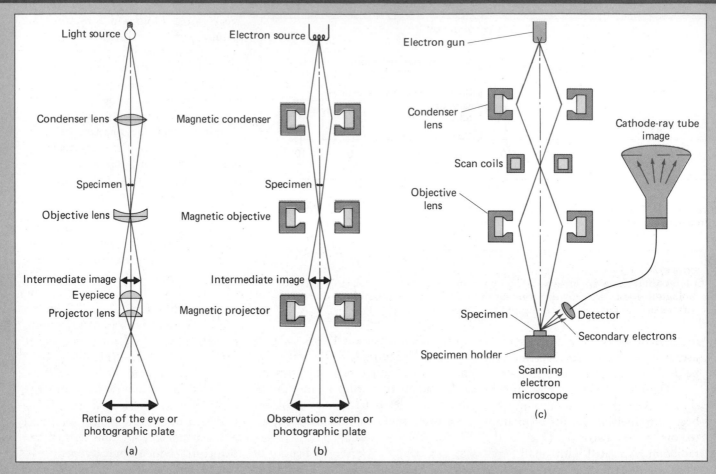

FIGURE **2–G**
Comparison of Optics of Three Microscopes. (a) The light microscope. (b) The transmission electron microscope. The essential difference between the two is the lenses: the electron microscope uses electromagnetic lenses to focus the electron beam while the light microscope uses glass lenses to focus the light beam. (c) The scanning electron microscope. In the scanning instrument the direction of the electron beam is controlled in a regular way by magnetic scanning coils. The beam is focused on the specimen, which is covered with a thin layer of a heavy metal such as gold. The metal scatters secondary electrons onto a detector, which generates a dramatic three-dimensional image via a cathode ray tube.

still developing. We are approaching a point where the optical approach is yielding results which can be interpreted in molecular terms. The success of this venture depends not only on the development of microscopes, but also on the elaboration of a large repertoire of techniques for manipulating the biological material studied.

Preparation of Material for Optical Study

Under the light microscope, living cells are transparent and devoid of contrast. The problem of microscopy has thus been to render visible or "bring out" structures in the cell that would otherwise remain unnoticed. The assumption inherent in this approach is that the manipulations used to prepare specimens do no more than render existing structures visible. The alternative possibility is that they are creating *artifacts* (i.e., effects characteristic of the particular treatment rather than reflecting cell structure).

Distinguishing artifacts from enhanced but existing structures is a subtle problem which has been debated ever since microscopy began. It has been suggested than an artifact is that which a competent microscopist calls an artifact. The justification for this seemingly subjective definition is that a competent microscopist will have examined a large number of specimens of a given material after a variety of different treatments. An experienced microscopist can therefore judge whether the appearance of a structure is constant or variable. Varying the type of microscope and the conditions of treatment should change the number and kinds of artifacts but should not alter the characteristics of an underlying structure.

Chapter 2 Cell Structure— The Organizational Basis of Cell Function

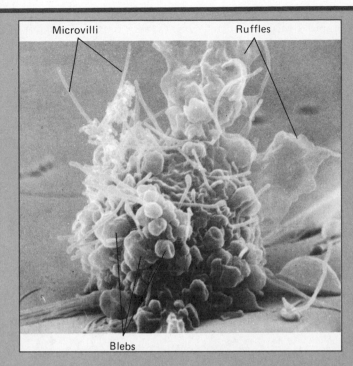

FIGURE 2-H
Scanning Electron Micrograph of a Neuroblastoma Cell in Culture.
The cell surface shows a variety of shapes: ruffles, blebs, and microvilli.
× 10,000. (Courtesy of K. R. Porter.)

After a series of such experiments the microscopist can determine whether a certain structure appears constant, or whether it appears different in relation to a particular treatment. Normally, the microscopist concludes that if the same structure appears under a variety of experimental conditions, it is likely to be real (that is, present in the living cell) rather than artifactual (created by the experimental procedure). Conversely, once a structure has been judged as real, it can be used to evaluate various experimental procedures. This circularity in the experimental approach may appear self-defeating to the logician, but there is much historical evidence to show that progress can be made in this fashion.

There are two main categories of cell manipulations that make structures visible: (1) treatments that maintain cells or organelles in a living or functional state, and (2) fixing and staining cells, frequently followed by additional treatment.

Living cells are homogeneous and transparent in their appearance under the light microscope. Never-

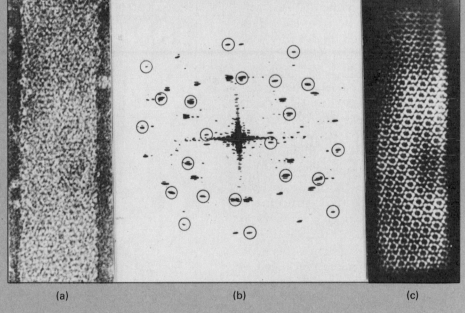

(a)   (b)   (c)

FIGURE 2-I
Optical Reconstruction by Diffraction and Image Filtering. A mutant form of T4 bacteriophage fails to form the normal spherical "head" and instead forms a long cylindrical structure of repetitive elements. (a) Electron microscope image of the cylindrical structure, lying flat on the carbon film of the microscope grid. (b) Optical diffraction pattern of the image in (a). The information in the diffraction pattern can be filtered before producing a reconstructed image. The information to be used is indicated by the circles drawn around one set of spots (diffraction peaks). A mask is constructed that allows only the circled diffraction peaks to pass through. (c) Optically filtered image. This image shows the repetitive features of only one side of the flattened mutant "head." One can see clearly the individual molecules, arranged in rings of six. (Courtesy of D. J. DeRosier and A. Klug.)

theless, it is possible to manipulate them in a variety of ways to obtain evidence of internal substructure.

- A number of stains are known which can enter the cell and become concentrated in a given structure while the cell survives. Examples of such *vital stains* include *neutral red,* which accumulates in the vacuoles of some plant cells and stains cytoplasmic granules, and *Janus Green,* which stains functioning mitochondria.

- Another approach which provides evidence for structural differentiation within the cell is *micromanipulation.* Microneedles moved with great precision by hydraulic systems can be used to free a membrane-bound vacuole from the cytoplasm or to study the fluidity of membranes. Micropipettes are used to inject dyes which would not otherwise penetrate the cell membrane, or to remove from cells such organelles as nuclei. Highly focused beams of ultraviolet radiation or lasers can be used to damage or destroy structures within the living cell and allow study of the consequences.

- A technique to investigate the biochemistry of cell structures is to visualize enzyme activities in living or recently disrupted cells. There exist substances that will form an insoluble precipitate as the result of the action of one or another cellular enzyme. The precipitate tends to form in the immediate vicinity of the enzyme. By bathing freshly sectioned tissues in one of these substances, it is possible to *localize the enzyme* by detecting the precipitate near a particular cellular structure (Figure 2–J).

By far the most widely utilized manipulations to study cells are *fixing* and *staining.* Fixation is used to inter-rupt cell processes as rapidly as possible and with as little alteration in structure as possible. The main effect of fixation is to "denature" the cell proteins and render them insoluble in water or organic solvents. Denatured proteins will thus precipitate at their normal location within the cell. Another function of fixation is to inacti-vate enzymes, thus preventing *autolysis,* the process of self-digestion that occurs after a cell dies, which can destroy structures to be studied. The more finely grained the precipitate formed by the fixative, the fewer will be the fixation artifacts. Very fine-grained precipitates are produced by using fixatives containing osmic acid,

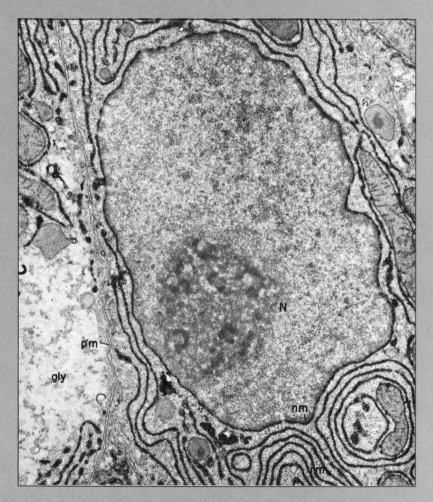

FIGURE 2–J
Localization of Enzyme Glucose-6-phosphatase in a Liver Cell. Thin slices of liver tissue were incubated in the presence of lead with the substrate glucose-6-phosphate, then fixed and sectioned for electron microscopy. The cellular glucose-6-phosphatase produced at the sites of enzyme activity is detected by its product, inorganic phosphate, which in the presence of lead forms a highly insoluble precipitate of lead phosphate. The electron micrograph shows that the deposition of lead phosphate, indicated by very dark staining, is concentrated at the outer nuclear membrane (nm) and at the membranes of the endoplasmic reticulum (rm), indicating the presence of the enzyme there. There is no precipitate at the plasma membrane (pm), in the mitochondria (m), or in the areas of glycogen storage (gly), which are therefore only lightly stained. × 20,000. (Courtesy of A. Leskes, P. Siekevitz, and G. E. Palade.)

neutralized formalin, mercuric chloride, or various combinations of these. Lastly, fixatives cross-link cell constituents, thereby preventing them from changing their relative positions during preparation for microscopy.

In electron microscopy, the use of a fixative giving fine-grained aggregates of the proteins and nucleic acids of the cell is even more important. Modern fixatives act by cross-linking macromolecules into an insoluble complex, rather than merely denaturing them. Among the most widely used fixatives are formaldehyde and, especially, glutaraldehyde. Adjusting pH turns out to be a very important feature of the fixation process, and one could claim that the success of cytological electron microscopy rests on the observation by K. R. Porter and G. E. Palade that fixatives should be buffered at pH 7.2 to 7.8 for best results.

Staining is used to achieve contrast between the constituents of the cell. *Hematoxylin* and *eosin* have been used traditionally to distinguish between nucleus (purple) and cytoplasm (pink). There are other stains that can bring out the structure of chromosomes, mitochondria, and the Golgi complex. Stains can also be used to obtain information regarding the chemical nature of the structure being stained. *Basic dyes* like methyl green, in which the colored ion is positively charged, will react with negatively charged substances such as nucleic acids (DNA and RNA); *acidic dyes* like eosin, in which the colored ion is negative, will react with positively charged proteins such as the histones in the chromosomes.

What constitutes an effective stain depends on the radiation utilized by a particular microscope. In light microscopy a stain is a substance which appears colored to the eye. In ultraviolet microscopy a stain may appear colorless to the eye, but strongly absorb ultraviolet light. In electron microscopy, compounds containing heavy atoms such as *osmium tetroxide* are used as stains because they absorb electrons most effectively. Of course, to be detected, even a stained complex must still be large enough to be resolved by the optical method used.

A variant of the staining technique is *autoradiography*. This often involves growing cells (or organisms) in the presence of a precursor compound containing a radioactive isotope. (A *precursor* is any substance that is chemically transformed into or incorporated into a cellular constituent or an intermediate in the cell's metabolism.) When the radioactive compound has been incorporated into the structure of the cell, its location can be detected (after fixing the cell and washing away unincorporated precursor) by coating the biological material with a photographic emulsion. This preparation is then stored in a light-free place for hours, days, or even weeks. The radiation emitted by the radioactive structure "exposes" the immediately adjacent photographic emulsion. After developing, it is possible to examine the preparation by microscopy and to identify the structure that appears to be most closely associated with the darkened photographic grains. An example of this is the localization of DNA in the chromosomes of cells grown in the presence of ^{3}H-thymidine, a precursor of DNA (Figure 2–K). Another use of autoradiography is to localize radioactive hormones or antibodies when they are bound to specific cellular receptors or antigens, respectively.

Antibodies can be visualized also by covalently coupling them to a fluorescent molecule and binding this complex to the antigen, a technique known as *direct immunofluorescence*. Under the ultraviolet microscope the location of the fluorescent antigen-antibody complex can be easily seen. The fluorescent marker need not even be attached directly to the antibody to

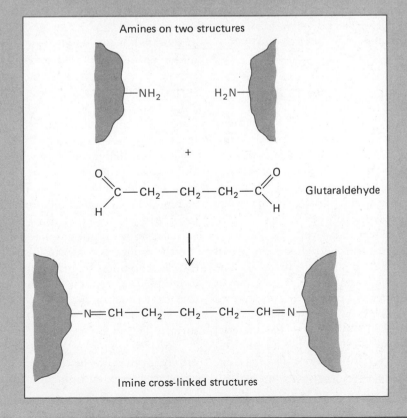

Amines on two structures

$-NH_2$ H_2N-

+

Glutaraldehyde

Imine cross-linked structures

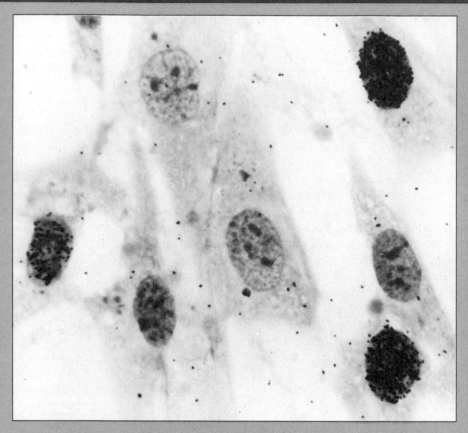

Autoradiography.   The micrograph shows hamster cell nuclei labeled with
^{3}H-thymidine, a precursor of DNA synthesis. The darkened grains in the
photographic emulsion are the sites of radioactive decay. The labeled nuclei are
those that synthesized DNA after the ^{3}H-thymidine was added to the cell culture.
× 4,500. (Courtesy of D. M. Prescott.)

creates a dry, foamlike structure that can be embedded and sectioned or even sectioned directly.

Sectioning of biological material prior to microscopic examination is necessary to provide specimens not much thicker than the microscope's depth of focus. If the material is too thick, the light entering and leaving the focal plane is scattered by objects above and below, reducing the clarity of the image. The higher the resolution and the smaller the depth of focus, the thinner the sections must be. In light microscopy sections of 2–15 μm are acceptable, but in electron microscopy they must usually be 0.05–0.10 μm thick.

An alternative to sectioning for electron microscopy is *freeze-fracturing* (freeze-cleaving) followed by *freeze-etching* (see Figure 7–6). Freeze-fracturing involves rapidly freezing a sample, then fracturing or cleaving it. If the fracture plane is appropriately oriented, it is possible to visualize the inner surface of the cell membrane. Freeze-etching is the removal of water from the surface of a specimen by sublimation (water molecules are evaporated from the frozen surface under vacuum). Freeze-etching will expose the cytoplasmic and outer surfaces of the cell membrane (Figure 2–La). An interesting consequence of freeze-cleaving is that membranes are frequently split lengthwise, separating the two halves of the membrane and exposing the center of each of the halves. Following the cleaving and etching, a *replica* of the exposed surfaces is made by condensing carbon and platinum vapor onto the surfaces. It is the replica that is examined under the electron microscope. Electron micrographs of replicas of the inner surfaces of membranes reveal numerous small particles lying within the surface (Figure 2–Lb).

Shadow casting is a method for bringing out with the electron microscope the size and shape of small

make it visible. A widely used technique to localize specific cellular antigens is *indirect immunofluorescence* (see Figure 7–18). In this method the fluorescent molecule is covalently attached to a "secondary" antibody whose antigen is another ("primary") antibody. Scarce antibodies that are specific to the cellular structure can thus be detected by the fluorescent anti-antibodies which, once prepared, can be used against all appropriate primary antibodies.

Frequent adjuncts to fixation and staining are embedding and sectioning. During *embedding* the biological material is surrounded and, fre-
quently, penetrated by a relatively hard substance which provides support for the soft cellular material during the sectioning process. For light microscopy a mixture of paraffin/plastic or glycol methacrylate plastic is widely used. In electron microscopy, however, the need for thinner sectioning requires the use of a much harder plastic material. The specimen is soaked in a solution of monomer which is then polymerized by the addition of a catalyst. Freezing is another way to give the material sufficient rigidity to permit sectioning. A variant is very rapid freezing, followed by freeze-drying, which

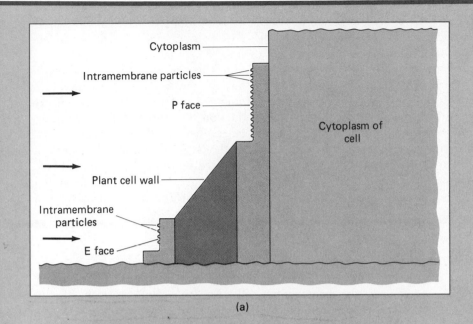

(a)

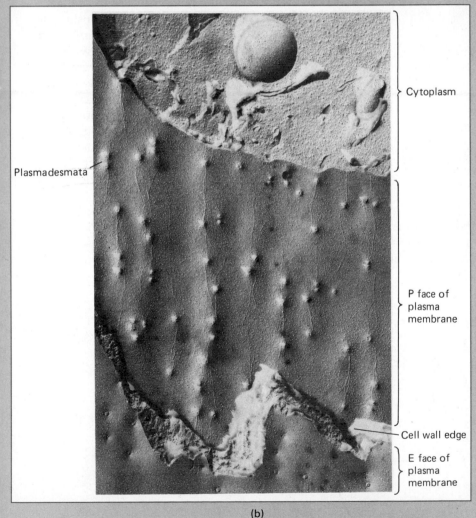

(b)

FIGURE 2–L
Freeze-fractured Tissue from Onion Root Tip. (a) Sketch of the cross-section of material seen in the micrograph. Leaders indicate the direction from which the electron micrograph is viewed. The internal surfaces of two fractured cell membranes can be seen, one on each side of the plant cell wall structure. When the cell membrane is fractured, the two exposed inner faces are the E face (outer half of the cell membrane) and the P face (inner half of the cell membrane). (b) Freeze-fractured membranes of two cells. The E face of the membrane of one cell and the P face of the membrane of the adjacent cell are separated by the cellulose cell wall. Numerous intramembrane particles are present in both internal membrane faces. In plants, connections between the cytoplasm of adjacent cells (plasmadesmata) are brought out clearly by the freeze-fracturing process. × 65,000. (Courtesy of D. Branton.)

independent objects. A beam of metal atoms (gold) falls at an angle onto a plane surface supporting the particles. A portion of the particle and of the grid (support) surfaces is left uncovered (shadowed). This produces a three-dimensional effect that not only makes the structure more visible, but allows one, knowing the shadow angle, to calculate the height of the structure. A variation of shadow casting, *rotary shadowing,* enhances the contrast of the image by rotating the specimen during the shadow casting. This procedure also enlarges very small structures, by the layer of metal deposited.

A recently developed method makes visible in a very dramatic way even the finest filaments of the cell's cytoskeleton. This method, originated by T. Reese and J. Heuser, combines the techniques of quick freezing, deep-etching, and rotary shadowing of the replica. (Deep-etching is simply a variant of the etching technique described above; a thicker layer of water is removed by sublimation.) Using these techniques one can detect filaments as thin as 3 nm in cross section. The three-dimensional structure of the cytoskeleton can be enhanced even further by stereoscopic photography of the images.

Another alternative to sectioning can be employed if the objects to be examined are very small, such as macromolecules, viruses, or some free-floating fractionated cell constituents (e.g., ribosomes). It is possible to fix these objects and examine them directly. In electron microscopy, the contrast of such structures is frequently brought out by *negative staining*. In this process an electron-dense stain is applied so that it surrounds but does not penetrate the object. Contrast is enhanced because the object appears lighter than its surroundings. An example of using this method to examine mitochondria can be seen in Figure 8–1, while Figure 13–18 shows similar images of ribosomes.

This description of optical methods for the study of cell structure is intended to provide a general introduction to a subject of great importance. We shall encounter optical methods and the results they provide repeatedly throughout this book, thereby greatly enlarging our understanding of this basic approach.

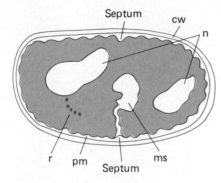

FIGURE 2–3
Diagram of Bacterium *Bacillus subtilis*. Among the structures illustrated are the cell wall (cw), inner membrane (plasma membrane, pm), nucleoids (n), ribosomes (r), and mesosome (ms), forming at the septum.

The cell wall of a gram-positive bacterium is composed of a *peptidoglycan,* which is a network formed by *polysaccharide* chains cross-linked covalently to each other by short *peptide* chains. One can think of this structure as a single, huge sac-like molecule enveloping the procaryotic cell. An important function of this structure is to maintain the size and shape of the cell. This can be demonstrated by digesting the cell wall with the enzyme *lysozyme*. If the water concentration inside the bacterium is lower than that of the environment, water moves into the bacterium and will cause it to swell and burst once the cell wall has been removed (Figure 2–4). This can be prevented from happening by adding sucrose to the external solution, reducing the water concentration of the environment to that of the cell. Without their walls, bacterial cells form spherical *protoplasts* (Figure 2–4) and can be used very effectively for studies of membrane permeability (Chapter 7). All gram-negative bacteria also have an *outer membrane* that lies outside and surrounds the cell wall.

In many cases, the plasma membrane is connected to a highly folded membrane structure in the interior of the cell called the *mesosome.* The mesosome is closely connected with the *septum* or cross wall which forms when

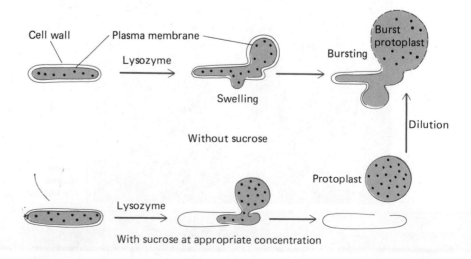

FIGURE 2–4
Protoplast Formation. The enzyme lysozyme, which digests away the cell wall, is used to induce protoplast formation or lysis in a bacterium. Bacterial protoplasts can be used to study the permeability properties of cell membranes. Lysis of protoplasts is a convenient procedure for releasing the cell contents.

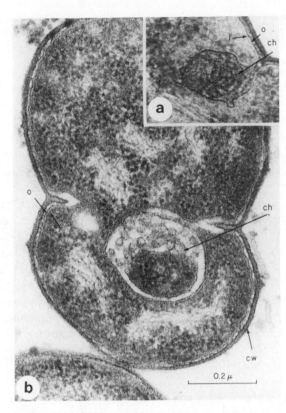

FIGURE **2–5**
A Mesosome in the Bacterium *Diplococcus pneumoniae.*
Indicated are the outer membrane (o), inner membrane (i),
cell wall (cw), septum (s), and the mesosome (ch) within the
cell. × 200,000. (Courtesy of A. Tomasz.)

and function of bacterial flagella and the mechanisms of
bacterial movement in response to substances in the
environment (chemotaxis). The bacterial flagellum consists of a single fiber, composed mainly of the protein
flagellin. The motion of the flagellum is a *rotation* brought
about at a basal structure lodged in the cell wall and
membrane (Figure 15–24).

Archaebacteria, while they have many distinctive
functions and macromolecular structures, resemble the
eubacteria in their general morphology and are discussed
in Chapter 15.

Cyanobacteria

Cyanobacteria (blue-green algae) and some photosynthetic bacteria are thought to be among the most
primitive autotrophs. They differ from bacteria mainly in
having organized structures to carry out photosynthesis
(Figure 2–6). These structures consist of flattened vesicles

the bacterium divides; it is also connected to the nuclear
region and believed, therefore, to be involved both in
DNA duplication and in cell division (Figures 2–3, 2–5;
see also Figure 1–2a).

Bacteria, depending on growth conditions, have one
or more nuclear regions. Each of these contains a copy of
the same genetic information, usually in the form of a
single circular molecule of DNA. DNA is an acidic
molecule carrying numerous negative charges. In procaryotes these charges are neutralized by Mg^{2+} ions and
positive organic ions, whereas in eucaryotes proteins are
included among the counter ions and, with the DNA,
form elaborate structures—chromosomes (see Chapter
11).

Ribosomes, the protein-synthesizing structures of
the cell, are somewhat smaller in procaryotes than in
eucaryotes. We shall study them in detail in Chapters 13
and 15.

Many bacteria are motile and many of the motile
forms have numerous flagella (Chapter 15). A great deal
of progress has been made in understanding the structure

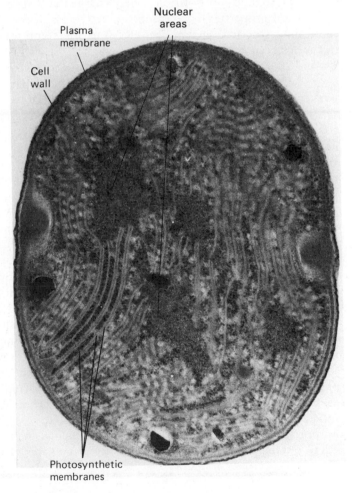

FIGURE **2–6**
A Blue-green Algae, *Gleocapsa alpicula.*   The
photosynthetic membranes or lamellae are shown × 72,000.
(Courtesy of M. M. Allen.)

The Biochemical Approach to Structure

It is possible to break cells with an appropriate *homogenizer* and to centrifuge or otherwise to fractionate the fragments *(homogenate)* under a variety of conditions, thereby separating various cellular structures from each other. The isolated structures in a quasi-living state can be studied by microscopy and their independent existence thereby confirmed. But also, the separation of cellular organelles allows one to study in isolation their biochemical properties (e.g., composition and associated enzyme activities) as has been done with mitochondria and chloroplasts (see Chapter 8).

Fractionation by centrifugation is successful because the various organelles within a cell are greatly different in size and density. The nucleus is by far the largest organelle in the cell; mitochondria are intermediate in size and density; and nucleoli and ribosomes are the smallest and densest. Various membranes, as **vesicles,** are the lightest because of their relatively high lipid content. These characteristics are used to separate one organelle from the rest. In general, there are two centrifugation methods one can use in these separations.

One is *differential* or *velocity centrifugation.* A homogenate is first spun at a low speed, forming a *pellet* and a *supernatant;* the supernatant is then spun at a higher speed, to sediment a second pellet and leave a second supernatant; and so on (Figure 2–Ma). If appropriate speeds are used, the first pellet contains the nuclei, while the

second contains mitochondria. Since both of these pellets are contaminated by other cellular constituents, they must be further purified, usually by *washing* (resuspending in fresh buffer) and resedimentation. The homogenizing medium may be a salt solution, but it is usually a sucrose solution since sucrose is an uncharged molecule and very soluble. Both of these characteristics are important: charged salts tend to harm organelles in various ways, while concentrated sucrose solutions can provide the high densities that in many cases are required for adequate separations. To accomplish differential centrifugation of most cell organelles, only a normal centrifuge is needed; separation of microsomes and ribosomes (see Chapter 13) requires the higher speeds of an ultracentrifuge.

The other method used to fractionate cell organelles is *density gradient centrifugation.* A *gradient of density* is made in the centrifuge tube and the homogenate is then carefully layered on top of the gradient (Figure 2–Mb). In many cases impure fractions obtained from the first method are further purified by layering them on top of a suitable density gradient. Upon centrifugation, the denser particles spin down to lower levels in the gradient than the less dense ones do. If the speed is high and the time of centrifugation prolonged, the particles will come to rest in *equilibrium zones* equivalent to their own densities; hence, this method has also been

called *zonal* or *equilibrium centrifugation.* This method has been used to separate mitochondria, lysosomes, and peroxisomes, all of which sediment together during velocity centrifugation. (It is important to realize that while the organelles come to rest in equilibrium with the surrounding density, the density gradient itself is not in equilibrium in gradient centrifugation. We shall discuss equilibrium density gradient centrifugation—for example, of DNA molecules—in Chapter 10.)

Sometimes a modification of this method is used: solutions of various densities are layered, one on top of the other, with the densest on the bottom and with the homogenate or subcellular fraction on top in a solution of the lightest density. Upon completion of centrifugation, the various organelles will come to rest at the interfaces between the density layers; particles lighter than the density of the solution beneath them will not penetrate into that solution and will usually have passed fully through the even lighter solution above.

In general, the velocity centrifugation method separates organelles by virtue of differences in their size, whereas density gradient centrifugation separates them through their differences in density. Mitochondria, for example, are usually obtained by the first method in the form of a tan colored pellet and are washed several times, by resuspension and resedimentation, to purify them further.

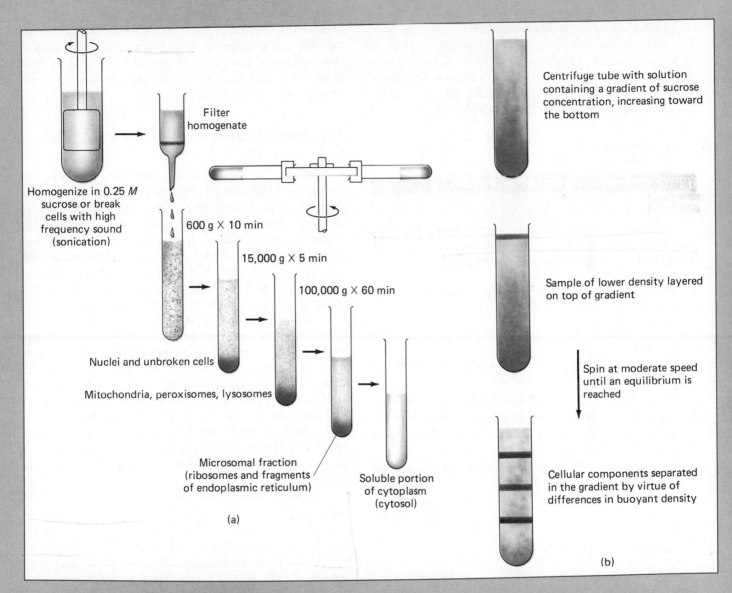

Filter homogenate

Homogenize in 0.25 *M* sucrose or break cells with high frequency sound (sonication)

600 g × 10 min

15,000 g × 5 min

100,000 g × 60 min

Nuclei and unbroken cells

Mitochondria, peroxisomes, lysosomes

Microsomal fraction (ribosomes and fragments of endoplasmic reticulum)

Soluble portion of cytoplasm (cytosol)

(a)

Centrifuge tube with solution containing a gradient of sucrose concentration, increasing toward the bottom

Sample of lower density layered on top of gradient

Spin at moderate speed until an equilibrium is reached

Cellular components separated in the gradient by virtue of differences in buoyant density

(b)

FIGURE **2–M**
Separation of Cell Components.   (a) Differential centrifugation. Higher rotational speeds generate increasing gravitational (g) forces which separate various components, as illustrated, mainly by their differences in size. (b) Density gradient centrifugation. To separate mitochondria from lysosomes and peroxisomes, differences in density can be exploited.

called *photosynthetic lamellae* arranged in a parallel array. Chlorophyll and carotenoid pigments are associated with the photosynthetic lamellae, as well as spherical granules called *cyanosomes* containing red and blue pigments (phycocyanin and phycoerythrin), which give these organisms their bluish or sometimes reddish hue. The outermost layer of photosynthetic lamellae may be continuous with the plasma membrane.

Outside the cell wall, which is similar to that of bacteria, there is a gelatinous sheath. Cyanobacteria engage in gliding, rotatory, or vibrational motion. However, they do not possess flagella, and the mechanisms of their movement is still unexplained.

The Eucaryotic Cell of Animals and Plants

All microorganisms, plants, and animals, other than bacteria and cyanobacteria, are eucaryotes. The main distinguishing characteristics of eucaryotic cells are:

1. When not dividing, their genetic material is surrounded by a membrane.
2. Their size (diameter 5–50 μm and volume 65–65,000 μm^3) is large.
3. Their DNA is associated with a large amount of protein.
4. They have a mitotic apparatus for nuclear division.
5. They have a number of organelles surrounded by membranes and supported by a cytoskeleton composed of filamentous elements.
6. They have a highly developed system of intracellular membranes.
7. They have mechanisms for motility that include complex flagella and elaborate dynamic structures for bringing about relative movements of the cell's interior components.

Compartments and Assemblies

We describe the internal architecture of the eucaryotic cell according to two general features. One is that of *compartmentalization.* The eucaryotic cell is divided into a number of compartments, differing in content and structure, each responsible for one or more specialized biological functions. These compartments or *organelles* are usually surrounded by a membrane which serves to regulate the movement of molecules in and out. Examples of such compartments are the *nucleus, chloroplasts, mitochondria, Golgi complex,* **endoplasmic reticulum,** and **lysosomes.**

Another feature of internal eucaryotic cell structure is the pervasive presence of certain *macromolecular assemblies.* These are structures composed of specific macromolecules organized in a specific three-dimensional array. Macromolecular assemblies can occur in the membranes of the cell, in organelles, or in the cytoplasmic matrix surrounding them. Examples of macromolecular assemblies found in eucaryotic cells are the **filaments** and **tubules** comprising the cytoskeleton, and the nuclear chromosomes. Examples of assemblies found in both procaryotic and eucaryotic cells are ribosomes, membranes, and multienzyme complexes.

The Overall Organization of Plant and Animal Cells

Figures 2–7a and 2–8a are generalized drawings of animal and plant cells based mostly on evidence obtained from electron microscopy. These drawings represent idealized versions of cells and contain structures found in a wide range of unspecialized cells. Since the size and relative abundance of these structures vary greatly from cell type to cell type, the student should not consider those shown as necessarily representative of all cells. It is unlikely that an electron micrograph can be made of any one cell showing all these structures simultaneously. We must, therefore, content ourselves with Figures 2–7b and 2–8b, which show the appearance by thin section transmission electron microscopy, and with Figure 2–7c, which shows the appearance by freeze-fracture transmission electron microscopy, of a number of the structures illustrated in the drawings. Even in the best electron micrographs a conscious effort must be made to discern clearly the appearance of the various cell components, a process of observation that is helped by extensive experience.

To summarize this introduction to cell structure: the eucaryotic cell is composed of (1) *a system of membranes* that surround the cell and each of its organelles, and are also found dispersed in the cell interior, (2) a *cytoplasmic matrix* that supports the organelles as well as macromolecular assemblies and food storage products, and (3) a number of *cell organelles.* Let us take each of these in turn.

The Membrane System

Topologically all membrane systems are closed bags—that is, they have an inner and an outer surface but no edges—which separate an interior space from its surroundings. In the eucaryotic cell, the membrane system consists of the *plasma membrane* and its modifications which surround the cell, and *internal membrane systems* which are either separate or form permanent or transient connections with each other and with the plasma membrane.

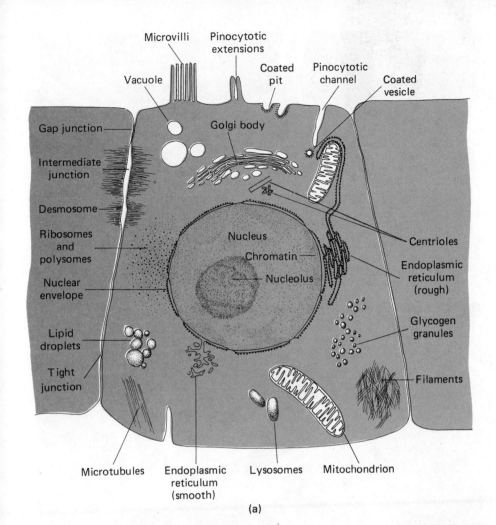

Microvilli Pinocytotic extensions
Vacuole Coated pit Pinocytotic channel Coated vesicle
Gap junction
Golgi body
Intermediate junction
Desmosome
Ribosomes and polysomes
Nuclear envelope
Nucleus
Chromatin
Nucleolus
Centrioles
Endoplasmic reticulum (rough)
Glycogen granules
Lipid droplets
Tight junction
Filaments
Microtubules Endoplasmic reticulum (smooth) Lysosomes Mitochondrion

(a)

FIGURE 2–7

Animal Cells. (a) Drawing of generalized animal cell, showing stylized representations of intracellular structures. (b) Electron micrograph of a liver cell. The picture indicates the nucleus (N), nucleolus (NS), nuclear membrane (nm), rough endoplasmic reticulum (RER), Golgi vesicles (G), the glycogen area (gly), and numerous cytoplasmic mitochondria (M). × 26,000. (Courtesy of G. E. Palade.) (c) Freeze-fracture electron micrograph of a pituitary tumor cell grown in vitro. The entire cell is visible. Nuclear pores appear as pits within the nucleus; the small bumps on the plasma membrane are artifacts caused by ice crystal formation during the freezing process. × 11,000. (Courtesy of J. I. Goldhaber.)

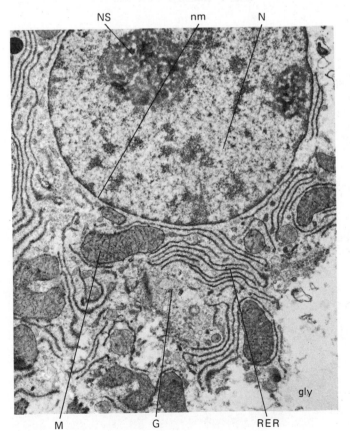

NS nm N

M G RER gly

(b)

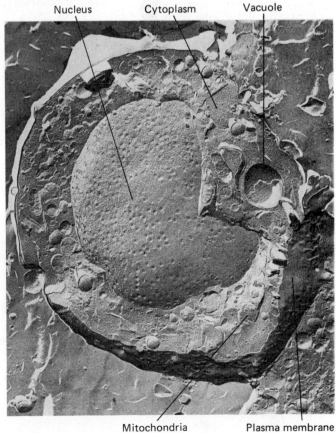

Nucleus Cytoplasm Vacuole

Mitochondria Plasma membrane

(c)

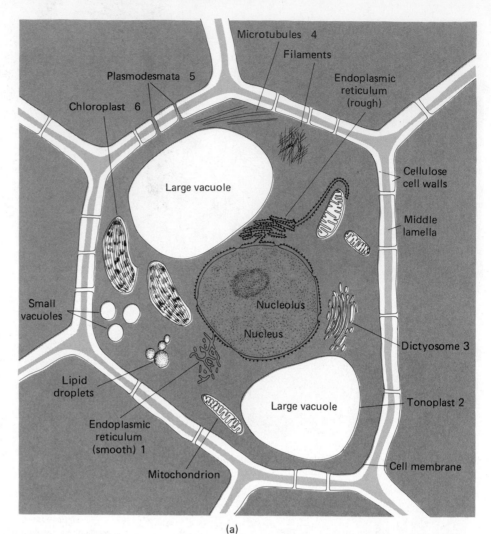

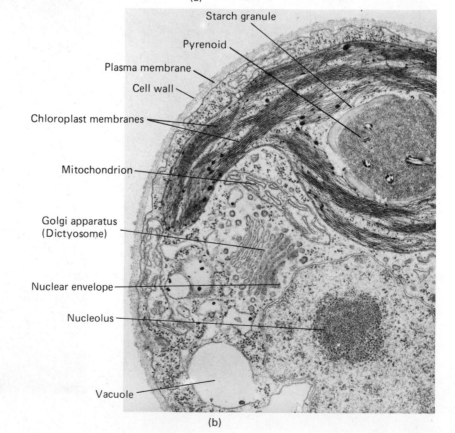

FIGURE 2–8
Plant Cells. (a) Drawing of generalized plant cell. Most elements found in animal cells (cf. Figure 2–7a) are also in plant cells. In addition, a plant cell contains cellulose cell walls, chloroplasts, vacuoles which regulate turgor, plasmodesmata connecting adjacent cells, and a Golgi apparatus (called a dictyosome) near the nucleus. (1) The smooth endoplasmic reticulum is less differentiated from rough endoplasmic reticulum than in animal cells; it is usually confined to cortical cytoplasm. (2) The vacuolar membrane or tonoplast is responsible for turgor and expansion. (3) The Golgi apparatus in plants is divided into "dictyosomes." There is no polar Golgi zone, as in most animal cells. (4) Microtubules in interphase cells are mostly confined to the cell cortex where they lie parallel to the adjacent cellulose microfibrils. (5) Plasmodesmata are unique to higher eucaryotic plant cells and are present in large numbers (thousands/mm²). They are believed to play an important role in cell-to-cell communication and polarity. (6) Chloroplasts usually are apposed to the plasma membrane. In some chloroplasts, fibrils run parallel to microtubules. (b) Electron micrograph of a section of a plant cell, the single-celled alga, *Chlamydomonas reinhardtii.* In this cell there is one large chloroplast occupying about half the cell volume. × 30,000. (Courtesy of G. E. Palade.)

The Plasma Membrane. The surfaces of both procaryotic and eucaryotic cells are delimited by a very sharply defined "skin," the plasma membrane. As we shall see in Chapter 7, this membrane slows the rate of molecular movement in to and out of the cell in a differential or selective manner. Thus, the plasma membrane helps to determine which molecules are allowed into the cell and which are excluded; it also determines which molecules are kept in the cell and which are allowed to escape. This *permeability barrier* function made cell physiologists fully aware of the plasma membrane decades before it was visually revealed by the electron microscope. We now know that the plasma membrane is not only a barrier to passive diffusion, but also contains catalytically active regions and the machinery that utilizes energy to carry out transport work. Through the electron microscope the plasma membrane appears deceptively simple. It is about 8–10 nm (80–100 Å) thick and is composed of three layers that differ in their electron density and in their behavior toward stains; the central region is less electron-dense and stains more weakly than the two external layers (Figure 2–9).

We shall study the molecular structure and function of membranes in considerable detail in Chapter 7. Here we simply point out that: (1) membranes are thin fluid structures not miscible with their aqueous surroundings; (2) they owe these properties to their lipid content; (3) they also contain proteins and protein assemblies or particles (Figure 7–7), which provide them with many of their functional characteristics; and (4) membranes are asymmetric structures, their outer and inner surfaces having different protein and lipid compositions. The important biological roles associated with the plasma membrane include not only regulating the movement of materials in and out of the cell, but also recognizing specifically hormones, neurotransmitters, viruses, and other cells.

Although we usually think of the plasma membrane as the outermost portion of the cell, most cells extrude some material that forms a "fuzzy" layer outside the plasma membrane. This layer can be a coating of protein and carbohydrate like that found outside many animal cells, or between them in the intercellular space. This *extracellular matrix*, sometimes called the *glycocalyx*, will be discussed in greater detail in Chapter 17. It was first described at the electron microscope level in epithelial tissue as a series of branched filaments (2.5–5 nm diameter) in close juxtaposition with the plasma membrane (Figure 2–10a, b). In other tissues, a *basement membrane*, or *basal lamina*, can be observed. It is composed of a thin translucent layer surrounded by a thick electron-dense layer (Figure 2–10c). The composition of the extracellular matrix is unusual in that it contains filamentous structural glycoproteins, such as *collagen*, *elastin*, and *fibronectin*. These proteins are major bodily constituents; collagen alone accounts for more than half the total protein in an adult mammal.

In most procaryotic cells and in most plant cells the extracellular layer is a complex, often multilayered structure called the *cell wall* (Figure 2–11; also see Figure 2–8b, which shows a cross section of the wall). In plant cells this wall, composed mainly of cellulose fibers, is a thin, but very strong, elastic structure built by the cell with the aid of special enzymes that link smaller molecular units into a continuous sac-like envelope.

Before the advent of the electron microscope we used to think of the plasma membrane as a thin coat, stretched tightly over the cell, as it indeed appears to be in the case of a few cells, such as the erythrocyte. Now we know that this is often not the case. The plasma membrane surface area is frequently amplified by numerous elongated slender projections called *microvilli* (Figure 2–12). These are found in cells active in the transport of material, such as absorptive cells in the intestinal epithelium, or secretory cells. Remarkably, microvilli have a very similar size and shape regardless of the cell type displaying them.

Besides the stable microvilli, the plasma membrane also engages in transitory evaginations and invaginations. Thus, the cell extends its surface to form folds or protrusions, cross sections of which can be readily seen under the electron microscope and which in some cells are large enough to be observed even under the light microscope. These protrusions often fuse on their outer margins, thus entrapping some liquid that is taken into the cell in the form of a membrane-bound *vesicle* or *vacuole*. Such cellular importation, which involves bulk transport of the external solution, is called **endocytosis**

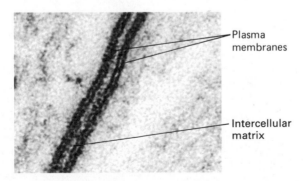

Plasma membranes

Intercellular matrix

FIGURE 2–9
Plasma Membranes. The boundary of two glial cells of the annelid *Aphrodite* is shown. Notice that each plasma membrane appears as three layers, the central region staining less strongly than the two external layers. × 260,000. (Courtesy of D. W. Fawcett.) See Figure 2–12b for another clear representation of trilaminar structure.

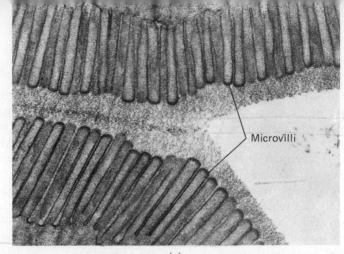

Microvilli

(a)

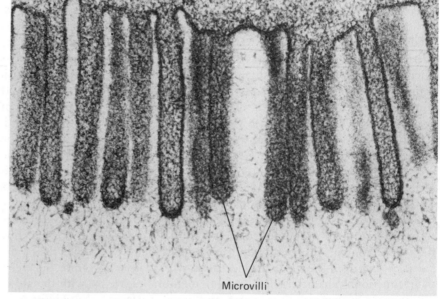

Microvilli

(b)

Intermediate
filaments

Cell membrane

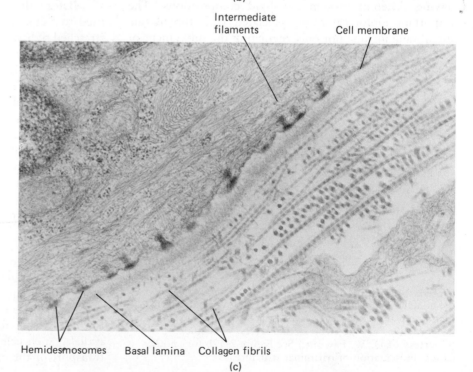

Hemidesmosomes   Basal lamina   Collagen fibrils

(c)

Flap of tissue rolled
up in sectioning
preparation

Helical thickening

Primary cell wall

FIGURE **2–11**
Plant Cell Wall.   Electron micrograph of a direct carbon
replica of a plant cell wall showing the multilayered structure.
S_1, S_2, S_3 = successive layers of secondary wall with cellulose
fibers running at different angles. × 3,550.
(Courtesy of W. A. Cote, Jr.)

(Figures 2–13 and 2–14). Endocytosis is measured by the
ingestion of an iron-containing protein, ferritin, which
can be seen in the electron microscope and is therefore
used for the study of this phenomenon. *Pinocytosis* refers
to the "drinking" of molecules in solution while *phagocytosis* refers to the ingestion of large particles such as
bacteria. The bulk transport of materials into cells
through topological changes in the membrane (see
Chapter 7) has attracted widespread attention in recent
years, though it was first observed by W. Lewis in 1935.
The phagocytic role of endocytosis will be discussed later
in this chapter (and in Chapter 7) when the lysosome is
presented (Figures 2–44, 2–45, 2–46). The plasma membrane is also involved in *secretion,* or **exocytosis,** of
materials from the cell into the extracellular space (see
below and in Chapters 7 and 13).

Finally, when cells are in contact with each other, the
plasma membrane forms structures which are involved in
cell-to-cell attachment and interactions, for example in
forming the tissues and organs of multicellular organisms. Although attachment between cells probably occurs all along the surface membrane, certain structures
provide additional special connections thereby anchoring
the cells to each other on a more permanent basis.
Furthermore, some membrane structures are believed to
be involved in cell-to-cell interactions that integrate the
behavior of cells (as in the case of heart muscle cells) or in
the growth and differentiation of a developing embryo.
The structures and functions of cell-cell contacts will be
discussed in Chapter 17.

The Endoplasmic Reticulum (ER).   As we have
already mentioned, the eucaryotic cell normally contains
extensive internal membrane systems. One of these is the
endoplasmic reticulum, a system of membrane-limited

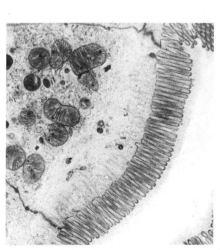

(a)

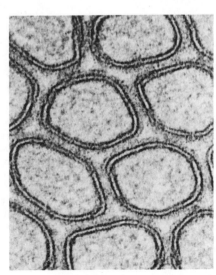

(b)

FIGURE **2–12**
Microvilli.   (a) Longitudinal section
of intestinal microvilli of the cat.
× 25,000. (Courtesy of S. Ito.)
(b) Transverse section of intestinal
microvilli. The triple-layered
appearance of the plasma membrane
can be clearly seen. × 230,000.
(Courtesy of D. W. Fawcett.)

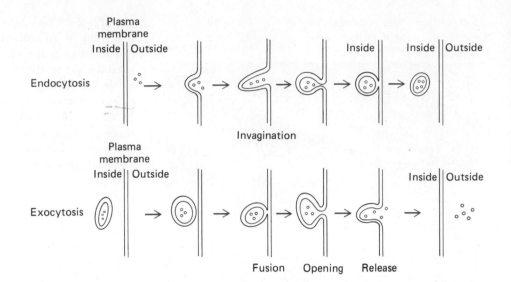

Plasma membrane

Inside || Outside

Endocytosis

Invagination

Plasma membrane

Inside || Outside

Exocytosis

Fusion   Opening   Release

FIGURE 2–13
Endocytosis and Exocytosis.
Diagrammatic representation of the processes.

channels distributed throughout the cell. These channels were first discovered with the electron microscope in thinly spread tissue culture cells in 1945 and were named by K. R. Porter in 1953. Although proof of the existence of the ER had to await the development of electron microscopy, the light microscope had already provided some evidence for it. In 1942, G. W. Scarth concluded

that there is a filmy membrane system in the cytoplasm of plant cells, connecting with the membrane of chloroplasts and of the nucleus (Figure 2–15). (Despite working at the limits of resolution of the light microscope, Scarth even recognized the nuclear membrane as being double in nature!) Scarth described this membrane system as having fluid properties and being "prominently lipoidal, though

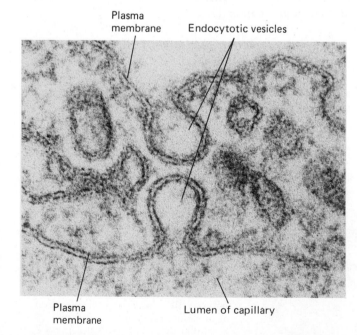

Plasma membrane   Endocytotic vesicles

Plasma membrane   Lumen of capillary

FIGURE 2–14
Endocytotic Vesicles.   Electron micrograph showing endocytotic (pinocytotic) vesicles in the process of forming on both sides of a blood capillary cell. × 315,000. (Courtesy of G. E. Palade.)

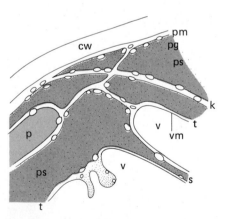

FIGURE 2–15
Diagrammatic Cross Section of the Cytoplasm of a Plant Cell.   Shown are plasma membrane (pm), vacuolar membrane (vm), and membrane around plastid (p) being interconnected by strands of "kinoplasm" (k). Kinoplasm was, according to Scarth, a fluid system which, like membranes, is composed of lipid and protein. This description anticipated the endoplasmic reticulum subsequently discovered by electron microscopy. Other structures are: plasmasol (ps), plasmagel (pg), plant cell wall (cw), and transvacuolar strand (s). (After G. W. Scarth, 1942.)

no doubt proteins also enter into its composition." He concluded that it was similar (though probably not identical) to the plasma membrane and the membranes that surround chloroplasts and nuclei. What is even more remarkable, Scarth described these membranes as a flowing dynamic system, constantly making and breaking interconnections, a property of the ER which has only recently been fully recognized. We mention this early description of the endoplasmic reticulum because it is an interesting example of a "premature" discovery —that is, one which depended on extraordinary powers of observation on the part of the experimenter, and was therefore not easily accessible to other workers. The advent of the electron microscope broadened the study of cytology, making available for observation phenomena that previously were seen only by the most skilled and imaginative investigators.

By the use of ultrathin sections and improved fixation techniques developed by Palade and Porter (1954), it was finally recognized that the endoplasmic reticulum represents cavities or channels of a great variety of shapes and dimensions, surrounded by membranes, all connected with one another (Figure 2–16). Subsequent work soon showed that the endoplasmic reticulum is found almost universally in eucaryotic cells. Since membranes can be clearly resolved in the electron microscope, the first impression one obtains from an electron micrograph of a thin section of a cell is that an extensive intracellular membrane system exists (Figures 2–17 and

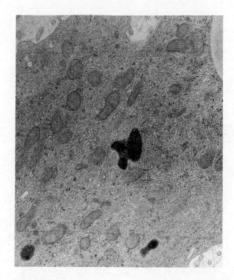

FIGURE **2–17**
Endoplasmic Reticulum of Testis Cell.   Some of the membranes have ribosomal particles attached to them (see Figure 2–18)—hence, they are called rough ER—while most ER membranes in this picture are devoid of particles, and are called smooth ER. The membrane system seems to "wrap around" the numerous mitochondria. × 35,000. (Courtesy of S. Ito.)

2–18). By filling the interior of the ER with an electron-dense material it is easy to demonstrate that there is an inside and an outside to this membrane system (Figure 2–19); i.e., that it forms a series of interconnected

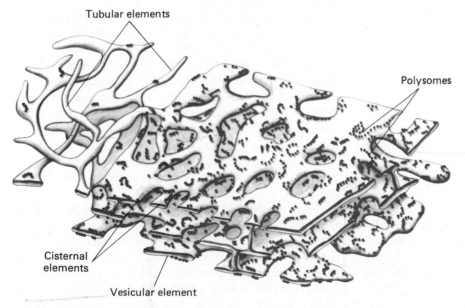

FIGURE **2–16**
Elements of the Endoplasmic Reticulum.   Three-dimensional drawing of different elements (cisternal, tubular, and vesicular) of the ER, showing the interconnections between them. (See also Figure 2–22.) (Redrawn from D. W. Fawcett and W. Bloom.)

FIGURE 2–18
Endoplasmic Reticulum of Pancreatic Acinar Cell.
The ER is formed here as a concentric array of a continuous
network of membrane-bound channels. Ribosomal particles R
are attached to that side of the membrane facing the
cytoplasm (c), but not on the inner side facing the channels.
× 65,000. (Courtesy of S. Ito.)

channels—vesicles (spherical, circular in cross section),
tubules (cylindrical), or *saccules* (*cisternae* = flattened
channels)—arranged in a loose array or parallel (Figure
2–17), or in concentric rows (Figure 2–18).

In some cells the endoplasmic reticulum is greatly
distended, taking up a large portion of the cell volume
(Figure 2–20). In other cells it becomes packed with
granules, droplets of lipid, and even with protein crystals.
In the muscle cells a similarly appearing structure occurs,
called the *sarcoplasmic reticulum,* which surrounds the
myofibrils (Figure 2–21) and is very intimately connected
with the process of contraction (see Chapter 16). These
numerous variations in structure and content of the
endoplasmic reticulum occur not only from one cell type
to another, but also in different developmental and
physiological stages of the same cell type, thus giving the
impression that it is a labile system capable of rapid
modifications in structure and function.

The membrane of the endoplasmic reticulum appears to be slightly thinner (5–6 nm, 50–60 Å) than the
plasma membrane. As we shall see in Chapter 7, studies
of the chemical composition of these two membrane
systems reveal some significant differences.

We know of a variety of structural roles played by
the endoplasmic reticulum though it is likely that the list
is far from complete. In general, it increases the amount
of membrane surface in contact with the cell interior and
divides the cell into two compartments—the cytoplasmic
ground substance or matrix, and the ER interior. The
consequences of these two general structural features are:

1. To provide a large amount of surface for membrane-
   bound enzyme systems synthesizing a number of
   substances; for example, steroid hormones.
2. To provide a system of channels which can transport
   synthetic products from one part of the cell to another
   or segregate them in a particular portion of the cell.
   The ER at times forms intermittent connections with
   the Golgi complex—small vesicles bud off the ER and
   fuse with the Golgi membranes. The ER also forms
   more persistent connections with the nuclear envelope. In both cases channels are opened between
   intracellular compartments.

There are two structurally dissimilar forms (see
Figure 2–16) of the ER: the *rough ER,* in which ribosomes
line the cytoplasmic (outer) face of the membranes
(Figure 2–18), and the *smooth ER,* which is free of
ribosomes (Figure 2–17).

A number of cell types are known, such as the acinar
cells of the pancreas, in which large quantities of protein
are synthesized for export. These cells contain large
amounts of rough ER arranged as flattened vesicles
(Figure 2–18). Tangential sections of the membrane face
show most of the ribosomes are arranged in short rows,
rings, or rosettes. These ribosomes are joined together in
a structure called a **polysome** or polyribosome (Figure
2–22). We shall see in Chapter 13 that such a linear
arrangement of ribosomes means that they are connected
by a strand of messenger RNA and involved in protein
synthesis.

The synthesized proteins are thought to be threaded
through the membrane of the ER into its interior where,
after being processed by enzymes located there, they
finish folding into their three-dimensional structure.
Once in the ER, the proteins move near the region of the
Golgi complex and are packaged into *intermediate vesicles*
which, in turn, transfer the proteins into the Golgi
complex for further concentration and packaging (see
Golgi Complex, below). The eventual fate of the proteins
is secretion out of the cell by exocytosis into the lumen of
a duct, as shown in Figure 2–23. Further details of this
process are described in Chapter 13.

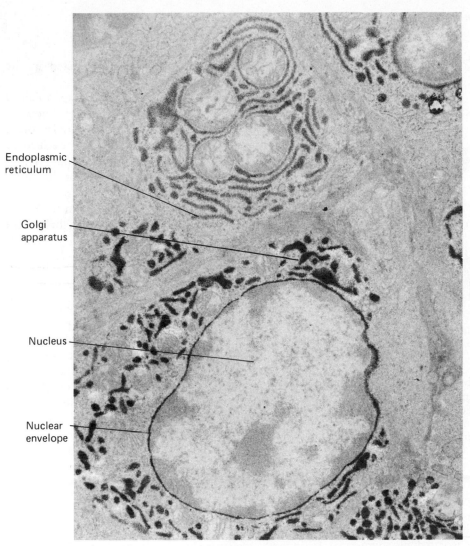

Endoplasmic reticulum

Golgi apparatus

Nucleus

Nuclear envelope

FIGURE 2–19
Endoplasmic Reticulum Composed of Enclosed Vesicles. A rat was first immunized by injection with the protein horseradish peroxidase. Plasma cells, which produce antibodies to the peroxidase, were then obtained from the rat. The inside of the ER of plasma cells accumulated these antiperoxidase antibodies. This could be shown by first incubating thin slices of the cells in a solution containing peroxidase. Then, the peroxidase activity that was bound to its antibody was localized by allowing it to catalyze a reaction that deposits electron-dense material in the vicinity of the enzyme. This electron-dense material within the ER indicates the presence of the antibody within these structures, as well as within the Golgi apparatus and the nuclear envelope. × 7,500. (Courtesy of E. H. Leduc.)

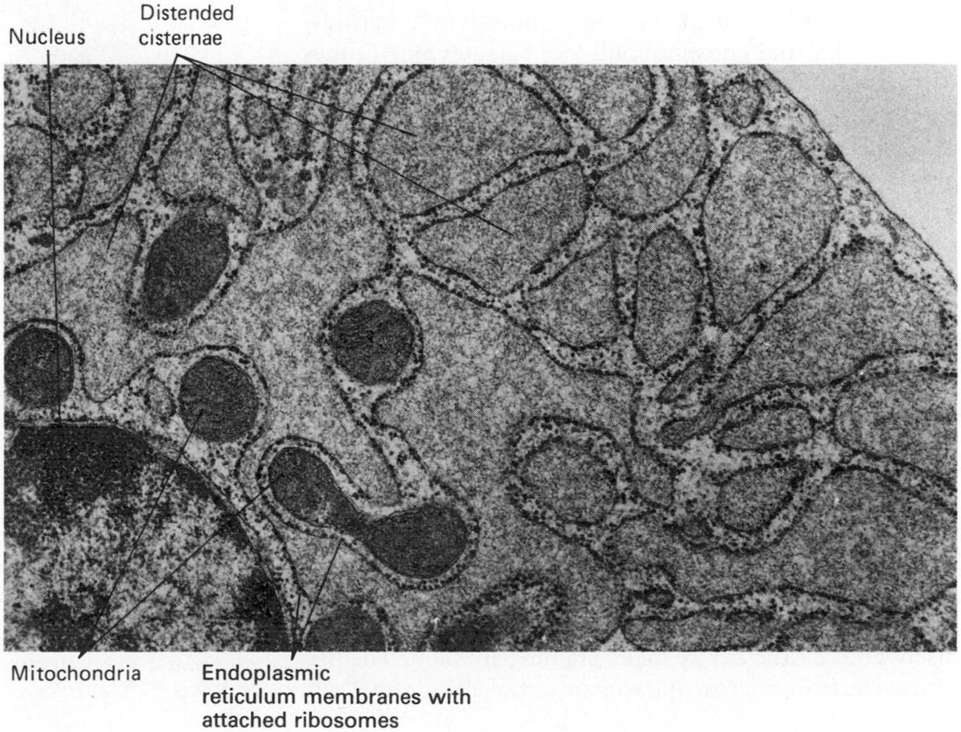

FIGURE 2–20
Endoplasmic Reticulum From a Plasma Cell in Bone Marrow. Due to the presence of large amounts of protein the cisternae of the ER (the darker gray areas) are greatly distended. The lighter gray areas are the spaces of the cytoplasmic matrix, as can be seen by the presence of mitochondria within these spaces. × 26,000. (Courtesy of D. W. Fawcett.)

Nucleus

Distended cisternae

Mitochondria

Endoplasmic reticulum membranes with attached ribosomes

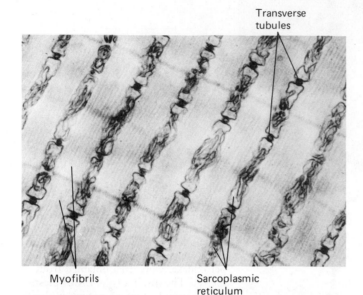

Transverse
tubules

Myofibrils

Sarcoplasmic
reticulum

FIGURE 2–21
**The Two Types of Membrane Systems Found in
Striated Muscle Cells.**   The sarcoplasmic reticulum, named
because it resembles the endoplasmic reticulum, stores Ca^{2+}
and releases it to the myofibrils. The slender cylindrical
transverse tubules are known to be continuous with the
plasma membrane. × 30,000. (Courtesy of D. W. Fawcett
and J. P. Revel.)

The ER is also a principal site of lipid synthesis in the
eucaryotic cell. Biochemical study of isolated ER membranes, both rough and smooth, show that they contain
the enzymes involved in the synthesis of steroids,
triglycerides, and phospholipids (see Chapter 3). At times
it is possible to see the accumulation of triglyceride lipid
droplets inside the ER. Steroid biosynthesis seems to
occur in connection with the smooth ER.

The Golgi Complex (GC).   The Golgi complex is
an internal membrane system composed of membrane-
bound cisternae, sufficiently characteristic in organization, location, and function to warrant classification as a
separate cell organelle (Figure 2–24). This structure was
discovered in 1898 by C. Golgi, who used a silver
impregnation technique to stain it. In the ensuing years
there was considerable disagreement about the existence
of this organelle, many cytologists believing that it was
an artifact of fixation or staining procedures. However,
studies with the phase contrast microscope in the early
1940s showed that there was a region of different
refractive index in living cells, usually close to the
nucleus, near the area in which the Golgi complex was
discovered by the heavy metal staining method. The use
of the electron microscope to study thin sections of cells

finally proved that a special organelle existed in the
precise location assigned to it by Golgi.

Figure 2–24b shows an electron micrograph of a cell
that was previously stained with osmium. It shows a
series of tightly packed smooth vesicles in parallel, often
semicircular, array. The outer vesicles alone are heavily
impregnated with electron-dense osmium stain. Thus, it
appears that Golgi saw only the outer part of the Golgi
complex.

The shape and arrangement of vesicles in the Golgi
complex vary a great deal from cell type to cell type and
even at different stages of the life history of a given cell.
The curved, disc-like vesicles, flattened at the center and
distended at the rims, are interconnected with each other.
Those vesicles at the convex side (also known as the *cis*
side) are very narrow, whereas the ones at the concave
side (also known as the *trans* side) are much wider,
especially when the Golgi complex is metabolically
active. At the convex side of the complex are numerous
small vesicles (30–80 nm, 300–800 Å). At the concave
side, especially in secretory cells, one observes enlarged
spherical structures at the ends of the Golgi cisternae with
varying amounts of secretory products *(condensing vacuoles)* and mature *secretory (zymogen) granules,* presumably
on their way to the apical portion of the secretory cell
(Figure 2–25). (See Chapter 13 for more details.)

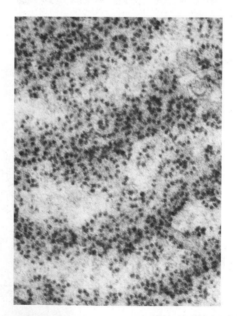

FIGURE 2–22
Rough Endoplasmic Reticulum.   Tangential section
showing cytoplasmic side of ER membrane with attached
ribosomes, many in curved clusters of polyribosomes.
× 100,000. (Courtesy of D. W. Fawcett.)

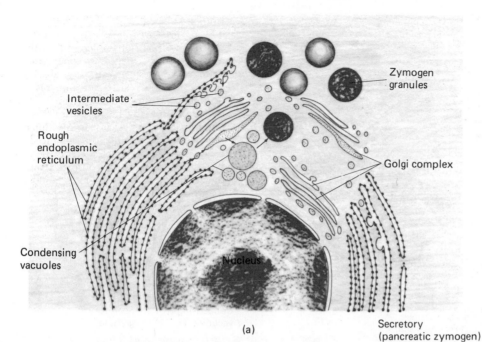

(a)

FIGURE 2–23

Secretion of Proteins. (a) The diagram shows the pathway of secretory protein transport, the rough ER ribosomes synthesizing proteins which are then discharged into the lumen or cisternae of the ER. From there the proteins are transported via intermediate or transition vesicles to the Golgi apparatus where the proteins are concentrated and packaged into the condensing vacuoles that, upon loss of water, finally form the secretory (zymogen) granules. Eventually the secretory granules discharge their contents into the lumen of the duct of the gland. (After D. W. Fawcett.) (b) Electron micrograph of pancreatic acinar cell, showing the secretory granules containing proteins packaged for export. × 25,000. (Courtesy of S. Ito and D. W. Fawcett.)

Membranes of the Golgi complex are smooth, that is, free of ribosomes. Their thickness is usually greater than 8 nm (80 Å), making them comparable to the plasma membrane. In some cells one sees thinner, 6 nm (60 Å) membranes more like the ER membranes, at the convex side of the Golgi complex.

A number of functions of the Golgi complex are known, although the list is probably far from complete. In plant cells, the Golgi complex is involved in the synthesis of polysaccharides used to build the cell wall surrounding the cell. In procaryotes, by contrast, the plasma membrane surrounding the cell probably is responsible for the synthesis and assembly of the cell wall. In eucaryotic plant cells, this process is presumably made more efficient by the complex, which not only synthesizes the polysaccharides but also assembles them into macromolecular structures, such as **fibrils** or

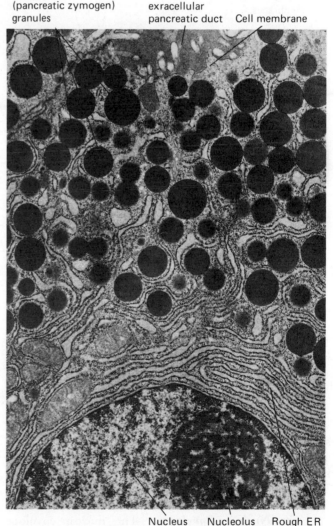

(b)

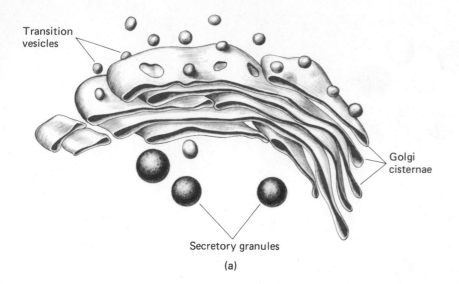

Transition vesicles

Golgi cisternae

Secretory granules

(a)

FIGURE 2-24
The Golgi Complex. As shown in this three-dimensional diagram, on the outside of this structure of concentric flattened vesicles one can often observe small transition vesicles, presumably originating from the endoplasmic reticulum and later fusing with the outermost Golgi vesicles. In glandular cells, secretory granules are produced by the innermost vesicles. (After D. W. Fawcett.) (b) Golgi complex from the epididymus of a mouse. × 50,000. (Courtesy of D. S. Friend.)

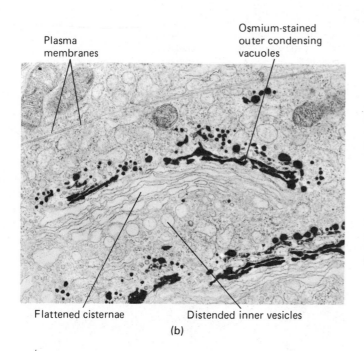

Plasma membranes

Osmium-stained outer condensing vacuoles

Flattened cisternae

Distended inner vesicles

(b)

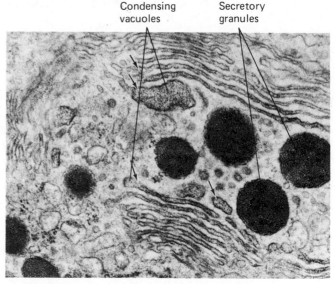

Condensing vacuoles

Secretory granules

FIGURE 2-25
The Golgi Apparatus of the Epithelial Cell of Brunners's Gland of a Mouse. Arrows point to the enlarged ends of the Golgi vesicles that are filling with secretory products. These condensing vacuoles eventually pinch off to form the electron-dense secretory granules. × 54,000. (Courtesy of D. S. Friend.)

networks, which are then transported to the surface and assembled further into the complex multilayered cell wall. In animal cells the Golgi complex is involved in the synthesis or modification of polysaccharide chains which are attached to proteins or lipids. The most carefully studied function is the processing of secretory proteins. This is discussed in detail in Chapter 13.

The Nuclear Envelope. The nuclear envelope consists of two membranes surrounding the nucleus of the eucaryotic cell (Figure 2–26). During most of nuclear division in higher eucaryotes, when the chromosomes go through the complicated dance that results in the equal distribution of hereditary material into the two daughter cells, the nucleus is not surrounded by a membrane system. During telophase, the last stage of nuclear division, a number of flat vesicles or sacs of the endoplasmic reticulum arrange themselves around the chromosomes and fuse extensively at their edges to form the envelope.

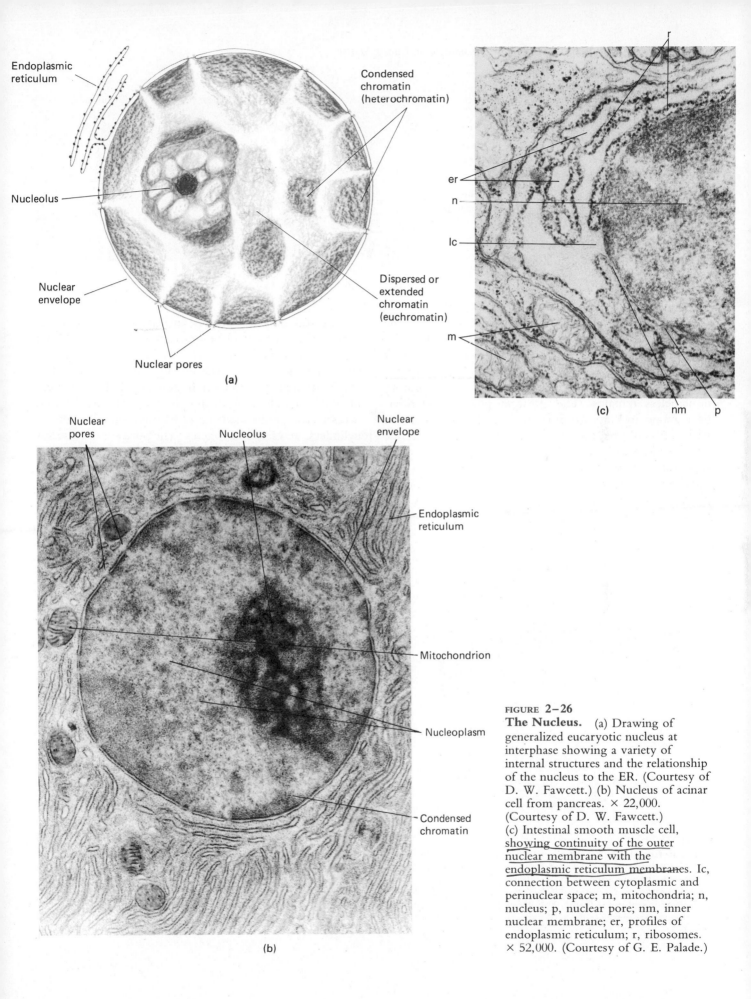

Endoplasmic reticulum

Condensed chromatin (heterochromatin)

Nucleolus

Nuclear envelope

Dispersed or extended chromatin (euchromatin)

Nuclear pores

(a)

r

er

n

lc

m

nm   p

(c)

Nuclear pores

Nucleolus

Nuclear envelope

Endoplasmic reticulum

Mitochondrion

Nucleoplasm

Condensed chromatin

(b)

FIGURE 2–26

The Nucleus. (a) Drawing of generalized eucaryotic nucleus at interphase showing a variety of internal structures and the relationship of the nucleus to the ER. (Courtesy of D. W. Fawcett.) (b) Nucleus of acinar cell from pancreas. × 22,000. (Courtesy of D. W. Fawcett.) (c) Intestinal smooth muscle cell, showing continuity of the outer nuclear membrane with the endoplasmic reticulum membranes. Ic, connection between cytoplasmic and perinuclear space; m, mitochondria; n, nucleus; p, nuclear pore; nm, inner nuclear membrane; er, profiles of endoplasmic reticulum; r, ribosomes. × 52,000. (Courtesy of G. E. Palade.)

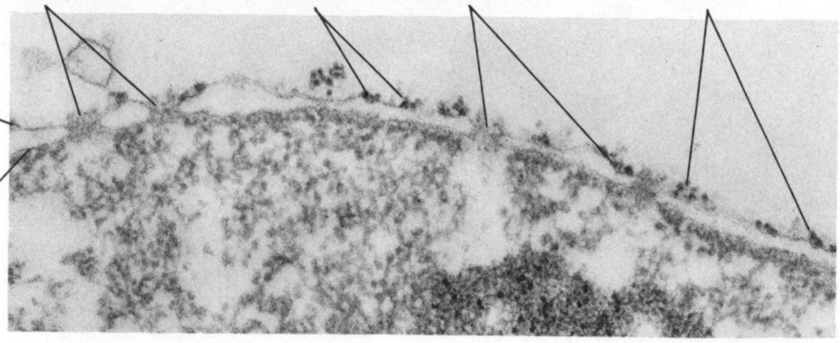

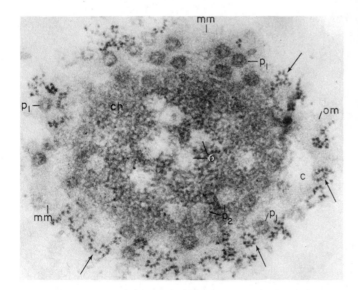

Nuclear pores   Ribosomes   Nuclear pores   Ribosomes

Outer membrane

Inner membrane

FIGURE 2–27
Nuclear Envelope. Higher magnification of nuclear envelope area in a liver cell, showing outer membrane, inner membrane, pores, and ribosomes (arrows) attached to outer membrane. × 70,000. (Courtesy of G. E. Palade.)

Included in this developing nuclear envelope are the distinctive structures called "pores" (Figures 2–26, 2–27, 2–28, 2–29). This name was given when it was thought they were actually holes. We now know they have a structure more akin to plugs, but the name "pores" nonetheless remains. The apparent pore diameter has been found to range from 60 to over 100 nm, but this variation could be due to the various methods of tissue preparation which have been used. The pores seem to be randomly arranged around the surface of the nucleus; their density in various cells ranges from 15–50 pores/μm^2, which comes to several thousand per nucleus.

FIGURE 2–28
Nuclear Envelope. This *en face* view is seen from the cytoplasm looking in toward the nucleus. Shown are pores (p_1, p_2) in the envelope, polysomal arrays of ribosomes (arrows) attached to outer membrane (om), overlap of inner and outer membranes (mm), and channels (o) cut through peripheral chromatin (ch). × 77,000. (Courtesy of G. E. Palade.)

Surrounding the pore is an *annulus,* or ring, composed of eight particulate aggregates arranged in exact radial symmetry around the pore periphery. These structures can be easily seen in a preparation of detached and isolated pore complexes (Figure 2–30). The structure of these annular aggregates is thought, by various investigators, to be made up of filament bundles, microcylinders, or a series of spheres. The pore complex seems to be supported by a peripheral protein layer called the *lamina,* which stabilizes the interaction between the pore complex and the nuclear envelope; the protein lamina with its pore complex can even be isolated free of the nuclear envelope. The chemical composition of this whole complex is mostly protein, with some carbohydrate and some RNA in the form of ribonucleoprotein.

It is almost certain that the function of the pore complex is involved with transfer of macromolecules between nucleus and cytoplasm. By using molecules of known size and shape, researchers found that the pores are actually sieves, allowing the easy passage of particles of up to 4.5 nm in diameter and 4,500 Da in mass, and the slow penetration of particles of 12,000–67,000 Da. The outer membrane of the nuclear envelope is continuous with the membranes of the endoplasmic reticulum. It sometimes is lined with ribosomes and polysomes (Figure 2–28). The inner membrane is lined with, and probably supported by, a filamentous layer or *fibrous lamina* and is intimately associated with chromatin (DNA-containing) material.

Annulate Lamellae. In a variety of cells, especially in germ cells of invertebrates and vertebrates, one can observe a highly organized system of membranes reminiscent of the nuclear envelope and, indeed, believed to be derived from it. This organelle consists of parallel arrays of flat vesicles interrupted by circular pores of uniform size (Figure 2–31). What is so striking about the appearance of annulate lamellae is that they are organized in register, the pores being stacked in a precise linear

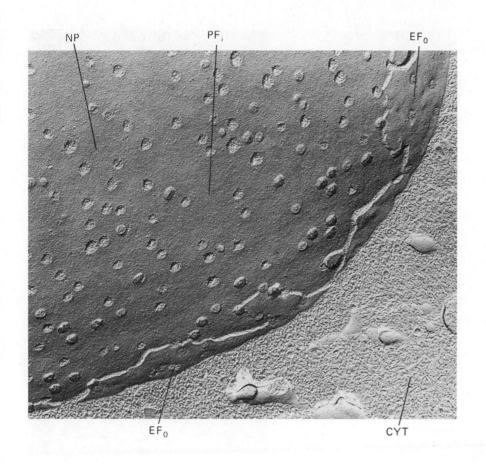

FIGURE 2-29
Nuclear Pores as Visualized by Freeze-fracture Electron Microscopy.   An *en face* view of the nuclear envelope, as seen from the cytoplasm, with the pores cross-fractured and appearing as flattopped elevations or shallow craters. The predominant visible surface is the P face (= protoplasmic = towards the protoplasm) of the inner membrane (PF_i) of the nuclear envelope, although the E face (= external = towards the outside) of the outer membrane (EF_0) is present near the rim. Beyond this is the cell cytoplasm (CYT). Studded across the face of the nucleus are many nuclear pore complexes (NP). $\times$ 54,000. (Courtesy of J. I. Goldhaber.)

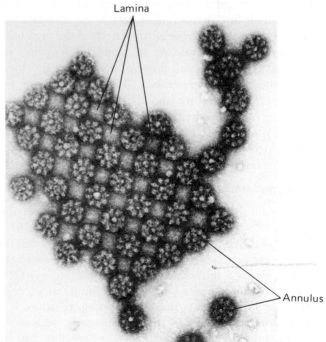

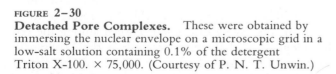

FIGURE 2-30
Detached Pore Complexes.   These were obtained by immersing the nuclear envelope on a microscopic grid in a low-salt solution containing 0.1% of the detergent Triton X-100. $\times$ 75,000. (Courtesy of P. N. T. Unwin.)

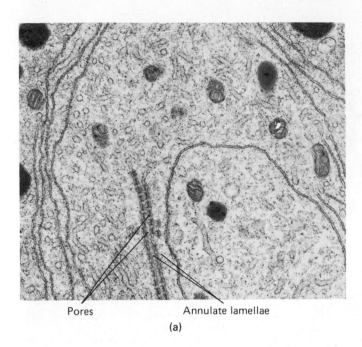

Pores   Annulate lamellae

(a)

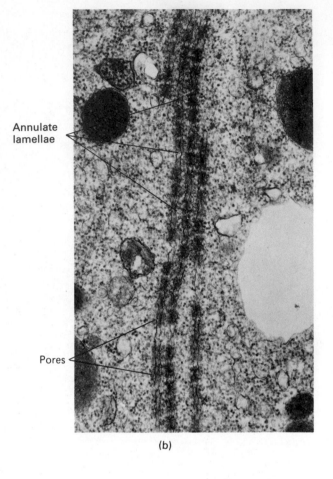

Annulate
lamellae

Pores

(b)

FIGURE 2–31

Annulate Lamellae in a Sea Urchin Egg.   The lamellae
are single or double rows of membrane-bound narrow
cisternae, with the dense differentiations lined up in register,
resembling nuclear pores. (a) = × 10,000; (b) = × 50,000.
(Courtesy of S. Ito.)

array. This is apparently achieved by the presence of
annuli which line the pore and which are composed of
fibrils running through the entire length of the organelle.
Furthermore, the pores of the annulate lamellae are
frequently in register with those of the nuclear envelope,
which they closely resemble. The functions of the
annulate lamellae are not known, but they might represent a way by which the cell can achieve an increase in the
amount of nuclear envelope material, thereby increasing
some biochemical function which is specific to the
nuclear envelope.

The Cytoplasmic Matrix

The development of the electron microscope permitted us for the first time to see the numerous membrane
systems of the cell, while it is only in recent years that the
cytoplasmic matrix or *ground substance* has come into the
focus of electron microscopy. The cytoplasmic matrix is
an aqueous phase containing large amounts of proteins
and nucleoproteins in various states of aggregation. It is
the supporting medium of the cell organelles such as the
nucleus, mitochondria, chloroplasts, ribosomes, and lysosomes. The ground substance has an unusual property:

it is capable both of viscous flow like a liquid and elastic
deformation like a solid. Furthermore, depending on the
particular physiological or developmental state of the cell,
the fluidity and solidity of the ground substance varies.
The cytoplasmic matrix near the plasma membrane,
often referred to as the **cortex** or *ectoplasm*, generally
behaves more like a solid, whereas the matrix in the
interior of the cell, or *endoplasm*, is generally in a more
fluid state.

The viscous properties of the ground substance were
studied extensively in the 1930s by (1) micromanipulation, (2) measuring in various portions of the cell the rate
of *Brownian movement* of particles (the random motion of
microscopic particles caused by the impacts of molecules
in the surrounding solution), (3) measuring the rate of
movement of cytoplasmic particles in the gravitational
field developed by centrifugation, and (4) other ingenious
techniques such as observing the movement of small
particles of iron in a magnetic field. As a result of these
measurements, lively discussions ensued among the cell
physiologists of that period regarding the "colloidal"
properties of the ground substance. Most workers agreed
that although the ground substance appeared uniform in

the light microscope, it must nevertheless contain a submicroscopic skeleton responsible for its elastic properties. These early workers conceived a "cytoskeleton" of highly elongated particles interacting with each other to form a "brush heap" or gel. In 1950 Francis Crick wrote about cytoplasm being like "mother's sewing basket," filled with spools of thread, buttons, and knitting needles in untidy array.

At first it looked as if no cytoskeletal "knitting needles" or elongated particles were likely to be found in the ground substance, except for the very obvious fibrillar organization of special cells, such as striated muscle cells, or special cellular structures, like the spindle, the centrioles, or the cilia. Then, as techniques improved, electron microscopists discovered **microfilaments, microtubules**, and then **intermediate filaments** in a large variety of cell types (see Table 2–3). It is now clear that these elongated rods are a universal component of the cytoplasmic matrix. They are generally arranged in a regular network, where they comprise a cytoskeleton that defines cell shape, cell movement, and intracellular compartmentalization (see Chapter 16 for details).

Microfilaments of Various Kinds.   Cytoplasmic microfilaments are rods of indefinite length and 5–7 nm in thickness. Their presence is, of course, most obvious when they are numerous and well organized, as the thin filaments in striated muscle (Figure 2–32), or in the axons of nerve cells. Once cytoplasmic filaments were recognized in their abundant and more organized manifestations, they were observed even in less specialized cells such as the amoeba, or undifferentiated plant cells.

The *thin filaments* (5–7 nm, 50–70 Å) in striated muscle are composed of *actin, tropomyosin,* and *troponin,* proteins involved in contractile processes (see Chapter 16 for a detailed discussion of these filaments and their functions). More recently such *5–7 nm microfilaments* have been found almost universally in cells and in many cases are shown to contain actin. Ingenious techniques were developed to demonstrate that this is so. One of these (developed by H. Ishikawa, following up earlier work by H. E. Huxley) utilizes heavy meromyosin, a digestion product of myosin protein from striated muscle. As we shall see in Chapter 16, actin and myosin bind together in a specific interaction during muscle contraction. When heavy meromyosin is added to the 7 nm microfilaments of nonmuscle cells, a distinctive "arrowhead" pattern is observed in negatively stained electron micrographs; the microfilaments are said to be "decorated" by heavy meromyosin. This shows that the 7 nm microfilaments contain a protein very closely related to muscle actin. Another method, improved by E. Lazarides and described above, uses antiactin antibodies and the indirect

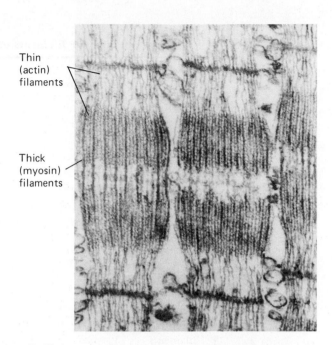

Thin (actin) filaments

Thick (myosin) filaments

FIGURE 2–32
Striated Rabbit Muscle.   The electron micrograph shows the highly organized filaments found in muscle cells. The thin and thick filaments are clearly seen. × 30,000. (Courtesy of H. E. Huxley.)

immunofluorescence method, making it possible to visualize bundles of actin in cells examined under the fluorescent microscope.

Actin microfilaments appear to fall into two groups. *Lattice filaments,* which are disrupted after treatment with the plant alkaloid *cytochalasin D,* are arranged in a loose network that is frequently in contact with the plasma membrane. *Sheath filaments,* also affected by cytochalasin D, are arranged in compact longitudinal bundles at places of cell-substratum contact. These are sometimes referred to as *stress fibers,* but this is somewhat of a misnomer since they seem to be formed mostly during maintenance of the cell in vitro. Evidence has accumulated showing that cytoskeletal elements interact with the cell membrane. This interaction is brought about by proteins such as *spectrin,* which lie at the inner surface of the membrane (Figure 2–33).

Intermediate Filaments.   These represent another class of cytoplasmic filaments and are also found in many cell types. They are 10–12 nm in diameter and have a composition different from that of microfilaments. From their appearance and interconnections it is suspected that their main function is to provide a cellular architecture. Unlike actin or microtubules, whose function is likely related to movement (see Chapter 16 for actin's role in muscle and Chapter 11 for the role of microtubules in

TABLE 2–3
Cytoskeleton Fibrillar Elements of the Cytoplasmic Matrix

Class	Size	Subunit Mol. Wt.	Characteristics
Microfilaments	5–7 nm diam.	45,000 (actin)	An example is actin, which binds to heavy meromyosin in muscle cells; also found in many other cell types as part of cytoskeleton. See Chapter 15. Disrupted by cytochalasin D.
a. "lattice" filaments			Arranged in loose network close to plasma membrane.
b. "sheath" filaments (stress fibers)			Arranged in longitudinal compact bundles at places of cell-substratum contact.
Intermediate filaments	10–12 nm diam.	various	Made up of a number of distinct proteins, found in various cell types. Discussed fully in Chapter 16.
Thick filaments	15 nm diam.	220,000 (myosin)	The most prominent is myosin, found in muscle cells and probably in other cell types. Discussed more fully in Chapter 16.
"Superfine" filaments	3–4 nm diam.	——	Interconnecting other filaments, microtubules, and also to some membranes.
Microtubules MAPs	24–25 nm diam.	56,000; 59,000; >200,000; ~20,000	Found in many cell types, and in cilia and flagella. The core is 14 nm diam., made up of 13 filaments and composed of two subunits: α-tubulin, MW 56,000, and β-tubulin, MW 59,000, and of microtubule-associated proteins (MAPs) of very high (>200,000) or low (~20,000) MW proteins.
a. Stable			Persists at low temperature and in colchicine. Found in centrioles, basal bodies, cilia, and flagella.
b. Labile			Disrupted in cold and in colchicine. Found in spindles and in cytoplasm of many cells.

Spectrin

FIGURE 2–33
Protein Association in the Erythrocyte Membrane Skeleton. Red blood cells were attached to polylysine-treated grids and then negatively stained with 2% uranyl acetate. The electron micrograph shows spectrin molecules (long white rods) on the inner surface of the membrane, converging with each other to form junctional complexes. × 200,000. (Courtesy of D. Branton.)

mitosis), the intermediate filaments probably provide the framework of the cytoskeleton—sometimes in cooperation with actin and microtubules (Figures 2–34, 2–35). A more detailed description of the filaments making up the cytoskeleton and of the structure of the cytoskeleton will be found in Chapter 16.

Thick Filaments. Striated muscle cells also contain thick filaments (15 nm, 150 Å), filaments which consist mostly of the very large protein myosin (Figure 2–32; see also Chapter 16). Less specialized cells, however, though they contain large amounts of actin (often as much as 25% of their protein), usually contain only small amounts of myosin. Thick myosin filaments have therefore been reported infrequently in nonmuscle cells. In these cells myosin may act in an unaggregated form or as very small aggregates. On the other hand, improved methods of preparation may yet reveal myosin filaments (they have, in fact, recently been observed in the cortex of nonmuscle cells).

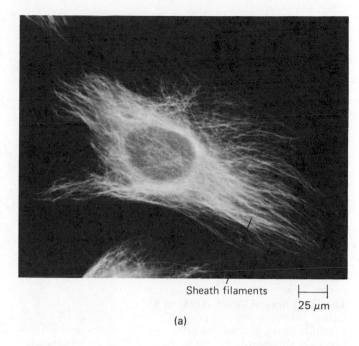

Sheath filaments

25 μm

(a)

Lattice filaments

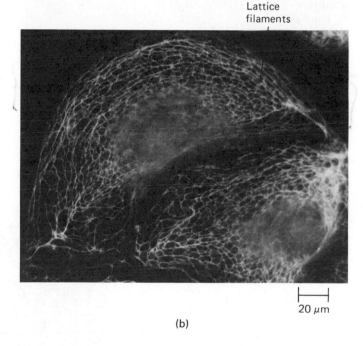

20 μm

(b)

FIGURE 2–34
Intermediate Filaments and Microtubules as Part of an Intracellular Network. (a) Microtubules as demonstrated by immunofluorescence microscopy, using an antibody to tubulin coupled to a fluorescent dye. × 3,000.
(b) Intermediate lattice filament pattern demonstrated by immunofluorescence microscopy, using the same technique but using the antibody to the filamentous protein, prokeratin. Two cells are shown; the nuclei are located in the fuzzy areas. × 4,000. (Courtesy of E. R. McBeath and K. Fujiwara.)

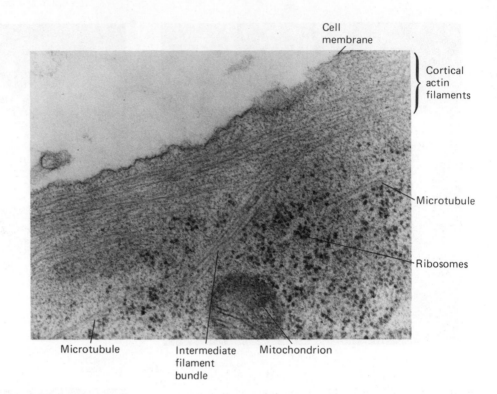

Cell
membrane

Cortical
actin
filaments

Microtubule

Ribosomes

Microtubule

Intermediate
filament
bundle

Mitochondrion

FIGURE 2–35
The Periphery (Cortex Area) of a Cell. Microtubules, actin microfilaments, and intermediate filaments are shown. × 70,000. (Courtesy of E. R. McBeath and K. Fujiwara.)

Microtubules. Microtubules, the other major component of the cytoplasmic ground substance, are straight and very long structures which can be seen clearly with the electron microscope (Figures 2–35, 2–36, 2–37). They are cylindrical structures, about 24 nm in diameter, with an electron-dense wall some 6 nm in thickness and usually composed of 13 globular subunits which are themselves dimers of the protein *tublin.* The

details of the structure and assembly of microtubules will be considered in Chapter 15; here we shall discuss the distribution and function of these structures.

Microtubules differ from each other both in longevity and function. Some, such as those found in *flagella,* are very stable while others, such as those in the *mitotic spindle*, are less stable and exist for only part of the cell cycle. These more labile microtubules can be depolymer-

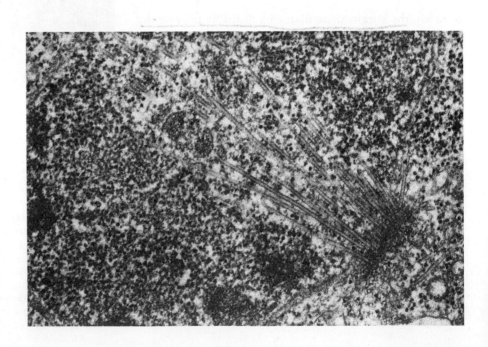

FIGURE 2–36
Microtubules. The long tubules are seen fanning out from part of the spindle in the metaphase part of cell division in *Dictyostelium.* Cell division, or mitosis, will be fully discussed in Chapter 11. × 81,000. (Courtesy of R. McIntosh and U.-R. Roos.)

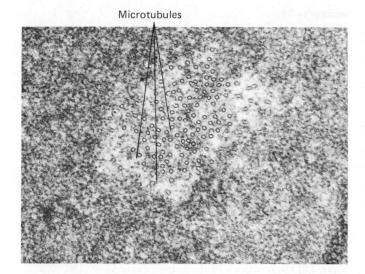

Microtubules

FIGURE 2-37
Microtubules. The microtubules are shown in cross-section during metaphase in *Dictyostelium*. × 81,000. (Courtesy of R. McIntosh and K. McDonald.)

ized by low temperatures, high pressures, or by the drugs *colchicine* or *colcemide*. Some microtubules, such as those in the outer edge of the avian erythrocyte, provide the stiffness necessary to give the cell its disc-like shape, while others, such as the flagellar and mitotic spindle microtubules, are involved in the machinery of motion. As we shall discuss in Chapter 16, microtubules are also involved in intracellular motion; "motor proteins" bring about movement along microtubules in one direction or the other—for example, in the *axonal transport* associated with neurons.

Table 2–3 summarizes some of the properties of the various fibrillar elements populating the cytoplasmic matrix. No doubt we shall learn a great deal more in the next few years about the chemical nature and function of these interesting cell constituents.

Superfine Filaments and the Microtrabecular Lattice. Use of the high-resolution electron microscope (100 kV) has provided evidence for a new class of superfine (2–3 nm) filaments, seen in cells extracted with non-ionic detergents (see Figure 16–45). These structures can also be seen when cells are rapidly frozen and the cell surface is etched by subliming away the ice (see Figure 16–A). The superfine filaments appear as relatively short "connectors" of the various cytoskeletal elements, joining them to each other and to membranes.

Keith Porter, one of the founders of cell electron microscopy, has proposed a dynamic interconnecting network of cytoskeletal filaments to which membrane-bound vesicles as well as ribosomes are attached (Figure

2–38). Porter sees this *microtrabecular lattice* as a protein-rich phase of the cytoplasm; many protein molecules, hitherto assumed to be freely dissolved, are proposed to be associated with this network of fibers. Since the system is dynamic and therefore labile, the methods used to preserve it for high-resolution microscopy have yielded variable and in some cases controversial results. Nevertheless, the observations of superfine connector filaments lend credibility to the concept that the microtubules and various filaments of the cytoskeleton are parts of an interconnected cytoplasmic matrix. Further experimentation is necessary to obtain a more precise picture of such a complex and dynamic intracellular network.

Microtubules are frequently found dispersed in the cytoplasm. At other times, however, they are present in high concentration and in special arrangements, thereby defining identifiable structures with specialized functions. Let us now examine a number of structures which are composed mainly of microtubules and which are involved in a variety of mechanochemical activities of the eucaryotic cell.

The Spindle. The spindle (Figure 2–39) is a structure formed during nuclear division—meiosis or mitosis—and is part of the machinery responsible for distribution of hereditary material into the two daughter cells. It is sufficiently cohesive to be separated from the rest of the cytoplasmic matrix by differential centrifugation. Its structure has been studied intensively by S. Inoue, who utilized a polarizing microscope to reveal the presence of the spindle's organized submicroscopic elements (Figure 2–39). Electron microscopic studies have established that these elements are microtubules lying in parallel array (Figures 2–36, 2–37). As we shall see later, movement of the chromosomes probably results from the interaction of special regions on the chromosomes (kinetochores) with the microtubules and microfilaments of the spindle.

Centrioles. Located near the nucleus of animal cells, or at each pole of the spindle during mitosis, are two centrioles. These are cylindrical bodies (Figure 2–40a), usually lying at right angles to each other, about 150 nm in diameter and 300–500 nm in length. Centrioles are probably self-replicating structures, perhaps even containing DNA, which seem to be involved with the organization of the spindle. The internal structure of the centriole is precise and universal in character, composed of nine groups of triplet microtubules evenly spaced around the circumference of a cylinder and oriented at a characteristic 30° angle of inclination (Figure 2–40b). High-resolution studies suggest that the innermost tubule

FIGURE 2–38
Microtrabecular Lattice. (a) High voltage electron micrograph of a thick section of a kidney cell showing a fine network of 4–6 nm diameter strands interlaced to form a network or lattice. Within this network small membrane-bound vesicles are apparently attached to the strands. This network has been found in all types of cells so far examined. K. Porter and his co-workers believe that, except for mitochondria, all membrane-bound vesicles as well as ribosomes are attached to it. × 128,000. (Courtesy of K. R. Porter.) (b) A model of the microtrabecular lattice showing its relationship to other cell organelles. This model was developed by studying many stereo images of a variety of cell types prepared in a number of different ways and examined by high voltage electron microscopy. The lattice filaments are depicted as continuous with the proteins (c) underlying the plasma membrane (pm), the surfaces of the endoplasmic reticulum (er), the microtubules (mt), the actin filaments of the stress fibers (sf), and the polysomes (r) located at the junctions of the lattice filaments. × 150,000. (Courtesy of K. R. Porter.)

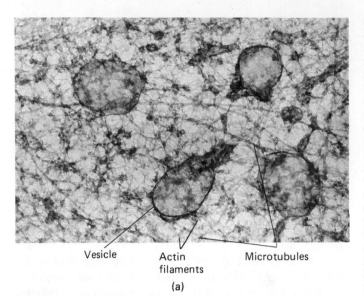

Vesicle   Actin filaments   Microtubules

(a)

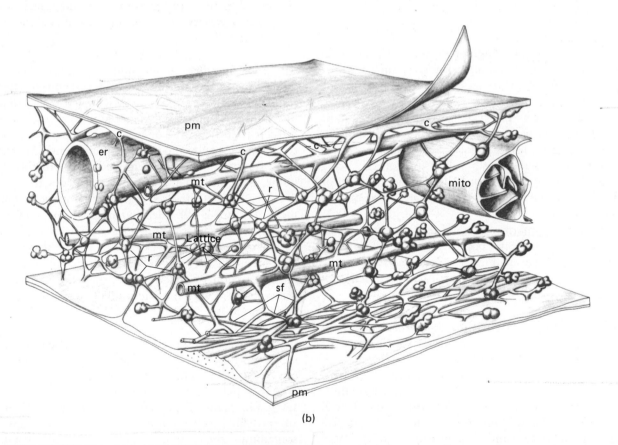

(b)

has two arms, one extended toward the center and the other toward the tubule of the neighboring group. The centriole is closed at one end (Figure 2–40a) and thus is not a symmetrical cylinder in the longitudinal direction; the centriole also does not have radial symmetry since the microtubules are inclined at an angle in a clockwise direction (Figure 2–40b) when one looks at them from the open end.

Centrioles in animal cells during mitosis are often surrounded by a structure that is visible in the light microscope and known as an *aster* because of its star-like appearance. The aster consists of spherical bodies and

FIGURE 2–39
Mitotic Spindle, as Seen in the Polarizing Microscope.
The mitotic spindle fibers that pull the chromosomes apart
are seen as bright streaks. Oocyte of marine worm,
Chaetopterus pergamentaceous. × 800. (Courtesy of S. Inoue.)

microtubules that radiate out from the centriole (Figure
2–40b). Even in nondividing cells there are microtubules
that radiate outward from the centrioles. This observation has given rise to the notion of a *cell center*, also called
the *centrosome*.

Basal Bodies, Cilia, and Flagella. *Cilia* and
flagella in eucaryotic cells differ from each other only in
their length and number; some cells have one or two long
flagella, others many shorter cilia. Since they have the
same internal structure, we shall refer mainly to cilia in
this section.

Cilia occur at all stages of evolution from the most
primitive animals and plants to a variety of cell types in
man. It is remarkable that they are the same in all
eucaryotic cells, not only in their general organization,
but even down to their last detail of fine structure and
dimension. This constitutes a dramatic piece of visual
evidence for the unity of life at the eucaryotic cell level. It
would seem that nature has been satisfied to use the

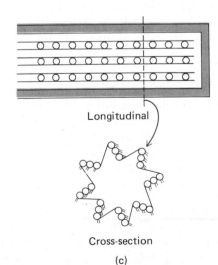

(a)

Centrioles

Longitudinal

Cross-section

(c)

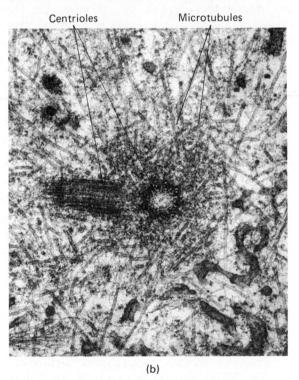

Centrioles   Microtubules

(b)

FIGURE 2–40
Centrioles. (a) Embryonic chick epithelium, showing a
longitudinal section of two centrioles at right angles to each
other. × 41,000. (Courtesy of S. Sorokin.) (b) Two
centrioles, at right angles to each other in the spindle pole of
metaphase cell. One is cut longitudinally and the other is in
cross section. Microtubules radiate out from the structure.
× 40,000. (Courtesy of R. McIntosh and K. McDonald.)
(c) Diagram of centriole structure, showing longitudinal and
cross-sectional views.

cilium essentially unmodified since it evolved several billion years ago in green algae.

Basal bodies, which apparently govern the assembly of cilia, have a microtubular structure similar to that of centrioles, being made up of nine bundles, with each bundle composed of three fused microtubules. One can frequently observe very long fibers extending from the basal body into the cytoplasm, which may have the function of anchoring the basal body. The transition from basal body to cilium occurs near the cell surface. The most obvious changes are the appearance of a pair of microtubules at the center of the cilium and the elimination of one microtubule from each triplet at the periphery (Figure 2–41). Thus, the 9 + 0 structure of the basal body changes into the 9 + 2 structure of the cilium. Ciliary microtubules are similar in dimensions and structure to cytoplasmic microtubules. Other structures which appear in the cilium are arms made of the protein *dynein,* a central sheath, *nexin bridges,* and radial links. The movement of the cilium is believed to be caused by the sliding of the microtubules against one another, the sliding motion resulting from the interaction of the proteins dynein and tubulin (see Chapter 16).

Cilia can be formed de novo around dense spherical bodies *(procentriole organizers)* composed of clusters of fibrous granules which develop near the centrioles, assemble into centrioles, and move to the cell surface. At the cell surface they become the basal body from which the cilia develop.

Ribosomes and Polysomes. These are the organelles on which protein synthesis occurs and will be discussed fully in Chapter 13. As well as being attached to some membranes of the ER and to the outer membrane of the nuclear envelope, ribosomes and polysomes can be found free in the cytoplasmic matrix. Both membrane-bound and free eucaryotic ribosomes are a little larger (80S) than procaryotic ribosomes (70S). They are found in the ground substance, either singly or attached to a form of RNA, called messenger RNA, like beads on a string (Figures 2–22, 13–18). The presence of polysomes in the ground substance can be detected either with the EM or by the use of ultracentrifugation. High-resolution studies are able to distinguish two subunits of different size that comprise the ribosome (Figures 13–24, 13–25).

Storage Products. Cells store energy for their metabolism by accumulating a variety of products. In order to be able to store large quantities of these products, cells convert them into macromolecules (polysaccharides and proteins) or into lipids. This has the virtue of allowing the cell to accumulate large amounts of mass without raising unduly the molar concentration of products. By keeping the total molar solute concentration

relatively fixed, the cell avoids the entry of water (**osmosis**) from its surroundings. In the case of polysaccharides and protein storage products, it is their high molecular weight which accounts for their low molar concentration; in the case of lipids, it is their insolubility in water. Some storage products are sequestered in the cell by a membrane, and these will be discussed in the next section devoted to membrane-bound cellular organelles. Other materials exist free in the cytoplasmic ground substance.

Polysaccharides are convenient storage products because they can be readily converted to monosaccharides by the cell's metabolism. Animal and plant cells differ in the particular polysaccharide they use. Animal cells store glycogen and plants store *starch;* both are polymers of the monosaccharide glucose.

Glycogen is stored as irregularly shaped particles of 15–30 nm in diameter. In some cells these particles are arranged in clusters of various sizes (Figure 2–42; see also Chapter 3).

Starch is found dispersed as granules in the chloroplasts of plant cells. In algae there are special structures called *pyrenoids* which are situated in the chloroplasts and are probably the sites of starch synthesis (see Figure 2–8b). Starch granules are found in many sizes. Sometimes they are so large that they can be seen readily in the light microscope as bright objects which stain purple with iodine and which have a characteristic appearance in polarized light.

Lipids are stored in cells, usually as tiny spherical droplets of oil. Some cells which specialize in fat storage, such as adipose tissue cells, store huge quantities of oil, which at times may constitute 99% of all the cell's organic matter.

Melanin, a dark pigment, is found widely distributed in the animal kingdom. Although not an energy storage product, it is nevertheless localized in dark granules which are initially surrounded by a membrane; mature melanin granules appear to have lost their membrane. Melanin functions as a pigment to absorb light, for example in skin cells.

Membrane-Bound Cell Organelles

The cell contains a number of bodies which are separated from the cytoplasmic matrix by one or two membranes. These bodies include the nucleus, mitochondria, and chloroplasts with two membranes, and lysosomes and peroxisomes with one.

The Nucleus. The nucleus, as the name suggests, is a central or crucial cell organelle. It is the site of storage and replication of most of the cell's hereditary material. Its immediate importance in the metabolic events of the cell is demonstrated by the instantaneous disappearance

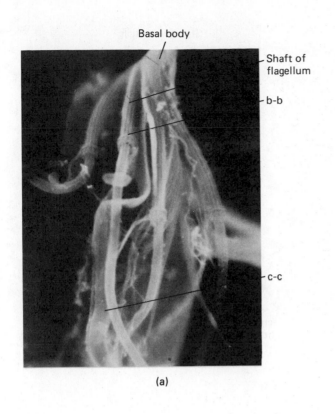

(a)

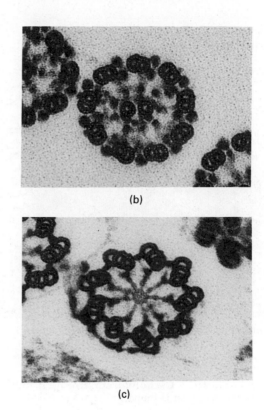

(b)

(c)

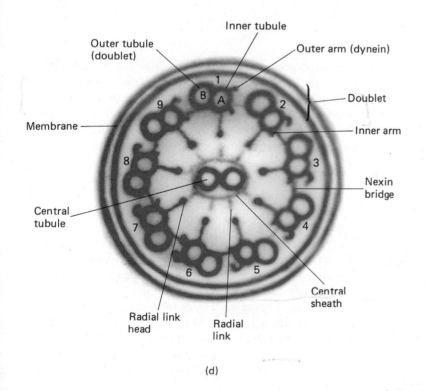

(d)

FIGURE **2–41**
The Structure of Cilia and Flagella, One of the Most Universal Structures of the Eucaryotic World.
(a) Flagella from the protozoan *Saccinobaculus,* as seen by scanning electron microscopy following removal of the cell membranes. The shafts of the flagella emanate from basal bodies (b–b, unusually long in this organism) in the cell cytoplasm. × 14,000.
(b) Cross section of a flagellar axoneme from the protozoan *Trichonympha,* in position c–c of part (a). × 130,000.
(c) Cross section of basal body. These sections were fixed with glutaraldehyde and tannic acid to show the subunit structure of the microtubule walls. × 150,000. (Courtesy of D. W. Woodrum and R. W. Linck.) (d) Diagram of a cross section of a cilium as seen at high resolution in the electron microscope. The structure consists of nine doublets (each made of two fused microtubules) and two central tubules. Two sets of arms are attached periodically to the inner tubule. Bridges made of the protein nexin connect the nine doublets to each other and to a central sheath.

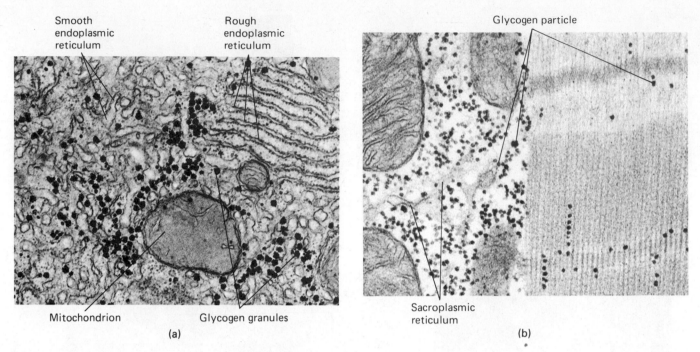

Smooth endoplasmic reticulum

Rough endoplasmic reticulum

Glycogen particle

Mitochondrion

Glycogen granules

(a)

Sacroplasmic reticulum

(b)

FIGURE 2–42
Glycogen Granules, a Carbohydrate Storage Product.   (a) In the liver of a salamander the glycogen is found in the smooth endoplasmic reticulum area. × 24,000. (b) In the muscle fibrils of the myocardium of a cat the glycogen is mainly associated with the sacroplasmic reticulum. × 120,000. (Courtesy of S. Ito.)

of important activities, such as organized cytoplasmic motion, when the nucleus is microsurgically removed from the cell of an amoeba. Cells lacking nuclei, either naturally (erythrocytes, platelets, and lens fibers) or after removal (enucleation), stop synthesizing protein and their life span becomes severely curtailed. Egg cells, which store a great deal of the protein-synthesizing machinery, are often capable of undergoing several cell divisions after experimental enucleation, but the ability of such enucleated cells to differentiate completely is generally lost, because as we shall see in later chapters, changes in nuclear function bring about cell differentiation.

The major components of the nucleus (Figure 2–26) include the *chromosomes,* one or more *nucleoli,* and the *nuclear envelope* (which we have already discussed in connection with the cell's membrane system). A number of different granules are also present.

The hereditary material present in the nucleus is located in *chromatin,* a complex of DNA, protein, and RNA, which is found in an extended state in the nondividing *(interphase)* cell and becomes highly condensed into *chromosomes* during nuclear division *(mitosis).* Each species has a characteristic number of chromosomes in its gametes, usually twice that number in the fertilized

egg (or zygote), and in all the cells of the organism that develops from it. We believe that each chromosome contains a single gigantic molecule of DNA which can be hundreds of billions of daltons in molecular mass. In the extended chromatin of the resting cell, the DNA is packaged in such a way that its length is probably condensed some 25 times. In the dividing nucleus the DNA is condensed at least 1,000 times, a degree and precision of packaging which we have yet to understand (see Chapter 12). Even in the resting cell there are two different degrees of chromatin-packing density that are visible by microscopy. The bulk of the chromatin is loosely packed and is called *euchromatin;* more densely staining, because of more highly condensed packing, is the *heterochromatin* (see Figure 2–26b).

The packaging of DNA into extended chromatin threads or fibers is believed to be partially the result of negatively charged DNA interacting with positively charged *histones,* basic proteins which carry a positive charge at the pH of the cell (see Chapters 10 and 11 for details).

Nuclear division (mitosis) is an ingenious piece of choreography whereby the eucaryotic cell manages to separate physically its previously duplicated hereditary material into two exactly equal parts. Nuclear division is generally correlated with the division of the entire cell. We shall consider the phenomenon of cell division in Chapter 11.

The *nucleolus,* which forms during the last stage of mitosis, does so in association with certain portions of particular chromosomes—the *nucleolar organizer regions.* They can be seen clearly in stained sections because they bind basic dyes. In the electron microscope (Figure 2–43) they appear to contain three components: (1) a shell of heterochromatin which sends projections into the interior of the nucleolus, (2) a core of RNA-containing fibrillar elements, and (3) RNA-containing granules roughly 15 nm in diameter. The arrangement of these three components varies from cell type to cell type and even varies with the physiological state of the cell.

Nucleoli are particularly large and prominent in cells that are actively synthesizing protein. There is much evidence that the nucleolus is the site of *ribosomal RNA* synthesis and that the nucleolus-associated chromatin contains the genes which code for the synthesis of ribosomal RNA. The nucleolar granules are believed to represent early stages of both ribosomal subunits. (The details of the synthesis and processing of ribosomal RNA and the formation of ribosomes will be discussed in Chapters 12 and 13.)

The chromosomes and nucleolus are supported in a "nuclear matrix" *(nucleoplasm),* which seems to contain granules of various sizes and densities and is of unknown composition and function. During mitosis the nucleoplasm is of course continuous with the cytoplasmic ground substance, since the nuclear membrane no longer separates them.

Mitochondria and *chloroplasts,* though they are membrane-bound organelles, will be discussed in Chapter 8, since structure in both is intimately linked to function.

The Lysosomes. *Lysosomes* are densely staining bodies that contain a large number of different digestive enzymes. These enzymes, when active, break down biological macromolecules (proteins, carbohydrates, nucleic acids, and phospholipids) by hydrolysis. Since lysosomal enzymes have maximal activity in acidic conditions, they are called *acid hydrolases.*

Lysosomes are spherical or nearly spherical, and can range from 50 nm to a few μm in diameter. They vary greatly in appearance and content from cell type to cell type and also in relation to the physiological state of a particular cell. There is a single, thin (5–6 nm) three-layered membrane surrounding the lysosome.

Surprisingly, lysosomes were discovered initially not with the electron microscope, but by differential centrifugation by C. de Duve. When homogenized cells were centrifuged, a certain fraction was found that

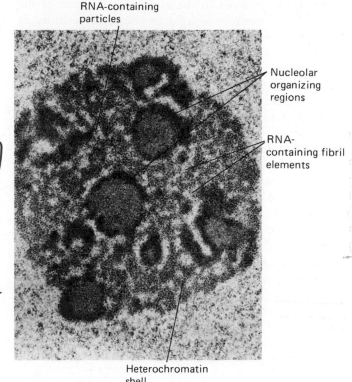

FIGURE 2–43
Electron Micrograph of a Nucleolus.
× 50,000. (Courtesy of D. Phillips.)

sedimented more slowly than mitochondria and that contained a number of hydrolytic enzymes. The activity of these enzymes became enhanced after severe mechanical disruption, which allowed the enzymes to attack substrates. This finding suggested that the enzymes were surrounded by a membrane. Careful electron microscopic and cytochemical studies on these fractions identified a number of membrane-bound bodies now known as lysosomes (see Figure 2–7a).

Lysosomes are widely distributed organelles, being present in most animal cells (mammalian red blood cells, which also lack other organelles, are an exception) and in a number of plant cells. At first the function of lysosomes was thought to be *autolysis* (self-digestion); that is, digesting the contents of a dead cell. We now know that lysosomes are part of a normal cellular digestion apparatus relating the process of endocytosis to the processes of intracellular synthesis and transport. We have alluded to this in our discussion of the different membrane systems of the cell and shall now draw this information together.

Scavenging is not the only lysosome activity. In certain cells they play a normal and important physiological role. In the thyroid gland, for instance, the protein *thyroglobulin* is stored in the extracellular lumen of the gland. This protein is brought into secretory cells by an endocytotic mechanism. The endocytotic vacuoles thus formed fuse with lysosomes, after which the thyroglobulin is hydrolyzed, releasing the hormone thyroxine as one of the digestion products. Thyroxine is then released by the secretory cells into blood capillaries.

In certain instances, lysosomes themselves appear to act as storage centers in which proteins are sequestered in large amounts. At times such proteins have even been known to form crystals (Figure 2–44b). In other instances, lysosomes contain small membrane-bound vesicles (Figure 2–44a) whose function is not yet understood. Lysosomes in the secretory cells of endocrine glands have also been observed to take up secretory granules, a process that could play a role in the control of secretion.

As scavengers, lysosomes are active not only during cell injury or death, but also participate in the digestion of cell material derived from the programmed destruction of other cells. Many developmental phenomena, such as the resorption of the tadpole's tail, the resorption of bone during normal modeling of its final shape, or the regression of the mammary gland after weaning, involve the *programmed death* of certain cell types. Some of the surviving cells in adjoining regions have been shown to ingest chunks of material from the dead cells (phagocytosis) or macromolecules in solution (pinocytosis). Lysosomes fuse with endocytotic vacuoles, after which the contents are partially or totally digested by lysosomal enzymes (Figure 2–45). Lysosomes containing the debris of cell organelles—at times even partially digested mitochondria can be discerned in them (Figure 2–46)—are called *heterophagic vacuoles.* It is not clear what cells do with the indigestible contents of heterophagic vacuoles, but it has been suggested that such material leaves by exocytosis.

The scavenging function of lysosomes is also believed to play a role in the controlled degradation of organelles within otherwise healthy cells. In this instance, *autophagic vacuoles* are thought to form around the organelles, after which lysosomes can be observed to fuse with the vacuoles. The precise mechanism by which a membrane forms around cell particulates is not yet understood, but it is likely that enfolding by endoplasmic reticulum vesicles is involved (Figure 2–46a).

What is the origin of the *primary lysosome;* that is, the

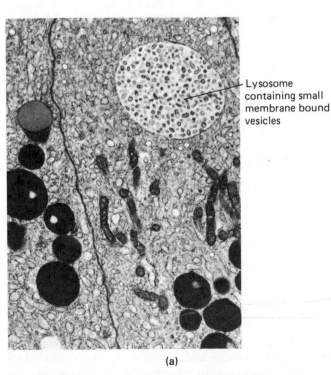

(a)

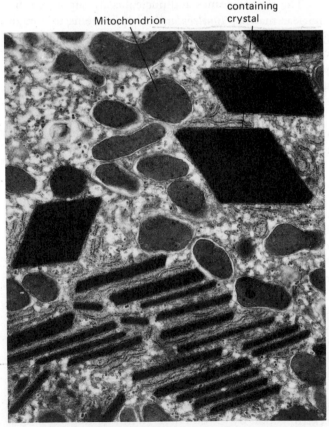

Mitochondrion

Lysosome containing crystal

(b)

FIGURE 2–44
Lysosomes. (a) Containing small membrane-bound vesicles from human epididymal epithelium. × 19,000. (Courtesy of D. W. Fawcett and A. Hoffer.) (b) Containing protein crystals. × 19,000. (Courtesy of D. W. Fawcett.)

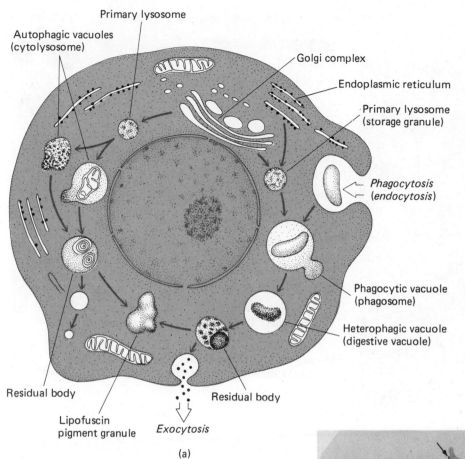

Primary lysosome

Autophagic vacuoles
(cytolysosome)

Golgi complex

Endoplasmic reticulum

Primary lysosome
(storage granule)

Phagocytosis
(endocytosis)

Phagocytic vacuole
(phagosome)

Heterophagic vacuole
(digestive vacuole)

Residual body

Residual body

Lipofuscin
pigment granule

Exocytosis

(a)

FIGURE **2–45**
Cellular Digestion by Lysosomes.
(a) Drawing of the role of lysosomes in the digestion of materials originating outside the cell (right, clockwise) and inside the cell (upper left). (Courtesy of D. W. Fawcett.) (b) An electron micrograph showing organelles which participate in endocytosis in a macrophage cell. The cell was exposed for two hours to colloidal thorium dioxide. The electron-dense colloidal particles adsorb in small amounts to the plasma membrane (arrows). Small pinocytotic vesicles (pv) containing small numbers of the electron-dense particles, arise from the plasma membrane. Larger vacuoles, called endosomes (end) probably arise by the fusion of pinocytotic vesicles. The bulk of the associated colloid is found in lysosomes (ly) which also contain an endogenous amorphous content. Note that no deposit is found in the Golgi apparatus. × 19,000. (Courtesy of R. Steinman.)

(b)

vesicle which contains the hydrolytic enzymes and which fuses with the heterophagic or autophagic vacuole? It is thought that the digestive enzymes of the lysosome are synthesized by ribosomes attached to the endoplasmic reticulum and then transported within the ER to the Golgi apparatus. There they are packaged and pinched off to form the primary lysosome vesicle. (See a more detailed discussion in Chapter 13.)

We have already discovered a large and varied list of activities associated with lysosomes. It is likely that this list of functions will increase in number and scope. There is, however, one common thread in all of the lysosome activities and that is the sequestering of the activity of

digestive enzymes inside membrane-bound vesicles where they cannot be disruptive to other cell functions.

The Peroxisomes. Peroxisomes are very similar to lysosomes in structure; in particular, they are both bound by only a single membrane. But peroxisomes differ from lysosomes in that they also contain enzymes which produce hydrogen peroxide—for example, D-amino acid oxidases—and catalase, an enzyme which catalyzes the breakdown of hydrogen peroxide to water and molecular oxygen. Peroxisomes are widely distributed in plants, where they play a role that relates to the processes of both photosynthesis and respiration. This explains why

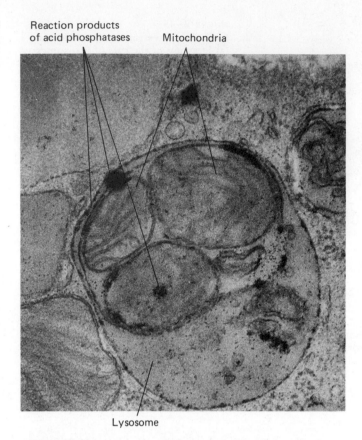

Reaction products of acid phosphatases   Mitochondria

Lysosome

FIGURE 2–46
Electron Micrograph of a Lysosome in Kidney Proximal Tubule Epithelium.   The organelle contains hemoglobin (dense, homogeneous background) from digested red blood cells, reaction products of acid phosphatases (dense, needle-like deposits) from digested cellular material, and mitochondria in the process of being digested (autophagy). × 50,000. (Courtesy of F. Miller and G. E. Palade.)

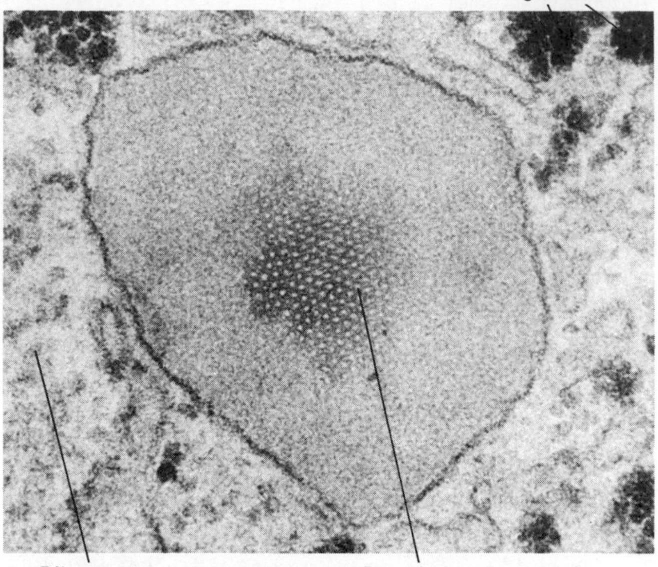

Glycogen granules

Ribosomes   Peroxisome

FIGURE 2–47
Peroxisome in Liver Cell.   The crystalline array of proteins is shown. × 186,000. (Courtesy of G. E. Palade.)

mitochondria, chloroplasts, and peroxisomes are closely associated in green leaves.

In animals the role of peroxisomes is still poorly understood, but the fact that they contain the enzyme catalase suggests that at least part of the function of this organelle is to protect the cell from the damaging effects of the hydrogen peroxide generated during metabolism. One frequent characteristic of peroxisomes is the presence within them of beautiful arrays of protein crystals (Figure 2–47).

SUMMARY

This concludes our electron microscopic tour of the cell. We have discussed the optical methods by which we investigate cells and have shown how the study of cell structure at resolutions finer than possible with the light microscope can be used to give important insights into cell function. We shall return to many of the topics discussed in this chapter after we have learned about the structure and function of the macromolecules of the cell and about the use of genetic techniques. This will permit us to discuss many of these topics in greater depth and in a more dynamic way; that is, in a way which relates the structure and function of the cell's parts at a deeper molecular level.

KEY WORDS

Angstrom unit (Å), cortex, cross-link, dalton (Da), electron microscope, endocytosis, endoplasmic reticulum, exocytosis, fibril, filament, intermediate filament (IF), lysosome, microfilament, micrometer (μm), microtubule, osmosis, polysome, resolution, tubule, vesicles

PROBLEMS

1. Towards the end of the nineteenth century, many biologists believed there was a certain "vital momentum" *(elan vitale)* necessary for life to go on. Cite an experiment which suggests that no such "momentum" is necessary.

2. On the average, eucaryotic cells are one thousand times larger than procaryotic cells. How does this difference in size appear to the microscopist?

3. Why can one see smaller objects when using blue light for microscopy rather than red light?

4. A globular virus is visualized in the electron microscope by the shadow-casting technique. The angle of shadow casting with respect to the surface supporting the specimen is θ; the distance between the center of the virus and the end of the shadow is Y. What is the approximate height (X) of the virus?

5. In microscopy, what is the purpose of fixation?

6. What is the advantage of high-voltage electron microscopy?

7. If raising the accelerating voltage of an electron microscope increases the resolving power, then why does a 200 kV microscope have no significant advantage over a 50 kV microscope in resolving images?

8. How would you use the stain neutral red to determine whether a cell in the epidermis of an onion scale is alive or dead? Give some additional experiments to demonstrate that your determination is correct.

9. What is the difference between a stain for the light microscope and a stain for the electron microscope?

10. Speculate about possible reasons why eucaryotic cells are so much larger than procaryotic cells.

SELECTED READINGS

Bainton, D. F. (1981) The discovery of lysosomes. *Journal of Cell Biology* 91:66s–76s.

de Duve, C. (1975) Exploring cells with a centrifuge. *Science* 189:186–94.

———. (1983) Microbodies in the living cell. *Scientific American* 248(5):74–84.

de Duve, C., and Beaufay, H. (1981) A short history of tissue fractionation. *Journal of Cell Biology* 91:293s–99s.

Diener, T. O. (1981) Viroids. *Scientific American* 244(1):58–65.

Fawcett, D. W. (1981) *The cell.* 2d ed. Philadelphia: W. B. Saunders.

Franke, W. W., Scheer, V., Krohne, G., and Jarasch, E. D. (1981) The nuclear envelope and the architecture of the nuclear periphery. *Journal of Cell Biology* 91:39s–50s.

Newport, J. W., and Forbes, D. J. (1987) The nucleus: Structure, function, and dynamics. *Annual Review of Biochemistry* 56:535–65.

Pease, D. C., and Porter, K. R. (1981) Electron microscopy and ultramicrotomy. *Journal of Cell Biology* 91:287s–92s.

Tolbert, N. E., and Essner, E. (1981) Microbodies: Peroxisomes and glyoxysomes. *Journal of Cell Biology* 91:271s–83s.

Computer-generated model of partially
liganded T-state hemoglobin, stabilized
by inositol hexaphosphate (center of
view); the hemes of the deoxy α-chains
(blue) contain manganese instead of
iron so they cannot bind the carbon
monoxide ligand. (Courtesy of A.
Arnone and G.E.O. Borgstahl.)

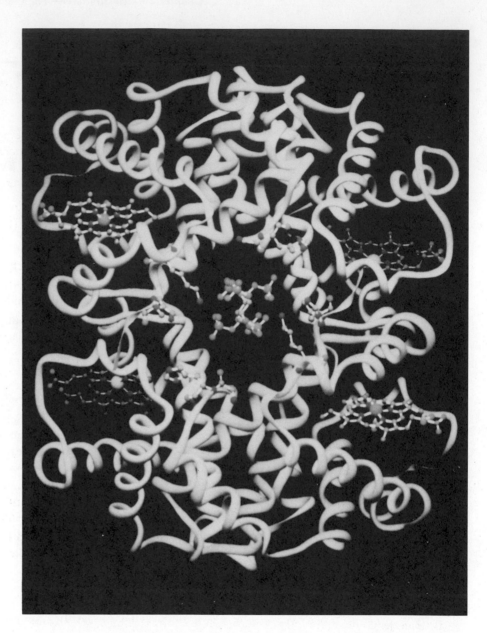

We will build an understanding of atoms and small molecules, and also of proteins, the large molecules that actively control the chemical and physical interactions of all the components of living matter. Our study of proteins lays the foundation of all that follows—the catalytic properties of enzymes, the metabolic processes that enzymes control, the movement of materials through membranes, and the transformation and utilization of energy.

This part begins to explain how the cell is the scene of numerous and complex interactions which are capable of transforming energy, and how the cell uses energy to maintain its highly organized state.

ATOMS, MOLECULES, AND THE FLOWS OF MASS AND ENERGY

Atoms and Small
Molecules in Cells

3

In the previous chapter we surveyed the complex cellular structures and organelles that can be detected by microscopy. Now we begin our discussion of the molecular basis for cellular structure and function. It is true that the systematic description of molecules and their behavior is called "chemistry." This should not dismay the student of cell biology. After all, molecules make up living systems and if we are to understand how cells work we first need to know their component parts. Before giving an account of the large molecules and assemblies that make up cell structures, it is helpful to study the smaller molecules of the cell. Their structures, functions, and interactions illustrate general principles singly rather than in bewildering combinations. Moreover, these small molecules play critical roles as intermediates and precursors in the synthesis of more complex molecular forms.

Elementary Biology

We begin by considering the various atoms found in cellular molecules. Later, two of the particles which make up atoms—the electron and the proton—will be discussed in their biological and chemical contexts. Various kinds of atoms exist; each kind is called an element.

The Relative Abundance of Elements

Living matter is selective in its relationship to the environment. An example of this selectivity is the considerable difference between the abundance of elements available on the earth and those found in living organisms (Table 3–1). The earth's crust is made mostly of oxygen, silicon, aluminum, sodium, calcium, iron, magnesium, and potassium; the remaining elements constitute much less than one percent of the total. Living organisms, on the other hand, are made mostly of hydrogen, oxygen, carbon, and nitrogen, with the remaining elements constituting little more than one percent of the total.

TABLE 3–1

Relative Abundance of the Major Elements in the Earth's Crust and the Human Body

Element	Atomic Number	Relative Abundance (Atoms %)	
		Earth's Crust	Human Body
Hydrogen	1	....	60.3
Carbon	6	....	10.5
Nitrogen	7	....	2.4
Oxygen	8	62.54	25.5
Sodium	11	2.64	....
Magnesium	12	1.84	....
Aluminum	13	6.47	....
Silicon	14	21.22	....
Phosphorus	15	....	....
Sulphur	16	....	....
Chlorine	17	....	....
Potassium	19	1.42	....
Calcium	20	1.93	....
Iron	26	1.92	....
Total		99.98	98.7

Note: Data recalculated from J. T. Edsall and J. Wyman.

Since it is likely that life began in the sea and not on land, it is necessary to consider how the composition of sea water might have affected the subset of elements which have become part of the structure of living matter. In fact, the relative abundance of elements in the earth's crust and in sea water differ markedly. Aluminum, for instance, is the third most abundant element in the earth's crust but is found only in concentrations of 4×10^{-8} M in sea water. There are, however, two problems connected with the assumption that the concentration of elements in sea water played a decisive role in what elements are included in living matter. One problem is that we do not know precisely the composition of sea water several billion years ago, when primitive cells first evolved. The other problem is that of deciding the threshold value below which the concentration of an element is too low to be available for the cellular processes. Partial answers to both these problems can be gleaned from the biological data.

Let us first consider in detail the elements utilized by living matter. Although living cells often contain traces of all the elements found in the environment, only some 27 elements are believed to be essential for most organisms. Eighteen of these have universal importance, seven others are necessary for some living groups but apparently not for others, while the remaining two are suspected of being required, although convincing proof is still lacking.

Table 3–2 shows that the component elements of cells are found in a wide range of concentrations. As we have already noted, the major constituents are H, C, N, and O, which together with S and P are used to build the organic compounds of the cell. The *trace elements* are the next most abundant. They are often used as *counterions* to neutralize, with their opposite electrostatic charge, the charged groups of organic molecules (—COO$^-$, —NH$_3^+$, etc.). They are also used as *enzyme cofactors:* nonprotein parts of enzymes that are necessary for catalytic activity. The least abundant constituents are the *ultratrace elements,* which are either constituents of organic compounds used in trace amounts or cofactors of certain special enzymes. The relation of these biologically important elements to others is shown in Table 3–3, a modified version of the periodic table.

The only way to determine whether or not a particular element is required for life is to exclude it completely from an organism's environment and observe the result. The nutritional necessity of some ultratrace elements was not suspected until their near absence in certain soils caused the appearance of diseases and abnormalities in plants and animals. The absence of copper from certain regions in Australia, for example, caused a disease in sheep. A deficiency of boron in the soil causes conditions known as "heart rot" in beets, "cracked stems" in celery, "internal cork" in apples, and a host of abnormalities in other plants. The nutritional requirements of ultratrace elements such as boron can best be shown by curing the deficiency disease through the addition of these elements to the soil. One hundredth of a part per billion parts (by weight) of boron in the soil

TABLE 3–2

Elements Used in Living Cells

Category	Element	Symbol	Atomic Number	Requirement
Major constituents, 2–60 atoms percent	Hydrogen	H	1	Required universally
	Carbon	C	6	" "
	Nitrogen	N	7	" "
	Oxygen	O	8	" "
	Sodium	Na	11	" "
	Magnesium	Mg	12	" "
Trace elements, 0.02–0.10 atoms percent	Phosphorus	P	15	" "
	Sulphur	S	16	" "
	Chlorine	Cl	17	" "
	Potassium	K	19	" "
	Calcium	Ca	20	" "
	Boron	B	5	Required in plants
Ultratrace elements, less than 0.001 atoms percent	Fluorine	F	9	Required in higher animals
	Silicon	Si	14	Required in some lower organisms
	Vanadium	V	23	Required universally
	Chromium	Cr	24	Required in higher animals
	Manganese	Mn	25	Required universally
	Iron	Fe	26	" "
	Cobalt	Co	27	" "
	Nickel	Ni	28	Required in higher animals and by some bacteria
	Copper	Cu	29	Required universally
	Zinc	Zn	30	" "
	Selenium	Se	34	Believed to be required in some bacteria and some higher animals
	Bromine	Br	35	Required in some organisms
	Molybdenum	Mo	42	Required universally
	Tin	Sn	50	Believed to be required in some organisms
	Iodine	I	53	Required in higher animals

will cause the disease, one tenth of a part per billion will cure it, and one part per billion is a sufficiently high concentration to poison the plant.

The ionic composition of a number of organisms of different evolutionary types is given in Table 3–4, along with the ionic composition of seawater. There are marked similarities in the ionic concentrations, normalized to that of the sodium ion, between living systems and seawater. A. B. Macallum was the first to conclude that this parallelism means that life originated in the sea and that subsequent evolution did little to change the relative ionic balance. After one billion years of evolution on land, and although our body fluids are less concentrated than seawater is today (perhaps even less concentrated than seawater was one billion years ago), we still carry the ionic balance of seawater in our body fluids! This argument assumes that the present-day forms of more primitive organisms have ionic compositions similar to those of our ancient ancestors. But since the morphology of these organisms has not changed appreciably, it is not far-fetched to assume that their physiology has also remained fairly constant.

The Fitness of Elements for Participation in the Processes of Life

It seems likely that of the elements with similar chemical properties—for example, Li^+, Na^+, K^+, Rb^+, and Cs^+—the ones used by cells are those with the highest concentrations in seawater. The fitness of these elements can be further appreciated when one considers the roles they play in cells. Most ultratrace elements serve as cofactors for certain enzymes (Table 3–5, p. 92 and Table 3–6, p. 93). Since enzymes are effective in very low concentrations, cofactors need only be present in low

TABLE 3–3

The Periodic Table of the Elements, Showing Their Different Requirements.

1A																	8A
1 Hydrogen **H** 1.0079	2A											3A	4A	5A	6A	7A	2 Helium **He** 4.0026
3 Lithium **Li** 6.941	4 Beryllium **Be** 9.0122											5 Boron **B** 10.81	6 Carbon **C** 12.011	7 Nitrogen **N** 14.0067	8 Oxygen **O** 15.9994	9 Fluorine **F** 18.9984	10 Neon **Ne** 20.179
11 Sodium **Na** 22.9898	12 Magnesium **Mg** 24.305	3B	4B	5B	6B	7B	8B	8B	8B	1B	2B	13 Aluminum **Al** 26.9815	14 Silicon **Si** 28.0855	15 Phosphorus **P** 30.9738	16 Sulfur **S** 32.06	17 Chlorine **Cl** 35.453	18 Argon **Ar** 39.948
19 Potassium **K** 39.0983	20 Calcium **Ca** 40.08	21 Scandium **Sc** 44.9559	22 Titanium **Ti** 47.88	23 Vanadium **V** 50.9415	24 Chromium **Cr** 51.996	25 Manganese **Mn** 54.9380	26 Iron **Fe** 55.847	27 Cobalt **Co** 58.9332	28 Nickel **Ni** ? 58.69	29 Copper **Cu** 63.546	30 Zinc **Zn** 65.38	31 Gallium **Ga** 69.72	32 Germanium **Ge** 72.59	33 Arsenic **As** 74.9216	34 Selenium **Se** ? 78.96	35 Bromine **Br** 79.904	36 Krypton **Kr** 83.80
37 Rubidium **Rb** 85.4678	38 Strontium **Sr** 87.62	39 Yttrium **Y** 88.9059	40 Zirconium **Zr** 91.22	41 Niobium **Nb** 92.9064	42 Molybdenum **Mo** 95.94	43 Technetium **Tc** (98)	44 Ruthenium **Ru** 101.07	45 Rhodium **Rh** 102.9055	46 Palladium **Pd** 106.42	47 Silver **Ag** 107.868	48 Cadmium **Cd** 112.41	49 Indium **In** 114.82	50 Tin **Sn** ? 118.69	51 Antimony **Sb** 121.75	52 Tellurium **Te** 127.60	53 Iodine **I** 126.9045	54 Xenon **Xe** 131.29
55 Cesium **Cs** 132.9054	56 Barium **Ba** 137.33	57 Lanthanum **La·** 138.9055	72 Hafnium **Hf** 178.49	73 Tantalum **Ta** 180.9479	74 Tungsten **W** 183.85	75 Rhenium **Re** 186.207	76 Osmium **Os** 190.2	77 Iridium **Ir** 192.22	78 Platinum **Pt** 195.08	79 Gold **Au** 196.9665	80 Mercury **Hg** 200.59	81 Thallium **Tl** 204.383	82 Lead **Pb** 207.2	83 Bismuth **Bi** 208.9804	84 Polonium **Po** (209)	85 Astatine **At** (210)	86 Radon **Rn** (222)
87 Francium **Fr** (223)	88 Radium **Ra** 226.0254	89 Actinium **Ac··** (227.0278)	104 Unnilquadium **Unq** (261)	105 Unnilpentium **Unp** (262)	106 Unnilhexium **Unh** (263)												

58 Cerium **Ce** 140.12	59 Praseodymium **Pr** 140.9077	60 Neodymium **Nd** 144.24	61 Promethium **Pm** (145)	62 Samarium **Sm** 150.36	63 Europium **Eu** 151.96	64 Gadolinium **Gd** 157.25	65 Terbium **Tb** 158.9254	66 Dysprosium **Dy** 162.50	67 Holmium **Ho** 164.9304	68 Erbium **Er** 167.26	69 Thulium **Tm** 168.9342	70 Ytterbium **Yb** 173.04	71 Lutetium **Lu** 174.967
90 Thorium **Th** 232.0381	91 Protactinium **Pa** 231.0359	92 Uranium **U** 238.0289	93 Neptunium **Np** 237.0482	94 Plutonium **Pu** (244)	95 Americium **Am** (243)	96 Curium **Cm** (247)	97 Berkelium **Bk** (247)	98 Californium **Cf** (251)	99 Einsteinium **Es** (252)	100 Fermium **Fm** (257)	101 Mendelevium **Md** (258)	102 Nobelium **No** (259)	103 Lawrencium **Lr** (260)

Required major constituents

Required trace elements

Required ultratrace elements

Not universally required ultratrace elements

? Suggested to be essential for higher animals

Present in concentrations above 5×10^{-9} M, but not required

Present in concentrations below 5×10^{-9} M, and not required

TABLE 3–4

Ionic Composition of Seawater and the Body Fluids of Several Species

	Na^+	K^+	Ca^{2+}	Mg^{2+}	Cl^-	SO_4^{2-}
VERTEBRATES						
Man	145	5.1	2.5	1.2	103	2.5
(mammal)	100	3.5	1.7	0.83	71	1.7
Rat	145	6.2	3.1	1.6	116	
(mammal)	100	4.2	2.1	1.1	80	
Frog	103	2.5	2.0	1.2	74	
(amphibian)	100	2.4	1.9	1.2	72	
Lophius	228	6.4	2.3	3.7	164	
(fish)	100	2.8	1.0	1.6	72	
INVERTEBRATES						
Hydrophilus	119	13	1.1	20	40	0.14
(insect)	100	11	0.93	17	34	0.13
Lobster	465	8.6	10.5	4.8	498	10
(arthropod)	100	1.9	2.3	1.0	110	2.2
Venus	438	7.4	9.5	25	514	26
(mollusk)	100	1.7	2.2	5.7	120	5.9
Sea cucumber	420	9.7	9.3	50	487	30
(echinoderm)	100	2.3	2.2	12	120	7.2
Seawater	417	9.1	9.4	50	483	30
	100	2.2	2.3	12	120	7.2

Note: Black numbers are expressed in millimoles per liter (mM); shaded numbers are relative, expressed in terms of 100 units of Na^+.

concentrations. A metal can be a cofactor by interacting directly with the protein of an enzyme or it can be part of a *coenzyme*, an organic compound which is essential for the activity of some enzymes. An example of the latter is cobalt, which is part of the coenzyme known as vitamin B_{12}. In some instances, ultratrace elements can be part of the structure of compounds such as *hormones*, which are also found in very low concentrations. For instance, iodine, which is not an essential element for lower forms, is part of the structure of the mammalian hormone thyroxine.

The roles of the trace elements are a little more varied than those of the ultratrace elements. Phosphorus and sulfur make important contributions to the structures of critical macromolecules, phosphorus being present in nucleic acids and sulfur in proteins. Sulfur and phosphorus are also found in negatively charged inorganic ions, sulfate and phosphate, which serve as counterions to such positively charged species as the ammonium group ($-NH_3^+$) and various metal ions.

Magnesium, calcium, and potassium also can act as cofactors for certain enzymes (Tables 3–5, 3–6), but the relatively high concentration of magnesium and potassium inside cells suggests that they play other important roles. With sodium, these elements comprise the major fraction of positive ions found in living systems. The total ionic concentration contributed by these species inside an average cell is equivalent to about 0.1–0.2 M. This is precisely the concentration of ions which many proteins require in order to stay in solution (see Chapter 4). Besides this general role of solubilizing proteins, these ions play specific roles in maintaining the structure of certain cell components. The magnesium ion, for instance, is necessary to maintain nucleic acids in their proper configurations (see Chapter 10) and it is likely that calcium ions are necessary for the integrity of the cell's membranes (see Chapter 7). The calcium ion is also a very important regulator of a number of cell functions. It is typically sequestered by special proteins or within membrane-bound compartments; its release is used to control a variety of cell processes (see Chapters 14, 16).

As for the major constituents of living matter—hydrogen, carbon, nitrogen, and oxygen—their role, together with phosphorus and sulfur, is to build the covalent compounds of the cell. Their utilization seems to depend both on their availability in seawater and on their chemical properties, upon which evolution has based the covalent chemistry of the cell.

TABLE 3–5

Some of the Elements Found in Cells and Some of the Functions of Those Elements

Prevalence	Element	Functions
Major	H C N O	Universally required for organic compounds
Trace	Na	Important counterion, as Na^+; found mostly outside the cell, but necessary for maintaining electrical polarity of membrane (see Chapter 7)
	Mg	Counterion, as Mg^{2+}, for nucleic acids; cofactor for enzymes such as phosphotransferases and phosphohydrolases
	P	Constituent of nucleic acids; used for energy transduction in "high energy compounds" (as phosphate, PO_4); negative phosphate ion
	S	Found in proteins; counterion, as SO_4^{2-}
	Cl	As Cl^-, major negative ion and counterion inside cells
	K	Major positive ion, as K^+, inside cells; cofactor for a few enzymes such as pyruvate kinase; maintains electrical polarity of membrane
	Ca	As Ca^{2+}, regulatory cofactor in muscle contraction, secretion, blood clotting, etc.; cofactor for a few enzymes such as translutaminase; stored in membrane-bound spaces*
Ultratrace	Mn	Cofactor, as Mn^{2+}, for enzymes such as arginase and phosphotransferase; required for O_2 production by photosynthesis
	Fe	Found (as Fe^{2+} and Fe^{3+}) in electron transfer proteins such as cytochromes, oxygen-carrying proteins such as hemoglobin and myoglobin; cofactor for enzymes such as peroxidase and catalase
	Cu	Cofactor for enzymes such as cytochrome oxidase, tyrosinase, as Cu^{2+}
	Zn	Cofactor for many enzymes, such as dehydrogenases, carbonic anhydrase, carboxypeptidase, as Zn^{2+}
	Mo	Cofactor for enzymes such as nitrate reductase, nitrogenase, as Mo^{2+}

*Calcium is also found in large amounts in certain extracellular structures—notably the bones of vertebrates—as a salt with phosphate.

Carbon, An Element with Unique Properties

Organic chemistry, the chemistry of the living world, is the chemistry of *carbon*. An outstanding property of carbon compounds associated with cells is that they are remarkably *inert*. Under physiological conditions of temperature, pressure, and acidity, they react only at very low rates—often hardly at all—with each other, with water, and with atmospheric oxygen. This is so despite the fact that large amounts of energy are often released when these compounds do react. The

wooden chair on which we may be sitting appears to be a very inert object at room temperature, but the cellulose of which it is mostly made releases enormous amounts of energy when heated in the presence of oxygen. In fact, the rate of the heat release is so great that a piece of wood, once heated under appropriate conditions to a certain point (kindled), will maintain its elevated temperature, thereby sustaining the oxidation process (burning) until it is completed.

The cell also "burns" carbohydrates such as glucose, but it does so in a controlled manner and under mild

TABLE 3–6

Some Metal Ion-requiring Enzymatic Reactions

Glucose + ATP	$\xrightarrow[\text{Mg}^{2+}]{\text{Hexokinase}}$	Glucose-6-P + ADP
Tryptophan + H_2O	$\xrightarrow[\text{K}^+ \text{ or Rb}^+]{\text{Tryptophanase}}$	Indole + pyruvate + NH_3
Soluble fibrin	$\xrightarrow[\text{Ca}^{2+}]{\text{Plasma transglutaminase}}$	Insoluble fibrin
Arginine + H_2O	$\xrightarrow[\text{Co}^{2+} \text{ or Mn}^{2+} \text{ or Ni}^{2+}]{\text{Arginase}}$	Urea + 2,5-diaminovaleric acid
Histidine	$\xrightarrow[\text{Fe}^{3+} \text{ or Al}^{3+}]{\text{Histidine decarboxylase}}$	Histamine + CO_2
Catechol	$\xrightarrow[\text{Cu}^{2+}]{\text{Phenol oxidase}}$	*o*-benzoquinone
Lactic acid + NAD^+	$\xrightarrow[\text{Zn}^{2+}]{\text{Lactic dehydrogenase}}$	Pyruvic acid + NADH + H^+
Nitrate + NADPH + H^+	$\xrightarrow[\text{Mo}^{2+}]{\text{Nitrate reductase}}$	Nitrite + $NADP^+$ + H_2O

conditions of temperature, pressure, and acidity. This is accomplished through the use of specific *enzyme catalysts* which speed up those reactions necessary for cellular metabolism so they can occur efficiently under cellular conditions. *The other reactions of which the cellular molecules are capable do not occur at appreciable rates because the appropriate catalysts are not present.* The cell exploits the usual sluggishness of reactions involving organic compounds to regulate its chemical activities. In later chapters we shall discuss in detail how the cell manages to control the several thousand reactions it utilizes for its activities. At present, we shall concern ourselves with the properties of carbon compounds which make biological control of organic reactions possible.

A second outstanding property of carbon is its *versatility*—its ability to form a practically unlimited number of compounds that vary widely in their properties. Both the inertness of carbon compounds and their diversity are due to the specific chemical properties of carbon.

First, carbon is a light element with the atomic number six. Like other elements in the second horizontal row (period) of the periodic table (Table 3–3), carbon cannot expand its octet of electrons. Thus, it does not

form complex ions or coordination compounds to any appreciable extent. Chemical reactions of elements in higher numbered rows, which have other shells of electrons, can be speeded up by forming coordination complexes. It is not surprising that most reactions of carbon compounds are relatively slower since they lack this capability.

Second, carbon can form four covalent bonds with other carbon atoms. This property permits the formation of a potentially complex *carbon skeleton*, the fundamental framework around which organic molecules are constructed. Furthermore, the strength of the C—C bond, comparable to that of the C—O, C—H, and C—N bonds (Table 3–7), ensures the stability of the basic carbon skeleton and permits the existence of organic compounds containing long chains of carbon atoms. Since a carbon atom can combine with 1, 2, 3, or 4 other carbon atoms, great structural variety is possible in the basic skeleton, which can include straight or branched chains, rings, networks, and combinations of these structures.

Third, located centrally in the periodic table, carbon is capable of reacting both with elements that can attract valence electrons (*electronegative* elements), such as oxygen,

TABLE 3–7

Average Values for Bond Energies Found in Organic Compounds, kJ/mole (kcal/mole)

C—C	C—H	C—O	C—N
344 (82)	415 (99)	344 (82)	272 (65)
C═C	O—H	C═O	N—H
609 (145)	461 (110)	713 (170)	432 (103)

Note: Bond energy is defined here as the energy (in kJ and kcal per mole) needed to break the bond and separate the atoms to a great distance. (100 kJ/mole = 23.9 kcal/mole and 100 kcal/mole = 418 kJ/mole.)

nitrogen, phosphorus, sulfur, and chlorine, and with hydrogen (an *electropositive* element), which can relatively easily share its valence electron. Thus, carbon compounds are known in which the carbon atom can be assigned formal oxidation states of −4, −2, 0, +2, and +4. This ability to react with a variety of elements is further expanded by the ability of carbon to form single, double, or triple bonds with carbon, and double bonds with oxygen and nitrogen. These properties of carbon make possible the enormous versatility of its compounds.

Another element in the same group (IV A) of the periodic table is silicon. It also is capable of significant versatility in its reactions. But silicon dioxide is a solid under physiological conditions, while carbon dioxide is a gas and is sufficiently soluble in water to distribute itself equally between the gas phase and aqueous solution. Because it is so freely available in solution, carbon dioxide is the primary source of carbon from which all of the living material on this earth is built.

In summary, carbon shares with other elements in the same horizontal row of the periodic table the property of forming compounds which in the absence of catalysts react very slowly; it also shares with elements in the same vertical column of the periodic table the property of versatility. Carbon is the only element which possesses this combination of slow reactivity and great versatility. One must conclude that the properties of carbon are truly unique and it is difficult to imagine that life based on a different element, such as silicon, would ever evolve in the universe.

Chemical Bonds and Weak Interactions

The elements are almost never found as such in biological systems. Instead, they are associated with one another to form compounds. These associations are of varying strengths and permanence. The strongest are the covalent chemical bonds.

Covalent Bonds

A general principle of chemical structure is that atoms will bond together when by doing so they are able to form the eight-electron "noble gas" configuration in each outer shell. There are a number of elements, such as Cl, H, O, N, S, and P, situated in the upper right-hand corner of the periodic table, which have the ability to achieve the eight-electron noble gas configuration by *sharing* pairs of electrons with each other. The electron density becomes greater in the region between the bonded atoms where their atomic orbitals overlap. Alternatively, one can think of the attractive force of the nuclei for the shared electrons as holding the molecule together. Thus we have:

$$:\ddot{C}l\cdot \ + \ \cdot\ddot{C}l: \ \rightleftarrows \ :\ddot{C}l:\ddot{C}l: \ \text{or} \ Cl\text{—}Cl.$$

A variation on the general rule of an eight-electron neon-like configuration is hydrogen which, with its one electron, can form a two-electron helium-like configuration:

$$H\cdot \ + \ \cdot H \ \rightleftarrows \ H:H \ \text{or} \ H\text{—}H.$$

Covalent bonds form not only between like atoms, but also between unlike atoms. In methane (CH_4) for instance, hydrogen, which has one electron, and carbon, which has an outer shell of four electrons, can form a compound in which four pairs of electrons are shared. By this sharing process hydrogen becomes helium-like and carbon becomes neon-like in their respective electronic configurations:

$$4\,H\cdot \ + \ \cdot\ddot{C}\cdot \ \rightleftarrows \ H:\overset{\displaystyle H}{\underset{\displaystyle \ddot{H}}{\ddot{C}}}:H \ \text{or} \ H\text{—}\overset{\displaystyle H}{\underset{\displaystyle H}{\overset{|}{\underset{|}{C}}}}\text{—}H.$$

In carbon dioxide (CO_2), all three atoms achieve octets, becoming neon-like, but in this instance *two* pairs of electrons are shared with each of the oxygen atoms, forming a pair of *double bonds:*

$$2\,:\ddot{O}: \ + \ \cdot\ddot{C}\cdot \ \rightleftarrows \ \ddot{O}::C::\ddot{O} \ \text{or} \ \ddot{O}\text{═}C\text{═}\ddot{O}$$

This sharing of electrons, the overlapping of their atomic orbitals, causes them to become part of *molecular orbitals,* which are common to the entire molecule. The geometry of molecular orbitals and, therefore, the structure of the molecule often differ from that of the atomic orbitals which overlap (Figure 3–1). This is because molecular orbitals are not the simple sum of two atomic orbitals; rather, they are more complex combinations of several atomic orbitals called *hybrid orbitals*. (The student who is unfamiliar with these notions should review the relevant chapters in a textbook of general chemistry.)

So far we have suggested that in covalent bonds electrons are shared equally between the atomic nuclei forming these bonds. This is true for compounds such as H_2 or O_2 in which the atoms involved are identical. But in a compound such as NH_3, we have covalent bonds formed between very different atoms which do not attract the shared electrons with equal strength. This is because nitrogen is relatively *electronegative*, which means that it attracts the shared electrons more strongly than does the relatively *electropositive* hydrogen atom. Because of this, the nitrogen atom bears a slight negative charge and the hydrogen atom a slight positive charge, which we symbolize as $N^{\delta-}$—$H^{\delta+}$. The N—H bonds are called *polar covalent bonds*, in contrast to symmetrical covalent bonds (as in Cl_2, for example) where neither atom has even a partial electrostatic charge.

Bond Energies, Bond Distances, and Bond Angles

The bonds that hold covalently linked compounds together can be described in a number of ways that help illuminate the structure and function of classes of organic

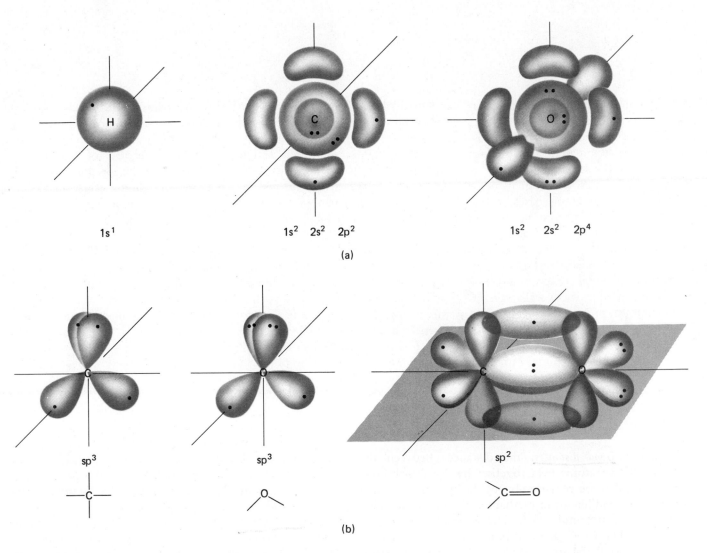

(a)

(b)

FIGURE **3–1**
Atomic and Molecular Orbitals.
(a) The atomic orbitals of hydrogen, carbon, and oxygen. (b) The molecular orbitals of carbon in methane, oxygen in water, and carbon-oxygen in formaldehyde.

Important Functional Groups of Bio-Organic Molecules

Though living systems are capable of synthesizing many esoteric, even bizarre, chemical compounds, the bulk of living matter consists of just a few basic structures. Thus, the cell generally uses just twenty amino acids, five nitrogenous bases, a few carbohydrates, and a few lipids. Organic molecules are a composite of a limited number of functional groups; if one knows the properties of each functional group in a compound, one can often predict its overall physical and chemical properties. Electrostatically charged groups are almost invariably *carboxyl, ammonium, guanidinium,* and *imidazole* groups in proteins

and lipids, and *phosphates* in nucleic acids and phospholipids. Oxygen-containing *hydroxyl (alcohol)* and *carbonyl* groups appear in proteins and in carbohydrates; the latter groups form *aldehydes, ketones, esters,* and *amides*. *Sulfhydryl* groups are found in proteins; *methyl* and other *alkyl (hydrocarbon)* groups are found in lipids and in a few amino acids.

The carbon backbones of the monomers utilized as building blocks for macromolecules are most frequently small, ranging from two to seven carbon atoms; lipids generally have backbones of 16–18 carbon atoms. These carbons can be found in

straight chains, branched chains, five- or six-membered cyclic compounds, or six-membered *aromatic rings*. One remarkable feature of bio-organic molecules is the preponderance of nitrogen-containing rings, such as the imidazole group in the amino acid histidine and the *purines* and *pyrimidines* of the nucleic acids. Even though the carbon backbones of the building blocks are simple, the order in which they are strung together into long chain polymers and the three-dimensional folding assumed by these chains result in a prodigious structural versatility.

molecules. *Bond distances,* the distances between the centers of two atoms held together by covalent bonds, have proved to be remarkably regular quantities. Thus, the C—H bond length in methane is 0.109 nm (1.09 Å); in benzene, 0.108 nm (1.08 Å); and in ethylene, 0.107 nm (1.07 Å). The C—C bond distance in either ethane or cyclohexane is 0.154 nm (1.54 Å). In ethylene it is 0.134 nm (1.34 Å); the smaller distance turns out to be due to the fact that in ethylene the carbon-carbon bond is double. Since covalent bond lengths are fairly constant, they are also additive. Each atom has a characteristic *covalent radius*

that is associated with its contribution to a covalent bond (Table 3–8). Thus, it is possible to obtain a pretty good estimate of bond distances in molecules by adding the covalent radii of the respective atoms.

Bond energies, listed in Table 3–7, also prove to be additive (bond energies are positive; heat must be transferred from the environment to the system to break the bonds of stable molecules). It is possible to estimate the energy released in forming an organic compound from its constituent elements—its *heat of formation*—by summing the respective bond energies. (The heat of

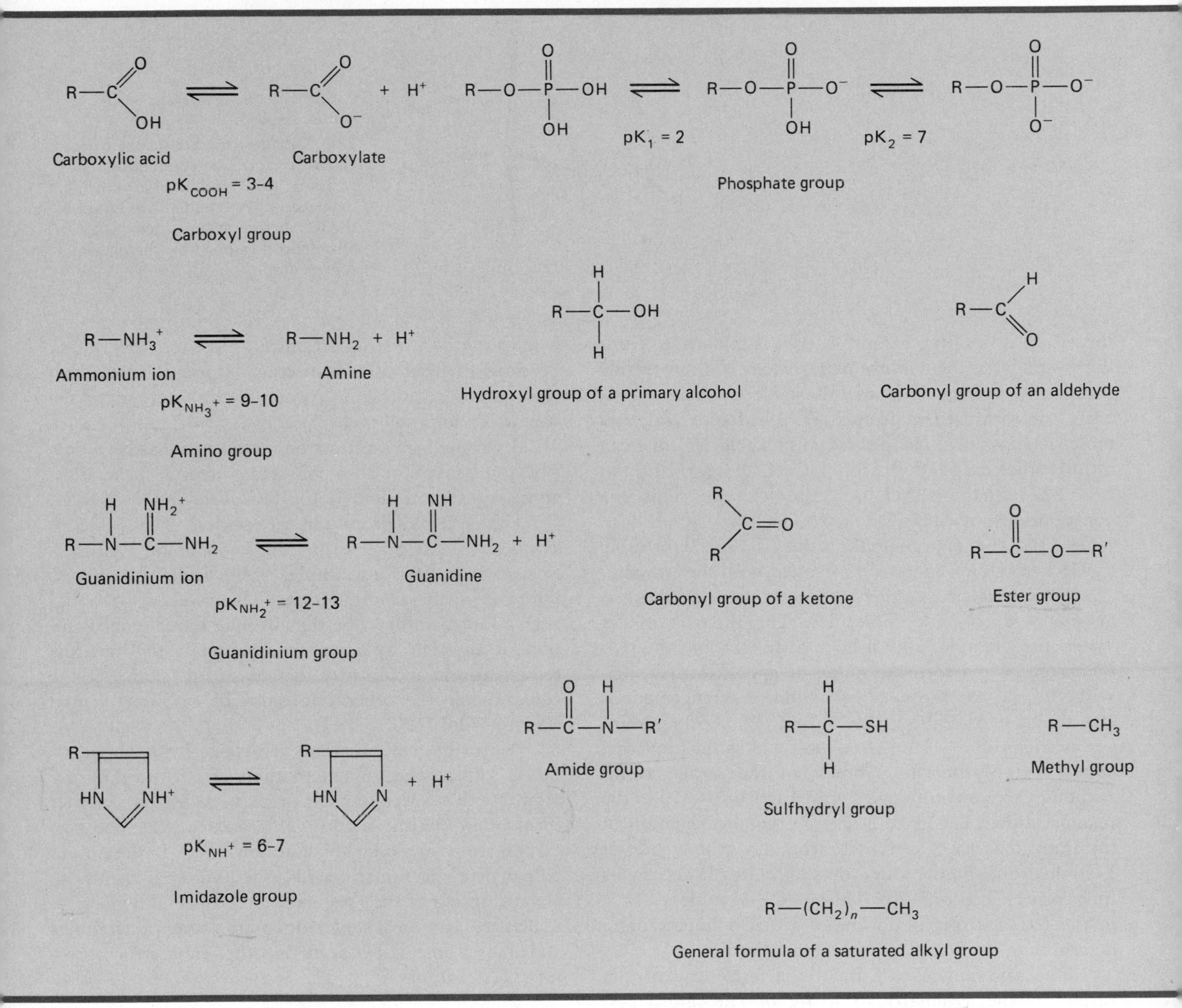

Carboxylic acid $\rightleftharpoons$ Carboxylate $+ H^+$

$pK_{COOH} = 3\text{-}4$

Carboxyl group

Phosphate group

$pK_1 = 2$ $pK_2 = 7$

Ammonium ion $\rightleftharpoons$ Amine $+ H^+$

$pK_{NH_3^+} = 9\text{-}10$

Amino group

Hydroxyl group of a primary alcohol

Carbonyl group of an aldehyde

Guanidinium ion $\rightleftharpoons$ Guanidine $+ H^+$

$pK_{NH_2^+} = 12\text{-}13$

Guanidinium group

Carbonyl group of a ketone

Ester group

Imidazole group

$pK_{NH^+} = 6\text{-}7$

Amide group

Sulfhydryl group

Methyl group

$R\text{—}(CH_2)_n\text{—}CH_3$

General formula of a saturated alkyl group

TABLE 3–8
Covalent Radii of Atoms Found in Organic Molecules, nm (Å)

Bond	H	C	N	O	P	S
Single	0.037 (0.37)	0.077 (0.77)	0.070 (0.70)	0.066 (0.66)	0.110 (1.10)	0.104 (1.04)
Double	...	0.067 (0.67)	0.060 (0.60)	0.055 (0.55)	...	...
Triple	...	0.060 (0.60)	0.055 (0.55)	...	...	...

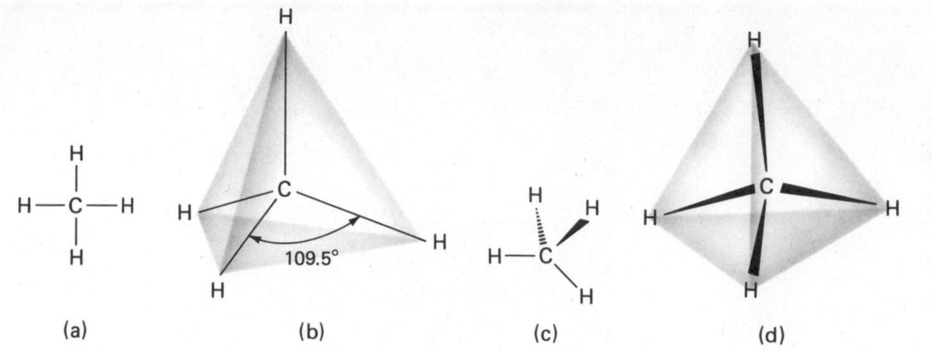

FIGURE 3–2
The Tetrahedral Structure of Carbon. (a) The conventional formula. (b) Tetrahedral arrangement in space. (c) A schematic diagram of the tetrahedral arrangement. (d) A tetrahedron implied by the planar representation in (a).

formation is negative because energy is transferred from the system to the environment when a more stable compound is formed from less stable elements.) This energy is measured in **joules (J)** or **calories (cal)** per mole. For instance, using the data in Table 3–7, one can estimate the heat of ethyl alcohol (C_2H_5OH) formation to be -3224 kJ/mole (-769 kcal/mole): one sums the contributions of 5(C—H) + 1(C—C) + 1(C—O) + 1(O—H). The experimental value is -3200 kJ/mole (-765 kcal/mole), in good agreement with the estimate.

Bond angles are another parameter that describes the geometry of organic molecules. The three atoms in water, for example, do not lie in a straight line; the two hydrogen atoms form an angle of approximately 105° with the oxygen atom. This nonlinear arrangement is also found in H_2S (93°) and NO_2 (134°). An important geometric property of organic molecules is the *tetrahedral arrangement* of the single bonds of the carbon atom. Although we conventionally write methane (CH_4) in the manner shown in Figure 3–2a, the actual orientation of the atoms in space is such that carbon lies at the center of a tetrahedron, and the four atoms of hydrogen occupy its four corners (Figure 3–2b). The H—C—H angle is 109.5° in the tetrahedron. Bond angles are also important in determining the structure of macromolecules.

One consequence of the tetrahedral structure of carbon is the *asymmetry* of many organic molecules. When a carbon atom in an organic molecule is bonded to four different atoms, two spatial arrangements are possible, one the mirror image of the other. The compounds have the same composition but different structures; they are *isomers* of one another (Figure 3–3). When plane-polarized light is passed through a solution of one such organic molecule, the plane of polarization will be rotated in one direction; the molecule is therefore said to show *optical activity*. The mirror image isomer will rotate the plane of light equally, but in the opposite direction. This is why two such isomers are called *optical isomers,* and why the carbon atom responsible for this lack of symmetry is called an *asymmetric carbon* atom. In cellular chemistry, specificity occurs via molecules coming into

contact, with their structures fitting close together. It will be clear after carefully inspecting Figure 3–3 that such close fitting requires the correct optical isomers of the contacting molecules.

By using x-ray diffraction methods, it has been possible to discover the absolute arrangement in space of the atoms of tartaric acid (COOHCHOHCHOHCOOH; see Figure 3–3). From tartaric acid it is possible to synthesize many other compounds containing asymmetric carbon atoms (for example, amino acids) and thus, to infer their spatial configurations. The convention that has been adopted to describe the absolute spatial configuration of an optically active compound is to draw the tetrahedron in such a way that two of its sides face the reader while the other two sides lie invisibly behind (Figure 3–2d).

Normally, carbon atoms can rotate freely about the C—C single bond. When rotation of two neighboring atoms with respect to each other is restricted, another form of isomerism results. This occurs when the two carbon atoms are linked by a double bond. In the isomer pair maleic and fumaric acids, the hydrogen atoms are always on either the *same* side of the C=C bond (*cis-*, maleic acid), or on *opposite* sides (*trans-*, fumaric acid). Cis and trans isomers also occur in other rigid structures—*rings,* for example.

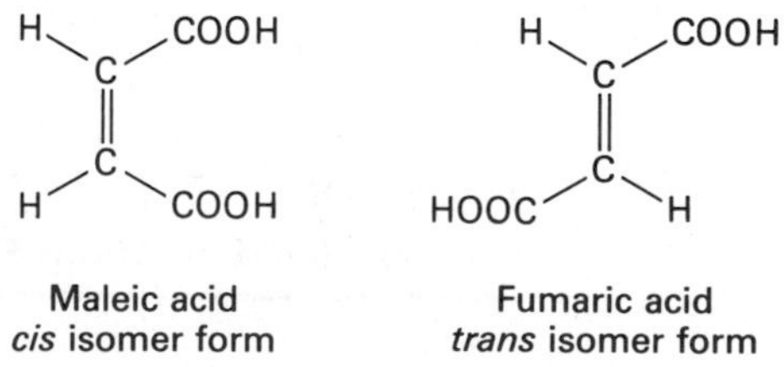

Maleic acid
cis isomer form

Fumaric acid
trans isomer form

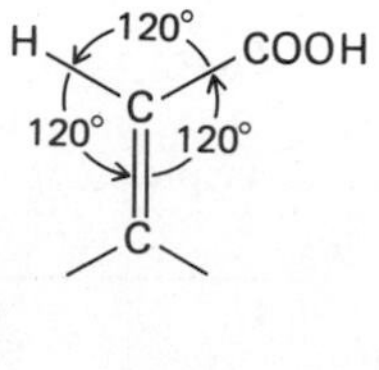

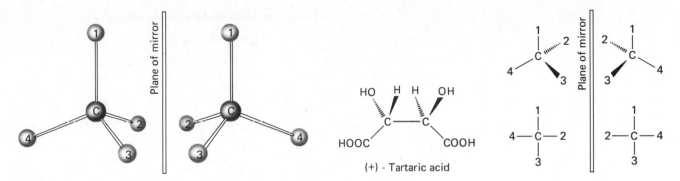

FIGURE 3–3

Mirror Symmetry. Shown are two nonsuperimposable forms of a molecule that is built from one carbon atom to which four different atoms or groups are bonded. The two forms are mirror images of one another. The absolute configuration of tartaric acid (+ form) is used to standardize the configuration of many other organic compounds.

Ribose,
a *cis*-2,3-hydroxy ring

Arabinose,
a *trans*-2,3-hydroxy ring

In a molecule with a double bond, the isomerism is due to a restriction to rotation, which "freezes" one arrangement or the other. In molecules that consist of a chain of three or more singly bonded carbon atoms, where rotation is possible around each bond, an immense number of arrangements or *conformations* is possible. Despite the existence of many possibilities, most molecules assume *preferred conformations* that are energetically more favorable than others, due to weak—but important—forces, which are discussed below. Because of these weak forces, the molecules comprising a so-called straight-chain hydrocarbon of the formula CH_3—$(CH_2)_x$—CH_3 adopt a *planar zigzag* conformation when packed in a crystalline solid. That is, the carbon atoms lie in a plane, and the C—C—C bond angles are 114°, rather than 180° (a truly straight chain) or 109.5° (the tetrahedral angle) (see Figure 3–4a).

Because the zigzag angle is not 120°, if the six singly bonded carbon atoms are joined into a ring, the ring will not lie flat but will twist, giving rise to still another kind of isomerism. As shown in Figure 3–4b, the twisted ring of cyclohexane can exist in two structures, called the *chair* and *boat* conformations. Normally, the chair form is more stable because it allows the hydrogen atoms to be separated as much as possible. If other groups are attached to the cyclohexane framework, the boat form

may occur because it allows the molecule to achieve a conformation that results in a lower total energy and is, therefore, more stable.

Ionic Bonds

Another simpler way for certain atoms to attain a complete outer electron shell is to donate or to accept completely one or two electrons. For example, an atom with one electron in its outer shell can donate it to another atom with seven electrons, thereby causing the former atom to be positively charged and the latter negatively charged:

$$Na\cdot \ + \ \cdot \ddot{\underset{..}{C}}l\text{:} \ \rightleftarrows \ Na\text{:}\ddot{\underset{..}{C}}l\text{:} \ \rightleftarrows \ Na^+ \ + \ \text{:}\ddot{\underset{..}{C}}l\text{:}^-.$$

In such *ionic compounds,* the atoms are held together by an electrostatic attraction, called an *ionic bond.* Ionic bonds can have large energies in appropriate environments, almost as large as covalent bond energies. In the presence of water, as we shall see later in this chapter, ionic bond energies are reduced significantly.

The Weak Interactions Between Organic Groups

As shown previously, a covalent interaction is a strong one, involving energies of about 420 kJ/mole (100 kcal/mole). There are also a number of weak interactions. When these occur between molecules, they are called *intermolecular interactions*—in contrast to the *intra*molecular covalent bonds which hold an organic molecule together. For all but the smallest molecules, however, the weak interactions can also be intramolecular; that is, they can operate between atoms or groups within a molecule. The important feature to recognize about weak interactions is that they are short-range interac-

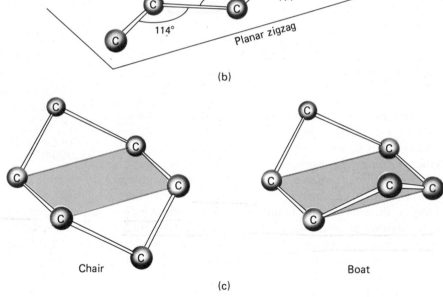

FIGURE 3–4
Conformations of Singly Bonded Carbon Atoms. (a) Different rotational conformations of ethane. The groups attached to a carbon atom can rotate with respect to a carbon atom attached by a single bond, but there are energy barriers at particular conformations. The energy of interaction is least favorable in the "eclipsed" conformation. (b) In a straight chain of carbon atoms, adjacent staggered conformations lead to a planar zigzag arrangement. (c) In a six-membered ring, the chair form is normally more stable than the boat form.

tions, significant only when the interacting groups are close together. Since biological structure, specificity, and reactions involve just such close contacts, weak interactions are of utmost importance to these phenomena. Even though weak individually, the overall effect of *summing* the various types of these interactions can lead to strong stable associations between, or within, large molecules.

Ionic Interactions

Biologically important molecules often include ions—groups with one or two electrostatic charges. These can interact by either attracting or repelling ions of opposite or the same charge, respectively. In a covalently linked protein, for example, *intra*molecular ionic interactions, like those between positive ammonium ($-NH_3^+$) and negative carboxylate ($-COO^-$), can help stabilize the biologically active conformation.

Van der Waals Interactions

The fact that many substances form liquids, under the appropriate conditions of temperature and pressure, is evidence that forces of attraction exist between molecules, even if they are uncharged. These interactions are collectively classified as *van der Waals interactions* because they can be evaluated from the corrections that the chemist van der Waals made to the ideal gas law (PV = RT) to account for gases at high pressures. The *intermolecular forces* responsible for the deviation from the ideal gas law differ from each other in intensity and by how much they decrease with increasing distance between molecules. Before we discuss them, we must introduce more specifically the concept of the *polarity* of molecules.

A molecule, although electrically neutral, may be *polar* if its center of negative charge does not lie at the same point as the center of positive charge. Such a molecule is said to have a *dipole moment*. The greater the polarity of the molecule, the greater its dipole moment and the more strongly it will tend to become oriented in an electric field, with its positive side toward the negative direction of the field and vice versa. If such a molecule has only two atoms, they are held together by a polar bond, as we discussed previously.

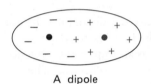

A dipole

The strongest of the van der Waals forces is the *ion-dipole* interaction, which is due to the interaction of an ion (such as Na^+) and a polar molecule such as H_2O (Figure 3–5a). This force is responsible for the hydration of ions, surrounding them with oriented water molecules (Figure 3–6).

The *ion-induced dipole* interaction is due to the effect of an ion on a nonpolar molecule (Figure 3–5b). Here, the charge of the ion *induces* a dipole in what would be, in the absence of the ion, a symmetrical nonpolar molecule. This can occur in certain molecules whose electrons are relatively free to respond to external electrical influences. The result is an interaction somewhat weaker than the ion–dipole one.

An attractive force may also occur between two dipoles (Figure 3–5c). This *dipole-dipole* interaction causes

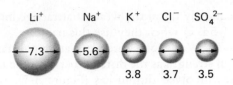

Hydration sphere diameters in Å

FIGURE 3–6
Hydration Spheres of Various Ions.
In an aqueous solution, ions are surrounded by oriented water molecules. The smaller the absolute size of the ion the greater the number of water molecules that are associated with it. Na^+ has a diameter of 0.10 nm (1.0Å) in a crystal, while K^+ has a diameter of 0.12 nm (1.2Å); Na^+ has approximately 4.5 water molecules associated with it, while K^+ has approximately 3. The figure indicates the relative sizes (in Å) of the hydration spheres, composed of ion plus water, for positive ions of lithium, sodium, and potassium, and negative ions chloride and sulfate.

molecules to orient with respect to each other. The hydrogen bond, discussed later in this chapter, is an important example of a dipole-dipole interaction.

A dipole may also induce a dipole in a nonpolar molecule. *Dipole-induced dipole* forces are among the weakest interactions (Figure 3–5d).

Finally, we discuss an interaction that can exist between almost any two molecules. Two nonpolar molecules are capable of mutual attraction because of *mutually induced dipoles* (Figure 3–5e). Although the electrons of a neutral group may be evenly distributed about the nuclei when viewed for a long period of time, at any single instant it is quite likely that the center of negative charge does not coincide with the center of positive charge. Consequently, there exists an instantaneous dipole that, however temporary it is, can induce a similar dipole in a nearby molecule. The dipoles thus formed can generate weak attractions, called *dispersion forces*, which explains the fact that nonpolar molecules of gases such as H_2, N_2, or He can form liquids under proper conditions of temperature and pressure.

It is unfortunately true that the term "van der Waals interaction" has become somewhat vague. Biologists

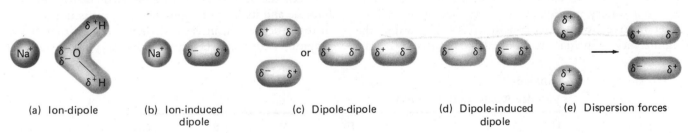

(a) Ion-dipole   (b) Ion-induced dipole   (c) Dipole-dipole   (d) Dipole-induced dipole   (e) Dispersion forces

FIGURE 3–5
Categories of van der Waals Interactions. These weak attractions between molecules are shown in order of decreasing strength.

most frequently mean the weakest attractive interactions (Figure 3–5d, e) when they use this name.

Van der Waals forces are weak at long distances, but increase in strength as the distance between the molecules or groups involved diminishes (Figure 3–7). Molecules that are not covalently bonded, however, cannot come too close without exerting strong repulsive forces on each other. (This is the origin of the concept of *steric hindrance;* nonbonded atoms cannot approach too closely and this can affect the rates of chemical reactions.) It is therefore possible to establish *van der Waals radii* for atoms, functional groups, and molecules. These radii represent the point of balance between the repulsive forces and the van der Waals attraction; the sum of two such radii defines the minimum expected distance between non-covalently bonded atoms (Table 3–9).

One particular type of van der Waals interaction calls for a more extended treatment. This is the special instance of dipole-dipole interaction known as the *hydrogen bond.* Hydrogen bonds are extremely important in maintaining the conformation of biological macromolecules. They are equally important in determining the properties of an *inorganic* substance, one which critically influences the organic chemistry of life. This substance is water. It is impossible to discuss the remarkable properties of water without reference to the hydrogen bond, and it is most convenient to discuss the hydrogen bond in the context of the water molecule. As we shall see, another weak interaction that greatly affects biological structure and function depends on the details of the behavior of water. This is the *hydrophobic interaction,* which plays an important part in stabilizing the structure of proteins, nucleic acids, and biological membranes. In the following section, therefore, we shall explore both the nature of the hydrogen bond and the structure of water. Before that, you may wish to consult Figure 3–8, which summarizes the weak interactions.

Water and Life

Water is the medium in which living matter is dispersed. Even land organisms that at first glance appear to thrive in a gaseous environment are seen actually to live in a watery medium when examined at the cellular level. Active living cells consist of 60–95% water; the significance of water can be further appreciated by the

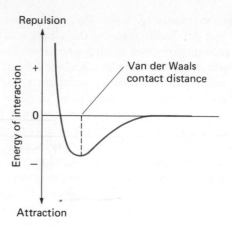

FIGURE 3–7
The Effect of Distance on the Interaction Between Two Atoms.   The distance of strongest attraction is called the van der Waals contact distance and is equal to the sum of the van der Waals radii of the two atoms.

observation that even dormant cells and tissues, such as spores and seeds, have water contents of 10–20%. This is because water has an important influence on the structure of biological macromolecules. The ubiquity of water should not distract us from examining its very special and unique properties. We shall first discuss the molecular properties of water and then mention how these have been exploited by living systems. The details of how water's properties influence the structure and function of macromolecules will be discussed further in Chapters 4, 10, and 11.

A Hydride of Unusual Properties

Water is a hydride of oxygen. It exhibits uniquely strong interactions between its molecules. Figure 3–9 shows the heat of vaporization of a number of hydrides. The heat of vaporization, which is a measure of the energy necessary to separate the molecules of a liquid from one another, is used here as a measure of the strength of intermolecular forces. In the case of the carbon series (lowest curve), there is a rough proportion between the atomic weight of the element forming the hydride and its heat of vaporization. This is not the case in the other series; in each of them the first member is atypical, with water being the most atypical of all. The

TABLE 3–9
Van der Waals Radii, nm (Å)

H	N	O	F	P	S	−CH$_3$
0.12	0.15	0.14	0.135	0.19	0.185	0.215
(1.2)	(1.5)	(1.4)	(1.35)	(1.9)	(1.85)	(2.15)

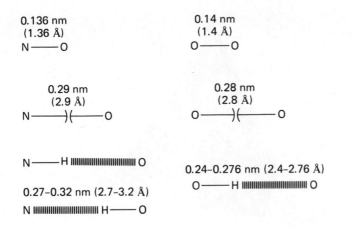

Covalent bond

Van der Waals radii

Hydrogen bond

FIGURE 3–8
Three Types of Interactions Between Two Electronegative Atoms. In the hydrogen bond the two electronegative atoms are closer than if they were separated by van der Waals distances, including the hydrogen atom.

reason for these unusually strong interactions is that oxygen, fluorine, and nitrogen form hydrides in which only some of their electrons are involved in covalent bonds with the hydrogen. The remaining "lone pair" electrons are not located symmetrically in the molecule which, as a result, has a _high electrical dipole moment_. This causes very strong forces between such molecules.

Let us consider the case of the water molecule. The more electronegative nucleus of oxygen exerts a strong attractive pull on its electrons, and the resulting unequal sharing leaves each hydrogen with a small positive charge and the oxygen with a small negative charge. The centers of negative charge, which include two lone electron pairs, are located in space away from the O—H bonds. The precise geometry of this arrangement is such that if the oxygen is placed in the center of a *tetrahedron* and the two hydrogens occupy two of the corners, then the centers of negative charge are concentrated in the direction of the two other corners (Figure 3–10). Thus, water is not only an *electrically polar* molecule, but also has a structure in which the two centers of positive charge and the two centers of negative charge are *tetrahedrally* arranged. When water freezes into crystalline ice the adjacent molecules are oriented to preserve and extend this tetrahedral arrangement (Figure 3–11).

In the liquid state, water molecules also exert orienting effects on each other; centers of negative charge attract centers of positive charge and vice versa. Each water molecule tends to attract four other molecules, which arrange themselves in ice-like tetrahedral fashion around it. In liquid water these tetrahedral arrangements are in a dynamic state; instantaneous clusters of oriented "crystalline" water are in equilibrium with more randomly distributed water molecules, and a single water molecule will be part of a crystal-like cluster at one instant and a less structured arrangement the next. The tetrahedral structure of water, in which each oxygen attracts two neighboring hydrogens and each hydrogen attracts the oxygen of a neighbor, is therefore a direct consequence of the tetrahedral geometry of the centers of positive and negative charge.

The Hydrogen Bond in Water

Using x-ray diffraction methods, it is possible to determine the exact dimensions of the ice crystal lattice. As shown in Figure 3–11, the distance of the O—H bond is 0.099 nm, and the oxygen-to-oxygen distance in ice is 0.276 nm. Therefore, the distance between protons and

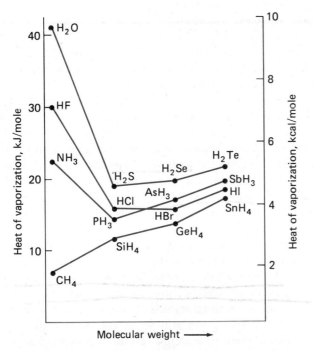

Molecular weight ⟶

FIGURE 3–9
Heats of Vaporization of Hydride Molecules. The lines join values for hydrides of elements in the same vertical column of the periodic table—that is, with the same number of valence electrons. The values for hydrides of N, F, and O are atypical, with water being the most atypical. (From General Chemistry, 2d ed. by Linus Pauling, W. H. Freeman and Co., 1953.)

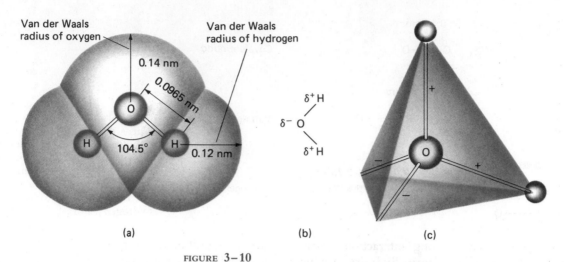

FIGURE **3–10**
The Structure of Water.   The figure illustrates the precise geometry of water as well as its polar properties. (a) The bond distances and bond angle of the water molecule. (b) Because oxygen is more electronegative than hydrogen, the oxygen has a partial negative electrostatic charge and the hydrogens have partial positive charges—that is, water has a dipole moment. (c) The hydrogen bonding orbitals and the orbitals occupied by the unshared pair of electrons are directed toward the corners of a tetrahedron.

oxygen atoms in neighboring water molecules is 0.177 nm which, while shorter than the sum of the van der Waals radii (0.26 nm), is longer than the covalent O—H bond distance. Since chemical bond lengths are inversely related to chemical bond energies, we expect the interaction between two water molecules also to be intermediate between a covalent bond and a weak van der Waals interaction. These intermediate interactions or linkages are called *hydrogen bonds* because the hydrogen atom holds together the two oxygen atoms.

Hydrogen bonds also form with nitrogen or fluorine, and this is why hydrogen fluoride and ammonia are also endowed with very high heats of vaporization (see Figure 3–9). In the case of ammonia and hydrogen fluoride, however, the geometry of the interactions is such that only chains and rings of molecules are formed. Even with branches, these structures are two-dimensional; it is impossible for them to form continuous, three-dimensional lattices. The properties of water are unique in that it has both *strong noncovalent interactions* and *the ability to form three-dimensional structures*.

Hydrogen bonds are fairly strong compared to the weakest van der Waals interactions but weak in comparison with covalent bonds. The energy needed to break a covalent bond lies between 200 and 400 kJ/mole (50 and 100 kcal/mole), whereas the energy to break hydrogen bonds ranges from 2 to 50 kJ/mole (0.5 to 12 kcal/mole). L. Pauling determined the strength of the hydrogen bonds in water by measuring the heat of sublimation of

ice and subtracting from this value the heat of sublimation of CH_4 (methane), which has similar general molecular properties (see Figure 3–9) but does not form hydrogen bonds. He obtained a value of 11.9 kJ/mole (4.5 kcal/mole) which is less than one-twentieth the strength of the covalent O—H bond.

In biological systems, hydrogen bonds can form by sharing a hydrogen between any combination of the two electronegative atoms O and N. The atom to which the H is covalently bound (when the hydrogen bond is not present) is called the *hydrogen donor* and the atom whose unshared electron pair bonds more weakly to the H is called the *hydrogen acceptor*. Since it is the H nucleus to which the unshared electron pair bonds, we speak also of *proton* donors and acceptors. The bond distances vary according to the identity of the hydrogen donor and acceptor atoms (Table 3–10).

The intermediate strength of hydrogen bonds has important consequences for biological structures that contain them. They are strong enough to stabilize many complex macromolecular structures, yet still weak enough for molecules containing them to undergo certain important changes in structure. These properties give hydrogen bonds the preeminent position they occupy in the structure and function of living matter. It is also important to recognize that, although an individual hydrogen bond is not as strong as a single covalent bond, biological macromolecules such as proteins and nucleic acids are capable of forming many hydrogen bonds simultaneously. The

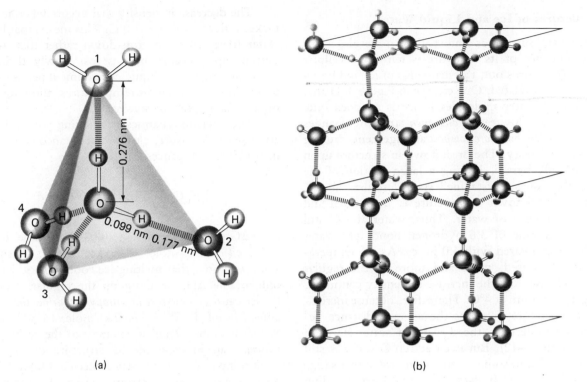

(a)   (b)

FIGURE **3–11**

The Structure of Ice, Form I(h). (a) Tetrahedral arrangement of water molecules in ice. Molecules 1 and 2, as well as the central H_2O molecule, lie entirely in the plane of the page. Molecule 3 lies above the page and molecule 4 below, so that oxygens 1–4 lie at the corners of a regular tetrahedron. (From Edsall and Wyman, 1958.) (b) The arrangement of water molecules in the ice crystal. The diagram shows how the water molecules are arranged in a relatively open network of adjacent tetrahedrons, forming a structure that is less tightly packed than water in the liquid state. (From General Chemistry, ed. by Linus Pauling, W. H. Freeman and Co., 1953.)

structure and interactions of these macromolecules are therefore the result of the *additive contributions* of all the hydrogen bonds.

Another crucially important property of hydrogen bonds is that the angle required for optimum bonding is relatively specific. The highest bonding energy of

O—H----O is at an angle of 180° and the energy falls off very rapidly when the bond is distorted even by a few degrees. The *directionality* of the hydrogen bond endows macromolecules with *geometrical specificity of interaction,* which is a most important property of the molecules of living systems.

TABLE **3–10**

A Variety of Hydrogen Bonds Occurring in Living Systems

Hydrogen Donor	Hydrogen Acceptor	Distance Between Electro-negative Atoms, nm (Å)
—O—H· · · · · · · · · · · · · · · · · · · ·	O—	0.276 (2.76)
—O—H· · · · · · · · · · · · · · · · · · · ·	N—	0.288 (2.88)
—N—H· · · · · · · · · · · · · · · · · · · ·	O—	0.304 (3.04)
—N—H· · · · · · · · · · · · · · · · · · · ·	N—	0.310 (3.10)
—O—H· · · · · · · · · · · · · · · · · · · ·	⁻O—	0.263 (2.63)
—N⁺—H· · · · · · · · · · · · · · · · · · · ·	O—	0.293 (2.93)

The Structures of Ice and Liquid Water

Since water surrounds almost all the molecules of the cell, the influence of its structure is felt in all cellular processes. As it turns out, the structures of ice and liquid water are closely related. One can see in Figure 3–11 that the packing of water molecules in ice is an open one. When ice melts, this *regular and open* structure changes into a *more random but more compact* arrangement, causing an increase in density. The breakdown in structure upon melting is only partial, leaving a large portion of the liquid in the "quasi-crystalline" state described above. This partial breakdown of structure is an unusual and interesting property of water. Thus, water at 0°C still contains an average of 3.6 hydrogen bonds per water molecule, as compared with 4.0 in ice. As the temperature is slowly raised, more and more of the regular structure collapses and the increase in density continues, reaching a maximum at 4°C. Thereafter, because increasing the temperature increases the average distance between water molecules, the density of water starts to decrease (Figure 3–12). But even at 100°C water is still hydrogen-bonded to some extent, as indicated by its high heat of vaporization. In fact some hydrogen bonding occurs even in water vapor and it is believed that these infrequent bits of structure are the loci for condensation when water vapor forms mist.

The decrease in density that occurs when ice freezes makes it float on water at 0°C. This means that bodies of water freeze from the top down rather than from the bottom up. Because heat is lost slowly through the insulating ice layer, a liquid environment persists at lower depths despite winter air temperatures, allowing resident organisms like fish to survive.

The thermodynamics of melting ice at 10°C and atmospheric pressure, clearly a spontaneous process, are instructive. The process

$$H_2O \text{ (solid)} \rightleftarrows H_2O \text{ (liquid)}$$

has a $\Delta G = -0.23$ kJ/mole (-0.054 kcal/mole) but ΔH is positive ($+6.4$ kJ/mole, $+1.53$ kcal/mole), not surprising since we know that melting ice cools drinks. Despite an unfavorable ΔH, breaking up the regular structure of the ice causes *an increase in entropy that drives the melting reaction forward* ($T \cdot \Delta S = +6.62$ kJ/mole, $+1.58$ kcal/mole). We shall discuss later an analogy to the reverse of this process, an increase in the structure of water from attempting to dissolve a fatty molecule. Often this would result in a decrease in entropy so large that the process is nonspontaneous.

It is reasonable to ask why water with so much internal structure is nevertheless so very fluid. The answer lies in the rapidity with which hydrogen bonds are made and broken. The ice-like structure in liquid water is *statistical*. The lifetime (half-life) of any given hydrogen bond is very short, possibly as little as 10^{-11} seconds. This means that water has a low viscosity and that it takes only a short time for a molecule to move a given distance through water by diffusion.

Other Unique Properties of Water

The structural features discussed so far are the cause of other unique properties of water, such as its unusually high melting point, boiling point, cohesion, enthalpy of fusion, enthalpy of vaporization, heat capacity, surface tension, and dielectric constant (Table 3–11). Some of these properties have a special significance for living matter. The hydrogen bonding between molecules in liquid water is responsible for its *cohesion*. Provided that continuous water structure is present, evaporation from the leaves of trees can "pull" water, by such cohesion, through the trunks to heights of several hundred feet. The high heat of vaporization provides the cooling capacity of evaporating water that is used by organisms to regulate their body temperatures. Water's high heat capacity acts as a "temperature buffer" to protect the labile structures of the cell from destruction by local, short-lived releases of heat.

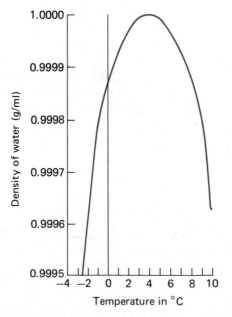

FIGURE 3–12
The Effect of Temperature on the Density of Water. Having a maximum of density while still a liquid (at 4°C) is a unique characteristic of water.

TABLE 3–11

Physical Properties of Water Compared with Other Liquids

	H_2O	Liquids		
		Ethanol	Hexane	Chloroform
Heat capacity (J/gram)	4.2	2.5	2.1	1.0
Enthalpy of vaporization (J/gram)	2500 at 100°C	1100 at 79°C	330 at 68°C	250 at 61°C
Enthalpy of fusion (J/gram)	334 at 0°C	1040 at −117.2°C	...	...
Surface tension (liquid-air, mJ/m²)	76	22	18	11
Dielectric constant	79	24	1.9	5

There are four other properties of water that are of fundamental importance to living systems and deserve more extensive discussion. They are (1) the solvent power of water, (2) the role of water in hydrophobic interactions, (3) the dissociation of water, and (4) the dissociating effects of water.

The Solvent Power of Water

Water is an excellent solvent for molecules ranging from polar organic compounds to salts that are completely dissociated into ions, even in the solid crystalline state. Molecules that dissolve well in water are called *hydrophilic* ("water-loving").

Hydrocarbons are notoriously insoluble in water, for reasons we shall discuss presently, but as soon as one includes a hydroxyl group (—OH), the resulting alcohol is much more soluble in water. This is because the hydroxyl group, like the water molecule, is strongly dipolar. The hydrogens of hydroxyl groups carry a partial positive charge and readily form hydrogen bonds with the oxygen atoms of water. There is also the possibility of sharing water hydrogens with the lone pairs of electrons of the hydroxyl oxygen. Water thus acts as a solvent for polar substances because *it forms hydrogen bonds* (dipole-dipole interactions) *with the solute molecules,* competing with the bonds which solute molecules form with each other. Other organic groups also enhance the solubility of compounds containing them by forming hydrogen bonds with water.

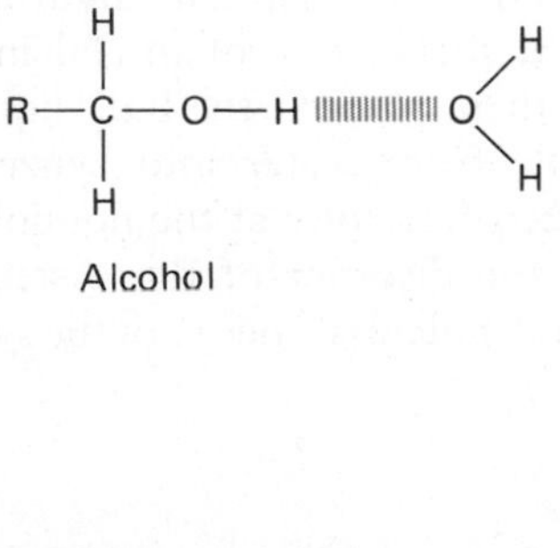

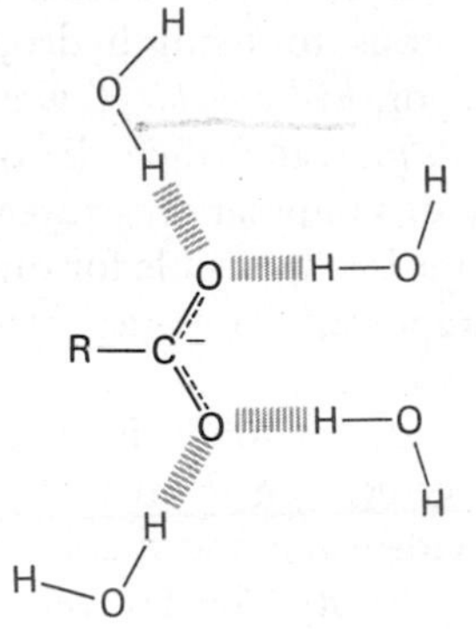

Carboxylate ion 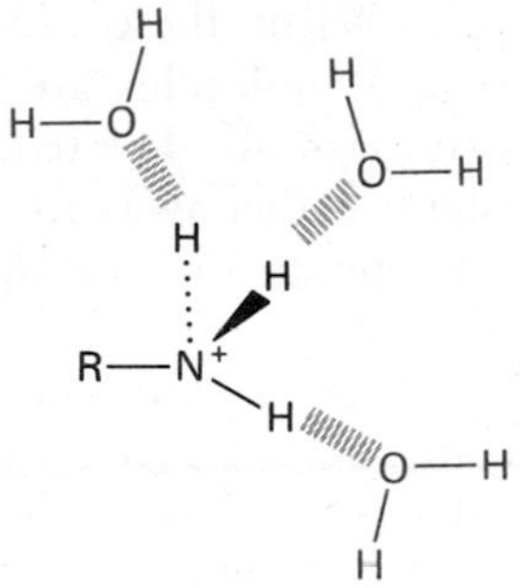   Quaternary ammonium ion

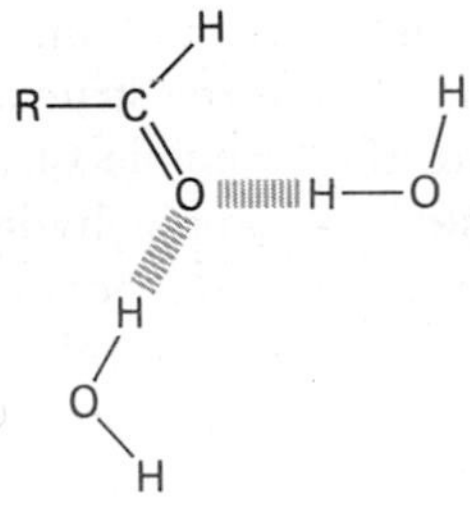

Keto group

Alcohol

Water is also a very good solvent for salts which are fully dissociated into positive and negative ions in the crystalline state. The strong attractions between these ions make salts insoluble in other solvents. But many salts will dissolve readily in water because water molecules tend to become packed and oriented around ions, forming an aqueous layer of special properties called the *hydration shell*. The hydration shell, which is stabilized by ion-dipole interactions, tends to screen out the electrostatic interactions between ions. This reduces their energy of interaction, compared with the interaction in a vacuum, by a factor of 80, the value of the *dielectric constant* of water. Ion-ion interactions are weakened within aqueous solutions, but in nonaqueous environments (the interior of certain proteins, for example) such interactions can be much stronger and thus help stabilize the conformation of a macromolecule.

The hydration shell enlarges the effective diameter of ions. Even though K^+ has a larger atomic number, Na^+ can be approached more closely, can therefore attract more water molecules, and is a larger hydrated ion. This is one reason why Na^+ penetrates the membranes of cells more slowly than K^+ (see further discussion of this point in Chapter 7).

The Hydrophobic Interactions of Water

The secret of water's great solvent power is its ability to form hydrogen bonds with polar solutes. But many biologically important molecules, or parts thereof, can neither accept nor donate protons to form hydrogen bonds. When these nonpolar or *hydrophobic* ("water-fearing") molecules are placed in water they dissolve poorly, if at all. The tendency of nonpolar structures to cluster together away from water is responsible for much of the energy that stabilizes important biological structures.

Nonpolar molecules do not dissolve in water, despite the fact that these molecules can form perfectly good van der Waals contacts with water and despite our general expectation that mixing should be favored because it results in less organization, and thus a higher entropy (see Chapter 1). Since many different molecular forms behave in a similar way with respect to water solubility, we infer that the molecular basis for this puzzling behavior lies in the structure of water itself.

To illustrate the principle of hydrophobic interactions, we consider a simple hydrophobic compound, ethane, that is free to move between different environments. We use the same symbols as for a chemical reaction to refer to this transfer process, the movement of ethane from water to benzene. The precursor state is ethane in water; the product state is ethane in benzene:

$$\text{ethane (water)} \rightleftarrows \text{ethane (benzene)}.$$

It has been found that under certain conditions this transfer process, as written from left to right, has an enthalpy change of $\Delta H = +9.2$ kJ/mole. Benzene is a nonpolar molecule and we might therefore expect it to be a more congenial environment for ethane; nevertheless, heat is taken up by the system when nonpolar ethane moves into benzene. This means that the product state has a higher potential energy than the precursor state. But the process as written is spontaneous at 300 K (27°C), with a free energy change of $\Delta G = -11.6$ kJ/mole. This occurs because there is a *large positive entropy change*:

$T \cdot \Delta S = +20.8$ kJ/mole. To understand why the *entropy increases* during this process we shall have to reconsider the structure of liquid water.

When ethane moves into water the entropy of the system is decreased because water is less able to form three-dimensional ice-like structures. Although it may at first seem paradoxical, an *increase* in the amount of these quasi-crystalline, ice-like structures in liquid water *reduces the degree of organization*. This is because of the dual participation of the water oxygen, as both a donor and an acceptor of hydrogens, in the formation of hydrogen bonds. At any instant, the member of a pair of hydrogen-bonded oxygen atoms that is closer to the hydrogen can be considered to be the donor and the other the acceptor. But by a small relative motion of the hydrogen (which is very light), the roles of donor and acceptor are reversed. This happens even in crystalline ice and is happening all the time in liquid water as well, along with individual water molecules moving out of the ice-like structure into more random orientations with respect to their neighbors.

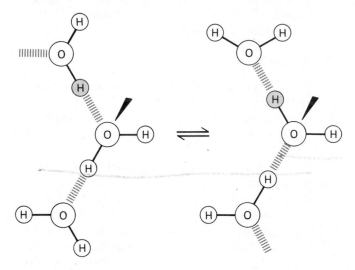

Hydrogen nuclei (protons) are thus *delocalized* in the ice structure in the same sense that electrons are delocalized in molecular orbitals—for example, of benzene. The important point is that the hydrogens are delocalized more in a water molecule that is part of a three-dimensional quasi-crystalline ice structure than they are in a water molecule adjacent to an ethane molecule. The water molecules next to the ethane cannot participate in ice-like structures to the same extent as they could were the ethane not present because the water molecules adjacent to the ethane cannot make hydrogen bonds with it. The water at the surface of the ethane is, therefore, *more organized* than water in the bulk liquid. Moving an ethane molecule from water into benzene reduces the number of water molecules at the nonpolar surfaces and thus increases the disorder of the system. The change actually raises the potential energy of the system ($\Delta H > 0$), but the increase in entropy ($\Delta S = +69 \text{ J K}^{-1} \text{ mole}^{-1}$) more than compensates and the process is spontaneous.

A simple example of this phenomenon can be experienced on a more homely scale when the oil of salad dressing must be remixed with the aqueous phase (vinegar). After the homogenate stands for a relatively short time the nonpolar oil is squeezed out as the water tries to achieve its maximum entropy.

The same idea explains why water has a high *surface tension* (measured as the force necessary to increase the surface area). The water molecules at the surface cannot make hydrogen bonds with the adjacent gas molecules, so the protons of those water molecules are more localized. *The structure of the water at the surface layer therefore has a lower entropy* than the structure of bulk water. Since the system tends spontaneously to increase its entropy, it exerts a force against any attempt to increase the area of surface at an air-water interface; as a result the surface area is minimized.

The term *hydrophobic interaction* is the name given to the tendency whereby nonpolar substances, to minimize their contact with water, come together. Although there are usually good van der Waals interactions in the resulting oily pockets, the main driving force for forming them is the increase in entropy of the surrounding water that occurs when the hydrophobic structures are withdrawn. Hydrophobic interactions play a very important role in maintaining the biologically active structures of proteins, nucleic acids, and membranes. (The term hydrophobic interaction is preferred over the older "hydrophobic bond" since there is neither the specificity nor the definite length and angle one associates with a chemical bond.)

Some biologically important molecules are nonpolar over most of their length but have a polar or even a charged group at one end (Figure 3–13a). Such *amphiphilic* ("both-loving") or *amphipathic* ("both-feeling") molecules will dissolve in water but behave in a very special way. The tendency of the water to maximize the delocalization of the protons involved in hydrogen bonds produces a segregation of the nonpolar parts of amphipathic molecules. The result is the formation of *micelles*: small symmetrical structures with hydrophobic interiors and hydrophilic surfaces. The simplest micelle formed by fatty acids, for example, is a small sphere in which the hydrophobic tails of the molecule are all in the interior and the charged groups are located on the outside surface, where they form hydrogen bonds with the surrounding water (Figure 3–13b). Hydrophobic molecules can be dissolved in the interior of micelles formed by an amphiphilic detergent, while the micelles themselves are dissolved in water. This accounts for the solubilizing power of these amphiphilic molecules—it also explains how household detergents can clean oily dirt that water alone won't wash out.

The Dissociation of Water

Water not only acts as a solvent for charged substances, but is itself capable of dissociating to a slight degree. Hydrogen-bonded protons manage at infrequent intervals to shift completely to the oxygen of an adjacent water molecule, thus leaving a negatively charged hydroxyl ion (OH^-) and forming a positively charged *hydronium ion* (H_3O^+). At 25°C this event occurs on the average in one out of 5.5×10^8 molecules of water. (This probability of dissociation, $10^{-8.7}$, times the concentration of water, approximately 55 molar, yields the concentration of hydronium ions: 10^{-7} molar.)

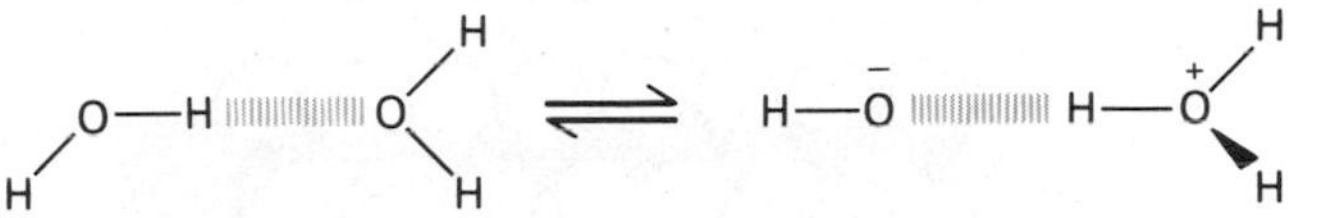

By dissociating into ions, water acts both as an acid and a base. This can best be understood in terms of the Brønsted-Lowry theory of dissociation, which defines an acid as a substance that donates a proton to a base. *A Brønsted acid is a proton donor; its conjugate base is a proton acceptor*. Water acts as both a base and an acid, one molecule accepting a proton, the other donating a proton:

$$H_2O + H_2O \rightleftharpoons H_3O^+ + OH^-.$$

The dissociation of water can be expressed quantitatively by the equilibrium constant of the reaction. The true equilibrium constant, independent of concentrations, is determined by measuring *thermodynamic activities*. Luckily, this is not necessary in most biological systems because the concentrations of the molecules and ions in solution are low, and the temperature of the systems studied does not vary greatly. Under these conditions the *molar concentrations* of solute ions and molecules are numerically very close to their thermodynamic activities. According to the principles of chemical equilibrium, the equilibrium constant is

$$K_A = \frac{[H^+][OH^-]}{[H_2O]}$$

where the symbol [ ] stands for concentration in moles per liter. (Although we normally write H^+, we really mean it as a shorthand for H_3O^+; extremely few free protons are found in aqueous solutions.)

Since the concentration of water (~55 M) is so much larger than that of the ions H^+ and OH^-, it is for all practical purposes constant with respect to any likely changes in their concentrations. It is therefore customary

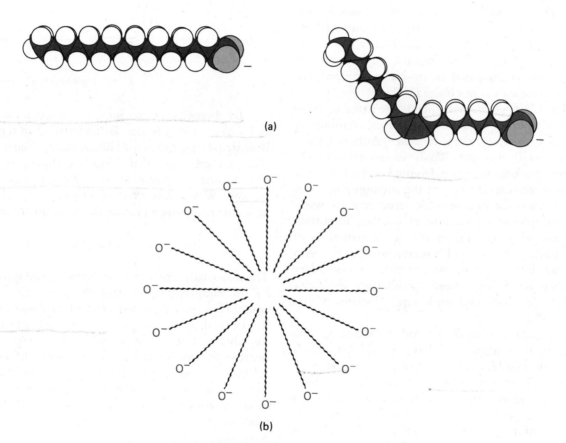

FIGURE 3–13

Hydrophobic Interactions and Micelle Formation.   (a) Two fatty acids, showing the hydrophilic carboxylate groups (colored) and the long hydrophobic hydrocarbon groups. (b) Formation of a soap micelle. The high negative charge density of the surface of the micelle (due to the −COO⁻ groups) repels other micelles, thus making it unlikely they will coalesce with one another. This prevents the constituent molecules from coming out of solution and forming a separate phase, thus making the micelle system stable.

to combine the equilibrium constant K_A with the concentration of water and obtain a new constant K_{WA}, the ion product of water:

$$K_{WA} = [H^+][OH^-].$$

K_{WA} has been determined by a variety of methods and turns out to be 1.0×10^{-14} at 25°C. In pure water the → $[H^+] = [OH^-] = 1.0 \times 10^{-7}$ molar.

According to the above equation, if one increases the concentration of H^+ the concentration of OH^- decreases in such a way that the product of the concentration of → both ions is 10^{-14}. Thus, if a dilute solution of HCl has a $[H^+] = 10^{-3}$ M then the $[OH^-] = 10^{-11}$ M. Conversely, if a dilute solution of NaOH has a $[OH^-] = 10^{-4}$ M then the $[H^+]$ is 10^{-10} M.

The acidity of a solution is dependent on $[H^+]$. Since the exponential character of the above quantities is often inconvenient, the *pH scale* has been defined as a more useful way of expressing acidity:

$$pH = -\log[H^+] = \log \frac{1}{[H^+]},$$

where the logarithms are taken to the base ten.

It is important that students familiarize themselves with this definition. First, it must be recognized that the pH scale is *inversely related* to the hydrogen ion concentration $[H^+]$; that is, the lower the pH the higher the hydrogen ion concentration. Secondly, the pH scale is *exponential* rather than linear. This means that when a solution changes from pH 7 to pH 6, the hydrogen ion concentration increases tenfold, and when it goes from pH 7 to pH 5, it increases a hundredfold. One convenient property of the pH scale is that for solutions of strong acids which dissociate completely, such as HCl, the pH is approximately equal to the logarithm of the reciprocal of the acid concentration. Thus, the pH of a 0.001 M HCl solution is close to 3.

The Dissociating Effects of Water

Besides being itself capable of dissociating, water is able to enhance the dissociation of protons from certain other substances—those that we call *weak electrolytes*. When pure, they are largely undissociated, but they become more and more dissociated as they are increasingly diluted in water. For example, a weak acid, such as acetic acid, participates in the following equilibrium:

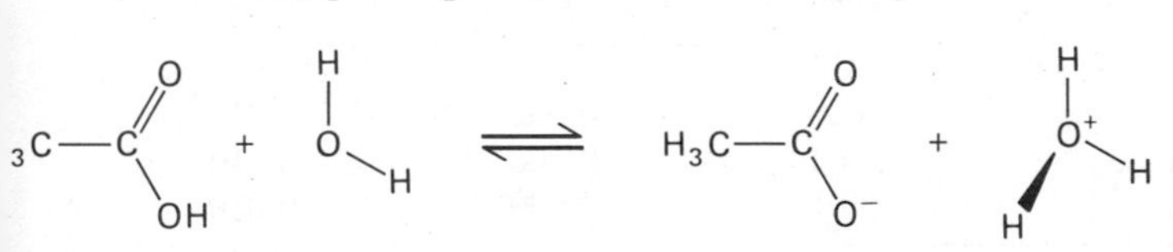

This shows how water competes with the acetate ion for the proton of the carboxyl group. Thus, we write in shorthand fashion:

$$CH_3COOH \rightleftharpoons CH_3COO^- + H^+.$$

The equilibrium constant for this reaction at 25°C is:

$$K_A = \frac{[CH_3COO^-][H^+]}{[CH_3COOH]} = 1.78 \times 10^{-5} = 10^{-4.75}.$$

The contribution of $[H_2O]$ is again included in the value of K_A.

In the above equation, K_A is really the *apparent dissociation constant* because it is based on measurements of concentrations. The true dissociation constant is based on the thermodynamic activities of the ions and does not vary with concentration or the presence of other ions, as does the apparent K_A. However, most conditions relevant to cells involve sufficiently low concentrations of ions to make the difference between the true K_A and the apparent K_A of little significance. We shall, therefore, keep on referring to K_A simply as the dissociation constant, but the student should remember the special conditions that make this appropriate.

Weak acids are those with small dissociation constants. Here again the exponential form is inconvenient, and it is useful to express the dissociation of the weak acid in terms of the negative logarithm of the equilibrium constant.

$$pK_A \text{ (acetic acid)} = -\log K_A = \log \frac{1}{K_A} = 4.75.$$

All weak electrolytes have one or more pK_A values, and a convention of subscripts is used to distinguish them: the most acid dissociation is pK_1, the next one pK_2, and so forth. Most authors simply use the term pK to mean pK_A, and we shall also do so from now on. Table 3–12, p. 114, lists a number of acids and their conjugate bases, as well as their pK values.

As we shall see in succeeding chapters, the macromolecules that have structural and functional importance in the cell are weak polyelectrolytes; that is, they contain several weak acidic and/or basic groups. They owe their state of dissociation, and hence many of their physical characteristics, to the presence of water or dilute salt solutions. Water, therefore, is not merely the dispersion medium of the cell, but also a major influence on the molecules it disperses. The properties of water that we have discussed have innumerable connections with the functioning of the living machine. It is not any one but

(text continued on p. 114)

Buffers

A *buffer system* has the property of resisting pH changes despite additions of acid or base. Since cellular metabolism is constantly producing and consuming protons, the pH of the cell in the absence of buffers would swing wildly between low and high levels. Large changes in pH alter the structure of macromolecules, affect the rates of chemical reactions, and generally wreak havoc on cellular processes. Buffers, then, are an important way for the cell to maintain constant or smoothly changing conditions.

To see the properties of a buffer system in detail, consider adding a strong base to a solution of a weak acid, a process we call titration (Figure 3–A). The midpoint of the titration curve—that is, the point where the pH changes least for a given added amount of NaOH—occurs at a pH value that is numerically equal to the pK_A value of the acetic acid. This can be demonstrated by applying the principles of chemical equilibrium to the dissociation of a weak acid.

A weak Brønsted acid, HA, dissociates as:

$$HA \rightleftarrows A^- + H^+.$$

At equilibrium,

$$K_A = \frac{[A^-][H^+]}{[HA]} . \qquad (3-1)$$

We also know from the *law of conservation of mass* that the total amount of the weak acid, both in its undissociated form (HA) and in its conjugate base form (A^-), must remain constant during the addition of base:

$$[HA] + [A^-] = [HA]_0, \quad (3-2)$$

where $[HA]_0$ is the amount present before any base was added. (All amounts have been divided by the volume to give molar concentrations.)

At any point during the addition of sodium hydroxide, we know the sodium ion concentration. This is because NaOH is a strong base that will dissociate completely to sodium and hydroxide ions. By the *law of conservation of electrostatic charge* we know that *the total number of positive ions must equal the total number of negative ions*:

$$[Na^+] + [H^+] = [A^-] + [OH^-]. \qquad (3-3)$$

This is because all the compounds originally added are electrostatically neutral. We also know that the ion product of water is a constant:

$$[H^+][OH^-] = K_{WA}. \qquad (3-4)$$

By listing the *equilibrium relations* for the dissociation of the weak acid and water, and the *conservation relations* of mass and electrostatic charge, we have found *four independent simultaneous equations* in four unknowns: $[A^-]$, $[H^+]$, $[OH^-]$, and $[HA]$. The rules of algebra guarantee an explicit solution for $[H^+]$ as a function of $[Na^+]$, which should agree with the experimental results depicted in Figure 3–A. In fact, equilibrium and conservation relations like these can *always* be used to solve numerical problems involving strong or weak acids, bases, buffers, titrations, etc.,

even though the mathematics involved may be cumbersome. By using chemical intuition, however, we can reduce the algebraic agony. Opportunities to do this arise when a sum includes terms of very small relative magnitude. *Omitting very small terms in a sum does not significantly alter the value of the sum.*

As an example, consider the point during the titration at which base has been added in an amount equivalent to one-half the weak acid. We shall assume the total initial mass of acetic acid to have been 0.010 moles and the volume at the time of completing the addition of 0.005 moles of sodium hydroxide to be one liter: $[HA]_0 = 0.010$ M; $[Na^+] = 0.005$ M. Since there are approximately equal amounts of $[HA]$ and $[A^-]$ after adding one-half of the equivalent of the base, and since the solution will still be acidic, we expect $[A^-] \gg [OH^-]$. Equation 3–3 can thus be approximated by $[Na^+] + [H^+] \cong [A^-]$. This also removes the need to use Equation 3–4, because $[OH^-]$ no longer needs to be eliminated. There are now three equations in three unknowns. The student can go through the algebra necessary to find that $[H^+]^2 + ([Na^+] + K_A) \cdot [H^+] - K_A \cdot ([HA]_0 - [Na^+]) = 0$.

In general, there are two solutions to a quadratic equation like this,

TABLE 3–A

Conservation of Mass During Titration

	HA	A^-	HA + A^-
Before adding base	0.010	0	0.010=HA_0
After adding base	0.008	0.002	0.010
After adding more base	0.004	0.006	0.010
Near end of titration	0.001	0.009	0.010

Note: Amount is expressed in moles.

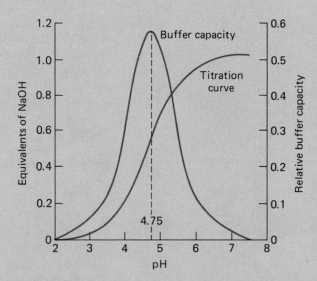

FIGURE 3-A
Titration Curve of Acetic Acid. These data are obtained by adding known quantities of NaOH to a known quantity of acetic acid, and measuring the resulting pH values. The buffer capacity can be derived by computing the change in pH per change in NaOH added—that is, the derivative of the titration curve. This curve has a maximum at the midpoint of the titration curve, the point at which the addition of base has the least effect on pH. The pH at which this occurs is numerically equal to the pK of the acid. It is thus the pH at which the system has the maximum buffering capacity.

but only one solution will make sense chemically. Furthermore, it is not always necessary to solve a quadratic equation. Here, for example, we can see that with half the weak acid neutralized, the concentration of $[H^+]$ will be much less than $[Na^+] = 0.005$ M. Thus, Equation 3–3 can be approximated further to $[Na^+] = [A^-]$. By algebra, it is possible to solve the three equations to give

$$H^+ = K_A \left[\frac{[HA]_0}{[Na^+]} - 1 \right].$$

Substitution of the numerical values gives $[H^+] = K_A = 1.78 \times 10^{-5}$ M. It should be noted that this implies $[OH^-]$ is roughly 10^{-9} M, which justifies our ignoring it in Equation 3–3 relative to $[A^-] = 5 \times 10^{-3}$ M.

EXAMPLE 3–1

What equilibrium $[H^+]$ would result from mixing 100 ml of a 100-mM solution of the weak base "Tris" ($K_A = 1.0 \times 10^{-8}$) with 100 ml of an 80-mM solution of strong acid HCl?

Solution

The dissociation reaction of Tris is

$$Tris \cdot H^+ \rightleftharpoons Tris + H^+$$

The equations expressing the equilibrium relations are thus:

$$\frac{[Tris][H^+]}{[Tris \cdot H^+]} = 1.0 \times 10^{-8} \quad (3\text{–}A)$$

$$\frac{[H^+][Cl^-]}{[HCl]} \gg 10^0 \text{ (strong acid) } (3\text{–}B)$$

$$[H^+][OH^-] = 10^{-14} \quad (3\text{–}C)$$

The equations expressing the conservation relations are:

$$[H^+] + [Tris \cdot H^+] =$$
$$[Cl^-] + [OH^-] \text{ (charge)} \quad (3\text{–}D)$$

$$[Cl^-] + [HCl] =$$
$$40 \text{ mM (mass, Cl)} \quad (3\text{–}E)$$

$$[Tris \cdot H^+] + [Tris] =$$
$$50 \text{ mM (mass, Tris)} \quad (3\text{–}F)$$

The numerical values assigned to the last two equations arise because the final volume will be 200 ml and, as a consequence, both total Tris and chloride will be diluted one-half from the solutions added: 100 ml / 200 ml.

There are six equations in six unknowns: $[H^+]$, $[OH^-]$, $[Tris \cdot H^+]$, $[Tris]$, $[HCl]$, $[Cl^-]$. Chemistry can be used to simplify the set of simultaneous equations. Acids like HCl are "strong" because they dissociate completely into protons and the conjugate base. Thus one can ignore the concentration of HCl in Equation 3–E: $[Cl^-] = 40$ mM.

Since less HCl acid was added than Tris base, the final concentration will be basic, and the hydrogen ion concentration will be less than 10^{-7}. This means that we can ignore $[H^+]$ in Equation 3–D in comparison with $[Tris \cdot H^+]$, which will be of the order of millimolar: $[Tris \cdot H^+] = [Cl^-] + [OH^-]$. But if $[Cl^-]$ and $[OH^-]$ were about the same value, 40×10^{-3} M, then the equilibrium $[H^+]$ would, by Equation 3–C, be $10^{-14}/40 \times 10^{-3} = 2.5 \times 10^{-12}$ or pH = 11.6. This is unlikely, since a comparable amount of acid (final $[Cl^-] = 40$ mM) and base (final total Tris = 50 mM) were added. This argument justifies ignoring $[OH^-]$ in relation to $[Cl^-]$.

Buffers (continued)

As a first approximation, assume Equation 3–D can be stated as $[\text{Tris} \cdot \text{H}^+] = [\text{Cl}^-] = 40$ mM. From Equation 3–F, one can then calculate $[\text{Tris}] = 10$ mM. Substituting these values into Equation 3–A yields

$$[\text{H}^+] = \frac{(1.0 \times 10^{-8}) \times (40 \times 10^{-3})}{(10 \times 10^{-3})}$$

$$= 4.0 \times 10^{-8} \text{ M, or pH} = 7.4.$$

This answer is consistent with the approximations made above: $[\text{H}^+] \ll [\text{Tris} \cdot \text{H}^+]$ and $[\text{OH}^-] \ll [\text{Cl}^-]$. (Note: Solutions buffered with Tris (tris-hydroxymethyl amino methane) are often used in research on cell biology.)

The equilibrium relation describing the dissociation of a weak acid (Equation 3–1) can be restated as

$$K_A = \frac{[\text{conjugate base}]\,[\text{H}^+]}{[\text{acid}]} \quad \text{or}$$

$$\frac{1}{[\text{H}^+]} = \frac{1}{K_A} \cdot \frac{[\text{conjugate base}]}{[\text{acid}]}.$$

Taking logarithms,

$$\text{pH} = \text{p}K_A + \log \frac{[\text{conjugate base}]}{[\text{acid}]}$$

$$= \text{p}K_A + \log \frac{[\text{A}^-]}{[\text{HA}]}.$$

This form of the equilibrium equation is always true and is often called the *Henderson-Hasselbalch* equation. You will notice that under certain conditions in our titration, when one-half of the equivalent of the base has been added, the concentration of sodium ion is, numerically, approximately equal to the concentration of the conjugate base A^-. *Under these conditions,*

$$\text{pH} = \text{p}K_A + \log \frac{[\text{salt}]}{[\text{acid}]},$$

which is the usually given form of the Henderson-Hasselbalch equation. The buffer capacity, measured as the ability of the solution to minimize changes in pH due to addition of base, is strongest near the midpoint of a titration, when $[\text{A}^-] = [\text{HA}]$ and $\text{pH} = \text{p}K_A$ (Figure 3–A).

Buffers are found both in nature and in the laboratory where pH must be controlled. If one wants to maintain the pH of a solution at a given value, one chooses a buffer in which the weak acid component has a $\text{p}K_A$ that is as close as possible to the value of the desired pH. Thus, to keep the pH of a solution at 4.7, a buffer containing acetic acid is well suited for the purpose. An inspection of Figure 3–A shows that the buffering efficiency of such a system is high over the range from one pH unit above to one pH unit below the pK value of the weak acid.

---

the simultaneous presence of all these properties that makes water the unique solvent for the world of life.

Structural and Functional Properties of Organic Molecules

To prepare for a discussion of the molecules of biology, it is helpful to review some of the features common to all organic molecules. These include proper-

ties that they show in respect of their physical environment and that they demonstrate when they react with one another. An understanding of some of the physical properties of organic molecules not only helps us to understand important aspects of their behavior in the cellular environment but is also of value for developing effective methods, sometimes called "handles," by which these molecules can be manipulated and studied.

Interaction with Electromagnetic Radiation

Electromagnetic radiation can be viewed as a *wave* of oscillating electric and magnetic fields in which the *velocity* of propagation is related to *wavelength* and *frequency* by the equation

$$c = \lambda\, \nu$$

where c = velocity, λ = wavelength, and ν = frequency. Wavelength is commonly expressed in meters and frequency in hertz (Hz: 1 Hz = 1 sec^{-1}); the value of c, in a vacuum, is 2.99×10^8 ms^{-1}. Electromagnetic radiation can also be viewed as composed of particles, called

TABLE 3–12
Acids, Conjugate Bases, and their $\text{p}K_A$ Values

Acid	Conjugate Base	$\text{p}K_A$
CH_3COOH	CH_3COO^-	4.75
H_3PO_4	H_2PO_4^-	2.0 ($\text{p}K_1$)
H_2PO_4^-	HPO_4^{2-}	6.7 ($\text{p}K_2$)
HPO_4^{2-}	PO_4^{3-}	11.7 ($\text{p}K_3$)
NH_4^+	NH_3	9.3
H_2CO_3	HCO_3^-	6.4 ($\text{p}K_1$)
HCO_3^-	CO_3^{2-}	10.2 ($\text{p}K_2$)

EXAMPLE 3–2

The N-terminal amino group of a certain cellular protein has a pK_A of 7.3. If the cytoplasm pH is 7.6, what fraction of these amino groups is in the charged ($-NH_4^+$) form?

Solution

Assume that the association and dissociation of protons is at equilibrium with respect to the pH: $-NH_4^+ \rightleftharpoons -NH_3 + H^+$. (This is a reasonable asumption, since protons dissociate and reassociate very rapidly). The equations describing this equilibrium are:

$$K_A = \frac{[-NH_3]\,[H^+]}{[-NH_4^+]}$$

$$pH = pK_A + \log \frac{[-NH_3]}{[-NH_4^+]}.$$

Solve for the concentration ratio by substituting values for pH and pK_A:

$$7.6 = 7.3 + \log \frac{[-NH_3]}{[-NH_4^+]}$$

$$\log \frac{[-NH_3]}{[-NH_4^+]} = 0.3 \quad \text{and} \quad \frac{[-NH_3]}{[-NH_4^+]} = 2.$$

The *fraction* of amino groups in the charged form is

$$\frac{[-NH_4^+]}{[-NH_3] + [-NH_4^+]}$$

and the inverse of this fraction can be written

$$\frac{[-NH_3]}{[-NH_4^+]} + \frac{[-NH_4^+]}{[-NH_4^+]}.$$

From above, the first term has the value 2 and the second term has the value 1; their sum is 3. The fraction of charged amino groups is the inverse of the inverse, 1/3. (Note: A group with a pK_A comparable to the cellular pH will have roughly the same concentrations of charged and uncharged forms.)

---

photons, which carry different amounts of energy, depending on the wavelength. Einstein showed that

$$E = h\,\nu$$

where E = the energy of a photon and h = Planck's constant. If the energy is expressed in joules and the frequency is expressed in hertz, the value of Planck's constant is 6.624×10^{-34} J·sec. Thus, the energy of electromagnetic radiation is directly proportional to its frequency and inversely proportional to its wavelength. Photons with shorter wavelengths are more energetic than photons with longer wavelengths.

The electromagnetic spectrum of radiation ranges from the very short gamma rays to the very long radio waves (Figure 3–14). Visible light, defined as radiation detected by the human eye, occupies only a small portion (380–700 nm) of the entire spectrum. The molecules of the cell interact with various forms of electromagnetic radiation. This interaction can be: (1) important to life, as in photosynthesis, vision, or phototropism; (2) detrimental to life, as in the effects of ultraviolet light in producing mutations; and (3) of great importance to the investigator as a probe for the study of many kinds of biomolecules.

The Absorption of Light

The fluctuating electric and magnetic fields of visible and ultraviolet radiation bring about the oscillation, at the same frequency, primarily of the small, negatively charged electrons in the molecules being irradiated. The electrons of different molecules respond differently to the frequency of the radiation to which they are exposed. Maximum absorption of radiation occurs at certain specific frequencies, corresponding to the differences in energy between quantum states of the electrons in a given molecular species. Thus, different substances have different absorption spectra; that is, they absorb significant amounts of radiation in different regions of the electromagnetic spectrum.

Figure 3–15 shows an absorption spectrum of chlorophyll *a*, found in most green plants. Most of the light absorption occurs in the blue and red regions of the spectrum. The green portion of the spectrum is reflected or transmitted, and so chlorophyll *a* appears green to the

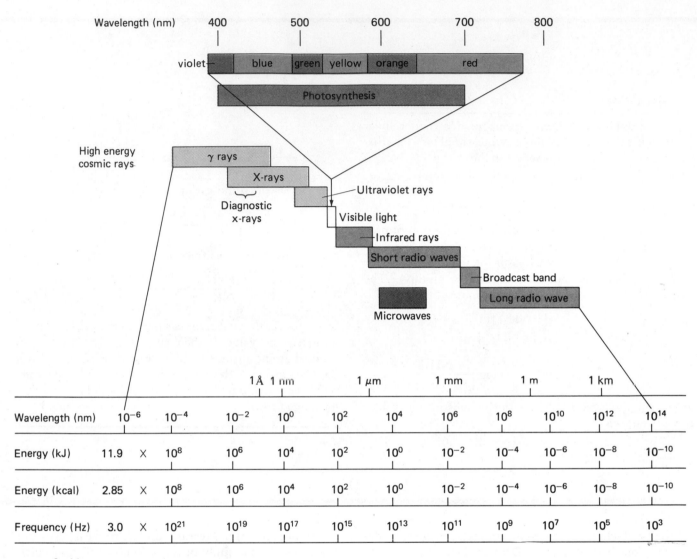

Wavelength (nm)	10^{-6}	10^{-4}	10^{-2}	10^{0}	10^{2}	10^{4}	10^{6}	10^{8}	10^{10}	10^{12}	10^{14}
Energy (kJ)	11.9 ×	10^{8}	10^{6}	10^{4}	10^{2}	10^{0}	10^{-2}	10^{-4}	10^{-6}	10^{-8}	10^{-10}
Energy (kcal)	2.85 ×	10^{8}	10^{6}	10^{4}	10^{2}	10^{0}	10^{-2}	10^{-4}	10^{-6}	10^{-8}	10^{-10}
Frequency (Hz)	3.0 ×	10^{21}	10^{19}	10^{17}	10^{15}	10^{13}	10^{11}	10^{9}	10^{7}	10^{5}	10^{3}

FIGURE 3–14
The Electromagnetic Spectrum.   The various bands of wavelengths are shown, as well as their energy content per mole of photons. Wavelength is inversely related to frequency and energy.

human eye. The energy of the absorbed photon changes the distribution of electrons of a molecule like chlorophyll *a*. We say the absorption of a photon results in chlorophyll being in an *excited state*. This excitation energy can be dissipated in a number of ways (see Chapter 8 for a more complete discussion). For biology, the most important way is by donating an excited electron to an acceptor molecule—which becomes reduced—as part of the process of photosynthesis.

The contents of most cells appear colorless and transparent to the human eye because proteins and nucleic acids do not absorb visible light. They do, however, absorb radiation with wavelengths below 300 nm, in the *near ultraviolet* region of the spectrum. Fortunately for cell biologists, the nitrogenous bases of the nucleic acids, such as adenine, absorb maximally at about 260 nm and the

proteins (owing to the presence of the amino acids tyrosine and tryptophan) absorb primarily at 280 nm (Figure 3–16). This means that we can distinguish by spectrophotometry between proteins and nucleic acids and even assess their relative amounts in a cell or solution.

Another pertinent region of the electromagnetic spectrum includes the *radio* frequencies, which can be used for the study of nuclear magnetic resonance (NMR). This technique can distinguish between different atoms of a given element in a particular molecule. The four protons (hydrogen nuclei) in methane (CH_4) are equivalent because of methane's symmetry, and will give only one NMR absorption peak. The protons in methyl alcohol (CH_3OH), however, exist in two different environments: three of them are bonded to a carbon atom and one is bonded to an oxygen. We are therefore able to

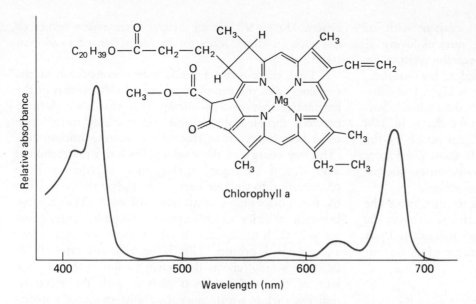

FIGURE **3–15**

The Absorption Spectrum of Chlorophyll *a*. Maximum absorption of light is in the blue (420 nm) and red (670 nm) regions of the spectrum. The molecular structure is also shown.

observe two absorption peaks. This property has made NMR the single most important analytical tool in the study of small organic molecules. In the last decade the technique has been applied, with considerable success, to the problem of distinguishing hydrogen atoms involved in more complicated biological molecules like transfer RNA. The magnetic properties of certain other atoms (^{13}C, ^{19}F, ^{31}P) are also suitable for analysis of the structures containing them by NMR.

Electrons in organic molecules are usually found in pairs; a molecule containing an unpaired electron absorbs radio frequency energy and is said to be *paramagnetic*. The analytical technique used to measure the absorption is known as *electron paramagnetic resonance* (EPR) or *electron spin resonance* (ESR) spectroscopy. It is especially useful in studying metal-containing proteins and molecules to which a special spin label group has been attached.

Reaction Mechanisms of Organic Molecules

Even the simplest of cells can carry out an astonishingly large number of chemical reactions. These reactions are typically catalyzed by specific enzymes. As we shall explain further in Chapter 5, in spite of the high degree of versatility and specificity exhibited by enzymes, most utilize one of two fundamental reaction mechanisms in their catalytic actions: *nucleophilic attack* and *acid-base catalysis*.

Nucleophilic Attack

Chemical reactivity can be usefully discussed in terms of interactions between electrically charged groups. The term *nucleophilic attack* describes the interaction

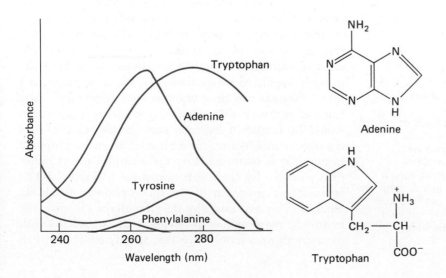

FIGURE **3–16**

Absorption Spectrum of Three Amino Acids and Adenine. All these species absorb in the ultraviolet region of the electromagnetic spectrum. The amount of absorption per mole of amino acid (extinction coefficient) is highest for tryptophan. The concentration of proteins, which contain tryptophan, can be estimated from the absorbance of light at 280 nm since this is the absorption maximum for tryptophan. Nucleic acids absorb light maximally at 260 nm, as illustrated by the absorption spectrum of nucleic acid component adenine.

between the "attacking" molecule, a reagent with an *unshared pair of electrons*, or *nucleophile* ("nucleus-loving"), and the molecule attacked, an *electron-deficient substrate*, or *electrophile* ("electron-loving"). Consider, for instance, the molecule methyl chloride (Figure 3–17a). The chlorine atom is more electronegative than the carbon atom. The carbon atom, with its partial positive charge (δ^+), is an electrophilic site, or *nuclear center*. An attack by the strongly nucleophilic hydroxide ion is shown in Figure 3–17b. This reaction is called S_N2, substitution *nucleophilic bimolecular* (that is, using *2* molecules).

Simply put, nucleophilic attack results from the attractive force between opposite electrical charges: the negative charge of the nucleophile, with its unshared pair of electrons, and the partial positive charge of the nuclear

center. However, several factors determine whether or not such an attack leads to the formation of a covalent bond.

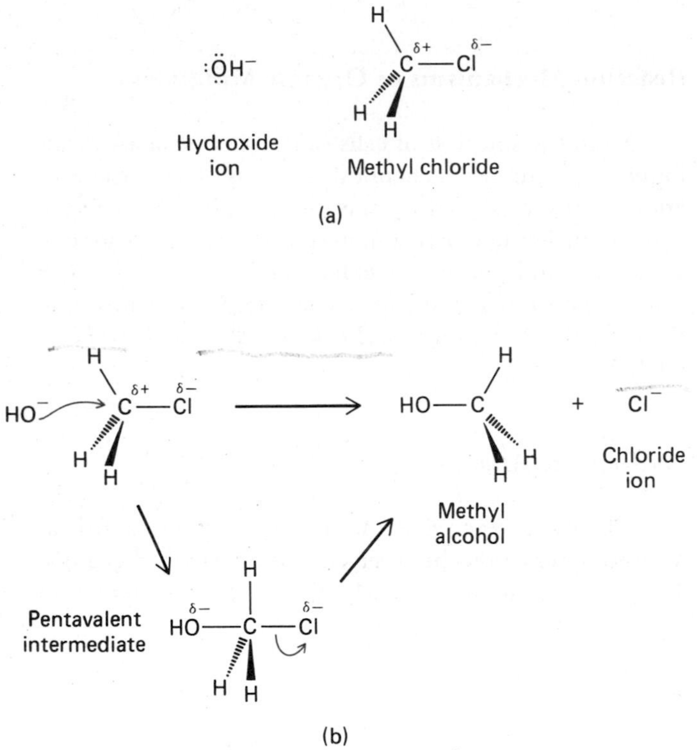

First, there is the thermodynamic consideration: the free energy of the products must be less than that of the reactants. This usually means that the bond formed between the electrophile and the attacking nucleophile must be stronger than that of the leaving nucleophile. (The free energy of the nucleophile-water interaction is very similar for most nucleophiles, so there is little contribution to the free energy change by the interaction of the nucleophiles with the solvent.) The relative bonding abilities of different nucleophiles correspond roughly with the degree to which they can act as bases and accept a proton. This is because the strength of covalent bonds reflects the extent to which the electron pair of the nucleophile is shared with the electron-deficient carbon atom. Since the ability to accept a proton is also a reflection of the ability to share an electron pair, *strong bases are in general strong nucleophiles*. In the reaction shown in Figure 3–17, the attacking and leaving nucleophiles—OH$^-$ and Cl$^-$—can be compared when each shares an electron pair with a hydrogen nucleus, to form HOH and HCl, respectively. The pK_A of the resulting acid is inversely related to the strength of the base: weak acid H_2O gives rise to strongly basic OH$^-$ while strong acid HCl gives rise to Cl$^-$, a very weak base.

Second, there is the kinetic consideration. If no reasonable pathway exists for a postulated molecular rearrangement, then the reaction cannot occur even if it would meet the thermodynamic requirements. For example, S_N2 attack (Figure 3–17) must occur at the *side opposite (rear attack)* the bond between the carbon atom and its departing group. This rear substitution requires a shift in the positions of the electrophilic carbon's other three bonds so that the tetrahedral bonding geometry of the saturated carbon atom (Figure 3–2) is maintained. (In fact, any optical isomerism existing at the nuclear center carbon before the attack is reversed when the reaction is complete, as shown in Figure 3–17.) The requirement for rear attack also places certain *steric restrictions* on the substrate molecule. (Steric restrictions result from the strongly repulsive interactions that occur between two atoms brought too close together—see Figure 3–7. Any reaction pathway that required too close an approach would be restricted by such *steric hindrance*.) The S_N2 reaction cannot occur, for example, at an electrophilic carbon if it is surrounded by three carbon atoms rather than partially by hydrogen atoms (as in Figure 3–17), because rear approach of the nucleophile is physically blocked. This explains why the strength of a nucleophile in an S_N2 reaction cannot be absolutely correlated with its strength as a base, since that is defined in terms of

(a)

(b)

FIGURE 3–17
The S_N2 Reaction Pathway. (a) The two reactants are the hydroxide ion (nucleophile) and methyl chloride (electrophile). (b) Hydroxide ion has an unshared pair of electrons and a full negative charge; it is attracted to the nuclear center carbon atom which has a partial positive charge. The color arrows represent the movement of electrons. A pentavalent intermediate is first formed, then it resolves into the products: methyl alcohol and chloride ion. The figure also illustrates inversion of the bond configuration about the nuclear center carbon atom; this would result in a change in optical activity if two of the hydrogens were substituted by different groups.

the simple proton transfer—a reaction not subject to steric restrictions.

The most common electrophile in cellular reactions is the <u>*carbonyl carbon*</u> atom. An example of such a reaction is base hydrolysis of an ester, as shown in Figure 3–18. Polarity is established by the greater attractive force of the oxygen for the electrons involved in the C=O bond. But the arrangement of the hybrid molecular orbitals of the carboxyl carbon (sp^2) is different from that of the saturated methyl chloride carbon (sp^3). The steric requirements of nucleophilic attack at these centers, therefore, are also different. The atoms of the carboxyl group lie *in a plane;* nucleophilic attack occurs perpendicular to the face of this plane and, therefore, with fewer restrictions than at nuclear centers containing only single bonds.

Acid-Base Catalysis

The degree to which a molecule is protonated (has accepted one or more protons) has a profound effect on the availability of that molecule's electrons and thus on its reactivity. There is first the general correlation between the strength of a nucleophile and its strength as a base, as discussed above. The degree to which a potential electrophile is protonated can also have an effect on the kinetics of a reaction. This is illustrated in the context of the acid-catalyzed *transesterification reaction* (Figure 3–19b). Adding a proton to the carbonyl oxygen makes it relatively electron-deficient; thus, donating an electron pair during nucleophilic attack to an even more electron-deficient carbon atom is encouraged. In low-pH (acidic) solutions, the reaction therefore proceeds at a significantly increased rate, as compared to neutral solutions.

Catalysis of nucleophilic reactions also occurs in strongly basic solutions. For example, proton transfer plays a critical role in the base hydrolysis reaction (Figure 3–17). An even clearer pathway that is catalyzed by base is the transesterification reaction shown in Figure 3–19c. When a proton is dissociated from an alcohol to combine

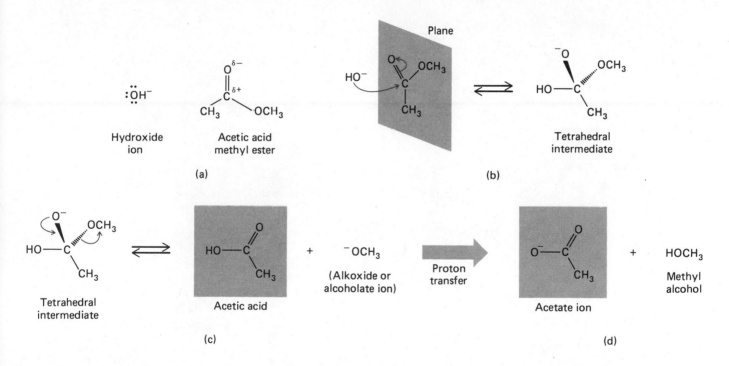

(a) (b)

(c) (d)

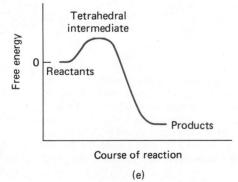

(e)

FIGURE 3–18
Hydrolysis of an Ester (Saponification Reaction).   (a) The two reactants: hydroxide ion and acetic acid methyl ester. (b) Nucleophilic attack by hydroxide occurs perpendicular to the plane that contains the nuclear center and its attachedgroups. A tetrahedral intermediate is formed. (c) The tetrahedral intermediate resolves into the first products: acetic acid and methyl alcoholate ion. (d) Proton transfer reactions conclude the pathway. Since the pH is high, a proton leaves the acetic acid to form an acetate ion; a proton becomes associated with the alcoholate ion to form methyl alcohol. (e) An energy diagram illustrates the thermodynamics of the reaction. The products acetate and methyl alcohol have a lower free energy under basic conditions than the reactants acetic acid methyl ester and hydroxide ion.

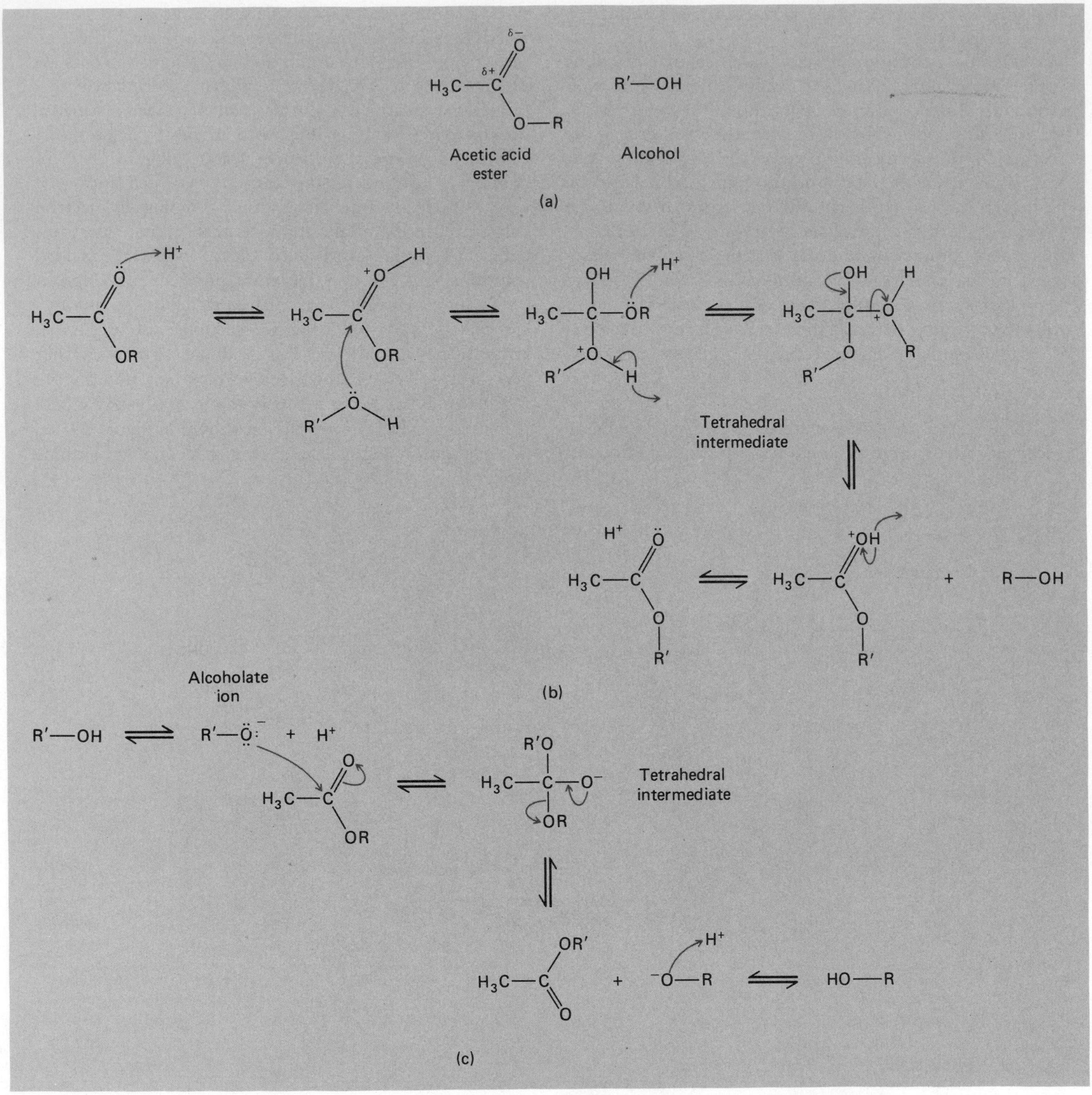

FIGURE 3–19
Transesterification, Catalyzed by Acid or by Base.   (a) The two reactants: an ester of acetic acid (from alcohol R—OH) and another alcohol (R'—OH). (b) In the presence of strong acid the carbonyl oxygen of the ester becomes protonated. This facilitates a nucleophilic attack by the unshared pair of electrons in R'—OH. A tetrahedral intermediate is formed which resolves into a protonated ester of the R' alcohol and the displaced alcohol R—OH. Under these conditions the reaction is reversible; it comes to an equilibrium that depends on the relative nucleophilic strength of R—OH and R'—OH. (c) In the presence of strong base, the R'—OH alcohol loses its proton, becoming an alcoholate ion (R—O⁻). This strong nucleophile attacks the carbonyl carbon nuclear center, forming a tetrahedral intermediate. When the carbonyl group reforms, the products are the transformed ester and the alcohol R—OH.

with hydroxide ions in the strongly basic solution, an alcoholate ion is formed. The electrons of alcoholate are more reactive than the electrons of hydroxide: the pK_A of water is 8.7, while that of methyl alcohol, for example, is 16. The strongly nucleophilic alcoholate ion attacks the carbonyl carbon and the final products formed are the results of an equilibrium between the two kinds of ester.

In summary, nucleophilic attack can be understood as an attraction between unlike electrical charges. Thermodynamic and kinetic factors determine whether or not such an attraction between a nucleophile and an electrophile can lead to formation of a covalent bond. Protonation can profoundly affect the distribution of electrons in a molecule, providing an explanation for the influence of pH in nucleophilic attacks. Many enzyme-catalyzed biological reactions exploit mechanisms of nucleophilic attack and acid-base catalysis. Detailed discussion of the reaction mechanisms of several enzymes is presented in Chapter 5.

Some Biologically Interesting Small Molecules and Some of Their Polymers

The full range of diversity found in small molecules and their many reactions is too much to catalogue here; it is a task which not even advanced textbooks in biochemistry manage to do. Instead, we shall give a brief overview of some of the principal carbohydrates and lipids found in the cell, leaving the proteins and nucleic acids until later chapters where we discuss them in detail.

Carbohydrates

Carbohydrates are a group of universally occurring compounds having the general formula $(CH_2O)_n$. They perform both functional and structural roles in the cell, serving as intermediates in metabolism, as storage products, and as structural materials in the walls of plant and microbial cells. Even the extracellular matrix of animal cells contains considerable amounts of carbohydrates (see Chapter 17). Carbohydrates are also components of nucleic acids and many proteins. As part of a protein in the cell membrane, carbohydrates face outward and can act as signals to other cells.

D-Glucose, a six-carbon *hexose,* is the most widespread sugar. It contains five hydroxyl groups and one aldehyde group. In solution, it is found mostly in the form of a six-membered ring, a form that is in equilibrium with a small amount of the open-chain form. In the ring *(pyranose)* form the oxygen of the aldehyde group is incorporated into the ring, and thus does not show its normal degree of reactivity.

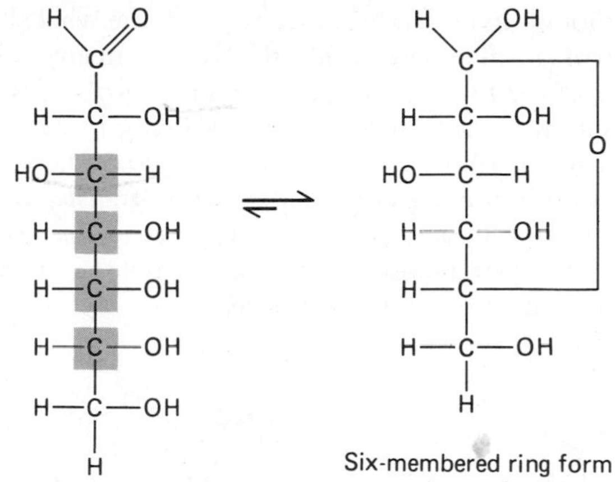

Open-chain form

Six-membered ring form of
α-D-glucose (pyranose)

D-Glucose

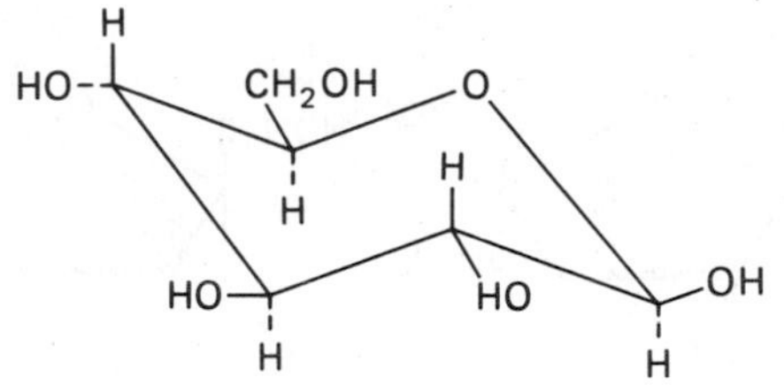

β-D-Glucose

The six-membered ring of glucose is found mostly in the chair conformation. Since this hexose has four asymmetric carbon atoms (the boxed carbon atoms in the structure shown above) and since each carbon's hydroxyl group can be found in one of two possible positions, $4^2 = 16$ different stereoisomers can in principle be formed; only four of these other sugars are found in nature. The short representation of D-glucose (shown below) provides a convention (suggested by the chemist Sir W. N. Haworth) whereby its particular isomeric structure can be specified. The heavy part of the ring plane is regarded as closest to the viewer, and the orientation of the hydroxyl groups is either up or down. The hydrogen atoms are often not shown but are assumed to be at the empty ends of the vertical lines.

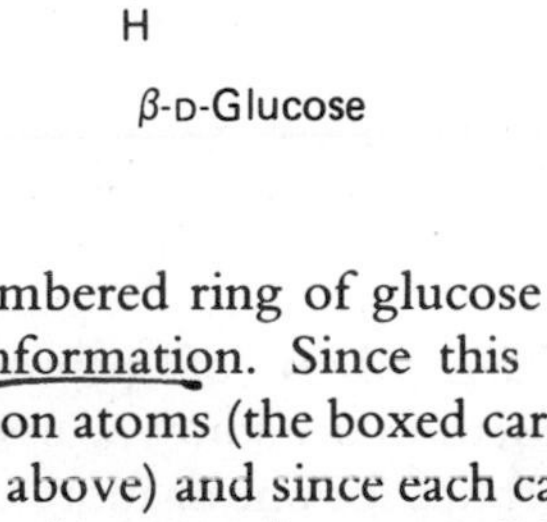

Short representation of D-glucose
(The convention for numbering carbon atoms is shown.)

Although five-membered rings are on the whole less stable than six-membered ones, the five-membered *furanose* ring of the five-carbon sugar *(pentose)* D-ribose is an important and universal constituent of living matter. As we shall see, D-ribose and its derivative deoxyribose are involved both in energy- and information-transfer reactions. Of the possible isomers, two other pentoses (D-xylose and L-arabinose) exist in nature, but they do not play as prominent a role as D-ribose.

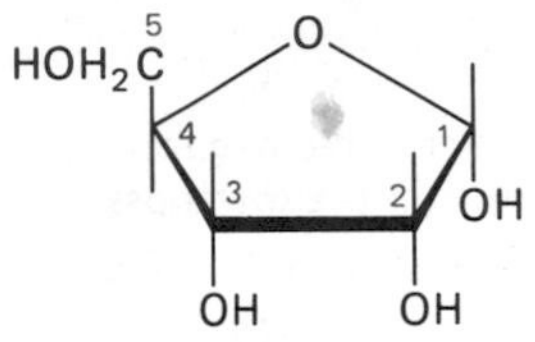

D-Ribose

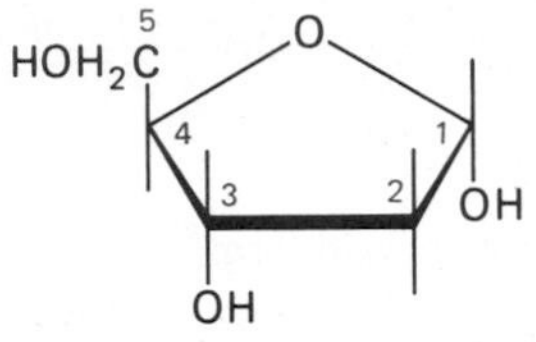

Deoxyribose

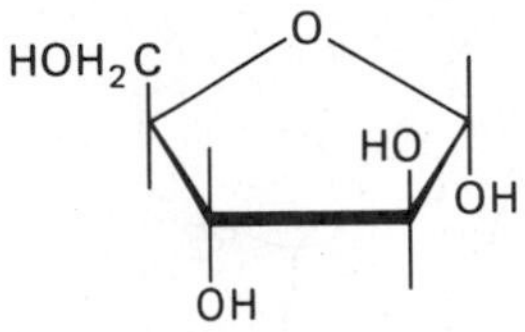

L-Arabinose

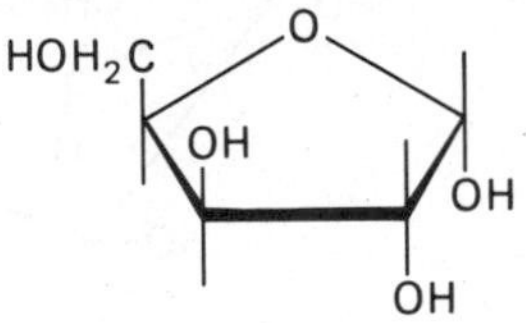

D-Xylose

Numerous derivatives of pentose and hexose *monosaccharide* structures are known, involving substitutions of various groups for one or more of the hydroxyls. As examples, we show below some forms of glucose which have been substituted by a carboxyl group (D-glucuronic acid), hydroxyl group (D-sorbitol), amino group (D-glucosamine), and phosphate group (glucose-6-phosphate). Also shown is the substitution of a methyl group, on the L-form of galactose, to form L-fucose.

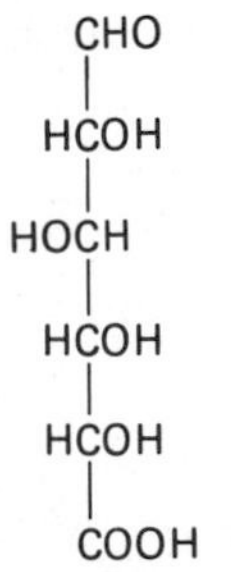

CHO	CH$_2$OH	CHO
HCOH	HCOH	HCNH$_2$
HOCH	HOCH	HOCH
HCOH	HCOH	HCOH
HCOH	HCOH	HCOH
COOH	CH$_2$OH	CH$_2$OH
D-Glucuronic acid	D-Sorbitol	D-Glucosamine
(a)	(b)	(c)

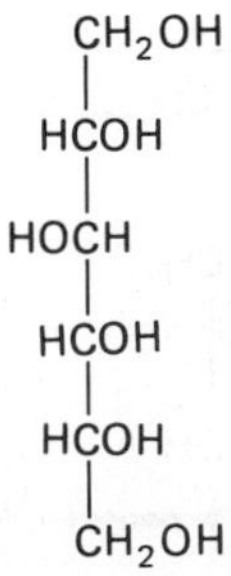

CHO	CHO	CHO
HCOH	HOCH	HOCH
HOCH	HCOH	HCOH
HCOH	HCOH	HCOH
HCOH	HOCH	HCOH
CH$_2$—O—P—O$^-$	CH$_3$	CH$_2$OH
Glucose-6-phosphate	L-Fucose	D-Galactose
(d)	(e)	(f)

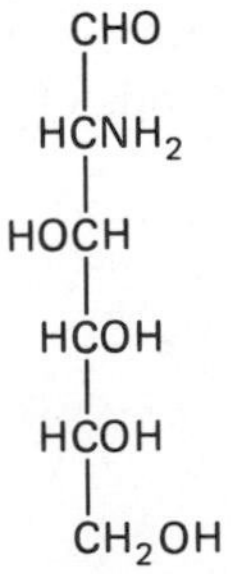

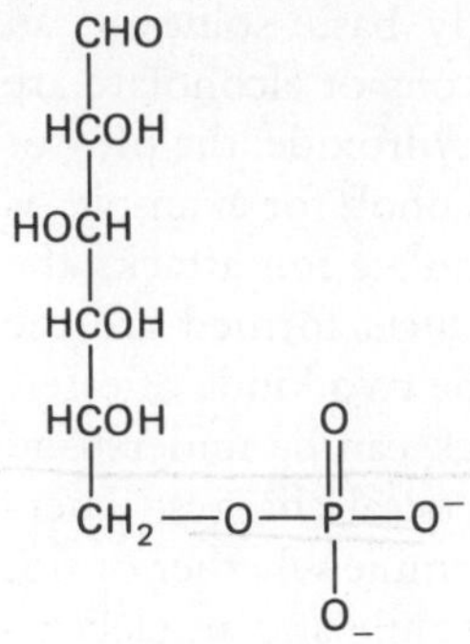

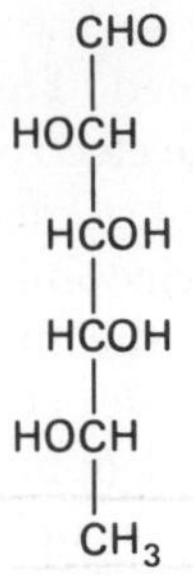

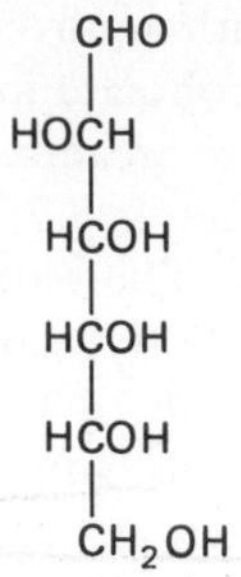

Monosaccharides can link together to form chains. A chain composed of between two and nine monosaccharide residues is called an *oligosaccharide*. The most important of these are *disaccharides* like sucrose. Sucrose is found in all plants that carry out photosynthesis and is a major low-molecular-weight food source for animals.

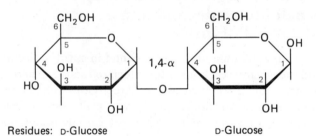

Residues: D-Glucose   D-Glucose

Maltose
(obtained when starch is degraded)

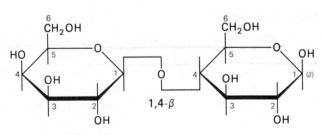

Residues: D-Galactose   D-Glucose

Lactose
(found in milk of mammals)

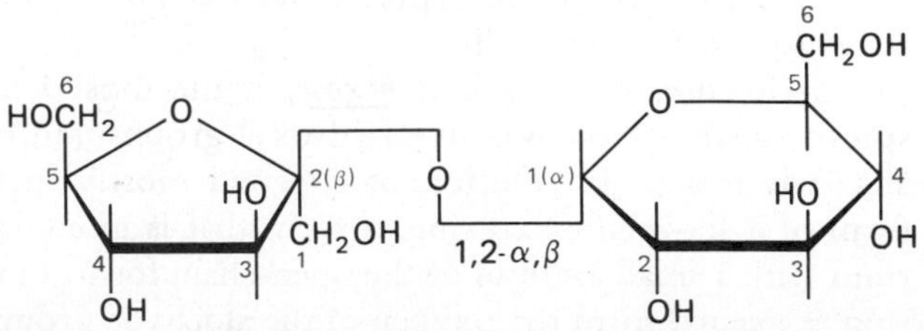

Residues: D-Fructose   D-Glucose

Sucrose

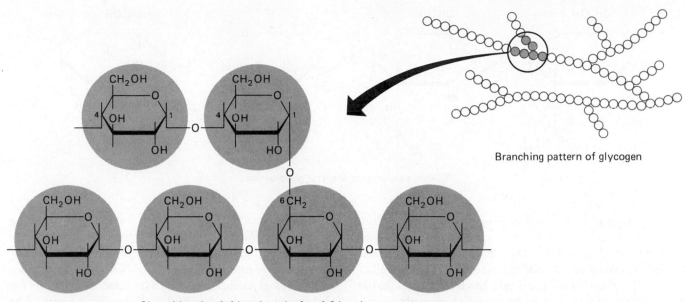

Branching pattern of glycogen

α-Glycoside using 1,4 bonds and a few 1,6 bonds

The *glycosidic bond* joining the residues of oligo- and polysaccharides can form in two ways, but only one of these isomers is usually found in nature. In the structures at the bottom of the previous page we designate these types of linkages as α, if the bond joining the C1 carbon to the connecting oxygen is down, or β, if the bond is up. For a detailed understanding of the conventions used, the student should consult a biochemistry textbook.

Polysaccharides are chains consisting of more than nine monosaccharide residues. The cell does not usually stop with as few as 10 residues, but synthesizes gigantic molecules that serve either food storage or structural functions in the organism. *Glycogen* and *starch* are the animal and plant kingdoms' major food storage products, respectively.

Liver glycogen is a polymer of about 30,000 glucose residues (α-glycosidic linkage) and contains long branched chains. The chain length between branch points is usually 10–14 glucose residues. The whole structure has a "bushy" appearance and forms huge water-soluble molecules (molecular weight = 5×10^6) having the general shape of a flattened ellipsoid.

Starch is deposited as microscopically visible granules in the plant cell. The starch granule is made up of two types of molecules: an unbranched chain of 1,4-α-glycoside linkages joining 250–300 glucose residues, usually arranged in the form of a helix; and a molecule much like glycogen but smaller (1000 residues) and less frequently branched (on the average, every 30 residues). The precise arrangement of these two types of molecules in the starch grain is not yet understood. However, a strong rotation of the plane of polarized light by starch grains suggests that highly ordered structures are involved.

Cellulose is the major structural component of plants; it is present in the cell wall of all plants from green algae to oak trees, either by itself or with other compounds such as *lignin*. Wood is essentially a collection of empty cells, with walls consisting of cellulose and lignin. The cellulose molecule is a long, unbranched chain of 1,4-β-glycoside linkages, joining some 14,000 glucose residues, with a molecular weight of 2.3×10^6. This gigantically long thread is folded, producing fibrils that are visible under the electron microscope. The folded molecule is insoluble because the hydroxyl groups interact with one another to cross-link the molecule, rather than dissolve by forming hydrogen bonds with the water.

Portion of cellulose molecule

Chitin, a polymer of *N*-acetyl glucosamine

The secondary cell wall of plants is made of cellulose fibrils arranged in alternate layers lying at different angles to each other (Figure 2–11), the whole structure interlaced by lignin, acting apparently as a cement. The cell wall, a sac-like envelope of the plant cell, has tremendous tensile strength, exceeding even that of the highest quality steel.

Chitin is a structural substance of the cell wall of fungi and the exoskeleton of arthropods. It also is a linear, unbranched molecule held together by 1,4-β-linkages, but in this case the sugar residue is *N*-acetylglucosamine.

The structural materials that form the cell walls of bacteria are among the most complex macromolecular edifices found in nature. The bacterial cell wall is constructed of long chains that are covalently interconnected by shorter side branches. It can therefore be argued that the bacterial cell wall is a single, gigantic,

bag-shaped macromolecule. The fundamental building block of the long polysaccharide chain is a disaccharide consisting of glucose derivatives *N*-acetylglucosamine (NAG) and *N*-acetylmuramic acid (NAM), held together by a 1,4-β-glycosidic linkage. The disaccharide units are also linked to each other by 1,4-β-glycosidic bonds. The bonds between C1 of NAM and C4 of NAG are susceptible to hydrolysis by the enzyme lysozyme, a phenomenon that both bacteriophages and higher animals use in their predatory or defensive interactions with bacteria (see Chapter 5).

The polysaccharide chains in the bacterial cell wall are cross-linked by short runs of peptide chains, which are polymers of amino acids. Because it contains both peptide and polysaccharide parts, the structure is known as a *peptidoglycan*. One of several structures that

The polysaccharide structure of the bacterial cell wall

we know to occur is shown at the bottom of page 124. There is a great deal of variation in the details of these structures, especially among different types of bacteria. Even within a given bacterium the cell wall appears to be composed of multiple layers, varying in their structural details.

Lipids

Fats play an important role in the cell as storers of chemical energy. Because they have long hydrocarbon chains, their overall state of oxidation is very low. They have much further to go than carbohydrates and proteins, therefore, to yield CO_2 and H_2O during oxidation by O_2, and they release more energy per unit of weight than carbohydrates and proteins.

Lipids are defined only by the *common physical property of solubility in nonpolar, organic solvents*. Not all classes of lipids, therefore, are related to each other in chemical structure. Several types of compounds belong to the lipids, the most important of which are fatty acids, neutral fats, phospholipids (glycerophosphatides), sphingolipids, and steroids.

Fatty acids are compounds consisting of a long hydrocarbon chain with a carboxylate group at one end. The general formula is:

$$CH_3-(CH_2)_n-C\overset{\displaystyle O}{\underset{\displaystyle O}{\Bigg\langle}}\ -.$$

Since fatty acids are synthesized from 2-carbon units (derived from acetic acid, CH_3COOH), all naturally occurring fatty acids have an even number of carbon atoms, the most frequent being 16 or 18. Fatty acids can

be either *saturated*, when their hydrocarbon chain contains only single bonds, or *unsaturated*, when one or more double bonds are present. Typical saturated fatty acids are:

$$CH_3(CH_2)_{14}\,COO^-\ \text{palmitic acid}\ (C_{16})$$

$$CH_3(CH_2)_{16}\,COO^-\ \text{stearic acid}\quad (C_{18}).$$

Typical unsaturated fatty acids are:

$$CH_3(CH_2)_7CH{=}CH(CH_2)_7\,COO^-$$
$$\text{oleic acid}\ (C_{18}\Delta^9)$$

$$CH_3(CH_2)_4CH{=}CHCH_2CH{=}CH(CH_2)_7\,COO^-$$
$$\text{linoleic acid}\ (C_{18}\Delta^{9,12}).$$

(The shorthand $C_{18}\,\Delta^9$ means an 18-carbon fatty acid with a double bond between carbon number 9 and carbon number 10, counting from the carboxyl carbon.)

Fatty acids have mutually contradictory parts; they consist of a polar, water-soluble "head" combined with an extremely nonpolar, water-insoluble "tail." Their schizophrenia is beautifully demonstrated by their interaction with water: they both do and do not dissolve in water—that is, they form a *monomolecular layer* at the surface of the water, with the negatively charged carboxyl groups sticking into the water and the hydrocarbon chains waving above (Figure 3–20). Or, as we have seen in Figure 3–13, fatty acids can form stable spherical micelles in the interior of the liquid water.

The monomolecular layer of fatty acids at the surface lowers the surface tension of water, an effect that increases the "wetting" power of water. This helps explain the cleaning action of soaps, which are sodium or potassium salts of fatty acids.

Neutral fats are esters of glycerol and fatty acids.

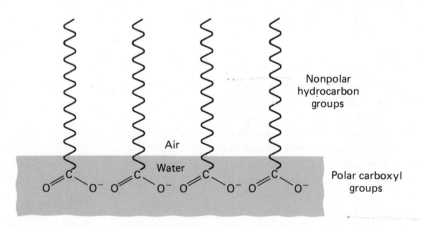

FIGURE 3–20
Orientation of Fatty Acid Chains at a Water-Air Interface. Fatty acids consist of a hydrophilic carboxyl group and a hydrophobic hydrocarbon chain. At a water-air interface they form a monomolecular layer in which the carboxyl groups lie preferentially in the water and hydrocarbon chains lie in the air.

$$CH_2O\ H \qquad HO\ OC(CH_2)_n\ \ CH_3$$
$$CH_2O\ H \quad + \quad HO\ OC(CH_2)_n{}'\ CH_3 \quad \rightleftharpoons$$
$$CH_2O\ H \qquad HO\ OC(CH_2)_n{}''\ CH_3$$

Glycerol Fatty acids

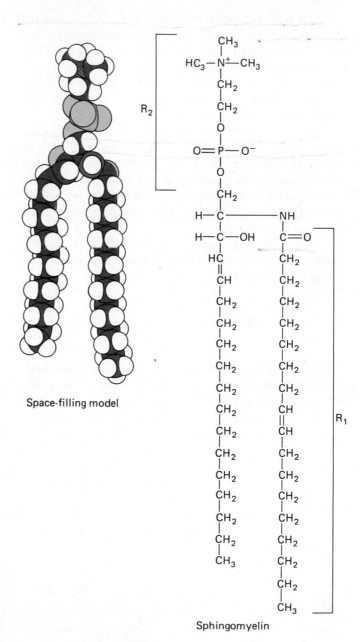

$$
\begin{array}{c}
\quad\quad H \quad\quad\quad O \\
H-C-O-C-(CH_2)_n\ \ CH_3 \\
\quad\quad\quad\quad\quad\quad O \\
H-C-O-C-(CH_2)_n{}'\ CH_3\ +\ 3H_2O \\
\quad\quad\quad\quad\quad\quad O \\
H-C-O-C-(CH_2)_n{}''CH_3 \\
\quad\quad H
\end{array}
$$

Fat

Space-filling model

As the formula above indicates, the fatty acids need not be identical—a fat may contain one, two, or three different fatty acids. A neutral fat is hydrolyzed (saponified) to fatty acids and glycerol by heating it in strong alkali. This is how for thousands of years our ancestors converted animal fat to soap.

The melting point of a fat depends on how unsaturated its fatty acids are: the higher the degree of unsaturation, the lower the melting point. When a fat has a sufficiently high degree of unsaturation to be a liquid at room temperature, it is called an *oil*. Oils can be converted to "hard" fats by saturating the fatty acids—that is, adding hydrogen across the double bonds, a process also known as hydrogenation. This is how margarine is manufactured from vegetable oils.

Phospholipids (glycerophosphatides) and *sphingolipids* comprise a rich variety of compounds that play important structural and functional roles in membranes (Figure 3–21a). The chemical structures of a large number of these compounds have been elucidated, but the details of how they all participate in membrane systems are not yet completely understood. When added to water the free energy of the system is reduced by increasing the area of the phospholipid as much as possible, which brings about the formation of *bimolecular leaflets* (Figure 3–21b). When proteins are added to phospholipids to create artificial membranes, their appearance in the electron microscope is similar to that of natural membranes. We shall discuss the details of the structure of biological membranes in Chapter 7.

Sphingomyelin

(a)

FIGURE 3–21
Sphingomyelin. This sphingolipid has polar and nonpolar properties that endow it with the ability to form a bilayer membrane. (a) There are two hydrophobic "tails" and a strongly hydrophilic "head." The latter is called a zwitterion, because it carries both a negatively charged group

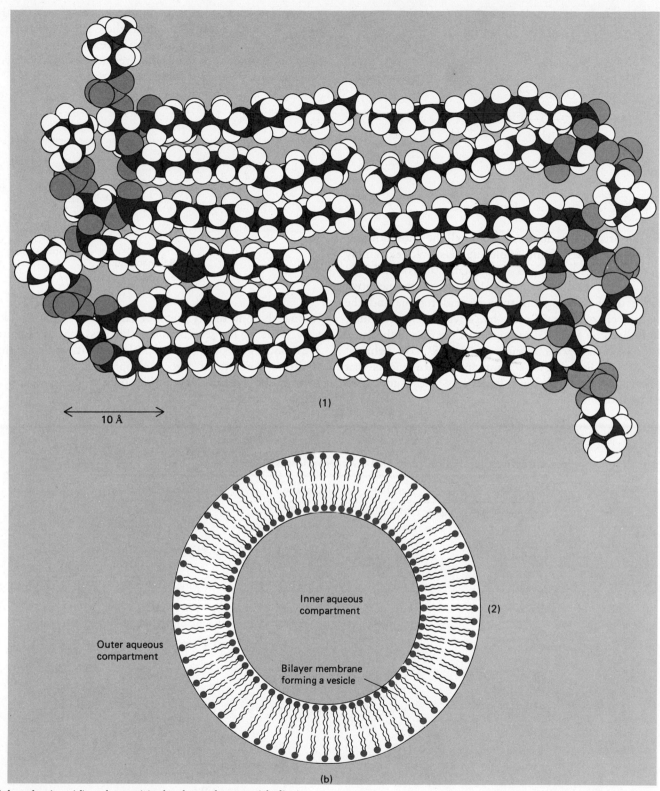

(1)

←——→ 10 Å

Outer aqueous
compartment

Inner aqueous
compartment

Bilayer membrane
forming a vesicle

(2)

(b)

(phosphoric acid) and a positively charged group (choline).
(b) The tendency of a phospholipid when in contact with
water to form a bilayer, a bimolecular surface ("leaflet"):
(1) Space-filling model of a small section of bilayer. (2)
Leaflets curving on themselves to form a closed spherical
vesicle.

Phospholipids are all phosphate esters of glycerol, with the following general formula:

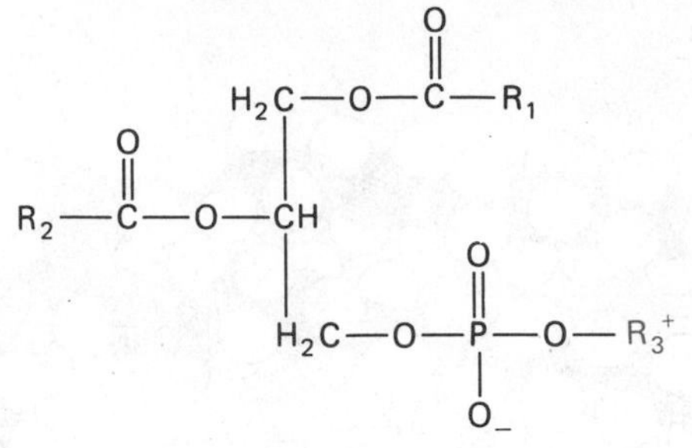

Phospholipid

The important common property of all the phospholipids is that one end of the molecule has two extremely hydrophobic tails (R_1 and R_2 are fatty acids) while the other end is very hydrophilic because of the negative charge of the strong phosphoric acid residue and the positive charge contributed by a variety of residues, symbolized by R_3. (Having opposite charges in the same structure makes this a *zwitterion*.) We have brought together in Table 3–13 a number of components that show the complexity and variety of these important compounds.

Sphingolipids comprise another category of lipids found in membranes. Here again, there is a great variety of complex compounds. The structure common to all of the compounds is a *sphingosine* residue on which two other residues are substituted as follows:

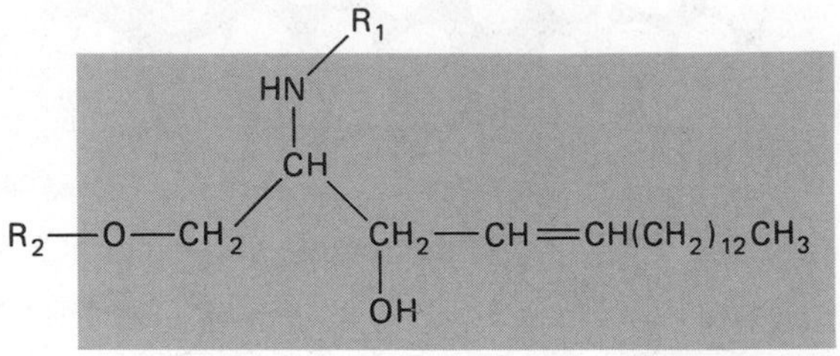

Sphingolipid

Sphingosine

R_1 may be one of a variety of fatty acids while R_2 can vary widely. Rather than list the possibilities, Figure 3–21a shows a typical example. Notice that here again the structure consists of two hydrophobic tails tied to a hydrophilic portion containing both a positive and a negative charge.

TABLE 3–13

A Partial List of Phospholipids (Glycerophosphatides)

Name	Residue 1	Residue 2	Residue 3
Phosphatidic acids	Fatty acid	Fatty acid	—H
Lecithins	Fatty acid	Fatty acid	Choline residue
Cephalins	Fatty acid	Fatty acid	Ethanolamine residue / Serine residue
Inositides	Fatty acid	Fatty acid	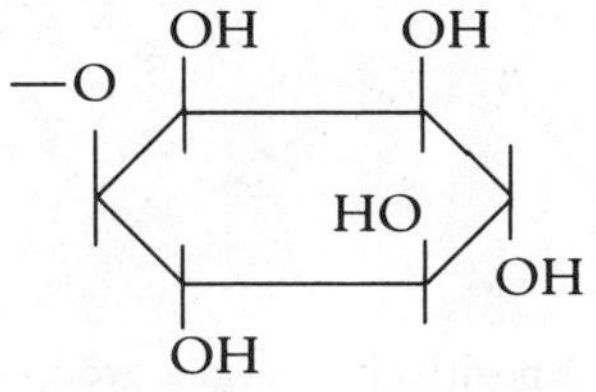 Inositol residue
Plasmalogens	Enol ether of fatty acid	Fatty acid	Ethanolamine or choline

Finally, there is an entirely different class of lipids synthesized from a basic *isoprene* skeleton:

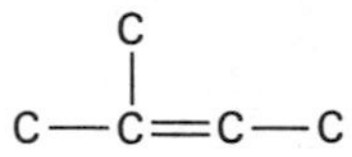

C—C=C—C
(with C above central carbon)

Isoprene skeleton

Cholesterol

Among the most widely occurring of these compounds are the *steroids*, many of which act as intercellular regulators, for example, hormones like testosterone in multicellular animals. It is also likely that steroids, especially cholesterol, play an important role in many cell membranes.

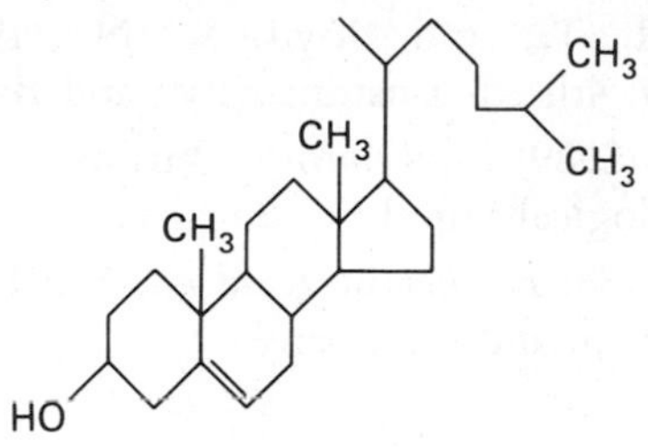

Carotenoids are another group of isoprene derivatives. They are broadly distributed in the living world, but only procaryotes and plants can synthesize them. Animals require them, however, for many physiological functions, such as vision. β-carotene, one of the best-known compounds in this class, is shown below.

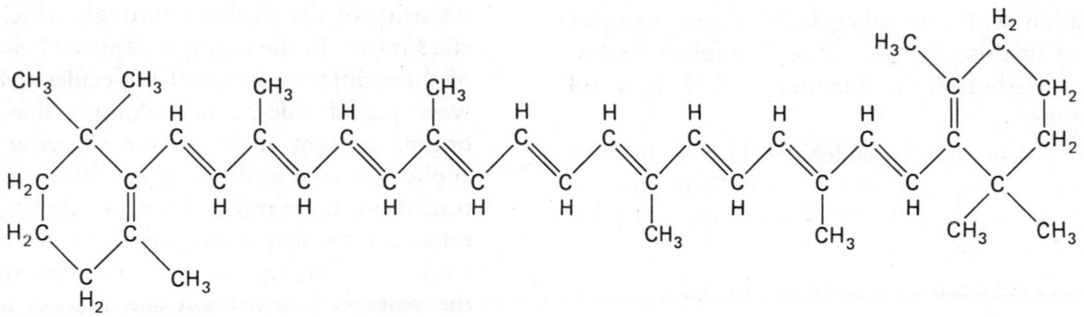

β-Carotene, a precursor to vitamin A

SUMMARY

The above discussion includes but a limited sample of the immense variety of small organic compounds found in nature. We already know of about a thousand compounds involved in the metabolism of the bacterium *E. coli*. In this simple cell there are probably some 500 or 1000 additional compounds that we must still isolate from the complex network of its metabolism. In the cells of higher organisms there are probably many thousands, possibly tens of thousands, of reactions still to be identified. Our goal is not to discuss systematically all those reactions and compounds that are currently known. Rather, we hope to identify general principles of metabolism and general kinds of molecular structures that are found in all cells. To do this we shall first describe the structure of proteins (in Chapter 4) and then discuss the action of protein enzymes in catalyzing the chemical reactions that go on inside cells (in Chapter 5).

KEY WORDS

Calorie (cal), cofactor, conformation, energy, equilibrium, hydrogen bond, joule (J), phospholipid, solute, ultraviolet radiation

SELECTED READINGS

Brown, T. L., and Lemay, H. E. (1985) *Chemistry: The central science*. 3d ed. Englewood Cliffs, N. J.: Prentice-Hall.

Edwards, N. A., and Hassall, K. A. (1980) *Biochemistry and physiology of the cell: An introductory text*. 2d ed. New York: McGraw-Hill Book Company.

Morrison, R. T., and Boyd, R. N. (1983) *Organic chemistry*. 4th ed. Boston: Allyn and Bacon, Inc.

Rawn, J. D. (1989) *Biochemistry*. Burlington, NC: Carolina Biological Supply Company.

Stryer, L. (1988) *Biochemistry*. 3d ed. San Francisco: W. H. Freeman and Company.

Ucko, D. A. (1982) *Basics for chemistry*. New York: Academic Press.

Zubay, G. (1983) *Biochemistry*. Reading, MA: Addison-Wesley Publishing Company.

PROBLEMS

1. (a) Name two trace elements used in cells as counterions for carboxylate (—COO⁻). (b) Name a trace element used, as the element alone, in cells as a counterion for ammonium (—NH₃⁺). (c) Name a trace element that when combined with oxygen is used in cells as a counterion for ammonium.

2. Carbon monoxide (CO) is a compound of carbon in combination with one other atom. From examples illustrated in this chapter, name examples of compounds of carbon in combination with 2, 3, and 4 other atoms.

3. (a) In the compound formaldehyde ($H_2C{=}O$), what will be the partial electrostatic charge, if any, on the carbon atom and why? (b) What about the molecular oxygen compound (O_2)?

4. Distinguish between a covalent radius and a van der Waals radius. Which is larger?

5. Does the compound shown below (glycine, an amino acid) have optical activity? Why or why not?

H   O

| ‖

H_2N—C—C

| \\

H   OH

6. What is the angle between the H—O—H covalent bonds in water and the O—H—O hydrogen bond? Which is the better defined?

7. There are two ways of making up a pH 5, 0.5 M sodium acetate buffer solution. The first way starts with 0.5 moles of acetic acid, adds NaOH to get the correct pH, then dilutes to one liter. The second way starts with 0.5 moles of sodium acetate, adds acetic acid to get the correct pH, then dilutes to one liter. For each method calculate (a) the final concentration of sodium ions and (b) the final concentration of total acetic acid plus acetate ions.

8. How many other isomeric forms are there for the lipid called linoleic acid?

9. Which of the various functional groups in organic molecules would be effective in buffering cells at their physiological pH (approximately 7.0)?

10. Electrophoresis is the movement of charged particles through a fluid, or other supporting medium, in response to an electric field. A common supporting medium is a polyacrylamide gel soaked with a buffer-salt solution. After the voltage is applied between the ends of a slab of gel, suspended molecules will move at a rate determined by their size and net average electrostatic charge; the rate of movement is also determined by the temperature and viscosity of the medium through which the molecules move. In the example depicted below, samples of three different kinds of molecules (M_1, M_2, M_3) were placed side-by-side along a line, called the origin, at right angles to the electrical field to be applied, as seen in drawing (a). When the voltage is turned on, these molecules move through the gel at relative rates determined *solely* by the net average electrostatic charge on the molecules. After a time, the voltage is turned off and the gel is stained to reveal the place(s) to which various molecules have migrated; the locations of M_1 and M_3 molecules at the end of the electrophoresis are indicated in drawing (b).

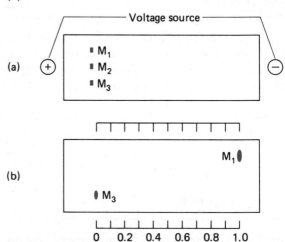

Each molecule is a weak base whose conjugate acid can undergo the following dissociation reactions under the conditions of the experiment:

$$M_1H^+ \rightleftharpoons M_1 + H^+$$

$$M_2H^+ \rightleftharpoons M_2 + H^+$$

$$M_3H^+ \rightleftharpoons M_3 + H^+.$$

The buffer solution in which the gel was soaked has a pH such that:

$$pK_1 - pH = +2.0$$

$$pK_2 - pH = -0.5$$

$$pK_3 - pH = -2.0.$$

Indicate on drawing (b) your best estimate of the distribution of molecules of type 2 and indicate how you arrived at your answer.

11. Consider another set of three molecules (M_1, M_2, M_3), slightly different from those in the previous problem: when subjected to electrophoresis, these molecules also move at rates determined solely by the net average electrostatic charge on the molecules, but these molecules have different dissociation constants. After electrophoresis, M_1 and M_3 are found as shown in drawing (b), in positions 0.0 and 1.0 respectively, and M_2 is located at position 0.7. If pK_1 for molecule M_1 is 10.0, pK_3 for molecule M_3 is 6.0, and the electrophoresis was carried out at pH 8.0, what is pK_2 for M_2?

Proteins—Agents of Biological Specificity

4

The word *protein* was derived from a Greek word meaning "primary." As Table 4–1 shows, proteins constitute the major component of the dry weight of an actively growing cell. What is so remarkable about proteins is that they are not only the main building material of the cell, but they also regulate almost all its activities. Most biological processes would not occur under the conditions of temperature, pressure, pH, etc. found in cells if not for the existence of enzymes to speed up the chemical reactions involved. Almost all enzymes discovered so far in cells are proteins. Protein enzymes are endowed with *specificity*, the ability to distinguish among different molecules. This property allows enzymes to catalyze only appro-

TABLE 4–1

Typical Analytical Results Obtained upon Fractionation of Rapidly Growing Cells

Material	Criteria Used for Fraction	Percent Dry Weight
Small molecules "acid-soluble fraction"	Solubility in cold 5% trichloroacetic acid	2–3
Lipids "organic solvent-soluble fraction"	Solubility in alcohol-ether at 50°C	10–15
Nucleic acids "hot acid-soluble fraction"	Soluble after 30 min treatment in 5% trichloroacetic acid at 90°C	10–20
Proteins "hot acid-precipitable fraction"	Insoluble after 30 min treatment in 5% trichloroacetic acid at 90°C	55–85

Note: These data are characteristic for cells lacking polysaccharide cell walls or large amounts of other structural and storage materials.

priate reactions, and not all those of which the small organic molecules found inside cells are capable. Specificity, more than any other characteristic, is emblematic of life itself.

As we have stressed repeatedly, it is an abiding law of biology that structure and function are intimately related. Thus, the key to understanding how proteins behave is to know in detail how they are put together. In the past 40 years an astounding series of developments has provided detailed insights into the exact structure of proteins. We shall attempt here to give an account of these developments in order to build in the student's mind a vivid picture of the architecture of the protein molecule. We begin at a necessary first step for the investigation of structure—how to isolate and purify a particular protein. Next, there is a discussion of the gross morphology of proteins, their size, and general shape. The complexity of protein structure is then considered in terms of four levels of structure: primary (the sequence of amino acids), secondary (regular arrangements of sequence), tertiary (complex folding of regular arrangements), and quaternary (the association of separate protein chains). Finally, we introduce the idea that alterations can be made in the structure of a protein to change its function.

Purification of Proteins

The number of different proteins in a given cell is extremely large. In the smallest and simplest bacteria there may be as few as one or two thousand, but human cells may be able to make as many as 100,000 different proteins, though probably only about 10,000 of them are made in any one kind of cell. In order to study a given type of protein molecule, one must prepare it in pure form—that is, one must discard all other proteins and increase the concentration of the particular protein in which one is interested. The techniques used for the purification of proteins are the craft of a highly refined and still developing art, which we shall describe only briefly.

The first notable success in the purification of proteins was achieved by J. Sumner (1926), who crystallized the protein urease from the tissue of the jack bean. This important achievement marked the end of an era during which biologists came to regard proteins with an awe that precluded the use of straightforward chemical approaches to the study of these complicated compounds. Indeed, many biologists gave Sumner's discovery little credence for a number of years. By now, however, hundreds of different proteins have been crystallized, and a much larger number have been prepared in highly purified form.

The methods by which proteins can be separated from each other depend on the very same properties which form the basis for their biological activity. Proteins have a highly specific architecture that endows them not only with their individual catalytic or structural properties, but often also with very specific physical properties that affect their solubility in different solutions or their binding at various interfaces. These differences in solubility and binding properties, under a wide variety of conditions, have been used to separate proteins from one another.

One must be able to distinguish between the protein one wants to purify and all the other proteins to be discarded, and one should be able to express this distinction in quantitative terms. If the protein to be purified is an enzyme, it is possible to devise a test that utilizes the rate of a chemical reaction, catalyzed specifically by the enzyme, that transforms a *substrate* into a *product*. The rate of the reaction at very high substrate concentration is a measure of the *enzyme activity* and is proportional to the enzyme's concentration (see Chapter 5). The ratio of the enzyme activity to the mass of total protein present is the *specific activity* of the enzyme preparation which is, of course, related to its state of purity. The higher the specific activity reached during the purification of a given enzyme, the purer it is.

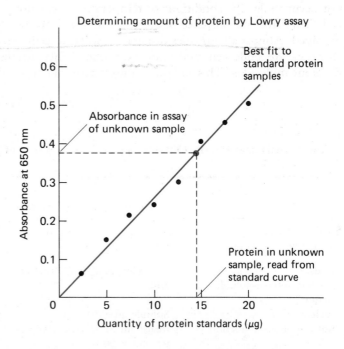

Determining amount of protein by Lowry assay

Best fit to standard protein samples

Absorbance in assay of unknown sample

Absorbance at 650 nm

Protein in unknown sample, read from standard curve

Quantity of protein standards (μg)

The total mass of protein can be measured by a variety of specific chemical reactions. One of the most widely used is the Lowry assay, in which cupric ions react with peptides in alkaline solution to give a blue-colored product. The intensity of the blue color is then deter-minced in a spectrophotometer and related to a *standard curve,* which graphs the intensity of color produced by known quantities of protein.

If the protein to be purified is not an enzyme, some other test endowed with sufficient specificity to distinguish the desired product from other proteins must be utilized. In the case of a protein hormone, a biological assay capable of measuring its hormonal activity is appropriate. To assay the purification of a protein with an unknown function, it is necessary to use a test based on some structural property; for example, the mobility of the protein during electrophoresis on a polyacrylamide gel (see page 146).

After purification, immunological methods can assay the amount of the protein of interest in a complex mixture (see the discussion in Chapter 17). Specific **antibodies** are elicited by injecting the purified protein as an **antigen** into a vertebrate animal, often a rabbit. The antibodies will react only with the protein which stimulated their synthesis. When the antibodies are added to a mixture of proteins they will therefore precipitate only the ones whose presence one wishes to establish. Here again biological specificity is critical. By exploiting the specificity inherent in protein structure, biochemists have developed a variety of methods for determining the relative amount of a given protein in a complex mixture.

In order to purify a given protein, it is necessary to use some physical method to increase its concentration with respect to the other proteins ("impurities") in the mixture. One of the most widely used methods to achieve this is to change the medium in which the protein mixture is dissolved to bring about the precipitation of some of the proteins. If this is done correctly, it leaves most of the desired protein either in the solution (the *supernatant*) or in the *precipitate.* Which of these contains the desired protein can be determined by measuring, for example, the specific enzyme activity in the two fractions. If one of them has significantly higher specific activity, it is kept and the other is discarded, after which the "active" fraction is subjected to another precipitation. It is possible to eliminate systematically more and more of the unwanted proteins by varying the conditions of precipitation.

Some proteins can be *salted out* of solution by adding large concentrations of a salt such as ammonium sulfate. The positive and negative ions of the salt tie up water molecules in *hydration shells* around themselves and decrease the effective concentration (or activity) of the water around the protein, causing the protein to precipitate.

Some proteins, called *globulins,* become insoluble at low ion concentrations. It is therefore possible to separate them from the water-soluble proteins known as *albumins* by adding distilled water to a mixture of proteins or by

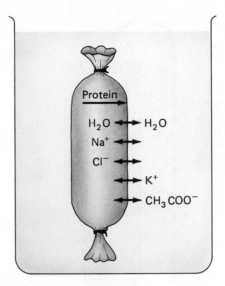

FIGURE 4–1

Dialysis. This process allows one to add small molecules and ions to, or remove them from, solutions of macromolecules. Above, a protein solution containing sodium chloride is dialyzed against a solution of potassium acetate. Large protein molecules (mass at least 10,000 Da) cannot pass through the membrane of the dialysis bag; molecules or ions less than 1,000 Da move easily through pores in the membrane. The small molecules thus equilibrate in the entire volume while the macromolecules remain inside the bag. After the first equilibration, the dialysis bag is placed in a fresh volume of potassium acetate solution and the sodium chloride remaining in the bag is again allowed to equilibrate. By changing the dialysis bag several times, the residual sodium chloride is reduced to a negligible concentration and the macromolecules are in essentially the desired environment.

dialyzing it against distilled water. **Dialysis** is a process that separates large molecules (above 10,000 Da) from small ones (in this case, the unwanted ions). The macromolecules are enclosed in a dialysis bag made of a material containing small pores which do not allow the macromolecules to pass through. The dialysis bag is surrounded by a solution which has the particular composition of small molecules that one desires to achieve inside the bag (Figure 4–1). Since the volume of the external solution is many times that of the bag and since it is possible to change the external solution several times, one can produce any type of small molecule environment one desires for the macromolecules without substantially increasing the volume of the solution containing the macromolecules.

Hydrogen ion concentration has a profound influence on protein solubility. As we learned in Chapter 3, the amount of dissociation of a weak acid is determined by the pH. The acid or its conjugate base will have an electric charge, so the pH determines what fraction of the acid is

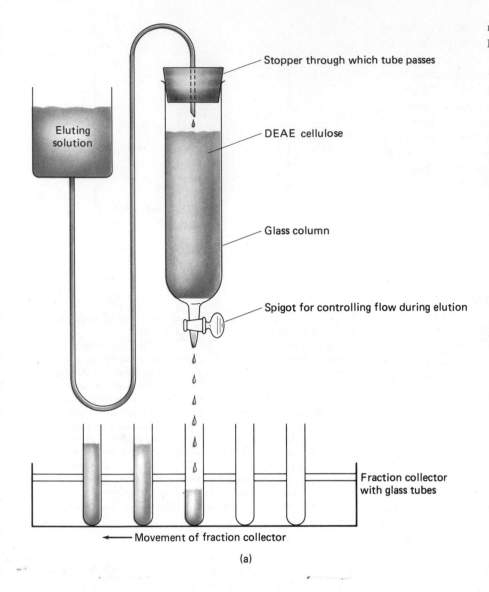

Stopper through which tube passes

Eluting solution

DEAE cellulose

Glass column

Spigot for controlling flow during elution

Fraction collector with glass tubes

← Movement of fraction collector

(a)

FIGURE 4-2

Ion Exchange Chromatography.
(a) Organic groups bearing an ionic charge, in this case positively charged *di*ethyl*a*mino*e*thyl (DEAE), are covalently bound to cellulose. A glass column is packed with a dense suspension of this DEAE-cellulose, then pre-equilibrated by slowly percolating through it a solution containing a low concentration of a salt, NaCl. The chloride ions form ionic bonds with the DEAE groups. A protein, dissolved in a buffer containing the same low concentration of salt, is then percolated through the column. Negatively charged groups on the protein make stronger ionic bonds with the positively charged groups on the DEAE cellulose, exchanging with the chloride ions. This binds the protein to the column packing. To remove the protein (elute it), increasingly concentrated salt solutions are percolated through the column. A particular protein will be eluted in a narrow range of salt concentrations, allowing it to be separated into a few fractions. Each fraction can be assayed for total protein (by absorbance at 280 nm, for example) and for enzyme activity. (b) DEAE cellulose chromatography of a human blood plasma fraction containing plasma transglutaminase activity. The peak of absorbance labeled 4 is the only one containing the enzyme activity (data not shown). The immunological test in the center of the diagram (Ouchterlony gel diffusion), however, reveals that peak 4 also contains a protein related to protein in peak 3. (For an explanation of gel diffusion methods, see Figure 4–4.) In particular, peak 3 (fractions 58 and 62) show an antigen-antibody precipitation zone that is continuous with a zone from peak 4 (fraction 93, 98, 100). Other experiments showed that the peak 3 protein is an inactive subunit of the peak 4 active enzyme. Thus, a subunit of the enzyme can be distinguished from the active enzyme by chromatographic as well as immunological tests. (Courtesy of A. G. Loewy.)

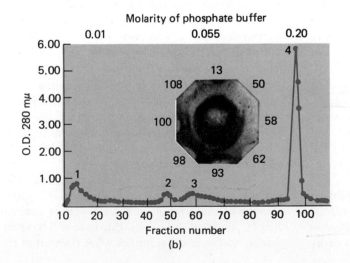

Molarity of phosphate buffer

0.01   0.055   0.20

O.D. 280 mμ

Fraction number

(b)

in a charged form. Proteins contain many weakly acidic and basic groups and the pH, therefore, determines the distribution of electrical charges on a protein. Some of these charges will be positive, if a proton associated with a weakly basic group, and some will be negative, if a proton dissociated from a weakly acidic group. The solubility of a protein is lowest at the *isoelectric point (pI),* the pH at which the net electrostatic charge on the protein is zero. When the pH is equal to the pI of a particular protein, there is no electrostatic repulsive force holding the molecules apart; they tend to precipitate preferentially. Other proteins, with different pIs, will still have net charges, will repel one another, and will tend to stay in solution (see also Figure 4–16).

These and other methods for precipitating proteins are useful for the early stages of purification, when protein impurities might outnumber the desired protein by a factor of 100 or even 1000. As the specific activity of the desired molecule increases, further purification often involves a *chromatographic method* which distinguishes among proteins by the differences in their ability to bind at a given solid-liquid interface. Thus, a solid material such as diethyl-aminoethyl cellulose (DEAE cellulose), when packed into a *column,* will bind different proteins to different extents. By pouring the protein solution into the top of the column *(loading),* *eluting* with different buffers, and *collecting* sequential samples *(fractions),* it is possible to separate many of the proteins from one another (Figure 4–2). The total amount of protein in each fraction is measured by a general test; for example, the absorbance (optical density) at near ultraviolet wavelengths (280 nm), and a specific test such as enzyme activity, are used to measure the amount of the protein to be purified.

In recent years, a refinement called *affinity chromatography* has come more and more into use. In this method, the solid phase contains chemical groups similar to the substrate of the enzyme one wants to purify. As a result, only this enzyme binds to the solid phase in the column, and all other proteins pass through. By using the substrate or by changing the pH or the salt concentration so that the desired enzyme no longer binds, it can be eluted from the column in very pure form.

There are a number of other methods that have been used successfully for the purification of proteins. They include *electrophoresis,* which separates on the basis of electric charge differences among proteins (see pages 138, 146); *molecular sieving,* which discriminates among differences in size (see page 145); *zone sedimentation* in stabilizing **sucrose gradients,** which depends on size, shape, and density differences among proteins (see page 143); and finally *crystallization.*

Table 4–2 illustrates the results of a typical fractionation of the enzyme plasma transglutaminase in which a variety of different methods have been used: salting out,

TABLE 4–2

Fractionation of Plasma Transglutaminase

Fraction	Conditions of Fractionation	Yield Percent	Specific Activity	Purification
Plasma	Starting material	100	2.1	1
1	Precipitate: 20% ammonium sulfate, pH 7.0	100	87	42
2	Precipitate: 16% ammonium sulfate, pH 5.4	80	190	91
3	Precipitate: 16% ammonium sulfate, pH 7.0	80	290	138
4	Supernatant: 56°C for 3 min	72	4,600	2200
5	Precipitate: 36% ammonium sulfate, pH 7.0	70	13,000	6200
6	Peak 4: DEAE cellulose chromatography eluted at 0.20 M phosphate buffer, pH 7.0	67	16,800	8000

Note: (Recalculated from Loewy, et al.) The success of this purification depends on the fact that plasma transglutaminase seems to be attached to fibrinogen and thus follows fibrinogen in fractions 1, 2, and 3. The heat treatment that follows denatures and precipitates fibrinogen, leaving active plasma transglutaminase in the supernatant. The enzyme, now in the absence of fibrinogen, shows entirely different solubility characteristics so that the impurities not removed in the earlier fractionations can be eliminated. This allows the high degree of purification while maintaining a high yield. The extent of purification is calculated as the ratio of specific activity at any stage to that at the initial stage of purification.

heat precipitation of impurities, and chromatography. The effectiveness of the fractionation procedure is determined by keeping a quantitative accounting of the progress of the fractionation. A successful fractionation maximizes purification and yield while keeping the number of steps to a minimum.

Criteria of Purity

When varying the conditions of fractionation produces no further increase in specific activity, it is appropriate to determine the purity of the protein. The formation of a crystal, the traditional test for small organic molecules, is not a good criterion of purity for a protein because proteins form loose crystals that can include in them considerable amounts of impurities. There are numerous other methods, however, which can often be used both on a preparative scale to purify proteins and on an analytical scale to determine their purity. These include:

- *column chromatography* (Figure 4–2) repeated under a variety of conditions and/or with a variety of chromatographic materials;

- polyacrylamide gel electrophoresis (PAGE), a powerful and widely used method which can determine the purity of a protein preparation and, if detergent is used to **denature** the protein, the molecular weights of its constituent polypeptide chains (see Figure 4–9, p. 146);

- *isoelectric focusing* (Figure 4–3), electrophoresis in a pH gradient, which resolves proteins according to their isoionic points; a commonly used technique is focusing in a polyacrylamide gel in one dimension, followed by electrophoresis in a denaturing detergent at right angles to the first dimension; the result is a two-dimensional array of proteins, separated by charge in the first dimension and by molecular weight in the second, which suffices to resolve virtually every protein in a cell extract; and

- *immunological gel-diffusion,* which uses the diffusion in an agarose gel of antibodies and antigens (proteins) towards each other to measure the purity of protein preparations used (Figure 4–4).

After a protein has been purified, it is possible to study it in a systematic manner. One can first study the overall size and shape of the molecule, then build a detailed picture of the complete molecule, step by step.

The Size and Shape of Protein Molecules

The very large size and relative fragility of proteins present the structural chemist with a number of problems that cannot be solved with the methods applied to small molecules. However, a variety of methods developed and refined during the last 35 years have made it possible to

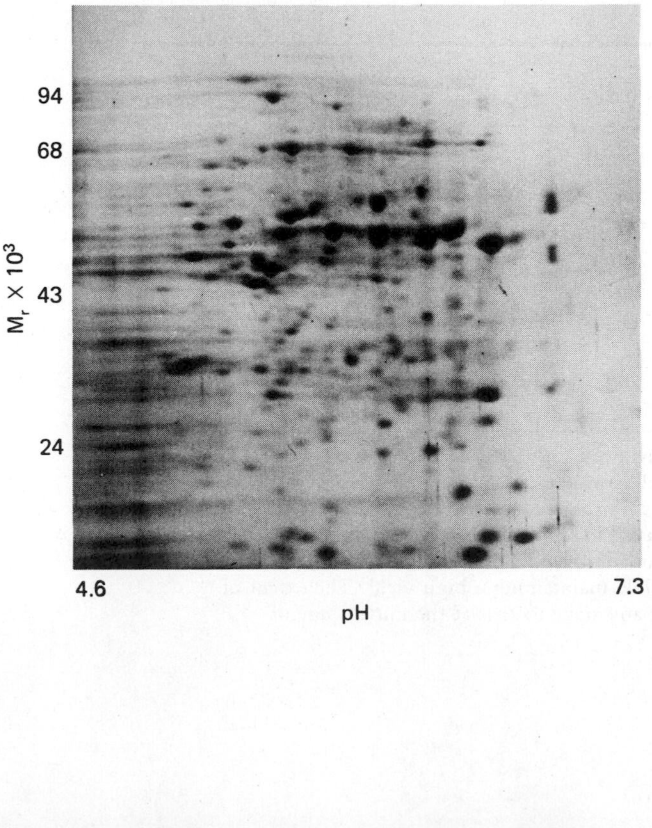

FIGURE 4–3

Two-dimensional (O'Farrell) Gel of *Tetrahymena* Proteins. Whole *Tetrahymena thermophila* cells were washed in buffer, then boiled in 1.5% sodium dodecyl sulfate (SDS, a strong detergent), 7.5% mercaptoethanol (to reduce disulfide bonds), and 0.75 mM phenylmethylsulfonyl fluoride (to inhibit proteases). Aliquots were subjected to isoelectric focussing (pH 4.6 to 7.3) in the horizontal direction in the presence of urea and a nonionic detergent. Then slab gel electrophoresis was performed in the vertical direction in the presence of SDS. The positions of the various proteins were determined by staining the gel with the dye Coomassie brilliant blue. (Courtesy of N. E. Williams.)

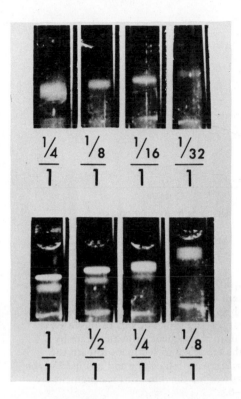

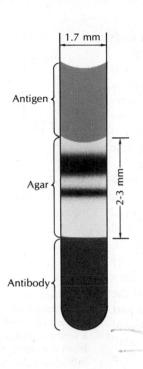

1.7 mm

Antigen

Agar

2-3 mm

Antibody

FIGURE 4-4

The Use of Immunodiffusion Methods to Determine the Purity of a Protein Preparation. A rabbit is immunized with a preparation of plasma transglutaminase. After immunization the serum contains antibodies against the enzyme as well as against impurities. Protein preparations and immune serum are both allowed to diffuse through agar (double diffusion). Where they meet, a precipitation zone or band is produced, one band for each type of antigen/antibody complex. By varying the relative concentrations of antigen and antiserum, one can obtain changes in the position and sharpness of the bands. Since each band detected represents a different complex, one can determine the number of different antigens in the preparation. (a) Double diffusion in tiny tubes (1.7 mm diameter): a plasma transglutaminase preparation and antiserum raised against the purified protein. Various dilutions (¼, ⅛. . .) of transglutaminase (antigen) are added to a constant amount of the antiserum. The presence of plasma transglutaminase (upper band) and its smaller subunit (lower band) was detected. This method has high sensitivity (10 μgm protein/ml) and precision. (b) The Ouchterlony method of gel diffusion permits the identification of related and identical proteins in different preparations. Circular wells are cut into a thin layer of agar on the surface of a microscope slide and are filled with antigen and antiserum, respectively. Each diffuses through the agar and precipitation bands form where antigen meets antibody. This experiment shows that peak 4 (isolated from the experiment shown in Figure 4-2) contains two proteins, and peak 3 contains only one protein. The single protein in peak 3 is related to one of the proteins in peak 4 because the two precipitation bands are fused. Precipitation bands of proteins that are not related cross each other rather than fuse because noncross-reacting antibodies can diffuse through the precipitation bands. (Courtesy of A. G. Loewy.)

determine with considerable precision the molecular mass and the shape of macromolecules.[1]

Because proteins have large molecular masses it is not as easy to get highly concentrated solutions as it is for smaller compounds. It is not generally possible, therefore, to use colligative properties like the freezing point depression (which depends on the molar concentration of solutes and which is a valuable method for the organic chemist) to measure the molecular mass of proteins.

One method that can directly yield the molecular mass of proteins is *sedimentation equilibrium*. If a solution containing a purified protein is spun at moderate speed (10,000 rpm) for sufficient time, an equilibrium will develop in which the tendency of the molecules to sediment away from the axis of rotation is exactly balanced by the tendency of the molecules to diffuse from the region of high concentration in the outer part of the tube to the region of low concentration in the inner part of the tube (Figure 4–5; by analogy with gravitational forces we usually refer to the outer part as the "bottom" and the inner part as the "top" of the tube). The position of the equilibrium will depend on both the molecular weight and the density of the protein. The density of the protein can be measured independently, and it is therefore possible to use this method to calculate the molecular weight.

Methods Sensitive to Shape that Determine Molecular Mass Indirectly

By centrifuging a solution containing protein at high speeds (50,000 rpm), it is possible to sediment the protein molecules toward the bottom of the tube (Figure 4–6). As the protein molecules move away from the axis of rotation, the region near the top is left free of protein. The *boundary* between this region and the solution containing protein is easy to detect and it moves with the same radial velocity as the protein molecules themselves. The measured rate of sedimentation (in cm/sec) will be faster if the speed of rotation is higher. To take account of this effect one uses the **sedimentation coefficient,** the ratio of the sedimentation rate to the acceleration resulting from the spinning:

[1]The student should recognize the distinction between *molecular mass* and *molecular weight ratio*. The value of molecular mass is expressed in grams for a mole, or in *daltons* for an individual molecule. A dalton is the same as the atomic mass unit used by chemists. The molecular weight ratio is a pure number, the ratio of the molecular mass to the unit molecular mass. It is symbolized as M_r and is usually, albeit incorrectly, referred to simply as "molecular weight."

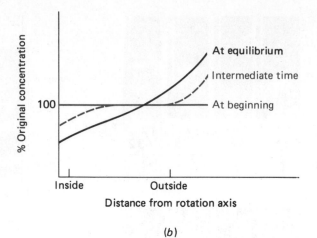

(b)

FIGURE 4–5

Sedimentation Equilibrium.   The graph shows the distribution of protein concentration in the rotor cell at various times after beginning the centrifugation. Starting with a uniform concentration, the protein tends to move away from the meniscus until the tendency to equilibrate by diffusion balances the tendency to sediment toward the bottom of the centrifuge cell. The rotor and cell used in the analytical ultracentrifuge for this kind of measurement are shown in Figure 4–6.

$$S = \frac{dx/dt}{\omega^2 x}.$$

The sedimentation coefficient has units of seconds; x is the distance of the boundary (Figure 4–6) from the axis of rotation, t is time (in seconds), dx/dt is the rate of sedimentation of the boundary, ω is the angular velocity (in radians per second), and $\omega^2 x$ is thus the centrifugal acceleration on the molecules at position x. Sedimentation coefficients of proteins range from 0.25×10^{-13} sec to 500×10^{-13} sec (Table 4–3). To simplify the manipulation of the numbers, a sedimentation coefficient of 1×10^{-13} sec is defined as the *Svedberg unit* or *S value*. Thus, a sedimentation coefficient (S) of 6×10^{-13} sec is denoted as 6S.

Nowadays this *sedimentation velocity* technique, which originally required the use of an analytical **ultracentrifuge,** has been adapted to the simpler and cheaper preparative ultracentrifuge. The *zone sedimentation* method, which is a common version of the sedimentation velocity technique, utilizes a sucrose **density gradient** in the centrifuge tube. The density gradient is a consequence of a smoothly increasing concentration of sucrose from the top to the bottom of the tube. A protein solution is layered on top of the sucrose solution. The sucrose gradient persists during the short time of the sedimentation because the equilibration of sucrose by diffusion is

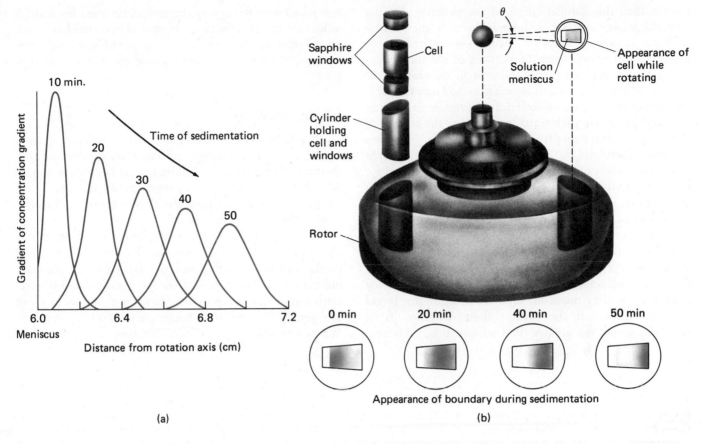

(a)

(b)

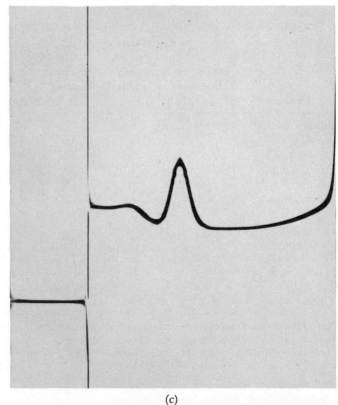

(c)

FIGURE **4–6**

The Determination of Molecular Weight Using the Sedimentation Rate Method.   (a) A high centrifugal force creates a boundary between the solution containing protein and the protein-free solution that is closer to the axis of rotation. The boundary moves through the solution as the sedimentation proceeds. In the diagram a Schlieren optical system displays the gradient of the concentration gradient (the change in the change in concentration with distance) at the boundary in the form of a peak. As the boundary leaves the meniscus, it is sharp; as it moves farther from the rotation axis, the boundary widens because of the diffusion of the macromolecules. The sedimentation coefficient for plasma transglutaminase is 9.9 S (see Figure 4–3). (b) Rotor and cell used in the analytical ultracentrifuge and the appearance of the cell while rotating. Light passes from a light source below the rotor, through the protein solution in the cell, and into an optical system that measures the concentration of protein at different points in the centrifuge cell. (c) Sedimentation in the ultracentrifuge as a criterion of purity. A solution of plasma transglutaminase (see Figure 4–2) was centrifuged at 59,780 rpm for 36 minutes. The large peak represents the enzyme plasma transglutaminase; the small peak to the left is caused by the subunit of the enzyme. (Courtesy of H. Schachman.)

slower than the sedimentation of the proteins. During centrifugation, proteins sediment through the sucrose solution as narrow zones or "bands"—stabilized by the density gradient—toward the bottom of the tube. Centrifugation is stopped before the protein has reached the bottom of the tube, a small hole is punched into the tube, and small fractions are collected. It is possible to measure the total protein in each fraction as well as the activity of the protein that is being studied (Figure 4–7, p. 143). This method, therefore, is extremely useful: it provides a criterion of purity (specific activity can be determined for each fraction) or, in the case of an impure protein preparation, it allows purification as well as estimation of the sedimentation rate.

The rate at which protein molecules move through a solution while spinning depends on their molecular mass (the more massive the molecules, the greater the effect of the centrifugal field), and on the frictional resistance they experience as they move through the solution (the larger the molecules and the greater their departure from spherical shape, the greater the friction). It is, therefore,

not possible to determine the molecular mass from the S value alone. By making another determination, such as *diffusion rate* or *viscosity*, it is possible to calculate molecular weight from any two of these three parameters.

Sieving Methods

Molecular sieving methods for separating macromolecules and measuring molecular weights are among the most common in the cell biological laboratory today. We shall first discuss *gel permeation* or *molecular sieve chromatography* (the method also is commonly, but incorrectly, called gel filtration).

It is possible to prepare polymers of carbohydrates (dextrans) or other materials in the form of spherical beads which contain pores or passages of relatively uniform size. Variations in the preparation yield beads with a characteristic pore size; beads with uniform pores are generally used after being packed into a column. When a mixture of molecules differing in size is passed

TABLE 4–3

Physical Constants of Some Proteins and Nucleoproteins and a Comparison of their Molecular Weights Calculated from Several Different Parameters

Protein	Diffusion Coefficient $(D_{20,\omega} \times 10^7)$	Sedimentation Coefficient $(S_{20,\omega})$	Partial Specific Volume $\bar{v}$	Molecular Weights						Intrinsic Viscosity (cm^3/g)	Frictional Ratio f/f_0
				S and D	Sed. Eqb.	Chemical Methods	Light Scattering	Osmotic Pressure			
Ribonuclease, cow	11.9	1.64	0.728	12,700	13,000	13,683*			3.3		
Ovalbumin	7.76	3.55	0.748	44,000	43,500		38,000	46,000	4.0		
Hemoglobin, horse	6.90	4.31	0.749	63,000		64,650*		67,000	3.6	1.16	
Serum albumin, cow	6.10	4.60	0.733	65,400	68,000	66,296*	70,000	69,000	3.7	1.29	
Hemocyanin (*Polynurus*)					450,000	453,000		461,000			
Tomato bushy stunt virus	1.15	13.2	0.74	10,600,000	7,600,000			9,000,000	4.0		
Tobacco mosaic virus	0.46	198	0.73		40,700,000			40,000,000	29	2.03	

Note: Partial specific volume ($\bar{v}$) is the increase in volume of a solution caused by adding more protein to the solution. It is expressed as liters per mole (although the experiments to determine the values use very much lower quantities). For further information concerning diffusion coefficients, see Chapter 7. S and D is the estimate of molecular weight based on both the sedimentation coefficient and the diffusion coefficient.

*We now know this value with an accuracy of five significant figures (provided we know the pH) because the complete amino acid sequence of the protein has been determined.

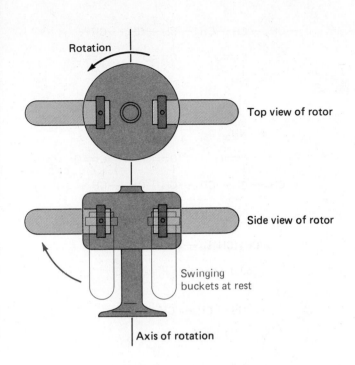

FIGURE 4–7
Determination of the Sedimentation Rate and Purity of a Protein by Zone Sedimentation through a Sucrose Density Gradient. A centrifuge tube is filled with a solution of varying sucrose concentration, starting with a high concentration at the bottom of the tube and ending with a low concentration near the top. Then a solution containing protein(s) is carefully layered on top of the sucrose gradient. A swinging bucket centrifuge rotor is used; the centrifugal force causes the centrifuge tubes to swing into a horizontal position. After spinning, a hole is punched in the bottom of the centrifuge tube, and drops are collected in separate fractions. One can measure both total protein (absorbance at 280 nm) and enzyme activity in each fraction. Since no complex optical system for measuring the concentration of protein during centrifugation is required, this is a simpler and cheaper procedure for measuring sedimentation coefficients.

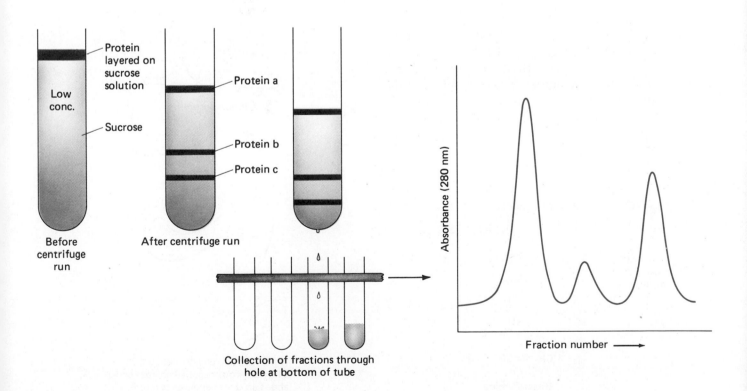

through the column they will separate by a *sieving process:* the very largest molecules will not be able to diffuse through the small pores into the beads and will therefore move relatively rapidly through the column by moving only in the liquid surrounding the beads. The smallest molecules will be able to diffuse into the beads readily and thus will move in the fluid both within and between the beads. They will therefore be slowed down in their passage through the column. Molecules of intermediate size will move at intermediate rates.

For spherical molecules, the rate at which such intermediate size molecules move through the column is proportional to their molecular mass, and by standardizing a column with spherical proteins of known molecular mass, it is possible to determine the molecular mass of an unknown protein (Figure 4–8). Caution, however, must be exercised in interpreting the results, since the method yields accurate molecular weights only when the unknown protein is spherical. Therefore, if one performs such measurements with a protein whose shape as well as size is unknown, one can only calculate an approximate molecular mass—the molecular mass the protein would have if it were an unhydrated sphere.

The most common method used in the laboratory today for quickly estimating molecular masses of proteins is to sieve them through a polyacrylamide gel. When polymerized, acrylamide forms a three-dimensional network with a mesh size that varies inversely according to the fraction of acrylamide in the gelling mixture: the higher the fraction the smaller the area of the mesh. The gel can be cast in a glass tube as a cylinder (tube gel) or between two glass plates separated by thin spacers (slab gel). To minimize the effects of molecular shape during sieving, the interactions between various groups on the protein are disrupted by heating briefly to 100°C in the presence of mercaptoethanol and a strong negatively charged detergent, sodium dodecyl sulfate (SDS).

Mercaptoethanol is present to reduce any disulfide bonds (see page 169); the protein will then be held

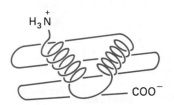

Polymerized acrylamide

$$CH_3(CH_2)_{11}-O-SO_3^-Na^+$$

Sodium dodecyl sulfate (SDS)

$$HS-CH_2-CH_2-OH$$

2-Mercaptoethanol

together only by the peptide bonds between amino acids and can unfold completely. Unfolding is aided and stabilized by the SDS, which coats completely the hydrophobic groups of the protein. Most proteins treated this way have a similar structure, called a *random coil,* which includes a surrounding layer of negatively charged detergent molecules.

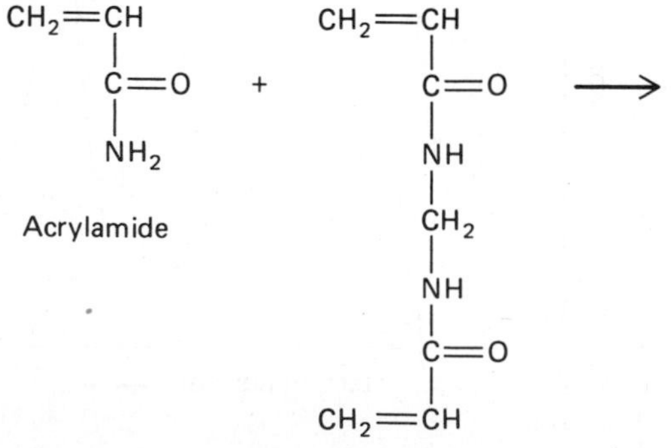

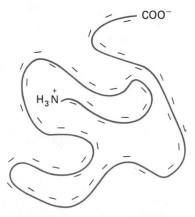

Globular protein

A polypeptide chain binding SDS. The numerous negative charges bring about an open random structure.

Acrylamide

$+$

Bisacrylamide

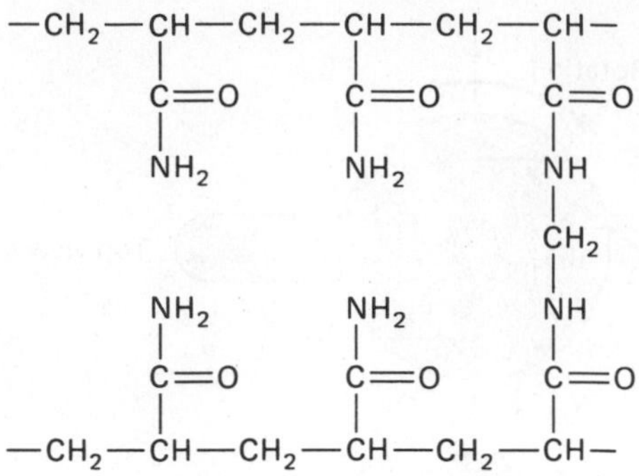

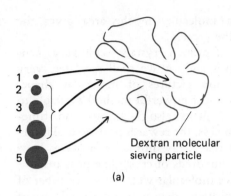

Dextran molecular
sieving particle

(a)

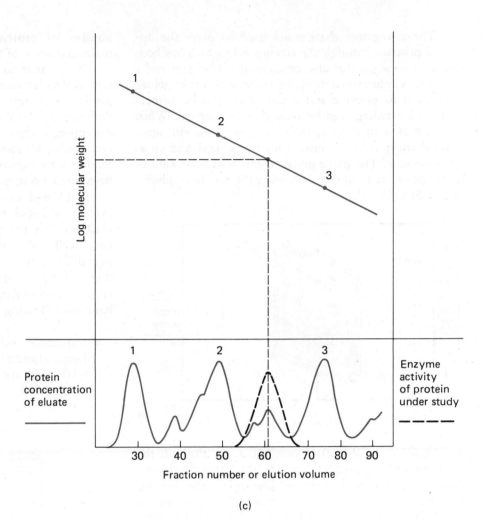

(c)

Protein mixture
applied at top
of column.

Proteins enter column.
Small protein able to
enter sieving particles;
large proteins excluded
from sieving particles.

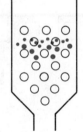

Small protein slowed
down by diffusion
into particles; large
protein is excluded
from particles and
therefore moves faster
down column.

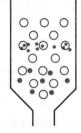

○   Molecular sieving particle

·   Small protein molecule

●   Large protein molecule

(b)

FIGURE 4–8

The Molecular Sieving Method for Measuring Molecular Masses of Proteins.   (a) Illustration of the molecular sieving principle. The particle made of dextran has a pore size which allows small molecules to diffuse freely into the particle, excludes large molecules completely, and permits intermediate molecules to enter the pores but at a reduced rate, proportional to the Stokes radius of these molecules. (b) Illustration of column separation of two proteins differing in size. The large protein molecule is completely excluded from the sieving particles. Fractions of equal volume are collected sequentially and their protein concentration determined. (c) Molecular weight determination by gel permeation of an enzyme in a complex mixture. Proteins 1, 2, and 3 are standards of known molecular weights which have been added to the mixture in sufficient quantity that the fractions containing them can be identified by their high protein concentrations. The enzyme can be identified by measuring its activity. It is found for the standard molecules (1, 2, 3) that the logarithm of the molecular weight, or more precisely, of the Stokes radius, is proportional to the elution volume.

These negative charges are used to drive the unfolded proteins through the sieving gel, which has been cast in a solution that also contains the detergent and a buffer (pH approximately 8). Opposite ends of the gel are connected to positive and negative electrodes, respectively, so a voltage can be applied across the gel. When the current is turned on, the coated proteins will move toward the positive electrode in the process known as *electrophoresis*. The name given to this particular kind of electrophoresis is SDS *polyacrylamide gel electrophoresis*, or SDS-PAGE.

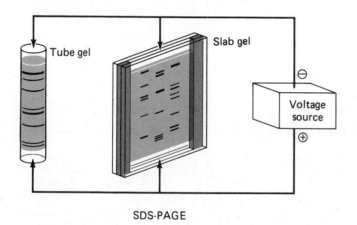

SDS-PAGE

Unfolded SDS-coated proteins move through the gel at rates determined by their size (the larger the slower) and by the mesh size of the gel (the tighter the slower). To determine the molecular mass of a protein, its electrophoresis rate is compared with the rates of a group of standard proteins whose molecular masses are known (Figure 4–9).

Methods Yielding Size and Shape Independently

There are a number of optical methods that can provide independent estimates of the size and shape of macromolecules. Direct visualization in the electron microscope has come into increasing use, especially with large molecules. Molecular mass can be determined quite accurately by a counting technique. For this purpose, one uses a homogeneous preparation of polystyrene beads, the exact dimensions of which are known. A mixture of a known mass of protein and a known number of beads, per volume, is sprayed on the specimen grid of an electron microscope. The electron microscopic image will show, in a given area, a certain number of protein molecules and a certain number of beads. One knows the concentration of beads in the sample, so the beads per area determine the volume of the solution represented by the area. Since one also knows the mass concentration of the protein in the sample, one can determine the number of grams of protein in the area. This mass divided by the

number of protein molecules in the area gives the molecular mass of the protein.

X-ray analysis of protein crystals yields very accurate molecular masses and, as we shall see later, very precise, high-resolution pictures of molecular shape. Although by far the most powerful method available to the protein chemist, it is also laborious and time-consuming. Nevertheless, the rewards in structural information are so great that x-ray diffraction analyses have been carried out on hundreds of crystalline proteins.

Table 4–3 shows molecular weights of a number of proteins and nucleoproteins, determined with a variety of methods. For the most part, the values obtained agree very well. We have seen that molecular weight and overall molecular dimensions of a protein can be determined if it is available in a pure state. To summarize a large amount of data on the size and shape of proteins, we have the following picture:

- Molecular weights of proteins vary over a wide range (insulin subunit, 6,000; snail hemocyanin, 6,700,000). It turns out, however, that large proteins are generally composed of subunits (see page 188).

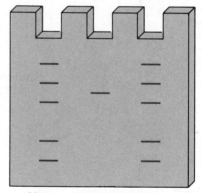

Markers Unknown Markers

SDS polyacrylamide gel electrophoresis
(unknown protein and markers)

FIGURE 4–9

Separation of Polypeptide Chains by Acrylamide Gel Electrophoresis. A protein sample is boiled briefly in the anionic detergent SDS and the disulfide-splitting reagent β-mercaptoethanol. The samples are then added to the wells at the top of the gel and a current is passed through the gel. The binding of SDS to the protein chains endows them with a negative charge which causes them to move toward the positive electrode positioned below the gel. The rate of migration of the protein chains is inversely proportional to the log of their molecular weight. By using polypeptide chains of known molecular weight (markers) one can establish a standard curve and thereby obtain estimates of the molecular weight of the unknown polypeptide chains.

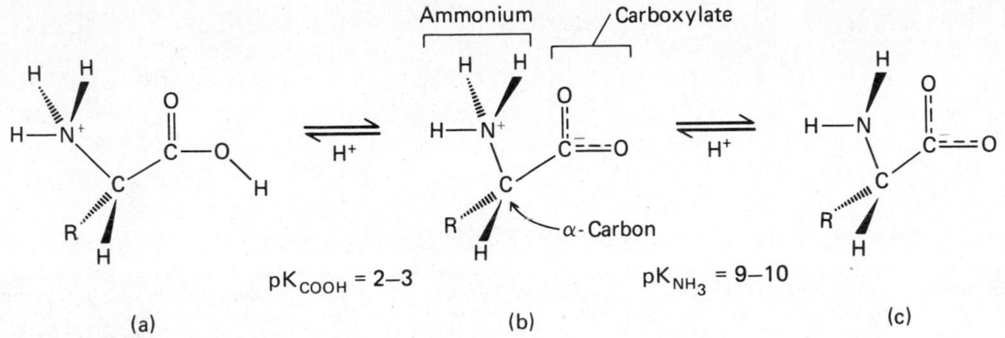

Ammonium Carboxylate

$pK_{COOH} = 2–3$

$pK_{NH_3} = 9–10$

(a)   (b)   (c)

FIGURE 4–10

Ionic Forms of Amino Acids. At physiological pH values (near pH 7), amino acids have a negatively charged α-carboxylate group and a positively charged α-ammonium group; it is thus called a zwitterion, or double ion (*b*). Lowering the pH tends to protonate the α-carboxylate group and leaves the molecule with a net positive charge (*a*). Raising the pH brings about the dissociation of a bound proton from the α-ammonium group and leaves the molecule with a net negative charge (*c*). The pH at which a given amino acid has equal numbers of positive and negative charges (that is, at which its net charge is zero) is called the isoelectric point of the amino acid. At this pH the amino acid will not migrate in an electric field.

There are relatively few instances of polypeptide chains of molecular weight larger than 100,000 occurring in nature.

- Shapes vary considerably from near-spherical to highly elongated, but the latter are frequently made of polymers of near-spherical subunits.

- Unlike many synthetic polymers, protein molecules appear to be defined particles of remarkably constant shape. As we shall see later in this chapter, certain subtle but important changes in **conformation** do occur in protein molecules.

Primary Structure—The Sequence of Amino Acids

Before discussing how the sequence of amino acids in protein molecules is determined, we must begin by describing the chemistry of amino acids in some detail.

The Twenty Amino Acids and Their Separation

When a protein molecule is heated in 6 M hydrochloric acid at 120°C in vacuo for several hours, the peptide bonds holding the monomers together are hydrolyzed and amino acids are released. The building blocks for protein synthesis are 20 different amino acids. Not all are used in every protein and some are modified after being incorporated. Figure 4–10 shows the basic structure of all the 20 amino acids except proline. It shows that at neutral pH the amino acid is a "zwitterion," containing simultaneously a negative α-carboxylate ($-COO^-$)

group and a positive α-ammonium ($-NH_3^+$) group. It also shows that in all amino acids except glycine, the α-carbon atom is asymmetrical since it has four different groups attached to it. The absolute spatial configuration of amino acids can be determined by x-ray diffraction (Figure 4–11). In all organisms tested so far all amino acids derived from proteins are L-amino acids, a fact that strongly suggests the common origin of all living matter on earth. The D-isomers of amino acids have on occasion been found in nature (e.g., in the cell walls of bacteria), but as far as we know they do not become incorporated into the structure of proteins.

The symbol R in Figure 4–10 represents the variable portion of the molecule, commonly known as the *R group*

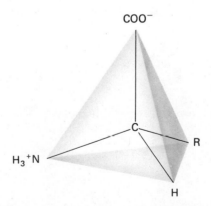

FIGURE 4–11

The Absolute Configuration in Space of L-amino Acids, Determined by X-ray Diffraction. All proteins synthesized on ribosomes are formed exclusively from L-amino acids.

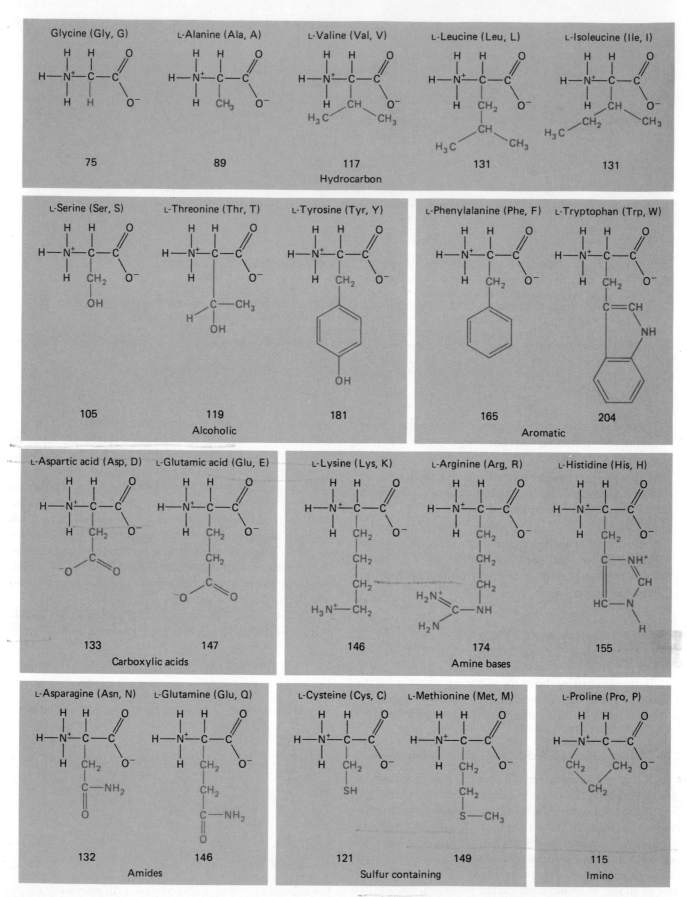

FIGURE 4-12

The Structures of the 20 Commonly Occurring Amino Acids. The student should memorize this basic 20-letter alphabet, from which the language of biological variability is built. Both three-letter and one-letter symbols are given. The side chains, which vary from one amino acid to another, are indicated in color. Below each structure is the molecular mass in daltons.

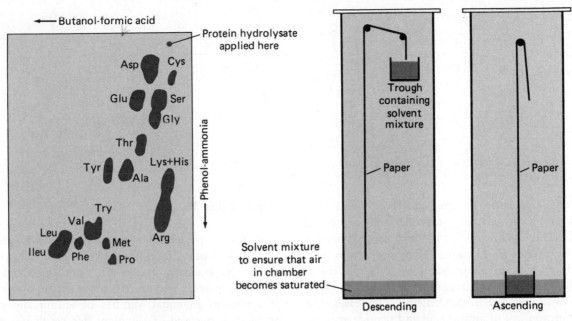

Two-dimensional chromatogram

Two types of chromatography apparatus

FIGURE 4–13

Two-dimensional Paper Chromatography of Amino Acids.   The amino acid mixture is applied in one corner and a mixture of solvents either ascends by capillary action or descends by capillary action and the influence of gravity. After drying, the position of each amino acid is determined by spraying the filter paper with ninhydrin, which reacts with amino acids to produce a purple color. Alternatively, the position of radioactive amino acids can be determined by placing the chromatogram on a sheet of x-ray film, which after development will show the location of the amino acids as darkened spots. (Glutamine and asparagine were deamidated by prior acid hydrolysis and converted to glutamic and aspartic acids, respectively; they therefore do not appear on the chromatogram.)

or *side chain*; it differentiates one amino acid from another. Figure 4–12 illustrates the 20 amino acids. The last one, L-proline, is atypical in that the α-amino group is not free, but part of a ring structure. The amino acids, with their distinct side chains, are the alphabet of protein structure and are ultimately responsible for the specificity and variability of living matter.

The separation of the 20 amino acids from each other was at first a formidable task, one to which the famous chemist E. Fischer (1852–1919) devoted many years of his life. In 1941, A. J. P. Martin and R. L. M. Synge proposed a new approach to the problem of purifying and assaying compounds, such as amino acids, that closely resemble each other. This method involved dissolving the amino acids in a mixture of solvents, some hydrophilic and some hydrophobic, and then allowing this mixture to percolate along filter paper strips or through columns. By using a second mixture of solvents after turning the paper through 90 degrees, it is possible to separate amino acids two-dimensionally over a large sheet of filter paper (Figure 4–13). Because these methods bear some resemblance to a technique applied many years earlier by M. Tswett to the separation of leaf pigments, Martin and Synge named the procedure *chromatography*. In the last 40 years, these methods have been extended

and refined so that chromatography is now a very widely used technique for analysis and purification.

S. Moore and W. Stein were the first to perfect column chromatography for the analysis of amino acids. They built an automatic device capable of taking the hydrolysate (the products formed after hydrolysis in acid) of one mg of protein and estimating the concentration of each one of its component amino acids to within a few percent. It utilizes a column of ion-exchange resins made of sulfonated polystyrene. The amino acid hydrolysate is placed on the column in a buffer at low pH and low ion concentration, which is a condition for maximum binding of the positively charged amino acids with the $—SO_3^{2-}$ groups on the resin. Buffer is then run through the column, and both the pH and the temperature of the column are raised. This process, called *elution*, causes the different amino acids to percolate down the column at different rates, eventually separating them into distinct bands. The product of elution, the *eluate* that emanates from the bottom of the column, is then processed by an automatic machine that detects amino acids through a colorimetric reaction and spectrophotometric recording of the color intensity.

Numerous modifications and improvements of the original machine have been made, including new separa-

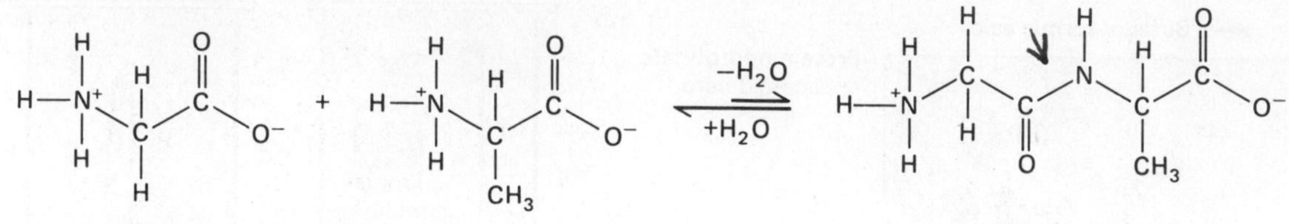

FIGURE 4–14

The Peptide Bond. Glycylalanine is a dipeptide made of two residues, the glycyl and alanyl residues.

tion methods and reagents that produce amino acid derivatives with fluorescent or ultraviolet light-absorbing groups. These new instruments are much more rapid and can detect as little as 10^{-11} moles (10 picomoles) of each amino acid and can run many samples consecutively at the rate of 12 or more chromatograms per day. The speed and degree of automation of these methods increases each year. The automation of amino acid and peptide analysis is not merely a laborsaving gimmick. Our experience of the past 25 years shows that the elucidation of the structure of a number of proteins would not have been feasible without this device. Like computer methods in physics, automated analytical methods in biology have made a qualitative difference in the nature of the problems that can be approached.

The Peptide Bond

Once a protein has been defined in terms of the composition of its building blocks, it is logical to examine how these building blocks fit together in the protein molecule. Fischer was able to show that upon hydrolysis of a protein an equal number of amino and carboxyl groups are released. To explain this result he suggested that the amino acids were linked to each other by an *amide linkage* called a *peptide bond* (Figure 4–14). The equilibrium of the reaction shown in Figure 4–14 is far to the hydrolysis side, and as we shall see later, the cell synthesizes the peptide bonds of the protein molecule by a mechanism entirely different from a mere reversal of hydrolysis. Fischer's peptide bond theory for protein

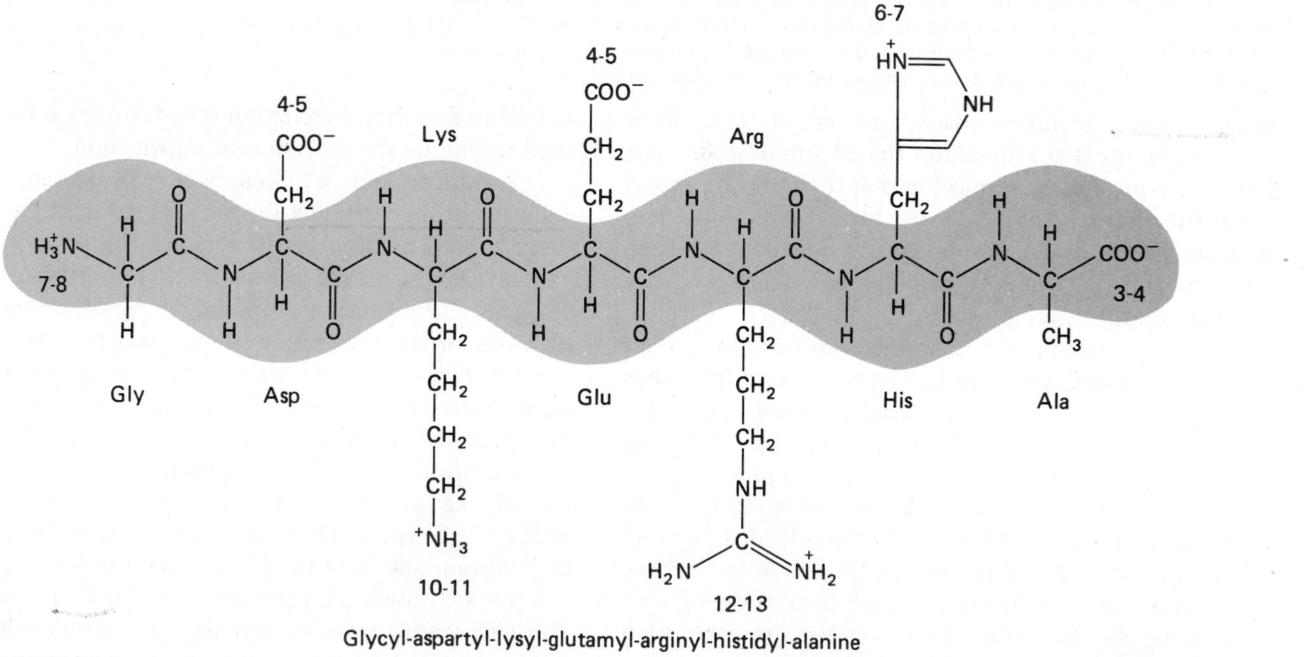

Glycyl-aspartyl-lysyl-glutamyl-arginyl-histidyl-alanine

FIGURE 4–15

A Hypothetical Polypeptide Containing All the Groups That Normally Contribute Positive and Negative Charges to Proteins. The numbers represent the pK range of each dissociating group. The charged groups consist of (1) an N-terminal ammonium and a C-terminal carboxylate; (2) negatively charged aspartyl and glutamyl side chains; and (3) positively charged histidyl, lysyl, and arginyl side chains. In a large polypeptide chain, the charge contributions of the N- and C-terminal groups are, of course, small compared with those of the more numerous charged side chains.

structure has since been validated by many separate lines of evidence. Proteolytic enzymes, for instance, which are known to break the peptide bond of synthetic peptides, are also capable of hydrolyzing proteins.

Figure 4–15 is a diagram of a hypothetical chain of amino acids, termed a *polypeptide*. Notice that except for the terminal carboxyl (**C-terminal**) and the terminal amino (**N-terminal**) groups, all the remaining α-carboxyl and α-amino groups are involved in the amide linkage. This linkage is not capable of donating or accepting protons and therefore does not contribute to the acid-base properties of the polypeptide. The electrochemical properties at physiological pH of the polypeptide (and hence, of the protein) are primarily determined by the R groups of the *acidic amino acids*, aspartic acid and glutamic acid; by the R groups of the *basic amino acids*, lysine and arginine; by histidine; and by the terminal amino and carboxyl groups. Histidine is important because it has, along with the N-terminal amino group, a pK in the region of physiological pH, and thus it can contribute to a change in charge on the protein caused by small changes in pH under physiological conditions. The role of this amino acid side chain in mechanisms of enzyme catalysis and cellular regulation will be discussed in later chapters.

The sequence of amino acids in polypeptides is generally written as in Figure 4–15, that is, with the α-amino or N-terminal group on the left and the α-carboxyl or C-terminal group on the right. One generally uses the three letter abbreviations, as shown,

but computer retrieval systems make use of a single letter code (Figure 4–12), so this particular peptide could be written

GDKERHA.

The term *residue* is used to denote the part of the amino acid found in the polypeptide chain. Amino acid residues are denoted with a "-yl" ending (for example, one refers to the alan*yl* residue of the amino acid alanine in the peptide chain).

The hypothetical polypeptide in Figure 4–15 shows all the R groups that contribute to the charge on a protein at physiological pH. It shows how a protein can be a *polyvalent ion,* containing a number of positive and negative charges. From the dissociation properties of the groups one can predict what should happen when acid or base is added to a protein. Thus, when acid is added, —COO^- will change to —COOH and therefore leave the protein with less negative charge; when base is added, —NH_3^+ will change to —NH_2, leaving the protein with less positive charge. At a certain intermediate pH the number of positive charges equals the number of negative charges (Figure 4–16). Two different conditions have been defined under which a protein can carry a zero net electrostatic charge: (1) The *isoionic point* is the pH at which the net charge on the protein is zero in a solution containing no ions other than the protein, H^+ and OH^-. A protein which has been dialyzed against distilled water is at its isoionic point. (2) The more often used property *isoelectric point* refers to the pH of the protein solution, in

FIGURE 4–16
The Effect of pH on the Charge of Proteins.   (a) At the isoelectric pH, even though there may be large numbers of individual positive and negative charges, the net charge on each molecule is zero. Given the absence of long-range electrostatic repulsive forces between the molecules, the solubility of the protein is at a minimum. It can therefore be precipitated by salt or organic solvents more easily than at other pH values. (b) As the pH decreases from the isoelectric point a protein begins to have a net positive charge—protons become associated with the conjugate bases of weakly acidic groups. Under these conditions the protein will migrate in an electric field toward the negative pole, or *cathode*—the protein behaves as a *cation*. If the pH increases from the isoelectric point the opposite changes occur—the protein behaves as an *anion*, migrating toward the positive pole *(anode)*.

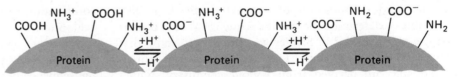

Acid pH	Isoelectric pH	Alkaline pH
Net charge is +	Net charge is zero	Net charge is −
Migration toward cathode	No migration in an electric field	Migration toward anode

(a)

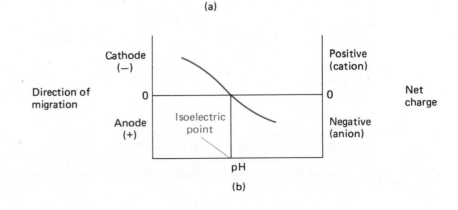

(b)

a given buffer and at a specified ion concentration, at which the net charge on the protein is zero. A protein exhibiting no electrophoretic migration in a given buffer is at its isoelectric point.

This property is used to separate proteins from one another, a technique known as *isoelectric focusing*. In a suitable supporting medium (polyacrylamide gels are typically used), a gradient is set up so the pH changes smoothly with distance. If an electric field is applied across the gradient, a protein will move until it is at a position where the pH is the same as the protein's isoelectric point. At that position there is no longer any electrical force on the protein, since it has no net electrostatic charge, so it remains focused there.

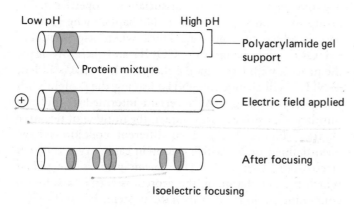

Isoelectric focusing

Determining the Sequence of Amino Acids in Polypeptide Chains

We have shown how it is possible to determine the amino acid composition of a polypeptide. Is it also possible to determine the specific sequence of amino acids in the polypeptide? F. Sanger was the first to show that this indeed could be done. The early traditions of protein chemistry are derived from colloid chemistry, which

regarded its materials as chemically indefinite, to be characterized by statistical rather than precise chemical parameters. Sanger challenged this tradition by raising a question that until 1945 was considered preposterous by most protein chemists. He asked: "Do proteins have a specific chemical composition, down to the *sequence* of the 20 building blocks?" To ask such a question and to be willing to invest 10 years of one's life in answering it implies a belief in the absolute accuracy by which biological macromolecules are replicated and synthesized. The extent to which we take this for granted today is a tribute to the magnitude of Sanger's achievement.

When Sanger began his epoch-making study on the protein insulin, the method of paper chromatography for the separation of amino acids had just been developed by Martin and Synge. To this Sanger added a technique of his own, the "labeling" of the amino end group (N-terminal group) of a peptide by combining it with the compound 2,4-*dinitrofluorobenzene* (DNFB) to give a yellow-colored *di*nitrophenyl (DNP) peptide (Figure 4–17). This combination is relatively stable during the acid hydrolysis of the peptide, and it is possible by the use of chromatographic separation after hydrolysis to identify the particular amino acid to which the DNP is attached. By determining the N-terminal amino acid, it was possible for Sanger to orient the peptide, that is, to distinguish one end from the other.

Sanger also developed a number of methods for *partially hydrolyzing* insulin into smaller peptides of various lengths. He developed chromatographic procedures for separating these peptides from each other, which permitted him to determine their amino acid composition and the identity of their N-terminal amino acid. Some of the longer peptides had to be hydrolyzed a second time, chromatographed, and again analyzed for end groups. As the data accumulated, more and more of the sequence became uniquely defined. Figure 4–18 is a

FIGURE 4–17

Sanger's Method for Determining the Amino-terminal Residue of a Peptide. The dinitrophenyl derivative of the terminal residue is relatively stable under the conditions used to hydrolyze peptide bonds. Because of its aromatic nature and yellow color, the derivative can readily be separated from other amino acids and identified.

	N-terminal end → C-terminal end
Peptides from acid hydrolyzates	Phe–Val Val–Asp Asp–Glu Phe–Val–Asp Glu–His Val–Asp–Glu Phe–Val–Asp–Glu His Leu–CySO₃H Ser–His Leu–Val Val–Glu Val–Glu–Ala Ala–Leu Tyr Leu–Val Leu–Val–Glu Leu–CySO₃H Val–CySO₃H Val–CySO₃H–Gly CySO₃H–Gly Gly–Glu Glu–Arg Glu–Arg Arg–Gly Gly–Phe Thr–Pro Pro–Lys–Ala Lys–Ala Ser–His–Leu Glu–His–Leu His–Leu–Val Tyr–Leu–Val–CySO₃H Leu–Val–CySO₃H–Gly His–Leu–CySO₃H Ser–His–Leu–Val Leu–Val–Glu–Ala Tyr–Leu–Val–CySO₃H–Gly Ser–His–Leu–Val–Glu Glu–His–Leu–CySO₃H His–Leu–Val–Glu–Ala Glu–His–Leu–CySO₃H–Gly Ser–His–Leu–Val–Glu–Ala Phe–Val–Asp–Glu–His Phe–Val–Asp–Glu–His–Leu–CySO₃H–Gly
Sequences deduced from the above peptides	Phe–Val–Asp–Glu–His–Leu–CySO₃H–Gly–Ser–His–Leu–Val–Glu–Ala–Leu–Tyr–Leu–Val–CySO₃H–Gly–Glu–Arg–Gly Thr–Pro–Lys–Ala
Peptides from pepsin hydrolyzate	Phe–Val–Asn–Gln–His–Leu–CySO₃H–Gly–Ser–His–Leu His–Leu–CySO₃H–Gly–Ser–His–Leu Val–Glu–Ala–Leu Tyr–Leu–Val–CySO₃H–Gly–Glu–Arg–Gly–Phe Tyr–Thr–Pro–Lys–Ala
Peptides from chymotrypsin hydrolyzate	Phe–Val–Asn–Gln–His–Leu–CySO₃H–Gly–Ser–His–Leu–Val–Glu–Ala–Leu–Tyr Leu–Val–CySO₃H–Gly–Glu–Arg–Gly–Phe–Phe Tyr–Thr–Pro–Lys–Ala
Peptides from trypsin hydrolyzate	Gly–Phe–Phe–Tyr–Thr–Pro–Lys Ala
Structure of phenylalanyl chain of oxidized insulin	Phe–Val–Asn–Gln–His–Leu–CySO₃H–Gly–Ser–His–Leu–Val–Glu–Ala–Leu–Tyr–Leu–Val–CySO₃H–Gly–Glu–Arg–Gly–Phe–Phe–Tyr–Thr–Pro–Lys–Ala

FIGURE 4–18

Determining the Sequence of Insulin. This shows a compilation of Sanger's analytical results that determined the final sequence of one of the polypeptide chains of insulin. Acid hydrolysis permitted the determination of the sequence of five short peptides (acid hydrolysis also converts Gln to Glu, Asn to Asp, and destroys Trp). Enzyme hydrolysis provided the overlap that established the overall sequence. The student can work through the logic of this procedure by putting the sequences on separate cards and deducing the overall sequence from them. The symbol CySO₃H refers to the cysteic acid produced from the amino acid cysteine by oxidizing the insulin. (Courtesy of E. O. P. Thompson and *Scientific American*.)

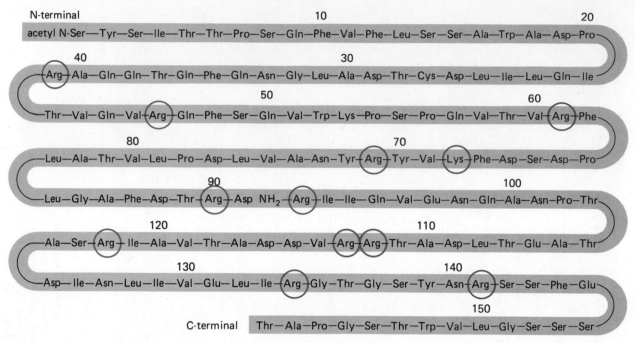

N-terminal ... 10 ... 20

acetyl N-Ser—Tyr—Ser—Ile—Thr—Thr—Pro—Ser—Gln—Phe—Val—Phe—Leu—Ser—Ser—Ala—Trp—Ala—Asp—Pro

40 ... 30

Arg—Ala—Gln—Gln—Thr—Gln—Phe—Gln—Asn—Gly—Leu—Ala—Asp—Thr—Cys—Asp—Leu—Ile—Leu—Gln—Ile

50 ... 60

Thr—Val—Gln—Val—Arg—Gln—Phe—Ser—Gln—Val—Trp—Lys—Pro—Ser—Pro—Gln—Val—Thr—Val—Arg—Phe

80 ... 70

Leu—Ala—Thr—Val—Leu—Pro—Asp—Leu—Val—Ala—Asn—Tyr—Arg—Tyr—Val—Lys—Phe—Asp—Ser—Asp—Pro

90 ... 100

Leu—Gly—Ala—Phe—Asp—Thr—Arg—Asp NH₂—Arg—Ile—Ile—Gln—Val—Glu—Asn—Gln—Ala—Asn—Pro—Thr

120 ... 110

Ala—Ser—Arg—Ile—Ala—Val—Thr—Ala—Asp—Asp—Val—Arg—Arg—Thr—Ala—Asp—Leu—Thr—Glu—Ala—Thr

130 ... 140

Asp—Ile—Asn—Leu—Ile—Val—Glu—Leu—Ile—Arg—Gly—Thr—Gly—Ser—Tyr—Asn—Arg—Ser—Ser—Phe—Glu

150

C-terminal Thr—Ala—Pro—Gly—Ser—Thr—Trp—Val—Leu—Gly—Ser—Ser—Ser

FIGURE 4–19

Primary Structure of the Protein of the Tobacco Mosaic Virus.   Circles are placed around the lysyl and arginyl residues on the N-terminal side of the peptide bond cleaved by trypsin. At the N-terminus the amino group is not free but occurs in an acylated form. (Courtesy of Virus Laboratory, University of California.)

FIGURE 4–20

The Specific Hydrolysis of a Polypeptide Chain by Trypsin.   This enzyme cleaves the chain on the carboxyl side of a positively charged lysyl or arginyl residue.

summary of the results that Sanger used in determining the amino acid sequence of one of the two polypeptide chains of the insulin molecule.

Using the above approach, Sanger was able to show that only *one sequence* would satisfy the data he had obtained. Thus, he demonstrated for the first time that, despite the complexity of the protein molecule, the cell is able to synthesize it in a reproducible and chemically precise manner.

Figure 4–19 shows the sequence or *primary structure* of a much larger protein (molecular weight 18,000) that is composed of a single chain. Determining the sequence of (**sequencing**) this larger polypeptide proved a much more difficult problem, requiring the precision of the automated column chromatographic method. Also required was the technique of *specific cleavage.* The proteolytic enzyme *trypsin,* for instance, cleaves (proteolyzes) only those peptide bonds that have the positively charged arginyl or lysyl residues donating the carboxyl group to the peptide bond (Figure 4–20). This procedure breaks a large polypeptide chain into smaller *specific peptides.* These can be separated by column chromatography, and each of their structures can be determined separately.

Since such specific enzyme hydrolysis produces "nonoverlapping" peptide sequences, it is necessary to

Tryptic peptides

Gly—Ala—Trp—Glu—Arg

Asp—Ala—Leu

Val—His—Lys

Chymotryptic peptides

Val—His—Lys—Gly—Ala—Trp

Glu—Arg—Asp—Ala—Leu

Overall sequence of polypeptide

Tryptic peptides

Val—His—Lys—Gly—Ala—Trp—Glu—Arg—Asp—Ala—Leu

Chymotryptic peptides

FIGURE 4–21
Overlap in the Determination of Sequences. Two proteolytic enzymes, trypsin and chymotrypsin, produce peptides with overlapping sequences. Lining up the sequences at the site of overlap allows one to fit the peptides into an overall sequence.

determine the order of the tryptic peptides by an additional technique. This consists of using either another enzyme or a chemical method capable of splitting polypeptide chains at other specific places. The enzyme *chymotrypsin,* though not as specific as trypsin, cleaves polypeptide chains on the carboxyl side of aromatic and other bulky nonpolar residues. Figure 4–21 shows how the *overlap* obtained from determining the sequence of ("sequencing") tryptic and chymotryptic peptides can be used to fit these into an overall sequence. A common chemical method for obtaining overlapping sequences is the use of *cyanogen bromide,* which converts methionine to homoserine lactone, cleaving the peptide bond on the carboxyl side of the methionyl residue (Figure 4–22). Most proteins have very few methionyl residues, so that this method usually yields a few relatively large peptides.

Other enzymes of great value for amino acid sequence determination are the *aminopeptidases* and the *carboxypeptidases.* The former hydrolyze the peptide bond closest to the N-terminal end of the peptide, and the latter hydrolyze the peptide bond closest to the C-terminal end. Since these enzymes digest peptides sequentially, they can often be used to establish the sequence of a few amino acids from either end of a chain. In Figure 4–23 the amino acid sequences of peptides following various types of digestion of the protein hormone glucagon are provided, to illustrate the logic of amino acid sequence analysis.

In recent years the sequencing of short peptides (40–50 residues or less) has been greatly simplified by the development of the *Edman degradation* method which is able to cleave sequentially the N-terminal amino acid

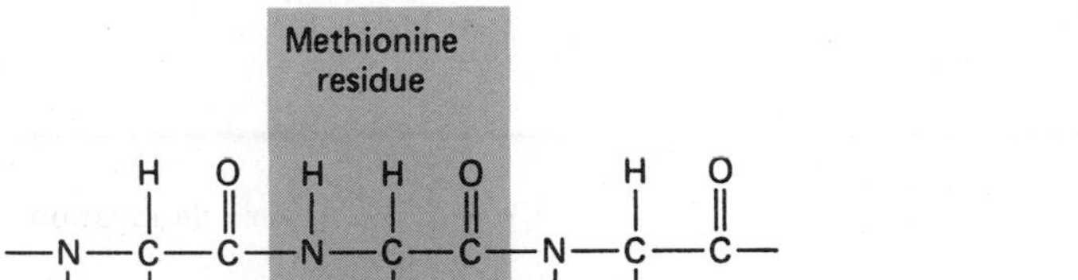

FIGURE 4–22
The Use of Cyanogen Bromide for the Specific Cleavage of a Polypeptide Chain. A methionyl residue is converted to homoserine lactone and the peptide bond on the carboxyl side of the methionyl residue is broken.

FIGURE 4–23

Sequences of Peptides Obtained From the Protein Glucagon After a Number of Different Enzyme and Acid Hydrolyses. The student should write these sequences on individual cards and try to deduce a unique sequence for the glucagon chain. It is important to demonstrate to your own satisfaction that all the data are consistent with a single sequence. The order of the amino acids within parentheses was not determined and the amino acids, therefore, are listed alphabetically.

Trypsin (2.25 hr.)	His (Asp, Glu, Gly, Lys, Phe, Ser, Ser, Ser, Thr, Thr, Tyr) Tyr (Arg, Asp, Leu, Ser) Arg (Ala, Asp, Asp, Glu, Glu, Leu, Met, Phe, Thr, Trp, Val) Ala (Asp, Asp, Glu, Glu, Leu, Met, Phe, Thr, Trp, Val) Arg
Chymotrypsin	Val (Gln, Trp) Ser (Lys, Tyr) Thr (Asp, Ser, Tyr) Leu (Asp, Met, Thr) His (Gln, Gly, Phe, Ser, Thr) Leu (Ala, Arg, Arg, Asp, Asp, Glu, Phe, Ser)
Carboxypeptidase action	Ala (Asp, Gln, Gln, Leu, Phe, Trp, Val) Met, Asn, Thr
Trypsin (50 hr.)	Arg Leu (Asp, Met, Thr) Tyr (Arg, Asp, Leu, Ser) His (Glu, Gly, Phe, Ser, Thr) Thr (Asp, Lys, Ser, Ser, Tyr) Ala (Asp, Glu, Glu, Phe, Trp, Val) Ala (Asp, Asp, Glu, Glu, Leu, Met, Phe, Thr, Trp, Val)
Subtilisin	Arg (Ala, Gln) Asn · Thr Lys · Tyr Asp · Phe His (Glu, Ser) Leu · Met Thr · Ser Leu (Arg, Asp, Ser) Val (Glu, Trp) Gly (Phe, Thr) Asp (Ser, Tyr)
Acid degradation	Thr His Asp · Ser Tyr Leu · Asp Asp · Tyr Glu · Gly Thr · Phe Ser · Lys Ser (Asp, Tyr) Ser · Asp Ser (Glu, Gly) Ser · Arg Tyr · Leu Tyr (Asp, Leu)

from the polypeptide chain (Figure 4–24). Here, also, a machine has been developed to perform the sequential operations automatically.

The sequencing strategies we have described thus far all lie in the province of protein chemistry; they demand that a protein be purified to homogeneity, and then subjected to amino acid sequence analysis by one or another (or a combination) of chemical methods. An entirely different strategy has emerged from the revolutionary developments in molecular genetics during the last decade.

It is now often possible to **clone** the gene coding for the protein of interest, and to determine the nucleotide sequence of the gene (see Chapter 9). In fact, cloning a gene is often easier than purifying a protein, and techniques of DNA sequence analysis have developed to the point where they are simpler than amino acid sequence analysis of a protein. When the nucleotide sequence of a gene is determined, the amino acid

sequence of the encoded protein can be specified directly via the genetic code (see Chapter 10). Most of the amino acid sequence data obtained today does not come from protein chemical methods, but rather from decoding the information contained in the DNA nucleotide sequence of a cloned gene. Here we have another example of the way in which biological research makes use of the biological information system itself.

Why We Study Primary Structure

There are two good reasons why an understanding of the primary structure of proteins is important. First, the primary structure of a protein is determined uniquely by the nucleotide sequence of the corresponding DNA gene, through the processes of transcription and translation. It is therefore necessary to know the primary structure of proteins if one is to understand the overall process whereby the hereditary materials control the

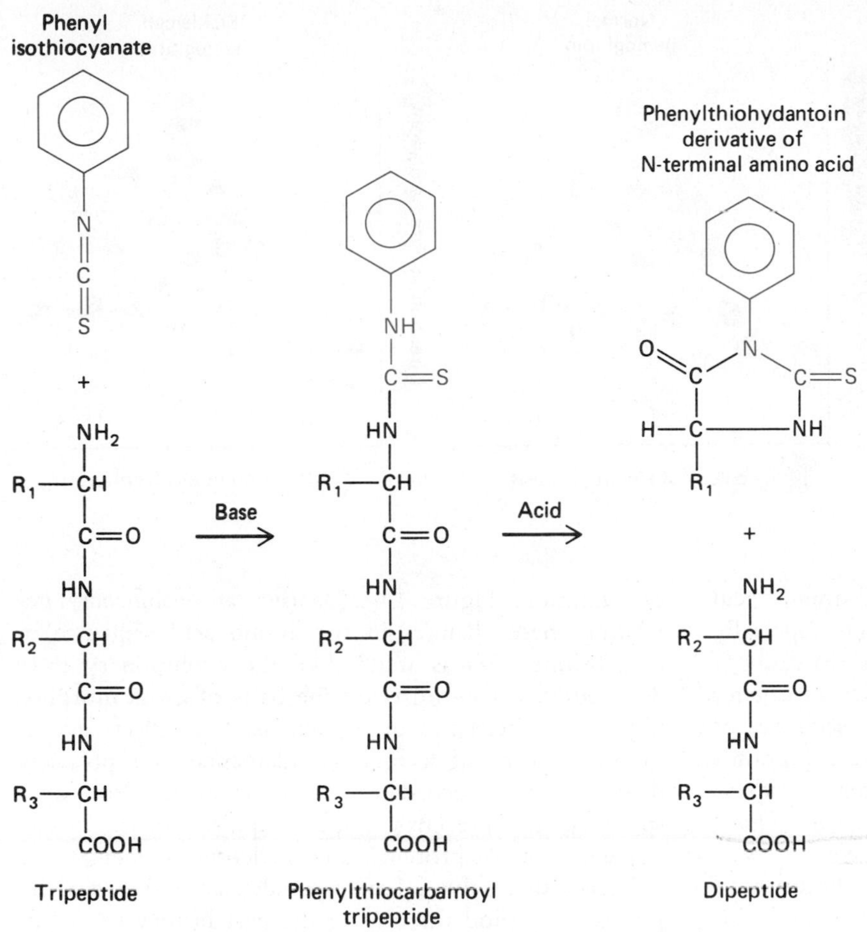

Phenyl
isothiocyanate

Phenylthiohydantoin
derivative of
N-terminal amino acid

FIGURE 4–24

The Edman Degradation. This reaction removes the N-terminal amino acid residue by forming a derivative and leaving the remaining peptide chain intact. The derivative can be identified chromatographically. By repeating the procedure one can determine the sequence of the peptide chain from the N-terminal end, until the resolution of the method becomes too low.

Tripeptide

Phenylthiocarbamoyl
tripeptide

Dipeptide

synthesis of proteins. Second, it is the primary structure of a protein that determines the final three-dimensional structure of the molecule and hence the physiological role it plays. *The primary structure is the fundamental link between the genetic material on the one hand and the physiological function on the other.*

The first discovery, by V. Ingram, relating a genetic mutation to a known molecular change in protein structure was made not in a microorganism, whose genetics were much more fully understood, but in human hemoglobin. There is a disease in humans known as sickle cell anemia, which is inherited in a simple Mendelian fashion as a single gene defect. The gene for hemoglobin produces a form of the protein which L. Pauling and H. Itano demonstrated to have an electrophoretic mobility different from that of normal hemoglobin. The abnormal hemoglobin also exhibits an impaired oxygen-binding ability and an altered solubility, so that under certain conditions it forms large aggregates that deform the red blood cell and give it a sickle shape. Deformed cells are removed preferentially from the circulation by the spleen and symptoms of

anemia result. To aid the analysis of sickle cell hemoglobin, Ingram developed a technique for comparing proteins, the *peptide map* (or *fingerprint*) method. This consists of hydrolyzing the hemoglobin with trypsin and separating the resulting peptides from each other chromatographically in one direction, then electrophoretically at right angles. Figure 4–25 shows a comparison of the maps of normal and abnormal hemoglobin. It shows that of the 27 spots, only one (in color) appeared in a different position, corresponding to a difference in electrophoretic migration rate and, therefore, most likely to a difference in a charged residue. Analysis of the amino acid content of this peptide indicated that a glutamyl residue (negative charge) was missing and a valyl residue (no charge) was substituted in the mutant form.

Human hemoglobin is made of four polypeptide chains, two α-chains and two β-chains. Ingram showed that the abnormal gene was responsible for the synthesis of a hemoglobin molecule in which the β-chains had a valyl residue substituted for the glutamyl residue in the sixth position from the N-terminal end. A number of other abnormal hemoglobin molecules have since been

FIGURE 4–25

Peptide Maps ("fingerprints").
These were derived from normal
hemoglobin (Hb A) and abnormal
(sickle cell anemia) hemoglobin
(Hb S). The position of one peptide,
identified in color, has changed because
a mutation has caused the substitution
of a valyl residue for a glutamyl
residue in this peptide. (Courtesy of C.
Baglioni.)

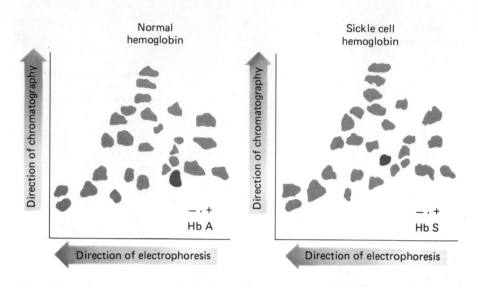

discovered (Figure 4–26). The study of amino acid substitutions resulting from genetic mutations, especially in microorganisms, was a major approach to the study of the coding problem, that is, the rules by which a sequence of nucleotides in the DNA gives rise to a sequence of amino acids in a protein (see Chapter 12). Ingram's finding was one of the most important discoveries in molecular biology, namely that *a genetic mutation can bring about a specific amino acid substitution in a protein.*

Another interesting biological result of the study of primary structure is the discovery that proteins with similar functions have similar primary structures even if they occur in quite unrelated organisms. For example, the important protein cytochrome *c,* which occurs in almost all cells, exhibits only minor modifications in its primary structure as one proceeds from the primitive yeast cell to much more recently evolved species such as mammals (Figure 4–27).

Thus, the primary structure of functionally similar proteins from different organisms reveals the succession of mutations in the gene for the protein during the course

of evolution. Figure 14–27 shows an *evolutionary tree* relating these changes in the amino acid sequence of cytochrome *c.* It is much like the evolutionary trees deduced by taxonomists on the basis of gross morphological characteristics of organisms, but reflects evolutionary history at a more fundamental and precisely defined level: it records the changing morphology on a specific gene. In recent years, the explosion of amino acid sequence data of proteins, and nucleotide sequence data on genes themselves, have provided an extraordinarily powerful method for tracing the past history of evolutionary change.

Secondary Structure—Elements of Regularity

A reasonable approach to studying the structure of large molecules is to identify regular elements, or principles of molecular architecture, which help in comprehending the overall structure of the molecule. By **secondary structure**, we mean regular associations of

Normal hemoglobin	(HbA)	$\overset{+}{N}H_3$-Val-His-Leu-Thr-Pro-$\overset{-}{Glu}$-$\overset{-}{Glu}$-$\overset{+}{Lys}$...
Sickle-cell hemoglobin	(HbS)	$\overset{+}{N}H_3$-Val-His-Leu-Thr-Pro-Val-$\overset{-}{Glu}$-$\overset{+}{Lys}$...
Mutant hemoglobin	(HbC)	$\overset{+}{N}H_3$-Val-His-Leu-Thr-Pro-$\overset{+}{Lys}$-$\overset{-}{Glu}$-$\overset{+}{Lys}$...
San José hemoglobin	(HbG)	$\overset{+}{N}H_3$-Val-His-Leu-Thr-Pro-$\overset{-}{Glu}$-Gly-$\overset{+}{Lys}$...

FIGURE 4–26

The Sequence of the N-terminal Tryptic Peptide from the β-chain of Normal Hemoglobin and of Three Mutant Forms.   Over 100 mutations in both the α- and β-chains of human hemoglobin have been identified as amino acid substitutions at certain points. The arrows represent the positions at which the proteolytic enzyme trypsin breaks the β-chain. (Courtesy of V. Ingram.)

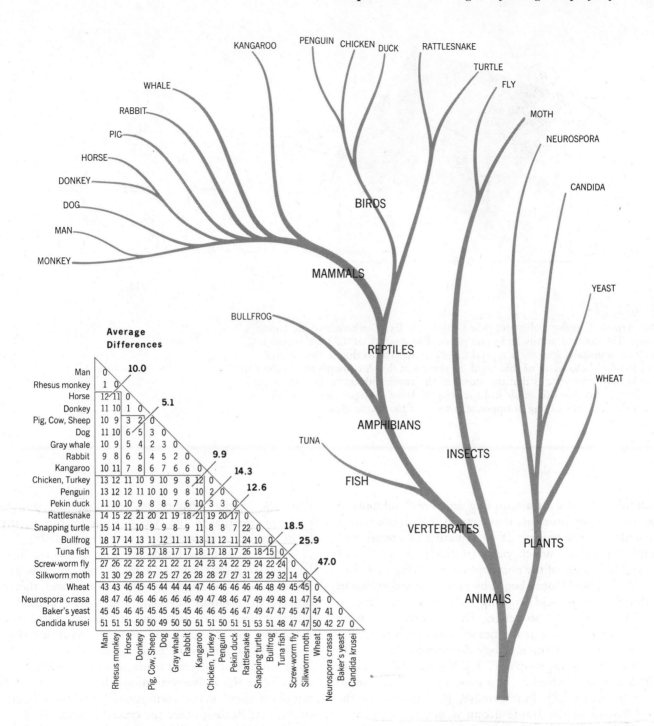

Average Differences

	Man	Rhesus monkey	Horse	Donkey	Pig, Cow, Sheep	Dog	Gray whale	Rabbit	Kangaroo	Chicken, Turkey	Penguin	Pekin duck	Rattlesnake	Snapping turtle	Bullfrog	Tuna fish	Screw-worm fly	Silkworm moth	Wheat	Neurospora crassa	Baker's yeast	Candida krusei
Man	0																					
Rhesus monkey	1	0																				
Horse	12	11	0																			
Donkey	11	10	1	0																		
Pig, Cow, Sheep	10	9	3	2	0																	
Dog	11	10	6	5	3	0																
Gray whale	10	9	5	4	2	3	0															
Rabbit	9	8	6	5	4	5	2	0														
Kangaroo	10	11	7	8	6	7	6	6	0													
Chicken, Turkey	13	12	11	10	9	10	9	8	12	0												
Penguin	13	12	12	11	10	10	9	8	10	2	0											
Pekin duck	11	10	10	9	8	8	7	6	10	3	3	0										
Rattlesnake	14	15	22	21	20	21	19	18	21	19	20	17	0									
Snapping turtle	15	14	11	10	9	9	8	9	11	8	8	7	22	0								
Bullfrog	18	17	14	13	11	12	11	11	13	11	12	11	24	10	0							
Tuna fish	21	21	19	18	17	18	17	17	18	17	18	17	26	18	15	0						
Screw-worm fly	27	26	22	22	22	21	22	21	24	23	24	22	29	24	22	24	0					
Silkworm moth	31	30	29	28	27	25	27	26	28	28	27	27	31	28	29	32	14	0				
Wheat	43	43	46	45	45	44	44	47	46	46	46	46	48	49	45	47	45	45	0			
Neurospora crassa	48	47	46	46	46	46	46	46	49	47	48	46	47	49	49	48	41	47	54	0		
Baker's yeast	45	45	46	45	45	45	45	45	46	46	45	46	47	49	47	47	45	47	47	41	0	
Candida krusei	51	51	51	50	50	49	50	50	51	51	50	51	51	53	51	48	47	47	50	42	27	0

Diagonal labels: 10.0, 5.1, 9.9, 14.3, 12.6, 18.5, 25.9, 47.0

amino acid residues that are near neighbors in the primary sequence and/or that combine over extensive lengths. The most important contributions to the elucidation of secondary structure were made by Pauling and his co-workers.

The α-Helix—A Hypothesis

In 1951, Pauling and R. Corey provided the first clue to understanding the secondary structure of protein

FIGURE 4–27
A Comparison of the Number of Amino Acid Differences between the Enzyme Cytochrome *c* of Humans and of Other Organisms. With an increase in phylogenetic distance, the number of amino acid differences also increases. By a detailed study of the pattern of substitutions, it is possible to deduce a phylogenetic tree resembling the phlyogenetic relationships that are based on generally accepted morphological grounds. (Courtesy of R. E. Dickerson and I. Geis.)

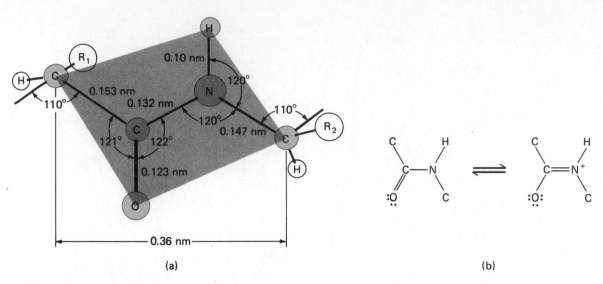

FIGURE 4–28

The Amide Linkage of a Peptide Bond. (a) Exact structure of the amide group. The colored atoms lie in one plane. The distance of 0.32 nm between C and N is unusually low for a typical C—N bond. This is due to the "partial double-bond" character of this bond, as shown in (b). A consequence of the partial double-bond character is that the atoms of the amide linkage lie in a single plane. (b) Notice also that the amide linkage is in the trans configuration, with the two asymmetric carbons lying at opposite corners of the amide group.

molecules. Instead of attempting an overall solution to the structure of proteins, they started their work with a meticulous x-ray analysis of the structure of a number of simple dipeptides, which yielded a precise description of the amide linkage of the polypeptide chain (Figure 4–28).

Pauling and Corey began by consciously deciding to ignore the amino acid R groups which give each amino acid its own specific properties. By doing this they were able to focus on those properties of amino acids which are common to all of them. They discovered that the six atoms of the amide group (CCONHC) are *coplanar,* lying within a few hundredths of a nanometer in a common plane (Figure 4–28). Furthermore, they found that the C—N bond of the amide group is unusually short. It is 0.1325 nm (1.325 Å) as compared with 0.1487 nm (1.487 Å) for the usual C—N bond. However, it is longer than a C=N bond, which is 0.127 nm (1.27 Å) in length. Pauling and Corey concluded that the C—N bond has a *partial double-bond character,* as does the C—O bond, which is somewhat longer than a true double bond between these two atoms. The bonding electrons are *delocalized,* occupying a molecular orbital that extends over all three atoms: O, C, and N. This explains not only the length of the bonds but also the coplanarity of the six atoms in the amide linkage. Pauling and Corey also found

that the amide group is arranged in the *trans configuration*—that is, the two asymmetric carbon atoms lie in opposite corners of the plane (Figure 4–28).

Even though the bonds forming the amide linkage lie in a plane, the bonds to the corner carbons of the amide group are single covalent bonds and therefore capable of rotation. Pauling and Corey argued that different degrees of rotation around this bond would determine the configuration of the polypeptide (Figure 4–29). By building scale models, they discovered that a large number of different configurations could be generated despite the above restrictions. It is here, in deciding which of these many configurations is most likely to occur, that Pauling made the crucial intuitive leap.

Pauling had already determined by x-ray diffraction analysis the structure of model compounds like the dipeptide glycylglycine. He found that in crystals adjacent molecules interacted strongly by the formation of hydrogen bonds between —NH groups and O=C— groups. Pauling predicted, therefore, that the preferred configuration of the polypeptide would be the one that favored maximum hydrogen bonding between the —NH and O=C— groups of the amide linkage. He then discovered that, if each amide group along the polypeptide experiences a small and equivalent rotation (Figure

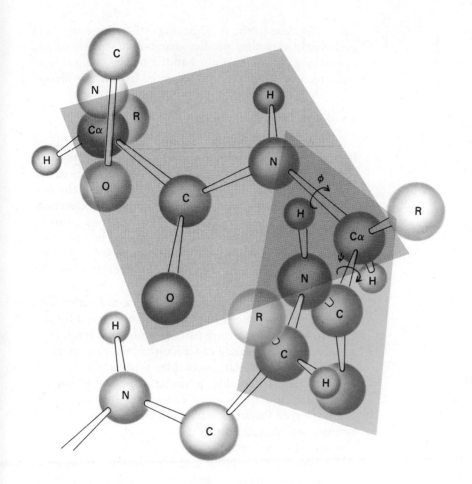

FIGURE 4–29
Rotation Around the Single Bonds of the Peptide Chain. Each asymmetric carbon atom has two single bonds and rotation around these puts each successive amide group into a new plane. If the rotation is the same for each asymmetric carbon, a helix is generated. The angle ψ measures rotation about the C_α—C single bond; the angle ϕ measures rotation about the C_α—N single bond.

4–30), a helix is generated that has the property of forming hydrogen bonds between the turns. This predicted structure for a polypeptide was called the *α-helix*.

The α-helix (Figure 4–30) is capable of forming hydrogen bonds between all of its amide groups, a fact that on a priori grounds should give it a great deal of stability. The structure can be described by stating the number of amino acid residues per turn (3.6), the pitch of the helix (0.54 nm = 5.4 Å), the diameter including the side chains (1.05 nm = 10.5 Å), and the number of amino acid residues in one hydrogen-bonded loop (3). The formula for the hydrogen-bonded loop of the α-helix is shown below.

The final step in the argument is to consider the effect of the R groups on the formation of the α-helix. Very bulky R groups are hard to pack into the structure. Of great importance, the cyclic structure of proline places the amino N in a position that is incapable of being in the middle of an α-helix.

It is, of course, possible to generate two kinds of helices, right-handed and left-handed (the one in Figure 4–30 is right-handed). These would be equivalent to each other were it not for the presence of the asymmetric carbon atom of the L-amino acid residues. In a left-handed α-helix generated by L-amino acids, the R groups

For the Pauling and Corey helix, *n* = 3

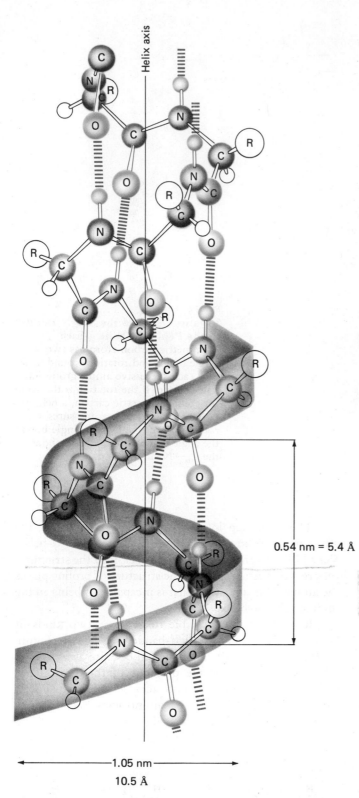

0.54 nm = 5.4 Å

←————1.05 nm————→

10.5 Å

FIGURE 4–30

**Model of the Right-handed α-helix, Proposed by
L. Pauling and R. Corey.** There are three amino acids in
one hydrogen-bonded loop, 3.6 amino acids per turn, a pitch
of 0.54 nm, and a diameter of 1.05 nm. Notice that all
—C=O and —NH groups form hydrogen bonds. In the
right-handed α-helix, all the R groups point away from the
helix. (Taken from B. Low and J. Edsall.)

come too close to the neighboring —C=O groups, and
one would therefore predict that L-amino acids would
normally give rise to right-handed helices. As we shall
see, this prediction has been validated by direct evidence.

Evidence for the α-Helix

In the years following Pauling and Corey's brilliant
hypothesis, the results of x-ray studies and other physical
evidence demonstrated that the α-helix indeed occurred
in protein molecules. M. Perutz, upon reading the first
Pauling and Corey article, pointed to clear evidence for
the presence of the α-helix in x-ray reflections given by
keratin, the structural protein of hair. The structural
protein of sheep hair (wool) was already studied in the
early 1930s by the great pioneer of x-ray crystallography
of biological macromolecules, W. T. Astbury. He
showed that a distinctive pattern of x-ray reflections,
which he named the α-keratin type, was given by hair
(Figure 4–31). This strong pattern could only result from
an underlying regularity of structure in the protein. Since
hair is an extracellular, insoluble, fibrous protein, students of cellular, soluble, globular proteins did not pay
much attention to Astbury's results. However, when
Perutz proved Pauling's assertion that the α-helix was the
primary structural motif of hair, and later when he
showed that α-keratin patterns also occurred in the
diffraction by myoglobin and hemoglobin, it began to
look as if the α-helix might well turn out to be an
important feature of protein architecture.

The picture which emerges from x-ray and other
optical studies is that the α-helical content of globular
proteins varies a great deal. On one end of the scale are
proteins such as muscle tropomyosin, which appears to
be 100% α-helical, consisting of two long α-helical rods
wrapped around each other to form a coiled coil (see
Chapter 16). On the other end of the scale are proteins
such as the immunoglobulins, which appear to have
hardly any α-helical content at all.

The β-Pleated Sheets

In Astbury's classical x-ray diffraction studies of
animal fibers, he noticed that they could be divided into
two major classes. One class, to which he gave the
general name of a α-keratins, resembled the keratin of
wool and could be found also in hair, fur, skin, nails,
claws, hooves, beaks, and horns. The structure of these is
dominated by long stretches of α-helix. The other class,
β-keratins, are found in silk, bird feathers, and a variety
of other nonmammalian fibers.

All α-keratins have x-ray diffraction patterns with a
prominent reflection corresponding to a repeated distance
of 0.54 nm (5.4 Å) which, as we have seen, is caused by

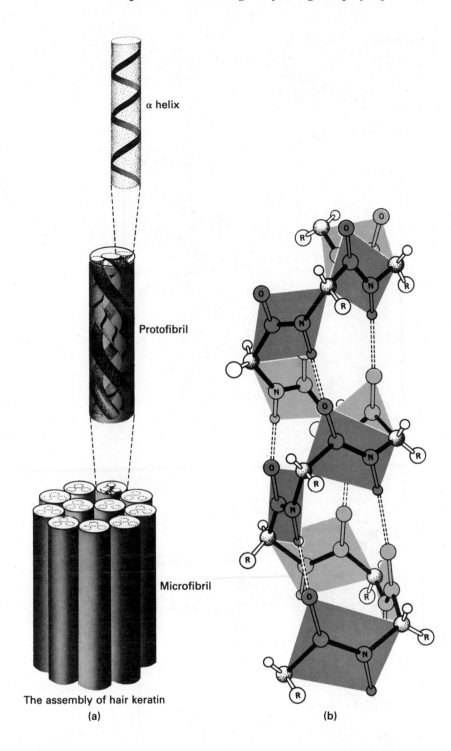

α helix

Protofibril

Microfibril

The assembly of hair keratin

(a)

(b)

FIGURE 4–31

The α-helix in Hair. (a) The organization of hair at a grosser level of structure is fairly complex and as yet not completely understood. At the level of fine structures, however, it is likely that the structural unit is a protofibril composed of three α-helices wrapped around each other in a supercoil. (b) α-helix, showing the relations of successive planes in which the six atoms of the amide linkage lie. (Courtesy of R. E. Dickerson and I. Geis.)

the pitch of the α-helix. In β-keratins, however, Astbury observed x-ray patterns corresponding to a 0.7 nm (7 Å) repeating structure. Astbury made the further interesting discovery that when α-keratins are treated with steam, they can be stretched to a 0.66 nm (6.6 Å) repeating structure, approximating that of the β-keratins. Another

difference between stretched α-keratins and β-keratins is that the former are not stable in the stretched state and will revert back to their native conformation upon exposure to moist heat.

To explain these observations, Pauling and Corey proposed that the β-keratin structure consists of extended

N-terminal

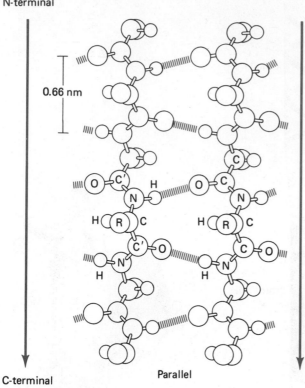

0.66 nm

C-terminal

Parallel

C-terminal

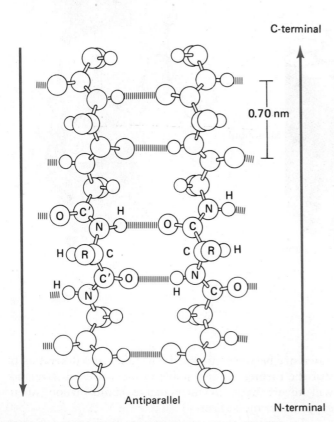

0.70 nm

Antiparallel

N-terminal

(a)

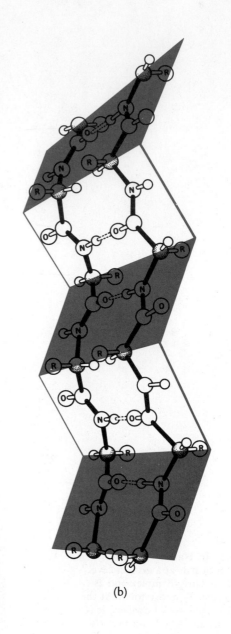

(b)

FIGURE 4–32

The β-pleated Sheet. (a) Extended polypeptide chains, running parallel or antiparallel to each other, can form hydrogen bonds between —CO and —NH groups in the amide linkages of their backbones. (b) Since the tetrahedral carbon bonds cannot be bent to form a 180° angle, the two-dimensional sheet formed by the aligning of two or more chains is riffled or pleated. Hydrogen bonds connect —CO and —NH groups of the polypeptide chain backbone. (c) Side chains (in the case of silk they are alanine and glycine) point up and down, thereby interconnecting successive layers of

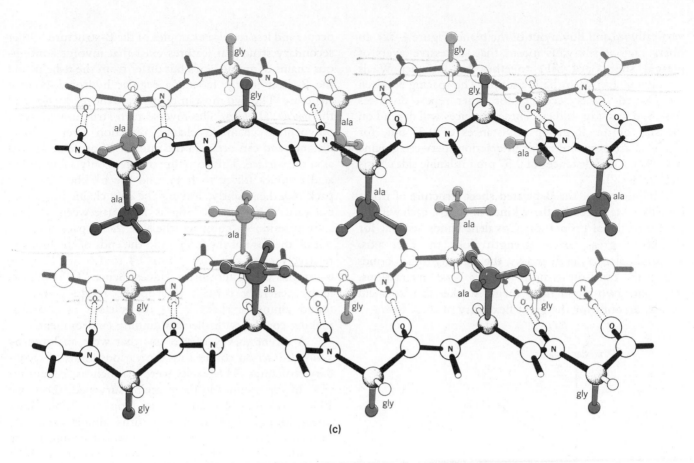

(c)

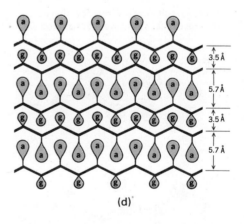

(d)

β-sheets. A side chain pointing up fits between two side chains pointing down. (d) The sheets in silk fibroin are arranged in such a manner that alanine packs against glycine forming spacings of 0.57 nm (5.7 Å) alternating with spacings of 0.35 nm (3.5 Å). (b–d courtesy of R. E. Dickerson and I. Geis.)

polypeptide chains lying next to each other and oriented in a manner that allows the —C=O and HN— groups of the amide linkages of neighboring polypeptide chains to interact with each other. When a number of polypeptide chains associate in this manner, two kinds of interactions occur: those in which the chains run in the same direction (parallel) and those in which the chains run in opposite directions (antiparallel, Figure 4–32). The resulting structure is a two-dimensional sheet. The sheet is not planar but is pleated or rippled because this conformation allows

the regular formation of hydrogen bonds (Figure 4–32). In parallel pleated sheets the hydrogen bonds must either be distorted or the sheets must be a little more pleated, or both. This is why silk fibroin, an antiparallel structure, has a 0.70 nm repeat, whereas stretched hair, which is distorted to a parallel β structure, shows only a 0.66 nm periodicity.

The pleated sheet configuration is such that if the hydrogen bonds point in a horizontal direction approximately in the plane of the sheet, the R groups will point vertically up and down out of the plane (Figure 4–32). In fibers such as silk, this means that successive layers of pleated sheets are held together by van der Waals interactions between the R groups. The spacing between the pleated sheets accounts for other repeat distances observed in x-ray studies. These distances will depend on the size of the R groups. Distances of 0.35 nm, for instance, are observed in the interactions between glycine side chains, and distances of 0.57 nm in alanine side chains (Figure 4–32).

In summary, the β-pleated sheet structure of fibers like silk is stabilized by three kinds of bonds, each with its own directional properties. Covalent bonds account for the fibers' great tensile strength along the fiber axis. Perpendicular to the fiber axis, there are hydrogen bonds and in the third dimension, van der Waals interactions. The latter two interactions, being weaker than covalent bonds, account for the great flexibility of these fibers.

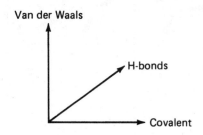

Forces stabilizing β-sheets forming a 3-dimensional structure like hair and silk

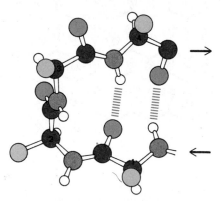

Structure of a β-turn. The CO group of residue 1 of the tetrapeptide shown here is hydrogen bonded to the NH group of residue 4, which results in a hairpin turn.

Is Secondary Structure Predictable?

So far, we have discussed two of the more important elements of secondary structure: the α-helix and the β-pleated sheet. There are other helical structures that

occur, and less regular examples of the β-structure. Other secondary structural features exist that involve continuous chains of amino acids but differ from the α-helix and the β-sheet in not having a regular hydrogen-bonded structure. The two most important are the *reverse turn* and the *omega (Ω) loop.* The reverse turn structure allows a polypeptide chain to change direction over a short distance and can occur in the interior of proteins as well as at the surface. This structure involves only a few amino acid residues (three to five), and the backbone groups pack together tightly, leaving the side chains projecting outward. The omega loop is large (between six and sixteen residues), compact (the side chains pack together within the core of the loop), and the ends of the loop are relatively close together (from 0.4 to 1.0 nm between α-carbons of the terminal residues, hence "Ω"). Omega loops generally contain hydrophilic residues and are found almost exclusively at the surface of proteins, making contact with the surrounding environment.

A recent survey of proteins that were analyzed by x-ray diffraction counted up the various kinds of secondary structures. The results were that helices account for 26% of the residues in the proteins surveyed, sheets for 19%, turns for 15%, and omega loops for 21%. (Since omega loops can contain reverse turns, almost a necessity to achieve a compact structure, certain corrections had to be made in the bookkeeping; we have given the adjusted values.) With the inclusion of the omega loop, very few residues in the known proteins are not in identified structures and are, therefore, thought to be present as random coils.

The important question is: what determines the assumption by a given polypeptide chain of a helix, a pleated sheet, turns, or loops? On present evidence, it appears that the interaction between certain neighboring amino acid residues determines which of these structures will be formed. For example, on the basis of both theoretical calculations and the behavior of model compounds, it is possible to classify the 20 amino acid residues into three categories: helix breakers, helix indifferent, and helix formers (Table 4–4). H. Scheraga and others have calculated the probability that a given amino acid residue is involved in a helical configuration and have compared the calculation with a number of proteins for which the detailed structure has been worked out by x-ray diffraction analysis. The results showed a surprisingly close correspondence between the predictions and the actual structures.

Theories that predict β-structures and extended structures are not so well developed, but empirical studies of a number of proteins have demonstrated that certain amino acid residues tend to occur in each of these structures with high frequencies. Thus, for instance, residues such as Gly, Ser, and Asp, which have a low

TABLE 4-4

Helix-Forming Tendencies of the Amino Acid Residues*

Helix Breaker	Helix Indifferent	Helix Former
Gly	Lys	Val
Ser	Tyr	Gln
Pro	Asp	Ile
Asn	Thr	His
	Arg	Ala
	Cys	Trp
	Phe	Met
		Leu
		Glu

*From Lewis, P. N., and Scheraga, H. A. (1971) *Arch. Biochem. Biophys* 144:576.

tendency for helix formation, have a strong tendency to form β-structures. Using such empirical generalizations, a number of workers have made successful predictions of these structures' occurrence in a number of test proteins.

The conclusion of these studies so far is that short-range interactions—interactions between neighboring amino acid residues—play a dominant role in determining whether a given amino acid residue is found in an α-helix, a β-pleated sheet, or other structures. One imagines that during the formation of the structure of a

given protein, special regions of the polypeptide chain assume a given structure, thus determining in a direct manner what the local secondary structure of the protein will be (Figure 4–33).

Tertiary Structure—The Conformation of the Polypeptide Chain

The soluble, globular proteins of the cell are compact, well-defined, but not rigid molecules. It is easy to see how α-helices and β-pleated sheets can form elongated fibers. But if these structures are to be part of compact, near-spherical molecules, there must be another (tertiary) level of protein structure to account for the property of compactness. The **tertiary structure** of proteins has to do with the way the regions of secondary structure are folded together.

As we have seen, the interactions among the side chains in the amino acid sequence determine the secondary structures formed. Likewise, these interactions are the basis for the folding of secondary structures to form the tertiary structure. This folding often brings into close proximity residues that are distant along the polypeptide chain. The specific three-dimensional structure, or *conformation,* of a protein gives it the ability to have a specific function. The existence of the high specificity of enzymes and antigen-antibody interactions makes it clear that the sequence-determined folding forms a uniquely characteristic structure for each given protein. *The sequence of amino acids determines the conformation of the protein, which determines its function.*

Forces Stabilizing Protein Conformation

Proteins exhibit the entire range of interactions available to organic compounds. These include covalent bonds and a variety of hydrophobic and ionic interactions. The main groups responsible for the interactions stabilizing the tertiary structure of proteins are the 20 different amino acid side chains. Figure 4–34 attempts to classify these into categories which describe the role they play in determining the conformation of polypeptide chains.

Covalent Interactions

Disulfide bonds are the most frequently encountered covalent linkages between side chains of the protein molecule. It was known for a long time that such bonds occur in proteins, but it was only when the structure of insulin was elucidated by Sanger, and ribonuclease by Moore, Stein, and their co-workers, that the precise structural chemistry of intrachain disulfide bonds was

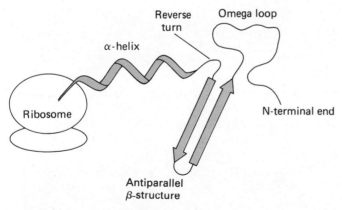

FIGURE 4-33

The Determination of Secondary Structure. This scheme illustrates how different portions of the polypeptide chain might form local areas of structure—determined by the local neighborhood of amino acids in the sequence—as the chain is synthesized, starting at the N-terminal end. These local areas may well retain their structure as the protein folds into a compact tertiary structure.

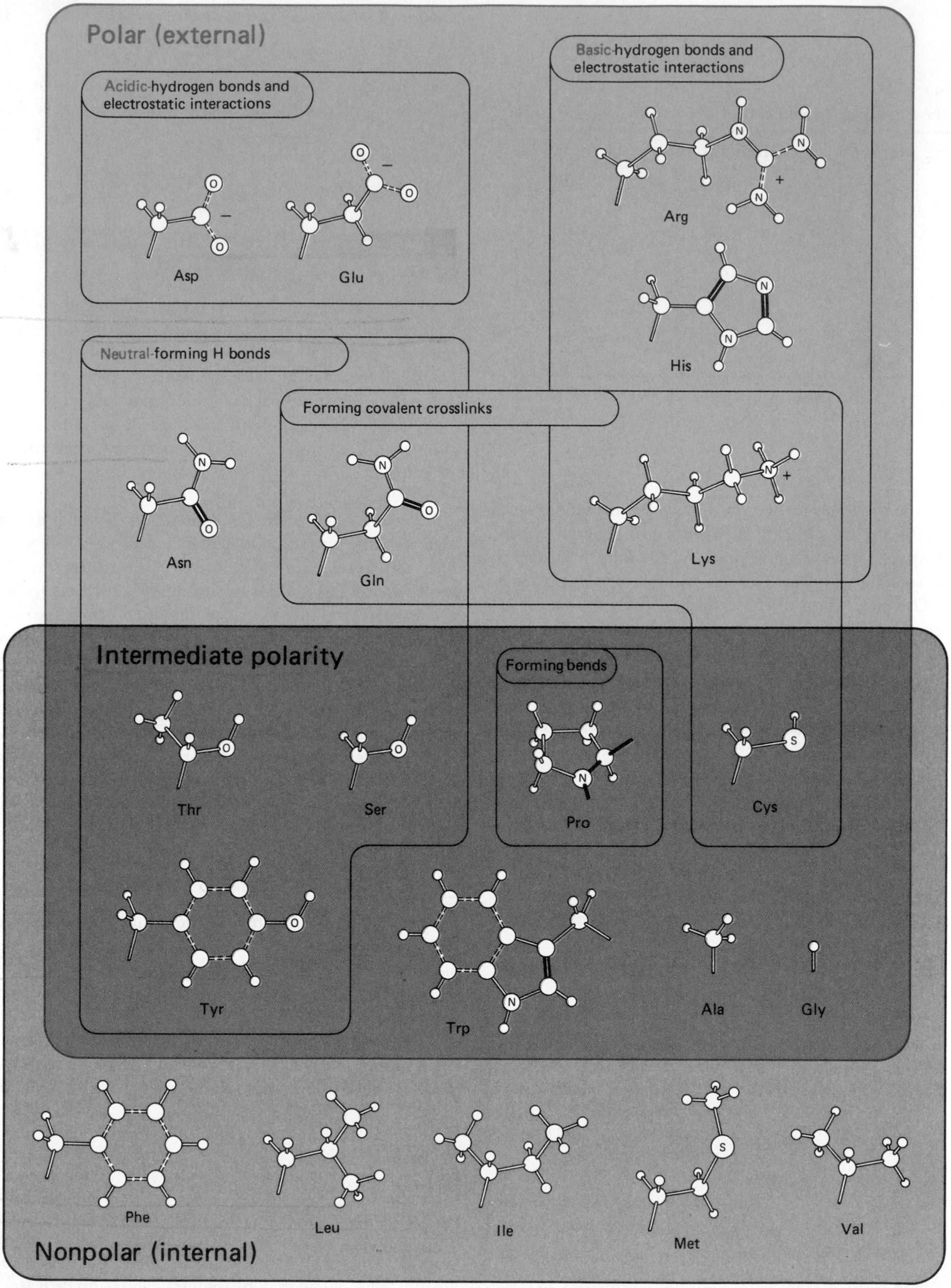

Polar (external)
Acidic-hydrogen bonds and electrostatic interactions
Asp
Glu
Basic-hydrogen bonds and electrostatic interactions
Arg
His
Neutral-forming H bonds
Forming covalent crosslinks
Asn
Gln
Lys
Intermediate polarity
Forming bends
Thr
Ser
Pro
Cys
Tyr
Trp
Ala
Gly
Phe
Leu
Ile
Met
Val
Nonpolar (internal)

FIGURE 4–34

Properties of Amino Acid Side Chains Responsible for Stabilizing Tertiary Structure of Protein Molecules. Five residues are of a nonpolar nature and contribute to hydrophobic interactions, while seven residues are polar. The remaining eight residues are of intermediate polarity. Some residues have more than one function; for instance, glutamine can interact with water at the surface, or form a hydrogen bond with another residue, or form a covalent (isopeptide) bond with lysine.

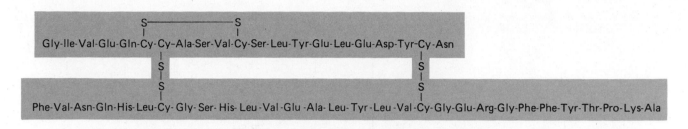

(a)

(b)

FIGURE **4–35**

Covalent Interactions in Bovine Insulin and Ribonuclease. (a) The amino acid sequence of the mature hormone insulin, showing interchain and intrachain disulfide bonds. Pancreatic islet cells synthesize a precursor of insulin, a single polypeptide chain from which the active protein (hormone) is formed by the action of proteolytic enzymes. During this processing an enzyme removes a length of peptide from the chain, thus producing the two-chained structure shown above. (b) The amino acid sequence of the enzyme ribonuclease, showing disulfide bonds cross-linking the peptide chain in four places. (Courtesy of D. G. Smyth, W. H. Stein, and S. Moore.)

documented in detail (Figure 4–35). Disulfide bonds can form spontaneously when two —SH groups of two cysteinyl residues are brought close to each other in the presence of an oxidant:

$$2 \text{—SH} + \frac{1}{2} O_2 \rightleftharpoons \text{—S——S—} + H_2O.$$

The protein structural chemist can break these bonds by a variety of *reducing agents* such as cysteine, mercaptoethanol, or dithiothreitol; by oxidizing agents such as performic or perchloric acid; or by sulfitolysis with sodium sulfite (Figure 4–36). Indeed, the disulfide bonds of a protein such as ribonuclease must be broken before

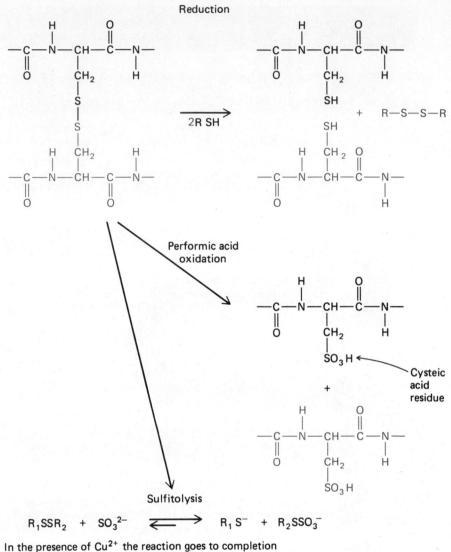

FIGURE 4-36
Splitting of Disulfide Bonds by Reduction, Oxidation, or Sulfitolysis.

its primary structure can be determined. Once the sequence of the amino acids in the polypeptide chain is worked out, it is possible to determine the exact location of the disulfide bonds by leaving them intact, digesting the protein with a variety of enzymes, and identifying the amino acids that are connected with the —S—S— linked cysteines (that is, disulfide-containing peptides).

A considerable number of proteins do not contain disulfide bonds (the proteins of the bacterium *E. coli,* for example, have very few), so one cannot claim that they are necessary to maintain protein structure. On the other hand, it has been possible to show in a number of proteins that do contain disulfide bonds that when they are broken by reduction, the protein either loses its specificity or becomes significantly less stable. Thus, disulfide bonds

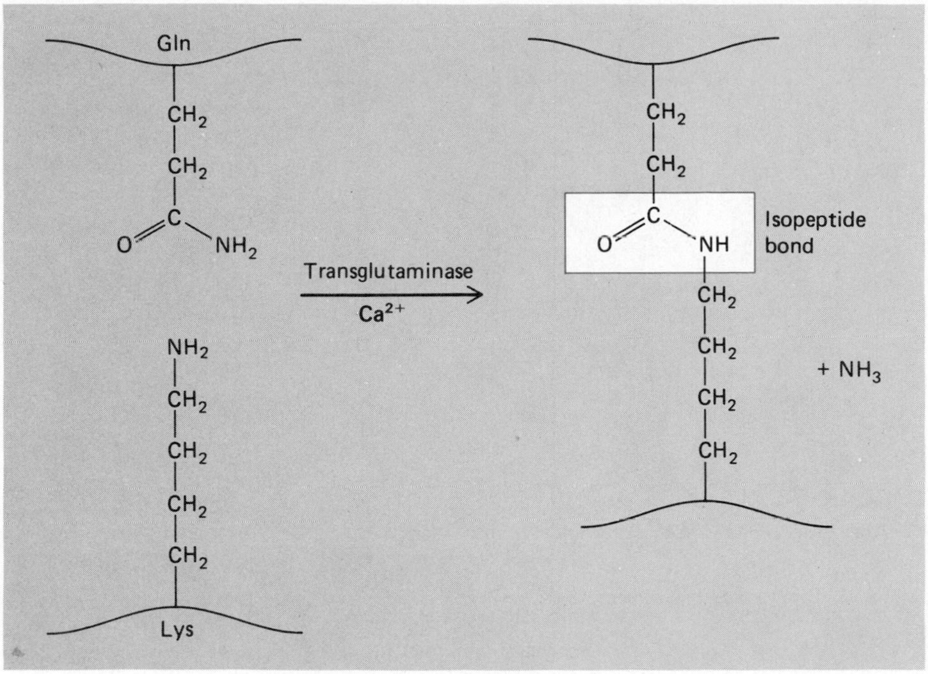

FIGURE 4–37
Formation of the Isopeptide Bond.
The side chains of glutamine and lysine
react to form this covalent bond,
splitting out ammonia. The reaction is
catalyzed by the enzyme known as
transglutaminase.

generally contribute to the stability of those proteins containing them.

For a long while there was very little, if any, direct evidence for other covalent bonds involved in the maintenance of protein structure. As will be seen in Chapter 15, it has been demonstrated that the transglutaminase enzyme found in blood plasma can cross-link fibrin molecules by forming "isopeptide" bonds between the γ-carboxyl groups of glutaminyl residues and the ε-amino groups of lysyl residues (Figure 4–37). Transglutaminase activity is also found inside cells. Assays for the presence of the isopeptide bond in cellular proteins indicate that even though this bond occurs in a wide range of cells, it is found at concentrations as low as one bond in 10^6 to 10^7 Da of proteins.

Nonpolar or Hydrophobic Interactions

The nonpolar or hydrophobic interactions involving nonpolar amino acid side chains (such as valine, leucine, isoleucine, phenylalanine, tryptophan, and methionine) are of great importance in stabilizing proteins. These interactions bring about an overall molecular conformation that buries the nonpolar amino acid side chains in the interior of the molecule, leaving the polar and ionic groups at the surface where they can interact with water. The driving force for this effect is primarily entropic. When hydrophobic groups are withdrawn from water, the number of hydrogen bonds is increased, while at the same time the structure of the water becomes more random—that is, its entropy is increased. Consequently,

hydrophobic groups tend to move away from the water and close to each other, thus forming hydrophobic regions inside the protein molecules.

Hydrophobic interactions are of crucial importance for the stabilization of macromolecules, as well as larger structures like membranes. Of the noncovalent interactions, hydrophobic interactions provide the largest share of the *energy* stabilizing the conformation of macromolecules. The protein molecule gains approximately 17 kJ/mole (4 kcal/mole) of free energy of stabilization for every hydrophobic side chain transferred from the aqueous environment to the nonpolar interior of the molecule. However, these noncovalent interactions lack the *specificity* exhibited by covalent bonds and hydrogen bonds.

Polar Interactions

Polar interactions can be divided into two closely related types of bonds: hydrogen bonds and ionic bonds. We have already encountered the hydrogen bonds which stabilize the α-helix and the β-pleated sheet. These hydrogen bonds form between —CO and —NH groups on the backbone of the polypeptide chains. But hydrogen bonds can also form between amino acid side chains: tyrosine and histidine, or serine and aspartic acid (Figure 4–38). When polar groups—the backbone —CO and —NH groups, as well as hydrophilic side chains—are buried in the hydrophobic interior of a globular protein molecule, the increase in energy that would occur by removing them from the aqueous solvent is compensated

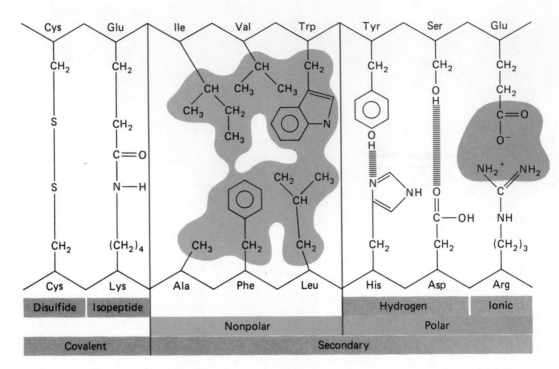

FIGURE 4–38
**Variety of Bonds or Interactions Stabilizing the Tertiary Structure
(conformation) of Protein Molecules.**   These are examples of general types
of interaction and not an exhaustive list. Although most charged side chains will
be oriented toward the aqueous solvent at the surface of the molecule, some polar
bonds might remain inside, presumably shielded from the water and salt by
nonpolar groups. Noncovalent interactions are indicated by color.

by their being involved in interior hydrogen bonding. A
number of physical studies have definitely identified such
interactions, especially in the case of tyrosine side chains.

Another form of polar interaction is the ionic
bonding that may occur between positively charged side
chains of such amino acids as lysine, arginine, or
histidine, and the negatively charged side chains of
glutamic or aspartic acids. An ionic bond between amino
acid side chains is an extremely strong form of interaction
if other ions are not available to compete and if the
surrounding environment is nonpolar. Since only certain
pairs of appropriately positioned atoms can form ionic
bonds, their presence inside a protein tends to favor
specific structures.

X-Ray Diffraction Analysis of Protein Conformation

The early students of protein chemistry conceived of
proteins as indeterminate "colloidal" molecules whose
various properties distributed themselves broadly around
statistical averages. The discovery that proteins can be
crystallized and the observation that enzymes and antibodies are highly specific in their functions represented
two types of early evidence suggesting that proteins, in
spite of their size and complexity, have a well-defined
structure. But it was not until methods for analyzing
large molecules by x-ray diffraction were developed,
largely in the laboratory of M. Perutz and J. Kendrew,
that it became clear that protein molecules exhibit a single
structure down to atomic dimensions; that is, less than
0.2 nm (2 Å).

Figure 4–39 shows a *precession photograph* of myoglobin, obtained by recording x-ray diffraction patterns
while moving a protein crystal, a mask, and the photographic plate simultaneously in a complex pattern. This
allows a regular display of the diffracted radiation on the
plate. The dots represent scattering maxima; the ones
close to the center represent large regularities in the
protein crystal, whereas the ones at the periphery represent regularities within the protein molecule at distances
as small as 0.1–0.2 nm.

With simpler structures one makes an educated guess
about the possible structure, calculating what kind of
scattering pattern one would obtain, and comparing it
with the experimental diffraction photographs. Then, if

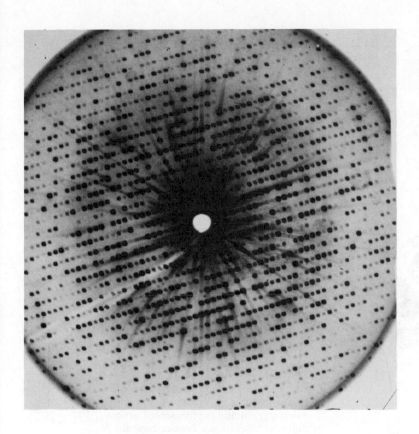

FIGURE **4–39**

Precession X-ray of Myoglobin.
This pattern is obtained when the crystal, a mask, and the recording film are moved simultaneously. The scattering maxima near the center represent long spacings in the crystal; the scattering maxima near the outer edge of the diagram represent short spacings in the crystal. Kendrew and his associates were eventually able to study myoglobin at 0.15 nm resolution, which involved the measurement of some 10,000 intensities and an equal number of complex calculations. X-ray diffraction data can now be collected from two-dimensional multiwire detectors that are more sensitive, more accurate, and faster than film. (Courtesy of F. H. C. Crick and J. Kendrew.)

need be, one refines the guess in an attempt to improve the agreement between the calculated and the observed diffraction patterns. Protein molecules are far too complex to permit the use of such a procedure. There is, however, a method that permits one to calculate the structure directly from the scattering maxima. This involves comparing the scattering by normal crystals of a given compound with the scattering by crystals in which a strongly scattering atom (such as a heavy metal) or a group of atoms has been substituted at a specific place in the molecule. If this creates only a change in the intensities (Figure 4–40) of the scattering maxima, and not in the spacings between them—that is, if the derivative is *isomorphous*—then from the change in inten-

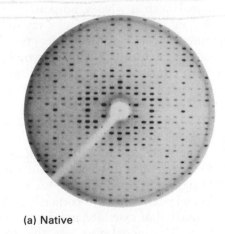

(a) Native

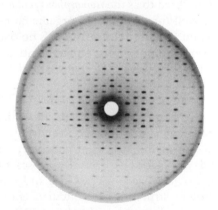

(b) Isomorphous replacement

FIGURE **4–40**

Comparison of X-ray Diffraction Patterns. The two examples were obtained from a normal crystal of horse heart cytochrome *c* and from an "isomorphous replacement" crystal, obtained by diffusing in the heavy metal derivative $PtCl_4$. The electron-dense platinum atoms bind to specific sites on the protein. Numerous differences in the intensity of the scattering maxima can be seen but the spacing between the maxima is unchanged. The measurement of intensity differences permits the determination of the platinum atom's position. From three or more such isomorphous crystals, the three-dimensional structure of the protein can be deduced. (Courtesy of R. E. Dickerson.)

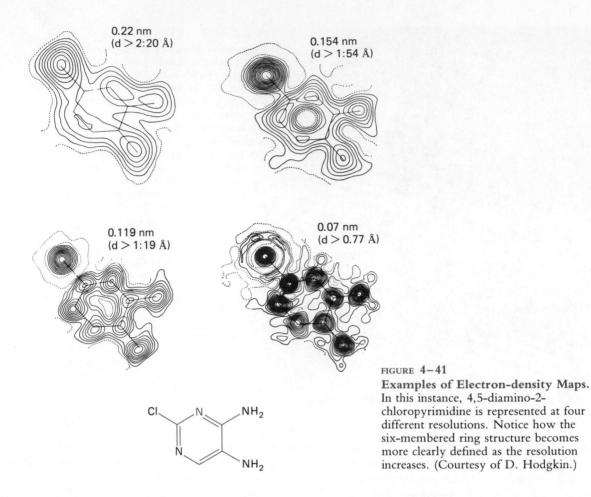

FIGURE **4–41**

Examples of Electron-density Maps.
In this instance, 4,5-diamino-2-
chloropyrimidine is represented at four
different resolutions. Notice how the
six-membered ring structure becomes
more clearly defined as the resolution
increases. (Courtesy of D. Hodgkin.)

sities it is possible to calculate the exact position of the heavy atom(s).

This procedure, referred to as the *isomorphous replacement method,* worked well for complicated organic compounds such as penicillin, but until the early 1950s no one succeeded in preparing isomorphous replacements of protein crystals. In 1956, however, H. Dintzis and G. Bodo, working in the laboratory of Kendrew and Perutz, were able to prepare several heavy metal derivatives of myoglobin. When x-ray diffraction measurements of the crystals showed that they were indeed isomorphous replacements, a new era in the structural chemistry of proteins began. Today, detailed structural analyses of more than 100 proteins have been carried out and our understanding of protein structure has solidified enormously. We shall not dwell on how Kendrew and his associates measured thousands of scattering intensities and, utilizing at that time new high-speed electronic computers, traced electron-density maps (Figure 4–41) at various levels, superimposed them (Figure 4–42), and built models of the myoglobin molecule, first at 0.6 nm (6 Å) resolution (Figure 4–43), and then at 0.15 nm (1.5 Å) resolution (Figure 4–45).

Myoglobin is an oxygen-storing protein found especially in the skeletal muscles of marine diving mammals such as whales, porpoises, and seals. It proved to be an ideal protein for x-ray diffraction analysis because it readily forms large crystals. As an added advantage, it turned out that myoglobin is sufficiently similar to the much more complex hemoglobin, that the elucidation of the former's structure helped in working out the detailed structure of the latter.

The big surprise of Kendrew's 0.6 nm resolution myoglobin model was that it provided direct evidence of the importance of the Pauling and Corey α-helix. Figure 4–43 shows how the myoglobin molecule is formed from eight α-helical segments that are folded around each other and interrupted by "corners" or "turns," consisting of nonhelical or extended polypeptide chains of different lengths. The model also was able to locate the heme group, which appeared to be nestled in a pocket surrounded by four portions of the polypeptide chain. Perhaps the biggest surprise of all was that the myoglobin molecule is an asymmetrical, irregular structure. As we shall see, this turns out to be true for globular protein molecules in general.

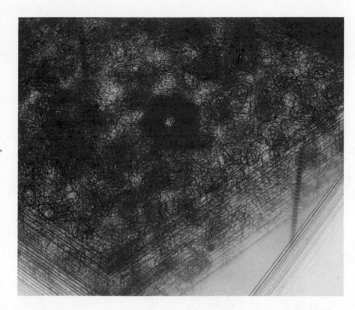

FIGURE **4–42**

A Three-dimensional Electron-density Map of Myoglobin. Electron densities at various levels in the crystal are plotted on lucite sheets; then the sheets are stacked over each other so the three-dimensional structure of the molecule can be visualized. Computer graphic techniques can now be used to plot electron density maps and to build the molecular model on a video display screen. (Courtesy of J. Kendrew.)

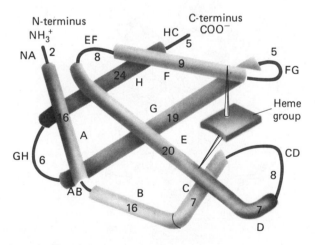

FIGURE **4–43**

The Conformation of the Myoglobin Chain at 0.6-nm Resolution. The original model built by Kendrew and his co-workers showed eight straight portions (later demonstrated to be α-helices) interrupted by seven nonhelical corners of different lengths. The heme group is shown resting in a pocket. The numbers represent the number of amino acids present in a given helix or corner. The region between the N-terminus and the A-helix is NA; HC is the region between the H-helix and the C-terminus.

Val-	Leu-	Ser-	Glu-	Gly-	Glu-	Trp-	Gln-	Leu-	Val-	10
NA1	NA2	A1	A2	A3	A4	A5	A6	A7	A8	
Leu-	His-	Val-	Trp-	Ala-	Lys-	Val-	Glu-	Ala-	Asp-	20
A9	A10	A11	A12	A13	A14	A15	A16	AB1	B1	
Val-	Ala-	Gly-	His-	Gly-	Gln-	Asp-	Ile-	Leu-	Ile-	30
B2	B3	B4	B5	B6	B7	B8	B9	B10	B11	
Arg-	Leu-	Phe-	Lys-	Ser-	His-	Pro-	Glu-	Thr-	Leu-	40
B12	B13	B14	B15	B16	C1	C2	C3	C4	C5	
Glu-	Lys-	Phe-	Asp-	Arg-	Phe-	Lys-	His-	Leu-	Lys-	50
C6	C7	CD1	CD2	CD3	CD4	CD5	CD6	CD7	CD8	
Thr-	Glu-	Ala-	Glu-	Met-	Lys-	Ala-	Ser-	Glu-	Asp-	60
D1	D2	D3	D4	D5	D6	D7	DE1	E1	E2	
Leu-	Lys-	Lys-	His-	Gly-	Val-	Thr-	Val-	Leu-	Thr-	70
E3	E4	E5	E6	E7	E8	E9	E10	E11	E12	
Ala-	Leu-	Gly-	Ala-	Ile-	Leu-	Lys-	Lys-	Lys-	Gly-	80
E13	E14	E15	E16	E17	E18	E19	EF1	EF2	EF3	
His-	His-	Glu-	Ala-	Glu-	Leu-	Lys-	Pro-	Leu-	Ala-	90
EF4	EF5	EF6	EF7	EF8	F1	F2	F3	F4	F5	
Gln-	Ser-	His-	Ala-	Thr-	Lys-	His-	Lys-	Ile-	Pro-	100
F6	F7	F8	F9	FG1	FG2	FG3	FG4	FG5	G1	
Ile-	Lys-	Tyr-	Leu-	Glu-	Phe-	Ile-	Ser-	Glu-	Ala-	110
G2	G3	G4	G5	G6	G7	G8	G9	G10	G11	
Ile-	Ile-	His-	Val-	Leu-	His-	Ser-	Arg-	His-	Pro-	120
G12	G13	G14	G15	G16	G17	G18	G19	GH1	GH2	
Gly-	Asn-	Phe-	Gly-	Ala-	Asp-	Ala-	Gln-	Gly-	Ala-	130
GH3	GH4	GH5	GH6	H1	H2	H3	H4	H5	H6	
Met-	Asn-	Lys-	Ala-	Leu-	Glu-	Leu-	Phe-	Arg-	Lys-	140
H7	H8	H9	H10	H11	H12	H13	H14	H15	H16	
Asp-	Ilc-	Ala-	Ala-	Lys-	Tyr-	Lys-	Glu-	Leu-	Gly-	150
H17	H18	H19	H20	H21	H22	H23	H24	HC1	HC2	
Tyr-	Gln-	Gly								153
HC3	HC4	HC5								

FIGURE **4–44**

Amino Acid Sequence of Sperm Whale Myoglobin. Residues lying in α-helices are labeled A to H (see Figure 4–45). Amino acids lying between helices—at the "corners"—are labeled with letters designating the helices on either side (e.g., GH1, GH2, etc.). (Courtesy of A. E. Edmundson and H. C. Watson.)

The higher-resolution maps provided detailed structural information at all levels. At the level of primary structure it was possible to identify directly a large number of the amino acid residues. By combining these data with the chemical sequencing work being done at the time, it was soon possible to work out the complete amino acid sequence of the myoglobin polypeptide (Figure 4–44). This permitted the building of detailed three-dimensional structures in which the exact conformation of the polypeptide, both helical and nonhelical

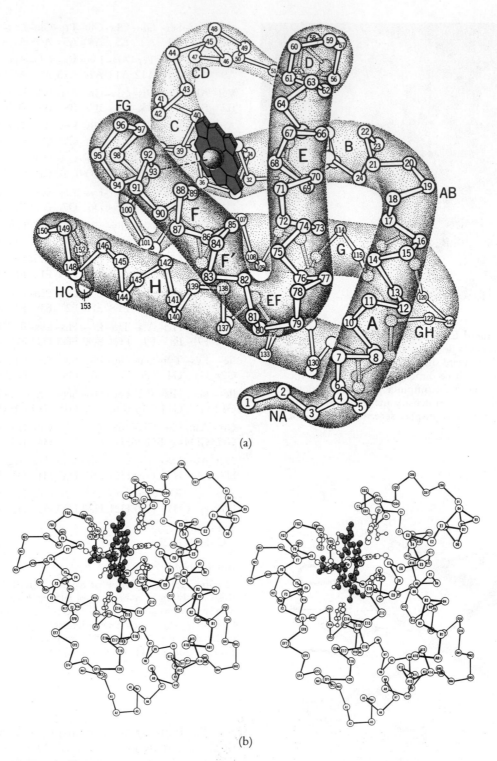

(a)

(b)

FIGURE 4–45

Tertiary Structure of Myoglobin.   (a) The structure of myoglobin at 0.15 nm resolution. The helical as well as the nonhelical portions of the molecule can be clearly seen. The heme group is nestled in a pocket and its nearness to Histidines F8(93) and E7(64) can be seen. (Courtesy of R. E. Dickerson and I. Geis.)
(b) Stereoviews of the myoglobin chain showing a few side chains that interact with the heme group. These include 39(Thr C4), 43(Phe CD1), 64(His E7), 68(Val E10), 71(Ala E13), 93(His F8), 97(His FG3), 99(Ile FG5), 104(Leu G5), 107(Ile G8), and 138(Phe H15). (Drawing courtesy of I. Geis.)

portions, was specified (Figure 4–45a, b). When this was done, Kendrew was able to analyze visually the interactions that presumably stabilized the myoglobin molecule. A number of generalizations emerged that, on the whole, confirmed the earlier speculation of the physical chemists regarding protein structure:

1. Myoglobin is a very compact molecule (4.5 × 3.5 × 2.5 nm) that contains fewer than four molecules of water in its interior (Figure 4–45c).

2. The molecule consists of eight right-handed α-helices (labeled A–H) including 78% (121 of the 153 residues) of the amino acids. There are seven nonhelical segments of the polypeptide chain, two at the C- and N-terminal ends and five between helices.

3. Since prolines can only be accommodated near the end of an α-helix, one would expect to find them in the nonhelical portions of the polypeptide chain. And this is indeed where one finds the four prolines of the myoglobin molecule. Since proline is not found at all the corners it is clear that its presence is not a *necessary* condition for helix termination.

4. Most of the polar side chains are on the surface of the molecule where they interact with water. Side chains which are partially polar, such as threonine, tyrosine, and tryptophan, are oriented to point the polar portion of the molecule towards the surface and the nonpolar portion towards the inside. A very few polar side chains are buried inside the molecule. Two of these are

histidines that play an important role in the oxygen-binding site of the molecule (Table 4–5).

5. The nonpolar side chains such as valine, leucine, isoleucine, methionine, and phenylalanine are in the interior of the molecule.

6. A number of polar interactions do appear to occur between various side chains, and between side chains and the polypeptide backbone, as evidenced by the closeness (0.2–0.3 nm) of these groups to each other (Table 4–5).

7. The heme group, which is the oxygen-binding site of the myoglobin molecule, is situated inside a pocket that provides a largely hydrophobic environment (Figure 4–46). The two polar propionic acid side chains of the heme group point out of the pocket into the aqueous environment. The iron of the heme group is octahedrally coordinated. Its six nearest neighbor atoms at the corners of the octahedron are:

(a) Four nitrogens in the porphyrin ring of the heme.

(b) A fifth nitrogen belonging to a histidine side chain (His F8), which is held rigidly by a hydrogen bond to a nearby leucyl (Leu F4) residue (Figure 4–46). (F8 refers to the eighth residue of the F helix, F4 to the fourth.)

(c) The bound oxygen molecule (occupying the sixth coordination position), which can also form a hydrogen bond with the nitrogen of another histidine (His E6) (Figure 4–46).

TABLE 4–5
Polar Interactions in Myoglobin

Residue	Total Number	Number on Surface Interacting with Solvent	Number "Buried" Inside Molecule	Number Involved in Strong Intramolecular Interactions	Partners
Lys	19	19	0	3	Glu
Arg	4	4	0	1	Asp
Glu or Gln	19	19	0		Lys, Try Chain NH
Asp or Asn	8	8	0	4	Arg, His
Ser	6	5	1	4	Chain CO Chain NH
Thr	5	3	1–2	3	Chain CO Chain NH
His	11	7–8	3	3	Chain CO Fe^{2+} (H$_2$O) Asp
Try	2	2	0	1	Glu
Tyr	3	3	0	1	Chain CO

Note: Courtesy of F. M. Richards.

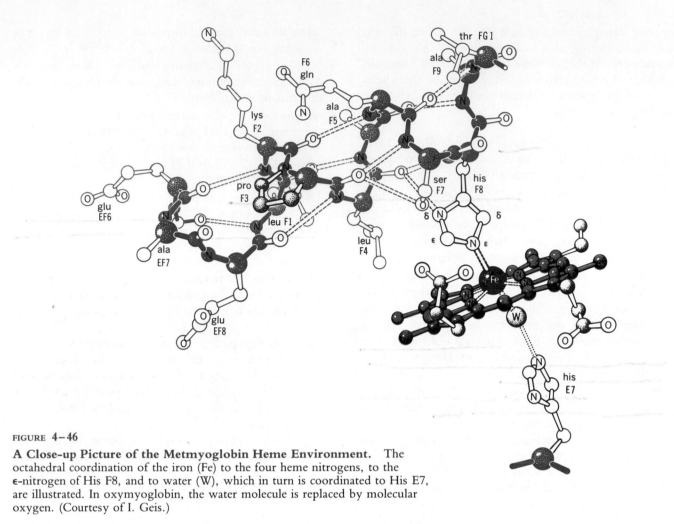

FIGURE **4—46**

A Close-up Picture of the Metmyoglobin Heme Environment. The octahedral coordination of the iron (Fe) to the four heme nitrogens, to the ε-nitrogen of His F8, and to water (W), which in turn is coordinated to His E7, are illustrated. In oxymyoglobin, the water molecule is replaced by molecular oxygen. (Courtesy of I. Geis.)

The nonpolar environment of the heme group in myoglobin seems to be necessary for its function in oxygen storage. Among other things, it prevents the oxidation of the iron atom from the ferrous to the ferric form, which cannot bind oxygen.

The polypeptide chain plays an important role in determining the specificity of a nonprotein group such as the heme group: different proteins use the heme group in different ways. The very same heme that is used in myoglobin to bind oxygen is found in cytochrome c, where it acts as an electron donor and acceptor, and in catalase where it catalyzes the splitting of hydrogen peroxide into water and oxygen. The hydrophobic environment of the heme in cytochrome c does not prevent oxidation of the iron atom like the environment in myoglobin. In fact, as we shall see in Chapter 8, the alternating reduction and oxidation of the Fe of cytochrome c is the basis of its function as an electron carrier.

The magnificent work on myoglobin by Kendrew and his co-workers supplied the first capstone of the structural work on proteins initiated by Sanger. The latter's hunch that the primary structure of proteins is highly specific for each type of protein molecule proved to be true also for the secondary and tertiary structures of proteins. And indeed, looking at it in retrospect, how could it be otherwise? How else could one account for the high degree of specificity exhibited by protein molecules such as enzymes and antibodies?

Soon after the structure of myoglobin was determined, the results of a structural study of the enzyme lysozyme were published by D. Phillips and his co-workers. Its structure exhibits the same general principles as those observed in myoglobin: helical and nonhelical regions, the presence of charged hydrophilic side chains outside and hydrophobic side chains inside the molecule, and hydrogen bonds (Figure 4–47). However, there are also some differences. The helices do not seem to be perfect α-helices but appear to be somewhat distorted. On the other hand, another conformation of the polypeptide chain predicted by Pauling and Corey is present, the antiparallel "pleated sheet" arrangement formed by a polypeptide chain that bends on itself (Figure 4–47).

Shortly after the elucidation of lysozyme, W. N. Lipscomb and his colleagues worked out the structure of

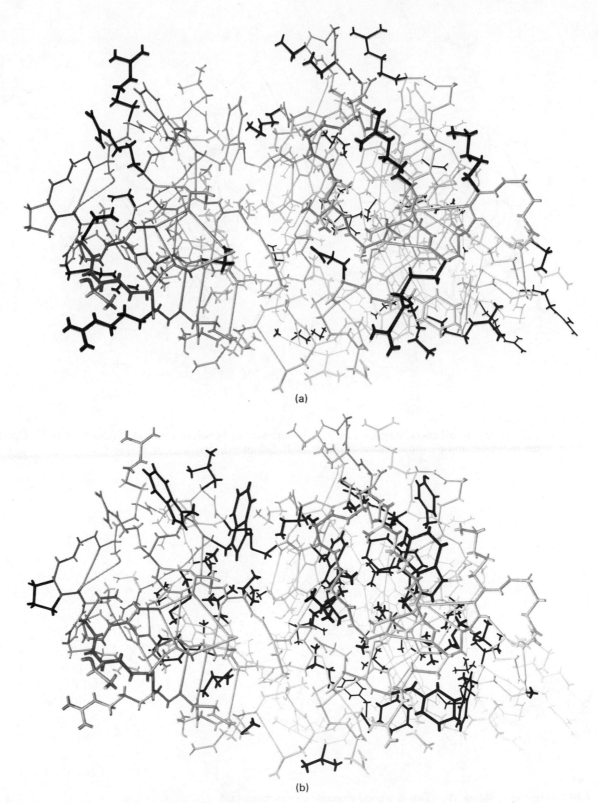

(a)

(b)

FIGURE 4–47

The Location of Various Side Chains of Hydrogen Bonds and Pleated Sheets in the Tertiary Structure of Egg-white Lysosyme. (a) The charged groups (Lys, Arg, Asp, Glu, N-terminal, and C-terminal), shown in color, are located on the outside of the molecule. (b) The hydrophobic groups (Trp, Tyr, Phe, Pro, Ile, Leu, Val, Ala, Cys-S-S-Cys, and Met), shown in color, are located inside the molecule. (c) Hydrogen bonds (shown in color) come in three categories: (1) those stabilizing β-sheets, at the lower left of the

(*illustration continued on p. 180*)

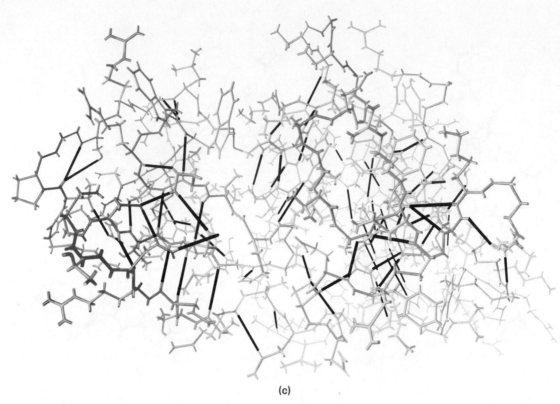

(c)

FIGURE   **4–47 (continued)**

molecule; (2) those stabilizing two α-helices, near the center and at the right of the molecule; and (3) those stablizing other interactions between side chains. (Courtesy of C. Cantor and P. Schimmel; drawings by I. Geis.)

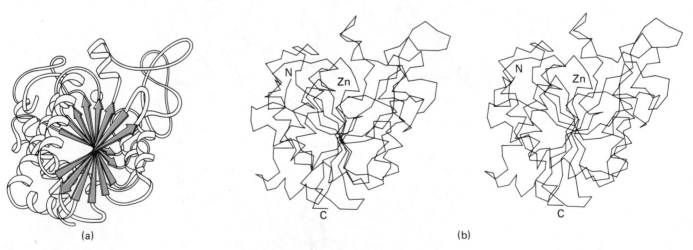

(a)   (b)

FIGURE   **4–48**

Structure of Carboxypeptidase A.   This is a good example of miscellaneous structural principles. (a) A graphic representation in which spiral ribbons represent α-helices, arrows represent strands of β-sheet, and thin lines represent turns and loops. The β-sheet twists, changing in orientation, and the chains run both in parallel and antiparallel senses. The sphere shows the location of the Zn^{2+} ion found in the molecule. (b) A stereo view of the polypeptide chain in which each angle represents an α-carbon atom. (Courtesy of J. S. Richardson.)

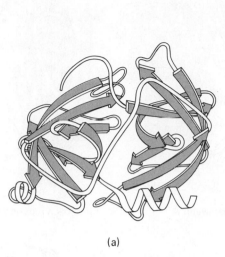

(a)

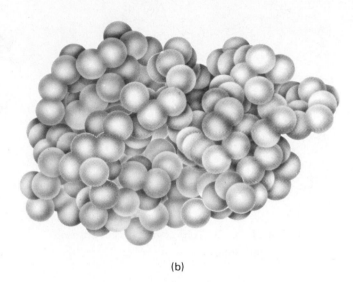

(b)

FIGURE **4–49**

The Presence of Domains in the Structure of Protein Molecules.
(a) Schematic drawing of the backbone of the elastase molecule, showing two similar domains composed of barrel-shaped β-structures. (Courtesy of J. S. Richardson) (b) Space-filling computer-drawn structure of the papain molecule showing two similar domains wrapping "arms" around each other. (From J. S. Richardson.)

the enzyme carboxypeptidase A (Figure 4–48). This molecule contains four major and four minor α-helical regions, constituting 30% of the molecule; 20% of the molecule consists of eight extended chains forming a pleated sheet that includes both antiparallel and parallel chains. The pleated sheet of carboxypeptidase A is a most interesting structure: it runs down the center of the molecule, changing the orientation of its plane so that the top chain is rotated 120 degrees with respect to the bottom chain (Figure 4–48).

As the structures of more and more protein molecules are worked out, a number of general conclusions have emerged. Perhaps most important is that no single architectural principle seems to dominate the structure of proteins. We expect to see α-helices, but other kinds of helices have also been observed. Both parallel and antiparallel pleated sheets play important structural roles in a variety of proteins. Extended chains allow for changes in the direction of the polypeptide backbone or endow the molecule with a certain amount of conformational flexibility. Compact mixtures of these elements can be identified as *domains* of structure, similar domains being found in several different proteins (Figure 4–49). Domains are not only identified by their structural

properties but often share similar functions and possibly evolutionary origins.

In spite of the tremendous potential for variability in protein structure, certain basic patterns of folding are found so often that proteins can be grouped into basic categories, as suggested by J. Richardson. Figure 4–50 shows examples of four of these major categories of folding: antiparallel α, parallel α/β, antiparallel β, and small disulfide-rich or metal-rich proteins. A good example of the folding of domains is the enzyme pyruvate kinase, which consists of three domains, each folding in a different pattern (Figure 4–51). It is not surprising that proteins closely related in function often show similarities in structure (Figure 4–52). This could be due to a common evolutionary origin, but other interpretations are possible. Curiously, some apparently unrelated proteins can exhibit striking similarities in structure (Figure 4–53).

In the normal cellular environment, protein molecules function in solution or in solid-phase systems containing more than one component. It is relevant to ask, therefore, whether the structure of proteins observed in crystals by x-ray diffraction is sufficiently similar to the structure of proteins in vivo. A number of studies

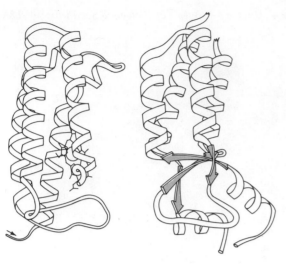

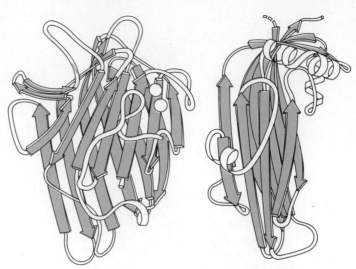

Cytochrome C1   Tobacco mosaic virus protein   Concanavalin A   Southern bean mosaic virus protein

(a) Antiparallel α   (c) Antiparallel β

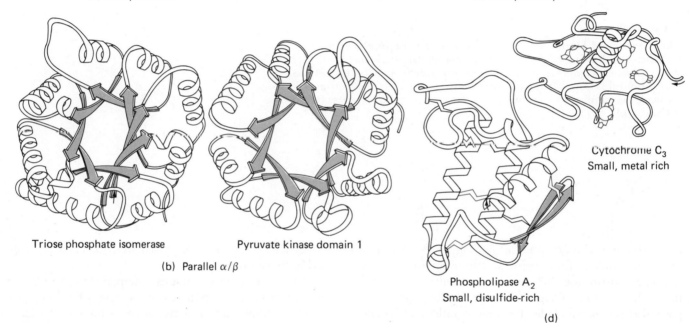

Triose phosphate isomerase   Pyruvate kinase domain 1

(b) Parallel α/β

Cytochrome C₃
Small, metal rich

Phospholipase A₂
Small, disulfide-rich

(d)

FIGURE 4–50

Four Major Categories of Protein Structure.
(a) Antiparallel α, in which two or more neighboring α-helices run in an antiparallel manner. (b) Parallel α/β, in which both α-helices and β-sheets are oriented in parallel, respectively. (c) Antiparallel β, in which β-sheets form a consistent antiparallel pattern. (d) Small proteins rich in disulfide bonds or metal ligands. Disulfide bonds and heme groups are in color. (Courtesy of J. S. Richardson.)

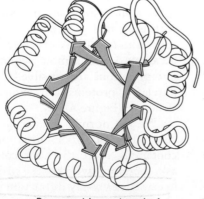

FIGURE 4–51
Three Different Domains Within the Same Molecule (Pyruvate Kinase) Fold in Entirely Different Ways. (Courtesy of J. S. Richardson.)

Pyruvate kinase domain 1   Pyruvate kinase domain 2   Pyruvate kinase domain 3

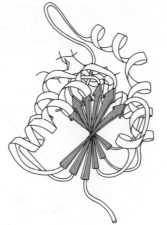
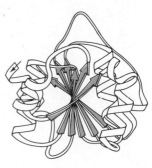

Alcohol dehydrogenase domain 2

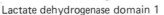

Lactate dehydrogenase domain 1

FIGURE 4–52
Proteins with Similar Functions Frequently Have Similar Structures. The reason for this may be the common evolutionary origin of the proteins or their closely related molecular function. (Courtesy of J. S. Richardson.)

FIGURE 4–53
Sometimes Proteins with Entirely Different Functions have Similar Structures. (a) Superoxide dismutase is an enzyme that protects cells from the damaging effects of superoxides. (b) The variable domain of the immunoglobulin G light chain (V_L) is a structural feature in one of the antibody proteins that circulate in the blood stream. The barrel-shaped structure composed of β-sheets is identical in both proteins. (Courtesy of D. Richardson.)

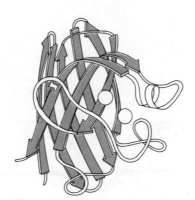

Cu, Zn Superoxide dismutase

(a)

Immunoglobulin V_L domain

(b)

performed with myoglobin and other proteins show that the structure of the protein in solution and in the crystalline state are not very different. These experiments include a detailed analysis of the absorption spectrum of the heme group, measurements of oxygen binding, and the determination of α-helical content. Also, crystals of certain enzymes demonstrate enzymatic activity similar to that found in solution. The introduction of extremely bright x-ray beams from synchrotrons has allowed detection and analysis of the changes in conformation associated with enzyme catalysis. Such results demonstrate that the forces which bind the protein molecules together in crystals do not necessarily distort the structure from that present in the cell.

Tertiary Structure—A Consequence of Amino Acid Sequence

As we have seen, it is possible to demonstrate that certain amino acid residues favor helix formation, that

others favor β-pleated sheets, and that others favor an extended structure. We must now ask whether the folding of the polypeptide chain, to form a compact three-dimensional molecule is also a consequence of the amino acid sequence of the polypeptide chain, and if so, whether folding can be predicted. The first question was answered affirmatively by studies of specific protein molecules. As for the second—though it is an article of faith of most molecular biologists that amino acid sequence determines both secondary and tertiary structure—we have so far made only modest progress in the important venture of predicting protein folding.

The Spontaneous Folding of Certain Proteins

It has been known since the classical studies by M. L. Anson and A. E. Mirsky (1945) that some proteins which have lost their biological properties after being exposed to conditions that *denature* the proteins can be brought back to a biologically active state by some appropriate treat-

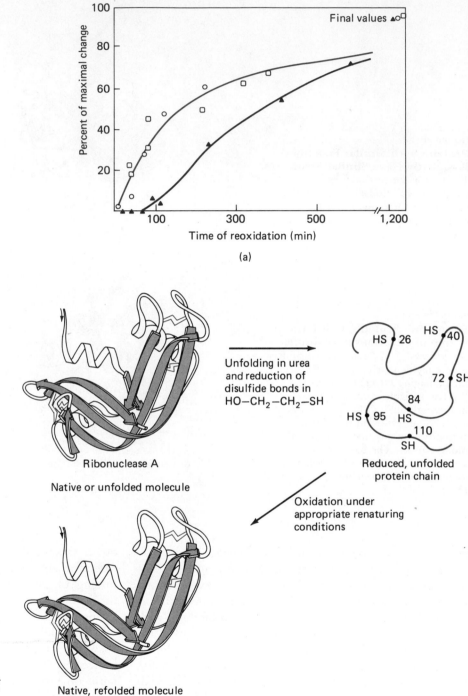

FIGURE 4–54
The Renaturation of Ribonuclease.
(a) Renaturation, which is stabilized by oxidation, can be detected by the return of enzyme activity (▲), restoration of the optical rotation of the native enzyme (□), and disappearance of the free SH groups of cysteine (○). According to all three criteria, almost 100% of the native state is reestablished. (After C. B. Anfinsen et al.) (b) A diagrammatic representation of the experiment demonstrating that renaturation of unfolded and reduced ribonuclease is accomplished by the reestablishment of the original disulfide bonds. (Courtesy of J. S. Richardson.)

ment. However, since the structure of proteins was not known at that time, it was not possible to determine whether the return of biological activity was accompanied by a return to the original protein structure. The dramatic work on ribonuclease in 1957–1962 by a number of investigators, including C. Anfinsen and M. Sela, demonstrated that when denatured ribonuclease was brought back to a biologically active state, it also assumed its original, native three-dimensional folding. As shown in Figure 4–35, ribonuclease consists of a single polypeptide chain, self-linked by four disulfide (—S—S—) bonds. When ribonuclease is treated with 8 M urea (to weaken secondary interactions between groups on the protein) in the presence of a disulfide reducing reagent such as mercaptoethanol, the chain becomes unfolded, the disulfide bonds are broken, and the enzyme loses its biologial activity. If the urea is now removed by dialysis

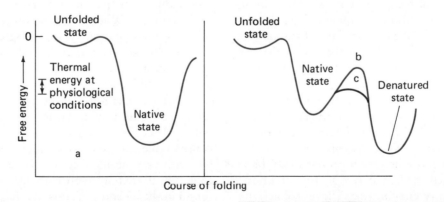

against mercaptoethanol, the secondary interactions will reform but the disulfide bonds will not. If the reducing agent is then removed by dialysis against a buffer solution, disulfide bonds will form again, being oxidized by the oxygen normally dissolved in water. The protein will have almost completely regained its biological activity (Figure 4–54).

To demonstrate the importance of the noncovalent interactions in determining the folding, one can attempt to **renature** the unfolded enzyme in a different way: reform the disulfide bonds by dialyzing against 8 M urea alone, then allow the secondary interactions to form by dialyzing against a physiological buffer solution. The result is a protein with almost no enzyme activity. This outcome is explained by hypothesizing that when the disulfide bonds reform in the presence of urea they combine in a variety of ways. In principle four disulfide bonds can form in 105 different ways[2] and, if the process is random, only 1/105 of the preparation will have the right combination of disulfide bonds. This hypothesis was in fact verified by performing the type of structural analysis of disulfide bonds illustrated in Figure 4–36. The results showed that the disulfide bonds of ribonuclease

which were reoxidized in urea were indeed scrambled rather than uniquely reconstituted. It can also be demonstrated that the renatured product with full enzyme activity has formed precisely the same disulfide bonds that were present in the original, native enzyme.

Similar results have been obtained with a number of other proteins and many workers in this field have concluded that, given the proper environmental conditions, a protein's native structure is the thermodynamically most stable conformation—that conformation with the lowest free energy, or at least an energy minimum along the most favorable path of refolding (Figure 4–55). In thinking about the renaturation of enzymes, one must take care in the definition of the native state to include the surrounding solution. As we have seen, the state of the water surrounding the protein contributes a great deal of free energy to the stabilization of the protein structure.

It seems clear that in the case of ribonuclease the native structure is also the lowest free energy structure, since it forms spontaneously from a completely unfolded random coil when the denaturing urea is slowly removed. The interactions that lead to the formation of this specific conformation are due to the specific sequence of amino acid side chains in the protein. This kind of experimental result is taken as proof that the sequence of amino acids is *sufficient* to produce the native tertiary

[2]Starting with one of the eight —SH groups, there are seven possible partners; for one of the six remaining, five possible partners; and for one of the four remaining, three possible partners. The remaining disulfide bond is made from the two remaining —SH groups. Thus, there are $7 \times 5 \times 3 = 105$ different combinations of —SH to form disulfide bonds in ribonuclease.

FIGURE 4–55

The Energetics of Protein Folding.   The free energy required to break the noncovalent interactions between amino acid side chains, so as to unfold the polypeptide into a random coil, is graphed against a pathway of folding the polypeptide chain. Three possiblities are illustrated. (a) The native conformation represents the lowest free energy state. (b) The native conformation is a metastable state—that is, it is

protected (or trapped) by a high activation energy barrier which guarantees the stability of the molecule even if it is not at its lowest free energy state. (c) The native state of the molecule is unstable because it is protected by only a small activation energy barrier. This protein molecule has a short half-life; thermal motions will relatively quickly "push" the molecule over the barrier to a denatured state, with loss of biological activity.

structure. It is even possible to demonstrate, in the case of ribonuclease, that not all the primary structure is necessary for an active tertiary structure. By brief exposure to subtilysin, a proteolytic enzyme, only a single peptide bond of ribonuclease is cleaved. This makes it possible to ask whether that peptide bond is *necessary* for the activity of ribonuclease, and the answer is no. One can remove the small, N-terminal fragment (S-peptide) from the large ribonuclease fragment (known as RNAse-S) by gel filtration chromatography in dilute acid, neutralize the pH, and then test again for enzymatic activity. Neither the S-peptide nor the ribonuclease-S are active alone. They can be recombined, however, with the result that nearly full enzymatic activity is restored despite the peptide bond remaining broken. Apparently the secondary interactions of the amino acid side chains are sufficient to reform the tertiary structure even in the absence of one of the covalent peptide bonds.

Whether the native structure of *all* proteins is also the thermodynamically most probable one remains to be seen. The possibility exists that the native structure of some proteins may represent a *metastable state:* a state that is not lowest in free energy but one in which the structure is "trapped" by energy barriers (Figure 4–55). A metastable state would have to be unique to explain the specificity of protein function but it could be achieved by guiding protein folding along a certain special pathway.

We know of one such pathway, the *pathway of synthesis* of the polypeptide on the ribosome which starts at the N-terminal amino acid and proceeds toward the C-terminal end. As the polypeptide chain emerges from the ribosome, the initial structure which forms will be that with the lowest free energy, but the free energy may no longer be lowest when the structure eventually becomes part of the completed protein. Cellular conditions will generally not unfold a protein, so this kinetically trapped structure is the one on which natural selection is likely to operate. Since the mechanism of synthesis was probably fixed at an early stage of evolution, most protein structures that we observe today must have evolved under the influence of a stepwise synthesis. Of course, even in this nonequilibrium interpretation of protein structure, the amino acid sequence still determines the native structure of the protein.

There is another sense in which proteins may be in metastable states when in their native form. We know, for example, that many enzymes have levels of activity that vary with the binding of a small molecule (**ligand**) to a location other than the active site responsible for catalysis (this is discussed in Chapters 5 and 14). The binding of a ligand brings about a functional effect by inducing a change in the structure of the protein. This means that what we refer to as the *native conformation* must be composed of a variety of structures, each with a different activity and separated by relatively small energy barriers. These barriers are high enough to give specificity to the various conformations but low enough that the binding of a ligand can shift the molecule to another conformation.

Finally, we must mention that in eucaryotic cells in unperturbed growth *most proteins are degraded* at a constant rate. Enzymes also often exhibit instability of their activity in the purified state. Both these observations may be due partially to the spontaneous relaxation of a protein to its lowest free energy conformation, one which is not native (i.e., enzymatically active) and which is more susceptible to the action of intracellular proteases (see further discussion in Chapter 13).

Can Folding Be Predicted?

One way to discover how amino acid sequences influence tertiary structures is to study different amino acid sequences which fold to form similar tertiary structures. This phenomenon has been observed at two levels. Proteins with precisely the same function (for example, cytochromes *c* from various organisms) have in the process of evolution experienced considerable change in their amino acid sequence (Figure 4–27). Yet x-ray diffraction analysis of the various cytochromes *c* shows they have not changed significantly in their folding pattern.

A more controlled example is the effect of point mutations, which substitute one amino acid for another in the sequence of a given protein, on the structure of the protein. C. Yanofsky and his co-workers showed that certain point mutations can destroy a protein's biological activity which, however, can be restored by a second point mutation. One can hypothesize that this will occur where there is a specific and crucial interaction between two residues. When a mutation changes one of the residues, thereby destroying biological activity, a second mutation in the position of the other residue could restore the proper interaction, thereby bringing about a return of biological activity.

Different examples of the influence of the amino acid sequence on tertiary structure are provided by proteins with related, or even unrelated, functions and with considerable differences in their amino acid sequences, but with very similar folded structures. The "myoglobin fold" illustrated in Figure 4–43, for instance, is also found in the α- and β-chains of adult hemoglobin, in the γ- and δ-chains of fetal hemoglobin, and even in heme-containing proteins of worms and insects (see Figure 4–56 for a comparison of the primary structures of myoglobin and hemoglobin). Another common folding structure has been found for a number of proteolytic enzymes such as chymotrypsin, trypsin, elastase, and thrombin, although they differ somewhat in their proteolytic properties.

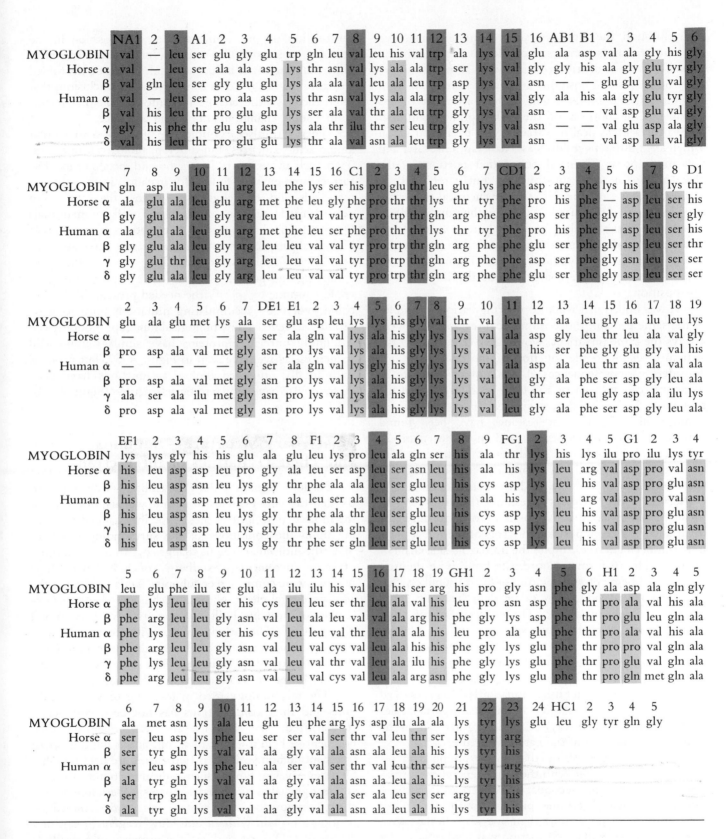

FIGURE 4—56

Primary Structure of a Number of Chains Belonging To the Myoglobin Family. It is remarkable that in spite of the many amino acid substitutions, all these chains have the myoglobin fold as their tertiary structure. Such comparisons allow one to identify the amino acid residues which are not essential for a particular tertiary structure—that is, ones that show high variability among several sequences, like B1–B5 and D1–D6. (Courtesy of R. E. Dickerson and I. Geis.)

Although the similarly shaped myoglobins have different primary structures, the differences in amino acid sequence are not distributed uniformly along the molecule. Some residue positions are highly conserved from one molecule to another, while others are more or less variable. One portion of the polypeptide chain that allows for a great deal of variability in position E13, in which such different residues as Thr, Asp, His, Gly, and Tyr may appear (Figure 4–56). Other portions of the chain allow for some substitutions, but it seems to be important that a residue with similar chemical properties be substituted. An example would be the position B8 in which Glu is uniformly present in human hemoglobin chains and which accepts the substitution of structurally similar Asp in the myoglobin chain (Figure 4–46). It appears that a negative charge is crucially important in this position.

It was hoped that when comparing various examples of the same protein, those amino acid residues that were always the same (invariant) would give clues to how the structure relates to the function of the protein. If one considers various sequences of myoglobin, there are 70 invariant residues (out of 153 total residues) among the sequences from 43 different species. In the β-chain of hemoglobin, 42 invariant residues have been found among 52 different sequences; and in the α-chain, 21 invariant residues among 49 different sequences. It is remarkable, however, that among all vertebrate proteins that form the myoglobin fold there are only 6 fully conserved ("superinvariant") residues (Figure 4–56). In most of these cases it is possible to rationalize the invariance of a given position in terms of the role it plays in the molecule. For example, the imidazole side chain of invariant His F8 forms a coordinate bond with the iron atom of the heme group (Figure 4–46). It is unlikely that any amino acid side chain other than histidine could perform this chemical task. The conservation of other invariant amino acids, however, is hard to explain. Consider Lys H9, which sticks into the aqueous medium away from everything else and could, one might guess, have been replaced by arginine (Figure 4–46). It is difficult to imagine why it has apparently been preserved for 500 million years! Could it be that its importance lies in the kinetics (or pathway) of folding rather than in the ultimate stabilization of the native structure?

These examples show that many amino acid sequences are compatible with the same general conformation. The important secondary and tertiary structural interactions that determine protein conformation can apparently be made in more than one way. In the above examples, the overall conformation was held constant by selection despite the variation in primary sequence due to mutations. More recently, use of modern genetic techniques (see Chapter 12) has allowed the systematic substitution of any amino acid at any position in the primary structure. This method promises a complete understanding of the influence of side chains on the folded structure. So far, the data confirm earlier conclusions that often there is little change in conformation that results from a small change in amino acid sequence.

There are also examples of proteins with similar amino acid sequences, but different biological functions. Two such proteins are lysozyme and lactalbumin. The former is an enzyme found, for instance, in the saliva of mammals and the eggs of chickens; the latter is a nonenzymatic protein found in the milk of mammals (Figure 4–56). It is believed that proteins with different biological functions but with similar structures arose by *gene duplication* in which another copy of the gene specifying the protein is made. One copy of the gene could continue to specify the needed protein. The other could evolve by accumulating mutations until its gene product achieved a new biological function. Not only can seemingly unrelated proteins have similar amino acid sequences, as we have seen in Figure 4–57, but proteins with totally unrelated biological functions have been discovered to have similar tertiary structures (Figure 4–53).

Quaternary Structure—Subunits and Their Dynamics of Interaction

As long ago as 1926 T. Svedberg, working with oxygen-carrying proteins of marine worms, noticed that the molecular weight of these proteins changed depending upon the conditions of the surrounding solution. Since the molecular weight changes occurred as multiples of large increments, Svedberg concluded that these proteins were composed of subunits. Today, by using the acrylamide gel electrophoresis technique, numerous examples of proteins built of subunits have been found (Table 4–6). The arrangement, with respect to one another, of polypeptide chains and subunits in a protein is called its **quaternary structure**. The subunits may be the same or different, as indicated in Table 4–6.

The terminology used in referring to various subunits and their combinations has not been completely standardized. We shall use the term *chain* for each polypeptide held together by a continuous linear array of true peptide bonds. Even if two chains are held together by other covalent bonds, such as disulfide bonds, they are still recognized as two separate chains. Operationally, the chain composition of a protein can be determined by treating a protein with an unfolding reagent (urea), or a detergent (SDS), or both, in the presence of a disulfide reducing reagent (mercaptoethanol), and subjecting the resulting mixture to acrylamide gel electrophoresis. The subunits dissociate from one another and travel separately

	1	2	3	4	5	6	7	8	9	10	11	12	13	14	15	16	17	18	19	20	21	22	23	24	25	26	27
Chicken egg lysozyme	lys	val	phe	gly	arg	cys	glu	leu	ala	ala	ala	met	lys	arg	his	gly	leu	asp	asn	tyr	arg	gly	tyr	ser	leu	gly	asn
Human salivary lysozyme	lys	val	phe	glu	arg	cys	glu	leu	ala	arg	thr	leu	lys	arg	leu	gly	met	asx	gly	tyr	arg	gly	ilu	ser	leu	ala	asx
α-Lact.	glu	gln	leu	thr	lys	cys	glu	val	phe	arg	glu	leu	lys	—	—	asp	leu	lys	gly	tyr	gly	gly	val	ser	leu	pro	glu

(cys at position 6 = 127)

	28	29	30	31	32	33	34	35	36	37	38	39	40	41	42	43	44	45	46	47	48	49	50	51	52
Chicken egg lysozyme	trp	val	cys	ala	ala	lys	phe	glu	— ser	asn	phe	asn	thr	gln	ala	thr	asn	arg	asn	thr	— asp	gly	ser	thr	asp
Human salivary lysozyme	trp	met	cys	leu	ala	lys	trp	glu	— ser	gly	tyr	asn	thr	arg	ala	asn	asx	tyr	asx	ala	gly — asx	arg	ser	thr	asp
α-Lact.	trp	val	cys	thr	thr	—	phe	his	thr ser	gly	tyr	asx	thr	glx	ala	ilu	val	glx	asx	—	— asx	glx	ser	thr	asx

(cys at position 30 = 115)

	53	54	55	56	57	58	59	60	61	62	63	64	65	66	67	68	69	70	71	72	73	74	75	76	77	78	79
Chicken egg lysozyme	tyr	gly	ilu	leu	gln	ilu	asn	ser	arg	trp	trp	cys	asn	asp	gly	arg	thr	pro	gly	ser	arg	asn	leu	cys	asn	ilu	pro
Human salivary lysozyme	tyr	gly	ilu	phe	gln	ilu	asx	ser	arg	tyr	trp	cys	asx	asx	gly	lys	thr	pro	gly	ala	val	asn	ala	cys	his	leu	ser
α-Lact.	tyr	gly	leu	phe	glx	ilu	asx	asx	lys	ilu	trp	cys	lys	asx	asx	glx	asx	pro	his	ser	ser	asn	ilu	cys	asn	ilu	ser

(cys at position 64 = 80; cys at position 76 = 94)

	80	81	82	83	84	85	86	87	88	89	90	91	92	93	94	95	96	97	98	99	100	101	102	103	104	105	106
Chicken egg lysozyme	cys	ser	ala	leu	leu	ser	ser	asp	ilu	thr	ala	ser	val	asn	cys	ala	lys	lys	ilu	val	ser	asp	gly	asp	gly	met	asn
Human salivary lysozyme	cys	ser	ala	leu	leu	glx	asx	asx	ilu	ala	ala	asx	val	ala	cys	ala	lys	arg	val	arg	asx	pro	—	glx	gly	ilu	arg
α-Lact.	cys	asp	lys	phe	leu	asx	asx	asx	leu	thr	asx	asx	ilu	met	cys	val	lys	lys	ilu	leu	asp	lys	—	val	gly	ilu	asn

(cys at position 80 = 64; cys at position 94 = 76)

	107	108	109	110	111	112	113	114	115	116	117	118	119	120	121	122	123	124	125	126	127	128	129
Chicken egg lysozyme	ala	trp	val	ala	trp	arg	asn	arg	cys	lys	gly	thr	asp	val	gln	ala	trp	ilu	arg	gly	cys	— arg	leu
Human salivary lysozyme	ala	trp	val	ala	trp	arg	asn	arg	asx	val	arg	gln	tyr	val	glx	—	—	—	—	gly	cys	— gly	val
α-Lact.	tyr	trp	leu	ala	his	lys	ala	leu	cys	ser	glu	lys	leu	asp	gln	trp	trp	leu	—	—	cys	glu	lys leu

(cys at position 115 = 30; cys at position 127 = 6)

FIGURE 4-57

Homology in Primary Structure Between Chicken and Human Lysozyme and Bovine α-lactalbumin, Proteins of Entirely Different Function. There does not appear to be any relationship in function between lysozyme, which hydrolyzes polysaccharides, and α-lactalbumin, which is a nonenzymatic protein found in milk. It is believed that proteins with similar primary structure have evolved from a common ancestral protein by gene duplication, followed by accumulation of different mutational changes. (Courtesy of R. E. Dickerson and I. Geis.)

on the gel. (An exception to this is the infrequent case when the polypeptide chains are held together by covalent bonds other than disulfide bonds. In such cases, it may be much more difficult to determine the chain composition of the protein.)

We shall use the term *subunit* for two or more chains held together by covalent bonds (usually —S——S— bonds). Operationally, a subunit travels as a single entity in SDS polyacrylamide gels and splits into more than one chain in SDS-mercaptoethanol gels.

An example for the terminology introduced so far is provided by insulin, which has a molecular mass of 11,466 Da and splits into two identical subunits upon the addition of SDS. Upon addition of mercaptoethanol to the SDS, each subunit splits into two nonidentical chains: α (21 amino acids) and β (30 amino acids). To describe

TABLE 4–6

The Quaternary Structure of a Few Proteins

Protein	Molecular Weight[*]	Formula[+]	Chain Weights	
Insulin	11,460	$(\alpha\beta)_2$	α	2,334
			β	3,396
α-Amylase	50,000	α_2	α	25,000
Aminoacyl transferase	186,000	α_3	α	62,000
Catalase	232,000	α_4	α	57,500
Lysine 2, 3-aminomutase	285,000	α_6	α	48,000
Yeast NAD-isocitrate dehydrogenase	300,000	α_8	α	39,000
Membrane ATP-ase	385,000	α_{12}	α	33,000
Erythrocurosin	3,000,000	α_{162}	α	18,500
Lactose synthetase	60,000	$\alpha\beta$	α	44,000
Bovine procarboxypeptidase A	88,000	$\alpha\beta_2$	α	40,000
			β	23,000
Hemoglobin	64,500	$\alpha_2\beta_2$	α	16,000
			β	16,000
Aspartokinase	116,000	$\alpha_2\beta_2$	α	47,000
			β	17,000
Histidine decarboxylase	190,000	$\alpha_5\beta_5$	α	29,500
			β	9,000
IgG immunoglobulin	150,000	$(\alpha_2\beta_2)$	α	50,000
			β	25,000
IgM immunoglobulin	900,000	$(\alpha_{10}\beta_{10}\gamma)$	α	70,000
			β	25,000
			γ	20,000
Fibrinogen	340,000	$(\alpha_2\beta_2\gamma_2)$	α	63,500
			β	56,000
			γ	47,000
Myosin	520,000	$(\alpha_2\beta\beta'\gamma_2)$	α	207,000
			β	16,000
			β'	21,000
			γ	18,000
Aspartate carbamoyl transferase	300,000	$\alpha_6\beta_6$	α	33,000
			β	17,000

Note: Modified from Klotz, Darnall, and Langerman. *The Proteins,* Vol. 1, 3d ed. Polypeptide chains that are not identical are designated by different Greek letters.

[*]Molecular weights and chain weights may be determined by different methods and therefore one may not get exactly the same values for molecular weights when one adds up the weights of the individual chains.
[+]Parentheses around chains in the formulae denote that they are held together by disulfide bonds.

the chain and subunit structure of insulin we can use the formula $(\alpha\beta)_2$ where α denotes one chain and β the other, the parentheses denote the subunit structure, and the subscript denotes the number of subunits (Table 4–6). In the case of fibrinogen, we use the formula $(\alpha_2\beta_2\gamma_2)$ because it is composed of one subunit of three dissimilar chains, each present in two copies.

A chain or a subunit can be associated with others noncovalently; the same secondary interactions that hold a single polypeptide chain in a globular conformation can hold together two chains or subunits. *Oligomer* is the term used to describe the whole protein molecule; an oligomer is composed of noncovalently associated chains and/or subunits. The oligomer may be composed of more than one identical *protomer*, which is the minimum number of chains and/or subunits that associate together. In aspartate transcarbamylase, the protomer is $\alpha\beta$ and the oligomer is $\alpha_6\beta_6$ (Table 4–6). In some cases the protomer is also the minimum structure that still retains full biological activity.

Quaternary structure thus concerns itself with the chain, subunit, and protomer structure of the protein molecule or oligomer. There is a wide range of possibilities for protein quarternary structure, but when one looks at the actual proteins found in nature, one finds that most oligomers consist of either two or four subunits and that only a very small percentage have nonidentical chains. It seems that the quaternary structure of proteins is usually fairly simple and not subject to great variation.

The Biological Function of Quaternary Structure

Few proteins have chains with molecular weights much larger than about 200,000, corresponding to about 2,000 amino acids. When a protein is larger than that, it usually is built from two or more chains. One can imagine a number of biological advantages to this strategy:

Reducing the amount of information storage in the DNA. Constructing an oligomer of four identical chains reduces the amount of information which must be stored in the DNA for this particular protein by a factor of four.

Reducing the damage caused by errors in synthesis. A simple kind of error during protein synthesis would result in the substitution in a polypeptide chain of one amino acid for another. Randomly occurring errors in protein synthesis are inevitable; we are assured of this by the second law of thermodynamics. Let us assume, therefore, that the probability of making an error (P_E) is the same for each addition of an amino acid to the polypeptide chain. The average probability of *not* making an error is then ($1 - P_E$) and we can estimate the probability of synthesizing an error-free polypeptide chain of length L as $(1 - P_E)^L$. Since P_E is approximately 1/1000 (see Chapter 13), a simple calculation will convince one that it is not very likely to synthesize error-free copies of very long (L > 1000) polypeptides. (For example, the total number of amino acids in IgM immunoglobulin is approximately 9500. If IgM were a single polypeptide chain, the probability of synthesizing an error-free copy would be $(0.999)^{9500}$, which is less than 10^{-4}.)

For this reason, if the function of a protein requires a structure that involves many amino acids there will be an advantage to constructing the protein from identical small chains. This is because each of the small chains can be synthesized error-free with an acceptable probability. The accurate chains can then combine to form an accurate protein. Furthermore, any error that does occur may prevent the erroneous chain from associating into an oligomer. A single error would in that case produce only a defective chain and not a defective protein.

Providing physiological flexibility. We shall see in Chapter 5 that the cell can construct, for instance, five slightly different enzymes of the same type (*isozymes*) by synthesizing two slightly different chains and using them in different combinations to make a four-chain oligomer. These enzymes presumably play different physiological roles in different tissues of the same organism.

Providing the opportunity for cooperative effects. Many different protein molecules, such as the oxygen-carrying hemoglobin or enzymes whose activity can be regulated, show cooperative interactions which adapt them to their biological roles. These cooperative properties appear to be related to their oligomeric structure, being caused by changes in the interactions between chains.

Providing additional molecular diversity for evolution. Combining two or more different chains, each with its own biological function, into an oligomer with new properties may well provide important opportunities for enzyme evolution. We can imagine, for instance, that a regulatable enzyme might arise through a new combination of an inhibitor chain with a catalytic chain. This could occur by a mutation in the catalytic chain that allows it to associate with an inhibitor chain already being used in conjunction with another enzyme. If the additional control over the catalytic chain activity has selective advantage, the mutation will become fixed. The independent evolution of control by the catalytic chain alone would require many more mutations.

In the remaining portion of this section, we shall illustrate some of these advantages by discussing an important nonenzyme protein molecule.

The Structure and Function of Hemoglobin

Hemoglobin, the oxygen carrier of mammalian blood, has functional properties that we can now relate to changes in its quaternary structure. The knowledge that made this possible was worked out by M. Perutz and his colleagues who labored for three decades to provide the first detailed description and explanation of the behavior of this important protein. Hemoglobin has a molecular mass of 64,500 Da and consists of two α-chains and two β-chains, differing slightly in their amino acid sequence.

In early studies, Perutz and his co-workers used a number of isomorphous replacements to work out, by x-ray diffraction methods, the structure of hemoglobin to a resolution of 0.55 nm (5.5 Å). To everyone's great surprise, the conformation of each of the four hemoglobin chains closely resembles that of the single myoglobin chain, a fact which allowed the hemoglobin workers to draw upon the results of the myoglobin structure. Even more surprising was the fact that the four hemoglobin chains fit closely together to form a compact, almost spherical, structure (6.4 × 5.5 × 5.0 nm; Figure 4–58).

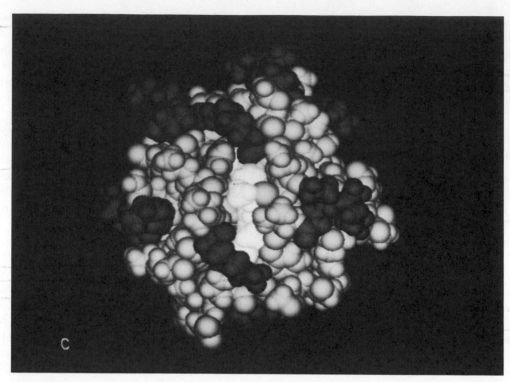

FIGURE 4–58

Computer-drawn Space-filling Model of Hemoglobin Showing How the Four Chains Fit Together to Form an Almost Spherical Molecule. The model also reveals the snug fit of the two heme groups, shown in contrasting shades. The $\alpha_1\beta_1$ protomer is light and the $\alpha_2\beta_2$ protomer is dark. (Courtesy of R. J. Feldman.)

Since the functional properties of hemoglobin are different from those of myoglobin, despite the conformational similarity of the individual chains, the quaternary structure must play a significant role in explaining the differences.

The hemoglobin molecule contains four pockets, open to the outside, in which the four heme groups fit snugly (Figures 4–58 and 4–59). The four chains of hemoglobin can, with some difficulty, be dissociated and reassociated by varying the pH and the ionic concentration of the surrounding medium. Clearly, one must be able to explain in terms of the amino acid sequence of these chains why four chains of hemoglobin associate and why four very similar chains of myoglobin do not.

The precise nature of the interaction between chains within the hemoglobin molecule became evident when the x-ray analysis was refined to 0.28 nm (2.8 Å) resolution. It turned out that there is very little interaction between like chains (i.e., α and α, or β and β). On the other hand, two types of close association are observed between the unlike chains, namely those between chains with neighboring hemes ($\alpha_1\beta_2$ or $\alpha_2\beta_1$) and those

between chains with widely separated hemes ($\alpha_1\beta_1$ and $\alpha_2\beta_2$) (Figure 4–59). As might be expected, the interactions are of two kinds: a few hydrogen bonds and charged-group interactions, and a much larger number of hydrophobic interactions.

The $\alpha_1\beta_1$ and $\alpha_2\beta_2$ interactions, involving 34 amino acid side chain contacts, are more extensive than those of $\alpha_1\beta_2$ and $\alpha_2\beta_1$, involving only 19. If one compares these interacting side chains with the residues in the same position in the myoglobin molecule, one finds that a great many of them are polar in myoglobin and nonpolar in hemoglobin. The nonpolar residues of hemoglobin are more stable when they interact with each rather than with the aqueous solvent, which accounts for the observed association of polypeptide chains in hemoglobin.

Let us now consider the special functional properties of the hemoglobin molecule and see if we can explain them in terms of its tertiary and quaternary structure. Hemoglobin is an oxygen *carrier* protein while myoglobin is an oxygen *storage* protein. Hemoglobin is adapted to its carrier function because: (1) it can pick up or release oxygen readily at partial pressures of oxygen normally

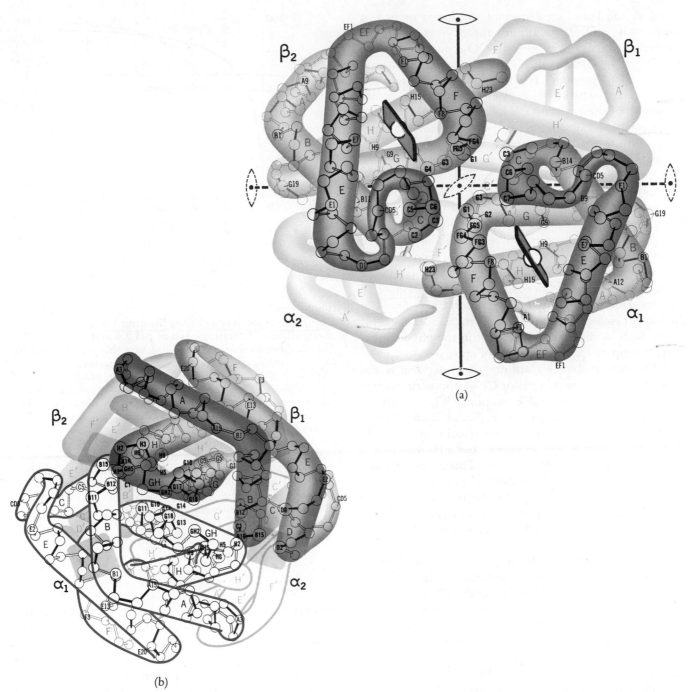

(a)

(b)

FIGURE 4-59

The Four Chains of Oxyhemoglobin.
The views are arbitrarily designated as
"front" and "right side." Only the α
carbons of the main chains are shown.
Those α carbons whose side chains are
involved in contacts between subunits
are given boldfaced numbers in large
circles. (a) The front view figure
shows the $\alpha_1\beta_2$ contacts. (b) The side
view figure shows the $\alpha_1\beta_1$ contacts.
(Courtesy of R. E. Dickerson and I.
Geis.)

found in the blood capillaries of the lungs or tissues, respectively; and (2) its affinity for oxygen changes in response to variations in pH and CO_2 concentration.

Figure 4–60 shows the O_2-binding curves for hemoglobin and myoglobin. The myoglobin curve is *hyperbolic,* a shape which we shall encounter again in our discussion of enzyme action, while the hemoglobin curve is *sigmoidal* (S-shaped). At low pO_2, hemoglobin has a lower affinity for oxygen than myoglobin, as seen by comparing the slopes of their two O_2-binding curves. The low O_2 affinity of hemoglobin allows it to release oxygen to cellular metabolism before O_2 concentrations become too low. Hemoglobin will also release O_2 to myoglobin.

At the partial pressures of O_2 found in peripheral blood capillaries, the O_2-binding curve for hemoglobin is relatively steep. This means that the affinity of the hemoglobin molecule for oxygen is very sensitive to oxygen concentration at those concentrations found in the blood capillaries. Hemoglobin is thus a sort of *oxygen buffer,* with its dissociation constant in the physiological range (compare Figure 4–60 with Figure 3–A).

The oxygen-binding affinity of myoglobin is not affected by the changes in pH or CO_2 concentrations that occur under physiological conditions. The affinity of hemoglobin for oxygen, however, is profoundly affected by small changes in pH and CO_2 concentration. As shown in Figure 4–61, lowering the pH reduces the affinity of hemoglobin for oxygen. This is reflected in a shift of the oxygen dissocation curve to the right. Lowering the pH will bring about a release of O_2 even while the oxygen concentration remains constant. Oxygen affinity is also lowered by increasing CO_2 concentration at constant pH.

These properties of hemoglobin are central to its physiological function. In the alveolar capillaries of the lungs, the pH is relatively high and the CO_2 concentration relatively low. Under these conditions the hemoglobin has a higher O_2 affinity and is thus very efficient at picking up oxygen. When the hemoglobin is then transported to a rapidly metabolizing tissue such as muscle, where the pH in the capillaries is lower and CO_2 concentration is higher, the oxygen affinity of the molecule falls, causing it to release more of its oxygen.

The influence of pH and CO_2 on the affinity of hemoglobin for O_2 is called the *Bohr effect.* Both the sigmoidal shape of the O_2-binding curve and the Bohr effect make hemoglobin a very efficient oxygen carrier under physiological conditions. We shall now attempt to explain these properties in terms of the tertiary and quaternary structure of hemoglobin.

The sigmoidal shape of the O_2-binding curve of hemoglobin reflects the combined oxygen-binding behavior of the four heme groups. The increasing slope of

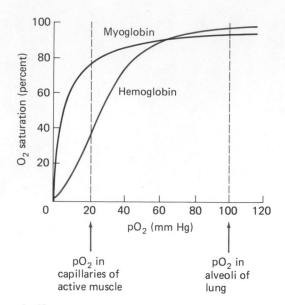

FIGURE **4–60**

Comparison Between the O_2 Binding Curves of Myoglobin and Hemoglobin at pH 7.4 and 38°C.
The equilibrium fraction of binding sites that are occupied by O_2 is plotted against increasing oxygen concentration (expressed as the partial pressure of oxygen, pO_2). Myoglobin is fully saturated at low oxygen concentrations, a property which is appropriate for its oxygen storage function. To carry out its transport function, hemoglobin must be able to pick up or release oxygen readily at the oxygen concentrations found in lungs or body tissues, respectively, and respond with great sensitivity to changes in oxygen concentration. The steepest portion of the hemoglobin curve (where the most effective release occurs) lies at an oxygen concentration just above that of capillaries in active muscle. (Courtesy of F. Daniels and R. A. Alberty.)

the curve at low pO_2 shows that the first oxygen is bound with a relatively low affinity, the next oxygen with a higher affinity, and so on. (The last oxygen is bound with the highest affinity, but the leveling off of the oxygen-binding curve—due to saturation of the hemoglobin—obscures this. Only at low oxygen concentrations is the slope of the binding curve proportional to the affinity of the binding.) The *slope* of the binding curve increases, at low pO_2, as though adding oxygen to hemoglobin were proportional to $[O_2]^n$, where the exponent n has a value greater than one. This implies either that more than one oxygen is involved in a binding event or that an already bound oxygen influences subsequent binding. The latter interpretation, that an occupied binding site can somehow cooperate to increase binding affinity at an unoccupied site, gives rise to the notion of a *positive cooperative effect.*

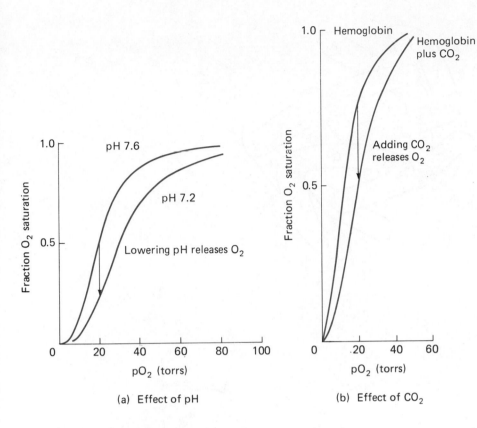

(a) Effect of pH   (b) Effect of CO_2

FIGURE **4–61**

The Effect of pH and CO_2 on the Oxygen-carrying Capacity of Hemoglobin. Oxygen concentration is measured by its partial pressure, expressed in torrs; one atmosphere pressure is 760 torrs. (a) A lowering of pH at constant oxygen concentration brings about the release of oxygen (Bohr effect). This is an adaptive property of hemoglobin: during muscle activity the pH of muscle blood drops, helping the hemoglobin to release needed oxygen. No such effect is observed in myoglobin. (Courtesy of J. V. Kilmartin.) (b) An increase of CO_2 at constant oxygen concentration brings about a release of oxygen. This is also an adaptive property of hemoglobin because during muscle activity the CO_2 concentration increases, thereby releasing oxygen to the muscle. Here again, no such effect is observed in myoglobin.

From a classical point of view in chemistry, this is a very unusual type of behavior. In the chlorination of methane, for example, each succeeding chlorine atom is added with *increasing difficulty*. We might therefore suspect some very unusual chemistry to be involved in the sigmoidal curve for hemoglobin. And there is!

The heme groups of the hemoglobin molecule are quite far from each other (2.5–4.0 nm). In order for the binding of O_2 at one heme group to affect the affinity for O_2 of another group, some sort of signal must be transmitted, since chemical forces cannot act directly over such large distances. The signal turns out to be a *conformational change* in the molecule.

That such a conformational change must occur seemed clear as long ago as 1937 when F. Haurowitz added oxygen to crystals of *deoxyhemoglobin* (hemoglobin lacking any O_2) and observed that they shatter. It has also been known for some time that reduced horse hemoglobin crystals, which have the form of hexagonal plates, change their crystal structure to needles when they become oxygenated. The oxygenation causes conformational changes of such magnitude that the crystal structure established by deoxyhemoglobin cannot accommodate the *oxyhemoglobin* molecule (hemoglobin with O_2). And, indeed, low resolution x-ray studies by Perutz demonstrated that major changes in the quaternary structure of the hemoglobin occur upon oxygenation.

The most obvious change is in the relative distances between some of the heme groups; the largest is between the hemes of the two β-chains, where their separation is increased by as much as 0.65 nm. This movement is achieved by sliding and rotational changes between the four chains of the molecule. Of special interest is the relative motion of the two types of contact between unlike chains: the motion between α_1 and β_1 (and between α_2 and β_2) is small because of the large number of "packing" contacts holding them together (Figure 4–62); the $\alpha_1\beta_2$ and $\alpha_2\beta_1$ "sliding" contacts are shifted enough to break some hydrogen bond pairs and make others. As Figure 4–62 demonstrates, the major motion of the chains, relative to each other, occurs as a rotation of the $\alpha_1\beta_1$ dimer with respect to the $\alpha_2\beta_2$ dimer.

But how then does oxygen binding at a particular heme group bring about sliding and rotational changes in the other chains? The answer must lie in *changes in the tertiary structure which bring about changes in the quaternary structure*. To obtain evidence for this, Perutz and his co-workers had to study oxy- and deoxy-hemoglobin crystals at a much higher resolution.

A number of high-resolution x-ray crystallographic studies have been carried out on deoxyhemoglobin and on hemoglobin with bound oxygen or other ligands. A

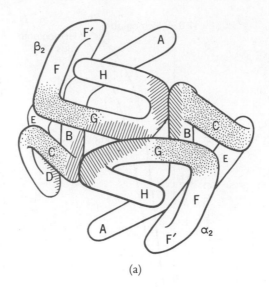

(a)

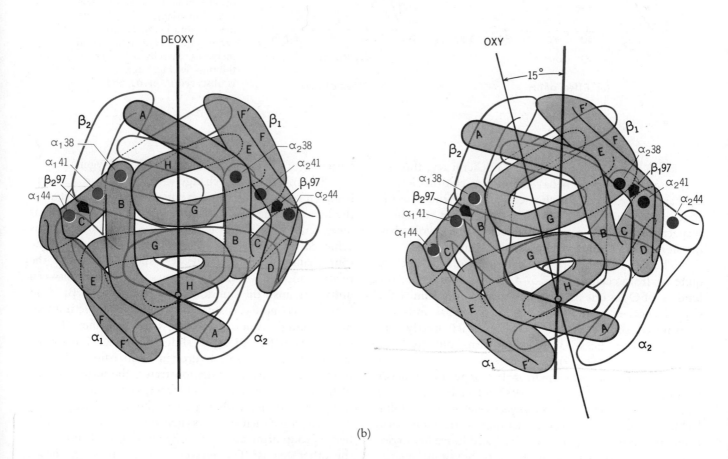

(b)

FIGURE 4–62

Subunit Motion in Hemoglobin Upon Oxygen Binding. (a) Side views of $\alpha_2\beta_2$ dimer showing *packing contacts* (cross-hatching) holding the two chains of the subunit together, and *sliding contacts* (stippling) representing surfaces involved in the rotational sliding motion between the $\alpha_2\beta_2$ subunit and the $\alpha_1\beta_1$ subunit.
(b) Sliding motion between the two subunits $\alpha_1\beta_1$ and $\alpha_2\beta_2$. The position of $\beta97$ moves from between $\alpha44$ and $\alpha41$ to between $\alpha41$ and $\alpha38$. (Courtesy of I. Geis.)

detailed picture has emerged, confirmed by independent physical evidence, which describes hemoglobin's little "levers, hooks, knobs, and pulleys." These mechanics relate the binding of the first oxygen molecule to the conformational changes that account for the cooperative properties of oxygen binding. The molecular dynamics of these changes are beautifully described in *Hemoglobin* by R. E. Dickerson and I. Geis (see Selected Readings). Briefly, histidine F8, which is bound to the iron atom (Figure 4–46), is tilted away from the perpendicular in the deoxy state; upon binding of oxygen, the histidine assumes a perpendicular position (Figure 4–63). In addition to this, the heme changes from a domed to a more planar configuration, an adjustment that allows a reduction in strain between heme, histidine, oxygen, and valine FG5 (see region marked "S" in Figure 4–63).

Figure 4–64 shows the conformational changes in the α subunit, particularly as it affects the F helix; similar changes can be demonstrated in the β subunit. These changes in subunit structure, subtle though they may appear, influence the hydrogen bonds and electrostatic interactions between subunits. The resultant rotation is shown in Figure 4–62. This rotation in turn has an influence on the heme groups that still lack oxygen, bringing about in them conditions more conducive to oxygen binding.

Much is also known about the molecular effects of other bound ligands on the binding of oxygen to hemoglobin. These ligands include not only the protons and CO_2 of the Bohr effect, but also chloride ions and *di*phosphoglycerate (DPG), all of which play important physiological roles in controlling oxygen transport by the blood of mammals. The binding of each of these ligands causes conformational changes in the subunits of hemoglobin. These changes have the effect of favoring the deoxy state—the conformation adopted by hemoglobin in the absence of oxygen. As this state has a low affinity

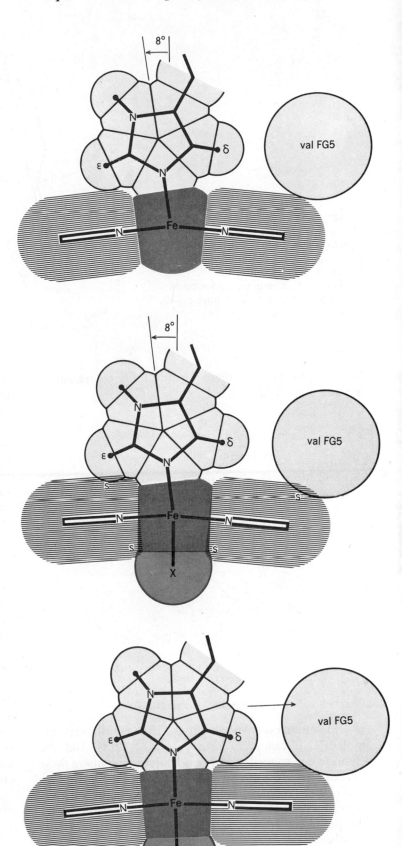

FIGURE 4–63

The Effect of O_2 Binding on the Heme and Its Immediate Environment. The figures show the space occupied by the heme group (shaded), histidine F8 (above), and valine FG5. (a) The heme in deoxyhemoglobin is domed, the histidine is tilted, the Fe^{2+} is out of the plane of the heme, and the Val FG5 side chain is in contact with the heme. (b) Upon O_2 binding, steric strain is generated in the region marked "s." (c) Steric strain is relieved by a decrease in the domed shape of the heme, movement of the Fe closer to the heme plane, and a change in the tilt of histidine. (Courtesy of I. Geis.)

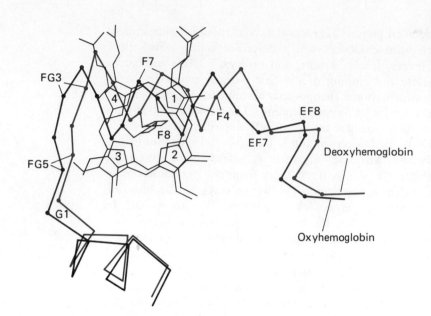

FIGURE 4–64

Helix and Heme in the α-subunit of Hemoglobin Following Oxygen Binding. The bold black lines represent the position of the oxyhemoglobin structure. Motion in the β-subunit is similar. (Courtesy of I. Geis.)

for oxygen, binding these ligands leads to release of oxygen.

Diphosphoglycerate (DPG)

In summary, hemoglobin is a marvelously versatile molecule; it can carry protons, CO_2, DPG, and the chloride ion, as well as oxygen. Moreover, binding these ligands affects the conformation of hemoglobin and vice versa. When oxygen binds in the lungs, protons and CO_2 are dissociated and when protons and CO_2 are bound in the peripheral tissues, oxygen is dissociated more efficiently.

We should not leave the discussion of hemoglobin without mentioning the substantial contributions made to our understanding of structure's relation to function by various mutant forms of the hemoglobin molecule. The mutations giving rise to these hemoglobins have arisen spontaneously in the human population over evolutionary time. They are discovered in patients who present themselves to hematology clinics throughout the world. In fact, most of these abnormal hemoglobins are named for their sites of discovery: Kansas, Köln, Sydney, Yakima, etc. Most of the hundreds that have been studied are due to a single amino acid substitution in the primary structure of either the α or the β chain of the molecule.

Some of these amino acid substitutions cause serious clinical problems because they tend to "lock" the hemoglobin molecule in either the deoxygenated or the oxygenated form. The position of these critical amino acids in the quaternary structure of hemoglobin gave important clues to the way in which the structure normally changes when oxygen is bound. The general principle can be imagined easily when one realizes that the kind of hydrogen bonds that can be made by an aspartyl residue, for example, cannot be made by a histidyl residue. Substitution of one for the other at the contact region between the β_2 and α_1 chains, which occurs in Hemoglobin Yakima, would destabilize the form of hemoglobin in which the hydrogen bond normally occurs (in this case, the deoxygenated form). As a result, patients with Hbg Yakima are ruddy with very highly oxygenated hemoglobin, which has a high oxygen affinity and from which it is more difficult to release oxygen to the tissues.

One reason for devoting so much space to describing hemoglobin is because it is the best understood example of a general principle which is of the utmost importance to understanding cellular phenomena. Proteins are not

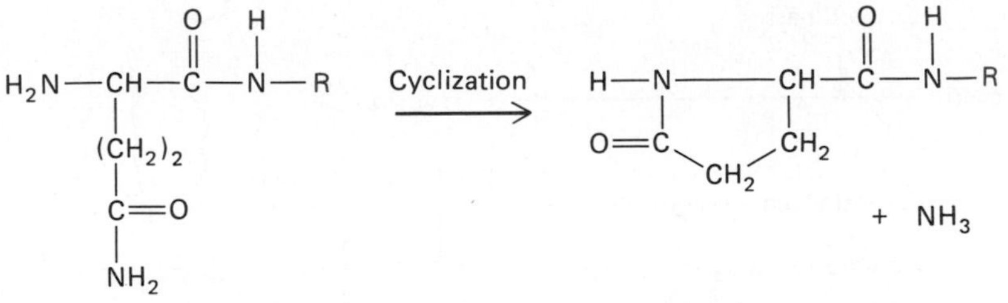

$$
\begin{array}{c}
\text{H}_2\text{N}-\text{CH}-\overset{\overset{\displaystyle O}{\|}}{\text{C}}-\overset{\overset{\displaystyle H}{|}}{\text{N}}-\text{R} \\
| \\
(\text{CH}_2)_2 \\
| \\
\text{C}=\text{O} \\
| \\
\text{NH}_2
\end{array}
\quad\xrightarrow{\text{Cyclization}}\quad
\begin{array}{c}
\text{H}-\text{N}-\text{CH}-\overset{\overset{\displaystyle O}{\|}}{\text{C}}-\overset{\overset{\displaystyle H}{|}}{\text{N}}-\text{R} \\
| \qquad | \\
\text{O}=\text{C} \quad \text{CH}_2 \\
\backslash \quad / \\
\text{CH}_2
\end{array}
\quad + \ \text{NH}_3
$$

A peptide with glutamine in the N-terminal position

A peptide with pyrrolidone carboxylic acid (pyroglutamic acid) in the N-terminal position

FIGURE 4–65

A Posttranslational Modification of an Amino Acid. N-terminal glutamine is cyclized to pyrrolidone carboxylic acid.

merely catalysts, they are also *sensors;* that is, they can perceive some aspect of the state of the environment and respond to it in a highly precise manner. Thus, hemoglobin can react with O_2, hydrogen ions, and CO_2, and respond appropriately by changing its conformation and thereby modifying its affinity for oxygen. The cell physiology of the future will concern itself in considerable measure with the conformational changes within, and interactions between, protein molecules. The story of the cooperativity among hemoglobin subunits is just a beginning.

Protein Modification—A Way to Alter Structure and Function

The biosynthesis of proteins might be thought to involve enzyme-mediated formation of covalent bonds only at the level of primary structure (i.e., the synthesis of the polypeptide chain). After that point, spontaneous processes such as the formation of polar interactions, nonpolar interactions, and disulfide bonds and disulfide exchange would take over. Many proteins, however, do not achieve biological activity until they interact with one or more enzymes which bring about some modification in the covalent primary structure of the protein. These changes are known as *post-translational modifications,* because they occur after the process of translation by which proteins are synthesized (see Chapter 13).

There is a large variety of biological functions which depend on post-translational modifications and the examples we cite will illustrate only a few of them, grouped according to the following categories: modification of translated structure; removal of part of translated structure; and addition to translated structure.

Modification of Protein Structures

We have already encountered the 20 amino acids which are coded in DNA and translated from RNA. Occasionally one finds some "rare" amino acids in proteins; these are not coded differently but instead owe their origin to enzyme-catalyzed modifications of one of the 20 amino acids used to synthesize the protein molecule. Glutamine, for example, when present at the N-terminal position, often becomes cyclized enzymatically to form pyrrolidone carboxylic acid (Figure 4–65). (When this occurs it is difficult to establish the chain composition of a protein by using N-terminal labels such as Sanger's reagent because those labels won't react with the cyclized form.) Two other examples of unusual amino acids derived from such modifications are hydroxyproline and hydroxylysine, commonly found in the collagen of connective tissue. The precise functions of each of these covalent modifications in the structure of a given protein are not yet understood, although hydroxylysine is known to be involved in the crosslinking of collagen chains in fibrous tissues (see Chapter 17).

Another mechanism of post-translational protein modification is the hydrolysis of specific peptide bonds by the action of certain proteases. There are a number of inactive enzyme precursors, or *zymogens,* which are activated by proteolytic enzymes. We have already mentioned in Chapter 2 how a zymogen is synthesized in the endoplasmic reticulum of acinar cells of the pancreas, and transported through the Golgi complex to form zymogen granules. These granules eventually discharge their contents into the pancreatic duct, whence the zymogens are transported to the duodenum where activation occurs. The enzyme enterokinase converts trypsinogen to trypsin, which then works on the other zymogens, chymotrypsinogen, proelastase, and procarboxypeptidase, converting them, respectively, to chymotrypsin, elastase, and carboxypeptidase (Figure 4–66).

The situation is more complex than this, however, since the activated enzyme can sometimes act on its own zymogen, as in the case of chymotrypsin. The zymogen chymotrypsinogen is a single chain molecule composed of 245 residues and 5 disulfide bonds. Trypsin specifically breaks one bond between Arg 15 and Ile 16, thereby generating full chymotrypsin activity. The short 15-residue peptide is not, however, liberated by the action of trypsin since it is attached to the major chain by a

FIGURE 4–66

The Activation of Pancreatic Enzymes for Digestion of Proteins in the Duodenum. The activation is coordinated by the proteolytic enzyme enterokinase which converts trypsinogen to trypsin. Trypsin then acts to convert all four zymogens (including trypsinogen) into the respective active enzymes.

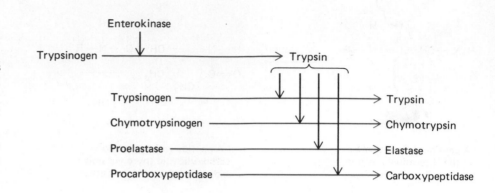

disulfide bond (Figure 4–67). Curiously, activated chymotrypsin then works on other molecules like itself and splits three more bonds, but this does not increase the activity of the molecule.

Removal of Peptides

Numerous cases have been discovered in which an inactive protein precursor is converted to an active molecule by the removal, through limited proteolysis, of one or more pieces of polypeptide chains. Examples of these include the removal of special sequences ("signal peptides") that allow proteins to penetrate membranes (discussed in Chapter 13), the conversion of inactive precursors into hormones (e.g., insulin), or the conversion of soluble precursors into insoluble structures (e.g., fibrin or T4 bacteriophage; see Chapter 15).

The case of insulin has an interesting history. Because insulin was the first protein to have its primary structure determined, it became fixed in our minds as the archetypical protein. It was therefore very puzzling to find that when insulin is unfolded and its two chains separated, the resulting mixture cannot to any significant extent be reconstituted into the native molecule. This was

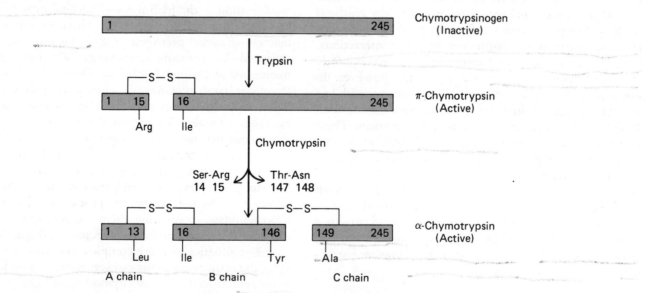

FIGURE 4–67

The Proteolytic Processing of Chymotrypsinogen. The zymogen is converted to the enzymatically active π-chymotrypsin by trypsin. Chymotrypsin then acts on itself to break three more peptide bonds, liberating Ser-Arg and Thr-Asn. The remaining three polypeptide chains of the resulting, and equally active, α-chymotrypsin remain associated, being held together at least by two interchain disulfide bonds.

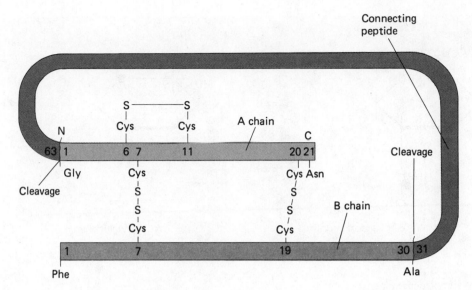

FIGURE 4-68

The Conversion of Proinsulin to the Active Hormone Insulin by the Proteolytic Removal of a Large Connecting Peptide.

an apparent exception to the frequent observation that the native conformation is the most probable (lowest free energy) conformation. The difficulty was resolved when it was discovered that insulin arises from **processing** a larger protein precursor, *proinsulin* (Figure 4-68).

The proinsulin molecule is a single-chain molecule, 84 residues long and containing three intrachain disulfide bonds. After synthesis and folding in the pancreas, a proteolytic enzyme cleaves specifically two peptide bonds, converting the molecule into the two-chain structure that Sanger first worked out, with two interchain disulfide bonds and one intrachain disulfide bond. In proinsulin molecules from all the various species studied so far, the 33-residue-long peptide thus removed has two positively charged residues (Arg-Arg and Lys-Arg) at either end of the chain, which suggests that this is a requirement for the proteolytic enzyme involved in activating proinsulin.

Additions to Protein Structures

There are numerous instances in which new functional groups of various dimensions are added to a protein by enzymatically controlled reactions. For instance casein, the primary protein in milk, is a *phosphoprotein* in which seryl residues are esterified by the addition of phosphate. The phosphorylation of proteins has become, in recent years, a very actively investigated area because it plays an important regulatory role in cellular processes. The most ubiquitous group of proteins to be modified by covalently linking to them another group of compounds, however, are the *glycoproteins*.

As techniques for identifying carbohydrates improved, it became increasingly clear that a very large number of proteins have carbohydrates linked to them. Most of the proteins of blood plasma, the ubiquitous structural protein collagen, protein constitutents of membranes, blood group substances, and many soluble enzymes are also glycoproteins. As our understanding of cell surfaces has progressed, we have become aware that glycoproteins play important roles in cell immunity, cell interaction and recognition, and possibly even in the initiation of cell division. We shall therefore consider these interesting protein conjugates briefly. A fuller discussion is given in Chapter 17.

First a few words about structure. There are only five amino acid residues to which the sugars are normally linked. They are asparagine, threonine, and serine, as well as the two modified residues, hydroxyproline and hydroxylysine. Of the many species of carbohydrates found in nature only seven hexoses and two pentoses are utilized in glycoproteins (Figure 4-69 shows some examples of the types of linkages found). Nevertheless, the number of structural possibilities offered by the utilization of glycoproteins is enormous. This is not only because the glycoproteins can form chains, both branched and linear, of various dimensions—the sugar content can vary from 1% (ovalbumin) to 85% (blood group substances)—but also because the exact composition of the carbohydrate moiety of a specific protein is not

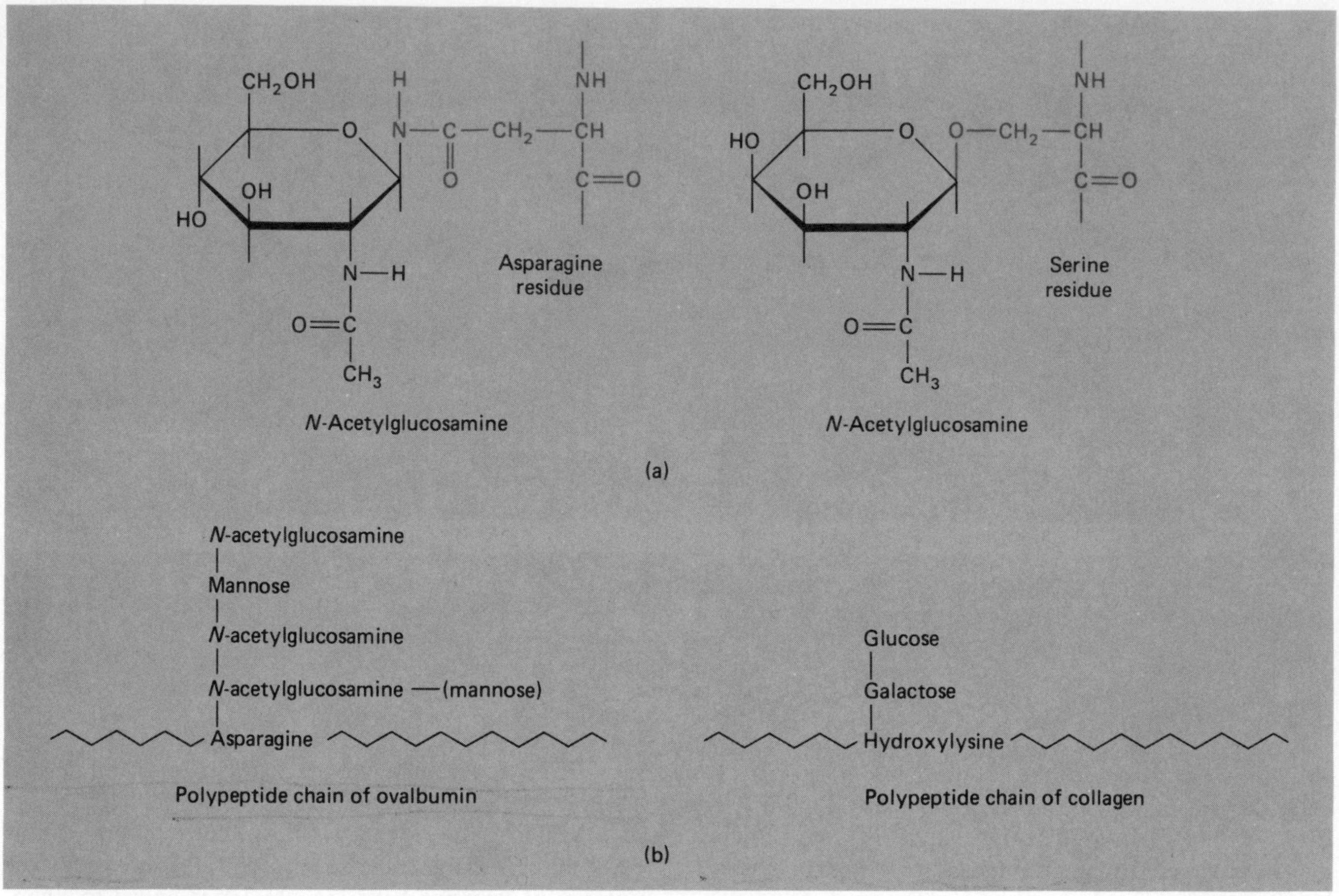

FIGURE 4-69

Examples of Glycoprotein Structures.   (a) The types of linkages found
between N-acetylglucosamine and asparaginyl or seryl residues. (b) Example of the
structure of the oligosaccharide part of two glycoproteins.

perfectly defined. Careful studies of a single glycoprotein
have demonstrated that there is some variation in the
number of amino acid residues to which the carbohydrate
is linked, and in the precise structure and composition of
the carbohydrate moiety. Unlike the protein portion,
which is synthesized first and under direct genetic
control, the carbohydrate part of the molecule is added
subsequently by a series of enzymatic reactions (see
Chapter 13). It is therefore only under indirect genetic
control and so is subject to some statistical variation.

SUMMARY

We have come a long way in this chapter in
providing a molecular basis for understanding the dynamics of the cell. We shall have to discuss genetics and
nucleic acids before we understand how proteins are
made. But to understand proteins is to understand
practically everything else about the cell, because proteins
are the building blocks, the catalysts, the regulators, the
sensors, the transmitters, and the transducers of the cell.
In a sense, the rest of this book is a study of all the things
proteins do.

KEY WORDS

Antibody, antigen, C-terminal end, clone, conformation, denature, density gradient, dialysis, ligand, N-terminal
end, primary structure, processing, quaternary structure, renature, secondary structure, sedimentation coefficient,
sequencing, sucrose gradient, tertiary structure, ultracentrifuge.

PROBLEMS

1. The pK_A values of amino acid side chains that dissociate protons are: Asp 3.9, Glu 4.3, His 6.0, Cys 8.3, Lys 10.8, Tyr 10.9, and Arg 12.5.
   (a) List the amino acid residues which contribute to the charge on a protein at pH 6.0.
   (b) If the isoelectric point of a protein is 6, what would be the net charge on the protein at pH 7?
   (c) During electrophoresis, toward which electrode would a basic protein (isoelectric point 8.5) migrate in a buffer maintained at pH 7?
   (d) Which amino acid residue would contribute to a change in net charge on a protein if its surrounding buffer were changed from pH 7.5 to 6.5?

2. Dowex 50, a resin frequently used to separate amino acids by chromatography, is made of polystyrene that contains hydrophobic benzene groups and negatively charged sulfonic acid ($-SO_2^-$) groups. A mixture of amino acids is placed on a column of Dowex 50 at pH 3.5 and eluted from the column with a buffer rising in concentration and pH. Give the order in which the following amino acids elute from the column and the reason for the order: Ala, Arg, Asp, Glu, Lys, Ser, Val.

3. Poly-L-glutamate is a synthetic peptide made only of Glu residues. It forms an α-helix at pH 3 and a random coil at pH 5. Explain this change in conformation.

4. A protein is cleaved by partial acid hydrolysis into a number of peptide fragments which are separated from each other by chromatography. Each peptide is reacted with fluorodinitrobenzene, after which the peptide is hydrolyzed and its N-terminal amino acid and amino acid composition are determined. The results are expressed below by listing the dinitrophenylated residue to the left and the other amino acids inside the parentheses, in alphabetical order. What is the amino acid sequence of the original protein?

Cys (Ala, Ser)	Cys (Ser)
Glu (Cys, Gln)	Gln (Cys)
Gly (Ile, Val)	Gln (Leu)
Leu (Asn, Glu)	Glu (Asn)
Leu (Gln, Tyr)	Gly (Ile)
Tyr (Asn, Cys)	Leu (Tyr)
Val (Cys, Ser)	Ser (Leu)
Asn (Tyr)	Ser (Val)
Cys (Ala)	Tyr (Cys)
Cys (Cys)	Val (Glu)

   (More information is provided than necessary to establish a unique sequence. The additional information can be used to test the sequence you have determined.)

5. In the following polypeptide, how many fragments would one obtain when using:
   (a) trypsin digestion
   (b) treatment with cyanogen bromide
   (c) reduction with β-mercaptoethanol
   (d) trypsin digestion and reduction with β-mercaptoethanol
   (e) treatment with cyanogen bromide and reduction with β-mercaptoethanol
   (f) all three treatments.

   ```
   N-Asn-Gly-Ala-Arg-Tyr-Cys-Asp-Met-Asp-Lys
                         |              |
                         S              |
                         |              |
                         S              |
                         |              |
   Val-Met-Ile-Lys-Phe-Cys-Ala-Met-Thr-Gln
   ```

6. Name the two structures depicted below and identify the atoms of which each is composed.

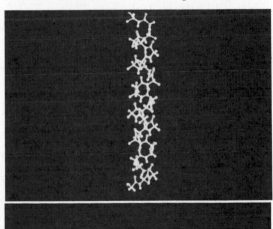

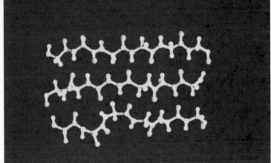

7. Of the following amino acids, which residues would you expect to find inside a folded protein and which at the surface?

Thr	Met	Glu	Gln
Asp	Ile	Ser	Arg
Lys	Val	His	Tyr

8. A certain protein is a dimer of two identical units which originated from a monomeric protein during the course of evolution. What kind of amino acid substitutions would you expect to have taken place to make possible the evolution from the monomer to the dimer?

Monomer form   Dimer form

9. If it takes 30 minutes to sediment a 20S molecule halfway down a centrifuge tube, how long will it take a molecule of 10S if you spin it twice as rapidly?

10. Angiotensin (a hormone that regulates blood pressure) from the horse has a unique amino acid sequence which you can deduce from the data given below. The peptides are tabulated with the N-terminal amino acids to the left and other amino acids separated by dashes. If the other amino acids in the peptide are of unknown sequence they are listed in alphabetical order within parentheses.
   (a) Peptide fragments resulting from trypsin digestion:
       Pro - Phe - His
       Leu - (Leu, Ser, Tyr, Val)
       Asp - Arg
       Val - Tyr - Ile - His
   (b) Peptide fragments resulting from chymotrypsin digestion:
       Ile - His - Pro - Phe
       Asp - (Arg, Tyr, Val)
       Ser
       His - Leu - Leu - Val - Tyr

11. Melanocyte-stimulating hormones from cows, pigs, horses, sheep, and monkeys all have the same amino acid sequence, which you are asked to deduce from the data given below. The sequences are written as usual with the N-terminal amino acid to the left. Amino acids whose position is not determined are listed alphabetically, separated by commas and enclosed by parentheses. (The native hormones have both the N-terminal amino and the C-terminal carboxyl blocked by covalently attached groups. These are assumed to have been removed in the data given below.)
   (a) Peptide fragments resulting from trypsin digestion:
       Pro - Val
       Ser - (Arg, Glu, His, Met, Phe, Ser, Tyr)
       Trp - (Gly, Lys)
   (b) Peptide fragments resulting from chymotrypsin digestion:
       Arg - Trp
       Gly - (Lys, Pro, Val)
       Ser - (Glu, His, Met, Phe)
       Ser - Tyr
   (c) Order of release of amino acids when hormone is digested with carboxypeptidase: Val first; Pro second.
   (d) Among the peptides released by acid hydrolysis:
       Met - Glu - His
       Ser - Tyr - Ser

12. Below are listed the amino acid sequences of four examples of fibrinopeptide A, from human, dog, rabbit, and cat. (Fibrinogen is a plasma protein of 340 kDa and contains three different polypeptide chains: A, B, and C. During blood clotting, the enzyme thrombin cleaves off fibrinopeptides A and B from the N-terminus of the A and B chains, respectively; polymerization of the resultant fibrin follows.) Fill in the table using your knowledge of protein structure.

Fibrinopeptide A Sequences

Human: Ala- Asp-Ser- Gly-Glu-Gly-Asp-Phe-Leu-Ala- Glu-Gly-Gly-Gly-Val- Arg
Dog: Thr-Asn-Ser- Lys-Glu-Gly-Glu- Phe-Ile- Ala- Glu-Gly-Gly-Gly-Val- Arg
Rabbit: Val- Asp-Pro-Gly-Glu-Ser- Thr-Phe-Ile- Asp-Glu-Gly-Ala- Thr-Gly-Arg
Cat: Gly-Asp-Val-Gln-Glu-Gly-Glu- Phe-Ile- Ala- Glu-Gly-Gly-Gly-Val- Arg

Result of experiment, or property	*Human*	*Dog*	*Rabbit*	*Cat*
(a) Rate of migration towards anode (positive electrode) during electrophoresis at pH 7. (#1 = fastest; #2 = next fastest, etc.)	____	____	____	____
(b) Rate of migration during chromatography in which the most hydrophobic peptide moves fastest. (#1 = fastest; #2 = next fastest, etc.)	____	____	____	____
(c) Likelihood of having α-helical structure in N-terminal half of peptide. (#1 = most likely; #2 = next most . . .)	____	____	____	____
(d) Makes most H-bonds with solvent water. (#1 = most H-bonds; #2 next most . . .)	____	____	____	____
(e) Number of subpeptides resulting from complete trypsin digestion.	____	____	____	____
(f) Number of subpeptides resulting from complete chymotrypsin digestion.	____	____	____	____

SELECTED READINGS

Anfinsen, C. B. (1973) Principles that govern the folding of protein chains. *Science* 181:223–30.

Dickerson, R. E., (1972) The structure and history of an ancient protein. *Scientific American* 226(4):58–72.

Dickerson, R. E., and Geis, I. (1983) *Hemoglobin: Structure, function, evolution and pathology.* Menlo Park: Benjamin/Cummings.

Dickerson, R. E., and Geis, I. (1982) *Proteins: Structure, function and evolution.* Menlo Park: Benjamin/Cummings.

Doolittle, R. F. (1985) Proteins. *Scientific American* 253:88–99.

Karplus, M., and McCammon, J. A. (1986) The dynamics of proteins. *Scientific American* 254:42–51.

Kendrew, J. (1961) The three-dimensional structure of a protein molecule. *Scientific American* 205:96–111.

Moore, S., and Stein, W. H. (1973) Chemical structures of pancreatic ribonuclease and deoxyribonuclease. *Science* 180:458–64.

Perutz, M. Haemoglobin: Genetic abnormalities. *New Scientist and Science Journal,* 24 June 1971, pp. 762–65.

Perutz, M. Haemoglobin: The molecular lung. *New Scientist and Science Journal,* 17 June 1971, pp. 676–79.

Enzyme Catalysis—
The Mechanism of
Biological Chemistry

5

Most organic compounds of the cell are remarkably stable at physiological temperatures, pressures, and hydrogen ion concentrations, as we pointed out in Chapter 3. If urea, for instance, is dissolved in water at standard conditions it will not react with its solvent at an appreciable rate even though the reaction would release a considerable amount of free energy.

$$\begin{array}{c} H_2N \\ \diagdown \\ C{=}O + H_2O \rightleftarrows CO_2 + 2\,NH_3 \qquad \Delta G' = -57\ kJ/mole^{[1]} \\ \diagup \\ H_2N \end{array}$$

In 1935, H. Eyring proposed that the urea and water do not react rapidly because the reaction has to pass through an *activated complex,* the formation of which takes a great deal of free energy (Figure 5–1). In order to form the activated complex, water and urea molecules must collide with a certain minimum amount of energy. As Figure 5–2 shows, only an infinitesimal fraction of the relevant molecules at room temperature have at least this minimum amount of free energy. When the temperature is raised, a larger proportion of molecules has this minimal free energy, and the rate of the reaction increases correspondingly. This is what an organic chemist does in the laboratory (along with varying the pressure and utilizing extremes of pH) in order to speed up the rate of organic reactions. The cell, however, carries out its reactions at mild temperatures, low pressures, and pH values close to neutrality, yet the reactions often proceed at considerable rates.

These remarkable properties of the cell are of course due to the specific biological catalyses carried out by enzymes. Speeding up the chemical reactions in the cell, however, is only half of the important role of enzymes. The other half is the regulation or control of the reactions. As we pointed out previously, this regulation is accomplished through *selectivity*—the cell can control which reactions are speeded up, and by how much. It is therefore

---

[1]The standard conditions signified by the prime (') in $\Delta G'$ are: one atmosphere pressure, 300 K, pH 7, and all solute concentrations (except H^+ ions) at 1.0 molal.

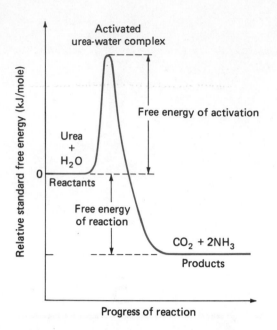

FIGURE 5–1
Energy Diagram of the Hydrolysis of Urea. The free energy of the urea-water system is plotted at various stages during the hydrolysis reaction. In order to react, a molecule of urea and water must have sufficient energy (the *activation energy*) to form an *activated complex*—a combination of urea and water that is capable of the hydrolysis reaction. At room temperature very few molecules have sufficient energy to form this complex. As a result, the rate of the reaction is exceedingly slow, even though the formation of products would result in a large negative free energy change.

essential that uncatalyzed organic reactions be so terribly sluggish; if they were not, it is hard to see how the cell could regulate them. We can see then that *the reactions going on inside cells represent a truly biological chemistry,* one that exists and is controlled by virtue of the presence and activity of specific enzyme catalysts.

Enzymes: Protein Catalysts of Cells

The concept of catalysis was proposed by the Swedish chemist J. J. Berzelius in 1853 to explain the speeding up of chemical reactions not only by acids, but also by extracts obtained from living tissues. He compared, for instance, the effect of sulfuric acid on the hydrolysis of starch with the much greater effect of an equal weight of the enzyme diastase. Berzelius even suggested that the so-called "vital force" of organisms resided in the unusual property of biological catalysts which could greatly speed up chemical reactions—a concept to which many of his contemporaries (including the great chemist J. F. von Liebig) objected fervently. Precise measurements of reaction rates by J. H. van't Hoff and W. Ostwald finally convinced most chemists that extracts from tissues could indeed increase the rate of certain chemical reactions, and led to the definition of catalysis as *speeding up a reaction without effect on the overall result of the reaction.*

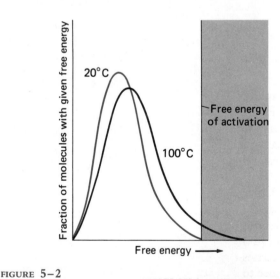

FIGURE 5–2
Free Energy Distribution of Urea in Water at Two Temperatures. The fraction of urea molecules with a certain free energy is graphed against the value of the free energy. Most urea molecules do not have sufficient free energy to react with water. The fraction with free energy at least as high as the activation energy is represented by the area under the curve in the shaded region. At 100°C a much higher fraction of molecules has free energies lying in the shaded region. This explains the considerable acceleration that raising the temperature has on the rate of this reaction.

Enzymes As Catalysts

This notion was at first thought to imply that the catalyst does not enter into the reaction, but subsequent work has made it clear that the catalyst does interact with the substrate and is regenerated at the end in its original form. This enables it to go through many cycles of reaction, and explains why it is effective in such very low concentrations. The reaction occurs at a special location on the enzyme, known as the *active site*. It is in binding substrate to the active site that selectivity is exerted. This binding also allows the catalytic activity of the enzyme to be expressed. The *enzyme-substrate activated complex* has a free energy of formation lower than that of the activated intermediates of the uncatalyzed reaction, thus permitting a larger number of molecules to react per unit of time at a given temperature (Figure 5–3).

It should be noted that since there are usually so few catalyst molecules, compared to the number of substrate molecules, the relative amount of substrate in catalyst-substrate intermediates is very small and so has essentially no effect on the equilibrium of the reaction. Catalysts are not thermodynamic magicians, capable of pushing reactions "uphill." They do not shift equilibria, but merely increase the *rate* at which the equilibrium is reached. In the case of urease—the enzyme that catalyzes the reaction of urea with water—the rate of the enzymatic reaction can be speeded up by a factor of 10^{14} over the uncatalyzed reaction. We call the absolute rate in optimum conditions the *turnover number* of the enzyme, the number of reactions catalyzed by the enzyme per second. Urease has a turnover number of 10^4/sec: one enzyme molecule will catalyze the hydrolysis of 10,000 urea molecules per second.

Biological catalysis by enzymes obeys the same general rules observed for nonenzymatic catalysis. That is, enzymes:

1. are effective in very small concentrations.
2. are unchanged by the reaction (when we compare their initial and final states).
3. exert their physiological effects by changing the rate at which equilibrium is reached, not by changing the equilibrium of the reaction.

Enzymes differ from other forms of catalysts chiefly in their greater effectiveness in lowering the activation energy barrier (see Figure 5–3) and in the extraordinary degree of specificity that they exhibit. Herein lies a clue to the delicate control of cell processes that enzymes exert. *By regulating the supply and activity of tiny amounts of highly specific enzymes, the cell can control the metabolic flux of the compounds within it.* But in order to understand their regulatory capacity we must first study the general properties of enzymes.

The Protein Nature of Most Enzymes

The general procedure we have just outlined was employed by W. H. Sumner, who in 1926 was the first to crystallize an enzyme—urease. This achievement was especially remarkable because it took place at a time when an atmosphere of mystery and awe surrounded the phenomenon of biological catalysis. The crystals of urease that Sumner obtained proved to be made of

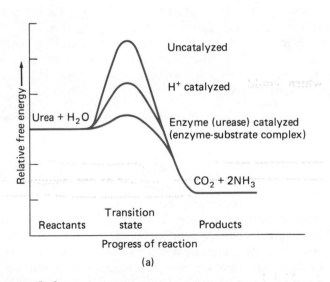

(a)

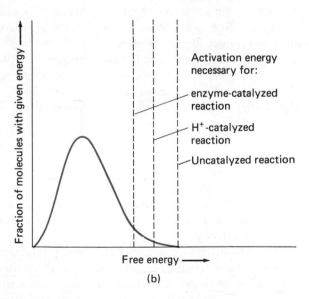

(b)

FIGURE 5–3
The Effect of Enzyme Catalysis on the Hydrolysis of Urea. (a) Free energy diagram. Enzyme catalysis is more effective than H^+ catalysis: the free energy of activation in the enzymatic reaction is much lower. (b) Free energy distribution of urea in water. The activation energy required for each case in panel (a) is shown. Lowering the activation energy requirement speeds up the rate, dramatically during enzyme catalysis, because a greater fraction of the urea molecules has sufficient energy to form the activated complex and can, therefore, react.

The Enzyme Assay

Modern biochemistry owes its success to a biological phenomenon that living systems have utilized from their inception; namely, the specificity of enzyme action. This property permits us to mix a substrate with enzyme preparations that might contain 1000 times as much impurity as the enzyme and obtain a quantitative assay of the amount of the particular enzyme present. The reason for this is that the enzyme can ignore all the extraneous compounds and catalyze specifically the reactions of its substrate. Equally important, its substrate is ignored by all the other enzymes present since they are specific for their substrates. For example, it is possible to grind up some jack beans and assay the enzyme urease found inside by adding this crude preparation to purified urea. The hydrolysis of the urea can be followed by measuring the production of ammonia with an appropriate chemical test. Figure 5–A, in which the amount of product formed is plotted against time, depicts the course of this reaction. A *tangent to the curve at zero time* has a *slope* which corresponds to the *initial rate of the reaction*.

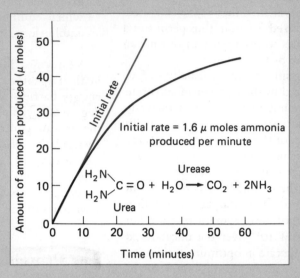

Initial rate = 1.6 μ moles ammonia produced per minute

$$H_2N\!-\!C\!=\!O + H_2O \xrightarrow{\text{Urease}} CO_2 + 2NH_3$$

Urea

FIGURE 5–A

A Typical Enzyme Assay. The product, ammonia, formed by the action of urease on urea and water, is plotted against the time of reaction. The initial rate is measured by drawing a tangent to the curve starting at zero time.

This initial rate is determined in part by the enzyme concentration. As we have described in Chapter 4, to estimate the purity of an enzyme preparation, *the initial rate is divided by the total amount of protein present*. This measure is called the *specific activity,* and it increases as successive steps of purification provide a cleaner enzyme preparation. When the specific activity cannot be increased beyond a certain limit, even though a wide range of purification procedures has been employed, it is provisionally assumed that the enzyme is pure.

An enzyme assay is not an absolute indicator of enzyme activity. The initial rate depends on a number of parameters, such as substrate concentration, pH, ionic strength, the kind of buffer used, and the temperature. Care must be taken that all of these are specified, and especially that the substrate concentration is high enough not to limit the measured activity.

protein, but owing to the prestige of the great chemist R. Willstätter, who opposed the notion that enzymes were "mere" proteins, it took a great deal of further experimentation before proteins were generally believed to have enzyme activity.

Over 500 different enzymes have been crystallized, and a great many more have been purified; all of them were found to be proteins. Perhaps the most convincing proof for the protein nature of enzymes is the often-repeated experiment of treating an enzyme with a protein-digesting enzyme (protease). The loss of enzyme activity parallels the disappearance of protein. It is possible also to demonstrate a parallelism between protein denaturation, as caused by a number of agents, and loss of enzyme activity. Since the same conditions which

affect protein structure also affect enzyme action, it has generally been concluded that enzymes are proteins. In spite of this, occasional claims are made that certain preparations containing no protein have nevertheless some enzymatic activity. Even the classic Watson and Crick article (see Appendix) on the structure of DNA contained the suggestion that DNA might carry the catalytic function for its own replication. Until very recently, none of these claims was fully confirmed or substantiated. Even though there may be one or two exceptions (we shall report in Chapter 12 on the behavior of catalysts made of RNA), we now have accumulated sufficient experience to be certain most of the enzymatic activity inside cells is due to proteins. Because of that experience and because it is not yet possible to describe the structures of other macromolecules in the same detail as proteins, we shall discuss catalysis here in terms of protein enzymes only.

The most important consequence of the protein nature of enzymes is that *the synthesis of the biological catalysts is under direct genetic control.* From this fact follows the uniquely biological property of the cell: *metabolism*—that is, cellular function—*can evolve.*

The fact that enzymes are proteins has a number of additional consequences for their properties. One of these is the effect of temperature on enzyme action. Figure 5–4 shows that this effect has two components: at low temperatures there is an increase, caused by the effect of temperature on a chemical reaction (see Figure 5–2), but at higher temperatures there is a steeply falling decrease, which is due to the effect of temperature in denaturing the enzyme. The marked effect of temperature on denaturation is a unique and characteristic property of proteins and nucleic acids, in part because their highly ordered structure "melts"—that is, becomes disordered relatively suddenly and throughout the whole molecule. It is therefore important that enzyme assays be carried out at temperatures at which no appreciably irreversible inactivation of the enzyme occurs.

Protein enzymes are polyvalent ions; their net charge depends on various groups that donate or accept protons. Enzyme activity is therefore highly pH-dependent; the pH at which maximum activity occurs varies considerably from one enzyme to another (Figure 5–5). The dependence of enzyme activity on pH is due to a number of effects, such as the dissociation of protons from the

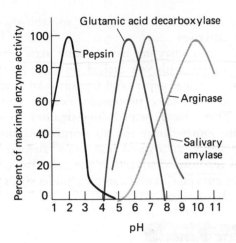

FIGURE 5–5

Effect of pH on the Activity of a Number of Enzymes.   The effect is a complex phenomenon, often exerted on the substrate as well as the enzyme. Digestive enzyme pepsin has an optimum activity at pH 2, consistent with the high activity of this enzyme in the very acidic stomach juice. (Redrawn with permission from J. S. Fruton and S. Simmons, *General Biochemistry,* New York, John Wiley & Sons, Inc., 1953.)

substrate, and from the side chains on the enzyme that directly interact with the substrate during catalysis. There is also, as we have discussed in Chapter 4, an effect of pH on the overall conformation of the enzyme molecule.

For a number of different reasons enzymes are not always found in an active state. Some enzymes, among them the proteolytic (protein-digesting) variety, are often found in the cell as inactive precursors, or *zymogens.* Trypsinogen, for instance, is inactive but can be activated to trypsin by a number of enzymes, including trypsin itself. The activation involves the removal of a short peptide, followed by some structural changes in the protein molecule that finally yield the active trypsin enzyme (see the discussion in Chapter 4). Other enzymes, as we have already pointed out in Chapter 3, require for activity certain divalent cations such as Ca^{2+}, Zn^{2+}, Mn^{2+}, Mg^{2+}, or Co^{2+}. Another category of enzymes requires a relatively small organic compound as a *coenzyme.* The divalent cation or the coenzyme, being small, can usually be separated from the enzyme by dialysis. Even when it is an organic molecule, a coenzyme is usually stable in boiling water, while the *apoenzyme*—the protein portion of the active enzyme—is inactivated by heat. Dialyzability and heat stability are therefore the usual tests for a coenzyme. Some enzymes, however, contain a nonprotein portion that is sufficiently attached that it cannot be removed by dialysis. This nonprotein portion, called the *prosthetic group,* plays an important role

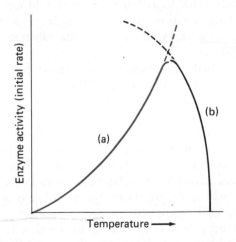

FIGURE 5–4

The Effect of Temperature on Enzyme Activity.   The observed curve is a composite of two processes: (a) is the exponential effect of temperature on an enzyme-catalyzed reaction; (b) is the effect of temperature on the denaturation of the enzyme. The abruptness of this latter curve is due to the very high temperature dependence of protein denaturation.

in the interaction between enzyme and substrate, as we shall presently see.

In the past, enzymes have usually been named after their substrates with -*ase* tacked on (though there are exceptions, such as pepsin and trypsin). An international commission has since proposed rules for naming enzymes. These have helped investigators to create a consistent nomenclature for the burgeoning number of enzymes being discovered and draw attention to the *classes* of reactions catalyzed in the cell. Table 5–1 lists groups of enzymes, with examples, under this system of enzyme classification.

Enzyme Kinetics

In general, the progress of a chemical reaction is followed by assaying either the disappearance of one or another of the reactants or the appearance of product(s). Since the reactions are usually done in a solution of constant volume, the change in concentration of reactant [R] or product [P] is the variable observed. A typical reaction might involve the combination of two reactants to give two products:

$$R_1 + R_2 \underset{k_2}{\overset{k_1}{\rightleftarrows}} P_1 + P_2.$$

We expect as the reaction progresses to see the product concentration increase with time—that is, [P] at time t_2 will be greater than [P] at an earlier time t_1. The rate of the reaction (usually symbolized as v, for velocity) is measured as the increase in [P] over the time interval $t_2 - t_1$:

$$v = \frac{[P(t_2)] - [P(t_1)]}{t_2 - t_1} = \frac{\Delta[P]}{\Delta t}.$$

The symbol Δ has been used to designate an increase in product concentration, $\Delta[P]$, or an increase in time, Δt.

It has been found experimentally that at the beginning of the reaction, the *initial velocity* (v_i), measured as the rate of product formation, is

$$v_i = k_1 [R_1] [R_2]. \qquad (5-1)$$

That is, the initial velocity or rate of the forward reaction is proportional to the concentration of each of the reactants. The constant of proportionality k_1 is called the *forward rate constant*.

It is reasonable that the forward rate of this bimolecular reaction should be proportional to the concentration of both reactants. To react, two reactant molecules must collide; that is, they must be in the same small volume in the same small interval of time. The concentrations are a measure of the probability that a molecule will be in a particular place at a particular time. The higher the concentration of R_1, the higher is the probability that, during a certain interval of time, a certain

small volume will contain a molecule of R_1, and likewise for R_2. Hence, the probability of collision is proportional to the product of the concentrations of R_1 and R_2.

At later times the concentration of products will no longer be negligible and we must consider also the backward reaction:

$$v = k_1 [R_1] [R_2] - k_2 [P_1] [P_2]. \qquad (5-2)$$

The constant of proportionality k_2 is called the *backward rate constant*. Since the velocity by definition is measured as an *increase* in product concentration, the backward reaction rate has a negative sign. At equilibrium, of course, $v = 0$ and the forward and backward reactions have equal rates. Since the equilibrium constant is defined as

$$K_{eq} = \frac{[P_1] [P_2]}{[R_1] [R_2]},$$

it is easy to see that the equilibrium constant is equal to k_1/k_2.

The Enzyme-Substrate Complex

Let us now consider a simpler case of a chemical reaction, one in which there is only one reactant. For uncatalyzed reactions of this type (an isomerization, for example) the forward rate is proportional only to the concentration of the reactant. The same reaction, however, may be catalyzed by the presence of an enzyme (e.g., an *isomerase*). In this case, the reactant is the substrate of the enzyme and we shall symbolize it as S. By analogy with Equation 5–1, we should expect $v_i = k_1 [S]$. This equation predicts a simple proportionality between the initial velocity v_i and the substrate concentration [S].

The usual way rates are measured for reactions catalyzed by enzymes is to add varying amounts of substrate to a constant amount of an enzyme preparation. If the initial rate of such a reaction is measured at a number of different substrate concentrations (Figure 5–6a) and if these initial rates of reaction are then plotted against substrate concentration, a curve like the one depicted in Figure 5–6b is obtained. Instead of increasing continuously as the substrate concentration is increased, the value of v_i approaches a *limiting maximum velocity*, V_{max}. Another way of describing the result is to say that at high substrate concentrations, the rate of the reaction becomes limited by something other than substrate. To explain this observation, V. Henri (1902) was the first to suggest that *the enzyme and substrate combine with each other*. At high substrate concentrations, therefore, the rate of the enzyme-catalyzed reaction becomes limited by the enzyme concentration, which is held constant in the various reactions. L. Michaelis and M. Menten (1913) were able to validate Henri's hunch in quantitative terms.

TABLE 5–1

Classification of Enzymes

Class of Enzyme

Substrate	Reaction Catalyzed	Specific Example

1. Oxido-reductases
  1.1 alcohol R—OH
	Oxidation-reduction	1.1.1.1 alcohol
		              dehydrogenase
		1.1.1.27 lactate
		              dehydrogenase

1.2 carbonyl R_1—C—R_2
                      $\parallel$
                      O
		1.2.1.12 glyceraldehyde-
		              phosphate
		              dehydrogenase
		1.2.3 luciferase

1.4 amine R—NH_2		1.4.3.2 L-amino acid oxidase
2. Transferases	Transfers of functional groups	

2.1 one-carbon groups		2.1 aspartyl
		              transcarbamoyl
		              synthetase
		              (aspartyl
		              transcarbamoylase:
		              ATCase)

2.3 acyl groups		2.3.1.6 choline acetyl transferase
2.4 glycosyl groups		2.4.1.1 phosphorylase A
2.6 amine groups		2.6.1.1 glutamic oxaloace- tic transaminase
2.7 phosphate groups		2.7.7.7 DNA polymerase
3. Hydrolases	Hydrolysis	

O
                   $\parallel$
  3.1 esters R—C—O—R'
		3.1.1.7 acetyl
		              cholinesterase

3.2 glycosidic bonds		3.1.4.22 ribonuclease 3.2.1.23 β-galactosidase

O
                   $\parallel$
  3.4 peptide bonds R—C—NH—R'
		3.4.1.21 carboxypeptidase A

3.6 acid anhydrides		3.6.1.1 inorganic pyrophosphatase

4. Lyases
  4.1 —C≡C— lyases
	Addition to double bonds	4.1.1.31 phosphoenolpyr-
		              uvate carboxylase
		4.1.2.13 aldolase

4.2 —C≡O lyases		4.2.1.2 fumarase

5. Isomerases
  5.1 racemases,
  epimerases
	Isomerization	5.1.3.1 ribulose-5-
		              phosphate
		              epimerase

5.3 intramolecular
  oxidoreductases
		5.3.1.1 triose-phosphate
		              isomerase

6. Ligases
  6.1 C—O bonds
	ATP-linked syntheses	6.1.1 amino acid–RNA
		              ligase (tRNA-acti-
		              vating enzymes)

6.2 C—S bonds		6.2.1.4 succinyl-Co A synthetase
6.3 C—N bonds		6.3.4.2 CTP synthetase

6.4 C—C bonds		6.4.1.3 propionyl-Co A
		              (CO_2 ligase)
		              carboxylase

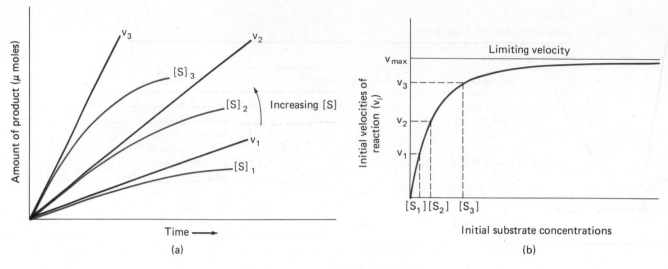

FIGURE **5–6**
The Kinetic Properties of Most Enzyme-Catalyzed Reactions. (a) The time course of product formation. Three reactions are shown, each with a different initial substrate concentration ($[S]_1$, $[S]_2$, $[S]_3$). The slope of the curve of product formation at early times is the initial velocity of the reaction (v_1, v_2, v_3). Higher substrate concentrations yield a higher initial rate of product formation. (b) The initial velocities measured at different substrate concentrations plotted against [S]. This curve has the shape of a rectangular hyperbola. The limiting velocity (V_{max}) is the initial velocity at very high substrate concentrations.

The Michaelis–Menten Equation

We shall model the involvement of the enzyme in the reaction by the following scheme:

$$E + S \underset{k_2}{\overset{k_1}{\rightleftharpoons}} E \cdot S \overset{k_3}{\longrightarrow} E + P. \qquad (5\text{–}3)$$

We can see from the graph in Figure 5–6a that early in the course of the reaction the rate of product formation is essentially constant. Since according to our model this rate is proportional only to the concentration of the enzyme-substrate complex, ES, we conclude that [ES] must also be constant. (Keep in mind that we are considering only the initial stage of the reaction, when there is so little product that it cannot react to any significant extent with the free enzyme, so that the reaction

$$ES \overset{k_3}{\longrightarrow} E + P$$

proceeds only in the forward direction.) One can also see that the higher the substrate concentration, the higher the initial rate of product formation, which means a higher value of [ES].

The unvarying level of [ES] is not the result of an equilibrium. Rather, it reflects the fact that ES is in a *steady state* (see Chapter 1). That is, the rate of formation of ES from E and S is equal to the sum of the rates of destruction of ES into E + S and E + P. We can write an equation for the rate of increase of ES,

$$\frac{\Delta[ES]}{\Delta t} = k_1 \cdot [E] \cdot [S] - k_2 \cdot [ES] - k_3 \cdot [ES] \qquad (5\text{–}4)$$

in which the positive first term represents the production of ES by the reaction of E + S (with rate constant k_1) and the two negative terms represent the destruction of ES to form E + S (with rate constant k_2) or to form E + P (with rate constant k_3). In the steady state, [ES] is unchanging, and so, $\Delta[ES]/\Delta t = 0$. The steady state will therefore occur when

$$k_1 \cdot [E] \cdot [S] = k_2 \cdot [ES] + k_3 \cdot [ES]. \qquad (5\text{–}5)$$

Besides the requirement that [ES] be in a steady state, there is another constraint on the variables of this formulation. Because of *conservation of mass, the total amount of E (E_t) must remain constant,* although the relative amounts of free E and enzyme-substrate complex ES can vary:

$$[E] + [ES] = \text{constant} = [E]_t.$$

Thus

$$[E] = [E]_t - [ES], \qquad (5\text{–}6)$$

and we can substitute this expression into Equation 5–5 for the steady state condition:

$$k_1 \cdot [S] \cdot ([E]_t - [ES]) = (k_2 + k_3) \cdot [ES].$$

This equation can then be solved for [ES]:

$$[ES] = \frac{k_1\,[S]\,[E]_t}{(k_2 + k_3) + k_1\,[S]} = \frac{[S]\,[E]_t}{\left[\dfrac{k_2 + k_3}{k_1}\right] + [S]}.$$

The first term in the denominator of the right-most fraction is composed only of rate constants and is given the special name *Michaelis constant,* K_m, to honor one of the first biochemists to develop this analysis. Note that K_m does not depend on concentrations; it measures an intrinsic property of the interaction between enzyme and substrate.

The rate of product formation is $k_3\,[ES]$, and so

$$v_i = \frac{k_3\,[E_t]\,[S]}{K_m + [S]} \qquad (5\text{–}7)$$

where the subscript in v_i reminds us that this equation is appropriate for *initial velocities,* when there is very little product (so we can ignore the reverse reaction), and when the concentration of substrate has not decreased significantly.

By examining Figure 5–6b, we see that at high substrate concentrations v_i reaches a limiting maximum value called V_{max}. Since $v_i = k_3\,[ES]$, $V_{max} = k_3\,[ES]_{max}$, which must be when essentially all the enzyme is in the ES form—that is, when $[ES] = [E]_t$. Replacing $k_3[E]_t$ in Equation 5–7 with V_{max}, we find the final form of the Michaelis–Menten equation:

$$v_i = \frac{V_{max}\,[S]}{K_m + [S]}. \qquad (5\text{–}8)$$

EXAMPLE 5–1

If the K_m for enzyme X-ase is 50 μM, at what substrate concentration will the initial velocity of the catalyzed reaction be 20% of V_{max}?

Solution

Assume the enzyme follows simple Michaelis–Menten kinetics:

$$v_i = \frac{V_{max}\,[S]}{K_m + [S]}.$$

Substituting the values given, $v_i = 0.20\,V_{max}$ and $K_m = 50\ \mu$M,

$$0.20\,V_{max} = \frac{V_{max}\,[S]}{50\ \mu\text{M} + [S]}.$$

After dividing both sides of the equation by V_{max} and multiplying both sides by 50 μM + [S],

$$(0.20 \times 50\ \mu\text{M}) + (0.20\,[S]) = [S].$$

Solving for [S] yields 10 μM = $(1.0 - 0.20)\,[S] = 0.80\,[S]$ or [S] = 13 μM. (Note: The equation evaluates to [S] = 12.5 μM, but the data given are appropriate for a precision of only two significant figures.)

Experimental Tests of the Michaelis-Menten Equation

If one plots the Michaelis-Menten equation, one obtains a curve (Figure 5–7) that has the same shape—a rectangular hyperbola—as the curve obtained experimentally for the dependence of initial reaction rate on substrate concentration (Figure 5–6b). Since the derivation of this relationship is based on assuming the formation of an ES complex, the agreement between the experimental and theoretical curves constitutes good, if indirect, evidence for the formation of an enzyme-substrate complex.

For enzymes which absorb ultraviolet or visible light, more direct evidence of an enzyme-substrate complex can be obtained by studying changes in their absorption spectra when the enzymes interact with their substrates. H. Theorell, B. Chance, and their co-workers perfected a *rapid mixing and continuous flow* technique (Figure 5–8) that enables one to measure the absorption spectra of colored enzymes, catalase, or peroxidase, even though the lifetime of the ES complexes is in the

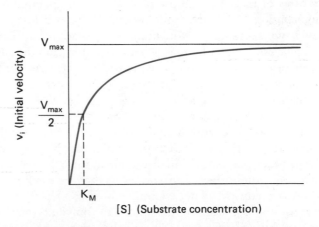

FIGURE 5–7

Curve Obtained by Plotting the Michaelis-Menten Equation. This curve is a rectangular hyperbola of the same general shape as the curve obtained experimentally in Figure 5–6b. By substituting ½ V_{max} for v_i in the Michaelis-Menten equation, one obtains $K_m = [S]$. Thus, the substrate concentration giving $v_i = V_{max}/2$ is numerically equal to the value of K_m.

Equilibrium and Steady State

Because they are so important as models, we wish to discuss again the differences between equilibrium and steady state. The complexity of cellular function has frustrated most attempts at perfect quantitative description. Cell biologists nevertheless make progress by the use of simplified conceptual models which try to account for complex observations. One quantitative model in common use is that of *equilibrium*, whose operational definition for our purposes can be taken as a condition in which, even though change is possible, nothing is observed to change no matter how long the period of observation.

An equilibrium results from a balance of forces. In the case of a chemical equilibrium, the opposing forces are represented by the tendency to form products and the reverse tendency to form reactants. At equilibrium these opposing tendencies are exactly balanced and the result is that the macroscopic concentrations of reactants and products are unchanging. At the microscopic level, we speak of this situation as a *dynamic equilibrium;* there is actually a finite rate of forming products, but it is balanced by the rate of the reverse process of forming reactants.

This can be observed by adding to a reaction at chemical equilibrium some radioactively labeled reactant (or product), but such a small amount that the equilibrium concentrations are not perturbed. The dynamic nature of the equilibrium can then be proved by isolating the radioactively labeled product (or, more dramatically for the reverse case, reactant) from the reaction. The labeling can occur because *the reaction pathway still exists at equilibrium*. (Were this not the case, it would be inappropriate to describe the system as being in equilibrium.) The mixing of the radioactivity throughout the population of molecules can be viewed as a spontaneous process of the type predicted by the second law of thermodynamics. It is the thermal motion of molecules that allows the collisions necessary to form some product, on the one hand, and some reactant, on the other. Thermal motion, of course, does not stop when equilibrium is reached.

In a steady state the observables also do not change, but this constancy depends on a *steady flow of material or energy,* or both, through the system. An example is the apparently smooth constant surface of a placid river. The height of the river does not vary, but only because the water flowing out of a particular location is exactly matched by water flowing in. Contrast this with the smooth surface of a pond in equilibrium on a humid day. Water is lost from the pond by evaporation but is replaced by condensation, and the surface level is constant. In the case of the pond, the water evaporates and condenses at equal rates; *the two processes are exact reversals of one another*. In the case of the river, water flows in by one path and flows out by a completely independent path; *the two processes are not the reverse of one another* (Figure 1–A).

The flowing river is analogous to our description of the steady concentration of ES; it is due to equal rates of formation (from E + S) and destruction (by breaking down either to form E + S or to form E + P). Since substrates are flowing through the ES complex on the way to becoming the product, the substrate concentration must eventually fall and the state will no longer be steady. In general, steady states do not remain unchanging forever since they depend essentially on a source of matter and/or energy to provide the net flow.

Because *living systems are open systems,* exchanging matter and energy with their environments, we are sure that the steady state will always be a more realistic model than equilibrium. Nevertheless, the equilibrium model may provide useful numbers with a relatively simple calculation. Therefore we shall often use the equilibrium approximation. As an example, we have seen that the steady state assumption gives a $K_m = [(k_2 + k_3)/k_1]$. It is possible to interpret the formation of the enzyme-substrate complex by assuming that it is in equilibrium with enzyme and substrate. This assumption leads to an equation similar in form to Equation 5–8, but with a different definition for K_m: that is, k_2/k_1. The K_m given by the steady state assumption will become the same as that given by the equilibrium assumption, under conditions of k_2 being much greater than k_3. Chapter 6 contains other examples; the concentrations of small molecules in the cell are assumed to be in equilibrium for the purpose of calculating changes in free energy. Although metabolism is at least as complicated as a steady state, the assumption of equilibrium gives useful values for comparing the changes in free energy associated with various metabolic reactions.

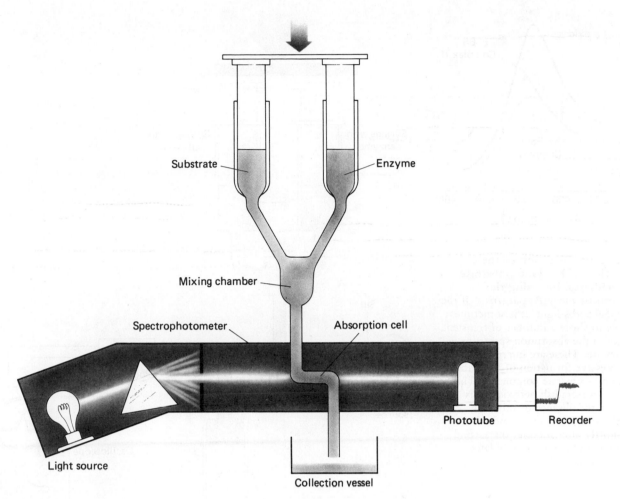

FIGURE 5–8
Continuous Flow Apparatus for Following Rapid Reactions. The enzyme and substrate, in separate syringes, are forced by pressure into the mixing chamber to begin the reaction. The reacting substances flow through a cell where changes in the absorption of light are measured and recorded by a spectrophotometer. The flow rate and the volume of tubing between the mixing chamber and the absorption cell determine the earliest time interval that can be investigated. The length of the interval is proportional to the length of the flow cell. A precisely controlled flow rate ensures measuring a well-defined interval.

millisecond range. They found in a number of instances that more than one ES intermediate occurs; in the case of peroxidase, for instance, the brown enzyme first turns green because of the formation of ES_1 and then turns red from the formation of ES_2 (Figure 5–9). Thus, one must write for the reaction scheme

$$E + S \rightleftharpoons ES_1 \rightleftharpoons ES_2 \rightleftharpoons E + P$$

and so on for additional intermediates if necessary. There is also good evidence for the existence of enzymes in at least two conformations: an "open" one from which reactants can be released readily and a "closed" one from which reactant release is very slow. A more realistic minimum mechanism would thus be:

$$E_{open} + S \rightleftharpoons (ES)_{open} \rightleftharpoons (ES)_{closed} \rightleftharpoons$$
$$(EP)_{closed} \rightleftharpoons (EP)_{open} \rightleftharpoons E_{open} + P.$$

It turns out that even when more than one intermediate is formed, Michaelis-Menten kinetics still apply.

A more convenient way to measure fast-changing chemical events is the _stopped flow_ method, which rapidly mixes enzyme and substrate and then observes the reaction after the flow has been stopped (Figure 5–10). This method has the advantage of requiring less careful regulation of the flow rate; it also uses smaller quantities of material, without any loss of resolution.

An ingenious method for measuring very fast reactions uses very rapid temperature changes ("jumps"). In

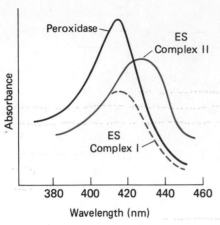

FIGURE 5–9

Direct Demonstration of the Formation of Enzyme-Substrate Intermediates, by Using the Continuous Flow Apparatus. If the enzyme absorbs light, it is sometimes possible to show a number of transient changes in the absorption spectrum of the enzyme. These are interpreted as the successive formation of a number of enzyme-substrate complexes. The enzyme, horseradish peroxidase, is brown in the absence of substrate. The first ES complex is green and can be seen shortly after mixing; the second one is red and is visible only later. (After B. Chance.)

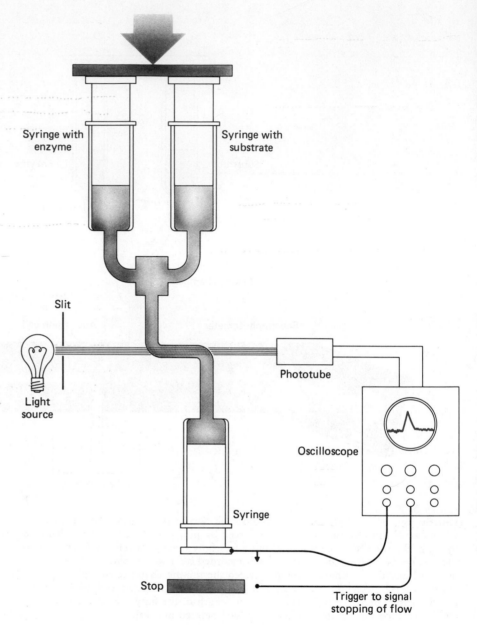

FIGURE 5–10

Stopped Flow Apparatus for Measuring the Time Course of Rapid Reactions. Enzyme and substrate are forced from syringes into a mixing chamber and then into a cell in which rapid changes in the absorption of light can be measured. The flow fills another syringe whose piston contacts a switch that stops the flow. Measurements of light absorption begin at the time the flow stops. Rapid mixing is essential but the flow rate need not be controlled precisely. This method uses significantly less material than the continuous flow apparatus.

this method, introduced by M. Eigen and his collaborators, a reaction at equilibrium is illuminated by an extremely brief, but very intense, flash of light. The temperature of the solution is raised suddenly and, since equilibrium constants depend on temperature, the concentrations of reactants, intermediates, and products change to reach a new equilibrium. By using equally rapid assays of light absorbance, the change in concentration of colored substances can be followed over time intervals as brief as 10 μsec. Temperature-jump methods are adding much new information on the nature of the ES complexes, on the kinetics of enzyme reactions in general, and on the structure of macromolecules.

Linearization of the Michaelis-Menten Equation

We turn now to the problem of determining by experiment the values of K_m and V_{max}. To see how this might be done let us examine the Michaelis-Menten equation and the curve it generates (Figure 5–7) a little more closely. Suppose one finds a substrate concentration that gives an initial velocity (v_i) equal to one-half the limiting velocity ($\frac{1}{2} V_{max}$). If one substitutes these values of the concentration and velocity into Equation 5–8, one sees that the substrate concentration is numerically equivalent to the Michaelis constant (K_m):

$$\text{If } v_i = \frac{V_{max}}{2} = \frac{V_{max}[S]}{K_m + [S]}, \text{ then } \frac{1}{2} = \frac{[S]}{K_m + [S]},$$

which implies that $[S] = K_m$ in these conditions.

EXAMPLE 5–2

From the data tabulated below, evaluate the K_m of the enzyme.

[S]	v_i
0.01 μM	0.05 μmoles/min/mg protein
0.02	0.10
0.04	0.20
0.10	0.48
0.20	0.91
0.40	1.7
1.0	3.3
2.0	5.0
4.0	6.7
10	8.3
20	9.1
40	9.5
100	9.8

Solution

As the substrate concentration increases to 100 μM, the observed initial velocity seems to be limiting at approximately 10 μmole/min/mg protein. One-half this apparent V_{max} is observed at $[S] = 2.0$ μM, so the value of the K_m is approximately 2.0 μM.

To use this approach to determine a value for K_m, one must be able to measure accurately the value of V_{max}. The limiting velocity, however, is hard to determine by inspection of experimental data plotted in the form of Figure 5–7. Further, deciding whether a given set of

experimental data can be fit to a particular Michaelis-Menten curve, such as Figure 5–7, is very difficult, probably because our minds do not recognize the exact shape of a rectangular hyperbola very easily. This problem comes up often in experimental analysis and the usual solution is to *linearize* the data; that is, to reformulate the mathematical equation that describes the model into the equation of a straight line. It can be done in more than one way for the Michaelis-Menten equation, but the most common method is that of H. Lineweaver and D. Burk (Figure 5–11): plotting the reciprocal of the initial velocity ($1/v_i$) against the reciprocal of the initial substrate concentration ($1/[S]$).

The linearized equation is

$$\frac{1}{v_i} = \frac{K_m}{V_{max}} \frac{1}{[S]} + \frac{1}{V_{max}},$$

which is the equation of a line

$$(y = mx + b)$$

with slope K_m/V_{max} and intercept (at $1/[S] = 0$) of $1/V_{max}$. Since the human mind is adept at recognizing deviations from a straight line, one can estimate quickly by inspection the best line for a given set of data on such a "double reciprocal plot." Of course, the mathematical technique of linear regression analysis can also be used to determine the line that fits the data best. The determination of V_{max} and K_m has turned out, in practice, to be useful for characterizing a given enzyme-substrate system.

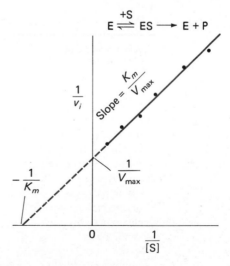

FIGURE 5–11

The Lineweaver-Burk Plot. The reciprocal of v_i is plotted against the reciprocal of [S]. An enzyme which follows Michaelis-Menten kinetics gives results that yield a straight line. On extrapolation, the value of V_{max} can be determined as the intercept with the Y-axis. Either the slope or the further extrapolation of the line to the X-axis determines the value of K_m.

The Importance of the Michaelis Constant

Let us examine further the meaning and uses of the Michaelis-Menten treatment. As shown in Tables 5–2 and 5–3, values of K_m of various enzymes differ widely. One should note that the K_m of a particular enzyme and substrate depends on conditions such as pH, type of buffer, ionic strength, and temperature. Care must always be taken to define these conditions and to ensure that they remain constant during the measurements.

An error sometimes made is to think uncritically of K_m as a measure of the equilibrium dissociation constant of the ES complex. Except in special cases, it is not. It is a property of the enzyme that describes the formation and decay of [ES] and that can be measured only during the *initial steady state* of the reaction shown in Equation 5–3. Only in cases where k_2 is much greater than k_3 does K_m approach the value of the equilibrium dissociation constant of the ES complex.

If the concentrations of E, S, and ES were measured at equilibrium, then the ratio $K_D = [E] \cdot [S]/[ES]$ would be the equilibrium dissociation constant for the reaction $ES \rightleftharpoons E + S$. The *affinity* of an enzyme for its substrate (how tightly the two bind together) can be measured by the equilibrium dissociation constant. The smaller the value of K_D, the greater the affinity. During the steady state formation of enzyme-substrate complex, the affinity of the enzyme for the substrate is measured by K_m, in the sense that the ratio $[E] \cdot [S]/[ES]$ has the value K_m in the steady state. The interested student can prove this by expressing [ES] as $[E_t] - [E]$, solving the steady state equation for [E], then using the expressions for [ES] and [E] to evaluate the above ratio. Analogous to the interpretation of K_D, the lower the value of K_m, the higher the affinity of the enzyme for its substrate.

Despite these limitations, the Michaelis-Menten formulation is useful for a number of reasons:

1. It allows one to calculate the degree of saturation of an enzyme at a given substrate concentration by the following equation. The fraction of sites filled is

$$\frac{v_i}{V_{max}} = \frac{[S]}{[S] + K_m} = \frac{1}{1 + K_m/[S]}.$$

Thus, at the concentration of substrate equal to K_m, the enzyme is half-saturated.

2. One can determine K_m and V_{max} for an impure enzyme preparation in which the concentration or even the amount of enzyme is not known. Typically, the K_m will be the same as for the pure enzyme but V_{max} can be expressed only in *specific enzyme units*—the number of micromoles of product formed per minute per milligram of protein.

3. If the molar concentration of catalytic sites is known (this is found by dividing the weight concentration of enzyme protein by its molecular weight and multiplying by the number of catalytic sites per enzyme), then one can calculate the *turnover number*—the number of substrate molecules changed into product per enzyme molecule per second—when the enzyme is fully saturated (Table 5–4): turnover number = k_3 = $V_{max}/([E] \cdot [\text{number of active sites per enzyme}])$. This formulation accounts for the case when the enzyme has more than one active site.

TABLE 5–2

K_m Values of a Number of Enzymes from Different Sources

Enzyme and Source	Substrate	K_m (M)
Alcohol dehydrogenase	Ethanol	
Yeast		1.3×10^{-2}
Horse liver		5.5×10^{-4}
Human liver		1.2×10^{-3}
Carbonic anhydrase	H_2CO_3	
Human erythrocyte C		6.8×10^{-2}
Bovine erythrocyte		9.6×10^{-3}
Chymotrypsin A	Benzoyl-L-phenylalanine ethyl ester	
Bovine pancreas		2.8×10^{-3}
Porcine pancreas		5.0×10^{-3}
Succinate dehydrogenase	Succinate	
Bovine heart		1.3×10^{-3}
Yeast		1.0×10^{-3}
Micrococcus lactilyticus		5.3×10^{-3}
Urease	Urea	
Jack bean		1.1×10^{-2}
Bacillus pasteurii		4×10^{-2}
Corynebacterium renale		3×10^{-2}

TABLE 5–3

K_m Values of a Number of Enzymes with a Variety of Substrates

Enzyme and Source	Substrate	K_m (M)
Aldehyde Oxidase		
Pig liver	Formaldehyde	3.80×10^{-1}
	Acetaldehyde	1.00×10^{-1}
	Propionaldehyde	3.00×10^{-2}
	Butylaldehyde	2.50×10^{-2}
	Valeraldehyde	1.25×10^{-3}
	Heptaldehyde	1.30×10^{-3}
Chymotrypsin A		
Bovine pancreas	Acetyl-L-tyrosine ethyl ester	$7 \quad \times 10^{-4}$
	Acetyl-L-tryptophan ethyl ester	$9 \quad \times 10^{-5}$
	Acetyl-L-phenylalanine ethyl ester	1.2×10^{-3}
	Acetyl-L-phenylalanine methyl ester	1.8×10^{-3}
	Nitrophenyl acetate	$4 \quad \times 10^{-5}$

4. It also permits the comparison of different enzyme preparations obtained from different species, or from different organs in the same species, with respect to a given substrate (Table 5–2). Or conversely, it is possible to compare the kinetics of a given enzyme with respect to a variety of substrates (Table 5–3).

5. It provides a formulation for the description of the phenomenon of enzyme inhibition which, as we shall see, is of importance for the study of the mechanism of enzyme action.

6. It helps identify regulatory enzymes which do not strictly obey classical Michaelis–Menten kinetics.

K_m and V_{max} are also important from the point of view of the cell. An enzyme cannot catalyze a reaction that uses a substrate whose concentration is very much smaller than the value of K_m. Cellular metabolism has presumably evolved to keep K_m in the proper relation to the substrates. Similarly, an enzyme cannot catalyze a reaction faster than V_{max}. When the reaction product is needed at a faster rate, more enzymes must be synthesized. We shall discuss the control of enzyme synthesis in Chapter 14.

Enzyme Inhibition

As we pointed out in Chapter 3, the inhibition of catalysis is a plausible explanation for the drastic effects that minute quantities of some poisons have on living systems. The Michaelis–Menten treatment of enzyme-substrate kinetics gives a simple descriptive framework for the phenomenon of enzyme inhibition.

Enzyme inhibition can occur in a variety of ways, and its study can reveal a great deal about the nature of

TABLE 5–4

K_3 (Turnover Numbers) of Different Enzymes

Enzyme and Source	Substrate	Turnover Number (per second)
Carbonic anhydrase		
Human erythrocyte	CO_2	620,000
Catalase		
Beef liver	Hydrogen peroxide	93,000
Chymotrypsin A		
Bovine pancreas	Acetyl-L-tyrosine ethyl ester	193
β-galactosidase		
E. coli	Lactose	100
Aldehyde oxidase		
Pig liver	Acetaldehyde	4.6
Succinate dehydrogenase		
Bovine heart	Succinate	1.0

the enzyme and the catalyzed reaction. Some inhibitors exert their effect on enzymes by reacting with them to form a covalently bonded derivative. Thus, diisopropyl-fluorophosphate (DFP) reacts with enzymes such as trypsin, chymotrypsin, and acetylcholinesterase to form an inactive product (see page 230). DFP labeled with ^{32}P has been used to study this reaction, and as we shall see, considerable information has been obtained about the mechanism of action of chymotrypsin and related enzymes by these means. Another example is the reaction of some enzyme sulfhydryl groups with an alkylating reagent such as iodoacetamide (see page 229). These types of inhibition are *irreversible;* only a stoichiometric amount of the inhibitor is required to inactivate all the enzyme. If the inactivation is not complete, a Michaelis-Menten analysis will reveal an unchanged K_m but a reduced V_{max}, because the concentration of active enzyme has been reduced.

There are also reversible forms of enzyme inhibition, in which the substrate concentration plays an important role. We refer to these inhibitions as competitive, uncompetitive, and noncompetitive. They differ from one another by the ways in which the reversible inhibitor can bind to the enzyme:

- A *competitive inhibitor* can bind only to free enzyme molecules (those not having already bound a substrate molecule).
- An *uncompetitive inhibitor* can bind only to an enzyme-substrate complex and not to free enzyme.
- A *noncompetitive inhibitor* can bind both to free enzyme and to the enzyme-substrate complex.

Competitive Inhibition

Let us examine the classical case of the oxidation of succinic acid to fumaric acid, an important reaction in the energy metabolism of aerobic cells. It has been known for a long time that malonic acid is an inhibitor of this reaction.

$$
\begin{array}{c}
\text{COO}^- \\
| \\
\text{CH}_2 \\
| \quad + \text{FAD} \\
\text{CH}_2 \\
| \\
\text{COO}^- \\
\text{Succinate}
\end{array}
\underset{\substack{\text{Inhibited} \\ \text{competitively} \\ \text{by malonate}}}{\overset{\substack{\text{Catalyzed by} \\ \text{succinate} \\ \text{dehydrogenase}}}{\rightleftharpoons}}
\begin{array}{c}
\text{COO}^- \\
| \\
\text{CH} \\
|| \quad + \text{FADH}_2 \\
\text{HC} \\
| \\
\text{COO}^- \\
\text{Fumarate}
\end{array}
$$

Malonate:
$$
\begin{array}{c}
\text{COO}^- \\
| \\
\text{CH}_2 \\
| \\
\text{COO}^-
\end{array}
$$

Like many other metabolic reactions catalyzed by enzymes, there are two reactants for the oxidation of succinate. The FAD is a co-substrate of succinate dehydrogenase. It accepts the electrons removed during the oxidation of succinate and becomes reduced to $FADH_2$ (see Chapter 8). We shall direct our attention in the following discussion to succinate alone. Experimentally, this requires that we keep the FAD (and enzyme) concentration constant while we vary only succinate.

When the rate of fumarate formation is measured, at constant concentration of malonate (and FAD), one finds that at low succinate concentrations the rate is less than that measured in the absence of malonate—that is, the rate is inhibited in the presence of malonate. The degree of inhibition becomes less at higher substrate concentrations, however, and the limiting velocity at very high succinate concentration seems not to be affected much by the presence of malonate. This behavior is easier to visualize on a double reciprocal (Lineweaver-Burk) plot of the data (Figure 5–12). As the initial substrate concentration is made higher and higher, and 1/[S] gets closer to zero, the initial velocity comes closer to that found in the absence of inhibitor. Extrapolated to infinite substrate concentrations (1/[S] = 0), the values of V_{max} are the same in the presence and absence of malonate, the competitive inhibitor. The inhibitor does alter the apparent K_m of the reaction, however; the slope of the double reciprocal plot is higher.

These effects can be explained by assuming that there is a *competition* for the enzyme between the inhibitor and the substrate. That is, either the substrate can bind to the enzyme, or the inhibitor can, but not both simulta-

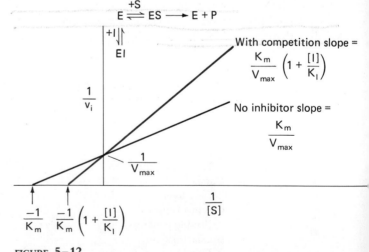

FIGURE 5–12
Competitive Inhibition. Shown are the reaction scheme and double reciprocal plots of the uninhibited and the competitively inhibited enzyme reactions. In competitive inhibition the value of V_{max} does not change but the apparent value of K_m increases.

neously. This competition takes place at the *active site* of the enzyme, the area which binds substrate during ES formation and where the catalytic action takes place. And indeed the explanation based on competition becomes entirely plausible when one examines the similarity in structures of malonic and succinic acids. The two molecules are approximately the same size and both have two carboxyl groups which are negatively charged at physiological pH. One can imagine the active site of the enzyme to have two positively charged pockets that can accommodate these groups. Furthermore, one can understand why malonate won't undergo the chemical change experienced by succinate, the real substrate—it is because one of the —CH$_2$ groups in succinate is missing in malonate, and an analogous oxidation, therefore, is impossible.

When they are structurally similar to substrates, reaction *products* can act as competitive inhibitors of the enzymes catalyzing their production. As the concentration of product rises, it slows the enzyme-catalyzed reaction by interacting with the active site in competition with substrate. (This effect should be distinguished from the influence of product concentration, by Le Châtelier's principle, on the equilibrium of the reaction.) In the cell, product inhibition is minimized when the product is utilized in another reaction.

Further quantitative analysis of competitive inhibition is instructive. With minimal algebra we can develop a model that will be useful not only in sharpening our appreciation of enzyme action but also in the context of membrane transport (see Chapter 7). In competitive inhibition the enzyme can combine with substrate, or with inhibitor, but not simultaneously with both:

$$E + I \underset{K_i}{\rightleftharpoons} E \cdot I$$

$$E + S \rightleftharpoons E \cdot S \longrightarrow E + P.$$

Since the inhibitor can only bind to, and not react with, the enzyme, it will come into equilibrium with the enzyme-inhibitor complex; the equilibrium constant for the dissociation of EI is

$$K_i = \frac{[E]\,[I]}{[EI]}$$

where [I] is the molar concentration of inhibitor. This equation can be rearranged to give [EI] as a function of [E] and [I]:

$$[EI] = [E]\,\frac{[I]}{K_i}.$$

The equation for the initial steady state rate of product formation is the same as Equation 5–2, but the equation expressing conservation of mass of enzyme is different because of the existence of EI:

$$[E] + [EI] + [ES] = [E]_t.$$

Substituting the expression for [EI] derived from the equilibrium constant equation above, the conservation equation can be written

$$[E] + [E]\,\frac{[I]}{K_i} + [ES] = [E]_t,$$

or

$$[E]\left[1 + \frac{[I]}{K_i}\right] + [ES] = [E]_t.$$

Solving for [E] then gives

$$[E] = \frac{[E]_t - [ES]}{\left[1 + \frac{[I]}{K_i}\right]}.$$

The derivation of the expression for v_i from this point onward follows that of the Michaelis-Menten equation: one starts with Equation 5–6. The final result is

$$v_i = \frac{V_{max}\,[S]}{K_m\left[1 + \frac{[I]}{K_i}\right] + [S]}. \tag{5–9}$$

This equation for the initial velocity has the same form as we found for the equation for the uninhibited case (Equation 5–8); the factor that is analogous to V_{max} is exactly the same, but the term that is analogous to K_m (called the apparent K_m, $K_m{}^{app}$) is larger:

$$K_m{}^{app} = K_m\left[1 + \frac{[I]}{K_i}\right].$$

The mathematical model reveals that the effect of the competitive inhibitor is to reduce the apparent affinity of the enzyme for its substrate. In double reciprocal form this equation is

$$\frac{1}{v_i} = \frac{1}{V_{max}} + \frac{K_m\left[1 + \frac{[I]}{K_i}\right]}{V_{max}} \cdot \frac{1}{[S]}.$$

It is easy to see by comparing Figure 5–11 with Figure 5–12 that the slope of the line on the Lineweaver-Burk plot is increased by a factor of

$$\left[1 + \frac{[I]}{K_i}\right]$$

over that measured in the absence of inhibitor. This generally happens when an inhibitor binds to the same form of the enzyme with which the substrate binds. The intercept that corresponds to indefinitely large substrate concentrations is unchanged because the inhibition can be overcome by sufficiently high substrate concentrations. This can be understood in terms of the competition between substrate and inhibitor. The active site is sampling the population of small molecules in its environment. As the concentration of substrate rises, the probability that an inhibitor will be the next small molecule to diffuse into the active site falls. Inhibition occurs only when the inhibitor is able to compete effectively for the active site; with the inhibitor bound, the substrate cannot gain access and the rate of product formation must decrease.

Another well-known example of competitive inhibition is the effect of sulfanilamide (a sulfa drug), which "fools" the enzyme normally acting on *p*-aminobenzoic acid, an important growth factor in many bacteria:

p-Aminobenzoic acid   Sulfanilamide

Note again the general structural similarities between the two compounds. There are numerous competitive inhibitors known; their use has contributed greatly to our understanding of the mechanism of enzyme action, to the study of metabolic sequences, and to the development of chemotherapy in medical practice.

Competitive inhibition was one of the earliest and most direct lines of evidence for the notion that an active site is the locus of enzyme-substrate binding and catalysis. As we shall see, the use of x-ray crystallography has provided an even more direct and detailed description of the active site of an enzyme, and its interaction with substrate and inhibitor. These studies have amply corroborated the earlier predictions based on kinetic studies. Some reversible inhibitors, however, seem to act competitively but bind at a location other than the active site of the enzyme. When bound, they apparently cause a distortion of the active site so that the substrate(s) cannot bind—thus, the appearance of competition.

Reaction schemes and kinetic equations for the cases of uncompetitive and noncompetitive inhibition are analogous to those of competitive inhibition and are discussed in texts of biochemistry. Most inhibitors behave as mixtures of these simple types. Detailed analysis of their behavior, however, gives information about the mechanism of catalysis that is not obtainable from structural analysis alone.

The biologically most important enzyme inhibitors are those which bind to the enzyme at sites other than the active catalytic site. This seemingly trivial variation is of fundamental importance since it allows the control of

metabolic chemistry by molecules with structures very different from the substrates and products of a single reaction. The use of such inhibitors and multi-subunit enzymes has allowed the evolution of subtle control mechanisms which are the envy of engineers. However, in order to do this subject justice, we should first discuss the topics of *enzyme specificity* and *mechanisms of enzyme action*.

Specificity of Enzyme Action

Specificity is perhaps the most characteristic property of living processes, and the enzyme-substrate interaction perhaps the most dramatic example of biological specificity. Many hundreds of different enzymes, each with its own catalytic properties and specificities, have been isolated from various tissues.

As we have seen, studies of competitive enzyme inhibition led to a precise notion of a special place on the enzyme, called the *active site,* which can interact in a specific manner with the substrate. In fact, it is possible to appreciate much of the specificity of the active site by observing the interaction of enzymes with their normal substrates, as E. Fischer did in 1894. His studies of the stereospecificity of the enzyme hydrolysis of glycosides led him to propose the *lock-and-key* theory of enzyme-substrate interaction which proclaimed, with a great deal of prophetic insight, the *complementary structural relationship of enzyme and substrate.* We shall cite a few instructive examples of enzyme specificity and draw some conclusions about a range of properties exhibited by the active sites of different enzymes.

Absolute specificity is found in an enzyme such as urease, which can utilize only urea as a substrate and will fail to act on even the slightest chemical modification of urea. This must mean that the active site of the enzyme urease is very "snug" and that it has very precise requirements for each of the three functional groups of the urea molecule.

Urease

Other enzymes are known to have a high degree of specificity only for one particular group, usually the group participating in the chemical reaction. Alcohol

V_{max}/K_m: A Parameter to Measure Specificity

A useful parameter for judging the specificity of an enzyme is the ratio of rates at which the enzyme catalyzes competing substrates that are present at equal concentrations. From a list of substrates and their relative rates under such conditions, one can identify the substrate with the highest rate as that for which the enzyme is most specific. The Michaelis-Menten scheme allows one to calculate the appropriate expression for two substrates, 1 and 2:

$$\frac{v_1}{v_2} = \frac{\dfrac{(V_{max})_1 [S]_1}{(K_m)_1 + [S]_1}}{\dfrac{(V_{max})_2 [S]_2}{(K_m)_2 + [S]_2}}$$

$$= \frac{(V_{max})_1 [S]_1}{(V_{max})_2 [S]_2} \frac{(K_m)_2 + [S]_2}{(K_m)_1 + [S]_1}.$$

Since the substrates are to be evaluated at equal concentrations, $[S]_1 = [S]_2$ and

$$\frac{v_1}{v_2} = \frac{(V_{max})_1}{(V_{max})_2} \frac{(K_m)_2 + [S]_2}{(K_m)_1 + [S]_1}.$$

At high concentrations S is very much larger than K_m and the ratio of rates is simply the ratio of maximum velocities. More interesting is the case when substrate concentrations are low. Then $K_m + S$ is essentially K_m and, as can be seen from the above equation,

$$\frac{v_1}{v_2} = \frac{(V_{max})_1}{(K_m)_1} \bigg/ \frac{(V_{max})_2}{(K_m)_2}.$$

From this analysis one can see that the parameter that is related to the discrimination between two substrates is V_{max}/K_m, and not merely K_m alone.

EXAMPLE 5-3

Calculate the ratio of v_i for two Michaelis-Menten enzymes operating at two different substrate concentrations:

- Enzyme 1: $V_{max} = 5$ μmole/min/mg protein, $K_m = 20$ μM
- Enzyme 2: $V_{max} = 20$ μmole/min/mg protein, $K_m = 200$ μM
- Case A, both substrates at $[S] = 10$ mM
- Case B, both substrates at $[S] = 1$ μM.

Solution

Case A:

$$\frac{v_1}{v_2} = \frac{5}{20} \frac{(200 + 10,000)}{(20 + 10,000)} \simeq 0.25$$

Case B:

$$\frac{v_1}{v_2} = \frac{5}{20} \frac{(200 + 1)}{(20 + 1)} \simeq 2.4$$

At high substrate concentrations ($[S] \gg K_m$), the ratio of initial velocities is approximately the ratio of maximum velocities: v_2 is greater than v_1 even though enzyme 1 has the lower K_m. At low substrate concentrations, the reaction catalyzed by enzyme 1 has a higher velocity despite having a lower V_{max}.

This is illustrated by the enzyme alcohol dehydrogenase derived from rat liver cells (Table 5–A). Listed are the values of V'_{max}, K_m, and the ratio of V'_{max}/K_m for a series of alcohols. Straight chain alcohols increase in their ability to be used by the enzyme as the length increases (compare ethanol, propanol, and butanol). The most effective substrate listed is methylbutanol; it has a V'_{max}/K_m ratio almost a factor of ten higher than the presumed natural substrate, ethanol. When the hydroxyl group is not on the terminal carbon, the substrate is not utilized as effectively; isopropanol has a V'_{max}/K_m ratio lower than propanol by a factor of over 100. Another hydroxyl group in the molecule also lowers the effectiveness of the substrate. The ratio of V_{max} to K_m for propanediol is lower than that for propanol by over 100-fold.

(*continued*)

TABLE 5–A

Kinetic Constants for Oxidation of Alcohols by Rat Liver Alcohol Dehydrogenase

Alcohol		V'_{max}	K_m (mM)	V'_{max}/K_m
Ethanol	CH_3CH_2-OH	1.0	0.64	1.6
Propanol	$CH_3CH_2CH_2-OH$	0.95	0.22	4.3
Butanol	$CH_3CH_2CH_2CH_2-OH$	1.3	0.14	9.3
Methylpropanol	$\overset{\displaystyle CH_3}{\overset{\displaystyle \mid}{CH_3CHCH_2}}-OH$	0.80	0.19	4.2
Methylbutanol	$\overset{\displaystyle CH_3}{\overset{\displaystyle \mid}{CH_3CHCH_2CH_2}}-OH$	0.54	0.042	13
Isopropanol	$\underset{\underset{\displaystyle OH}{\displaystyle \mid}}{CH_3\ CHCH_3}$	0.40	36	0.011
Isobutanol	$\underset{\underset{\displaystyle OH}{\displaystyle \mid}}{CH_3CH_2CHCH_3}$	0.48	12	0.040
Ethanediol	$\underset{\underset{\displaystyle OH}{\displaystyle \mid}}{CH_2CH_2-OH}$	0.99	740	0.0013
Propanediol	$\underset{\underset{\displaystyle OH}{\displaystyle \mid}}{CH_2CH_2CH_2-OH}$	0.75	24	0.031

Values of V'_{max} are given relative to V_{max} for ethanol. (Unpublished data courtesy of B. Plapp.)

dehydrogenase is an enzyme exhibiting such *group specificity*. There is a strong requirement for the —OH group, which is oxidized to an aldehyde.

$$CH_3CH_2OH + \overset{}{\underset{}{\text{Oxidized coenzyme}}} \underset{\xrightarrow{\hspace{2cm}}}{\overset{\text{Alcohol dehydrogenase}}{\xleftarrow{\hspace{2cm}}}}$$

$$CH_3CHO + \text{Reduced coenzyme}$$

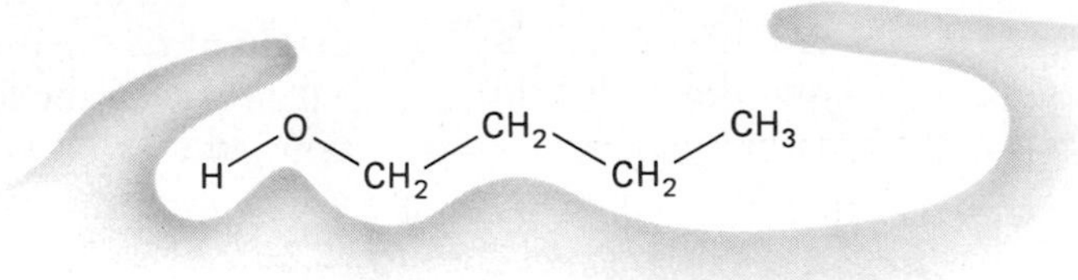

Although ethanol is probably the natural substrate, alcohol dehydrogenase can also act on straight chain alcohols of other lengths. We can make a good guess about the nature of the active site in such an instance. It is probably a cleft with some rather precise requirements for the oxidation of the —OH group at one end, with a hydrophobic region next to it to accommodate the —CH_2— group, and beyond this a region which can fit increasing lengths of a hydrophobic tail while not necessarily interacting with it very strongly.

Alcohol dehydrogenase

Apparently, broader specificity is exhibited by other enzymes which can catalyze reactions of more than one kind of functional group, provided they have similar structures. Such *relative group specificity* is found in the enzyme trypsin, which can hydrolyze an *ester* bond as well as a *peptide* bond, but requires a positively charged side chain of lysine or arginine on the carbonyl (C=O) side of the ester or peptide bond (Figure 5–13). The high degree of proteolytic specificity by trypsin has been used to great advantage in the sequence studies described in Chapter 4.

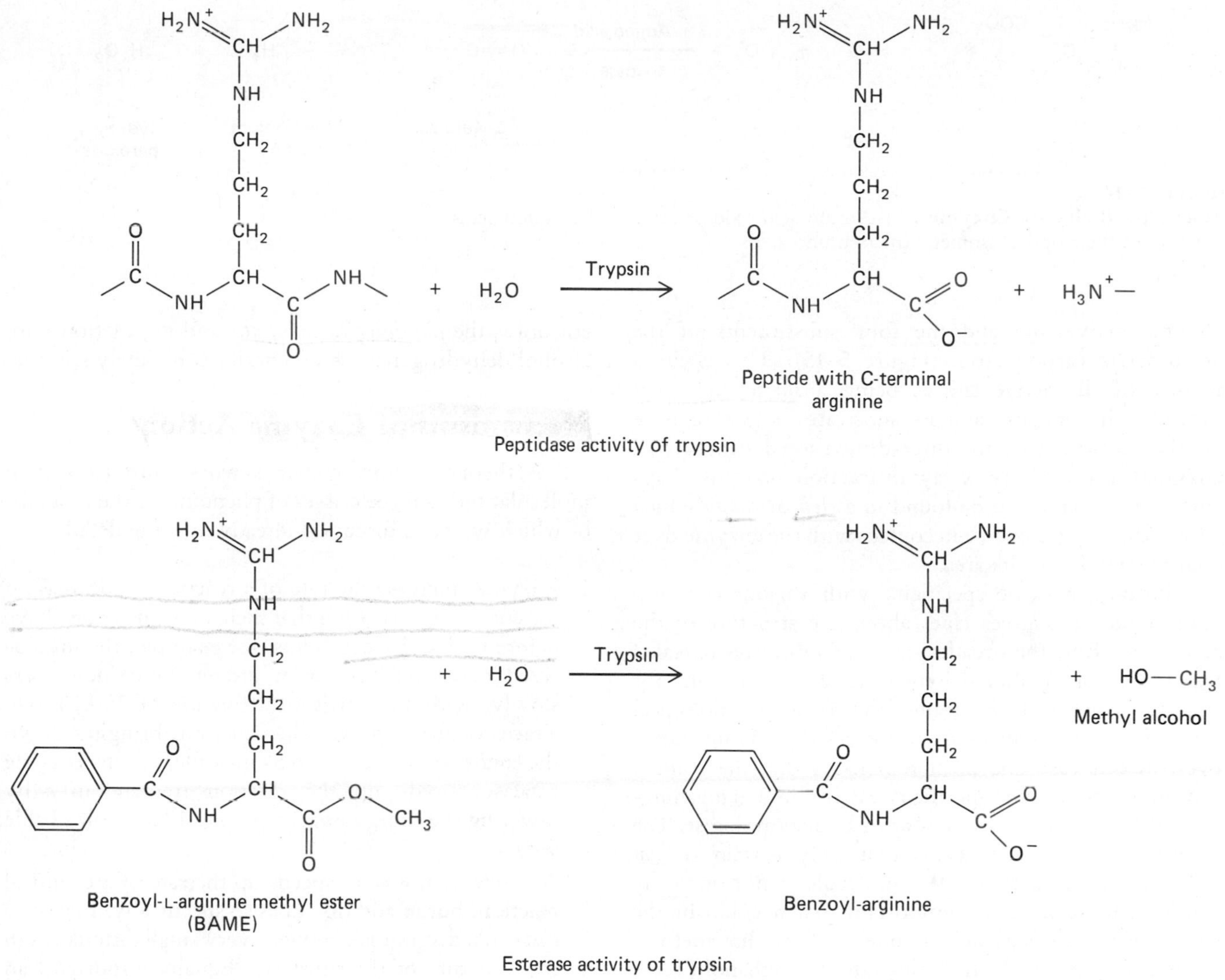

Peptidase activity of trypsin

Esterase activity of trypsin

FIGURE **5–13**
Relative Group Specificity by Enzymes.   Trypsin is an endopeptidase that
catalyzes the hydrolysis of a peptide bond (color) in which the —C═O group is
contributed by an arginyl (or lysyl) residue. Provided that one of these positively
charged residues is present, trypsin can also act as an esterase. This is illustrated by
the action of trypsin on the artificial substrate BAME, which contains an ester
bond (color).

Trypsin exhibits two principles of specificity which
we must distinguish from each other and which can occur
separately in different enzymes. First there is relative
group specificity with respect to the functional group
involved in the reaction, trypsin being able to catalyze the
splitting of an ester as well as an amide. Second, there is
the absolute requirement of a positively charged functional group situated at a precise distance away from the
site of catalytic action. The enzyme acetylcholinesterase
has similar relative and absolute specificities, and one can
imagine easily the structure of an active site to fit these
requirements.

Enzymes are able to distinguish between the D and L
forms of optical isomers (Figure 5–14). Such stereochemical specificity illustrates dramatically the remarkable
degree of specificity exhibited by biological systems.
Most reactions in the organic chemistry laboratory do not
have sufficient specificity to distinguish between optical
isomers. An example of enzymatic stereospecificity is
that of L-amino acid oxidase, which will not act on
D-amino acids. Furthermore, a D-amino acid oxidase is
known which will oxidize D-amino acids exclusively.

Stereospecificity can only be explained by assuming
that there is at least a *three-point contact* between the

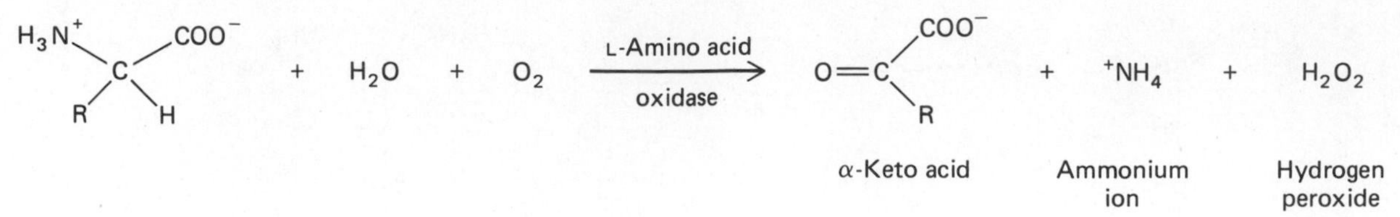

$$H_3N^+ \underset{\underset{R}{|}}{\overset{\overset{COO^-}{|}}{C}} H + H_2O + O_2 \xrightarrow[\text{oxidase}]{\text{L-Amino acid}} O=C \underset{R}{\overset{COO^-}{|}} + {}^+NH_4 + H_2O_2$$

α-Keto acid   Ammonium   Hydrogen
ion   peroxide

FIGURE **5–14**
Stereospecificity by Enzymes.   L–Amino acid oxidase can act on L-amino acids
but not on their optical isomers, the D-amino acids.

enzyme active site and the four substituents of the asymmetric carbon atom (Figure 5–15). This elicits a picture of the active site as being a *surface of contact* between the enzyme and its substrate, a picture now amply confirmed by the three-dimensional structure of enzymes revealed by x-ray diffraction analysis. This surface is generally to be found in a *cleft* or *cavity* which allows the substrate to be in contact with the enzyme over a larger fraction of its area.

Probing enzyme specificity with various chemical compounds thus gives clues about the structure of the active site. But, the broader range of substrates revealed by studies of alcohol dehydrogenase and trypsin, for example, appears to contradict the notion of biological specificity. Appearances are sometimes deceiving, however; in this case one must remember that the context within which enzyme specificity evolves and is meaningful is the cell, not the laboratory of an enzymologist. The significance of specificity is that only certain *cellular* molecules are substrates. If a molecule is not normally found in the cell, it is unimportant that it could be the substrate for an enzyme. Despite the fact that methylbutanol is a more effective substrate than ethanol (Table 5–A), there is essentially no formation of methylbutanaldehyde in normal cells. Under physiological conditions

ethanol is the *physiological substrate,* and its reaction with alcohol dehydrogenase is practically completely specific.

Mechanisms of Enzyme Action

A theory of how enzymes work must explain in molecular terms three classes of phenomena, the first two of which we have discussed already in some detail:

1. Enzymes increase the rate of a reaction by decreasing the energy of activation that their substrates must have before they can react. Thus, for example, the uncatalyzed decomposition of hydrogen peroxide occurs slowly, with an energy of activation of 75 kJ/mole. Traces of iron catalyze this reaction, bringing down the energy of activation to 54 kJ/mole, but the enzyme catalase speeds up the reaction tremendously by lowering the energy of activation to 17 kJ/mole (Table 5–5).

2. An enzyme not only speeds up the rate of a chemical reaction, but in addition does so specifically. Enzymes can often distinguish between very small differences in the structure of the substrate. Equally important, an enzyme usually catalyzes only one specific reaction among the many of which the substrate is capable.

FIGURE **5–15**
Optical Specificity by Enzymes.
Demonstration that optical specificity
must be due to a minimum of three
points of contact. Two-point contact
cannot distinguish between the optical
isomers (a) and (b).

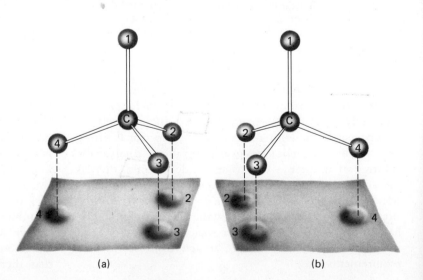

(a)   (b)

TABLE 5-5

The Effect of Enzymes on the Energies of Activation of Two Reactions

Reaction	Catalyst	Activation Energy (kJ/mole)
$2H_2O_2 \longrightarrow 2H_2 + O_2$	None	75
	Traces of Iron	54
	Enzyme catalase	17
$\begin{array}{c} H_2N \\ \diagdown \\ \quad C{=}O + H_2O \rightarrow CO_2 + 2NH_3 \\ \diagup \\ H_2N \end{array}$	Hydrogen ion	110
	Enzyme urease	54

3. The rate of enzymatic catalysis is capable of being regulated under biological conditions.

Significant progress in understanding the function of a few enzymes has been made in recent years by using the descriptions of protein structure made possible by x-ray crystallography. On the other hand, the large variety of other approaches used over a period of 40 years to probe the properties of active sites is also of interest. In particular, it is instructive to realize that one's view of "state-of-the-art" methodology changes as new techniques reveal properties that lie beyond the resolution of older assays. The working scientists of today can reliably expect their techniques to be superseded in the future. Considering the historical development of a field thus teaches us humility, which is useful to temper enthusiasm. We shall therefore proceed first by describing the general results from older methods of identifying functional groups and chemical reactivities in the active site, then discuss some specific results obtained by x-ray crystallography, and finally conclude with a brief introduction to the exciting approach of "protein engineering."

Experimental Approaches to the Study of the Active Site

From our studies of protein structure in Chapter 4, we can predict what the chemical forces at the active site are likely to be. Polypeptide chains have (1) 20 different amino acid residues, each with its characteristic side chain, (2) one N-terminal and one C-terminal group, and (3) a large number of peptide bonds forming the backbone of the chains. We expect covalent bonds, ionic interactions, van der Waals forces, and hydrophobic interactions to play a role in enzyme catalysis. And, in fact, all of these interactions have been implicated by more or less direct means. Let us begin by describing

studies which demonstrated that certain functional groups are present in the active site of an enzyme.

A large number of enzymes are known in which certain of their cysteinyl residues must have free —SH groups in order to maintain activity. Alkylating reagents such as iodoacetamide or bromoacetic acid react with —SH groups (Figure 5–16). It is usually not possible to distinguish between an —SH group which is part of the active site—that is, directly involved in catalytic activity—and an —SH group situated elsewhere on the protein but nevertheless required to maintain the active site in its proper conformation.

There are some circumstances, however, when irreversible alkylating reagents can give more precise information about the catalytic site. Such a case is the enzyme ribonuclease. This enzyme was examined by E. A. Barnard and W. H. Stein, who discovered that it has a histidine imidazole group, located in position 119, that is unusually reactive toward bromoacetic acid. (The position of the alkylated residue can be determined by sequence analysis, after using a radiolabeled alkyl group as a marker; see Chapter 4.) Bromoacetic acid does not react as readily with the other three histidines in the molecule, nor does it react readily with histidine in solution. Most significantly, the special reactivity of histidine 119 is lost if ribonuclease is denatured; from this observation it can be concluded that the special reactivity of histidine 119 depends on the active conformation of the enzyme. Since the alkylation of this imadazole by bromoacetic acid results in complete loss of enzymatic activity, the simplest hypothesis to explain all the data is that histidine 119 lies within the active site. Further kinetic studies of ribonuclease using a variety of inhibitors also implicated histidine 12 and lysines 7 and 41 in the active site of the enzyme.

These studies were verified in a dramatic manner when D. Harker and his co-workers succeeded in working out the three-dimensional structure of ribonu-

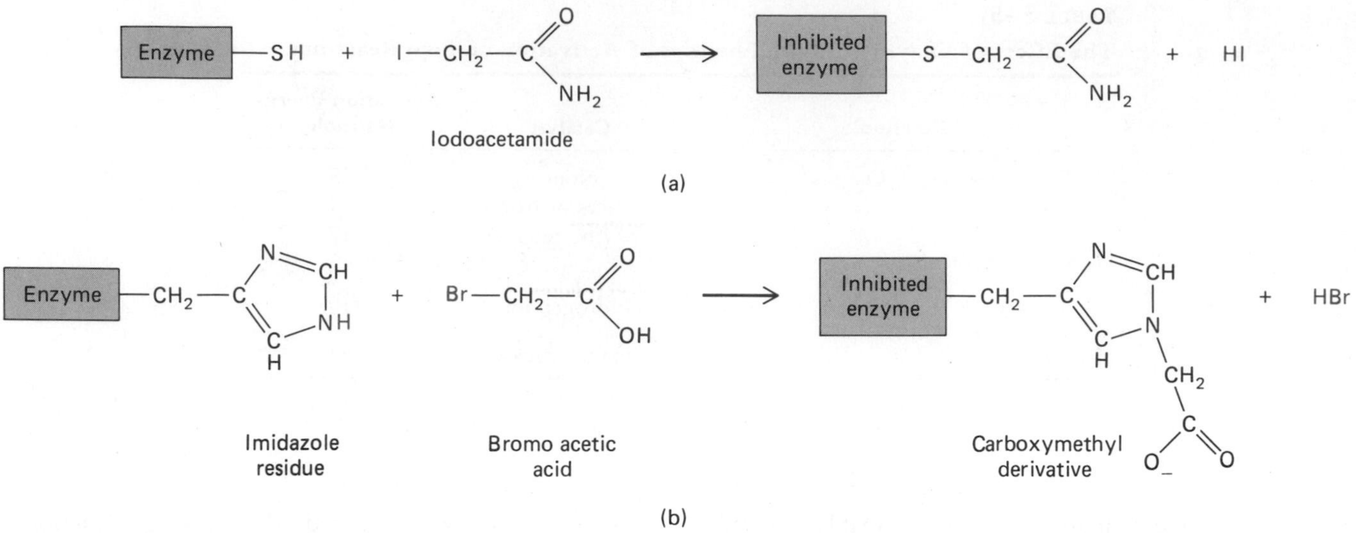

(a)

(b)

FIGURE 5–16

Alkylation of an Enzyme.   (a) Reaction of an —SH group with iodoacetamide.
(b) Reaction of an imidazole residue with bromoacetic acid. These spontaneous
reactions often lead to an irreversibly inhibited enzyme.

cleave to 0.35 nm (3.5 Å) resolution using the x-ray
diffraction method. According to this structure, histidines 119 and 12 and lysine 41, though separated by
many peptide bonds, are all in the same vicinity, brought
together by the complex folding of the protein's tertiary
structure. The histidine 119 group, which reacts with
bromoacetic acid at a rate several hundred times faster
than histidine in solution, does so *by virtue of its location in
the active site.* We emerge with the finding, which we shall
have to explain later, that functional groups in the active
site can experience greatly enhanced reactivities.

A similar case is that of diisopropylfluorophosphate
(DFP), which reacts stoichiometrically with a number of
esterase and protease enzymes to form a diisopropylphos-
phoryl (DP) derivative. DFP is a *nerve gas* that poisons
because it reacts with acetylcholinesterase, an enzyme of
critical importance in the functioning of the nervous
system. Partial hydrolysis of the derivatized enzyme has
shown that in each case the hydroxyl group of a particular
serine side chain is involved in this reaction (Figure 5–17).
Partial hydrolysis of a number of DFP-sensitive enzymes
and recovery of the DP-peptides showed that in many
cases glycyl or alanyl residues are found next to the
reactive serine (Table 5–6). When structural features like
this are *conserved* among several examples, one begins
to suspect there is a functional significance or an evolutionary relationship, or both. The fact that other serine
side chains in the enzyme molecules are not equally reactive, and that in each of these enzymes only one serine
reacts rapidly and stoichiometrically, constitutes strong

presumptive evidence that this particular serine is part of
the active site. The involvement of a serine hydroxyl
group in such a wide range of enzymatic reactions
suggests that it plays a role in the bond-breaking function
of catalysis, rather than in helping to determine substrate
specificity. The latter function must be the consequence
of additional side chains in the active sites of the enzymes
listed in Table 5–6.

The use of small, essentially nonspecific alkylating
reagents to obtain specific structural information, as
described above, depends on enhanced chemical reactivity of the target groups to provide the specificity. A more
general approach, closely related to the above techniques,
is *affinity labeling*. The strategy is to synthesize a compound whose structure is sufficiently similar to the
substrate to bind specifically to the enzyme, presumably
at the active site. In addition, the compound carries a
functional group which can react covalently with an
amino acid residue at, or close to, the active site. If this
bond is sufficiently stable to survive hydrolysis of the
protein, it is often possible to identify the amino acid
which has been labeled. This method, of course, requires
a great deal of initial trial and error to find the proper
functional group which will react with a given amino acid
residue located in a given region of the active site, but
once one has succeeded in devising an effective label for a
given side chain, one can interpret the results obtained
with a great deal of precision. Figure 5–18 shows an
affinity label, tosyl-L-phenylalanine chloromethyl ketone
(TPCK), which has been synthesized to have a structure

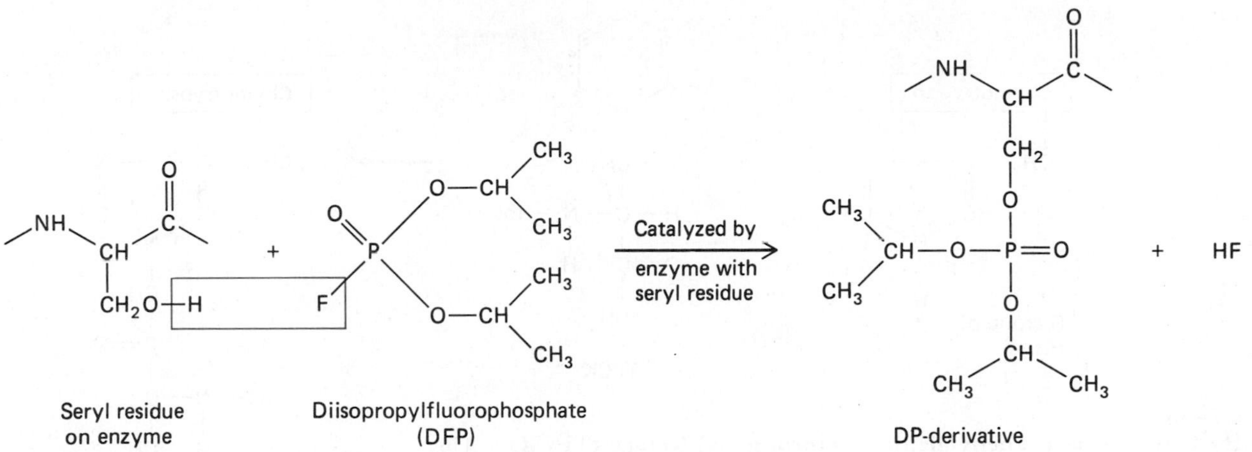

FIGURE 5-17
The Reaction of Nerve Gas Diisopropylfluorophosphate (DFP).   A number of proteases and esterases catalyze the addition of this irreversible inhibitor to a special seryl residue on the enzyme.

similar to one of the substrates of chymotrypsin. TPCK contains a functional group which reacts rapidly with a histidyl residue in the chymotrypsin molecule. When chymotrypsin is treated with TPCK, histidine residue 57 becomes alkylated. Chymotrypsin is also one of the enzymes with a serine (residue 195) that reacts with DFP, so alkylation experiments reveal two amino acids that are

likely to be within the active site. X-ray crystallographic studies by D. M. Blow and T. A. Steitz beautifully confirmed this earlier work (see Figure 5–21b).

The ultimate approach to understanding catalysis is to synthesize chemically a model of the active site and then determine whether it has catalytic properties. Early attempts to achieve this goal tried to model enzymes with

TABLE 5-6

Amino Acid Sequence Around the DFP-Reactive Serine Residue of Some Esterases, Proteases, and Other Enzymes

Sequence	Enzyme
Val Ser Ser Cys Met Gly Asp *Ser* Gly Gly Pro Leu Val Cys Lys	Chymotrypsin
NH₂ (subscript)	
Asp Ser Cys Glu Gly Glu Asp *Ser* Gly Pro Val Cys Ser Gly Lys	Trypsin
Asp *Ser* Gly	Thrombin
Asp *Ser* Gly	Elastase
Phe Gly Glu *Ser* Ala Gly (Ala, Ala, Ser)	Butyryl cholinesterase
Glu *Ser* Ala	(Eel) acetylcholinesterase
Gly Glu *Ser* Ala Gly Gly	Liver aliesterase (horse)
NH₂	
Asp Gly Thr *Ser* Met Ala Ser Pro His	Subtilisin *(B. subtilis)*
Thr *Ser* Met Ala	Mold protease *(Aspergillus oryzae)*
Thr Ala *Ser* His Asp	Phosphoglucomutase
NH₂	
Lys Glu Ile *Ser* Val Arg	Phosphorylase
Pro Asp Tyr Val Thr Asp *Ser* Ala Ala Ser Ala	Alkaline phosphatase *(E. coli)*

Note: Courtesy of R. A. Oosterbaan and J. A. Cohen.

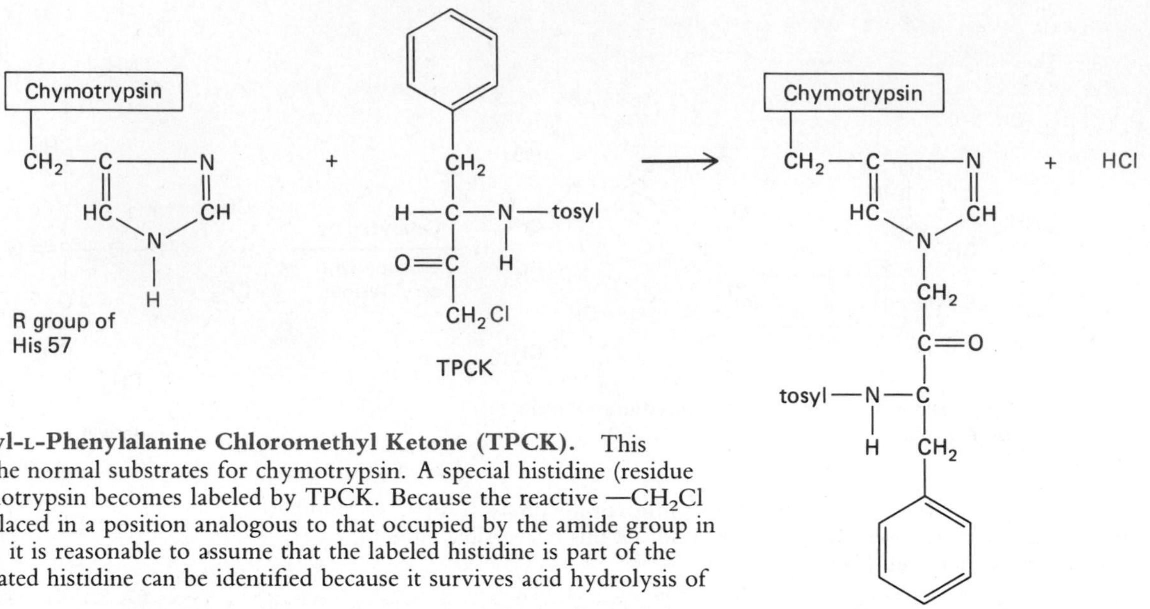

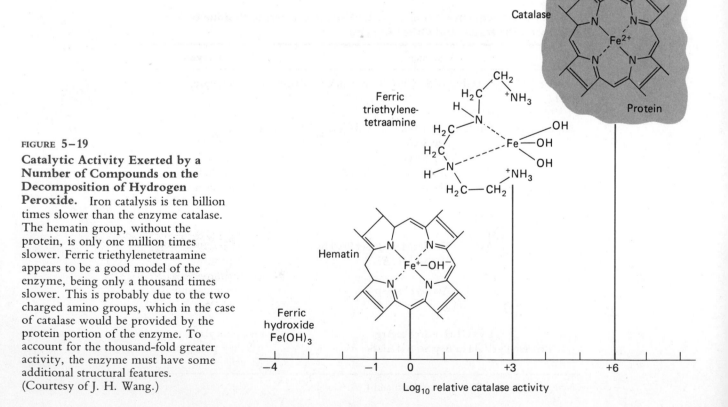

FIGURE **5–18**

Affinity Label Tosyl-L-Phenylalanine Chloromethyl Ketone (TPCK).   This analogue resembles the normal substrates for chymotrypsin. A special histidine (residue number 57) on chymotrypsin becomes labeled by TPCK. Because the reactive —CH_2Cl group on TPCK is placed in a position analogous to that occupied by the amide group in the normal substrate, it is reasonable to assume that the labeled histidine is part of the active site. The alkylated histidine can be identified because it survives acid hydrolysis of the protein.

a prosthetic group, because the known structure of the prosthetic group provided a good beginning for mimicking the active site. J. H. Wang carried out an interesting study on catalase, the enzyme catalyzing the decomposition of hydrogen peroxide into oxygen and water. Catalase has a prosthetic group composed of an iron atom

and a heme group (called hematin). It is possible to measure the catalytic effects of the iron alone (weak) and of hematin alone (less weak). The conclusion is that the protein portion of the enzyme does play a crucial role in catalysis (Figure 5–19): it is very likely that the hematin group is located on the enzyme in such a manner that the

FIGURE **5–19**

Catalytic Activity Exerted by a Number of Compounds on the Decomposition of Hydrogen Peroxide.   Iron catalysis is ten billion times slower than the enzyme catalase. The hematin group, without the protein, is only one million times slower. Ferric triethylenetetraamine appears to be a good model of the enzyme, being only a thousand times slower. This is probably due to the two charged amino groups, which in the case of catalase would be provided by the protein portion of the enzyme. To account for the thousand-fold greater activity, the enzyme must have some additional structural features. (Courtesy of J. H. Wang.)

substrate will interact with portions of the protein molecule as well as with the prosthetic group. An alternative hypothesis assumes that the interaction of the protein and the hematin modifies the properties of the latter in such a way as to increase its catalytic activity.

In other studies, Wang was able to synthesize a compound (ferric triethylenetetraamine) that is a much more effective catalyst than hematin. In fact, legend has it that when Wang first added his model compound to hydrogen peroxide, he observed an explosive release of oxygen. This is not because the model compound is a better catalyst than the enzyme. As Figure 5–19 shows, it is less effective by three orders of magnitude. But since catalase has a high molecular weight, it is not possible to use it in concentrations of more than 10^{-6} M, whereas triethylenetetraamine can be dissolved in concentrations of 10^{-1} M, thus allowing a rate of oxygen release 100 times greater than that of the enzyme. By placing two —NH_3^+ groups close to the iron, Wang synthesized an effective compound for catalyzing the conversion of hydrogen peroxide into water, and it is tempting to speculate that the presence of these —NH_3^+ groups simulates the situation achieved in the protein by the appropriate positioning of histidyl residues in the active site. The use of "protein engineering" techniques (see discussion later in this chapter) offers the possibility of extending the reconstruction approach to build from scratch an entire enzyme, one with entirely novel but predictable catalytic properties.

Another useful approach to the study of the active site is to use the substrate itself as a chemical probe. There are a number of enzymes which actually form covalent enzyme-substrate intermediates, which are known as *enzyme-substrate compounds*. The relatively stable covalent bond between enzyme and substrate provides a reliable handle for identifying the functional groups actually involved in the catalytic reaction.

Chymotrypsin is a good example of an enzyme forming a covalent intermediate with its substrate. Evidence for this first appeared from studies which demonstrated the accumulation of appreciable amounts of the ES compound at low pH. By using the substrate trimethyl acetic acid it was even possible to crystallize the acyl-enzyme compound and to demonstrate that the hyperreactive serine 195 is actually involved (Figure 5–20).

These and other studies of chymotrypsin led to the proposal of an overall mechanism for the action of chymotrypsin shown in Figure 5–21. The revelation by x-ray diffraction analysis that serine 195 and histidine 57 are in fact brought close to each other in the tertiary structure of the enzyme was a dramatic confirmation of the earlier chemical research.

Many studies, such as the ones we have described, have been carried out with a large variety of enzymes. From these and also from the results of x-ray crystallography, there has emerged a series of concepts attempting to explain the high efficiency and specificity of enzyme action.

Theories Accounting for the Efficiency and Specificity of Enzyme Action

The foregoing account of research on enzymes suggests that their specificity has to do with the precision of fit between enzyme and substrate, while the efficiency

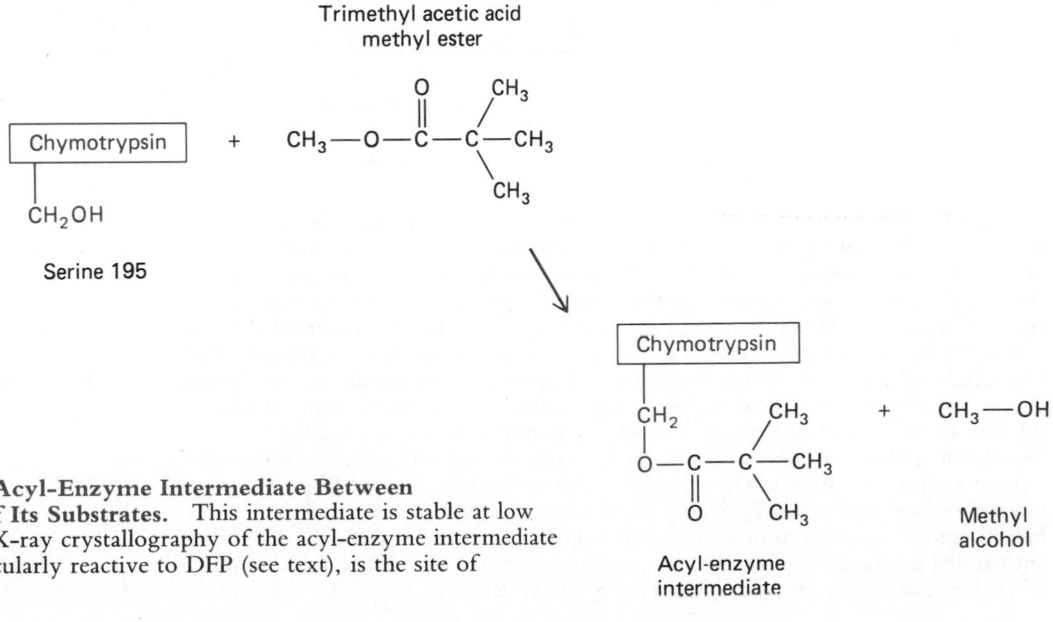

FIGURE 5–20
Formation of a Covalent Acyl-Enzyme Intermediate Between Chymotrypsin and One of Its Substrates. This intermediate is stable at low pH and can be crystallized. X-ray crystallography of the acyl-enzyme intermediate reveals that serine 195, particularly reactive to DFP (see text), is the site of attachment.

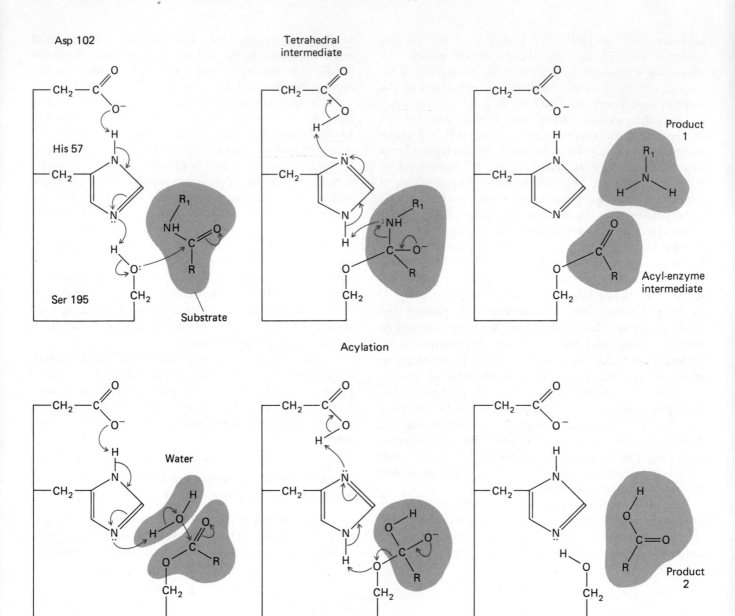

Acylation

Deacylation

(a)

FIGURE **5–21**
The Structure and Function of Chymotrypsin. (a) Proposed reaction
mechanism. The attack of the serine hydroxyl group on the carbonyl carbon atom
of the substrate is aided by both general base and acid catalysis, a proton being
alternately accepted and donated by the imidazole ring of histidine 57. The side
chain of aspartate 102 may temporarily store the proton. In a symmetrical fashion,
the decomposition of the tetrahedral intermediate also proceeds via a general acid-
base catalyzed pathway. A water molecule takes the place of the amino part of the
original substrate during deacylation, which is essentially the reverse of acylation.
(b) Older style model of the structure of chymotrypsin at 0.2 nm resolution,
depicted as ribbons mounted on a board by rods, two of which represent twofold
axes of symmetry (dyads). The stability of the molecule is derived from interchain
interactions—disulfide bonds (color) and hydrogen bonds. There is only one short
helical region. Chains can be seen to run next to each other in parallel or
antiparallel β-pleated sheets. The close juxtaposition of serine 195 and histidine
57 (color) can readily be seen. (Courtesy of D. M. Blow.)

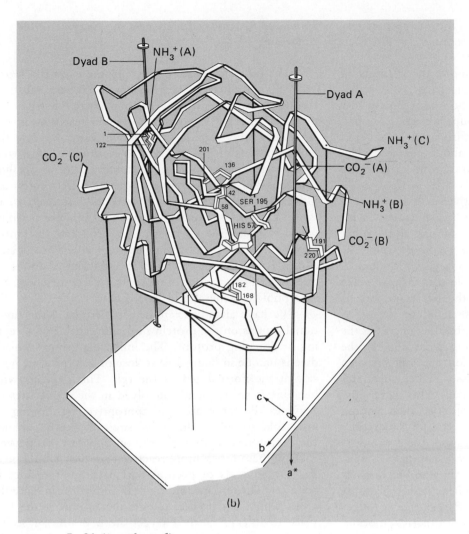

(b)

FIGURE **5–21 (continued)**

of catalysis depends on lowering the energy of activation (and thus the enhancement of reactivity) brought about by the interaction between enzyme and substrate. We shall therefore group our explanations into these two broad categories although, as we shall see, these categories overlap.

Specificity of Binding Between Enzyme and Substrate

The substrate binds to the enzyme with great precision. This binding involves at least three points of contact with the enzyme. The same forces responsible for the tertiary folding of proteins are believed to operate here as well:

- *Ionic interactions* between oppositely charged groups which, though weak in water, are extremely strong when water is excluded.

- *Hydrogen bonds,* which are directional and therefore important for aligning the enzyme and substrate. The effect of such hydrogen bonds is weakened by competition from water, but greatly enhanced when water is excluded.

- *Van der Waals interactions,* which are the weakest interactions but can be numerous and therefore add significant binding energy. Repulsive interactions of the van der Waals type, which give rise to steric hindrance, are very important in restricting the position and orientation of the substrate in the active site.

- *Hydrophobic effects,* which involve the same groups as van der Waals interactions. In excluding water, these interactions provide a large entropic contribution to the free energy of binding (see Chapter 3).

Covalent bonds occur in some enzyme-substrate interactions and, once formed, add to the strength of binding.

These interactions, however, are more appropriately considered to enhance chemical reactivity, as discussed below.

To organize our thinking about the relation of structure and function, it is convenient to classify structural groups involving the interaction between enzyme and substrate into three classes: groups providing binding energy, groups bringing about conformational changes in the enzyme, and groups bringing about enhanced reactivity. Of course, the same structural group may belong to more than one class.

The tertiary structure of the active site has evolved so that these basic interactions can be made strongly with only the correct substrate molecule, among all the other molecular structures that are found within the cell. The active site can be considered as stationary, while potential substrates collide with it. Those molecules that can make some interactions with groups in the active site will be bound to the extent that their interaction is strong, as indicated by the *binding energy*. The binding energy is the sum of all the interaction energies and is measured by the value of the equilibrium dissociation constant between enzyme and substrate. As we discussed earlier, substrates with low Michaelis constant values also have low dissociation constant values and, therefore, high binding energies. It is typically found that the physiologically correct substrate molecule shows the strongest interactions—that is, among possible competitors, the correct substrate has the tightest binding and the highest binding energy. Tight binding implies that the substrate will remain associated with the active site for a longer time, despite the disruptive influence of thermal motion. The delay allows catalysis to occur.

Tight binding may also have one or more of the following effects:

1. It *increases the effective concentration* of the substrate in the precise region of the reactive groups on the enzyme. There can be increases in reaction rates of greater than 10^5-fold due to this effect.

2. It *aligns the reactive groups* of the substrate with great precision not only to increase its concentration but also to bring about its optimum orientation. The restriction of possible rotations may bring about a 10^3-fold enhancement of rate. (The first two effects of tight binding reduce the entropy of activation.)

3. It may help create a *hydrophobic environment* in which the interactions of certain structural groups on both enzyme and substrate become enhanced by *removing water of solvation*. This can in special cases increase the reaction rate by more than 10^5-fold.

Enhanced Reactivity Induced by the Enzyme in the Substrate

The other influence of the enzyme is on the *kinetic pathways* by which catalyzed reactions take the substrate through greatly reduced activation energy barriers. In order to increase the rate of organic reactions in the laboratory, the organic chemist will often increase the concentrations of H^+ or OH^- ions. Proteins, however, are not generally stable in extremely acid or alkaline solutions. Instead, enzymatic catalysis frequently involves a process known as *general acid-base catalysis,* in which various functional groups can act as *proton donors* or *proton acceptors* (see Chapter 3). It turns out that most organic reactions occurring in the cell, whether the addition or removal of water, the splitting of amide or ester bonds, or a variety of molecular rearrangements, are susceptible to general acid-base catalysis.

We have already examined in Chapter 4 the functional groups on the protein molecule capable of accepting or donating protons. The imidazole group of histidine is unique in being able to lose or gain protons with ease at the normal pH of the cell. This explains why histidine is so frequently involved in the active sites of enzymes. Furthermore, the appropriate positioning of imidazole groups in the active site can greatly enhance their proton-accepting or proton-donating properties, making them much more reactive than they would be as free amino acids or even as part of simple peptides in solution. The tertiary folding of the enzyme can place the atom which donates or accepts the proton at the best place to interact with a target atom of the substrate, which in turn is held in the active site by binding to the protein's tertiary structure.

A number of enzymes form *covalent compounds* with their substrates, making use of serine, cysteine, histidine, or lysine residues as *nucleophilic groups* to attack the relevant nuclear center of the substrate (Figure 5–22). (Nucleophilic attack is discussed in Chapter 3.)

It should be understood that the enzyme's simply providing a nucleophilic group does not by itself enhance the rate of the reaction. It is necessary that the formation and breakdown of the covalent enzyme-substrate intermediate involve lower energies of activation than the uncatalyzed reaction. We have already noted this phenomenon in our discussion of functional groups on the enzyme that exhibit enhanced reactivities to alkylating probes—for example, the reactivity of the serine 195 in chymotrypsin, relative to serine free in solution.

Let us pause a moment to reconsider the structural properties of the enzyme and substrate during the

Nucleophile residue	Enzyme	Intermediate
Seryl — O: H	Serine proteases phosphoglucomutase	Acyl-enzyme phosphoryl enzyme
Cysteinyl — S: H	Thiol proteases, Glyceraldehyde-3-phosphate dehydrogenase	Acyl-enzyme
Histidyl — C=CH / HN, N: / C / H	Phosphoglycerate mutase Histone phosphokinase	Phosphoryl enzyme
Aspartyl — C — O: ‖ H / O	Pepsin $Na^+ - K^+$ ATPase	Acyl-enzyme Phosphoryl enzyme
Tyrosyl — O: H	Glutamine synthetase	Adenylenzyme

(a)

Uncatalyzed reaction:

$$RX \ + \ H_2O \ \xrightarrow{\text{slow}} \ ROH \ + \ X^- \ + \ H^+$$

Catalysis by a nucleophilic reagent Y:

$$RX \ + \ Y \ \xrightarrow{\text{fast}} \ RY \ + \ X^-$$

$$RY \ + \ H_2O \ \xrightarrow{\text{fast}} \ ROH \ + \ Y \ + \ H^+$$

Resulting reaction:

$$RX \ + \ H_2O \ \xrightarrow{\text{fast}} \ ROH \ + \ X^- \ + \ H^+$$

(b)

FIGURE 5–22
Nucleophilic Catalysis. This mechanism provides a pathway in which the formation of a covalent ES compound lowers the overall energy of activation. (a) Some nucleophilic amino acid residues in proteins. (b) How the formation of an ES compound through the action of a nucleophilic group in the enzyme active site increases the reaction rate. Even though participating in a covalent bond, the nucleophilic group is restored unchanged at the end of the reaction.

reaction. We previously introduced the idea that these molecules must have a certain activation energy in order to form an activated complex. During the reaction, these molecules are said to be in a **transition state,** a term evoking the image of a system that is midway between the two more stable structures we call reactant and product. The transition state is conceived as an interme-diate in the reaction mechanism and as having a specific structure—tetrahedral intermediates, etc. (see Chapter 3 and the following discussion for examples). In the case of an enzyme-catalyzed reaction, the structure of a transition state could in principle include a different conformation of the enzyme as well as a different structure of the substrate.

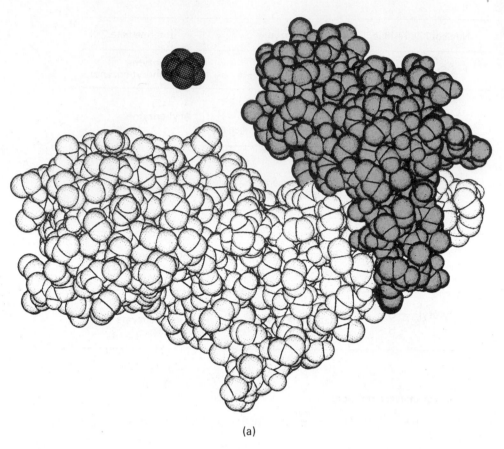

(a)

FIGURE 5–23

Change in the Conformation of Hexokinase Upon the Binding of Glucose.
(a) Computer-generated space-filling drawings of free hexokinase and glucose.
(b) The hexokinase-glucose complex. Note how the cleft between the large lobe (lightly shaded) and the small lobe (more darkly shaded) narrows, increasing specific contacts with glucose (cross-hatched shading) and reducing accessibility of the solvent. (Courtesy of W. S. Bennett and T. A. Steitz.)

It was originally suggested by J. B. S. Haldane and L. Pauling that the enzyme enhances reactivity in the substrate by *putting the substrate under strain*. The suggestion was that tight binding might bring about a stretching or change in angle of some relevant bond in the substrate, thereby lowering the reaction's energy of activation. In visualizing this strain, enzymologists think of the active site's surface as more complementary to the structure of the transition state than to the structure of the substrate in solution. Binding the substrate would not achieve the maximum interaction energy until the transition state is reached. Such a mechanism suggests that enzymes might be rather stiff molecules. According to this view, the secondary and tertiary structural levels of the polypeptide chain are there simply to bring the right residues

(frequently widely separated in the primary structure of the peptide) into proper juxtaposition to produce an active site in which the functional groups are accurately positioned and rigidly held.

There is, on the other hand, optical and other evidence that *enzymes change their conformation* when they combine with substrate (Figure 5–23). This *induced fit*, first suggested as a general mechanism of enzyme action by D. Koshland, may bring a certain functional group into the proper position with respect to the substrate, especially in a situation in which the group may have to be out of the way to permit binding the substrate in the first place.

It may seem that the concepts of strain and induced fit contradict each other—that an enzyme cannot be both

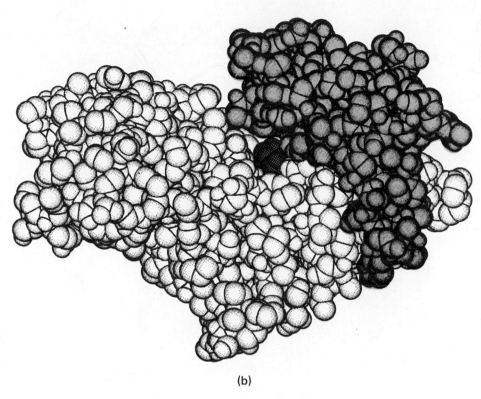

(b)

FIGURE 5–23 (continued)

rigid and flexible at the same time. This is not necessarily true, for one can imagine that some parts of the active site are relatively rigid and that binding substrate to those portions brings about a stress in the substrate, while other parts of the active site become distorted by the binding of the substrate, thereby bringing about a conformational change in the active site that accelerates the reaction.

The energy required to bring about conformational changes in the enzyme and strain in the substrate comes from the energy of binding. W. P. Jencks proposed that this energy is much larger than some previously thought. The energy of binding, being maximized after a change in the conformation of the substrate and enzyme, might be primarily responsible for the great increase in the rate of reaction of the correct substrate. This idea predicts that if an analogue of the substrate were used—one which already had a structure similar to that of the transition state—the binding observed should be tighter. The prediction of tighter binding by such *transition-state analogues* has been borne out by experiment.

Another way of testing the idea is to bind to an enzyme a substrate analogue that has a smaller size than normal (a "partial substrate") and measure the catalyzed rate of transfer of some group to another partial substrate that is also present. W. J. Ray, Jr., and his colleagues have

measured the effect of binding various partial substrates of phosphoglucomutase on the "phosphorylation" of water. Phosphoglucomutase catalyzes the isomerization of glucose-1-phosphate to glucose-6-phosphate (Figure 5–24) and has an active serine residue which donates and accepts covalently bound phosphate groups during catalysis. Glucose-1,6-diphosphate is an intermediate during the reaction; it ordinarily returns a phosphate group to the enzyme, yielding glucose-6-phosphate. The isolated phospho-enzyme intermediate can transfer the phosphate (phosphorylate) to many different acceptors, including water, but the rate of transfer to the normal substrate, glucose-1-phosphate, is 10^{10} times as rapid as the transfer to water. The startling discovery from Ray's group emerged when they measured the rate of transfer of phosphate from phospho-enzyme to water when the active site was also occupied by xylose-1-phosphate, a smaller sugar whose other hydroxyls are not placed appropriately to accept a phosphate. The rate was increased over 10^5-fold! The presence of xylose-1-phosphate could hardly have affected the chemical reactivity of the accepting group (the water hydroxyl). Rather, the energy of binding xylose-1-phosphate must have been partly used to change the structure of the enzyme to one more closely resembling the transition state and thus

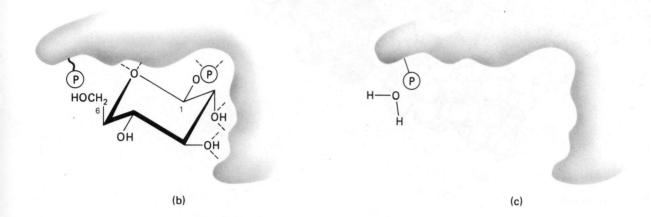

Glucose-1-phosphate   Glucase-6-phosphate

(a)

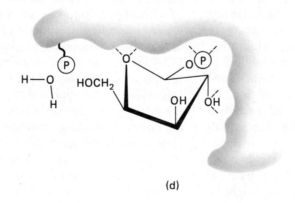

(b)   (c)

(d)

<small>FIGURE</small> **5−24**
Effect of Substrate Binding on Catalytic Activity. (a) The reaction catalyzed by phosphoglucomutase. Glucose-1-phosphate is isomerized to glucose-6-phosphate. (b) The reaction pathway includes the compound glucose-1,6-diphosphate and an enzyme-bound phosphate. (c) The phospho-enzyme can transfer its phosphate to water, a reaction equivalent to hydrolyzing the phosphate from the enzyme. This reaction is much slower, by a factor of 10^{10}, than the transfer to glucose-1-phosphate. (d) Effect on enzyme activity of binding a "partial" substrate, xylose-1-phosphate. By its binding, this transition-state analogue is thought to change the shape of the enzyme to promote catalysis. As the result of binding, the rate of donating phosphate to water is enhanced by a factor of 10^5. Xylose-phosphate, being smaller, cannot itself accept the phosphate.

increased the rate of reaction (Figure 5–24d). This result thus provides striking evidence for the *reduction of activation energy barriers by binding of a "substrate."*

The Structure and Function of Lysozyme

Lysozyme is an enzyme found in human saliva and tears that can bring about the rupture of bacteria by hydrolyzing a specific linkage in the polysaccharide chains of bacterial cell walls (Figure 5–25). The lysozyme molecule is composed of a single chain of 129 amino acids linked in four places by disulfide bonds (Figure 5–26). The amino acid sequence and position of the disulfide pairs were worked out independently by P. Jolles and R. E. Canfield and their respective co-workers. In 1960 the first isomorphous replacements were prepared (see Chapter 4), and in 1962 a low-resolution image of the molecule was worked out by D. C. Phillips and co-workers.

In 1965 a high-resolution image of lysozyme (0.2 nm = 2 Å) was completed. This gave a reasonably detailed picture of the conformation of the polypeptide chain (Figures 4–55 and 5–27). The helical content of lysozyme is less than one-half of the 75% observed in myoglobin, and some of the helices are different from the classical α-helix of Pauling and Corey: the —C=O and —NH groups are not as close as they are in the α-helix. Furthermore, the lysozyme molecule contains a structure, the "antiparallel pleated sheet," which Pauling and Corey predicted to occur in certain fibrous proteins. The remainder of the structure does not appear to have any obvious regularity except that, just as in myoglobin, the polar side chains are on the outside of the molecule interacting with the water, whereas the nonpolar side chains are buried in the interior. One other interesting feature of the molecule is that the first 40 residues from the amino-terminal end form a compact helical structure.

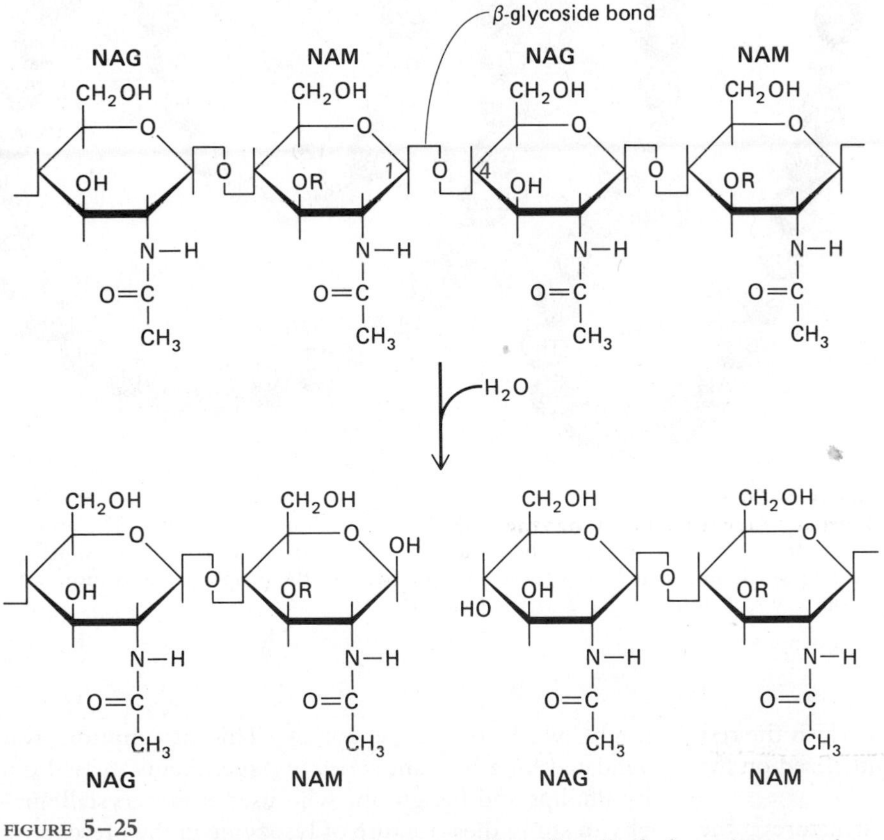

FIGURE 5–25
The Reaction Catalyzed by Lysozyme. The cell wall of a bacterium contains long polysaccharide chains, consisting of *N*-acetylglucosamine (NAG) and *N*-acetyl muramic acid (NAM), linked to each other as shown. Lysozyme catalyzes the hydrolysis of the bond between C-1 of NAM and C-4 of NAG, but not between C-1 of NAG and C-4 of NAM.

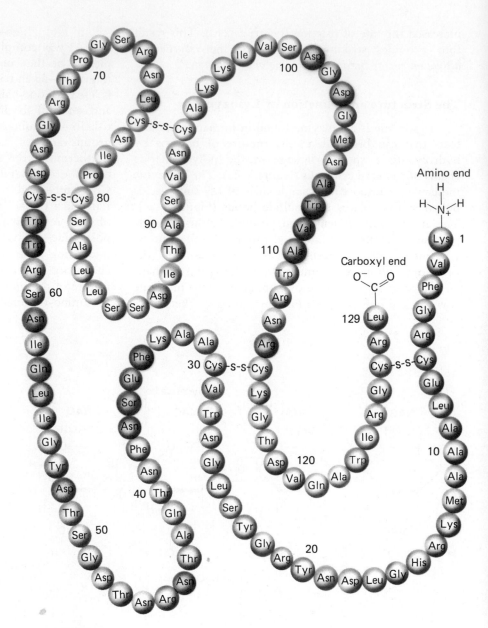

FIGURE 5–26
Primary Structure of Lysozyme.   Four disulfide bonds are shown. The amino acids forming the active site of the molecule are in color. The three-dimensional folding of the molecule brings these amino acids into the proper juxtaposition. (Courtesy of J. Jolles, J. Jarequi-Adell, and P. Jolles.)

This could act as a "substructure" around which the rest of the molecule is folded as it is being synthesized on the ribosome.

An intriguing aspect of lysozyme's structure is the *cleft* that runs down the middle of the molecule (Figure 5–27). It was tempting to assume that this cleft represented the binding site for the polysaccharide chain which is the enzyme's normal substrate (there was at the time very little notion from solution chemistry how catalysis

might work in this enzyme). This assumption was validated by a brilliant series of experiments carried out by Phillips and his group, who used x-ray crystallography to study the structure of lysozyme in the presence of a number of competitive inhibitors, such as tri-N-acetyl glucosamine, that closely resemble the normal substrates (Figure 5–28a). These studies showed the precise location of the competitive inhibitors in the binding site and provided the basis for a model of the interactions between

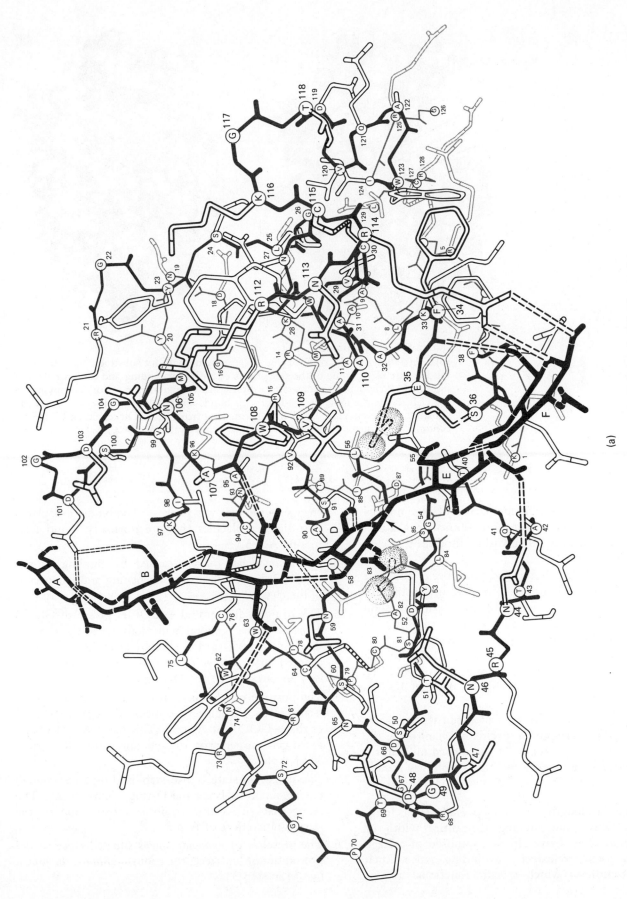

(a)

FIGURE 5–27
Three-Dimensional Structure of Lysozyme. (a) The precise location of the substrate on the active site of the enzyme. The amino acid side chains involved in the interaction of substrate with enzyme are shown, as well as the position of the four disulfide bonds. The four spheres represent oxygen atoms belonging to the carboxyl groups of a glutamyl and an aspartyl residue; they are involved in the mechanism of catalysis. (Courtesy of M. O. Dayhoff and R. V. Eck; drawn by Geis.)

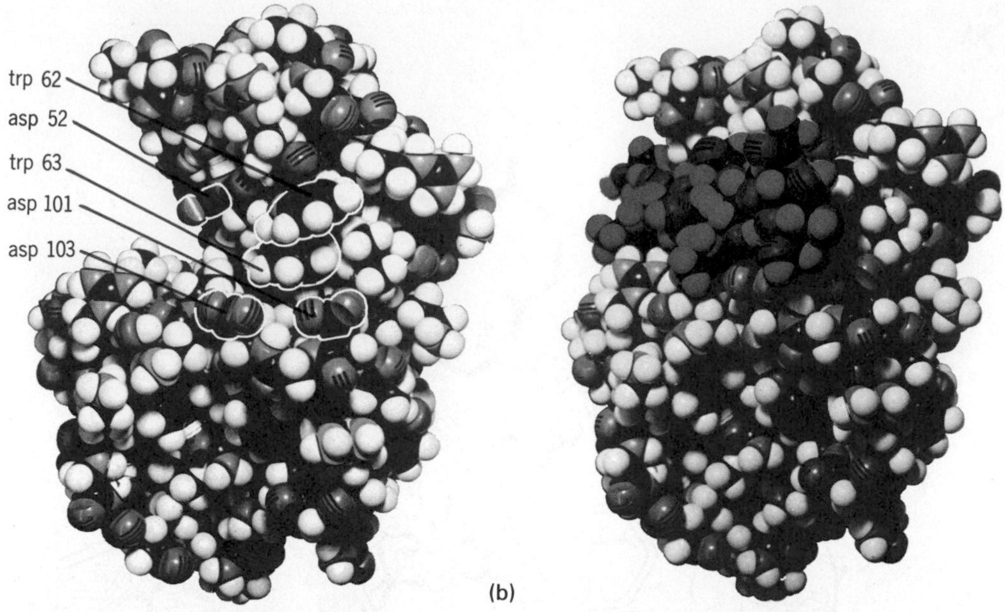

(b)

FIGURE **5–27 (continued)**
(b) Space-filling model of lysozyme. *Left:* enzyme without substrate, showing binding site cleft. Some of the side chains essential for binding and catalysis are indicated. *Right:* enzyme-substrate complex showing substrate (color) occupying the cleft. Note the close fit, via hydrogen bonds and van der Waals contacts, of the substrate to the enzyme. Used with permission from R. E. Dickerson and I. Geis, *Structure and Action of Proteins* (Harper & Row, New York, 1969.)

enzyme and the polysaccharide chain of the normal substrate. (Model building from three-dimensional structural information is now a standard method for guessing the mechanism of catalysis.)

Some six hydrogen bonds and a number of nonpolar interactions seem to be involved in binding substrate to the active site cleft (Figure 5–28b). Furthermore, upon binding of the substrate, a small change in conformation occurs to cause the cleft to narrow and deepen slightly. This, then, was the first direct confirmation of the induced fit theory of Koshland, although, admittedly, the extent of the conformational change induced by the substrate is very slight.

It is not possible to discuss in detail the evidence which led Phillips to suggest a specific mechanism for the enzymatic action of lysozyme. The proposed mechanism itself is illustrated in Figure 5–29 and involves several features:

1. A substantial length of *substrate binds* to a long, cleft-like active site running the entire width of the enzyme. The active site is composed of a large region devoted exclusively to binding and a much smaller region in which specific functional groups

engage in the catalytic process. The binding displaces two water molecules that are normally found near residue Asp 52.

2. A *conformational change,* caused by binding of the substrate, brings about the deepening and narrowing of the cleft.

3. The strategically located —COOH group of Glu 35 *donates a proton* (general acid catalysis) to the oxygen that joins rings D and E, thereby creating a positively charged *carbonium ion* (= *carboxonium ion* = carbocation) on carbon 1 of the D-ring.

4. The positive carbonium ion of the transition state is *stabilized* by the closely positioned negative charge of the —COO$^-$ of Asp 52. This has been calculated to lower the energy of activation by more than 30 kJ/mole.

5. The transition state is also stabilized by a *distortion of the substrate,* involving the D-ring (Figure 5–30). The stress that elicits this strain is mostly due to the electrostatic effect of Asp 52.

6. The presence of *nonpolar groups* situated close to the proton donor and acceptor groups *facilitates the proton transfer* in step 3.

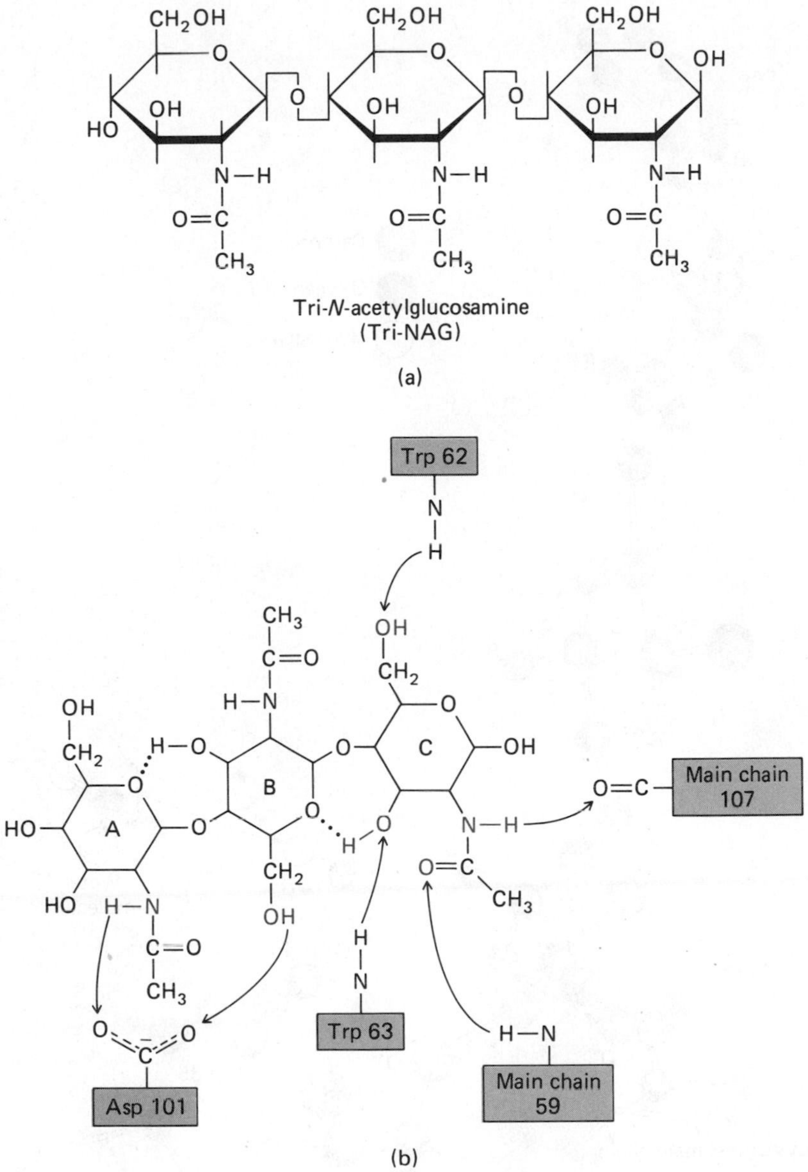

Tri-*N*-acetylglucosamine
(Tri-NAG)

(a)

(b)

FIGURE 5–28
Tri-N-Acetylglucosamine (tri-NAG), a Competitive Inhibitor of Lysozyme. (a) The structural formula of tri-NAG. This inhibitor is one of the probes used to explore the active site of lysozyme. (b) Interaction involving six hydrogen bonds between tri-NAG and lysozyme. Tri-NAG inhibits because, although it closely resembles the normal substrate, it occupies only half the cleft. The normal substrate is cleaved at the bond between rings D and E (see Fig. 5–29), which bind in the other half of the cleft. Using the example of the binding interactions between the enzyme and rings A, B, and C of tri-NAG, it was possible to build a model for the binding of rings D, E, and F of the normal substrate. The D ring is probably distorted relative to its normal chair conformation in the carbonium ion transition stage.

7. The ES complex partly breaks up by the release of the first reaction product (ring E).
8. *Water supplies an OH group,* to combine with the carbonium ion, *and an H⁺,* to protonate the —COO⁻ of Glu 35.
9. The second product (rings A through D) is released.

We mentioned previously the probable relationship between catalysis and the energy of binding substrate. The results of two studies involving lysozyme demonstrate this relationship very clearly. Substrate analogues containing four saccharide rings (A–D) but missing the hydroxymethyl group of ring D (R_1 in Figure 5–29) bind to lysozyme with a free energy that is over 20 kJ/mole more favorable than an analogue containing the group. It appears that the hydroxymethyl group of ring D N-acetylglucosamine bears on a part of lysozyme that can only give way by having the whole enzyme conformation change. The presence of the hydroxymethyl group stabilizes the enzyme-substrate conformation that is necessary for catalysis. Another analogue of four rings, one containing a bond arrangement around carbon atom 1 of the D-ring that is planar (and thus resembling the proposed carbonium ion intermediate), binds more tightly than the normal four-ring analogue by a factor of 6×10^{10} (equivalent to 22 kJ/mole more favorable free energy of binding). These data are in agreement with the hypothesis that the energy of binding substrate is of great importance in speeding the reaction rate, especially when

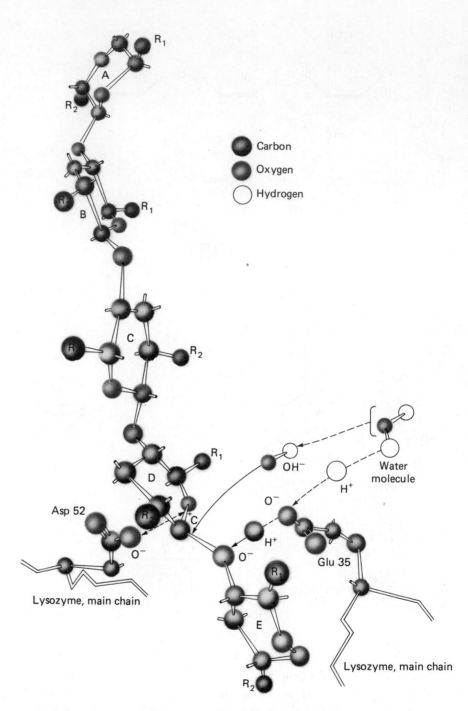

FIGURE 5–29

Proposed Mechanism of Action of Lysozyme. Residues Glu 35 and Asp 52 are directly involved in splitting the C—O bond in the polysaccharide chain. A hydrogen ion dissociates from the —OH group of Glu 35 and associates with the oxygen atom that joins rings D and E, thus breaking the bond between the two rings. This leaves carbon atom 1 of the D ring with a positive charge, a form known as a carbonium ion. The positive atom is stabilized by the negatively charged side chain of residue Asp 52 (arrow). The surrounding water supplies an OH⁻ to combine with the carbonium ion and an H⁺ ion to replace the one lost by residue 35. The two parts of the substrate then fall away, leaving the enzyme free to cleave another polysaccharide chain. (From D. C. Phillips: 1966. Reprinted with permission, *Scientific American*, W. H. Freeman and Co.)

the energy is used to favor the formation of an appropriate transition state.

Although we have not described all the quantitative details that would account for the velocity of the catalyzed reaction, the x-ray diffraction study of lysozyme offers us one of the first high-resolution models in which the structure and function of the enzyme have been related to each other. Interestingly, most of the concepts formulated prior to this important work received some independent and direct support. These include: (1) the binding of the substrate to an active site on the enzyme, (2) the induction by the substrate of a conformational change in the active site, (3) the distortion of some bonds in the substrate as a result of its binding to the enzyme, and (4) the presence of some specially reactive side chains in the neighborhood of the bond to be split.

There are now hundreds of enzymes whose catalytic mechanisms are understood in part and whose crystal structures have been solved at nearly atomic resolution.

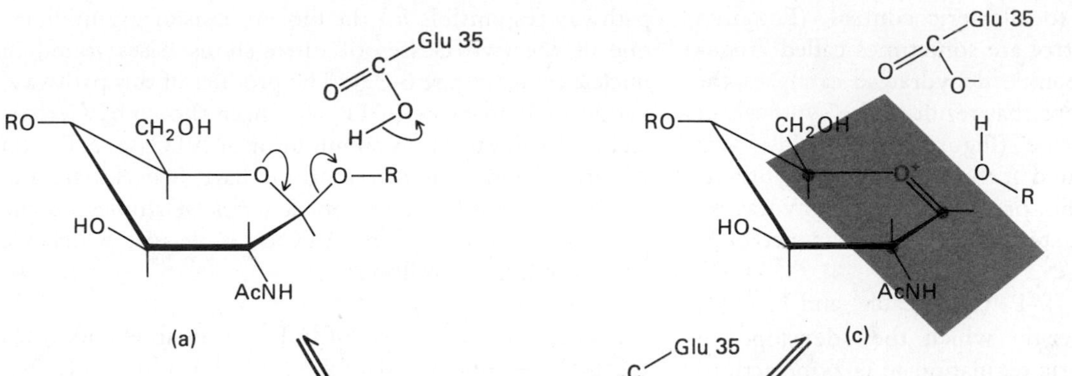

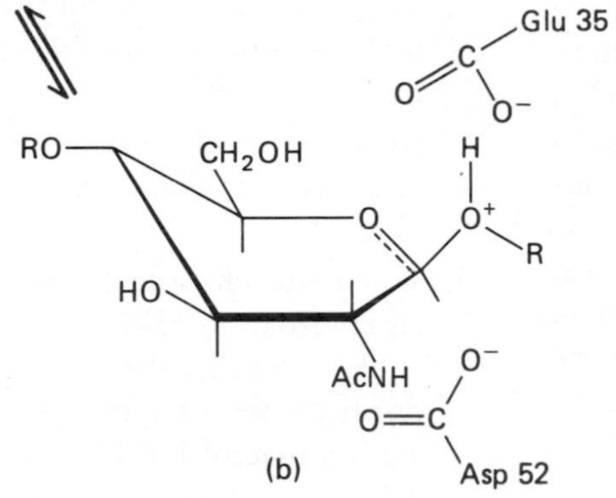

FIGURE 5–30

Distortion of the Substrate D-Ring in the Transition State of Lysozyme. (a) The sugar residue in its usual chair conformation, before binding to enzyme. (b) Upon binding, the oxygen atom and carbon 5 move into the same plane as carbons 1 and 2. This conformation is known as a half-chair or "sofa." (c) The distortion stabilizes the planar carbonium ion, a transition state in the catalytic mechanism. (Based on D. C. Phillips, 1966. *Scientific American*, W. H. Freeman and Co.)

Lysozyme was merely the first to be studied in this way. In reviewing the structure and mechanism of this enzyme our purpose is to provide examples of the strategy used in this type of investigation rather than to look systematically at all the methods and results. There are references in the Selected Readings at the end of this chapter to more comprehensive treatments of the subject.

Allosteric Regulation of Enzyme Activity

As we have seen, the activity of an enzyme is greatly affected by factors such as pH and the concentration of other ions and coenzymes. In the cell, however, neither the pH nor ion concentrations vary to any great extent. Therefore, in order to regulate the activity of enzymes, the cell relies on more subtle interactions. These can be grouped into two general categories:

1. *Covalent modulation* of enzyme activity, brought about by adding a ligand (small molecule) to or removing it from the enzyme by covalent bond formation or breakage. We have already examined such mechanisms as applying to proteins in general in Chapter 4 and we shall have more to say in Chapter 14 about the matter, including the control of the enzymes that catalyze the formation and removal of the covalent modifications.

2. *Noncovalent control,* brought about by adding or removing a ligand through noncovalent interactions. This form of regulation can produce *graded and rapidly reversible responses* in the rate of catalysis, rather than the more abrupt and longer-lived response typically observed in covalent modulation.

In the last 25 years, an increasing number of enzymes have been discovered that are specifically inhibited by compounds whose structure resembles neither the substrate nor the product of the inhibited enzyme. Molecules that do not resemble the substrate but bind to the enzyme and exert an effect are called **allosteric** (allo-steric = other-shape). The difference in structure led to the supposition that such inhibitors probably bind at a different site on the enzyme from the active site; this location is sometimes called the "allosteric site." The hypothesis was confirmed by later x-ray diffraction studies. Enzymes that are subject to allosteric control usually do not display simple Michaelis-Menten kinetics. We shall discuss the metabolic role of allosteric enzymes in Chapter 6 and examine them again in Chapter 14 within the context of cellular regulation in general. Here we shall discuss only the molecular mechanism of allosteric regulation.

L-threonine dehydratase (commonly known as threonine deaminase) was discovered by H. E. Umbarger and B. Brown in 1958. It was one of the earliest enzymes studied that is subject to allosteric control. (Enzymes subject to allosteric control are sometimes called "regulatory enzymes.") Threonine dehydratase catalyzes the first of a sequence of steps that result in the synthesis of the amino acid isoleucine (Figure 5–31). Note that isoleucine, which is located at the end of its biosynthetic pathway, acts as an inhibitor of threonine dehydratase, which catalyzes the first step of this particular metabolic pathway.

In 1963, J. Monod, J.-P. Changeaux, and F. Jacob wrote a prophetic paper in which they developed a general theory of allosteric regulation of enzyme action. They felt that the balance of evidence favored a model in which the catalytic and regulatory sites were located in different places on the enzyme. Furthermore, they suggested that another kind of regulatory site, one involving the action of *allosteric activators,* was also present in certain regulated enzymes. In the case of threonine deaminase they concluded that there are two sites for allosteric **effectors:** one to react with isoleucine, causing *inhibition;* and the other to react with valine, bringing about *activation.*

When Monod, Changeaux, and Jacob first proposed these general ideas, the evidence supporting them was fragmentary and inconclusive. Later, J. Gerhart and H. Schachman provided, in a beautiful series of studies, direct confirmation that the catalytic and regulatory sites can be located on different portions of an enzyme. They also showed that binding of the substrate induces in the enzyme a conformational change that is reversed by binding the inhibitor.

The enzyme they studied is *a*spartate *t*ranscarbamoyl synthet*ase* (ATCase), which catalyzes the first step of the pathway responsible for the biosynthesis of pyrimidines, one of the two classes of nitrogenous bases found in nucleic acids (Figure 5–32). The product of this pathway, cytidine triphosphate (CTP), has been shown by Gerhart and A. Pardee to act as an inhibitor of ATCase. ATP, on the other hand, is an activator. Gerhart, Schachman, and W. N. Lipscomb carried out a series of studies on the physical properties of the ATCase molecule, which we can summarize as follows:

1. The molecular mass of ATCase is about 300 kDa (kDa = one kilodalton = 1000 Da of molecular mass).

2. Acrylamide gel electrophoresis shows the presence of two different chains, of molecular mass 33 kDa and 17 kDa.

3. When the enzyme is treated with *p*-hydroxymercuribenzoate (PHMB), a reagent which reacts with —SH groups, the enzyme dissociates into two distinct types of subunits, with molecular masses of approximately 100 kDa and 34 kDa (Figure 5–33).

4. The large subunit of 100 kDa contains only the large polypeptide chains of 33 kDa and, therefore, three of them.

5. The small subunit of 34 kDa contains only the small polypeptide chains of 17 kDa and, therefore, two of them.

6. The large subunit binds three substrate molecules, one per chain.

7. The small subunit binds two CTP molecules, one per chain, and contains two Zn^{2+} ions, one per chain.

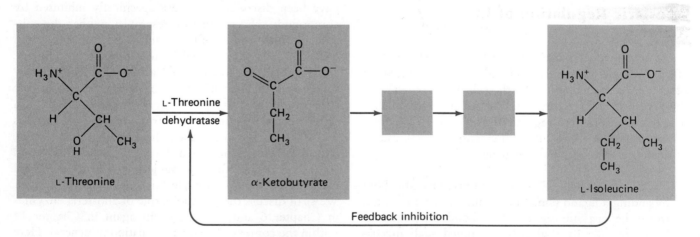

FIGURE 5–31
Biosynthetic Pathway for the Synthesis of Isoleucine. This pathway is controlled by feedback inhibition. Isoleucine, synthesized after a sequence of four enzymatic steps, regulates the rate of the first step by inhibiting the enzyme threonine dehydratase (threonine deaminase). An enzyme thus controlled is sometimes called a "regulatory enzyme." (From H. E. Umbarger and B. Brown.)

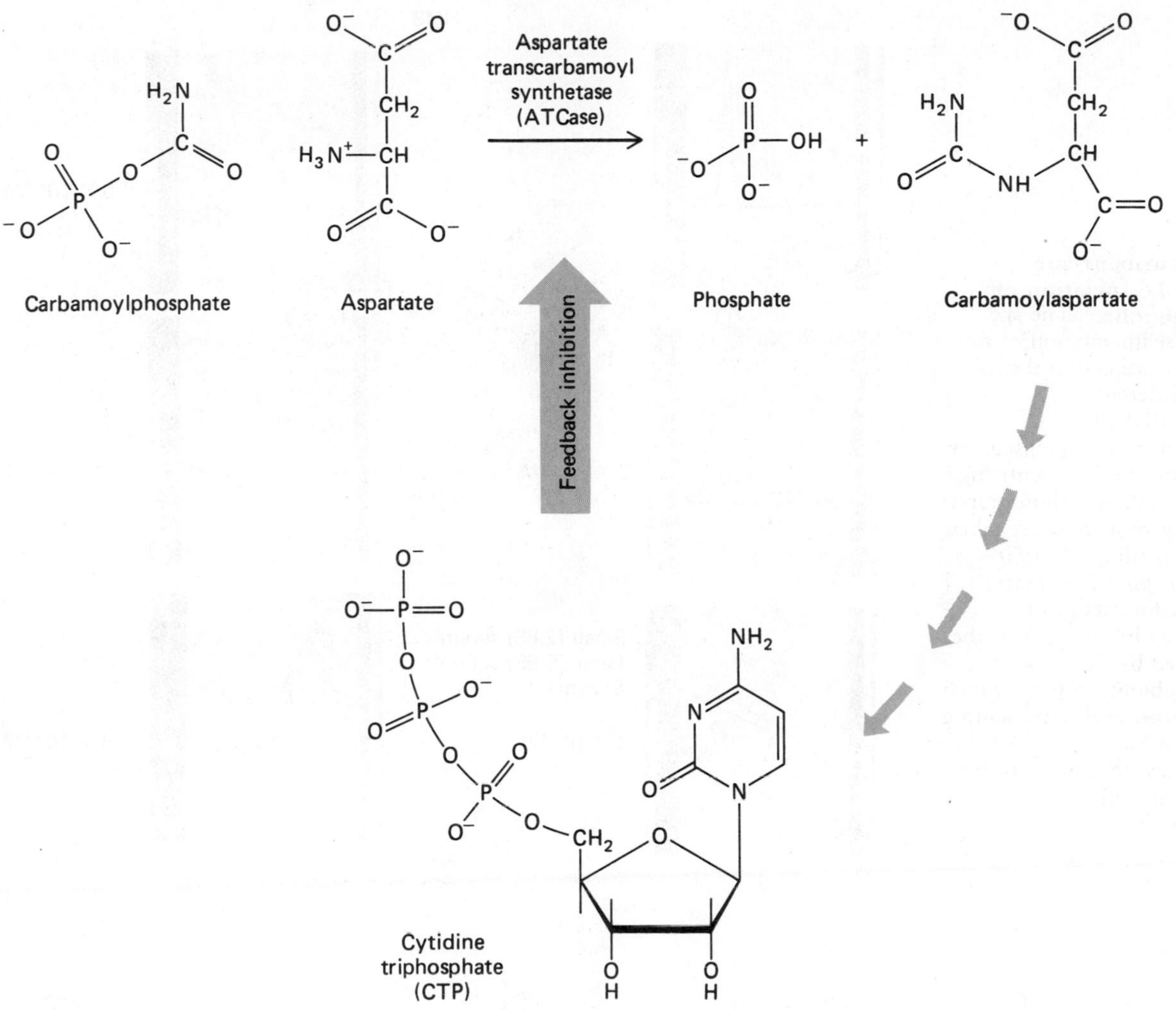

FIGURE 5–32
Biosynthetic Pathway for the Synthesis of Cytidine Triphosphate (CTP).
This pathway is controlled by feedback inhibition. CTP, formed at the end of a
sequence of enzymatic steps, regulates the rate of the first enzyme in the pathway:
aspartate transcarbamoyl synthetase (ATCase). CTP bears little structural
resemblance to the substrates or the product of the reaction catalyzed by ATCase.
(From J. Gerhart and A. Pardee.)

8. After separating the two kinds of subunits by zone
sedimentation in a sucrose gradient, Gerhart and
Schachman showed that *only the large subunit has
catalytic activity*. Moreover, it is no longer inhibited
by CTP (Figure 5–34).

9. Upon mixing the two subunits, however, one can
regenerate the inhibitory effect of CTP, presumably
by recombining the two subunits into a regulatory
enzyme (Figure 5–34).

10. The reconstituted molecule has the same molecular
mass as the native ATCase.

From these studies we can conclude: (a) there are two
kinds of subunits in the enzyme ATCase—a large
catalytic (C) subunit and a small regulatory (R) subunit;
(b) the stoichiometry of the molecule is like that diagrammed in Figure 5–35.

Lipscomb and his co-workers performed an ambi-
tious x-ray crystallographic study of this large and
complex enzyme molecule. Their 0.26 nm (= 2.6 Å)
resolution electron density map reveals a molecule in
which the regulatory chains are arranged peripherally and
the catalytic chains are contacting one another centrally

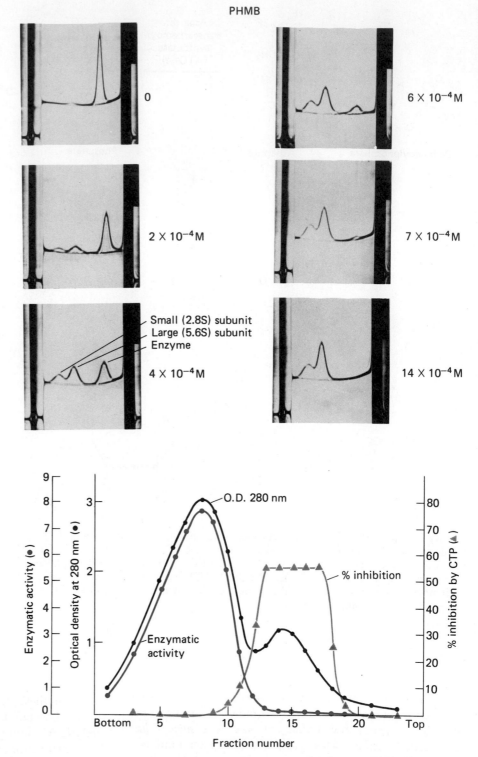

PHMB

FIGURE 5–33
The Effect of
_p_-Hydroxymercuribenzoate
(PHMB) on the Dissociation of
ATCase into Subunits.   The six
figures represent sedimentation of the
enzyme in the absence and in the
presence of five different
concentrations of PHMB.
Sedimentation is from the meniscus on
the left to the bottom of the centrifuge
cell, on the right. Shlieren photographs
of the sedimenting proteins were taken
in an analytical centrifuge after 50
minutes at 60,000 rpm. The intact
enzyme, which sediments most
rapidly, is indicated by the peak at the
right; it is followed by the peak
indicating large subunits (5.6 S), which
in turn is trailed by a peak representing
small subunits (2.8 S). At 14×10^{-4}
M PHMB, the enzyme appears to be
fully dissociated into subunits.
(Courtesy of J. Gerhart and
H. Schachman.)

FIGURE 5–34
Experiment Demonstrating the Catalytic Role of
Large Subunits and the Regulatory Role of Small
Subunits of ATCase.   After treating the enzyme with
_p_-hydroxymercuribenzoate, the dissociated subunits were
separated by sucrose density gradient centrifugation. Total
protein in the various fractions was estimated by the absorbance at 280 nm (●). The larger subunits, appearing in the
earlier fractions, contained all the enzyme activity (●); the
smaller subunits retained the property of binding CTP
(data not shown). When pooled material from the peak
containing enzyme activity was combined successively with
material from all the fractions and tested for inhibition by
CTP (▲), only the later fractions were active. This proves
that the subunits in the later fractions have the regulatory
role. (Courtesy of J. Gerhart and H. Schachman.)

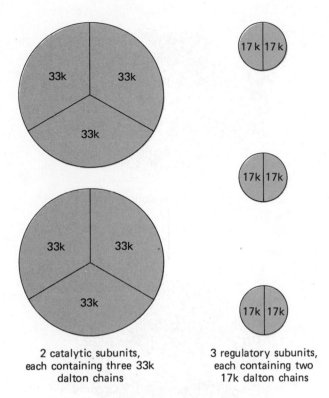

2 catalytic subunits,
each containing three 33k
dalton chains

3 regulatory subunits,
each containing two
17k dalton chains

2 × (3 × 33k) daltons + 3 × (2 × 17k) daltons ≈ 300k daltons

FIGURE 5–35
The Aspartate Transcarbamoyl Synthetase Molecule.
The 300 kDa molecule is composed of two catalytic
subunits, each containing three 33 kDa chains, and three
regulatory subunits, each containing two 17 kDa chains.

encountered in Chapter 4. There is a *cooperative effect* exerted by successive binding of substrate molecules to a number of catalytic sites on the enzyme. What we mean by this is simply that as the multiple catalytic sites on the enzyme become successively filled by the substrate, the affinity of the enzyme for the substrate increases, producing the accelerating effect shown by the part of the sigmoid curve at low concentrations of aspartate.

Since the multiple catalytic sites are situated on separate polypeptide chains, it follows that the effect of filling one site on others yet to be filled must be *transmitted by a conformational change* in the molecule. Gerhart and Schachman demonstrated that such a conformational change does occur in the ATCase molecule; binding succinate, a substrate analogue, caused the sedimentation coefficient to change about 3%. W. Lipscomb, R. Kantrowitz, and their co-workers found that binding a transition state analogue caused the catalytic subunits to move apart by 1.2 nm and turn by 12°, and the regulatory dimers to rotate 15° about their twofold axis (Figure 5–38).

The interaction between ATCase and its six substrate molecules is still cooperative when the reaction is inhibited by the presence of CTP (Figure 5–37). The inhibitor does not change the sigmoid shape of the

(Figure 5–36). One interesting feature of this model is that at the center of the molecule there is a large, central aqueous cavity, made accessible from the outside by a number of channels.

The series of studies by Gerhart and Schachman on ATCase brought direct evidence to bear on the theory of Monod, Changeaux, and Jacob that the catalytic and inhibition sites are on separate portions of the enzyme. It also confirmed their prediction that these sites could be located on distinct subunits of the enzyme molecule. And this leads us to the important question about how allosteric effectors activate or inhibit enzyme activity.

If one plots the initial velocity of the reaction catalyzed by ATCase against concentration of substrate (aspartate) in the presence and absence of the allosteric inhibitor (CTP) and the activator (ATP), one obtains a set of curves illustrated in Figure 5–37. Unlike the usual Michaelis-Menten hyperbolic curve, an "S-shaped" or *sigmoidal* curve is obtained in the absence of allosteric effectors. This is reminiscent of the relationship observed for the binding of oxygen with hemoglobin, which we

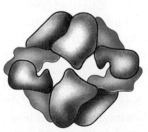

FIGURE 5–36
The Arrangement of Chains and Subunits in Aspartate Transcarbamoyl Synthetase. The molecule is composed of two catalytic subunits (colored), each consisting of three chains. Separating these two catalytic subunits are three regulatory subunits, each consisting of two chains. Note the large central cavity, an unusual aspect of the quaternary structure of this protein molecule.

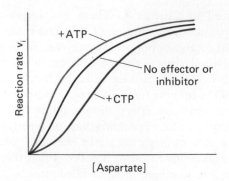

FIGURE 5–37

The Kinetics of the Enzyme Aspartate Transcarbamoyl Synthetase. The reaction rate is plotted against aspartate concentration in the presence and absence of the allosteric inhibitor CTP and the allosteric activator ATP. Notice the sigmoidal (S-shaped) curve in the case of the uninhibited and inhibited enzyme, in contrast to the usual hyperbolic curve in the presence of ATP. A simple analysis of sigmoidal kinetics cannot yield a true K_m, but it is possible to obtain an apparent K_m: the aspartate concentration when v_i is equal to one-half V_{max}.

reaction curve but merely makes the enzyme less responsive to substrate (as reflected by the shifting of the curve to the right). Thus, the cooperative kinetics are a property of the holoenzyme—the combination of the catalytic and regulatory subunits. It can be shown that the presence of the regulatory subunits is necessary for the cooperativity because the catalytic subunits by themselves do not exhibit sigmoid kinetics. It is also possible to "desensitize" the complete enzyme (make it unresponsive or insensitive to inhibitors like CTP) by heat treatment (Figure 5–39), by enzyme poisons such as mercurials, or by the effect of certain mutations. In these cases the kinetics are more like those expected from a simple Michaelis-Menten enzyme.

Allosteric activators like ATP also bind to the regulatory subunits. But the effect of activators is to make the enzyme kinetics less sigmoid and more Michaelis-like. As seen in Figure 5–37, this means the cooperativity is reduced and enzyme activity is more responsive to substrate. Since the allosteric effectors interact with different chains than the substrate molecules, they also must transmit their effect on the catalytic sites via some conformational changes in the molecule.

From such studies we emerge with a concept of the ATCase molecule carrying a number of catalytic and inhibitory sites which interact cooperatively and can "communicate" with one another via conformational changes in the molecule. The significance of these interactions is to adjust the binding and catalytic activities of the enzyme so they are appropriate to local cellular conditions.

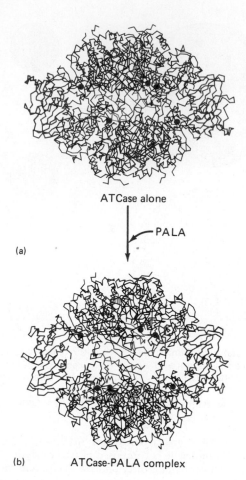

ATCase alone

PALA

(a)

(b)   ATCase-PALA complex

FIGURE 5–38

Stereoviews of the T state (CTP-liganded) and the R state (PALA-liganded) of ATCase. The transition from the T state to the R state separates the poles of the molecule by 12 Å and brings about a 5° rotation of the regulatory subunits. These changes in the quaternary structure are accompanied by conformational changes in the tertiary structure of each chain. (After K. E. Krause; K. W. Volz, and W. N. Libscomb.)

Cooperativity arises in general from the influence that binding a substrate has on the affinity of other subunits for substrate. If one subunit is unaffected by the presence of substrate on other subunits, there is no cooperativity. Binding the first substrate to a tetrameric enzyme may cause the affinities of all the subunits to increase to the maximum, giving high positive cooperativity. Or, the affinities may be increased by binding a single substrate, but not to maximal levels; binding subsequent substrates may further increase the affinity of the subunits.

The analysis of many enzymes reveals a variety of behaviors that include the two extremes but more frequently show an intermediate degree of cooperativity.

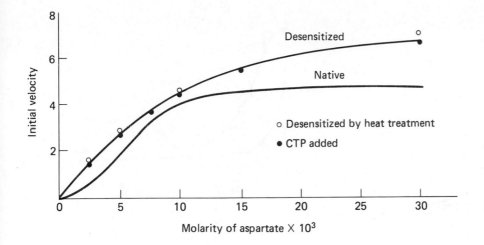

FIGURE **5–39**
Kinetics of Native and Desensitized Aspartate Transcarbamoyl Synthetase.   In the native form, the kinetics are *S*-shaped. After appropriate heat treatment, the kinetics become the usual Michaelis–Menten variety (○). The altered enzyme is also insensitive to the inhibitor CTP (●). (Courtesy of J. Gerhart and A. Pardee.)

There are also enzymes that respond to allosteric effectors, but which consist of only a single polypeptide chain and which exhibit ordinary Michaelis-Menten kinetics— for example, simple competitive inhibition. Furthermore, *negative cooperativity* has been encountered in some enzymes. In these cases the binding of a substrate molecule *decreases* the affinity with which subsequent substrate molecules are bound. One way this may occur is for oligomeric enzymes to be composed of polypeptide chains with identical amino acid sequences, but with tertiary structures that differ because of the substrates being bound to one but not to the other(s). When such chains occur in pairs (asymmetric dimers), they can give rise to the phenomenon of *half-of-the-sites reactivity,* in which only one of the two identical polypeptide chains functions catalytically.

It is appropriate to speculate a bit on the possible biological functions of such a large variety of regulatory arrangements. A little thought convinces one that the kinetic behavior of a Michaelis-Menten enzyme is often inappropriate to cell metabolism. Michaelis-Menten enzymes respond linearly to even very low concentrations of substrates, transforming them rapidly into products. If all enzymes responded to physiological substrate concentrations with kinetics like those of Figure 5–6b, it would be difficult to build up any concentration of metabolic intermediates. Now consider the behavior of a regulatory enzyme with substrate-induced positive cooperativity (Figures 5–37, 5–39 and, for comparison, 4–58). A certain minimum concentration of substrate must be present before the activity of the enzyme is significantly increased. This can be seen by noting that a given rate of reaction by ATCase requires a much greater concentration of aspartate in the absence of ATP than in its presence, when ATCase displays essentially Michaelis-Menten kinetics (Figure 5–37).

Equally nonadaptive, the effect on a Michaelis-Menten enzyme of all three types of reversible inhibitors that we have mentioned—competitive, uncompetitive, and noncompetitive—is to reduce the velocity relative to uninhibited conditions in the same way (Figure 5–40a). A small concentration of inhibitor, relative to the K_i, inhibits the enzyme substantially. If end-product inhibitors behaved as shown in Figure 5–40a, their biosynthetic pathways would be inhibited before much end product could accumulate. Since most end products (like isoleucine) are used elsewhere in metabolism (in protein synthesis, for example), this could be a serious disadvantage. Regulatory enzymes with allosteric control avoid this, as shown in Figure 5–40b. A certain *threshold concentration* of isoleucine can build up before significant inhibition of threonine deaminase occurs.

Negative cooperativity, shown in Figure 5–41, allows an enzyme to remain responsive to substrate over a much wider range of concentrations. While a Michaelis-Menten enzyme can be brought from 10 to 90% of its maximum velocity by an 81-fold increase in substrate concentration, the same increase in rate of a negatively cooperative enzyme may require changing substrate concentration over a 6,000-fold range.

Why do many regulatory enzymes consist of more than one type of chain? It may well be that the answer to this question is not some requirement of the molecular machinery of regulation, but evidence of the pathway by which most regulatory enzymes have evolved. In the first stage in the evolution of metabolic regulation the cell might have contained two separate proteins, one with catalytic function and the other acting as a specific inhibitor of the catalyst. Inhibition would have occurred whenever the inhibitor protein was synthesized. This might have been followed by a stage in which these two proteins were always present in the cell, but with the

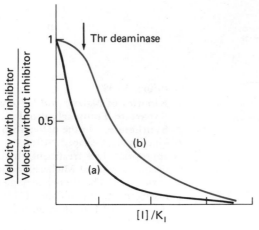

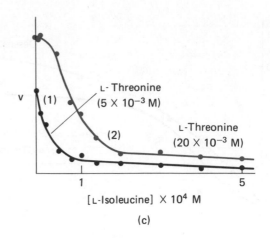

FIGURE 5–40

Response of Enzymes to Concentration of Inhibitor. (a) Michaelis-Menten enzyme. Relatively little inhibitor is required for substantial inhibition.
(b) Allosteric enzyme. There is relatively little inhibition until a threshold-like concentration of inhibitor is reached; then a slight increase in [I] produces a large inhibition. A feedback inhibitor would be expected to act this way, with its normal cellular concentration indicated by the arrow. (c) The activity of enzyme threonine dehydratase in response to isoleucine. This is the first enzyme in the pathway for synthesizing isoleucine, which acts as a feedback inhibitor. At low substrate (threonine) concentration, the inhibition is like that of a Michaelis-Menten enzyme. At higher threonine concentrations the response to inhibitor is like that of an allosteric enzyme.

inhibitory interaction modulated by an appropriate small molecule. The active "inhibitor" would then be the small molecule acting on what came to be a regulatory subunit protein. More precise regulation, permitted by the cooperativity brought about by subunit interaction, may be a further step in the evolution of these enzymes.

The role of small molecules in bringing about conformational changes in proteins has become one of the

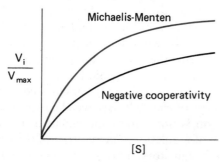

FIGURE 5–41

Negative Cooperativity Extends the Range of Response by an Enzyme to Substrate Concentration.
A Michaelis-Menten enzyme increases from 10 to 90% of its maximal activity over an 81-fold range of substrate concentration. An enzyme showing negative cooperativity, however, may require an over 6000-fold increase in substrate concentration to change from 10 to 90% of V_{max}.

most fruitful areas of research in cell biology. In Chapter 14 we will note how allosteric interactions can be used by the cell to regulate the functioning of metabolic pathways. But other important physiological phenomena (such as mechanochemical, bioelectric, and osmotic transductions), membrane receptor action, and the regulation of the cell cycle also involve conformational changes in proteins brought about by their interaction with each other and with small molecules. The amazing versatility of allosteric effects provides such a seductive way of explaining the most complex physiological phenomena that we must, in our enthusiasm, guard against overstepping the limits set by experimental evidence.

The foregoing discussion of enzyme action shows that we are well on the way to understanding the general features of the mechanisms whereby enzymes increase and regulate the rate of chemical reactions in a highly specific manner. To do this, enzymes use the entire variety of structural and functional principles found in the repertoire of protein molecules. They may use the rigidity and precision of the folded structure of proteins to impose a strain on the substrate molecule; on the other hand, they may use the conformational flexibility that proteins sometimes exhibit to allow a functional group to close in on the substrate or to seal a hydrophobic pocket from access by water molecules. Functional groups on enzymes may use charge interactions to polarize atoms in the substrate, they may use proton donors or proton acceptors to promote the formation of intermediate states, they may use nucleophiles to form covalent intermediates, or they may use hydrophobic groups to enhance the reactivity of the above interactions.

The intermediate states observed in catalytic events represent pathways allowing the reaction to proceed with lower energies of activation than that of the uncatalyzed reaction. The energy to do this seems often to come from the attractive interactions expressed when substrate binds snugly to enzyme. (It should not be a complete surprise that a complementary surface that can tightly bind a specific substrate, put it under strain, and thus catalyze a chemical reaction might be constructed from materials other than amino acids. Organic chemists are building such surfaces in the laboratory out of carbohydrates. As we see in Chapter 12, the cell has evolved RNA structures that can catalyze certain metabolic reactions.)

The goals of enzymology include assigning to each factor in a catalytic scheme its precise contribution to the velocity constant. A recent development that will have a tremendous impact on these goals is the specific substitution of amino acid residues (site specific mutagenesis by protein engineering methods; see box). This is an area of intense research, but we are as yet far from a complete quantitative account of the observed rate of catalysis. For instance, the rate of 6×10^5 molecules per second of carbonic acid being converted to CO_2 and H_2O by a single molecule of carbonic anhydrase C has certainly not yet been explained in detail. The direction in which progress will take place seems to be increasingly clear; but then the enzyme may yet have a few surprises in store for us.

Protein Engineering

The revolution in biological research brought about by the concepts of molecular genetics and the techniques of recombinant DNA (see discussion in Chapter 9 and later) have also been applied to the study of enzymes and their catalysis. Some call this effort protein engineering, but others believe that term to be more enthusiastic than deserved by the current state of the art.

The details of the experimental methods will become more clear after studying the descriptions in later chapters, but the general approach can be easily grasped: it is essentially to alter an amino acid residue thought to be important in specificity or catalysis of an existing enzyme (the result is sometimes called a "factitious" protein) and to see whether the predicted consequences are observed. Alternatively, the method offers the promise of constructing a new amino acid sequence (a fictitious protein) which will fold up to give an enzyme with desired and novel catalytic properties. The methods in each case depend on the determination of amino acid sequences by genetic nu-cleic acid sequences and our growing ability to alter or synthesize the appropriate nucleic acid sequences.

What kinds of insights can be produced by tinkering with proteins? One can exchange one amino acid residue—for example, negatively charged aspartic acid—for another with the same shape but different function—for example, uncharged asparagine. A more extensive change was made in trypsin, an enzyme that hydrolyzes peptides after a residue bearing a positive charge (see Chapter 4). A negatively charged aspartic acid believed to lie in a "pocket" and to confer specificity for positively charged substrates by electrostatic attraction was changed to a positively charged lysine by W. J. Rutter's group. The expected result was a modified trypsin that would catalyze the hydrolysis of peptide bonds in which the residue donating the carbonyl group is negatively charged. The observed result, however, was a trypsin that no longer accepted positively charged substrates, but instead favored *neutral* substrates. Apparently, the lysine side chain extends out of the pocket and does not play the expected role in determining specificity. As well as exemplifying the strategies of protein engineering, this illustrates the difficulties of the approach. The enzyme structure has evolved in response to natural selection, not in response to the simple modeling by enzymologists, and we will have to deal with disappointments along the road. Luckily, the specificity of other enzymes has been altered in more expected ways by changes in amino acid residues.

Trypsin has a group of residues in its active site similar to the ones in chymotrypsin, shown in Figure 5–21a: Asp 102, His 57, and Ser 195. When Asp 102 was changed to a neutral asparagine residue by C. S. Craik and colleagues, the catalytic activity fell much more than expected, by 10^4-fold, and so did the ability of Ser 195 to be alkylated by the enzyme poison DFP (see Figure 5–17). The ability of His 57 to be modified by the appropriate tosyl chloromethyl ketone (see Figure 5–18), however, dropped by only a factor of five. There was also a

(continued)

greatly enhanced activity at basic pH. When the structure of the enzyme with the altered amino acid was examined by x-ray diffraction at pH 6, the plane of the ring in His 57 was altered in some molecules, implying a change in the catalytic reaction mechanism. Again, the engineering produced results that were partly expected, and the surprises produced an improved understanding of the enzyme design complexity.

The bacterial virus T4 elicits, in infected cells, the synthesis of a lysozyme with catalytic activity and active site structure similar to egg white lysozyme. Numerous alterations have been made by B. W. Matthews and his colleagues in residues of the T4 lysozyme to test hypotheses about the catalytic mechanism and the stability and folding of the tertiary structure. Many alterations produced an enzyme that was easier to denature at a high temperature; analysis of these gave information about the architectural principles that lead to protein stability. One interesting result is that most of the structural changes elicited by the alterations, assessed by x-ray diffraction studies, were very local in their effects. That is, the change in the enzyme structure was confined to the region of the alteration and not propagated throughout the molecule. This is a reassuring observation, since it would be quite difficult to predict the results if changing one residue gave rise to significant changes in tertiary structure far from the altered site. It must be mentioned that there are other results suggesting some alterations of amino acid residues do cause structural changes far from the site of alteration.

The synthesis of new proteins, as opposed to the alteration of natural products, is just getting underway. An intermediate stage in this process, a "chymohelizyme" with a simplified chymotrypsin active site, has recently been constructed by J. M. Stewart and his colleagues, via chemical rather than genetic techniques. The structure is composed of parallel amphiphilic α helices, covalently attached at their C-terminal ends and having N-terminal residues found in the active site of chymotrypsin: serine, histidine, and aspartic acid (Figure 5–B). The chymohelizyme structure was designed using computers and rules derived from the structure-function relations of known proteins. It bears no resemblance to the amino acid sequence or general three-dimensional conformation of chymotrypsin, but it nevertheless can catalyze the hydrolysis of acetyltyrosine ethyl ester with approximately 0.03% the activity of native enzyme. This is 10^5 times the spontaneous hydrolysis rate. There is no activity toward trypsin substrates. Further refinements are expected to improve on this. It seems likely that continuing effort will be devoted to such experiments; the scientific (and commercial) payoff could be immense.

Ac-E-E-A-E-E-K-A-K-R-L-L-E-E-L-K-K-A—
(1)   (17)
Ac-H-E-E-A-K-K-K-A-E-K-L-L-E-E-L-K-K-L-K
(57)   (75)
Ac-D-E-A-G-K-K-A-E-E-L-K-K-L-L-E-E-L-K-K-K - Orn-amide
(102)   (123)
Ac-S-E-K-A-K-K-L-L-E-E-L-K-K-L-A—
(195)   (209)

FIGURE 5–B
The Chymohelizyme. Amino acid sequence of the four constituent peptides. Single-letter symbols for the amino acids are used. Ac is an N-terminal acetyl group. The C-terminal carboxyl groups either form isopeptide bonds with amino groups on the side chains of lysine (K) or ornithine (Orn), or have a terminal amide group. (Courtesy of K. W. Hahn, W. A. Klis, and J. M. Stewart.)

SUMMARY

One of the most important functions of proteins in cells is to act as enzymes. Models that describe catalysis and inhibition—for example, the Michaelis-Menten equation—can be used to describe other cellular processes. Protein structural features are ultimately responsible for catalysis, but the details are not yet understood. The use of the genetic technique known as protein engineering offers a systematic way to analyze enzyme function. Control of catalysis is exerted by binding small molecules to enzymes, which alters their shape. Cells contain ensembles of enzymes, working together in the network of processes called metabolism, the topic of the next chapter.

KEY WORDS

Allosteric, coenzyme, conformation, denature, effector, energy, enzyme, equilibrium, free energy, ligand, steady state, substrate, tertiary structure, transition state, x-ray diffraction.

PROBLEMS

1. Discuss in a few words: (a) why organic chemical reactions are slow under mild conditions at room temperature; (b) how the organic chemist can speed them up in the laboratory; (c) why the same methods would not be appropriate for increasing the rates of cellular reactions; (d) what general method is used to speed up reactions in cells; and (e) what structural features permit the increase in rates of specific reactions.

2. One hundred ml of an 8 μg/ml solution of pure urease from jack beans consumes 10^{-3} moles of urea per minute. If under the conditions of the experiment the turnover number of urease is 10^4 and there is one catalytic site per urease molecule, estimate the molecular weight of the enzyme.

3. Rewrite Equation 5–4 (p. 214) for the case when there is a significant amount of product present—that is, later during the reaction, rather than during the initial conditions.

4. Which molecule, enzyme or substrate, forms the activated complex? Which the active site? What other properties distinguish one from the other?

5. Alcohol dehydrogenase from which tissue (yeast, horse liver, human liver) will best be able to catalyze the oxidation of ethanol when its concentration is 10^{-3} molar?

6. Respond "true" or "false," and briefly explain your reason for the following: Increasing the succinate concentration from 1×10^{-6} M to 3×10^{-6} M will triple the initial velocity of oxidation when catalyzed by succinate dehydrogenase from yeast. What about an increase from 1×10^{-3} to 3×10^{-3} M?

7. After treating chymotrypsin A from bovine pancreas with a certain amount of DFP, it was found that the V_{max} of the enzyme preparation was only 10% of that before treatment. What is the K_m of the enzyme preparation, after treatment, with substrate nitrophenyl acetate?

8. When the competitive inhibitor **C** was added at 2×10^{-3} M to a series of enzyme assays with varying [S], the apparent Michaelis constant was measured at 2.7×10^{-3} M, as compared with 1.5×10^{-3} M in the absence of the inhibitor. What was the inhibition constant, K_i, for the inhibitor?

9. Consider the following data from an enzyme assay carried out in the presence of the inhibitor **U** at

1×10^{-3} M. Calculate the Michaelis constant in the presence and absence of inhibitor. Is **U** a competitive inhibitor? Why or why not?

[S]	v_i (−U)	v_i (+U)
1.0×10^{-4} M	9.1	8.3 units
2.0	16	14
3.0	23	19
5.0	33	25
7.0	41	29
10	50	33
20	67	40
30	75	43
50	83	45

10. The table below gives the initial velocities of an enzyme-catalyzed reaction run at different initial substrate concentrations. The reactions were also run at two different pHs, all other parameters having been held constant.

Initial Substrate Concentration	Initial Velocity (min^{-1} × 10^3)	
	pH 8.5	pH 7.5
1.0×10^{-5} M	3.3	1.8
2.0	5.7	3.3
3.0	7.5	4.6
4.0	8.9	5.7
6.0	11	7.5
8.0	12	8.9
10	13	10
15	15	12
20	16	13
30	17	15
50	18	17
100	19	18
200	20	19
300	20	19

(a) Determine the apparent Michaelis constant for each pH.
(b) Propose a simple specific hypothesis, consistent with the data, to explain the difference in behavior of the enzyme at the two different pH conditions. Suggest an experiment that can be done in a reasonable period of time (less than one month) and that will test your hypothesis.

SELECTED READINGS

Blackburn, P., and Moore, S. (1971) "Pancreatic ribonuclease," in *The Enzymes*. Vol. 4. P. Boyer, ed. New York: Academic Press.

Fersht, A. (1985) *Enzyme structure and mechanism*. 2d ed. New York: W. H. Freeman and Company.

Jencks, W. P. (1975) Binding energy, specificity, and enzymatic catalysis: The Circe effect. *Advances in Enzymology* 43:219–410. (See also *Catalysis in chemistry and enzymology*. New York: Dover, 1987.)

Knowles, J. R. (1987) Tinkering with enzymes: What are we learning? *Science* 236:1252–58.

Phillips, D. C. (1966) The three-dimensional structure of an enzyme molecule. *Scientific American* 215(5):78.

Shaw, W. V. (1987) Protein engineering: The design, synthesis and characterization of factitious proteins. *Biochemical Journal* 246:1–17.

Stryer, L. *Biochemistry*. (1988) 3d ed. San Francisco: W. H. Freeman and Company.

Voet, D., and Voet, J. G. (1990) *Biochemistry*. New York: John Wiley & Sons.

Metabolic Pathways

6

Thus far we have discussed enzymes and how they act on substrates but we have said little about the substrates themselves. We now take a close, hard look at these molecules: where they come from, what they do, and what becomes of them. A useful way of organizing one's thoughts about this complicated chemistry we call metabolism is to consider again the three flows that characterize the cell's activities: *flows of mass, energy, and information*.

The *mass* contained in the small substrate molecules originates in the foodstuffs taken in by organisms or cells. The foodstuffs may be carbohydrates, fats, proteins, nucleic acids, various other constituents, or combinations of any or all of them. For plants and certain bacteria the foodstuffs may all be inorganic: water, carbon dioxide, ions like ammonium, phosphate, and sulfate, and so on. For humans and other animals, foods are of animal or plant origin, made up of large and complex molecules first processed by extracellular enzymes in the saliva, and then in the gastrointestinal tract. These enzymes break the large molecules (e.g., proteins and polysaccharides) into smaller molecules (e.g., amino acids and monosaccharides) able to be absorbed more easily into cells. The breakdown products are then transported into the bloodstream and circulated to the cells of the body, where intracellular enzymes may split them into still smaller molecules, transform them through oxidation, reduction, or other chemical reactions into other substances, or combine small molecules into larger ones. In different cells the predominant mode of use or metabolism of substrates varies according to the activity of the enzymes present. These activities in turn are responsive to the physiological state of the cell or organism. In the liver cells of a recently fed mammal, for example, certain enzymes synthesize esterified fatty acids which circulate in the blood while others convert glucose into its storage form, glycogen, from which glucose can later be recovered for use in all the body's cells (Figure 6–1).

The *information* required for this complicated network of chemistry is embodied in the enzymes that catalyze the reactions. The activity of enzymes can be affected by small molecules, *allosteric effectors,* which bear information

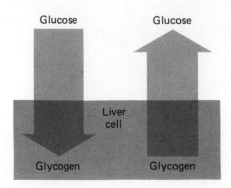

FIGURE 6–1
Glycogen-glucose Interconversion. Liver cells in mammals can convert excess glucose into glycogen, and convert glycogen back into glucose when it is needed. The glycogen is stored inside the liver cell.

about the processes that produced them. By controlling the activities and amounts of the enzymes, the cell regulates its metabolism. These matters have been mentioned already in Chapter 5 and will be explained further in Chapters 13 and 14.

Energy is involved in all the chemical transformations that produce the molecules of the cell. Clearly *all metabolic reactions must be spontaneous processes* ($\Delta G < 0$) under physiological conditions; otherwise they would not take place. This seems at first glance to involve a paradox. Since the breakdown of many cellular molecules is spontaneous, the reverse reaction—the synthesis of those cellular molecules from their breakdown products—must be nonspontaneous. As intimated in Chapter 1, the metabolic solution to the thermodynamic problem of synthesizing molecules whose breakdown is spontaneous is to use *different reaction pathways* for synthesis and breakdown (Figure 6–2). The synthetic path involves the formation of different intermediates, ones with high internal energy that can be released in a spontaneous reaction to form the needed products. One can think of this as combining, via the new pathway, a spontaneous process—the one whose negative free energy change is used to form the new intermediates—with a nonspontaneous process—the reverse of the breakdown—so the combination is spontaneous.

In this chapter we shall discuss how that combination is brought about—that is, how the energy released from one reaction can be used to "drive" another. Then we shall consider the common features of metabolic pathways and some specific examples that illustrate the common features. Among these examples are two pathways that play a central role in the metabolism of cells: glycolysis and the Krebs cycle. These illustrate how energy is derived from molecules to operate cellular processes.

Free Energy Change: The Nernst-Planck Equation

We wish now to make our understanding of the free energy change more quantitative. This will allow us to use the concept of free energy to see how cells synthesize molecules whose breakdown is spontaneous.

The free energy change that occurs during a chemical reaction has two aspects, the difference in intrinsic free energy between reactants and products, and the difference in concentrations between reactants and products. For a simple reaction like an isomerization—for example, $A \rightleftharpoons B$—we believe the free energy associated with product B is generally different from that associated with reactant A. Changing one molecule into another will thus involve an energy change independent of their concentration. But we also know, by the principle of Le Châtelier, that the more concentrated A is, the more B is produced; increasing the concentration of A shifts the equilibrium of the reaction to form more product B. Another way to express this is to recall that to dilute a substance is a spontaneous process, and to concentrate a substance is nonspontaneous.

These qualitative considerations will now be formalized by considering the free energy change that accompanies the reaction $A \rightleftharpoons B$ to be composed of two parts: (a) the free energy change that would occur under a standard set of conditions, including concentrations, and (b) the free energy change that would occur by diluting (or concentrating) the reactant(s) and product(s) from standard to cellular conditions. We will merely assert the particular form of these two parts; the rigorous derivation we leave to more advanced texts on thermodynamics.

The *Nernst-Planck equation* says that for $A \rightleftharpoons B$,

$$\Delta G = \Delta G^0 + R T \ln \frac{[B]/[B]_0}{[A]/[A]_0}. \qquad (6-1)$$

The first term on the equation's right side (ΔG^0) is known as a *standard free energy change*. It is designated by the superscript ^{0} and refers to the free energy change for the reaction when it occurs under *standard conditions*. The standard free energy change can also be understood as the free energy released when forming product B at the standard state from its constituent elements, minus the free energy released when forming reactant A at the standard state from its constituent elements. Usually, such a *standard free energy of formation*, G_f^0, is negative—that is, the formation of a compound from its elements is a spontaneous process. In this context the elements are arbitrarily assigned free energies of formation of zero under standard conditions. Thus, for the reaction $A \rightleftharpoons B$, we can write

$$\Delta G_{A \to B}^0 = G_f^0(B) - G_f^0(A).$$

FIGURE 6-2
Spontaneous Pathways Are Not Reversible. The pathway for synthesis of an energy-rich cellular molecule is usually not the exact reversal of the pathway by which it is broken down (although some steps may be common to both pathways). The breakdown (catabolic) pathway is symbolized by the ski run. Like the descent of the skier, the breakdown of the molecule is energetically "downhill" (i.e., it has a negative ΔG and so proceeds spontaneously). The synthetic (anabolic) pathway is symbolized by the ski lift. Since the synthesis is energetically "uphill" (and thus nonspontaneous), the process must be coupled to a source of energy, represented by the motor that drives the ski lift. In cells, the energy source is a very spontaneous reaction, such as the hydrolysis of ATP.

The arbitrary definition of the *standard state* represented by superscript ^{0} is one atmosphere pressure (1.01×10^5 N/m²), 25°C (298 K), and with all reactants and products (except water) at the same standard concentration of 1.0 molal (1.0 mole/kg solvent).[1] The free energy change is measured in kJ/mole of product formed. To precisely measure the value one must arrange the conditions so the *concentrations are held constant during the reaction*.

The second term of Equation 6–1 contains the factor R, the gas constant (8.314 J/mole·K = 1.987 cal/mole·K), and T, the absolute temperature measured in Kelvins (K = °C + 273). This term involves only the concentra-

tions of reactants and products. It represents the *energy that would be released by changing the concentration* of product B from the standard state, $[B]_0$, to the reaction concentration, $[B]$, minus the energy that would be released by changing the concentration of reactant A from the standard state, $[A]_0$, to the reaction concentration, $[A]$. Equation 6–1 models this, recalling the rule for calculating the logarithm of a ratio ($\ln (y/z) = \ln y - \ln z$): thus, Equation 6–1 can be rewritten as

$$\Delta G = \Delta G^0 + R\,T \ln \frac{[B]}{[B]_0} - R\,T \ln \frac{[A]}{[A]_0}.$$

Since we arbitrarily set the numerical value of the standard concentrations as 1.0, the Nernst-Planck equation can also be written, for $A \rightleftharpoons B$,

$$\Delta G = \Delta G^0 + R\,T \ln \frac{[B]}{[A]}. \tag{6–2}$$

[1] Actually, the standard state is with all reactants and products (except water) at *unit chemical activity*. For our purposes the concentrations of 1.0 *molal* and 1.0 *molar* are a sufficiently close approximation. The important parameter is the *ratio* of the concentration under the reaction conditions to the concentration under standard conditions.

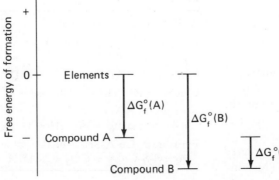

EXAMPLE 6-1

Inside a certain mammalian red blood cell at 37°C the following reaction occurs:

$$\text{dihydroxyacetone phosphate} \rightleftharpoons \text{glyceraldehyde phosphate}.$$

Calculate the change in free energy if the concentrations of dihydroxyacetone phosphate and glyceraldehyde phosphate are 300 μM and 15 μM, respectively, and the ΔG^0 for the reaction is 7.70 kJ/mole.

Solution

$$\Delta G = \Delta G^0 + R\,T\ln\frac{[\text{glyceraldehyde phosphate}]}{[\text{dihydroxyacetone phosphate}]}$$

The standard free energy change is given in kJ/mole, so the value of the gas constant R must be in _kilo_joules per mole · Kelvin:

$$\Delta G = 7.70 + \frac{8.314 \times 10^{-3}\text{kJ}}{\text{mole} \cdot \text{K}}\,(310\text{ K})\left(\ln\frac{15}{300}\right)$$

$$= \frac{-0.02\text{ kJ}}{\text{mole}}.$$

(Note: Even though the standard free energy change is positive, the reaction is nevertheless spontaneous under cellular conditions because the concentration of the reactant is higher than that of the product.)

We note that at lower concentrations of B, or at higher concentrations of A, the value of ΔG becomes more negative. This indicates that the reaction becomes more spontaneous and is in agreement with Le Châtelier's principle.

This can be clearly seen by examining a simple dilution reaction:

$$A_{(\text{concentrated})} \rightleftharpoons A_{(\text{dilute})}.$$

Since the product is the same molecule as the reactant, there is no difference in the standard free energy of formation and $\Delta G^0 = 0$. Therefore,

$$\Delta G = R\,T\ln\frac{[A]_d}{[A]_c}.$$

The lower concentration of the product state, relative to the reactant state, ensures that the free energy change is negative, in agreement with our experience about the spontaneity of dilution. As we mentioned in Chapter 1, this spontaneous process is driven by the increase in entropy of the system as it changes from a more ordered (concentrated) to a less ordered (dilute) state. (This kind

of equation will come up again in the next chapter when we consider the transport of a molecule from one place to another.)

EXAMPLE 6-2

Calculate the free energy change for a potassium ion flowing from inside a cell, where $[K^+]_i = 150$ mM, to the surrounding medium, where $[K^+]_o = 5$ mM, when $T = 37°C$.

Solution

This reaction can be symbolized by

$$K_i^+ \rightleftharpoons K_o^+.$$

The standard free energy change is zero since the same substance is both reactant and product. Therefore,

$$\Delta G = R\,T\ln\frac{[K^+]_o}{[K^+]_i}$$

$$= \left(\frac{8.314 \times 10^{-3}\text{ kJ}}{\text{mole} \cdot \text{K}}\right)(310\text{ K})\left(\ln\frac{5}{150}\right)$$

$$= \frac{-8.8\text{ kJ}}{\text{mole}}.$$

(Note: Potassium ions would leave the cell spontaneously, but there is a barrier that prevents this process—the cell membrane.)

Values of $\Delta G'$ have been tabulated for many reactions that occur in cells (Table 6–1). With such a table it is possible to use the Nernst-Planck equation to calculate the free energy change for the reaction under cellular conditions. The tabulated values were not often actually measured by mixing elements together to make compounds. Rather, the reactants and products were mixed

TABLE 6-1

Some Standard Free Energy Changes

Reaction	$\Delta G'$ (kJ/mole)
$ATP + H_2O \rightleftharpoons ADP + P_i$	−34
Phosphoenolpyruvate + ADP $\rightleftharpoons$ pyruvate + ATP	−31
Pyruvate + NADH + H^+ $\rightleftharpoons$ NAD^+ + Lactate	−25
Glucose-6-P + H_2O $\rightleftharpoons$ Glucose + P_i	−14
Isocitrate + NAD^+ $\rightleftharpoons$ α-ketoglutarate + NADH + H^+ + CO_2	−21
Pyruvate + Co A-SH + NAD^+ $\rightleftharpoons$ Acetyl-S-Co A + CO_2 + NADH + H^+	−33

Standard Concentrations

Equation 6–1 is more complicated than the form of the Nernst-Planck equation usually given in chemistry textbooks (Equation 6–2). The preceding discussion shows how the arbitrary definition of standard states leads from one equation to the other. Despite its greater complexity, the student is urged to memorize the form of Equation 6–1, because it is less likely to be misinterpreted. Consider, for example, the following type of reaction:

$$C \rightleftharpoons D + E$$

$$\Delta G = \Delta G^0 + \qquad (6\text{–}A)$$
$$R\,T\,\ln\frac{[D]/[D]_0 \cdot [E]/[E]_0}{[C]/[C]_0}.$$

Each concentration under the reaction conditions is divided by the concentration under standard conditions. This has two advantages.

First, one is not limited to a single definition of standard concentrations. Provided all concentrations are expressed in the same units, one can change the numerical value of some standard concentrations and still use Equation 6–1 to calculate the free energy change. This property allows cell biologists to use a standard state designated by a superscript prime (') in which the standard concentration of the hydrogen ion is 10^{-7} molal while all other substances have standard concentrations of 1.0 molal.

For the reaction in which a proton dissociates from an acid,

$$HA \rightleftharpoons A^- + H^+$$

$$\Delta G = \Delta G' + \qquad (6\text{–}B)$$
$$R\,T\,\ln\frac{[A^-]/[A^-]' \cdot [H^+]/[H^+]'}{[HA]/[HA]'}$$

which yields

$$\Delta G = \Delta G' +$$
$$R\,T\,\ln\frac{[A^-] \cdot [H^+]/10^{-7}}{[HA]}.$$

The values of ΔG for the dissociation must be the same under a constant set of conditions—that is, ΔG for Equation 6–B must be the same as ΔG for an equation like 6–A. The values of ΔG^0 and $\Delta G'$ will not, therefore, be the same for any reaction that either produces or consumes protons. To see the relation between these two standard free energy changes, imagine Equation 6–A to be describing the same acid dissociation reaction as Equation 6–B:

$$\Delta G = \Delta G^0 +$$

$$R\,T\,\ln\frac{[A^-]/[A^-]_0 \cdot [H^+]/[H^+]_0}{[HA]/[HA]_0}$$

$$= \Delta G^0 + R\,T\,\ln\frac{[A^-] \cdot [H^+]}{[HA]},$$

since $[HA]_0$, $[A^-]_0$, and $[H^+]_0$ each equals one. Because both equations must describe the same free energy change for the same reaction, one can solve for $\Delta G'$ in terms of ΔG^0:

$$\Delta G' = \Delta G^0 + R\,T\,\ln 10^{-7}$$
$$= \Delta G^0 - 40\ \text{kJ/mole}.$$

This means that at pH 7 the dissociation of the acid is more spontaneous than at pH 0. The $\Delta G'$ is more negative by 40 kJ/mole than ΔG^0 because 10^{-7} molal is a lower H^+ concentration than 1.0 molal; thus, the equilibrium at the ' (prime) standard state is displaced toward the dissociation. For our purposes we shall use the biological standard conditions, pH = 7, and values of $\Delta G'$ as standard free energy changes.

The second advantage to remembering the more complex Nernst-Planck equation can be seen by examining Equation 6–A. Recall that concentrations are not just numbers; they include units of measurement. Only a pure number, however, can have a logarithm. In Equation 6–A, despite the fact that there are two products and but one reactant, the units of measurement all cancel out and one is left with the natural logarithm of a pure number. Whether the number of reactants is equal to the number of products or not, one always ends up with the logarithm of a pure number when concentrations are normalized in this way.

---

and allowed to come to equilibrium. We have already learned that at equilibrium the free energy change is zero (see Chapter 1). For our reaction $A \rightleftharpoons B$ *at equilibrium*, the Nernst-Planck equation is

$$\Delta G = 0 = \Delta G' + R\,T\,\ln\frac{[B]_{eq}}{[A]_{eq}},$$

where the subscript$_{eq}$ signifies the concentrations at equilibrium. This equation can be solved for $\Delta G'$:

$$\Delta G' = -\,R\,T\,\ln\frac{[B]_{eq}}{[A]_{eq}}.$$

Since at low concentrations the ratio of the product of the equilibrium concentrations of the reaction products to the

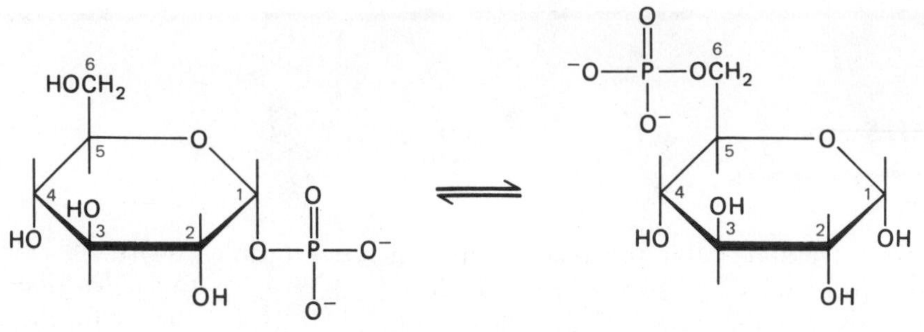

Glucose-1-phosphate   Glucose-6-phosphate

product of the equilibrium concentrations of the reactants is the equilibrium constant, K_{eq}, we obtain

$$\Delta G' = - R T \ln K_{eq}. \qquad (6–3)$$

Analytical chemistry will reveal the various concentrations at equilibrium, and Equation 6–3 can then be used to calculate the value of $\Delta G'$. (See figure above.)

EXAMPLE 6–3

Consider the isomerization of glucose-1-phosphate to glucose-6-phosphate and calculate $\Delta G'$. The equilibrium of this reaction has been studied at 25°C (298 K) and pH 7, and K_{eq} has been found to be 17.

Solution

Substituting into Equation 6–3, we have

$$\Delta G' = - (8.31) (298) (\ln 17) = -7.0 \text{ kJ/mole}$$
$$= -1.7 \text{ kcal/mole}.$$

Since the standard free energy change is negative, this reaction (a transphosphorylation) will occur spontaneously under standard conditions, a fact which we shall have the opportunity to observe again later in this chapter.

Although $\Delta G'$ is the free energy change expected at standard conditions, we often use it as a rough guide to the free energy change under cellular conditions. This is helpful because we know accurately the cellular concentrations of relatively few molecules. If the number of reactants and products is identical and if the concentrations of the reactants and products are all roughly the same, then the concentration term of the Nernst-Planck equation will not be very large and ΔG will be dominated by $\Delta G'$.

Concentrations of Small Molecules

If one wished to make spontaneous a reaction with a positive $\Delta G'$, it is only necessary to raise the concentration of the reactants and lower the concentration of the products. Equation 6–1 would then predict a lowering of ΔG until a negative, spontaneous value was achieved. Unfortunately, this is not a practical method for most metabolic reactions because the reactions inside cells are only rarely independent of one another. Usually the product of one reaction is the reactant of another. If we reduce the concentration of the product (glucose-6-phosphate) of reaction A, to lower its ΔG, we would also reduce the concentration of the reactant of reaction B, thus making *its* ΔG more positive. (See figure on top of p. 265.)

There is another reason why small molecules must all be at a low concentration in the cell: there are too many of them to fit if they are at a high concentration. It has been estimated that there may be as many as 10^4 different kinds of small molecules in a cell. (A small molecule is one with a molecular weight of a few hundred.) If they were dissolved in the cell at a concentration of one millimolar, each would represent a mass density of 10^{-3} (moles/l) $\cdot$ 10^2 g/mole, or 0.1 g/l. If there were 10^4 of them, they would yield a solution with a total mass concentration of 10^3 g/l, which is the same as the density of water—except that there wouldn't be room for much water in a "solution" like that. The viscosity and osmotic pressure of such a cytoplasm would be incompatible with the operation of cells as we know them. This argument (about which you can read more in the monograph by D. E. Atkinson listed in the Selected Readings) suggests that the small molecules of the cell must on the average be present in concentrations much less than 1.0 mM.

Among the consequences of these low concentrations is the slow intrinsic rate of metabolic chemistry, as we discussed in the previous chapter. This is compensated by the presence of enzyme catalysts which speed up the cell's reaction rates. Another consequence is that metabolism must be arranged so there are very few reactions with large values, either positive or negative, of $\Delta G'$. Reactions with large negative $\Delta G'$ would tend to lead to high concentrations of products. Reactions with

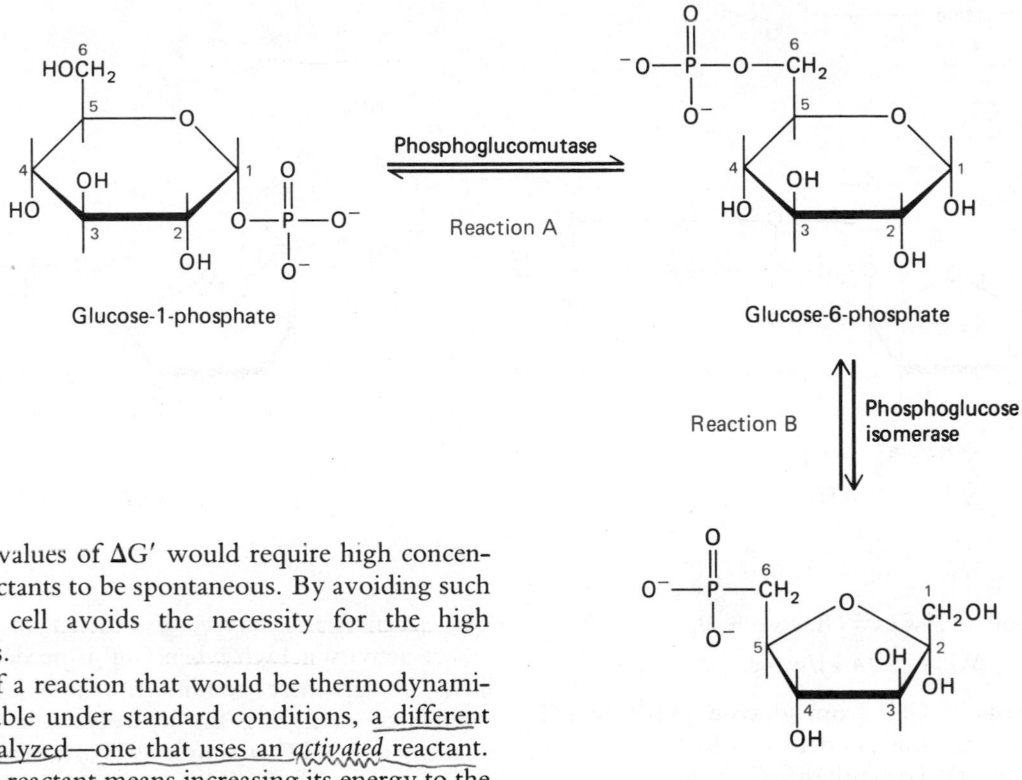

Glucose-1-phosphate

Phosphoglucomutase

Reaction A

Glucose-6-phosphate

Reaction B — Phosphoglucose isomerase

Fructose-6-phosphate

high positive values of $\Delta G'$ would require high concentrations of reactants to be spontaneous. By avoiding such reactions the cell avoids the necessity for the high concentrations.

Instead of a reaction that would be thermodynamically unfavorable under standard conditions, a different reaction is catalyzed—one that uses an *activated* reactant. Activating the reactant means increasing its energy to the extent that the desired reaction proceeds with $\Delta G'$ negative or close to zero. A reactant is activated by changing its structure; for example, by transferring to it a phosphate group from an even more energetic donor such as ATP. During participation in the favorable reaction, the added group is often removed. If not, the later removal, under cellular conditions, is usually spontaneous as well. The free energy change associated with the removal may contribute significantly to the overall free energy change of a sequence of reactions.

Energy Coupling and Common Intermediates

The synthesis of an unstable molecule requires combining a nonspontaneous reaction with a spontaneous one so that the combination is spontaneous. In some cases this can be done directly, without the need for any intermediates.

Combining Reactions

A typical molecule found in the cellular metabolism of carbohydrates is glucose-6-phosphate (G6P). This is an ester of phosphoric acid and the sugar glucose. It is unstable in the sense that its hydrolysis to glucose and inorganic phosphate (P_i, the ionized forms of phosphoric acid that exist at pH 7) has a negative standard free energy change:

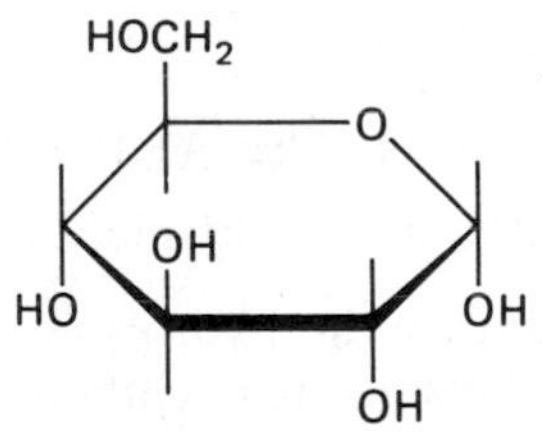

Glucose

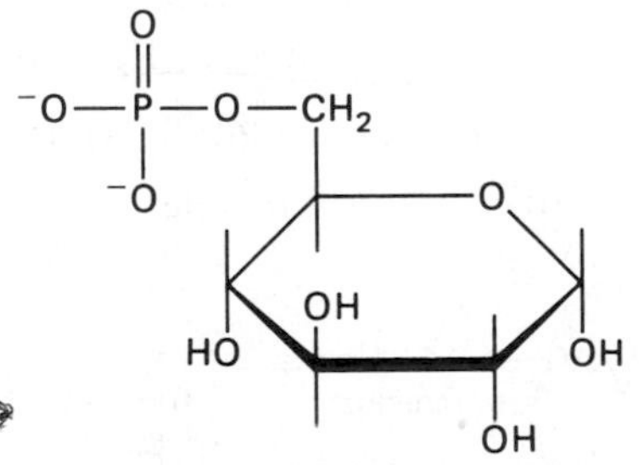

Glucose-6-phosphate (G6P)

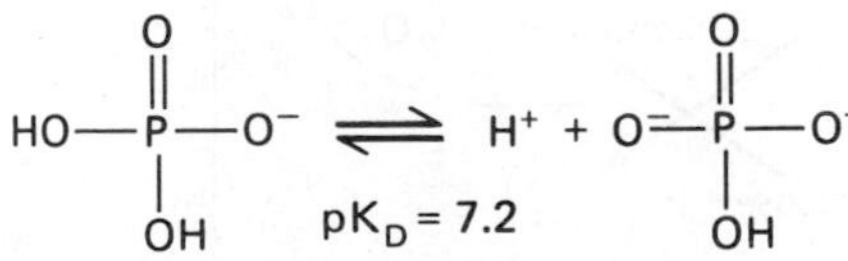

$$pK_D = 7.2$$

Inorganic phosphate (P_i)

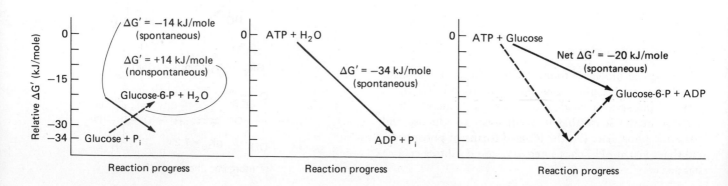

$$G6P + H_2O \rightleftharpoons Glucose + P_i$$
$$\Delta G' = -14 \text{ kJ/mole}. \tag{6-4}$$

The synthesis of G6P from glucose and P_i would thus be nonspontaneous under standard conditions ($\Delta G' = +14 \text{ kJ/mole}$). To synthesize G6P requires energy from a spontaneous process with a standard free energy change more negative than -14 kJ/mole. The most celebrated spontaneous reaction in cells is the hydrolysis of ATP (adenosine triphosphate) to give ADP and phosphate:

$$ATP + H_2O \rightleftharpoons ADP + P_i$$
$$\Delta G' = -34 \text{ kJ/mole}. \tag{6-5}$$

The energy released by the hydrolysis of ATP is more than sufficient to allow for the synthesis of G6P: $-34 + 14 = -20 \text{ kJ/mole}$ (see Table 6–1).

In cells the synthesis of G6P proceeds by a pathway that avoids very high concentrations of intermediates; that is, it is not simply the reverse of Reaction 6–4:

$$ATP + Glucose \rightleftharpoons G6P + ADP$$
$$\Delta G' = -20 \text{ kJ/mole}. \tag{6-6}$$

This reaction is catalyzed by an enzyme (hexokinase) whose activity is high when G6P is needed for metabolism. Under these conditions, the enzyme activities capable of catalyzing the hydrolysis of G6P (a phosphatase) and ATP (a simple ATPase) would be low or absent. (Indeed, since ATP is needed for many reactions in metabolism, a simple ATPase activity would be counterproductive and we would not expect to find it in cells.) It is in the existence of the pathway for Reaction 6–6 that one finds the combination of the nonspontaneous reaction (phosphorylation of glucose, the reverse of Reaction 6–4) with the spontaneous reaction (hydrolysis of ATP, Reaction 6–5). In Reaction 6–6, mass flows through a spontaneous process into G6P because (1) the free energy change is negative and (2) an enzyme with a high activity is available to catalyze the process. The existence of hexokinase activity allows part of the energy that would be released by ATP hydrolysis to "drive" the synthesis of glucose-6-phosphate (see below).

It is important to always make sure that the energy bookkeeping is consistent. One way to see this is to suppose that Reactions 6–6 and 6–4 occur in series. The net products formed would be:

	$\Delta G'$ (kJ/mole)
ATP + Glucose $\rightleftharpoons$ ADP + G6P	-20
G6P + H_2O $\rightleftharpoons$ Glucose + P_i	-14
ATP + H_2O $\rightleftharpoons$ ADP + P_i	-34

Since the additive combination of Reactions 6–6 and 6–4 is Reaction 6–5, the standard free energy change should be −34 kJ/mole, which is in fact the case.

Expressing these relations in a different way, we can think of part of the standard free energy of ATP hydrolysis (−34 kJ/mole) as being "stored" during the synthesis of G6P. We note, however, that Reaction 6–4 has a standard free energy change of −14 kJ/mole. All the free energy available from the hydrolysis of ATP has not in fact been stored. If it had been—for example, by synthesizing some compound X-P with a standard free energy of hydrolysis of −34 kJ/mole—then the reaction ATP + X $\rightleftharpoons$ ADP + X-P would have a standard free energy change of zero, and so would be at equilibrium. It is *necessary* that free energy *not be stored completely*: some free energy must be dissipated (as heat) so that each reaction has a negative ΔG and proceeds spontaneously.

Coupling Reactions

In most cases where a spontaneous reaction is used to drive a nonspontaneous one, a new pathway is required—one that typically contains more than one metabolic reaction. **Coupling** these reactions requires one or more *common intermediates*—that is, the product of one reaction is a precursor of the next (Figure 6–3).

An example of this is the synthesis of sucrose, common table sugar. Sucrose is an unstable compound; its hydrolysis proceeds with a negative standard free energy change:

$$\text{Sucrose} + H_2O \rightleftharpoons \text{Glucose} + \text{Fructose}$$
$$\Delta G' = -27 \text{ kJ/mole}. \qquad (6\text{–}7)$$

Here, as in the synthesis of G6P, the hydrolysis of ATP ($\Delta G' = -34$ kJ/mole) could provide enough energy to drive the synthesis of sucrose. The problem for the cell is

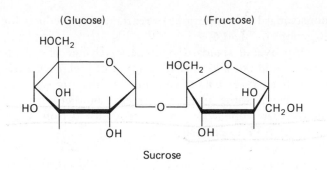

(Glucose)   (Fructose)

Sucrose

how to exploit that energy. In certain bacterial cells the trick is done by the following sequence of reactions:

	$\Delta G'$ (kJ/mole)	
ATP + Glucose $\rightleftharpoons$ ADP + G6P	-20	(6–6)
G6P $\rightleftharpoons$ G1P	$+7$	(6–8)
G1P + Fructose $\rightleftharpoons$ Sucrose + P_i	$+6$	(6–9)
ATP + Glucose + Fructose $\rightleftharpoons$ ADP + P_i + Sucrose	-7	(6–10)

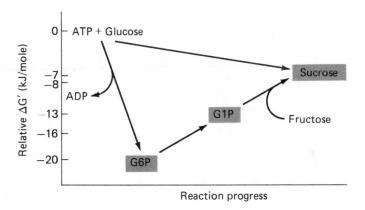

The three reactions in this pathway are coupled by common intermediates: the product G6P of Reaction 6–6 is a precursor of Reaction 6–8; the product G1P of Reaction 6–8 is a precursor of Reaction 6–9. The energy coupling occurs by means of the concentrations of the common intermediates. Thus, the negative standard free energy change of Reaction 6–6 implies that the equilibrium constant of that reaction is large, so the reaction will produce high concentrations of G6P. This is what can drive Reaction 6–8; its small positive standard free energy change can be overcome by a high precursor concentration. Similarly, the production of G1P in sufficiently high concentration can lead to the spontaneity of Reaction 6–9. This discussion focuses our attention again on the differences between standard conditions and cellular conditions. Each reaction (6–6, 6–8, and 6–9) must be spontaneous to synthesize sucrose, and this will require

(1)   A   +   B $\rightleftharpoons$ H   +   Ⓘ

(2)   Ⓘ   +   J $\rightleftharpoons$ K   +   L

FIGURE 6–3
Common Intermediate. Reaction 1 is coupled to Reaction 2 because I is a common intermediate: a product of Reaction 1 and a substrate of Reaction 2.

nonstandard concentrations of reactants and products for Reactions 6–8 and 6–9.

An overall standard free energy change of −7 kJ/mole would not allow the formation of high concentrations of sucrose, like those found in the sugarcane or sugar beet. Some plant tissues use another activated intermediate, one which has energy stored from more than one ATP hydrolysis:

	$\Delta G'$ (kJ/mole)	
UTP + G1P $\rightleftharpoons$ UDP-Glucose + PP$_i$	+ 4	(6–11)
PP$_i$ + H$_2$O $\rightleftharpoons$ 2 P$_i$	−33	(6–12)
UDP-Glucose $\rightleftharpoons$ Sucrose + UDP + Fructose	+ 1	(6–13)
UTP + G1P + Fructose + H$_2$O $\rightleftharpoons$ UDP + 2 P$_i$ + Sucrose	−28	(6–14)

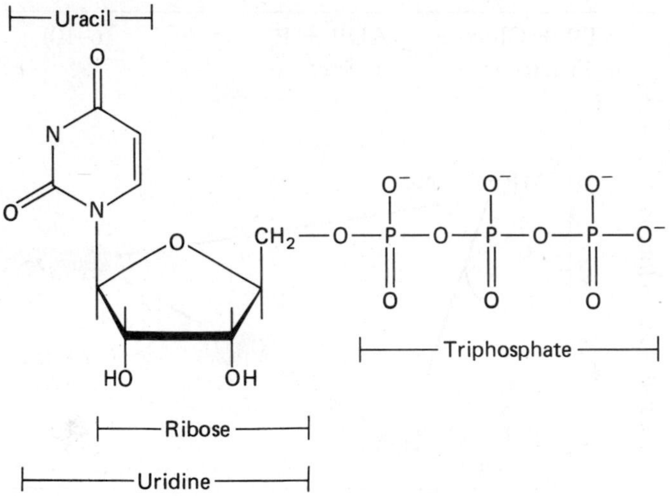

UTP

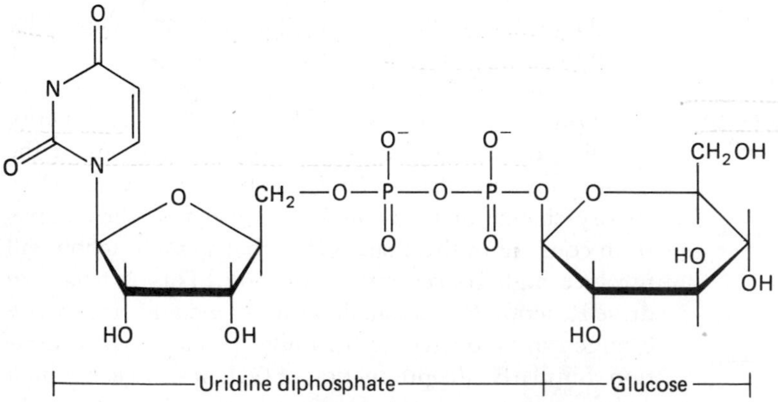

UDP–Glucose

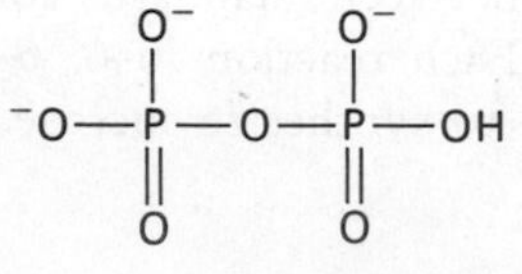

Inorganic pyrophosphate (PP$_i$)

The compound UTP has a standard free energy of hydrolysis similar to ATP. This pathway (Reactions 6–6, 6–8, 6–11, 6–12, and 6–13) thus involves the energy equivalent of two ATP hydrolyses to synthesize one molecule of sucrose: one to form the activated intermediate G1P and the other to form activated intermediate UDP-glucose. Reaction 6–11 is made more spontaneous by a low cellular concentration of inorganic pyrophosphate (PP$_i$), whose hydrolysis (catalyzed by a pyrophosphatase, Reaction 6–12) is very spontaneous. This tends to increase the concentration of UDP-glucose, which in turn tends to make Reaction 6–13 more spontaneous. In this sense, Reaction 6–12 is the driving force of the overall reaction yielding sucrose. The overall free energy change under standard conditions for this synthesis of sucrose is −41 kJ/mole (a value that can be verified by adding the $\Delta G'$s for Reactions 6–6, 6–8, 6–11, 6–12, and 6–13).

Generally in metabolism, one finds that energy-yielding reactions, like the hydrolysis of ATP, are coupled to the synthesis of unstable molecules to allow their spontaneous synthesis in appropriate amounts. Molecules that can participate in very spontaneous reactions ($\Delta G' \ll 0$) are termed *high-energy compounds*. A number of these are commonly found in cells and we pause here to discuss the most typical of them.

High-Energy Phosphate Compounds

Organic phosphate esters are involved in almost every aspect of cell function; they are important in the storage and transfer of genetic information, in the structure of membranes, in the synthesis of macromolecules, in the storage of metabolic energy, and in the various types of energy transduction. Many classes of biomolecules—such as carbohydrates, lipids, proteins, and nucleosides—form phosphate derivatives, and we shall encounter numerous examples in succeeding chapters. We begin with a discussion of the "high-energy" phosphates that the cell utilizes for most energy transfer reactions.

Cells transfer usable energy from the environment to cellular synthetic chemistry primarily by the synthesis and hydrolysis of high-energy phosphate compounds. These molecules can release a large amount of energy in the form of heat when the phosphate is removed by hydrolysis. Among these substances are phosphocreatine, phosphoenolpyruvate, uridine diphosphate-glucose (UDP-glucose), and the triphosphates of adenosine (ATP), uridine (UTP), cytidine (CTP), and guanosine (GTP). Of all these, the most important energy mediator is adenosine triphosphate (ATP). An example of the extraordinary potential energy locked in these compounds is that the hydrolysis of ATP, at standard conditions, releases about 37,000 joules of heat per mole. The value of $\Delta H'$ is about twice as great as for the equivalent reaction of other phosphate compounds such as glucose-6-phosphate.

An even more relevant measure of the high-energy content of a compound is its *free energy of hydrolysis, ΔG.** The various phosphate compounds can be arranged in order of their free energies of hydrolysis. Table 6–2 gives such a list of biologically important compounds. This list is useful because it allows one to determine the direction in which the phosphate group can be transferred, each compound being capable of doing so spontaneously under standard conditions, to the dephosphorylated compounds below it. The free energies of hydrolysis vary over a wide range. Because cells use them to form activated intermediates, the molecules in the upper part of Table 6–2, ATP and those above, are classified as *high-energy phosphate compounds*. These all have very large negative standard free energies of hydrolysis, as much as −62 kJ/mole in the case of phosphoenolpyruvate.

You may recall from our discussion in Chapter 1 that the change in free energy involves not only a change in enthalpy (heat flow at constant pressure) but also a change in entropy. Put another way, the concentrations of reactants and products affect the free energy of the reaction, but not the heat flow. The hydrolysis of a mole of ATP will always produce 37 kJ of heat, but if this

TABLE 6–2

Standard Free Energy of Hydrolysis of Biologically Important Phosphate Compounds

Phosphate Compound	$\Delta G'$	
	kJ/mole	kcal/mole
Phosphoenolpyruvate	−62	−14.7
1,3-Diphosphoglycerate	−49	−11.8
Acetyl phosphate	−47	−11.3
Phosphocreatine	−43	−10.2
Phosphoarginine	−38	−9.1
ATP	−34	−8.2
Pyrophosphate	−33	−7.9
Glucose-1-phosphate	−21	−5.0
Fructose-6-phosphate	−16	−3.8
Glucose-6-phosphate	−14	−3.3
Glycerol-1-phosphate	−9	−2.2

reaction takes place near its equilibrium it will have almost no free energy to supply to the cell. To be useful in driving metabolism, the relative concentrations of ATP, ADP, and P_i must be kept far from equilibrium. As we shall see later, the cell's regulatory mechanisms do just that, by maintaining low concentrations of ADP and high concentrations of ATP.

Let us consider more precisely some implications of the term "high-energy compound." Because the phosphate linkage is important in the free energy of "high-energy" phosphate hydrolysis, one often encounters references to the "high-energy phosphate bonds" in such compounds. This is, strictly speaking, a misnomer; bond energy is properly defined as the energy necessary to break a bond between two atoms, and there is really nothing special about the bonds holding the atoms together in a high-energy phosphate compound. They are completely ordinary covalent linkages, strong enough to keep the atoms from flying apart. The phosphate linkage is, however, weak or labile in the sense that the phosphate is a *good leaving group* in a displacement reaction. In such a reaction, bonds between atoms are not permanently broken; instead, they are redistributed to form product molecules whose stabilities, in the cases we are discussing, are much greater than those of the reactants (Figure 6–4). The energy liberated during the

$$R_{he}\!-\!O\!-\!P_i \;+\; R_{ac}\!-\!OH \;\rightleftharpoons\; R_{he}\!-\!OH \;+\; R_{ac}\!-\!O\!-\!P_i$$

FIGURE 6–4

Redistribution of Bonds During a Displacement Reaction Involving a Phosphate Group. The high-energy phosphate compound (R_{HE}—O—P_i) donates its phosphate to an acceptor (R_{AC}—OH). No chemical bonds are permanently broken; the new configuration, however, may have much lower energy. In the hydrolysis of ATP, for example, R_{HE} is ADP, and the acceptor R_{AC}—OH is water.

hydrolysis of the P—O bond is thus unusually high, not because anything is extraordinary about the bond itself but because of the large difference in free energy between the reactants and the products. It is convenient to use F. Lipmann's "squiggle" ($\sim$) to identify the bond which features so prominently in the high energy of hydrolysis of some phosphate esters—for example, ATP = ADP$\sim$P.

Although the detailed structural basis for the "high-energy" nature of these compounds has not been completely explained, it is fundamentally due to differences between the compounds and the products of their hydrolysis. The most important differences involve their *ability to be stabilized by resonance*, their *ionizations*, and their *intramolecular electrostatic repulsions*.

First, it was noted by F. Lipmann and H. Kalckar in 1941 that a prominent feature of several high-energy compounds is that they are anhydrides of phosphoric acid

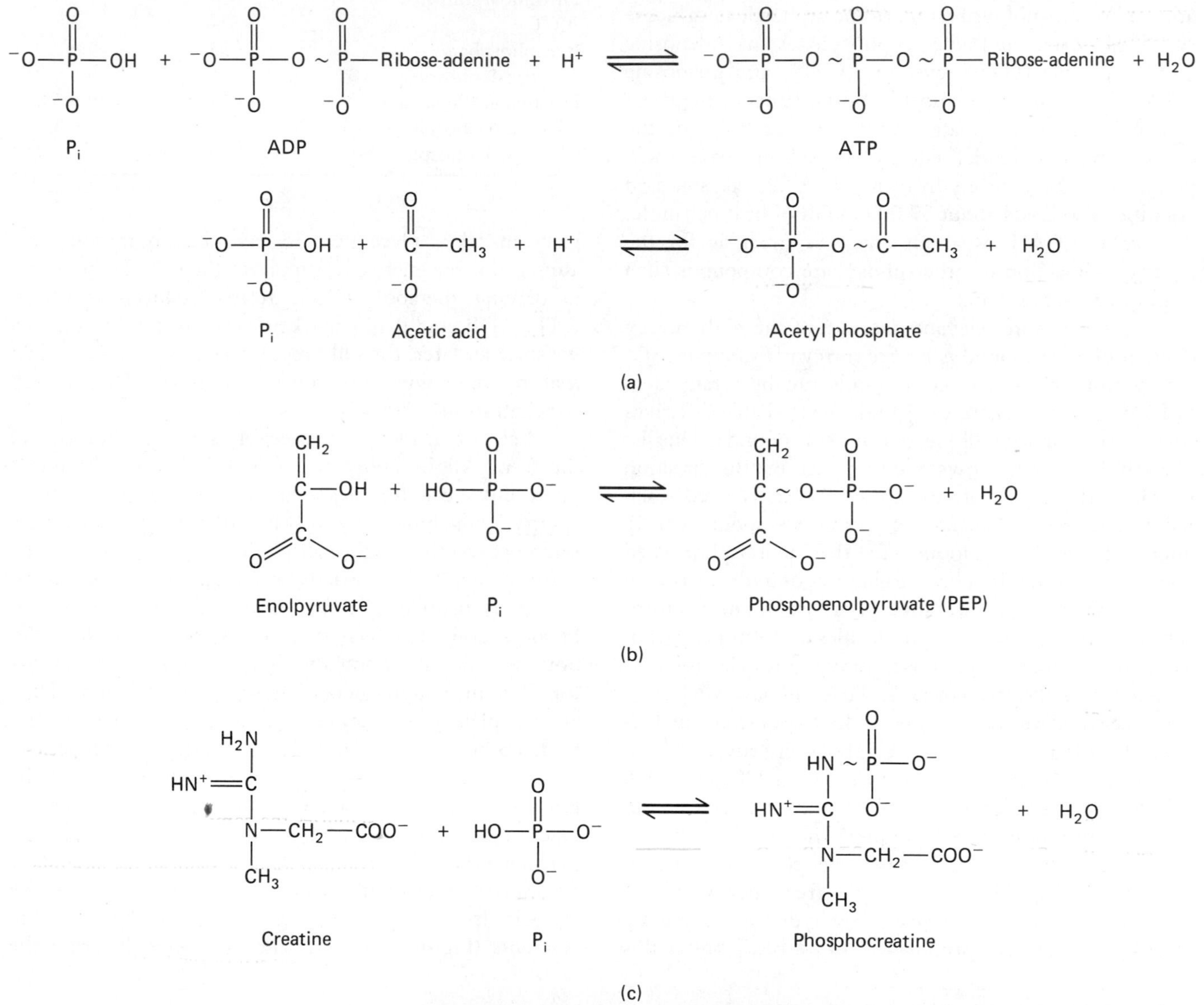

FIGURE 6–5
Formation of Various High-Energy Phosphate Compounds. (a) Phosphoric acid and another acid can combine, with removal of water, in an acid anhydride linkage to form compounds such as ATP and acetyl phosphate. (b) Phosphoric acid in an ester linkage with the hydroxyl of enolpyruvate, forming phosphoenolpyruvate (PEP). (c) Anhydride linkage of phosphoric acid with a basic nitrogen compound, creatine, to form phosphocreatine (creatine phosphate). Anhydride bonds are shown by squiggly lines ($\sim$).

with a second acid. The second acid can be a substituted phosphoric acid, to form the nucleoside di- and triphosphates (e.g., ATP); or a carboxylic acid (e.g., acetic acid), to form acetyl phosphate (Figure 6–5a). Other high-energy phosphates involve esters of phosphoric acid with an enol to form phosphoenolpyruvate (Figure 6–5b), or with a basic nitrogen compound to form phosphocreatine (Figure 6–5c). In all these cases, the formation of the phosphate linkage reduces the number of resonance forms available to the molecule (Figure 6–6a). We know that the thermodynamic stability of a substance increases directly with the number of resonance forms that are accessible to it. Since the high-energy compounds have fewer accessible resonance forms, relative to their hydrolysis products, they are less stable and their hydrolysis is therefore favored.

The dissociation and reassociation of protons from phosphoric acid is rapid. The forms of phosphoric acid that exist under cellular conditions—that is, inorganic phosphate, HPO_4^{2-}, and $H_2PO_4^-$—contain three oxygen atoms with which the protons might associate. There are, therefore, several different arrangements (corresponding to a higher entropy) of the inorganic phosphate forms (Figure 6–6b). When the phosphate is combined in an anhydride or ester linkage, there are fewer of these protonated forms available because one of the oxygen atoms must be involved in the covalent linkage. Again, the compounds containing the phosphate linkage are less stable and higher in energy.

Finally, most high-energy compounds contain multiple negative charges in close proximity. When a phosphate is removed, the density of negative charge is

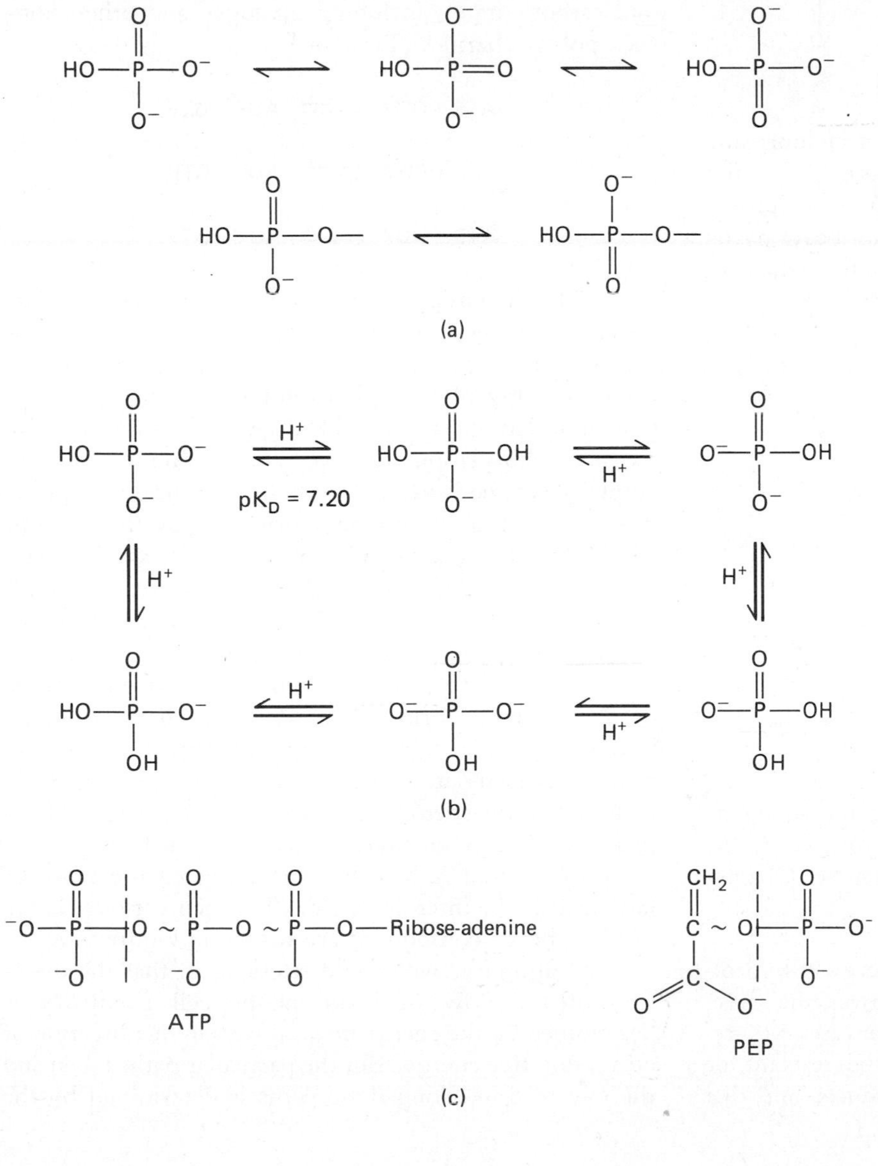

(a)

(b)

ATP

PEP

(c)

FIGURE 6–6

Structures of Phosphoric Acid and High-Energy Phosphates.
(a) Resonance forms of one species of phosphoric acid, HPO_4^{2-} (above). The existence of these forms stabilizes the compound and reduces its energy. When the group is incorporated into a compound such as ATP (below), one of the oxygen atoms becomes involved in a covalent linkage and the number of resonance forms is reduced, raising the energy content. (b) Ionized species of phosphoric acid. As with resonance, the existence of these various species stabilizes the compound. In phosphate compounds the number of possible configurations is reduced because one of the oxygens is again involved in a covalent bond. Consequently, the energy of the molecule is increased. (Note that each of the protonated species will have several resonance forms, as shown in (a) above.) (c) High-energy phosphate compounds, such as ATP and PEP, have several closely spaced negative charges which repel each other electrostatically. Removal of a phosphate group (at dashed lines) decreases the concentration of charge, lowering the energy of the molecule.

reduced, with a consequent increase in the stability of the products (see Figures 6–5 and 6–6c).

ATP is by far the most important high-energy compound in cell chemistry—so important that it has been called the "energy coin of the biological realm." The several ways it is formed by the cell are described later in this chapter and in Chapter 8. ATP functions in various ways as an energy donor:

- As a *phosphorylating agent,* transferring inorganic phosphate to a suitable acceptor compound, as in the hexokinase reaction to form glucose-6-phosphate (see Reaction 6–6).

Phosphoribosyl pyrophosphate

- As a *pyrophosphorylating agent,* transferring inorganic pyrophosphate to a suitable acceptor, as in the formation of the compound ribosyl pyrophosphate-5′-phosphate (also known as phosphoribosyl pyrophosphate or PRPP), an important intermediate in the biosynthesis of several compounds.

Amino acid adenylate

- As an *adenylating agent,* transferring the adenylate (AMP) moiety to a suitable acceptor, as in the formation of the amino acid adenylates (see Chapter 13).

In all these cases, the *standard* free energy of hydrolysis of the synthesized product may be lower, the same, or even higher than that of ATP. The latter cases require, of course, that the concentrations of the reactant be sufficiently higher than those of the products, and that

under cellular conditions the overall ΔG of the phosphate transfer be negative.

In the cell, ATP functions as a mediator between strongly exergonic reactions that would release—as heat—large amounts of free energy, were it not stored by synthesizing ATP, and reactions that would otherwise be nonspontaneous. These endergonic reactions are avoided by using a different pathway, one that involves ATP. The mediation is possible because ATP is a *common intermediate* of both types of reactions.

Other nucleoside diphosphates—GDP, CDP, and UDP—can be phosphorylated by ATP to give the corresponding triphosphates. These reactions are important because all of the nucleoside triphosphates are instrumental in synthetic reactions: ATP is involved in the synthesis of fatty acids, proteins, and nucleic acids; GTP, in the synthesis of carbohydrates, proteins, and nucleic acids; CTP, in the synthesis of phospholipids and nucleic acids; and UTP, in the synthesis of nucleic acids and carbohydrates, including glycogen and other complex polysaccharides (Table 6–3).

$$ATP + GDP \rightleftharpoons ADP + GTP$$

$$ATP + CDP \rightleftharpoons ADP + CTP$$

$$ATP + UDP \rightleftharpoons ADP + UTP$$

ATP is regarded as the ultimate energy supplier for the synthetic needs of the cell, since it is by far the most common high-energy compound formed in the strongly exergonic glycolytic, photosynthetic, and oxidative phosphorylations. From ATP the energy is funneled by transphosphorylations to other energy donors. For example, glycogen is made in animals from uridine diphosphate-glucose, a compound synthesized by the reaction of UTP with glucose-1-phosphate (see Reaction 6–11). The UTP is formed from ATP through the reaction ATP + UDP $\rightleftharpoons$ ADP + UTP, catalyzed by a ubiquitous enzyme, nucleoside diphosphate kinase. In addition, ATP can phosphorylate creatine, forming the phosphocreatine used by muscle fibrils. It can also phosphorylate acetic acid (in bacteria), giving acetyl phosphate, or it can transfer its adenylate moiety to acetic acid (in eucaryotes), forming acetyl-adenylate. Both products, acetyl phosphate and acetyl-adenylate, can be transformed into acetyl-coenzyme A, which is the activated precursor of fatty acids. The interconversions between various classes of phosphorus compounds are shown in Figure 6–7.

In summary, we should stress again that the direction of the flow of mass in the cell chemistry is determined by the energy flow: always in the direction of decreasing free energy. But the particular paths taken and the rate of flow along these paths is determined by the

TABLE 6-3
Nucleoside Triphosphates in Biosynthetic Pathways

Energy Donors	Synthesis of			
	Proteins	Carbohydrates	Phospholipids	Nucleic Acids
ATP	X	X	X	X
UTP		X		X
GTP	X		X	X
CTP			X	X

availability and activity of enzyme catalysts. These enzymes catalyze the transfer of high-energy phosphate through a series of reactions which result not only in the ultimate release of the phosphate into the solution but also in the synthesis of some other molecule of use to the cell. The reactions are influenced by one another because the product of one is the substrate of the next. *The presence of common intermediates acts to couple the flows of energy as well as mass through a metabolic pathway.*

The Common Features of Metabolic Pathways

Metabolism has been studied for many years and the data are extensive and varied. Metabolic maps have been drawn that are more than a square meter in area but nevertheless require magnification for all the reactions to be clearly seen. These maps show how the chemical transformations that proceed inside cells are interrelated to one another. Luckily, this complicated functional network can be approached by a study of *unbranched linear segments* which have many features in common:

$$A \longrightarrow B \longrightarrow C \longrightarrow D \longrightarrow E. \qquad (6\text{-}15)$$

Molecules A and E may be the substrates or products of several reactions but the route from A to E is specifically controlled by the activity of the enzymes producing B, C, D, and E. The intermediates B, C, and D undergo *no* reactions other than the ones indicated, because there are no other enzymes that will accept B, C,

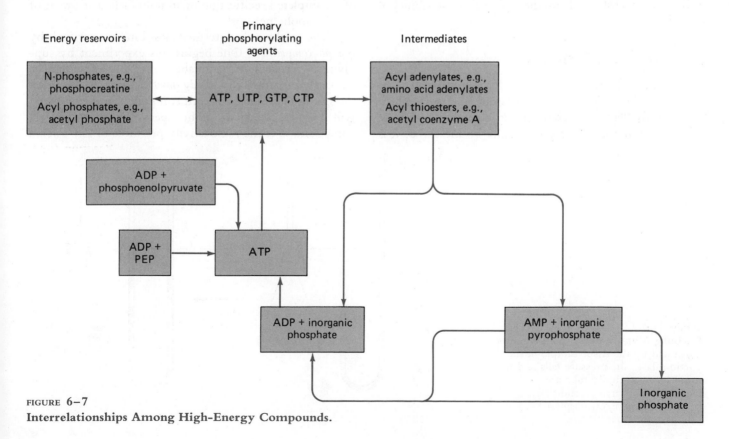

FIGURE 6-7
Interrelationships Among High-Energy Compounds.

or D as substrates; this restriction of chemical reactivity is essentially why cellular metabolic pathways exist. Metabolic networks can be built up from combinations of these unbranched linear segments.

The arrows of our schematic pathway are shown pointing in one direction only. This is untrue chemically, of course, since enzymes don't control the direction of mass flow—only the rate. But it seems to be a general rule in cells that mass travels in only one direction along a metabolic pathway. In each pathway there are usually one or more reactions that are *practically irreversible under cellular conditions* and that prevent the backward movement of material. These reactions will have large negative free energy changes under cellular conditions; often they involve the hydrolysis of a phosphate from an activated intermediate. Such reactions, called *pacemakers,* are often found as the first or last steps in a segment of the metabolic network and are typically under feedback control (see Chapters 5 and 14). In Pathway 6–15 the reaction from A to B and/or from D to E would be expected to perform this pacemaker function.

It is often possible to feed a cell substance E (for example) and find that substance A is made. In such cases, however, the mass usually follows a different pathway from E to A, one that is not simply the reversal of the A-to-E pathway. The paths of synthesis and degradation of a cellular molecule like a protein are almost always different. This follows from the practical irreversibility of pathways.

How Pathways Are Discovered

You may be wondering how all this complexity was first worked out. It is a long and fascinating story which we can only begin to sketch here. For a start, the flow of mass through a pathway can actually be measured. It is possible to assay, for example, the amount of gas liberated or consumed by a metabolic pathway. O. Warburg used a manometer to measure (by the reduction in pressure) the consumption of oxygen during respiration by tissue slices. Different molecules, found in the slices and identified by analytical organic chemistry, can be added to see whether they enhance or inhibit oxygen consumption by the slices (Figure 6–8). Alternatively, the added molecules may themselves undergo a chemical modification which results in the enhanced production or consumption of another, more easily measured, compound.

Radioactive Tracers

Until the availability of radiolabeled compounds it was necessary, and often tedious, to measure the amounts of all the various compounds in a cell to see which, if any, were increasing. But by using radiolabeled substrates, it is possible to detect the flow of labeled material into a form whose absolute concentration remains unchanged. In Pathway 6–15, for example, after the addition of labeled A the first compound in which radioactivity can be detected will be the product of the first enzymatic step of the pathway, B (Figure 6–9). In such experiments the radioactivity is used as a *tracer* of the flow of mass through the pathway. This type of analysis is capable, in principle, of a complete specification of an unbranched segment of the metabolic network.

The pathway's intermediates can be verified by *isotopic competition*. One begins this experiment by supplying an excess of radiolabeled A and allowing the pathway to come into a *steady state*. Under this condition the *specific radioactivity* of each compound in the pathway will become constant. (The specific radioactivity of compound Z is the radioactivity per mole of Z.) Then a

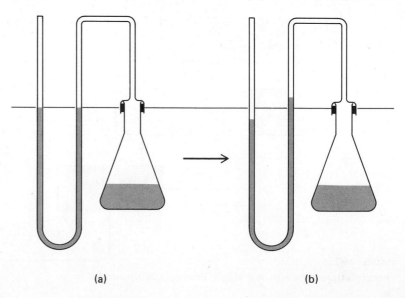

FIGURE 6–8
Warburg Manometer.   As oxygen is consumed by the tissue sample in the reaction flask, the pressure falls and the liquid levels in the U-tube are displaced. (a) Starting conditions. (b) After oxygen consumption. Compare the liquid levels.

(a)

(b)

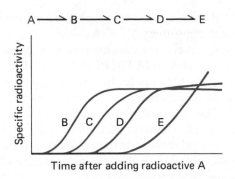

FIGURE 6–9
Specific Radioactivity of Pathway Intermediates. The specific radioactivity (radioactivity per μmole) of each intermediate in the pathway increases after radiolabeled A is added.

suspected intermediate, unlabeled, is added to the tissue slice. If the unlabeled compound is an intermediate between A and E, the specific radioactivity of E will diminish (Figure 6–10). This isotopic dilution is a useful strategy because it is easier to synthesize suspected intermediates without also having to radiolabel them.

Inhibitors of Pathways

A very general approach to analyzing pathways is to use *specific inhibition* of the enzymes in the pathway. Consider a chemical inhibitor that blocks the enzyme catalyzing the C-to-D reaction while the pathway is in a steady state. The earliest events that can be detected after addition of the inhibitor should include a rise in the concentration of precursor C and a fall in the concentration of D (Figure 6–11), called a "crossover." The interpretation of this crossover is that intermediate C lies upstream of intermediate D in the pathway. Next, one may observe a rise in the concentration of B and a fall in E, etc. Each intermediate can be assigned a position in the pathway, either upstream or downstream of the inhibited enzyme. The inhibition strategy has also been used to investigate pathways in intact cells.

The scarcity of known chemical inhibitors is a drawback to the application of the inhibition strategy to most pathways. But geneticists have rescued their biochemist colleagues by discovering that the phenotype of many simple mutations can be explained as a reduction in the activity of a single specific enzyme. To see how this represents a significant advance both in the study of metabolic pathways and in our understanding of the flow of genetic information, consider a mutation that inactivates the enzyme catalyzing the B-to-C reaction. This is detected as a cell that can no longer synthesize E when supplied with A in its diet. G. Beadle and E. Tatum used this strategy to isolate large numbers of mutant *Neurospora*, a bread mold, that could not grow unless one or another molecule was added to its usual diet, and the

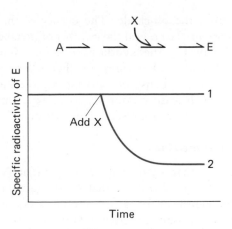

FIGURE 6–10
Isotope Competition. Radiolabeled A is added to the cells and the specific radioactivity of product E is measured over time. If no additions are made, the specific radioactivity does not change (case 1). When unlabeled compound X is added, the specific radioactivity falls (case 2). This shows that unlabeled X can compete with an intermediate for a reaction in the pathway from A to E. If X can also be isolated from the cells, it is likely that it is itself an intermediate in the pathway.

same approach was subsequently exploited with bacteria. The mutants fell into various classes, depending on which molecule (E) would permit the mutant to grow. Each blocked pathway can be analyzed chemically in a mutant cell by observing the increase in concentration of molecules (for example, B) before the block, due to the inactive mutant enzyme. Alternatively, one can identify other molecules (C, D) by trial and error that can substitute for the end product, because they can enter the

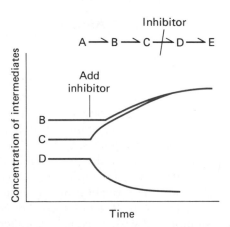

FIGURE 6–11
Effect of Pathway Inhibitor. A block in a pathway, whether caused by an inhibitor or a mutation in an enzyme that lowers its activity, leads to an accumulation of intermediates in front of the block (B and C) and a reduction in concentration of intermediates beyond the block (D).

pathway after the blockade. The beauty of the method lies in its potential to reveal the number of metabolic steps in a pathway for the synthesis of—for example—an amino acid, without knowing any chemical details. This benefit is at the expense of isolating a large number of mutants, which arise randomly, to be sure that all the various enzymes are represented at least once.

One Gene–One Enzyme

The fact that such analyses of mutant *Neurospora* could be performed implies that most mutations which block the synthesis of a metabolite are due to the inactivation of one enzymatic step in the synthetic pathway. The mutations are like specific enzymatic inhibitors. This finding was generalized in 1945 by Beadle and Tatum into the *one gene–one enzyme* hypothesis, an ideological precursor to the later central dogma of molecular biology, namely, that DNA information flows through RNA into protein. We shall discuss this in more detail in Chapters 10–13. The Beadle and Tatum one gene–one enzyme hypothesis was all the more remarkable for being made when our knowledge of the molecular nature of the gene was almost nil.

Examples of Metabolic Pathways

The following examples illustrate the general properties of pathways and are chosen from the many available because of their instructive value and also because they involve enzymes and/or metabolic intermediates of some clinical or social interest. The details of these pathways are found in the figure legends; the discussion in the text focuses on more general aspects.

Biosynthesis of Aromatic Amino Acids

Figure 6–12 shows a path that is found in all plants and microorganisms capable of synthesizing the amino acids phenylalanine and tyrosine. The sequence of reactions begins with two phosphorylated (activated) intermediates, erythrose-4-phosphate and phosphoenolpyruvate. Their condensation (Reaction 1) proceeds with the loss of one of the phosphates, which makes this a pacemaker step and practically irreversible in cellular conditions. Other reactions with large negative free energy changes are the activation of shikimate by ATP (Reaction 5, analogous to Reaction 6–6) and the condensation of shikimic-5-P with activated phosphoenolpyruvate (PEP, Reaction 6). An example of a pacemaker reaction at the end of an unbranched linear segment is the hydrolysis of P_i (Reaction 7) from enolpyruvyl shikimate phosphate, the precursor of chorismate. Pacemakers can also involve the irreversible decarboxylation of CO_2, as

shown by the conversion of prephenate to phenylpyruvate (Reaction 9).

Although there are cosubstrates that participate in some of these reactions (ATP, PEP, NAD^+, NADH, and NADPH), the pathways leading to shikimate and from shikimate to chorismate are both unbranched linear segments. The intermediates in those segments undergo only the reactions shown. Shikimate itself is a branch point. It is the precursor for the complex polymer *lignin*, which makes up part of the woody structure of plants,

FIGURE **6–12**

An Abbreviated Pathway, Found in Microorganisms and Plants, for the Synthesis of Aromatic Amino Acids.
The flow of atoms through the pathway can be followed by adding radiolabeled precursors. Two such atoms are traced in the figure, indicated by an open circle and a solid circle.

The first reaction is a condensation of erythrose-4-phosphate with phosphoenolpyruvate (PEP), both compounds used and/or produced by other pathways in cells. Reaction 1 is catalyzed by DAHP synthase and releases the phosphate that was bound to high-energy PEP; as the result, this reaction is practically irreversible in cellular conditions. Next comes the cyclization step (2), to form dehydroquinate, which involves another release of P_i. This reaction, probably multistep, results in the oxidation of one carbon (from CH—OH to C=O) and the reduction of its neighbor (from CH—O—P_i to CH_2). Since no net change in the oxidation state occurs there are no electrons carried off by NAD^+, even though this compound is necessary for the reaction. There follows a dehydration (Reaction 3, catalyzed by a dehydratase) and a reduction, catalyzed by a dehydrogenase (Reaction 4, the electrons for which are supplied by NADPH) to form shikimate, the end of the first segment.

The second segment begins (Reaction 5) with the activation of shikimate by ATP, catalyzed by a kinase, to form shikimate phosphate. This then condenses (Reaction 6, catalyzed by a synthase) with activated PEP to form enolpyruvyl shikimate phosphate, releasing P_i in the process. The following dehydration (7) is also accompanied by release of P_i. Chorismate is another branch point in the metabolic network. One set of reactions leads to synthesis of the amino acid tryptophan. The other proceeds via chorismate mutase (Reaction 8) to form the branch compound prephenate. Both branches are driven by the loss of (radiolabeled) CO_2. The branch leading to phenylalanine (Reaction 9, catalyzed by a dehydratase) also involves a dehydration, while the branch leading to tyrosine (Reaction 10, catalyzed by a dehydrogenase) also involves an oxidation, the electrons from which are carried away by NADH.

Phenyl pyruvate and hydroxyphenyl pyruvate are the keto-acid forms of amino acids phenylalanine and tyrosine, respectively. Keto-acids and amino acids can exchange roles via transamination (Reactions 11, 12), catalyzed by a transaminase. Glutamate, converted to α-keto-glutarate, usually provides the needed amino group, but other amino acids can be used. These reactions are almost at equilibrium under cellular conditions.

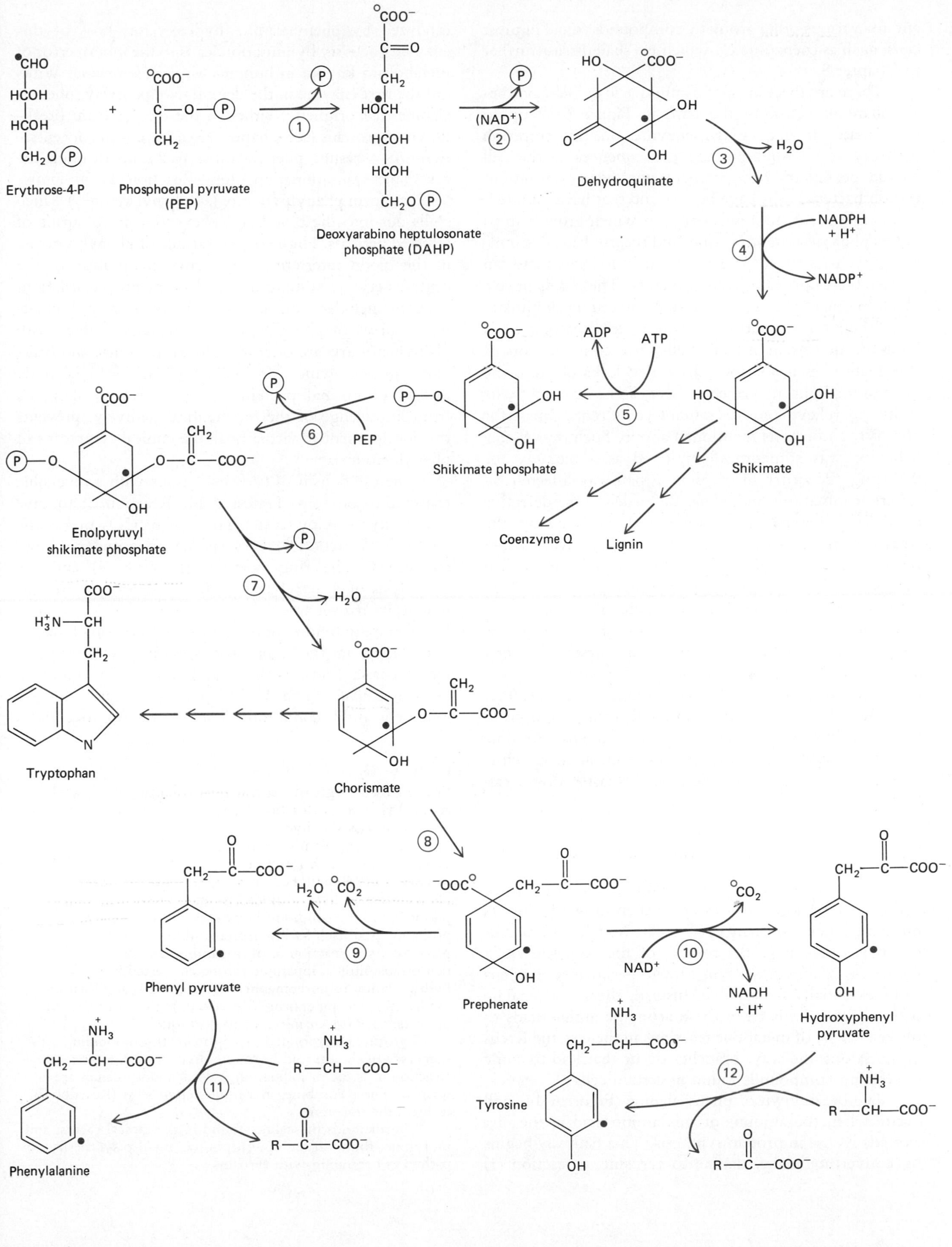

Erythrose-4-P
Phosphoenol pyruvate (PEP)
Deoxyarabino heptulosonate phosphate (DAHP)
(NAD⁺)
Dehydroquinate
H₂O
NADPH + H⁺
NADP⁺
ADP
ATP
Shikimate phosphate
Shikimate
Coenzyme Q
Lignin
PEP
Enolpyruvyl shikimate phosphate
H₂O
Tryptophan
Chorismate
Phenyl pyruvate
Prephenate
NAD⁺
NADH + H⁺
Hydroxyphenyl pyruvate
Phenylalanine
Tyrosine

and for various other aromatic compounds found in most cells, such as coenzyme Q, which we shall discuss further in Chapter 8.

There are three aromatic amino acids whose syntheses share in common the path in Figure 6–12 up to chorismate. If one of the enzymes in this common pathway were inhibited, the consequences to the cell should be severe. In agreement with this expectation, certain bacterial cells have been found that have mutated, in a single event, to a phenotype in which growth stops unless phenylalanine, tyrosine, and tryptophan are simultaneously present in the growth medium. Shikimate can replace their nutritional requirements. These cells have a block in one of the enzymatic steps preceding shikimate.

When the activation of shikimate (Reaction 5) is blocked, in certain mutant cells, the concentration of shikimate rises inside the cells and it leaks out into the growth medium. Its presence there can be detected by its ability to relieve another mutant cell's requirement for phe, tyr, and trp, as mentioned above. Such *cross-feeding* behavior was sometimes employed as a bioassay for deducing the order of steps in a pathway affected by different mutations and, thus, the order of the defective enzymes. This could be done without knowing the structure of the various intermediates involved. More recently, mutant cells with a blocked pathway, which causes them to excrete an intermediate, have been exploited as commercial sources of chemicals.

The position of tyrosine in the metabolic network is worth noting (Figure 6–13). It is a precursor of several compounds of great importance to organisms. Among these compounds are the neurotransmitters DOPA, dopamine, norepinephrine, and epinephrine; the thyroid hormone thyroxine; the psychotropic alkaloids mescaline and morphine; and the dark pigment melanin, which is found in many cells. Mutations in this last pathway can lead to albinism.

Degradation of Phenylalanine and Tyrosine

Although not all cells can synthesize phenylalanine and tyrosine, most cells can break them down for use as building blocks or energy. The pathway for the breakdown is not simply the reverse of the path shown in Figure 6–12, since it is arranged with pacemaker reactions to be essentially irreversible. Instead, there is a separate pathway that leads from these aromatic amino acids to the central set of metabolic reactions in the cell, the Krebs cycle. Along the way, branches occur that lead to some interesting compounds found in certain cells.

The breakdown of phenylalanine (Figure 6–14) will occur when the amount of this amino acid in the diet exceeds its use in protein synthesis. This pathway begins by converting phenylalanine to tyrosine, a reaction (1)

catalyzed by phenylalanine hydroxylase. Loss of this enzyme's activity by mutation leads to the inborn error of metabolism known in humans as *phenylketonuria*. Without the first enzyme in the degradative pathway, phenylalanine that originates either in the diet or from breakdown of proteins (see Chapter 13) increases in concentration. As a result, phenylalanine backs up through the reversible transamination step (Reaction 11 in Figure 6–12) to form phenylpyruvate (a "phenyl ketone") which spills out into the blood and is excreted in the urine of afflicted patients. High concentrations of phenylpyruvate in the blood interfere with normal maturation of the central nervous system and lead to mental retardation. Therapy includes reducing (though not entirely eliminating) sources of phenylalanine in the diet. Babies with phenylketonuria are often fed special formulae and fruits like bananas, which are very low in phenylalanine, instead of normal proteins. This treatment, designed from knowledge of the biochemical pathway, prevents the development of virtually all the clinical symptoms of phenylketonuria.

The catabolism of tyrosine begins with a reversible transamination step (Figure 6–14, Reaction 2) to give *p*-hydroxyphenylpyruvate. The loss of CO_2 makes the next step (Reaction 3) very spontaneous under cellular conditions. The ring opening (Reaction 4) and the hydrolysis of fumaryl-acetoacetate (Reaction 6) are also practically irreversible.

Failure to oxidize homogentisate (Reaction 4), due to a mutation in the gene that specifies homogentisate dioxygenase, leads to the inborn error of metabolism known as *alkaptonuria*. Unlike phenylketonuria, this is a relatively mild condition. The excess homogentisate

FIGURE 6–13

Some Physiologically Important Compounds that Arise from Tyrosine.   Reaction 1 shows the conversion by tyrosinase (tyrosine hydroxylase) to DOPA, dihydroxyphenylalanine. A branch to the dark pigment melanin begins with Reaction 2, also catalyzed by tyrosinase, to form DOPA quinone. This is converted to melanin. Any block in this pathway can lead to underproduction of the pigment and the condition known as albinism. DOPA is probably used itself as a neurotransmitter. It is also the precursor, via Reaction 3, of dopamine, a known neurotransmitter. Dopamine can be converted by hydroxylation to norepinephrine (Reaction 4) and then by methylation to epinephrine (Reaction 5). Both of these are important neurotransmitters and hormones.

Tyrosine incorporated in a protein, thyroglobulin, is the precursor of the thyroid hormone thyroxine. The first step, Reaction 6, is the di-iodination; two di-iodotyrosines are formed. These combine, in a multistep process (Reaction 7), to form the hormone.

The alkaloids mescaline, found in the mescal cactus, and morphine, found in the opium poppy, are also synthesized by pathways beginning with tyrosine.

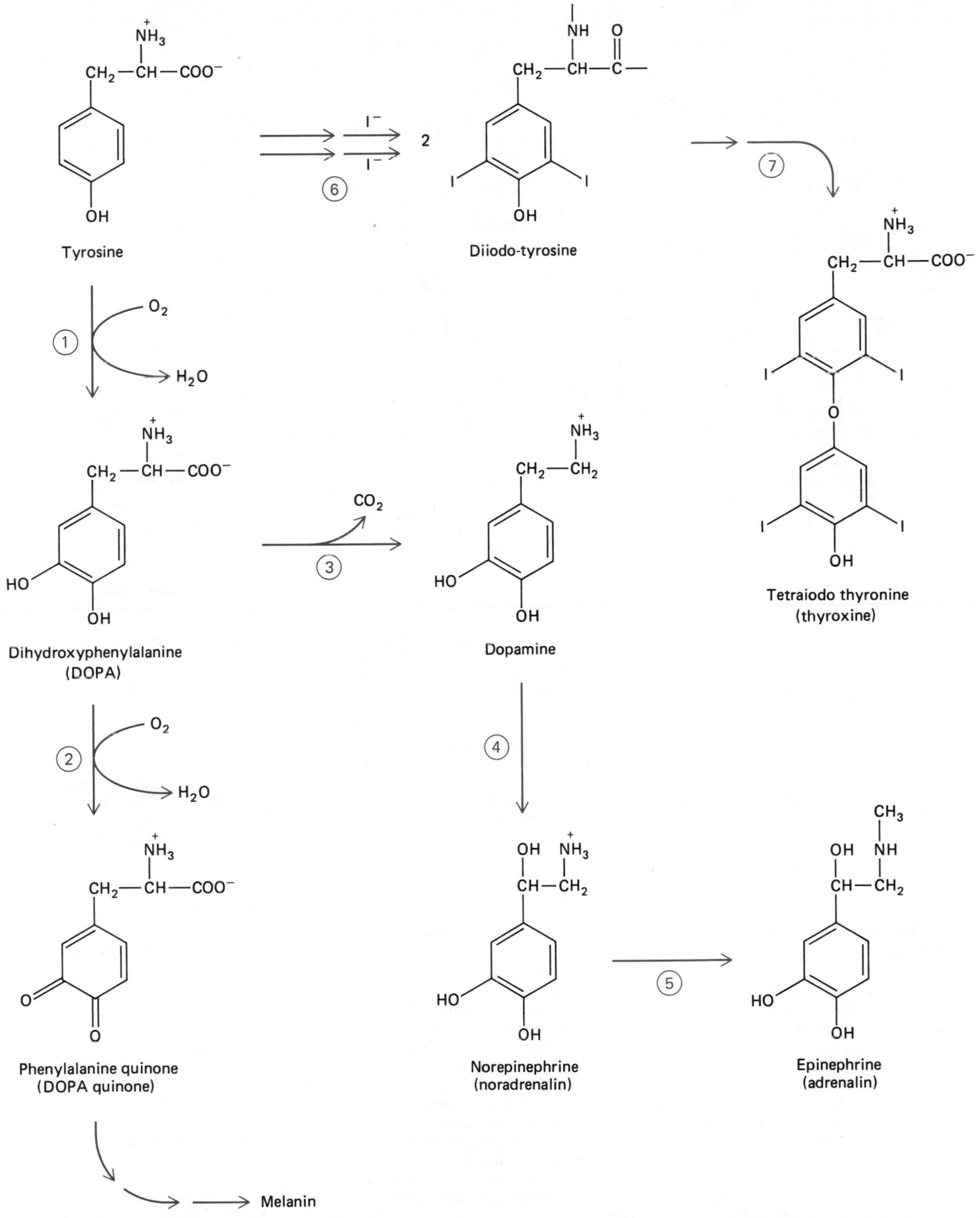

Tyrosine

Diiodo-tyrosine

Tetraiodo thyronine
(thyroxine)

Dihydroxyphenylalanine
(DOPA)

Dopamine

Phenylalanine quinone
(DOPA quinone)

Norepinephrine
(noradrenalin)

Epinephrine
(adrenalin)

Melanin

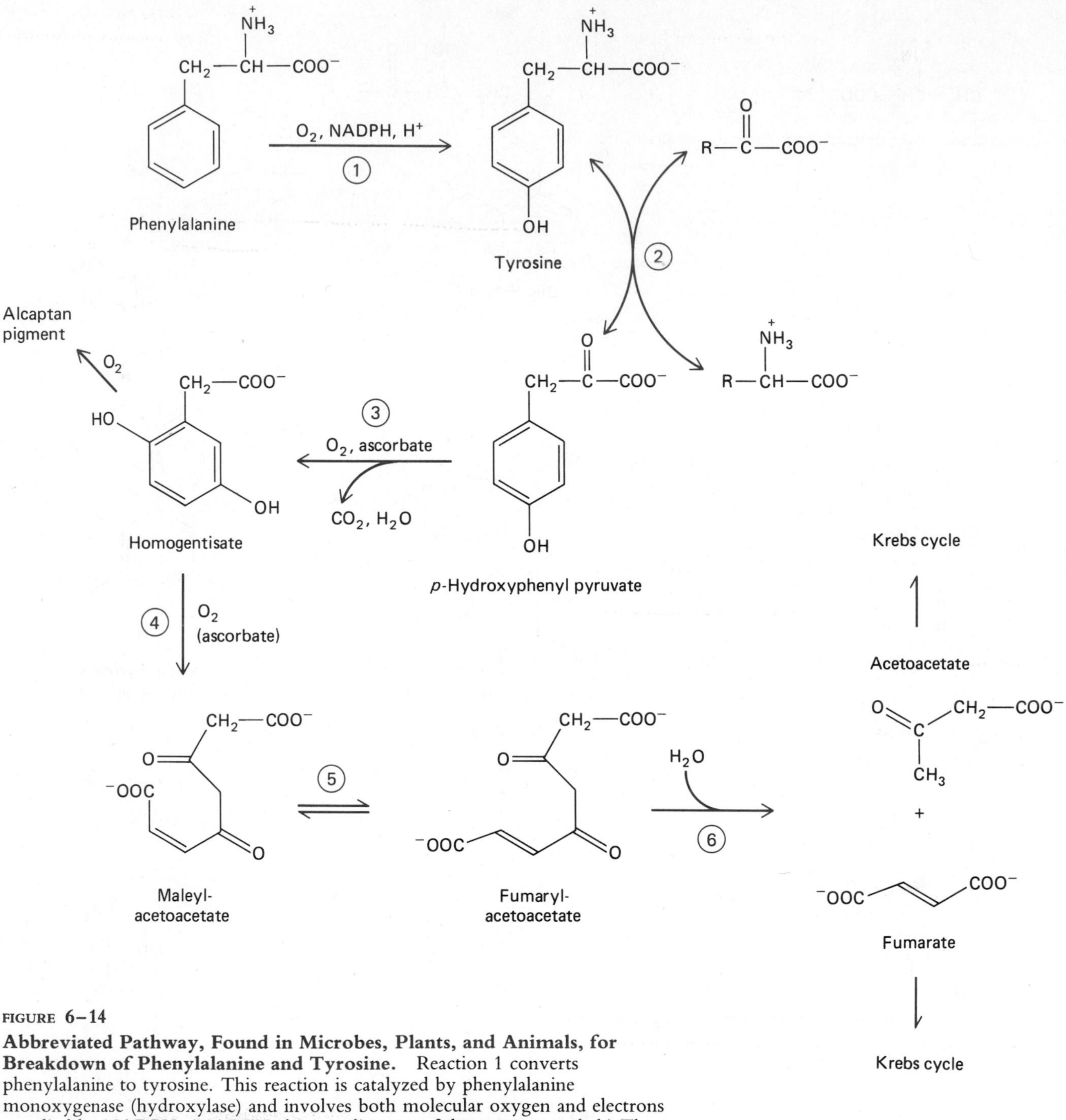

FIGURE 6–14

Abbreviated Pathway, Found in Microbes, Plants, and Animals, for Breakdown of Phenylalanine and Tyrosine. Reaction 1 converts phenylalanine to tyrosine. This reaction is catalyzed by phenylalanine monoxygenase (hydroxylase) and involves both molecular oxygen and electrons supplied by NADPH. (NADPH also supplies one of the protons needed.) The further common breakdown pathway for phenylalanine and tyrosine starts with a reversible transamination (Reaction 2) to yield hydroxyphenylpyruvate. There follows a complex reaction (3) involving an oxidation of the aromatic ring, a decarboxylation releasing CO_2, and a migration of the side chain to yield homogentisate. This is catalyzed by an enzyme that uses molecular oxygen as the oxidant. Ascorbic acid (vitamin C) is also involved. Homogentisic acid is attacked by another dioxygenase, stimulated by ascorbate, which breaks the aromatic ring to yield maleyl-acetoacetate (Reaction 4). This isomerizes to fumaryl-acetoacetate (Reaction 5). Finally, this compound is cleaved by a hydrolase to yield acetoacetate and fumarate (Reaction 6), both of which can enter the Krebs cycle to be processed into almost any of the metabolites found in cells.

spontaneously oxidizes to form an ochre-colored alkapton pigment which darkens the urine and connective tissue and is associated with arthritis.

The characteristics of these pathways are typical of most others found in cells. The synthetic pathway is different from the degradative pathway; both usually have irreversible steps at the beginning and end of their respective segments. Intermediates which are blocked from being utilized by defects in the enzymes using them as substrates have nowhere to go and when concentrations rise are excreted or react at appreciable rates spontaneously with a ubiquitous cosubstrate like molecular oxygen. This illustrates again the value of using enzyme catalysts; the low concentrations of intermediates required for an enzyme-catalyzed reaction tend to reduce very much the rate of any uncatalyzed and potentially deleterious reaction.

Cellular Energy Supply

There are only two ways that cells can obtain energy from their surroundings. They can oxidize molecules found in the environment or they can absorb light. We shall discuss photosynthesis and the use of light energy to generate ATP (photophosphorylation) in Chapter 8 after learning more about the structure and functions of the membranes on which those processes depend. We begin the story of oxidation metabolism here, but it also will be completed in Chapter 8 since membranes also play an important role in the generation of ATP by mitochondria, the process known as oxidative phosphorylation.

Fermentations

In most cellular systems there is essentially no capacity for the storage of naked electrons. When a molecule is oxidized, the electrons removed must be incorporated into the structure of another molecule, which thereby becomes reduced. In this chapter we shall consider oxidation-reduction reactions that together form a fermentation pathway. In a *fermentation*, electrons are removed from an *organic* molecule (reductant) found in the environment, and are used ultimately to reduce another *organic* molecule (oxidant) that is released to the environment. Coupling these spontaneous fermentation oxidations to the synthesis of ATP is how many cells make an energetic living. *Anaerobic* cells—ones that don't use oxygen—can exploit the energy available from oxidation only by fermentations.

The environmental molecule most commonly used as a reductant in cellular fermentations is glucose. As it turns out, the pathway used in fermenting glucose is found in most cells, even those that do not complete the fermentation. We shall first describe *glycolysis,* the breakdown pathway from glucose to pyruvate, and then show how pyruvate is utilized to complete various fermentations.

Glycolysis

Our knowledge of the glycolytic pathway (Figure 6–15) is based on the work both of G. and C. Cori, and O. Myerhoff and G. Embden, who elucidated the energy requirements of the muscle cell. Indeed, this series of reactions is also called the *Embden-Myerhoff pathway*.

Most foods can be digested either to yield glucose directly or to yield metabolites that can be transformed into glucose. Glucose enters cells directly from their environment, or it may arise from the breakdown of intracellular storage molecules such as glycogen or starch. To enter glycolysis, glucose must first be converted to glucose-6-phosphate (Figure 6–15, Reaction 1) which, because of the charge on the phosphate group, is more easily retained in the cell (see also the earlier discussion of Reaction 6–6).

The first important energy business of glycolysis occurs during Reaction 6, the oxidation of glyceraldehyde phosphate (GAP), catalyzed by its dehydrogenase. The electrons are incorporated into the structure of coenzyme NAD^+, as shown in Figure 6–16. The incorporation of inorganic phosphate also occurs during the oxidation, to yield 1,3-diphosphoglycerate (DPG). If the oxidation of the aldehyde to an acid were to occur without the incorporation of P_i, it would be very spontaneous; a free energy change under standard conditions has been calculated as -42 kJ/mole. This would lead to a very large concentration of the acid product. With the formation of the activated DPG, however, $\Delta G' = +5$ kJ/mole and Reaction 6 is close to equilibrium under standard conditions. DPG is sufficiently activated to transfer its phosphate to ADP with a $\Delta G' = -20$ kJ/mole (Reaction 7). Thus, the overall oxidation and generation of ATP has $\Delta G' = -15$ kJ/mole, is quite spontaneous at standard conditions, avoids the buildup of high concentrations of intermediates, and stores part of the energy of oxidation as ATP.

An audit of ATP at this point shows no net energy gain; two ATPs were required to form six-carbon fructose diphosphate, and two ATPs were formed from oxidizing the two three-carbon glyceraldehyde phosphates. But glycolysis is not yet finished. An isomerization (Reaction 8) followed by a dehydration (Reaction 9) yields phosphoenolpyruvate, another activated intermediate. It is at this point that the energetic payoff of glycolysis occurs: the transfer of the phosphate from phosphoenolpyruvate to ADP to yield pyruvate and ATP (Reaction 10). This reaction is very spontaneous ($\Delta G' = -31$ kJ/mole), and is an essentially irreversible pacemaker reaction under cellular conditions. At this

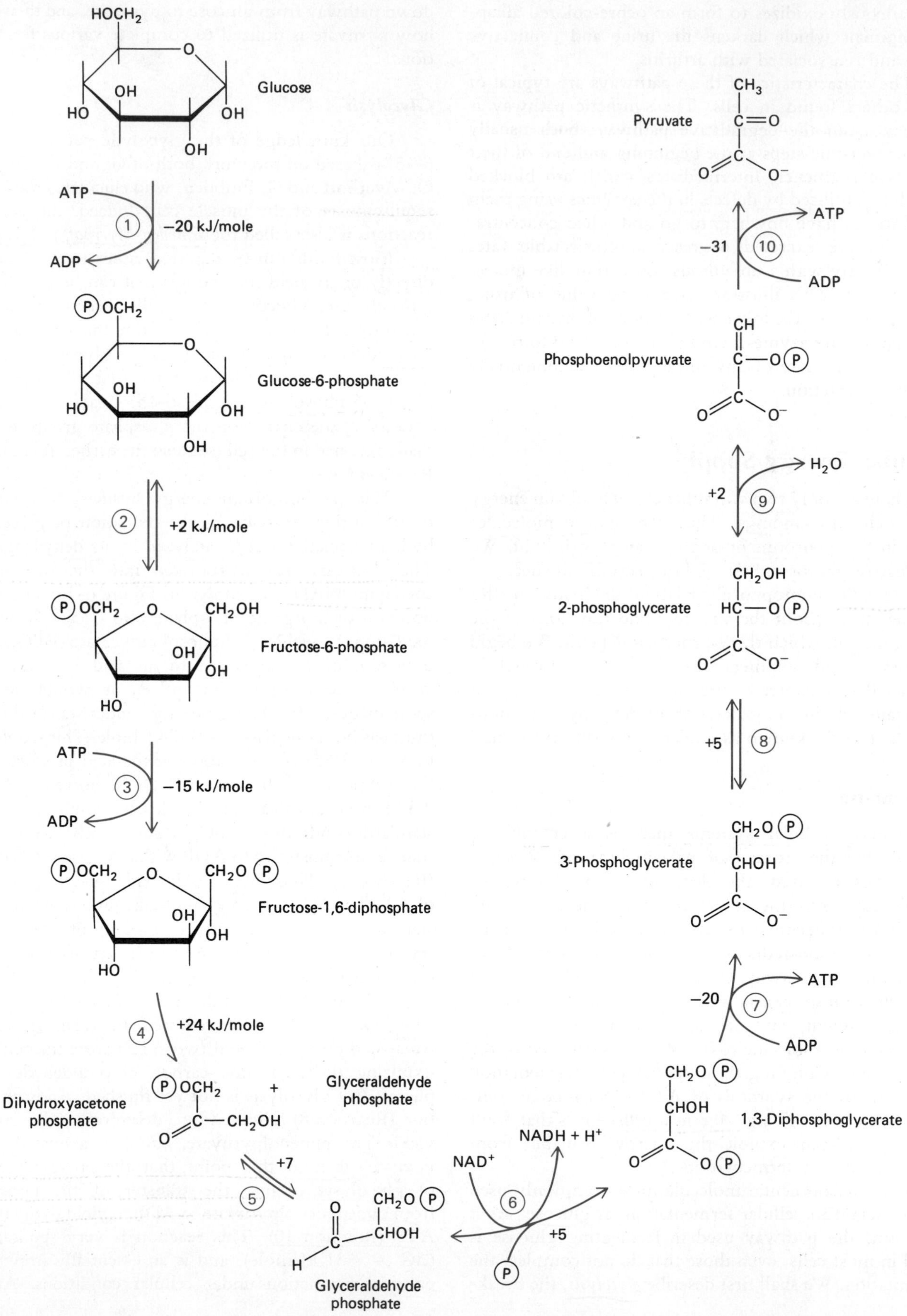

HOCH₂

Glucose

ATP

① −20 kJ/mole

ADP

(P)OCH₂

Glucose-6-phosphate

② +2 kJ/mole

(P)OCH₂   CH₂OH

Fructose-6-phosphate

ATP

③ −15 kJ/mole

ADP

(P)OCH₂   CH₂O(P)

Fructose-1,6-diphosphate

④ +24 kJ/mole

Dihydroxyacetone
phosphate

(P)OCH₂
  |
  C—CH₂OH
  ‖
  O

+   Glyceraldehyde
    phosphate

⑤ +7

    O   CH₂O(P)
    ‖ |
    C—CHOH
   /
  H

Glyceraldehyde
phosphate

NAD⁺   NADH + H⁺

⑥ +5

(P)

Pyruvate

CH₃
 |
 C=O
 |
 C
// \
O   O⁻

ATP

−31 ⑩

ADP

Phosphoenolpyruvate

CH
‖
C—O(P)
 |
 C
// \
O   O⁻

+2 ⑨   H₂O

2-phosphoglycerate

CH₂OH
 |
HC—O(P)
 |
 C
// \
O   O⁻

+5 ⑧

3-Phosphoglycerate

CH₂O(P)
 |
 CHOH
 |
 C
// \
O   O⁻

−20 ⑦   ATP

ADP

1,3-Diphosphoglycerate

CH₂O(P)
 |
 CHOH
 |
 C
// \
O   O(P)

FIGURE **6–15**

Glycolysis, the Breakdown of Glucose to Pyruvate. In Reaction 1, glucose is phosphorylated by ATP to give glucose-6-phosphate (G6P). This reaction is catalyzed by hexokinase and has a standard free energy change ($\Delta G'$) of -20 kJ/mole, as indicated (see the earlier discussion of Reaction 6–6). G6P next undergoes an isomerization (Reaction 2), catalyzed by phosphoglucoisomerase, to fructose-6-phosphate (F6P). In Reaction 3, F6P is phosphorylated, again by ATP, to yield fructose diphosphate (FDP); the reaction is catalyzed by phosphofructokinase. This reaction is an important control point in glycolysis. The breaking of a 6-carbon hexose into two 3-carbon molecules occurs in Reaction 4, catalyzed by the lyase enzyme, aldolase. The products dihydroxyacetone phosphate (DAP) and glyceraldehyde phosphate (GAP) are interconvertible (Reaction 5) via triosephosphate isomerase. The sole oxidation step of glycolysis (Reaction 6) occurs next and converts glyceraldehyde phosphate to 1,3-diphosphoglycerate (DPG). The extra phosphate in the acid anhydride product comes from inorganic phosphate in the cytoplasm; the electrons removed during the oxidation from GAP by its dehydrogenase reduce the coenzyme NAD^+ to $NADH + H^+$ (see Figure 6–16). High energy DPG can phosphorylate ADP to give ATP (Reaction 7) and 3-phosphoglycerate (3PG), catalyzed by phosphoglycerate kinase. 3PG then undergoes isomerization to 2PG, 2-phosphoglycerate, via phosphoglyceromutase (Reaction 8). The dehydration (Reaction 9) of 2PG to phosphoenolpyruvate (PEP) is catalyzed by enolase. Finally, the phosphate on PEP is used to phosphorylate ADP to ATP, yielding pyruvate as the end product (Reaction 10).

point the cell has a net gain of two ATPs from the catabolism of one glucose, and glycolysis is complete.

Glycolysis, like other pathways, is a series of coupled reactions, coupled because the product of one reaction is the substrate of the next. We also say that glycolysis is *tightly coupled* to the phosphorylation of ADP to form ATP. The coupling exists because ADP is a *cosubstrate* in two reactions of the pathway (7 and 10). Even if glucose is present in excess, glycolysis cannot proceed if there is no ADP. Stimulation of the glycolysis pathway by added

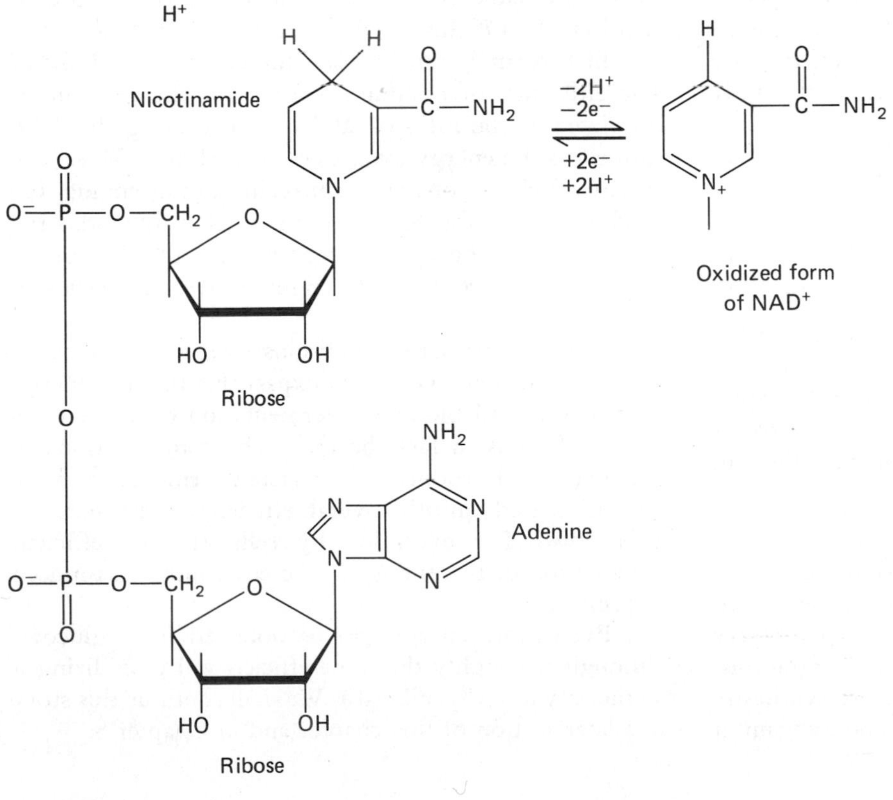

FIGURE **6–16**

Reduced NADH and Oxidized NAD$^+$ (*nicotinamide adenine dinucleotide*). Two electrons are carried by the structure, at the colored hydrogen atoms. The positive charge is carried either by the nicotinamide (in NAD$^+$) or by an unbound hydrogen ion (in NADH). NADPH and NADP^+ have analogous structures but contain a phosphate esterified at the colored oxygen.

ADP, the *ADP effect*, was one of the early indications of this coupling. There is a similarly explained *phosphate effect*, because of the use of P_i in Reaction 6. Since glycolysis is tightly coupled to phosphorylation, we expect that the operation of any process which uses energy from the conversion of ATP to ADP will stimulate glycolysis if the concentration of ADP is rate-limiting (we shall see an example of this in Chapter 9).

The arsenate ion (AsO_4) *uncouples* the phosphorylation of ADP from glycolysis. This ion, which resembles P_i, can substitute for it in Reaction 6 to form arseno-phosphoglycerate. This is an unstable compound: the arsenate is removed by hydrolysis spontaneously (that is, no enzyme is needed to catalyze the reaction). The effect of the formation and hydrolysis of arseno-phosphoglycerate is to provide an *alternative pathway* between GAP and 3PG, one without any formation of ATP. Glycolysis can proceed to pyruvate, but no net ATP is obtained. It is in this sense that we regard arsenate as an uncoupler of the phosphorylation of ADP.

Arsenate   Arseno-phosphoglycerate

All the enzymatic steps of glycolysis are performed by soluble enzymes located in the cell sap. These enzymes were isolated and the whole pathway was reconstructed in vitro by biochemists in the 1930s and 1940s. It was a triumph of technique and gave great impetus to the study of the molecules of the cell.

The overall reaction of glycolysis is:

Glucose + 2 NAD$^+$ + 2 ADP + 2 P_i $\longrightarrow$

$$2 \text{ Pyruvate} + 2 \text{ H}^+ + 2 \text{ NADH} + 2 \text{ H}^+ + 2 \text{ H}_2\text{O} + 2 \text{ ATP}.$$

In aerobic cells, pyruvate and NADH are oxidized further in the mitochondria via the Krebs cycle (discussed later in this chapter) and the electron transport chain (see Chapter 8). The two net ATPs formed can be used, in appropriately coupled reactions, to run the metabolism of the cell.

Fermentation End Products

Glycolysis is not by itself a fermentation because the electrons removed from glucose can only be stored temporarily in NADH. After all, NAD$^+$ is a coenzyme and is present in cells at relatively low concentrations. Also, it is energetically expensive for a cell to synthesize NAD$^+$, and there would thus be no energy gain if

NADH were constantly released to the environment. For these reasons there must be some way to regenerate NAD$^+$ from NADH. Cells using fermentation for energy vary a great deal in how they dispose of the electrons stored in NADH. The human red blood cell (which contains no mitochondria) uses these electrons to reduce pyruvate, forming lactate, which is transported into the blood (Figure 6–17). The anaerobic yeast cell decarboxylates pyruvate to form acetaldehyde which, in turn, is reduced by NADH to form ethanol. Both the CO_2 and the ethanol are excreted into the environment, to the everlasting gratitude of brewers, vintners, bakers, and those who consume their products. Certain procaryotes can transform pyruvate into a variety of fermentation end products: lactate, acetate, isopropanol, acetone, butyrate, etc.

Energy Efficiency of Fermentation

In the mammalian red blood cell, the energy necessary to operate its metabolism comes from the fermentation of glucose to lactate via pyruvate. It is useful to evaluate the efficiency with which the energy is produced and to compare that process with man-made machines that also perform energy transductions. The sum of all the standard free energy changes during the fermentation is -130 kJ/mole (remember that all the steps after Reaction 5 in Figure 6–15 occur twice). To this must be added the additional $\Delta G'$ that would result from the hydrolysis of the two ATPs formed: -68 kJ/mole. This total of -198 kJ/mole represents the overall free energy change available from the fermentation of glucose to two lactates. Of this, -68 kJ/mole is "stored" as ATP, in which form it can be used for cellular metabolism. The efficiency of obtaining energy for the cell can be calculated as the ratio of $\Delta G'$ of hydrolyzing the ATP formed to the energy available: $-68/-198 = 34\%$. The energy efficiency of the steam reciprocating engine, the machine responsible for powering much of the industrial revolution, has been estimated as only 16%. Gasoline-driven internal combustion engines have efficiencies of around 25%.

Because standard conditions are not necessarily cellular conditions, we might expect that the true energy efficiency of red blood cell fermentation could be even higher. That is in fact the case. The concentrations of precursors and products of the lactate fermentation have been measured, and the overall efficiency is estimated at over 50%! The evolution by cells of such efficient exploitation of the energy in the environment is indeed impressive.

Even more energy can be obtained from glucose, although at roughly the same efficiency, by oxidizing it all the way to CO_2 and H_2O. We shall continue this story in a later section of this chapter and in Chapter 8.

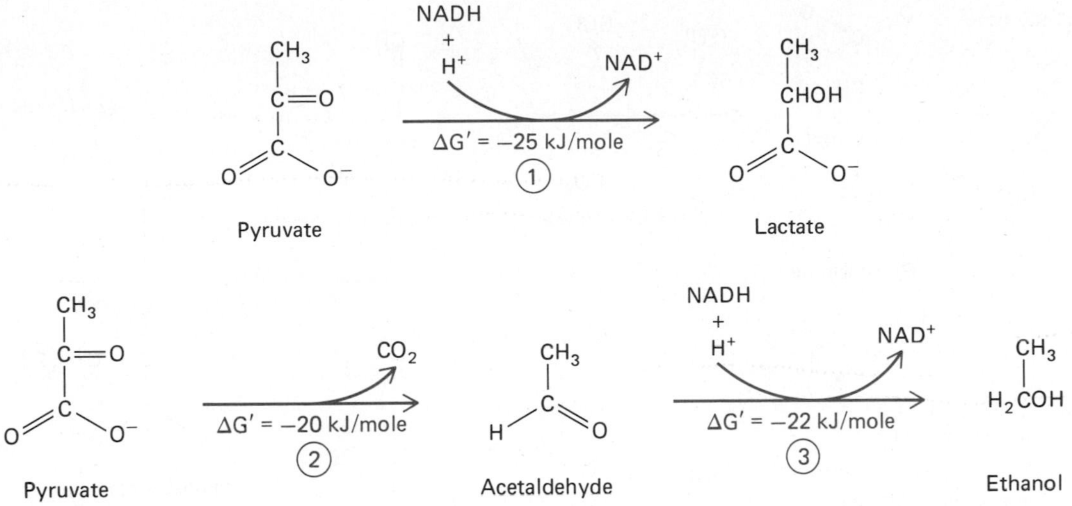

FIGURE 6–17

The Completion of Two Fermentations. The pathway by which mammalian red blood cells (and certain bacteria during food fermentation) transform glycolytic pyruvate is shown in Reaction 1. Pyruvate is reduced, via lactate dehydrogenase, to lactate. The electrons from NADH are used and NAD^+ is regenerated for further participation in glycolysis. The end product, lactic acid, is excreted into the blood (or into yogurt). Fermenting yeast use a different pathway. Pyruvate is first decarboxylated to form acetaldehyde, the reaction (2) being catalyzed by pyruvate decarboxylase. It is the acetaldehyde that is reduced with the electrons in NADH, to give ethanol. This reaction (3) is catalyzed by alcohol dehydrogenase. The end products are released into the environment.

Gluconeogenesis

Pyruvate is formed by other reactions of metabolism and, when it is present in excess in mammalian liver cells, can follow a pathway to glucose and onward to glycogen (Figure 6–18). Although it shares some steps in common, this path (*gluconeogenesis*) is not merely the reverse of glycolysis. The differences between the two constitute another example of the practical irreversibility of most pathways.

The first hurdle to overcome is the conversion of pyruvate to phosphoenolpyruvate (PEP). This happens in a round-about way because the $\Delta G'$ of the PEP-to-pyruvate reaction is -31 kJ/mole. Excess pyruvate is usually generated in the liver mitochondria. There it becomes carboxylated with CO_2 to form oxaloacetate, a reaction that requires the hydrolysis of an ATP to ADP and phosphate. Once outside the mitochondrion, oxaloacetate is decarboxylated and phosphorylated, by GTP, to give phosphoenolpyruvate. This complex path thus requires the investment of two high-energy phosphates to get around the essentially irreversible step of glycolysis.

The glycolytic pathway is followed backwards through the formation of fructose diphosphate. Here the extra phosphate is removed by hydrolysis to yield fructose-6-phosphate, a reaction with a much more favorable free energy change ($\Delta G' = -17$ kJ/mole) than the reversal of the glycolytic step ($\Delta G' = +14$ kJ/mole).

The glycolytic path is then followed backwards to glucose-6-phosphate (G6P).

Glucose can be formed by hydrolyzing G6P (see the earlier discussion of Reaction 6–4). Glycogen is formed by isomerizing glucose-6-phosphate to glucose-1-phosphate (Reaction 6–8) and then converting G1P to UDP-glucose (Reaction 6–10). Uridine diphosphate-glucose is the donor of glucose to the polysaccharide structure of glycogen:

$$\text{Glycogen}_n + \text{UDP-Glucose} \longrightarrow \text{Glycogen}_{n+1} + \text{UDP}.$$

Since glycolysis and gluconeogenesis share several reactions in common, the cell must carefully regulate the flow of material along these paths. The problem is more clearly seen by considering the consequences of both synthesizing fructose diphosphate from fructose-6-phosphate and ATP, and breaking down fructose diphosphate to fructose-6-phosphate and P_i. If they operated simultaneously, these two reactions would constitute a *futile cycle*, in which ATP was hydrolyzed with no apparent benefit to the cell. The enzymes involved must be tightly controlled to prevent such waste.

The Krebs Cycle

The evolution of metabolism can be regarded as conservative since cells often use a common pathway both for the production of energy and as the starting

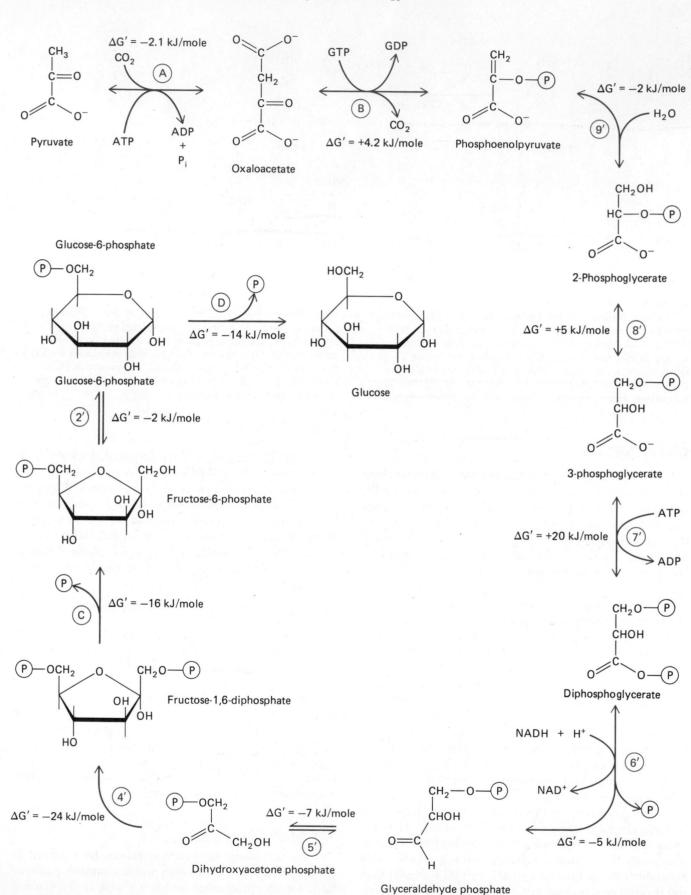

Pyruvate

$\Delta G' = -2.1$ kJ/mole

Oxaloacetate

$\Delta G' = +4.2$ kJ/mole

Phosphoenolpyruvate

$\Delta G' = -2$ kJ/mole

2-Phosphoglycerate

Glucose-6-phosphate

Glucose-6-phosphate

$\Delta G' = -14$ kJ/mole

Glucose

$\Delta G' = +5$ kJ/mole

3-phosphoglycerate

$\Delta G' = -2$ kJ/mole

Fructose-6-phosphate

$\Delta G' = +20$ kJ/mole

$\Delta G' = -16$ kJ/mole

Diphosphoglycerate

Fructose-1,6-diphosphate

$\Delta G' = -24$ kJ/mole

Dihydroxyacetone phosphate

$\Delta G' = -7$ kJ/mole

Glyceraldehyde phosphate

$\Delta G' = -5$ kJ/mole

FIGURE **6-18**

Gluconeogenesis, the Conversion of Pyruvate to Glucose. This pathway uses many of the same steps, catalyzed by the same enzymes, as glycolysis. The shared reactions are numbered the same way as in Figure 6–15, but with a prime ('). The steps that are not the reverse of glycolytic reactions are designated by circled capital letters. The first of these (Reaction A) is the carboxylation of pyruvate to oxaloacetate. This reaction also involves the hydrolysis of ATP to ADP and P_i. The reaction is catalyzed by pyruvate carboxylase. In liver cells, this reaction occurs in the mitochondria where excess pyruvate is mostly found. The mitochondrial membrane will not permit oxaloacetate to pass directly through, so it is reduced to malate, which does penetrate the membrane, and re-oxidized to oxaloacetate in the cytoplasm. Reaction B, catalyzed by phosphoenolpyruvate (PEP) carboxykinase, converts oxaloacetate to PEP. The phosphate comes from GTP, which is converted to GDP, and CO_2 is given off. Two high-energy phosphates are needed to convert pyruvate to PEP; nevertheless, the reaction is practically at equilibrium under standard conditions ($\Delta G' = 2.1$ kJ/mole). The reactions 9', 8', 7', 6', 5', and 4' are the same as in glycolysis, but reversed, and lead to the formation of fructose-1,6-diphosphate (FDP). The next reaction (C) that is specific to gluconeogenesis is the hydrolysis of phosphate from FDP to give fructose-6-phosphate. This removal of phosphate is catalyzed by a phosphatase (fructose diphosphatase) and the enzyme's activity is under strict control (see Chapter 14). The reaction (D) that terminates the pathway is again catalyzed by a phosphatase, glucose-6-phosphatase, and converts G6P to glucose. This enzyme must also be carefully controlled.

point for the synthesis of precursors of large molecules. This sharing is a feature of several segments of the metabolic network and is convincingly illustrated by the details of the most famous of these, the **Krebs cycle**—named for its formulator (H. Krebs), and also known as the *tricarboxylic acid (TCA) cycle* and the *citric acid cycle*. In examining this cycle, we shall also learn a bit of how a biochemist works, and how these pathways came to be hypothesized and confirmed.

The Krebs cycle is shown in Figure 6–19. Essentially, it is a scheme by which two carbon atoms in a derivative of acetic acid are oxidized to carbon dioxide. The derivative used is actually acetyl-coenzyme A (acetyl-S-Co A), and it is formed during the breakdown of fat, certain amino acids, or carbohydrate foodstuffs. The end product of many of these breakdown steps is pyruvate, which is transported into mitochondria. There it is oxidized and decarboxylated to acetyl-S-Co A in a series of enzymatic reactions. This *activated acetate* can be oxidized via the Krebs cycle, or it can go through a series of reactions in which it successively condenses with itself or with a derivative of itself to form higher fatty acids—straight-chain compounds, 14–20 carbon atoms long. This is how carbohydrate becomes converted to lipid, and how sugar eaten ends up as body fat.

The first step in the oxidation of acetyl-coenzyme A is its condensation with oxaloacetate and water to form citrate. Citrate is then processed by an enzyme, aconitase

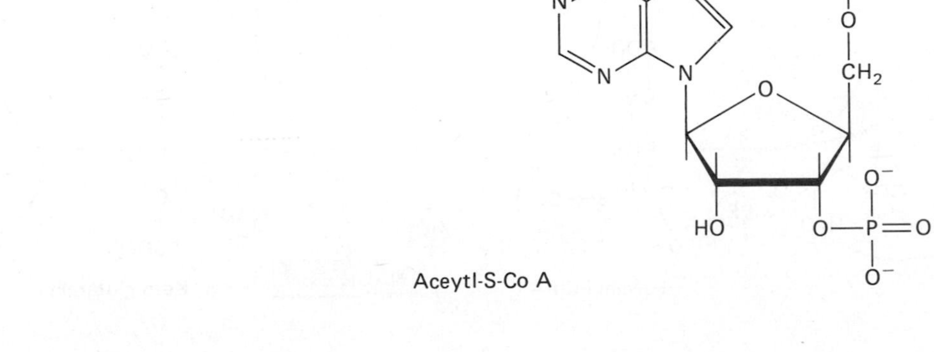

Aceytl

Aceytl-S-Co A

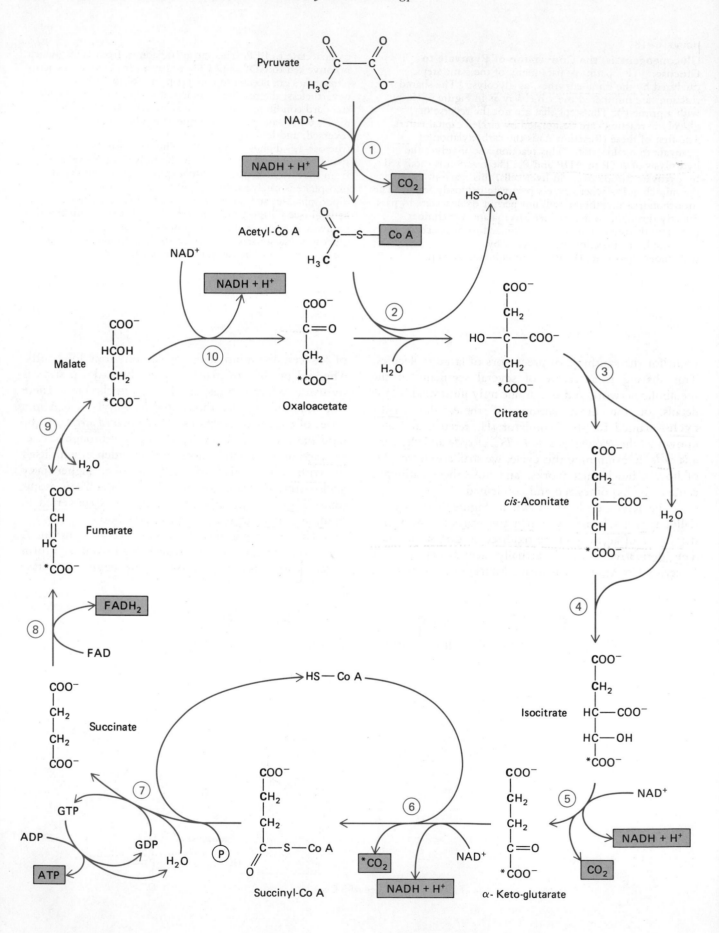

FIGURE 6–19
The Krebs Cycle. This sequence of reactions oxidizes acetyl-Co A. Several metabolic reactions can produce the activated 2-carbon fragment with which the cycle begins. Shown in Reaction 1 is the oxidative decarboxylation of pyruvate, the route by which carbohydrate carbons enter the cycle (see Figure 6–17). This reaction, essentially irreversible, has a complicated mechanism and is catalyzed by the multi-enzyme pyruvate dehydrogenase complex. The first reaction (2) of the cycle proper is the condensation of oxaloacetate with acetyl-Co A to form citrate. Water enters the cycle during this exergonic reaction, catalyzed by citrate synthase. Citrate is dehydrated (Reaction 3) to form *cis*-aconitate, which is rehydrated (Reaction 4) to form isocitrate. Both of these reactions are catalyzed by aconitase and the free energy change in both is small. The first oxidation reaction (5) in the cycle decarboxylates isocitrate as well and thus forms α-ketoglutarate; the electrons are carried off as NADH. This step, catalyzed by isocitrate dehydrogenase, is probably the slowest in the cycle and determines, therefore, the rate of all the other reactions. It is a step at which control over the rate of the cycle is exerted. The next reaction (6) is also an oxidative decarboxylation; coenzyme A is used and the product is succinyl-Co A. This is similar to Reaction 1; it also has a large negative free energy change and is catalyzed by a multi-enzyme complex, the α-ketoglutarate dehydrogenase complex. The $\Delta G'$ of hydrolysis of succinyl-Co A has practically the same value as the $\Delta G'$ for hydrolyzing ATP. Reaction 7, therefore, is practically at equilibrium as GTP is formed from GDP and P_i. This reaction, forming succinate, also involves the entry of water into the cycle and is catalyzed by succinyl-Co A synthetase. Reaction 8, catalyzed by succinate dehydrogenase, is another oxidation; the electrons removed from succinate to form fumarate are carried away by $FADH_2$, however. Fumarate is then hydrated to form malate (Reaction 9), a reaction catalyzed by fumarase and practically at equilibrium. The final oxidation of the cycle (Reaction 10) is catalyzed by malate dehydrogenase and the electrons are carried away by NADH. At standard conditions this reaction has a large positive free energy change, but under cellular conditions the reaction proceeds because of the very low concentrations of the products oxaloacetate and NADH.

The fate of the two boldfaced carbons of acetyl-Co A can be followed through the cycle as far as succinate. Because that molecule is symmetrical, the pairs of carbons cannot be distinguished in their further reactions. The net reaction of the cycle is to oxidize the acetyl-Co A carbons to CO_2, carrying off the eight electrons in the form of carriers NADH and $FADH_2$. One GTP is formed from GDP and P_i during each turn of the cycle. This GTP can be used to recharge ADP to ATP. Besides the oxidation of isocitrate (Reaction 5), the formation of citrate (Reaction 2) and the oxidation of α-ketoglutarate (Reaction 6) are important control points in the cycle.

(aconitate hydrase), that catalyzes the interconversion between citrate, enzyme-bound *cis*-aconitate, and isocitrate. At equilibrium, the ratio of these species would be citrate: *cis*-aconitate: isocitrate = 91: 3: 6. But during normal operation of the Krebs cycle, the direction of mass flow is toward isocitrate formation because of the presence of an enzyme (isocitrate dehydrogenase) that catalyzes the energetically favored oxidation of isocitrate, thus reducing its concentration. The overall effect is to displace the interconversion reactions from equilibrium so that more isocitrate is formed from citrate.

This situation—the displacement from an equilibrium because of a subsequent reaction of a product—is common in the cell. As a result of the coupling between reactions, many reactions that might themselves appear to be *thermodynamically reversible*, because there is very little change in free energy, are constantly being rendered *kinetically irreversible*—that is, in the steady state of metabolism they go almost entirely in one direction. We can demonstrate the thermodynamic reversibility of the reaction catalyzed by aconitase by increasing the concentration of isocitrate. This can be done experimentally by adding isocitrate and blocking its oxidation; the result will be the formation of citrate.

Following the two carbon atoms of acetyl-S-Co A in Figure 6–19 (boldface type) throughout the cycle reveals that they end up first in succinyl-S-Co A and then in the new oxaloacetate molecule. Thus, although oxaloacetate is resynthesized by the cycle, it is not precisely the same molecule as before. Two carbons of the original oxalo-acetate molecule were lost as carbon dioxide. The two lost atoms were replaced by the acetyl-S-Co A. This is why we say that the effect of the cycle is to oxidize the two acetyl carbon atoms to carbon dioxide.

Most of the electrons removed during the oxidation of the acetyl carbon atoms are stored in NADH. The subsequent oxidation of NADH by the mitochondrial electron transport chain is what leads to the synthesis of ATP. How this coupling is accomplished will be discussed in Chapter 8. The step of succinate dehydrogenation, historically a very important reaction for elucidating the whole cycle (see below), is a typical oxidation in which two hydrogen atoms are removed from succinate to form fumarate. The two electrons, together with their accompanying protons, are held on the enzyme succinate dehydrogenase as part of the reduced coenzyme flavine adenine dinucleotide ($FADH_2$). This electron carrier is also oxidized by the electron transport chain, but enters it at a place different from the entry point of NADH.

There is one step in the Krebs cycle that provides energy directly, rather than via NADH or $FADH_2$ oxidation. This is the enzymatic hydrolysis of succinyl-S-Co A, which also forms guanosine triphosphate (GTP) from guanosine diphosphate (GPD) and inorganic phosphate. We describe this kind of direct phosphorylation as occurring at the *substrate level* (recall our earlier discussion of the oxidation of glyceraldehyde phosphate). The GTP formed can react with ADP to form GDP and ATP, the more generally used high-energy compound.

It is interesting to consider the bookkeeping of hydrogen atoms in the cycle (Table 6–4). The number entering with acetyl-S-Co A is three, plus two more from the water involved in the condensation with oxaloacetate to form citrate. One hydrogen leaves as *H*S-Co A, so the net gain on entering the cycle is four. There is no net gain or loss as citrate is isomerized to isocitrate. During the oxidation of isocitrate, however, two hydrogen atoms leave as NAD*H* plus a proton. The next entry of hydrogen is when *H*S-Co A reacts with α-ketoglutarate to form succinyl-S-Co A. Two hydrogens leave during this oxidation as NAD*H* and a proton. The net number of hydrogens in the cycle is now one and this rises to two

Flavine adenine dinucleotide (FAD)

TABLE 6–4

A Balance Sheet for the Krebs Cycle

Atom	In		Out	
Carbon	2	{from acetyl-Co A}	2	{2CO$_2$}
Oxygen	4	{1 from acetyl-Co A / 3 from 3H$_2$O}	4	{2CO$_2$}
Hydrogen	9	{3 from acetyl-Co A / 6 from 3H$_2$O}	9	{8 to reduced coenzymes / 1 to recycle HS–Co A (step 2)}

when H_2O is used to hydrolyze succinyl-S-Co A to yield HS-Co A and succinate. Two more hydrogens leave during the oxidation of succinate to fumarate; they are carried away by FADH_2. Water supplies two more hydrogens during the hydration of fumarate and they are lost, during the subsequent oxidation of malate, as NADH and a proton. Hydrogen atoms and their electrons enter the cycle from acetyl-S-Co A and from water but leave as NADH (or FADH_2) and protons. Thus, in a sense, the cycle *splits water* to provide oxygens for the CO_2 that is released by the cycle and some of the electrons for later oxidation by the mitochondrial electron transport chain.

The Krebs cycle is central to the oxidation of fats and carbohydrates; most of the human body's fat and a good deal of its carbohydrates are oxidized via this cycle. Since α-ketoglutarate, oxaloacetate, and pyruvate can be formed very easily by transamination from amino acids (glutamate, aspartate, and alanine, respectively), the cycle is also directly responsible for the oxidation of some of the breakdown products of proteins. Many of the other amino acids can be broken down to cycle intermediates: acetyl-S-Co A, α-ketoglutarate, oxaloacetate, succinate, or fumarate. It is probable that oxidation of *most* substrates' carbon atoms taken in by the aerobically metabolizing cell proceeds through this cycle. Moreover, the Krebs cycle is found in many cells, including most kinds in our body, as well as plant, yeast, and bacterial cells.

Thus far we have focused on the operation of the Krebs cycle in the breakdown of its intermediates to CO_2 and NADH. The Krebs cycle is also involved in supplying intermediates that lead to the formation of larger molecules. By the easily reversible *transamination* reaction, for example, many amino acids are formed from the keto acids of the cycle: glutamate from α-ketoglutarate, aspartate from oxaloacetate, and other amino acids are formed from these. In turn, some of the amino acids are precursors of purines and pyrimidines (and thus of nucleic acids), and are also precursors of the

porphyrins found in the electron transport chain. Oxaloacetate is also on the pathway for the synthesis of hexose and pentose sugars (gluconeogenesis); pentoses are the sugar part of nucleic acids.

Transamination

The reactions of the Krebs cycle continue, even though intermediates are permanently removed to act as precursors for amino acids (Figure 6–20) or other metabolites. This means that the cycle intermediates must be replaced. The major reaction by which this occurs in mammals is the carboxylation of pyruvate to form oxaloacetate (Figure 6–20). Other reactions that feed in cycle intermediates in other cells are the carboxylations of PEP to form malate and transaminations of certain amino acids—for example, of glutamate to form α-ketoglutarate. Thus, the incorporation of CO_2 into organic molecules is found not only in photosynthetic cells, but even in our own bodies. Of course, we must supply the energy for such carbon dioxide incorporation out of oxidative metabolism; the energy to fix CO_2 in photosynthetic cells comes mostly from light.

You can see, even from this very brief survey, that the Krebs cycle has a central role in cellular metabolism. Through the cycle pass most of the chemical intermediates—most of the carbon atoms, to be exact—of the cell. The cycle is the hub of a great many pathways involved in the breakdown and synthesis of cellular constituents.

FIGURE 6–20

Reactions that Use Krebs Cycle Intermediates or Add Intermediates to the Cycle.   Reactions 1 and 2 are transaminations; they form the amino acids corresponding to the α-keto acids oxaloacetate and α-ketoglutarate, aspartate, and glutamate, respectively. Succinyl-Co A is a precursor of the synthesis of porphyrins, complex components of the mitochondrial electron transport chain. Reaction 3, catalyzed by pyruvate carboxylase, is the major pathway by which intermediates can be added to the Krebs cycle. This is the carboxylation of pyruvate to form oxaloacetate, a reaction that requires the hydrolysis of ATP to form ADP and P_i but still has a ΔG′ near zero. Phosphoenolpyruvate carboxykinase, the enzyme that catalyzes Reaction 4, also can help load up the cycle; its product is malate. This reaction is associated with the phosphorylation of GDP to form GTP. A similar reaction can lead in some cells to the formation of succinate from CO_2 and propionate. Also shown is the position of the reaction blocked by the enzyme inhibitor malonate.

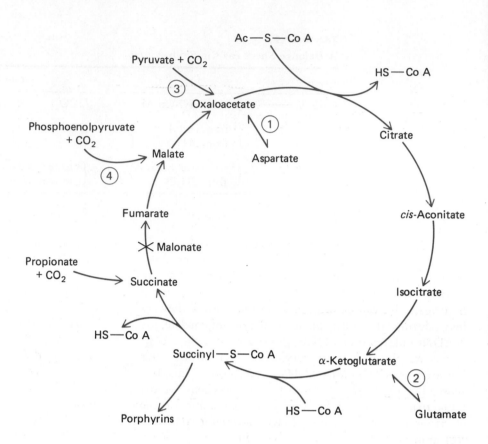

How the Krebs Cycle Was Verified

The experiments which demonstrate that something like the Krebs cycle actually happens within the individual cell or within the whole animal are examples of those discussed earlier. Specific atoms of chemical compounds like pyruvic acid can be labeled with radioactive tracers such as carbon 14. The labeled pyruvic acid can then be given to a yeast cell or to a whole organism like a rat, and various chemical compounds—glucose, fat, and amino acids—can be isolated. By knowing which carbon atom of the pyruvic acid was labeled, by determining which carbon atoms of isolated compounds are labeled, and by knowing the individual biochemical reactions involved, we can infer what happened to the individual carbon atoms of the pyruvic acid. On the basis of many such experiments, investigators have come to the conclusions that (1) the Krebs cycle is operative in the cell as well as in the test tube; (2) there are alternative metabolic pathways for many of the intermediates of fat, carbohydrate, and amino acid metabolism; and (3) these pathways intersect each other at the level of the Krebs cycle. The reactions of the Krebs cycle can explain fully the means whereby the carbon atoms of pyruvate end up in various other compounds.

Using minced tissue, such as pigeon breast muscle or liver, Krebs observed that in the presence of excess pyruvate the addition of only a very small amount of citrate catalyzed a large uptake of oxygen, larger than required to oxidize the added citrate. It was also found that citrate could be synthesized when oxaloacetate was added, and that added citrate, isocitrate, cis-aconitate, and α-ketoglutarate were all rapidly oxidized to CO_2 by these minced tissues.

The key finding in Krebs' early laboratory work involved the use of malonate, which specifically inhibits the enzyme succinate dehydrogenase and thus prevents the conversion of succinate to fumarate (see discussion in Chapter 5). In the presence of malonate, succinate could nevertheless be synthesized when fumarate or oxaloacetate was added. Figure 6–20 shows that succinate could not have been formed by the direct reduction of fumarate, for this enzyme was blocked by malonate. Krebs deduced that there must be another pathway—he called it a "back" reaction—by which succinate could be formed via the breakdown of α-ketoglutarate or isocitrate, or the preceding intermediates.

Krebs thus postulated a cycle that involved the tricarboxylic acids, citrate, isocitrate, and *cis*-aconitate, and the dicarboxylic acids, succinate, malate, and oxaloacetate. He was able to verify this cyclic pathway by demonstrating that although malonate blocked the utilization of pyruvate, the block could be relieved by added oxaloacetate and one oxaloacetate molecule was used for each molecule of pyruvate oxidized. It was noted earlier by Szent-Györgyi that in an uninhibited preparation, added oxaloacetate stimulated much more pyruvate oxidation and oxygen uptake than could be explained by the oxidation of the oxaloacetate.

Later evidence made use of radioactive carbon dioxide, added to minced liver in the presence of malonate and excess pyruvate. Under these conditions, carbon dioxide can be incorporated into oxaloacetate by carboxylating pyruvate (as we discussed previously). There was an increase in the concentration of succinate but it contained no radioactivity. The radioactive label was found first in oxaloacetate (see the carbon with the asterisk in Figure 6–19), but, because of the block by malonate, the labeled carbon had to go around the cycle clockwise. Citrate, isocitrate, and α-ketoglutarate were labeled, and the radioactivity in the isolated α-ketoglutarate was in the carboxyl carbon next to the carbonyl group. The labeled carboxyl was removed by α-ketoglutarate dehydrogenase and thus could not produce radioactive succinate.

Examining the cycle, you can see that radioactivity should be found in fumarate and malate, moving counterclockwise from radioactive oxaloacetate. It was; and since no radioactivity was present in succinate, it is apparent that the succinate did not arise from fumarate but must have been formed via another reaction, just as Krebs deduced. When malonate was omitted, all the dicarboxylic acids, including succinate, contained radioactivity, showing the interconvertibility of the compounds in question.

Blocking succinate oxidation by malonate allows one to test compounds for their ability to form succinate. During malonate blockade, any succinate formed will accumulate; it can be extracted and its amount can be estimated. Using malonate, Krebs found earlier that all of the intermediates of the cycle can form succinate.

That the cycle is important for the economy of the cell is illustrated by the finding that ingesting fluoroacetate can kill an animal. Fluoroacetate is converted in the body to fluoroacetyl-Co A, which condenses with oxaloacetate to form fluorocitrate, a powerful inhibitor of the enzyme aconitase. When the cycle stops the cell is in serious trouble. This is an example of a relatively innocuous compound, fluoroacetate, being converted into a poison by a *lethal synthesis,* not an uncommon process in cells.

Metabolic Networks: An Example of Their Meaning

Although many metabolites are substrates for only one enzyme, in other cases the same metabolite can interact with more than one enzyme. One enzyme may catalyze the oxidation of the metabolite, another its reduction, still another the attachment of some group or compound. Branch points in the metabolic network arise where an intermediate is produced or utilized by more than one enzyme-catalyzed reaction. The same substance, for example, can follow one pathway to produce chemical energy, or can be routed onto another pathway to synthesize larger molecules of the cell, such as proteins and nucleic acids.

A good example of this is the fate of glucose-6-phosphate in the liver cell (Figure 6–21). It can be formed in various ways: from glycogen, smaller intermediates of fermentation, or free glucose. Once it is formed, it can be acted on by at least four different enzymes; that is, there are four directions its metabolism can take: (1) to glucose, to replenish its supply in the blood; (2) to glycogen, to build up that energy storehouse; (3) to pyruvate, to provide energy and intermediates for the synthesis of fats and proteins; and (4) to phosphogluconate, to provide intermediates for the synthesis of nucleic acids.

The simplified diagram of Figure 6–21 does not show all the alternative pathways known in this section of the metabolic network. No single small diagram could. Our goal, however, is not a systematic and comprehensive discussion of metabolism, but rather to illustrate the basic features by presenting a few examples of related pathways.

One such example is the synthesis of nucleic acids, a requisite for cell growth. A constituent of these complicated compounds is the 5-carbon sugar, ribose. The actual precursor of nucleic acid monomers, ribose-5-phosphate, is synthesized via many pathways. Three lead to its formation from the precursor glucose-6-phosphate (Figure 6–22).

The *oxidative pentose phosphate pathway* leads to ribose-5-phosphate in five enzymatic steps. It is also a prime source of the coenzyme NADPH, which is reduced during the oxidation of both glucose-6-phosphate and 6-phosphogluconic acid. The electrons carried by NADPH (though not, usually, those carried by NADH) are used to reduce other substrates during the synthesis of fatty acids and steroids, for example, and for this reason NADPH is often referred to as having a *reducing potential.*

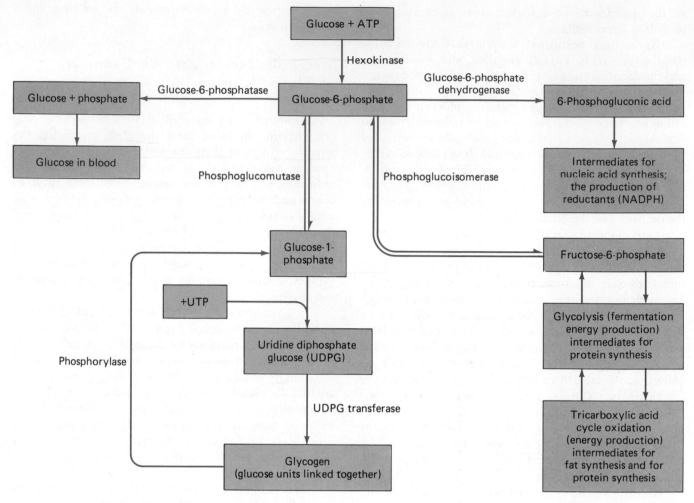

FIGURE 6–21
Part of the Metabolic Network.   The glucose-6-phosphate branch points in liver cells.

The *nonoxidative pentose phosphate pathway* begins with the conversion of glucose-6-phosphate to fructose-6-phosphate. However, instead of continuing on the glycolytic road (see Figure 6–15), the carbon atoms of fructose-6-phosphate go through a series of condensation and transfer reactions, leading, via the enzymes transaldolase and transketolase, to xylulose-5-phosphate and ribose-5-phosphate.

Xylulose-5-phosphate is also made via the oxidative pathway, and via still another route, called the *uronic acid pathway,* involving uridine diphosphate-glucose and eleven enzymatic reactions. This pathway is also known for its contribution to the synthesis of ascorbic acid (vitamin C) which occurs in plants and in the livers of most vertebrates (not, unfortunately, including man). Since the end product, xylulose-5-phosphate, can be converted to ribose-5-phosphate via the intermediate ribulose-5-phosphate, the uronic acid pathway is also a source of the ribose moiety of nucleic acids.

By using radioactive tracers, we have learned that all these pathways seem to be operating in the cell at the same time. The traffic on the path leading to ribose-5-phosphate via the uronic acids seems to be insignificant compared with that on the other two pathways. In liver and muscle cells, at least, much of the uridine diphosphate-glucose formed via the uronic acid pathway is funneled off into glycogen synthesis. We shall discuss in Chapter 14 the mechanisms by which the control necessary to partition mass flow among competing pathways can be achieved.

Is there a rationale to explain why these multiple sources of ribose exist? Since nucleic acids are essential for the cell's growth, even its very existence, the cell may have provided itself with more than one pathway for

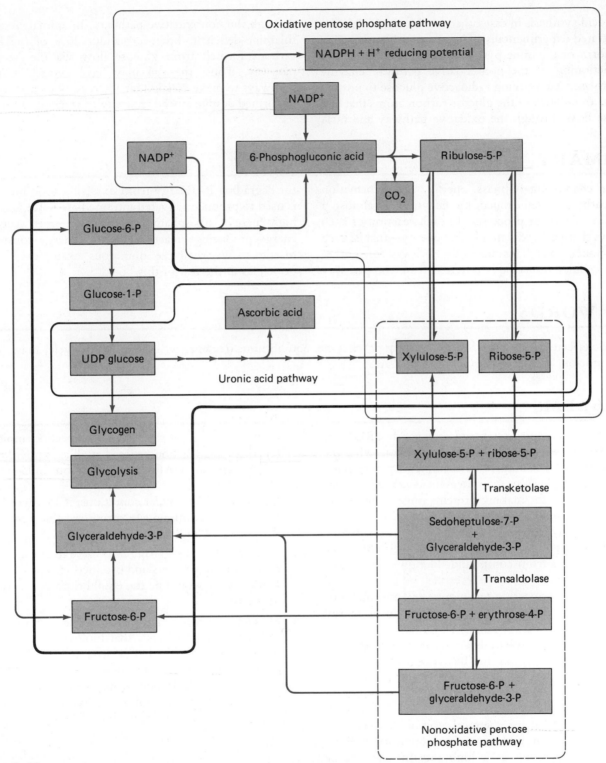

FIGURE 6-22
Pentose Phosphate Pathways. Not all enzymatic steps are indicated by arrows.
The solid black line shows how the glycolytic pathway and the pathway for
synthesis of glycogen are connected, via common intermediates, to the uronic acid
pathway and the two pentose phosphate pathways.

nucleic acid synthesis in case others are blocked. This can be illustrated experimentally. The vitamin thiamine (B_1), in the form of thiamine pyrophosphate, is a cofactor in the functioning of the nonoxidative pathway enzyme transketolase. After giving radioactive glucose to normal animals, about 40% of the glucose carbon atoms that end in ribose flow through the oxidative pathway and 60% through the nonoxidative pathway. In animals rendered thiamine-deficient, however, about 80% of the ribose carbons formed from glucose flow via the oxidative pathway. Thus, the cell may have evolved multiple pathways to make certain that a necessary compound can be formed despite any adverse environmental conditions.

SUMMARY

The channeling of mass, energy, and information along pathways determined by enzymes is almost a definition of cellular processes. In each reaction of such pathways there is a reduction in free energy—that is, each step is spontaneous. This means the pathway for synthesizing a cellular molecule is different from the pathway for degrading it. Energy from oxidizing food molecules is used to produce activated intermediates—for example, high-energy phosphate compounds—thus coupling energy-producing reactions with energy-consuming reactions. We shall see numerous examples of these principles in the rest of this book.

KEY WORDS

Allosteric effector, coenzyme, coupling, energy, enzyme, equilibrium, free energy, kinase, Krebs cycle, mass, mitochondrion, oxidation, reduction, substrate.

PROBLEMS

1. To the following statements, respond TRUE or FALSE. If you respond TRUE, explain why the statement is true and give an example; if you respond FALSE, explain why the statement is not true.
   (a) Since $\Delta G'$ for making proteins must be negative, to explain why their synthesis is spontaneous, $\Delta G'$ for the hydrolysis of proteins must be positive.
   (b) A phosphorylase enzyme removes inorganic phosphate from compounds like glucose-6-phosphate.
   (c) Genetic engineering techniques will eventually be able to yield a cell with an efficiency of energy storage of 100%.
   (d) Shikimate phosphate is an activated intermediate in the pathway for the biosynthesis of aromatic amino acids.
   (e) The ADP effect in glycolysis will not be seen if arsenate is present.
   (f) It is likely that any chemical reaction that a cellular molecule can undergo in a laboratory will be found operating at a high rate in one cell or another.

2. To a solution of 0.10 M glucose-6-phosphate you add an appropriate amount of phosphoglucomutase, which catalyzes the reaction:

$$\text{G-6-P} \underset{\text{phosphoglucomutase}}{\rightleftharpoons} \text{G-1-P.}$$

Assume the $\Delta G'$ of this reaction is +6.3 kJ/mole.
   (a) Does this reaction proceed at all as written, and if so, what are the final concentrations of G-6-P and G-1-P?
   (b) Under what cellular conditions, if any, would this reaction continuously produce glucose-1-phosphate at a high rate?

3. If the Reaction 6–10 (page 267) were to reach equilibrium with all reactants and products except sucrose at 1 mM, what would be the equilibrium concentration of sucrose?

4. Manned landings on Mars are expected any time now and the possibility exists that biological high-energy compounds other than ATP may be discovered. To prepare yourself for such news, imagine a Martian Unknown Cytoplasmic Kompound (MU—C—K) which can undergo the following hydrolytic reactions, analogous to but not identical to ATP:

	$\Delta G'$ (kJ/mole)
MU—C—K + H_2O $\rightleftharpoons$ MU—C + K	−44
MU—C + H_2O $\rightleftharpoons$ MU + C	− 5
C—K + H_2O $\rightleftharpoons$ C + K	−25

   (a) Using the first law of thermodynamics, as true on Mars as on Earth, predict the standard free energy of the reaction

$$MU—C—K + H_2O \rightleftharpoons MU + C—K$$

(b) We may expect that Martian cells will use MU—C—K to drive unfavorable metabolic reactions, such as the following:

	$\Delta G'$ (kJ/mole)
$A + B \rightleftharpoons D + H_2O$	+21
$D + E \rightleftharpoons F + H_2O$	+36

Assume two molecules of MU—C—K are hydrolyzed to two molecules of MU—C and K during the formation of one mole of F. Write out a thermodynamically defensible scheme by which these reactions might occur. Assume that none of the compounds A, B, D, E, F contain C or K groups initially. Give $\Delta G'$ values for each step in your scheme.

5. Two "high-energy" compounds found in some cells are phosphocreatine (PC) and phosphoarginine (PA).

			$\Delta G'$ (kJ/mole)
$PC + H_2O \rightleftharpoons$	C	$+ P_i$	−49
$PA + H_2O \rightleftharpoons$	A	$+ P_i$	−40
$ATP + H_2O \rightleftharpoons ADP$		$+ P_i$	−35
$PC + ADP \rightleftharpoons$	C	$+ ATP$	—
$PA + ADP \rightleftharpoons$	A	$+ ATP$	—

(a) From the data given, estimate the standard free energy change associated with the transfer of phosphate from PC to ADP, and from PA to ADP.

(b) Phosphocreatine is often used for energy *storage* when [ATP] is high. What is the minimum ratio of [ATP]/[ADP] that would have to exist in a cell to store energy in PC if the ratio of [PC]/[C] were 1.0?

6. Let $[A]_{eq}$ and $[B]_{eq}$ be the concentrations of A and B at equilibrium. Show that ΔG for the reaction $A \rightleftharpoons B$ measures how far the reactants and products are from equilibrium by showing that

$$\Delta G = R\,T\ln\frac{[B]}{[B]_{eq}} - R\,T\ln\frac{[A]}{[A]_{eq}}$$

7. Imagine that a biotechnology company has engineered a new cell, a Biologically Unusual Recombinant Procaryote (BURP). A government agency has asked your lab team to analyze the BURP, which is a facultative anaerobe that gets energy from a fermentation pathway. Your contribution is to compare the BURP pathway with lactate fermentation by red blood cells. The BURP pathway intermediates are not known and are symbolized by letters from U to Z; *P* means inorganic phosphate while NAD$^+$, NADH, ADP, and ATP have their usual meanings. (See figure below.)

(a) In lactate fermentation there is an "ADP effect." Is there a similar ADP effect in the BURP? Briefly explain your answer.

(b) Compare the net amount of NAD$^+$ consumed by the BURP fermentation with the net amount consumed in lactate fermentation. Briefly explain your answer.

(c) Compare the amount of ATP generated by processing one molecule of U in the BURP fermentation pathway with the amount generated by processing one molecule of glucose in the lactate fermentation pathway. Briefly explain your answer.

(d) Compare the following ratios of mass flow through intermediates in the lactate fermentation pathway of red blood cells with the BURP fermentation pathway (a mass flow could be measured, for example, as μmoles/min/10^8 cells):

$$\frac{\text{flow through glucose}}{\text{flow through PEP}} \quad versus \quad \frac{\text{flow through U}}{\text{flow through } P\text{-Y}}$$

Briefly explain your answer.

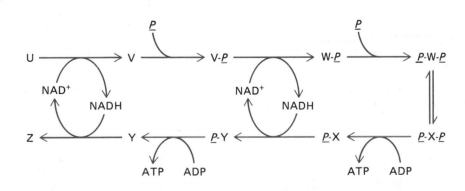

(e) If BURP cells are exposed to compound U that is uniformly labeled with ^{14}C to a specific activity of 100 μCi/μmole and if this U is utilized by the cells at a rate of 10 μmoles/min, at what initial rate will the ^{14}C specific activity of NADH increase? (^{14}C specific activity is measured as radioactivity/mole of compound. Assume that Z is removed rapidly so it cannot label other components of BURP metabolism.) Briefly explain your answer.

8. Use the table of standard free energies given below.

Enzyme-catalyzed Reaction	$\Delta G'$ (kJ/mole)
$NADP^+ + NADH + ATP + H_2O \rightleftharpoons$ $NADPH + NAD^+ + ADP + P_i$	-31.3
fructose-6-P + $H_2O \rightleftharpoons$ fructose + P_i	-15.9
glucose-6-P + $H_2O \rightleftharpoons$ glucose + P_i	-13.8
$NADP^+ + NADH \rightleftharpoons$ $NADPH + NAD^+$	-0.8
glucose-6-P $\rightleftharpoons$ fructose-6-P	$+1.7$

(a) Calculate the equilibrium ratio of [glucose-6-P] to [fructose-6-P].

(b) Calculate the equilibrium ratio of [glucose] to [fructose].

(c) In mammalian semen, carbohydrate is present predominately as fructose. This sugar is synthesized from glucose in the seminal vesicle cells by the reactions:

1. glucose + NADPH + $H^+ \rightleftharpoons$
 sorbitol + $NADP^+$

2. sorbitol + $NAD^+ \rightleftharpoons$
 fructose + NADH + H^+

Are these reactions coupled? Briefly explain your answer.

(d) Using the data in this question, propose a thermodynamically and physiologically reasonable way the seminal vesicle cells can convert, essentially quantitatively, glucose to fructose.

SELECTED READINGS

Atkinson, D. E. (1977) *Cellular Energy Metabolism and its Regulation*. New York: Academic Press.

Rawn, J. D. (1989) *Biochemistry*. Burlington, NC: Carolina Biological Supply Company.

Stryer, L. (1988) *Biochemistry*. 3d ed. San Francisco: W. H. Freeman and Co.

Zubay, G. (1983) *Biochemistry*. Reading, MA: Addison-Wesley.

Cell Transport—The Transport of Materials Through Membranes

7

Inside the cell various properties are maintained relatively constant despite a drastically different and often-changing environment. The cell membrane lies at the interface of the controlled interior and the fluctuating exterior, and represents not only the *structural* but also the *functional boundary* of the cell. Compartments with different properties also exist inside the eucaryotic cell and these too are bounded by membranes. One of the important mechanisms by which a cell or an organelle achieves independence from its environment is regulation of the movement of mass across the surface of the membrane. Despite its delicate thickness (7.5 nm), a biological membrane can engage in two distinct and crucial activities:

1. It can select among different molecular species, slowing down the permeation of some materials while allowing others to pass almost unimpeded.
2. It can bring about the transport of material, either inward (accumulation) or outward (excretion, secretion), *against* both concentration and electrical potential gradients, by coupling the transport to energy-yielding reactions.

These two phenomena of *selective permeability* and *active transport,* respectively, make possible many essential cellular functions, such as irritability and the transmission of excitation along the cell surface. But the membranes of the cell and its organelles are also the sites of other functions:

- By virtue of the same type of coupling found in active transport, the movement of protons down an energy gradient is transformed into the synthesis of ATP. (This function is discussed in Chapter 8.)
- Membranes provide a structural framework for some enzymes.
- Adhesion or specialized junctions between cells in tissues and in other multicellular structures occur at the membrane (Chapter 17).

- Structures on or in the membrane serve as identifying markers that are specific for the cell, the tissue, the organism, or the species (Chapter 17).
- Information is transmitted to cells via receptors that are located on the cell membrane and that are sensitive to the presence of extremely low concentrations of substances in the environment (hormones, for example; see also Chapter 14).
- Certain cellular extensions (cilia), extracellular structures (flagella), and intracellular structures (the cytoskeleton) are associated with the plasma membrane (Chapter 16).
- The movement of a cell through its environment must be expressed at its surface and thus also involves the membrane (Chapter 16).

These functions are discussed in detail elsewhere. This chapter will explore the structure of the membrane and the various mechanisms that regulate the flow of mass across it. Since, in biological systems, both energy and information have a material basis, our discussion will also reveal how the cell exchanges energy and information with its environment.

Membrane Structure

Although the first microscopes could demonstrate the existence of cells, the membrane at the cell boundary could not be resolved because of its small size. For a time it was doubted if there was any special surface structure; the cytoplasm was thought to end at a well-defined place, like a gelatin dessert, without any qualitative change in properties. But in the early 1950s, when the electron microscope first revealed the silhouette of the plasma membrane, cell biologists were not surprised because its existence and approximate dimensions had been deduced by generations of physiologists who studied the *osmotic properties* of cells.

Osmosis and Osmotic Pressure

As early as 1885 H. M. de Vries demonstrated the selective permeability properties of plant cell vacuoles, and in 1897 W. F. P. Pfeffer proved the existence of a permeability barrier at the external surface of the cell. When a cell was immersed in a water-soluble dye, the interior did not become colored unless the outer layer of the cell was deliberately damaged. This work was later extended by microinjecting water-soluble dyes into intact cells, and again the dye could not cross the boundary of the cell even though it penetrated every part of the cell.

Pfeffer also showed that when a cell is immersed in a solution of certain solute molecules, its volume changes;

it increases or decreases depending on the concentration of the solution (Figure 7–1). Since cells swell when placed in distilled water, movement of water into the cell is the likely cause. This rapid net water movement can be prevented, however, by using solutions of particular concentrations. With experiments such as those illustrated in Figure 7–2, Pfeffer showed that no net movement of water occurs in a 0.30 M sucrose solution, a 0.15 M solution of NaCl, or a 0.10 M solution of CaCl$_2$. By analogy with his earlier studies on water movement through inorganic ferrocyanide films, Pfeffer proposed the existence of an invisible barrier that surrounds cells and permits the passage of water but not solutes like sucrose. This proposal was not only a triumph of inductive reasoning but also illustrates the useful interplay between concepts of structure and function in the study of cell biology.

The observed changes in cell volume are due to water movement in response to the reduced concentration of water (more accurately, reduction of its thermodynamic activity) caused by dissolved solute molecules and ions. If the concentration of water is different in two regions, there will be a negative free energy change associated with water movement from the higher concentration (i.e., where there is less solute) to the lower concentration (i.e., where there is more solute). In this respect water is no different from any other substance we have discussed; it moves or changes to lower its free energy. The movement of water toward regions where its concentration is lower is called *osmosis.*

This movement can be prevented if suitable hydrostatic pressure is applied to the region of lower water concentration. A concentrated solution of any soluble substance thus has a high **osmotic pressure,** meaning that the hydrostatic pressure needed to oppose the entry of additional water is relatively great. Figure 7–3 shows how osmotic pressure can be measured. The osmotic pressure (Π) of the solution in side 2 is equal to the hydrostatic pressure (ΔP) which, when applied to side 2, is sufficient to halt the osmotic flow of water. In this apparatus, water flowing into side 2 causes fluid to rise in the chimney until the weight of the solution column provides sufficient pressure.

The equilibrium osmotic pressure of a solution is related to both the temperature and the concentration of dissolved solutes; according to the Van't Hoff equation,

$$\Pi = R\,T\,C_s, \qquad (7-1)$$

where C_s is the summed molar concentrations of all *nonpermeating* solutes present (at equilibrium, all permeating solutes will have equal concentrations on both sides of the membrane). It is C_s, the **osmolar** concentration, that determines the osmotic pressure. When there are solutes

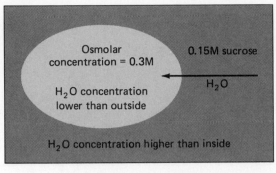

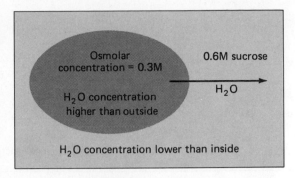

(a) Water enters the cell

(b) Water leaves the cell

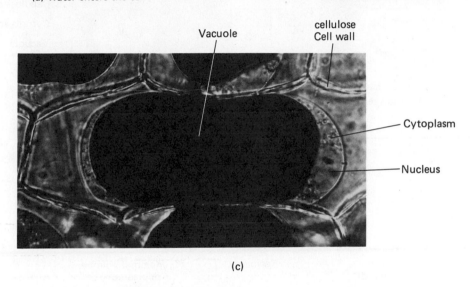

(c)

FIGURE 7-1
The cell as an osmometer. Because of the presence of different concentrations of solutes, the concentration (and thus the thermodynamic activity) of water may be different inside and outside a cell. The osmolar concentration (or osmolar activity) inside the cell is the sum of the concentrations of all the solute (nonwater) species, including ions. This sum is dominated by small molecules and ions since the osmolar concentration of macromolecules is very low. (a) When the cell is immersed in a hypotonic sucrose solution—one whose solute concentration is lower than the osmolar concentration inside the cell—water will diffuse into the cell because the concentration of the water is lower inside the cell due to the higher solute concentration. (b) Water diffuses from the cell when immersed in a hypertonic sucrose solution whose concentration is higher than the cell's osmolar concentration. (c) A cell from the inner epidermis of an onion. The cell has been plasmolyzed in a hypertonic sucrose solution; the cytoplasm has shrunk and pulled away from the cell wall. The very large central vacuole is stained by a dye (neutral red) which enters the cell and accumulates in the vacuole. × 800.

dissolved on both sides of the membrane, the net osmotic pressure is

$$\Delta\Pi = R\ T\ \Delta C_s,$$

where ΔC_s is the difference in concentration of nonpermeating solutes on opposite sides (see Appendix 7A).

We now can see the basis for Pfeffer's observation of the osmotic equivalence in certain cells of 0.3 M sucrose and 0.1 M $CaCl_2$ (Figure 7–2). The complete ionization of 0.1 M $CaCl_2$ produces a 0.3 molar solution of ions (0.1 M Ca^{2+} and 0.2 M Cl^-), all of which participate

equally in reducing the concentration of water. (Pfeffer's findings, which originally puzzled the great chemist S. Arrhenius, eventually were explained by the latter's theory of the dissociation of electrolytes—one instance in which a biological observation led to the formulation of an important physical theory.)

If two solutions at the same hydrostatic pressure are in osmotic equilibrium they must, according to the Van't Hoff equation (7–1), have the same total solute concentration. Such solutions are *isosmotic* and their concentrations are *isosmolar*. If a cell is placed in a solution whose total solute concentration is lower than the cell's—a

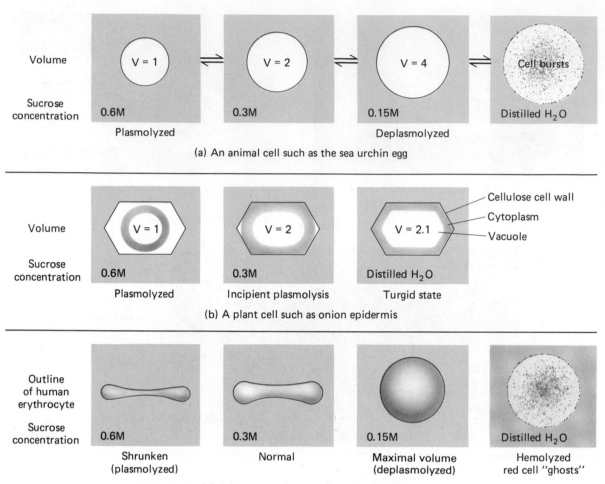

FIGURE 7-2
Osmotic behavior of plant and animal cells. The osmolar concentration under normal growth conditions is measured by determining the concentration of a solution of sucrose or other nonpenetrating solute that causes no volume change in a cell placed in this solution. (a) In the presence of a nonpenetrating solute, a deformable animal cell acts like an osmometer, swelling and shrinking in approximately inverse proportion to the molar concentration of sucrose. This volume change occurs because sucrose lowers the concentration of water, which moves across the membrane from regions of higher to lower concentration. (b) In plant cells placed in distilled water (or water of low solute concentration like soil water), osmotic pressure exerted by the water trying to enter the cell is counteracted by hydrostatic pressure. This pressure—the turgor pressure—is also exerted against the rigid cellulose wall; it gives nonwoody plants the rigidity they require to stand erect. (c) The normal human erythrocyte is a biconcave disk. In a solution of lower osmotic concentration, it expands with almost no change in surface area until it reaches a spherical shape (which has maximum volume per unit area); any further volume increase causes the membrane to become leaky, allowing hemoglobin molecules to diffuse out (hemolysis). Then the turbid red blood cell suspension suddenly becomes a clear hemoglobin solution of much higher optical absorbance, a very useful end point for permeability studies. The "ghosts" (red blood cells without hemoglobin) are almost pure membrane. After repeated washing by sedimentation from hypotonic solutions, ghosts are suitable for studies on the chemical and physiological properties of membranes.

hypoosmotic solution with a *hypoosmolar* concentration—water will flow into the cell and its volume will increase. This kind of solution is also *hypotonic* for that cell. If a cell is placed in a solution whose total solute concentration is higher than the cell's—a *hyperosmotic* solution with a *hyperosmolar* concentration—water will flow out of the cell and its volume will decrease, at least at first. Such a solution is *hypertonic* for that cell, for a while at least. When a cell is placed in a solution of equal solute concentration—an *isosmolar* solution—the cell may experience no water gain or loss; if so, the solution is also *isotonic* for that cell.

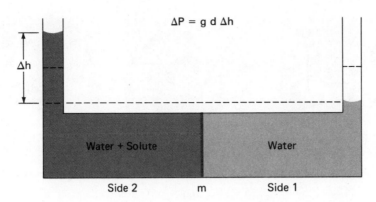

$$\Delta P = g\, d\, \Delta h$$

FIGURE **7–3**

An experiment to measure osmotic pressure. The two sides of the osmometer are separated by a semipermeable membrane (m) through which water can pass. Initially, the fluid levels are at the position marked by the upper dotted lines. Water flows from side 1 through the membrane to side 2, where the water concentration is lower. As it enters side 2 the fluid level rises in the chimney until eventually, at equilibrium, the pressure developed by the column of fluid blocks the further net movement of water. The pressure depends on the gravitational constant (g), the fluid's density (d), and the difference in height of the two fluid levels (Δh). At equilibrium, the measured hydrostatic pressure difference is numerically equal to the osmotic pressure difference between the two solutions. Osmotic pressure is defined so the *side to which the water flows* has the *higher* osmotic pressure. This is because that side has the *lower* water concentration (since it has the higher solute concentration).

You can see from the above definitions that _osmolarity_ refers to the summed molar concentrations of all the dissolved solutes in a solution, whereas _tonicity is determined by the cell's response to the osmotic properties of the bathing solution_. We described the isosmolar and the hyperosmolar cases conditionally because not all isosmolar solutions are isotonic, nor are all hyperosmolar solutions hypertonic. The reason for the distinction between osmolarity and tonicity is that some solutes can pass through the cell's membrane. When there is a permeability pathway and a negative free energy change driving the movement of a solute, it will flow through the membrane, reducing the water concentration in the cytoplasm. When solute enters the cell, water will enter also (to restore the osmotic equilibrium) and the cell's volume will increase (see Figure 7–4). In such cases, the solution, though isosmolar or hyperosmolar, will still be hypotonic.

How much water will enter? We can easily determine the ratio of solute to water entering if, as is usually the case, the water's permeation rate is very much higher than the solute's. Then the solution that enters will be almost isotonic. Of course, it cannot be exactly isotonic; if it were, there would be no difference in water concentration to force the water entry. But only a slightly higher osmotic pressure inside the cell, due to the solute entry, is enough to account for the water entry. An isotonic solution, one in which the cell volume is constant, can usually be found for any particular cell by trial and error. Animal cell membranes often have a very low permeability to sucrose, so one needs merely to find the molar concentration of sucrose that leads to zero volume change in the cells of interest. This is usually in the neighborhood of 0.30 M, or (since the water concentration is 55 M) a molar ratio of $0.30/55 \cong 0.0055 \cong 1/180$. Thus, for every millimole of permeating solute, approximately 180 millimoles of water will also enter to maintain a constant isotonic concentration of solute.

This property is exploited by animal cells to rid themselves of water; active pumping out of solutes, such as sodium ions, gives rise to a passive outward flow of water that helps to maintain the cell's volume. Many bacterial and plant cells, however, are in osmotic equilibrium even with very dilute bathing solutions. In this environment, water enters until a high internal hydrostatic pressure is created, which balances the tendency for further water entry. These cells suffer no structural distortion from the outward pressure because they are surrounded by a sturdy cell wall. Such an arrangement allows such cells to avoid osmotic swelling without resorting to the pumping schemes used by animal cells.

Understanding the physical basis for osmosis is important because the phenomenon is used extensively to investigate both the structural and functional properties of cell membranes. Many of the fundamental experiments were done before the beginning of this century, based on the volume changes of cells immersed in solutions (see Figure 7–4). These experiments depend on the high permeability of most cells' membranes to water. That being the case, the rate of volume increase of a cell

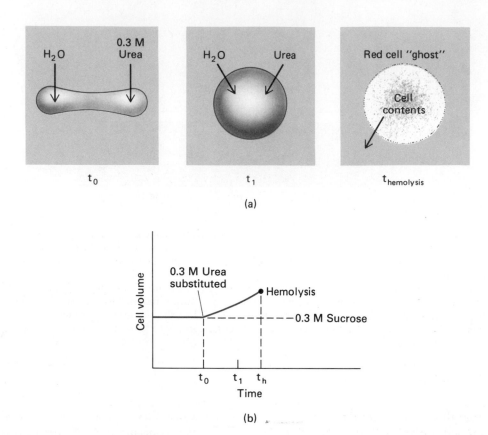

FIGURE **7–4**

Measurement of the human erythrocyte permeability to urea. The erythrocytes are initially in 0.3 M sucrose, where their volume is stable. At time t_0 the cells are centrifuged and resuspended in 0.3 M urea. This concentration is isosmolar with the cell's interior but it is not isotonic. At first there is no change in volume because the concentration of water is balanced on both sides, but since the cell membrane is slightly permeable to urea, it slowly enters the cell, moving to a region of lower urea concentration. The osmolar concentration of the cell's interior is increased, water infuses to equilibrate the osmolar concentrations, and the cell volume increases. The membrane is much more permeable to water than to urea so the volume increase follows very closely the urea entry. This is conveniently measured as the time taken to hemolyze. (a) Diagram of erythrocyte at different times. (b) Graph of cell volume versus time.

placed in an isosmolar but hypotonic solution will be determined by the rate of solute entry. The cell physiologist E. Overton did what must rank as one of the most thorough studies of this question, perhaps of any scientific question, by investigating the general properties of plant and animal cells for nearly a decade. After more than 10,000 experiments involving over 500 different chemical compounds he reached what may be deemed a satisfactory general view of these properties.

To understand Overton's results fully, we must develop a general framework for describing the forces that lead to flows of materials and how they are measured. This we shall do in the second half of this chapter. For our present purposes, to grasp the historical basis for the concept of membrane structure, we need only summarize Overton's extensive studies: *the more hydrophobic the solute, the faster it penetrates the cell membrane.* This was the first and weightiest evidence for the

notion that *lipids* play an important structural role in the cell membrane. A solute that cannot dissolve easily in lipid will pass through the membrane only very slowly; a solute that dissolves easily will pass through rapidly.

From the properties already mentioned we can infer not only that the membranes are composed of lipid, but also that they contain a large number of different proteins. These proteins must be present to endow membranes with their wide range of specific functions. Generally, cell surfaces also include an outside coating of variable thickness and chemical composition (secreted proteins and polysaccharides, mucopolysaccharides, and so forth) and a thin inside layer of gel-like cytoplasm.

History of Theories

While Pfeffer, Overton, and their successors studied the plasma membrane through its permeability and electrical characteristics, other biologists investigated the surfaces of cells more directly. Until the electron microscope revealed the presence of a finite membrane surrounding all cells, however, there was scant evidence that the structure at the surface and the one accounting for the permeability characteristics of cells were the same.

An early direct demonstration that the plasma membrane contained considerable amounts of lipid was carried out by E. Gorter and F. Grendel (1925). They extracted the lipid from red blood cells with acetone and placed the extract on a Langmuir trough (an apparatus for measuring the area of a monomolecular lipid film spread on water). Their results suggested there was enough lipid in the red blood cell to account for *two* unimolecular layers surrounding the cell. They concluded, therefore, that cells were covered with a *bimolecular layer of lipid.* (It is interesting that later experiments showed this conclusion, though true, was reached because of two compensating errors: they underestimated both the total amount of lipid per cell and the surface area of the erythrocyte.)

If the surface of the cell is oily, it should have a characteristic surface tension. E. N. Harvey, H. Davson, J. F. Danielli, and others performed many ingenious experiments to measure this property but to their surprise, the surface tension of cells was found to be considerably lower than that of oil/water interfaces. Since it was known that the addition of protein to oil (for example, egg white to mackerel oil) lowers the surface tension, Davson and Danielli guessed that protein is an integral part of the membrane. Their model interpreting the structure of the membrane as a three-layered (protein-lipid-protein) sandwich influenced cell biologists for over 20 years. In 1959, J. D. Robertson elevated the Davson-Danielli model to a "unit membrane" theory, which stated boldly that all membranes of the cell were constructed of the protein-lipid-protein sandwich. Indeed, the electron microscopy and x-ray diffraction evidence seemed to support these conclusions very well. Using $KMnO_4$ as a fixative, one could observe a layer approximately 7.5 nm thick. Two electron-dense lines, 2.0 nm wide, presumably of protein, were separated by a lighter layer, 3.5 nm thick, presumably of lipid (Figure 7–5).

Using the freeze-etching technique, which does not require any chemical fixative, D. Branton and others showed in the 1960s that the membrane does behave as a layered structure. The center seems to be weak, being frequently split by the fracturing process used to expose the interior of the cell (Figure 7–6). This is what one would expect if the middle layer were made mostly of a bimolecular leaflet of lipid, because such a layer would be held together only by the relatively weak van der Waals forces between the nonpolar hydrocarbon chains.

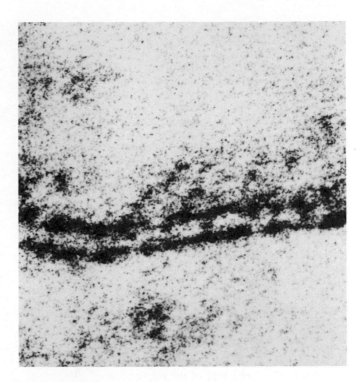

FIGURE 7–5
Electron micrograph of unit membrane. The "railroad track" appearance is illustrated. × 432,000. (Courtesy of G. E. Palade.)

Although many data could be interpreted as a simple protein-lipid-protein sandwich, workers in the field of membrane structure during the 1960s began to have second thoughts regarding the universal applicability of the unit membrane theory. Such a simple, static architecture seemed inadequate to account for the observed complexities of membrane function. Cell biologists now accept as a working model the *fluid mosaic* structure proposed by S. J. Singer and G. Nicolson (Figure 7–7).

The Fluid Mosaic Model

The fluid mosaic model shares with earlier proposals the primary structural motif of the lipid bilayer. The proteins, however, are much more variably disposed in the structure. Some completely *span* the bilayer and have part of their structures facing both inside and outside compartments. A few may be *inserted* into the bilayer from one side or the other without passing completely through. These two classes comprise the *integral* or *intrinsic* membrane proteins. They interact with the lipids primarily by hydrophobic forces and cannot be removed unless the bilayer is disrupted, by detergents, for example. Near the surface, integral proteins can also interact with the hydrophilic heads of the lipids.

The *peripheral* or *extrinsic* proteins are attached to the head groups of the lipids, or to the superficial portions of

FIGURE 7–6

Evidence for the existence of a three-layered sandwich structure. The vacuole membrane of the onion root tip, prepared by using the freeze-etching technique, is illustrated. This method uses no chemical fixatives. Instead, it involves rapidly freezing a specimen and fracturing it with a microtome. The newly exposed frozen surface is etched by briefly sublimating water from it, then it is shadowed with heavy metals, the tissue removed by acid treatment, and a replica constructed. Electron micrographs are taken of the replica. Application of this technique reveals that the membrane is composed of three layers, the middle layer being weakest and therefore easiest to fracture. (a) Electron micrograph of the replica. The three-layered membrane (m), at right, is in continuity with the fractured membrane, at left. The procedure leads to the appearance of a ridge (r) of one outer layer and the face (f) of a middle layer. (b) Explanatory diagram of the appearance of the ridge, face, and leaflet structure of a membrane. (Courtesy of D. Branton.)

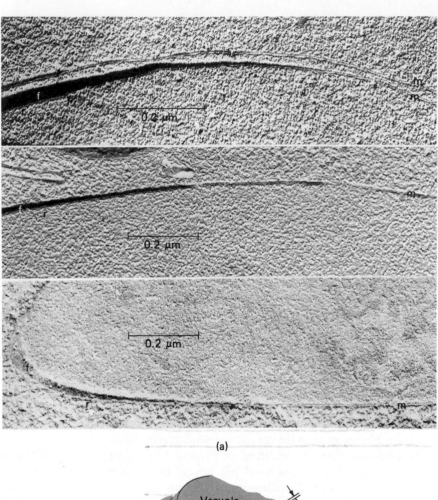

(a)

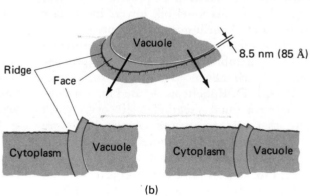

(b)

integral proteins. They can be dislodged by relatively mild treatments—for example, raising the ionic concentration of the solution surrounding the cells to screen electrostatic interactions.

The relative independence of the proteins from one another gives rise to the "mosaic" theory; the proteins are embedded in a matrix of lipid. However, it is now known that some intrinsic membrane proteins do interact with each other, or with peripheral proteins on the outside, or with cytoskeletal proteins and enzymes on the inside. The idea of fluidity comes from experiments showing how mobile the lipids of the bilayer are. As we

review the investigations that led to this modern model of membrane structure, the student should note the experiments that form the basis of the model and how they rule out the earlier picture of membrane structure.

Properties of Membranes

Cell biologists have become skillful at isolating and purifying various types of membranes. First, cells are broken open and the resulting homogenate is separated into various fractions by centrifugation (see Chapter 2). The plasma membrane and the endoplasmic reticulum

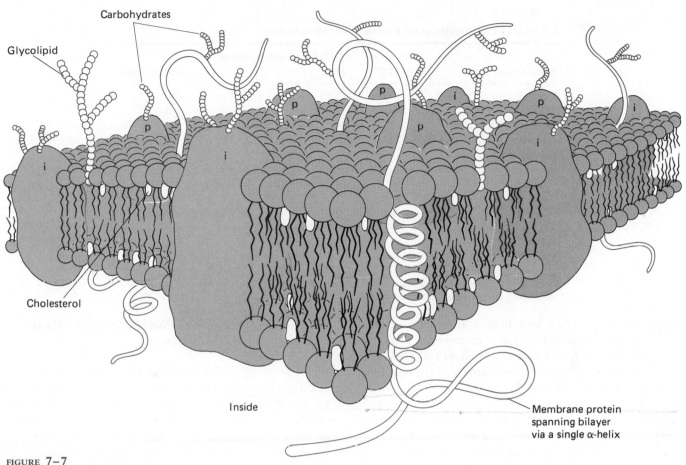

FIGURE 7–7
Fluid mosaic model of membrane structure. The bimolecular layer of lipid supports peripheral proteins (p), lying attached to one side or the other, and integral proteins (i), lying partially within the bilayer or completely spanning the membrane. Also shown are outward-facing carbohydrates on glycoproteins and glycolipids. Some of the integral proteins associate to form channels through the membrane, some span the membrane via a single α-helix. Cholesterol is found in the bilayer of animal cell mcmbranes.

membrane typically form vesicles. Some organelles— nuclei, mitochondria, and chloroplasts—can be isolated intact and their membranes converted to vesicles later. The outer vesicle surface can be washed by centrifugation; the insides are only exposed by breaking and reforming the vesicles, using more vigorous homogenization or ultrasonication. Dissolving pure vesicles in a detergent—Triton X-100, for example—allows for separation of the proteins and lipids. The mammalian red blood cell is particularly simple to study because, unlike most other eucaryotic cells, it has only a plasma membrane.

The Gross Chemical Composition of Membranes Varies

It is hard to reconcile the observed variability in the protein-lipid ratio and the carbohydrate content of membranes (Table 7–1) with any model requiring a uniform structure, such as the unit membrane theory. Membranes isolated from different organelles or structures also have vastly different enzyme activities. The inner mitochondrial membrane, for example, has an electron-transport chain (a series of electron carriers) quite unlike two other electron-transport chains found in

Triton X-100

TABLE 7–1

Chemical Composition of Various Cell Membranes

Membrane Type	Protein (%)	Lipid (%)	Carbohydrate (%)	Protein/Lipid
Myelin	18	79	3	0.23
Mitochondria, inner	76	24	(0)	3.2
Mitochondria, outer	52	48	(2–4)	1.1
Chloroplast lamellae	70	30	(0)	2.3
Plasma membranes				
Liver	52	48	(5–10)	1.1
Red blood cell	49	43	8	1.1
Cultured cells	60	40	(2–10)	1.5
Nuclear envelope	59	35	3	1.6
Gram-positive bacteria	75	25	(0)	3.0
Endoplasmic reticulum	60	40	(5)	1.5
Sarcoplasmic reticulum	67	33	(5)	2.0

Note: Values shown in parentheses are either estimates or so low that they may be influenced by contamination with other cell fractions.

endoplasmic reticulum membranes. Liver cell endoplasmic reticulum membranes have an enzyme, glucose-6-phosphatase, not found in any other liver cell membrane.

The precise chemical composition of the lipid component also differs extensively (Table 7–2). _Phospholipids are the most common component of membranes_, whereas neutral lipids (triglycerides and cholesterol esters) are usually found in very low quantities. The plasma membrane is an exception—in mammalian cells neutral lipids can make up as much as 25% of the plasma membrane.

Even among the phospholipids there is great variation among different intracellular membranes in the same cell type. For example, Table 7–3 shows that in rat liver cells, four different membranes have differing percentages of phospholipid species. Indeed, the presence of a large amount of _phosphatidylcholine_ is a diagnostic "marker" for endoplasmic reticulum membranes from all types of cells. Similarly, the presence of a relatively large amount of _sphingomyelin_ and _cholesterol_ is used as a marker for the plasma membrane, and _cardiolipin_ is a unique marker for the inner mitochondrial membrane.

The Bimolecular Layer of Phospholipids Is a Basic Structural Feature of Membranes

The x-ray diffraction patterns obtained from oriented stacks of phospholipid bilayers have been compared by M. Wilkins and his collaborators with the patterns obtained from similar stacks of plasma membranes from erythrocytes and bacteria. Both sets of patterns showed a spacing of 4.5–5.5 nm, the distance across the membrane between the layers of electron-rich phosphorus atoms, and a central disordered region that was attributed to the hydrophobic tails of the fatty acids.

Although most membranes contain complex mixtures of lipids, it is possible to force specific fatty acids into membranes by various techniques. Mycoplasma bacterial cells ordinarily require fatty acids in their diet since they cannot synthesize their own. By supplying them in the growth medium, it is possible to have the cells construct membranes from single types of fatty acids. Wilkins found that the x-ray diffraction patterns from stacked mycoplasma membranes (especially convenient because, unlike most bacteria, there is no cell wall to

TABLE 7–2

Lipid Composition of a Variety of Membranes*

| Lipid | Percentage in | | | |
	Myelin	Chloroplasts	Erythrocyte	Mitochondria
Phospholipids	32	10	55	95
Cholesterol	25	0	25	5
Sphingolipids	31	0	18	0
Glycolipids	0	41	0	0
Others	12	50	2	0

*Courtesy of D. Branton and R. B. Park.

TABLE 7–3

Lipid Composition of Rat Liver Membranes (As Percent of Membrane Lipid)

	Endoplasmic Reticulum	Plasma	Mitochondria Outer	Inner
Phosphatidylcholine	60	30	52	45
Phosphatidylethanolanine	15	16	25	27
Phosphatidylserine + phosphatidylinositol	10	12	14	4
Sphingomyelin	4	20	3	3
Cardiolipin	0	0	2	14
Cholesterol	8	17	3	2
Triglycerides	1	1	—	—
Miscellaneous	2	4	1	5

remove) showed greater spacings between the phosphates with longer fatty acids. This supported their interpretation of the diffraction pattern. The thickness of the membrane can be assessed when the x-ray beam is perpendicular to the cross sections of stacked membranes (Figure 7–8).

When the x-ray beam was perpendicular to the faces of the membranes, strong diffraction was detected at low (nonphysiological) temperatures. The maximum diffracted intensity was at an angle corresponding to a spacing of 0.42 nm. Since this is the same spacing observed between the stacked rod-like molecules in a phospholipid crystal, the diffraction was interpreted as due to similarly ordered arrays of phospholipids in the membrane. The degree of order observed in these diffraction patterns was altered by the temperature, being lower at higher temperatures and greater at lower temperatures. The structure changed from ordered to disordered over a *transition temperature* region, which varied with the fatty acids incorporated in the membrane.

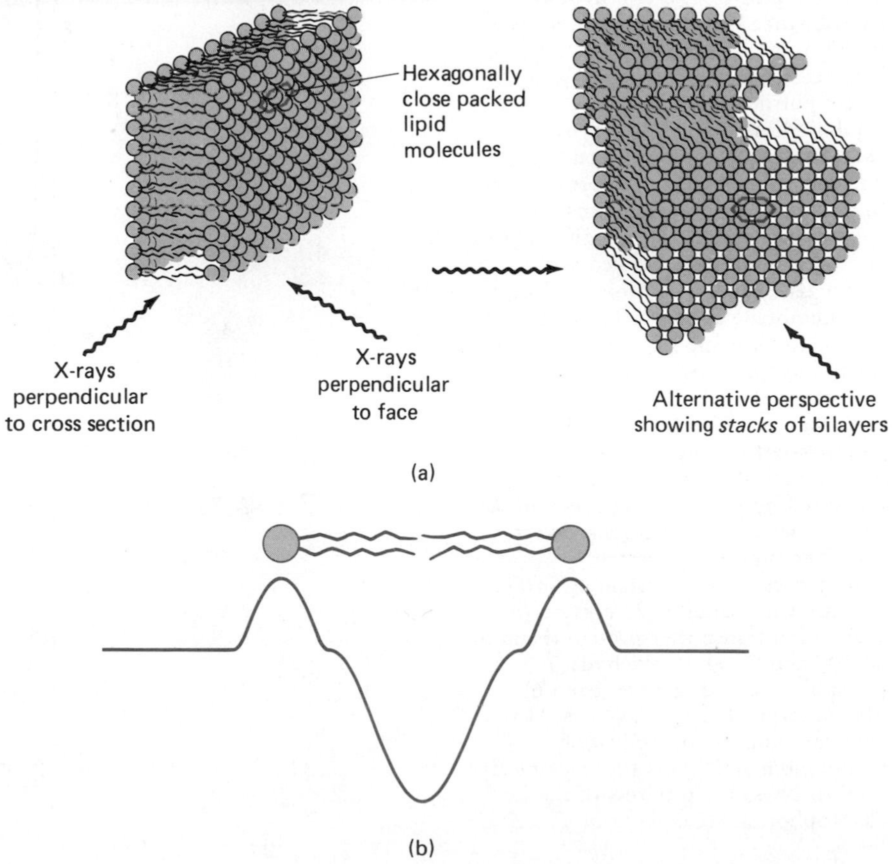

FIGURE 7–8

X-ray diffraction analysis of membranes. (a) Diagram showing stacks of *Mycoplasma* membranes and the hexagonally close-packed lipid molecules which give the x-ray patterns described in the text. X-rays fell perpendicular to the face or to the cross section of the lipid array. (b) Part of electron density pattern inferred from x-ray diffraction. Two phospholipid molecules are indicated above the electron density pattern as a reference. The peaks of electron density are due to the heavy phosphorus atoms in the head groups of the phospholipids. The trough in the center is due to the lighter and less ordered fatty acid tails.

The higher the melting point of the fatty acid, the higher the temperature needed to disorder the x-ray diffraction pattern.

All these data led to the conclusion that *lipid-lipid interactions are of great importance* in the structure of membranes. This is consistent with the previously mentioned observation that only treatments which dissociate these interactions (e.g., detergents) are successful at breaking down the integrity of the membrane.

Proteins May Be Associated with the Inside or the Outside Face, or May Completely Span the Membrane

The fluid mosaic model of membrane structure assigns the fundamental architectural role to the lipid bilayer. Further, it provides for nonsymmetrical placement of proteins within the structure. The experimental proof of this possibility involved the use of *nonpermeating probes*. If a probe is excluded from the inside, then it can react only with parts of proteins that lie exposed on the outer face of the membrane. A permeating probe, or a nonpermeating probe used on disrupted membranes, can get access to both sides. Any difference in the results of these two experimental modes can be provisionally ascribed to inward-facing structures.

Before considering the results of this strategy, we need to describe the assay system. One can prepare membranes from erythrocytes simply by exposing them to dilute (hypotonic) solutions, which causes the cells to lyse and release their hemoglobin, a process known as *hemolysis*. The membranous remains, lacking the intense red of normal cells, are known as *red cell ghosts*. Lipids can be removed by extracting them with organic solvents like chloroform or acetone; the proteins can be separated from one another on the basis of molecular weight by *electrophoresis in denaturing solvents*. The universal choice for separating membrane-associated proteins is electrophoresis in *polyacrylamide gels,* in solutions containing the strong detergent *sodium dodecylsulfate* (SDS-PAGE, see Figure 4–5). The proteins in the gel are visualized with

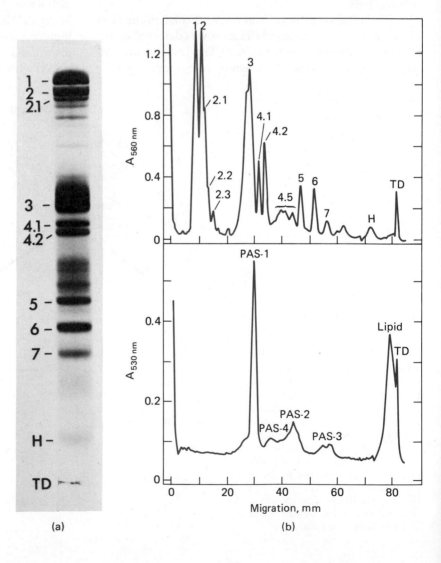

FIGURE 7–9

The proteins of the erythrocyte membrane.   Erythrocytes were lysed in dilute, mildly alkaline buffer and repeatedly washed to obtain "ghosts." These purified membranes were solubilized in sodium dodecylsulfate (SDS) and electrophoresed on polyacrylamide gels in the presence of SDS, which separates proteins on the basis of their molecular weight. (a) Gel after electrophoresis, stained with Coomassie Blue. The proteins from the cell membrane appear as bands; the numbering scheme for red blood cell membrane proteins is indicated. (Courtesy of T. L. Steck.) (b) Absorbance of the gels stained with Coomassie Blue (upper trace) or with periodic acid-Schiff (PAS, lower trace), measured by scanning the gels with a densitometer. The PAS stain detects carbohydrates and indicates glycoproteins. Deeply staining proteins are: spectrin (bands 1, 2) ankyrin (band 2.1), anion transporter protein (band 3), actin (band 5), glyceraldehyde-3-phosphate dehydrogenase (band 6), glycophorins (bands PAS-1, 2); H is a small amount of contaminating hemoglobin and TD is the tracking dye used to assess the progress of the electrophoresis. (Courtesy of T. L. Steck.)

(a)

(b)

stains diffused into the gel after electrophoresis, usually in an acetic acid solution that tends to *fix* the proteins so they do not diffuse out. *Coomasie blue* stains most proteins and the *periodic acid Schiff* stain is used to locate glycoproteins. Radioactive proteins can be located by autoradiography on a sheet of photographic film placed over the gel, or by slicing the gel and measuring the radioactivity of each slice in a scintillation counter.

When these methods are applied to the human red blood cell, only three bands on the gel account for one-half the total protein of the membrane (Figure 7–9). Much of the rest of the protein is represented by weakly staining bands on the gel, and there is a very large number of extremely weak bands. These last probably include many components of great functional importance; the sodium ion pump, for example, is present in only a few hundred copies per red blood cell. The stain for glycoproteins shows that only a few are present in significant amounts and one band is dominant.

The use of various nonpermeating probes to study the proteins in the human red blood cell membrane gave consistent results. One of the earliest used was the proteolytic enzyme pronase. It and other proteases attack proteins in the membrane that represent only two of the major bands on SDS gels, a glycoprotein known as *glycophorin* (or PAS-1) and the protein known as *band 3*. After pronase digestion of the membrane, these bands on the gel are reduced in amount, and new bands, with smaller molecular weights, appear. Another probe, introduced by N. M. Whitely and H. Berg, is *isethionyl acetimidate*. This reagent does not penetrate the membrane because of a strong negative charge (Figure 7–10) and can attach covalently to primary amines—that is, to lysine side chains. Since the amidine produced by this reaction is both small and positively charged like the original amine, the attached ligand should have minimal effects on the functional properties of proteins. In fact, Berg found that attaching isethionyl acetimidate to all the available external amines has only minor detectable effects on K^+ or anion or glucose permeation. Like the proteolytic enzymes, this probe labels the band 3 protein and glycophorin. Many other reagents which are thought to be nonpermeating also label these two proteins. Other proteins are too few to detect easily, or have no susceptible groups. We can conclude from these results that only two species, glycophorin and band 3 protein,

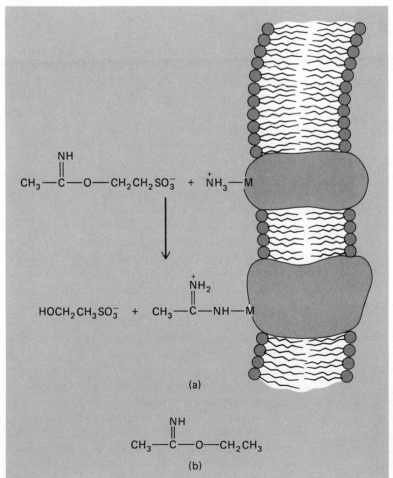

FIGURE 7–10
Probes of membrane surface structures.
(a) Isethionyl acetimidate reacting with a membrane-associated primary amine to form covalently attached amidine and an alcohol. (b) The membrane-permeable ethyl acetimidate. Its reaction with amines is the same as isethionyl acetimidate.

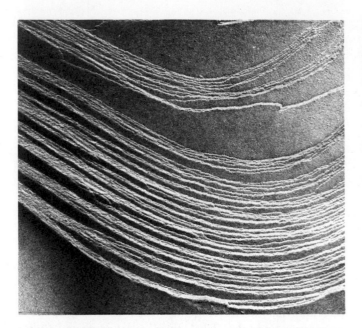

FIGURE 7–11
Comparison of membrane faces of myelin and chloroplast lamellae prepared by the freeze-etch technique.   Note the smooth faces of the myelin membranes (a) and the numerous globular subunits on the chloroplast membrane (b). × 80,000. (Courtesy of D. Branton and R. B. Park.)

account for the bulk of the externally facing proteins of the red blood cell.

To label or modify proteins exposed on the inner face, one needs either to disrupt the cell membrane or use a permeating probe. The permeant analogue of isethionyl acetimidate is *ethyl acetimidate;* it lacks the negative charge that makes isethionyl acetimidate nonpenetrating (Figure 7–10). An advantage of the mild and permeant ethyl acetimidate is that it is less likely to perturb the structure and therefore more likely to label the amines in their native conformation. While there are 10^6 sites in the exterior face that react with isethionyl acetimidate, there are more than 15 times as many on the interior face that are labeled with ethyl acetimidate. Of course, various structural features might block access of these reagents to some amines, so failure to label cannot be interpreted clearly. Nevertheless, because most of the SDS gel bands can be labeled, but only a few with the impermeant reagent, this suggests an important structural asymmetry in the distribution of membrane proteins. Studies on disrupted membranes with other probes, including proteolytic enzymes, are in agreement: all the easily detectable bands on SDS gels can be labeled or altered.

The band 3 protein can be digested by proteolytic enzymes or labeled by acetimidate from either side of the membrane. Thus, it must be accessible on both sides of the membrane: it *spans the bilayer*. By using ^{14}C-labeled isethionyl acetimidate and ^{3}H-labeled ethyl acetimidate, Whitely and Berg demonstrated four other polypeptides

that span the human erythrocyte membrane. The peptides resulting from proteolytic attack on the band 3 protein from the outside differ from the peptides resulting from attack on disrupted membranes. This means that the band 3 protein *is asymmetric* and *does not rotate within the bilayer*. Without these properties, the sites accessible on the outside would represent all the susceptible sites rather than a set different from those accessible from the inside.

In summary, our picture of the disposition of proteins in the red blood cell membrane is that some proteins are exposed on the outer surface, while others are exposed on the inner surface. Some completely span the bilayer and are exposed on both faces, in complete disagreement with the Davson-Danielli sandwich model. The sugar residues of glycoproteins are confined to the outer face of the erythrocyte membrane (but glycophorin itself spans the bilayer). Data are still being gathered from many other cell types but we should feel surprised if the fundamental feature of protein asymmetry in biological membranes were not universal.

Microscopically visible features of membranes that did not seem consistent with the Davson-Danielli model can be explained by the fluid mosaic model. For example, electron microscopy using the freeze-etch technique revealed evidence for embedded integral proteins. Myelin membranes are smooth faced, but chloroplast membranes contain particles (Figure 7–11). In the erythrocyte, both faces of the membrane contain numerous 8-nm

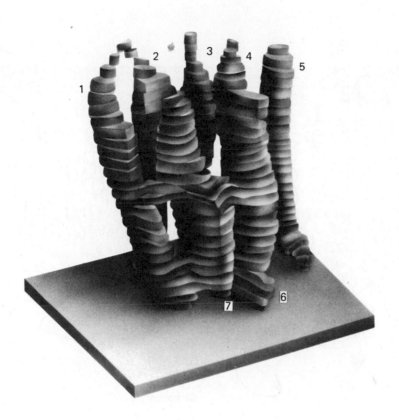

FIGURE **7–12**
Model of bacteriorhodopsin at resolution of 1.2 nm.   This 248-amino acid protein absorbs light and uses the energy to "pump" hydrogen ions outside the cell. Most versions of the structure show seven bent rods clustered together, roughly perpendicular to the plane of the membrane. The amino acid sequence can be interpreted as seven largely nonpolar α-helical membrane-crossing stretches, with polar loops between. The light-absorbing chromophore attaches to a lysine residue in the C-terminal membrane-crossing helix. (Courtesy of R. Henderson and D. Leifer.)

particles, about $2600/\mu m^2$ on the inner face and $1300/\mu m^2$ on the outer. Treatment of the membrane faces with proteolytic enzymes before freeze fracturing makes these particles undetectable, suggesting they are integral proteins.

Transport Proteins Span the Membrane Many Times

To be involved in transport of mass from one side of the membrane to the other, a protein must span the membrane. The polypeptide chain of such proteins lies folded back and forth across the bilayer several times. This can be demonstrated by the observation that a series of protease-sensitive or labeling sites lie on opposite sides of the membrane. For example, if the order of five such sites in the amino acid sequence is a, b, c, d, e, and if sites a, c, and e lie on one face while b and d lie on the other, the polypeptide chain must cross the membrane at least four times.

Until we have good three-dimensional data from high-resolution x-ray diffraction analysis, the exact structure of membrane proteins is uncertain. The first membrane protein for which significant structural information was obtained is bacteriorhodopsin, the purple membrane protein from the photosynthetic bacterium *Halobacterium halobium.* Electron diffraction and electron microscopic data from this protein have been analyzed to give a model with a cluster of seven rod-like segments,

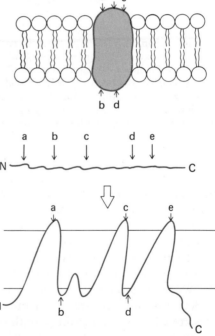

lying roughly at right angles to the plane of the membrane (Figure 7–12).

Polypeptide chains forming the rods of bacteriorhodopsin cross the membrane as an α-helix of approximately 21 residues—the number that forms a helix of length equal to the thickness of the nonpolar part of the bilayer (~3.2 nm). It is likely that *membrane-spanning α-helices* are a common structural motif of transport

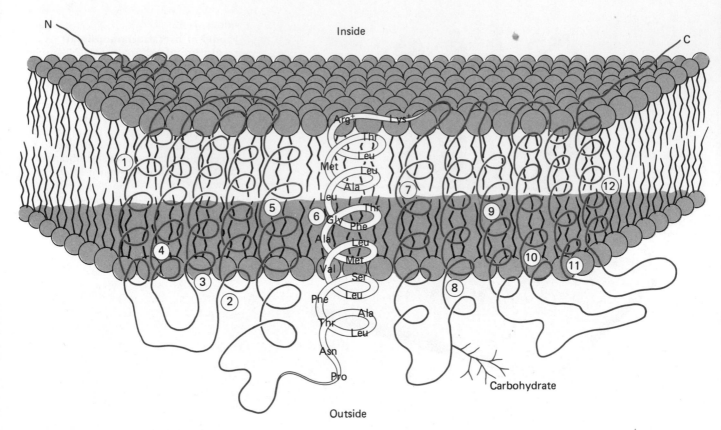

FIGURE 7–13

Scheme for orientation of band 3 protein in membrane.   The N-terminal half (~400 amino acids) faces the cytoplasm and interacts with cytoskeletal components and other proteins. The C-terminal half (~500 amino acids) is inserted into the lipid bilayer and has the anion exchange activity. The polypeptide chain crosses the membrane 12 times; each crossing is represented as an α-helix. Between the seventh and eighth crossing is the carbohydrate attachment site. The 21 amino acids of the sixth crossing are almost all hydrophobic and can interact strongly with the fatty acid chains of the lipid bilayer. Some helices are amphiphilic, with polar residues lying on one face of the helix where they could interact with adjacent amphiphilic helices. Such an arrangement is likely to be involved in forming the channel for anion exchange.

proteins. In support of this notion are features in the amino acid sequence of many membrane proteins; most of these sequences have been determined by analysis of the genes that specify the protein (see Chapter 10). There are stretches of largely nonpolar amino acids that could form an α-helix having outward-facing side chains in good van der Waals contacts with the fatty acid chains in the bilayer. Between the hypothesized membrane-crossing sequences are stretches of variable length containing polar and charged amino acids that are likely to interact with the phospholipid head groups and the adjacent aqueous layers.

Some of these α-helices are *amphiphilic;* they have both polar and nonpolar residues, arranged to have one side or face of the helix polar and the rest nonpolar. Amphiphilic helices in membranes can interact to form a cluster with the polar sides facing inward and having an exterior surface of nonpolar side chains that interact with

the fatty acids. The interior polar surface can act as a channel for the movement of polar substances from one side of the membrane to the other. (A protein arranged this way is inside out relative to typical soluble proteins, which have a polar exterior surface and a hydrophobic interior.) Band 3 protein from the mammalian red blood cell probably contains both hydrophobic and amphiphilic helices, crossing the membrane at least 12 times (Figure 7–13). Both bacteriorhodopsin and band 3 protein catalyze the transport of polar substances from one side of the membrane to the other: bacteriorhodopsin catalyzes the movement of H^+ ions outward, in response to light, and band 3 protein catalyzes the exchange of anions (Cl^- and HCO_3^-) across the membrane.

The α-helix is not the only protein secondary structure that can span a membrane. *E. coli* has a protein called *porin*, which as a trimer forms three channels through the bacterial outer membrane. A similar protein is found in the outer membrane of mitochondria. About two-thirds of the bacterial porin is in the form of antiparallel pleated strands, β-structures, arranged perpendicular to the plane of the membrane and probably making extensive hydrogen bonds with one another. They are about 10–12 amino acids long, significantly smaller than the number required to form a membrane-spanning α-helix. Porin has many polar residues and lacks large nonpolar domains. Unlike other transport proteins, porin is relatively nonspecific for what kinds of molecules it conducts across the outer membrane; most small molecules of less than 10,000 molecular weight penetrate relatively freely. A similarly nonspecific pore is found in the nuclear envelope of eucaryotic cells. A 10-nm aqueous channel allows the rapid movement of molecules up to 40,000 molecular weight between the cytoplasm and the nucleus (the nuclear pore also has an active transport function, using ATP for larger structures).

The Composition of Lipids in the Bilayer Is Also Asymmetric

A bilayer structure can be formed with any phospholipid and can admit a variety of other hydrophobic molecules, such as cholesterol and glycosphingolipids, etc. Is there a random distribution of these molecules in membranes, or might the two portions of the lipid bilayer show specificity of composition? Measurements designed to answer this question leave little doubt about the asymmetrical distribution of phospholipids in the membranes of at least some cells.

The strategy of the experiments is essentially the same one described earlier to assess the asymmetry of proteins in the membrane. Instead of a proteolytic enzyme, a phospholipase is used as a nonpermeating probe of the lipids on the outer face of the bilayer. Phospholipase A_2 cleaves the fatty acid residue from the middle position of the glycerol of a phospholipid. The venom of the cobra *(Naja naja)* is often used as a source for this enzyme. When added to osmotically balanced intact cells, phospholipase is first likely to degrade phospholipids that face outwards. In the case of the

human erythrocyte, 68% of the phosphatidylcholine is degraded under these conditions. If a sphingomyelinase is also present, 76% of the phosphatidylcholine is degraded and 82% of the sphingomyelin, but only 20% of the phosphatidylethanolamine and none of the phosphatidylserine. The degraded species account for 48% of the total present, enough for one leaflet of the bilayer. The phospholipids of disrupted erythrocyte ghosts, by contrast, are completely degraded by this combination of enzymes. These data, and many other experiments of a similar sort, suggest that the outer face of the human erythrocyte is rich in phosphatidylcholine and sphingomyelin while the inner face contains most of the amino lipids, phosphatidylethanolamine, and phosphatidylserine (Figure 7–14).

Radioactive or fluorescent labeling probes can also be used to explore the bilayer composition. One of the first such reagents was formylmethionyl sulfone methylphosphate, which M. Bretscher used to label amino lipids. He found that phosphatidylethanolamine could be easily labeled in disrupted ghosts of the human erythrocyte, but not in intact cells. Whitely and Berg confirmed this result with the finding that roughly a hundred times more amino lipids could be labeled with the permeant ethyl acetimidate than with the impermeant isethionyl acetimidate. Erythrocytes also show asymmetry in the position of the fatty acids within the phospholipid.

As we shall see, exchange of a lipid from one side of the bilayer to the other is not very frequent, but it is sufficiently common to randomize the distribution of lipids over the life of a cell. There are, therefore, probably specific mechanisms for establishing and maintaining the observed asymmetry. Protein–lipid interactions may preserve the asymmetry, but the details of the mechanism remain unknown. Asymmetry in the distribution of the many other nonprotein hydrophobic molecules found in membranes has also been demonstrated. X-ray diffraction evidence suggests that cholesterol is found preferentially in the outer portion of the bilayer. Glycolipids are all located in the outer portion, with their carbohydrate groups exposed to the external aqueous solution.

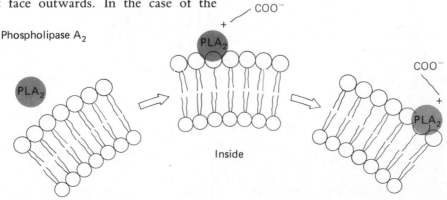

Phospholipase A_2

Inside

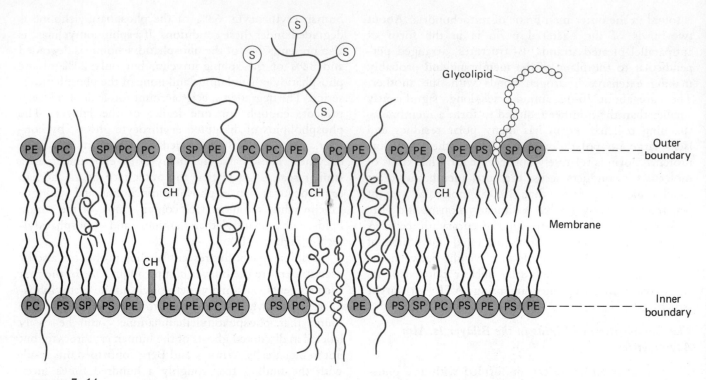

FIGURE 7–14
Asymmetric distribution of components in erythrocyte membrane. PE, phosphatidylethanolamine; PS, phosphatidylserine; SP, sphingomyelin; PC, phosphatidylcholine; CH, cholesterol; S, carbohydrate residues on protein, used to identify the main glycoprotein, glycophorin.

The Bilayer Acts Like a Two-Dimensional Fluid

One of the most startling departures from the old ideas about the structure of membranes was abandoning the notion of a static architecture. The experiments which led to our current view of a *fluid* membrane involved a wide range of disciplines and techniques, from physical chemistry to the visualization of surface structures in whole cells. The idea of fluidity implies motion, both of the molecules forming the fluid and other substances immersed in it. In a membrane, motion can be lateral (in the plane of the membrane) or vertical (from one side of the membrane to the other, a motion also called flip-flop).

An important difference between a solid and a fluid is the degree of order in the two types of structures. A

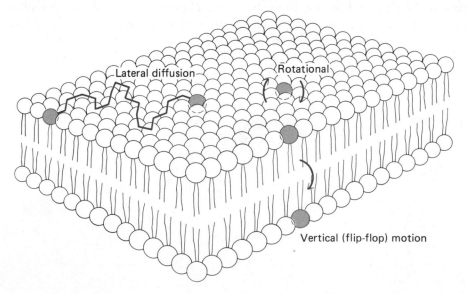

Lateral diffusion   Rotational

Vertical (flip-flop) motion

solid is characterized by the long-range, relatively constant ordering of its molecules; knowing the position of one molecule allows one to predict the position of nearby molecules with high precision. Fluids are more chaotic. To determine which state more accurately represents the bilayer one uses a technique that has minimal disruptive effects on its structure: spectroscopy—shining radiation into the bilayer and observing the properties of the emerging radiation. The characteristic time for the events to be measured must be carefully matched to the frequency of the radiation. The radiation of choice for this measurement is high-frequency radio waves (microwaves), which are perturbed by the spin of unpaired electrons; the technique is known as *electron spin resonance (ESR) spectroscopy*. Such radiation is perturbed very little by ordinary organic molecules and there is, therefore, no ESR signal to measure from the bilayer alone. This problem, a common one in cell biology, is solved by attaching to the bilayer a *spectroscopic probe* (or *reporter group*)—a molecule with enough absorbance to perturb the radiation but small enough to avoid perturbing the structure. H. M. McConnel used for this purpose a nitroxide-containing ligand attached to a fatty acid. Nitroxides have strong electron spin resonance signals; they are good *spin labels* (Figure 7–15). The fatty acid was incorporated into stacked multiple layers of phosphatidylcholine, or diffused into spherical vesicles whose walls were made from cellular membranes.

The details of the theory for interpreting these experiments lie beyond the scope of this discussion; essentially, the nature of the absorption spectrum gives information about the movement of the probe. Important data that can be obtained from analysis of the microwave spectra include the *nearest neighbor exchange rate* for the membrane phospholipids. This rate is extremely rapid in the plane of the membrane, over 10^6 per second. This means that a phospholipid in the membrane could diffuse the length of an *E. coli* cell, about one μm, in about one second. In contrast, the estimated rate of *flip-flop exchange* by a phospholipid, from one face of the bilayer to the other, was one per six hours. The lateral movement of a phospholipid molecule within the bilayer is thus some 10^{10}-fold faster than movement across the bilayer.

By attaching the nitroxide spin label to various carbon atoms in a fatty acid chain it was possible to assess in model membranes the degree of order in the bilayer structure at various distances from the charged head groups of the phospholipids. There is apparently a great deal of order near the charged head groups (i.e., at the membrane surface) and inward for the first nine carbon atoms. In the center of the bilayer the order is much lower. The disorder arises from rotations about the C—C single bonds in the fatty acid chains, which can create "kinks"—departures from a smooth linear arrangement

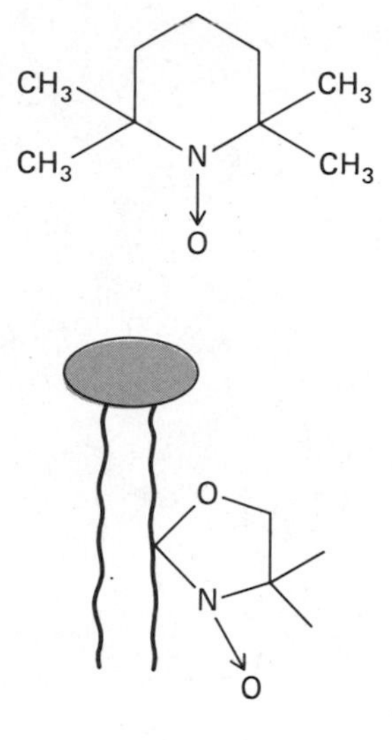

FIGURE 7–15

Probes for electron spin resonance spectroscopy.
(a) The spin label, 2,2,6,6-tetramethyl-piperidene-1-oxy, also known as TEMPO. (b) A TEMPO-like probe incorporated into a phospholipid is sensitive to the motion of the fatty acid chains to which it is attached.

of atoms. The resulting structural defects can propagate across the bilayer and accommodate the transport of very small molecules such as water.

Observations on the rapid movement of lipid molecules within a membrane monolayer have been repeated and extended. The picture that emerges is that of a *two-dimensional fluid* in which isolated objects can move relatively unimpeded from place to place in the membrane surface.

A more direct way to demonstrate the fluid nature of the bilayer is to label surface components and observe the results of their movement within the bilayer. One can obtain antibodies to red blood cells that will cause them to stick together (agglutinate). The structures to which the antibodies bind can be localized on the cell surface by electron microscopy when the antibodies are labeled with an electron-opaque molecule such as ferritin (an iron-protein complex). Figure 7–16a shows the image seen in the electron microscope after treating a red blood cell with trypsin (to expose the surface better), adding the ferritin-labeled agglutinating antibodies, incubating at 0°C, and fixing for microscopy. The structures to which the antibodies are bound are distributed uniformly on the cell surface. If the incubation is at 20°C (Figure 7–16b), the antibodies are found in patches. This is consistent with movement of the surface structures and their cross-linking, via the antibodies, when they come into contact.

For visual drama in cellular biology, the microscopic demonstration of membrane fluidity depicted in Figure

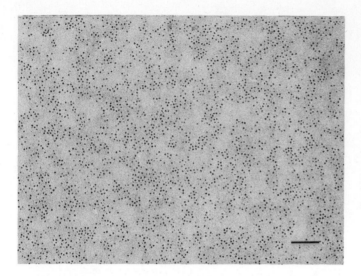
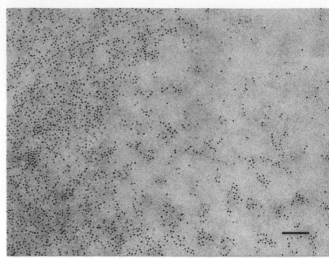

FIGURE **7–16**
Redistribution of cell surface proteins.   Red blood cells (treated with trypsin to expose the surface) were labeled at 0°C with an agglutinating antibody to which ferritin (an iron-protein complex viewable in the electron microscope) was chemically bound. (a) The cells were further incubated at 0°C and random distribution of the ferritin on the surface is shown. (b) The cells were further incubated at 20°C and much of the ferritin was aggregated. The results indicate movement at 20°C of the surface proteins to which antibody was attached, as monitored by the movement of the ferritin along the surface of the cells (bar indicates 0.1 μm). (Courtesy of G. Nicolson.)

7–18 has few equals. The basic strategy was to fuse two cells with different detectable surface features and follow the positions of these features after fusion. M. Edidin and his collaborators used an immunological labeling method to distinguish the surface features of cultured mouse cells and cultured human cells (Figure 7–17).

The cell fusion was accomplished with the mediation of inactivated *Sendai virus particles.* The fusion of two cells doubtless occurs by intimate contact between their lipid bilayers. (Today, such fusions are more often mediated by *polyethylene glycol* or by *electrofusion,* in which cells are brought together by a nonuniform alternating electric field and fused by a brief high voltage pulse.) After fusion the cells were incubated for varying times, then fixed and treated with the antibodies to locate the various surface structures.

At five minutes after fusion the surface antigens of each of the two original cells were confined to opposite hemispheres. The image in the ultraviolet microscope resembled the diagram in Figure 7–18b, two "half-rings." This was because the microscope was focused on a plane that both bisected the spherical cell and was perpendicular to the plane of fusion. (That the whole of each hemisphere was labeled could be demonstrated by focusing on the nearer or farther surface.) After ten minutes, a mixing of the red-labeled and green-labeled surface antigens could be observed. This was seen most

clearly by looking at the same view through a green or through a red filter to allow visualization of each fluorescent label without interference from the other. After a longer time (40 min) complete mixing had occurred and a diffuse intermingling of red and green fluorescence could be seen (Figure 7–18c).

The fraction of the population of fused cells that displayed these distributions of surface antigens varied with time after fusing the mouse and human cells. At first most of the cells showed separate half-rings, then most showed the partial mixing, and finally the whole population showed the mosaic pattern. Edidin and his group tried various methods to affect this time course. Among strategies that did *not* work were brief inhibition of cellular energy metabolism, brief and extended inhibition of protein synthesis, and other metabolic inhibitors. A method which *did* slow the mixing was lowering the temperature by ten to twenty degrees. The interpretation of these studies was that the two-dimensional movement of the surface antigens through the lipid bilayer fluid is spontaneous, requiring no input of metabolic mass or energy. Lowering the temperature affects the movement by raising the viscosity of the fluid bilayer and slowing the spontaneous mixing.

Studies of other cells reveal similar behavior. Proteins in the lymphocyte surface, in the myelin and the axonal surface of neurons, in bacterial membranes, even

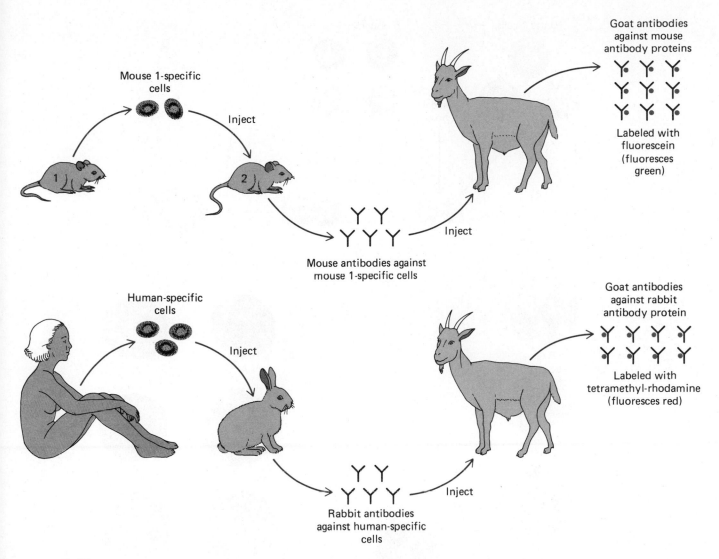

FIGURE **7–17**
The production of antibodies against mouse and human cells. Antibodies against mouse cell-specific antigens were raised in a different strain of mouse having different cell-surface antigens. These antibodies could be used to locate the mouse cell-surface antigens; microscopic visualization was accomplished by adding goat antibodies capable of reacting with the mouse antibody proteins. The goat antibodies were previously labeled with *fluorescein,* which fluoresces green in an ultraviolet microscope. This technique is known as *indirect immunofluorescence.* A similar set of antibodies was raised against human cell surface antigens in rabbits and anti-rabbit antibodies were raised in goats. The visualization was accomplished here by labeling the goat anti-antibodies with *tetramethyl-rhodamine,* which fluoresces red in the ultraviolet microscope. Thus, the human surface features would appear red and the mouse surface features would appear green.

the rhodopsin molecules in the highly ordered photoreceptor membrane of the eye's outer segment discs, are all capable of lateral movement. When the surface components of cells are exposed for a long time to fluorescent-labeled antibodies, there is movement within the fluid bilayer but it is not random. The surface antigens, cross-linked by the antibodies to form large "patches," all

tend to flow to one pole of the cell and collect there, a phenomenon known as "capping." One way to achieve capping is to reorganize the surface structures from underneath, via attachments to the cytoskeleton. Another is *membrane* or *lipid flow*—a movement that would result from addition of lipid (or membrane, for example, by exocytosis) to the membrane at one point on the cell

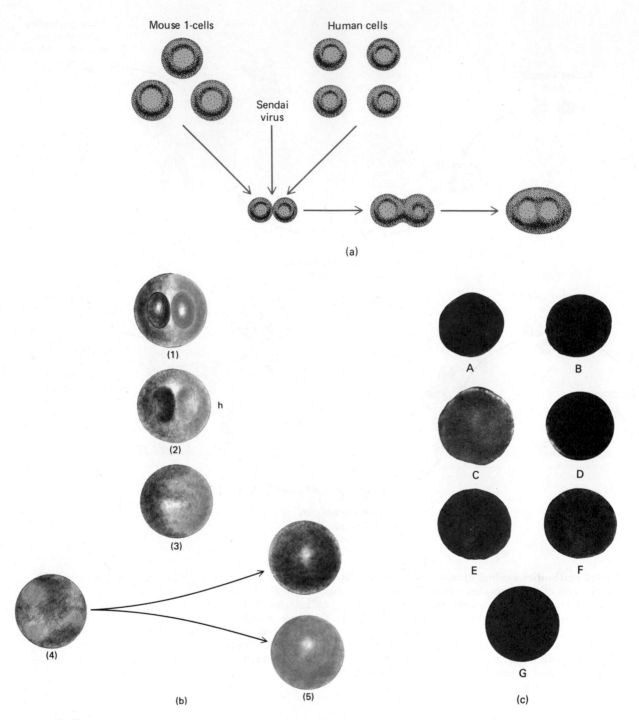

FIGURE 7–18
Lateral movement of cell surface components.   (a) Diagram of fusion of mouse and human cells. (b) Localization of surface antigens by indirect fluorescence of fused cells. At different times after fusion the cells were fixed and treated with the various antibodies. (1) Appearance of surface labeling shortly after fusion, with mouse antigens appearing green and human antigens appearing red. (2) Appearance of cells when microscope is focused on equator; h = half-rings (corresponding to 3 in c). (3) Appearance of fused cells at ten minutes. (4) Appearance of fused cells at forty minutes. (5) Visualization of fluorescent labels when viewed through separate filters (corresponding to E and F in c). (c) The appearance of the fused cells, as viewed in an ultraviolet microscope, showing the spread of mouse (green label) and human (red label) antigens on the cell surface.
A = stable hybrid cell line; appearance through green filter shows spread of mouse antigens on surface of cell. B = stable hybrid cell line; appearance of same cell through red filter shows human antigen on cell surface. C = cell was doubly exposed to record both antigens (red and green). D = appearance of another hybrid cell through green filter shows small amount of spread of mouse antigens. E = appearance of same cell through red filter shows large amount of spread of human antigens. F = appearance of another hybrid cell shows large spread of mouse antigens. G = same cell as in F shows large spread of human antigens.

FIGURE 7–19

"Capping" in lymphocytes.

(a) Capping can be visualized by labeling the surface with fluorescent-labeled antibodies raised against surface antigens. At first the fluorescence is seen all over the cell surface; then it collects in "patches," due to cross-linking of the surface antigens by the antibodies; eventually the patches congregate at a pole of the cell, forming the cap.
(b) Membrane or lipid flow in the lymphocyte. This is one way to explain capping. Materials enter the membrane structure via a source at one end of the cell and flow through the membrane to a "sink" at the other end. Colored arrows indicate the direction of flow. Large antigen-antibody complexes (patches) on the outer surface of the membrane are carried along by the flow and end up in a cap near the sink. (c) Pulling from underneath by the cytoskeleton. Fibrous structures in the interior of the cell can move and exert force. Connections between the cytoskeleton and some of the surface structures could move cross-linked aggregates towards the pole.

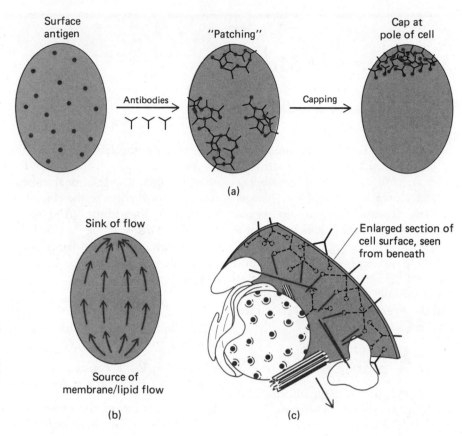

surface and removal (by endocytosis) near where the cap forms (Figure 7–19).

If all the proteins in a membrane were free to move independently of one another, the result in the absence of directed flow would be complete mixing, as seems to be the case with the surface antigens of fused mouse and human cells. But we know that cells in many tissues maintain special locations for membrane-associated functions. Such cells are described as *polarized*. In an exocrine gland, for example, only the cell surface that faces the duct is involved in secretion. Among the most dramatic examples of polarized cells are the free-living protozoa, like *Paramecium,* which have very asymmetric surface structures (Figure 1–5). How is this asymmetry or heterogeneity brought about? We shall discuss the mechanisms of inserting proteins into membranes later, in Chapter 13, but the basic problem is not yet satisfactorily explained. Part of the answer to the question of how the cell achieves heterogeneity in its membrane probably lies in the attachment of some surface components to the underlying cytoskeleton. This prevents the free and independent movement of those components. How the architecture of the cytoskeleton is laid down is a fundamental problem of morphogenesis at the cellular level and awaits the intellectual breakthrough that one of today's cell biology students may provide.

The Membrane Constituents Turn Over

Besides the fluid nature of the membrane there is another sense in which the lipids and proteins of the membrane are dynamic. They are continually being removed and replaced with newly synthesized components, and each component has its own particular rate of turnover.

Ever since R. Schoenheimer demonstrated that the constituents of the body are not stable but undergo continual destruction and resynthesis, it has been a tenet of biochemistry that there is constant replacement going on in the cell. The experiments to demonstrate this are fairly straightforward but can only be done by using radioactive tracers to follow the molecules or complexes in question. A radioactive precursor of the molecule or structure under study is injected into a series of animals or cells that are not growing any more and are thus in a steady state. At various times afterwards, the molecule is isolated and purified, and its radioactivity is determined. The precursor must contain radioactive atoms that are relatively long-lived; it must also be one that is incorporated quickly, and the incorporation must be able to be stopped quickly by dilution with unlabeled precursor molecules. In a whole animal such a "pulse" of radioactivity is terminated by the unlabeled food the animal eats.

Mathematical Model for Turnover

Suppose there are N molecules of a given kind (such as a particular protein), and the molecules are being broken down in a completely *random* fashion. Random breakdown means that the probability per unit time of a single molecule being broken down is constant; we represent this probability by P. The number of molecules broken down per unit time is $P \cdot N$ and the overall rate of breakdown for molecules of this type is

$$-\frac{\Delta N}{\Delta t} = P\,N, \qquad (7\text{-}A)$$

where the minus sign means that $-\Delta N/\Delta t$ represents a *decrease* in N with time. This is the form of a first-order chemical rate equation. To reveal the solution, one can rearrange the equation and obtain

$$\left[\frac{\Delta N}{N}\right] = -P\,\Delta t.$$

The term on the left represents the *fractional change* in N ($\Delta N/N$). For small changes, the fractional change in a quantity is equal to the change in its natural logarithm: $(\Delta N/N) = \Delta \ln N$. (This can be verified by trying a few values on a hand calculator.) We can then rewrite Equation 7-A as

$$\Delta \ln N = -P\,\Delta t.$$

With the aid of calculus, this can be solved to give

$$\ln N = -P\,t + \text{constant, or}$$
$$\log N = -P\,(\log e)\,t + \text{constant.}$$

At the beginning of the experiment $t = 0$, so the constant must be $\log N_0$, where N_0 is the initial number of molecules. Rearranging the equation and substituting the value of $\log e$,

$$\log \frac{N}{N_0} = -0.434\,P\,t. \qquad (7\text{-}B)$$

EXAMPLE 7–1

Equations with logarithms, like Equation 7–B, can be put in exponential form:

$$\ln \frac{N}{N_0} = -P\,t \rightarrow \frac{N}{N_0} = e^{(-P\,t)}.$$

Show how analyzing the graph of the exponential equation can give a value

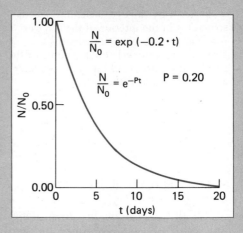

$$\frac{N}{N_0} = \exp(-0.2 \cdot t)$$
$$\frac{N}{N_0} = e^{-Pt} \qquad P = 0.20$$

If the experiment is done in cell culture the pulse must be followed by an added excess of unlabeled precursor, the "chase" (Figure 7–20).

If the isolated molecule is stable, the incorporated radioactivity will still be there days after the initial injection. But if it is unstable and broken down, radioactivity will be lost and the *specific radioactivity* (the amount of radioactivity per mass of isolated component) will decline. Since the animals or cells chosen are in a steady state, with no loss or gain of mass, there is no *net* decline in any of their components. Loss of radioactivity can therefore occur only through the replacement of radioactive components by nonradioactive ones. Such replacement implies a continual breakdown and synthesis of components. A protein, for example, that is synthesized from radioactive amino acids (and so is radioactive) is eventually hydrolyzed to its amino acids. The total

amount of the protein remains constant since the rates of breakdown and synthesis are approximately equal. During the "chase" phase, however, the protein is synthesized from nonradioactive amino acids which have diluted out the radioactivity in both the initial pulse and the labeled amino acids released by protein hydrolysis.

Membrane components also show turnover. What was surprising, however, was the discovery that the membrane does not break down as such; instead, its component parts are constantly being broken down and synthesized, each at a characteristic rate. Figure 7–21 indicates the results of experiments measuring the turnover of certain components of the liver endoplasmic reticulum membranes. It is clear that the rates of radioactivity loss, indicated by the slopes of the lines plotted on the graph, are different for each component: their *half-lives* (the time it takes for one-half of the original

for P, the probability per unit time of a molecule breaking down.

Solution

The diagram to the left shows a graph of N versus time, and the table gives for various values of t the values of N, where N_0 has been set to 1. Also tabulated are $\Delta N/N$, and $(\Delta N/N)/\Delta t$ for $\Delta t = 0.08$ days.

t	N	$-\Delta N$	$-\Delta N/N$	$-(\Delta N/N)/\Delta t$
1	0.819	0.0131	0.0160	0.200
2	0.670	0.0107	0.0160	0.200
3	0.549	0.00878	0.0160	0.200
5	0.368	0.00589	0.0160	0.200
8	0.202	0.00323	0.0160	0.200
12	0.091	0.00145	0.0160	0.200
15	0.050	0.00080	0.016	0.20
20	0.018	0.00029	0.016	0.20

By inspecting the table, one can see that as the value of N decreases, the value of ΔN (for the fixed Δt) also decreases, so the ratio $\Delta N/N$ remains constant. This is in agreement with Equation 7–A. This equation also allows one to determine the value of $P = -(\Delta N/N)/\Delta t = 0.20$ per day.

For a pulse-chase experiment of the sort described in the text, in which radioactively labeled molecules are broken down and resynthesized from nonradioactive precursors, N/N_0 represents the specific radioactivity left at time t. If Equation 7–B is graphed with the specific radioactivity (N/N_0) on a logarithmic scale and the time on a linear scale, as in Figure 7–21, the result is a straight line with a negative slope proportional to P, the probability of breakdown for a single molecule. The half-life for molecules of this type is the time when one-half the specific radioactivity remains (i.e., when $N/N_0 = 1/2$). From Equation 7–B, therefore,

$$t_{1/2} = -\frac{\log (1/2)}{0.434 \, P}$$

$$= \frac{\log 2}{0.434 \, P} = \frac{0.693}{P}.$$

The half-life is inversely proportional to P, but independent of the number of molecules originally present; it is always the same, regardless of sample size. In Figure 7–21 one can see that the half-life of the fatty acids in phospholipids is about six days. This means the probability per unit time of a fatty acid's leaving the membrane phospholipid is $P = 0.693/(6 \text{ days}) = 0.12$ per day or 12% per day.

Since the experimental data do in fact give a straight-line graph, we can adopt *random breakdown* as a *provisional model to explain how molecules participate in turnover.*

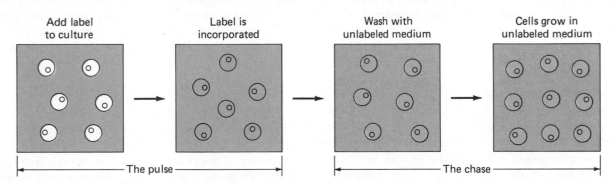

FIGURE 7–20
A pulse-chase labeling experiment used to determine turnover of cellular constituents. A radioactive compound used by the cells is added to a culture (the pulse); the compound is incorporated into various specific cellular molecules. The culture is then incubated in the presence of a large excess of unlabeled compound (the chase); the compound is incorporated into the same cellular molecules. The unlabeled molecules gradually replace the labeled ones, and thus the specific radioactivity (radioactivity/molecule) constantly diminishes. Measurements of this loss give values for the lifetime of these molecules in the cell; the quantity usually used is the "half-life," that time at which one-half of the radioactive molecules have been replaced by unlabeled ones.

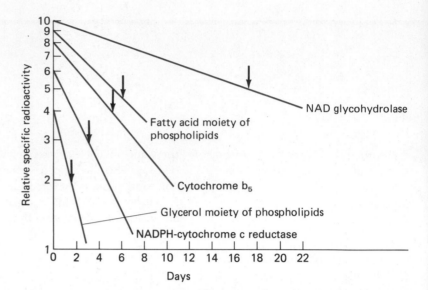

FIGURE 7–21
Turnover of constituents of liver endoplasmic reticulum membranes.   The logarithms of the specific radioactivities, relative to each other, are plotted against the days after terminating the radioactive pulse. The vertical arrows indicate the calculated half-lives. (Data from Y. Y. Omura, X. X. Bock, G. E. Palade, and P. Siekevitz.)

radioactivity to be lost, indicated by the vertical arrows) are different. Despite the continual replacement of its protein and lipid components, the *membrane* as a structure is stable.

In Figure 7–21 the horizontal time scale is linear, but the vertical scale, indicating the specific radioactivity, is logarithmic. (This is known as a *semilog graph*.) The fact that the plots for all the membrane components are straight lines indicates that each replacement process is indistinguishable from a random one (see box). Protein molecules in the membrane are not broken down according to age; each molecule has the same probability of being replaced, regardless if it is several days old or just synthesized. There is no experimentally demonstrable "old" or "new" protein in these membranes.

Although membrane proteins turn over by being broken down into their amino acid monomers, the phospholipids turn over by an exchange process. To demonstrate this, mitochondria with radioactive lipids (prepared by injecting a radioactive precursor into animals) were isolated and mixed in vitro with nonradioactive microsome membranes. After a suitable incubation period, the mitochondria and microsomes were reisolated. It was found that the mitochondria had lost, and the microsome membranes had gained, radioactive lipids. There was no net decrease or increase in lipids in either of the fractions, but there was an exchange of lipids between the two compartments. Subsequently, it was discovered that whole lipid molecules are exchanged continually between the various membranes of living cells, mediated by specific lipid-binding protein carriers (PLEPs—*phospholipid exchange proteins*).

Lipid-Protein Interactions Affect Membrane Function

The transport functions of the membrane are almost all mediated by proteins, as we shall discuss later. But do the membrane lipids play any role in transport or do they merely provide a two-dimensional fluid in which the transport proteins are dissolved? Part of the answer to this question was worked out by F. Fox and his collaborators. They studied the lactose permease system in *E. coli;* by using mutant bacteria, ones which could not synthesize fatty acids, they were able to control the cell membrane lipid composition. For example, they could grow cells in medium supplemented with linoleic acid (an 18-carbon fatty acid with two double bonds; melting temperature −5°C) or with oleic acid (an 18-carbon fatty acid with one double bond; melting temperature 13.4°C). They could then compare the temperature dependence of the lactose transport system in linoleic- and oleic-enriched cells. (See figure to right.)

As the temperature is lowered, the velocity of an enzyme-catalyzed reaction generally decreases. If one graphs the logarithm of the maximum velocity against the reciprocal of the absolute temperature (an Arrhenius plot), one usually sees a straight line with a negative slope. This is because chemical reactions generally have rate constants that follow the equation

$$\ln k = -\frac{E_a}{R}\frac{1}{T} + \text{constant}$$

where *ln k* is the natural logarithm of the rate (velocity) constant, E_a is the *activation energy* (see Chapter 5), R is the universal gas constant, and T is the absolute temperature. When the temperature range explored was large, however, Fox and his colleagues found that the Arrhenius plot of the transport rate for β-galactosides showed two straight-line sections, with a break or transition between them (Figure 7–22). The temperature at which the break occurred depended on the fatty acid present during cell growth. It can be seen in Figure 7–22 that the transition temperature was lower for the cells supplemented with linoleic acid (the fatty acid with the lower melting

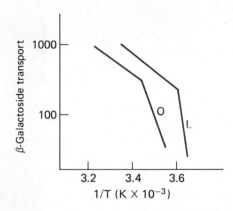

FIGURE 7–22
Arrhenius plot of transport.   The natural logarithm of the rate of lactose transport into *E. coli* is plotted against the reciprocal of the absolute temperature. The bacterial cells were grown to incorporate either linoleic acid (L) or oleic acid (O) in their membranes.

Linoleic acid   Oleic acid

temperature) than for the cells supplemented with oleic acid (the one with the higher melting temperature). Fox and his group studied a large series of different fatty acid supplements and found that the correlation between the melting temperature and the transition temperature is very strong indeed.

The transition is interpreted as a point at which the activation energy of the reaction abruptly rises as the temperature is lowered further. This almost surely happens because of a phase transition: a sort of freezing, characterized by a sudden rise in the viscosity of the lipid bilayer. These experimental results, therefore, strongly suggest that the lactose transport protein is structurally flexible and sensitive to the viscosity of the lipid that surrounds it. As lipid–lipid interactions change during the transition, the lipid–protein interactions change also.

When membrane proteins with enzymatic activity are isolated using a mild detergent, they usually remain associated with a certain amount of membrane lipid. If the solubilizing detergents are sufficiently strong to remove all the lipid, the protein's catalytic function is usually lost as well. This observation has led to the notion

of a *boundary layer of lipid* that surrounds membrane proteins and is only slowly exchangeable with the freely moving lipid outside the layer.

Components of the Cell Plasma Membrane Act as Recognition Markers

Evidence is accumulating that some of the outward-facing proteins of the plasma membrane are responsible for various recognition functions, including cell-to-cell adhesion to form tissues and organs. Many of these proteins are glycoproteins (cf. Chapter 4). The main body of the protein may be immersed in the membrane, while the small fraction of the protein containing the sugar residues is exposed on the outer surface (cf. Figure 7–14). In many cases it is the kind and amount of the sugar moieties on the outside of the cell that are responsible for this recognition. For example, glycoproteins and glycolipids form the structures of the blood group substances on erythrocytes, and it is the carbohydrate of these substances that is responsible for their specificities.

The sugar residues on glycoproteins in the environment also act as signals, for recognition by cells. There must be a recognition and binding of environmental proteins before they are taken into the cell by endocytosis (see further discussion later in this chapter). It has been found in some cases that modification of the sugar residues of an environmental glycoprotein destroys its ability to be taken in by endocytosing fibroblast cells. Albumin and lysozyme are not glycoproteins and normally are not taken up by rat liver cells. If appropriate oligosaccharide residues are covalently attached to these proteins, however, they are then taken up by the same cells.

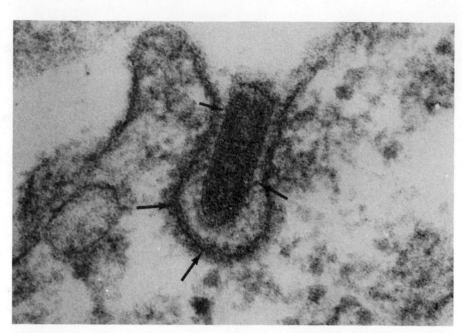

FIGURE 7–23

Early interaction between vesicular stomatitus virus and a mouse fibroblast cell.   The cell culture was sampled a few minutes after infection. The virus has spikes which are attached to the cell membrane (short arrows); the invaginated part of the cell membrane shows the characteristic bristly appearance of a coated pit (long arrows). × 200,000. (Courtesy of S. Dales.)

Another kind of recognition function is displayed in the infection of cells by viruses. The cell-virus interaction is rather specific, depending on the nature of the glycoproteins in the virus envelope and on the cell surface. (This does not hold for all viruses; in some cases, subviral particles or viral cores, lacking an envelope, can infect cells.) For example, a mucopolysaccharide covering of the plasma membrane is necessary for the attachment of influenza or polio viruses. In other cases (Figure 7–23) the virus has surface "spikes" that are probably glycoprotein in nature and used in attaching to the cell surface. When the cells are pretreated with the enzyme neuraminidase, they lose their capacity to adsorb virus, presumably because this enzyme attacks neuraminic acid residues on the cell surface glycoproteins that are responsible for the attachment. These cells become capable of infection upon further culturing, probably through the synthesis and insertion of new glycoprotein residues into the membrane.

The glycoproteins of the virus envelope are specified by the viral genome and synthesized by the infected cell; in some cases these proteins are inserted into certain sites in the plasma membrane of the host cells. In the process of extrusion of the virus particles from the cell, patches of cell membrane containing the viral glycoproteins are wrapped around the exiting virus particle, to become the new envelope of the virus.

Signaling molecules with polar structures, like polypeptide hormones, exert their effects on cells through interactions at the cell surface. Early evidence for plasma membrane interaction came from work with insulin. Using highly radioactive insulin and either intact cells or isolated membrane preparations, it was found that insulin binds with high affinity to receptors at the cell surface;

binding occurs even at concentrations as low as 10^{-10} M, lower than the normal concentration of circulating insulin. The conclusion that insulin binds to surface protein receptors also rests on experiments showing that brief exposures of sensitive cells to proteolytic enzymes, made even less permeating than usual by being bound to an insoluble polymer, can destroy the cell's insulin-binding properties. Finally, P. Cuatrecasas showed that when insulin is attached to an insoluble matrix (e.g., the polymer agarose) and cannot enter the cell, it still binds to the cell surface and exerts some of its typical hormonal effects.

The receptor is quite specific since modified forms of the hormone, insulin whose —S—S— bonds have been broken by reduction to —SH groups, are neither biologically active in vivo nor able to bind to isolated membranes in vitro. The receptor seems to be a uniformly distributed surface glycoprotein; some galactose residues are critical for recognition, since galactosidase destroys the hormone-binding capacity. Binding to the receptor protein does not alter insulin; it can be dissociated without losing its biological activity. From the specific radioactivity of labeled insulin, one can estimate that 10^4 molecules are bound per intact fat cell, indicating that the same number of receptor molecules lie in the surface membrane. The receptor is an integral membrane protein and can only be solubilized by detergents.

This detergent-solubilized receptor protein has been highly purified from fat cell membranes; it has a molecular weight of 300,000 and still retains the same binding properties towards very low concentrations of insulin as when it was part of the membrane. The difficulty of purification can be appreciated when it is estimated that only one part in 500,000 of the proteins in receptive cells is a receptor. That it can be purified at all is due to its very high and specific affinity for insulin. Cuatrecasas made use of this property by covalently linking insulin to insoluble polymers, such as agarose, and pouring detergent-solubilized cellular material through a column of this insulin-agarose. The receptor binds very tightly and specifically to these *affinity chromatography* columns and a purification of some 5,000-fold can be achieved in one passage through the column.

Our present knowledge of membrane structure can be summarized thus: protein and lipids are complexed together through mostly hydrophobic interactions to present hydrophilic faces on the inner and outer faces of the membrane, with hydrophobic parts buried in the interior. Protein can extend into and through the membrane. The various membranes of the cell are specific with regard to the proteins they contain; only certain enzymes are found in certain membranes and other enzymes are found in other membranes. Whether, as once believed, there are "structural" membrane proteins, which have no physiological function but to act as assemblers of other proteins and lipids, is not known at present. The final assembled membrane is fluid; its constituent molecules can exhibit lateral motion in the plane of the membrane, and some of the lipids and proteins can be removed without altering the visible structure. We will now relate these structural properties to the central function of membranes—controlling the flow of mass, energy, and information between compartments that are separated by membranes.

Movement of Materials Across Membranes

There are two aspects of the flow of material that we must consider: the thermodynamic driving force and the pathway taken by the material. These two ideas are of great utility; they can also be applied to flows of electrons, heat, and fluids. To make these concepts easier to grasp we shall treat them first in a general way, then apply them to cells.

Why Flows of Materials Occur

The physical basis for the movement of mass through space is the same as for the flow of mass through a metabolic pathway. Flow of material across a membrane occurs because of a negative free energy change. If the free energy of a substance on one side of a membrane is lower than on the other, there is a force that can drive a *flow to the region of lower free energy*.

In the formal vocabulary of physics, a change in energy with a change in distance is the definition of a *force*. We can therefore speak of a *concentration force* (or, more loosely, an *osmotic force*) that moves materials from

place to place within a cell and across the plasma membrane. In general we measure such forces as $-\Delta G/\Delta x$, where x measures distance in the direction of interest and ΔG is the change in free energy between the destination and the origin. (Since Δ always means destination *minus* origin, the negative sign is necessary to give a positive force in the direction of increasing x.)

The Nernst-Planck equation (Equation 6–1) can again be used to calculate ΔG from measurable quantities. Since the *standard* free energy of formation does not change between the destination and origin, $\Delta G^0 = 0$. Thus, for an uncharged molecule, A, moving across a cell boundary from outside to inside at constant pressure:

$$\Delta G = R\,T \ln \frac{[A]_{in}}{[A]_{out}}.$$

If A were an ion and had a net electrostatic charge, we would also need to consider the influence of any electrical potential across the cell membrane. By convention, the electrical potentials of cells, measured in volts, are always measured as $\Delta E = E_{inside} - E_{outside}$. For an ion B with net electrostatic charge z,

$$\Delta G = R\,T \ln \frac{[B]_{in}}{[B]_{out}} + z\,F\,\Delta E_m \qquad (7\text{-}2)$$

where F is a proportionality constant, the Faraday constant, equal to 96.5 kJ/volt · equivalent (23.1 kcal/volt · equivalent). (An *equivalent* of an ion is the amount that has one mole of charge.) The energy driving the movement of a mole of calcium ion with net charge $z = +2$ equivalents/mole into a cell with -0.1 volt electrical potential inside would be $2 \times 96.5 \times (-0.1) = -19.3$ kJ (or -4.6 kcal), provided the concentrations of Ca^{2+} were kept the same at the origin and the destination ($[Ca^{2+}]_{in}/[Ca^{2+}]_{out}$ would then equal 1, and the logarithm of 1 is zero).

EXAMPLE 7-2

Calculate the free energy change that results from moving a chloride ion from the surrounding fluid to the cytoplasm of a cell at 37°C. Assume $[Cl^-]_i = 61$ mM, $[Cl^-]_o = 130$ mM, and the electrical potential difference across the cell membrane is 20 mV, inside negative.

Solution

This process can be represented by the symbols

$$Cl_{out}^- \rightleftharpoons Cl_{in}^-.$$

Since the substance is the same inside or outside the cell, there is no standard free energy change and Equation 7–2 describes the process:

$$\Delta G = R\,T \ln \frac{Cl_{in}^-}{Cl_{out}^-} + z\,F\,\Delta E_m.$$

The Physical Basis for Fick's First Law

A discussion of the physical rationale for Fick's first law shows how the free energy change influences the flow of materials into and within cells. The driving force for mass flow is the decrease in free energy with distance. If this force is experienced by a substance at high concentration, there will naturally be a larger flow than if the concentration were lower. We expect, therefore, that the flux density will be proportional to both the driving force and the concentration:

$$J = - U C \frac{\Delta G}{\Delta x} \qquad (7\text{-}C)$$

where $-\Delta G/\Delta x$ is the osmotic force on the substance, C is the concentration of the substance, and U is a constant of proportionality known as the *mobility* of the substance. The mobility constant (usually determined by experiment) is a function of the size and shape of the substance, and of the viscosity of the medium through which movement occurs.

TABLE 7–4

A List of Substances in Decreasing Order of Their Diffusion Constant, $D^{\circ}_{20,w}$ (Measured in Water at 20°C, Extrapolated to Infinite Dilution)

Species	Molecular Weight	$D^{\circ}_{20,w}$ (cm²/sec)	Average Transit Time (milliseconds) for 1 μm Movement
NH$_3$	17	180×10^{-7}	0.28
KCl	74	160	0.31
Urea	60	109	0.46
H$_3$O$^+$	19	94	0.54
Acetic acid	60	90	0.56
Sucrose	342	29	1.7
RNAse	13700	12	4.2
Lysozyme	14000	10	4.8
Chymotrypsinogen	23200	9.5	5.3
Hemoglobin	68000	6.9	7.2
Tobacco mosaic virus (a short rod)	50×10^6	0.30	170
DNA (a long random coil)	6×10^6	0.13	390

Note: The transit time is the average time for each substance to diffuse a distance of 1 μm, about the length of an *E. coli* cell. The effect of size and shape on the diffusion constant and on the transit time can be seen. The student will find it instructive to calculate the average distance traveled in one millisecond, both in units of μm and in units of a molecular diameter.

The temperature T = (37 + 273) K; R = 8.314 × 10^{-3} kJ/(mole · K). The negative electrostatic charge on chloride ions means z = −1 equivalent/mole. The membrane potential must be expressed in volts: −0.050, negative since the destination state of ΔE_m is inside. The Faraday constant F = 96.5 kJ/(volt · equivalent).

$$\Delta G = (8.314 \times 10^{-3})(310) \ln \frac{61}{130} + (-1)(96.5)(-0.020) = -0.02$$

The forces leading to the movement of a chloride ion across the cell membrane are practically in equilibrium under these conditions.

Flow Measurement

To describe quantitatively the pathway taken by a material as it passes through a membrane, we must be able to measure precisely the flow. We can use familiar concepts here—a measure of mass and a measure of time. The flow, or *flux, of substance S* is measured as the number of moles (n) of S that enter the region of interest in unit time (one second). Clearly, in a flow stream the amount entering a region depends on the area of the entry channel—the larger the area, the larger the flow. To study those properties of flows that are independent of the channel area, one measures the flow divided by the area, the *flux density (J)*. If the flux is given by $\Delta n/\Delta t$, then the flux density J = $(\Delta n/\Delta t)/A$, where Δn is the increase in

Equation 7–C states that the greater the concentration of the substance, the greater the flow at constant force and mobility, and the greater the mobility of the substance, the greater the flow at constant force and concentration. We can use the Nernst-Planck equation to evaluate the osmotic force:

$$\frac{\Delta G}{\Delta x} = \frac{\Delta G^0}{\Delta x} + R T \frac{\Delta \ln C}{\Delta x}. \quad (7\text{-}D)$$

Since the standard free energy of formation of a substance does not depend on position x, $\Delta G^0/\Delta x = 0$. In the limit of very small ΔC,

$$\Delta \ln C = \frac{\Delta C}{C}. \quad (7\text{-}E)$$

Substituting Equation 7–E into Equation 7–D gives

$$\frac{\Delta G}{\Delta x} = R T \frac{\Delta C}{C \cdot \Delta x}$$

and substituting this expression into Equation 7–C for the flux density gives

$$J = - U R T \frac{\Delta C}{\Delta x}.$$

It is convenient to combine U, R, and T into a single *diffusion constant*, D, which takes account of the variation in diffusion both with the mobility of the substance and the temperature: $D = U R T$. We can therefore express the flux density as:

$$J = - D \frac{\Delta C}{\Delta x}.$$

The usual units of D are cm^2/sec; some values of D, for diffusion through pure water, are shown in Table 7–4.

EXAMPLE 7–3

Calculate the flux of sucrose moving across a 1-mm thick, freely permeable membrane of a 1-cm^2 area that separates a 1.0-M solution from a 0.1-M solution.

Solution

Use Fick's first law (Equation 7–3): $J = - D \Delta C/\Delta x$. From Table 7–4, one can get the diffusion constant of sucrose in water at 20°C, the best approximation available: $D = 29 \times 10^{-7}$ cm^2/sec. ΔC is given as $0.1 - 1.0 = -0.9$ M (the negative sign is the result of the destination being the lower concentration) and the membrane thickness is given as 0.1 cm. Therefore,

$$J = -(29 \times 10^{-7}) \frac{-0.9}{0.1}$$

$$= 2.6 \times 10^{-5} \frac{molar \cdot cm}{sec}.$$

The area is given as 1 cm^2 and the flux is the area times the flux density:

$$\Phi = 2.6 \times 10^{-5} \left(\frac{moles}{liter}\right) \left(\frac{cm^3}{sec}\right).$$

Since $cm^3 = 10^{-3}$ liter,

$$\Phi = 2.6 \times 10^{-8} \frac{moles}{sec}$$

$$= 2.6 \times 10^{-2} \frac{\mu moles}{sec}.$$

moles of substance S during the time interval Δt, and A is the area of the entry channel.

The flux density for a substance is proportional to the difference in its concentration, ΔC, between two regions separated by distance Δx:

$$J = - D \frac{\Delta C}{\Delta x}. \quad (7\text{-}3)$$

The constant of proportionality, D, is the *diffusion constant* (units are typically cm^2/sec; see box). This equation is known as *Fick's first law of diffusion* and was originally established as a summary of numerous experimental results.

At the molecular level, the physical reason why particles diffuse is the *constant bombardment* by adjacent molecules of fluid. Under physiological conditions these collisions occur at $\sim 10^{15}$ per second and their rate depends only on the temperature. The *effectiveness* of the collisions in producing motion depends on the viscosity of the surrounding fluid, and the size and shape of the diffusing particle. These properties are lumped together in the mobility constant.

The thermal motion of molecules is random in the sense that the probability of a particle's being hit on one side in an interval of time is the same as the probability of being hit on the opposite side. The mean number of collisions on opposite sides during a long interval is therefore equal. But this does not mean that the particle does not move. The statistics of this kind of movement, called a *random walk,* were worked out near the beginning of this century by A. Einstein. For movement along one dimension, x,

$$\overline{x^2} = 2\,D\,t, \qquad (7\text{-}4)$$

where $\overline{x^2}$ is the mean squared distance from the starting point at the time of interest, t is the time since starting, and D is the diffusion constant. We now can appreciate that the macroscopic diffusion of molecules is due to the statistics of microscopic random walking. In the absence of electrical effects, the probability of movement in the two directions is the same. But since there are more molecules of the substance in the region of higher concentration, more molecules move from the region of higher concentration to the region of lower concentration than in the reverse direction. The concentration dependence is embodied in the Nernst-Planck equation.

Equation 7–4 for the random walk can be used to give some insight into the physiological effectiveness of diffusion inside a cell. Assuming the values of D measured in water are appropriate to the cytoplasm, we can calculate the time it would take a small molecule to diffuse a typical cellular distance of one micrometer. This corresponds to the distance that a substrate molecule, for example, might have to travel to arrive in the vicinity of an enzyme. Table 7–4 lists several molecules and the transit times required to diffuse 1 μm. The table shows that enzymes with small substrates (molecular weight approximately 100) and turnover numbers of 10^3 substrates per second need no special mechanisms to guarantee adequate supplies of substrate molecules.

The Permeability Constant

We can now use the concept of diffusion to explore the flow of materials through membranes into cells. The general model for diffusion includes explicitly the distance x. To apply the models to cells means dealing with the troubling fact that we do not know exactly the thickness of the cell membrane. Moreover, there is next to each face of the membrane an *unstirred layer* of fluid which interferes with an accurate knowledge of the concentrations of substances in the immediate vicinity of the membrane. Despite these difficulties, it is possible to integrate Fick's first law—that is, to sum the contributions (using the methods of calculus) of the small changes in C in going from one side of the membrane to the other—and obtain the following equation:

$$J = -\frac{D}{X}\,\Delta C$$

where ΔC is the difference in concentration between the two solutions bathing the membrane and X is the *effective*

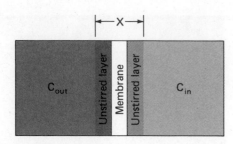

FIGURE 7–24
Measuring the permeability constant. In the equation $J = -\,P\,\Delta C$, the difference in concentration, $\Delta C = C_{in} - C_{out}$, refers to the concentrations of a diffusing substance (S), in the bulk solution bathing the membrane. These concentrations can be set by the experimenter or analyzed after sampling. The concentrations of S in the unstirred layers (drawn larger than scale) immediately adjacent to the membrane are not accessible for measurement. These unstirred layers are included in the effective thickness (X) of the membrane.

thickness of the membrane and adjacent unstirred layers (see Figure 7–24). Because the effective membrane thickness cannot be determined accurately, it is convenient to rewrite this equation introducing a new quantity, the *permeability constant,* $P = D/X$:

$$J = -\,P\,\Delta C.$$

The permeability constant (typical dimensions in cm/sec) measures the ability of a solute molecule to penetrate cells. It allows the comparison of different cell types, different solutes, and different mechanisms of transport.

To determine the permeability constant of substance *i,* it is necessary to measure both the difference in [i] between the inside and outside of the cell, and the *net flux density.* This is now done almost exclusively with radioactive tracer methods, which assess *unidirectional fluxes* (Figure 7–25). To measure the *exit flux density* (J_{out}), for example, a cell is first equilibrated with labeled compound *i* of known specific radioactivity, until the inside and outside concentrations are equal. Then the cell is rapidly washed with unlabeled substance *i*. At various times thereafter, the cells are filtered and the radioactivity inside is measured by a suitable detection device. The unidirectional *entry flux density* (J_{in}) can be measured in a very similar way. The cell is equilibrated with unlabeled *i,* filtered, and resuspended in radiolabeled *i*. The initial gain of radioactivity with time gives the desired data. The *net* entry flux density is simply $J_{in} - J_{out}$. It is then straightforward to calculate that

$$P = -\,\frac{J_{in} - J_{out}}{C_{in} - C_{out}}.$$

Movement of Materials Into Cells

A cell achieves its individuality by being separated from its environment by a permeability barrier. But to

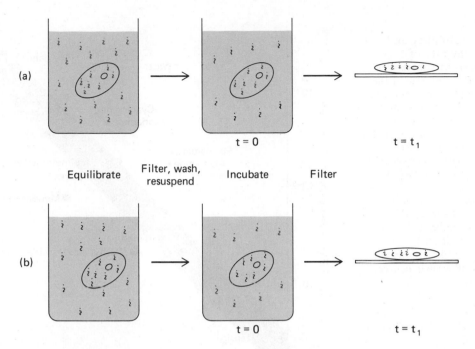

FIGURE **7–25**
Measurement of unidirectional fluxes. (a) Unidirectional efflux. Cells are first equilibrated in a solution of radiolabeled compound *i* at concentration 1. This determines the intracellular concentration for the experiment. After rapidly filtering and washing to remove radiolabeled *i* from the outside of the cells, they are resuspended in unlabeled *i* at concentration 2. Radiolabeled *i* moves out of the cells from concentration 1 to concentration 2. The radioactivity loss rate (determined by filtering the cells and counting the radioactivity in them), divided by the specific activity of *i* measures the efflux of *i*. Division by the cell surface area gives the outward flux density. (b) Unidirectional influx. Cells are equilibrated in a solution of unlabeled *i* at concentration 1. After rapidly filtering and washing the cells, they are resuspended in the presence of radiolabeled *i* at concentration 2. The radioactivity gain rate can be converted to an inward flux density by calculations analogous to those described above.

stay alive, a cell must communicate with its environment via the passage of mass, energy, and information through the permeability barrier. There are five different processes by which molecules can penetrate the cell membrane:

- *Simple diffusion* through the structure of the membrane by random molecular motion after dissolving in the membrane material.
- *Facilitated diffusion,* or catalyzed diffusion; the permeating molecule combines specifically with a transporter structure, which provides a pathway through the membrane. Simple diffusion and facilitated diffusion will allow net entry of molecules only if the ΔG for entry is negative; these two processes are, therefore, referred to as "passive."
- *Active transport,* in which the permeating molecule combines with a special membrane component, their interaction being *coupled to some energy-yielding chemical reaction or flow*. Net transport of the molecule can occur despite a positive ΔG for the molecule alone—that is, against gradients of concentration or electrical potential or both—when the coupled process has a combined $\Delta G < 0$.

- *Group translocation,* in which the cellular concentration of a permeating molecule is kept low at the destination by chemically transforming it to another species (usually at the expense of metabolic energy) or combining it with some binding agent, so that the molecule can continue to move passively down a favorable concentration gradient.
- *Bulk transport* (i.e., *endocytosis*), in which a small area of membrane invaginates and buds off within the cell, forming a vesicle that contains a sample of the extracellular fluid, its constituents, and any species bound to the surface of the membrane making up the vesicle.

To communicate effectively with its environment a cell must have pathways for movement into and out of the cell. With the possible exception of group translocation, the same kinds of processes that allow entry can be used for exit.

Simple Diffusion Through the Membrane

Using a variety of plant cells, and an experimental design similar to that depicted in Figure 7–2b, Overton

FIGURE 7–26

The effect of lipid solubility and molecular size on the rate of penetration of different substances into the alga *Chara*. The size of the points is roughly proportional to the size of the molecule. On the whole, small molecules (colored points) lie above the line, and large molecules lie on or below it, indicating the sieve-like nature of the membrane. The correlation between lipid solubility and permeability can easily be seen by comparing homologous molecules—for example, urea, methyl-urea, ethyl-urea, and dimethyl-urea. The membrane behaves like a solvent for nonpolar molecules such as ethyl-urea, and like a molecular sieve for polar molecules such as urea. (After R. Collander.)

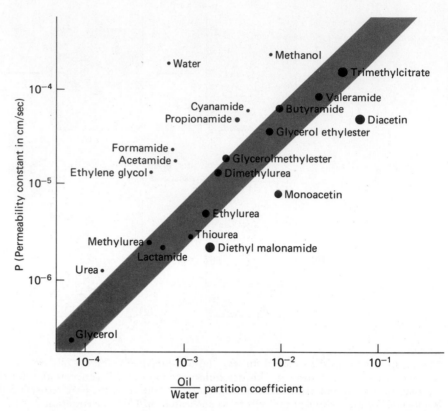

studied the permeability of a large number of substances (see Figure 7–26). He noticed that, in general, the rate of penetration of a substance was related to its lipid solubility, which he measured by determining the *partition coefficient* of that substance between olive oil and water. That is, after shaking the substance in an olive oil-water mixture, he determined the concentration of the substance in each phase:

$$\text{partition coefficient} = \frac{\text{concentration in oil}}{\text{concentration in water}}.$$

Overton concluded from the correlation between lipid solubility and cell permeability that the cell membrane was constructed of a thin film of lipid. To penetrate the membrane would require the *solubilization of the substance in the lipid, diffusion through the membrane, and solubilization again in the aqueous cytoplasm* (Figure 7–27). Although this generalization has been subjected to continued modification in the intervening years, the basic conclusions of Overton's theory, namely that the membrane is thin and that lipid is a major component of it, have held up remarkably well.

Overton's successors performed additional studies that confirmed his results, but they also discovered that Overton's lipid solubility rules seemed to break down for very small molecules such as water, methanol, formamide, and so forth (Figure 7–26). Such small molecules penetrate the cell much more rapidly than can be explained on the basis of their bulk solubility in lipid.

To explain these results cell physiologists first hypothesized that Overton's lipid membrane was interrupted by *small aqueous pores* that permitted the unexpectedly rapid penetration of small polar substances such as water or methanol. Measurements of water's permeability coefficient fall in the 10^{-5} to 10^{-2} cm/sec range, equivalent to 10^{-3} to 10^{-6} of the rate at which water

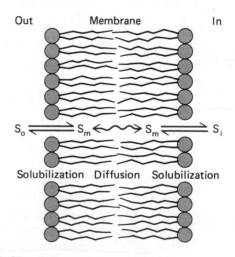

FIGURE 7–27

The solubilization-diffusion pathway. Substance S solubilizes within the membrane structure and diffuses (S_m) to the other side. S_o signifies substance S outside the cell; S_i signifies substance S inside the cell. The subscripts denote the solubilization state as well as the location of S.

diffuses through a water layer of thickness similar to the membrane. Thus, we can conclude that if special pores for water penetration exist in the cell membrane, they represent a very small fraction of the cell surface area. The most recent evidence suggests that any nonlipid paths ("pores") for small molecules (MW < 80) are probably associated with the proteins in the membrane.

The simple *solubilization-diffusion* mechanism may in any case be able to explain the high permeability of water: the reason why water penetrates the membrane so rapidly may be attributable to its very small size. A tiny molecule could wiggle through the spaces generated by the molecular motion of the lipid fatty acid chains. One would not expect many water molecules to be present in the lipid (the solubility is after all quite low), but not many have to be present at one time since the layer of lipid is quite thin.

The data in Figure 7–26 give some idea of how very isolated a cell is from its environment. Urea, for example, which is considered a readily permeant substance, penetrates 100 times more slowly than water, and some ions penetrate cells 1,000,000 times more slowly than urea. Nevertheless, we shall see that these rates of penetration, slow as they are, serve definite physiological functions in the life of the cell. We shall discuss the permeation of ions through membranes at greater length later; suffice it to say here that ions are among the most impermeant of substances. This is to be expected from the solubilization-diffusion model of transport, because of the large free energy required to solubilize a small charged ion in a lipid medium. The same interactions that keep amino acids with charged side chains on the surface of a soluble protein and those with hydrophobic side chains in the interior (see Chapter 4) retard the movement of ions through a membrane.

Facilitated Diffusion Through the Membrane

Consideration of simple diffusion leads to the conclusion that the cell is surrounded by a thin lipid membrane which has the dual properties of being a solvent for nonpolar molecules and a molecular sieve for polar molecules. Since the lipid is present in substantial excess, relative to the number of molecules that must pass through it under physiological conditions, simple diffusion should follow Fick's first law: The flux density J should be proportional to the concentration difference ΔC. In other words, a graph of J versus ΔC should be a straight line (of slope P). This is what is found for many small molecules, but not always for those of particular interest to cell biologists.

Consider, for instance, glycerol's rate of penetration into a plant cell (Figure 7–28). Glycerol, being a highly polar substance that forms hydrogen bonds with water,

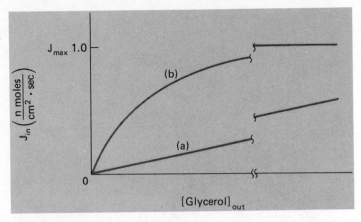

FIGURE **7–28**

Unidirectional entry flux of glycerol into the cell at various external concentrations of glycerol. (a) By simple diffusion. (b) By facilitated diffusion. The entry by facilitated diffusion is faster and, more important, it shows saturation at high external glycerol concentrations. At the maximum flux density (J_{max}), the entry rate is limited by some parameter other than the glycerol concentration, analogous to the V_{max} limit in enzyme-catalyzed chemical reactions.

has in plant cells a very low P of 2×10^{-7} cm/sec. In the human red blood cell, however, the P of glycerol is one hundred times greater, about 2×10^{-5} cm/sec. Since both membranes have similar basic structures—lipid bilayers—some special mechanism must be involved in the penetration of glycerol into the human erythrocyte. In such cases, when the substance penetrates with unusual rapidity, its permeability has properties that are familiar to us from our study of enzyme catalysis. That is, the permeation shows *substrate saturation kinetics, competitive inhibition* by structurally similar compounds, *inactivation* by trace concentrations of certain compounds, and *great pH dependence*.

Let us take these up in turn. Figure 7–28 shows the concentration dependence of the initial entry flux density of glycerol in two cells, measured by the isotopic tracer method. In case b, the entry rate becomes essentially *saturated* at high glycerol concentrations—that is, the rate responds very much less to a given increase in glycerol concentration. Similar observations led Michaelis, Menten, and Henri to postulate the existence of an enzyme-substrate complex (Chapter 5). We interpret such saturation of initial flux densities in a similar way (Figure 7–29): the transported molecule interacts with a specific transporter structure and the rate of movement is then determined by the properties of the transporter-substrate complex. When all the transporters are occupied, increasing the concentration of the substrate cannot increase the rate of the transport reaction.

The structural properties of the transporter can be modeled by assuming a membrane-spanning protein

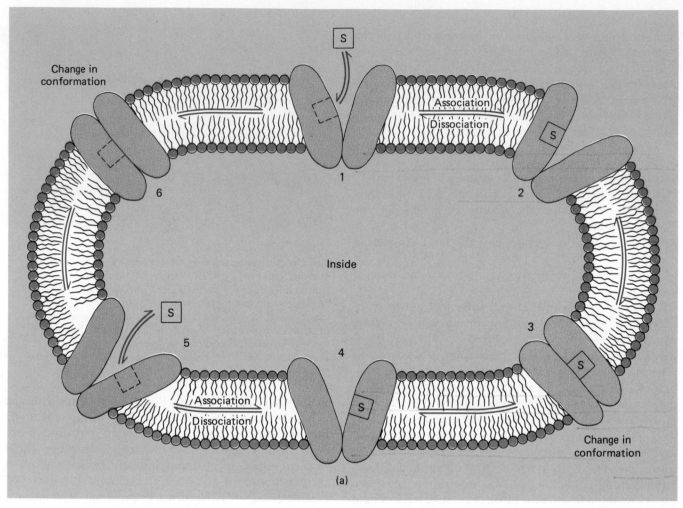

(a)

$$S_{out} + Transporter_{out} \xrightleftharpoons{1 \to 2} S \cdot T_{out}$$

$$S \cdot T_{out} \xrightleftharpoons{2 \to 4} S \cdot T_{in}$$

$$S \cdot T_{in} \xrightleftharpoons{4 \to 5} T_{in} + S_{in}$$

$$T_{in} \xrightleftharpoons{5 \to 1} T_{out}$$

$$S_{out} \rightleftharpoons S_{in}$$

(b)

FIGURE 7–29

Generalized scheme for transport of a substrate (S) across a membrane, via a specific transporter (T). (a) The scheme involves six states: (1) the empty transporter faces outward; (2) the outward facing transporter with bound substrate; (3) an occluded state in which the bound substrate is not accessible to either side; (4) the inward facing transporter with bound substrate; (5) the empty transporter faces inward; (6) an occluded state in which substrate does not have access to the transporter. Between states 1 and 2, and between 4 and 5, the substrate *associates or dissociates;* the binding equilibrium is determined by the concentration of S in the adjacent solution. Between states 2 and 4, and between 5 and 1, there is a *change in conformation* of the transporter—*translocation*—which results in changing the accessibility of its substrate binding site from one to the other adjacent solution. If the transporter cannot undergo the 5-to-1 translocation, *net transport* cannot occur; only *exchange* of S between the two adjacent solutions. In the absence of energy input, the direction of net transport will depend only on the difference of concentrations (more strictly, chemical activities) between the solutions separated by the membrane. Transport need not require energy; thermal motion may be sufficient to bring about transitions between adjacent states of the transporter. (b) Chemical reaction scheme for transport. The net result of all six steps in (a) is the transport of S from outside to inside the cell (sum of reactions, below line).

structure that undergoes changes in conformation. In its simplest form (Figure 7–29), such a transporter consists of (1) a specific combining site for the substrate; (2) a set of conformational changes which translocates the substrate across the membrane; and (3) the ability to reverse

the conformational changes without substrate being bound. Without the third step, only a catalyzed *exchange* of substrate could occur.

The mathematical modeling of the kinetics of this process gives equations which are the same as those we used for simple enzyme catalysis. Transport of a molecule implies *direction*, however, so the mathematical symbols are a bit more complex than for a simple chemical reaction (see Equation 5–8). For the *unidirectional* inward transport of substance S:

$$J_{in} = \frac{J_{imax}\,[S]_o}{K_{in} + [S]_o},$$

where J_{in} is the inward flux density, J_{imax} is the maximum inward flux density, and $[S]_o$ is the outside (external) concentration of S. K_{in} is analogous to the Michaelis constant. It characterizes the interaction on the outward face of the membrane between the substrate S and the transporter structure. One can measure K_{in} as that $[S]_o$ which gives $J_{in} = \frac{1}{2}J_{imax}$. There is a similar unidirectional outward or exit flux density, described by an analogous mathematical model. In the study of enzyme kinetics we also examine essentially unidirectional reactions because we ignore the backward reactions at early times when not much product has formed. That is why the same equations model unidirectional fluxes. Even if substance S normally is found on both sides of the membrane, it is easy to measure its unidirectional flux using small (tracer) amounts of radiolabeled S (Figure 7–25).

When the transporter behaves in the same way on both sides of the membrane—that is, when J_{max} and K have the same values for unidirectional entry and exit—and when the transporter is not saturated—that is, [S] << K—the equation for net flow reduces to Fick's first law:

$$J = \frac{J_{max}}{K}\,([S]_o - [S]_i) = -\,P\,\Delta C.$$

The *competitive inhibition* of glycerol penetration into human erythrocytes is another property typical of facilitated diffusion, and similar to the competitive inhibition of enzyme-catalyzed reactions. Thus, the presence of extracellular ethylene glycol markedly reduces the measured entry of glycerol, although ethylene glycol has no effect on the transport of sugars like glucose and galactose. This reduction of entry rate can be quantified in exactly the same way as we did in Chapter 5 for competitive inhibitors of enzyme catalysis. Like enzymes, the structures facilitating diffusion show specificity in the species with which they will combine.

$$\underset{\text{Glycerol}}{\overset{\displaystyle \overset{OH}{\underset{|}{}}\ \overset{OH}{\underset{|}{}}\ \overset{OH}{\underset{|}{}}}{CH_2\!-\!CH\!-\!CH_2}} \qquad \underset{\text{Ethylene glycol}}{\overset{\displaystyle \overset{OH}{\underset{|}{}}\ \overset{OH}{\underset{|}{}}}{CH_2\!-\!CH_2}}$$

When one adds Cu^{2+} ion to the extracellular solution, the facilitated diffusion of glycerol into the human erythrocyte is blocked (Figure 7–30). Cu^{2+} ions form complexes with imidazole groups like those on the amino acid histidine. The data suggest that a protein containing histidine, and therefore inhibitable by Cu^{2+}, may be very important in facilitating the entry of glycerol. When one measures the *pH dependence* of the glycerol entry (Figure 7–30), one sees a sharp transition in the permeability constant at about pH 6.2. The pH at the midpoint of the transition is close to the pK of the imidazole group on the side chain of the amino acid histidine. This seems to corroborate the hypothesis that a protein in which histidine occupies a key location plays a crucial role in the facilitated entry of glycerol, and that the catalytic function of this protein requires an unprotonated histidine side chain.

The phenomenon of catalyzed permeability adds a new element to our picture of membrane structure: the presence in membranes of specific catalytic entities made of protein. The catalytic structures can, like enzymes, interact specifically with given substrates. Also, like enzymes, they do not by themselves change the equilibrium of a reaction but merely increase the rate at which equilibrium is reached. By this we mean that the catalysis of permeability does not by itself allow the cell to

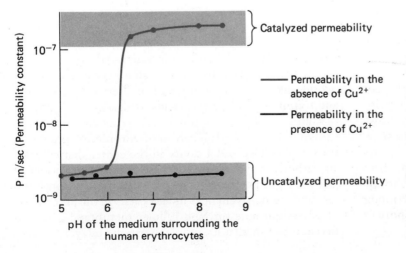

FIGURE 7–30

The effect of pH and Cu^{2+} on the catalyzed penetration of glycerol into human erythrocytes. The pH of blood plasma is normally 7.2–7.3. Lowering the pH to 6.0 decreases the glycerol permeability one hundred-fold. Adding the Cu^{2+} ion in very small quantities has the same effect. The response to changing pH resembles the titration curve for a weak acid, one with a pK_A similar to that of the imidazole group of histidine. Imidazole also complexes with Cu^{2+} ions. The data suggest that the unprotonated (uncharged) form of imidazole is essential for the catalysis of glycerol transport. (From a class experiment at Haverford College.)

L-Lactate (L)   L-Valine (LV)   D-Hydroxy-isovalerate (H)   D-Valine (DV)

(a)

(b)

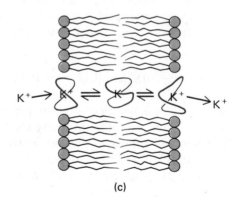

(c)

FIGURE **7–31**
Schematic model of valinomycin.
(a) The cyclic peptide antibiotic is made up of three tetramers of L–lactate (L), L–valine (LV), D–hydroxyisolvalerate (H), and D–valine (DV). (b) The coordination of the molecule with a K^+ ion. The ion is coordinated to six centrally directed carbonyl oxygens. The methyl and isopropyl side chains extend on the outer surface of the peptide, providing hydrophobic residues that allow the molecule to dissolve in the lipid environment of a cell membrane. (c) Transport mechanism. A potassium ion binds to valinomycin at one face of the membrane; the valinomycin · K^+ complex diffuses through the bilayer; the potassium ion dissociates from valinomycin at the other face.

accumulate or extrude molecules against concentration gradients. *Facilitated diffusion only allows more rapid net flow of material in the direction of decreasing free energy.*

Most transporters are probably proteins, with general structural features similar to those of bacteriorhodopsin and band 3 protein. General features, however, are not a sufficient guide to how specific catalysis of transport works. The structural details even of nonspecific porin are not yet known. On the other hand, there are examples of model compounds that give clues to the structural basis of specific facilitated transport. Among these are the *ionophores,* which mediate the transport of ions across cell membranes.

The ionophore for which we have the most detailed structural data is the antibiotic *valinomycin,* excreted by certain *Streptomyces* bacteria. Valinomycin is a cyclic peptide-like molecule composed of alternating pairs of D and L amino- and hydroxy-acids: [D-hydroxyisovalarate, D-valine, L-lactate, L-valine]₃. A schematic model is shown in Figure 7–31. There are hydrophobic groups on the outer surface and a cavity in the center that contains six carbonyl oxygen atoms. These oxygens are located at just the right positions to bind coordinately a potassium ion, and valinomycin does indeed catalyze the permeation of potassium ions through cellular membranes. Valinomycin is too small a structure to span the bilayer; instead, it acts as a *carrier,* picking up potassium ions on one side of the membrane and ferrying them across.

One imagines that the structure, the backbone of which resembles the seam on a tennis ball, opens like a set of false teeth to admit the potassium ion, then closes before moving through the lipid phase of the membrane. The exact structural basis of how a hydrophobic molecule can approach the surface of the membrane close enough to engulf or disgorge a hydrophilic ion is still a mystery. The model of a mobile carrier was prominent for many years among transport researchers, but it is more likely that transporters are nonmobile structures, embedded stably in the membrane but able to undergo conformational changes to bring about transport.

Examples of *channel* structures can also be found among the ionophore antibiotics. One of the best understood is *gramicidin A,* which is also a peptide antibiotic, synthesized by the bacterium *Bacillus brevis.* Two gramicidin molecules interact to form a narrow (0.2-nm radius) channel by head-to-head association. The fact that two molecules must come into contact to form the conducting channel means that catalysis of permeability should increase as the second power of the gramicidin concentration, and this is the case. For a carrier such as valinomycin, by contrast, one expects a first power concentration-dependence since each molecule can by itself catalyze potassium ion permeability. Gramicidin and other channel-forming ionophores show much less specificity than valinomycin or natural transporters.

The relatively low specificity is also conveyed by the term "channel," which suggests the image of a small tube through which various ions and water can move. This image is consistent with the high fluxes observed for the gramicidin channel, 10^7 cesium ions per second, a thousand times higher than valinomycin. Although a relatively low resistance pathway, the narrowness of the gramicidin opening means that two ions or water molecules cannot pass one another within the channel. Thus, *single-file transport* occurs rather than simple bulk flow. The *colicin E1* protein is another channel-forming antibiotic, secreted by certain strains of *E. coli.* Because *E. coli* has such well-developed genetics, the sophisticated panoply of genetic engineering techniques (see Chapter 10) can be brought to bear on the problem of structure-function relations in colicin E1.

Active Transport

Active transport must be involved when the free energy change calculated for the observed movement of a substance across a membrane is positive. The laws of thermodynamics are not disobeyed, of course; instead, the observed transport is *coupled* to another process whose ΔG is negative enough to allow the combination of the two processes to occur spontaneously. The thermodynamic arrangement is the same as for the coupling of energy-yielding reactions in metabolism to energy-requiring reactions. The fundamental problem of active transport is to identify the molecular details of the coupling systems, which are not nearly so well understood as they are in metabolic pathways. First, we consider some of the basic observations that point to the existence of active transport.

Table 7–5 gives the concentrations of various ions in the human erythrocyte and in the blood plasma surrounding it. It shows that the human erythrocyte, like most cells, is much higher in $[K^+]$ and much lower in $[Na^+]$ than the surrounding fluid. The dramatic differences in these ion concentrations can be eradicated, reversibly, by a number of treatments:

- On cooling to 2°C, erythrocytes release K^+ and pick up Na^+ until equilibration of the ions occurs. This process can be reversed if the cells are restored to 37°C.
- Treatment with certain poisons of energy metabolism, such as iodoacetate, also brings about an equilibration of K^+ and Na^+.
- When erythrocytes are stored for a while in plasma at 37°C, a time comes when K^+ begins to leak out of and Na^+ begins to leak into the cells. Upon addition of glucose, the cells will again resume the extrusion of Na^+ and accumulation of K^+.

TABLE 7–5
Ionic Concentrations in the Erythrocyte and the Surrounding Blood Plasma

	Concentrations in Milliequivalents per Liter			
	K^+	Na^+	Cl^-	Ca^{2+}
Erythrocyte	150	25	75	$\leq 10^{-4}$
Blood plasma	5	145	110	3

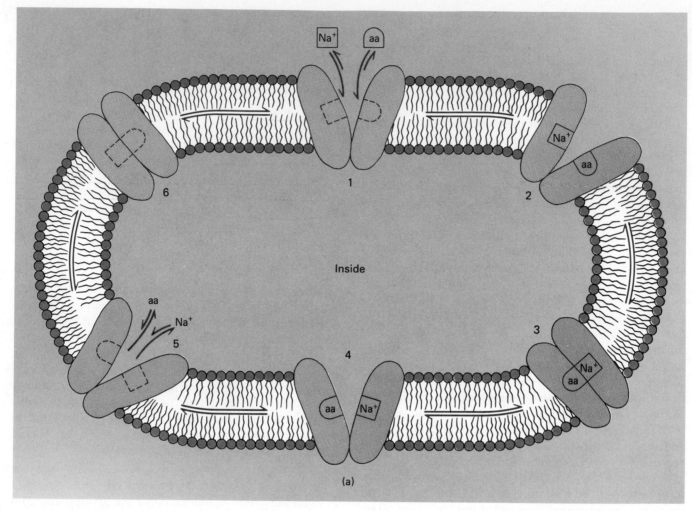

(a)

FIGURE 7–32

Cotransport of Na⁺ and an amino acid (aa) across a cell membrane. (a) Diagrammatic model. The six states are identical to the scheme depicted in Figure 7–29. The transporter differs, however, in having *two binding sites* and in being *unable to translocate if only one of its binding sites is occupied.* The flow of Na⁺ and the amino acid are thus coupled. Accumulation of the amino acid can occur against a concentration gradient, provided there is a negative free energy gradient for Na⁺. (b) Chemical reaction scheme for cotransport. The net result of all six steps in (a) is the linked transport of the Na⁺ and the amino acid (sum of reactions, below line).

$$Na^+_{out} + aa_{out} + Transporter_{out} \xrightleftharpoons{1 \to 2} Na^+ \cdot T \cdot aa_{out}$$

$$Na^+ \cdot T \cdot aa_{out} \xrightleftharpoons{2 \to 4} Na^+ \cdot T \cdot aa_{in}$$

$$Na^+ \cdot T \cdot aa_{in} \xrightleftharpoons{4 \to 5} Transporter_{in} + Na^+_{in} + aa_{in}$$

$$Transporter_{in} \xrightleftharpoons{5 \to 1} Transporter_{out}$$

$$\overline{\phantom{Na^+_{out} + aa_{out} \xrightleftharpoons{} Na^+_{in} + aa_{in}}}$$

$$Na^+_{out} + aa_{out} \xrightleftharpoons{} Na^+_{in} + aa_{in}$$

(b)

These and other experiments show that the erythrocyte accumulates K⁺ and extrudes Na⁺ by an active process requiring metabolic energy. This active transport balances permeability pathways that allow Na⁺ to enter and K⁺ to leave the cell spontaneously, following their free energy gradients.

Active Transport by Coupling of Flows

Some of the permeability pathways through which ions like Na⁺ and H⁺ flow across cell membranes along free energy gradients involve the <u>*cotransport*</u> of other molecules, such as amino acids and sugars. For example, the transport of some amino acids depends on the simultaneous transport of Na⁺. We can model the phenomenon with a mechanism like that depicted in Figure 7–32. <u>The transporter has *two* binding sites: one for Na⁺ and one for a specific amino acid.</u> The transporter can bind both species on either side of the membrane. Further, it can translocate across the membrane whether fully loaded or empty. In the simplest case, the cotransporter is *not* able to translocate if only one

The Lactose Permease of E. coli

A different example of facilitated sugar cotransport is found in the bacterium *E. coli*. As we shall learn in Chapter 14, when placed in a medium containing the sugar lactose these cells form the enzyme β-galactosidase, which catalyzes the hydrolysis of lactose to galactose and glucose. After synthesizing the enzyme, the bacteria can grow on the sugar lactose, using it as the sole carbon source for all cellular material. Some ingenious studies by G. Cohen, H. Rickenberg, and J. Monod showed the bacterial cell membrane contains catalysts, which they named "permeases," that play a specific role in mediating the penetration of certain compounds. As with many enzymes inside bacteria, the synthesis of these membrane permeases is stimulated by the presence of the penetrating substance. Subsequent work showed that lactose permeation in *E. coli* is catalyzed by one of these permeases.

In *E. coli,* the lactose is bound to a specific membrane protein (called the "M" or *lacY* protein) and cotransported with H^+. The energy for lactose transport is a *transmembrane proton gradient*. The mechanism is analogous to the Na^+-linked uptake of sugars and amino acids in the kidney and intestine. The H^+ gradient is maintained by a H^+ pump (see Chapter 8).

It is worth pausing here to stress the advantages of combining physiological studies of transport with genetics. This can now be done conveniently only in the transport systems of microorganisms. Favorite objects of study include the bacteria *Streptococcus faecalis* and *Escherichia coli,* and the baker's yeast *Saccharomyces cerevisiae*. The ease of obtaining mutations in these organisms and the armamentarium of techniques for recombining the mutations allow a study of structure-function relationships which has not yet been approached in higher cells. New ways of isolating and manipulating genes, using recombinant DNA methods, promise to extend such studies to other organisms. To exploit these opportunities requires that serious students of cell structure and function also study genetics.

of its binding sites is occupied. It is this last feature which provides coupling between the movement of Na^+ and the amino acid, much like a bisubstrate enzyme couples the flow of mass of both its reactants.

An influx of neutral amino acid can occur *against a concentration gradient,* provided the required energy exists in the free energy gradient of Na^+ influx. These two transports can be imagined to occur by separate pathways, one with $\Delta G_{AA} > 0$, the other with $\Delta G_{Na} < 0$. The transporter structure (Figure 7–32) models an obligatory coupling between the two flows, and the actual combined influx process therefore has a $\Delta G = \Delta G_{AA} + \Delta G_{Na}$. Since most cells have a negative electrostatic potential inside, there is both a chemical and electrical energy gradient favoring the influx of Na^+ and, hence, the influx of amino acid.

The energy gradient for Na^+ is maintained by the Na^+ pump, whose energy comes from ATP (see later discussion in this chapter). Given this fact, it is easy to see how placing amino acids in the medium surrounding a cell would lead to an increased hydrolysis of ATP. The result of adding amino acids is in effect to unmask an additional pathway for Na^+ entry. The entry of Na^+ is in turn counteracted by an increase in the rate of the Na^+ pump, at the expense of metabolic energy in the form of ATP.

The coupling of Na^+ influx with other flows is essential for the absorption of both sugars and amino acids in the intestinal epithelium and the kidney. In mitochondria, as we shall see in Chapter 8, the transport of many organic molecules is driven by a *H^+ gradient* set up by electron flow. (Cotransport flow in mitochondria is said to occur through a *symport*.) Proton gradients also drive some of the transport across the membranes of procaryotic cells.

Active secretion can be accomplished by coupling the efflux of the unwanted material with the influx of some other substance. For such *antitransport,* the model transporter must have different properties—namely, the equivalent of a single binding site which can be occupied by one substrate or the other, but *not* by both at the same time. Secretion will occur if, for example, on the outside of the membrane Na^+ is more likely to bind, since it is in relatively high concentration, and on the inside the

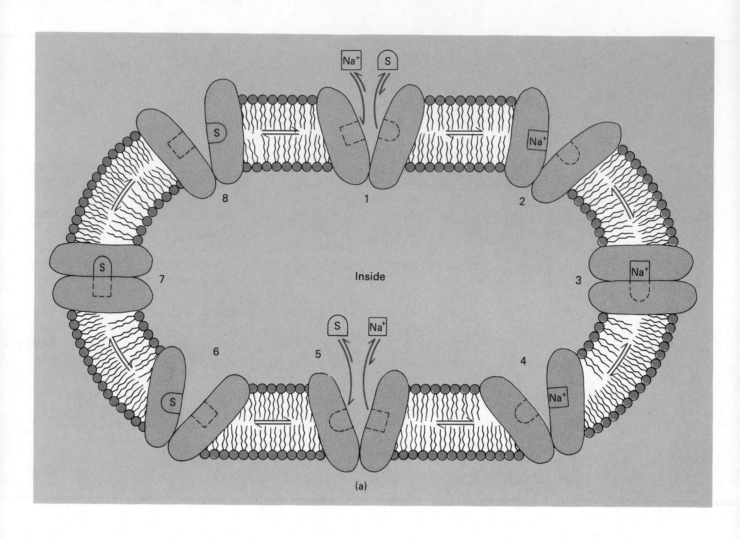

(a)

$$Na^+_{out} + Transporter_{out} \rightleftharpoons Na^+ \cdot T_{out}$$

$$Na^+ \cdot T_{out} \rightleftharpoons Na^+ \cdot T_{in}$$

$$Na^+ \cdot T_{in} \rightleftharpoons T_{in} + Na^+_{in}$$

$$S_{in} + T_{in} \rightleftharpoons S \cdot T_{in}$$

$$S \cdot T_{in} \rightleftharpoons S \cdot T_{out}$$

$$S \cdot T_{out} \rightleftharpoons T_{out} + S_{out}$$

$$\overline{\phantom{Na^+_{out} + S_{in} \rightleftharpoons Na^+_{in} + S_{out}}}$$

$$Na^+_{out} + S_{in} \rightleftharpoons Na^+_{in} + S_{out}$$

(b)

FIGURE 7–33

Antitransport of Na^+ and a compound S across a cell membrane. (a) Diagrammatic model. The various states are analogous to the schemes depicted in Figures 7–29 and 7–32. The transporter differs, however, in having *two binding sites* and in being able to *translocate only if one or the other (not both) site is occupied.* Opposite flows of Na^+ and S are thus coupled. Secretion of S can occur against a concentration gradient, provided there is a negative free energy gradient for Na^+. (b) Chemical reaction scheme for antitransport. The net result of the eight steps in (a) is the linkage of oppositely directed flows of Na^+ and S (sum of reactions, below line).

substrate to be secreted is likely to bind. The easiest way to imagine this for two structures so disparate as an ion and a small organic molecule would be to have two different but complementary sites, only one of which could be occupied at a time. As shown in Figure 7–33, the influx of Na^+ along its free energy gradient will drive the

secretion of substrate against a lower energy gradient. Coupling will occur to the extent that an empty transporter cannot translocate across the membrane. (In mitochondria, the Na^+/H^+ _antiport_ catalyzes the antitransport of Na^+, driven by the free energy gradient of H^+.)

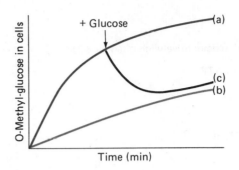

FIGURE 7–34
Countertransport.   (a) Cells were exposed to radiolabeled O-methyl-glucose (OMG) and the influx was measured. (b) The same measurement done in the presence of an excess of unlabeled glucose. (c) The experiment as in (a), except that the glucose was added at the time indicated. After glucose addition, radiolabeled OMG flowed out of the cells, against its concentration gradient. This occurs because the influx of OMG is sharply reduced by competition from the externally added glucose, while its efflux is initially unaffected.

A particularly simple example of this is *countertransport,* exhibited by molecules that cross the membrane along facilitated diffusion pathways. The basic observations are illustrated in Figure 7–34. Heart tissue slices were bathed in a solution containing radiolabeled O-methyl-glucose (OMG), a treatment that led to the increase in radioactivity in the cells shown by curve a. If glucose was also present in the bathing medium, however, the increase in radioactivity followed curve b. The close structural resemblance of glucose to OMG suggests that the reduction in the rate of radioactivity uptake with glucose present was caused by glucose acting as a competitive inhibitor of OMG transport.

In another experiment, glucose was added to the bathing medium containing radioactive OMG at the time indicated by the arrow (Figure 7–34). In the absence of glucose, the cells would have continued to accumulate radioactive OMG (curve A), indicating that the concentration of OMG was lower inside the cells than outside at the time of adding glucose. Yet upon addition of glucose, some OMG actually flowed *out of* the cells (curve c), against its concentration gradient! No external sources of energy are required to demonstrate this phenomenon. Since the efflux of OMG, which would be expected to be responsive only to the concentration of molecules inside, is enhanced by adding glucose to the outside, this phenomenon is sometimes described as *transstimulation.*

The added glucose competes with the influx more than with the efflux of radioactive OMG, because initially there is very little glucose inside the cells. Let us assume that the carriers move in both directions across the membrane at the same rate. Immediately after the addition of glucose outside the cells, the carriers are just as likely to transport radioactive OMG outward as they were before, but less likely to transport it inward because of the competition (see Appendix B). Of course, the carriers are more likely to move unlabeled glucose inward and unlikely to move it outward, since there is yet hardly any glucose inside the cells. Thus, the concentration gradient of the glucose drives the efflux of OMG. The ability to demonstrate countertransport is considered one of the strongest experimental tests of facilitated diffusion.

The Na$^+$ Pump

Experiments on red blood cells by E. J. Harris, starting in the early 1950s, showed that accumulation of K$^+$ and extrusion of Na$^+$ are linked. Perhaps the most convincing experiments were those by I. M. Glynn and R. L. Post, who studied red blood cells returned to 37°C after they had equilibrated their K$^+$ and Na$^+$ during storage at 2°C. They found that reducing the *external K$^+$* concentration also reduced the active *Na$^+$ efflux*. They also showed that the influx of K$^+$ against a concentration gradient would not occur in Na$^+$-free cells (obtained by equilibrating them with Na$^+$-free solutions at 2°C). Thus, external K$^+$ stimulates the Na$^+$ efflux and internal Na$^+$ stimulates the K$^+$ influx. The active pump process apparently couples the two movements of ions, but in opposite directions. The exact stoichiometry of the pump was worked out by these investigators and found to be three Na$^+$ extruded for every two K$^+$ accumulated. The fluxes depend on ion concentrations in a complex way, as might be expected, but the dependence is also consistent with the 3 Na$^+$: 2 K$^+$ stoichiometry.

When erythrocytes are gently hemolyzed in hypotonic solutions of defined pH, they lose their hemoglobin and gain an intracellular composition similar to that of the lysing solution. Under appropriate conditions, the membrane reseals itself ("reverse hemolysis") and the permeability barrier of the resulting red cell ghost is reestablished. By incorporating various energy sources in the lysing solution, it was shown that ATP is the only molecule capable of driving the active Na$^+$ pump.

In the red blood cell this ATP is normally produced by glycolysis. Glycolysis, in turn, cannot continue without a steady supply of ADP as substrate, and much of this ADP is produced by the hydrolysis of ATP that powers the pump. Thus, the production of lactate (the end product of fermentation in the erythrocyte) depends partly on the operation of the pump, and poisons of the pump, specifically *ouabain,* also inhibit lactate production. If the membrane is intact and the operation of the pump is limited by low intracellular [Na$^+$], movement of Na$^+$ into the cell is *tightly coupled* to glucose fermentation: Na$^+$ influx speeds up the pump, by raising the concentration of Na$^+$ (a substrate of the pump), and the pump

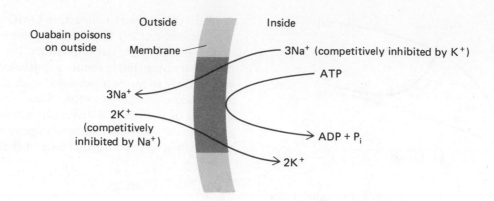

Outside   Inside

Ouabain poisons
on outside

Membrane

3Na⁺ (competitively inhibited by K⁺)

ATP

3Na⁺

2K⁺
(competitively
inhibited by Na⁺)

ADP + Pᵢ

2K⁺

FIGURE **7–35**

Stoichiometry and localization of active transport of Na⁺ and K⁺ across the membrane of the human erythrocyte.
The hydrolysis of one ATP phosphate bond provides energy for the linked transport of two K⁺ ions inward and three Na⁺ ions outward. The tinted region in the membrane symbolizes the Mg^{2+}-activated (Na⁺ + K⁺)-ATPase enzyme. This enzyme activity is poisoned by ouabain, which also inhibits active transport of Na⁺ and K⁺. (After A. K. Sen and R. L. Post.)

produces the common intermediate ADP, which is needed for fermentation. Thus, any perturbation to the cell that leads to altered inward movement of Na⁺ will tend to alter the cell's production of lactate.

A. K. Sen and Post demonstrated in a series of ingenious experiments that the extrusion of three Na⁺ and the accumulation of two K⁺ were linked to the hydrolysis of one ATP molecule (Figure 7–35). The enzymatic activity of the pump is called an *ATPase* because during normal operation it hydrolyzes ATP to ADP and inorganic phosphate. One can study this activity on membrane fragments, where pumping cannot be observed because no permeability barrier exists, and so it is common, if somewhat misleading, to speak of the ATPase as though it were separate from the pump. By loading red blood cell ghosts with various solutions, it was found that intracellular Mg^{2+} is required for optimum operation of the pump; the isolated ATPase is also activated by Mg^{2+}. Many experiments convince us that the Mg^{2+}-activated (Na⁺ + K⁺)-ATPase is indeed the source of the active transport.

An important feature that had to be demonstrated to link the ATPase activity with the pump is *asymmetry*. A chemical reaction that is symmetrical in space cannot give rise to the preferential movement of mass in a particular direction. The membrane location of the ATPase provided the possibility of asymmetry, but its actual existence had to be experimentally proven. Originally, functional tests showed that the pump is driven by ATP and activated by Mg^{2+} on the inside, but not the outside; activated by Na⁺ on the inside; and activated by K⁺ on the outside (Figure 7–35). More recently, impermeable membrane probes have confirmed that the Na⁺ pump structure is asymmetric.

The (Na⁺ + K⁺)-stimulated ATPase activity, which was originally discovered by J. C. Skou in crab nerve, occurs in all mammalian tissues, especially the brain and

kidney, and in many other organisms. Extremely active pumping tissues such as the electric organ of the electric eel, the rectal gland of the dogfish, and the salt gland of the herring gull are especially rich sources of this enzyme system.

Quite a bit is known about the biochemistry of the ATPase. It is usually composed of two major polypeptide chains: a species-specific glycoprotein, called β, of around 50,000 molecular weight and a larger protein, called α, of 110,000 molecular weight which can accept the terminal phosphate of ATP on an aspartic acid side chain to form a phosphoprotein. The phosphate transfer reaction is stimulated by Na⁺; the later dephosphorylation (removal of the phosphate from the protein) is stimulated by K⁺. This enzyme activity is inhibited by low concentrations of *ouabain* (its I_{50}, the concentration that gives 50% inhibition, is only 1 μM). The enzyme also requires some lipids to be active, consistent with its location in the cell membrane. An αβ dimer is functional but there may be two of each polypeptide subunit joined to form the native enzyme.

If the hydrolysis of ATP is coupled to the transport of Na⁺ and K⁺, then forcing the flows of ions to reverse should result in the net synthesis of ATP from ADP and phosphate. This experimental proof of coupling was for a long time difficult to achieve because of the action of other enzymes on ATP. In 1970, V. L. Lew, following earlier work by Glynn, loaded guinea pig erythrocytes with high K⁺ and low Na⁺ concentrations (by equilibrating them in a glucose-free medium containing iodoacetamide, a poison of the glycolytic pathway), ADP, and P_i. After washing the cells thoroughly they were suspended in a high-Na⁺, K⁺-free medium also containing iodoacetamide, and incubated for 30 minutes. Lew was able to show a net increase in the ATP inside the cells, an increase blocked by adding ouabain to the resuspension medium (Figure 7–36). This experiment shows the pump is *fully*

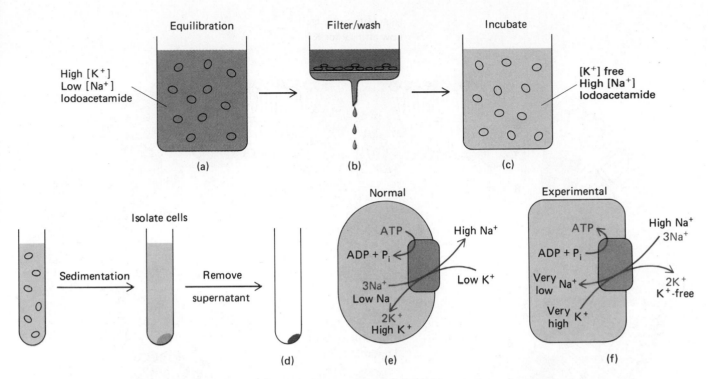

FIGURE 7–36

Experiment showing the reversibility of the Na⁺ pump and the synthesis of ATP.   Cells were equilibrated in a medium containing high $[K^+]$, low $[Na^+]$, and the glycolytic poison iodoacetamide (a). They were then washed with cold buffer (b), then resuspended and incubated in a high $[Na^+]$, $[K^+]$-free medium containing iodoacetamide (c). After collecting the cells by sedimentation and removing the supernatant (d), the cells were analyzed for their content of ATP. The normal and experimental conditions are diagrammed in (e) and (f).

reversible and that forcing Na⁺ influx and K⁺ efflux, along free energy gradients, can generate ATP.

In the case of Na⁺ and K⁺ transport we must assume a more complex carrier system than the ones depicted earlier. First, the carrier must translocate Na⁺ specifically out and K⁺ specifically in; that is, it must have affinities for the two ions that are different on the two sides of the membrane. The feature of asymmetric affinities for a substrate is a fundamental property of all active transport models. Second, the carrier must be coupled to an energy-releasing mechanism to allow it to move these ions against free energy gradients. The mechanism shown in Figure 7–37 is based on the ideas of R. W. Albers, A. Judah, R. L. Post, and T. Shaw. Na⁺ combines with high-affinity sites on the inward-facing α subunit of the pump. ATP phosphorylates the α subunit to yield a phosphoenzyme (sometimes called $E_1 \sim P$). A conformational change yields an occluded form in which bound Na⁺ ions cannot exchange easily with Na⁺ ions on either side of the membrane. Na⁺ ions are released from the α subunit to the outside. K⁺ combines with high-affinity sites on the outward-facing phosphorylated α subunit (also known as E_2-P), which stimulates the hydrolysis of the phosphate from the protein. The α

subunit then undergoes another conformational change to yield an occluded form, followed by release of the K⁺ on the inside. This completes the cycle. ATP is necessary to phosphorylate the α subunit; the breakdown of the ATP is what is observed as an ATPase.

The lower part of Figure 7–37b shows a formal kinetic scheme for the transport mechanism. This is, of course, not the only model that could be made. A specific alternative is one in which there is simultaneous, instead of sequential, movement of the two ions. In the near future we can expect more of the details to be determined, such as the nature and extent of the conformational changes and how they account for the observed changes in the pump's affinity for the two ions.

The presence of passive movements of Na⁺ and K⁺ can be shown in various ways, such as poisoning the active transport with ouabain or slowing it down by cooling to 2°C. Although the values obtained vary somewhat depending on the methods used and the membranes studied, it seems clear that the permeability to Na⁺ ($P \approx 10^{-9}$ cm/sec) is lower than that to K⁺ ($P \approx 10^{-7}$ cm/sec, about 10^{-5} the value in free solution). The difference between Na⁺ and K⁺ might be explained by the fact that Na⁺ is more hydrated and thus is effectively

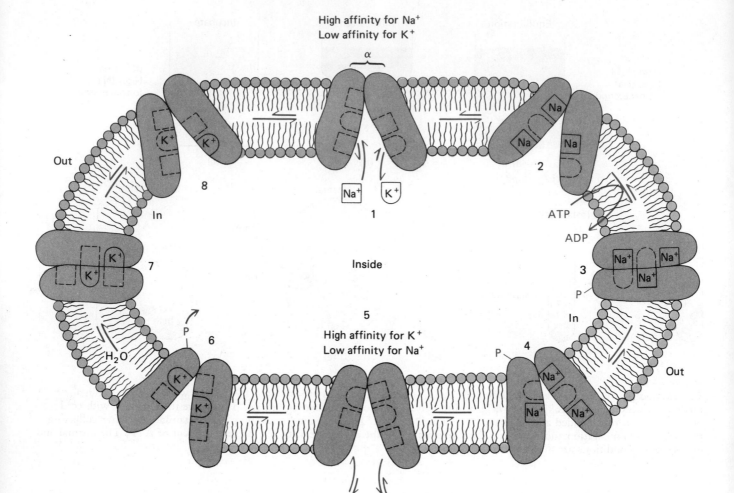

(a)

FIGURE 7–37

The Na⁺ pump.   (a) The scheme shown involves eight states: (1) the empty α subunit faces inward; (2) the inward-facing α subunit has three Na⁺ bound; (3) an ATP-linked phosphorylation of the α subunit is accompanied by a conformational change to an occluded state, in which the bound ions are not accessible to either bathing solution; (4) the phosphorylated α subunit faces outward with three Na⁺ bound; (5) the empty phosphorylated α subunit faces outward; (6) the phosphorylated outward-facing α subunit has two K⁺ bound; (7) the dephosphorylation of the α subunit is coupled with a conformational change to an occluded state, in which the bound ions are not accessible to either bathing solution; (8) the inward-facing α subunit has two K⁺ bound. The α subunit in states 4, 5, and 6 (phosphorylated) has a high affinity for K⁺ and a low affinity for Na⁺; in states 8, 1, and 2 (dephosphorylated), it has a high affinity for Na⁺ and a low affinity for K⁺. Coupling exists because translocation of the α subunit requires phosphorylation or dephosphorylation and occurs only if there are three Na⁺ or two K⁺ bound. (b) The equivalent chemical scheme shows how the transitions between states result in the ATP-linked transport of three Na⁺ outward and two K⁺ inward.

$$3Na_{in}^+ + \alpha_{in} \rightleftharpoons Na_3^+ \cdot \alpha_{in}$$

$$Na_3^+ \cdot \alpha_{in} + ATP \rightleftharpoons Na_3^+ \cdot \alpha\text{-}P_{occluded} + ADP$$

$$Na_3^+ \cdot \alpha\text{-}P_{occ} \rightleftharpoons Na_3^+ \cdot \alpha\text{-}P_{out}$$

$$Na_3^+ \cdot \alpha\text{-}P_{out} \rightleftharpoons \alpha\text{-}P_{out} + 3Na_{out}^+$$

$$2K_{out}^+ + \alpha\text{-}P_{out} \rightleftharpoons K_2^+ \cdot \alpha\text{-}P_{out}$$

$$K_2^+ \cdot \alpha\text{-}P_{out} + H_2O \rightleftharpoons K_2^+ \cdot \alpha_{occ} + P$$

$$K_2^+ \cdot \alpha_{occ} \rightleftharpoons K_2^+ \cdot \alpha_{in}$$

$$K_2^+ \cdot \alpha_{in} \rightleftharpoons \alpha_{in} + 2K_{in}^+$$

$$3Na_{in}^+ + ATP + H_2O + 2K_{out}^+ \rightleftharpoons 3Na_{out}^+ + 2K_{in}^+ + ADP + P$$

(b)

a larger ion (diameter of Na^+ = 0.51 nm, and of K^+ = 0.40 nm). Indeed, the permeation of Li^+ (the most hydrated of the alkali metal ions) is slower than Na^+, and Rb^+ (the least hydrated) is faster than K^+.

This very low permeability to cations is physiologically important for the erythrocyte: a great deal of energy is saved that the cell would otherwise have to spend retaining the ions it has accumulated or to keep out the ones it has extruded. One can estimate the number of pumps on an erythrocyte membrane by binding [^{3}H]-ouabain to the cells (assuming very tight binding, so all the pumps are labeled, and very high specificity of binding, so nothing else is labeled). These experiments suggest there are approximately 250 Na^+ pump sites per human erythrocyte, a density of the order of 1–2/μm^2. (In many other cell membranes the density of pump sites is of the order of $10^3/\mu m^2$.) Quantitative flow measurements with tracers have estimated a maximum rate of 100 K^+ ions per second per pump site. Thus, it can be calculated that if K^+ had a permeability constant like that of urea (8 × 10^{-5} cm/sec), the work necessary to maintain the K^+ inside the cell would be 7.5 × 10^6 J/kg · hr, whereas the work actually necessary to maintain K^+ with its own permeability constant of 1 × 10^{-9} cm/sec is only 55 J/kg · hr. From the glucose consumption of the human erythrocyte, assuming 100% efficiency in fermentation, one can calculate that the maximum amount of work the cell can do is equivalent to 170 J/kg · hr. The passive permeability to ions is just low enough to make it energetically feasible for the human erythrocyte to maintain constant its cellular concentrations of cations.

It has been estimated that approximately 30% of the ATP utilized in an adult human goes into the operation of the Na^+ pump. A reasonable question to ask, therefore, is why a less energetically expensive way of controlling [Na^+] and [K^+] has not evolved—for example, the cellular equivalent of a brick wall. Some understanding of this problem can be gained by considering certain problems the cell faces and how the expenditure of cellular energy though the pump can solve them. The details of controlling the pumping rate are still not entirely clear but the effect of this control is to maintain constant the intracellular ion concentrations. Many physiological experiments suggest that the normal operation of the cell requires such constant intracellular concentrations. That is why cell physiologists devoted to the Na^+ pump often refer to passive flows of these ions across the membrane as "leaks." When Na^+ enters or K^+ exits, the pump must operate to counteract these flows.

The cell is an open system and must exchange mass and energy with its environment in order to live. The concept of a leak, therefore, is relative. Many passive pathways for ion movement are important to cellular functions even though their operation perturbs the normal Na^+ and K^+ concentrations. In fact, the exchange of certain molecules between the cell and its environment is directly linked to the flow of Na^+ ions down their energy gradient. This observation rationalizes both the pumping of sodium ions to maintain their low intracellular concentration and the flow of Na^+ into the cell.

One universally important function in animal cells is the maintenance of *osmotic balance*. A cell in a steady state has equal influxes and effluxes for all species, including water. If the pump were blocked, there would be a tendency for unbalanced fluxes of cations. This is because the membranes of most cells have different permeability pathways for K^+ and Na^+, and because the forces on the ions that tend to make them move also differ. Generally, the difference in free energy for movement of an ion involves not only its concentration but, because it has an electrostatic charge, also the electrical potential across the cell membrane. When the Na^+ pump of the sheep red blood cell is inhibited by ouabain, Na^+ and K^+ continue to move along their free energy gradients and the net intracellular cation concentration rises. The volume of the cell increases at a proportional rate because water enters to maintain osmotic balance. The net inward flow of cations is the result of several physiologically important processes like the cotransport and antitransport processes described earlier. In the physiological steady state, the combined effect of the pump and the "leaks" keeps the ion concentrations constant, and the osmotic forces remain balanced.

As a summary of active transport, all mechanisms require a transporter structure that can specifically bind the substrate, the ability of the transporter to translocate, and energy inputs to change the behavior of the transporter from high affinity for the substrate on one side of the membrane to low affinity on the other. We expect that all cellular transporters will be found to involve proteins, for proteins are the only molecules which combine specificity of structure (necessary for a binding site) with plasticity of structure (required for conformational changes that can lead to affinity changes and translocation). The energy input for active transport can involve ATP, ion gradients, proton gradients (discussed in Chapter 8), or the transport of electrons. Cells use all these strategies, and may use more than one, to transport the same substance by different systems under different conditions.

Anion Exchange Pathways

The low permeability of the cell membrane to Na^+ and K^+ does not extend to small anions. In the red blood cell the half-life for exchange of an interior for an exterior Cl^- is only 0.25 sec, some million times faster than for a cation. In fact, in many cells the small anions appear to be

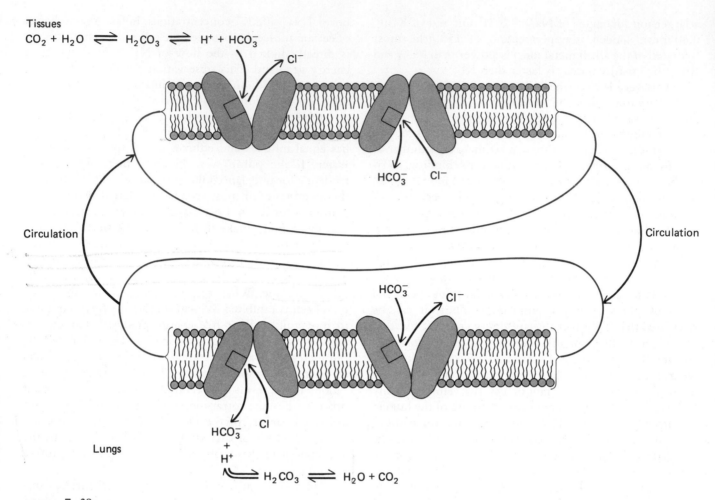

FIGURE 7–38

The HCO_3^-/Cl^- exchange pathway found in mammalian red blood cells.   The band 3 protein catalyzes this exchange. In the lungs HCO_3^- inside the cell is exchanged for Cl^- outside. The HCO_3^- becomes protonated and the resulting H_2CO_3 is dehydrated to form CO_2, which is exhaled. In other tissues Cl^- inside is exchanged for HCO_3^- outside, which arises from hydration of tissue CO_2 to form H_2CO_3 (catalyzed by carbonic anhydrase) and dissociation of a proton. A Cl^- outside can exchange for a Cl^- inside by two successive exchanges for HCO_3^-.

in equilibrium across the membrane. This means that net transport processes must be rapid enough to allow Cl^-, for example, to respond quickly to changes in concentration or electrical potential. It turns out in many red blood cells that Cl^- exchange is much more rapid than net Cl^- transport, perhaps by as much as a factor of 10^5. Anion exchange occurs by a facilitated diffusion pathway in which the protein transporter very seldom translocates when empty. This is the same property observed for antitransport proteins, but the relative lack of specificity of the anion exchange pathway—many anions can use it—leads us to discuss it separately.

What mostly happens in the red blood cell is a rapid exchange of Cl^- for HCO_3^-. This is of great physiological significance to animals since it allows rapid uptake in the tissues, or efflux in the lungs, of CO_2 (which combines with water to give H_2CO_3, then dissociates to give H^+ and HCO_3^-). Since the HCO_3^-/Cl^- exchange can occur in either direction, two such events will result in inside-outside Cl^- exchange (Figure 7–38).

The study of anion permeability in the red blood cell has been aided by the availability of potent inhibitors of the process. *Phloretin,* or the more commonly used analogue *phlorizin,* is an example of a reversible inhibitor; it is effective only if applied to the outside of the membrane. This asymmetry of action is exactly what you would expect from an asymmetric membrane protein. Certain chemical reagents that can attach by covalent bonds to structures in the membrane can also block anion movement. Reagents that attack amino groups typically produce large inhibitions of Cl^- and SO_4^{2-} exchange. One of the most potent of these is known as DIDS (4,4′-diisothiocyanostilbene-2,2′-disulfonic acid), which can form covalent linkages with membrane structures and which also is available in tritium-labeled form. There is a linear correspondence between DIDS molecules bound and transport inhibited, which allows one to titrate the number of sensitive sites and thus count them. Using this radioactive affinity label, it has been estimated that there are about 3×10^5 anion exchange sites per human red blood cell. Knowing the number of sites allows one to calculate, from the measured anion exchange rates, the exchange rate per site: approximately 7×10^5 ions per site per second. This turnover number is quite large, almost as high as the fastest enzymes known. It is much larger than both the rate of the $(Na^+ + K^+)$-ATPase (100 ions per site per second) and the rate of the mobile carrier of K^+ ions, valinomycin (80 ions per site per second).

But the biggest payoff for the use of radiolabeled DIDS was identification of the structure to which it binds, which is presumably the structure that catalyzes anion exchange. After covalently attaching tritiated DIDS, membrane proteins were solubilized in SDS and separated from one another by electrophoresis on polyacrylamide gels. The radioactivity traveled primarily in a single peak, corresponding to a molecular mass of 95,000 Da, at the position of the previously identified *band 3 protein*. As discussed earlier, this protein spans the lipid bilayer, consistent with its involvement in a transport function. Several irreversible inhibitors of anion exchange, like DIDS, have been found to attach to one or another residue of the band 3 protein, and the extent of attachment by each is proportional to the extent of inhibition of the anion exchange. Finally, purified preparations of this protein have been inserted into lipid bilayers and the anion exchange function has thereby been reconstituted. It thus seems very likely that the molecule responsible for this transport function has been identified. Even though the amino acid sequence of band 3 protein is known, our knowledge of its structure is still insufficient to account for the anion transport activity. Equally mysterious is how band 3 protein might form pores for the movement of water through the human red blood cell membrane.

Group Translocation

The only formal example of group translocation is the *phosphotransferase system* found in some bacteria, which can transport various sugars, discovered by S. Roseman and his collaborators. The reactions that modify the sugar are as follows:

$$\text{Phosphoenolpyruvate} + \text{HPr} \xrightarrow{\text{EnzI}} \qquad (7\text{-}5)$$
$$\text{(PEP)} \qquad \text{(heat-stable protein)}$$
$$\text{pyruvate} + \text{phospho-HPr}$$

$$\text{Phospho-HPr} + \text{sugar} \xrightarrow{\text{EnzIII}} \qquad (7\text{-}6)$$
$$\text{sugar-phosphate} + \text{HPr}$$

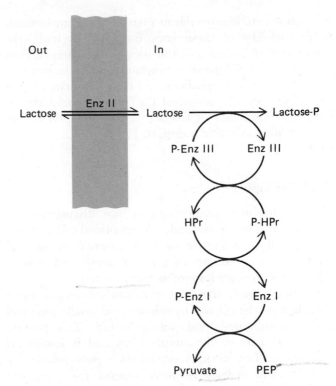

FIGURE **7–39**

Lactose transport in *Staphylococcus aureus* by the phosphotransferase system. Phosphoenolpyruvate (PEP) acts as the source of energy and phosphate for a linked series of protein phosphorylations. Finally, lactose is phosphorylated to lactose-phosphate. The value of ΔG for this step is so negative that the intracellular lactose concentration is kept low. Lactose can therefore enter the cell passively via a specific transporter called EnzII.

The two enzymes (EnzI and EnzIII) have been purified and shown to catalyze reactions 5 and 6, respectively. HPr and EnzI are soluble proteins, while EnzII is an integral membrane protein. Figure 7–39 shows how this system could operate in *S. aureus* to transport lactose. The sugar is phosphorylated via a series of phosphorylated proteins and the energy for the process (as well as the phosphate group) comes from the "high energy" phosphate in PEP. The phosphorylated sugar is trapped inside the cell; the charged phosphate group makes the sugar unable to permeate the membrane. Since the phosphorylation reactions proceed with large negative free energy changes, the concentration of free intracellular lactose is kept low and lactose spontaneously travels through the membrane, presumably via the specific EnzII transporter.

Although group translocation of the PEP-linked type is relatively rare, there are other examples of apparent active transport which have an analogous explanation. One instance is the accumulation of calcium ions in cellular compartments that already contain large amounts of Ca^{2+}. This is sometimes due to the binding of calcium ions to *impermeable sites* within the compartment. If the affinity of these sites for Ca^{2+} is high, the concentration of free calcium ions in the compartment will be low and further influx can occur spontaneously along a free energy gradient. If a clear distinction is not made between the *amount* of Ca^{2+} in the compartment and the *concentration* of free Ca^{2+}, this situation can be misidentified as active transport.

Bulk Transport

The transport phenomena we have discussed so far involve small single molecules. Most animal cells can take up or deliver a sample of their external or internal environment by means of a membrane-bound vesicle. These processes are known as *bulk transport.*

Inward bulk transport involves forming a small infolding of the cellular membrane and finally pinching off a membrane-bound vesicle inside. This process *internalizes* part of the environment and is known as *endocytosis.* The vesicles—called *endocytotic vesicles* or *vacuoles,* and *endosomes*—move inward from the cell surface, meeting and fusing with other vesicles, and increasing in size.

In some cells, the endocytosis process is called *pinocytosis* ("cell drinking"); the vesicle seems to contain only the liquid part of the environment. Small (100-nm diameter) vesicles move inward where they fuse with other vesicles. This process can be seen easily by examining amoebae, in which it happens on a large scale; this is probably how amoebae obtain much of their food. When grown in a medium containing glucose, an amoeba can take in this sugar by means of pinocytotic vesicles; using radioactive glucose, as did H. Holter, one can determine by autoradiography what is happening. The process is induced by the presence of certain proteins in the medium, for without them no pinocytosis takes place. Once inside the vesicles, the radioactive glucose moves into the general cytoplasmic milieu. It now seems likely that small molecules entering the cell via pinocytotic vesicles pass into the cytoplasm through specific membrane permeation pathways.

When particulate matter from the environment is taken up by an invagination of the cell membrane, the process is called *phagocytosis* ("cell eating"). The infolding and the resultant vesicles are larger than in the case of pinocytosis—phagocytotic vesicles may be micrometers in size.

There are several aspects of the endocytotic mechanism we must consider: firstly, the internalization process itself (invagination of the membrane and budding off as a vesicle inside), while known to require energy, still has mysterious details. Secondly, material is brought into the cell by bulk transport without having to pass through the plasma membrane, and once the material is in the cell it is still surrounded by a membrane. The question arises as to what happens next. We still do not have a complete answer for all bulk transport processes, but one process we do know about is that in which endocytotic vesicles fuse with lysosomes to form *digestion vacuoles* (Figure 7–40). There, in the acid environment of the lysosome, one or more of its three dozen degradative enzymes break down the particles or macromolecules into small molecules capable of diffusing (probably via facilitated pathways) through the vesicle membrane. It is not entirely clear what happens to other endosomes, but probably they also eventually fuse with one of the internal membrane systems of the cell. Endocytosis thus acts as a device by which heterotrophic cells acquire metabolites in concentrated form (as particles or macromolecules).

Pinocytosis and phagocytosis represent so-called "bulk phase" endocytosis; they appear to bring in representative samples of the external environment. Even in bulk transport there is some selectivity, however; not all compounds are moved with equal probability. This comes about because the cell membrane has *receptors* that specifically bind certain compounds and not others. When the membrane invaginates, the bound extracellular compounds and their receptors enter together. Receptors are integral proteins that span the membrane. Their outward-facing portion has a high-affinity binding site (dissociation constant $< 10^{-8}$ M) for the substrate or *ligand.*

These general ideas have in recent years received detailed support in studies of *absorptive or receptor-mediated endocytosis* by J. L. Goldstein and M. S. Brown, for which they received the Nobel Prize in 1985. The prototypic example of this process is the bulk transport by animal cells of cholesterol, which is synthesized mainly in the liver and transported by the blood (Figure 7–41). In the circulation, cholesterol is carried as an ester in a large (MW 3×10^6) complex structure associated with fatty acids, phospholipids, and special proteins: the *low-density lipoprotein (LDL).* Many cells have on their surfaces special *LDL receptors.* Shortly after binding to the receptors (less than 10 minutes), the LDL is internalized by invagination of the membrane. The region of the cell membrane that contains the receptors has a special structure, visualizable in the electron microscope, called a *coated pit* (Figure 7–42). After the internalization has occurred the LDL is found in *coated vesicles,* which can be distinguished from the usual vesicles by a fuzzy coat on the cytoplasmic side of the membrane. The coated vesicles lose their fuzzy coat and eventually fuse with lysosomes. Within these fused *secondary lysosomes,* hydrolysis of the esters occurs, which frees the cholesterol and fatty acids for membrane synthesis (Figure 7–41).

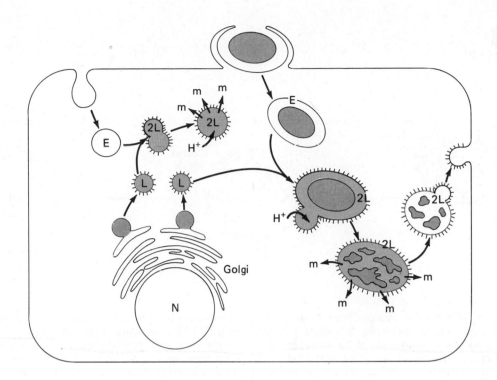

FIGURE 7-40

Pathway of degradation by lysosomes. Primary lysosomes (L), containing numerous degradative enzymes, arise by budding off from the Golgi membranes, at which time they are likely to be associated with a clathrin coat. Small pinocytotic vesicles arise by budding inward from the plasma membrane and then fuse to form an endosome. In some cases a large particle, such as a bacterium, is engulfed and the resultant vesicle-enclosed sac is called a phagosome. The result of fusing a primary lysosome with an endosome or with a phagosome is called a secondary lysosome (2L) and they can be quite different in size. Both endosomes and secondary lysosomes have an acid pH. Digestion of the contents of the secondary lysosome (also known as a digestion vacuole) by acid hydrolases is enhanced by the acid environment inside, created by a H^+ pump in the lysosomal membrane. Small molecules (m) produced by the digestion in the secondary lysosome enter the cytoplasm. Endosomes, and to an extent, secondary lysosomes, bud off small vesicles which fuse with the plasma membrane, delivering their contents to the environment by exocytosis and returning membrane area to the cell surface. The reminder of the secondary lysosome persists, sometimes loaded with debris, in the cytoplasm. Lysosomes can also fuse with vesicles surrounding other cellular organelles—for example, mitrochondria (not shown). This process is called autophagy. *Note:* Electron micrographs of some lysosomes can be seen in Figure 2–44, p. 80 and Figure 2–45, p. 81.

The large, smooth endosome vesicles that result after the removal of the fuzzy coats contain both receptor and ligand. In most cases the ligand dissociates from the receptor, which is recycled to the plasma membrane rather than being exposed to lysosomal enzymes. This dissociation is enhanced by the accumulation of hydrogen ions by the endosome via an ATP-linked pump, producing a significantly lower pH (~5). The receptors are then sequestered in a tube-like region which separates from the remainder of the endosome. The receptors return to the cell surface in their special vesicles, which fuse with the plasma membrane; the remainder of the endosome goes on to fuse with a lysosome. The mechanism by which receptors are separated into special vesicular structures is still unknown.

Coated vesicles appear to be membrane vesicles surrounded by a basket made of an unusual protein, clathrin. They have been isolated and studied in relatively pure form by B. M. F. Pearse and her collaborators (Figure 7–42). Coated vesicles have also been implicated in the movement of enzymes from the Golgi region to lysosomes, the secretion of enzymes (see Chapter 13), and the internalization of excess membrane resulting from the fusion of presynaptic vesicles of neurons with the synaptic membrane (Chapter 17). The polymerization of the clathrin subunit to form the basket structure may drive the invagination of a coated pit to form a coated vesicle.

Other proteins are probably also involved in receptor-mediated endocytosis. Among those implicated are

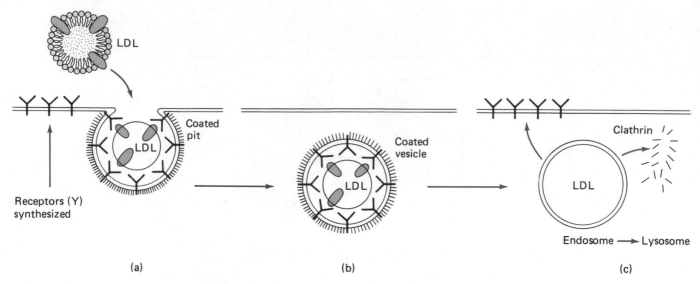

FIGURE 7-41

Model for the internalization of extracellular molecules by receptor-mediated endocytosis.   The example illustrated is the internalization of cholesterol-carrying low density lipoproteins (LDL) by mammalian cells. The LDL is a spherical structure with a monolayer of phospholipids on its outer surface and approximately 1500 cholesterols, esterified to fatty acids, in the interior. There are also specific proteins embedded in the surface with some of their structures facing outward. (a) LDL receptor proteins are synthesized in the cytoplasm and inserted at random in the plasma membrane. The receptors cluster in coated pits, specialized regions of the plasma membrane with a high density of underlying clathrin molecules (marked by their bristly appearance). (b) The receptors are internalized, together with bound LDL, by invagination of the plasma membrane and formation of a clathrin-coated vesicle. (c) After becoming uncoated, the vesicle is known as an endosome. Recycling of both clathrin to the cytoplasm and the receptor to the surface occurs. The endosome fuses with a primary lysosome to form a secondary lysosome, whose contents become acidified by the action of a H^+ pump, hydrolysis occurs, and the cholesterol is released to the cytoplasm. (Redrawn from R. G. W. Anderson, J. L. Goldstein, and M. G. Brown.)

egg yolk proteins, the iron-carrying transferrins, growth factors, certain immunoglobulins, and protein hormones such as insulin. The receptors seem to have two active sites, one to recognize the specific protein ligand and the other to interact with a cytoplasmic system involved both in the clustering of the receptors into coated pits and in the internalization process.

In the beautiful example of the very thin endothelial cells that line blood capillaries (Figure 7-43), it seems that materials are transferred from the blood into the interstitial space by literally being ferried across the cell in small vesicles. These form at one border, travel across the narrow cell, fuse with the membrane at the other side, and thus discharge their contents outside the cell but on the other side.

Bulk transport proceeds not only into or across the cell, but also frequently out of the cell (the process called *exocytosis*). In animals, a number of secretory (glandular) cells are known which manufacture products that are transported out of the cell by exocytosis. Examples include the release of digestive enzyme precursors by the acinar cells of the pancreas (Figure 7-44), and the release of the hormone prolactin by a pituitary gland cell into the circulation. (These processes will be discussed in detail in Chapter 13 when we consider the secretion of proteins.)

All of these cases indicate that intracellular membranes are not static—the cell structure itself, even in the tightly restricted cells of a multicellular organism, is dynamic. Coated pits often account for two percent of an animal cell membrane and the receptors are continually internalized, sequestered, and recycled to the surface, each cycle taking approximately a quarter of an hour. Observations of tissue culture cells also show very clearly a constant movement of materials—even large bodies like mitochondria—within these cells. We are thus led by this discussion of bulk transport to the phenomena of mechanical movements of large macromolecular complexes. This subject will be treated in detail in Chapter 16.

Membrane Electrical Potential and the Action Potential

Thus far we have considered the ability of membranes to bring about and maintain a difference in concentration of diffusible substances between the cell and its surroundings, or between different parts of the cell. There are two apparently different functions of the cell membrane that depend on this same ability: the establishment and maintenance of a relatively constant

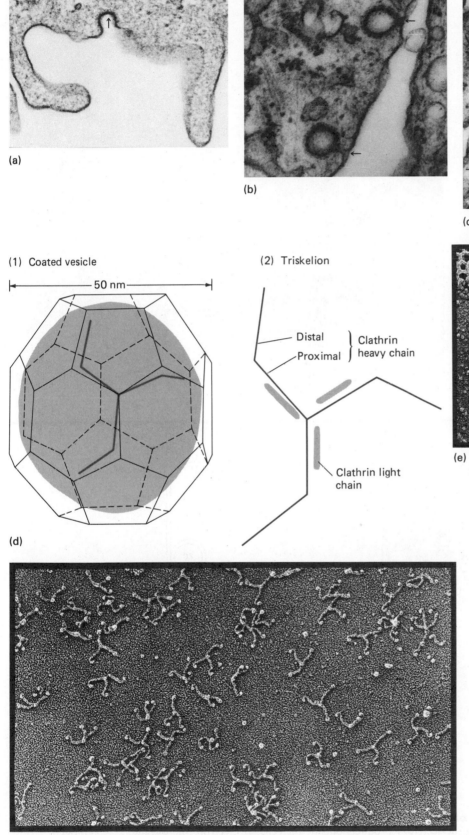

(a)

(b)

(c)

(1) Coated vesicle

— 50 nm —

(d)

(2) Triskelion

Distal
Proximal } Clathrin
heavy chain

Clathrin light
chain

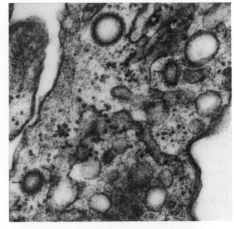

(e)

(f)

FIGURE 7–42

Coated vesicles and coated pits.
Electron micrographs of dermal cells from human skin. (a) An invagination of part of the plasma membrane. The membrane appears thicker at this area and a series of protein bristles appears on the cytoplasmic surface. The structure is called a coated pit (arrow). (b) The coated pit begins to become enclosed within the cell. Two such pits (arrows) are shown, each with an opening to the extracellular space. (c) Two coated vesicles (arrows), with spikes on their outer surfaces. a–c, × 150,000. (Courtesy of N. Romani and G. Schules.) (d) Diagram (1) shows a coated vesicle with a clathrin coat surrounding it; (2) shows how the heavy and light clathrin chains intersect to form the triskelion repeating unit of the clathrin coat. (Based on R. A. Crowther and B. M. F. Pearse.) (e) Flat and curved clathrin lattices on the inner surface of the cell membrane of cultured fibroblasts. The samples were quickly frozen, etched, and rotary replicated. (Courtesy of J. Heuser.) (f) Purified clathrin triskelions from bovine brain. e–f, × 176,800. (Courtesy of J. Heuser.)

FIGURE 7–43

Transfer of material across an endothelial cell of a blood capillary by sequential endocytosis and exocytosis. l = lumen of capillary; f = injected ferritin particles in lumen and extracellular space; ec = endothelial cell lining capillary. Long arrow on the left indicates the uptake of ferritin particles into a pinocytotic vesicle and the presumed transport of ferritin particles (arrow on capillary vesicle) to the other side of the capillary cell (short arrows). × 130,000. (Courtesy of G. E. Palade and R. Bruns.)

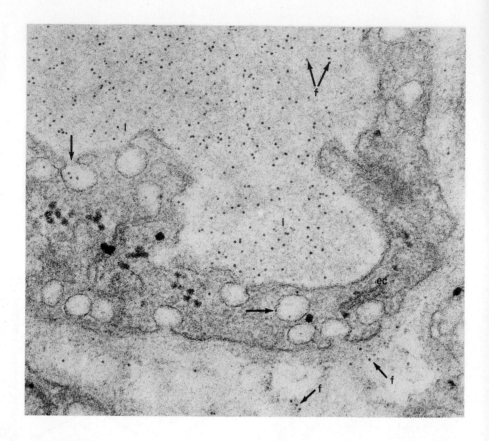

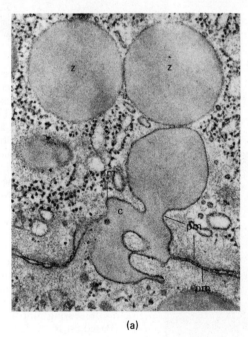

(a)

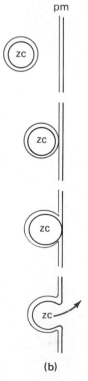

(b)

FIGURE 7–44

Zymogen granules and the discharge of their contents into the duct lumen. The membrane of the zymogen granules fuses with the plasma membrane, leaving the secretory products outside the cell. (a) Electron micrograph. z = zymogen granules; c = contents of zymogen granule spilling out; pm = plasma membrane fusing with zymogen granule membrane. × 87,000. (Courtesy of G. E. Palade.) (b) Diagram showing the fusion process. zc = contents of zymogen granule; pm = plasma membrane.

electrical potential difference between the inside and outside of the cell (the *resting potential*) and the rapid variation of this transmembrane electrical potential that is used for signaling (the *action potential*). Both of these functions are due to properties of the cell membrane that we have discussed already; the electrical effects arise when the transported or excluded substances have an electrical charge—that is, when they are ions. The general mechanisms of these functions have familiar features: the ability to move ions by energy-linked processes and special permeability pathways. A new function necessary for the action potential is the rapid opening and closing of permeability pathways through the membrane. Although the action potential has its most specialized expression in nerve cells, it can be found in attenuated form in some other cells.

In this chapter we shall briefly consider first the resting potential, and then the action potential and its propagation in a single cell system such as the giant axon of the squid. Synaptic transmission and related topics will be discussed further in Chapter 17. For more information the student may also wish to consult a truly inspiring little book by B. Katz, *Nerve, Muscle and Synapse* (see Selected Readings).

The Resting Potential

The phenomena we wish to describe can all be studied using a segment of the giant axon of the squid. This ideal experimental system, discovered by J. Z. Young, is a part of the cylindrical portion of a nerve cell. The cell body has been removed, as well as the synaptic contacts to other nerves found at one end and the motor nerve endings at the other (Figure 7–45a). This apparently

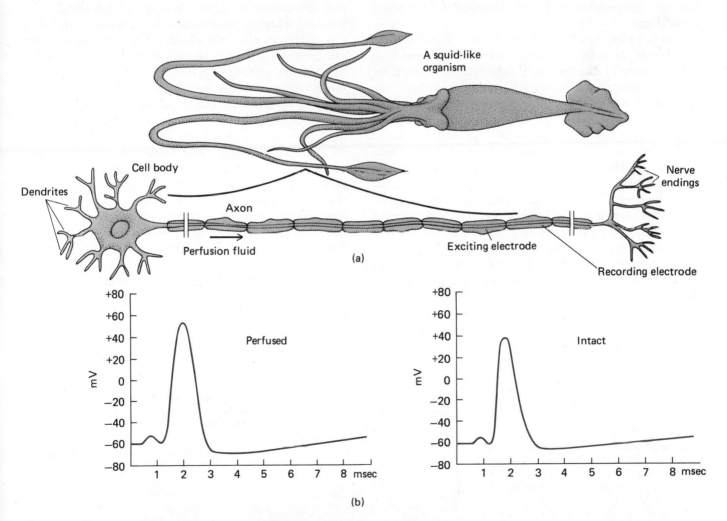

FIGURE **7–45**
Preparation of the squid giant axon. (a) The axon is dissected free from the squid's arm, cut at both ends and then perfused with 0.15 M KCl solution to replace the axoplasm inside. (b) When recording electrodes are placed inside the axon, the observed action potential is very similar to that of the intact axon, both in amplitude and time course.

TABLE 7–6

Ionic Concentrations in the Squid Axon and the Surrounding Body Fluid

				Concentrations in Milliequivalents per Liter
	K^+	Na^+	Cl^-	Organic Anions (For Example, Aspartate, Isothionate)
Squid axon	400	50	40–100	345
Body fluid	10	460	540	—

mutilated specimen can function properly for several hours when placed in seawater and can conduct hundreds of thousands of nerve impulses. Its giant dimensions (up to 1 mm in diameter) make it easy to insert electrodes or remove material (axoplasm) by perfusion. This can be done without irreversible injury to the function, as shown by comparing an action potential generated by the perfused preparation with one generated by the intact axon (Figure 7–45b).

If one measures the concentrations of ions inside the squid axon and in the body fluid surrounding it, one finds a distribution (Table 7–6) that, although not identical to, is reminiscent of that observed in human erythrocytes. Numerous experiments have established for the squid axon, for other nerve cells, and for muscle cells as well, that the distribution of ions can be explained in terms of (1) the presence of a differentially permeable membrane surrounding the cell and (2) a pumping mechanism which extrudes Na^+ and accumulates K^+. These properties could be demonstrated very conveniently with the squid axon. Because of its size, direct chemical and electrical manipulation is particularly easy. The squid axon is very similar to the erythrocyte in its ion-transport behavior. There is, however, a consequence of ion movements that we have not yet discussed—namely, the separation of positive and negative charges that occurs across the plasma membrane. If one takes a pair of electrodes, carefully inserts one of them into the squid axon, and places the other one in the bathing seawater, it is possible to show that there is a potential difference of 60-90

millivolts (mV) between the inside and the outside of the axon membrane, the inside being negative with respect to the outside (Figure 7–46). This *steady-state potential difference* is called the *resting potential*.

The data in Table 7–7 show that the cell membranes can possess a considerable resistance to the flow of ions. Because of this high resistance and because it separates two ion-containing (and, therefore, conducting) solutions, a biological membrane has a *capacitance*—that is, it can collect an electrical charge. The magnitude of the capacitance is given by the ratio of the charge collected to the potential difference across the membrane. The measured values for many membranes are low, lying in the range of 1–5 $\mu F/cm^2$. This means that a resting potential difference of 100 mV would lead to a charge distributed on 1 cm^2 of membrane area of 0.1–0.5 μC, or 1–5 $\times$ 10^{-12} moles of charge. The thinner the membrane and the less it conducts, the higher the capacitance. From these capacitance measurements it is possible to calculate that the thickness of the membrane is 5.0 nm—again, a good confirmation of the existence of a thin membrane and of its controlling influence on the movement of ions. The quantitative electrical measurements were extended by direct measurements of permeability using isotopic tracers. The data in Table 7–8 show that for the nerve membrane, K^+ permeability is much greater than Na^+ permeability.

Although the resting potential has been studied most extensively in nerve cells, we know that it is a widespread phenomenon among cells of widely differing origin,

FIGURE 7–46

Measurement of the resting potential of the squid axon.   Modern instrumentation can measure the potential difference without drawing appreciable current, thus without disturbing the ionic steady state being measured. The convention is to express the potential difference as inside minus outside. For most cells, the inside is negative with respect to the outside.

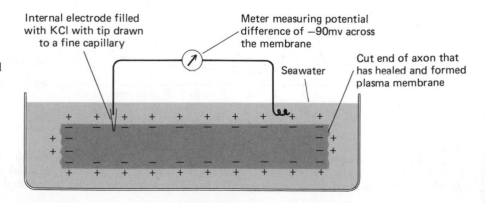

Internal electrode filled with KCl with tip drawn to a fine capillary

Meter measuring potential difference of −90mv across the membrane

Seawater

Cut end of axon that has healed and formed plasma membrane

TABLE 7–7

Electrical Properties of Some Nerve and Muscle Cells★

	Cell Diameter (μm)	Membrane Resistance (ohm-cm^2)	Membrane Capacitance (μF/cm^2)	Resistivity of Cell Interior (ohm-cm)	Resistivity of Medium (ohm-cm)
Squid axon	500	700	1	30	22
Lobster nerve	75	2000	1	60	22
Frog muscle	75	4000	2.5	200	87

★After B. Katz, 1966.

including the red blood cell. In nerve cells, however, it is especially interesting because it plays a role in impulse transmission, which, after all, is the main biological function of nerve tissue. What is the mechanism that produces the resting potential? It turns out that this important phenomenon has a simple explanation: the resting potential is primarily a consequence of *differential movement of ions*. It is important to realize that the resting potential is *not* determined primarily by the Na$^+$ pump. This is easy to show in the squid axon. I. Tasaki and his co-workers found that it is possible to insert pipettes longitudinally into the axon, withdraw the axoplasm, and replace it with solutions of proper ionic concentrations, but without energy sources such as ATP. This preparation could maintain a stable resting potential for hours and was still able to propagate thousands of nerve impulses, despite not having an active Na$^+$ pump.

The reason the resting potential can be maintained without the action of the pump in the squid axon preparation is because of its large volume and the relatively high concentrations of ions inside, relative to the rate at which they move through the membrane. If the permeation of ions is very slow, the relatively high internal concentrations do not change much and, therefore, neither does the electrical potential difference across the membrane. In the red blood cell (and in small axons) the volume is much lower, the surface/volume ratio is higher, and the movement of ions across the cell membrane perturbs significantly the interior concentrations. Ultimately, the Na$^+$ pump is essential to maintain

a constant concentration of ions and, therefore, a constant resting potential.

An Equilibrium Model for the Membrane Electrical Potential

The simplest way to attempt an explanation for a constant electrical potential difference is to assume that some ion or other is at equilibrium across the cell membrane. Recall that Equation 7–2 gives the free energy change for moving an ion across a membrane:

$$\Delta G = R\,T \ln \frac{[B]_{in}}{[B]_{out}} + z\,F\,\Delta E_m,$$

where $\Delta E_m = E_{in} - E_{out}$. This equation is only applicable to an ion capable of moving across the membrane—that is, capable of coming into equilibrium. When equilibrium is reached, $\Delta G = 0$ and

$$\Delta E_m{}^{eq} = -\frac{R\,T}{z\,F} \ln \frac{[B]_{in}{}^{eq}}{[B]_{out}{}^{eq}}. \qquad (7\text{-}7)$$

This is also called a *Nernst equilibrium equation;* the concentrations are those that occur at equilibrium and ΔE_m is $E_{in} - E_{out}$.

But at equilibrium, how can the concentrations be different if ions are free to move across the membrane? This situation can arise when there are *other* ions present that do *not* penetrate the membrane and when the concentration of those ions differs between the inside and outside of the cell.

TABLE 7–8

Permeability Values for K$^+$, Na$^+$ and Cl$^-$ in the Squid Axon★

	Permeability (cm/sec)
K$_{out}{}^+$	6.2×10^{-7}
K$_{in}{}^+$	5.8×10^{-7}
Na$_{in}{}^+$	7.9×10^{-9}
Cl$^-$	1.0×10^{-8}

★After Katz, 1966, and Brinley and Mullins, 1965.

The Gibbs-Donnan Equilibrium

To see how the distortion due to impermeable ions works, we first recall that the membrane potential must be in equilibrium with both K^+ and Cl^-, since both are free to move across the membrane. We can, therefore, express the condition for equilibrium in another way. By applying Equation 7–7 to both K^+ and Cl^- we can write

$$E_m = -\frac{RT}{z_K F} \ln \frac{[K]_{in}}{[K]_{out}}$$

$$= -\frac{RT}{z_{Cl} F} \ln \frac{[Cl]_{in}}{[Cl]_{out}}.$$

Keeping in mind that $z_K = +1$, that $z_{Cl} = -1$, and that numbers whose logarithms are equal must themselves be equal, we can manipulate the second equality to give as a condition of equilibrium that

$$\frac{[K]_{in}}{[K]_{out}} = \frac{[Cl]_{out}}{[Cl]_{in}}. \qquad (7\text{–}F)$$

The other fact needed to describe this equilibrium is that, to a very good approximation, there are the same number of positive and negative charges on the left side of the membrane, and the right-side solution is electrically neutral also. (Obviously, this cannot be exactly true since the observed voltage implies an excess of negative electrical charge inside the cell. It turns out, however, that only minute differences in the number of charges are necessary to produce the observed electrical potentials, differences less than the total concentrations of ions by orders of magnitude. This is a consequence of the low capacitances of biological membranes.) The assumption of *electroneutrality* allows us to write two simple equations to express the *conservation of electrical charge*:

$$[K^+]_{in} = [Cl^-]_{in} + [P^-]_{in} \qquad (7\text{–}G)$$

and

$$[K^+]_{out} = [Cl^-]_{out}.$$

Equations 7–F and 7–G must be simultaneously true at equilibrium. This set of three simultaneous equations can be solved by algebra to yield

$$[Cl]_{out}^2 = [Cl]_{in}^2 + [Cl]_{in} [P]_{in}$$

and the electrical potential can be calculated by solving this equation for $[Cl]_{in}/[Cl]_{out}$ and substituting it in Equation 7–7:

$$E_m = \frac{RT}{F} \ln \left[1 + \frac{[P]_{in}}{[Cl]_{in}} \right]^{1/2}. \qquad (7\text{–}H)$$

EXAMPLE 7–4

Calculate the electrical potential difference at 37°C across a chloride-permeable membrane that is due to the Gibbs-Donnan equilibrium when there is an excess of 50 mM negatively charged, impermeant protein ions inside the cell and $[Cl^-]_{in} = 65$ mM. Recalculate the

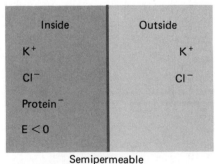

Inside	Outside
K^+	K^+
Cl^-	Cl^-
Protein$^-$	
E < 0	

Semipermeable membrane

FIGURE 7–47

The Gibbs-Donnan equilibrium. This simple membrane is freely premeable to K^+ and Cl^- ions, but not to the negatively charged proteins found on one side. The impermeable charged molecules cause unequal concentrations of K^+ and Cl^-, and an electrical potential difference, to develop across the membrane.

The effect of large, impermeable ions on membrane potential is illustrated by the experimental setup shown in Figure 7–47. This preparation models a cell that is placed in a salt solution. On the left side of the membrane is a solution of permeable ions, say K^+ and Cl^-, and impermeable ions that taken together have a net negative charge. The set of impermeable ions is usually dominated by proteins, and typical intracellular proteins have a net negative charge under physiological conditions; we therefore symbolize them as P^-. On the right side of the membrane is a solution of KCl alone. At equilibrium in the *absence* of the negative macromolecules, the K^+ and Cl^- would be distributed so that each side would have equal concentrations of both. The presence of the negative macromolecules distorts this equilibration ten-

membrane potential for $[Cl^-]_{in} = 100$ mM.

Solution

Use Equation 7–H to find the answer:

$$E_m = \frac{R\,T}{F} \ln \left[1 + \frac{[P^-]_{in}}{[Cl^-]_{in}} \right]^{1/2}.$$

Substituting the known values for R, T, F, and the concentrations:

$$E_m =$$

$$\frac{(8.314 \times 10^{-3})\,(310)}{(96.5)} \ln \left[1 + \frac{50}{65} \right]^{1/2}$$

$$= 0.0076 \text{ V} = 7.6 \text{ mV}.$$

For $[Cl^-]_{in} = 100$ mM, $E_m = 5.4$ mV. Comparing this value with that determined for $[Cl^-]_{in} = 65$ mM illustrates how a higher concentration of permeable ion can reduce the Gibbs-Donnan equilibrium potential.

Since all cell membranes separate impermeable macromolecules, the Gibbs-Donnan equilibrium is a universal mechanism for part of observed cellular membrane potentials. How much of E_m is due to the impermeable macromolecules depends, as can be seen from Equation 7–H, on the *net* concentration of impermeable ionic charges—that is, on the difference between the protein charges inside and outside the cell. For most cells the Donnan potential is probably less than 10 mV. Equation 7–H also reveals that increasing the salt concentration (and therefore the chloride concentration) tends to reduce the Donnan potential.

The Gibbs-Donnan equilibrium equations also point up another property of cells containing trapped macromolecules: they are not in osmotic equilibrium if the pressures on the two sides of the membrane are equal. The student can prove this by noting that for the simplified setup in Figure 7–47, the total ion concentration outside, $[T]_{out} = 2\,[Cl]_{out}$, and the total ion concentration inside, $[T]_{in} = 2\,([Cl]_{in} + [P]_{in})$ are different. The equations can be manipulated to show that $[T]_{in}/[T]_{out} = [Cl]_{out}/[Cl]_{in}$, where the right-hand side must be greater than unity if the membrane potential is negative. Animal cells with deformable cell membranes avoid, in many cases, the influx of water that inevitably follows the osmotic gradient by actively pumping a net flux of positive ions outward; the water follows. This is another function of the Na^+ pump. Bacteria and plant cells can withstand the internal hydrostatic pressure that brings them into osmotic equilibrium because of their strong cell walls.

dency (see box above) and the result is an equilibrium electrical potential across the membrane. This kind of equilibrium is called a *Gibbs-Donnan equilibrium* and the electrical potential is known as the *Donnan potential.*

A Steady-State Model for the Resting Potential

As a model for the electrical properties of cell membranes the Gibbs-Donnan equilibrium is unsatisfactory. First, it cannot generally account for the size of the observed resting potentials of many cells; it certainly cannot explain the potential observed in the squid axon after all the protein has been removed. Second, to change the potential to propagate signals would mean allowing proteins to flow across the membrane. This has not been observed to occur, and in any case would be very expensive for the cell, since proteins require a great deal of ATP for their synthesis (see Chapter 13).

The equilibrium model assumed that some ions did not permeate the cell membrane at all. The steady-state model assumes that small ions like Na^+ and K^+ do permeate the membrane but, given equal driving forces, at *different rates*. Let's look again at the steady-state concentrations of ions inside and outside the squid axon in Table 7–6. The squid nerve membrane, as we have learned, differs markedly in its permeability to these ions. K^+ permeates most easily; Na^+ permeates very much less easily (and the organic anions inside the cell are virtually unable to leave). At the observed steady-state concentrations, and in the absence of a membrane potential, there would be a tendency for K^+ to diffuse outwards and Na^+

An Electric Current Model for the Resting Potential

We shall try to account for the steady-state resting potential in terms of passive *flows of Na⁺ and K⁺*. The *two ion concentrations will be assumed to remain constant*. This is the result under physiological conditions of an active Na^+ pump, but in a perfused squid axon, without an active pump, the concentrations can be kept constant by the experimenter. For this model, therefore, we can ignore the action of the pump. We shall first express the flow of each positive ion as an electric current, one that responds to forces on the ion. Since the resting potential is constant in value, *the condition for the steady state is that the sum of the inward currents of the two ions is zero;* if it were not, there would be an accumulation of positive ions inside and a rising membrane potential.

From our earlier discussion of Fick's first law we have a mathematical model (Equation 7–C) for the flux density of an ion:

$$J = - U C \frac{\Delta G}{\Delta x}.$$

Equation 7–2 shows that an ion responds to both osmotic and electrical forces; rewriting for the case of Na^+ gives:

$$\Delta G = R T \ln \frac{[Na]_{in}}{[Na]_{out}} + F \Delta E_m \quad (7\text{-}I)$$

(the charge sign on the sodium ion has been omitted; $z = +1$).

In our model *the osmotic force on the ion will be constant* because it depends only on the concentrations of ions, which are assumed to be constant. For an electric current model, it is convenient to have both kinds of force in electrical units. We can express the constant osmotic force as an *equilibrium potential for the ion* by using Equation 7–4:

$$\Delta E_{Na}{}^{eq} = - \frac{R T}{z F} \ln \frac{[Na]_{in}}{[Na]_{out}}. \quad (7\text{-}J)$$

This is the equation for the *sodium equilibrium potential*, which is usually symbolized E_{Na}; there is an analogous equation for the *potassium equilibrium potential*, E_K.

Both forces are now modeled in the form $-\Delta E/\Delta x$. The methods of calculus can be used to sum up the contributions to the total force, along a path from outside to inside the cell. The result gives an equation for the sodium flux density.

$$J_{Na} = U C z F (E_{Na} - E_m). \quad (7\text{-}K)$$

Electrical current (I) is proportional to the flux density:

$$I_{Na} = J_{Na} A F$$

where A is the area and F is the Faraday constant. The various parameters that determine how the ion current responds to the force on the ion are generally lumped together into an *ion conductance*, g_{Na} or g_K. This yields the desired equations for

inwards, due solely to their concentration gradients. But, since K^+ moves out more readily than Na^+ moves in, there would under those conditions be a net efflux of positive charges, leaving the inside of the cell negative with respect to the outside. Eventually a limit would be placed on the movement of K^+; at a certain point the electrical potential generated by the separation of charges would slow the concentration-driven efflux of K^+ to the rate of Na^+ influx. This would be a steady state, with balanced fluxes of ions and a stable electrical potential. This simple qualitative argument works very well indeed

when expressed in quantitative terms. (See box on the electric current model for the resting potential.)

In order to see how well the current model (Equation 7–M) works as a model for the resting potential, its predictions can be compared with other data. Values for E_{Na} and E_K can be calculated from the data in Table 7–6, using Equation 7–7:

$$E_{Na} = +58 \text{ mV} \quad \text{and} \quad E_K = -96 \text{ mV}.$$

The observed resting potential (-60 to -90 mV) has the same sign as E_K and the opposite sign from E_{Na}; we

the inward electrical currents carried by the two ions:

$$\frac{I_{Na}}{g_{Na}} = E_{Na} - E_m \qquad (7\text{–}L)$$

and

$$\frac{I_K}{g_K} = E_K - E_m.$$

(These current equations are sometimes called Hodgkin-Horowitz equations.)[1] I_{Na} represents the current carried by sodium ions (due to the flux of Na^+) and I_K has a similar significance. Values of the conductances are determined experimentally by the use of these equations: the current is measured by the flow of a radiolabeled ion under conditions of constant concentrations, and E_m is measured by a suitable pair of electrodes. It is unnecessary to include a current carried by chloride ions because they are practically at equilibrium across the membrane of the squid axon.

At the steady state (resting condition) there is no *net* inward current flow of positive ions: $I_{Na} + I_K = 0$. Combining this criterion with the equations for currents of the two ions (7–L) yields

$$g_{Na}\,(E_{Na} - E_m) + $$
$$g_K\,(E_K - E_m) = 0.$$

Rearranged, this gives the resting membrane potential:

$$E_m = \frac{g_{Na}}{g_{Na} + g_K}\,E_{Na} + \qquad (7\text{–}M)$$

$$\frac{g_K}{g_{Na} + g_K}\,E_K.$$

EXAMPLE 7–5

Calculate the steady-state resting potential across a membrane with a potassium ion conductance of 0.96 milliSiemens/cm^2, a sodium ion conductance of 0.14 milliSiemens/cm^2, a potassium ion equilibrium potential of −90 mV, and a sodium ion equilibrium potential of +60 mV.

Solution

Equation 7–M can be used to calculate the resting potential for this membrane:

$$E_m = \frac{g_{Na}}{g_{Na} + g_K}\,E_{Na} + \frac{g_K}{g_{Na} + g_K}\,E_K.$$

Substituting the values given for the various conductances and equilibrium potentials yields:

$$E_m = \frac{0.14}{1.1}\,60\text{ mV} + \frac{0.96}{1.1}\,(-90\text{ mV})$$

$$= -71\text{ mV}.$$

(Note: The unit of electrical conductance is the Siemen or mho [the reciprocal of the unit of electrical resistance, the ohm].)

[1]The student should be aware that these equations are sometimes derived with a different convention of signs. Equation 7–I shows that a positive inward current will be enhanced by a negative membrane potential and by a positive value of the equilibrium potential—equivalent to a concentration ratio (in/out) with value less than one.

conclude that the term in Equation 7–M containing E_K has a larger absolute magnitude than the other term. This must be partly due to a larger value for g_K, since if g_K were equal to g_{Na} the membrane potential would be only −19 mV. A higher potassium conductance (permeability) is in agreement with tracer measurements and with our previous arguments. If the sodium conductance were very much smaller, the current carried by Na^+ would be almost zero and the resting membrane potential would, according to Equation 7–J, become equal to E_K. This would be analogous to a Gibbs-Donnan equilibrium (in our earlier example, proteins had a zero conductance).

One way to demonstrate the relative contributions of the two ion currents to the resting potential is to alter the ion concentrations in the medium bathing the cell. Changing $[K^+]_{out}$, which alters E_K, has much more effect than changing $[Na^+]_{out}$, which alters E_{Na}. The external concentration of the more permeant ion has a greater influence on the resting potential.

In summary, the resting potential can be modeled as the result of steady-state ion currents. The current carried by an ion is determined by its conductance, which is

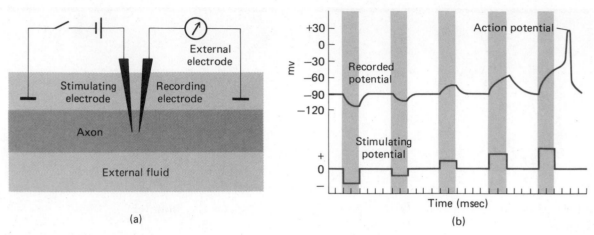

FIGURE 7–48

Effect of direction and strength of electrical stimulation on the potential difference across the membrane of a single nerve axon. (a) Assembly showing stimulating and recording electrodes. These are located close to each other so that subthreshold perturbations can be recorded. (b) Demonstration that the stimulating potential must be in the right direction and of sufficient intensity to produce an action potential. (Modified from B. Katz.)

affected by its concentration and mobility, and by the driving force on the ion, a function of its concentration ratio. In many cells the differential flows of Na^+ and K^+ account for most of the observed resting potential.

The Action Potential

The nerve fiber is not constructed like a copper wire, which has a very small resistance and can support the flow of electric currents over long distances without loss. The interior of the nerve cell is an aqueous solution of ions with a considerable resistance to the passage of currents, which become greatly attenuated after even a few millimeters of travel. Yet we observe that signals can travel along nerve fibers for several hundred centimeters without becoming smaller. How then do nerves, which are such poor conductors of electric currents, manage to be such excellent transmitters of electric signals?

Let's begin by describing the electrical changes observed when a nerve is stimulated. Figure 7–48a illustrates the arrangement used for stimulating a nerve electrically and measuring the electrical consequences. Many types of stimuli, including chemical and physical ones, will elicit a response from a nerve like the squid axon, but electrical stimulation is easiest to control and measure. A nerve preparation shows a resting potential of 60–90 mV as soon as the recording electrode has penetrated the membrane. When a resting potential is observed, the nerve membrane is said to be *polarized*. One can now stimulate the nerve by applying with the stimulating electrode a current pulse—typically one millisecond in duration. If positive current flows outward during the pulse, the membrane becomes locally even

more polarized *(hyperpolarized)*, creating a temporary local "disturbance" that can be observed with the recording electrode (Figure 7–48b). If, however, positive current flows inward, the membrane will be *depolarized*.

Should the stimulus become large enough to depolarize the membrane by some 50 mV (that is, to change locally the potential from −90 mV to −40 mV), then a *threshold* is reached. Each time this occurs the nerve cell responds, independently of the continued presence of stimulus current, with a characteristic change in potential difference that at its peak even reverses the polarity of the membrane (Figures 7–48b, 7–49); this is the *action potential*. The change in potential travels from the stimulating electrode down the entire length of the nerve fiber without suffering attenuation. Increasing the intensity and duration of the stimulus beyond the minimum value does not affect the height, shape, or velocity of propagation of the action potential (Figure 7–49c). This set of properties is one of the classical examples of an *all-or-none response*.

Another interesting property of neurons is that numerous action potentials can be generated in close succession, but when the interval between them becomes less than a few milliseconds, one observes a *refractory period*, during which the cell will not respond to the usual stimulation. This is how the action potential works:

1. The negative resting potential (membrane polarization) is the consequence of the differing intracellular concentrations of Na^+ and K^+, and the differential permeabilities of the axon membrane to Na^+ and K^+, which lead to different flows of these two ions through the membrane.

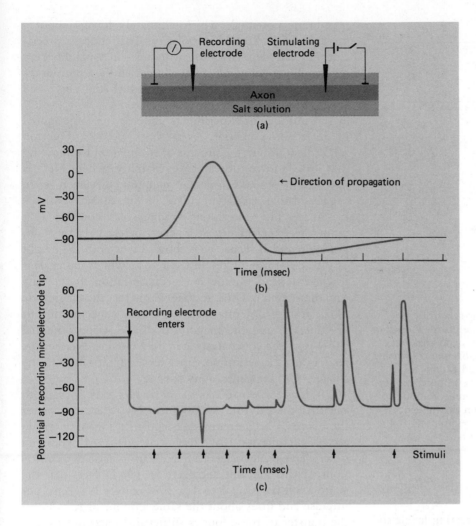

(a)

(b)

(c)

FIGURE **7–49**

The shape of the action potential and the all-or-none response.
(a) Assembly showing stimulating and recording electrodes. These are located sufficiently far apart to record the shape of the self-propagating action potential. (b) Amplitude and time course of action potential. (c) Demonstration that the nature of the action potential does not depend on strength of stimulus, displayed here as sharp spikes and indicated by vertical arrows, provided it is above the threshold level. The three stimuli at the right, although different in size, elicit the same action potential. (Modified from B. Katz.)

2. In response to a partial depolarizing stimulus (a change in the membrane potential to less negative values), the Na^+ conductance (g_{Na}) increases. The result is an increased influx of Na^+, which tends to depolarize even further the membrane potential (raising it toward the Na^+ equilibrium potential, E_{Na}; see Equation 7–M).

3. At subthreshold stimulations, K^+ efflux (and, to a lesser extent, Cl^- influx) can balance the entry of Na^+, preventing further increase of the sodium conductance and restoring the resting potential when the stimulus is removed.

4. When the stimulus exceeds the threshold level, the increasing membrane potential (E_m) raises the sodium conductance to the point where the resulting rapid influx of Na^+ overwhelms the K^+ efflux. The large net inward flow of positive charge changes the membrane potential still more, until it almost reaches the sodium equilibrium potential. It is this positive feedback which gives the "all" of the all-or-none response. Unless the threshold is reached the feedback does not occur (the "none") and no action potential is observed. Since the Na^+ equilibrium potential is positive, the action potential of the membrane "overshoots" the zero volt level (Figure 7–49b).

5. At first the sodium conductance rises when the membrane potential rises. But after a short interval (about 1 msec), the sodium conductance falls again due to a process called *inactivation*. These specific changes in the Na^+ permeability are attributed to specific Na^+ channels in the axon membrane, each controlled by a *gate*. Depolarizing the axon opens the Na^+ gates; inactivation blocks the Na^+ channels. For a certain interval after inactivation has begun the Na^+ channels cannot conduct. This interval represents an *absolute refractory period* during which no stimulation, however great, can elicit an action potential.

6. During the opening of the Na^+ gate, a slower increase in K^+ permeability develops. More K^+ gates open even while the Na^+ channels are inactivated. The K^+ channels are completely *voltage-dependent;* at any particular membrane potential the potassium conductance has a well-defined steady-state value and there is no inactivation. The efflux of K^+ tends to make the

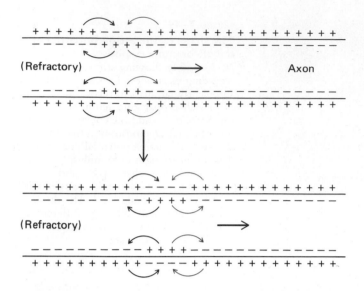

FIGURE 7–50
Propagation of the nerve impulse.   The schematic diagram of the nerve axon shows membrane polarization as separation of + and − charges. A region of depolarization has a reversed distribution of charges. This region induces current flows, indicated by color arrows, which tend to induce depolarization in adjacent regions. The nerve impulse moves from left to right so the region to the left cannot respond, being in its refractory period. The region on the right does respond and becomes depolarized. The process repeats and a wave of depolarization moves along the axon.

membrane potential more negative (see Equation 7–M) and thus *repolarizes* the axon.

7. When the sodium conductance is at resting levels and the potassium conductance is higher than normal, the membrane potential becomes more negative than the resting potential (undershoot, Figure 7–49b). During this hyperpolarization phase there is a *relative refractory period* in which a stimulus stronger than the usual threshold value is necessary to elicit another action potential. Hyperpolarization causes the K^+ gates to close, thus returning E_m to the resting state.

8. Propagation of the action potential is due to the influence of one depolarized region of the neuron on another. The effect of depolarization is to induce current flows from the adjacent polarized regions, which tends to depolarize them (Figure 7–50). As Na^+ gates open in the adjoining region, the depolarization becomes self-sustaining through the positive feedback described in (4) above. This process repeats itself over and over again in successive adjacent regions, thus propagating the wave of membrane polarity reversal.

To summarize this explanation of the action potential and its propagation: the permeability of the membrane to Na^+ is briefly increased by an increase of the membrane potential, which ultimately brings about a brief localized reversal of polarity. This effect is propagated by the depolarization it induces in a neighboring region. The resting potential is restored by a temporary increase in K^+ permeability after which the nerve cell again becomes susceptible to stimulation.

A number of experiments underlie the ionic model of the action potential. As far back as 1902, Overton implicated Na^+ in the process of stimulation by showing that frog muscle loses its excitability when Na^+ is removed from the outside medium. Overton even suggested that excitation might be due to Na^+ and K^+ exchange. This interesting finding was confirmed in numerous experiments with nerve, particularly by A. L. Hodgkin and B. Katz, who showed that both the height of the action potential and the rate at which the peak is reached depend on the Na^+ concentration in the surrounding region. Thus, it was realized that the magnitude of the resting potential varies with the concentration of K^+ outside, and the magnitude of the action potential varies with the concentration of Na^+ outside. Neither the resting nor the action potential varies significantly if one changes Cl^- concentrations outside.

Using the postulates of the model it is possible to calculate the minimum amount of Na^+ that must enter the nerve to account for the depolarization of a certain area of membrane: 10^{-12} moles/cm². R. D. Keynes and others have used tracers to make very precise determinations of this amount and have found, in fact, that the squid axon takes up 3 to 4 × 10^{-12} moles Na^+/cm² per impulse and loses about the same amount of K^+. Thus, the transfer of these ions is sufficient to account for the magnitude of the electric currents observed. It should also be noted that one impulse produces only a very minute change in ion concentration: in the squid axon only one-millionth of the K^+ is lost. These lost ions are replaced by metabolic pumping.

H. J. Curtis and K. S. Cole did a number of remarkable studies of the electrical properties of the squid axon and demonstrated that during excitation there is a large and transitory decrease in resistance of the membrane but no change in capacitance. The specific resistance of the membrane falls from 1000 to about 20 ohm-cm² while the specific capacitance remains at 1 μF/cm². This means that the major change occurs not in the vast lipid area of the membrane, which would be expected to alter the capacitance, but only in the small portion of the membrane involved in the passage of ions. In fact, one can calculate that less than one percent of the area of the membrane is involved in ion flows.

The most convincing evidence for the movement of ions being responsible for the action potential was provided by the *voltage clamp* technique developed by G. Marmont, and used to great advantage by A. L.

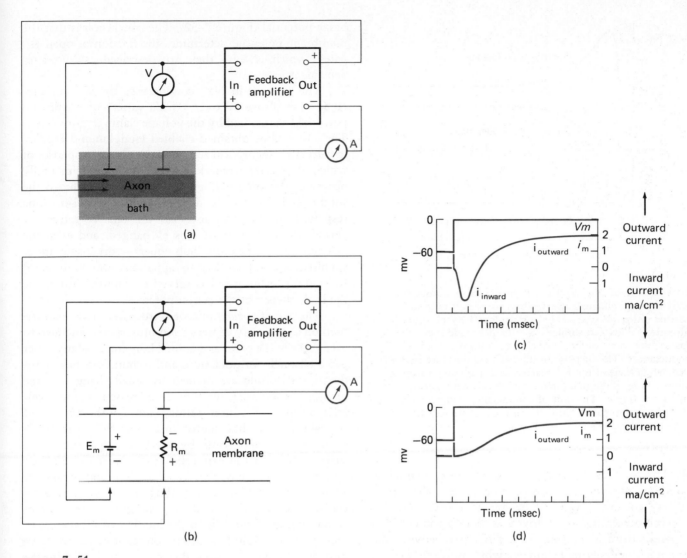

FIGURE **7–51**
The voltage clamp.   This apparatus allows one to study the ionic conductance of the nerve membrane at a fixed potential difference across the membrane. (a) Diagram showing the disposition of two sets of electrodes, voltage (black) and current (color), and their connections to a feedback amplifier. (b) Simplified electrical circuit showing clamping to zero volts across the membrane. Any voltage detected by the voltage electrodes is amplified and fed back through the current electrodes. Current flowing through the ammeter (A) and the membrane (shown as a resistance R_m) will create a voltage opposite in sign to the membrane voltage (shown as a battery E_m). The result is to force ("clamp") the actual potential across the membrane to a very small value. (c) Typical measurement obtained by clamping to zero volts across a squid axon membrane. The current required to maintain the voltage at zero varies, first flowing inward then outward, eventually reaching a steady value. (d) The current required to keep the membrane clamped to zero volts when the outside $[Na^+]$ was made equal to inside $[Na^+]$ by partial replacement with membrane-impermeable choline ions. Under these conditions only an outward current is found, suggesting that the inward current normally observed is carried by sodium ions.

Hodgkin, A. F. Huxley, and B. Katz. This technique can prevent the "explosive" electrical events of the action potential by maintaining (that is, "clamping") a desired potential difference on the membrane with a feedback amplifier connected to two electrodes on each side of the membrane (Figure 7–51). By combining this technique with the use of tracer methods and by replacing the outside Na^+ with the monovalent cation choline, it was possible to disentangle the changes in conductance due to Na^+ influx and K^+ efflux. For instance, the complicated current flow observed when E_m was clamped to 0 volts (Figure 7–51c) was seen to be simpler if external $[Na^+]$ was made equal to internal $[Na^+]$ by partially replacing the external Na^+ with choline (Figure 7–51d). When

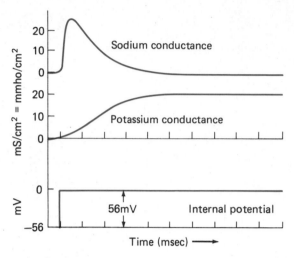

FIGURE 7–52

Time course of Na$^+$ and K$^+$ conductances computed from voltage clamp experiments in which external sodium chloride was varied or replaced by choline chloride.   This and similar experiments made it possible to separate the total conductance changes into Na$^+$ and K$^+$ conductances. The increase in the Na$^+$ conductance is short-lived, while that of the K$^+$ reaches and maintains a steady value as long as the membrane potential is held constant, in this case at 0 mV. The unit of conductance is the Siemen, or mho (the reciprocal of ohm, the unit of resistance). (From A. L. Hodgkin, as shown in B. Katz.)

[Na$^+$]$_{out}$ = [Na$^+$]$_{in}$ and the membrane potential is maintained at zero, there is no driving force at all on Na$^+$ and any current flow must be due to other ions. The observed outward current flow was therefore interpreted as arising from a K$^+$ flux. The difference between the current in this experiment and one where normal [Na$^+$]$_{out}$ is present is thus attributable to the Na$^+$ flux. Since the membrane potential was held constant by the clamp's feedback circuits, the changes in ion currents (fluxes) must have been the result of *changes in ion conductances*.

Another strategy used in conjunction with the voltage clamp to eliminate a particular ion flux is the blockade of channels. Tetraethylammonium (TEA) can block the K$^+$ channels fairly well and in its presence a stimulus will elicit mostly an inward current carried by Na$^+$ ions. When the voltage clamp was used to raise the membrane potential beyond the threshold, at which point the nerve would ordinarily fire, the Na$^+$ influx was short in duration. Since the membrane potential was constant, as were the sodium concentrations, the changing Na$^+$ current must have been due to a changing Na$^+$ conductance—the Na$^+$ gate opened, then shut as a consequence of the inactivation process. When Na$^+$ channels were blocked and the membrane potential was clamped to a constant value above threshold, the K$^+$ flux rose to, and remained constant at, a level determined by the mem-

brane potential (Figure 7–52). The conclusion is that the membrane potential determines the fraction of open K$^+$ gates, which is why they are described as "voltage-dependent gates."

The changes in Na$^+$ conductance and K$^+$ conductance were measured over a wide variety of membrane potentials maintained by the voltage clamp (Figure 7–53). The curves thus obtained enabled Hodgkin and Huxley to calculate a theoretical action potential, the properties of which duplicated remarkably well the experimentally observed phenomena (Figure 7–54). These included the subthreshold events, the development of the all-or-none response, the polarity reversal, the relative refractory period, the quantities of ions exchanged, and even the velocity of propagation. Subsequent work using more sophisticated voltage clamp apparatus and high-speed computing techniques has served to confirm and refine the ionic theory of the action potential.

The use of special microelectrodes has extended the method to a variety of nerve and muscle cells and also to other cells in the plant and animal kingdoms. Many other gated channels for potassium and sodium ions have been discovered—some are opened by small changes in the membrane potential, some by an increase in the intracellular calcium ion concentration, some by the binding of special molecules like neurotransmitters (see Chapter 17) or hormones, and some by mechanical perturbations, such as those in the hair cells of the inner ear. In fact, recent studies suggest that gated sodium and potassium channels play broader signaling roles in nonneural cells—for example, in mediating the growth response of certain white blood cells to the binding of substances to their surface. Minor variations on the themes we have introduced have also been discovered, such as different rates of response by gates or channels, as well as major variations, such as the use of different ions (Ca^{2+} in the barnacle, and even Cl$^-$ in *Nitella*). The basic principles of ion pumping, selective leakage, and specific transitory changes in ionic conductances of the membrane, however, are involved in all bioelectric phenomena.

One approach to elucidating the structural basis of the Na$^+$ channel has been to measure its limiting size by trying organic ions of different shapes. B. Hille and his collaborators found in the voltage-clamped frog neuron that only ions of cross-section 0.3 × 0.5 nm, or smaller, could fit through the Na$^+$ channel. Apparently, there is a hydrogen bond acceptor within this narrow channel, since ions like CH$_3$—NH$_3^+$ will not pass through while NH$_2$—NH$_3^+$ will, presumably because the methyl group cannot form a hydrogen bond with the postulated acceptor. The H-bonds that allow specific association of the channel with the penetrating ion are weak, so it can pass through by breaking and making a succession of these bonds.

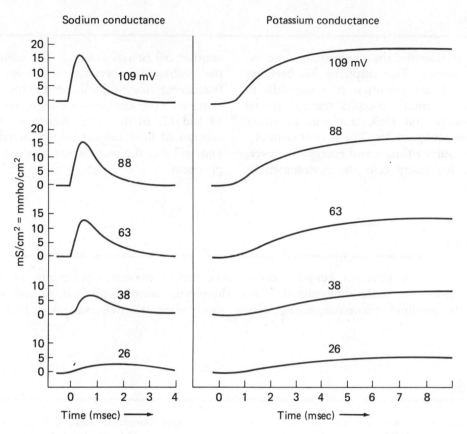

FIGURE 7–53
The effect of five different perturbations of membrane potential on the Na$^+$ and K$^+$ conductances. The membrane potentials were increased from the resting level by the indicated amounts, held constant by the voltage clamp method, and the conductances were measured. (From A. L. Hodgkin, as shown in B. Katz.)

Much productive research on the Na$^+$ channel involves the use of specific biological inhibitors, *tetrodotoxin* and *saxitoxin*. The former is found in large quantity in the Japanese fugu fish (*Spheroides porphyreus,* a puffer fish) and the latter in the plankton (such as *Gonyaulax* and *Gymnodinium*) that cause the notorious "red tides." Both of these inhibitors are potent toxins to man and other animals. The specificity of these inhibitors and their high affinity of binding to the Na$^+$ gate make it possible to use radiolabeled toxin molecules in binding assays to count the number of Na$^+$ channels in a neuron. Their density in the squid axon has been measured at $500/\mu m^2$, which is more than ten times as high as densities found in other neurons. An affinity-labeling analogue of tetrodotoxin that can be covalently attached to structures in or near the sodium channel allows it to be labeled and followed during isolation and purification procedures.

This strategy was used to purify a tetrodotoxin-binding protein from the electric organ of the electric eel *Electrophorus*. It is a large molecule (230,000 MW) and its amino acid sequence has been determined, but we shall have to wait a bit to learn the detailed molecular basis of voltage-dependent ion-specific channels.

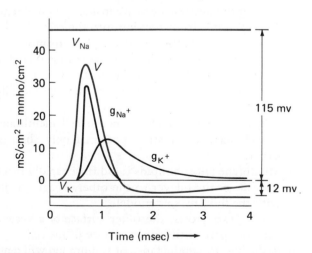

FIGURE 7–54
Theoretical reconstruction of changes in conductances due to Na$^+$ gate (g_{Na^+}) and K$^+$ gate (g_{K^+}) during the action potential. (From A. L. Hodgkin and A. F. Huxley, as shown in B. Katz.)

SUMMARY

This chapter has described the structure and function of biological membranes. The emphasis has been on those functions which are common to single cells in isolation: transport of small molecules through special permeability pathways and bulk transport of larger molecular complexes and particles. The environment of a cell is more than a source of mass and energy, however. That is is because for many cells the environment is another cell or cell product. The membrane also acts as the ambassador between sovereign cell states. This fascinating story of cellular diplomacy—how cells get along with one another—we begin to explore in Chapters 14 and 17. In the next chapter we take up again the account of how cellular energy metabolism works. As you will see, the membranes play a crucial role in these processes.

KEY WORDS

Active transport, carrier, conformation, coupling, endocytosis, energy, enzyme, equilibrium, exocytosis, feedback, free energy, ligand, lysosome, macromolecule, mass, mitochondrion, osmolar, osmosis, osmotic equilibrium, osmotic pressure, phospholipid, plasma membrane, solute, steady state, substrate, vacuole, vesicle.

PROBLEMS

1. To the following statements, respond TRUE or FALSE. If you respond TRUE, explain why the statement is true and give an example; if you respond FALSE, explain why the statement is not true.

   (a) You read a report that galactose is accumulated by *E. coli* cells: $[\text{gal}]_{\text{cell}}/[\text{gal}]_{\text{medium}} = 100,000$. This is consistent with one ATP hydrolyzed per galactose accumulated, assuming the ΔG for ATP + H_2O → ADP + P_i is -33 kJ/mole.

   (b) Raising the osmotic pressure inside a cell will tend to keep water from flowing into the cell.

   (c) A hypertonic solution is always hyperosmotic.

   (d) The part of the intracellular electrical potential due to the Gibbs-Donnan equilibrium will not be perturbed by increasing the potassium permeability of the membrane.

   (e) Valinomycin applied to a nerve cell membrane would make the resting potential only slightly larger.

   (f) Membrane phospholipids can diffuse rapidly from one side of the bilayer to the other, in agreement with the fluid mosaic model.

   (g) During exocytosis, the outer surface of a vesicle becomes part of the inner surface of the cell.

   (h) Reducing the conductance of sodium ion will tend to reduce the size of the nerve cell resting potential.

   (i) When placed in 0.3 M glycerol, the volume of a certain cell increases; from this observation one can conclude that 0.3 M glycerol cannot be isosmotic with the cell contents.

   (j) A coated pit is the seed of a peach from which the fruit has been incompletely removed.

2. A certain red blood cell membrane has been analyzed for its protein components. The cells were broken open and the "total membrane" fraction was isolated by centrifugation, solubilized by sodium dodecyl sulfate, and the proteins separated by electrophoresis in polyacrylamide gels (in the presence of sodium dodecyl sulfate). Only four major proteins (I–IV) were detected by staining with Coomassie Blue (see diagram).

   On the basis of the results seen in the diagrams, identify the topographic location of each of the major membrane proteins.

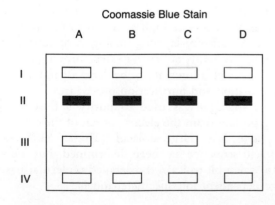

Coomassie Blue Stain

Autoradiography

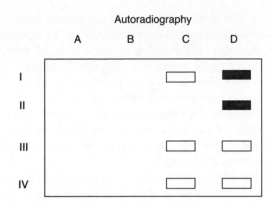

The material placed on the gel was:

Lane A: total membranes.
Lane B: total membranes, after washing membrane preparation three times in buffered 200-mM NaCl.
Lane C: total membranes, after labeling the cells with radioactive impermeable probe isethionyl acetimidate. The radioactivity was detected by autoradiography (see diagram).
Lane D: total membranes, after labeling the cells with radioactive permeable probe ethyl acetimidate; autoradiography was used to detect the radiolabeled proteins.

3. The Biologically Unusual Recombinant Procaryote (BURP) has been analyzed for membrane transport properties. The graph below records the relation between the results of two kinds of experiment: (a) a determination of the permeability of the BURP membrane to certain compounds, including two isomers of a 5-carbon sugar; (b) the partition coefficient (oil/water) of the same compounds. Discuss the relative positions on the graph of each of the compounds graphed in terms of your general knowledge of membrane structure and function.

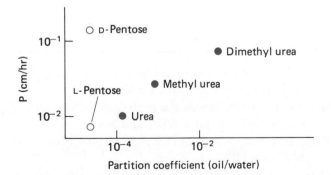

4. Prove that the mathematical model for the net flux in a facilitated diffusion is the same as Fick's first law when the transporter is symmetrical—that is, has the same affinity for substrate on both sides of the membrane and the same maximum velocity—and is very far from saturated with substrate.

5. How, during receptor-mediated endocytosis of LDL, does cholesterol get from the secondary lysosome into the cytoplasm of the cell?

6. Imagine the BURP to have a nonionic substrate (S) that is brought into the cell by cotransport with sodium ions, one S per Na^+.

	[Substrate]	[Na]	ΔE_m
Outside	0.5 mM	100 mM	0 mV
Inside	?	10	−50

(a) Given the tabulated distribution of *concentrations only* of substrate and sodium ion, estimate the maximum intracellular concentration against which net inward transport could still occur.
(b) Taking account *also* of the tabulated membrane potential, assumed to be maintained constant, again estimate the maximum intracellular substrate concentration against which net inward transport could still occur.
(c) Would your answer to part (b) be different if the substrate was an amino acid, like alanine?
(d) Would your answer to part (b) be different if the substrate was the amino acid arginine?
Assume the utilization of the substrate during metabolism can be blocked without perturbing the transport system. Each of the lines in the double reciprocal graph below illustrates the kinetics of inward transport of S as a function of $[Na^+]$ in the bathing medium.

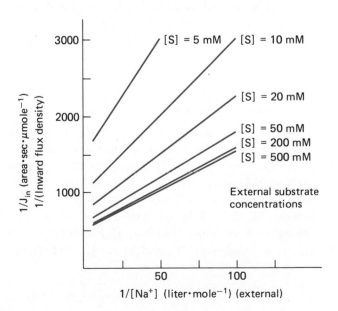

(e) What is the apparent dissociation constant of the cotransporter for sodium ion? Briefly explain your reasoning.

(f) What is the apparent dissociation constant of the cotransporter for substrate? Briefly explain your reasoning.

7. The BURP cell membrane contains a glycoprotein which can be broken with cyanogen bromide into three peptides, each of which contains the amino acid tyrosine. The carbohydrate is confined exclusively to peptide #1. All the sialic acid can be removed from the carbohydrate by treating it with the enzyme neuraminidase. When the cells are incubated with lactoperoxidase and radioactive iodine, a treatment that labels tyrosine residues, only peptide #1 is labeled; when membrane fragments are treated, both peptides #1 and #3 are labeled. Draw a model of the glycoprotein lying in the BURP cell membrane and explain how your drawing accounts for these results. Discuss the amino acid composition that you would expect to find for each peptide.

8. Animal cells are not the only ones with excitable membranes. Consider and discuss the ion distributions, shown below, in a fresh water alga from the genus *Chara*.

Ion	Pond (mM)	Cell Sap (mM)	E^{eq} (mV)
K^+	0.06	65	——
Na^+	0.10	65	——
Cl^-	0.06	110	——
protein		——	

The measured membrane potential is −180 mV.

(a) On the basis of the data presented in the table, are there ionic forms not listed for the pond water which must be present? If so, what could they be? Briefly explain.

(b) A 300 mM solution of impermeant sucrose has been determined experimentally to be isotonic for this alga. How much of what kind of protein is likely to be inside the cell? Briefly explain. How does *Chara* survive in pond water of such low osmotic pressure? Briefly explain.

(c) Calculate the equilibrium membrane potentials for each of the three small ions listed in the table, using the sign appropriate for the intracellular electrical potential.

(d) Sir William of Occam recommended that to explain any set of data, one ought first to use the simplest hypothesis. Therefore, we make the hypothesis that the ions are in electrochemical equilibrium across the *Chara* cell membrane. Considering the observed resting potential and *only* the K^+ distribution, can you rule out this equilibrium hypothesis? Briefly explain.

(e) Considering *also* the Na^+ distribution, can you rule out the equilibrium hypothesis? Briefly explain.

(f) Considering *also* the Cl^- distribution, can you rule out the equilibrium hypothesis? Briefly explain.

(g) If necessary, propose a specific alternative hypothesis to explain the observed ion distributions and the observed resting potential.

(h) Rank in decreasing value the resting ion conductances for the three ions: K^+, Na^+, Cl^-. Briefly explain your ordering.

(i) *Chara,* when stimulated, shows an action potential: depolarization of the membrane followed by slower repolarization. Compare and contrast your best qualitative guess of how an action potential might be generated in this cell with the mechanism of the action potential in the giant squid axon. Suggest an experiment to test your proposed mechanism.

SELECTED READINGS

Barnes, C. D., and Kircher C., eds. (1968) *Readings in Neurophysiology.* New York: John Wiley and Sons.

Branton, D., and Park, R. B., eds. (1968) *Papers on Biological Membrane Structure.* Boston: Little, Brown and Company.

Bretscher, M. S. (1985) The molecules of the cell membrane. *Scientific American* 253:100–108.

Finean, J. B., Coleman, R., and Michell, R. H. (1984)

Membranes and Their Cellular Functions. 3d ed. Oxford: Blackwell Scientific Publications.

Jain, M. K. (1988) *Introduction to Biological Membranes.* 2d ed. New York: John Wiley and Sons.

Katz, B. (1966) *Nerve, Muscle and Synapse.* New York: McGraw-Hill.

Ostro, M. J. (1987) Liposomes. *Scientific American* 256:102–11.

Osmosis and the Nernst-Planck Equation

APPENDIX 7–A

The form of the Van't Hoff equation (7–1) is not obviously related to the Nernst-Planck equation (6–2), yet it is a change in the free energy of water that provides the driving force for osmosis. Here we show how to calculate the ΔG of water and explore the properties of osmotic equilibrium. For water movement from side 1 to side 2 in the apparatus shown in Figure 7–3, the Nernst-Planck equation is written as

$$\Delta G_w = \Delta G_w{}^0 + R\, T \ln \frac{C_2}{C_1} + \overline{v}_w\, \Delta P. \qquad (A7\text{-}1)$$

ΔG_w is the free energy change for the process of moving a mole of water from side 1 (pure water) to side 2 (the solution). $\Delta G_w{}^0$ is the change in standard free energy, but since the energy released in forming water from its elements at standard conditions does not depend on location, $\Delta G_w{}^0$ is zero. C_2 and C_1 are the concentrations of water on the two sides, relative to the water concentration in the standard state. The last term of Equation A7–1 describes the influence of a difference in hydrostatic pressure ($\Delta P = P_2 - P_1$) on the change in free energy.

For incompressible fluids like water, the free energy at any location is proportional to the hydrostatic pressure at that location and $\Delta G > 0$ for movement into regions of higher pressure. The constant of proportionality, $\overline{v}_w$, is the *partial (molar) specific volume* of water, defined as the ratio of the increase in volume with increase in amount (measured in moles) of water, at constant temperature and pressure, and with the concentrations of other components also held constant. For solutions composed of several components, the products of partial molar specific volume ($\overline{v}_i$) times the number of moles (n_i) sums to the total volume: $V = \Sigma \overline{v}_i\, n_i$. For dilute aqueous solutions this sum is dominated by the partial volume of water (over 55 moles/l) and $V \approx \overline{v}_w \cdot n_w$.

At equilibrium, $\Delta G = 0$ and Equation A7–1 can be solved for the hydrostatic pressure difference, ΔP^{eq}, which depends on the concentration of water in the two regions separated by the semipermeable membrane:

$$\Delta P^{eq} = -\frac{R\,T}{\overline{v}_w}\ln\frac{C_2^{eq}}{C_1^{eq}}. \qquad (A7\text{-}2)$$

Equation A7–2 shows that ΔP is positive when the concentration of water is lower in side 2 than in side 1—in other words, the tendency of water to flow into a region of lower water concentration will be balanced by a hydrostatic pressure difference. We now show how this statement is related to the concentration of dissolved solutes. For this purpose it is convenient to measure the water concentration as a *mole fraction,* defined as the number of moles of water divided by the total number of moles of all substances: $n_w/(n_w + n_s)$, where n_s represents the number of moles of all dissolved solutes. The standard state appropriate to this concentration measure is a mole fraction of one—that is, pure water. Thus, we can set $C_1^{eq} = 1$ and $C_2^{eq} = n_w/(n_w + n_s)$. But, since the mole fractions of solutes and water must add up to one, $C_2 = 1 - n_s/(n_w + n_s)$ and

$$\Delta P^{eq} = -\frac{R\,T}{\overline{v}_w}\ln\left[1 - \frac{n_s}{n_w + n_s}\right].$$

For dilute aqueous solutions the number of moles of solute(s), n_s, is numerically small relative to n_w; thus, $1 - n_s/(n_w + n_s)$ is approximately equal to $[1 - n_s/n_w]$. It can be verified by a hand calculator or a table of natural logarithms that $\ln[1 - n_s/n_w]$ is approximately equal to $-(n_s/n_w)$. This allows us to approximate Equation A7–2 as

$$\Delta P^{eq} = R\,T\,\frac{n_s}{n_w\overline{v}_w}.$$

Since $n_w\,\overline{v}_w$ is approximately equal to the total volume V, in liters, $n_s/(n_w\,\overline{v}_w)$ is approximately equal to n_s/V, the *molar concentration* of all the dissolved solutes: C_s.

By analogy with macroscopic water flow that is driven by hydrostatic pressure, we express the driving force that causes water to move by osmosis as due to an *osmotic pressure,* $\Delta\Pi$. The osmotic pressure is equal to the hydrostatic pressure at equilibrium:

$$\Delta P^{eq} = \Delta\Pi = R\,T\,C_s, \qquad (A7\text{-}3)$$

where C_s is the summed molar concentrations of all dissolved solutes. The Van't Hoff relation shows how the osmotic pressure, the "pressure" that would exist in equilibrium with the hydrostatic pressure when flow ceases, depends on the concentration of dissolved solutes.

Since $\overline{v}_w \cdot \Delta\Pi$ is equal to the second term on the right-hand side of Equation A7–1, the free energy change for water movement is

$$\Delta G_w = \overline{v}_w\,\Delta P - \overline{v}_w\,\Delta\Pi.$$

It is common to speak of a concentrated solution of constant composition as having a high osmotic pressure. This equation shows that water will spontaneously enter a solution with an osmotic pressure higher than its surroundings unless there is a compensating hydrostatic pressure. When there are impermeable solutes on both sides of the membrane, the Van't Hoff equation is written $\Delta\Pi = R\,T\,\Delta C_s$, where ΔC_s is the difference in concentration of all impermeable solutes, [inside] − [outside]. If this difference in concentration is one millimolar, $\Delta\Pi = R\,T\,10^{-3} = 0.0821$ (liter atmospheres mole^{-1} K^{-1}) $\cdot$ 300 (K) $\cdot$ 10^{-3} (moles liter^{-1}) = 0.025 atmospheres = 19 mm Hg.

EXAMPLE 7–6

Calculate the osmotic pressure that exists in a bacterial cell with internal concentration of 300 milli-osmolar when it is growing in a 37°C laboratory growth medium of concentration 175 milliosmolar.

Solution

One can use the Van't Hoff equation: $\Delta\Pi = R\,T\,\Delta C_s$. Substituting the given values:

$$\Delta\Pi = 0.0821\ (\text{liter} \cdot \text{atmospheres})/\text{mole} \cdot \text{K}\ \times \\ 310\ \text{K} \times 0.125\ \text{molar}.$$

$$\Delta\Pi = 3.18\ \text{atmospheres} = 2420\ \text{mm Hg}.$$

To keep the cell from lysing, this very high pressure difference must be resisted by the bacterial cell wall.

APPENDIX 7–B

Countertransport Kinetics

Quantitatively, countertransport can be described by a mathematical model that describes simple competitive inhibition of an enzyme, similar to Equation 5–9. For this case it is easy to satisfy the condition that the enzyme (transporter) can combine with the substrate (OMG) or the inhibitor (glucose) but not both, because there is assumed to be only one binding site on the transporter. The major difference from the model for competitive inhibition of enzymes is that the model for countertransport must keep track of both inside and outside concentrations and account for two unidirectional fluxes. If we assume the Michaelis-like constants are the same for both influx and efflux of OMG, and that the "inhibition" constant for glucose is the same on the inside and outside surfaces of the membrane, then the equation analogous to 5–9 for the net influx ($J_{in} - J_{out}$) is:

$$J = \frac{J_{max} [O]_o}{K_O \left[1 + \dfrac{[G]_o}{K_G} \right] + [O]_o} - \frac{J_{max} [O]_i}{K_O \left[1 + \dfrac{[G]_i}{K_G} \right] + [O]_i}$$

where [G] represents the concentration of glucose, [O] the concentration of radioactive OMG, and the subscripts i and o represent inside and outside the cell, respectively. K_O is the Michaelis-like constant for OMG and K_G the inhibition constant for glucose.

By examining this equation carefully one can see that when glucose is added to the outside, the inhibition factor

$$\left[1 + \frac{[G]}{K_G} \right]$$

is very small on the inside. The competitive inhibition by glucose thus reduces the influx of OMG (first term of the equation) much more than the efflux (second term), and the net flux J of OMG becomes outward. Alternatively, one notes that the inhibition factor times the K_O is a measure of the *effective affinity* of the transport system for OMG. The equation shows that the effective affinity is different on the two sides of the membrane, a general feature of active transport mechanisms.

Mitochondria, Chloroplasts, and the Fixation of Energy

8

We have learned that cells can derive useful energy from fermentation processes like glycolysis plus the reduction of pyruvate to lactic acid. Most cells, however, oxidize their organic foodstuffs completely to CO_2 and H_2O. By doing so, they extract much more of the available energy from each food molecule. The oxidation of organic molecules in eucaryotic cells is completed in the mitochondrion, an organelle to which we will turn our attention presently. Procaryotic cells carry out the analogous reactions within their cytoplasm and in their plasma membranes—one of the reasons why mitochondria are thought to have evolved from procaryotes that established a symbiotic relationship with a primitive eucaryotic cell.

The complete oxidation of organic molecules yields much energy, part of which cells conserve in the form of ATP. A corresponding amount of energy must therefore be supplied when these large molecules are synthesized from CO_2. The organic molecules we see around us in such profusion could not have been formed without the input of nonbiological energy: the solar energy of visible light harnessed by the process of photosynthesis. This process occurs in eucaryotic cells in organelles called chloroplasts, which we shall discuss later in this chapter. Photosynthetic procaryotes are thought to have been the progenitors of the chloroplast. Before turning to these fascinating organelles, however, we first review the basic nomenclature and chemical concepts necessary to complete the discussion of cellular energy metabolism.

Electrons and Energy

Carbon and Energy Metabolism

Cells may be categorized on the basis of the source of the energy needed during metabolism and the carbon atoms utilized for synthesis. The *heterotrophs* use reduced organic compounds, like glucose, as a carbon source. In contrast, the *autotrophs* can use inorganic and oxidized CO_2 as a source of

TABLE 8–1

Standard Reducing Potentials (at pH 7) of Various Reactions Important in Cellular Energy Metabolism

Reduction Reaction	Standard Reducing Potential E′ (pH 7, V)
Succinate + CO_2 + $2e^-$ + $2H^+$ ⇌ α-ketoglutarate	−0.67
H^+ + e^- ⇌ 1/2 H_2	−0.42
Pyruvate + CO_2 + $2e^-$ + 2H ⇌ malate	−0.33
NAD^+ + $2e^-$ + $2H^+$ ⇌ NADH + H^+	−0.32
1,3-diphosphoglycerate + $2e^-$ + 2H ⇌ glyceraldehyde-phosphate	−0.29
Pyruvate + $2e^-$ + $2H^+$ ⇌ lactate	−0.19
Fumarate + $2e^-$ + $2H^+$ ⇌ succinate	−0.03
NO_3^- + $2e^-$ + $2H^+$ ⇌ NO_2^- + H_2O	+0.42
1/2 O_2 + $2e^-$ + $2H^+$ ⇌ H_2O	+0.82

Note: All the reactions are written as reductions. Any pair would result in one reaction being a reduction and the other being an oxidation:

$$NADH + H^+ + 1/2\ O_2 \rightleftharpoons NAD^+ + H_2O$$

which would have a standard potential of $+0.82 - (-0.32) = +1.14$ V. Since $\Delta G' = -2\ F\Delta E'$, the free energy change of this redox reaction pair at standard conditions (pH 7) would be -220 kJ/mole, illustrating both the spontaneous tendency of electrons to participate in reductions with large positive standard reducing potentials and the large amount of energy to be obtained by the oxidation by oxygen of the NADH produced in metabolism. (F is the Faraday constant, 96.7 kJ/volt-equivalent; the factor -2 is used because two electrons participate per mole of products formed.)

carbon for growth. The *phototrophs* can use electromagnetic energy in the form of visible light to generate ATP and other high-energy compounds for their metabolism, while the *chemotrophs* use the free energy released during oxidation reactions. Of course, photoautotrophs are very common (e.g., the green plants), but there are also *chemoautotrophs,* like the hydrogen bacteria that oxidize hydrogen gas with oxygen while obtaining their carbon atoms from CO_2.

The formal definition of *oxidation* is *the removal of electrons* (e^-) from a compound and of *reduction, the adding of electrons* to a compound. Thus, the reaction $Fe^{2+} \rightleftharpoons Fe^{3+} + e^-$ is an oxidation when read from left to right and a reduction when read from right to left. Organic molecules are rarely oxidized by the removal of an electron alone; usually the electron is accompanied by a proton to give a net loss of a hydrogen atom. It is nevertheless important to keep track of the electrons and protons separately, as we shall see later.

Chemotrophic cells are distinguished by the type of electron donor (**reductant**) and electron acceptor (**oxidant**) they can use. If an inorganic electron donor can be used (e.g., H_2 gas by the hydrogen bacteria), the cell is a *chemolithotroph.* If the electron donor is an organic molecule like glucose, the cell is a *chemorganotroph.* When the ultimate electron acceptor is an inorganic molecule

like O_2, the cell gets its energy from *respiration.* When the ultimate electron acceptor is an organic molecule like pyruvic acid in the red blood cell, the cell gets its energy from *fermentation.*

Oxidations and Reductions

Oxidation of organic molecules by molecular oxygen is spontaneous and gives off much energy. How can one measure the tendency to be oxidized? Cell biologists concerned with respiration or photosynthesis always measure this tendency, from the point of view of the electrons, in terms of a *reduction potential.* The electrons in an oxidizable substrate are considered to be at a particular reduction potential. They will tend to flow spontaneously to a molecule of a more positive reduction potential, thus reducing the destination molecule (B) and oxidizing the origin molecule (A):

$$AH_2 + B \rightleftharpoons A + BH_2 \tag{8–1}$$

This paired reduction-oxidation *(redox)* reaction can be written as two *half-reactions* to make it more obvious what is happening:

oxidation (of A) $\quad AH_2 \rightleftharpoons A + 2\ e^- + 2\ H^+ \tag{8–2}$

reduction (of B) $\quad B + 2\ e^- + 2\ H^+ \rightleftharpoons BH_2 \tag{8–3}$

To compare the reduction potential of two half-reactions, they are both written as reductions; Reaction 8-2 becomes

$$A + 2\,e^- + 2\,H^+ \rightleftharpoons AH_2 \qquad (8\text{-}4)$$

If the reduction potential of the half-reaction (8-3) is more positive than the reduction potential of the half-reaction (8-4), then the pair of half-reactions written and summed in Reaction 8-1 will proceed spontaneously from left to right. Each half-reaction represents an environment for the other, an environment which may accept or donate electrons. The definition of reduction potential is formulated in this way to agree with our expectation that a negative electron flows spontaneously away from a less positive environment, thus participating in an oxidation, and towards a more positive environment, participating in the reduction there.

As an example, the standard reduction potential (see Table 8-1 and compare with Table 6-1) for converting pyruvate to lactate is -0.19 V, while the standard reduction potential for the reverse of the oxidation of NADH to NAD^+ (the reduction of NAD^+ to NADH) is -0.32 V. Since the pyruvate-lactate standard reduction potential is more positive, pyruvate will accept electrons at the expense of NADH to give the redox reaction

$$\text{pyruvate} + \text{NADH} + H^+ \longrightarrow \text{lactate} + NAD^+.$$

The net standard potential for this reaction is $(-0.19) - (-0.32) = 0.13$ V.[1] We can then apply the conversion from electrical potential to free energy $\Delta G = z\,F\,\Delta E$ (see Equation 7-2), noting that for electrons, $z = -1$ and that there are two electrons involved in this redox reaction. This gives $\Delta G = -25$ kJ/mole (see Table 6-1).

The numerical values of reduction potentials depend both on the concentrations of the oxidized and reduced species, and on intrinsic features of the molecular structure of the reduced and oxidized form. Thus, an equation like the Nernst-Planck equation can be written to model the potential. It turns out that the free energies of redox reactions are among the most accurately known. This is because, by a suitable separation of the oxidation from the reduction reaction, the electrons can be forced to travel from the reductant to the oxidant through a wire, where their flow can be measured easily. This is, of course, another way of describing a battery, whose potential can be measured by a voltmeter.

The advantage of complete oxidation to CO_2 and H_2O of an organic compound over its fermentation is that *inorganic oxidants have very high positive reducing potentials,* much higher than those of organic oxidants (Table 8-1). Since the free energy of oxidation is proportional to the total rise in reduction potential, this means that more energy is available per food molecule oxidized. The oxidation of NADH by oxygen, for example, yields 220 kJ/mole of NADH, almost nine times the energy available from oxidizing NADH by pyruvate. The mitochondrion breaks up the large release of free energy into a series of steps and couples some of those steps ultimately to the synthesis of ATP.

The Mitochondrion and Respiration

Mitochondria: Preparation and Appearance

In 1910 O. Warburg, the inventor of the Warburg respirometer used to measure oxygen consumption by tissues, found that the oxidative reactions which take place in most tissues are concentrated in a small part of the cells. By grinding up the tissue, he made what we call a *homogenate* of the tissue; today we make a homogenate with a revolving pestle that fits snugly inside a glass tube into which tissue slices and a suitable medium have been placed. The homogenate can be spun in a centrifuge, and if this is done at successively increasing speeds, a fractionation or separation of the cell constituents can be obtained (see Chapter 2). When Warburg did this, he found that most of the enzymes responsible for the oxidation of the acids now known to be intermediates in the Krebs cycle were contained in a fraction called "large granules," because it could be spun down at relatively low speeds. Later, this fraction was called mitochondria (mitos = thread; chondrion = granule).

Cytologists knew about the existence of mitochondria for a long time. They saw these microscopically small objects in most cells of the human body and knew that mitochondria contained enzymes that reacted with certain dyes. After some 40 years of work with mitochondria, biochemists now know quite a bit more about them. They are the important constituents of Warburg's large granule fractions; all the enzymes of the Krebs cycle are found in these bodies, and they are therefore responsible for most of the energy transformation of aerobic cells. They have been called "powerhouses of the cell," converting foodstuffs—"input fuels"—by oxidation into ATP—the "working fuel" of the cell.

Mitochondria have been observed as slender, short threads with the light microscope since the middle of the last century, but only since differential centrifugation (see Chapter 2) was used to separate them from other cell constituents has their biochemical function been clearly identified. The application of electron microscopy to fixed, stained, and thin-sectioned material allowed the fine structure of mitochondria to become apparent. The mitochondrion is a structure found in most eucaryotic cells, including plant cells, algae, and protozoans (exceptions include mature mammalian red blood cells). In all

---

[1] We subtract the standard reduction potential of NAD^+/NADH (-0.32 V) because NADH is oxidized in this redox reaction. For redox calculations z is often represented as $-n$, where n is the number of electrons.

FIGURE 8-1
Mitochondria. (a) Three-dimensional drawing of a generalized mitochondrion, showing smooth outer membrane and inner membrane, thrown into folds (cristae), with particles attached. (b) Mitochondria in pancreas cell, showing outer and inner cristae membranes. × 70,000. (Courtesy of G. E. Palade.) (c) Mitochondria in pancreas cell surrounded by layers of rough endoplasmic reticulum to which ribosomes are attached. × 120,000. (Courtesy of S. Ito.)

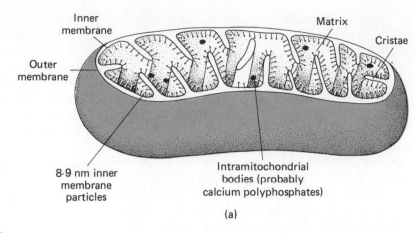

(a)

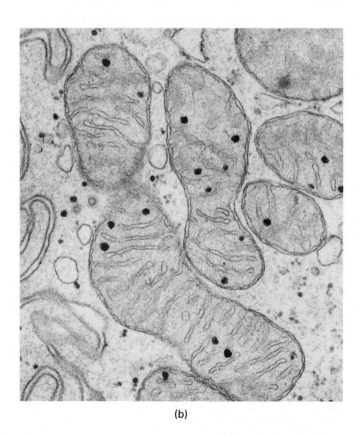

(b)

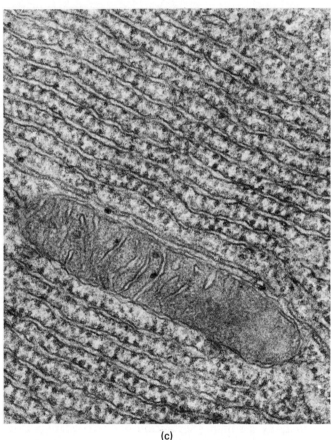

(c)

these cells its appearance is so characteristic that it is easy to recognize a mitochondrion simply by its morphology.

Mitochondria are composed of two membranes, each 5 to 6 nm thick, and thinner than the plasma membrane. The outer membrane covers the organelle and the inner membrane is thrown into folds, the *cristae,* which penetrate the interior (Figure 8–1). In lower forms such as protozoa (and, surprisingly, in the adrenal cortex of mammals), the folds of the inner membrane form tubules (Figure 8–2), whereas in the cells of other animals and plants they form flattened vesicles (Figure 8–1c). The

inner membrane encloses the amorphous material known as the *matrix* and a variety of electron-dense granules. Mitochondria contain DNA, which is consistent with the fact that they are partially self-duplicating organelles. The DNA of the mitochondrion contains only a small part of the hereditary information necessary for replication and growth; the remainder is found in the nuclear DNA (see Chapter 15).

Negative staining techniques show that the surface of the inner membrane adjoining the interior of the organelle is lined with particles attached to the mem-

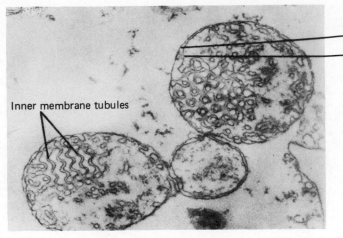

Outer mitochondrial
membrane

Inner mitochondrial
membrane

Inner membrane tubules

FIGURE 8–2
Mitochondria in an amoeba. One
mitochondrion shows the inner
membrane tubules in cross section. the
other in longitudinal section. × 50,000.
(Courtesy of G. Pappas.)

branes by slender stalks (Figure 8–3). The particles, called
coupling factors, together with the stalk and the portion of
the membrane to which it is attached, are where ATP is
synthesized. (In positive-stained electron micrographs
these particles are not seen, perhaps because they lie
buried and invisible in the matrix; alternatively, they may
be destroyed by the fixation technique. While modern
negative staining and freeze-etch methods show the
particles and stalks, there is still a possibility that their
appearance may be an artifact of the drastic dehydration
step used in negative staining.)

The disposition of the mitochondria in the cell
constitutes a visual documentation of their function and
the general relationship between cell structure and function. Mitochondria are often found in close juxtaposition
to the cellular machincry that they supply with ATP. In
muscle, for instance, mitochondria are neatly sandwiched
between the contractile elements of the muscle cell

(Figure 8–4); in the sperm tail, they are seen surrounding
the longitudinal fibrils responsible for the motion of the
sperm flagellum; in epithelia engaged in energy-requiring
active transport, mitochondria are in close proximity to
the enlarged, highly folded surface membrane (Figure
8–5); in cells that actively synthesize proteins, ribosome-
lined endoplasmic reticulum can often be seen closely
wrapped around the mitochondria; finally, in other cells,
mitochondria are found next to droplets of lipid, one of
the input fuels from which much energy can be obtained
by oxidation.

Enzymes in Mitochondria

Even when isolated, mitochondria retain their typical appearance in the electron microscope, although the
process of isolating them often damages their membranes, making their outlines less sharp. Mitochondria

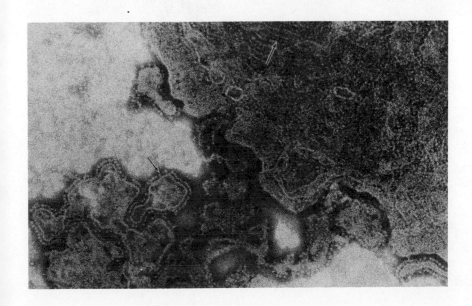

FIGURE 8–3

**Inner membrane particles of
mitochondria.** Isolated and
disrupted beef heart mitochondria were
negatively stained with
phosphotungstic acid so that the
background is dark. The picture clearly
shows the inner membrane particles or
coupling factors (arrows) attached to
what we know from other pictures are
the cristae (inner membranes) of the
mitochondria. × 135,000. (Courtesy of
S. Fleischer and A. Saito.)

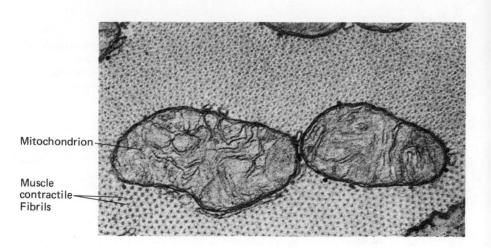

FIGURE 8–4

Mitochondria in a cardiac muscle cell. The mitochondria are embedded between the muscle contractile elements, shown in cross section. × 65,000. (Courtesy of D. W. Fawcett.)

from most cells almost always have the same enzymatic activities. Certain enzymes in the cell seem to be found solely in the mitochondria. Table 8–2 shows the results of a typical experiment in which a liver homogenate was separated into its various subcellular fractions by differential centrifugation. Of the total succinate dehydrogenase (SDH) activity and cytochrome oxidase activity in the liver cell, some 60 and 80%, respectively, were found

in the mitochondrial fraction. The presence of these enzymes in the other cell fractions is thought to be due to contamination by whole or broken mitochondria. The interpretation that both enzymes are mitochondrial is strengthened by the observation that not only are large fractions of the activities found there, but also that the activities are most concentrated there; the amount of enzymatic activity per milligram of protein, the *specific*

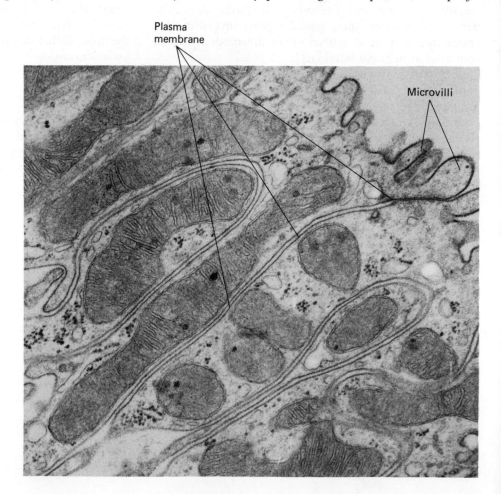

FIGURE 8–5

Mitochondria in proximity to surface membrane of distal convoluted tubule of kidney. × 45,000. (Courtesy of M. Farquhar and G. E. Palade.)

TABLE 8–2

Distribution of Succinate Dehydrogenase and Cytochrome Oxidase Activities in Fractions of Mouse Liver Cells[*]

Fraction	Succinate Dehydrogenase		Cytochrome Oxidase	
	Activity	Activity/mg Protein	Activity	Activity/mg Protein
Homogenate	4.25	1.34	6.86	2.06
Nuclei, cell debris[1]	0.84	1.65	1.36	2.44
Mitochondria[2]	2.40	3.18	5.39	6.46
Mitochondrial supernatant[3]	0.18	0.09		
Small particles[4]			0.29	0.35
Final supernatant[5]			0.00	0.00
Mitochondria plus supernatant	3.15			
Mitochondria plus nuclei, cell debris plus mitochondrial supernatant	4.18			

[*]Data from G. H. Hogeboom and W.C. Schneider; enzyme activities measured in abitrary units.
[1]Pellet of first centrifugation.
[2]Pellet of second centrifugation.
[3]Supernatant of second centrifugation.
[4]Pellet of third centrifugation.
[5]Supernatant of third centrifugation.

activity, is greater in the mitochondria than in any other fraction, or in the whole cell homogenate.

Sometimes there is a cofactor for a mitochondrial enzymatic activity in some other cell fraction. Although little or no SDH activity as such resides in the supernatant fraction, there is something there that when added to mitochondria increases their succinate dehydrogenase activity.

Experiments like these have permitted us to determine the cellular site of many enzymes and cofactors. In just this way it was found that the enzymes of the Krebs cycle are localized in the mitochondria. This was also inferred from the early observation that mitochondria can oxidize pyruvate completely to carbon dioxide and water. Since this happens only via the Krebs cycle, we can say that all the enzymes of the cycle must reside in the mitochondria. For some of these enzymes there is a large portion of the total cellular activity that is found outside the mitochondria, but the Krebs cycle itself is solely a mitochondrial process.

As methods for the preparation of mitochondria improved, it became possible to isolate various mitochondrial subfractions and demonstrate that certain enzyme activities were associated with one or another of them. Within the matrix, for example, are found not only the Krebs cycle enzymes but also DNA, RNA, and ribosomes. Electron carriers such as cytochromes are integral proteins of the inner membrane (an exception is cytochrome *c,* which is a peripheral protein), as are the flavoproteins involved in the oxidation of certain substrates—for example, α-ketoglutarate dehydrogenase. Also found in the inner membrane are a special phospholipid known as cardiolipin and several transport proteins that catalyze the movement of substrates into and out of the matrix. Between the inner and outer membranes enzymes are found that catalyze the exchange of phosphates among various nucleotides—for example, adenylate kinase, which catalyzes $ADP + ADP \rightleftharpoons ATP + AMP$. The outer membrane contains *porin,* which provides an aqueous channel through the membrane (see Chapter 7). This allows the movement of substrates and products with molecular weights less than 10^4 between the cytoplasm and the intermembrane space. Also in the outer membrane are found various oxidase enzymes that help to degrade certain molecules—monoamine oxidase, for example, whose substrate is the hormone epinephrine.

Energy Transduction and Phosphate Metabolism

Biological energy conversion is closely related to inorganic phosphate metabolism. As long ago as 1907, the biochemists A. Harden and W. Young found that inorganic phosphate disappeared during yeast cell-free fermentation. They also found that enzymes in yeast esterify this phosphate into organic forms, such as the hexose monophosphates and hexose diphosphates, which they later isolated from yeast cells. In the period from 1930 to 1938 it was found that during the anaerobic oxidation of carbohydrate, inorganic phosphate is also converted into an organic form, which in this case is 1,3-diphosphoglyceric acid (see Figure 6–15). We now know that these enzymatic steps are part of the glycolytic scheme in which glucose is broken down anaerobically to lactate and during which energy is made available. But respiration-dependent phosphorylation was not observed until 1930 by V. Engelhardt and 1937 by H. Kalckar.

In 1939 substantial progress was made in determining the role of phosphates in the oxidation of what we now call the Krebs cycle substrates. Many laboratories, almost simultaneously, observed that inorganic phosphate was *necessary* for the cellular oxidation of citrate, glutamate, fumarate, malate, and pyruvate; without the addition of phosphate to cell extracts, very little oxidation was observed. Hence, disappearance of inorganic phosphate and oxidation of substrate are *coupled* to each other. In kidney, liver, and muscle tissues the inorganic phosphate that disappeared was found in combination with glucose, fructose, or adenylic acid; in muscle it was also found in combination with creatine.

One object of research in those days was to find how many molecules of inorganic phosphate are taken up into organic form (as ATP) per atom of oxygen consumed during the oxidation of substrate: the *P/O ratio*. This ratio measures the conversion of oxidation energy into biologically useful ATP. The P/O ratio was usually found to be equal to two or three. Since each O atom can accept only the two electrons removed by a single oxidation, P/O ratios greater than one imply there must be for each pair of electrons more than one enzymatic step that is coupled to the synthesis of ATP.

Similarly, these high P/O ratios mean that much more inorganic phosphate disappeared than could be accounted for by the then-known step of glycolytic phosphorylation: oxidation of glyceraldehyde-3-phosphate to 1,3-diphosphoglyceric acid (see Chapter 6). Hence, it was proposed that the remainder of the phosphorylations occurred during the transfer of electrons from the Krebs cycle substrates to oxygen. Later on, as the Krebs cycle was more fully worked out, these proposals were confirmed by experiment. For example, during the oxidation of α- ketoglutarate to succinate, one inorganic phosphate becomes esterified into guanosine triphosphate; this was an easily observed result during early work on the Krebs cycle. But as techniques improved, much more ATP was found to be synthesized as the result of oxidizing α-ketoglutarate to succinate: not one, but up to four moles of inorganic phosphate are esterified for every atom of oxygen consumed, or for

every mole of α-ketoglutarate oxidized. Thus, three more phosphate esterifications had to be accounted for. They turned out to be the consequence of oxidizing the NADH that was reduced when α-ketoglutarate was oxidized. In general, Krebs cycle oxidations result in the reduction of coenzymes, NAD^+, and *flavin adenine dinucleotide (FAD)*; most of the formation of organic phosphate occurs as the result of oxidizing these coenzymes, ultimately by molecular oxygen.

The Electron Transport Chain

The term *electron transport chain* is a shorthand way of describing a series of enzyme-catalyzed reactions, in each of which a donor transfers electrons to an acceptor, the donor becoming oxidized and the acceptor becoming reduced in the process. The electron transport chain in the mitochondrion provides a *common pathway for the oxidation of reduced coenzymes*. These coenzymes become reduced during the oxidation of various intermediates, such as pyruvate, and the electrons removed during oxidation of the coenzymes are passed along the electron transport chain, ultimately to reduce O_2 to H_2O. Pyruvate, for example, is oxidized by an enzyme that transfers a pair of electrons to the coenzyme NAD^+, to form NADH. (Actually, the NAD^+ "coenzyme" should be considered as a substrate, one that acts by accepting electrons, thus becoming reduced.) NADH then dissociates from that enzyme and combines with another enzyme to become oxidized. The oxidation of other substrates like α-ketoglutarate is accompanied by the reduction of coenzyme FAD to form $FADH_2$. Some of the electron transport components have been known for a long time. One of the best known is cytochrome *c*, found in so many different cells that it serves as a marker of the evolution of various organisms. Figure 8–6 illustrates the general idea of the electron transport chain, from oxidizable substrate to reducible oxygen.

Many substrates are oxidized by enzymes known as *dehydrogenases,* so-called because they remove two electrons and two hydrogen ions from the substrate. Like most enzymes, dehydrogenases are specific for their particular substrate. In the mitochondrion there are dehydrogenases specific for many of the Krebs cycle intermediates: α-ketoglutarate, malate, succinate, and so on. In some cases the dehydrogenase is a *flavoprotein,* consisting of a protein plus a flavin prosthetic group; both together constitute the active enzyme. (A prosthetic group is a tightly bound, specific, nonpeptide molecule required for enzyme activity.) The electron acceptor for some dehydrogenases is NAD^+; others use $NADP^+$ specifically; and still others can use either.

The first step in the mitochondrial electron transport chain (Figure 8–7) is the oxidation of NADH by an NADH dehydrogenase; this enzyme is a flavoprotein in

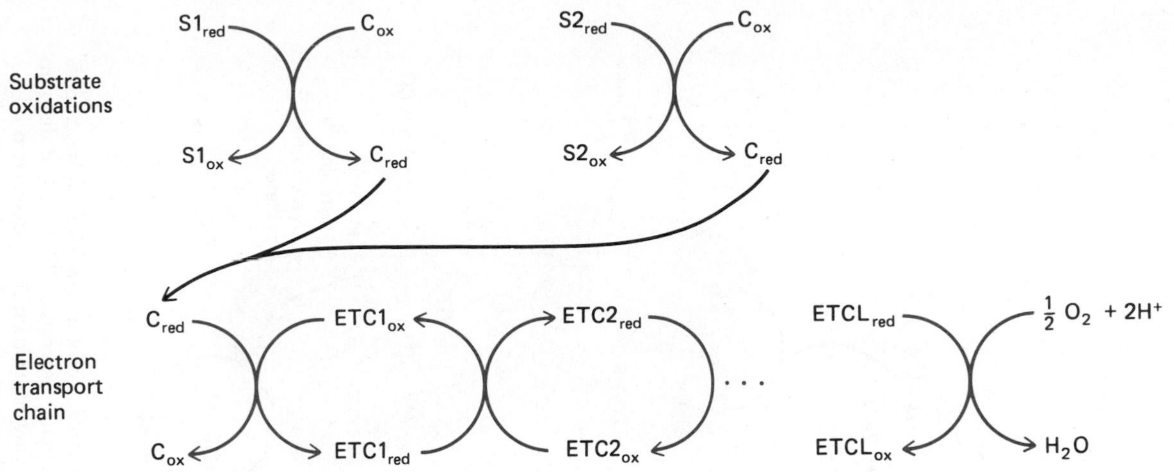

FIGURE 8–6
Simplified view of the electron transport chain in mitochondria.   Various substrates (S1, S2. . .) are oxidized, and a electron carrier (C_{ox}) common to all of them becomes subsequently reduced (C_{red}). The reduced carriers transfer their electrons to the common pathway of the electron transport chain. Components of this chain, ETC1, ETC2. . .) become successively reduced and oxidized. The last component of the chain ($ETCL_{red}$) transfers electrons to molecular oxygen. The reduced oxygen combines with hydrogen ions, also formed during substrate oxidations, to give H_2O.

which the prosthetic group *flavin mononucleotide (FMN)* is reduced to $FMNH_2$, then reoxidized by iron-sulfur complexes ($Fe \cdot S$) in the protein. Various iron-sulfur complexes can be distinguished by electron spin resonance spectroscopy; other proteins containing such complexes are found elsewhere in the electron transport chain. The NADH dehydrogenase complex gives up its electrons to the next carrier in the chain, *ubiquinone (or coenzyme Q)*, which then becomes reduced. Coenzyme Q also can accept electrons from another donor, the covalently bound *flavin adenine dinucleotide (FADH_2)* of flavoproteins like succinate dehydrogenase. The electrons from $FADH_2$ are transferred to coenzyme Q via an iron-sulfur protein. There are thus two entry points to the electron transport chain, one for NADH electrons and another for flavoproteins. Reduced ubiquinone (QH_2) transfers its electrons to cytochrome *b*, of which there are two types in the mitochondrion, distinguishable by their different absorption of visible light. Cytochrome c_1, cytochrome *c*, cytochrome *a*, and cytochrome a_3 are reduced and then oxidized in turn as electrons pass along the chain. Beyond coenzyme Q, only electrons are passed down the electron transport chain; the protons follow a separate pathway, as will be discussed below. Finally, four protons and four electrons from cytochrome a_3 reduce O_2 to two H_2O. See Fig. 8–8, p. 384, for the structures of some of the electron carriers.

Cytochromes are *heme proteins* in which the iron in the center of the heme becomes alternately oxidized and reduced. It is the heme group that absorbs visible light so strongly, thus giving the cytochromes their characteristic

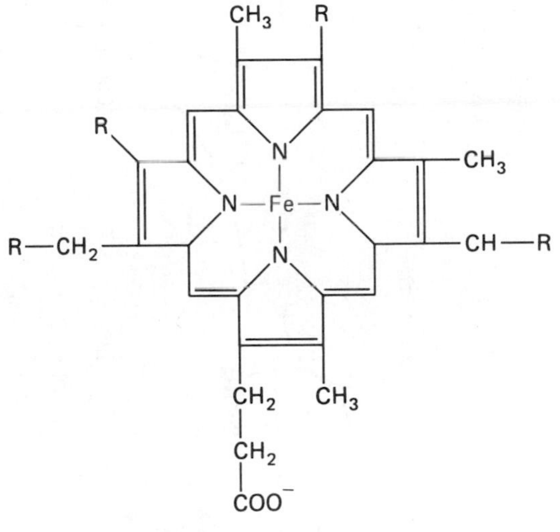

Generalized heme

color ("chromes"). Since each cytochrome has a distinctive absorption spectrum and since the spectrum changes when the iron atom is reduced from the Fe^{3+} to the Fe^{2+} state (Figure 8–9, p. 385), one can determine by spectrophotometry the fraction of each cytochrome that is reduced under various conditions. This analytical method was quite useful in working out the details of the electron transport chain pathway. The structure of the heme group is slightly different in cytochromes *a* and a_3, which together with other protein subunits form the *cytochrome oxidase* complex. Cytochrome oxidase complexes also contain copper atoms that are alternatively reduced and oxidized from Cu^+ to Cu^{2+} during electron transfer to oxygen.

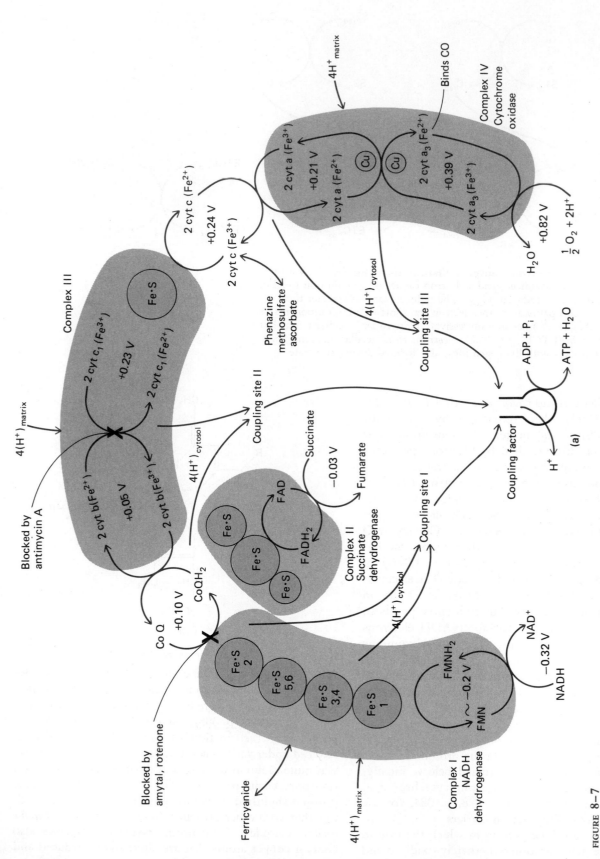

FIGURE 8–7

Electron transport chain of mitochondria. (a) The four complexes, the three sites of coupling to proton translocation (4H⁺ₘₐₜᵣᵢₓ →
4H⁺_cytosol), the common pathway for coupling the proton gradient with ATP synthesis, and the sites of inhibition of electron carriers by exogenous compounds are shown. Fe · S denotes iron-sulfur complexes. Exogenous compounds can donate electrons to or receive electrons from compounds in the chain. The numerical values associated with the various electron carriers (e.g., –0.25 V) are their standard reducing potentials. (b) Enlarged diagram of the coupling factor structure, showing inner membrane-associated F₀ and matrix-associated F₁. These are the sites where proton flow from the cytosol to the matrix is coupled to ATP synthesis. Note the site of the oligomycin block. (See further discussion beginning on p. 395.)

In the figure:

Blocked by amytal, rotenone

Ferricyanide

Complex I NADH dehydrogenase

FMN ~ –0.2 V $FMNH_2$

–0.32 V

NAD^+ / NADH

Fe · S 2, Fe · S 5,6, Fe · S 3,4, Fe · S 1

$4(H^+)_{matrix}$

$4(H^+)_{cytosol}$

Coupling site I

Co Q +0.10 V $CoQH_2$

Blocked by antimycin A

Complex III

$4(H^+)_{matrix}$

2 cyt b(Fe^{2+}) +0.05 V 2 cyt b(Fe^{3+})

2 cyt c_1 (Fe^{3+}) +0.23 V 2 cyt c_1 (Fe^{2+})

Fe · S

$4(H^+)_{cytosol}$

Coupling site II

Complex II Succinate dehydrogenase

Fe · S, Fe · S, Fe · S

FAD / $FADH_2$

Succinate –0.03 V Fumarate

Phenazine methosulfate ascorbate

2 cyt c (Fe^{2+}) +0.24 V 2 cyt c (Fe^{3+})

$4(H^+)_{cytosol}$

Coupling site III

2 cyt a (Fe^{3+}) +0.21 V 2 cyt a (Fe^{2+})

Cu   Cu

2 cyt a_3 (Fe^{2+}) +0.39 V 2 cyt a_3 (Fe^{3+})

$4H^+_{matrix}$

Binds CO

Complex IV Cytochrome oxidase

H_2O +0.82 V $\frac{1}{2} O_2 + 2H^+$

Coupling factor

$ADP + P_i$ → $ATP + H_2O$

H^+

(a)

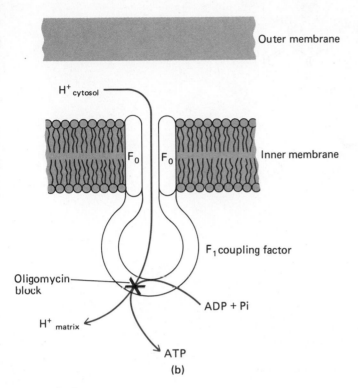

FIGURE 8–7
(continued)

There are three segments of the electron transport chain with sufficient energy differences under standard conditions for energy storage in the form of ATP: the oxidation of NADH by coenzyme Q, the oxidation of reduced cytochrome b by cytochrome c_1, and the oxidation of reduced cytochrome c by oxygen. Figure 8–7 shows that when a substrate is oxidized by NAD^+—that is, by an NAD^+-linked dehydrogenase—all three segments are traversed by the electrons and three molecules of ATP can be formed for each oxygen atom that is reduced to H_2O. The P/O ratio for these substrates is thus expected to be three. In the case of electrons from succinate oxidation, glycerol-phosphate oxidation, and for some of the electrons from fatty acid oxidation, the first segment of oxidizing NADH is bypassed; only two ATPs can be synthesized per pair of electrons and a P/O ratio of two would be expected.

As mentioned earlier, there is also a *substrate level phosphorylation* that occurs during the Krebs cycle in the mitochondrion. This is the NAD^+-linked oxidation of α-ketoglutarate to form succinyl-S-CoA, plus the synthesis of GTP from GDP + P_i that accompanies the release of succinate from coenzyme A. A transphosphorylation then occurs, forming ATP: GTP + ADP → GDP + ATP. In the case of α-ketoglutarate oxidation, therefore, the P/O ratio could be four.

The transfer of electrons from the substrate to oxygen takes place because in each transfer the electrons go from a compound with a more negative reducing potential to one with a more positive reducing potential. It is as if the electron transport chain were made up of a series of "waterfalls": the electrons start out at a high energy level, in the substrate, and end up at a lower energy level, in oxygen. This is normally a one-way flow. As we shall see, however, the flow of electrons can be reversed under certain conditions by putting energy into the system, just as water can be made to go uphill if sufficient energy is supplied.

Coupling of Oxidation to Phosphorylation

The experimental observation of oxidative phosphorylation and the early determination of P/O ratios led naturally to the idea of *coupling* between the transfer of electrons from substrate to oxygen and the synthesis of ATP. Coupling of these two processes is, in fact, directly observable. In respiring mitochondria, oxidation of Krebs cycle substrates is quite low without concomitant phosphorylation. Carefully isolated mitochondria oxidize α-ketoglutarate or pyruvate very slowly in the absence of inorganic phosphate and ADP. In the presence of inorganic phosphate only, without ADP, there is also little oxidation. As soon as ADP is added, however, the substrate is oxidized, inorganic phosphate disappears, and ATP is formed. Coupling of phosphorylation to oxidation is therefore mandatory—that is, normally the electrons are not transported unless synthesis of ATP also takes place. ADP is the necessary acceptor for inorganic phosphate; addition of ATP only has very little effect on the oxidation rate. If ATP is broken down to ADP, however, oxidation ensues and the ADP is then rephosphorylated to ATP. This can easily be demonstrated by adding glucose, ATP, and the enzyme hexokinase, which catalyzes the transfer of phosphate from ATP to glucose to form glucose-6-phosphate and ADP—the ADP necessary for oxidation to take place.

Oxidative Phosphorylation

It is helpful to ask why the cell employs this complicated means of capturing the energy in reduced substrates. The reason is that by breaking the large difference in energy potential between substrate and oxygen into several steps, it is possible to conserve more of the energy. Each step with a sufficient difference in energy potential to meet the requirements can be used for storage. Any energy in excess of the requirements will be transferred to the environment as heat. By providing more such steps, the mitochondrion can store more of the energy released per pair of electrons passing from substrate to oxygen. The ultimate form of energy stored by the mitochondrion is ATP.

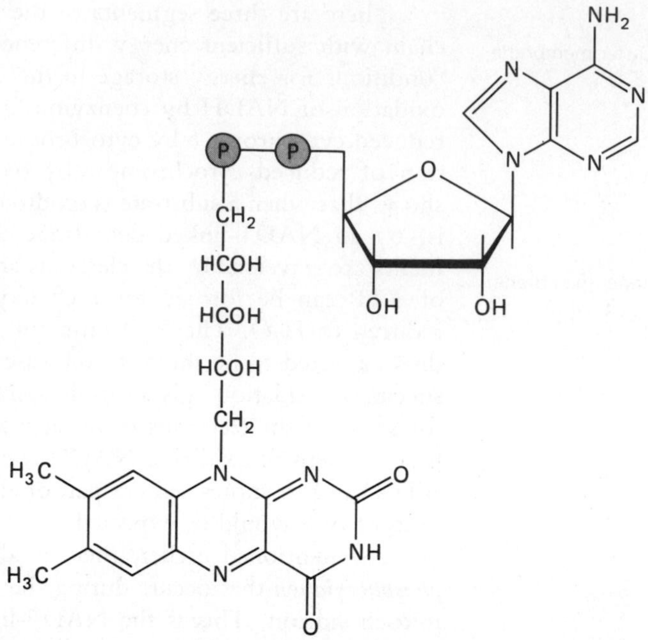

Flavin adenine dinucleotide
(FAD)

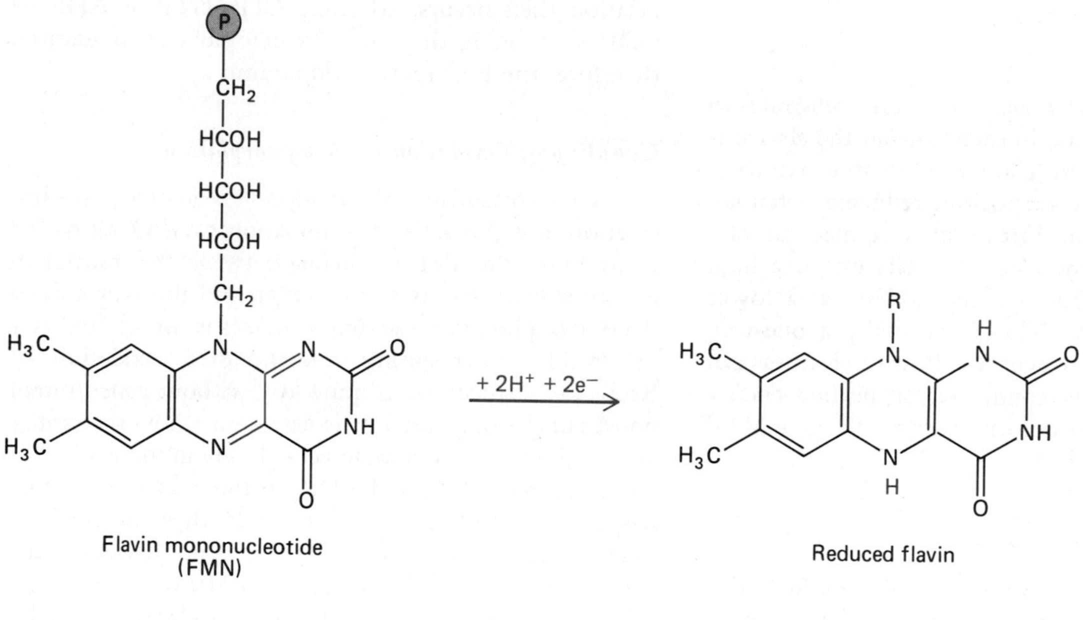

Flavin mononucleotide
(FMN)

$+ 2H^+ + 2e^-$

Reduced flavin

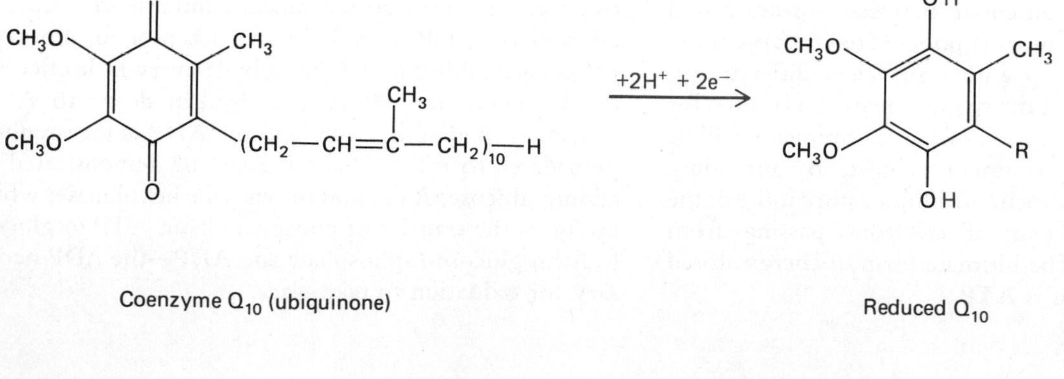

Coenzyme Q_{10} (ubiquinone)

$+2H^+ + 2e^-$

Reduced Q_{10}

FIGURE 8–8
Carriers in the electron transport chain. The structure of FAD, FMN, and coenzyme Q are shown, the latter in both oxidized and reduced forms.

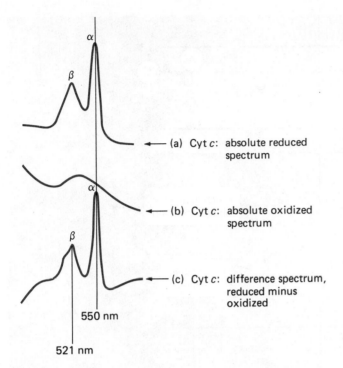

←— (a) Cyt *c*: absolute reduced
 spectrum

←— (b) Cyt *c*: absolute oxidized
 spectrum

←— (c) Cyt *c*: difference spectrum,
 reduced minus
 oxidized

550 nm

521 nm

FIGURE 8–9

Spectra of purified cytochrome *c*. The absorbance of light by this electron transport chain component changes with its oxidation state. When fully reduced (a) by an agent like dithionite, the spectrum shows two peaks (α at 550 nm and β at 521 nm). When fully oxidized (b) by an agent like ferricyanide, the peaks disappear. The *difference spectrum* (c) is the result of subtracting spectrum (b) from spectrum (a). The spectrum of a sample of cytochrome *c* can be analyzed, knowing the spectra (a) and (b), to determine the fraction of the sample in the reduced state.

Table 8–3 shows a typical experiment in which the addition of hexokinase, in the presence of ATP and glucose, causes an increase in the oxidation of pyruvate or fumarate. In intact tissues, the *respiratory control ratio*—the ratio of the oxygen consumption rate (or substrate oxidation rate) in the presence of ADP to the rate in the absence of ADP—can be as high as 100. The response to ADP of increasing the rate of oxidation is called the "phosphate acceptor" effect. Similar experiments indicated the necessity of inorganic phosphate for oxidation; the reason is precisely the same as in the case of ADP: both act as cosubstrates for oxidative phosphorylation. Experiments such as these firmly established the existence, although not the mechanism, of coupling between phosphorylation and the oxidation of substrates in mitochondria. In Chapter 14 the possible meaning of this for the regulation of certain aspects of cellular metabolism will be discussed.

In mitochondria from heart and skeletal muscle the rate of electron transport is affected by the activity of one form of *creatine kinase,* found in the inner membrane of the mitochondria. The ATP formed by mitochondrial phosphorylation is used by creatine kinase to phosphorylate creatine; the ADP formed is then available to the phosphorylation system (Figure 8–10). In the presence of creatine this process keeps generating ADP and thus provides a phosphate acceptor to ensure a high rate of oxidation. Of course, not all of the ATP produced by muscle mitochondria is used by creatine kinase, but it has sufficient influence to keep oxidative phosphorylation running until the creatine pool is fully charged to creatine phosphate. The rest of the ATP produced is transported through the mitochondrial membrane, via the ADP-ATP exchange porter system shown in Figure 8–10.

Because phosphorylation normally is tightly coupled to oxidation, an agent that inhibits either process affects both. If mitochondria are isolated with their membranes damaged, however, the coupling will be imperfect (this is discussed in more detail later in this chapter). Under these conditions, an inhibitory substance can be evaluated for its primary effect on oxidation or phosphorylation. Thus, the antibiotic *oligomycin* was found to exert its primary

TABLE 8–3

Effect of Phosphate Acceptors on Oxidative Rate of Liver Mitochondria

	Oxidative Rate (μl O_2/10 min/5 mg protein)	
Additions	**Pyruvate as Substrate**	**Glutamate as Substrate**
ATP, inorganic phosphate	12	8
Inorganic phosphate	18	12
ADP, inorganic phosphate	75	101
ATP, inorganic phosphate, hexokinase, glucose	80	116

Note: The rates of substrate oxidation, measured by O_2 uptake, are stimulated by the presence of ADP, whether added directly or produced during the phosphorylation by ATP of glucose to glucose-6-P. (ATP + glucose $\xrightarrow{hexokinase}$ ADP + glucose-6-phosphate.) (Data from P. Siekevitz.)

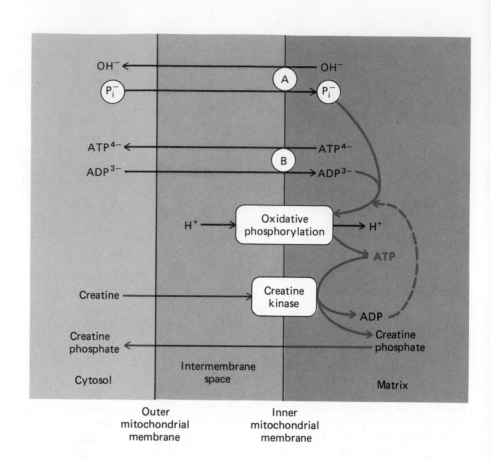

FIGURE **8-10**
Function of creatine kinase in oxidative phosphorylation by muscle mitochondria. The energy for oxidative phosphorylation is produced by the flow of protons from the intermembrane space to the matrix. Inorganic phosphate can enter the mitochondria via an hydroxyl-phosphate symport (A). Likewise, there is an ADP-ATP antiporter for the adenine nucleotides (B). Creatine, coming into the mitochondrial matrix, is phosphorylated by creatine kinase and the ATP generated through oxidative phosphorylation. The ADP thus formed within the mitochondrion can continue to be used as a "phosphate acceptor"; the creatine phosphate is transported to the cytoplasm to provide energy for the muscle contractile system. This continues as long as there is both substrate available to be oxidized via the electron transport chain and a supply of creatine.

effect on phosphorylation; it will also inhibit electron flow to the extent that tight coupling exists. The poison *rotenone,* on the other hand, blocks electron transfer from NADH dehydrogenase to coenzyme Q; its effect on phosphorylation is secondary (Figure 8–7). These inhibitors were among those used to identify the components of the electron transport pathway, using strategies for pathway analysis similar to those discussed in Chapter 6.

The Search for Coupling Sites

After the existence of coupling between oxidation and phosphorylation was fully documented, the attention of biochemists was directed to establishing where along the electron transport chain the coupling to ATP generation occurs, the *coupling sites.* The existence of such sites was confidently expected, by analogy with the sites found in the glycolysis pathway: the oxidation of glyceraldehyde phosphate and the transfer of phosphate from phosphoenolpyruvate (see Chapter 6). But the analogy with glycolysis was false, and misled biochemists for an embarrassingly long time.

One way of approaching this problem is to isolate separate segments of the electron transport chain, assaying each segment for its P/O ratio. Thus, the oxidation of one mole of exogenously added, reduced cytochrome *c*

gives one mole of ATP, the formation of which must be coupled to reactions occurring between cytochrome *c* and oxygen, *coupling site III* (Figure 8–7). The oxidation of succinate gives two ATPs, the formation of the additional one being coupled to a reaction occurring somewhere between the succinate-linked flavoprotein and cytochrome *c*—either during the transfer of electrons from coenzyme Q to cytochrome *b,* or from cytochrome *b* to cytochrome c_1, *coupling site II.* The oxidation of NADH gives three ATPs, the third one being coupled to a reaction somewhere between NADH and coenzyme Q, *coupling site I.*

Elegant work by B. Chance and his colleagues made possible a detailed kinetic study of the electron transport chain in mitochondria, first in isolated whole cells and later even in various animal tissues in situ. By using micromethods—oxygen electrodes to measure oxygen consumption and rapid-flow dual wavelength spectrophotometry to measure simultaneously the reduced and oxidized states of the electron carriers—the sequence of electron carriers along the chain was established. Instead of measuring respiration by its last reaction, oxygen consumption, they measured the reduction of the intermediate electron carriers using the characteristic absorption spectra of the oxidized and reduced states (see Figure 8–9).

An example of their work is a set of experiments involving known electron transport inhibitors such as *antimycin A* and *rotenone* (a major commercial insecticide). These react with components of the chain as shown in Figure 8–7 and lead, in the presence of NADH and oxygen, to reduction of some components and oxidation of others. This means there is a *crossover point* that identifies the position in the pathway where electron flow has been inhibited (see discussion of pathway analysis in Chapter 6). In the presence of antimycin A, for example, cytochrome b and coenzyme Q are reduced while cytochromes c_1, c, and a/a_3 are oxidized (Figure 8–11). This crossover of the oxidation-reduction state allows identification of the inhibition point of antimycin A and also confirms the order of components of the pathway of the electron transport chain: ubiquinone and cytochrome b must be closer to NADH than cytochromes c_1, c, or a/a_3. As for the glycolytic pathways described in Chapter 6, the use of specific inhibitors immensely aided the analysis of the order of electron carriers from NADH to oxygen.

The action of one of these inhibitors isolates functionally two segments of the electron transport chain. Thus, if one treats intact respiring mitochondria with antimycin A, little ATP can be synthesized after the electron carriers upstream of the block become fully reduced and those downstream become fully oxidized. Adding excess reduced ascorbate, which can donate its electrons to cytochrome c, allows the synthesis of one ATP per pair of electrons (Figure 8–7). This experiment shows that coupling site III lies after cytochrome c.

Another interesting aspect of the work of B. Chance and G. R. Williams was an attempt to identify the coupling sites using a radically different experimental approach. Simply put, they varied the concentrations of substrate and ADP, and measured the states of oxidation and reduction of various components of the electron transport chain. If mitochondria are starved of substrate

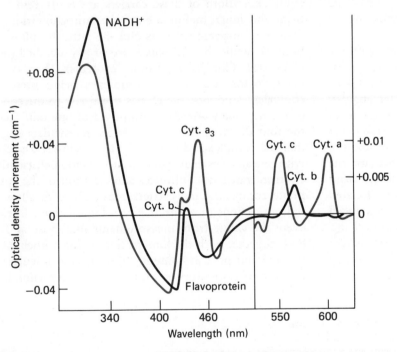

FIGURE 8–11
Difference spectra in rat liver mitochondria. The spectrum was obtained by using as the reference a sample of mitochondria with electron carriers in the fully oxidized state. The spectrum of the fully reduced (anaerobic) state was compared with the reference and the difference is plotted as the color line. This technique also cancels out turbidity and light scattering by mitochondria. Bands of wavelength(s) in the spectrum where the absorbance is dominated by one of the electron carriers are indicated by their names. The black line shows the spectrum, relative to the reference, when antimycin A was added in the presence of oxidizable substrate. Cytochromes a/a_3 and c are fully oxidized in the presence of antimycin A, but cytochrome b is fully reduced. This shows that antimycin A blocks electron transport between cytochrome b (which therefore becomes reduced) and cytochrome c (which becomes oxidized; see Figure 8–7).

in the presence of oxygen, all the electron carriers become oxidized. They are partially reduced during steady-state substrate oxidation and ATP synthesis. If substrate and only a small amount of ADP are added to previously starved mitochondria, respiration proceeds until all the ADP is phosphorylated to ATP, at which time the respiration rate becomes strongly inhibited because of the tight coupling we discussed earlier. There are changes in the oxidized-reduced ratio of the various electron transport chain components during the onset of respiration inhibition. Upstream of the block caused by ADP starvation, electrons flow into carriers, reducing them. Downstream of the block, electrons tend to flow out of the carriers, oxidizing them. In other words, crossover points can be observed. These are identified as sites where electron transport is dependent on the presence of ADP. Crossover points can also be observed during recovery following the addition of ADP. Under such recovery conditions there occurs, for example, a relative oxidation of cytochrome c and a relative reduction of cytochrome a/a_3. Adding ADP allows electrons to leave cytochrome c and flow to cytochrome a/a_3. This implicates a site between c and oxygen (site III) as one of those where coupling can take place.

Other crossover points could be associated with the NADH-to-coenzyme Q segment of the electron transport chain (coupling site I) and the coenzyme Q-to-cytochrome c segment (coupling site II). This analysis was aided by experimentally isolating larger segments of the electron transport chain using *artificial electron donors or acceptors* that can interact directly with mitochondrial electron carriers (Figure 8–7). Oxidized ascorbate and the dye phenazine methosulfate can accept electrons from cytochrome c; for example, electrons that can be donated to the electron transport chain by exogenous reduced ubiquinone. In this partial electron transport chain, starvation for the phosphate acceptor leads to a crossover of the oxidation-reduction state between cytochrome b and cytochrome c_1. These artificial electron donors and acceptors also provide useful ways to test the activity of reconstituted segments of the electron transport chain. In fact, all three coupling sites have been reconstituted in lipid vesicles, a critical test of their proposed compositions.

Respiratory Complexes

Analytical experiments, including the use of spectrophotometric methods, have provided measurement of the carriers in the chain. Some seem to exist in distinct stoichiometric relationships to each other, as seen in Table 8–4. When isolated under certain conditions, as shown by D. E. Green and his colleagues, some carriers tend to remain associated and are separated from the others as complexes: complex I, containing NADH dehydrogenase; complex II, containing succinate dehydrogenase; complex III, containing cytochromes b and c_1; and complex IV (cytochrome oxidase), containing cytochromes a and a_3. The other two carriers, coenzyme Q and cytochrome c, are present in numbers larger than the complexes. Both of these carriers are easily removable from the inner membrane of the mitochondrion (the former by organic solvents and the latter by phosphate buffers), while the complexes are firmly attached (being composed of integral membrane proteins). There is some evidence for each of the complexes being present in equivalent amounts, which has given rise to the concept of the *respiratory assembly*, composed of one unit each of the four different complexes. More recent data suggest that the complexes do not have to be closely associated with one another in an assembly to function efficiently in electron transport. Diffusion and collision of the appropriate complexes and carriers may suffice to allow the redox reactions to occur at physiological rates. In either event the values from measurements like those in Table 8–4, together with the known figures for mitochondrial volume and protein content, allow one to calculate the number of respiratory complexes for a mitochondrion.

TABLE 8–4

Stoichiometry of Electron Chain Components in Mitochondria*

Component	Micromoles per g Mitochondrial Protein
Cytochrome b	1
Cytochrome c_1	0.5
Cytochrome c	1
Cytochrome a	1
Cytochrome a_3	1
Ubiquinone (coenzyme Q)	~10
NAD(P)	~10
NADH-Flavoprotein	~0.1
Succinate-Flavoprotein	~0.2

*Data obtained from B. Chance and D. E. Green.

This amounts to somewhere between 5,000 and 20,000 sets of electron carrier chains, although the number will vary depending on the origin of the mitochondrion.

The principle that all chemical reactions and pathways are reversible microscopically suggests that, regardless of how they are coupled, both electron transport and the associated phosphorylation should also be reversible. Indeed, L. Ernster in 1957 discovered that when succinate and ATP are both added to a tightly coupled mitochondrion, NAD^+ is *reduced* to NADH. This can be explained by the model shown in Figure 8–12: after the oxidation of succinate by its flavoprotein, the electrons flow first to ubiquinone. Then, instead of electron flow to cytochrome *b*, energy released at coupling site I by the added ATP causes a *flow reversal* and the electrons go toward the NADH dehydrogenase, reducing NAD^+.

Mechanism of Electron Transport Coupling to Phosphorylation in Mitochondria, Chloroplasts, and Bacteria

As we shall see later in this chapter, the synthesis of ATP by illuminated chloroplasts is also associated with the flow of electrons in a chain with many similarities to that of the mitochondrion. We now believe that the fundamental mechanism of phosphorylation in these two organelles, as well as in bacteria, is probably the same. The complete answer to the very difficult and most important problem of how electron transport is coupled to ATP synthesis (and to ion transport, discussed below), is not known. However, the *chemiosmotic hypothesis*, proposed by P. Mitchell in 1961, has stood the test of extensive experimentation and is now accepted by almost everyone as the most likely basis for the mechanism of electron flow-linked phosphorylation. The chemiosmotic hypothesis proposes that the primary product of electron transport is the formation of an *electrochemical proton gradient* across the inner mitochondrial membrane, or across the thylakoid membrane of the chloroplast, or across the bacterial cell membrane. ATP synthesis results from the spontaneous flow of protons down this gradient; the path leads through the membrane via a complex of proteins known as the *coupling factor*. The gradient is established by an *obligatory movement of protons* during respiration and is preserved by the presence of a *membrane with generally low permeability to protons*, except at special

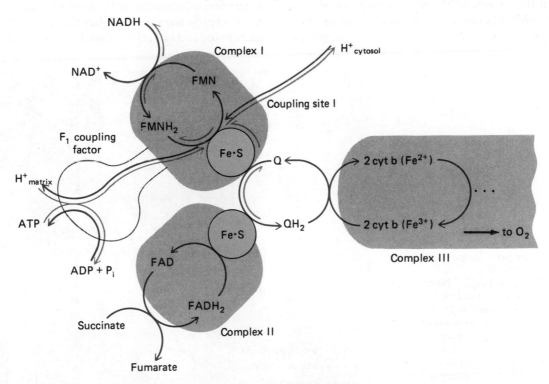

FIGURE 8–12
Reversal of electron flow in the mitochondrial electron transport chain. Normally, electrons resulting from the oxidation of succinate flow through complex II to complex III, and then to O_2 (darker color arrows). In the presence of high concentrations of ATP and low concentrations of ADP, a proton gradient is built up and electron flow through complex III is blocked. Instead, the electrons flow "backwards" through complex I to reduce NAD^+ (shown by the lighter color arrows).

locations where the energy of the proton gradient is harnessed to do cellular work. The critical proton barrier is the inner mitochondrial membrane, the thylakoid membrane of the chloroplast, or the cellular membrane of the bacterium.

Let us see how this might work. Mitchell proposed the idea of *loops* of electron transport that extend from one side of the inner mitochondrial membrane (the *matrix* or *M-side*) to the side facing the intermembrane space (known as the *C-side,* since that space is in relatively free communication with the cytoplasm). Along one segment of a loop, protons and electrons travel together from the M- to the C-side; for example, on the $FMNH_2$ of the NADH-dehydrogenase (Figure 8–13). But on the return segment of the loop, the Fe · S proteins can carry only electrons. Thus, the protons, once translocating the membrane, are forced to remain on the C-side. The action of such loops of electron transport would be to move protons to the C-side and thus create a proton gradient across the inner membrane.

Essential to the net movement of protons from one side of a membrane to another is the *asymmetric location* of various components of the electron transport chain. The electrons and protons from NADH can only be accepted by FMN on the M-side of the inner membrane and the electrons of $FMNH_2$ can only be donated to the Fe S centers on the C-side. Were the asymmetry not present

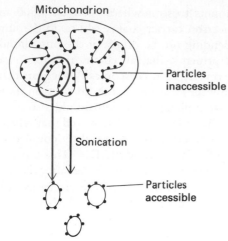

there would be an equal flow of protons in both directions across the inner membrane.

The experimental testing of the location of various electron transport components has not been completed, but the appropriate sidedness has been demonstrated for cytochromes *a, c,* and c_1 (on the C-side), and cytochrome a_3 (on the M-side) (Figure 8–14a). The methods to determine sidedness, similar to those described in Chapter 7, exploit the use of probes that cannot penetrate the membrane and can affect, therefore, only those structures

FIGURE **8–13**

An electron transport loop, according to Mitchell's chemiosmotic hypothesis. Note the paths of protons and electrons, sometimes separate, sometimes merged. The asymmetric location of different components, one on the matrix side of the inner mitochondrial membrane, another on the cytosolic side, makes possible the "pumping" of protons across the membrane. A black interrupted line shows a path taken by electrons; a color interrupted line shows a path taken by protons. An alternating black and color line indicates that protons and electrons travel together.

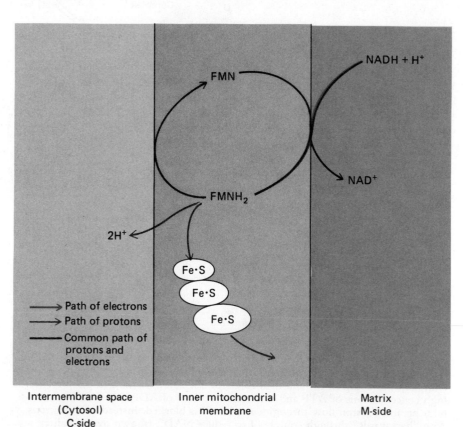

accessible on the exposed (C-side) surface. The experiments included exposure of intact mitochondria to inhibitory antibodies specific for the various components, chemical modification by nonpenetrating labeling reagents, and exposure to impermeant electron donors or acceptors. To test the location of M-side components the mitochondria are broken open by sonication. Under these conditions the folded cristae of the inner membranes fuse into spherical vesicles which are topologically inside out; that is, the M-side faces outward. This can be seen easily in the electron microscope; the knobs and stalks of the particles that normally line the cristae in intact mitochondria face outward (compare to Figures 8–1 and 8–3). These *submitochondrial particles* or *vesicles* can then be used with the reagents listed above to identify the M-side components.

In the scheme originally formulated by Mitchell, there were three such loops, each accounting for the translocation of at least two protons outwards for a pair of electrons flowing along the loop. The existence and exact locations of other loops and the stoichiometry of the proton/electron ratio are all in dispute. Instead of loops there may be special *proton pumps,* linked somehow to the flow of electrons. There is general agreement, however, that electron transport is accompanied by directional proton transport and that the proton gradient

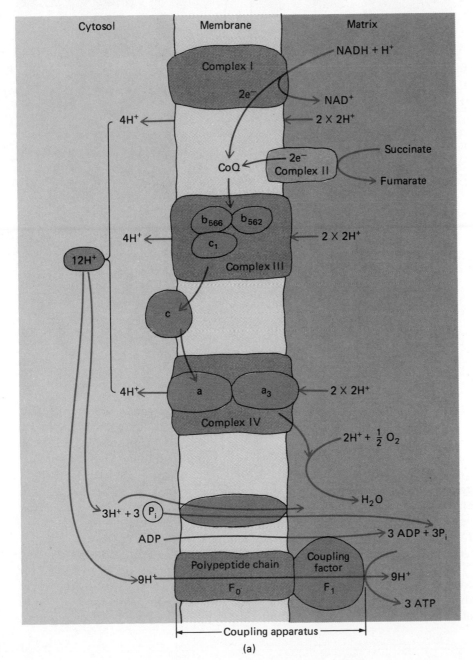

FIGURE **8–14**
Topography of components in energy-generating membranes.
(a) Topography of the inner mitochondrial membrane. The diagram shows the cytochromes (a, a_3, c, c_1, and b), the four electron transport complexes, the flow of electrons between them, the flow of protons across the membrane, and the coupling system (consisting of F_0, F_1, and hydrophobic proteins). This scheme is based on observed interactions of the components, on reconstitution experiments, and on the availability of some of the cytochromes on one or other side of the membrane to combine with antibodies against these specific cytochromes. The actual number of protons transported (shown as 12 in the figure) is still in dispute.

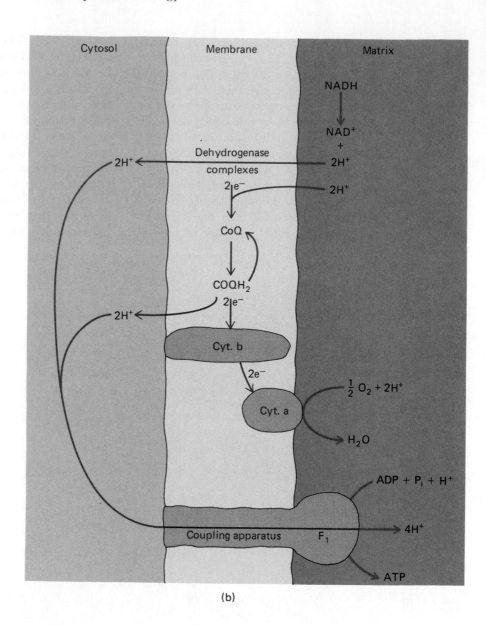

FIGURE **8–14**
(continued) (b) Schematic flow of electrons and protons through the carriers and coupling factors of the bacterial membrane. The postulated intramembrane sites of the electron transport carriers are shown.
(c) Schematic flow of electrons and protons through the carriers and coupling factors of the chloroplast membrane.

(b)

thus formed provides intermediate storage of the energy released during the oxidation of NADH by oxygen. This general scheme applies to many bacterial membranes as well, except that there probably are only two loops/ pumps, since only four protons seem to translocate per pair of electrons flowing from NADH to oxygen (Figure 8–14b). In chloroplasts the orientation of the proton translocation elements in the membrane is different and four protons are translocated *inward* (Figure 8–14c). Establishment of the proton gradient in chloroplasts is discussed later in this chapter.

How is the energy of oxidation stored? As protons are transported from the M-side to the C-side of the mitochondrial membrane, a gradient of proton concen-

tration is established. This represents an *energy gradient* with both an *osmotic* (concentration) and an *electrical* component. The concentration of protons is greater on the C-side, and this concentration gradient or pH gradient (about 0.5 pH units in a respiring mitochondrion) represents a source of free energy. Because protons have an electrostatic charge, there is also an electric potential gradient (of about 150 mV) across the inner mitochondrial membrane.

These two aspects of the energy gradient are in fact distinguishable from one another. It is possible, for example, to estimate the electrical potential by the use of a *permeable ion,* like tetraphenylphosphonium (TPP$^+$). In a respiring mitochondrion under steady-state conditions,

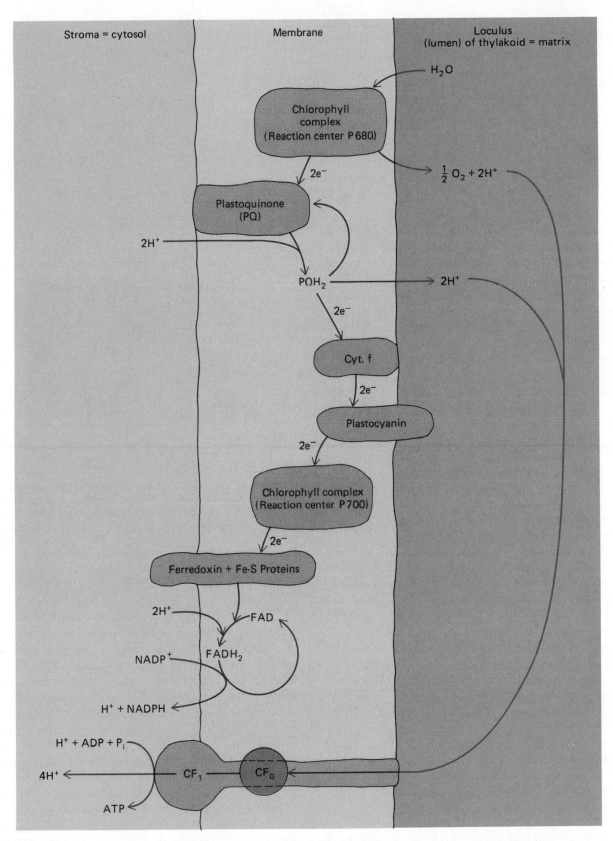

(c)

TPP^+ will accumulate preferentially on the more negative side of the inner membrane (Figure 8–15). If the permeable ion is radiolabeled, it can be detected easily and the ratio of its steady-state concentrations on the two sides of the inner membrane can be assessed. By assuming that the TPP^+ is in equilibrium with the electrical potential, one can use the Nernst-Planck equation to calculate the electrical component of the proton energy gradient:

$$\Delta E = - (RT/F) \ln ([TPP^+]_M/[TPP^+]_C).$$

The same measurements can, of course, be made on chloroplasts and bacteria membranes.

A similar strategy can be used to measure the pH gradient. In this case one measures the distribution across the appropriate membrane of a weak acid, like benzoic acid, whose acidic neutral form can penetrate the membrane easily but whose dissociated charged form cannot (Figure 8–16). The weak acid itself will equilibrate across the membrane, reaching equal concentrations. When the energy gradient is in a steady state, the weak acid's ion will accumulate on the more basic side of the membrane. The dissociation of the acid comes into equilibrium with the local proton concentration. If the benzoic acid is radiolabeled, one can measure the total concentration, in both associated and dissociated forms, on either side of the membrane. Along with the value of the dissociation constant for the acid, this gives sufficient information to determine the ratio of proton concentrations.

Another way to show the importance of each component of the energy gradient is to reduce one or the other and observe the effect on the synthesis of ATP. It is possible to partially collapse the electrical gradient, without changing the pH gradient much, by adding to isolated mitochondria in the presence of K^+ ions the ionophoretic antibiotic *valinomycin*. This carrier molecule provides a conductance pathway for K^+ ions (see Chapter 7), and they will flow in response to the electric potential gradient to the M-side of the inner membrane. During the flow of the K^+ current, less electrical potential energy is available to do the work of synthesizing ATP. In a respiring mitochondrion the rate of ATP production is, therefore, transiently lowered. Eventually, however, a concentration gradient of K^+ is reached which halts any further net influx of K^+. As further respiration pumps protons across the inner membrane, the pH gradient increases and ATP synthesis resumes.

It is possible to collapse the proton concentration gradient by providing a conduction pathway for hydrogen ions themselves. The molecules that do this, many of which are weak organic acids, were among the first reagents discovered that *uncouple oxidative phosphorylation.* The classical example is *dinitrophenol (DNP),* which can act as a carrier of H^+ by dissolving in a lipid membrane

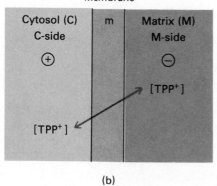

Tetraphenylphosphonium (TPP^+)

(a)

Membrane

Cytosol (C) C-side	m	Matrix (M) M-side
⊕		⊖

[TPP⁺]

[TPP⁺]

(b)

FIGURE **8–15**

Measuring the electrical component of the energy gradient due to proton displacement. A permeable ion, such as TPP^+ (a), which can be radioactively labeled, will accumulate on the side with opposite electrostatic charge. In the steady state, the distribution of TPP^+ is assumed to be in equilibrium with the electrical potential difference (b). The amount of TPP^+ can be estimated by rapidly filtering the mitochondria, then measuring the radioactivity on the filter.

in either its protonated or unprotonated form (Figure 8–17). By collapsing both the proton concentration and the electrical gradients, these molecules remove the energy source for the phosphorylation of ADP to ATP without blocking the movement of electrons along the electron transport chain. The *short-circuiting* thus breaks the coupling between electron flow and ATP generation. Since most cellular energy comes from oxidative phosphorylation, uncouplers like DNP, FCCP, and SFG847 (Figure 8–17) are powerful poisons of energy metabolism. The channel-forming ionophoretic antibiotics like gramicidin A (see Chapter 7) can also uncouple oxidative phosphorylation by providing conduction pathways for protons, as well as other positive ions that can respond to the electrical potential gradient. In the presence of uncouplers the energy of oxidizing NADH that does not go into ATP is given off as heat. Uncoupling occurs

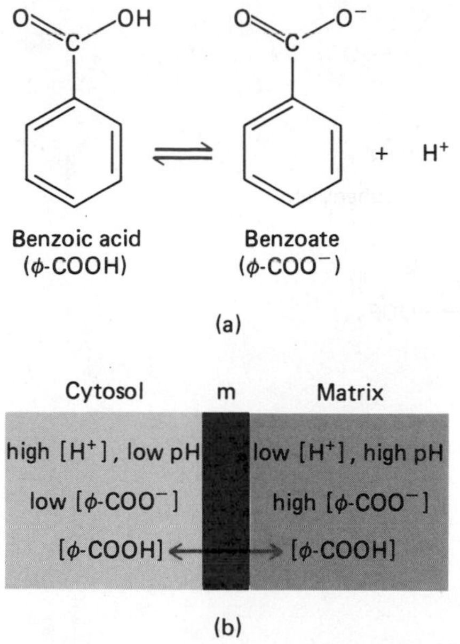

(a)

(b)

FIGURE 8–16

Measuring the osmotic (pH) component of the energy gradient due to proton displacement. A weak permeable acid, such as benzoic acid (a), will equilibrate across the membrane but will dissociate more on the side with the higher pH (lower $[H^+]$). The concentration of the benzoate ion, therefore, will be higher on that side (b). The sum of the concentrations of undissociated and dissociated acid can be measured using radioactive benzoic acid; thus, the amount of radioactivity will be higher on the side with the higher pH. By using equations for the equilibrium dissociation of benzoic acid, one can solve for the ratio of proton concentrations.

normally in some animals; for example, those that hibernate. Natural uncouplers (e.g., catacholamines) act to produce heat in mitochondria of the *brown fat tissue* found in these animals.

One can also demonstrate that the electrochemical gradient of protons is an intermediate form of energy storage at one or another of the three sites of phosphorylation. By using reduced ubiquinone as the electron donor and ferricyanide ions $[Fe(CN)_6^{3-}]$ as the electron acceptor, one can isolate the activity of site II (Figure 8–7). In lipid vesicles assembled with the components of site II (from ubiquinone to cytochrome *c*), oxidation of reduced ubiquinone and reduction of ferricyanide [to ferrocyanide, $Fe(CN)_6^{4-}$] is accompanied by the movement of protons outside the vesicle, an effect abolished by addition of the FCCP uncoupler.

Exactly how the electrochemical gradient of protons, the *proton-motive force (pmf)*, as Mitchell first called it, is used to phosphorylate ADP remains a central mystery of energy metabolism. We are nevertheless confident that the pmf is capable of driving the formation

of ATP because artificially induced proton gradients have been shown to do so. The first experiments of this type were done in chloroplasts by A. Jagendorf and E. Uribe in 1966. The direction of proton translocation in chloroplasts is inward, opposite to that of mitochondria and bacteria (compare Figures 8–14a and c). Therefore, the chloroplasts were first equilibrated in a pH 4 solution to acidify the inner space. Then they were rapidly washed and suspended in a solution of pH 8 to create a proton gradient across the thylakoid membrane. The movement of protons down this artificially induced gradient was accompanied by the net formation of ATP (Figure 8–18). Since then, appropriate proton gradients have been shown to generate ATP in both mitochondria and bacterial membrane vesicles.

Submitochondrial Particles

One reason why so much difficulty was encountered in elucidating the mechanism of oxidative phosphorylation lies in the fact that the respiratory enzymes with their associated coupling factors are a part of the insoluble lipoprotein membranes of the mitochondria. A. Lehninger, D. E. Green, and E. Racker were among the first to attempt to fragment these complex multienzyme units into workable pieces. After breaking up mitochondria by sonication, one can isolate closed vesicles of membrane fragments, which turn out to behave like miniature mitochondria. It thus appears that the respiratory function is fairly evenly spread over the membranes of intact mitochondria. The idea that the inner membranes of the mitochondria are the sites of respiration was first gleaned from the observation that heart mitochondria, which have more cristae per mitochondrion than liver mitochondria, also have a higher content of the respiratory pigments and a higher oxidative rate. Since the cristae contain, like other membrane structures, lipids and proteins, the respiratory components were suspected of being enzyme-lipid complexes. The miniature mitochondria were called *electron transport particles (ETP)*, or *submitochondrial particles*.

When prepared by ultrasonication, submitochondrial particles are usually *inside out*—that is, the knobs and stalks seen in electron micrographs extend outward from the submitochondrial particles into the surrounding fluid; in intact mitochondria the knobs and stalks extend inward to the matrix. This inside-out topology simplifies biochemical experimentation, since the substrates for oxidation and ADP + P_i can be added directly to the M-side of the inner membrane without having to be transported across.

Coupling Factors

If the mitochondrial phosphorylation system were run backwards it should behave like what might be called

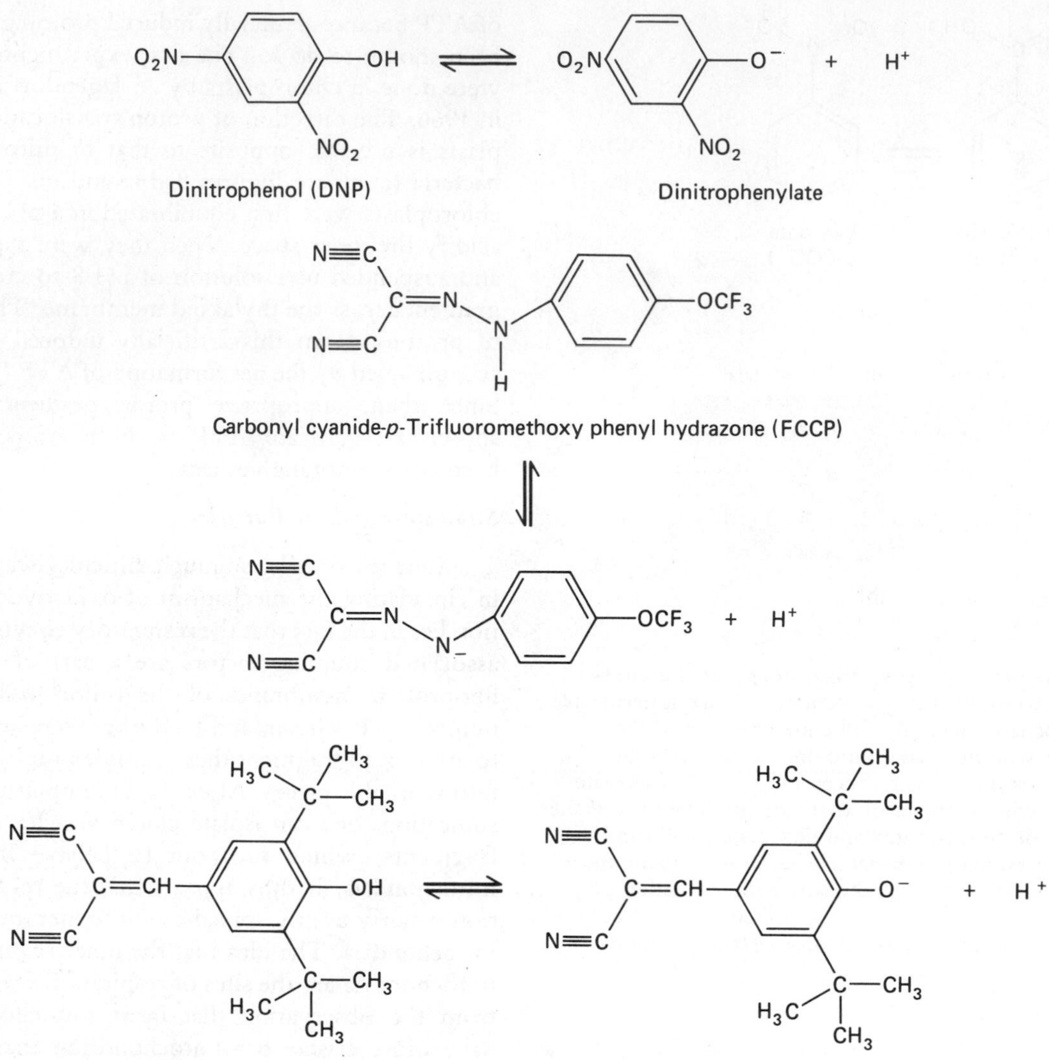

Dinitrophenol (DNP) ⇌ Dinitrophenylate + H⁺

Carbonyl cyanide-*p*-Trifluoromethoxy phenyl hydrazone (FCCP)

SF6847

FIGURE 8–17
Molecular formulae of three proton ionophoretic antibiotics.   These molecules uncouple oxidative phosphorylation and are known, therefore, as "uncouplers." FCCP is commonly used to collapse the proton gradient in preparations of isolated mitochondria. The uncoupler SF6847 is so efficient that it can completely abolish ATP production at a concentration only one-fifth that of the coupling systems in the mitochondria.

an "H^+-dependent ATPase." In mitochondria, chloroplasts, and bacterial cells, structures have been found with this property—hydrolyzing ATP in the presence of H^+ ions: the *coupling factors*. These are a set of proteins which include those making up the stalks and knobs seen in electron micrographs of energy-transducing membranes of mitochondria (Figures 8–3, 8–19), chloroplasts, and bacteria (see also Figure 8–7b).

Mitochondrial membranes capable of electron transport and ATP synthesis (Figure 8–19a) can be stripped of the stalks and knobs to obtain a particle-free fraction, having only electron transport activity (Figure 8–19b),

and a particulate fraction (Figure 8–19c), having only ATPase activity. When these two fractions are suitably mixed, one can obtain membranes with bound particles—the knobs and stalks (Figure 8–19d, compare with Figure 8–3). After adding the appropriate soluble fractions, the reconstituted membranes show activity for both electron transport and ATP synthesis.

The knob is known as coupling factor F_1 and is composed of at least five different polypeptides. It is attached to the membrane by a different set of integral proteins, known collectively as F_0, which form a channel in the membrane. In addition, there are other compo-

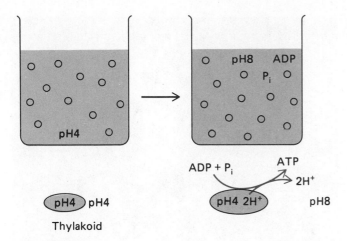

FIGURE 8–18
The Jagendorf-Uribe jump experiment. Isolated chloroplasts were equilibrated to pH 4 by a long incubation in a weak buffer. A strong pH 8 buffer containing ADP and P_i was then added to the chloroplast suspension, setting up a H^+ gradient between the thylakoid and stroma. As protons moved down their concentration gradient, from pH 4 to pH 8, a net synthesis of ATP was observed, concomitant with a flow of protons from inside to outside (see Figure 8–14c).

nents: the *oligomycin-sensitivity conferring protein (OSCP),* one called F_6, etc. The experiments that led to our present knowledge of the coupling system began with the isolation of a soluble protein fraction (F_1) in 1960 by the group led by Racker, described by him with verve in the delightful monograph *A New Look at Mechanisms in Bioenergetics* (see Selected Readings).

There are striking morphological similarities among the coupling systems from chloroplasts, mitochondria, and bacterial membranes, suggesting a possible evolutionary conservation in these structures (see Figures 8–19 and 8–20). In mitochondria, the search for the several components involved was aided by the use of an inhibitor specific for the ATPase, *oligomycin.* Oligomycin blocks ATP formation, and also blocks oxidation of NADH, because of the tight coupling we discussed above. It was known for a long time that uncouplers added to oligomycin-blocked mitochondria allow electron flow to resume and, therefore, that oligomycin must act directly on the ATP generating system. We know now that the sensitivity to oligomycin is provided by the oligomycin-sensitivity conferring protein and probably by other proteins, to one of which this antibiotic must bind to exert its inhibitory effect. Sensitivity to this drug also provided a useful test of success for attempts to reconstruct the coupling system in membrane vesicles. Only the correct ATPase would be inhibited by oligomycin.

There are many reasons to believe F_1 couples proton movement through the inner membrane with ATP synthesis. F_1 is capable itself of ATPase activity. Specific removal of F_1 both stops the formation of ATP and allows the proton gradient to equilibrate. The proton leak can be blocked by adding back F_1 or by adding certain inhibitors of phosphorylation, such as oligomycin or *dicyclohexyl carbodimide (DCCD).*

Although the components of the phosphorylation machine have been largely identified, the exact chemical mechanism involved is still in dispute. One possible mechanism is the *conformational coupling hypothesis* put forward by P. Boyer (Figure 8–21, p. 400). Protons passing into the F_1 knobs would bind and alter the conformation of proteins there, allowing the release of a tightly bound ATP. The binding site is imagined to be separated from water. Under such conditions, the phosphorylation of ADP by inorganic phosphate should be near equilibrium, but the product ATP is too tightly bound to be released until a proton-induced conformational change occurs. Although the mechanism is not yet established, there have been measurements of the stoichiometry of protons per ATP. The best estimates at present suggest three protons must flow from the C-side to the M-side to synthesize one ATP. As we shall discuss below, other protons may be needed to import P_i and ADP, or to export the ATP product. There are also uncertainties about the number of protons translocated during electron flow. Taken together, these complexities make accurate calculations of the P/O ratio very difficult.

Reconstructing Energy-linked Phosphorylation

Many techniques have been developed to reconstruct the energy-transducing systems of chloroplasts, mitochondria, and bacteria. The electron transport chain of mitochondria can be solubilized by bile salt detergents and, as shown by D. E. Green and his collaborators, can be resolved into three lipid-containing protein complexes: one each containing the segments for NADH-CoQ reductase, succinate-cytochrome c reductase, and cytochrome oxidase (see Figure 8–7). By lowering the detergent concentration (by dilution or dialysis) and adding phospholipids, these three solubilized complexes can be reassociated in the form of spherical membrane vesicles. That each complex can reassemble into a membrane is not surprising since each contains lipid. What did surprise researchers at the time was that each complex could show both proton transport and respiratory control. After mixing the three complexes with each other, and adding back the easily solubilized coenzyme Q and cytochrome c, the resulting membranous structure mimicked all the properties of the mitochondrial inner membrane.

Rather than describe further the detailed history of this effort (see Racker's monograph for an account by one

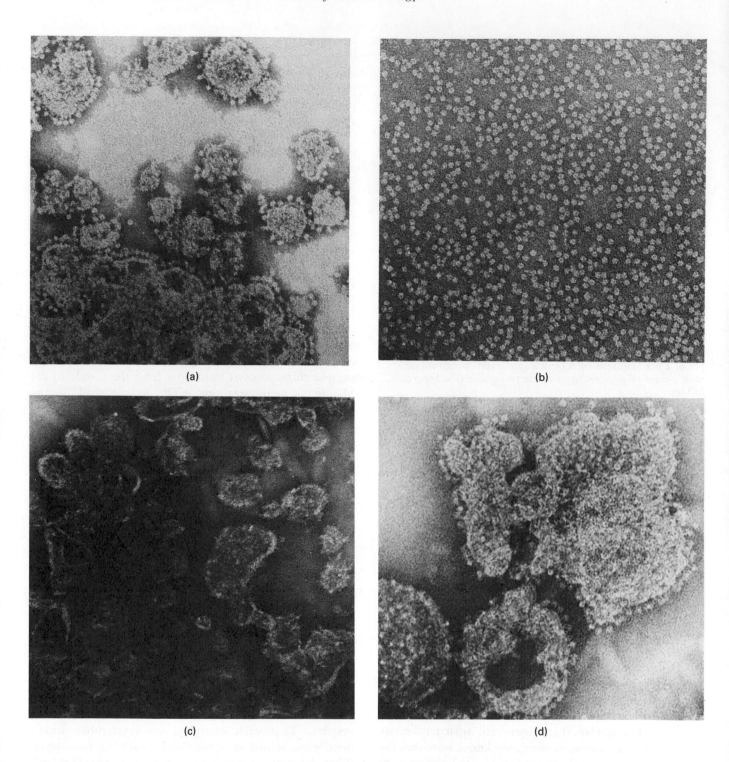

(a)

(b)

(c)

(d)

FIGURE **8–19**

Mitochondrial subfractions; reconstitution of inner membrane structure and function.   (a) Intact inner mitochondrial membranes lined with spherical particles. (b) Isolated inner membrane particles. (c) Isolated inner membrane devoid of particles. (d) Reconstituted inner membrane with attached particles, achieved by the appropriate mixing of fractions illustrated in (b) and (c). All the fractions were negatively stained with phosphotungstic acid, so that the background is dark (cf. Figure 8–3), in contrast to the positive staining shown in Figures 8–1b, 8–4, 8–5. ×120,000.
(Courtesy of E. Racker.)

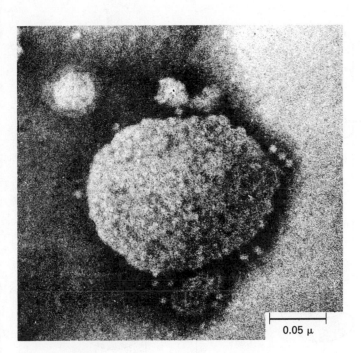

FIGURE 8–20

Isolated photosynthetic membranes (chromatophores) from the bacterium, *R. rubrum*. The preparation has been negatively stained. The particles attached to the outer surface are the same size as those shown in Figure 8–19a, from mitochondria, and are also similar to those found on the chloroplast thylakoid membranes. × 120,000. (Courtesy of S. Simon.)

personally involved), we shall conclude with a particularly dramatic example, the reconstruction of the purple membrane proton pump. The halophilic bacterium *Halobacterium halobium* grows in natural brines as a facultative anaerobe (i.e., tolerant of oxygen but not using it). By virtue of the pigment *bacteriorhodopsin,* found in a purple patch on the cell membrane, it can use light energy to form ATP. Light causes protons to be extruded from the bacterial cell. Bacteriorhodopsin acts as a light-driven proton pump and the proton gradient formed on illumination drives the phosphorylation of ADP. During light-stimulated formation of ATP, respiration is inhibited.

These ideas were confirmed brilliantly by E. Racker and W. Stoeckenius, who not only were successful at incorporating partially purified bacteriorhodopsin into lipid vesicles, but managed also to include the oligomycin-sensitive ATPase of bovine mitochondria. During illumination in the absence of oligomycin, these inside-out vesicles would accumulate protons and then synthesize ATP. The ATP synthesis was also fully sensitive to uncouplers of oxidative phosphorylation. The question was, what provided the coupling between the light absorbed by bacteriorhodopsin and the bovine ATPase? One could imagine, as an alternative to the chemiosmotic hypothesis, that some light-induced conformational

change in bacteriorhodopsin might somehow be linked mechanically to the phosphorylation of ADP by the mitochondrial ATPase. But the possibility of such a direct structural linkage between components from such evolutionarily divergent organisms as a cow and a halophilic bacterium seems very unlikely. The success of this improbable marriage thus stands as an especially dramatic confirmation of the chemiosmotic coupling theory of P. Mitchell.

Proton-Flow–linked Transport

The electrochemical gradient of protons is not used solely to drive the formation of ATP. In mitochondria the active transport of several ions is also driven directly by the energy gradient produced by oxidation of NADH. These transports may be driven by either the pH gradient or the electrical potential gradient, or by both. An example of the first is the proton-linked transport of Na^+ ions out of the mitochondrion (Figure 8–22). The system responsible behaves like a Na^+-H^+ exchange transporter, traveling inward to the M-side with a proton and outward to the C-side with Na^+. We discussed this type of process in Chapter 7, where it was described as antitransport. In the mitochondrion, the structure that catalyzes such a process is termed an *antiport*. Since both Na^+ and H^+ are positively charged, their exchange is indifferent to the electrical potential difference across the inner membrane.

A different mechanism is believed to be responsible for moving phosphate from the C-side into the matrix, where it is used for ATP synthesis (Figure 8–22). In this case the carrier catalyzes the cotransport of $H_2PO_4^-$ with H^1 to the M-side. In the mitochondrion such a cotransport process is catalyzed by a structure called a *symport*. The $H_2PO_4^-$/H^+ cotransport results in no net electrical charge transfer, so it responds only to the pH gradient. (This case can also be understood formally as a OH^-/$H_2PO_4^-$ antiport; see Figure 8–10.)

An example of a transport responsive to the electrical potential is the accumulation of Ca^{2+} ions by the mitochondrial matrix (Figure 8–22). The inner membrane contains a simple Ca^{2+}-specific transporter and calcium is accumulated until the free Ca^{2+} in the mitochondrion is in equilibrium with the electrical potential across the membrane. Much of the imported Ca^{2+} ends up, however, as insoluble calcium polyphosphate salts which can be seen when the mitochondria are examined by electron microscopy (Figure 8–1). The sudden influx of calcium ions into the cytoplasm during physiological functions like muscle cell contraction and secretion by other cells can be buffered by Ca^{2+} being taken up into mitochondria (and the endoplasmic reticulum).

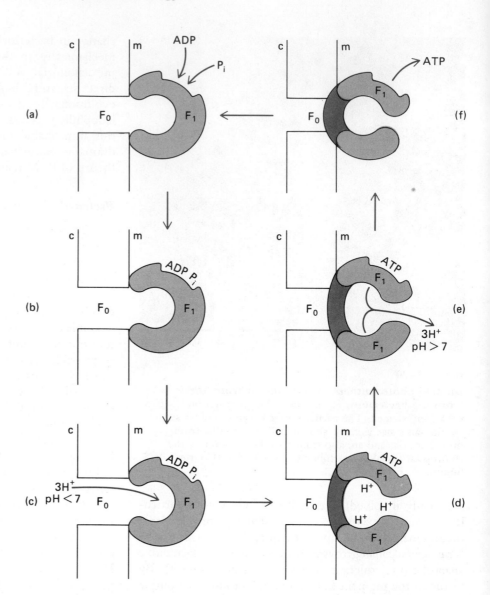

FIGURE 8–21
Conformational coupling hypothesis to explain ATP synthesis. F_1 can exist in two conformations, one with low-affinity catalytic and proton-binding sites, and the other with high-affinity catalytic and proton-binding sites. In going from (a) to (b), ADP and P_i bind to the low-affinity form, which also has its proton-binding site connected to F_0. In (c), protons bind to the low-affinity site, resulting in a conformational change in F_1 (d), an increase in the affinity of the catalytic and proton-binding sites, and disconnection from F_0. Because of the change in the configuration of F_1, the formation of ATP from ADP and P_i is practically at equilibrium (d). If the concentration of protons in the matrix is low, protons can be released from the high-affinity form of F_1 to the matrix (e), changing the conformation back to the low-affinity form, resulting in the release of ATP (f).

Another transport system responsive to the electrical portion of the pmf is the ATP^{4-}/ADP^{3-} antiport shown in Figure 8–10. When ATP^{4-} flows out of the matrix and ADP^{3-} flows in, there is a net flow of negative charge outward, driven by the electrical potential difference across the inner mitochondrial membrane.

Any transport that requires inward proton (or electrical current) flow, will compete with ATP synthesis for the energy in the electrochemical proton gradient. Thus, one must be careful to take account of all the transport processes when measuring in intact mitochondria the number of protons required to synthesize an ATP.

Although chloroplasts have not yet been shown to have any proton gradient-driven transport processes, bacteria have many. An example of one responsive to both components of the proton-motive force is the accumulation of the sugar lactose, discussed earlier in

Chapter 7 (Figure 8–23). Artificial proton gradients have been shown to drive the accumulation of lactose in *E. coli,* and the lactose flux into the cell is accompanied by a proton flux. Lactose is uncharged, so the cotransport of protons is driven by both the osmotic and electrical components of the H^+ gradient. Unlike mitochondria, the excretion of Ca^{2+} in *E. coli* is driven by an electrically neutral exchange with two protons. *E. coli* also harnesses the proton gradient across its cell membrane to drive the rotary motion of its flagella, truly an astonishing example of the versatility of the energy transductions deriving from the proton-motive force (see Chapter 16).

Physiological Behavior of Mitochondria

The mitochondrial membranes are similar to others in being semipermeable—that is, some substances (notably H^+ and K^+) have difficulty gaining entrance to the

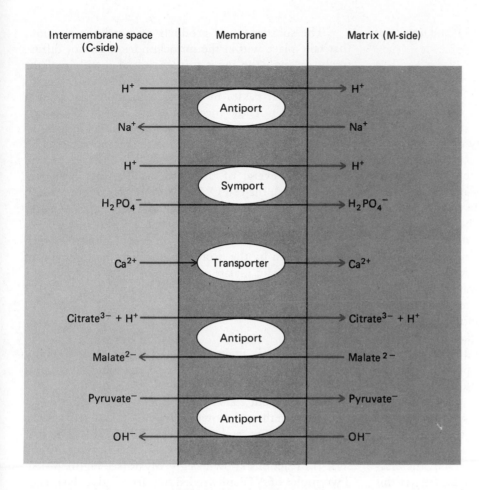

FIGURE 8–22

Symport, antiport, and transporter pathways in the inner membrane of mitochondria. When the transported species is moved in the same direction as protons, the transporter pathway is a symport; when in the opposite direction, an antiport. Both the symport and antiports shown respond only to the concentration gradient of protons, since neither results in a net flow of electrostatic charge. The movement of Ca^{2+} through the membrane is facilitated by a simple transporter but since a net charge is transferred, the transport is driven by the electrical potential gradient.

matrix while others pass through easily. The permeability barrier is associated largely with the inner membrane. Being bounded by a selectively permeable structure,

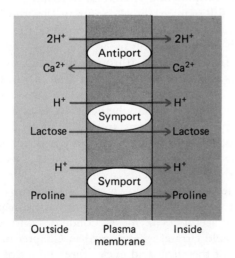

FIGURE 8–23

Symport and antiport pathways in the plasma membrane of *E. coli* bacteria. The lactose and proline symports respond to both the osmotic (pH) and electrical components of the proton motive force, the Ca^{2+} antiport only to the osmotic component.

mitochondria behave like osmometers: they take in water and swell when placed in hypotonic solutions. This property has been exploited by physiologists as a way to enlarge the interior volume of functioning mitochondria. Under more severe conditions of swelling they lose some of their low-molecular-weight soluble components, like the adenine nucleotides and the coenzymes NAD^+ and $NADP^+$, and hence lose their oxidative ability. A. Lehninger found that certain kinds of swelling—that caused by the addition to the bathing medium of phosphate ions, for example—are reversible; a very good reversing agent is ATP. Mitochondria swollen under these conditions will extrude water and contract upon addition of ATP. Indeed, it has been observed that during respiration (and hence when phosphorylation takes place), there are distinct changes in the shape—and presumably the volume—of the mitochondrion. What this alternate swelling and contraction of mitochondria has to do with the functioning of this organelle is not certain. We remember, however, that the function of the mitochondria is not just the manufacture of ATP but also its secretion to other parts of the cell. Thus, it is possible that these changes in the water movements across the mitochondrial membrane, as indicated by the volume changes undergone by the mitochondrion, may have something to do with the secretion of ATP and uptake of ADP or P_i.

Substrate Transport into Mitochondria

A perplexing problem concerning the function of mitochondria was the observation that they are impermeable to NADH. Since this coenzyme is produced extensively by glycolysis in the extramitochondrial cytoplasm (Figure 6–15) and since its oxidation by mitochondria could provide energy for the synthesis of ATP, biochemists pondered for a long time the ways and means of how NADH becomes available to the mitochondrial enzymes. The solution to this problem emerged from the observations that certain enzymes are found both in the mitochondria and the extramitochondrial cytoplasm. One of these enzymes is α-glycerophosphate dehydrogenase; in the cytoplasm the enzyme reduces dihydroxyacetone phosphate (DHAP) to α-glycerophosphate (αGP), using electrons from NADH. The α-glycerophosphate, which now contains the hydrogens and electrons from the extramitochondrial NADH, is transported across the mitochondrial membrane (Figure 8–24a). In the matrix a similar dehydrogenase oxidizes αGP back again to DHAP, and the electrons removed are used to reduce a flavoprotein, thus entering the electron transport chain. The dihydroxyacetone phosphate passes out of the mitochondrion and into the cytoplasm. Thus, the electrons originating in cytoplasmic NADH are fed via this *α-glycerol phosphate shuttle* into the electron transport chain and used to produce energy. Note, however, that they enter the chain at a point farther along than do electrons from mitochondrial NADH. For this reason there are only two ATPs generated for each pair of electrons from cytoplasmic NADH.

Another example of such an electron shuttle system is provided by the existence of two malate dehydrogenases and two glutamate-oxaloacetate transaminases, one each in the mitochondrion and the extramitochondrial cytoplasm. Together these two systems act to shuttle malate, a carrier of electrons originating in extramitochondrial NADH, into the mitochondrion (Figure 8–24b). In the cytoplasm, oxaloacetate is reduced by NADH, forming malate, in a reaction catalyzed by malate dehydrogenase. The malate then enters the mitochondrion, where the same reaction occurs in reverse: malate is oxidized to oxaloacetate while NAD^+ is reduced to NADH, which can then donate its electrons to the electron transport chain. The oxaloacetate, meanwhile, is shuttled out of the mitochondrion by means of a transaminase activity and other transporters. In all these shuttles the electrons are accompanied by protons so that whole hydrogen atoms are donated and accepted by the various redox systems.

The substrates and products of metabolic reactions that take place within the mitochondrion do not diffuse freely through the inner mitochondrial membrane (the outer membrane, as we have mentioned before, contains a channel that passes molecules with molecular weight less than 10^4). These metabolites are transported across the inner membrane by transporters, which in mitochondria are often called simply *porters* (a term introduced by P. Mitchell and J. B. Chappel). We have already mentioned several of these: the ATP/ADP antiporter (Figure 8–10), the $H_2PO_4^-/H^+$ symporter (Figure 8–22), and the Glu/Asp and α-ketoglutarate/malate exchange porters (Figure 8–24). Others include a pyruvate$^-$/OH$^-$ antiport and a (citrate $^{3-}$ + H^+)/malate^{2-} antiport (which can be driven by the proton gradient in the direction of citrate entry and malate exit).

Efficiency of Phosphorylation as an Energy Transducer

Let us look again at what happens to the energy available in a glucose molecule. If this glucose is completely burned in a test tube, it will give, under standard conditions, about 2900 kJ (690 kcal) of energy per mole, all as heat. In the cell, however, some of this energy is not lost as heat but is retained in the form of ATP. The first process that glucose undergoes in a cell is typically the series of reactions known as glycolysis (see Figure 6–15). Two moles of ATP are needed to start off glycolysis and, during the subsequent oxidation reaction, these two moles of ATP are regained. Two additional moles of ATP are formed in the subsequent conversion of two moles of phosphoenolpyruvate to pyruvate.

The following steps all occur in the mitochondria (see Figure 6–19). The NADH formed as a result of the oxidation during glycolysis is itself oxidized via the electron transport chain (through the mediation of the malate-oxaloacetate or some other shuttle, as shown in Figure 8–24). This process forms two or three more ATP molecules, depending on whether NADH or a flavoprotein receives the electrons from the shuttle. The end product of the glycolytic reactions is pyruvic acid, which enters the mitochondria and is oxidized by NAD^+. The NADH thus formed yields, overall, three moles of ATP as its electrons pass down the respiratory electron transport chain. The compound formed from pyruvate oxidation, acetyl-coenzyme A, condenses with oxaloacetate to yield citrate, as explained in Chapter 6. During one turn of the citric acid cycle, three moles of NADH are formed: from the oxidation of isocitrate to α-ketoglutarate, α-ketoglutarate to succinyl-CoA, and malate to oxaloacetate. These three moles of NADH are oxidized via the electron transport chain, yielding nine moles of ATP. In addition, the oxidation of succinate to

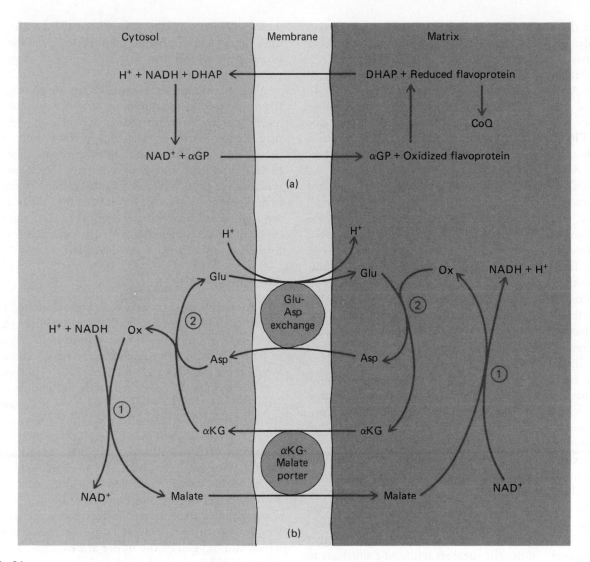

FIGURE 8-24
Mechanisms for transporting reducing equivalents across the mitochondrial membrane. (a) α-glycerophosphate shuttle. Dihydroxyacetone phosphate (DHAP) is reduced, by NADH and the α-glycerophosphate (αGP) dehydrogenase in the cytosol, to produce αGP; the αGP traverses the mitochondrial membrane where it is reoxidized to DHAP, providing electrons (reducing equivalents) to a flavoprotein. Thus, αGP is the carrier of reducing potential, gaining electrons in the cytosol and releasing them in the mitochondrial matrix. This can occur because of the presence of the same dehydrogenase in both locations. (b) In the glutamate-aspartate cycle, malate serves essentially the same purpose as αGP in (a). In the cytosol, oxaloacetate (OX) is reduced, by NADH and malate dehydrogenase, to form malate, which enters the mitochondria in exchange for α-ketoglutarate (αKG). In the mitochondria, malate is reoxidized by malate dehydrogenase (1) to form OX, generating intramitochondrial NADH. OX is transaminated with glutamate (Glu) to form aspartate (Asp) and αKG, which is then transported out in exchange with malate. The αKG is then transaminated (2) with cytosolic Asp to give Glu and OX. The Glu and Asp are exchange-transported in and out, driven by a proton-Glu symport system.

fumarate (by the FAD of succinate dehydrogenase) gives two moles of ATP, and the conversion of succinyl-CoA to succinate yields one mole of ATP, via GTP.

Adding up all these phosphorylations gives 17 or 18 moles of ATP formed from the oxidation of pyruvate to CO_2 and water; but, since there are two pyruvate molecules formed from one mole of glucose and one extra ATP formed per pyruvate during glycolysis, this one mole of glucose, after glycolysis and oxidation, gives 36 or 38 ATP molecules.

When ATP is hydrolyzed to ADP at standard conditions, the free energy released is about 34 kJ (8 kcal) per mole. Multiplying all the ATPs formed from the metabolism of glucose gives 36 or 38 × 34 kJ or about 1250 kJ/mole. This is about 40% of the total free energy available from oxidizing glucose under standard conditions; in other words, the process is 40% efficient—a high value. This efficiency, which might actually be higher under cellular conditions where metabolites are not at standard concentrations, can be compared favorably to the efficiency of large diesel engines (about 40%) or to gasoline internal combustion engines (about 22%).

The energy efficiency of lactic acid fermentation is approximately the same. The more complete utilization of reduced carbon compounds by respiring mitochondria, however, gives aerobic cells a distinct advantage in environments with limited food resources. Per mole of glucose oxidized, lactic acid fermentation yields two ATP while respiration with oxygen yields 38. Probably about 90% of the energy liberated during the oxidation of foodstuffs takes place during the process of electron transport. Thus, we can see that in heterotrophic cells most of the energy conversion—the conversion of that energy inherent in the chemical structure of a substrate to the energy inherent in the configuration of ATP—takes place in mitochondria.

The cell membrane has many functions. In the previous chapter we learned how the membrane is organized and later (Chapter 15) we shall read how it is formed. The energy-transducing membranes of mitochondria are a fine example of nature's conjoining of design and function to produce a structure that provides high efficiency.

Photosynthesis

Life would not be possible without radiant energy from the sun. Living forms have evolved to their present state in part because the process called photosynthesis has been available to provide usable energy for a long time. *Photosynthesis* may be defined as a series of steps in which *light energy is converted to chemical energy* that can be utilized for cellular work. This chemical energy is used in large part immediately to perform biosyntheses that are the primary source of *all* organic matter. The importance of

these processes may be appreciated from estimates that the annual yield of organic matter formed as the result of photosynthesis is about 9×10^{13} kg (10^{11} short tons), a truly astonishing figure.

The formation of organic compounds using energy from photosynthesis involves the *fixation* of carbon: highly oxidized carbon atoms in CO_2 are reduced and incorporated into sugars. This carbohydrate synthesis uses ATP as the immediate source of chemical energy and NADPH as the source of energetic electrons for the reduction of CO_2. Carbon fixation can in principle occur in darkness so long as ATP and NADPH are available. As we shall see, however, it is absorption of light by the pigment systems of photosynthetic cells that leads to the formation of both ATP and NADPH.

The Energy of Light

Although radiant energy, as we mentioned in Chapter 3, has both a particle and a wave nature, photosynthesis is best understood if we regard light as composed of particles, or *photons*. Each photon has an energy of $h\nu$, where h is Planck's constant, 6.6×10^{-34} J · second, and ν is the frequency associated with the photon, in hertz (Hz). The frequency is related to the wavelength, λ, by the expression $c = \nu\lambda$, where c is the velocity of light, 3×10^8 m/sec in vacuum. Thus, 400 nm blue light has a frequency of 7.5×10^{14} Hz, and 800 nm red light has a frequency of 3.8×10^{14} Hz. Of the estimated 5.2×10^{21} kJ of light energy received by our planet and its atmosphere from the sun each year, only 2.1×10^{21} kJ reaches the earth's surface. Ultraviolet light with wavelengths less than 350 nm is largely screened out by the ozone and oxygen of the upper atmosphere. This acts to protect cells from the harmful effects of energetic ultraviolet photons, which are strong enough to break covalent bonds in organic molecules. Roughly half the solar photons reaching the earth's surface are visible, with wavelengths between 400 and 800 nm. Photons have energies that depend inversely on their wavelengths: the 400 nm blue photons have an energy content of 300 kJ (72 kcal) per mole and the 800 nm red photons an energy of 150 kJ (36 kcal) per mole. Approximately 3×10^{18} kJ per year, or about 0.3%, of the incident visible light energy is stored by plants and other photosynthetic organisms, of which 1% is used by humans as food. This should be compared with the approximately 10^{18} kJ of energy per year used by humans as electricity, oil, and so forth. These numbers explain the current interest in harnessing a greater fraction of the solar energy flux for human uses.

Stoichiometry and Energy Requirements of Photosynthesis

The observed reaction of photosynthesis in green plants is

$$CO_2 + H_2O \xrightarrow{h\nu} (CH_2O) + O_2 \quad (8-5)$$

where (CH_2O) stands for carbohydrate. This reaction was reasonably well known by the end of the eighteenth century. As we have seen, the stoichiometry of Reaction 8–5 appears to be the reverse of respiration in mitochondria (except, of course, that no light is emitted). Appearances are sometimes deceptive, however, and even before isotope-labeling experiments in the early 1940s suggested that the oxygen evolved by chloroplasts came from water, C. B. van Niel had shown that certain bacteria could fix CO_2, using light as an energy source but without the evolution of oxygen. In the 1930s he measured the metabolism of sulfur compounds in anaerobic bacteria that live in brackish ponds and sulfur springs. He discovered that as hydrogen sulfide is oxidized to sulfate (in both purple and green sulfur bacteria) a proportional amount of carbon dioxide is taken up by the cells. The overall reaction in photosynthetic sulfur bacteria is

$$2\,CO_2 + H_2S + 2\,H_2O \xrightarrow{h\nu} \quad (8-6)$$
$$2\,(CH_2O) + H_2SO_4.$$

This reaction proceeds through two steps in which elemental sulfur is an intermediate product:

$$CO_2 + 2\,H_2S \xrightarrow{h\nu} (CH_2O) + H_2O + 2\,S. \quad (8-7)$$

$$3\,CO_2 + 2\,S + 5\,H_2O \xrightarrow{h\nu} \quad (8-8)$$
$$3\,(CH_2O) + 2H_2SO_4.$$

Van Niel realized that Reaction 8–7 closely resembles Reaction 8–5, the observed reaction of photosynthesis in green plants. He concluded that photosynthesis in all organisms has the same general stoichiometry:

$$CO_2 + 2\,H_2A \xrightarrow{h\nu} (CH_2O) + H_2O + 2\,A. \quad (8-9)$$

Van Niel proposed, therefore, that photosynthesis is an oxidation-reduction reaction; CO_2 is reduced by transfer of electrons and protons from donor molecules, designated as H_2A, which are oxidized in the light. The water produced in Reaction 8–9 gets its oxygen from CO_2. In green plants *the electron donor is water, from which oxygen is released:*

$$CO_2 + 2\,H_2O \xrightarrow{h\nu} (CH_2O) + H_2O + O_2. \quad (8-10)$$

This was shown in plants by using $H_2^{18}O$ and finding that the evolved gas was $^{18}O_2$. When $C^{18}O_2$ was used, water became labeled with the ^{18}O; as Reaction 8–9 indicates, the water formed by Reaction 8–10 is not one of those used as an electron donor, but a new molecule. In anaerobic sulfur bacteria, no oxygen is evolved because water is not split in either Reaction 8–7 or 8–8; the ultimate electron donor is H_2S. Among other phototrophic bacteria there is a wide variety of donor molecules, ranging from inorganic H_2 gas to organic molecules such as isopropanol and lactic acid. Although our discussion of photosynthesis thus far has been applicable to photosynthetic bacteria, most of the examples we shall give from now on are from green plants or algae.

A substantial amount of energy is needed to drive the reactions of photosynthesis in green plants. The energy requirement derives primarily from the use of water as the reductant. The standard potential for CO_2 reduction is:

$$CO_2 + 4\,H^+ + 4\,e^- \rightleftharpoons (CH_2O) \qquad E' = -0.4\ V$$

As can be seen from Table 8–1, the standard reduction potential for hydrogen is almost the same as for CO_2, so reduction by hydrogen would almost be spontaneous under standard conditions. When water is the reductant, however, the high reducing potential of oxygen must be overcome and the $\Delta E'$ is -1.2 V for the reduction of CO_2, equivalent to the standard free energy change of 460 kJ (110 kcal) per mole. This is the energy that must be supplied by light.

Chlorophylls

The mechanisms by which light energy is harvested and converted to chemical energy in photosynthesis are quite complex, involving the interaction of many compounds. The process starts by the absorption of a light photon by, and the consequent excitation of, a specific molecule. In photosynthesis today this role is played by *chlorophylls.*

Chlorophylls absorb the blue and red regions of the visible spectrum and therefore appear green. There are six principal types of chlorophyll in nature, differing very little in molecular structure. Chlorophyll *a* and chlorophyll *b* are present in green plants, algae, and certain protozoa; chlorophyll *c* is found in brown algae, diatoms, and dinoflagellates; chlorophyll *d* is a minor chlorophyll species in marine red algae; finally, bacteriochlorophylls and chlorobium chlorophylls occur in purple and green sulfur bacteria, respectively. All chlorophylls are made of four pyrrole rings bound to each other to form a *porphyrin* (Figure 8–25). An *extra ring,* a long *phytyl side chain,* and an atom of *magnesium* complexed to the rings are unique features of the chlorophyll molecules. These give them properties unlike those of other porphyrin-containing compounds found in hemoglobin and cytochromes, which possess an atom of iron instead of magnesium and do not usually contain long lipid-soluble side chains.

The useful property of all chlorophylls is the large number of double bonds in the molecular structure. There is extensive delocalization of electrons into molecular orbitals that differ only a little in energy level. For this reason, only a relatively small amount of energy is required to excite an electron from such a bonding orbital in the molecule to an unoccupied orbital with a higher energy. This process is fast, taking about 10^{-15} sec. Once

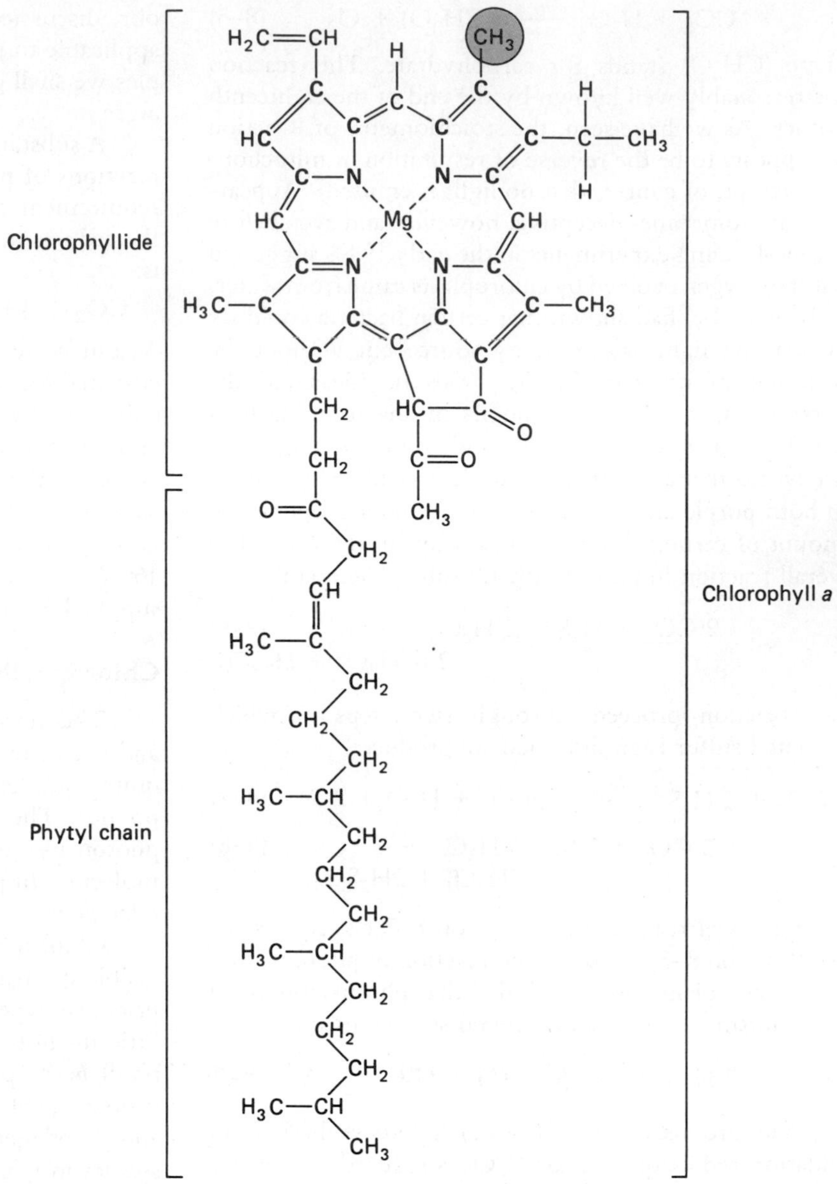

Chlorophyllide

Chlorophyll *a*

Phytyl chain

FIGURE 8–25

Structure of chlorophyll *a*.
Chlorophyll *b* differs in having a
—CHO group in place of the —CH₃
group enclosed by the circle. The four
pyrrole rings are indicated in bold
type.

in an excited state, an electron in the chlorophyll molecule can remain there for relatively "long" periods of time (10^{-8} sec) before the excitation is released, by direct transfer of the energy to another chlorophyll molecule or by the electron's falling back to a lower energy state. The latter process of de-excitation results in the emission of light energy observed as *fluorescence.*

During photosynthesis, however, the excited electron becomes completely dissociated. This occurs where a chlorophyll molecule is complexed with special proteins to form a *reaction center.* After excitation, electrons are transferred from the reaction center chlorophylls to specific compounds. The electron acceptors are reduced and the chlorophyll molecules of the reaction centers are oxidized. This is the important step in the photosynthetic

electron transport pathway; it is the process by which *light energy is converted to chemical energy.*

The oxidation of the reaction center cannot take place in the dark. It is only the light-excited forms of the reaction centers that can act as reductants. In a sense the fundamental problem of photosynthesis is to understand exactly how the absorption of photons causes the loss of the electrons to an acceptor from which they can return only infrequently.

The Photosynthetic Unit

In Reactions 8–5 to 8–9 the amount of light energy needed is not explicitly stated. This energy can be measured by experiments analogous to the enzyme assay, in which the photons absorbed are treated as though they were a reactant. Figure 8–26 shows a typical response of the rate of oxygen evolution to varying light intensities in a green alga. At low intensities the photosynthetic rate (assayed by O_2 evolution, which is easier to measure than CO_2 fixation) is linear with light intensity. At higher intensities the photosynthesis is *saturated* and some reactant other than light must be rate-limiting. This other limiting reactant might be needed anywhere along the coupled set of reactions that comprise the photosynthetic pathway.

If one measures light intensity in *photons absorbed per second* and O_2 production in molecules per second, the slope of the linear portion of the curve in Figure 8–26 is the *quantum yield*. This is now known to be one-eighth O_2 per photon or perhaps a bit less; the *quantum requirement* for O_2 evolution (the reciprocal of the quantum yield) is thus at least eight.

Evidence for postulating the existence of special reaction centers was first provided by R. Emerson and W. Arnold in 1932. They measured the oxygen evolution by *Chlorella* algae cells following illumination with very short flashes of light of saturating intensity. The interval length between flashes must be sufficiently long to warrant that no additional O_2 is evolved by making it longer (this interval is about 0.04 sec). Under these conditions the maximum amount of photochemistry that could be done would be done, but the yield of O_2 was found to be only one per 2500 chlorophyll molecules. This same ratio was found for *Chlorella* cultures grown to contain different amounts of chlorophyll. Since the measured quantum requirement is eight photons per O_2 evolved, there seems to be about 300 times as much chlorophyll as necessary. Emerson and Arnold used the term *photosynthetic unit*, proposed by H. Gaffron and K. Wohl, to describe collections of what we now know to be 300–500 chlorophyll molecules that act together in the absorption of light. One or two of these chlorophyll molecules are in a special environment, probably complexed to proteins, and can consequently help to perform the photochemistry; this is the *reaction center chlorophyll*.

It is now known that reaction center chlorophylls are molecules (perhaps dimers) of chlorophyll *a*. But if less than 1% of the chlorophyll is responsible for photosyn-

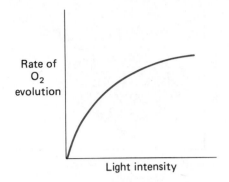

Rate of O_2 evolution

Light intensity

FIGURE 8–26

The activity of photosynthesis as a function of light intensity. Photosynthetic activity in a green alga was measured by the rate of oxygen gas evolution. The rate is proportional to the intensity at low light levels but becomes saturated at higher light intensities.

thetic reactions, what is the role of the other chlorophyll molecules in the photosynthetic unit? As it happens, in moderate sunlight an individual chlorophyll molecule can absorb, on the average, only about one photon per second. We have mentioned above that a dark interval of 0.04 sec suffices to perform the remaining steps leading to O_2 evolution. Thus, if each chlorophyll were connected to a separate enzyme system, the enzymes would have no substrate more than 95% of the time. As W. Sistrom has put it, "Unemployed enzymes are frowned on by natural selection." By connecting a large number of chlorophylls to a single enzyme system, the most efficient use is made of the metabolically expensive enzymes.

Typical photosynthetic units contain not only various chlorophylls but also *accessory pigments*, such as carotenoids and phycobilins. The combination of chlorophyll *a* with other chlorophylls and with the accessory pigments absorbs light over a wider range of wavelengths than chlorophyll *a* alone. Thus, the photosynthetic unit also maximizes the number of photons that can be absorbed. The ensemble of chlorophylls, other pigments, and their associated proteins not actually at the reaction

Lutein, a xanthophyll accessory pigment of leaves, green algae
(β, ϵ-carotene-3,3'-diol)

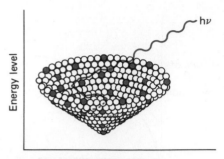

- 🔴 Initial excited molecule
- ⚪ Chlorophyll
- ⚫ Accessory pigment
- 🔴 Reaction center chlorophyll

FIGURE 8–27
The photosynthetic unit or antenna complex.   A light photon excites one molecule and the excitation moves within the densely packed chlorophyll and accessory pigments. Eventually, the energy is "trapped" in the molecule with the lowest energy, the reaction center chlorophyll.

center can be considered as an *antenna complex* that absorbs photons and transmits the excitation to the reaction center (Figure 8–27).

The lifetime of the excited state of the photosynthetic unit is estimated at 10^{-9} sec, during which the excitation that results from the absorption of a photon bounces among the various pigment molecules, gradually losing energy. The transfer of excitation does not involve radiation, but it does require that the antenna molecules be packed closely together, as they are in the photosynthetic unit. Eventually the reaction center traps the excitation, because it is then at the lowest energy state in the antenna. Although the various pigments can absorb light of higher energy, each absorbed photon's excitation ends up at a common reaction center and is thus of equal effect as a substrate for the attached enzyme system.

The Red Drop and Enhancement

R. Emerson and C. Lewis made measurements in the early 1940s of the quantum yield of oxygen evolution (O_2 evolved per photon absorbed) for light of different wavelengths. The graphic display of the results at a series of wavelengths is called an *action spectrum* and is shown in Figure 8–28 for the green alga *Chlorella*. If, as one expects for a simple photochemical reaction, all the light capable of being absorbed by the pigments in the photosynthetic unit were equally able to cause evolution of oxygen, one would expect an action spectrum that is a horizontal line, of height equal to one-eighth, wherever light is *absorbed*.

What Emerson and Lewis found, however, was a precipitous drop in the quantum yield at far-red wavelengths, at about 700 nm, even though isolated chlorophyll *a* has an appreciable ability to absorb light of those wavelengths (Figure 8–28). This decline in the photosynthetic efficiency of long wavelength light was generally observed in photosynthetic cells using water as an electron donor, and the phenomenon became known as the *red drop*.

Fourteen years later, Emerson discovered that he could eliminate the red drop by illuminating with blue (or green) light at the same time as with light above 700 nm in wavelength (Figure 8–28). The rate of oxygen evolution that occurred on simultaneous illumination with both colors was *greater than the sum* of the rates that occurred with the two colors individually. These experiments have been widely repeated on a variety of photosynthetic cells that show a red drop, and all show this *Emerson enhancement effect*. It was also found that the two colors of light do not need to be given simultaneously to observe the enhancement. It is possible to

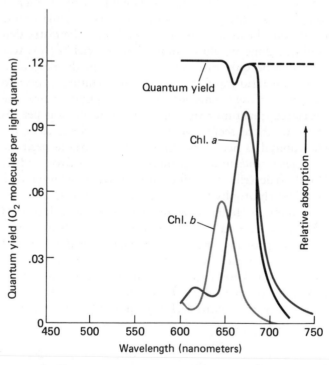

FIGURE 8–28
The quantum yield of oxygen evolution as a function of illumination wavelength.   Shown are the quantum yields and the absorption curves of chlorophylls *a* and *b*. The sharp reduction in quantum yield at about 700 nm is known as the "red drop." The high quantum yield is restored (broken line) when blue or green light is shone simultaneously with the far red light. (Data after R. Emerson.)

separate a flash of red light from a later flash of green light by as much as 10 sec and still find enhancement of the O_2 evolution rate.

On the basis of these results Emerson suggested that there are two distinct light-absorbing systems, only one of which can use red light of wavelength greater than 700 nm. This one was named *photosystem I (PS I)* and the other, which requires light of shorter wavelengths, was named *photosystem II (PS II)*. When a plant is exposed to far-red light, above 700 nm, the rate of oxygen evolution is limited by PS II, since it is getting very little usable light even though PS I may be saturated with photons. Adding some green light enables PS II to run faster and more O_2 is evolved. One or more of the products of PS I must be long-lived in the dark to explain how a delayed exposure to green light gives enhancement.

The Z-scheme

The observation of enhancement requires that there not only be two photosystems but also that they interact in some way. In 1960, R. Hill and F. Bendall proposed a theoretical model that connected the two photosystems by electron flow through an intermediate carrier system. This carrier was originally thought to be cytochrome *f,* a cytochrome of the *c*-type found in the leaves of green plants. We now know that other carriers are also involved, and a more modern rendering of the Z-scheme is shown in Figure 8–29.

The essential feature of the Hill and Bendall model has not, however, been changed by more recent data. PS I oxidizes cytochrome *f,* while PS II reduces it. In other words, electrons given up by PS II pass into and out of cytochrome *f* on their way to PS I. This suggestion was given quick experimental support in 1961 by L. N. M. Duysens and others. When the red alga *Porphyridium* was illuminated with red light, which is absorbed primarily by chlorophyll *a,* they detected an absorption spectrum characteristic of oxidized cytochrome *f,* while green light, absorbed primarily by an accessory pigment in PS II, led to a spectrum characteristic of reduced cytochrome *f.* Similar "push-pull" absorption changes have been seen for other components of the connecting pathway between PS II and PS I (e.g., *plastocyanin* and *plastoquinone*). There is now sufficient evidence to convince us that the two photosystems operate in series, as proposed in the Z-scheme of Hill and Bendall.

The ideas of enhancement and reaction centers were brought together by the work of B. Kok, who identified a special kind of chlorophyll *a* that has an absorbance at 700 nm. Kok surmised that this so-called *pigment 700 (P700)* might be the chlorophyll at the reaction center of PS I because the absorbance was diminished *(bleached)* by exposure to light capable of driving photosynthesis by PS

I (Figure 8–30). If a chlorophyll electron has been removed by an acceptor, that chlorophyll cannot absorb a photon with energy appropriate to excite the electron until it has been replaced. When exposed to light that preferentially activated PS II, the P700 regained its absorbance. Exciting PS II allows electrons to be removed to reduce cytochrome *f,* which in turn can reduce PS I and thus supply the electrons needed to absorb 700-nm photons.

The absorbance change can also be brought about by placing isolated P700 molecules in an oxidizing environment. A substance like P700 can be oxidized by an electron acceptor like ferricyanide if it is in sufficiently high concentration. The fraction of oxidized P700 is 50% at the midpoint of the bleaching response, and this level can be obtained experimentally by adjusting the ratio of concentrations of ferri- and ferrocyanide. This ratio and the known standard reduction potential for the ferricyanide ion can then be used to calculate the standard reduction potential of P700 (+0.43 V). These experiments allowed the placement of P700 on the Z-scheme potential scale (Figure 8–29). Similar methods were used to locate the standard reduction potentials of many other components in the Z-scheme. Later, another pigment system, *P680,* was implicated as the reaction center of PS II. There is some evidence to suggest that both P700 and P680 may be dimers of chlorophyll *a,* but the reason they have different absorbance maxima, though doubtless related to their association with proteins, is not yet understood in detail.

Photosystem I

Let us now follow electrons as they move through the Z-scheme. We shall begin with PS I since an analogous system is found in photosynthetic bacteria that do not evolve O_2, as well as in green plants that do. Shortly (10^{-15} sec) after P700 has absorbed a photon from the antenna of PS I, it becomes excited (P700* in Figure 8–29). P700* has a very negative reducing potential; it can reduce dyes such as methylviologen with standard reducing potentials of -0.5 to -0.7 V. This makes P700* a much stronger reducing agent than unexcited P700. The time required for the reduction to occur is about 10^{-11} sec, which is much shorter than the fluorescent lifetime mentioned earlier. Understanding the molecular basis for the speed of the reduction of the primary electron acceptor is one of the fundamental problems of photosynthesis.

The *primary electron acceptor* of PS I is not known, but there is some molecule whose absorbance at 430 nm is reduced (bleached) by light that excites PS I and which kinetics suggests is the first to be reduced. The spectrum of this X_{430} is somewhat different from the known

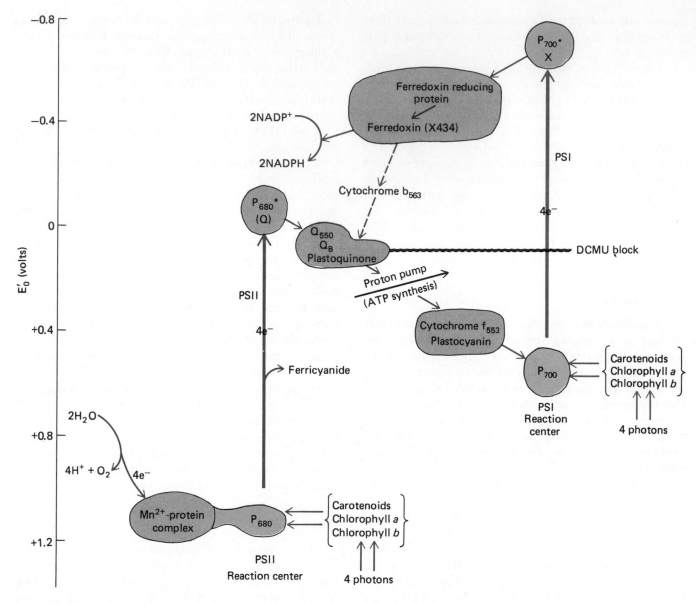

FIGURE 8–29

The Z-scheme: present-day concept of chlorophyll activation and electron flow during photosynthesis in plants. P680* and P700* are excited forms of reaction centers P680 and P700. Many of the components exist as complexes, such as the Mn^{2+}-protein complex, the Q_{550}-Q_B-plastoquinone complex, the cytochrome f_{553}-plastocyanin complex (which also contains an iron-sulfur protein), and the ferredoxin-reducing protein-ferredoxin complex. Between the plastoquinone and the cytochrome f_{553} complex is the position where a proton pump is coupled to ATP synthesis (see Figure 8–14c). The diagram indicates the eight photons necessary for the evolution of an O_2 molecule, and the electron flow among the components. The dotted line indicates a pathway for cyclic electron flow in the presence of cytochrome b_{563}, involving electrons only in the PS I pathway. The herbicide DCMU binds to the Q_B protein and blocks electron flow from PS II. This allows one to study the PS I system independent of PS II activity. Ferricyanide, when added, can be reduced by the electrons from PS II, allowing the study of PS II independent of PS I activity.

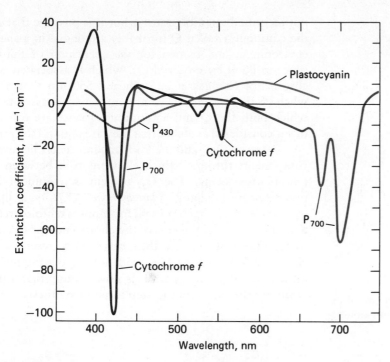

FIGURE 8–30
Difference spectra for various components of the Z-scheme. Using a low intensity source, the absorption by dark-adapted chloroplasts of light at different wavelengths was measured. The oxidation or reduction was accomplished by intense flashes of light. Immediately after an intense flash of light with a wavelength capable of driving PS I, the spectrum was found to be different; there was much less absorbance at 700 nm, for example. A similar spectrum could be obtained if the chloroplasts were treated with strong oxidants. Plotted in the figure is the difference in absorbance (oxidized minus reduced). The points in the spectrum dominated by the various electron carriers are indicated. The negative value of 700 nm means that the oxidized absorption is less than the reduced; thus, there is a "bleaching" response to light capable of exciting PS I. An intense flash with a wavelength capable of exciting PS II gives another spectrum; this treatment tends to reduce P700. (From T. Hiyama.)

spectra of ferredoxins, iron-sulfur proteins that are found in all photosynthetic cells, and it is suspected that a special protein-bound form of ferredoxin may comprise the X_{430}, perhaps in combination with other unknown molecules. After soluble ferredoxin has been removed, the bound form can still be detected in chloroplasts by its characteristic electron paramagnetic resonance spectrum. The bound form can be reduced after light excitation of PS I at temperatures too low to permit chemical reductions. This observation also suggests that the primary electron acceptor may be a bound ferredoxin.

The plant version of the ubiquitous ferredoxin has a molecular weight of about 12,000 and contains two iron atoms bound to two sulfur atoms (Figure 8–31); there is no heme group. Its standard reducing potential is 0.1 V

more negative than $NADP^+$-NADPH, and it is ferredoxin that delivers the two electrons to $NADP^+$, electrons used subsequently to reduce, albeit indirectly, carbon dioxide. The reaction is catalyzed by ferredoxin-$NADP^+$ reductase.

Some simple calculations can show the contribution by light to the energy requirements. Since unexcited P700 has a standard reducing potential of +0.43 V, its reduction of X_{430} would under standard conditions require a free energy change of $\Delta G = - F \, \Delta E = - F (-0.4 - 0.43) = +80$ kJ/mole, a very nonspontaneous process. P700*, on the other hand, has a standard reducing potential of −0.7 V, and the process of reducing X_{430} has a free energy change of $\Delta G = - F (-0.4 + 0.7) = -30$ kJ/mole. The light absorbed by

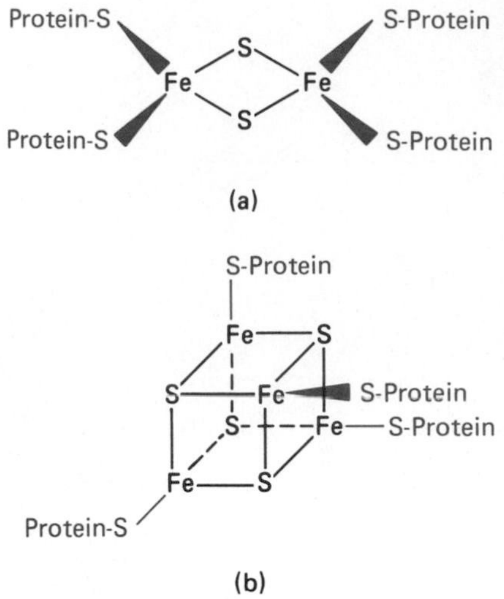

FIGURE 8–31

Two types of iron-sulfur complexes of the sort found in soluble and bound ferredoxin proteins. The iron atoms are also bound to the sulfur of cysteine in proteins. (a) 2Fe-2S. (b) 4Fe-4S.

P700, with energy = hc/700 nm = 170 kJ/mole, supplies the energy required to reduce X_{430}. The energy efficiency of using 700-nm light, the longest wavelength that can excite PS I, is 110/170, or about 65%—an overestimate, since more energetic photons will still lead to only 110 kJ/mole of conserved energy, but impressive nonetheless.

To generate a continuing supply of electrons for carbon fixation, photosynthetic bacteria, with only one photosystem, must use reduced chemical compounds in their environment—H_2S and succinate are examples. Some photosynthetic bacteria use the coenzyme NADH, instead of the NADPH used by green plants, to reduce CO_2. Also, the reduction of NAD^+ or $NADP^+$ in these bacteria occurs by a pathway whose energy is supplied by the proton gradient or by ATP, rather than directly by light. The main use that these bacteria make of their single photosystem, however, is to generate ATP (see the discussion of cyclic photophosphorylation later in this chapter).

The success of green plants and algae is partly attributable to their ability to use the electrons in water, which is so readily available on earth, for the fixation of carbon dioxide. The problem with this source of electrons is that the reduction potential of water is about +0.82 V. The solution to bridging the large energy gap (some 110 kJ/mole) between water and $NADP^+$ was to develop a second photosystem and connect it to the first by an electron transport chain.

Photosystem II

The absorption of a photon by PS II excites the P680 to the equivalent of a reducing potential less than zero volts. The primary electron acceptor from P680* is still

not identified unambiguously, but it is possible that it is the compound called Q (probably a quinone in a special environment and named for the *q*uenching of fluorescence in PS II by oxidized Q), which is associated with absorbance changes at 325 nm and 550 nm that occur when PS II is excited. Algal mutants have been isolated in which both the Q and 550-nm absorbance are missing; this is considered evidence that they are one and the same.

Between Q_{550} and the PS I reaction center lie several components through which electrons pass between the two photosystems. The Q_B protein is a quinone-like two-electron acceptor. *Plastoquinone (PQ)* is a lipid-soluble electron carrier present in amounts tenfold greater than the other components of the photosynthetic electron transport chain. It can thus serve as a reservoir of electrons from PS II. The oxidation of plastoquinone, ultimately by photosystem I, is slow enough that it may represent the rate-limiting step in photosynthesis.

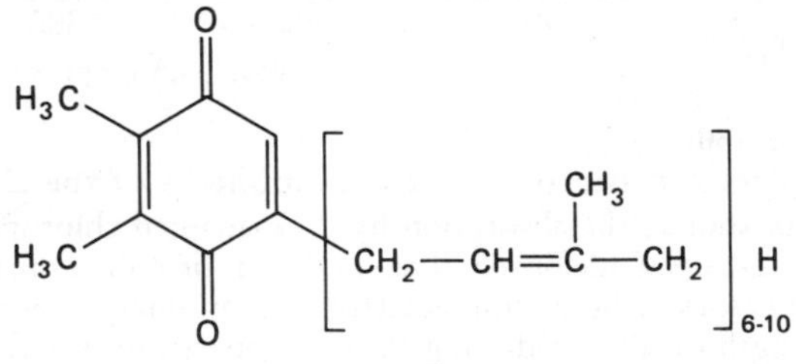

Plastoquinone

The reduction of PQ is blocked by the herbicide *DCMU* [3-(3,4-dichlorophenyl)-1,1-dimethyl urea]. In the presence of DCMU, to block electron flow from PS II, and with reduced ascorbate and a lipid-soluble dye to act as a source of electrons, it is possible to study the properties of PS I alone. Alternatively, if ferricyanide plus a lipid-soluble quinone oxidant are added as an electron acceptor, the activity of PS II can be studied, unlimited by the behavior of PS I. The process that results in the evolution of O_2 when using exogenous electron acceptors, such as ferricyanide, is called the *Hill reaction,* after its discoverer, R. Hill.

Two other electron carriers have been identified in the chain between PS II and PS I: cytochrome *f,* a cytochrome with an absorbance maximum at 553 nm (and thus sometimes called cytochrome b_{553}) and *plastocyanin,* a copper-containing protein of blue color. There have been several other components proposed for this segment of the Z-scheme, and for other segments, too. The student should be aware, therefore, of the provisional nature of the Z-scheme in Figure 8–29; future research may alter some of the details.

The functions we have been discussing all take place in the chloroplast. Before leaving this subject, we return to the yellow-orange pigments called carotenoids. They are a major component of the chloroplast; there are about

20 carotenoids for every 100 chlorophylls. The major carotenoid is β-carotene, which has several conjugated double bonds in its structure. Carotenoids absorb light maximally between 450 and 470 nm and transfer light energy to the reaction center rather inefficiently. Therefore, the main function of these pigments seems not to be in light harvesting. Instead, they probably act to protect light-sensitive structures in the chloroplast. This property can be explained by noting that chlorophyll is a "photosensitizer": in the light it can become excited to a "triplet" state and transfer the energy of that excitation to molecular oxygen, thus converting it to a reactive "singlet" state. It is the singlet state oxygen that destroys cellular components like the double bonds in membrane lipids and amino acid side chains in membrane proteins. Carotenoids, particularly if they contain 10 or more conjugated double bonds, can directly quench the triplet state chlorophyll and the singlet state oxygen, thus preventing their destructive effects.

β-Carotene, a carotenoid

Chloroplast Structure

In plant cells there is a class of organelles called *plastids,* which all appear to develop from a single type of precursor, the *proplastid.* Two of these are the *leucoplasts,* the plastids concerned with the transformation of glucose to starch, and the *chromoplasts,* the site of synthesis and storage of many plant pigments. But the most important of the plastids is the *chloroplast,* the organelle in which photosynthesis takes place (Figure 8–32). Chloroplasts are medium- to large-sized organelles which also vary a great deal in shape from species to species. Algae have an especially large variety of differently shaped chloroplasts, typically containing pyrenoids and pyrenoid sacs—bodies that store and possibly synthesize starch grains. In higher plants, each leaf cell contains a fairly large number of ovoid or disc-like chloroplasts, 2–4 μm in thickness and 5–10 μm in diameter.

The photosynthetic electron transport of plant cells, including the light-driven excitation of the photosystems, takes place in the membranes of the chloroplast, just as the respiratory electron transport system of animal cells is found in the membranes of the mitochondria. Even in photosynthetic bacteria (Figure 8–33), electron transport is found in the membranes of a network of vesicles spread throughout the cell. In the procaryote *Cyanobacteria* (blue-green algae), whose ancestral forms were the likely progenitors of chloroplasts, these vesicles are much enlarged and flattened. Such flattened vesicles are called *thylakoids.* In the eucaryotic green algae (Figure 8–34, p. 416), and in the chloroplasts of higher plants (Figure 8–32), the thylakoids are surrounded by an *inner* and *outer* membrane, which form the envelope of the chloroplast. The space enclosed by a thylakoid vesicle is called the *loculus* and the space within the inner membrane that surrounds the thylakoids is the *stroma.*

The groups of thylakoid membranes (sometimes called *lamellae*) in the chloroplasts of higher plants are more complicated: extensive fusion of the outer leaflet of these lamellar membranes has occurred. Electron microscopic examination of the chloroplasts reveals many of these fused lamellae piled one on top of the other, resembling layered discs of about 0.3-μm diameter. The stacks are called *grana* (Figure 8–32, 8–34). There are sometimes long, solitary membranous connections, called *stroma lamellae,* between the discs in adjacent grana. The stroma contains the enzymes of the carbon-fixation cycle, and probably all the enzymes involved in the assembly of chlorophylls and other large (molecular weight $\sim 10^3$) molecules. Also in the stroma are several structures unrelated to photosynthesis: the protein-synthesizing ribosomes, globules which are probably lipid droplets, starch granules, and in some, algae, pyrenoid bodies, and "eye-spots" whose function is not exactly known. A diagram of the current idea concerning this structure is given in Figure 8–35, p. 417.

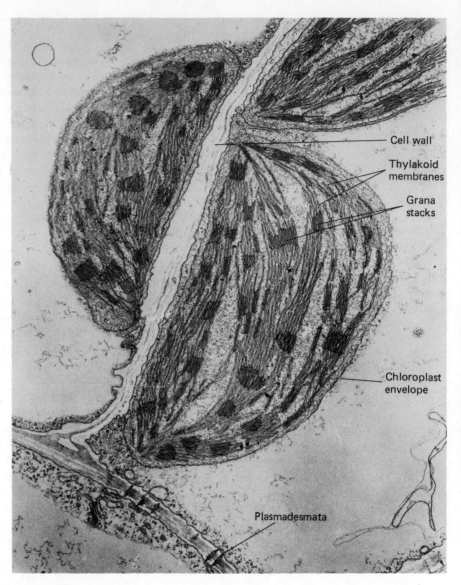

FIGURE **8–32**
Chloroplasts. Leaf mesophyll cell from *Leptochloa,* showing chloroplasts adjacent to cell membrane. The thylakoid membranes are all interconnected to form one membrane-enclosed space. × 150,000. (Courtesy of H. Mollenhauer.)

The thylakoid membrane is where the bulk of the chlorophyll is found and where most of the photochemical events take place. The stacking of these membranes into lamellae is consistent with the photosynthetic function. The chlorophylls and other components of the photosynthetic unit are concentrated into a small volume. Concentration of the pigment molecules allows rapid transfer of the light-induced excitation to the reaction center. The thylakoid membrane also separates the inside loculus compartment from the proteinaceous stroma, providing a permeability barrier that prevents discharg-

ing the electrical and pH gradient generated by photosynthetic electron flow. Freeze-etch preparations of chloroplast membranes (Figure 8–36, p. 418) reveal the presence of large and small particles embedded in the membrane. Some of these are probably the *coupling factors*—called CF_1, similar to the F_1 coupling factors in mitochondria—that transform the proton gradient into ATP.

The development of the grana is an interesting process. If young leaf tissue is examined, the chloroplasts are colorless and look quite different; instead of grana, all they contain are isolated vesicles. When this tissue is

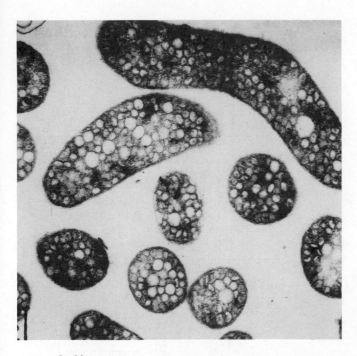

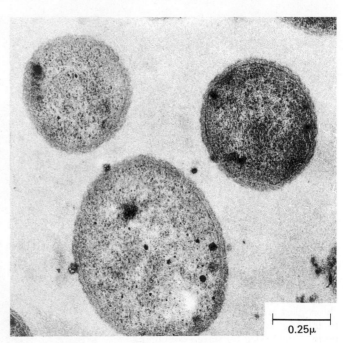

0.25μ

FIGURE **8–33**

Electron micrograph of *Rhodospirillum rubrum*, a photosynthetic bacterium.
(a) These cells have been grown anaerobically in the light. Under such conditions, the plasma membrane invaginates to form vesicles inside the cell, vesicles bound by the photosynthetic membranes. (b) When placed in the dark, these vesicles disappear, and the cell appears similar to nonphotosynthetic bacteria. × 150,000. (Courtesy of S. R. Simon.)

illuminated, some remarkable changes take place, culminating not only in a proliferation of membranes, but also in the typical appearance of the grana architecture. Thus, light seems to be necessary both for the synthesis of chlorophyll (which cannot occur in the dark in the chloroplasts of higher plants) and for the synthesis of the chlorophyll-containing membranes. The newly formed membranes come together to form the typical grana structure. The same kind of process, a light-induced formation of photosynthetic membranes, seems also to occur in the photosynthetic bacteria (cf. Figure 8–33). In Chapter 15 we will go further into the formation of the chloroplast.

Electron Transport Chain Function

The electron transport chain connecting PS II with PS I has two principal functions. First, it serves as a source of electrons for the reduction of PS I so a continued supply of NADPH can be maintained to fix carbon dioxide. Second, it serves as a generator of ATP, also a necessary reagent in the fixation of carbon dioxide (see Figure 8–38, p. 420). The series of electron carriers lie along a decreasing potential energy gradient, from lower to higher reduction potential, so electrons flow spontaneously from P680* to the oxidized reaction center of PS I or from H_2O to the oxidized reaction center of PS II. Movement of electrons along these paths is coupled to the phosphorylation of ADP in a way quite similar to that which we have discussed for mitochondria. The removal of electrons from water by oxidized P680 yields oxygen.

Generation of ATP

As in the case of mitochondria, the tightness of the coupling of phosphorylation to electron transport can be demonstrated by reversing the electron flow. Upon addition of ATP to chloroplasts that have been stimulated with PS I-specific light, there is a change in the absorbance of Q, the primary electron acceptor of PS II, that indicates its reduction. This demonstrates a reversal of electron flow along the path joining PS II to PS I, driven by the free energy of ATP hydrolysis. Thus, the ATPase and the photosynthetic electron transport chain are fully reversible. Also, as in the case of mitochondria (Table 8–3), there is a phosphate acceptor effect in chloroplasts

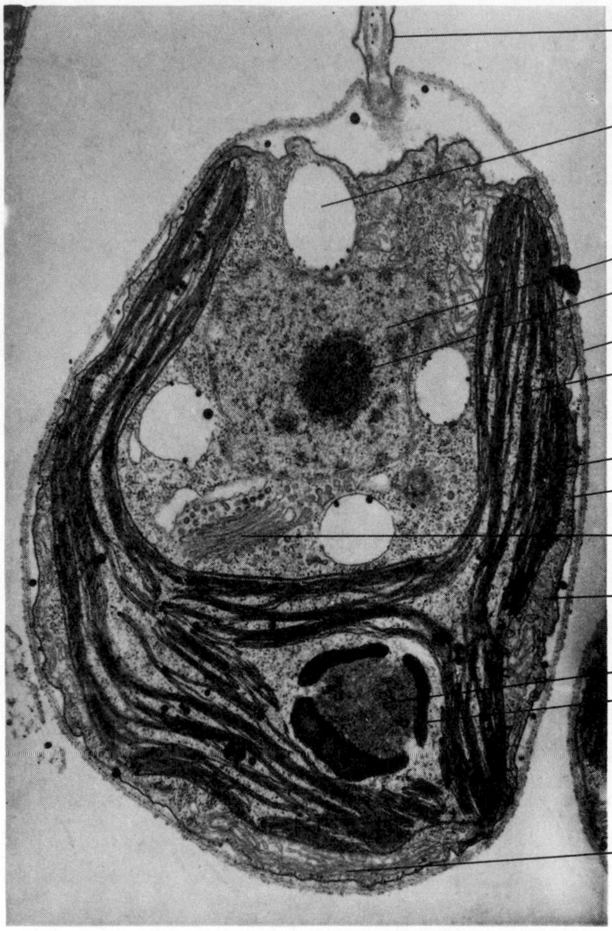
Flagella
Vacuole
Nucleus
Nucleolus
Cell wall
Thylakoid (chloroplast inner) membranes
Chloroplast membrane
Cell membrane
Dichtyosome (Golgi body)
Mitochondrion
Pyrenoid body
Starch granules
Endoplasmic reticulum
(a)

FIGURE 8–34
The alga *Chlamydomonas reinhardtii*, showing the single, cup-shaped giant chloroplast. (a) The cell is surrounded by a wall and a membrane, which encase the chlorplast, mitochondria, fat granules, and other cytoplasmic inclusions. Within the chloroplast are the thylakoid membranes and a central pyrenoid body, surrounded by starch-containing vesicles. × 21,000. (b) A higher magnification of part of the chloroplast. Shown are the cell wall, cell membrane, and double membrane around the chloroplast. Some of thylakoid membranes are fused to form grana stacks. The large dark granules are probably mineral deposits, while the smaller ones, occurring both within the chloroplast and in the cytoplasm, are ribosomes. × 82,000. (Courtesy of G. E. Palade.)

(Table 8–5). When various cosubstrates of phosphorylation are added, the rate of photosynthesis increases, due to tight coupling of electron flow with phosphorylation.

The electron carriers are disposed asymmetrically on the two sides of the thylakoid membrane (see Figure 8–14c). For each pair of electrons passing from water to NADPH, three protons are taken up from the stromal space (outside the thylakoids) and four protons are released into the loculus of the thylakoids. The stroma has a pH of around 8 and there is a gradient of about 3.5 pH units across the thylakoid membrane. The proton energy gradient has almost no electrical component. This is because the thylakoid membrane is fairly permeable to ions like Mg^{2+} and Cl^-. As protons are pumped into the

TABLE 8–5
Effect of Phosphate Acceptors on Photosynthetic Rate*

pH	Addition	NADP Reduced per mg Chlorophyll
7.5	Ferredoxin (Fd)	70
7.5	Fd, ADP, P, Mg^{2+}	205
8.0	Fd	130
8.0	Fd, ADP, P, Mg^{2+}	235
8.5	Fd	125
8.5	Fd, ADP, P, Mg^{2+}	195

*Data are from H. E. Davenport.

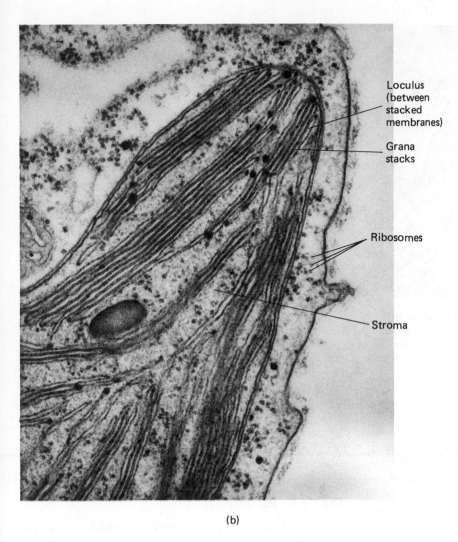

(b)

FIGURE 8–34
(continued)

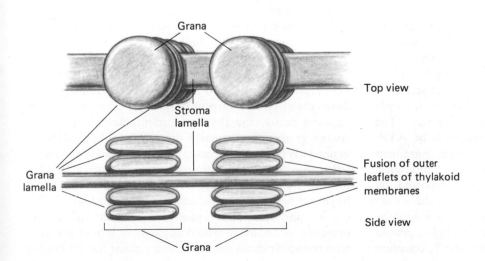

FIGURE 8–35
Diagram of lamellar structure of
the chloroplast.

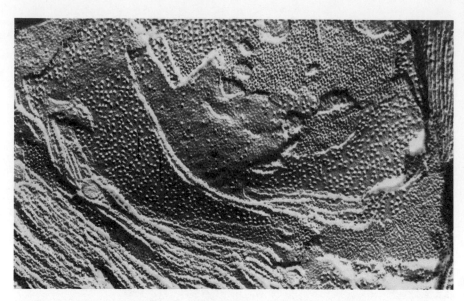

FIGURE **8–36**
Freeze-etch preparation of plant chloroplast thylakoid.   The picture shows the inner surface of the lamellae, with particles (arrows) protruding from the surface. × 80,000. (Courtesy of D. Branton.)

loculus space, these ions also move and no electrical gradient can develop. The protons can pass outward again through the CF_1–F_0 coupling system with the generation of ATP, at an estimated stoichiometry of three protons per ATP. The ATP usually remains within the chloroplast, where it is used in the biosynthetic reactions of carbon fixation, described below. The direction of the proton gradient in the chloroplast is different from that in mitochondria and there is no morphological equivalent of the loculus in mitochondria (compare Figure 8–14c with 8–14a). The disposition of the coupling factors is the same—the F_1 coupling factor is in the stroma and receives protons through an attached F_0 protein channel in the thylakoid membrane.

The chloroplast was in fact one of the first energy transduction systems that could be shown to behave as predicted by Mitchell's chemiosmotic theory. The experiment, as we mentioned earlier, was done by A. Jagendorf, who equilibrated isolated chloroplasts in solutions of low pH and then transferred them suddenly to high pH in the presence of ADP and inorganic phosphate. The resulting pH gradient drove a net synthesis of ATP. Another striking demonstration that a proton gradient can be generated by the action of light in a biological membrane, also mentioned earlier, was reported by Racker and Stoeckenius. They reconstituted the bacteriorhodopsin of the purple membrane of a halophilic bacterium in a lipid vesicle. On exposure to light a proton gradient could be measured. And when the F_1 coupling

factors from mitochondria were included in the reconstitution mixture, a light-dependent synthesis of ATP was observed!

The Cyclic Pathway

The Z-scheme is frequently described as *noncyclic photosynthetic electron transport* to differentiate it from a secondary pathway of electron flow. It was noted in 1954 by D. Arnon that isolated chloroplasts from spinach can under certain conditions synthesize ATP in the light *without* concomitant oxygen evolution; that is, without involving PS II. Since that time, it has been established that this photosynthetic electron transport involves only the reaction center of PS I and some of the intermediary electron carriers: plastoquinone, cytochrome b_{563}, cytochrome f, and plastocyanin. The electrons from excited $P700^*$ flow back to the electron transport chain, probably via cytochrome b_{563}, and return to P700. As they travel down the electron transport chain, they generate a proton gradient across the thylakoid membrane. The proton energy gradient is dissipated through coupling factors to generate ATP; this process is called *cyclic photophosphorylation*. No NADPH synthesis occurs during cyclic photosynthetic electron transport since $NADP^+$-reductase is bypassed to return the electrons to P700 (Figure 8–29). Cyclic photophosphorylation can be studied in the presence of DCMU, which apparently blocks the electron transport chain above the entry point for the cycling electrons, probably at the Q_B protein site (Figure 8–29). Such studies are needed to determine what controls the flow of electrons into the cyclic and noncyclic pathways. Many photosynthetic bacteria use only cyclic photophosphorylation to generate ATP; some of that ATP is used as an energy source to reduce NAD^+, the required electrons coming from compounds in the environment.

The Oxidation of Water

The story of photosynthesis is not complete without describing our current conception of how the electrons are removed from water to reduce PS II. Oxidized PS II is thought to have a very high reduction potential, around +1.2 V. Electrons can flow spontaneously, therefore, from water ($E' = +0.82$ V) to oxidized PS II (Figure 8–29). Carriers containing *manganese ions* are believed to be involved; one of the carriers is the so-called *Z-compound,* proposed to be tyrosine residues in the polypeptides associated with PS II. But the mechanism is more subtle than a general description of the pathway might suggest. This was discovered independently by P. Joliot and by B. Kok, who investigated the O_2 evolved by chloroplasts illuminated with very short pulses of light (Figure 8–37). After dark adaptation, the first short (1–10 µsec) pulse yielded no oxygen. The response to subsequent light pulses oscillated, with every fourth pulse producing a maximum amount of O_2. The release of protons followed a similar pattern. The model Kok proposed to explain this has a carrier ("S") that successively accumulates four positive charges during its oxidation by Z, and perhaps by other compounds. In the fully oxidized state, S can accept four electrons from two H_2O molecules and thus cause the evolution of a single O_2.

Photosynthesis—The Dark Reactions

The electron transport events described above are just part of the process of photosynthesis; we have been describing only the so-called *light reactions.* The remainder of the process consists of the *dark reactions,* so named not because they can only occur in the dark, but because in principle they do not require light for the chemistry to take place. In the dark reactions, the NADPH and ATP generated during the light reactions are used to reduce carbon dioxide, eventually to carbohydrate.

The enzymatic reactions whereby carbohydrate is formed from CO_2 in many plants and algae is shown in Figure 8–38. This scheme is the result mostly of the work of M. Calvin, J. Bassham, A. Benson, and their collaborators who, by using radioactive CO_2 and very short exposures to light, discovered and identified the initial steps in the reduction of CO_2. Later on, other workers

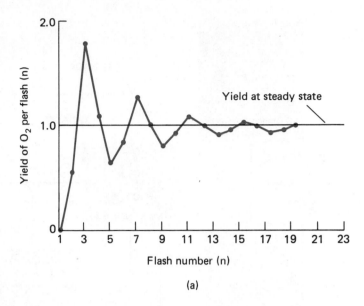

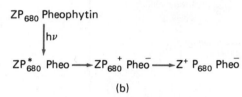

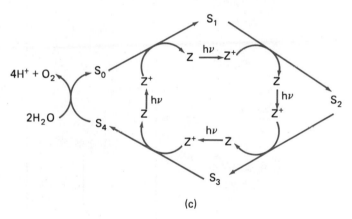

FIGURE 8–37
Generation of oxygen by PS II. (a) The relative yield of O_2, in response to very short pulses of light. The yield of O_2 by dark-adapted spinach chloroplasts is plotted relative to the steady state value. (b) Production of Z^+ compound. Excitation generates a positive and negative ion pair in a complex between the reaction center chlorophyll (P_{680}), the Z-compound (probably tyrosine residues in polypeptides associated with PS II), and pheophytin, an accessory pigment. (c) the Kok-Joliot charge accumulator. S_0–S_4 represent states of the Mn^{2+}-protein complex illustrated in Figure 8–29. Positive charges are accumulated successively from Z^+, which gives up its electron to P_{680}. (Data from Kok, Farbush, and McClain.)

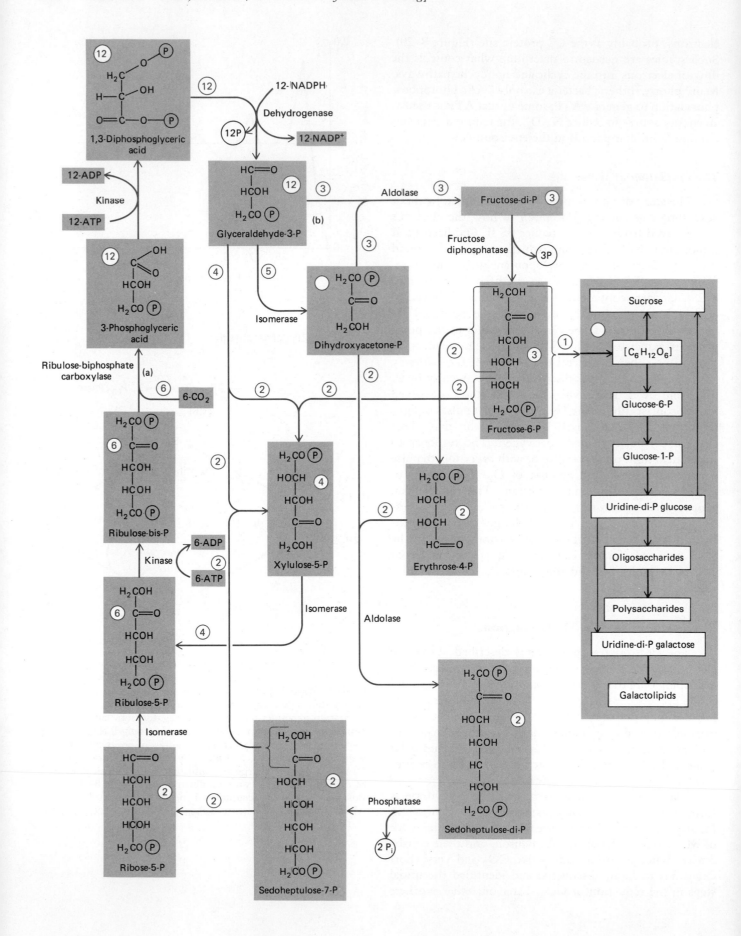

FIGURE 8–38

The Calvin-Benson (reductive pentose) cycle. Shown are the steps involved in CO_2 fixation by chloroplasts. Indicated are the enzymes involved, the CO_2 fixation step, and the use of ATP. The numbers in circles give the number of molecules of each intermediate generated by one turn of the cycle. For example (a), 6 molecules of CO_2 combine with 6 molecules of the five-carbon sugar, ribulose-bisphosphate, to give 12 molecules of the three-carbon compound, 3-phosphoglyceric acid. Another example (b) is the fate of the 12 molecules of glyceraldehyde-3-phosphate. Three of these molecules go to form 3 molecules of fructose-diphosphate by combining with 3 molecules of dihydroxyacetone phosphate. Five molecules also go to form this latter compound. Four molecules go to form 4 molecules of xylulose-5-phosphate, 2 by combining with a two-carbon fragment of fructose-6-phosphate, and 2 by combining with a two-carbon fragment of sedoheptulose-7-phosphate. All these, and other reactions shown, combine to regenerate the acceptor of CO_2, ribulose-bisphosphate.

such as B. Horecker and E. Racker began to identify and isolate the various enzyme activities involved so that at present the scheme is fully documented.

In the initial steps in this pathway, the addition of CO_2 to a 5-carbon compound, ribulose-bisphosphate *(RuBP)*, forms a short-lived 6-carbon compound, which immediately breaks down to 3-phosphoglyceric acid. When radioactive CO_2 was used, it was found that the first compound to be labeled (in the first few seconds) was 3-phosphoglyceric acid. The central position of this 3-carbon compound in the pathway has led to its being called the *C₃ pathway*, as well as the *reductive pentose cycle* and the *Calvin-Benson cycle*. The enzyme that catalyzes the addition of CO_2 is called *ribulose-1,5-bisphosphate carboxylase* ("RuBisCo"). It is a very large enzyme (molecular weight 550,000) and contains many subunits. It is an allosteric enzyme and its activity is controlled by Mg^{2+}, ATP, CO_2, and fructose-diphosphate. RuBisCo is probably the most abundant enzyme on earth; it represents about 15% of the mass of the chloroplast and has been estimated to contribute over half of all the protein mass on this planet.

Overall, 6 CO_2 molecules are incorporated into a molecule of hexose. For this process to take place, RuBP must be continually regenerated. The most important function of the complicated enzyme reactions that make up the cycle is the maintenance of a continuous supply of RuBP to ensure a continuous fixation of CO_2. Figure 8–38 shows how this is accomplished. Six molecules of CO_2 combine with 6 molecules of RuBP to give 12 molecules of a 3-carbon compound, 3-phosphoglyceric acid. This latter compound is phosphorylated by ATP to 1,3-diphosphoglyceric acid, which then is reduced by NADPH to give 12 molecules of glyceraldehyde-3-

phosphate. Of the 12 molecules formed, three are transformed into fructose-6-phosphate and one of these goes on to form the hexose sugars. Thus, each complete turn of the cycle requires 6 molecules of CO_2, 12 molecules of NADPH, and 18 molecules of ATP. One-third of the latter are required to phosphorylate ribulose-5-phosphate to RuBP. For every CO_2 fixed, 4 electrons (in the form of two molecules of NADPH) are required to reduce the carbon to the state represented in the general formula $C_6H_{12}O_6$.

The other parts of the cycle are a continuation of the process that regenerates RuBP. In the scheme shown in Figure 8–38 there are three such pathways leading from glyceraldehyde-3-phosphate. Some of the intermediates are useful not only in CO_2 fixation, but also as substrates in their own right; ribose-5-phosphate is one of the substrates for nucleic acid synthesis and erythrose-4-phosphate is one of the precursors for the synthesis of aromatic amino acids (see Figure 6–12). Fructose-6-phosphate is also used to form sucrose, the oligosaccharides, and polysaccharides, such as starch and the glycolipids, the last being found abundantly in the chloroplast membranes.

Other Syntheses

Other compounds are detected in rapid radiolabeling experiments in the light, in addition to the early formation of the compounds indicated in Figure 8–38. These include acids such as malate, fumarate, and phosphoenolpyruvate, and amino acids such as alanine, serine, and aspartate. The relatively rapid formation of compounds like these would indicate that some of the carbon atoms of the six molecules of CO_2 are neither directly converted to hexose nor do they regenerate RuBP; instead, they are sent through some bypass reactions for the synthesis of these compounds. For example, we know of enzymatic mechanisms that can form phosphoenolpyruvate from phosphoglycerate; once PEP is formed, it is possible that malate is formed from it by carboxylation and that alanine is formed by dephosphorylation to pyruvate and transamination. For those syntheses requiring reductions, it is probable that the NADPH generated in the light is used as the immediate electron donor. It is likely that acetyl-CoA is formed from one of the compounds in the carbon reduction cycle; in any case, acetyl-CoA can be formed from pyruvate. Once this compound is available, carbon can move onto pathways toward the synthesis of fatty acids, steroids, and carotenoids. Another metabolic connection is that one of the by-products of the carbon reduction cycle is the formation of the porphyrins, precursors to the chlorophylls, via a condensation of glycine with succinate, both of which arise from the cycle.

Photorespiration

Even during illumination, photosynthesizing cells using the C_3 pathway are also taking up O_2 and giving off CO_2. Some of this is due to the action of mitochondria but most of it is not inhibited by mitochondrial poisons; it is the result of an entirely different process called _photorespiration_. Glycolic acid is the substrate oxidized by photorespiration, and the glycolate oxidase enzyme is located in the peroxisomes of the plant cell. We shall not describe further the oxidative pathway, which varies in different cells, but instead describe where the glycolic acid arises and why it represents a drain on photosynthesis. When O_2 is present at high and CO_2 at low concentrations, the carboxylation reaction catalyzed by ribulose bisphosphate carboxylase is inhibited. This inhibition is partly understandable in terms of a reduction in the enzyme's activity. But recent research has revealed that O_2 can substitute for CO_2 as a _substrate,_ in which case the enzyme catalyzes an _oxygenation_ of RuBP. The products are phosphoglycolic acid and 3-phosphoglyceric acid. NADPH can reduce the latter to glyceraldehyde phosphate. The phosphoglycolate is dephosphorylated to glycolic acid, which then acts as a substrate for photorespiration in peroxisomes.

The photorespiration rate is high in bright light (five times the respiration rate in the dark), uses up a reduced form of carbon (RuBP), and generates no ATP. Cells using the C_3 (Calvin-Benson) pathway have a lower rate of photosynthetic carbohydrate production when photorespiration occurs. It is thus a surprisingly inefficient reaction to have survived so long. The investigation of apparent evolutionary paradoxes is a good approach for learning something novel about cell physiology and we await the outcome of further study of photorespiration with much curiosity.

The Hatch-Slack Pathway and C_4 Cells

There exists a group of plants, including sugarcane and maize, which are less hampered by photorespiration and which, therefore, have much greater potential effectiveness of their photosynthetic operations. These plants tend to be found in dry, sunny ecosystems. They were noticed when studies involving the incorporation of radioactive CO_2 showed that the first labeled molecules were 4-carbon compounds like oxaloacetate (OAA), malate, and aspartate. Much of the working out of this alternate CO_2 fixation scheme was done in the laboratories of M. D. Hatch and C. R. Slack, and the pathway is now known as the C_4 or _Hatch-Slack pathway_ (Figure 8-39). The cells using this pathway are named C_4 cells, as opposed to the C_3 cells using the reductive-pentose pathway described above.

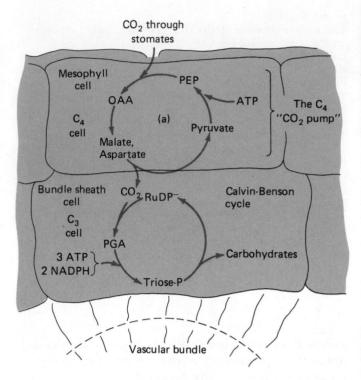

FIGURE 8-39

The C_4 pathway of carbon fixation. Two types of cells in these plants play a role in the pathway. The mesophyll (C_4) cell just under the leaf surface absorbs CO_2 through the openings, or stomates, in the leaf. The CO_2 is eventually delivered, after the reactions shown in (a), to an underlying bundle sheath (C_3) cell. There it gets fixed by the reductive-pentose, or Calvin-Benson, cycle. OAA = oxaloacetate; PEP = phosphoenolpyruvate; PGA = phosphoglycerate; RuDP = ribulose-bisphosphate.

The primary CO_2 assimilation is catalyzed by phosphoenolpyruvate (PEP) carboxylase:

$$PEP + CO_2 \rightleftharpoons OAA + P_i \qquad \Delta G' = -27 \text{ kJ/mole}$$

This enzyme has a much higher affinity for CO_2 than does RuBP carboxylase. The reaction is strongly exergonic and helps to drive the whole pathway forward. The remainder of the reaction pathway is shown in Figure 8-39, where its compartmentalization into two cell types is also illustrated. Oxaloacetate gives rise to malate and aspartate in the C_4 mesophyll cell. These compounds are taken up by the C_3 bundle sheath cell where they are decarboxylated; the CO_2 is then fixed into sugar by the Calvin-Benson cycle illustrated in Figure 8-38.

There is a cost in the C_4 cycle of two additional ATP high-energy phosphates for each CO_2 fixed. This extravagance is paid for, however, by keeping the CO_2 levels low in the mesophyll cells, so that CO_2 can diffuse in at useful rates, even when openings to the outside (the stomates) are closed to prevent water loss in arid climates where C_4 plants are found. Moreover, the CO_2 levels in the bundle sheath cells are kept high, which tends to reduce the deleterious photorespiration. The result of this biochemical adaptation is higher rates of hexose synthesis by C_4 plants for equal leaf areas. This difference between C_3 and C_4 plants is of such economic importance that we can expect much research to be devoted to understanding it and designing improvements in the efficiencies of crop plants.

Bioluminescence

We close our discussion of energy metabolism with a brief description of a performance by some cells which humans might well exploit to ease their energy budgets. This is the phenomenon of *bioluminescence,* the production of light from the chemical energy of cells. There are many cell types that can do this, ranging from simple marine bacteria (including those found as symbionts in the light organs of fish) to the cells of glow-worms and the lantern organ of the firefly.

Essentially, the free energy of an oxidation reaction is used to excite a molecule. When it returns to its ground state, a photon is emitted. (This can be seen at a low level even with chlorophylls, but as with other pathways, the emission of light by bioluminescent cells is not the reverse of chlorophyll excitation.) There are at least two

components involved, a smaller molecular weight heterocyclic ring structure called *luciferin* (Figure 8–40) and an enzyme that catalyzes the activation and oxidation of luciferin, *luciferase*. In fireflies, as shown by W. D. McElroy, the reduced luciferin (LH_2) becomes activated by ATP, a reaction catalyzed by the luciferase (L'ase), to form an enzyme-bound luciferyl adenylate:

$$LH_2 + ATP + L'ase \rightleftharpoons L'ase—LH_2—AMP + PP_i.$$

The oxidant is molecular oxygen:

$$L'ase—LH_2—AMP + O_2 \rightleftharpoons L^* + H_2O + L'ase + AMP + CO_2.$$

The CO_2 comes from decarboxylating the reduced luciferin, and L^* is an excited form of *oxyluciferin*. Light is emitted as the excited L^* decays to oxyluciferin (L). The firefly *(Photinus)* luciferin/luciferase system is used as a sensitive assay for ATP, because of the convenience of instruments that measure light rather than chemical reactions.

Luminescent marine bacteria use a different kind of luciferin and do not involve ATP in the reaction pathway. A reduced flavin, $FMNH_2$, interacts with the luciferase and the combination is oxidized by molecular oxygen. A long-chain fatty aldehyde is involved but the emitting molecule is probably either an altered flavin structure, or a recently reported blue fluorescing protein, or a combination of the two. The use of mutant bacteria

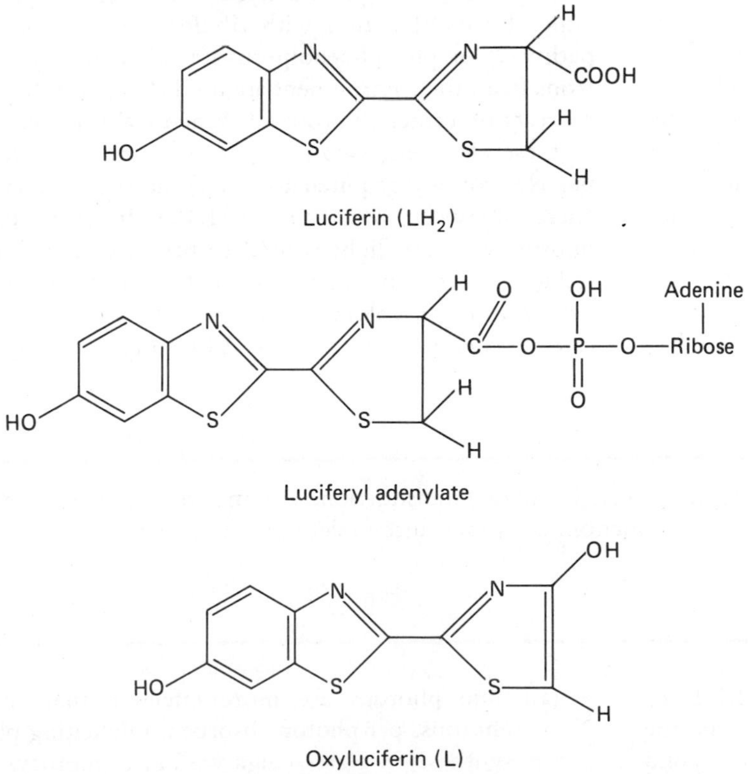

Luciferin (LH$_2$)

Luciferyl adenylate

Oxyluciferin (L)

FIGURE 8–40
Formulae of luciferin, luciferyl adenylate, and oxyluciferin.

with genetic lesions in the luciferase will help to work out the reaction pathway.

World Implications

At a time when exhaustible supplies of liquid fossil fuels are becoming cost-prohibitive and ecologically questionable, attention has naturally shifted to alternative sources of energy. Among the top contenders for future research and development are rapid conversion of biomass or more direct conversion of the solar energy flux to convenient fuels. Of course, mankind has for centuries been exploiting the biological conversion of the free energy in light to biochemical free energy; peat, coal, tars, and oils are the fossil remains of this ancient process. But now attention is turning to faster and possibly more efficient ways of trapping the sun's photons in a storage form.

Among the methods being assessed are those used by the nitrogen- fixing photosynthetic algae and bacteria (all photosynthetic bacteria also convert atmospheric N_2 to ammonia). The production of hydrogen gas occurs as a by-product of nitrogen fixation in vitro, but efforts are now underway to coax intact cells to devote more light energy to H_2 formation. Hydrogen gas in combination with the O_2 released during photosynthesis could be used to generate electricity via fuel cells. By genetic engineering techniques (see Chapter 11), it may be possible to transfer the required enzymatic apparatus to more conveniently manageable bacterial cells. (For discussion of the mechanism of nitrogen fixation and its enzymology, see the Selected Readings at the end of this chapter.)

We are confident that the community of cell biologists will play an important role in the effort to manage the worldwide need for energy. Their special knowledge of the pathways of biological energy flow, fine-tuned for maximum efficiency over eons of evolution, will enhance the human search for ecologically compatible energy sources.

SUMMARY

Biological energy comes either from oxidizing food molecules or from light. In the process called respiration, electrons removed during oxidation are used ultimately to reduce an inorganic molecule—for example, molecular oxygen—and as a result, more energy is available to the cell than in fermentations. In mitochondria, electrons are removed from reduced NADH and then passed along an electron transport chain of carriers to the oxidant O_2. At some steps, there is a coupled transport of protons from one side of the inner mitochondrial membrane to the other. The resulting proton gradient is dissipated either in driving transport processes or in the coupled synthesis of ATP. The mechanism of coupling ATP phosphorylation to electron transport via a proton gradient is the same in bacteria, mitochondria, and chloroplasts.

During photosynthesis in green plants, energy from visible light is absorbed by chlorophylls and secondary pigments, then funneled to a special reaction center chlorophyll. When excited, electrons from these molecules first reduce the primary electron acceptor, then are passed along an electron transport chain, ultimately to reduce $NADP^+$. Both ATP and NADPH are necessary to reduce CO_2 to glyceraldehyde phosphate. CO_2 is fixed either by combination with ribulose bisphosphate (C_3 pathway) or phosphoenolpyruvate (C_4 pathway). Electrons from the environment are used to replace those lost by reaction center chlorophyll. When water is the source of these electrons, two photosystems and two photons per electron are required to overcome the difference in energy between water and NADPH. In cyclic photophosphorylation, light is used to produce ATP but the reducing electrons come from the environment by pathways not involving photosystems. Some organisms can use chemical energy from ATP to produce light.

KEY WORDS

Carrier, chloroplast, coenzyme, coupling, cytochrome, energy, energy transduction, enzyme, free energy, Krebs cycle, mitochondrion, oxidant, oxidation, reductant, reduction, substrate, ultraviolet radiation, vesicle.

PROBLEMS

1. To the following statements, respond TRUE or FALSE. If you respond TRUE, explain why the statement is true and give an example; if you respond FALSE, explain why the statement is not true.

   (a) Blue photons are more effective than yellow photons, per photon absorbed, in eliciting photosynthesis in a green alga with two photosystems.

   (b) Action of the proton-$H_2PO_4^-$ symport of mitochondria tends to reduce the magnitude of ΔE across the inner mitochondrial membrane.

(c) Provided a pathway for the electrons was available, the oxidation of succinate to fumarate ($E' = -0.03$ V) could drive the reduction of pyruvate to lactate ($E' = -0.19$ V) at standard conditions. (E' is the standard reducing potential.)

(d) In the presence of rotenone and oligomycin, electrons can still flow from reduced ascorbate/phenazine methosulfate to oxygen by using part of the electron transport chain in a tightly coupled mitochondrion.

(e) In illuminated chloroplasts, adding dinitrophenol will inhibit oxygen production.

(f) If the reduced form of cytochrome b (Fe^{2+}) is in higher concentration than the oxidized form (Fe^{3+}), the reducing potential of cytochrome b will be higher than the standard reducing potential.

2. Imagine a cell that can carry out photosynthesis of glucose when illuminated in the presence of CO_2 and H_2O, but only if an organic acid ion (butyrate: $CH_3CH_2CH_2COO^-$) is also present in the growth medium. In the course of photosynthesis, butyrate is converted to crotonate ($CH_3CH{=}CHCOO^-$).

$$\text{crotonate} + 2\,e^- + 2\,H^+ \rightleftharpoons \text{butyrate}$$
$$E' = +0.25\ V$$

Answer the following with TRUE or FALSE; if you answer FALSE, explain why the statement is not true.

(a) The cell *must* have two photosystems, analogous to green plant PS I and PS II.

(b) CO_2 uptake must be accompanied by evolution of O_2.

(c) In the course of photosynthesis using isotopically labeled $C^{18}O_2$, some ^{18}O is incorporated into glucose.

(d) In the course of photosynthesis using isotopically labeled $C^{18}O_2$, some ^{18}O is incorporated into water.

(e) Most of the ATP generated photosynthetically by this organism is produced by noncyclic photophosphorylation.

3. Tetraphenyl phosphonium ion (TPP^+) has been used to estimate ΔE across the mitochondrial membrane. Consider the following experiment. Mitochondria were suspended in a medium supporting respiration and containing radiolabeled TPP^+. After equilibration, the mitochondria were rapidly filtered and washed so no TPP^+ remained outside. The radioactivity retained on the filter was a measure of the *amount* of TPP^+ in the matrix, TPP_m^+. To obtain the *concentration* of TPP^+ in the matrix, $[TPP^+]_m$, it is also necessary to have an estimate of matrix volume, V_m. This was done by equilibrating labeled TPP^+ with

mitochondria poisoned with antimycin A, under which conditions no ΔE was present. Assuming no binding of TPP^+ to a mitochondrial structure, the radioactivity retained by filtering such mitochondria estimates V_m. Data from such an experiment are shown in the graph below. (The specific activity of TPP^+ was 2000 cpm/μmole.)

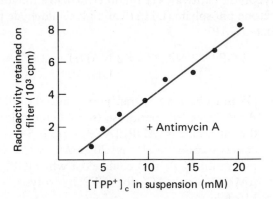

(a) How much radioactivity was retained on the filter when $[TPP^+]_c$ was 10 mM?

(b) Which of the following does this radioactivity measure: $[TPP^+]_c$, $[TPP^+]_m$, TPP_c^+, or TPP_m^+?

(c) In the conditions of this experiment, what is the most likely value for $[TPP^+]_m$ when $[TPP^+]_c = 10$ mM?

(d) Calculate the matrix volume (V_m) of the mitochondria using these data. Explain your method. The data graphed below were obtained from similar suspensions of mitochondria but in the absence of antimycin A, so the mitochondria were actively respiring.

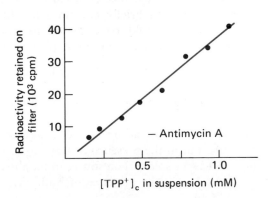

(e) How much radioactivity was retained on the filter when $[TPP^+]_c$ was 1.0 mM? Explain any differences between these results and those shown on the previous graph.

(f) What is the best estimate for $[TPP^+]_m$ when $[TPP^+]_c = 1.0$ mM? Explain your method of calculation.

(g) Evaluate ΔE, the electrical potential difference in mV, across the inner membrane of the actively respiring mitochondria used in these experiments.

4. Some cells have a variety of means of obtaining energy from their environment. Imagine a cell that can grow aerobically on glycerol phosphate (GP) as the sole source of carbon and energy. GP can center the glycolytic pathway via the intermediates dihydroxyacetone phosphate (DHAP) and glyceraldehyde phosphate (GAP):

$$2\text{ GP} + 2\text{ NAD}^+ \rightleftharpoons 2\text{ NADH} + 2\text{ H}^+ + 2\text{ DHAP} \rightleftharpoons 2\text{ GAP}$$

(a) GP must be actively transported into the cell by an ATP-dependent permease (see scheme below). If the intracellular [ADP]/[ATP] ratio is 5.0 and the concentration of P_i is 20 mM, what is the maximum $[GP]_i$ that could exist when $[GP]_o$ is 10 μM? (Assume $\Delta G'$ for ATP hydrolysis is -31 kJ/mole.)

$$2\text{ GP}_o + \text{ATP} + \text{H}_2\text{O} \rightleftharpoons 2\text{ GP}_i + \text{ADP} + P_i.$$

(b) In some environments this cell synthesizes nitrate reductase, which catalyzes the transfer of electrons from cytochrome b to nitrate under anaerobic conditions:

$$\text{NO}_3^- + 2\text{ H}^+ + 2\text{ e}^- \rightleftharpoons \text{NO}_2^- + \text{H}_2\text{O}$$
$$E' = +0.42\text{ V}$$

What is the $\Delta G'$ for the transfer of electrons from NADH to NO_3^-?

(c) Given the information from parts (a) and (b), predict the growth of this cell under the following conditions. State whether each culture would be able to grow and briefly explain why. Identify which culture would be expected to have the highest growth rate and briefly explain why.

  (1) GP as the sole carbon source in the presence of O_2.
  (2) GP as the sole carbon source in the absence of O_2.
  (3) GP as the sole carbon source in the absence of O_2, but in the presence of nitrate.
  (4) GP as the sole carbon source in the absence of O_2, but in the presence of both nitrate and HCN.

5. Space probes may bring back unusual cells, perhaps some with features similar to both procaryotic and eucaryotic terrestrial cells. Imagine such a cell with a Na^+-K^+ pump, inhibitable by ouabain but with a stoichiometry of 2 Na^+ and 2 K^+ per ATP.
(a) Will either the rate or the equilibrium point of the

Na^+-K^+ pump be influenced by the cell's interior electrical potential? Briefly explain your answer.
(b) The movement of K^+ into this cell is found to be stimulated by lowering the pH of the medium surrounding the cell. This result is also observed if the cell is treated with ouabain. Explain these observations and propose a mechanism for K^+ movement into the cell in the presence of ouabain.
(c) The movement of Na^+ out of the cell is also found to be stimulated by lowering the pH of the bathing medium. Furthermore, if the cell is equilibrated with a high Na^+ concentration at low temperature (5°C), then placed suddenly into a weakly buffered solution of low $[\text{Na}^+]$ at a higher temperature (30°C), the pH of the bathing medium rises. These results are observed if the cell is also treated with ouabain, or with antimycin A (which blocks electron transport from NADH to oxygen), or with both. Discuss these results and propose a mechanism for active Na^+ transport out of the cell.
(d) Careful measurements reveal the cell to have steady-state intracellular concentrations of 3.2 mM Na^+, 80 mM K^+, and an intracellular pH of 8 when growing in a medium of pH 6.5, $[\text{Na}^+] = 100$ mM, $[\text{K}^+] = 2.5$ mM. Can this distribution of protons, Na^+, and K^+ be distinguished from an equilibrium? Explain your answer.
(e) Another interesting feature of this cell is that it grows on nonfermentable carbon sources, such as succinate, but it has no F_1 ATPase. Assuming that ATP is a common cosubstrate in this cell's enzyme-catalyzed reactions, explain how its ATP can be generated by respiration.
(f) What ratio of intracellular [ATP]/ [ADP] would be expected in this cell if the intracellular $[P_i]$ is 20 mM? Explain your answer.

6. Imagine the crew of a space shuttle captures an asteroid containing spores of a novel life-form that is eventually shown to perform oxidative phosphorylation. Cell-free extracts of the organism oxidize NADH with oxygen, and the maximum yield of ATP is four per NADH oxidized. This ATP yield is found even when the ratio of [ATP]/[ADP] is maintained at 100/1. O_2 uptake and phosphorylation are tightly coupled in the extracts, similar to terrestrial mitochondria. The electron transport chain of the novel organism is found to contain *only* the following components:

$$a^{2+} + e^- \rightleftharpoons a^+$$
$$b + 2\text{ H}^+ + 2\text{ e}^- \rightleftharpoons bH_2$$
$$c + 2\text{ H}^+ + 2\text{ e}^- \rightleftharpoons cH_2$$
$$d^{2+} + e^- \rightleftharpoons d^+$$

Exp.	Conditions	O_2 Uptake	ATP per NADH Oxidized	Oxidation-Reduction Levels
1	O_2 present	+	4	a, b, c partially reduced
2	No O_2	−	0	a, b, c, d fully reduced
3	O_2 + inhibitor Q	−	0	a oxidized; rest reduced
4	O_2 + Q + uncoupler	−	0	a oxidized; rest reduced
5	O_2 + inhibitor R	−	0	a, b oxidized; c, d reduced
6	O_2 + R + uncoupler	−	0	a, b oxidized; c, d reduced
7	No O_2 + oxidized a	−	2	b, c, d partially reduced
8	No O_2 + oxidized c	−	2	a, b reduced; d partially reduced
9	No O_2 + oxidized d	−	0	a, b, c reduced

The extent of reduction of each component can be determined by spectrophotometry. Components a, c, and d can be obtained pure and soluble; when added to cell-free extracts in the absence of oxygen, the oxidized forms of these components support the oxidation of NADH. Inhibitors have been found which block O_2 uptake in the presence of NADH. Additional data are shown above.

(a) Arrange the components of the electron transport chain in order of increasing standard reduction potential:

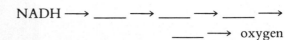

(b) Indicate the sites of inhibitors Q and R.
(c) Indicate the probable sites of ATP phosphorylation.
(d) Propose the minimum reducing potential difference between the two components that constitute each coupling site.

SELECTED READINGS

Bogorad, L. (1981) Chloroplasts. *Journal of Cell Biology* 91:256S–70S.

Boyer, P. D., Chance, B., Ernster, L., Mitchell, P., Racker, E., and Slater, E. C. (1977) Oxidative phosphorylation and photophosphorylation. *Annual Review of Biochemistry* 46:955–66.

Ernster, L., and Schatz, G. (1985) Mitochondria: A historical review. *Journal of Cell Biology* 91:227S–55S.

Hoober, J. K. (1984) *Chloroplasts.* New York: Plenum Press.

Racker, E. (1976) *A New Look at Mechanisms in Bioenergetics.* New York: Academic Press.

Racker, E. (1980) From Pasteur to Mitchell: A hundred years of bioenergetics. *Federation Proceedings (FASEB Journal)* 39:210–15.

Sprent, J. I. (1979) *The Biology of Nitrogen-Fixing Organisms.* New York: McGraw-Hill.

Tzagoloff, A. (1982) *Mitochondria.* New York: Plenum Press.

Youvan, D. C., and Marrs, B. L. (1987) Molecular mechanisms of photosynthesis. *Scientific American* 256(6):42–48.

Scanning electron micrograph of HeLa cells during mitosis: approximately late prophase, anaphase, early telophase (all ×3750), and late telophase (×3300). (Courtesy of R. G. Kessel and C. Y. Shih.)

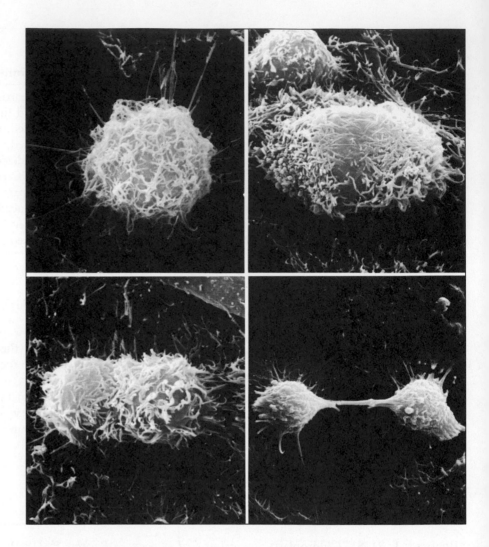

We discover that the cell, in order to continue and extend its organization, must maintain a fund of information built into the structure of its DNA molecules, the functions of which are to replicate and control the synthesis of RNA and proteins.

We find that this flow of information from DNA through the synthesis of specific proteins helps to regulate the flow of materials and energy, which ultimately control the integration of cell behavior.

PART

3

NUCLEIC ACIDS, PROTEINS, AND THE FLOW OF INFORMATION

Cell Continuity
Through Time

9

The preceding chapters have emphasized cell structure and function in terms of three-dimensional space. We come now to a remarkable feature of living things that has to do with the time dimension. Let us consider, for example, the history of a horseshoe crab (*Limulus polyphemus*). We take up our crab's history at a point where it is merely a tiny fertilized egg. As we watch over a period of weeks, this insignificant object undergoes an unfolding sequence of events which transforms it into something completely different: the crab's larval form (Figure 9–1a). This in turn transforms itself into still another structure, the adult horseshoe crab (Figure 9–1b). An enormously complicated series of morphological and biochemical changes have taken place in about a month. These events are exquisitely regulated: if we follow a large number of crabs through the time dimension in the same way, we find that the sequence of events is virtually identical in each of them. There is evidently a program, somewhere in the fertilized *Limulus* egg, which ensures its orderly development into a *Limulus* rather than, say, a llama.

Let us now expand the time scale to several *Limulus* lifetimes. The adult crabs die, but usually not before producing a new crop of fertilized eggs which recapitulate the whole developmental sequence to yield a second generation of crabs that resemble their parents in great detail. The second generation gives rise to a third, which gives rise to a fourth, and so on. The developmental process is reenacted each time with utmost fidelity. In fact, it is possible to expand the time scale by a factor of millions in the case of *Limulus*. Paleontologists have discovered *Limulus* fossils which are 80 million years old, and which indicate that the horseshoe crabs of that prehistoric time are almost indistinguishable from their distant descendants today.

What is remarkable in this is the element of *continuity*. This is the fundamental fact (or problem) of the science of heredity. To explain the continuity, biologists have hypothesized a program which ensures that a *Limulus* egg develops into an adult *Limulus*. This program has evidently been preserved intact from one generation to the next, for millions of generations.

Mendel, The Father of Formal Genetics

The Program

Among the early hypotheses devised to explain biological continuity, the most significant was the *preformation theory*. The idea was that the egg of each organism contained a perfect miniature of the adult form, or *homunculus,* which then simply grew in size. This theory, although plausible in principle, founders when one considers the problem of continuity from generation to generation. Suppose that an egg contains a homunculus of the adult organism: where do that adult's progeny come from? In terms of the theory, the homunculus must contain a miniature egg cell which contains even tinier homunculi of the next generation, which must in turn contain even more microscopic homunculi for the third generation, the fourth, and so on. A set of Chinese boxes of this sort is inconceivable for a run of many generations (e.g., our horseshoe crab).

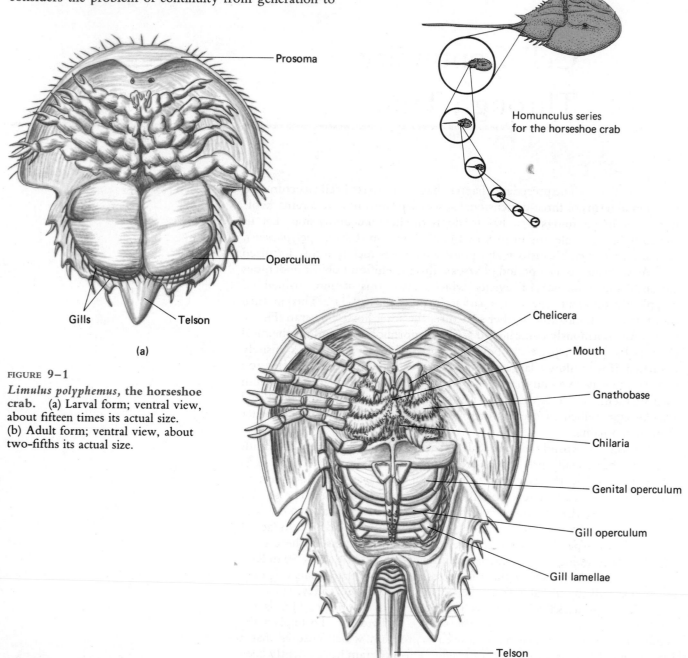

FIGURE 9-1

Limulus polyphemus, the horseshoe crab. (a) Larval form; ventral view, about fifteen times its actual size. (b) Adult form; ventral view, about two-fifths its actual size.

An alternative theory, proposed by A. Weismann in the 1880s, may have been inspired by the industrial revolution of the nineteenth century. A ship, for example, is not made by expanding a tiny scale model—which would be analogous to the homunculus theory. Instead, a ship is constructed by reading out the information that specifies its structure from an abstract coded form, a *blueprint.* The blueprint itself is not consumed in the process, but remains intact, available for the fabrication of unlimited "generations" of ships.

Here, clearly, was a practicable solution to the continuity problem: a blueprint for the organism's structure, passed on in unaltered form from one generation to the next. Weismann proposed that the blueprint is encoded in the molecular structure of a substance he called *germ plasm,* passed from one generation to the next through the sex cells. Since one organism may have many progeny, germ plasm must have one other property besides that of storing coded information: it must be able to produce additional copies of itself; in other words, to replicate.

Just as continuity is the basic datum of the science of heredity, the continuity of germ plasm is its primary working hypothesis. It took more than 50 years for the chemical identity of germ plasm to be discovered (see Chapter 10). In the meantime, however, fundamental discoveries about the way germ plasm is organized, and passed from generation to generation, were made without knowledge of its actual composition.

Statistical Relations

These discoveries, and the basic approach of genetics to this day, flowed from a revolutionary new technique in nineteenth-century biology: *counting.* From our vantage point it seems incredible, but for hundreds of years biologists simply did not count. Instead, they observed, they classified, they dissected. Even the simplest mathematical reasoning was scarcely used at all, and statistical considerations were unheard of. As a result, when the greatest biological thinkers of the time (including Darwin) wrestled with the problem of inheritance, the problem won.

It remained for one G. Mendel, an obscure country priest in the Czech part of the Austrian Empire, to join statistics to biology and, in so doing, to found the science of heredity. But essentially none of the biologists of that period understood his work, and the classic monograph he published in 1865 was universally ignored. It was, as one of its rediscoverers put it a generation later, "too beautiful for its time."

Mendel's astoundingly foresighted work—he invented the science of genetics 20 years before Weismann proposed the idea of germ plasm—rests on a deceptively simple procedure. He crossed strains of the common garden pea which differed in discrete either/or characteristics, such as round versus wrinkled seeds, green versus yellow seeds, and so forth, to produce **hybrid** plants. Farmers had been doing this for several thousand years, and Mendel, son of a peasant family, was adept at horticulture. Mendel was also educated (at the University of Vienna) in physics and mathematics. This training led him to introduce two new approaches, which were so foreign to the scientific world of the 1860s that they were not understood properly until 35 years later. First, he examined the hybrids and their second generation progeny, and the third generation as well, *counting and recording the number of plants in each generation* which displayed one

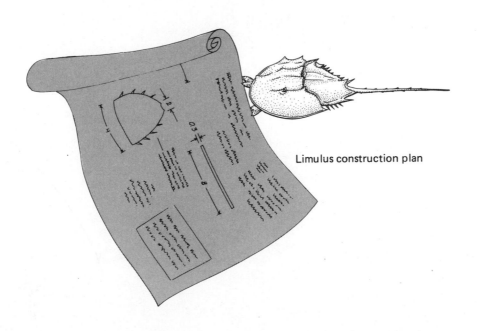

Limulus construction plan

or the other of the original parental characteristics. His intention, in his own words, was to "determine the number of forms under which the offspring of hybrids appear" and "to ascertain their statistical relations." Second, he used the statistical relations he had so meticulously recorded to devise a theory of inheritance.

We shall illustrate this gigantic series of experiments with data from one of them, the cross of yellow seed color by green seed color. In such experiments each parent is chosen from a stock that, when crossed with another member of the same stock, always shows the same color—that is, each parent is *true-breeding*. In the

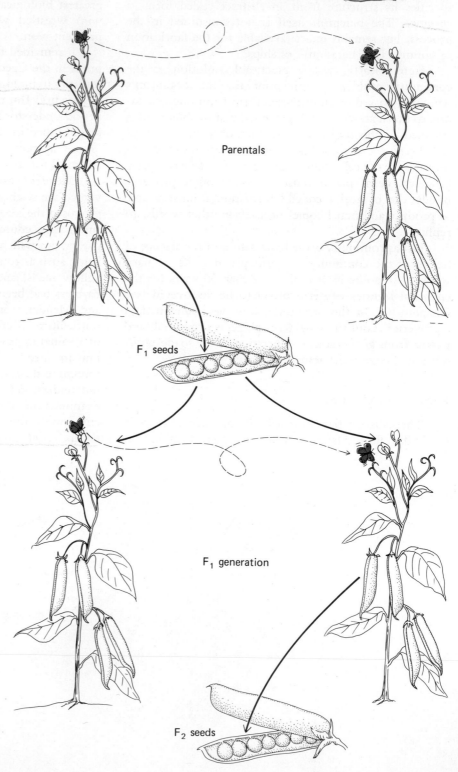

Parentals

F_1 seeds

F_1 generation

F_2 seeds

first hybrid generation, or F_1, all the seeds were yellow. Thus, yellow was the **dominant** property. In fact, in each of the seven factor pairs which Mendel studied, one of the two alternative parental types proved dominant in the F_1 hybrid. The parental character that was not observed in the F_1 he termed the **recessive** character.

Mendel then allowed the F_1 hybrids to pollinate themselves to produce a second hybrid generation (F_2). Table 9–1 shows the distribution of types he found in the offspring of the first ten F_1 plants, as well as the overall results from the entire set of 258 plants. The enormous number of plants Mendel worked with is of crucial importance, for this enabled him to obtain statistically meaningful results. As he put it in his monograph: "These . . . experiments are important for the determination of the average ratios, because with a smaller number of experimental plants they show that very considerable fluctuations may occur." From the data of Table 9–1, and the similar data obtained in the other six series of experiments, Mendel inferred the basic mechanism of inheritance. The steps in his reasoning can be retraced as follows.

First, although the F_1 generation all displayed the dominant yellow character, every F_1 plant produced some F_2 seeds of the recessive green type (Table 9–1). Thus, the information for green seed color was not lost in the F_1 generation, but only masked. This is what Mendel meant by recessive. It follows that the F_1 plants contained more than one set of heritable information: in fact, the data suggest two sets, one for the dominant yellow color which they displayed, and one for the green color which they did not display but nonetheless passed on to some of their progeny. Moreover, the hybrids are perfectly respectable pea plants; they are special only in that an experimenter selected their parents. Therefore, all pea plants, and presumably all organisms, must have a double set of hereditary information.

Second, from this Mendel drew attention to the fact that the F_2 generation contained no intermediate types, but only the original parental characters again—yellow and green. This meant that the yellow and green characters *did not mix* (to give a greenish-yellow color) during their joint residence in the F_1 generation, but retained their individual properties and simply *reassorted* in the F_2. It follows that the information for yellow and green color, and by implication all hereditary information, must be contained in indivisible units, like the elementary atoms of chemistry. This was a profound insight, and also a revolutionary one, for Mendel's contemporaries thought of inheritance in hybrids as some sort of blending process. Today, we refer to these elementary particles of inheritance as *genes*. Mendel's reasoning showed that each individual's germ plasm consists of a double set of genes for each character. The double set of genes for seed color in the parental line with yellow seeds might thus be represented *YY*, and the double set in the parent with green seeds as *gg* (using lower case letters to denote the recessive behavior).

The third step in his reasoning was to answer the question: where might any individual obtain its double set of genes? Obviously from its parents, of whom there are two—one set is derived from each parent. Thus, the "two-ness" of genetic constitution is related to the two-ness of sexual reproduction. The process of sexual reproduction represents a mechanism by which each parent transmits one of its double set of genetic factors to its progeny, and hybrids must inherit one factor set from each parent. The cross of YY plants by gg plants produces F_1 progeny whose genetic constitution is Yg, having inherited one Y gene from the YY parent and one g gene from the gg parent.

Since the Y character is dominant, these hybrid plants outwardly resemble their YY parent, but their genetic constitution is different. Nowadays, we call the

TABLE 9–1

Seed Coloration in the F_2 Generation

Individual Plants	Yellow	Green
1	25	11
2	32	7
3	14	5
4	70	27
5	24	13
6	20	6
7	32	13
8	44	9
9	50	14
10	44	18
Total of 258 plants	6022	2011

appearance the **phenotype** and the genetic constitution the **genotype.** Given the phenomenon of dominance, the genotype cannot always be inferred from the phenotype. It can, however, be inferred from an individual's progeny. The difference between the hybrid Yg plants of the F_1 generation and the parental YY strain is revealed in the fact that the Yg hybrids produce some green seed progeny while the pure-breeding YY parents do not.

From this, his fourth step was to observe, therefore, that the F_2 generation reflects the *reassortment* of the Y and g genes through self-pollination, which is equivalent to a cross of Yg by Yg. In fact, Mendel deliberately cross-pollinated some F_1 hybrids, and showed that the results were the same as those seen when the hybrids pollinated themselves. The assortment of genes in a cross of Yg by Yg is easy to predict, given an elementary knowledge of gambling. In a coin flip, for example, the chance of turning up heads is 1/2. In a double coin flip, the chance of turning up two heads is $1/2 \times 1/2 = 1/4$. The essence of Mendel's analysis is that heredity is just a double coin flip (Figure 9–2). The chance that an F_2 individual will inherit the g gene from one parent is 1/2, and the chance that the same individual will inherit the g gene from the other parent is also 1/2. Therefore, the chance that any given F_2 individual will inherit the g gene from both parents, and thus end up with the gg phenotype and exhibit the recessive green seed color, is $1/2 \times 1/2 = 1/4$. This is the frequency Mendel found, on average. (The phrase "on average" is important here; the frequency of 1/4 green will be approached closely only if the assortment is perfectly random and the sample is very large.) This idea of *random assortment* of genes explains, at one stroke, both the variation in ratio that Mendel observed in small samples and the overall average ratio (Table 9–1).

In his fifth step, Mendel concluded that this analysis makes a crucial prediction: the yellow seed individuals among the F_2 generation should be of two different genotypes, YY and Yg. The theory further predicts both the proportions of these two genotypes and that they can be distinguished by examining their descendants in another round of self-pollination, to produce an F_3 generation.

The predictions of the theory follow simply from the idea of the double flip. The chance that an F_2 individual inherited a Y gene from each of its two parents must be $1/2 \times 1/2 = 1/4$, exactly equal to the chance of ending up with a gg genotype. Thus, 1/4 of the F_2 generation should be of the YY genotype, and will therefore breed true for yellow seed color, just like their YY grandparents.

F_2 individuals of the Yg genotype can be formed by either of two routes, and one must take account of this in calculating their expected frequency (Figure 9–2). The chance of inheriting the Y gene from the parental egg cells is 1/2, and the chance of inheriting the g gene from the parental pollen cells is 1/2, so the chance of forming a Yg genotype by this route is $1/2 \times 1/2 = 1/4$. But the same genotype can be produced by inheriting the Y gene from the parental pollen cells and the g gene from the parental egg cells, and the chance of this occurring is also 1/4. Therefore, the overall probability of constituting a Yg genotype in the F_2 is $1/4 + 1/4 = 1/2$.

Thus, the theory predicts that the F_2 genotypes should be 1/4 gg (as Table 9–1 already confirms), 1/4 YY, and 1/2 Yg. Among the 3/4 of the F_2 which display the dominant yellow seed color, then, 1/3 should be YY and true breeding; the remaining 2/3 should be Yg, and these should show their hybrid genotype by producing 1/4 green seed offspring in the next generation.

Mendel raised 519 plants from yellow seeds of the F_2 generation. Of these, 166, almost exactly 1/3, yielded exclusively yellow seeds. The other 353 yielded yellow and green seeds in the predicted proportions of three to one. Similar results were obtained in the other six sets of crosses.

Further Tests of Mendel's Theory

The essence of Mendel's experimental discovery lies in the distinction between the YY and Yg genotypes. Let us generalize this distinction by means of a few definitions. The YY type and gg type are both true-breeding because both genetic elements are identical in these genotypes. This situation is called homozygosity and genotypes of this kind are called **homozygous.** YY would be termed *homozygous dominant* and gg *homozygous recessive*. The hybrid type, which is not true-breeding because it contains one genetic element of each type, is called **heterozygous.**

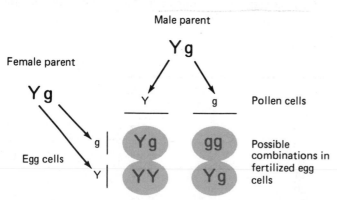

FIGURE **9–2**
Random assortment.   A graphical representation of all combinations of fertilized egg cells that result from crossing egg cells from a Yg parent and pollen cells from a Yg parent. When the probabilities of forming these combinations are all equal, the assortment is described as random.

The Y and g elements (genes) are distinguishable, but they are not different kinds of genes: they both affect seed color. In other words, they appear to be alternative forms of one genetic element whose domain is seed color. We denote this by calling them alternative **alleles** of one seed color gene. The concept of alleles is of critical importance, for it reveals that the element of randomness has to do with the assortment of alleles, not really the distribution of genes. The genetic process ensures that each individual inherits exactly two genes for seed color, not zero or one or three. The element of randomness has to do with the assortment of alleles in their pairwise combinations.

Let us therefore denote both alleles of the seed color gene by using upper case and lower case forms of the same letter: C for the dominant allele and c for the recessive. We can now abbreviate the homozygous dominant genotype as CC; the homozygous recessive as cc; and the heterozygous as Cc. Mendel's basic discovery was, under this nomenclature, that the cross Cc by Cc produced offspring which were, on the average, 1/4 cc, 1/2 Cc, and 1/4 CC. These frequencies, as he understood, simply reflect the random combination in pairs of the two parental alleles.

A further test of his theory can be carried out by crossing a Cc hybrid individual with a homozygous recessive cc individual. Such a cross is called a *back-cross* or a *test-cross*. In this case, each of the offspring must inherit one c allele from the homozygous recessive parent, which can pass on no other type. But from the heterozygous parent each of the offspring has a 1/2 chance of inheriting the C allele and a 1/2 chance of inheriting the c allele. Thus, Mendel's theory predicts that a back-cross of Cc by cc will yield 50% Cc and 50% cc. This is precisely what has been observed, not only with the seed color gene, but with a vast number of other genes as well.

The Mendelian principles of inheritance have been tested in an immense variety of plants and animals, from the corn smut to the carrier pigeon. In every case, with some very special exceptions we shall discuss later, they have been confirmed. That is, the frequencies observed in self-crosses, back-crosses, and more complicated breeding experiments have agreed with the simple frequencies we have just described and certain combinations of them. In reality, the most important aspect of a scientific test is not with what theory the results are consistent, but with what theory they are *inconsistent*, and thus rule out of consideration. For example, when Mendel discovered the segregation of the original parental characters in the F_2 generation, he ruled out *all* hypotheses based on the blending of parental attributes. The frequencies he obtained are consistent with the idea of a double set of genes and, more importantly, they are inconsistent with the idea of a single set, or a triple set, or 13 sets. The widespread verification of Mendelian principles therefore leads to the conclusion that most plants and animals have a double set of genes. We refer to such organisms as **diploid**.

Independent Assortment of Alleles of Several Genes

Mendel extended his statistical analysis to the inheritance of two and even three allele pairs simultaneously. For example, he performed a cross of a true-breeding parental strain with yellow, round seeds by one with green, wrinkled seeds. In single-factor crosses of round by wrinkled, seed shape had shown the 3:1 distribution of dominant:recessive phenotype in the F_2, the round property being dominant. In our modern terminology, the homozygous round shape would be designated SS, the homozygous wrinkled type ss, and the heterozygous hybrid Ss. Mendel's two-factor cross can thus be represented: SS CC × ss cc. The F_1 generation inherited, from each parent, one allele for each of the two genes and were thus doubly heterozygous: Ss Cc. Needless to say, they displayed the dominant character for each of the two genes, and had round yellow seeds. In the F_2 generation we would expect the results of two simultaneous random assortments of the two pairs of alleles if they assort *independently*—that is, if the assortment of one allele has no influence on the assortment of the alleles of the other pair. Let us first calculate what frequencies we would expect on the assumption of such *independent assortment*, and then compare these predictions with Mendel's actual data.

If the assortment is independent, we expect each allele pair to assort into the usual 3/4 to 1/4 ratio. Thus, we expect 3/4 of the F_2 to be yellow and 1/4 to be green; and 3/4 to be round and 1/4 to be wrinkled. The products of these individual frequencies, since each is assumed to be independent, will then give the probabilities of the four possible combinations of phenotypes.

$$\text{yellow, round:} \quad \frac{3}{4} \times \frac{3}{4} = \frac{9}{16} = 56.25\%$$

$$\text{yellow, wrinkled:} \quad \frac{3}{4} \times \frac{1}{4} = \frac{3}{16} = 18.75\%$$

$$\text{green, round:} \quad \frac{3}{4} \times \frac{1}{4} = \frac{3}{16} = 18.75\%$$

$$\text{green, wrinkled:} \quad \frac{1}{4} \times \frac{1}{4} = \frac{1}{16} = 6.25\%$$

Mendel examined 556 seeds bred from 15 of the F_1 plants. The distribution of phenotypes he found is shown in Table 9–2 together with the expected frequencies

TABLE 9–2
Mendel's Two-Factor Cross

Phenotype	Number	%	% Expected by Independent Probability
Yellow, round	315	56.7	56.25
Yellow, wrinkled	101	18.2	18.75
Green, round	108	19.4	18.75
Green, wrinkled	32	5.8	6.25

which we have just calculated. It can be seen that the agreement between predicted and observed frequencies is very good indeed. Mendel found that each of the seven characters (or genes) that he studied segregated independently of the others by this test, and were thus borne on separate particles of germ plasm.

The Organization of Genes

The rule of *independent assortment* is not as general as Mendel's first principle of random assortment. Independent assortment applies only to genes located on separate chromosomes, which are the physical units of germ plasm. Before we consider the way in which genes located on the same chromosome are transmitted from one generation to the next, it is convenient to describe the behavior of chromosomes.

Meiosis

The germ plasm blueprints are transmitted from one generation to the next via the sex cells: egg and sperm (these are also called **gametes**). Weismann discussed this transmission process as occurring through a succession of special cells, the *germ line*. During the late nineteenth and early twentieth centuries, cytologists discovered the chromosomes and traced their behavior during the production of sex cells, which occurs by means of a special type of nuclear division termed **meiosis.** Two aspects of meiosis are central to genetics.

The chromosomes differ from one another somewhat in size and shape. Careful inspection of the chromosomes of somatic cells reveals that there are *two* of each particular morphological type (Figure 9–3); structures like chromosomes that have a similar structure and origin are said to be **homologous.** This is made especially obvious in meiosis, for the members of each pair associate with one another, as if to call attention to their homologous nature (Figure 9–3). The existence of homologous pairs of chromosomes, visible through the microscope, reveals to the eye exactly what Mendel's statistical analysis had earlier inferred: germ plasm is organized into a double set of elements. It was this

cytological discovery, made after Mendel's death, that prepared biologists to understand, at last, the significance of Mendel's analysis.

The next step in meiosis is the movement of the members of the homologous chromosome pairs to opposite poles of the nucleus; nuclear and cell division follow, leaving one member of each homologous pair in each of the two daughter cells (see meiosis 1 of Figure 9–4). This first division of meiosis thus reduces the total chromosome number by a factor of two, and is consequently called a *reductional* division.

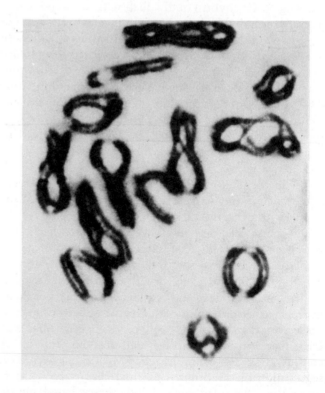

FIGURE 9–3
Homologous chromosomes paired in the first division of meiosis. The micrograph shows a spermatocyte (a cell in which male gametes are produced) of the salamander *Oedipina uniformis.* ×1,500. (Courtesy of J. Kezer.)

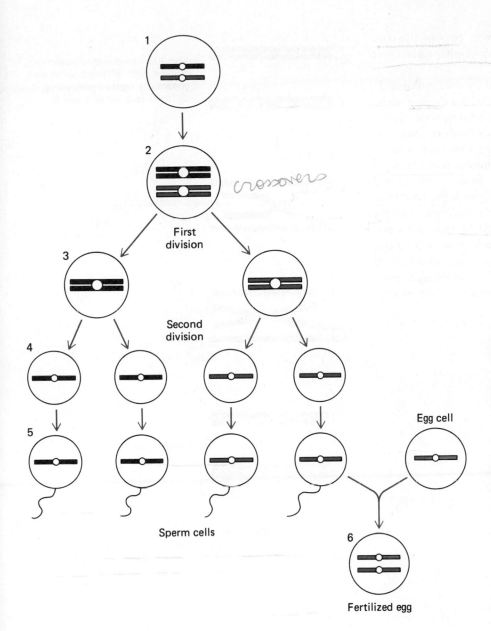

crossovers

1

2

First
division

3

Second
division

4

5

Sperm cells

Egg cell

6

Fertilized egg

FIGURE 9–4
Meiosis. The figure illustrates the meiotic divisions leading to the production of male gametes (sperm) in diagrammatic form with one pair of homologous chromosomes. (1) The ancestral premeiotic cell before chromosome replication. (2) Each chromosome replicates to produce a pair of chromatids, held together by the centromere (the blob in the middle). The homologues pair with each other at the beginning of the first division. (3) The homologues are segregated into two daughter cells at the first division. (4) Each chromatid segregates into a daughter cell at the second division, producing four haploid cells. (5) The four products of the two meiotic divisions develop into sperm. (6) The union of one of the sperm with an egg generates a diploid zygote.

Each chromosome itself consists of two replicas or *chromatids*, because DNA replication occurs before the first division of meiosis. The chromatids are separated from one another in a second meiotic division. Thus, the two successive divisions of meiosis yield four specialized germ line cells, or *gametes,* each with half the usual chromosome complement. Such cells are termed **haploid,** in contrast to all other cells, which contain pairs of homologues and are therefore diploid. The process of meiosis is thus an exact counterpart of the distribution of one of each pair of genetic elements from parent to its progeny, as Mendel had concluded.

Mendel's principle of independent assortment demands that separate pairs of alleles segregate into sex cells independently of one another. The same is true of the chromosome pairs in meiosis. This was demonstrated in 1913 by E. Carothers, an American graduate student, who took advantage of a strain of grasshopper in which the two members of a certain homologous chromosome pair are slightly different from one another in appearance. They could therefore be distinguished from one another under the microscope, and Carothers showed that they segregated randomly with respect to another marked chromosome during meiosis.

When the homologous chromosomes form their pairs in the first meiotic division, each pair is composed of *four* chromatids because each chromosome has already duplicated. These structures are called **tetrads** and they are crucial to the genetics of all higher organisms. Sometimes the tetrads have the appearance of four parallel strands. More often, however, two of the four strands cross over one another in an X-shape structure like that shown in Figure 9–5.

This structure was first observed by the cytologist F. Janssens in the early 1900s, and he called it a *chiasma*, from the Greek word for cross. Janssens suggested that a chiasma reflects a point at which two strands of the tetrad have become entangled in such a way that they actually exchange parts on either side of the junction of the X, as shown in Figure 9–6. Since then, more sophisticated analysis has confirmed that Janssens' hunch (he could not have actually seen the exchange process) was exactly right. The crossover of strands in chiasmata is not a rare event, but a common one. Most tetrads show at least one chiasma, and some more than one. The tendency of homologous chromosome strands to exchange parts during the first division of meiosis underlies the process

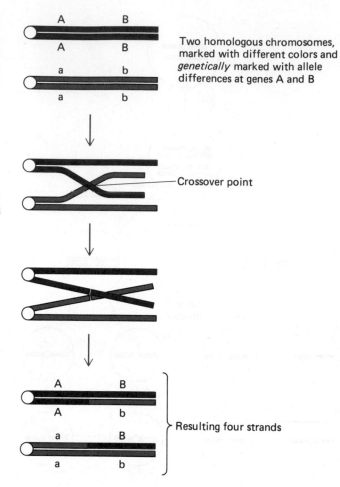

Two homologous chromosomes, marked with different colors and *genetically* marked with allele differences at genes A and B

Crossover point

Resulting four strands

FIGURE 9–6

Crossing-over in the first division of meiosis. The two homologous chromosomes are distinguished by different shading, and are *genetically* marked by allelic differences at two genes: A B in one and a b in the other. A crossover is indicated between the sites of the A/a and B/b genes.

of genetic **recombination** which, as we shall see, has provided geneticists with a way to map the positions of genes on chromosomes.

The Chromosome Theory of Inheritance

The similar behavior of Mendelian genes and chromosomes is suggestive, but offers only circumstantial evidence for the idea that genes are located on chromosomes. Decisive evidence came from **cytogenetics,** a subdiscipline which unites Mendel's statistical analysis of inheritance with direct cytological examination of the chromosomes. Most of this evidence was provided by the work of the American biologist T. H. Morgan and his many students and colleagues, who studied flies of the genus *Drosophila*.

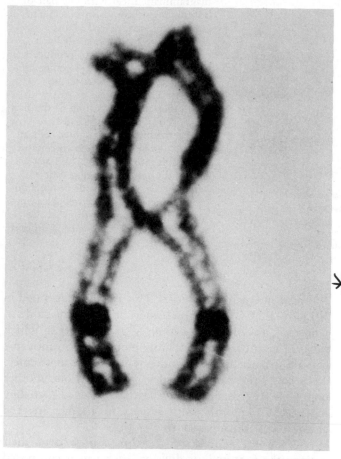

FIGURE 9–5

A chiasma. Two of the chromatids of paired homologous chromosomes undergo crossing over with one another. From a spermatocyte cell of *Oedipina uniformis* in the first division of meiosis, as in Figure 9–3. (Courtesy of J. Kezer.)

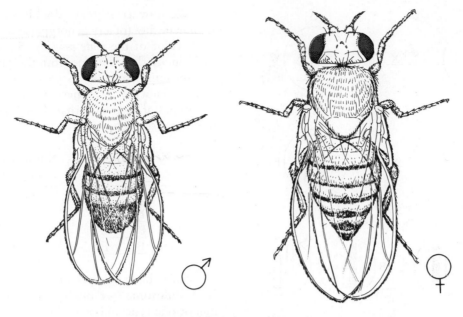

Drosophila melanogaster

In *Drosophila,* as in many other higher organisms including mammals, one pair of homologous chromosomes consists of partners that are visibly different under the microscope and, more importantly, differ in the two sexes. All somatic cells of female *Drosophila* contain two identical sex chromosomes termed X, whereas the somatic cells of males contain only one X chromosome, and a morphologically distinct homologue termed Y. All crosses between males and females can thus be represented, in regard to the sex chromosomes, as X/X by X/Y. The haploid egg cells produced in the female germ line all contain a single X chromosome, while the haploid sperm cells of males are of two different kinds: half contain an X chromosome and half contain a Y chromosome. Those progeny that inherit their father's Y chromosome thus have the XY constitution, which determines development into males. Those that inherit an X chromosome from their father as well as their mother have the XX chromosome constitution, and develop into females.

In 1910, Morgan reported a series of crosses involving a gene for eye color in *Drosophila:* the normal (or **wildtype**) allele dictates red eyes, and a rare mutant allele dictates white eyes. Morgan demonstrated that *male flies* always inherit the allele of this gene from their mother, and not their father, just as if the gene were located on the X chromosome with no counterpart on the Y (Figure 9–7).

Subsequently, Morgan and his associates discovered several other genes that showed the same *sex-linked* pattern of inheritance. Moreover, in certain cases where there were rare departures from the rule of mother-to-son transmission, these were shown to be due to aberrations in the transmission of the X chromosome, such that sons inherited an X chromosome from their father. This correlation of exceptional gene inheritance with exceptional chromosome inheritance provided strong evidence that the two processes were related.

The association of hereditary factors with a particular chromosome proved beyond doubt what many biologists had already come to suspect: the genes, hitherto abstract entities inferred from Mendel's statistical method, were in fact *physical components of specific chromosomes*.

Recombination

Morgan and his group then traced by *two-factor* Mendelian crosses the transmission of two genes, both borne on the X chromosome. We shall illustrate one of these crosses using the genetic terminology introduced earlier. Let us denote the dominant allele for tan *body* color as B, and its recessive allele (yellow body color) as b; the dominant allele for red *eye* color is denoted E and its recessive allele (white eye color) e.

We start by crossing a female homozygous for b and E by a male whose one X chromosome carries the alleles B and e. The *female* progeny of this cross inherit one X chromosome from their mother and one from their father. Their X chromosome genetic constitution can therefore be written:

$$\frac{b \qquad E}{B \qquad e}$$

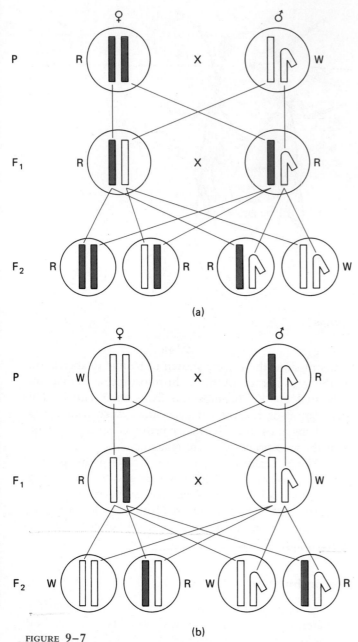

(a)

(b)

FIGURE 9–7
Sex-linked inheritance.   The results of two crosses of
Drosophila are shown. (a) A true-breeding red-eyed female is
crossed with a white-eyed male. The red eye color is
dominant and all the F_1 female progeny have red eyes. There
is no allele for eye color on the Y chromosome, so all the F_1
male progeny also have red eyes. When the F_1 flies are
crossed, the female F_2 flies are all red-eyed. Half of the male
F_2 flies are red-eyed and the other half are white- eyed. The
sex and eye color do not segregate independently. (b) A true-
breeding white-eyed female is crossed with a red-eyed male.
In the F_1, the females are all red-eyed but the males are all
white-eyed. The males have inherited eye color only from the
mother. This observation is consistent with the notion that
sex and eye color are linked, and that the males have only
one determinant for eye color. This hypothesis is borne out
in the F_2 generation, where the observed 2:2 segregation of
eye color is consistent with determination only by the
heterozygous red-eyed mother.

We now cross these doubly heterozygous females
with male flies (of any genotype) and examine the eye and
body color of the *male progeny*. Since the males inherit
their one X chromosome from their mother, and their Y
chromosome contains no genes for eye or body color,
their phenotype for these traits directly reveals the
genotypes of their maternal X chromosome. The results
of one set of such crosses were:

X Chromosome Genotype		Number Male Progeny	%
b	E	2302	51
B	e	2146	48
b	e	43	1.0
B	E	22	0.5

Since the female parents had one <u>b   E</u> and one <u>B   e</u>
X chromosome, we might expect 50% of their sons to
inherit one type and 50% the other. In other words, we
might expect the parental combination of alleles (which
geneticists call the coupling relationship) to be maintained, since the two genes are located together on the X
chromosome. By and large this is true, confirming that
the two genes show physical **linkage** to one another.

In a small minority of the male progeny, however,
the alleles have *recombined*. This result, typical of all genes
on the same chromosome, was termed *partial linkage*.
Partial linkage had been discovered several years earlier,
in plant crosses, and it seemed very puzzling at the time.
If the genes are physical components of the chromosomes, why is linkage partial and not complete?

Janssens' inspired guess that homologues can exchange physical parts while paired during the first
division of meiosis supplies the only plausible explanation. In a few of the meiotic cells of the female parents,
crossing-over somewhere between the locations of the
B/b and E/e genes on the X chromosome must have
generated the recombined X chromosomes, as diagrammed below:

b   E   b   e

.   .   .   .

B   e   B   E

Morgan's group performed two-factor crosses involving a great number of genes that exhibited partial
linkage. They found that *recombinant chromosomes* were
formed at frequencies characteristic of each gene pair
studied. Morgan attributed these characteristic recombination frequencies to the spatial separation of the genes
along the chromosome. For genes located close together
(like the eye and body color genes we have just
considered), the chance of a crossover happening to fall

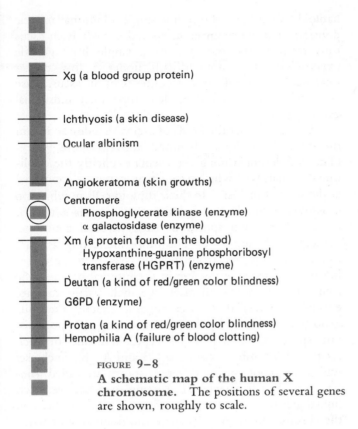

Xg (a blood group protein)

Ichthyosis (a skin disease)

Ocular albinism

Angiokeratoma (skin growths)

Centromere
Phosphoglycerate kinase (enzyme)
α galactosidase (enzyme)
Xm (a protein found in the blood)
Hypoxanthine-guanine phosphoribosyl
transferase (HGPRT) (enzyme)
Deutan (a kind of red/green color blindness)
G6PD (enzyme)
Protan (a kind of red/green color blindness)
Hemophilia A (failure of blood clotting)

FIGURE 9–8

A schematic map of the human X chromosome. The positions of several genes are shown, roughly to scale.

between them is small; for genes located farther apart, the chance is correspondingly greater.

One evening in 1911, A. H. Sturtevant, an undergraduate student in Morgan's lab, extended this analysis to the form in which it has been used ever since. Sturtevant devised a simple algebraic treatment of recombination frequencies from multiple two-factor crosses which defined the location of five X-linked genes in a linear array along the X chromosome. (Sturtevant recalled that he neglected his undergraduate homework that night, for which he can be forgiven, in order to establish

the first comprehensive **genetic map.**) Since that night, the same analysis has delineated the arrangement of genes in microorganisms, molds, flies, mice, chickens, plants, and people. For example, Figure 9–8 shows the organization of several genes on the X chromosome of the human species.

Genes that display linkage with one another define a *linkage group,* which is the genetic counterpart of the chromosome. Moreover, the transformation of recombination frequencies for linked genes into topological information reveals that the genes are located in linear arrays, corresponding to the linear structure of the chromosome. In some organisms, especially *Drosophila,* cytogenetic analysis has demonstrated the colinearity of the chromosome and its linkage group in extraordinarily fine detail. The chromosomes of some specialized *Drosophila* cells show a detailed morphology in the form of a varying thickness and a characteristic pattern of transverse bands (see Figure 9–9). Many genetic variants have been obtained in which particular regions of one or another of the fly's four chromosomes are visibly inverted, or duplicated, or even deleted altogether. (These deletion mutants must generally be maintained in the heterozygous condition, where one chromosome carries a deletion and the other is normal, so that one copy of the series of missing genes is present and thus enables the fly to survive.) Genetic mapping with these cytologically variant strains has demonstrated, in every case, that a particular block of genes shows the same rearrangement, inversion, duplication, or deletion, as the case may be, as the banding pattern of the chromosome. In this way, the genetic map of each linkage group has been associated with the cytological map of the chromosome to which it corresponds. Figure 9–9 illustrates one example.

FIGURE 9–9

Cytogenetics of a portion of the *Drosophila melanogaster* X chromosome and the corresponding genetic map. The lower picture is a micrograph of the salivary gland chromosome, revealing the characteristic banding pattern. Above it is an interpretative drawing which "straightens out" the spacing of the bands to correct for artifacts generated by squashing the cells for staining and microscopy. The numbers and letters refer to a coordinate system which cytogenticists use for identifying chromosome segments. The top line shows the genetic linkage map of genes located on this portion of the X chromosome; the numbers in the top line indicate map units (one map unit = 1% recombination) of distance from the left end of the linkage group. (Courtesy of G. LeFevre.)

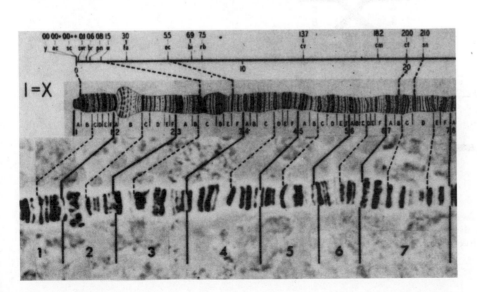

The Formation of Recombinant Chromatids

The Janssens-Morgan theory of recombination posits that crossing-over occurs when two of the four strands of a meiotic tetrad exchange parts to produce reciprocally recombinant chromatids (Figure 9–10). How is one to test this theory? Its first prediction is that *reciprocally recombinant chromatids*—those which arise in the same event (aB and Ab in Figure 9–10)—should be detected in equal frequencies. This prediction has been confirmed many times, by and large. In the example we showed earlier, the reciprocal recombinant types (b e) and (B E) were found in roughly equal numbers. (Their frequencies were not exactly equal due simply to sampling variation in the random distribution of chromatids to the gametes.) In higher organisms, one cannot critically test the theory that reciprocally recombinant chromatids are produced in the same event, because the descendants of individual meiotic cells cannot be recovered together. Happily, there is one group of lower organisms in which this can be done.

Yeasts and molds normally exist as single cells or a tangle of filaments, consisting of strings of single cells which reproduce vegetatively through mitotic divisions. Among these simple organisms, sexual reproduction consists of the fusion of two haploid cells of opposite mating type. The resulting diploid (zygote) cell, which can be identified under the microscope and isolated, then goes through meiosis to produce haploid *spore* cells. The process of sporulation might have been invented for the benefit of geneticists, for the four descendants of each isolated meiotic cell remain together in a four-spored *ascus* or spore sac. (In some molds, each of the four haploid spores goes through a single additional mitotic division to produce an eight-spored ascus.) Each ascus thus contains the outcome of a single little genetic experiment. It is not difficult under a low power microscope to break up the contents of an ascus, tease apart the spores, and then determine their individual genotypes.

Let us consider the kinds of asci which descend from the cross A B by a b posed above (Figure 9–11). (The double underlining represents explicitly the duplicate chromatids of which each chromosome is composed at the start of meiosis.) In those diploid cells in which no crossover occurs in the region between genes A and B, there will be two spores of each parental genotype, namely A B and a b. The more interesting asci are those which descend from meiotic cells in which crossing-over occurred anywhere in the interval between the two genes. In nearly all cases (we shall discuss the rare exceptions later) these asci contain exactly the four genotypes predicted by the Janssens-Morgan hypothesis: one spore of each parental type, and one of each reciprocal recombinant type (a B and A b). Such asci with four genotypically different spores are called *tetratype* asci. The frequency of crossing-over between two linked genes increases with the distance between them on the chromosome, and of course the frequency of tetratype asci increases in proportion. The enumeration of tetratype asci thus provides one way to obtain mapping information.

Since the four strands of individual meiotic tetrads are recovered together in asci, their genetic analysis is conventionally and properly referred to as *tetrad analysis* (Figure 9–12). We do not have space to go into tetrad

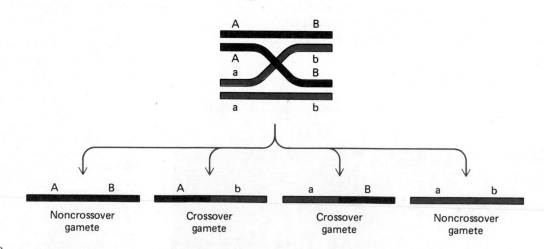

FIGURE 9–10

The consequences of crossing-over between two chromatids of paired homologous chromosomes at the first division of meiosis. The outcome is two noncrossover gametes and two crossover (or recombinant) gametes. (See also Figure 9–6.)

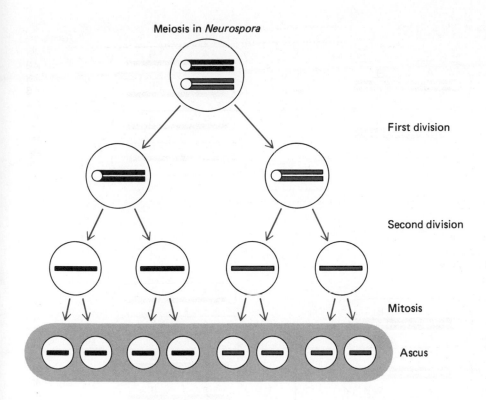

Meiosis in *Neurospora*

First division

Second division

Mitosis

Ascus

FIGURE 9–11
Meiosis in *Neurospora*. The first and second divisions are as in Figure 9–4. One subsequent mitotic division produces eight rather than four descendant cells. These are then wrapped up together in an ascus, shown in the bottom line of the figure.

analysis in detail, but some of its useful results are as follows:

1. Sophisticated analysis of asci in which multiple crossing-over events have occurred permits one to disentangle the behavior of each individual strand. For example, one can determine whether crossing-over in a given region between one pair of strands affects the chance of another crossover in an adjacent region of the same two strands, or in the other two.

2. In some mold species, the order of the spores within the ascus reflects the orderly segregation of strands as they are drawn apart by their centromeres (Figure 9–13). The process of meiosis thus makes the centromere a ghostly reference point, and permits one to measure the genetic distance separating a given gene from its centromere (Figure 9–14).

Structure Within the Gene

The methods of analysis we have discussed lead to a simple picture of genes arranged in series along the chromosome, like beads on a string. But what about the internal structure of the genes? In the following section, we describe how genetic analysis can be refined to discover the fine structure of individual genes.

Multiple Alleles

The first clue to the fine structure of genes came from the phenomenon of <u>multiple alleles</u>—a single gene having more than two forms. We have already mentioned Morgan's analysis of the sex-linked eye color gene in *Drosophila*, defined initially by its *white-eye* allele. In 1911, Morgan discovered another mutant strain with yellowish-pink (or eosin) eyes. When eosin-eyed females were crossed with white-eyed males, all the F_1 progeny had eosin eyes. Since the females of the F_1 had inherited one white allele from their fathers and one eosin allele from their mothers, one may conclude that eosin is dominant to white. In the F_2 there were 1147 eosin-eyed flies and 344 white-eyed flies, in fair agreement with the 3:1 ratio expected for a dominant and recessive pair of alleles. So eosin and white are alleles of the same gene. But they are also alleles of the wildtype form, which determines red eyes and is dominant to both of these mutant alleles. In short, eosin and white are different mutant alleles of the same gene.

Subsequent investigations turned up a remarkable variety of mutant alleles of this same gene, leading to apricot-colored eyes, purple eyes, or orange eyes. All are alleles of one another, by the criterion of Mendelian 3:1 segregation in the F_2, as illustrated above; and all map to the same position on the X chromosome. (We now know

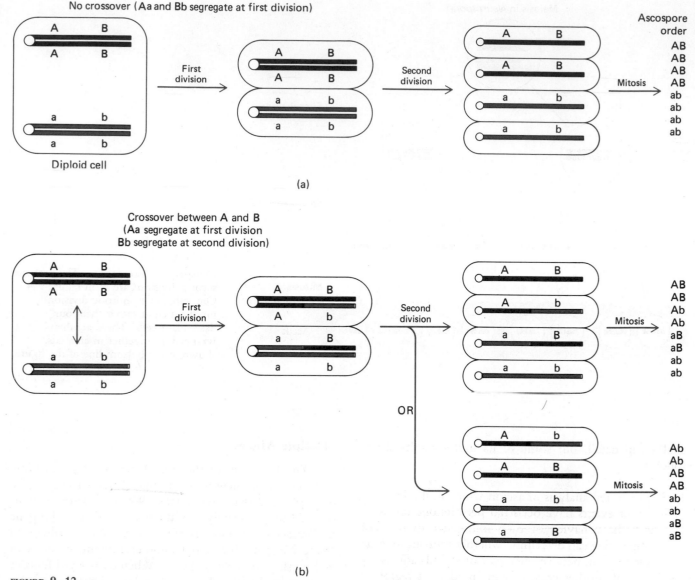

No crossover (Aa and Bb segregate at first division)

(a)

Crossover between **A** and **B**
(Aa segregate at first division
Bb segregate at second division)

(b)

FIGURE 9–12
Segregation patterns during meiosis.   (a) No crossing-over. (b) Crossing-over between genes A/a and B/b.

that this gene regulates the production of pigments during development of the *Drosophila* eye.) There are, it appears, many ways in which a single gene may be altered, leading to different phenotypic consequences (see Figure 9–15 for examples from another genetic locus).

More commonly, different mutant alleles produce the same or very similar phenotypes, but are sometimes distinguishable by their dominance relationships. For example, there are *Drosophila* mutants called *star-eyed*, in which the size and shape of the eye is affected. Two different mutant alleles that lead to the same phenotype can be distinguished, in that one, abbreviated S, is dominant to wildtype, whereas another, termed s, is recessive to wildtype.

Given the basic notion that germ plasm encodes genetic information through a specific chemical structure, one might expect different allelic variants of the same gene to reflect changes in different parts of the gene's internal structure. This notion implies that the order of mutable sites in a single gene might be determined by analyzing recombination *within* the gene by exactly the same methods used to map the order of discrete genes.

Intragenic Recombination

A pioneering step in this direction was reported by E. Lewis in 1941. He crossed the dominant S star allele and recessive s star allele to produce a hybrid F$_1$ of the

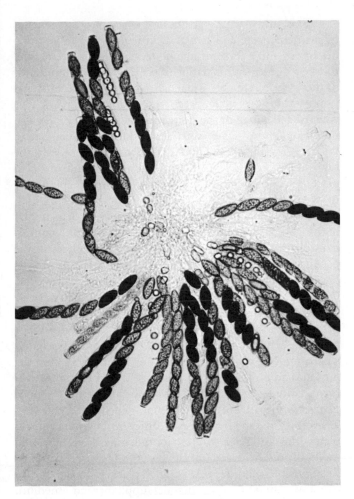

FIGURE **9–15**

Multiple alleles of the sd gene of *Drosophila melanogaster*, which governs wing size and shape. (Courtesy of A. Chovnick.)

FIGURE **9–13**

Asci of *Neurospora*. Alleles of a gene governing spore color are segregating. Different asci which show first or second division segregation patterns are evident. × 250. (Courtesy of D. Stadler.)

Gene to centromere distances in linkage group three of *Neurospora crassa*

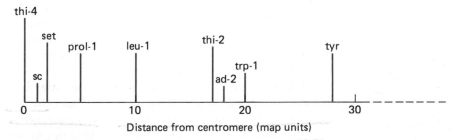

FIGURE **9–14**

Centromere map of one chromosome of *Neurospora*. Position "0" is the centromere, and the various genes are located in relation to their distance from the centromere. For example, gene thi-4 never shows second division segregation, meaning that it is extremely close to the centromere, as shown; gene trp-1 is 20 map units from the centromere, as determined from the proportion of second division segregations for this gene.

genotype S/s. Then, he backcrossed (i.e., crossed the hybrid with a parent) these heterozygous flies with homozygous s/s flies. If the S and s alleles occupied the very same position within the gene, then obviously only star-eyed progeny could result (S/s or s/s). But suppose that the two mutations giving rise to these alleles occupied different sites within the gene. The genotype of the F_1 hybrid could then be represented as $\frac{S\ +}{+\ s}$. (A "+"—which is conventionally used to denote the wild-type allele of any gene—is also employed to designate the normal form of a site *within* a gene.) In this case, crossing-over within the gene could produce gametes that had a wildtype eye gene, leading to the occurrence of +/+ progeny with normal eyes. Among a total of 31,110 progeny flies, Lewis found four with normal eyes. He had succeeded in splitting the gene.

In fact, Lewis rigged the cross so that mapping information on the location of the S and s alleles came out of the results. The F_1 females he used were the end result of a complicated breeding program that was arranged to yield a particular genotype. (Experimental geneticists refer to this program as constructing a genotype.) These F_1 females were constructed to be heterozygous for two other genes closely flanking the star-eye gene. These genes (the particular phenotype of which need not concern us) were called al and ho. The complete genotype of the F_1 females was

$$\frac{\text{al}\quad S\quad \text{ho}}{+\quad s\quad +}.$$

The four progeny with normal eyes (neither S nor s) all were recombinant for the flanking genes, having developed from eggs which were + + ho. This experiment shows that recombination was indeed responsible for the normal genotype at the star-eye gene. To give the observed progeny a crossover event must have occurred between the S and s mutations. Moreover, these results reveal the order of the S and s mutations relative to the outer markers. The genotype of the F_1 females must have been

$$\frac{\text{al}\ :\ S\ +\ :\ \text{ho}}{+\ :\ +\ s\ :\ +},$$

where the dotted lines bracket the internal structure of the star-eye gene. It would be more correct to symbolize the results of the single crossover between the two mutant sites inside the star-eye gene as yielding a recombinant + ++ ho genotype. (If the order of the star-eye mutations were the reverse—s S—then the F_1 females would have had the genotype

$$\frac{\text{al}\ :\ +\ S\ :\ \text{ho}}{+\ :\ s\ +\ :\ +},$$

and the flies with normal eyes would have been al ++.)

This study was a pathfinding entry into the genetics of the gene itself, but is was a difficult act to follow. To discover a mere four intragenic recombinants, Lewis had to carry out the heroic labor of scoring tens of thousands of flies. Genes are very small, it appears, and recombination within the gene is correspondingly rare. If we constructed a map of the **genome** (the entire set of genes) of *Drosophila* to scale, separating the two mutant sites within the star-eye gene by the distance used in the representation above (~1 cm), then we would have to place al some 45 cm to the left, and ho more than a meter to the right.

If intragenic recombinants could only be detected by hunting them out among a vast excess of nonrecombinant progeny, as Lewis did, then detailed analysis of the gene's fine structure would be out of the question. In the case of microorganisms and bacterial viruses, however, there is an easier way. Many of the genes studied in these creatures determine whether or not the organism can grow under a particular condition of cultivation. For example, there are mutants of the mold *Aspergillus nidulans* which cannot grow at all on medium lacking the nucleic acid building block adenine. Wildtype spores will form colonies on adenine-less solid medium in petri plates, while ad mutants (adenine dependent) will not. Therefore, if one crosses two different ad mutants and plates the progeny spores on adenine-deficient solid medium, only the rare intragenic ++ recombinants will form visible colonies. A parallel plating of the spores (suitably diluted) on adenine-supplemented medium serves to count the far more numerous nonrecombinant spores. The point is that one does not have to hunt for the rare wildtype spores *among* the others; one *selects them out* on the appropriate medium.

In 1955, the British geneticist R. H. Pritchard used selective plating of this sort in crosses involving seven different allelic ad mutants of *Aspergillus*. These two factor crosses, analyzed in exactly the way we discussed earlier, pointed to a linear sequence of the seven mutant alleles. Moreover, Pritchard made use of two flanking genes, in the same way that Lewis had done with the star-eye gene in *Drosophila*. The results established a unique order of the seven ad alleles between the two flanking markers.

It began to appear, as a result of these investigations and similar ones done about the same time, that the fine structure of the gene was very simple—a linear array of mutable sites disposed in order between flanking genes, although very much closer together. The image of genes as beads on a string began to disappear with these findings. Or rather, the beads disappeared, leaving only the string, of which each gene appeared to be nothing more than a very short segment that encodes the information for a particular phenotype. Needless to say, this picture of the organization of germ plasm, based on strictly genetic reasoning, agreed nicely with the discovery that the material carrier of germ plasm information was DNA, a linear molecule composed of a great long string of (nucleotide) subunits.

Mapping the Fine Structure of the Gene

The idea of identifying gene fine structure with the molecular structure of DNA was taken to its limit by S. Benzer in a classical analysis of the r_{II} gene (actually two contiguous genes) of bacterial virus T4. This virus grows on the bacterium *E. coli,* lysing each infected cell to produce a crop of a few hundred virus particles which can then in turn infect new *E. coli* victims. Individual T4 virus particles are easily enumerated by pouring them in soft agar together with about 10^8 *E. coli* cells onto a petri plate in which the bacteria can grow. After about 12 hours, the bacteria have grown into a thick, confluent lawn punctuated by holes (or plaques) where single virus particles initially landed (Figure 9–16). Each of these virus particles infected a single bacterial cell on the plate; its offspring then lysed more bacteria in the vicinity, their progeny in turn lysed still more bacteria, and these successive cycles of infection devoured a sufficiently large portion of the bacterial lawn to form the plaque. There is further discussion of bacteriophage T4 in Chapter 15.

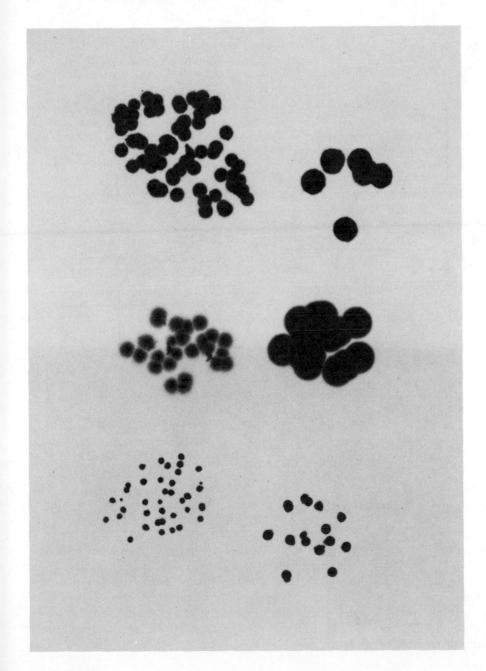

FIGURE 9–16
Plaques of bacterial virus T4 in a lawn of its host bacterium, *Escherichia coli.* The panels show a few plaques of each of six variants of T4 that differ by single mutations in genes which determine plaque morphology. × 6. (Courtesy of G. Doermann.)

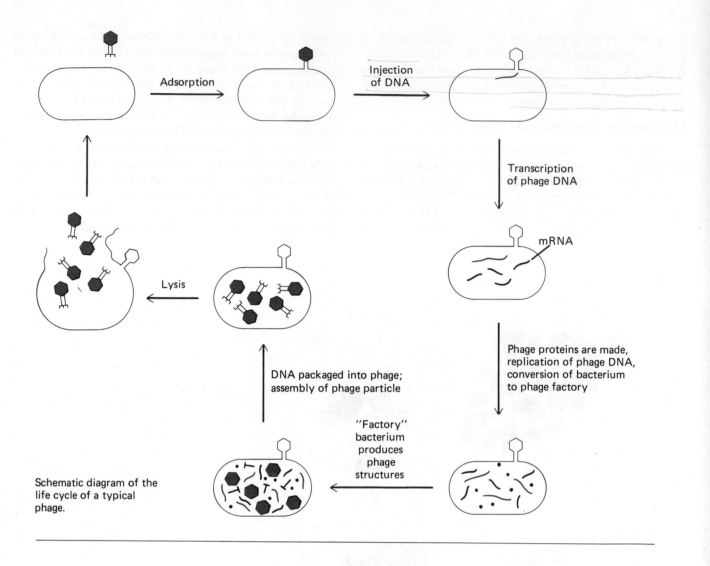

Schematic diagram of the life cycle of a typical phage.

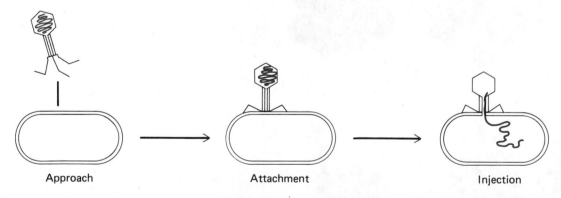

Approach   Attachment   Injection

The r_{II} mutants grow perfectly well on bacterial host strain B, but their plaques are morphologically distinct from those caused by wildtype viruses, making it easy to collect independent r_{II} mutants. To perform a two-factor cross of two of these mutants, one simply infects susceptible B cells jointly with stocks of the two mutant viruses. The techniques of virology make it possible to perform about 40 such crosses in patches of bacteria on a single petri plate, and score the results the next day. Once again, the use of microorganisms offers technical advantages over garden peas or fruit flies, which would have delighted Mendel or Morgan.

On another host strain, K(λ), r_{II} mutants cannot grow at all and therefore produce no plaques. This

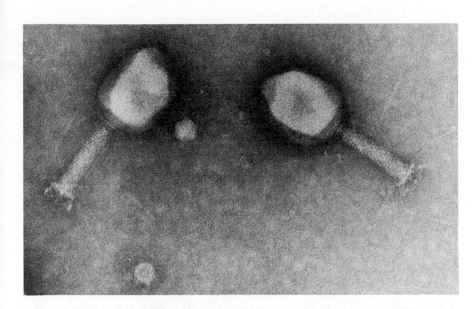

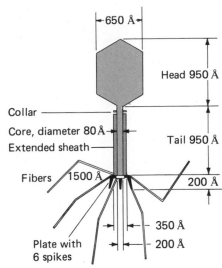

provides a stringent selection system analogous to the screen mentioned above in the case of *Aspergillus* nutritional mutants. One plates the progeny of a cross done in host strain B on K(λ); only the rare wildtype recombinants will form plaques on this host. The number of these recombinants, divided by the total progeny of the cross (enumerated, of course, on strain B), measures the frequency of recombination between the r_{II} mutant sites in the gene. Benzer began by performing two-factor crosses among eight independent r_{II} mutants, and found that the results ordered the mutant sites into a linear sequence. (Ordering mutations onto a linear map is called *mapping* them.) Then he collected an enormous number (about 3000) of new r_{II} mutants for a daring project: he would map the gene "into the ground," as he put it; that is, right down to the DNA nucleotide pairs of which the gene is composed.

To see what Benzer meant, suppose that the DNA of the r_{II} gene spans 1000 nucleotide pairs. If all of these are used to encode the structure of the protein product, then there would be 1000 separable genetic sites at which different mutants could map. If one collected 1001 independent mutants, then at least two of them would have to fall at the same site, and could not give any wildtype recombinants when crossed with each other.

Alternatively, suppose only 100 of the 1000 nucleotides were used to encode the gene product, and the remaining 900 nucleotides were structural filler, without genetic information. Then there would be only 100 distinct genetic sites of mutation. Moreover, the genetic map of these sites would indicate how they were distributed with reference to the remaining nongenetic 900 nucleotide pairs.

In principle, statistical analysis of the number of sites occupied by one, two, three, and so on, independently

isolated mutants in a large collection would reveal the total number of sites in the gene at which mutations could fall. (If the mutations occur at all sites with equal probability, the number of sites with no mutations can be estimated—because the more sites with high numbers of examples, the fewer the sites that will have been missed.) The size, in number of nucleotide pairs, of the DNA of the r_{II} gene could then be calculated from the total number of nucleotide pairs in the T4 virus chromosome, which is a single DNA molecule, times the fraction of the total linkage group occupied by the r_{II} gene.

Intragenic mapping on this scale might also be expected to turn up the lower limit of recombination frequencies, that found in a cross between mutants in two adjacent nucleotide pairs. Thus, Benzer hoped that the smallest non-zero recombination frequency he could detect, in many, many crosses, would define immediately adjacent mutable sites. Moreover, the ratio of the largest recombination frequency observed (between sites at either end of the gene) to the smallest (between adjacent nucleotide pairs) would provide another method for approximating the total number of mutable sites within the gene, and relating it to the number of nucleotide pairs composing the gene.

In order to reduce the number of two-factor crosses to a manageable number, Benzer made use of an ingenious shortcut method. Many of the mutants he isolated were *deletions* of a stretch of genetic material within the r_{II} gene (Figure 9–17). The genetic criterion which identifies a deletion mutant is that it cannot produce wildtype recombinants with two or more other mutants that do recombine with one another. For example, suppose we identify two mutations at different sites as *a* and *b*. A cross of these two mutants yields wildtype recombinants in proportion to the distance

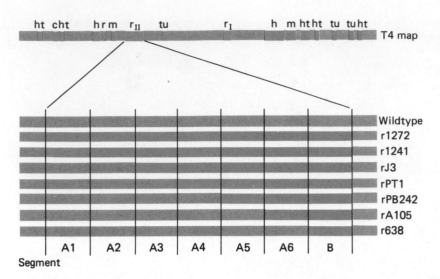

FIGURE 9–17

Overlapping deletions in the r_{II} gene of bacterial virus T4.. and their use in mapping point mutations.   The region *missing* in each deletion is shown in black. Thus, a point mutant which fails to give wildtype recombinants when crossed with deletion mutant r1272, but does with all the others, is identified as lying in segment A1. A point mutant which fails to give wildtype recombinants when crossed with deletions r1272 or r1241, but does with all the others, is identified as lying in segment A2, and so forth. The genetic map of the region around the r_{II} gene is shown above.

between sites *a* and *b*. But a deletion which lacks both sites cannot give wildtype recombinants when crossed with either. (This deletion mapping method has since been applied to many other viruses and cells.)

To map very large numbers of mutants quickly, Benzer crossed each new mutant with a standard set of deletions by a simple scoring procedure. A drop of a liquid stock of each mutant virus to be tested was spotted, together with a drop of one or another deletion mutant, on the restrictive host bacterium in which only wildtype recombinants could grow. If recombinants were produced, no matter how few, they would multiply and lyse the bacterial lawn; if the mutant fell into the span of genetic material absent in the deletion, then no recombinants could be produced, and the bacterial lawn would not be lysed. In this way, a new mutant could be mapped against dozens of standard deletions on a single petri plate, and its location assigned to a small segment of the gene's genetic map.

Once a mutation had been assigned to one of these segments, it remained only to cross it with other mutants which mapped to the same small segment. Here again, the spot testing procedure reveals whether two mutations lie at the same location—and thus cannot generate wildtype recombinants in a cross—or at different locations, in which case some wildtype recombinants are produced. The test has the absolute decisiveness of a

computer's binary logic in which there are only two alternatives, yes and no.

The application of these methods over several years enabled Benzer to map the r_{II} gene in unparalleled detail (Figure 9–18). He also performed a statistical analysis of the frequency distribution of mutations at identical sites—identified as those which did not yield any wildtype recombinants when crossed with each other—in relation to the recombination frequencies between nonidentical sites and the known size of the T4 genome, measured in numbers of DNA base pairs. This analysis suggested that in the region of T4 DNA identified as the r_{II} gene *all* the base pairs could be sites of mutation. Thus, the elementary unit of mutation, which Benzer dubbed the *muton,* is a single base pair in DNA. The elementary unit of recombination, or *recon,* is the frequency of recombination between adjacent base pairs; Benzer reckoned this to be about 0.01%, corresponding to the lowest nonzero recombination frequency he detected between very closely linked sites.

The simple picture which emerged is that a gene is just a continuous stretch of DNA, demarcated by *start* and *stop* symbols for the gene's expression. Later analysis showed that Benzer's statistical treatment was oversimplified, as is often the case with important clarifying ideas in science. Nonetheless, at its time, Benzer's analysis had the crucial effect of demystifying the concept of the gene,

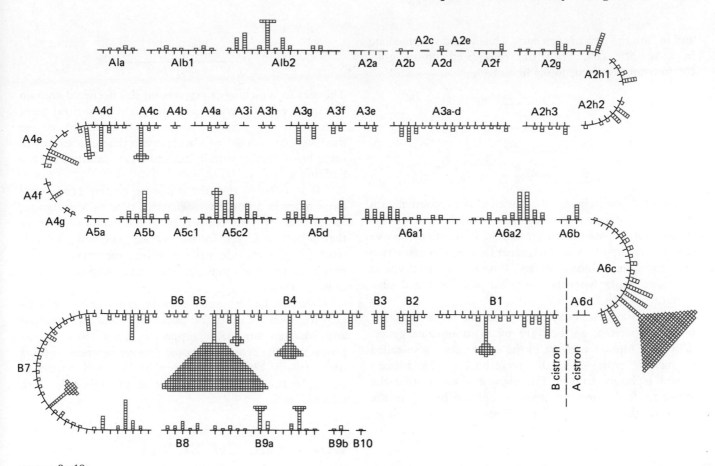

FIGURE 9-18

Fine structure map of the r$_{II}$ A and B genes. Each square represents the independent occurrence of a mutation in Benzer's gigantic collection of spontaneous mutants. The location of each mutant was determined by the deletion mapping method. Squares at the same position represent independent mutants which mapped to the very same site; they do not yield wildtype recombinants when crossed with one another. The letters and numbers below the line indicate a coordinate system defined by the various overlapping deletions (see Figure 9-17).

by superposing the genetic methods of Mendel and Morgan onto the structure of DNA in intimate detail.

With the further development of molecular genetics, this picture was modified in one regard. The genes of eucaryotic organisms generally contain noncoding DNA segments interspersed amongst the coding segments. The projection of the gene's structure onto DNA structure, exemplified by Benzer's analysis, solved one mystery and paved the way for deeper analyses into the structure of the gene. The later discovery of noncoding *intervening sequences* in higher organisms, which we describe in Chapter 11, thus illustrates the way in which science proceeds: mysteries are dispelled at one level of analysis, permitting more refined conceptual models and methods of investigation, which then encounter new mysteries.

Reciprocal and Nonreciprocal Recombination

Our discussion of recombination has so far implied that it is a simple exchange of parts between chromosomes, or DNA molecules, in which the genetic information is reshuffled but not altered. At a gross level of analysis, this is perfectly correct. At the finer level afforded by recombination between closely linked sites within a gene, it breaks down. This anomaly is best illustrated by returning to the tetrads of molds like yeast, where all four chromosomal strands of a single meiotic cell can be recovered together.

Consider a cross between two strains which harbor different mutations within the same gene. We may symbolize the cross as: $\underline{m_1 \quad +}$ by $\underline{+ \quad m_2}$, where m$_1$

and m_2 are the two mutant loci. The diploid intermediate in a meiotic cell, with the duplicated homologous chromosomes paired, looks like this:

$$
\begin{array}{cc}
m_1 & + \\
\hline
m_1 & + \\
\hline
+ & m_2 \\
\hline
+ & m_2 \\
\end{array}
$$

The dotted line indicates the position of a recombination exchange between two of the four strands. Our discussion of recombination so far would lead one to believe that the *tetratype* ascus thus generated will contain two *reciprocally* recombinant spores. One will be wildtype at both positions, whereas the other will be the double mutant $m_1\, m_2$.

This is almost always true when the mutant sites are widely separated, as they are when in separate genes. When the mutant sites are in the same gene—a so-called *heteroallelic* cross—then the reciprocal double mutant strand is rarely found. The typical result is that the expected double-mutant spore is replaced by one or the other of the parental combinations, as shown below.

$$
\begin{array}{cc}
m_1 & + \\
\hline
m_1 & + \\
\hline
+ & m_2 \\
\hline
+ & m_2 \\
\end{array}
\xrightarrow{\text{recombination}}
\begin{array}{cc}
m_1 & + \\
\hline
+ & + \\
\hline
+ & m_2 \\
\hline
+ & m_2 \\
\end{array}
$$

Clearly, this outcome cannot be generated by a simple exchange of parts. If all the genetic information in the chromosomes before recombination were conserved without alteration, then there would be two m_1 alleles and two m_1^+ alleles in the ascus, a 2:2 segregation. Instead, the genetic information at the m_1 site shows a 3:1 segregation of wildtype and mutant information, as if one of the m_1 mutant sites has been converted to its wildtype counterpart. In fact, this phenomenon is termed *gene conversion*.

If one examines very large numbers of products from a one-factor cross, only rarely can one find asci in which gene conversion produces a 3:1 (or 1:3) segregation for the mutant allele. In a heteroallelic cross involving two closely linked mutant sites, however, 3:1 segregation predominates. This observation suggests that gene conversion mostly occurs very close to the position of crossing-over.

The association of gene conversion with crossing-over can be revealed when the regions on either side of the heteroallelic gene are distinguishable genetically—that is, marked by allelic differences in two other genes, as illustrated:

$$
\begin{array}{cccc}
A & m_1 & + & B \\
\hline
a & + & m_2 & b \\
\end{array}
$$

The asci in which gene conversion has occurred contain one spore with the wildtype genotype at the central gene. Such spores are often recombinant (i.e., $\underline{A\ \ b}$) for the flanking genes. Also, another spore in the same ascus will often be of the reciprocally recombinant genotype for the flanking genes: $\underline{a\ \ B}$.

It is possible that the wildtype central gene could have arisen in an otherwise parental type by a mutation, rather than by gene conversion. To obtain a type spore that is both wildtype for the central gene and recombinant for the outside marker genes, however, would require two mutations in the same region—a very unlikely event.

We are led by such data to the conclusion that crossing-over is a more complicated process than Janssens and Morgan originally supposed. Very close to the position of the crossover, some "funny business" occurs that actually alters the genotype. We will return to an explanation of this anomaly at the DNA level in Chapter 10.

Genetics and Cell Biology

The pertinence of genetics to cell biology lies in both of the approaches we emphasize in this book, the structural and the functional. In this chapter, we have stressed the use of genetics as a means of structural analysis: the construction of genetic maps reveals the topological organization of genes along the chromosome, and of mutable sites within the gene. The correlation of this type of structural analysis with direct cytological and biochemical analysis of DNA, the bearer of genetic information, will be discussed in Chapter 10.

Genetics also provides an invaluable strategy for experimental studies on the function of individual cell components, and on the functional relationships between cell components that give rise to the cell's integrated behavior. To study a *function* by experiment means making that function the sole experimental variable: one compares cell activity in the presence and absence of a single, discrete functional activity carried out by a specific macromolecule. In principle, one could inhibit a functional activity by a specific chemical poison. Chemical intervention in a cellular process, however, is rarely specific enough to pinpoint a single type of macromolecule as the target. Experiments based on chemical inhibitors are therefore often difficult to interpret.

Luckily, all macromolecules in the cell are ultimately the products of information encoded in the genes. A mutation in the gene coding for a particular macromolecule affects that macromolecule with perfect surgical precision, due to the very precision of the process of gene expression itself (see Chapters 10–13). The effect of the mutation can be seen clearly in haploid cells, such as *E. coli,* where there are no other alleles present. Thus, one can tease out the functional consequences of a specific gene product's activity by comparing the behavior of cells which produce the normal gene product with cells differing only by the presence of a single mutant allele in the gene for the macromolecule in question.

A crucial feature of this approach is the use of *conditional mutants.* Many gene products are indispensable for cell growth and survival. Mutations in such genes would be expected to be lethal in haploid cells and, if so, could not be recovered or used for experimentation. Fortunately, many mutations in indispensable genes are only conditionally lethal—the gene product is defective under one *restrictive* condition but normal under another *permissive* condition. The most useful mutations of this type simply make the gene product *temperature sensitive;* the restrictive condition is high temperature, the permissive condition is low temperature. Such temperature-sensitive mutants can be propagated at the permissive (low) temperature, and then the gene product's function can be shut off at a defined time by shifting to the restrictive (high) temperature. In Chapter 14 we describe the use of such mutants to study the ordered sequence of steps through which the cell progresses in its division cycle. In Chapter 15, temperature-sensitive mutants will turn up again as the key to analyzing the assembly of a complex macromolecular structure, an entire bacterial virus. These examples of the use of temperature-sensitive mutants are but two out of a vast number.

Mendel's distinction between dominant and recessive alleles of genes will reappear in Chapter 14 as well. The regulatory processes of the cell depend on regulatory genes, which control the expression of other genes. The dominant or recessive character of mutant alleles of these regulatory genes provides, as we will see, information about the way they function. Regulatory processes control the expression of genes at specific sites within

them or adjacent to them. The discovery of these control sites has depended heavily on genetic analysis, especially deletion mapping. Finally, the decoding of the language of genes (Chapter 13)—a feat of cryptography that is unique in science—proceeded through a many-leveled interplay between genetic and biochemical approaches.

This sort of interplay between different approaches is emblematic of cell biology and is also, in a sense, the basic theme of this book. Cytology, biochemistry, and genetics are conventionally distinguished as different academic disciplines. But the cell does not follow an academic curriculum. Its every activity is at once cytological, biochemical, and genetic.

A Preface to Molecular Genetics

The conceptual structure of classical or formal genetics depends much more on analysis than on experimental manipulation. The elementary mutational and recombinational events of interest occur at random, owing nothing to the experimenter's hand; the positions of individual events are inferred by analysis, and the linear arrangement of genes on chromosomes and of mutable sites within genes reveals itself through this analysis.

For three generations, the dream of geneticists was to be able to perform precise operations on the genetic material and then examine the consequences. In the early 1970s, this dream came true. Discoveries in an esoteric area of bacterial genetics provided a remarkable set of techniques for cutting DNA at defined sites, and then manipulating these fragments virtually at will to produce new combinations of genes, genes fused to one another, or even genes hooked up to or containing artificially constructed DNA sequences. This sorcerer's bag of tricks is popularly termed "genetic engineering"; its amalgamation with classical genetics and DNA biochemistry has generated the new molecular genetics which we will recount in the next chapter, after first considering the chemistry of the genetic material in some detail.

KEY WORDS

Allele, chromatid, chromosome, deoxyribonucleic acid, diploid, dominant, gamete, genetic map, genome, genotype, haploid, heterozygous, homologous, homozygous, hybrid, information, linkage, meiosis, phenotype, recessive, recombinant, recombination, segregation, virus, wildtype.

PROBLEMS

1. The table below reports the results of pairwise crosses between *deletion* mutants in a gene of bacterial virus T4. A "+" indicates that some wildtype recombinants are produced; a "0" indicates that *no* wildtype recombinants are produced.

```
        A  B  C  D
      ┌─────────────
  A   │  0  0  +  0
  B   │  0  0  0  +
  C   │  +  0  0  +
  D   │  0  +  +  0
```

(a) Draw an approximate map showing the order of the deletions.

(b) Point mutant m63 yields a few wildtype recombinants when crossed with deletion B, several times more wildtype recombinants when crossed with deletion D, but none when crossed with deletion A. Place mutant m63 on your genetic map.

(c) Is it possible for a point mutant to yield wildtype recombinants when crossed with each of the four deletion mutants?

(d) Is it possible for a point mutant to fail to yield any wildtype recombinants when crossed with each of the four deletion mutants?

2. Three codominant alleles control flower color in violets. Red, yellow, and blue are pure breeding. Red × yellow → orange; yellow × blue → green; and red × blue → purple. What is the phenotypic ratio expected from a cross of orange × green?

3. A *Drosophila* male with a stubby body was crossed with a normal female. The male offspring were all normal, but the female offspring were all stubby. Is the gene controlling this trait autosomal or sex-linked? Is the stubby allele dominant or recessive?

4. The crossover frequency between linked genes A and B is 40%; between B and C, 20%; between C and D, 10%; between C and A, 20%; between D and B, 10%. What is the sequence of the genes on the chromosome?

5. I

   ☐ hemophilia—x-linked recessive

   II

   ? ☐ ○ normal

(a) The woman in line II is concerned that her unborn child may be a hemophiliac. Amniocentesis is performed and the fetus is a male. What is the chance that the fetus has hemophilia?

(b) What is the chance that her next child will have hemophilia?

6. Genes A, B, and C are linked with recombination values of AB 20%, AC 5%, and BC 15%.

(a) What is the order of these genes on the chromosome?

(b) If the F1 from a cross of an aaBB × AAbb is backcrossed to an aabb, what is the expected proportion of genotypes among the progeny?

7. Two alleles of one gene determine the character difference of *purple* versus *green* stems, and two alleles of a separate gene on another chromosome determine the character difference of *cut* versus *potato* leaves. From the results of the following crosses:

	Parental Phenotypes	Number of Progeny			
		Purple Cut	Purple Potato	Green Cut	Green Potato
1	purple cut × green cut	321	101	310	107
2	purple cut × purple potato	219	207	64	71

(a) Determine which alleles are dominant.

(b) What are the most probable genotypes for the parents in each of the two crosses?

8. A recessive allele on the X chromosome is responsible for red-green color blindness in humans. A normal woman whose father is color-blind marries a color-blind man. What is the probability that this couple's son will be color-blind?

A. 0   D. 3/4
B. 1/4   E. 1
C. 1/2

9. A man who carries an X-linked allele will pass it on to:

A. all of his daughters   D. half of his sons
B. half of his daughters   E. all of his children
C. all of his sons

10. Sexual reproduction provides a source of variation for a population through the:

A. crossing over of chromosomes
B. union of genetically different gametes at fertilization
C. independent assortment of nonhomologous chromosomes
D. only A and B are correct
E. all three answers (A, B, and C) are correct

Use the following to answer questions 11 through 13. In a certain plant the allele for large leaves (L) is dominant over the allele for small leaves (l), and the allele for yellow flowers (Y) is dominant over the allele for white flowers (y). The genes determining these two traits are unlinked.

11. A plant that is heterozygous for both traits is crossed with a plant that has small leaves and is heterozygous for flower color. What fraction of the offspring should have large leaves and white flowers?
    A. 1/4   D. 3/8
    B. 1/2   E. 3/4
    C. 1/8

12. Two plants that are heterozygous for both traits are crossed. What fraction of the offspring will have the genotype Llyy?
    A. 1/8   D. 1/2
    B. 3/8   E. 1/4
    C. 1/16

13. A plant with large leaves and white flowers was crossed with a plant with small leaves and yellow flowers. Many offspring were examined and it was found that one-half of the offspring had large leaves and yellow flowers and one-half had small leaves and yellow flowers. The genotypes of the parents must have been
    A. LlYy and llYY   D. LLyy and llYy
    B. Llyy and llYy   E. Llyy and llYY
    C. LlYy and LlYy

14. Albinism results from the failure to make the pigment melanin, which is produced from tyrosine. Like other enzyme deficiency diseases, it is transmitted as a recessive allele. In 1952 it was reported that two albinos, who met at a school for the partially sighted, married and had three children, all of whom had normal pigmentation. Assuming the children are not illegitimate, how do you explain that the children are normal?

15. In *Drosophila melanogaster,* there is a dominant gene for gray body color and another dominant gene for normal wings. The recessive alleles of these two genes result in black body color and vestigial wings, respectively. Flies homozygous for gray body and normal wings were crossed with flies that had black bodies and vestigial wings. The F1 progeny were then test-crossed, with the following results:

Gray body, normal wings	227
Black body, vestigial wings	253
Gray body, vestigial wings	54
Black body, normal wings	66

    Would you say that these two genes are linked?
    _____
    If so, how many units apart are they on the chromosome? _____

16. If an individual has the genotype AaBBDd, what fraction of the gametes produced will be AbD if the genes are unlinked?
    A. 0   D. 1/2
    B. 1/4   E. 1/8
    C. 1/16

17. If an individual with the genotype AaBb produces all four possible types of gametes with equal frequencies, then one can conclude that:
    A. The genes are definitely unlinked.
    B. The genes are definitely linked.
    C. If the genes are linked, then they are very close together on the chromosome.
    D. If the genes are unlinked, then they are on the same chromosome.
    E. If the genes are linked, then they are very far apart on the chromosome.

SELECTED READINGS

Beadle, George W. (1981) The genes of men and molds. In *Genetics: Readings from* Scientific American, ed. Cedric I. Davern. San Francisco: W. H. Freeman.

Benzer, Seymour. (1966) Adventures in the R$_{II}$ Region. In *Phage and the origins of molecular biology,* ed. John Cairns, Gunther Stent, and James Watson, Cold Spring Harbor, NY.

————. (1981) The fine structure of the gene. In *Genetics: Readings from* Scientific American, ed. Cedric I. Davern. San Francisco: W. H. Freeman.

Edgar, R. S., and Epstein, R. H. (1981) The genetics of a bacterial virus. In *Genetics: Readings from* Scientific American, ed. Cedric I. Davern. San Francisco: W. H. Freeman.

Gonick, Larry, and Wheelis, Mark. (1983) *The cartoon guide to genetics.* New York: Barnes and Noble.

Mendel, Gregor. (1981) Experiments in plant hybridization. In *Genetics: Readings from* Scientific American, ed. Cedric I. Davern. San Francisco: W. H. Freeman.

DNA Structure and Biological Information Storage

10

In 1868 F. Miescher, a 22-year-old Swiss physician, extracted from the nuclei of pus cells a macromolecular substance that he knew was not protein, which he called _nuclein._ At that time the significance of the nucleus was barely understood; not until several years later, in 1876, did O. Hertwig demonstrate that fertilization of the sea urchin egg involved the fusion of two nuclei, one from the egg and one from the sperm. Nevertheless, the influence of Miescher's work on nuclein was such that by 1884 Hertwig wrote that "nuclein is the substance that is responsible not only for fertilization but also for the transmission of hereditary characteristics. . . ." Toward the end of his life Miescher appears to have doubted that this was true, and most biologists did also, for the concept that nuclein is the hereditary substance sank into oblivion toward the end of the century, not to be resurrected until the early 1940s.

Even if Miescher was not aware that the substance he isolated was the hereditary material, his approach was a half-century ahead of his contemporaries. To begin with, he chose a biological material well suited for the isolation of nuclei: the pus cells (white blood cells) obtained from the discarded bandages of wounded soldiers, made abundantly available by the Franco-Prussian War.[1] Second, he used an amazingly advanced technique to disrupt cells: digestion with the enzyme pepsin in the presence of HCl—a condition that prevents most enzymatic activity—followed by ether extraction, which layered the nuclei at the ether-water interface. Thus, without access to the modern method of differential centrifugation, Miescher isolated a cellular organelle for the first time. Third, Miescher not only proceeded to study the nuclei under the microscope and identify them by a number of cytological techniques, but he also embarked on an extended chemical study of the materials found in them. This simultaneous cytochemical and biochemical approach to the study of cell organelles was truly visionary, for it was not to be used again by cell biologists until the 1940s.

[1]Human red blood cells, though very useful objects for the study of other aspects of cell biology—membranes (Chapter 7) and cytoskeleton (Chapter 16)—have no nuclei and do not contain DNA.

Miescher dissolved most of the material obtained from his nuclei in a salt solution. Upon acidification he obtained a precipitate that had a property hitherto observed only in some lipid fractions—it contained large amounts of phosphorus. The material Miescher called nuclein was later renamed *deoxyribonucleic acid (DNA)* and is now recognized as the chemical structure that stores the cell's hereditary information.

Miescher followed his discovery with many years of careful experimentation. By fractionating sperm heads from Rhine salmon at low temperature, he showed that the nuclein was a material of high molecular weight associated with an unusually basic protein which he called *protamine*. His measurement of the phosphorus content of nuclein (9.95%) agrees remarkably well with our modern values for DNA (9.22–9.24%) and is a tribute to the excellence of his preparations. When Miescher died, he bequeathed to science the firm foundations of an entirely new field of biology which he, a single individual, initiated. Thus, our understanding of the molecular basis of heredity was born just at the time when Mendel was discovering some of its biological manifestations.

Chemistry and Biology of Nucleic Acids

The Organic Chemistry of the Nucleic Acids

During the forty years following Miescher's death, the organic chemistry of the nucleic acids was clarified. It became apparent that there were two classes of these macromolecules. Deoxyribonucleic acid (DNA) was found to be composed of (1) the purine bases *adenine* and *guanine,* (2) the pyrimidine bases *cytosine* and *thymine,* (3) the pentose sugar *deoxyribose,* and (4) *phosphoric acid* (Figure 10–1). *Ribonucleic acid (RNA)* was found to be composed of the same building blocks, except that the pyrimidine *uracil* is substituted for *thymine* and the pentose *ribose* for deoxyribose (Figure 10–1).

(Additional nucleotide components have been found, in small amounts, in both DNA and RNA. These include a variety of methylated bases, sulfur-containing bases, the compound pseudouridine, in which an abnormal base-sugar linkage occurs, and other unusual sugars and bases (Figures 10–2 and 12–8). These *modified* bases are not a caprice of nature, for as we shall see in this and

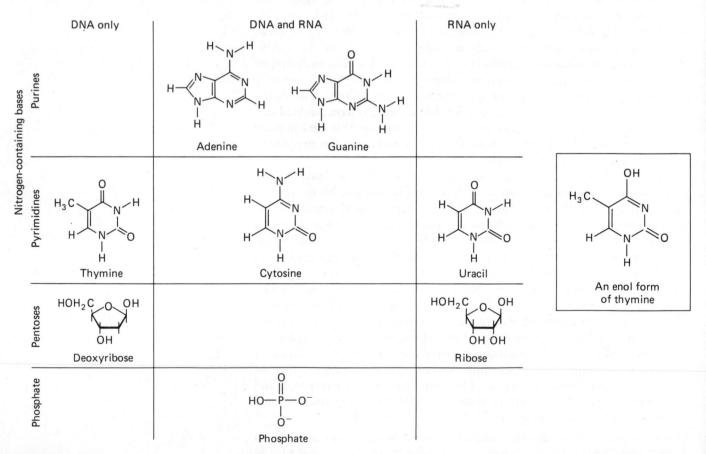

FIGURE 10–1
The Building Blocks of DNA and RNA.

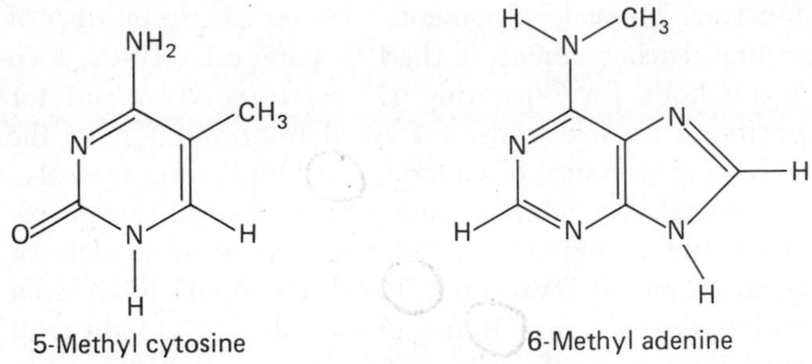

5-Methyl cytosine   6-Methyl adenine

FIGURE **10-2**
Modified bases found in DNA. The two bases shown are the most common. The methyl groups are added enzymatically after the normal bases are installed in the DNA.

later chapters, they play an important part in the function of nucleic acids.)

With the help of special enzymes called **nucleases,** which can split nucleic acids at various specific points in their macromolecular structure, it has been possible to identify the building blocks of DNA and RNA, the *nucleotides*. In both DNA and RNA the purine and pyrimidine nitrogen bases are linked to the deoxyribose or ribose at the 1'-carbon atom to form *nucleosides*, whereas phosphate is attached to the sugars at the 5'-carbon atoms, to form nucleotides (Figure 10-3).

How, then, are the nucleotides tied together in the polymer? Again, from enzymatic degradation studies

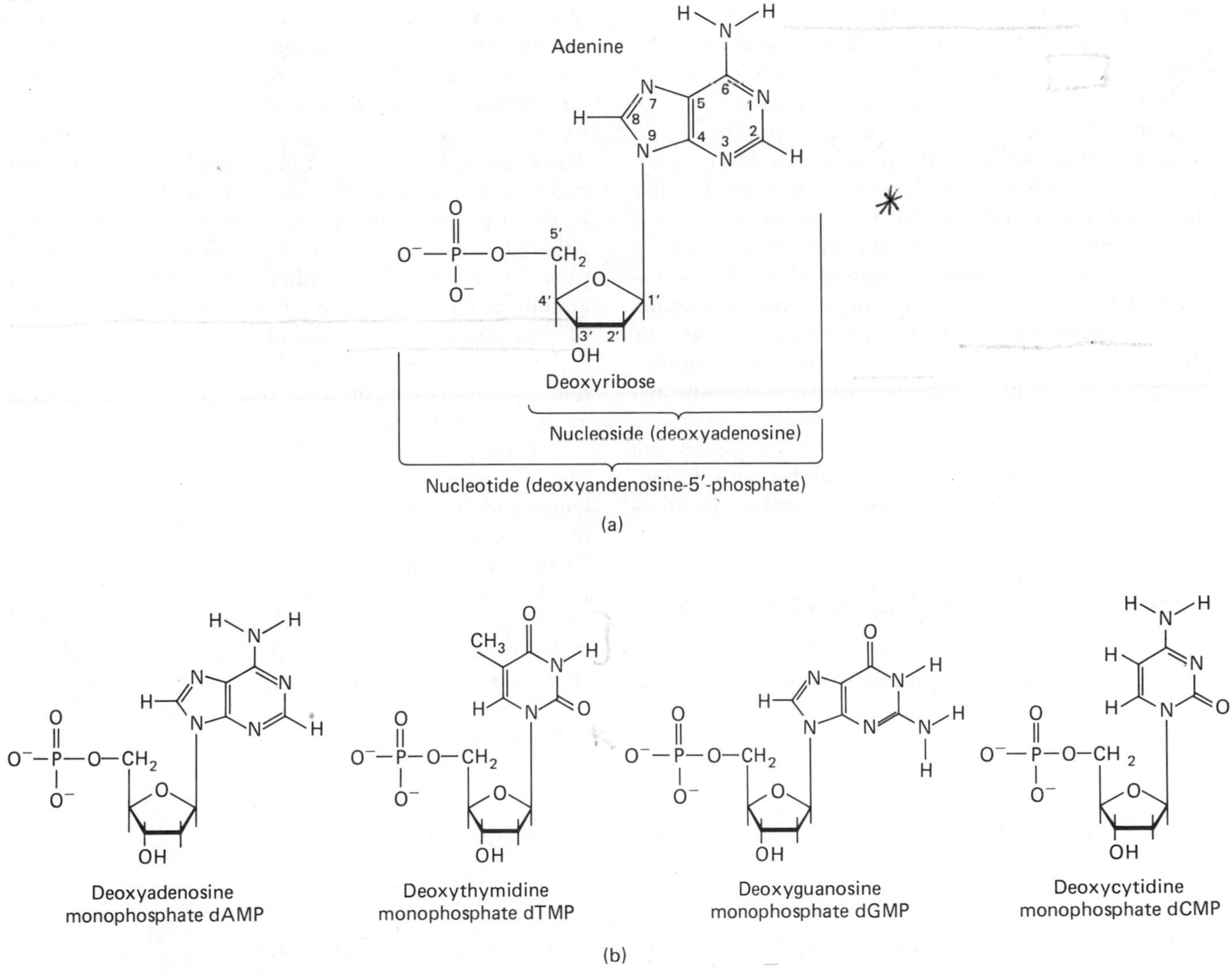

Adenine

Deoxyribose

Nucleoside (deoxyadenosine)

Nucleotide (deoxyandenosine-5'-phosphate)

(a)

Deoxyadenosine monophosphate dAMP

Deoxythymidine monophosphate dTMP

Deoxyguanosine monophosphate dGMP

Deoxycytidine monophosphate dCMP

(b)

FIGURE **10-3**
Monomers Found in DNA. (a) The relationship between the structure of a nucleoside and the corresponding nucleotide. (b) The four nucleotides of DNA. The bases are attached to the 1'-carbon atom of the ribose, and the phosphate is attached to the 5'-carbon. The nucleotides of RNA have the same structure, except that ribose rather than deoxyribose is the carbohydrate part of all the nucleotides, and uracil replaces thymine.

with specific nucleases A. R. Todd and his co-workers found that the phosphates are linked to both the 3' and 5' carbons of ribose. In DNA, where there is no possibility of linkage with the 2'-carbon (since the oxygen is missing), one could predict on purely structural grounds that DNA must be a *linear unbranched polymer*. This was found to be indeed the case when it became possible to observe DNA directly under the electron microscope (see Figure 10–22). The unbranched character of RNA was later established through ingenious enzymatic and chemical studies and electron microscopy. Masses of subsequent work have confirmed the notion that DNA and RNA of all biological systems are linear polymers in which successive nucleoside residues are joined to one another with *3', 5'-phosphodiester linkages.*

Figure 10–4 shows that the DNA and RNA polymers are acids; one phosphoric acid proton is still available for dissociation. In some viruses and in procaryotic cells the negative charges on the DNA molecule—resulting from the dissociation of the phosphate proton—are counterbalanced by *divalent cations* like Mg^{2+} and *basic polyamines*. In other viruses and in the chromosomes of eucaryotic cells the *amino groups* of proteins act as the *counterions,* bringing about the formation of even larger and more complex macromolecules called **nucleoproteins.** The chromosomes of eucaryotic cells, for instance, contain *histones*, which, because they are especially rich in *basic amino acids* (such as lysine and arginine), are well suited to act as counterions to the DNA of the chromosomes. (It was such a nucleoprotein complex that Miescher dissociated with his salt solution, thus permitting the DNA to be precipitated by acidification.)

Indications of Biological Function of the Nucleic Acids

While these painstaking investigations on the organic chemistry of the nucleic acids were in progress, several important developments helped clarify their biological function. These developments were based on a number of technical achievements in the 1930s and early 1940s, such as methods for separating DNA from RNA and for purifying nucleic acids. Of equal importance were the following technical discoveries: (1) DNA, after hydrolysis, reacts with Schiff's reagent to give a brilliant purple color (the Feulgen reaction used to stain nuclei for microscopic observation); (2) both DNA and RNA react with basic dyes; and (3) they absorb ultraviolet light with great intensity in the 240–280-nm region, the absorption maximum being at about 260 nm (Figure 10–5).

A large number of ingenious methods based on these discoveries were developed to localize the DNA and RNA in specific regions of the cell and to follow their relative abundance during various phases of cell growth and development. For example, cytochemists such as T. Caspersson and J. Brachet in the 1940s observed numerous instances of correlation between the abundance of RNA in cells and their protein-synthesizing activity. These classical studies have been supplemented by autoradiographic methods. We now know that (1) DNA is localized primarily in the chromosomes of the nucleus, although DNA is also found in mitochondria, chloroplasts, and possibly in other cell organelles; (2) the amount of DNA per haploid nucleus in any given species of organism is constant; and (3) the bulk of the RNA is found in the cytoplasm although some is found in the nucleus—primarily in the nucleolus, with smaller amounts in the chromosomes and even in the nuclear sap.

These generalizations, based on studies with eucaryotic cells, might in themselves not have been so enlightening had it not been for the work of a number of microbial geneticists and microbiologists. By the early 1950s several dramatic experiments with microorganisms firmly established two important generalizations: (1) DNA (or, rarely, RNA) is the hereditary material, and (2) RNA is in some way involved in protein synthesis. (Evidence for the missing connection between these two important generalizations—that is, that DNA is involved in RNA synthesis—became available only a decade later.)

$$\overset{\oplus}{H_3N}—CH_2—CH_2—CH_2—CH_2—\overset{\oplus}{NH_3}$$

Putrescine

$$\overset{\oplus}{H_3N}—CH_2—CH_2—CH_2—CH_2—\overset{\oplus}{NH_2}—CH_2—CH_2—CH_2—CH_2—\overset{\oplus}{NH_3}$$

Spermidine

$$\overset{\oplus}{H_3N}—CH_2—CH_2—CH_2—\overset{\oplus}{NH_2}—CH_2—CH_2—CH_2—CH_2—\overset{\oplus}{NH_2}—CH_2—CH_2—CH_2—\overset{\oplus}{NH_3}$$

Spermine

Basic (positively charged) polyamines

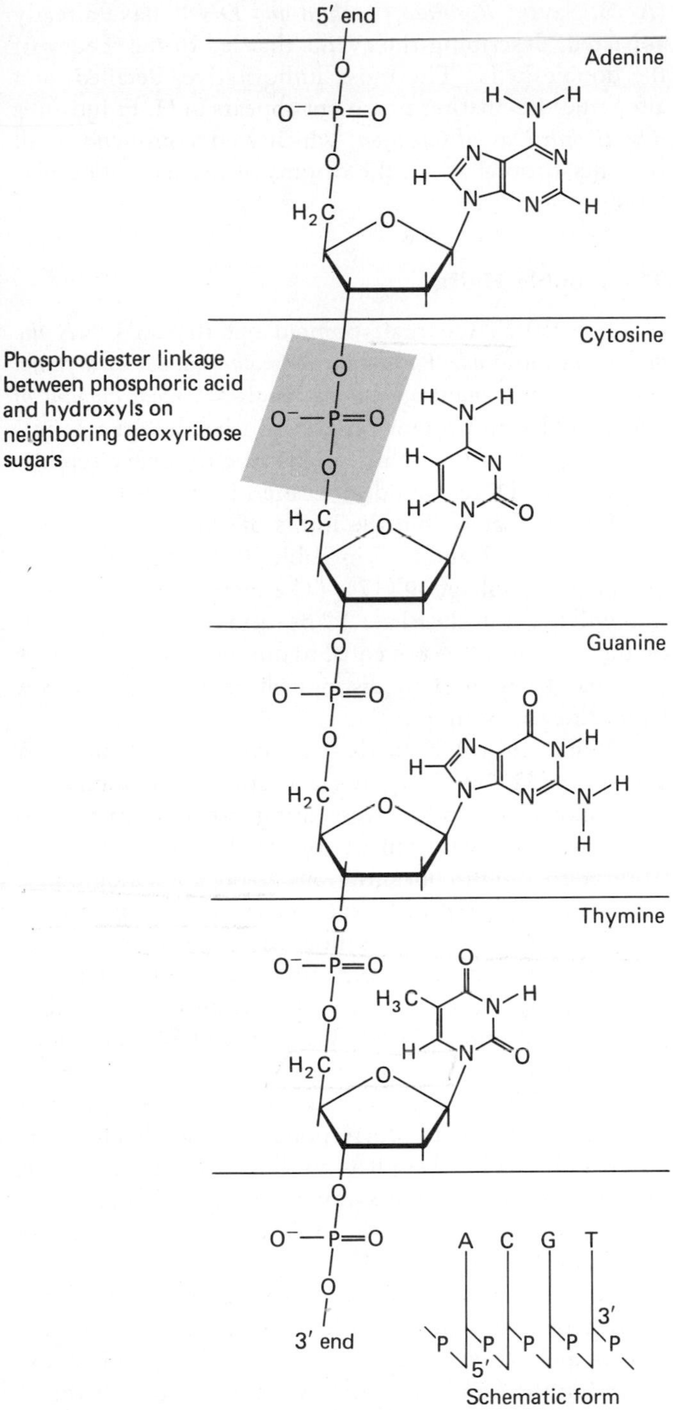

5' end

Adenine

Phosphodiester linkage
between phosphoric acid
and hydroxyls on
neighboring deoxyribose
sugars

Cytosine

Guanine

Thymine

3' end

A C G T

5'

3'

Schematic form

FIGURE **10–4**
The Linear Structure of DNA. A segment of a DNA
polynucleotide, showing how the phosphates form
3'-5'-phosphodiester bonds between the deoxyribose
nucleosides. The course of the deoxyribose-phosphate
backbone is also shown. The RNA polynucleotide chain has
the same structure except that ribose and uracil are present in
place of deoxyribose and thymine. Two alternative schematic
forms are illustrated at the bottom. These represent the same
DNA segment with simple symbols.

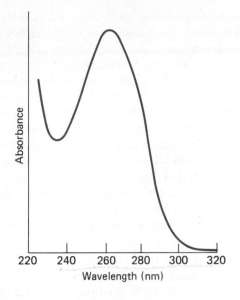

FIGURE **10–5**
Absorption Spectrum of Nucleic Acids. There is a
maximum absorbance near 260 nm. Proteins exhibit a lower
absorbance than nucleic acids in the 260-nm region, and a
peak of absorption at 280 nm. As a result, protein
contamination of a nucleic acid preparation makes the
absorption peak broader and shifts it to a longer
wavelength. The proportion of nucleic acids and protein in
a mixture can thus be estimated from the shape of the
absorption spectrum.

The critical experiments included:

- the isolation and purification of many viruses and the
  demonstration that these self-replicating systems
  contain only nucleic acid and protein.
- the work of O. Avery, C. MacLeod, and M.
  McCarty in 1944 on *bacterial transformation*. It was
  noted by F. Griffith in 1928 that a small fraction of
  pneumococcal cells of the R (rough) type, when
  exposed to dead cells of the pathogenic S (smooth)
  type, were transformed to the S type permanently—that is, the change was inherited. Avery,
  MacLeod, and McCarty did painstaking chemistry
  on the various components of the dead bacterial cells
  to see which molecules were responsible for the
  transformation. Their demonstration that the DNA,
  and not the protein, was capable of transforming
  bacterial cells surprised their contemporaries—
  nucleic acids were considered too dull to be involved
  in inheritance. Also of great significance was the
  demonstration that transformation by naked DNA
  actually worked. This technique is still in constant
  use by genetic engineers.
- the "Waring Blender Experiment" by A. Hershey
  and M. Chase in 1952. They utilized the radioactive
  isotope ^{32}P to label the DNA and ^{35}S to label the

protein of the bacterial virus T2. Under appropriately chosen conditions the virus attaches to the host cell but the infection cannot be completed. By careful agitation with the Waring blender, they could largely remove the ^{35}S-labeled protein and separate it from the cells. The ^{32}P-labeled DNA, injected into the cells during the early phase of the infection, remained there. Hershey and Chase then found the infection proceeded almost normally and mature virus, containing both DNA and protein, were produced. This demonstrated that the information necessary to carry the infection through to completion, including the synthesis of new viral protein and DNA, is most likely contained in the DNA of the infecting virus and not in its protein.

- the work on tobacco mosaic virus (TMV) carried out independently by H. Fraenkel-Conrat and by G. Schramm (1957) showed that the RNA of the virus (the protein having been stripped away from it) can infect tobacco plants and produce complete viruses composed of both RNA and protein. Fraenkel-Conrat also showed that in the case of two different strains of TMV, which could be chemically dissociated into RNA and protein, and then reconstituted to switch their respective proteins and nucleic acids, it was the RNA of the "hybrid" virus rather than the protein that upon infection determined the character of the progeny.

These and other experiments convinced biologists by the mid-1950s that DNA is the hereditary material of the cell (with a few exceptions, such as certain viruses in which RNA acts as the hereditary material). This generalization today seems self-evident since it has, in the intervening years, become part of our general scientific knowledge. However, the student need only read the literature of the late 1940s and even the early 1950s to discover how widespread the belief was that genes were in reality proteins. The 60 intervening years had indeed buried Miescher's brilliant intuitions about the nature of the hereditary material!

The Structure of DNA

Probably the single most important structural impetus for the concept that DNA is the hereditary material was provided by J. Watson and F. Crick, who in 1953 proposed a model for the structure of DNA. Although the model as such did not "prove" anything about the function of DNA, the structure did provide a useful conceptual framework for understanding DNA replication and protein synthesis. Interesting accounts, both personal (J. D. Watson, *The Double Helix*), detached (R. Olby, *The Path to the Double Helix*), and controversial

(A. B. Sayre, *Rosalind Franklin and DNA*) have already appeared, describing the events that led to the theory of the double helix. The most authoritative, detailed, and altogether fascinating treatment appears in H. F. Judson's *The Eighth Day of Creation*, which we recommend to all students curious about the origins of modern molecular biology.

The Double Helix

In 1950 E. Chargaff pointed out that in DNA *the molar amount of adenine equals the molar amount of thymine and the molar amount of guanine equals the molar amount of cytosine*. This important observation has become a universal biological law which holds true for every replicative form of DNA found in nature (Table 10–1).

Notice that within the limits of experimental error the ratios of A/T and G/C in Table 10–1 are equal to one, except for coliphage ΦX174. (The reason for this exception will become clear later.) The ratio (A + T)/(G + C), on the other hand, is not equal to one and the value of this ratio has been used to distinguish, crudely, the DNA from different sources.

While the work on the base composition of DNA was proceeding in Chargaff's laboratory, R. Franklin and M. Wilkins in London were using x-ray diffraction to obtain precise measurements of DNA pulled into fibers from gels of high concentration. X-ray crystallography, which we described in Chapter 4, is a powerful method for analyzing fine macromolecular structure. Earlier work by Astbury on DNA fibers showed a 0.34-nm "repeat"; that is, a regularity in the molecule that has a period of 0.34 nm. To this Franklin and Wilkins added the discovery of a 3.4-nm repeat, which was intriguing because nothing in the chemical structure of the polynucleotide appeared to correspond to this length—the distance from one phosphate to the next, for instance, is only about 0.7 nm. The x-ray diagrams of Franklin and Wilkins also showed that the molecule is a very long, thin rod about 2 nm in diameter.

An important question at that time concerned the number of chains in the molecule; there seemed to be a general inclination both in L. Pauling's group in California and in the London group favoring the number three. It was also important to decide whether the bases pointed toward the outside or toward each other in the center of the molecule. Pauling suggested that they stick out, but Franklin felt she had evidence that the phosphates are outside and that the bases consequently are in the center.

This was the situation in 1951 when J. Watson, a 23-year-old American postdoctoral fellow, arrived in Cambridge, England, and met F. Crick, a physicist working for his Ph.D. in biophysics. Although ostensibly they were meant to work on other problems, they

TABLE **10-1**

Base Composition of DNA from Various Sources

Species	Source	A	G	C	T	$\dfrac{A + T}{G + C}$	$\dfrac{A}{T}$	$\dfrac{G}{C}$
Calf	Thymus	29.0	21.2	21.2	28.5	1.35	1.01	1.00
Hen	Erythrocyte	28.8	20.5	21.5	29.2	1.39	0.99	0.95
Salmon	Sperm	29.7	20.8	20.4	29.1	1.42	1.02	1.02
Marine crab	All tissues	47.3	2.7	2.7	47.3	8.75	1.00	1.00
Euglena gracilis	Cell DNA	22.6	27.7	25.8	24.4	0.88	0.93	1.07
Euglena gracilis	Chloroplast	38.2	12.3	11.3	38.1	3.23	1.00	1.09
Escherichia coli W		24.7	26.0	25.7	23.6	0.93	1.04	1.01
Coliphage T2		32.5	18.2	16.8	32.5	1.88	1.00	1.12
Coliphage ΦX174	Replicative form	26.3	22.3	22.3	26.4	1.18	1.00	1.00
Coliphage ΦX174	Isolated virus	24.6	24.1	18.5	32.7	1.34	0.75	1.30

decided to collaborate on the study of DNA. Crick and Watson had two things in common: a deep interest in DNA, which they were convinced was the genetic material, and a great respect for Pauling and his approach to the study of macromolecules.

Pauling's approach combined a detailed knowledge of the stereochemistry of the relevant small molecules with the building of precise atomic models of these small molecules. This enabled him and his associates to build atomic models of macromolecules that were consistent with the laws of stereochemistry. These models turned out, in Pauling's hands, to be more than mere speculation for, by following certain rules—keeping atoms separated by at least their van der Waals radii, maximizing hydrogen bonding, etc.—the building of a model became a stereochemical experiment in space, often yielding as much information as a direct chemical experiment.

It turned out that Watson and Crick had more to offer than simply combining Pauling's approach with Franklin's data. Crick was fascinated by helices and together with W. Cochran and V. Vand worked out a theory regarding the scattering of x-rays by helical structures. According to this theory, helical structures produce a cross pattern (Figure 10-6). To the great excitement of Watson and Crick such a pattern actually appeared in Franklin's most recent x-ray diagram (which, though unpublished, had been briefly shown by Wilkins to Watson). Another contribution, this time by Watson, was that of _twoness_. As a biologist Watson knew that things related to sex come in twos. Whether he had Chargaff's data regarding base equality in mind or

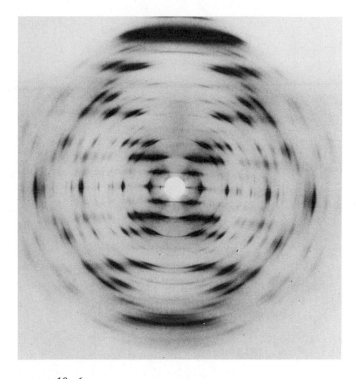

FIGURE **10-6**

X-ray Diffraction Pattern from a Crystalline Fiber of the Lithium Salt of DNA. The cross pattern of prominent spots (reflections), going from 11 o'clock to 5 o'clock and from 1 o'clock to 7 o'clock, provided Watson and Crick with evidence for the helical arrangement of DNA. The elongated pattern at the very top and bottom of the pattern provide evidence for the regular stacking of purine and pyrimidine bases at 0.334 nm separation, perpendicular to the axis of the DNA molecule. (Courtesy of M. H. F. Wilkins.)

whether Chargaff's important information was fitted into place when it was all over, we may never know. In any case, it seems clear that Watson preferred to build a model with two strands twisting around each other and with the bases pointing toward the center.

But even though Watson now had an almost complete model, not much happened until he was reminded by J. Donohue that the DNA bases, some of which had been shown in their enol form in major textbooks, really occur in their keto form (Figures 10–1 and 10–7). At this point Watson cut out models of the keto forms from cardboard and pushed them around each other on a flat surface; the solution to the structure of the molecule "fell out." Watson discovered that it was possible to form hydrogen bonds between A and T and between G and C in such a manner that *the two base pairs had very similar external dimensions* (Figure 10–7). If such **base pairing** leads to similar dimensions, then any sequence of bases can be paired with an appropriately hydrogen-bonded sequence to form a regular molecular

structure. This single structure, therefore, can contain information of all the types necessary to explain genetics. Thus, all at once Watson laid bare the principle which explains the stereochemical stabilization of the double helix as well as gene duplication.

Crick had already pointed out that some of Franklin's x-ray data required the strands to run in opposite directions (Figure 10–8), a feature also determined by the base pairing. He then calculated that many other properties of the structure were consistent with Franklin's x-ray photographs. Thus, the model confirmed Franklin's suspicion that the 0.34-nm repeat (now known to be 0.334 nm) represents the stacking of bases along the helical axis, and the 3.4-nm repeat (now known to be 3.34 nm) represents the distance along the axis required for a single complete twist of the double helix (Figure 10–9).

The model of DNA that Watson and Crick built was examined with understandable skepticism by x-ray

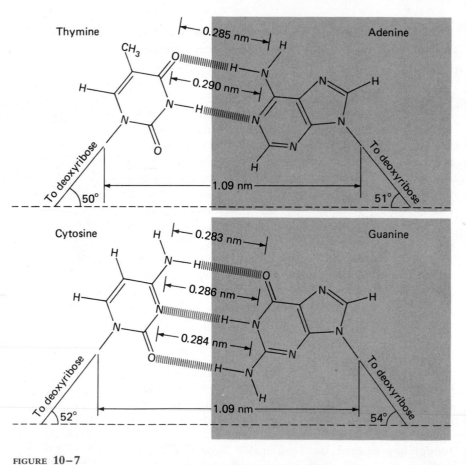

FIGURE **10–7**

Hydrogen Bonding between Adenine-thymine and Guanine-cytosine Bases in DNA.   The A–T base pair has two hydrogen bonds and the G–C base pair has three. The two pairs are very similar in external dimensions, allowing the formation of a smooth helical structure in DNA.

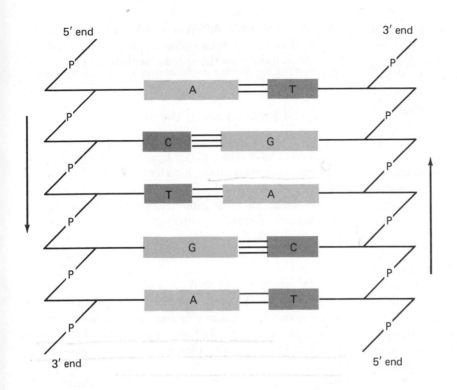

FIGURE **10–8**
Diagram of an Untwisted Portion of the DNA Double Helix. The sugar-phosphate backbones have opposite polarity. The diagram also shows how the similarity in the dimensions of the A–T and G–C base pairs ensures that the sequence of the bases does not place any strain on the structure.

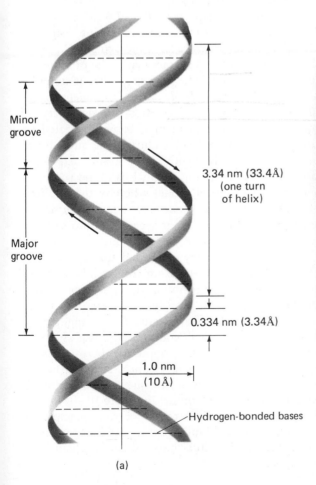

(a)

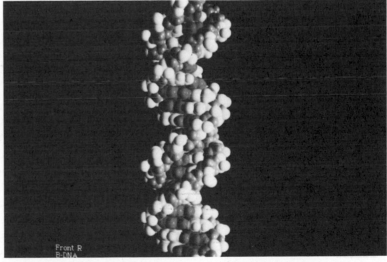

FIGURE **10–9**
The Structure of DNA. (a) The double helix. Open model of the DNA double helix showing hydrogen bonding of A–T and G–C base pairs at the center of the molecule, the dimensions of a turn of the helix (3.34 nm), and the distance between bases (0.334 nm). (b) Space-filling atomic model of the DNA molecule. The interior of the molecule is filled by the bases. Note the presence of a narrow ("minor") and a wide ("major") groove in the helix. (Courtesy of R. Feldmann.) (See Appendix for a reprint of the *Nature* paper by Watson and Crick.)

crystallographers and structural organic chemists. But it was soon apparent to all concerned that the "essential features" of DNA structure had been discovered. And so, the April 25, 1953, issue of the British magazine *Nature* carried a brief article entitled "A Structure for Deoxyribose Nucleic Acid" that began: "We wish to suggest a structure for the salt of deoxyribose nucleic acid (D.N.A.). This structure has novel features which are of considerable biological interest" (see Appendix).

We have indulged in this long historical description not only because of the enormous intrinsic importance of the events, but also because it dramatically illustrates two major features of the process of scientific discovery:

- The first has to do with the interdependence of the scientific community. It shows how much a theory is dependent on the work and thought of a large group of scholars, both past and present. X-ray crystallography from Bragg to Crick, the organic chemistry of DNA from Miescher to Todd, and the identification of the biological role of DNA from Griffith to Watson—all these thoughts and activities formed the background of the Watson-Crick theory.

- The other has to do with recognition of the importance of the relationship between structure and function in biological macromolecules. Here we must give full credit to Watson and Crick, who addressed the problem of DNA structure primarily because of their conviction that it must be related to DNA's biological function. It is in highlighting this relationship that the Watson-Crick theory had its greatest impact and rightly became the keystone of a new biology.

Figure 10–9 shows the double helix of DNA in detail; the molecule is indeed 2 nm in diameter with two polynucleotide chains twisted around each other, phosphate backbone on the outside, bases on the inside. The model also shows that the chains twist around each other to create a *major groove* and a *minor groove*. Note also that the strands run in opposite directions. The convention has been established of writing the order of bases of a single nucleic acid chain by starting with the *5′ phosphate on the left*.

One of the most important aspects of the model is that it allows tight *stacking* of the bases. The space-filling model (Figure 10–9b) demonstrates clearly the internal compactness of the double helix. The stacking of these hydrophobic residues away from the aqueous solvent contributes a substantial fraction of the energy that stabilizes the structure. Another consequence of the similarity in dimensions of the complementary base pairs is that the structure places *no restrictions on the sequence* in which the base pairs follow each other. Thus, it is possible to construct different DNA molecules that are almost identical in their general architecture and are distinguished only by the specific sequence of their base pairs.

Biological Significance of the Structure of DNA

It was suggested long ago that the genetic material of the cell must have two separate functions: self-duplication and encoding information for elements of cell structure and function. As a result of work in biochemical genetics, it is now clear that most genes encode information for the manufacture of specific proteins. *DNA must therefore be capable of providing the necessary information both for its own duplication and for the synthesis of proteins.* Let us take each in turn.

The structure of DNA, as realized by Watson and Crick, provides a convenient scheme whereby a molecule with a particular sequence of base pairs could be duplicated. Thus, each strand of the molecule could determine the laying down of a "complementary strand," resulting in the formation of two identical molecules. The significant property of the hydrogen bond formation between the complementary bases A=T and G≡C is that these interactions are highly specific. A sequence of 5′-ATGC-3′ in one strand implies a sequence of 3′-TACG-5′ in the complementary strand. Before this important insight it was difficult to imagine how a large molecule could serve as a biological *template* for the synthesis of another molecule just like itself. The answer is that it doesn't; instead, *each half of the molecule acts as a template for the synthesis of its complementary half.* As synthesis proceeds, the template and its complement become associated to form the new DNA molecule. During the subsequent replication, each newly synthesized strand acts as a template for a strand like the original half, and so on.

A precise geometrical model whereby the duplication of a DNA molecule could occur in a continuous manner, without requiring prior separation into isolated single strands, was proposed by C. Levinthal and H. R. Crane (Figure 10–10). They suggested that duplication could begin at one end, separating the strands, and the free energy released in forming the polymer would provide the energy for the rotation of the two lengthening daughter strands as well as the shortening parent strand. They concluded that enough energy was available to overcome the viscous drag opposing these rotations and that there was enough mechanical strength in the helix to withstand the necessary torque without seriously stretching the bonds. (The model in Figure 10–10 should be taken simply as an illustration of the principle of complementarity in the duplication of DNA. As we shall see in Chapter 11, the actual biosynthetic mechanism is more complicated.)

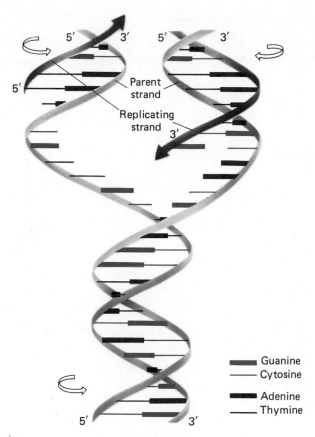

FIGURE 10–10

Model of DNA near the growing point of replication. As the new strands are laid down, the parent strands unwind. The model shows complementary base pairing between the parent and daughter strands. The enzyme structures and primers involved in replication are not shown.

If this picture of DNA duplication involving pushing apart the strands of the original molecule by the laying down of complementary strands is correct, then it should be possible to obtain direct evidence for the existence of Y-shaped growing points in duplicating DNA. And, indeed, many electron micrographic studies of procaryotic and eucaryotic DNA have demonstrated the presence of Y-shaped growing points (Figure 10–11). However, to everyone's surprise, it was found that growing points come in pairs because the growth is usually initiated in the middle of a DNA duplex and moves in both directions at once. This *bidirectionality of DNA replication* was established both by genetic and autoradiographic methods. If one transfers bacteria from a medium containing a low concentration of radioactive tritiated (^{3}H) thymine to a medium of high concentration, one observes that a stretch of DNA showing a low density of grains is always bracketed by *two* stretches, of equal length, of high grain density (Figure 10–12).

The second biological function of DNA is the transmission of information eventually to be used for the synthesis of proteins. The linear array of the four bases can be thought of as "information" based on a language of *4 letters* (bases), which finally becomes translated into a protein language of *20 letters* (amino acids). Thus, according to Crick, if one were to imagine the pairs of bases as analogous to the dots, dashes, and spaces of the Morse code, there is enough DNA in the human cell to encode 1000 large textbooks. We suggested in Chapter 1 that a signal characteristic of the living machine is its capability for microminiaturization. Here, indeed, is a striking example of this cellular property: the ability to store all the hereditary information responsible for the development of a complex individual in 10^{-18} kg of material. It is even more amazing to realize that the hereditary material which codes for *all* the inherited characteristics of *all* people alive today would weigh much less than 1 mg, and would easily fit onto the head of a pin. The explanation for this remarkable phenomenon of microminiaturization is that the "bits" of information of the language of the nucleic acids are the nucleotide residues—tiny pieces of a molecule—which are many times smaller than anything humans have so far invented for storing information.

Subsequent Evidence Supporting the Double Helix Hypothesis

In the years since 1953 a great deal of evidence accumulated that helped establish the Watson-Crick hypothesis on the structure of DNA as one of the most useful and secure generalizations in cell biology. Several very distinct lines of evidence are summarized here.

X-ray diffraction has certain limitations. Deriving a three-dimensional structure by pure deduction from the data obtained from a single crystal is usually difficult. For a large molecule it is in fact impossible. Moreover, while the DNA fiber used in the original experiments by Franklin was oriented along one axis, it did not form a three-dimensional crystal. Luckily, once an experimenter has a hypothesis about the three-dimensional structure of a molecule, he or she can compute the diffraction pattern such a structure would produce and then determine whether these predictions agree with the experimental results. This is what happened with DNA. Once there was a structural hypothesis, it was possible to predict certain x-ray scattering properties based on spacings such as 2 nm (diameter of the double helix), 0.334 nm (distance between bases), and 3.34 nm (distance for a complete turn of the double helix). These spacings were confirmed in the x-ray diagrams and so were a number of others, all of which corroborated the double helix hypothesis. In fact, x-ray studies by Wilkins and his co-workers even refined the original DNA model by showing that the planes of the bases were not exactly perpendicular to the helix axis.

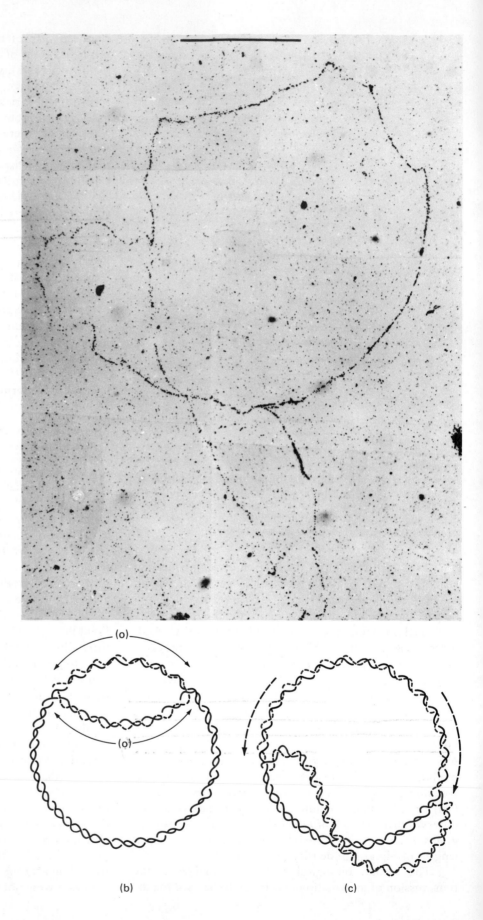

FIGURE **10–11**

The presence of Y- shaped growing points in replicating DNA.

(a) Autoradiograph showing replicating circular chromosome of *E. coli*. This experiment demonstrates the presence of two Y-shaped growing points. The DNA is rendered visible by growing the cells in tritium (^{3}H)-labelled thymidine. When a photographic emulsion is placed against the radioactive DNA, the radiation produces blacked grains which are seen after development of the emulsion. Bar = 100 μm. (b) Model of circular chromosome going through bidirectional replication; growing forks move in opposite directions (arrows) from the origin (O). The two complementary strands are shown in black and color, and newly replicated strands are shown dashed. This replication intermediate resembles the Greek letter Θ (theta) and is called a Θ form. (c) A model of a Θ form in an alternative appearance. In this case, each growing point has progressed a bit further down the circle than in (b), shown by arrows; and during preparation for autoradiography one of the replicated arms has accidentally flipped over the unreplicated portion of the circle. Note the close resemblance of this model to the actual radioautogram shown in (a).

Direct visualization and precise measurements of DNA with the electron microscope have also confirmed the dimensions predicted by the double helix model. If one knows the molecular weight of a certain DNA molecule and assumes the structure of the model, one can predict the length of the molecule. In a number of studies the measured and predicted lengths agreed extremely well. This result also ruled out fixation artifacts in the microscopy, allowing many subsequent, accurate estimates of the DNA length.

Numerous physical-chemical studies of the properties of DNA in solution, such as light scattering, viscosity, sedimentation, and diffusion have yielded results consistent with the predictions of the model, and inconsistent with alternatives. DNA is a somewhat stiff, unbranched rod and not a highly flexible polymer, due to the lower free energy that results from stacking together the hydrogen-bonded base pairs. To fold this kind of molecule into a compact chromosome requires special structural features—for example, the histone proteins (see discussion in Chapter 11).

The properties of a flexible polymer are, however, assumed by DNA when it is "denatured" by elevated temperatures. P. Doty and his co-workers found that when solutions of DNA are heated, there is a marked decrease in viscosity that occurs over a narrow temperature range (Figure 10–13). This phenomenon, reminiscent of the sharp melting of a crystal, is due to the cooperative breaking of the hydrogen bonds and hydrophobic interactions holding the complementary strands together, thus allowing each strand to separate and form a flexible, more compact molecule, which leads to a much lower viscosity. This simple method of heating has been often used to separate the two strands of DNA, preparatory to isolating them for further study.

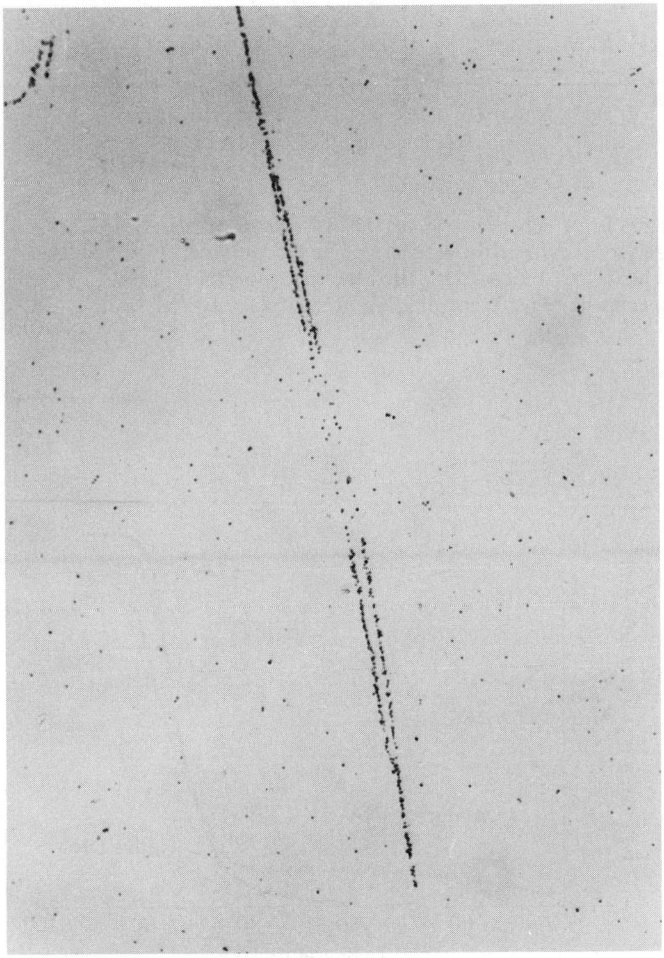

FIGURE **10–12**
Proof that DNA replicates bidirectionally. *E. coli* cells were labeled with ^{3}H-thymine for several minutes after the beginning of the DNA replication cycle, and then for a further 8 minutes with ^{3}H-thymidine at ten times higher specific radioactivity. DNA was extracted, spread on slides, and covered with a photographic emulsion. The emulsions were developed and photographed after prolonged exposure to the radioactive DNA. There are *two* regions of high grain density (corresponding to incorporation of the high specific activity thymidine) at either end of the region with low grain density (corresponding to incorporation of the low specific activity thymine). The two replicated strands in both regions are sufficiently far apart to be distinguishable. The length of the entire grain pattern is 370 μm. (Courtesy of P. Kuempel.)

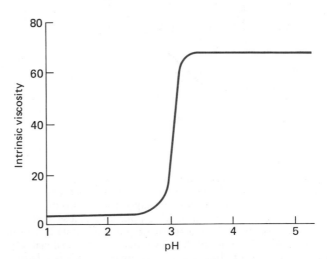

FIGURE **10–13**
Melting of DNA. The change in viscosity of DNA over a narrow pH range is explained by the cooperative nature of the structural transition. The stiff double helix collapses over a narrow pH range because each separation of base pairs facilitates the breaking of the remaining base pairs. (From P. Doty and C. Thomas.)

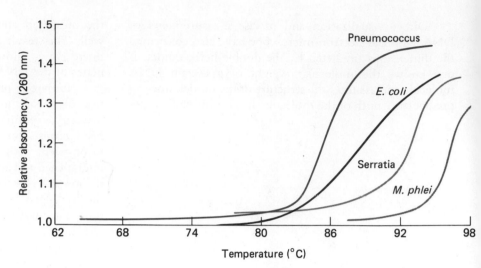

<small>FIGURE</small> **10–14**

The Increase in Ultraviolet Absorption of DNA Over a Narrow Temperature Range (hyperchromic effect). The temperature at which half the change in optical absorbance occurs is called the melting point (T_m). The DNAs of different species have different melting points. (From J. Marmur and P. Doty.)

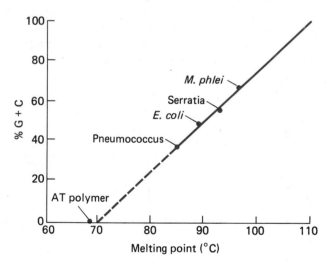

<small>FIGURE</small> **10–15**

Relationship between Melting Point and (G + C) Content of DNA Preparations From a Number of Bacteria. The curve can be extrapolated to zero (G + C) content, which predicts a melting point quite close to that observed for poly (AT), a synthetic polynucleotide consisting only of A and T nucleotides. The increased thermal stability conferred by G–C base pairs, as compared with A–T base pairs, reflects the presence of three hydrogen bonds in the former as compared with two hydrogen bonds in the latter.

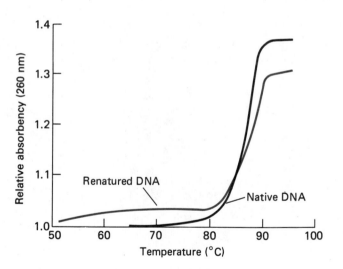

<small>FIGURE</small> **10–16**

Melting Curves of Native DNA and Annealing of Renatured DNAs. The latter is obtained by slow cooling (annealing) DNA that was denatured by heating above the melting point. The fact that the two curves are not identical shows that some of the DNA does not return to its native state. The residual denatured DNA can be removed by treatment with an enzyme (phosphodiesterase) which does not hydrolyze native DNA, after which one obtains a renatured DNA preparation with properties just like the native material. (From J. Marmur, C. L. Schildkraut, P. Doty: Biological and Physical Chemical Aspects of Reversible Denaturation of Deoxyribonucleic Acids. In *The Molecular Basis of Neoplasia,* University of Texas Press, Austin, Texas, 1962.)

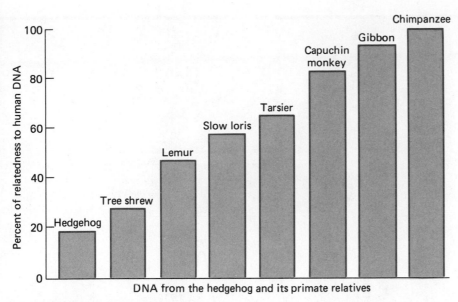

FIGURE 10–17
The use of DNA hybridization to study phylogenetic relationships.
Different species in the Primate order are compared. The technique employed in
these experiments was to take human DNA—melted, fast-cooled, and immobilized
in agar—and to incubate it with both ^{14}C-labeled fragments of melted human
DNA and unlabeled fragments of DNA from another species. The fragments of
labeled human DNA become immobilized because they hybridize with the DNA
trapped in the agar, an effect which is reduced if related DNA from a different
species is also present. Interestingly, this method cannot distinguish between
human and chimpanzee DNA; the latter inhibits hybridization between the DNA-
agar and labeled human DNA fragments as efficiently as does unlabeled human
DNA fragments. The degree of relatedness of the other species in this diagram
parallels the relatedness established on morphological grounds. The hedgehog,
which shows the least relatedness to humans, is not a primate but is thought to be
derived from a more generalized insectivore that is ancestral to the primates.
(Redrawn from Hoyer and Roberts.)

The decrease in viscosity of heated DNA solutions is
correlated with other properties such as an increase in
ultraviolet absorption and a decrease in optical rotation.
The change in ultraviolet absorption (called the hyperchromic effect) is the most convenient property to
measure (Figure 10–14). The abruptness of the change in
these properties of DNA is explained by the cooperative
nature of the effect. By this we mean that when some
secondary bonds are broken, the bonds in neighboring
bases are put under more strain; these bonds then break,
leading to a progressively rapid "melting" in large
regions of the molecule. Similar transitions occur when
solutions of DNA are exposed to pH extremes, certain
nonaqueous solvents, or a number of ionic species.

Doty and his co-workers noted that the melting
point of DNAs from different sources varies according to
their G + C content (Figure 10–15). This can be
explained both by the presence of three hydrogen bonds
between G and C (as opposed to only two between A and
T) and by the stronger stacking interactions between base
pairs involving G and C.

By very slow cooling solutions of "melted" DNA,
it is possible to *anneal* it; that is, to reestablish the
double-helical structure (Figure 10–16). When the DNAs
of two closely related species are mixed and melted, the
double-helical content after slow cooling is high, whereas
when the DNAs of unrelated species are used, the
double-helical content is low. The explanation of these
observations, of course, is that the greater the similarity
of sequence, the more readily do complementary strands
form "hybrid" double-stranded molecules. Such DNA
hybridization experiments have become a useful tool to
study taxonomic relationships between different species
of organisms (Figure 10–17).

Not only is the melting point of DNA related to its
G + C content, but so also is its density, a fact which
again points to the tightness of the G≡C bonding and the
stacking interactions of adjoining G≡C base pairs.

An ingenious and widely used technique, first
devised by J. Vinograd, allows one to measure the
density of DNA and to separate from one another DNA
molecules that differ in density (see box).

Equilibrium Density Gradient Centrifugation

When a concentrated solution of a dense salt, cesium chloride (CsCl) for example, is spun at high speed in an ultracentrifuge, its molecules will redistribute themselves. Like the molecules in our atmosphere, they will come to an *equilibrium* in which the tendency of the molecules to be sedimented away from the axis of rotation is balanced by their tendency to equilibrate, by diffusion, throughout the tube. The result is a smoothly changing concentration of CsCl, lower where closer to the axis of rotation and higher where farther away (the "bottom" of the tube).[2]

If DNA molecules are also present in the centrifuge tube, they will come to equilibrium by floating to certain regions of the tube where the density of the DNA molecules will be identical to the density of the CsCl solution (Figure 10–A). The width of the "band" thus formed is determined by thermal diffusion of the DNA molecules. The result is a bell-shaped distribution of concentration of molecules about the point where their buoyant density is equal

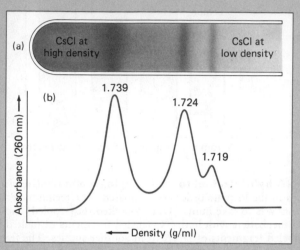

FIGURE 10–A
The "Banding" of Three Types of DNA in a Density Gradient of Cesium Chloride. The gradient was established by centrifuging the material at 354,000 rpm for 20 hours. The high centrifugal field causes a redistribution of the CsCl, which brings about the banding of the DNA molecules. (a) Diagram of the appearance of the centrifuge tube. (b) Absorbance measurements obtained from fractions removed sequentially from the bottom of the centrifuge tube. (Another way of obtaining such curves is to photograph the bands while the centrifuge rotor is still in motion and to perform densitometer tracings of the photographs.) The density of the CsCl solution at the center of the bands is given in g/ml.

[2]The equilibrium in the spinning tube distinguishes this method from the other density gradient centrifugations discussed in Chapter 4. When a density gradient of sucrose, for example, is preformed in a centrifuge tube, it is not at equilibrium, even when the tube is being spun at the rotational velocities used in such experiments. The sucrose density gradient will eventually disappear as diffusion equilibrates the sucrose concentration. Luckily, this takes much longer than the time required to separate the organelles or molecules that are floating in the gradient.

to that of the salt solution. When a mixture of different DNA molecules is present in the tube, bands will generally form in different positions. The exact position of a band can be determined in a variety of ways. In an analytical ultracentrifuge, it is possible for an optical system to make visible the rotating sample and to determine where the bands occur. More often, however, a preparative ultracentrifuge is used. After equilibrium is reached, the rotor is gently stopped, the tube is removed from

the centrifuge, and a small hole is punched into the bottom of the tube. As the tube slowly empties from the bottom, fractions are collected and transferred sequentially into a series of containers. The DNA in each fraction can be measured spectrophotometrically by the absorbance at 260 nm, and the density of each fraction can be assayed from the refractive index, which changes with the CsCl concentration. The results are usually presented graphically (Figure 10–A).

It is possible to estimate the G + C percent from the buoyant density of a DNA sample, and this value agrees reasonably well with the value obtained from direct chemical analysis. The **equilibrium density gradient centrifugation** method has also been used to separate helical (native) and nonhelical (e.g., heat-denatured) DNA. Native DNA is less dense because the double-stranded structure excludes more water than denatured DNA. Figure 10–18 shows how the method was used in a study of the denaturation and renaturation (annealing) of DNA. In this particular study, phosphodiesterase from *E. coli* was used to remove the portion of the DNA that remained denatured, demonstrating that it is possible to anneal denatured DNA so that its density distribution

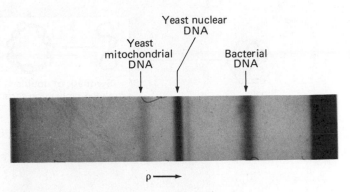

FIGURE **10–19**
Separation of Nuclear and Mitochondrial DNA. Total DNA of yeast was banded in a CsCl density gradient (cf. Figure 10–17). The centrifuge sample cell was photographed with ultraviolet light, revealing the positions of DNA bands as dark stripes of high UV absorption. The density of CsCl increases from left to right. The right-most band, at the position of greatest density, is a bacterial DNA preparation used as a marker. The left-most band, of least density, is yeast mitochondrial DNA. The dark band between these two is yeast nuclear DNA. Just to its right, there is a faint additional band which contains nuclear DNA fragments containing the ribosomal RNA genes; they have a slightly higher (G + C) content, and thus band at a slightly higher density than other nuclear DNA fragments. (Courtesy of E. Sana.)

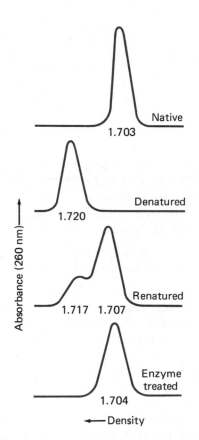

FIGURE **10–18**
Renaturation of DNA. Use of the density gradient centrifugation method to show that a substantial portion of heat-denatured DNA can be renatured to give DNA of the same density as native DNA. The enzyme phospodiesterase was used to hydrolyze residual denatured DNA. Since the density of the denatured DNA was essentially the same as native DNA, this experiment also provided evidence that heat denaturation of DNA is indeed reversible. (From C. L. Schildkraut, J. Marmur, and P. Doty.)

returns almost exactly to that of the original native material. This study exploited the much greater resistance to enzyme degradation of native DNA. Figure 10–19 shows a separation of nuclear and mitochondrial DNAs, illustrating in this instance that the mitochondria of a cell have circular DNA of a different density from that of linear DNA in the nucleus of the same cell.

Apparent exceptions can often help to understand the rule. A number of small bacterial viruses in which the rules A = T and G = C do not hold (see Table 10–1) have been discovered. Physical tests such as melting point and hyperchromicity studies demonstrated that the DNA packaged in these viruses is in a single-stranded form. The interesting question arises if single-stranded DNA, even with the help of enzymes, can replicate itself. What in fact happens is that when the "+" strand infects a bacterium, it acts as a template for the synthesis of a "−" strand, thus forming a conventional double helix structure, the so-called *replicative form*. This double-stranded form can replicate itself to produce more double-helical structures. Eventually, though (and this *is* an exception to the general rule), the replicative form will begin to produce only "+" strands, which, after becoming packaged with protein, emerge from the bacterium as new

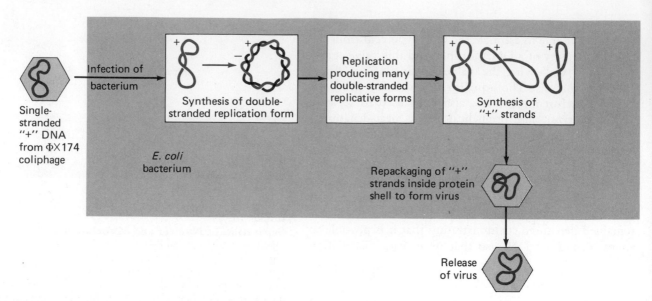

FIGURE 10–20
The Bacterial Virus ΦX174, Which Contains a Single Strand of DNA.
(a) Simplified account of the replication cycle of ΦX174. DNA replication still
occurs in the double-stranded form. More information about ΦX174 replication is
given in Figure 10–23. (Courtesy of A. Kornberg.)

virus particles (Figure 10–20). The requirement for a double-helical structure in the replication of even single-stranded DNA is a dramatic confirmation of the generality of the double-stranded system.

From these data we can conclude that the double-helical model of DNA is one of the most thoroughly validated hypotheses in biology and, as we shall see from the remaining portion of this book, conceptually one of the most useful ones.

There are now three structural forms of DNA known. The one we have discussed thus far is known as the *B form;* it exists in solutions of purified DNA. The *A form,* also a right-handed helix, occurs when DNA fibers are dried to about 75% relative humidity (Figure 10–21a). The A form has also been detected in double-stranded RNA molecules (see Chapter 12), and in single crystals of double-stranded oligonucleotides, such as

$$\text{dGdGdCdCdGdGdCdC}$$
$$\text{dĊdĊdĠdĠdĊdĊdĠdĠ}$$

and the mixed DNA/RNA oligonucleotides

$$\text{G C GdTdAdTdAdCdGdC}$$
$$\text{dĊdĠdĊdȦdṪdȦdṪ Ġ Ċ Ġ}$$

(The x-ray analysis of single crystals has significant advantages over the study of fibers that occurred earlier. The dimensions of the structural forms are easier to measure precisely and the influence of the base sequence

on the tertiary structure is easy to discern—in fibers, sequence effects are averaged out).

The third form of DNA, the *Z form,* is a left-handed helix (Figure 10–21c); its dimensions were measured by x-ray diffraction analysis of single crystals of

$$\text{dCdGdCdGdCdG}$$
$$\text{dĠdĊdĠdĊdĠdĊ}$$

The Z form appears to be favored by repeating dC–dG runs. In fact, the structure is actually formed by a repeating dimer of dC–dG. Methylation of cytosine at the 5′-carbon, a relatively common modification of eucaryotic DNA, also favors transition to the Z form. It has proved possible to raise antibodies against Z form DNA and, by their use, to detect this structure in portions of chromosomes.

It seems likely that different DNA sequences favor different helical forms. It is also possible that differences in the structural form of DNA may aid recognition of sequence information by proteins, without having to separate the strands of the double helix.

Some Overall Properties of DNA Molecules

The purification of homogeneous DNA molecules from bacterial viruses, a difficult feat of biochemistry, was accomplished in the 1950s. This allowed attempts to determine the molecular weight of the DNA in bacte-

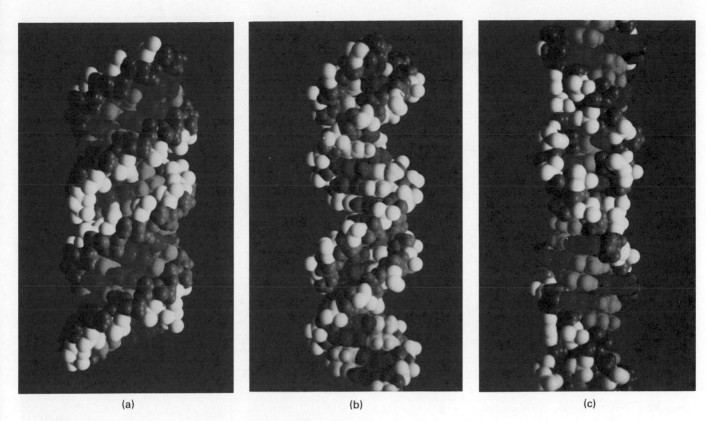

(a)　　　　　　　　　　　　(b)　　　　　　　　　　　　(c)

FIGURE **10–21**

Three structural forms of DNA. The dimensions are based on x-ray diffraction analyses of single crystals of small oligonucleotides. (a) The A form. This is a double-stranded molecule in the form of a right-handed helix. The two strands are antiparallel in their 5′-3′ directions. There are 10.9 base pairs per turn, the base pairs are tilted with respect to the helix axis (13°), and the distance along the axis per base pair is 0.292 nm. The major groove is deeper than in B form DNA and the minor groove is shallower. (b) The B form. (c) The Z form. This is a double-stranded molecule in the form of a left-handed helix. The two strands are antiparallel in their 5′-3′ directions. There are 12.0 base pairs per turn, the base pairs are tilted with respect to the helix axis (8.8°), and the distance along the axis per dC–dG dimer of base pairs is 0.765 nm—0.352 nm from dG to dC, and 0.413 nm from dC to dG. The riboses along a single chain alternate the direction that their oxygens point. This causes the zigzag course of the backbone, which gives the name of Z form to this structure. The minor groove is deep and narrow and the major groove is practically missing. (Courtesy of R. Feldman.)

riophage T2. Different workers obtained different values, ranging from hundreds of thousands to several millions. It was soon discovered that the more carefully one handled the DNA solutions, the higher the molecular weight values one measured. This led finally to the remarkable discovery that very large DNA molecules would easily break into two approximately equal fragments when subjected to the mild shearing forces in flowing liquid. At larger shear forces, the halves would break again, and so on. Therefore, depending on the shear force, one can obtain preparations of different, but reasonably narrowly distributed, molecular weights.

One T2 bacteriophage contains DNA with a mass of 1.2×10^8 Da. Was it possible that all of the DNA in T2 is present in a single giant molecule? It soon turned out that conventional physical-chemical methods for measuring molecular weight are not suited to the handling of such giant forms, but at that point electron microscopists came to the aid of biochemists. J. Cairns worked out an autoradiographic technique. He grew *E. coli* infected with T2 phage in the presence of tritium-labeled thymidine and thus labeled T2 DNA in the thymine residues. He then purified the phage particles and ruptured them delicately by placing them in a drop of distilled water on a copper grid. The grids were then placed against a photographic emulsion sensitive to the beta emissions of tritium, to produce blackening of grains in the emulsion. This autoradiographic method provides a reasonably high-resolution (100-nm) reproduction of the radioactive source. Cairns was able to show that all the DNA of T2 was indeed present in a single giant molecule of 1.2×10^8 molecular weight.

A. K. Kleinschmidt further succeeded in rupturing T2 phages on electron microscope grids, obtaining pictures of the DNA itself. These pictures had such dramatic impact on the biological community that one particular example has been reproduced in dozens of textbooks and has become a symbol of modern molecular biology (Figure 10–22). Looking at the photograph, one is awed by the molecular process that packages this gigantic molecule into the head of the phage particle.

Using the techniques of autoradiography and negative staining in the electron microscope, the DNAs of many viruses have been studied (Table 10–2). Two generalizations have emerged from these experiments:

- Each DNA virus particle usually carries a single DNA molecule.
- The DNA molecule is often circular.

There is some variation in the phenomenon of circularity. In some viruses, such as ΦX174 coliphage (Figure 10–23), the DNA is always a covalently bonded circle. In other viruses such as λ bacteriophage, it is linear inside the virus and circular in the host cell, while in a few other viruses such as T7 bacteriophage, circular forms have not been observed. It may well be that for purposes of intracellular replication in bacteria, DNA must be circular as well as double-stranded. The circles of DNA we frequently see on the grid of the electron microscope (Figure 10–23) are stretched out, open structures, many times larger than the host cell into which they must fit during DNA replication. Increasingly, we find that the DNA inside the cell is present as a *supercoil*, especially when it is in the process of replication.

The supercoiled structure is thought to be a consequence of replicating DNA in the form of a circle held together by covalent bonds. (To visualize this, we suggest a simple experiment with strings. To model a DNA circle, knot together two strings twisted around each other. Now try to separate the strands in a certain portion for purposes of "replication" and you will find that other parts of the "molecule" will form a twisted supercoil.) A single break in the covalent continuity of one of the strands permits the supercoil to relax and form an untwisted circle (Figure 10–24), a process that probably occurs often during DNA replication (see Chapter 11). It should be noted that evidence for circularity of

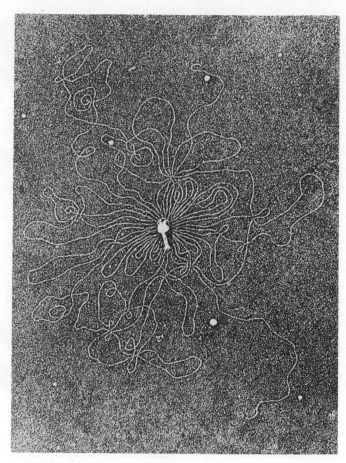

FIGURE 10–22

A single molecule of DNA released from one T2 bacteriophage particle. This molecule has a mass of 1.1×10^8 Da, is 57 nm long, and consists of 1.7×10^5 base pairs. ×81,000. (Courtesy of A. K. Kleinschmidt.)

viral DNA first emerged from genetic studies so that the appearance of circular molecules in the electron microscope did not come entirely as a surprise.

In procaryotic cells, studies of DNA structure have provided us with much the same picture as in viruses, with the exception that the molecules are much larger (Table 10–2). The bacterium *E. coli* has one or more genomes depending on the growth conditions. The amount of DNA in each genome would produce a DNA molecule with a mass of about 2×10^9 Da. Can it be that even this huge package of DNA exists as a single chromosome? Utilizing his autoradiographic technique, Cairns showed that in *E. coli* one really finds a single circular DNA molecule of 2×10^9 Da, its extended length being 1.2 mm! It seems likely that in procaryotes the circularity of DNA is also a widespread phenomenon.

TABLE 10–2

Size Distribution of DNA in a Variety of Biological Systems

Viruses	Molecular Mass (Millions of Da)	Length		Number of Base pairs
ΦX174 bacteriophage (single-stranded)	1.7	0.88	μm	
Polyoma virus	3.1	1.6	μm	4.6×10^3
T3,T7 bacteriophage	25	13	μm	3.8×10^4
Coliphage λ	31	16	μm	4.6×10^4
Herpes simplex virus	100	52	μm	1.5×10^5
T2,T4,T6 bacteriophage	110	57	μm	1.7×10^5
Vaccinia	160	83	μm	2.4×10^5
Procaryotes				
Mycoplasma	500	260	μm	7.5×10^5
E. coli	2,300	1.2	mm	3.5×10^6
Eucaryotes				
Yeast (average per chromosome)	540	280	μm	8.1×10^5
Drosophila (largest chromosome)	43,000	22	mm	6.5×10^7
(total haploid genome)	130,000	6.7	cm	2.0×10^8
Man (smallest chromosome)	33,000	1.7	mm	5.0×10^7
(largest chromosome)	180,000	9.2	cm	2.7×10^8
(total haploid genome)	2,000,000	1	m	3.0×10^9

Note: If the DNA of the haploid complement of human chromosomes was to be laid end to end it would form a "rope" 1 m long yet only 2×10^{-5} m (2 nm) wide!

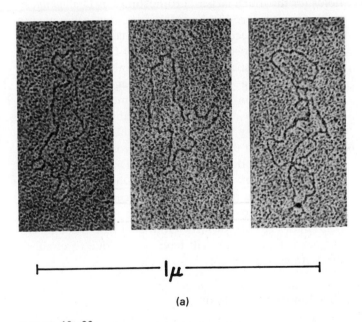

(a)

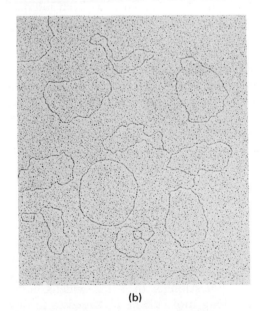

(b)

FIGURE 10–23

Electron micrographs of the circular DNA molecule (chromosome) of bacterial phage ΦX174. The infective form of the virus contains, within its protein coat, a *single-stranded* circular DNA molecule 0.88 nm long consisting of 5.1×10^3 nucleotides. Three such circular DNA molecules, at a magnification of ×90,000, are shown in (a). Within its host cell, the ΦX174 single DNA strand replicates semiconservatively, laying down a complementary strand, to produce a double-stranded circular DNA molecule termed the *replicative form*. A collection of such replicative double-stranded DNA circles, at a magnification of ×58,000, is shown in (b). Note that the double-stranded circles are "rounder" and less kinky than the single-stranded forms; this is because double-stranded DNA is stiffer and less prone to bend.

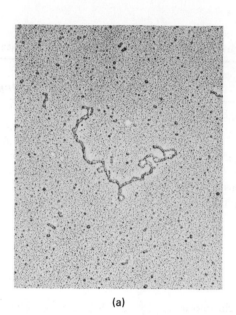
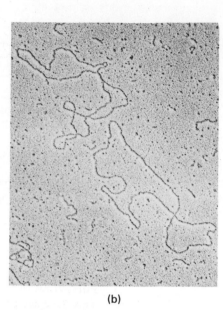

(a)

(b)

FIGURE **10-24**

Supercoiled and relaxed circular DNA. Human mitochondrial DNA is shown, a circle of 16,500 bases or 16.5 *kilobases* (kb), photographed in the electron microscope. ×38,000. (a) Supercoiled. (b) The relaxed or "open" circle, produced from (a) by a single nick. (Courtesy of D. Clayton.)

In eucaryotic cells, two distinct types of DNA are present:

- The DNA of self-replicating organelles such as mitochondria and chloroplasts very much resembles that of procaryotes (a) in its relatively smaller size, (b) in not being found in combination with histone proteins and, most importantly, (c) in its circular form (Figure 10-23). This has been interpreted by some workers as consistent with the endosymbiont hypothesis for the evolutionary origin of mitochondria and chloroplasts: an "infection" by a procaryote of another procaryotic cell, leading to a symbiotic relationship that eventually resulted in the development of the eucaryotic cell. (See also Chapters 2 and 8.)

- The nuclear DNA of eucaryotes differs fundamentally from other DNA in that it is complexed with certain proteins called *histones,* with *nonhistone proteins,* and with *small amounts of RNA* to form the structures called *chromosomes.*

The largest human chromosome (we shall discuss the details of chromosome structure and function in Chapter 11) has 2.7×10^8 base pairs. A single molecule of DNA with that many base pairs would be 10 cm long, 500 million times longer than it is wide! Could such a gigantically long molecule be packaged to form a chromosome? Work with the smaller chromosomes of yeast and *Drosophila* suggests very strongly that this in fact is the case. In yeast, the 17 chromosomes are very small, each chromosome containing on the average 5.4×10^8 Da (280 μm in length) of DNA. It is still possible to use electron microscopic techniques to determine the size of this DNA. By tracing the length of the image in composite electron micrographs of DNA molecules, one obtains an average length of 165 μm. This corresponds to more than half the mass of the average chromosomal DNA and suggests, therefore, that the number of DNA molecules per chromosome is probably one.

Using an ingenious technique which measures the *viscoelasticity* of DNA solutions, B. Zimm and his associates showed that the same single molecule structure holds for the much larger chromosomes of *Drosophila.* In a solution of heterogeneous molecules, this technique can measure the molecular weight of the largest. By using a variety of chromosomal aberrations of *Drosophila melanogaster* and other species, they showed that the molecular weight of the largest DNA molecule agreed closely with the DNA content of the chromosomes as determined independently by chemical means (Table 10-3).

It is too early to state with certainty that the above results apply also to the largest of chromosomes, which can be 100 times the size of those of *Drosophila* and 1000 times the size of those of yeast. One would like to conclude that the principle of *one chromosome—one DNA molecule* is a general biological rule, at least for the nonreplicating chromosome from anaphase through G_1, but much more work will have to be done to establish this with certainty.

Heteroduplex Analysis

As we mentioned earlier, the single-stranded chains of denatured DNA can be separated from undenatured double helices on the basis of their different densities. Skeptics initially questioned whether, upon melting, the strands actually separate completely from one another or whether they might remain stuck together in one or more places, thus helping to initiate later annealing. It turns out that the single-stranded components of most DNA

TABLE 10–3

Comparison of Chromosome Size with Results of Viscoelastic Measurements of *Drosophila* Lysates

Drosophila	Chromosomes	τ,hr $\left(\begin{array}{c}\text{Number of}\\\text{Measurements}\end{array}\right)$	Molecular Weight of Largest DNA	DNA Content of Largest Chromosome
melanogaster Wildtype		1.67 ± 0.23 (44)	41 ± 3 × 10⁹	43 × 10⁹
Inversion		1.77 ± 0.30 (9)	42 ± 4 × 10⁹	43 × 10⁹
Translocation		2.98 ± 0.54 (10)	58 ± 6 × 10⁹	59 × 10⁹
hydei Wildtype		1.62 ± 0.27 (8)	40 ± 4 × 10⁹	not reported
Deletion		0.69 ± 0.17 (8)	24 ± 4 × 10⁹	not reported
virilis		2.12 ± 0.28 (5)	47 ± 4 × 10⁹	not reported
americana		5.00 ± 1.1 (3)	79 ± 10 × 10⁹	not reported

Note: For each *Drosophila*, the idealized appearance of the chromosomes at metaphase is shown in the second column, with the largest chromosomes emphasized. The molecular weights given in the third column are based on the viscoelastic properties of the cell lysates, properties which are sensitive to the largest molecules in solution, in this instance the DNA of the largest chromosomes.

molecules differ from each other in overall base composition, and can be resolved from each other by equilibrium density gradient centrifugation or by chromatographic techniques. It is therefore a simple matter to isolate the individual single-strand components of any homogeneous preparation of a single type of DNA molecule, such as a virus chromosome.

When such purified, complementary single strands are brought together again under conditions which favor hydrogen bonding between complementary base pairs, the double helix reforms as efficiently as if the single strands had not been purified. This process of renaturation or reannealing has been used extensively to study the physical chemistry of the base-pairing interaction (see Chapter 11). It has also been used to relate the genetic map directly to the physical structure of chromosomal DNA.

Consider a deletion mutant lacking a sizeable stretch of DNA in a region identified genetically by crosses. If a preparation of one strand from such a mutant is renatured with a preparation of the complementary strand from the corresponding wildtype DNA, then complementarity between the two single strands will be interrupted in the region of the deletion: a stretch of DNA in the wildtype strand will find no complementary sequence within the region of the other strand defined by the deletion, and will necessarily *loop out* of the duplex structure of the renatured DNA molecule, as illustrated schematically in Figure 10–25a. A DNA molecule with a region of noncomplementarity between the two strands is termed a *heteroduplex*.

Direct electron microscopy of such heteroduplex DNA molecules reveals the position of the looped-out single-strand segment (Figure 10–25b). A heteroduplex

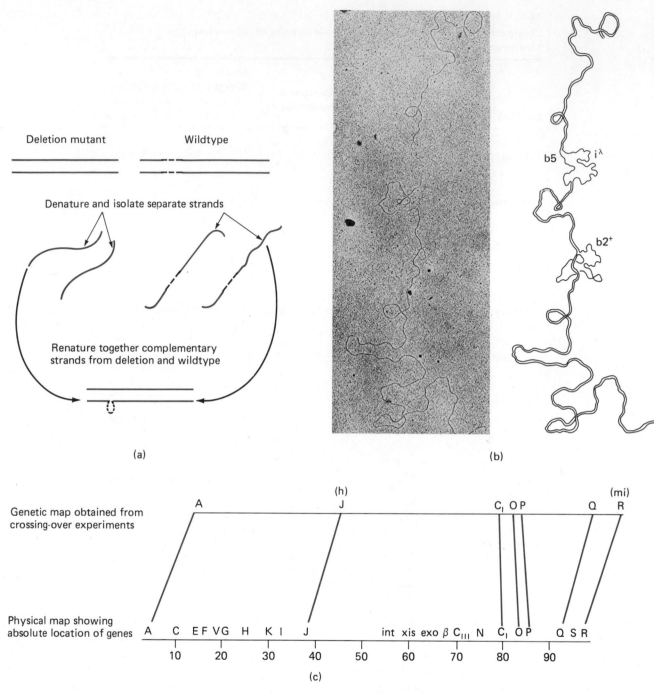

(a)

(b)

(c)

FIGURE **10–25**

Heteroduplex DNA construction, visualization, and use in physical mapping. (a) Diagrammatic representation of the construction of heteroduplex DNA. The two complementary strands of the DNA molecules are shown in black and color. The region in the wildtype molecule, which is missing in the deletion mutant, is shown as a dashed line. (b) Electron micrograph and interpretive drawing of a heteroduplex DNA molecule of bacterial virus lambda. One strand is from wildtype lambda, the other is from a double mutant: b2, b5. The b2 mutation is a deletion, resulting in the single-stranded loop at that position as in (a). The b5 alteration is a long nonhomologous sequence which cannot pair with its counterpart (called i$^\lambda$) in the wildtype lambda strain. As a result, this region of the heteroduplex molecule shows both strands looping out, owing to their inability to pair with one another. Double-stranded regions, which appear thick in the electron micrograph, are shown as a double line in the interpretive drawing. (Courtesy of W. Szybalski.) (c) Comparison of the physical map, obtained by electron microscopy of heteroduplex DNA molecules, with the genetic map, obtained by genetic crosses, of the chromosome of bacterial virus lambda. The numbers along the physical map indicate the percent of the distance from one end of the DNA molecule to the other. The letter symbols indicate the positions of genes.

prepared with one strand from another deletion mutant, located in a different gene, will exhibit the single-stranded loop in a different region of the molecule. In each case, the position of the loop is indexed by distance to one or the other end of the whole DNA molecule. Such heteroduplex mapping thus reveals the positions of different genes, and the distances between them, along the linear continuity of the chromosomal DNA. Reassuringly, such physical maps turn out to be colinear with the genetic map, as shown in Figure 10–25c.

The Sequence of Bases in DNA

The determination of the base sequence in DNA—that is, of the physical basis for the information encoded in the genes—is a spectacular example of the merging of biochemistry, cytology, and genetics, the historical roots of modern cell biology. Progress in the molecular approach to cell biology depends in a crucial manner on the development of techniques. Though technique is only a means to an end, it is the *only* means to an end in biology, a field in which theorizing is always closely tied to the presence of solid data. A student of the growth of modern biology cannot help but be impressed with the fact that all advances have depended on developments of techniques, such as the genetic cross, the enzyme assay, high-resolution optics, immunoassays, and especially on techniques of separation such as chromatography, gel permeation, density gradient centrifugation, and gel electrophoresis.

Enzymes that Degrade DNA

Since genetic information resides in the structure of DNA, it is necessary to find ways of analyzing the structural features that encode the information. The only way such coding can occur in a linear polymer like DNA is in the sequence of nucleotide bases. As for the determination of amino acid sequences in proteins, digestion by enzymes has played a critical role in the development of methods to sequence DNA. There are two general classes of enzymes that catalyze the hydrolysis of the phosphodiester bonds: **exonucleases,** which work inward from one end of the polynucleotide chain or the other, and **endonucleases,** which cleave internally. An example of an exonuclease is (rattlesnake) *venom phosphodiesterase,* which chews inward from the 3′ end and releases 5′-deoxyribonucleotides (5′-dXMP). *Spleen phosphodiesterase,* on the other hand, works inward from the 5′ end and releases 3′-dXMP. Neither of these enzymes shows specificity for particular sequences at the ends and both work on single polynucleotide chains. Double-stranded DNA is used as a substrate by *λ-exonuclease,* which is found in *E. coli* that are infected by

λ bacteriophage; it works from the 5′ end of a DNA strand and releases 3′-dXMP. Other exonuclease activities are found associated with the synthesis of DNA and will be discussed in Chapter 11.

dAp + dTpdApdGpdCpdT
(3′-dAMP)

↑ Spleen phosphodiesterase

dApdTpdApdGpdCpdT

↓ Venom phosphodiesterase

dApdTpdApdGpdC + pdT
(5′-dTMP)

The endonuclease most commonly used in the laboratory to digest DNA to small oligonucleotides is *pancreatic DNAse I.* It works on both double- and single-stranded DNA, splits off 5′-phosphate-terminated products, and has a preference for AT-rich sequences in double-stranded DNA substrates. From *Aspergillus* comes the *S1 endonuclease,* used by genetic engineers to trim the ends of broken DNA molecules because of its high specificity for single-stranded polynucleotide chains. Neither of the latter two enzymes shows specificity for particular DNA sequences. Without such specificity, using nucleases to determine the base sequence of a DNA is much more difficult. Luckily, a peculiarity of bacterial physiology provides the needed tools.

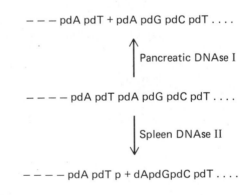

– – – pdA pdT + pdA pdG pdC pdT . . . .

↑ Pancreatic DNAse I

– – – – pdA pdT pdA pdG pdC pdT . . . .

↓ Spleen DNAse II

– – – – pdA pdT p + dApdGpdC pdT . . . .

Restriction Enzymes: Scalpels for DNA Surgery

Bacterial cells do not take kindly to the introduction of foreign DNA. Each bacterial species produces an array of *restriction endonucleases* which cut foreign DNA into fragments, the first step in the complete degradation of alien genetic material that might be introduced, for example, by a viral infection. The cell's own DNA is protected from this digestion system by a *specific pattern of methylation* on certain nucleotide residues, which flags the DNA as "self" rather than "foreign," and blocks the activities of the cell's particular restriction enzymes.

Thus, *E. coli* DNA transferred into *E. coli* cells is immune to hydrolysis by *E. coli* restriction enzymes, but DNA from a different species, methylated at a different set of positions, is hydrolyzed.

Clearly, both the methylation enzymes and the restriction endonucleases must be exceedingly specific; if they were not, there would be some chance that a bacterium could degrade its own DNA, hardly a useful activity. The specificity of this intricate system (termed the *restriction/modification system*) depends on the nucleotide sequence. The restriction endonucleases bind to DNA at particular sequences, while the species-specific methylating enzymes methylate nucleotide residues at or near these sites.

The sequence specificity of the restriction enzymes provides a tool which has revolutionized genetics. Table 10–4 summarizes the sequence specificity of a number of restriction enzymes from various bacterial species. Each enzyme cuts DNA *only* at the sites of its four-to-six base target sequence. These sites are relatively infrequent, as you can appreciate by a simple probability calculation. If the abundance of each base in a DNA strand is, for example, 0.25, then the average frequency of any particular six-base sequence will be $(0.25)^6 = 2.4 \times 10^{-4}$. A restriction enzyme which cuts at one particular six-base site will generate DNA fragments with an average size of $1/(2.4 \times 10^{-4}) = 4096$ nucleotide pairs (np). Of course, this is the mean of a distribution of fragment sizes, representing the distribution of distances separating target sequences in the DNA. When a DNA preparation from a given organism is cleaved with a single, purified restriction enzyme, the result is a collection of *discrete fragments* whose endpoints correspond to the positions in the DNA of the restriction enzyme's target sites (see Figure 10–26). These fragments can then be separated from one another by electrophoretic procedures which resolve DNA fragments according to molecular weight.

Now, think back to Chapter 9. Genetic analysis is simply the mapping of unique sites in the linear topology of the genetic material, DNA. Formal genetics gains this end by statistical inference, transforming the chaos of random sites of recombination into the order of a genetic map. Restriction enzymes circumvent the need for statistical inference, for their cutting sites are surgically specific. They thus provide the means for a new kind of genetic mapping.

TABLE 10–4
Restriction Enzyme Target Sites

Enzyme	Source	Target Site					
<u>Ava</u> I	*Anabaena variabilis*	C G	Py ↑ Pu	C G	G C	Pu Py	↓ G C
<u>Bam</u> HI	*Bacillus amyloliquifaciens*	G C	G ↑ C	A T	T A	C G	↓ C G
<u>Eco</u> RI	*Escherichia coli*	G C	A ↑ T	A T	T A	T A	↓ C G
<u>Hind</u> III	*Hemophilus influenza*	A T	A ↑ T	G C	C G	T A	↓ T A
<u>Hinf</u> I	*Haemophilus influenza*	G C	A ↑ T	N N	T A	↓ C G	
<u>Hpa</u> II	*Hemophilus parainfluenze*	C G	C ↑ G	G ↓ C	G C		
<u>Sma</u> I	*Serratia marcescens*	C G	C G	C ↓ G ↑	G C	G C	G C

Note: The target sequences are given with the top line running 5′-to-3′ left-to-right, and the bottom line, of course, the reverse. Pu stands for any purine, and Py stands for any pyrimidine. N stands for *any* nucleotide, N′ for its complement. The arrows indicate the phosphodiester bond cleaved. Note that most restriction enzymes cleave at staggered positions on the two complementary strands, but some (e.g., <u>Sma</u> I in this set of examples) do not. The restriction enzymes listed here are only a few of more than 400 known.

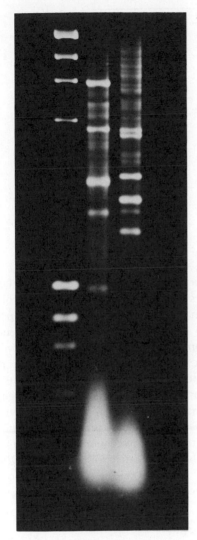

A B C

FIGURE **10–26**
Electrophoretic Separation of Restriction Fragments. DNA was cut with restriction enzymes, and the resulting fragments separated by gel electrophoresis. After electrophoresis, the gel was stained with ethidium bromide, a fluorescent molecule which binds to DNA, and the fluorescence of the tagged fragments was photographed. Lane A contains fragments cut from the DNA of two small bacterial viruses (phage lambda DNA, cut with <u>Hind</u> III, and phage ΦX174 replicative form DNA, cut by <u>Hae</u> III). The small genome size of these viruses means that there are only a few restriction sites of any given type, and so only a few distinct fragments are produced. Since the complete DNA sequence of these viruses is known, the molecular weight of each fragment is known, and a digest of this type is conventionally used to provide molecular weight markers. Lane B contains fragments cut from total yeast DNA by one restriction enzyme, and Lane C contains fragments cut from total yeast DNA by another restriction enzyme. The yeast genome is several hundred times larger than that of the bacterial viruses. As a result, there are many more restriction sites of any type, and many more fragments are produced. Note that the pattern of fragments is different in Lanes B and C, because different restriction enzymes were used to cut up the DNA. Eucaryotic organisms from yeast on up the phylogenetic scale contain certain highly repeated DNA sequences which are present in many copies distributed about the genome. A restriction fragment characteristic of one of these repeated sequences will of course be cut out of every copy of the sequence in question, and will thus be present in much greater abundance than those cut out of single-copy sequences. Fragments cut from these repeated sequences are obvious as intensely fluorescent bands on the gel. (Courtesy of R. Sclafani and J. Gallant.)

Restriction Mapping

Consider a linear fragment of DNA with ends A and D, and two internal sites B and C, which are specific for a particular restriction enzyme. Our problem is to deduce the order of the internal sites: is it A–B–C–D or A–C–B–D?

We do so in two steps. First, we digest the DNA to completion with the restriction enzyme and separate the resultant fragments A–B, B–C, and C–D by gel electrophoresis (Figure 10–27). The molecular weights (which is to say the lengths) of these fragments can be estimated very accurately from their electrophoretic mobilities.

In a parallel incubation, we reduce the concentration of the restriction enzyme and/or the time of hydrolysis, *so that many of the DNA molecules will be cut only once,* at either site B or site C. The fragments produced by this *partial* digest will now include two new ones: A–C and

B–D (Figure 10–27). If the order is A–B–C–D, as shown, then the length of fragment A–C will necessarily equal the sum of the lengths of A–B and B–C; and the length of B–D will equal the sum of B–C plus C–D. The perfect additivity of subfragment lengths thus provides us with an ordering principle, a molecular equivalent of genetic mapping.

This ordering procedure is based on the possibility of rigging conditions so that cuts are made in some cases at either B or C, but not at both. A variant of the technique can be employed where B and C are sites for *different* restriction enzymes—call them I and II. The complete digestion to produce fragments A–B, B–C, and C–D is carried out with both restriction enzymes together. A partial digestion to produce only fragments A–B and B–D is carried out with enzyme I alone; the other digestion to produce only fragments A–C and C–D is carried out with enzyme II alone (Figure 10–28). Again,

FIGURE 10-27
Mapping of Restriction Sites by Partial and Complete Digestion. The DNA molecule A–B–C–D, 3500 base pairs (bp) long, with the positions of restriction sites B and C, is shown approximately to scale. Gel electrophoresis (see Figure 10–26) is shown below, a complete digest at left, and a partial digest at right. The complete digest cuts every molecule at *both* positions B and C, resulting in the three fragments A–B, B–C, and C–D; the positions of these fragments on the gel and their lengths in base pairs are indicated. The partial digest cuts some molecules at both positions B and C, some at only B, some at only C; the positions of these fragments and their lengths are indicated.

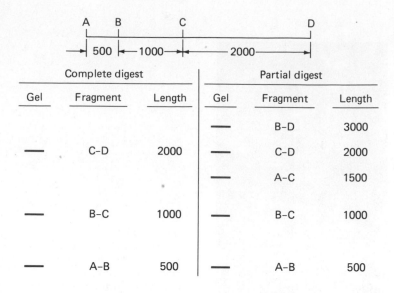

	Complete digest			Partial digest	
Gel	Fragment	Length	Gel	Fragment	Length
			—	B–D	3000
—	C–D	2000	—	C–D	2000
			—	A–C	1500
—	B–C	1000	—	B–C	1000
—	A–B	500	—	A–B	500

the simple additivity of lengths permits one to establish which of the larger fragments from a partial digest contains a given pair of adjacent smaller ones from a complete digest, and thus to order the sites.

Where does our initial A–D molecule come from? It comes from specific hydrolysis of still larger DNA (perhaps whole genome DNA) with a restriction enzyme which cuts at sites A and D. Thus, restriction mapping determines the order of various restriction enzyme target sites.

Transformation by Restriction Fragments

In conventional genetics, the sites ordered are genes, or mutant sites within genes. This approach can be combined with restriction mapping to identify genes carried on specific restriction fragments. Many microorganisms, and mammalian cells in tissue culture, can be induced to take up DNA molecules, a phenomenon termed *DNA transformation* (also called "transfection"). The imported DNA can then be incorporated into the resident genome by homologous recombination. Thus, if a particular restriction fragment carries the wildtype gene X^+, DNA transformation of X^- mutant cells will yield X^+ recombinants. The molecular equivalent of genetic linkage is demonstrated when two genes—call them X^+ and Y^+—reside on the same restriction fragment: DNA transformation of $\underline{X^- Y^-}$ cells with the fragment would then yield $\underline{X^+ Y^+}$ recombinants.

FIGURE 10-28
Mapping of Restriction Sites with Two Different Restriction Enzymes. The DNA molecule A–B–C–D is the same one shown in Figure 10–27. In this case, however, restriction enzyme I cuts only at position B, shown above the schematic representation of the DNA molecule; restriction enzyme II cuts only at position C, shown below the DNA molecule. The lines flanked by arrows above and below the DNA molecule indicate the fragments produced by each restriction enzyme; their position in gel electrophoresis and length are shown in the lower part of the figure. Complete digestion with both enzymes will produce fragments A–B (500 bp), B–C (1000 bp), and C–D (2000 bp). These will migrate in gel electrophoresis to the positions shown for them in Figure 10–27.

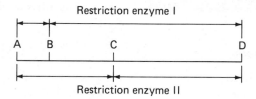

	Enzyme I digest			Enzyme II digest	
Gel	Fragment	Length	Gel	Fragment	Length
—	B–D	3000			
			—	C–D	2000
			—	A–C	1500
—	A–B	500			

In the last few years, DNA transformation methods have been extended from microorganisms and cultured cells to whole organisms. The method depends on the fact that all higher organisms start out as single cells, namely fertilized egg cells. Techniques of *micromanipulation* permit the isolation of fertilized egg cells and their transformation by specific DNA fragments. For example, one can isolate freshly fertilized mouse egg cells and *microinject* specific DNA fragments into their nuclei. A proportion of the egg cells survive this manipulation, and, after implantation in the oviduct of a surrogate mother mouse, develop into proper mice which are ordinary in all respects save one: some of them inherit the injected DNA fragment through its incorporation into the genome by DNA transformation during the early stages of embryonic development. These *transgenic* mice are of two kinds. In some, the injected DNA fragment is incorporated sufficiently early in embryonic development to become part of the animal's germ plasm, and these extraordinary mice transmit the incorporated gene to their progeny. Conventional genetic crosses (see Chapter 9) identify them as animals in which genetic engineering has produced a stable change in the genotype. Other transgenic mice contain the alien DNA fragment only in certain somatic tissues, presumably due to genetic incorporation at later stages of embryonic development, and do not transmit it to their progeny.

Transgenic organisms have also been produced by injecting specific DNA fragments into early embryonic cells of other species, such as *Drosophila*. The analysis of these genetically engineered animals will undoubtedly reveal much about gene expression during development.

Plasmids, Cloning, and Libraries

Some of the techniques of genetic engineering, including those we have just described, require sizeable quantities of the appropriate DNA sequences. These are not easily obtained by purification, especially in the case of unique genes which exist as only two copies (one on each homologous chromosome) per diploid genome. Fortunately, another component of modern DNA technology has provided a way to *amplify* any restriction fragment by many orders of magnitude, through a simple and elegant biological procedure. The procedure, which is called *cloning,* makes use of another exotic aspect of bacterial genetics.

Many bacterial species harbor genetic elements called *plasmids* that are distinct from chromosomes. They are circular DNA molecules, ranging from a few thousand to a few hundred thousand base pairs in size. They carry genetic information for their own autonomous replication and segregation, and, in some cases, for transfer from one cell to another. Plasmids are capable of "picking up" additional stretches of DNA from the chromosome, through nonhomologous recombinational interactions that are not well understood. These interactions are rare, but the recombinant plasmids they generate are subject to the classic principle of natural selection.

During the last forty years or so, medical practice has emphasized the use of antibiotics to treat bacterial infections. This provides a "natural" (albeit man-made) environment for selection. Plasmids that have picked up antibiotic-resistance genes confer a selective advantage on their host bacterial cells. Recombinant plasmids containing antibiotic-resistance genes have thus become abundant in those bacteria which maintain themselves in animals subject to antibiotic treatment, namely ourselves and our domestic animals.

Figure 10–29 shows the gene and restriction site map of drug-resistance plasmid pBR322, which carries genes for resistance to the widely used antibiotics tetracycline and ampicillin. If one treats bacterial cells with pBR322 DNA under appropriate conditions, some cells take up the plasmid, and can be selected on media containing either ampicillin or tetracycline. Such cells also express other genes contained in the plasmid. *The artificial insertion of restriction fragments into the plasmid* is the basic strategem of gene cloning.

The operation is remarkably simple, due to the special way in which most restriction enzymes cut DNA. The arrows in Table 10–4 indicate the positions of the phosphodiester bonds cut on each strand of the various restriction target sites. Note that in most cases the cuts are not at the same position on each strand, but *staggered*. As a result, the restriction fragments contain single-stranded "tails" at each end. The crucial point is that these single-stranded tails carry base sequences which are complementary to one another, precisely because they are cut from double-stranded DNA. For example, one end of *all* fragments generated by Bam HI will be

$$GATCC-----3'$$
$$G-----5'$$

while the other end will always be

$$5'-----G$$
$$3'-----CCTAG.$$

The complementarity of the two tail sequences means that all fragments, from whatever source, generated by a given restriction enzyme can form fairly stable end-to-end associations, due to hydrogen bonding between their single-stranded tails, as illustrated above and in Figure 10–30a.

It remains only to link up the strands at each end of such complexes, which can be done by means of *polynucleotide ligase,* a bacterial enzyme which reforges the missing phosphodiester bonds in DNA. This enzyme's

FIGURE 10–29

A Restriction and Gene Map of the Commonly Used Cloning Vehicle, Plasmid pBR322. Cleavage sites for various different restriction enzymes are shown outside the circle. The circular molecule contains about 4400 bp. The single Eco R1 site at the top of the circle is arbitrarily assigned position "0" to establish a coordinate system: clockwise distances from this position, at intervals of 1000 bp, are indicated inside the circle. The positions of three pBR322 genes are indicated inside the circle as arcs with an arrow at one end: Tet is the gene for resistance to antibiotic tetracycline, Amp is the gene for resistance to antibiotic ampicillin, Ori is the origin of replication. (Courtesy of Bethesda Research Laboratories.)

FIGURE 10–30

Splicing of Restriction Fragments. (a) In the first line, two different DNA sequences, indicated in different colors, are shown at left and right. The dashes indicate phosphodiester bonds connecting adjacent nucleotide residues. In the example, the two sequences share a common Bam H1 restriction site, flanked on either side by entirely different sequences (represented by dashes with no bases specified). Bam H1 cuts the two sequences at the positions on each strand shown by the arrows. Bam H1 digestion cuts the left and right DNA molecule into two fragments each. The second line shows the fragments of the left molecule and the fragment of the right molecule which have complementary ("sticky") single-stranded ends. When the two DNA preparations are annealed together, these complementary single-stranded ends will associate by base pairing, shown in the third line. Treatment with DNA ligase then forges phosphodiester bonds between the ends on each strand, indicated by squares. The result is the recombinant DNA molecule shown in the fourth line.

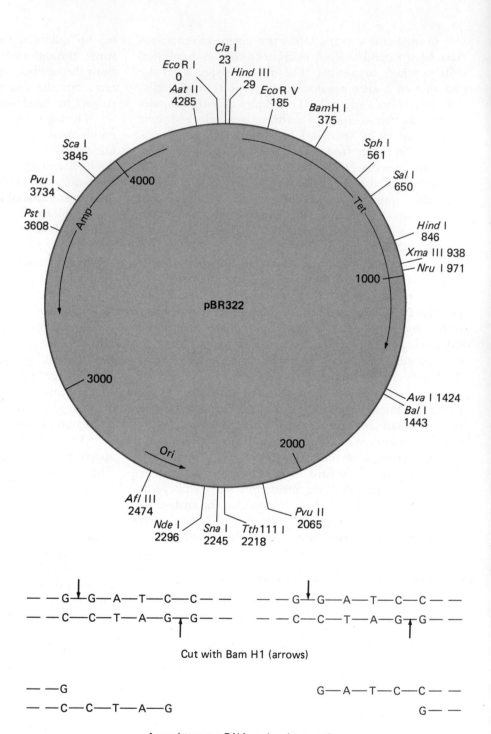

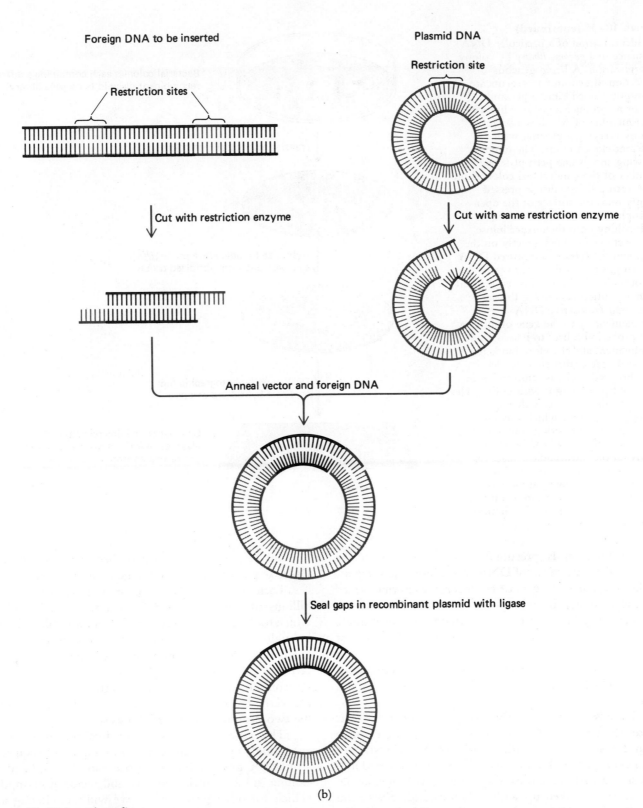

(b)

FIGURE 10–30 **(continued)**

(b) Insertion of a restriction fragment into a plasmid vector. Plasmid DNA is cut with a restriction enzyme specific for a unique site in the plasmid genome, such as the single <u>Bam</u> H1 site in pBR322. The foreign DNA is cut with the same restriction enzyme, so that its single-stranded ends will base-pair with the complementary "sticky ends" of the cut plasmid DNA, as in Figure 10–30a. After annealing of the two DNA preparations together, the gaps are sealed with *polynucleotide ligase* to yield the recombinant plasmid vector shown at the bottom.

FIGURE 10–30 (continued)

(c) Identification of a particular DNA sequence in a *genome library* by colony hybridization. A heterogeneous collection of genomic fragments are "shotgun cloned" into a plasmid vector to produce a genome library, each member of which is a bacterial colony carrying a plasmid with a different cloned insert. The upper drawing shows one petri plate with a number of these individual colonies on it. A nitrocellulose disc is pressed lightly onto the surface of the open petri plate, transferring a fraction of each colony onto the nitrocellulose. The bacteria are lysed directly on the disc, and the released denatured DNA at each position is fixed onto the nitrocellulose. The disc is then flooded with a solution containing the *probe:* denatured radioactive DNA complementary to the gene of interest. The probe hybridizes to its complement and therefore binds to the nitrocellulose at that spot, while unhybridized probe is removed from the disc by washing it with buffer. The disc is finally overlaid with a photographic film, which becomes darkened at the position corresponding to the hybridized radioactive probe DNA on the nitrocellulose disc. (In practice, a complete genome library requires several thousand individual colonies, so this procedure is carried out on a few dozen to a few hundred petri plates.)

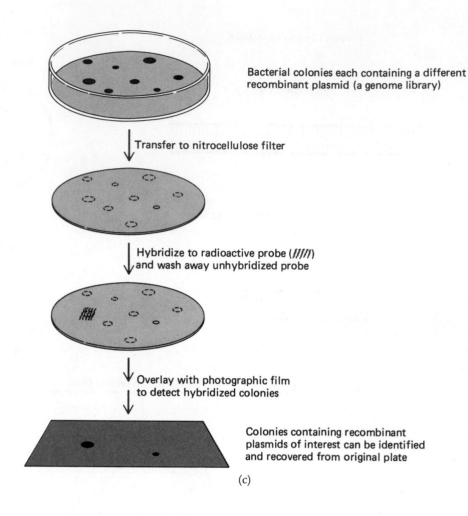

Bacterial colonies each containing a different recombinant plasmid (a genome library)

Transfer to nitrocellulose filter

Hybridize to radioactive probe (*/////*) and wash away unhybridized probe

Overlay with photographic film to detect hybridized colonies

Colonies containing recombinant plasmids of interest can be identified and recovered from original plate

(c)

natural function is presumably to close gaps in the intermediate structures of DNA recombination or repair. In the laboratory of a genetic engineer, the enzyme is used for the final step in linking one restriction fragment to another, or *splicing* an isolated restriction fragment into a plasmid cut by the same restriction enzyme. The result of the latter operation (Figure 10–30b) is the reformation of a circular *recombinant plasmid* (made of **recombinant DNA**) which contains a specific sequence inserted at a specific site.

Here is a typical, widely used procedure of this sort. Note (Figure 10–29) that pBR322 contains a single <u>Bam</u> HI site located within the tetracycline-resistance (tet^r) gene. A <u>Bam</u> HI fragment of interest is inserted into this site in pBR322 cut by the same restriction enzyme. The ends are linked up with polynucleotide ligase, and the plasmid DNA is used to transform *E. coli* cells. The cells are spread on petri plates containing ampicillin, so that only cells which have taken up plasmid DNA will form colonies. These colonies are then tested for tetracycline resistance: those which harbor a recombinant plas-

mid will be tetracycline-sensitive, because the insertion of DNA sequence within the tet^r gene inactivates it.

Each of these colonies is an independent *clone* of billions of cells descended from a single ancestral cell which took up one recombinant DNA molecule. Cloning individual lines of descent in this way isolates single DNA fragments and simultaneously amplifies them by a factor of 10^{11} or so. Plasmid DNA is easily isolated from a recombinant clone, and the inserted fragment can be recovered by cutting again with <u>Bam</u> HI and separating the two products by electrophoresis.

The cutting-cloning-cutting method provides an exceedingly useful strategem for moving DNA sequences around. A given DNA sequence can be amplified by cloning in one vehicle, cut out, and recloned in another vehicle for other purposes. As we will see, cloning into bacterial virus M13 is the first step in a widely used method for DNA sequence analysis.

Cloning also provides a way to move DNA sequences from species in which genetic analysis is difficult into the well-characterized genetic system of *E. coli* and its plasmids. For example, M. Silverman and co-workers have analyzed the genetics of bioluminescence in the marine bacterium *Vibrio fischerei* by cloning the genes into *E. coli* plasmids. Their constructions produced *E. coli* colonies that glow in the dark! Further genetic analyses could then be carried out with these clones.

Geneticists use the term "shotgun cloning" to describe the random ligation of genomic fragments into a vector. In this procedure, a suitable vector is first treated with a restriction enzyme that cuts at only a single site in the vector's DNA sequence—for example, the single Bam HI site in pBR322 (Figure 10–29). Total genomic DNA is then cut by the same restriction enzyme and the entire heterogeneous collection of fragments is ligated into the vector under conditions of a low ratio of genomic DNA to vector DNA, so that each vector molecule incorporates only a single genomic fragment. The recombinant vectors are then used to transform bacterial cells, selecting transformants on agar plates containing the antibiotic for which the vector carries a resistance gene, ampicillin in our example. The colonies which grow up on the selective agar plates all carry a different recombinant plasmid into which one or another genomic fragment has been spliced; they thus constitute a *library* of the entire genome under investigation. (The geneticist S. Brenner has pointed out this collection is not indexed in any way and, therefore, is better termed a "heap" or "pool" than a library.)

If one can select for a given gene, then it can be identified among the many clones of the library, provided that one screens a sufficient number of clones. For example, the genome of *E. coli* contains about 3×10^6 bp (base pairs) of DNA and the average Bam HI fragment contains about 4×10^3 bp. On the average, therefore, a collection of $(3 \times 10^6)/(4 \times 10^3) = 750$ random clones is likely to contain one fragment containing any gene of interest. In practice, screenings of about three times this number of clones provide near certainty of identifying the right fragment. If the genome of the organism under investigation is ten times longer, then one must screen ten times more clones.

How does one screen a library for a particular gene of interest? It is easy when the activity of the gene's product can be selected for, or detected. For example, if X^- mutant cells cannot grow in the absence of some particular nutritional supplement, then any recombinant plasmid which carries the X^+ gene will endow a clone of cells with the ability to grow in the absence of that nutritional supplement, and can be selected on media lacking it. If the expression of a cloned gene can be detected—for example, the *Vibrio fischerei* bioluminescence genes which enable *E. coli* colonies to glow in the dark—then these clones can be identified by appropriate tests of gene expression.

Many eucaryotic genes are poorly expressed in *E. coli;* they are identified by a different technique which detects the DNA itself, rather than gene expression. The method depends on the specificity of hybridization between complementary DNA strands. Suppose that we have purified a eucaryotic protein to homogeneity and now wish to identify those rare plasmids in a library that carry the gene for our protein. We begin by obtaining partial amino acid sequence data for the protein—for example, a short stretch of the N-terminal region by Edman degradation. This amino acid sequence reveals, with certain ambiguities, the nucleotide sequence of the corresponding region of the gene, via the genetic code (Chapter 13). Using this information, we can synthesize a DNA oligonucleotide which is complementary to that portion of the gene. This synthetic oligonucleotide is then tagged with a radioactive label and used as a *probe* to identify the clones in a library that carries the complementary DNA sequence.

In one identification procedure, termed *colony hybridization* (Figure 10–30c), an array of bacterial clones carrying different fragments of a library is transferred to a nitrocellulose sheet. The sheet is then incubated at high pH and baked at 80°C; this treatment lyses the bacterial cells on the sheet, denatures the DNA they release, and immobilizes the single-stranded DNA in the nitrocellulose at the position occupied by the colony. The sheet is then incubated with the radioactive probe under conditions which favor hybridization of complementary DNA sequences; our probe will hybridize only at positions on the sheet where colonies carrying its complement were located. After washing away unhybridized probe DNA, we detect the positions at which the probe hybridized by autoradiography. In this way, the probe identifies for us the clones in the library which carry the gene sought.

The expression of cloned eucaryotic genes in *E. coli* has enormous practical consequences for the pharmaceutical industry. Genes for medically important mammalian proteins, such as insulin, have been successfully cloned in *E. coli*. The poor expression of unmodified eucaryotic genes in *E. coli* can be alleviated by fusing them to the control region of an *E. coli* gene. For example, the coding sequence of the mammalian insulin gene can be inserted into an *expression vehicle* just downstream of the *promoter-operator* sequence of the *E. coli lac* genes which determines their efficient expression (see Chapter 14). In such a construct, expression of the mammalian gene is both efficient and subject to the well-understood *E. coli lac* control system, which the experimenter can turn on and off at will. In the past, insulin could only be produced by slaughtering great numbers of pigs and isolating the protein from their pancreatic tissue, a laborious, inhumane, and very expensive process. Now, the protein can be isolated far more rapidly and economically from cultures of *E. coli* strains that harbor a recombinant plasmid containing an insert of the mammalian insulin gene.

In some cases, the plasmid-mediated movement of genes from one species to another can be rerouted from *E. coli* into simple eucaryotic cells. Derivatives of pBR322 have been constructed which contain yeast DNA sequences that permit autonomous replication and regular segregation in yeast cells, as well as in the bacterial cells. These *shuttle vectors,* in which yeast and *E. coli* genes can be propagated in either organism, have been widely used to study the molecular genetics of yeast and, to some extent, related ascomycetes.

Restriction Fragments as Hybridization Probes

An isolated restriction fragment can be used as a reagent to detect identical or closely related sequences elsewhere in the genome, or in a collection of DNA or RNA molecules. This use of restriction fragments as "probes" follows from the specificity of DNA:DNA or DNA:RNA hybridization. A method commonly employed involves labeling a restriction fragment at one end with a radioactive isotope. The labeled restriction fragment can then be used to tag any complementary sequence, in either DNA or RNA, to which it will hybridize.

In one method, termed *in situ hybridization,* the labeled probe is hybridized to carefully prepared nuclei in which the chromosomes are not disrupted and can be examined by light microscopy. The giant salivary gland chromosomes of *Drosophila* consist of many identical DNA duplexes aligned side by side. These multistranded chromosomes exhibit a complex pattern of bulges and bands (see Figure 9–9) which can be associated with the locations of particular genes by cytogenetic techniques

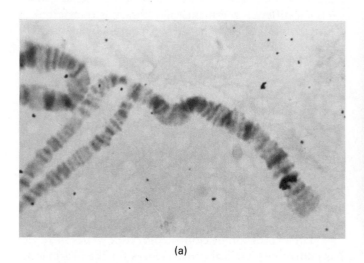

(a)

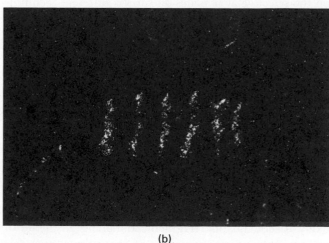

(b)

FIGURE **10–31**

Two Kinds of in Situ Hybridization.   (a) A preparation of *Drosophila* salivary gland chromosomes is incubated with a radioactively labeled DNA probe. The probe hybridizes to this complementary sequence on a particular chromosome. After autoradiography (see Figure 10–11), the position of the radioactive probe is visualized by microscopy. ×2,000. (Courtesy of R. Garber.) (b) In situ hybridization can also be used to detect cells or tissues in which a particular gene is expressed. In this technique, the radioactive DNA probe hybridizes to the RNA transcript of a gene which is vigorously expressed. The DNA probe in the illustration is a fragment containing the *engrailed* gene of *Drosophila.* Genetic studies of this very interesting gene suggest that its expression indicates the commitment of embryonic cells to differentiate into the tissues of distinct segments of the adult fly. The photograph shows in situ hybridization to groups of cells in a very early fly embryo—long before the adult fly is constructed. The bright bands (the photograph is a negative) show the positions of embryonic cells which express the *engrailed* gene and therefore contain high levels of mRNA which hybridize to the *engrailed* DNA probe. These groups of cells alternate from one end of the embryo to the other, implying an early commitment to gene expression which might assign cells, and their descendents, to differentiation into distinct segments of the adult fly. ×240. (Courtesy of T. Kornberg.)

(Chapter 9). Hybridization of a radioactive probe to a particular band can be visualized by autoradiography, overlaying the chromosome preparation with a photographic emulsion; local radioactive disintegration will then deposit silver grains in the emulsion directly over the position of the radioactive probe, and these grains are seen as dark blobs upon examination under the microscope (Figure 10–31). Thus, a given radioactive restriction fragment will mark its gene of origin. It will also mark any repeats of the same gene, or genes with closely related sequences, wherever they occur. In this way, in situ hybridization provides a means for identifying the chromosomal locations of *gene families* which share identical or nearly identical sequences.

Radioactive probes can also be used to tag identical or related sequences in a heterogeneous collection of restriction fragments. In this approach, genomic DNA is cut up by one or another restriction enzyme and separated according to size by gel electrophoresis. The various DNA fragments in the gel are then transferred to a strip of treated paper which immobilizes each fragment at a position corresponding to its location in the gel. The method, devised by E. Southern, is called the *Southern transfer* and its resulting pattern on the treated paper, the *Southern blot*. The paper containing the Southern blot is then treated with the radioactive probe, which hybridizes to complementary DNA fragments, excess radioactive probe is washed away, and the paper is submitted to autoradiography in order to visualize the positions tagged by the probe (Figure 10–32). These positions identify all sequences which are identical or nearly identical in base sequence to that of the probe, and thus locate members of a gene family among any heterogeneous collection of restriction fragments.

(The same approach can be brought to bear on a heterogeneous collection of RNA molecules that have been resolved by gel electrophoresis. This method is commonly termed a *Northern blot,* in whimsical reference to the Southern blot. The Northern blot permits one to identify RNA transcripts of specific genes, or members of the same gene family. Northern blots are of great value in the analysis of gene transcription during various stages of differentiation in a variety of organisms.)

The number of cloned probes for the human genome increases every day, due to the frenzied research activity in this field, and is currently over 4000. This is an average of nearly 200 probes per chromosome and, indeed, probes are available for each of the 23 different human chromosomes. The use of DNA probes has given rise to a spectacular advance in genetic analysis which combines the principles of classical genetics with the technique of the Southern blot.

The genome of all organisms is much more variable than most of us realize, a phenomenon called *genetic*

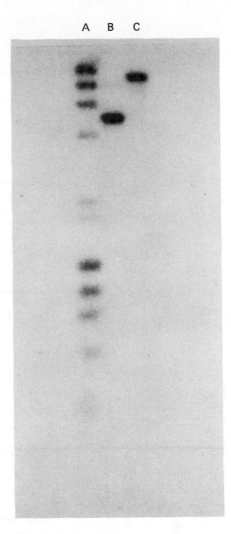

FIGURE **10–32**
Identification of Specific DNA Sequences by Southern Blotting.   DNA was subjected to gel electrophoresis in order to separate restriction fragments. Lane A contains specific fragments of bacterial virus DNA (from lambda and ΦX174 replicative form), to serve as molecular weight markers. Lane B contains total yeast DNA cut by restriction enzymes Hind III and Bam H1. Lane C contains total yeast DNA cut by restriction enzyme Eco R1. After electrophoresis, DNA in the gel was transferred to treated paper by the Southern blot procedure. The paper was then incubated with a radioactive DNA probe, under appropriate conditions for DNA:DNA hybridization. The probe contained labeled bacterial virus DNA identical to that in Lane A, plus a cloned yeast sequence containing the cdc7 gene. After hybridization and washing away excess radioactive probe, the paper was exposed to photographic film. The autoradiography locates the expected viral DNA fragments in Lane A. They are not labeled in Lanes B and C, because no viral DNA was present in these lanes. The yeast probe does identify restriction fragments in Lanes B and C which carry the cdc7 gene. Note that they are of different molecular weight, because different restriction enzymes were used. (Courtesy of R. Sclafani.)

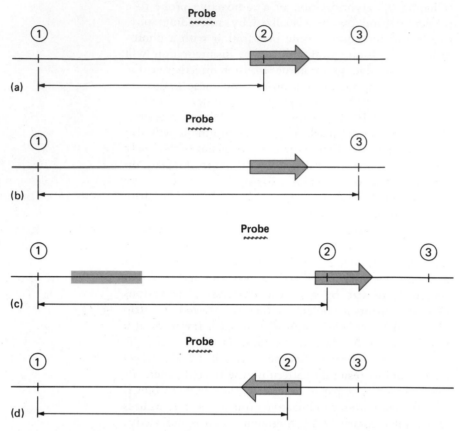

FIGURE 10-33

The Genetic Basis of Restriction Fragment Length Polymorphisms. Each panel shows a different variant or polymorphic form of the same stretch of DNA. The DNA is digested with a restriction enzyme which cuts at sites 1, 2, and 3. The fragments are then separated according to size by gel electrophoresis and transferred to treated paper by the Southern blot procedure (Figure 10–32). The same radioactive probe is employed in each case to mark the restriction fragments carrying the DNA sequence to which it is complementary, shown lying within the probe. The length of the restriction fragment identified by the probe is indicated. (a) The probe marks a restriction fragment extending from site (1) to site (2). (b) In this variant, a single base pair mutation in restriction site (2) alters it so the restriction enzyme does not cut at that position. Accordingly, the fragment marked by the probe extends from site (1) to site (3). (c) In this variant, an insertion or duplication (color box) increases the distance between sites (1) and (2). (d) In this variant, an inversion of a stretch of DNA surrounding site (2), indicated by a color arrow in panels (a) and (d), increases the distance between sites (1) and (2). Even though the probe hydrizes to the same complementary DNA sequence in each case, the restriction fragments are of different lengths. A particular probe may not detect RFLP variation of this sort in DNA cut with a *single* restriction enzyme. There is, however, sufficient polymorphism in DNA sequences, and a sufficient number of different restriction enzymes (over 100), that virtually all probes detect RFLPs in genomic DNA cut with one or another of a collection of restriction enzymes.

polymorphism. Consider a specific stretch of DNA along a particular chromosome in two different individuals: on average, the two DNA sequences will be different at least once in a stretch of roughly 1000 base pairs. In many cases, the difference will affect the length of a restriction fragment encompassing the region; the difference can be detected by Southern blot analysis with a cloned probe for the region of interest (Figure 10–33).

DNA sequence differences identified in this manner are called *restriction fragment length polymorphisms,* or

RFLPs for short. RFLPs are detected right at the DNA level, unlike the phenotypic variation in gene products which provided the markers of classical genetics. Nonetheless, the inheritance of RFLPs obeys the rules of classical genetics. For example, if one form of an RFLP is on the X chromosome, then it follows the sex-linked pattern of inheritance, just like the white eye mutant allele of *Drosophila:* fathers transmit it to all their daughters and none of their sons, and mothers transmit it to half of their daughters and all of their sons. If one form of an RFLP is closely linked to a gene for an inherited disease, then it will generally (except for the rare cases of crossing-over) be inherited together with the disease gene. The discovery of RFLP sites closely linked to genes for inherited diseases is a hot topic in human genetics, for it permits early, even prenatal, identification of individuals or fetuses cursed with a linked disease gene.

The very large number of cloned probes for the human genome, and the RFLP sites which they have revealed, raises the possibility that a complete restriction map of each human chromosome may be constructed within the next few years. Once that task is accomplished, fragments of the human genome carrying virtually any gene of interest will be available for study and its genetic position located and indexed. Then we will have not merely a pool of genomic fragments, but a true library of the human species' genetic endowment.

The Chromosome Walk

Hybridization of restriction fragments to one another is also the basis of a molecular genetic technique for mapping the order of contiguous DNA fragments in their chromosome of origin. The method, termed *overlapping hybridization,* is illustrated in simplified form in Figure 10–34.

Consider a stretch of genomic DNA from points A to G, which contains target sites B and C for restriction enzyme I, and target sites D, E, and F for enzyme II (Figure 10–34). We begin by isolating fragment A–B

from an enzyme I digest, and labeling it with a radioactive isotope for use as a probe. Next, we cut the whole piece A–G, or for that matter total genomic DNA, with enzyme II, separate the fragments by electrophoresis, and do a Southern blot; the blot is then probed with our radioactive A–B to identify those enzyme II fragments which overlap it. The autoradiogram identifies fragments A–D and D–E; in the jargon of the field, these bands on the Southern blot "light up." We have learned that A–D and D–E are contiguous, and must overlap the probe.

The next steps are similar. We employ labeled D–E to probe a Southern blot of enzyme I fragments, and find that bands A–B and B–C light up. Then we probe a blot of enzyme II fragments with labeled B–C; this time D–E and E–F will light up. And so on. Note that we obtain genetic information: the order of restriction sites, and the order of adjacent restriction fragments. Moreover, the overlap method permits us to "walk" along the DNA, extending our genetic information in each step.

The chromosome walk has been widely used, in conjunction with classical genetics, to locate restriction fragments bearing genes of interest. For example, there are genes affecting the pattern of embryonic development in *Drosophila* which cannot be located on particular restriction fragments directly, for technical reasons. However, these genes have been mapped genetically, and other genes which map near them can be located on restriction fragments. Starting from these identified restriction fragments, molecular geneticists have walked along the chromosome by overlapping hybridization in order to grope their way to the developmental genes, or rather to the stretches of DNA that contain them. These stretches are then identified unambiguously through their altered restriction map in genetically characterized *Drosophila* strains which carry rearrangements of the developmental genes. Once identified in this way, the developmental genes can be subjected to direct DNA sequence analysis and experimental manipulation.

The chromosome walk illustrates the manner in which molecular genetics amalgamates classical genetics,

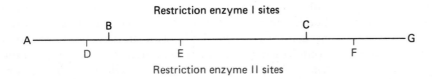

Restriction enzyme I sites

Restriction enzyme II sites

FIGURE 10–34
Overlapping hybridization. A hypothetical DNA sequence is shown schematically. The ends, A and G, are the sites at which a restriction enzyme cut this sequence out of a DNA preparation from whole cells; this unique fragment was isolated by gel electrophoresis, and perhaps amplified by cloning in a plasmid. The sequence contains unique target sites for restriction enzyme I (B and C, above the line) and for restriction enzyme II (D, E, and F, below the line).

cytogenetics, and recombinant DNA wizardry. Classical genetics and cytogenetics identify the genes, what they do, and the topological relationships within the chromosome that guide the chromosome walk; the DNA fragments are made by restriction enzyme digestion, amplified by cloning in plasmid vehicles, and analyzed by hybridization, and ultimately by DNA sequencing.

Reading the Information in Vitro—Sequencing DNA

It is the *primary structure,* or sequence of bases, which determines the major function of DNA; that is, information storage and information transfer. Can one determine the sequence of the bases in DNA by chemical means?

As recently as 1974 the prospect seemed bleak. The only bright element at that time was our increasing understanding of restriction endonucleases. The sequencing of amino acids in proteins became possible in the late 1940s, due to the work of F. Sanger, through the use of specific endoproteases. The sequencing of bases in RNA was achieved in the mid-1960s, again due to the work of Sanger and also R. Holley, through the use of specific endoribonucleases. The high specificity of these enzymes for only certain sequences of monomers in their respective substrates allows an unknown polymer to be broken into a small number of well-defined fragments. This is the first step in the analysis of the sequence of an informational macromolecule. The restriction endonucleases provided just such well-defined fragments of DNA, and thus allowed management of the problem of determining the otherwise dauntingly long sequence of bases in DNA.

The Sanger Method

The two major methods for determining the sequence of DNA are slightly different. We shall describe one of them in some detail, that of Sanger. His method exploits both the use of restriction endonucleases and the life cycle of a bacterial virus, M13, whose genomic DNA is normally in the form of a single strand. During replication inside a host *E. coli* bacterium, however, M13's DNA is in a double-stranded *circular replicative form*. This molecule can be isolated and purified for use in the determination of sequences (indeed, several variants of this replicative form are now available commercially). The first step is to open the circle by cutting it with one or another restriction endonuclease that attacks only a single site in the replicative form; one must use an enzyme that does not interrupt a gene specifying a vital function (Figure 10–35). The next step is to combine the opened circles with fragments of the DNA to be sequenced, that have been generated by cleavage with the same restriction endonuclease, and then let them join together by annealing. The same sticky ends are present in both the open replicative form and the fragments; many of the annealed products, therefore, will be closed circles containing a single fragment of the unknown DNA. Such a product is called *chimeric replicative form* or *recombinant DNA*. These noncovalently bonded circles are then sealed by the action of the ligase enzyme mentioned earlier. This procedure will yield a mixture of circular replicative forms, each with a different fragment of the test DNA.

The mixture is then separated by a biological, rather than a chemical, method. Naked DNA is taken into suitably prepared *E. coli* by the transformation process we discussed earlier. By choosing a sufficiently low ratio of DNA to cells, an experimenter can ensure that most cells receive no more than one chimeric replicative form. Once inside the *E. coli* and given appropriate conditions, the viral infection resumes and after a time M13 virus are secreted by the cell. If in the meantime the infected cells have been separated from one another, the secreted progeny virus can be isolated as pure populations, each containing information from only one of the original DNA fragments.

Now that a few pure viruses are available, their DNA can be obtained in larger quantities by going through another cycle of infection. Not much DNA is needed: as little as 50 μL of culture fluid from infected *E. coli* can yield sufficient viral DNA to determine the sequence of hundreds of nucleotides. After purifying the virus from the culture fluid and removing the viral coat protein, the single-stranded DNA is ready to be used for sequencing (Figure 10–35).

The choice by Sanger of a single-stranded DNA virus was ingenious because it allows another biological process to be exploited—DNA replication. The enzyme DNA polymerase I catalyzes the polymerization of deoxynucleotides into a single strand of DNA. The sequence of the product bases is determined by the sequence of bases in a necessary template DNA strand by means of the specific hydrogen bonding rules discussed earlier: A is placed opposite T, G opposite C, and vice versa. DNA polymerase I cannot initiate synthesis; it can only extend a preexisting single strand already hydrogen-bonded to the template. The experimenter supplies the needed *primer strand* DNA as a short sequence which is complementary to, and thus anneals to, a sequence in the viral DNA very near the site at which the restriction endonuclease cleavage occurs. (Providing suitable primers is now a growing business.) The DNA polymerase I then extends the primer through the cleavage site and into the unknown sequence of DNA. One or more of the precursor deoxynucleoside triphosphates is labeled with ^{32}P so the product DNA can be detected later by its radioactivity.

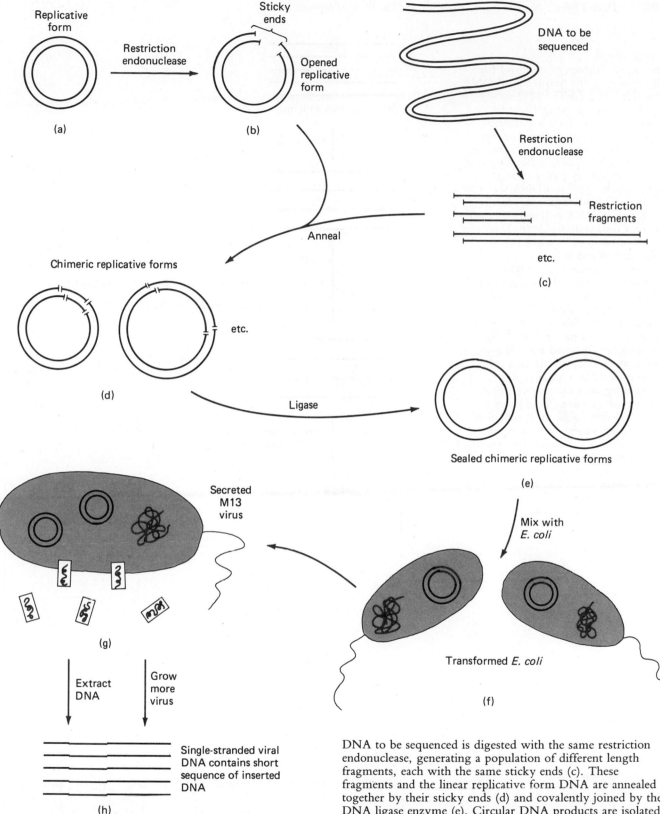

Replicative form

Restriction endonuclease

Sticky ends

Opened replicative form

(a)

(b)

DNA to be sequenced

Restriction endonuclease

Restriction fragments

etc.

(c)

Anneal

Chimeric replicative forms

etc.

(d)

Ligase

Sealed chimeric replicative forms

(e)

Mix with E. coli

Secreted M13 virus

Transformed E. coli

(f)

(g)

Extract DNA

Grow more virus

Single-stranded viral DNA contains short sequence of inserted DNA

(h)

FIGURE 10–35

Sanger's "dideoxy" method for determining the sequence of DNA. The two-stranded circular replicative form DNA of bacterial virus M13 (a) is isolated from infected *E. coli,* purified, and digested with a restriction endonuclease enzyme that makes a single staggered cut, generating a linear DNA molecule with "sticky" (complementary) ends (b). The DNA to be sequenced is digested with the same restriction endonuclease, generating a population of different length fragments, each with the same sticky ends (c). These fragments and the linear replicative form DNA are annealed together by their sticky ends (d) and covalently joined by the DNA ligase enzyme (e). Circular DNA products are isolated, purified, and mixed with suitably prepared *E. coli* at a density such that no more than one molecule of DNA enters a cell (f). Successful transformations yield a population of infected cells which are separated before they begin producing virus (g). Any virus from this growth cycle will have a single restriction fragment of the test DNA in its single strand of DNA. The virus can be grown as a pure population and the DNA isolated for further study (h).

FIGURE 10–36

Sanger's "Dideoxy" Method for Determining the Sequence of DNA (continued). (a) The M13 viral DNA containing the DNA to be sequenced (wavy line) is combined with a short single-stranded primer DNA. The primer is annealed to the viral DNA by hydrogen bonding to a complementary sequence close to the position where the test DNA was inserted. The primer is extended by the action of DNA polymerase I, using the four deoxynucleoside triphosphates as precursors (one or more being radiolabeled) and one dideoxynucleoside triphosphate, a precursor analogue. The radioactive product polymers are of different lengths because each has been terminated at a different point by the incorporation of the analogue. The products are separated according to length by electrophoresis on a gel made of agarose, typically, and detected by autoradiography. (b) A diagram of an autoradiograph of a ladder gel containing the products of five separate polymerization reactions, one containing a dideoxy analogue to each of the four normal DNA precursors and the fifth containing all four analogues. (c) The structure of the dideoxynucleoside triphosphate analogue. The color indicates the two deoxy positions.

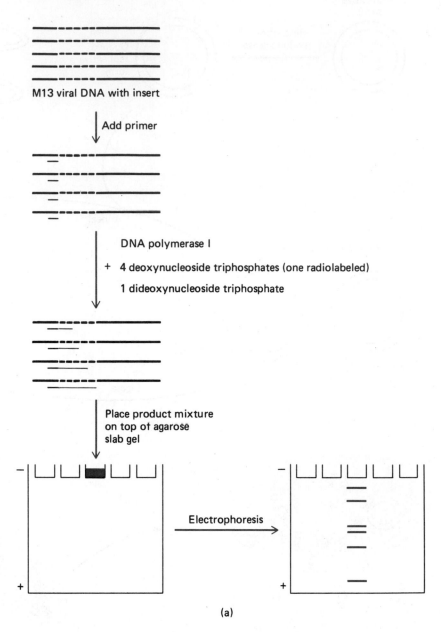

(a)

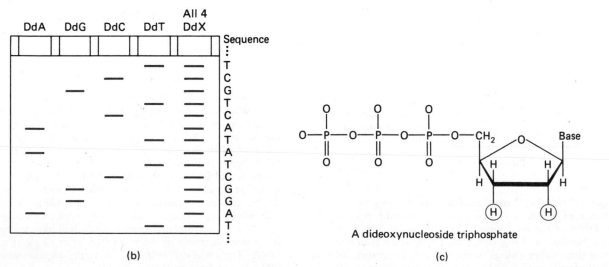

(b)

(c)

A dideoxynucleoside triphosphate

The principle of the final analysis of the DNA sequence is to produce fragments of the unknown DNA, each starting at the same place but each stopping at one or another of the points occupied by only one of the four possible bases. Sanger accomplished this by arranging to *terminate the polymerization,* with a certain probability, when one of the four deoxynucleotides was inserted. This was done by including in the polymerization reaction mixture a *dideoxynucleoside triphosphate* (Figure 10–36c). This precursor analogue has neither a 2′ nor a 3′ hydroxyl group (dideoxy); instead, it has only hydrogen at those positions of the ribose. As such, it can be added to a growing polynucleotide (through its 5′ phosphate) but it cannot accept another nucleotide at its 3′ position and DNA chain growth stops after the analogue has been inserted. A population of identical templates plus primers will, when replicated in the presence of the four deoxynucleoside triphosphates plus one dideoxynucleoside triphosphate, produce a set of radioactive product fragments. Each will be terminated with the dideoxynucleotide; some will have terminated early, others later (Figure 10–36a).

To separate the mixture of oligonucleotide fragments, Sanger and his group developed a refined slab gel electrophoresis technique which allowed them to separate chains differing in length by only one nucleotide. Agarose and acrylamide are substances which can be polymerized into a gel in the shape of a flat slab. The oligonucleotide mixture is applied to one end and is allowed to migrate through the slab in an electric field. Because the phosphates of the DNA are negatively charged, the fragments will migrate toward the positive electrode (anode). The rate at which a nucleic acid fragment will move through the gel will depend on the fragment's size; the larger it is, the more it is slowed down by the drag it experiences while passing through the gel. The position of the DNA in the gel is detected, after electrophoresis and drying, by placing the gel against a photographic film. After exposure for sufficient time (which may be only hours) and development, one obtains darkening of the film where the radiolabeled DNA bands are localized (Figures 10–36a, 10–37). Very long gels can be prepared which are capable of separating more than 300 oligonucleotides differing in length by one nucleotide each. These *ladder gels* (so named because the regular array of bands resembles the rungs of a ladder) form the basis of analysis of most methods of nucleic acid sequencing today.

Figure 10–37 shows a ladder gel of this sort which established the sequence of a DNA fragment from a human gene. Each lane of the gel corresponds to the incubation mixture that contained the indicated dideoxy analogue. Data of this kind contain sequence information in *positional* form: reading up or down the gel, the

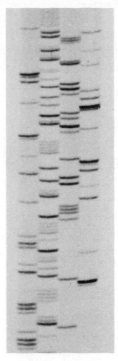

G A T C

FIGURE 10–37

A Ladder Gel, Illustrating the Sanger Dideoxy Sequencing Technique. Four DNA synthesis reactions are conducted; each one contains the dideoxynucleoside triphosphate of one of the four bases (G, A, T, or C), resulting in frequent termination of DNA chain growth at positions calling for the corresponding nucleotide. The reaction products of each of the four incubations are then resolved by gel electrophoresis, as in Figure 10–26. Each of the four incubation mixtures are run in separate lanes on the same gel, labeled in the figure according to the dideoxynucleoside triphosphate which was present. Autoradiography of the gel reveals the positions of end-labeled chains which terminated at positions calling for G, A, T, or C in each of the corresponding lanes.

position of successive bands in any of the four lanes locates positions calling for G, T, A, or C residues. The sequence can thus be read off by eye very easily. In fact, the sequencing can be automated by densitometer scanning of the autoradiographs, coupled to a computer that has been programmed to locate the positions of the bands in each lane and to record them in sequence.

We have explained how a length of DNA was sequenced from the cleavage product of a given restriction enzyme. Other portions of the DNA can be analyzed by using as the template other products of the same restriction enzyme, inserted in M13 viral DNA. Furthermore, one can get an entirely different set of cleavage products of the same DNA by using a different restriction enzyme. The *common* or *overlapping* sequences in these distinct cleavage products, generated by different restriction enzymes, permits longer sequences to be

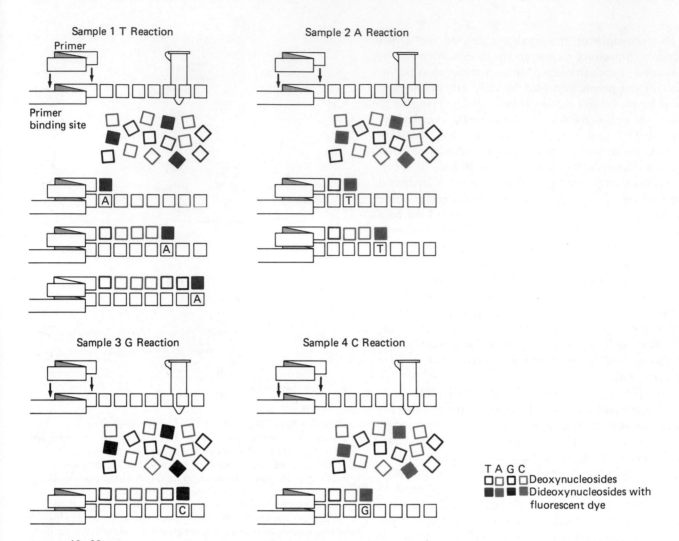

FIGURE 10–38

An automated version of the dideoxy method for sequencing DNA. In this variation of the method (see Figures 10–36 and 10–37) each of the four dideoxynucleotide triphosphates has been coupled to a different dye, so each terminated product fragment can be detected separately by the fluorescence of the specific dye. All four of the dideoxynucleotides are present in the polymerization reaction, together with the four deoxynucleotide triphosphates. The mixture of terminated fragments is resolved by electrophoresis through a single column of agarose gel, forming a ladder in the usual way. Electrophoresis continues so that each rung of the ladder passes completely through the column. As the rungs emerge from the column they are assayed by an optical system to determine which dye is attached and, therefore, which dideoxynucleotide terminated the fragment. The results are stored in a computer for further analysis.

reconstructed by connecting the fragmentary sequences at their positions of overlap. Moreover, the order of various restriction fragments along a length of DNA can be established by restriction mapping (as we discussed above). Thus, the relatively short sequences read from ladder gels (typically about 100 nucleotides) can be

hooked up to define very long sequences of genetic information.

Using these methods, Sanger's group determined the entire sequence of bacterial virus ΦX174 DNA (5386 nucleotides) in 1977, and the entire sequence of bacterial virus lambda DNA (48,502 nucleotides) in 1982. The most recent tour de force from this laboratory was the 1984 publication of the complete sequence of the genome of Epstein-Barr virus (172,282 nucleotides), which infects human cells. Complete sequence determination of mammalian chromosomes is technically within the realm of possibility and will undoubtedly be forthcoming in the next few years.

An automated version of the Sanger method has recently been developed by L. Hood and his colleagues. The four dideoxynucleotides are each coupled to a different colored (fluorescent) dye. Thus, each dideoxy-terminated fragment will have one of the four colors and each fragment terminated by the same dideoxynucleotide will have the same color. The four dideoxynucleotides are added to a single reaction mixture and the product fragments are all subjected to electrophoresis through a long tube containing an agarose gel, in which the usual ladder is formed (Figure 10–38). As a rung of this ladder emerges at the bottom of the column the color of the band is assayed by a spectrofluorimeter and recorded by a computer. This method offers greater convenience and allows more rapid analysis of sequences.

The Maxam-Gilbert Method

The strategy used by Sanger to determine DNA sequences requires replicating the DNA to establish the identity of the 3′ end of a fragment. Another way of accomplishing the identification would be to *break a DNA molecule preferentially at specific nucleotides*. The techniques were worked out by A. Maxam and W. Gilbert and their method is now in common use all over the world. We shall only give an outline of the logic of their approach:

1. DNA is broken into fragments by restriction endonucleases.
2. The ends of the fragments are labeled. First the terminal phosphates are enzymatically removed, usually with *alkaline phosphatase*. This is followed by ^{32}P labeling with *polynucleotide kinase,* which transfers the radioactive terminal phosphate from ATP to the DNA's 5′ ends.
3. After denaturation, the labeled single strands are separated by their differing electrophoretic mobilities on polyacrylamide gels.
4. A solution of one labeled single strand is subjected, in four separate reactions, to specific cleavages. The cleavages are done to produce low yields; roughly one in fifty susceptible sites are actually cleaved. This produces a series of labeled fragments of varying lengths.
5. The series is visualized by autoradiography of a ladder gel in which the different fragments have been separated by electrophoresis. The difference between two adjacent fragments in the ladder is the distance in the DNA sequence between the two susceptible bases.
6. The order of bands on adjacent lanes of the gel, corresponding to different cleavage regimes, allows the sequence to be read off.

The technology of determining DNA sequences is under continual development. The techniques are already so powerful that it is faster to determine the sequence of amino acids in a protein by analysis of its gene than by the conventional methods, provided, of course, that one has the DNA of that gene and that all the DNA information ends up in the protein. (As we shall see in Chapter 12, some of the sequence information in eucaryotic cell DNA is removed before being translated into amino acid sequence information.) Before turning to the interesting story of how the information is read out of the DNA we shall describe how the sequence of bases is replicated to provide the copies necessary for each daughter cell during cell division.

SUMMARY

DNA is the hereditary material, storing in a digital language the specifications for each and every living organism. In this chapter we have reviewed the gradual process of discovery and experiment which led to our present detailed knowledge of the structure of DNA, and the way in which this structure dictates its accurate replication, its variation, and the reading of its informational content in the expression of genes.

We conclude with a summary of modern *recombinant DNA* technology. This is a remarkable amalgam of biochemistry, genetics, and microbiology which makes it possible, even easy, to isolate a particular gene and determine its exact nucleotide sequence; to take it apart, put it back together, alter its sequence, splice it to a different gene, and transfer it from one species to another; methods, in short, which permit the manipulation of any gene virtually at will.

KEY WORDS

Base-pairing, endonuclease, equilibrium density gradient centrifugation, exonuclease, hybridization, ladder gel, nuclease, nucleoprotein, recombinant DNA.

PROBLEMS

1. Which of the two DNA fragments shown below has the higher melting temperature, and why?

GCAGTCACCTGG
CGTCAGTGGACC
Fragment 1

AATGTACATGTA
TTACATGTACAT
Fragment 2

2. For double-stranded DNA, assume the fraction of the adenine deoxynucleotides to be represented as "A," the fraction of guanine deoxynucleotides as "G," etc. If the quantity A + G = C + T = X, A + T = Y, and G + C = Z, then
   (a) What is the value of X?
   (b) What do the values of Y and Z signify, and how can they be used?

3. After melting by heating and reannealing by cooling a particular DNA preparation, it is found that some DNA remains single-stranded. What procedures can you use to remove the single-stranded DNA from this preparation?

4. The DNA of a certain virus has a mass of 4×10^6 Da. Of how many base pairs does this DNA molecule consist and how long is it?

5. (a) If the molecular weight of the DNA in an *E. coli* cell is 2×10^9 and the average molecular weight of an *E. coli* gene is 5×10^5, how many genes would be present in this procaryotic cell if we assume that all the DNA consists of genes (i.e., carries coding information).
   (b) Make the same calculation for a human cell in which the haploid DNA content has a molecular weight about 8×10^{10}.

6. Two samples of DNA, one from a wildtype virus and the other from a mutant strain, are mixed, heat-denatured, and annealed. Heteroduplex analysis shows the DNA from the mutant virus has two deletions (see diagram below). Lengths of the DNA segments are indicated in kilobases. Draw a map of the wildtype virus DNA.

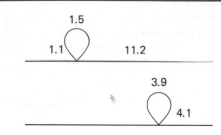

7. The autoradiogram diagrammed below is of a slab gel used to determine the sequence of a particular polynucleotide via the dideoxynucleotide method developed by Sanger.
   (a) What is the sequence of the DNA synthesized during the copying process?
   (b) What is the sequence of the desired polynucleotide?
   (c) What has happened to the primer DNA from which the copying process was initiated?
   (d) Why don't you see primer to which no nucleotides have been added?

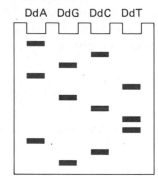

8. You have just cloned a DNA fragment which contains the gene for playing the clarinet and, of course, you want to obtain its restriction map. Complete digestion with the restriction enzyme <u>HaHa</u> II yields four fragments with the following molecular sizes (in ascending order): A (1.5 kb), B (3 kb), C (4 kb), and D (5 kb). Partial digestion with the same enzyme at a much lower concentration yields only traces of these fragments, but mostly five larger fragments: E (5.5 kb), F (6.5 kb), G (8.0 kb), H (9.5 kb), and I (10.5 kb). Draw a restriction map of

your cloned DNA, showing the locations of the <u>HaHa</u> II sites.

9. If, for a particular DNA, A + T = 66%, what is the probability of finding in a random sequence the recognition sequence of the following restriction enzymes:

   (a) <u>Bam</u> HI

   (b) <u>Hind</u> III

   (c) <u>Sma</u> I

   (d) <u>Hpa</u> II

10. You are using a probe to screen genome libraries, by colony hybridization, for a plasmid carrying the complementary stretch of DNA. Suppose you have one library made from genomic DNA cut with a restriction enzyme that recognizes a four-base target site, and another library made with a six-base cutter. Which library will permit you to screen the *smaller* number of colonies with a good chance of finding the gene?

11. If the (A+T)/(G+C) ratio is 0.39, what will be the average size of fragments produced by restriction enzyme:

    (a) <u>Taq</u> I

    (b) <u>Eco</u> RI

12. Why do different DNA fragments cut from genomic DNA with the same restriction enzyme have "sticky ends?"

13. Scientists in Seattle and Philadelphia cloned the gene for rat growth hormone and injected it into fertilized mouse eggs. A few of the resultant mice grew to about 60% larger than normal size (equivalent to a nine-foot human), thus putting the U.S. definitely ahead of Japan in the race to create a big mouse. Some of the big mice passed the trait of abnormal size on to their offspring, but most did not. Explain.

SELECTED READINGS

Abelson, J. and Butz, E., eds. (1980) Recombinant DNA. *Science* 209 (4463) [entire issue].

Bauer, W. R., Crick, F. H. C., and White, J. H. (1980) Supercoiled DNA. *Scientific American* 243:118–33.

Chargaff, E. (1978) *Heraclitean fire*. New York: Rockefeller University Press.

Dickerson, R. E. (1983) The DNA helix and how it is read. *Scientific American* 249: 94–111.

Judson, H. F. (1979) *The eighth day of creation*. New York: Simon and Schuster.

Maynard Smith, J. (1986) *The problems of biology*. London: Oxford University Press.

Olby, R. (1974) *The path to the double helix*. Seattle: University of Washington Press.

Sayer, A. B. (1978) *Rosalind Franklin and DNA*. New York: W. W. Norton.

Watson, J. D. (1980) *The double helix*. ed. G. S. Stent. New York: W. W. Norton.

Watson, J. D. and Crick, F. H. C. (1953) Molecular structure of nucleic acid: A structure for deoxyribose nucleic acid. *Nature* 171: 737–38.

Watson, J. D. and Tooze, J. (1981) *The DNA story: The documentary history of gene cloning*. San Francisco: W. H. Freeman.

The Replication and
Organization of
Genetic Information

11

Unlike the cytoplasm, which seems to have many and varied functions, the nucleus is mainly devoted to the preservation and reproduction of genetic information, and to the expression of this information to control the metabolic destiny of the cell. It is now well-established that the nucleus is the repository of the Mendelian genes. In eucaryotes these genes are located on structures called *chromosomes*. In both procaryotes and eucaryotes the genetic information is embodied in the chemical called DNA.

When the double-helical model for DNA was published by Watson and Crick it was immediately evident that the proposed structure neatly explained two basic requirements of genes: to contain information for control of the cell's activities and to be reproducible, guaranteeing that, after division, each daughter cell has the same information as the parent. Information is encoded in the *linear sequence of nucleotides in the DNA molecule,* just as the information in this text is encoded in the linear sequence of letters, spaces, punctuation marks, etc. The alphabet for the DNA code, however, contains only four symbols: A, G, T, and C.

In principle, information storage should require only a single strand of DNA. The other complementary strand found in cellular DNA is used for the *replication* (exact duplication) of the information. In fact, certain viruses in their mature forms have their genetic information encoded in only a single strand of nucleic acid. When that information is replicated during virus infection of cells, however, a complementary strand always seems to participate. During replication, one strand of DNA (or of RNA in the case of certain bacterial and plant viruses) acts as a *template* for the synthesis of a complementary strand.

DNA always plays a template role of this sort when its encoded information is *transcribed* into the other information molecule *ribonucleic acid* (RNA). (In any given gene, only one strand of the DNA is used as the template; however, one strand may be used in one gene and the other strand in another gene.) Transcription, described in the next chapter, is the first step in *gene expression*: the reading out of the genetic information into a form that can carry out the cell's metabolism. In the next step, much of the information undergoes *translation* into the form of proteins, including enzymes which actually exert the control by virtue of their ability to catalyze those chemical reactions that occur in the cell. Translation occurs on structures called *ribosomes*, located in the cytoplasm (see Chapter 13).

DNA Localization

There is a cytological stain that reacts specifically with DNA (the Feulgen stain). Early in this century it was shown by means of this stain that almost all the eucaryotic cell's DNA resides in the nucleus (Figure 11–1). More sensitive methods have since revealed that small amounts of DNA, probably less than 1%, are found in mitochondria and chloroplasts. Bacteria and other procaryotic cells have their DNA in a *nuclear region* (or *nucleoid*) not bounded by a membrane (Figure 11–1). Since viruses have genes and genes are made of nucleic acid, we should also expect, and have in fact found, DNA in viruses.

The amount of DNA per nucleus is generally the same for all somatic cells of an organism and is twice that found in the germ cells. We can go further and say that the quantity of DNA in any chromosome is generally a fixed amount, characteristic of the particular chromosome and the species. This amount of DNA does not fluctuate during different metabolic or nutritonal states in which the cell finds itself.

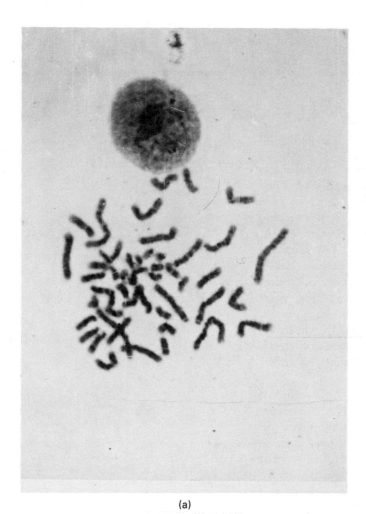

(a)

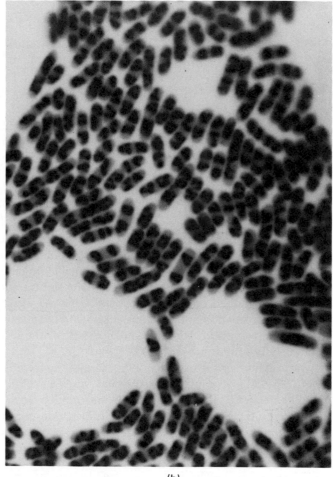

(b)

FIGURE 11–1

The Feulgen Technique for DNA.   (a) Human blood lymphocytes were obtained at metaphase arrest, hypotonically disrupted, and a spread made of the nuclei and chromosomes. The figure shows undisrupted interphase nuclei with DNA staining darkly and isolated metaphase chromosomes, showing the banding pattern of dark staining DNA, separated by negative regions. The Feulgen reaction utilizes the Schitt reagent to stain the DNA. (Courtesy of T. Rodman).

(b) Cells of the bacterium *Shigella dysenteriae* were fixed and subjected to the HCl-Giemsa procedure, which stains DNA specifically in a manner analogous to the Feulgen reaction. The bacterial *nucleoids,* which are not surrounded by a nuclear membrane, appear as darkly staining irregular bodies, generally one, two, or four per cell. (Magnification ×3900) (Courtesy of R. G. E. Murray).

A more stringent test of the apparent stability of the genes is whether any of the components of DNA might be synthesized and broken down at balanced rates, giving the illusion of constancy. Cell biologists do this by adding a radiolabeled precursor (e.g., $[^{32}P]$-phosphate), and examining the DNA for incorporated radioactivity. Nondividing cells become rapidly labeled in all sorts of other phosphorus-containing compounds, but very little radioactivity is found in the DNA. Once the DNA of the chromosome is synthesized in such cells, it is largely stable. This is precisely what we should expect of a molecule that carries the information necessary to specify a cell. Once the DNA is made its molecular pattern is largely set and it undergoes few additional metabolic transformations.

Nuclear Division and the Cell Cycle

This is not to say that DNA is completely inert, since when the cell is to divide, the DNA must be replicated and the chromosomes segregated into daughter cells. These latter events, visible under the microscope, constitute the process called *mitosis.* Mitosis, however, is only one phase—albeit the most dramatic—of a larger sequence of events known as the *cell cycle,* which includes many less easily observable processes that occur between successive cell divisions (see box on exponential growth).

Our understanding of the cell division cycle progresses as our analytical tools improve. Early microscopists could detect only two phases, mitosis and *interphase*. In the interphase nucleus—the nucleus of a nondividing or "resting" cell—the light microscope can resolve little but the dense unbounded area known as the *nucleolus* (see Chapter 2). Outside the nucleolus, in the *nucleoplasm,* lie less dense areas with a minimum of visible internal structure (Figure 2–31). During this period, the chromosomes are extended slender structures lying intertwined in the nucleus. They are usually too thin to be resolved and do not absorb much stain, so the nucleus appears, under the light microscope, to contain a fine mesh of slender threads, formerly known as the chromatin network. A few characteristic regions of the chromatin, however, are more compact and, therefore, stain more intensely at this stage. These regions are known as *heterochromatin*. The DNA of the extended regions, called *euchromatin,* is capable of being transcribed into RNA. The more condensed heterochromatin is believed not to be available for transcription. In the interphase nucleus, heterochromatin predominates in a region next to the nuclear membrane and also around the nucleolus (Figures 2–31 and 2–45). The relative amounts of heterochromatin and euchromatin in the interphase nucleus depend on the particular developmental stage a cell has reached in a developing organism. Some heterochromatin, however,

appears to be permanent, such as the heterochromatin of one of the X chromosomes of many female mammals, including the human.

During the early stages of mitosis, the structure of the nuclear material is progressively more visible. Proteins of the nucleoplasm, including histones and acidic proteins, participate in the *condensation* of DNA-containing chromatin threads into the chromosomes that are easily seen in the light microscope (Figure 11–2). Before the condensation occurs the DNA has already been

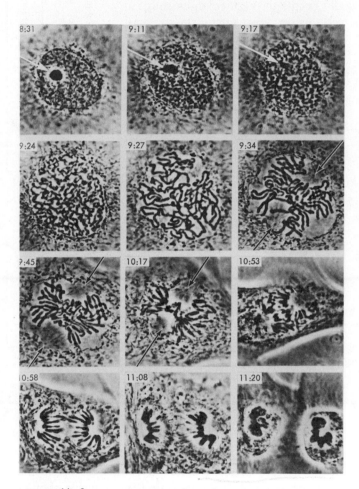

FIGURE **11–2**
Mitosis Sequence in an Animal Cell. Single salamander mesothelial cell in tissue culture was photographed at successive times during mitosis with phase contrast microscopy. ×800. The white arrows point to the nucleolus in the interphase nucleus; the black arrows point toward the centrioles. The numbers in the upper left indicate time by the clock. The cell was first photographed during interphase at 8:31. Prophase extended from about 9:20 (note that the nucleolus has disappeared and the chromosomes have begun to appear in the photograph taken at 9:24) to about 9:34. Metaphase extended from about 9:34 to about 10:30, and anaphase from about 10:30 to slightly after 11:08. The last photograph, taken at 11:20, shows the cell in telophase, with two newly formed daughter nuclei. (From *A Textbook of Histology,* by W. Bloom and D. Fawcett.)

Exponential Growth and the Doubling Time

Growth in the world of life usually occurs through enlargement of a cell to a certain characteristic size and its subsequent division into two daughter cells. This principle applies generally, whether the cells are part of a colony of microorganisms or a multicellular system like an oak tree or a man. Now, growth by division is exponential: 1, 2, 4, 8, 16, etc. After n generations, the number of cells will be 2^n. It thus takes only 20 generations, for example, for the population to exceed 10^6. The increase in numbers is so rapid that growth cannot remain in such an *exponential phase* for long. It is soon followed by a decelerating phase, with an eventual leveling off of cell numbers (Figure 11–A). During the exponential phase of growth, cells are in their most active metabolic condition; consequently, most experimental studies are performed with cells harvested in the exponential phase of

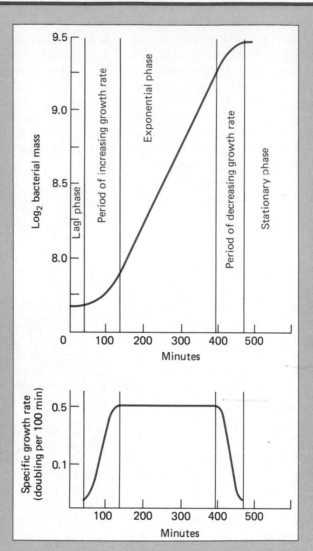

FIGURE **11–A**
The Bacterial Growth Curve. The upper part of the figure shows how bacterial biomass (which could be measured as protein per unit volume of culture, or as the number of cells per unit volume) increases with time. The logarithm (to the base 2) of the biomass is plotted against the time. In this semilogarithmic plot, exponential growth graphs as a straight line (this is often, though erroneously, called the "log phase" of growth). The lower part shows how the specific growth rate changes; it is constant only during the exponential phase of growth. (Courtesy of W. S. Sistrom, *Microbial Life.*)

replicated; chromosomes visible during mitosis at metaphase are doubled. After division is complete the chromosomes uncoil and become again invisible, though the DNA can still be demonstrated by the Feulgen stain or other chemical assay. After the Second World War, when radioisotopes became available, the incorporation of labeled precursors in DNA during interphase allowed identification of the S (synthesis) phase (Figure 11–3). The interval after mitosis and before S is called the first "gap" (G_1), and the interval between S and mitosis the second "gap" (G_2).

Although there are a number of variations on the main theme, the "mitotic cycle" of nuclear division correlates with the overall cycle of cell division in the following general way. Let us begin with the end of nuclear division. Two daughter nuclei have formed, and

the parent cell has divided into two daughter cells. There follows a period of growth when both the nucleus and cytoplasm enlarge until the cell has attained the same size as its parent just before division. Biochemically, it is a period of active protein and RNA synthesis and often represents some 30–40% of the generation time (doubling time) of dividing cells. This is the G_1 phase in Figure 11–3.

The G_1 phase is followed by the S phase, during which DNA and histone syntheses occur. At this stage the cell's hereditary material is duplicated in the form of two identical copies of each chromosome, which are called *chromatids*. In their slender extended form the chromatids cannot be separated from each other because all the chromosomes are extensively intertwined with each other. The S phase may take up 30–50% of the

TABLE 11–A
Exponential Growth

Time	Doublings	Cell Number
0	0	N_0
T_d	1	$N_0 \cdot 2$
$2 \cdot T_d$	2	$N_0 \cdot 2^2$
$3 \cdot T_d$	3	$N_0 \cdot 2^3$
.	.	.
.	.	.
.	.	.
$d \cdot T_d$	d	$N_0 \cdot 2^d$

Note: The data describe a culture of cells growing at a maximum rate, unlimited by resources. The initial number of cells, at time t = 0, is N_0. The cell number is tabulated after successive doublings, each of which takes T_d units of time.

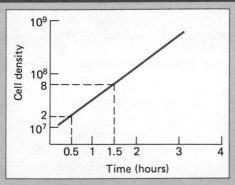

FIGURE 11–B
Time Course of Growth of a Bacterial Cell Culture. The cell density was measured at various times (by microscopic observation or turbidity, etc.) and plotted on semilogarithmic graph paper against the time in culture. Exponential growth is shown by the straight line graph. Since the culture increased from 2×10^7 to 8×10^7 (two doublings) in one hour, the doubling time must be 30 minutes for these cells.

their growth cycle. By making direct or indirect measurement of the growth kinetics of cultured cells or organisms, it is possible to calculate a "doubling time," which is the average time that has elapsed between successive cell doublings during the exponential phase of growth. The doubling time under favorable growth conditions for *E. coli* can be as low as 20 minutes, while for mammalian cells in tissue cultures it may be many hours.

Table 11–A shows how the cell number varies with the number of doublings and at what time during exponential growth that number describes the cell culture. If we take *t* as the time, then t = d · T_d or d = t/T_d. The cell number can then be written

$$\frac{N}{N_0} = 2^{t/T_d}$$

Taking logarithms of both sides of this equation, one obtains the equation of a straight line:

$$\log \frac{N}{N_0} = \frac{\log 2}{T_d} t.$$

The line relates log N/N_0 to the time t. The slope of the line is (log 2)/T_d. By plotting N/N_0 against t on semilogarithmic paper (Figure 11–B), the doubling time can easily be read from the graph.

generation time of the cell. It is followed by the G_2 phase; we do not yet understand all the metabolic significance of this phase, and thus we refer to it in our ignorance as "preparation for mitosis." The G_2 phase occupies some 10–20% of the generation time, followed by the M phase (mitosis) and the D phase (division of daughters), which normally lasts some 5–10% of the generation time.

DNA Replication

Having glanced briefly at the visible cellular events involved in the duplication of the genetic material, we now proceed to examine in some detail the molecular processes by which the DNA is replicated.

Enzymatic Synthesis of DNA

In 1957 A. Kornberg isolated an enzyme from the bacterium *E. coli* that could synthesize DNA. This earned Kornberg a share of the Nobel Prize for medicine in 1959. The enzyme was called *DNA polymerase* (now known as DNA polymerase I) and seemed capable at the time of explaining all that was necessary for the replication of DNA. During the intervening years many further observations have revealed a greater complexity and our ideas about the synthesis of DNA have evolved accordingly.

Even before the discovery of DNA polymerase, biochemists made some guesses as to its likely mode of action, guesses based on the structural features of DNA. The Watson-Crick model suggested that each chain of the double helix could act as a template for the alignment of

Mitosis

Because the condensed chromosomes in a dividing nucleus take up stains very strongly and can therefore be seen clearly in the light microscope, the structural events occurring during mitosis have been under intense scrutiny by light microscopists since the early years of this century. Their interpretations, in which the sequence of mitosis was pieced together from observations of dead cells, have been dramatically corroborated by studies of living material with the phase contrast microscope. By emphasizing differences in refractive index in the material to be observed, this microscope reveals beautifully the structure of the chromosomes of the living cell during mitosis.

The process of mitosis is traditionally divided into four main phases, which are arbitrary divisions of a continuous process (Figure 11–2).

We start with the *prophase* of an animal cell, in which the chromosomes gradually shorten and thicken —a phenomenon brought about by the extensive coiling of the two chromatids produced in the S phase (when the DNA was duplicated)—and still lie close and parallel to each other. The chromatids coil on themselves rather than around each other so that eventually they can separate. The double character of the chromosomes becomes increasingly obvious during prophase. At the same time one can also see the gradual disappearance of the nuclear boundary and the nucleolus (these last events are also considered to mark the beginning of a *prometaphase* stage in mitosis, which lies between prophase and metaphase). As the shortening and thickening proceeds, one can observe that the chromatids are held together at one particular point, called the *centromere.* Toward the end of prophase the chromosomes migrate to the *equatorial plane* of the dividing nucleus and, in animal cells, the *spindle* begins to be organized between the *centrioles* while they still lie to one side of the nucleus. As the centrioles move to opposite *poles* of the nucleus, the spindle enlarges and moves through the nuclear region. In plant cells centrioles appear to be absent, and the spindle is immediately organized to lie between the poles of the dividing nucleus.

At *metaphase* (Figure 11–C(4)) the spindle is fully developed, the chromosomes are poised at the equatorial plane, and the nuclear envelope and nucleoli have completely disappeared. At this point the chromosomes can easily be counted, and their size and shape determined. Thus, it can be shown that every species has a characteristic number of chromosomes. If one is observing a mitotic division during the diploid stage of a given species' life cycle, for every chromosome there is another *homologous chromosome* of similar morphology. Pairs of homologous chromosomes can often be distinguished from other pairs by the relative length of the "arms" on either side of the centromere, which is located in a constant position in each particular chromosome.

Anaphase begins when the chromatids of each chromosome start moving apart. The centromeres of each chromosome have split and the daughter centromeres appear to lead the rest of their chromosomes toward the poles. The use of micromanipulation techniques can show that the centromeres are attached to the spindle. We shall discuss the mechanism by which the chromosomes move in Chapter 16.

Once the chromosomes have reached the opposite poles of the spindle, *telophase* begins; it consists of the apparent reversal of all the changes occurring in prophase. The spindle disappears, the chromosomes extend by uncoiling, the nucleoli slowly appear again, and the nuclear envelope is once again laid down.

As these dramatic nuclear events unfold, the whole cell divides into two daughter cells *(cytokinesis),* in the

the nucleotide bases of a growing chain by specific hydrogen bond base-pairing as depicted in Figure 10–10. An enzyme would be necessary to test the fit of the incoming nucleotide base and to attach it to the growing chain by a phosphodiester bond. Since the depolymerization of polynucleotides is spontaneous, their synthesis should involve an activated molecule. Thus, the *mass* and *energy* required for the synthesis would both be contained in nucleoside triphosphate precursors and the *information* would be supplied by the DNA polymerase system and an already existing strand of DNA. These expectations were met in general and an overall scheme of DNA synthesis is shown in Figure 11–4.

DNA synthesis proceeds by the addition of monomeric units to a growing chain or **primer.** The deoxynucleoside triphosphate is joined to the 3'—OH of the

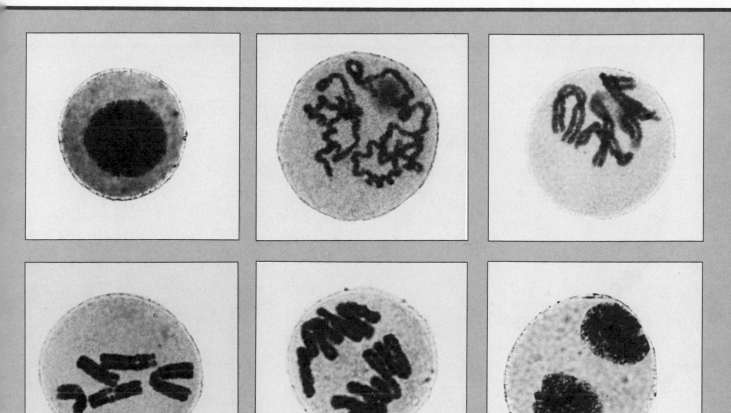

FIGURE 11–C
Mitosis Sequence in a Plant Cell.   Stages in the mitotic cycle of *Trillium erectum* microspores. ×800 (Courtesy of A. H. Sparrow and R. F. Smith.)

animal cell by a process of active constriction of the cell, and in the plant cell by the growth of a cellulose wall. Figure 11–C depicts the chromosomal changes during mitosis in a plant cell.

Mitosis is so characteristic of eucaryotic cells that it is easy to believe it evolved at the same time eucaryotes arose from procaryotes. Its function, clearly, is that of apportioning accurately the hereditary material, which in eucaryotic cells comes in several packages or linkage groups—that is, chromosomes. The visible result of cell division is the equal distribution of the daughter chromosomes to the two daughter cells. The shortening of chromosomes during the prophase of mitosis makes their disentanglement and migration possible.

primer by its α-phosphorous atom and inorganic pyrophosphate is split off. Thus, a new 3′—OH is in place on the primer and ready to join with the next precursor triphosphate. There is an absolute requirement for a template (parent) DNA strand to direct the order of nucleotides to be placed in the growing chain. The template strand and the growing strand are antiparallel, as shown in Figure 11–4, and the Watson-Crick pairing rules determine which of the four possible precursor deoxyribose triphosphates will be installed (adenine with thymine, guanine with cytosine).

The first attempts to show DNA synthesis in vitro were done with crude extracts. These contained so many nucleases that one observed a net degradation of the DNA in the reaction. The only way to prove that synthesis had in fact occurred was to use radioactive

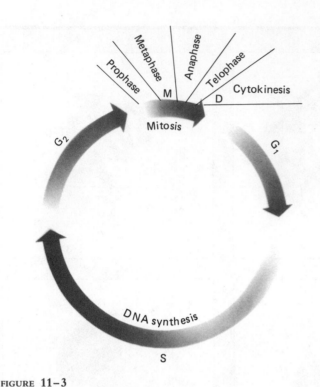

FIGURE **11–3**
The Cell Cycle.   The circle represents the cyclical behavior
of regularly dividing cells, with the clockwise arrow
representing the time direction; i.e., the order of the
successive phases in time. After cell division, there is a period
during which no DNA replication occurs (the G_1 phase).
After this, DNA replication occurs during the S phase. Once
DNA replication is complete, a phase (the G_2 phase) precedes
the condensation of the chromosomes at the beginning of
mitosis. Mitosis, the M phase, is subdivided into the
successive steps (prophase, metaphase, anaphase, telophase)
which reflect the behavior of the chromosomes. Nuclear and
cell division then ensues (the D phase, which is relatively
short) and the whole thing starts up again.

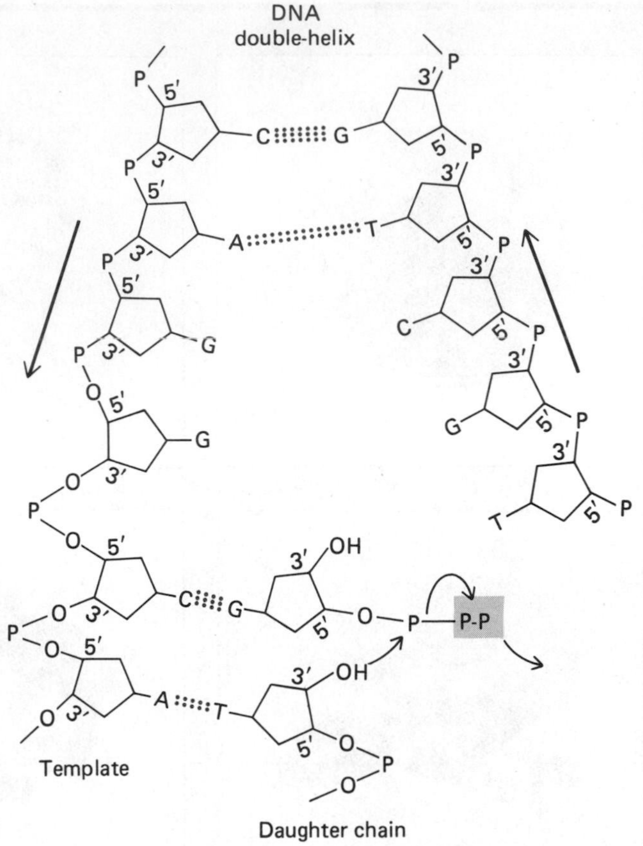

FIGURE **11–4**
Mechanism of Enzymatic Replication of DNA.   Two
parental (template) DNA strands are shown: the one on the
left runs downward from 5'-to-3' and the one on the right
runs in the opposite direction (antiparallel, see arrows). The
strands are unwound in the region shown and the left one is
being replicated by synthesis of a daughter strand in the
5'-to-3' direction, starting from the bottom. One base (T) of
the daughter strand is shown paired by two hydrogen bonds
to the A of the parent strand. The next nucleotide to be
added is a G (in the activated form of dGTP), which is
shown making three hydrogen bonds with the next base (C)
of the parent strand. The phosphodiester bond is made
between the thymidine monophosphate in the growing strand
and the α-phosphorus of the incoming dGTP with splitting
off of pyrophosphate (boxed). The later hydrolysis of the
pyrophosphate by inorganic pyrophosphatase keeps its
concentration low and drives the synthetic reaction forward.
(From A. Kornberg.)

precursors. Then it was found that radioactive nucleotides were in fact incorporated into large DNA molecules. This was shown by the conversion of an *acid-soluble*
precursor radioactivity into an *acid-insoluble* (because of
the size of the DNA) product. Using this crude system it
was possible to demonstrate a requirement for template
DNA and for all four deoxynucleoside triphosphates.
The substrates had to be the triphosphates and only
deoxyribonucleotides, not ribonucleotides, were active in
the system. Even though the enzyme was from bacteria,
Kornberg showed that the template DNA could be of
animal, plant, bacterial, or even viral origin.

Using a more highly purified enzyme preparation,
actual net synthesis of DNA was demonstrated: over
twenty times the amount added as template. This allowed
more rigorous testing of the product DNA in comparison
to the template. Since the physical properties of size,

sedimentation rate, viscosity, and optical properties were
those of double-strand DNA, it was hypothesized that
each of the single strands of the added DNA acted as a
template for the synthesis of complementary strands.
Chemical analyses verified the hypothesis: first, the
product had equal amounts of A and T and equal
amounts of G and C; second, the ratio of A + T to

G + C of the product was the same as that of the template. This can be seen from Table 11–1, and together these results could only be obtained if both strands of the added DNA functioned as templates. As also shown in the table, when the copolymer of adenine and thymine (poly dAT) was added as template, with all four deoxynucleoside triphosphates, only deoxyadenosine and deoxythymidine were found in the product.

Enzymatic replication of DNA was also able to verify the base-pairing theory by using as template a single-strand DNA from bacteriophage ΦX174. This DNA, being a single strand, need have no equivalence between adenine and thymine or between guanine and cytosine, and it does not. It was expected that, after a limited amount of synthesis in the presence of radioactive precursors, only enough DNA might be made to create double-strand molecules. In that case the newly synthesized chain should have radiolabeled deoxyadenosine equal not to the radioactive deoxythymidine but to the unlabeled deoxythymidine of the single-strand template. That was in fact what was observed.

The Meselson-Stahl Experiment

The first synthesis of DNA in vitro gave much information about the requirements for precursors, ions, template, etc., but could not answer a fundamental question posed by Watson and Crick. The question concerns the physical relation, as opposed to informational relation, between template and products. There were three main models to be distinguished: *conservative,* *semiconservative,* and *dispersive* replication. In conservative

replication one or both parental DNA strands would act as template for the synthesis of a new strand, but the two parental strands would reassociate, as would the two product strands, forming the two molecules that are distributed to daughter cells. Semiconservative replication is similar to conservative through the synthesis step, at which point each template strand associates with its own product strand to form two molecules. One can see that base pairing and antiparallel chain direction properties are shared by both of these models. In dispersive replication the template would disintegrate while (or after) forming two product molecules, leaving none of its structure to the daughter cells as highly polymerized DNA. To choose among these three possibilities required the use of a *physical label* so that the template and product DNA could be separated from one another.

M. Meselson and F. Stahl found that *E. coli* could be grown in a simple medium in which all the nitrogen came from nonradioactive but dense $^{15}NH_4Cl$. The DNA in such cells becomes heavily labeled with the ^{15}N and is more dense, from the additional neutron in ^{15}N, than DNA synthesized from precursors containing the normal isotope ^{14}N. To separate the two types of DNA, Meselson and Stahl exploited the technique of *equilibrium* *density gradient sedimentation* (see Figure 10–18). The band of DNA labeled with ^{15}N floats in the density gradient at a position well away from the band of DNA labeled with ^{14}N (see Figure 11–5a).

The experimental design that Meselson and Stahl employed was to grow a culture of *E. coli* for many generations in ^{15}N until essentially all the cells' DNA was "heavy." Then they added an excess of the normal

TABLE 11–1

Comparison of Base Composition of DNA Synthesized In Vitro with Template DNA

Source of Template	Adenine Plus Thymine/Guanine Plus Cytosine	
	Template DNA	Synthesized DNA
Micrococcus lysodeikticus	0.39	0.41
Mycobacterium phlei	0.49	0.48
Aerobacter aerogenes	0.82	0.80
Escherichia coli	0.97	1.01
Bacteriophage	1.06	1.00
Bacillus subtilis	1.29	1.26
Calf thymus	1.25	1.32
Hemophilus influenzae	1.64	1.62
Bacteriophage T2, T4, T6	1.84	1.76
dAT copolymer	>40	>250

Note: Data after A. Kornberg, *Enzymatic Synthesis of DNA,* John Wiley and Sons, 1961. The data of the table are explained in the text. In all cases there occurred a large net synthesis of DNA, using templates noted on the left; thus, the base composition of the DNA isolated at the end of the reaction was largely that of the DNA synthesized in the reaction.

The Nearest Neighbor Experiment

An elegant test of the Watson-Crick model made possible by the cell-free DNA synthesizing system was the "nearest neighbor" technique worked out by J. Josse, A. D. Kaiser, and Kornberg. To do the nearest neighbor experiment required each of the four deoxyribonucleoside triphosphates with ^{32}P in the alpha, or proximal, position (Figure 11–D). One of these radiolabeled precursors was added, with the other three unlabeled nucleoside triphosphates and DNA, to the DNA polymerase and allowed to react. After labeled DNA was synthesized in four such experiments, each with a different labeled precursor, the product DNAs were isolated, purified, and hydrolyzed enzymatically to yield single nucleotides. The enzymes used, micrococcal nuclease and bovine spleen phosphodiesterase, cleave DNA to give 3'-deoxyribonucleoside phosphates.

Since the ^{32}P label was incorporated from a 5'-triphosphate and was released as a 3'-monophosphate (Figure 11–D), the radioactivity was then attached to the labeled precursor's *nearest neighbor* in the DNA strand—but always the neighbor lying *in one direction* along the strand. Thus, if we call the direction from the 5' end to the 3' end of the strand the "downstream" direction (indicated by the arrows in Figure 11–E), the labeled phosphate is always transferred to the nearest neighbor on the "upstream" (5') side of the initially labeled nucleotide.

The Watson-Crick model specifies that the two complementary DNA strands run antiparallel to one another. One could imagine instead that the two strands were parallel. In Figure 11–E we show two possible ways of arranging the two DNA strands in a double helix: antiparallel and parallel. Suppose, for example, we used α-^{32}P-labeled dGTP in one of the reactions described above. Note that the left strand in Figure 11–4b contains the sequence 5'-ApG-3' (i.e., the labeled phosphate is on the 3' side of the A and on the 5' side of the G). Thus, after synthesis with the labeled dGTP and enzymatic digestion, we should expect to find some labeled dAMP, the exact amount reflecting the actual frequency of the 5'-ApG-3' sequence in the original molecule. It is clear that whenever and wherever the sequence AG occurs the sequence TC will occur in the complementary

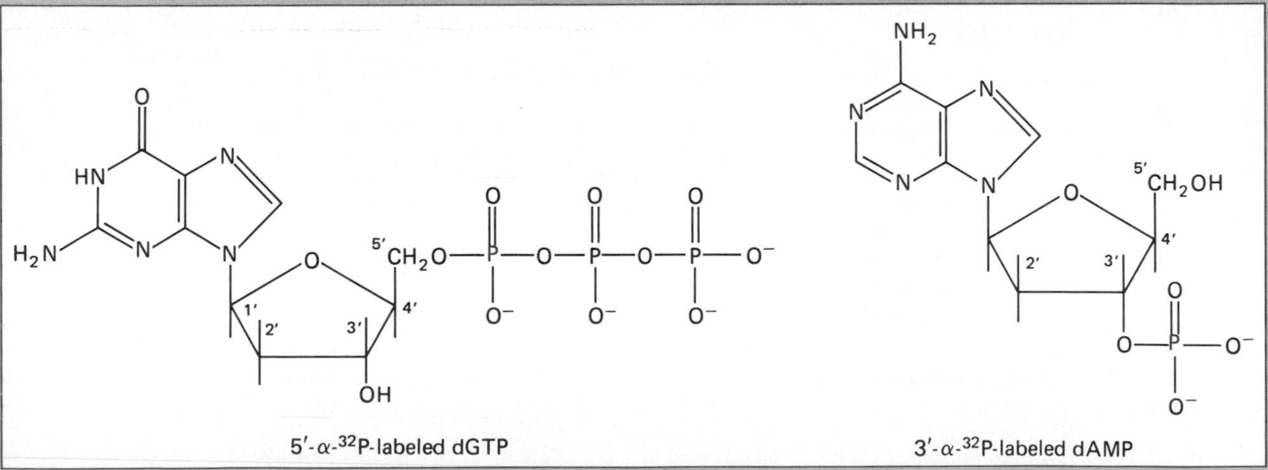

FIGURE 11–D

5'-α-^{32}P-labeled dGTP and 3'-α-^{32}P-labeled dAMP.   The α-^{32}P radioactivity in the dGTP will remain associated with the DNA after polymerization (replication). When the DNA is degraded with micrococcal nuclease and bovine spleen phosphodiesterase, the product 3'-deoxynucleotide phosphate will become labeled if that nucleotide was next to the dGMP in the DNA strand (see Figure 11–E).

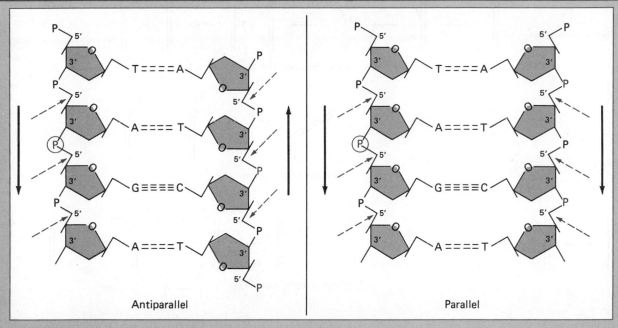

Antiparallel   Parallel

FIGURE 11–E

Use of "Nearest Neighbor" Analysis to Establish the Polarity of the Two Strands of DNA in the Double Helix. The colored Ps represent radioactive phosphorus. The dotted line with arrows shows the points of cleavage by the enzyme used to hydrolyze the DNA, giving 3'-deoxyribonucleoside monophosphates. The shaded heavy arrows show the direction of each of the strands, from 5'-to-3'.

strand. The antiparallel model predicts that opposite 5'-ApG-3' should always be 5'-CpT-3', while the parallel model predicts 5'-TpC-3' instead (Figure 11–E).

Suppose that we next repeat the experiment using α-^{32}P-dTTP labeling. If the strands are antiparallel, each occurrence of the 5'-CpT-3' sequence will give rise, after enzymatic digestion, to radioactive dCMP. Thus, the amount of radioactive dCMP recovered after dTTP labeling should be the same as the amount of radioactive dAMP recovered after dGTP labeling. This is, in fact, the case, providing an elegant confirmation of the antiparallel model.

The power of the nearest neighbor experiment is its ability to distinguish between the parallel and antiparallel models. One can see that there are 16 possible dinucleotides, and information about 4 of them can be obtained from an experiment using one radioactive precursor—if one incorporates α-^{32}P-labeled dGTP, the transfer, after enzymatic digestion, to nearest neighbors of all four possible kinds can be quantitated. The antiparallel model makes different predictions from the parallel model about the relative amounts of the 16 dinucleotide sequences. Table 11–B shows the data for all 16 possibilities from 4 such experiments. These data were taken to disprove the parallel model.

TABLE 11–B

Nearest Neighbor Frequencies

Labeled Triphosphate	Isolated 3'-deoxynucleotide			
	T	A	C	G
dATP	0.012	0.024	0.063	*0.065*
dTTP	0.026	0.031	**0.045**	0.060
dGTP	0.063	**0.045**	0.139	0.090
dCTP	*0.061*	0.064	0.090	0.122
Totals	0.162	0.164	0.337	0.337

Note: The values are the radioactivities found in each isolated 3'-nucleotide in four experiments, each using a different labeled α-^{32}P-triphosphate. The italic and boldface values indicate the two values which should be equal if the base pairing of the Watson-Crick model were obeyed and if the DNA strands were of opposite polarity. Notice that A = T and G = C in the sums. (Data from A. Kornberg, *Enyzmatic Synthesis of DNA*, John Wiley and Sons, 1961.)

Conservative replication (one mode)

(a)

Semiconservative replication

(b)

Dispersive replication

(c)

isotope ^{14}N to the growth medium to effectively stop the incorporation of heavy precursors. At various times after adding this "light" isotope, DNA was extracted (from an equal number of cells), and sedimented to equilibrium in the CsCl density gradient.

Immediately after the addition of ^{14}N, only one DNA band could be seen (by its absorbance of 260-nm light), corresponding to the ^{15}N-labeled DNA. As the cells grew and divided, the "heavy" band diminished in area and another band increased. The new band was not in the position expected for ^{14}N DNA but halfway between the positions of ^{15}N DNA and ^{14}N DNA (Figure 11–5b–e). After the cell number had exactly doubled, this intermediate (often called "hybrid") band was the only DNA observed. As the cells continued to divide, the intermediate band remained but a band corresponding to ^{14}N DNA began to increase and kept increasing in proportion to the number of cells (Figure 11–5e–k). Thus, the fraction of the DNA in the hybrid band diminished with further growth.

These results showed that the physical integrity of the DNA existing at the time of the isotope shift, the ^{15}N DNA, was maintained throughout the experiment. This contradicted the predictions of the dispersive model of DNA replication which could therefore be rejected immediately. The conclusion was further substantiated when Meselson and Stahl demonstrated that the DNA was composed of two subunits. They denatured the hybrid DNA by dissolving it in alkali, a treatment that breaks the hydrogen bonds holding the two strands together and allows them to separate, and sedimented it in the CsCl density gradient. The hybrid DNA gave rise

to two bands, corresponding to completely ^{15}N-labeled DNA and completely ^{14}N-labeled DNA that was similarly treated. The intermediate band thus must have been composed of two subunits, one of which was physically part of the template DNA and one of which was newly synthesized. The conservative replication model was, therefore, ruled out. Further semiconservative replication would maintain, as observed, a constant amount of hybrid molecules, corresponding to the ^{15}N-labeled DNA existing at the time when ^{14}N was added (Figure 11–6).

Meselson and Stahl were correctly cautious in their conclusion, referring to "subunits" of DNA. The similarity of the two subunits to the two strands of the Watson-Crick model, however, was too tempting for the community of biologists (especially writers of textbooks) to resist. Luckily, the easiest guess proved to be correct even though other interpretations existed. Only later was it shown that the two subunits are not joined end-to-end, a possibility not ruled out by Meselson and Stahl's experiment, and that the mass per length of even hybrid DNA is that to be expected of a two-strand DNA molecule and not something more complex. Numerous repetitions of the Meselson-Stahl experiment have established semiconservative as the dominant mode of cellular DNA replication.

That Other Strand

The properties we have discussed thus far were known in the late 1950s. They permitted a simple notion of how DNA was replicated: some enzyme like DNA

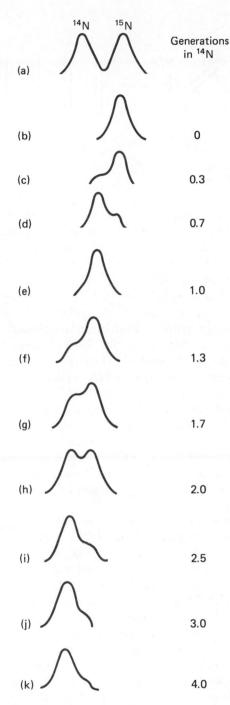

Generations
in ¹⁴N

(a)	
(b)	0
(c)	0.3
(d)	0.7
(e)	1.0
(f)	1.3
(g)	1.7
(h)	2.0
(i)	2.5
(j)	3.0
(k)	4.0

FIGURE **11–5**

The Meselson-Stahl Experiment, Showing Semiconservative Replication of DNA. The heights of the traces are proportional to the absorbance of 260 nm light and indicate the DNA concentration along the radial direction of the solution as it rotated in the analytical ultracentrifuge. The axis of rotation is to the left. Roughly equal amounts of DNA were sedimented in each example. Panel (a) shows the pattern that results from sedimenting a mixture of fully ¹⁴N- and fully ¹⁵N-labeled DNA. (Redrawn from M. Meselson and F. W. Stahl, Proc. Natl. Acad. Sci. USA, **44:**671, 1958.)

polymerase I would start at one end of a double-strand helix, unwind the two strands, catalyze the synthesis of one daughter strand by using one of the parent strands as template, then after reaching the end of the helix, catalyze the synthesis of the other daughter strand by using the other parent strand as template. In 1963 the applecart was badly tilted, if not upset, by J. Cairns' elegant autoradiographic visualization, in the electron microscope, of the replication of *E. coli* DNA. Viewing the replication of a complete genome was technically brilliant, but more startling still was the unmistakable fact that *E. coli*'s DNA was a circle. Worst of all, it was clear that the DNA was replicated on both template strands simultaneously from a common origin. Figure 10–10 shows the incorporation of label into two daughter DNA molecules that appear to be growing at a *replication fork*. The simple models of how replication began (e.g., at the end of the DNA molecule) were out and a fresh look was required. Let us first examine the problem of how both strands of the template DNA can be replicated together.

This problem arises because, although the strands are chemically antiparallel, only 5′-triphosphate precursors seem to be required and the chains grow only in the 5′-to-3′ direction. While the problem of replicating the other strand would have been neatly solved by the use of 3′-triphosphates or an enzyme with the opposite polarity, they have yet to be observed in nature.

Two discoveries pointed the way out of this dilemma. One was an enzyme, *DNA ligase,* that seals together, with a 3′-5′ phosphodiester bond, a butt joint of two DNA chains. The other was the observation that very recently synthesized DNA is found, after isolation, mainly as short unattached pieces. These were named *Okazaki fragments,* after their discoverer, and were observed after the introduction of tritiated thymidine into growing cells for very short times (or pulses). If the short pulse of label was followed by a chase—a period of time in unlabeled thymidine—the isolated labeled DNA was found to be much longer than the Okazaki fragments. The interpretation of these experiments is shown in Figure 11–7—a *discontinuous synthesis* of at least one of the DNA chains.

As illustrated, the template strand that runs from the 5′ end, to the replication fork, to the 3′ end could be replicated *continuously* by an enzyme like the Kornberg DNA polymerase. Continuous replication creates the *leading strand* of the product DNA. The other (3′-fork-5′) template strand must be replicated by *back filling* as new template strand sequences are exposed during movement of the replication fork down the template DNA (Figure 11–7). This discontinuous replication produces the *lagging strand*. The 3′-5′ phosphodiester linkages between the ends of the lagging strand fragments are sealed by the DNA ligase. This seemed likely from the finding that a

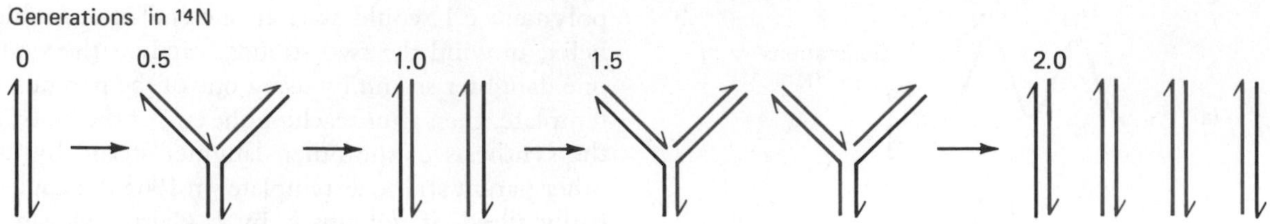

Generations in ^{14}N

FIGURE 11–6

Illustration of DNA Replication after an Isotope Shift from ^{15}N (black strands) to ^{14}N (color strands). As the number of DNA molecules increases, the *fraction* of strands labeled with ^{15}N decreases, but the *amount* of ^{15}N-labeled DNA is conserved.

mutant bacterial cell with depressed ligase activity had significantly longer lifetimes of Okazaki fragments. The role of DNA ligase has since been proved by using purified enzymes to synthesize DNA.

Primers and Origins of Replication

It was discovered early that DNA polymerase I cannot add nucleotides except to the end of a primer chain—that is, it cannot by itself initiate the synthesis of DNA strand, even in the presence of a template strand. This made it even more difficult to ignore the fact that the scheme for DNA synthesis, as elucidated by the study of isolated enzymes and labeling of product in vivo, did not explain how replication begins. The discovery of a circular DNA molecule in *E. coli* by Cairns made it clear that physical ends were not necessary for an origin of replication, since a circle has no ends. Moreover, discontinuous synthesis of DNA suggests that multiple initiations must occur.

The resolution of this problem revealed that the initiation of replication is a very complex process. First, none of the known DNA polymerases have an obvious initiating activity. Further, there is now convincing evidence that a short *RNA sequence* actually initiates the synthesis of an Okazaki fragment; the DNA strand is synthesized by addition to the 3' end of this primer. Later, the RNA is removed by a *5'-3' exonuclease activity*, possibly associated with a DNA polymerase; the gap is filled by a DNA polymerase, and DNA ligase seals the ends (Figure 11–8). The structures of some of the RNA priming molecules have been reported. During synthesis in vitro of DNA on bacteriophage T7 templates, the primers are pppApCpCpA or pppApCpCpC.[1] Covalent attachment of the RNA sequence to the DNA of the Okazaki fragment has been demonstrated in *E. coli* and

there is now little doubt that discontinuous synthesis of DNA from short RNA primers occurs in procaryotes.

Even this sophisticated scheme, though, does not answer the fundamental question of what physical property or sequence of bases in DNA is recognized as the place to begin replication. In *E. coli* there is persuasive evidence that the genome is replicated from a well-defined origin point, in both directions around the circle. This suggests that the origins for replication are reproducible and can be studied. The sequence of base pairs at and around the origin of DNA replication of bacteriophage lambda has been determined. As we shall see, eucaryotic cellular DNA has multiple sites at which DNA synthesis is initiated. There is also evidence that these sites may vary—for example, depending on the developmental stage of the organism. It seems likely that further analysis of the sequences of replication origins will yield additional clues to the mechanism of initiation of DNA synthesis.

Although our knowledge is still meagre about the details of initiating cellular DNA replication, greater progress has been made in studying the events of DNA synthesis in the icosahedral bacterial viruses, such as ΦX174. As shown in Figure 10–21, a circular single strand of DNA (the plus strand) is injected into the host *E. coli*. There, the complementary minus strand is synthesized to form a double-strand circular DNA. The initiation of minus strand synthesis uses only pre-existing cellular components; the enzymatic activity of forming the primer from ribo- and deoxyribonucleotides is probably due to the protein product of a bacterial gene (*dnaG*, also called *primase*) whose normal function is required for proper DNA replication. There are, however, at least a half-dozen other host cell proteins that act before the primer is synthesized. Notable among them is the *single-strand binding protein* (SSBP); SSBP has a high affinity for single-strand DNA and covers the plus strand as soon as it enters the cell. Initiation thus occurs on a protein-nucleic acid complex (sometimes known as the *primosome*). The role of some of the additional proteins may be to guide the primase protein to the correct starting point.

---

[1]RNA chains are initiated by adding a second nucleotide to a 5' ribonucleoside *tri*phosphate (see Chapter 12); that is why the primers begin with *ppp*Ap. . . .

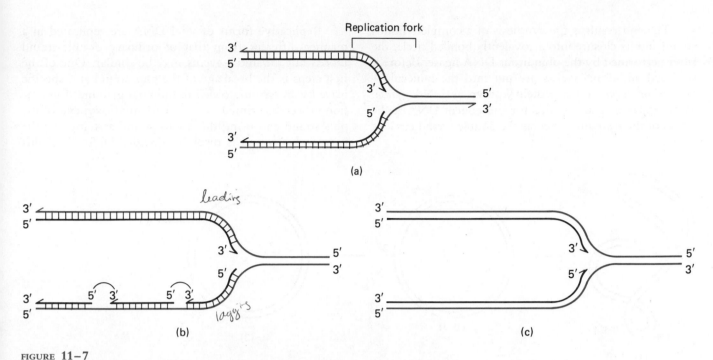

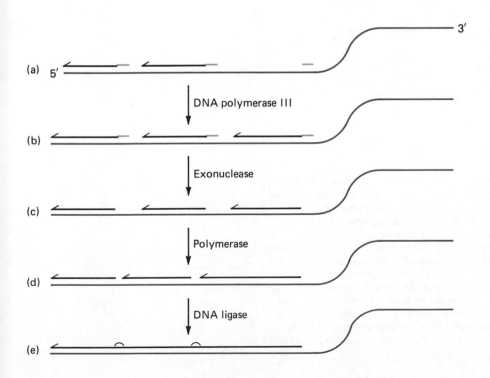

FIGURE **11–7**
Discontinuous Synthesis of DNA Chains. (a) Indicates the opposite polarity
(5′-to-3′ and 3′-to-5′) of the two DNA chains in the template helix (color lines),
and as a result, the necessary opposite polarity of the two new chains (black lines).
The helix has been straightened out for clarity. (b) Indicates the synthesis of the
two new chains (black lines). The lower chain starts on its template strand at many
discrete growing points, resulting in small pieces of newly synthesized DNA. (The
upper new DNA chain may also be synthesized discontinuously.) DNA ligase
eventually closes the gaps between these pieces by completing the phosphodiester
bond between the 5′-phosphate and 3′-OH ends to give the result shown in (c).

FIGURE **11–8**
**The Participation of an RNA
Primer in the Discontinuous
Synthesis of DNA.** (a) The short
RNA primer (represented as a light
color) is synthesized on the template
strand by a special enzyme activity. (b)
DNA polymerase III synthesizes
DNA, starting from the primer RNA,
until its progress is blocked by the
former product strand or its primer.
(c) An exonuclease activity, likely the
one associated with DNA polymerase
I, digests away the primer and possibly
some of the newly synthesized DNA
as well. (d) The gap is filled by a
DNA polymerase. (e) The butt ends
are sealed by DNA ligase to form a
continuous strand.

The end result is the synthesis of a complete minus strand and its closure into a covalently bonded circle, the latter performed by the ubiquitous DNA ligase. Before it is closed additional twists are put into the molecule—*superhelical turns*—by a protein known as *DNA gyrase*. These apparently are needed for subsequent DNA replication of the *replicative form*, as the double-strand circle is named.

Replicative forms of viral DNA are replicated in a manner different from that of ordinary double-strand DNA, but the initial events may be similar. One of the first steps is the breakage of the plus strand at a specific place by an enzyme coded in the viral genome. Replication proceeds, primed by the free 3'-hydroxyl end of the plus strand and using the minus strand as template; the parent plus strand is displaced (Figure 11–9). The plus

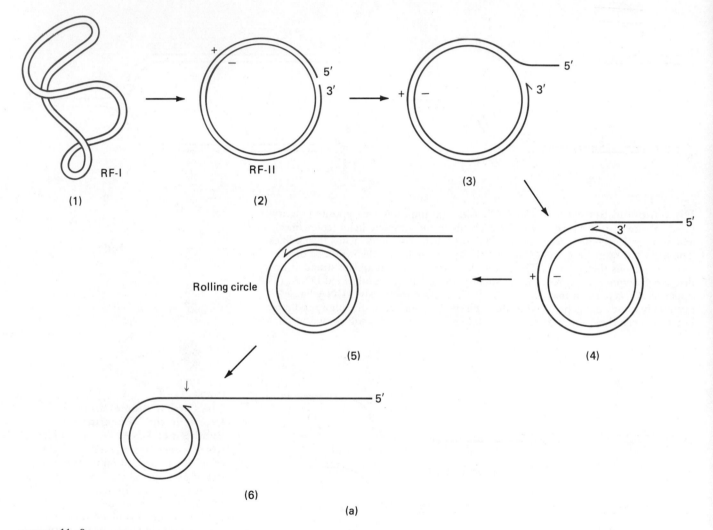

FIGURE 11–9
Replication of the ΦX174 Plus Strand of DNA. (a) The supercoiled double-strand RF-I (1) is nicked at the origin of replication to give a relaxed double-strand circular replicative form RF-II (2). Synthesis at the 3' end of the plus strand progressively displaces the 5' end in the rolling circle mode of replication (3–5). After the replication of the plus strand is complete, it may be freed to act as template for another minus strand, thus yielding RF-I, or it may become associated with bacteriophage-specified proteins and mature into an infective virus. The color arrow (6) marks the end of a complete plus strand. During massive plus strand synthesis, the tail of the rolling circle may be several plus strand lengths in size. (b) The cycle of events, and some electron microscopic images of intermediates, that lead from RF-I to RF-II (with viral protein DNA attached), to synthesis of single-strand circles, the coating of those strands with DNA-binding proteins (SSBP), and the synthesis of new RF-I. The participation of various host and viral enzymes is indicated. (Courtesy of J. Hurwitz.)

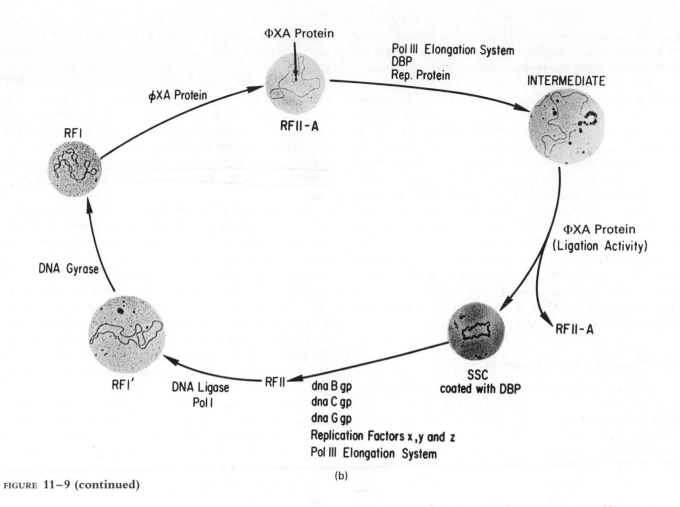

(b)

FIGURE **11–9 (continued)**

---

strand trails behind as the replication proceeds around this *rolling circle* (or the end may remain attached to some structure near the replication fork). This special synthesis requires proteins coded by the virus as well as proteins thought to be involved in host DNA synthesis. Although *rolling circle replication* has been observed mostly in viruses of procaryotes, it also occurs during mating in *E. coli* and perhaps during the amplification of ribosomal RNA genes in amphibian oocytes (see Chapter 12).

Strand Separation and Topological Problems

One of the host components required for ΦX174 DNA replication is the product of the *E. coli rep* gene, a protein of 65,000 molecular weight. This protein has an ATPase activity and acts to separate the strands of the parent DNA molecule. The *rep protein* binds to a section of single-strand DNA in the replicative form and moves along in a 3′-to-5′ direction in front of the polymerase, separating the double-strand parent DNA ahead. More recently another protein, named *helicase III,* has been

found that can move along the other parental strand in the 5′-to-3′ direction, with the same effect of separating the strands. Both of these proteins, called *helicases,* require ATP to effect strand separation and, in the case of the *rep* protein, estimates of two ATPs per base pair separated have been reported. The single DNA strand, once exposed, is coated by single-strand binding proteins, which prevent the reformation of the double-strand helix until the replication fork has passed.

Replication of DNA necessitates the unwinding of the template double helix. To do this for a circular DNA would quickly lead to problems, but even a linear molecule of the immense length of DNA would present enormous resistance to the unwinding rates observed. For *E. coli,* which can replicate its 4×10^6 base pair genome in 30 minutes, the replication forks move bidirectionally at roughly 67,000 base pairs/min. This means that the template double helix must unwind at 6700 rpm ahead of each replication fork. The strain of this twisting is undoubtedly relieved by single-strand breaks, or *nicks,* which provide a structure that can swivel. As the replication fork approaches one such nick, DNA ligase restores the integrity of the template double helix before its is replicated and another nick farther along the parental molecule provides the required swivel site.

The double-helical structure of DNA is generally stable if isolated in cellular conditions, and some perturbation is therefore necessary to achieve long-lived separation of the template strands during replication. On the other hand, it has been found that in solution the structure of DNA is dynamic: double helix regions open and close spontaneously in a process nicknamed "breathing." The open regions are of short length but occur transiently and randomly over the entire molecule. Single-strand DNA binding proteins will associate with these open regions, a process that tends to keep them open and thus unwind the double helix. When present in an in vitro DNA synthesizing system, these proteins enhance the replication by DNA polymerases. Such unwinding proteins have been isolated from both procaryotic and eucaryotic cells.

The "Real" DNA Polymerase

We have not yet answered a question implied earlier: why are there three (or more) DNA polymerases and what is the function of each? The details of the answer are still being worked out. At this point it appears that in *E. coli* DNA polymerase III is the enzyme primarily responsible for elongating the polynucleotide strands during replication. The enzyme discovered by Kornberg, DNA polymerase I, is mainly a *repair enzyme* that corrects mismatched base pairing caused by erroneous replication or external damage to DNA. This was first indicated when Cairns isolated a mutant *E. coli* having an altered polymerase I with almost no enzymatic activity and found nevertheless that such cells could replicate their DNA. This finding stimulated a search for other DNA polymerases, and they were not long in being discovered.

Although there are only three enzymatic activities in *E. coli* that can polymerize DNA, there are many other proteins involved. In the replication of ΦX174 there are probably a dozen. Many of the relevant proteins of *E. coli* have been identified by analysis of mutant cells that are either unable to replicate their DNA at high temperature or are unable to repair damaged DNA. Including the RNA polymerase (involved in primer synthesis), binding proteins, cofactors and so forth, there are probably at least a score of proteins required to replicate the genome of *E. coli*. Sorting out the roles of all these proteins now occupies the attention of several research groups spread throughout the world. In mammalian cell nuclei there are also three DNA polymerases, called α, β, and δ; a fourth (γ) polymerase is associated with the replication of mitochondrial DNA. The α-polymerase is likely to be the main enzyme associated with replication since it is a large and complex molecule, and its activity is greatest during S phase while least in differentiated nondividing cells. The β-polymerase is small and does not appear to be regulated during the cell cycle. The *unscheduled DNA synthesis*, observed outside the S phase of the cell cycle after ultraviolet light damage to DNA in differentiated cells, has been attributed to the β-polymerase (see the discussion of repair synthesis later in this chapter). The δ-polymerase has been shown to have a 3'-5' exonuclease activity, unlike the other DNA polymerases from animals.

In the case of *E. coli*, replicating DNA is now known to be associated with cell membrane fragments. This may indicate attachment to sites appropriate for later segregation of the product DNA molecules into the daughter cells. It also points up a lesson for aspiring cell biologists—do not assume that enzymes soluble in aqueous media are the only ones worth investigating; be sure to assay the activity of *all* the cell fractions. The "real" DNA polymerase may have been spun down to the bottom of the centrifuge tube with the cell "debris" and ignored for years before more systematic analysis discovered its importance.

The synthesis of DNA also includes the *postsynthetic modifications* to bases and sugars. Some of these modifications, methylated nucleotide bases, are known to play functional roles in procaryotes and are also important in eucaryotes (see Chapter 14). All such additions occur after the nucleotides have been synthesized into daughter strands.

Termination

The synthesis of DNA terminates in a controlled way. When two replication forks meet they do not pass one another and continue synthesizing DNA. This may merely reflect a steric hindrance of one replicating fork approaching another. In fact, the details (including the enzymology) of the termination process have been studied very little as yet.

Errors and Repair Synthesis

No transfer of information is perfect and DNA replication is no exception. There are at least two sources of errors in DNA: mistakes during replication (i.e., insertion of the wrong nucleotide), and damage inflicted during or after replication by some outside agent (such as x-rays, ultraviolet light, environmental mutagens, or the like). Errors during the replication process give rise to mutations, and the probabilities of mutation are known to be very low, perhaps of the order of 10^{-9} per nucleotide pair per generation. The physical forces involved in base pairing between the template strand and

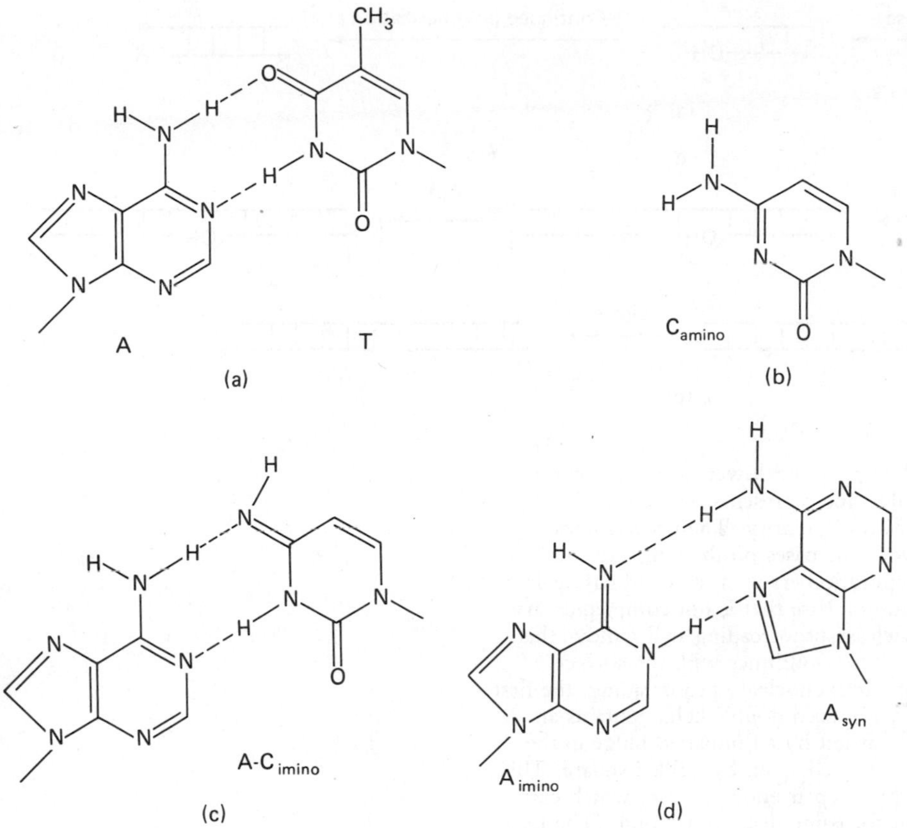

FIGURE **11–10**

Tautomeric Shifts in Nucleotide Bases Allow Incorrect Base Pairs to be Formed.   Illustrated are (a) the normal base pairing of adenine (A) with thymine (T); (b) the amino tautomer of cytosine (C) which predominates in physiological conditions; and (c) an erroneous pairing between adenine and the imino tautomer of cytosine (C_{imino}). (d) Isomerizations between the *syn* and *anti* forms of nucleosides, resulting from rotations about the glycosyl bond joining the base with the sugar, also allow mistakes to be made in base pairing. Purine-purine pairing between the imino tautomer of adenine (A_{imino}) and the unusual *syn* isomer of adenine (A_{syn}) is shown. (Redrawn from M. D. Topal and J. R. Fresco.)

the growing product strand have long been thought to be insufficient to explain this remarkable fidelity in DNA replication.

The two or three hydrogen bonds between the template nucleotide base and the complementary precursor nucleotide triphosphate are perhaps just strong enough to account for the insertion of the correct nucleotide. But the precursors are not always in the correct tautomeric or isomeric form to make the required hydrogen bonds (Figure 11–10). In fact, the wrong precursor can undergo a tautomeric shift often enough to be incorporated in DNA with a frequency much greater than the observed mutation rate. Alternatively, the template nucleotide base can undergo a tautomeric shift during the moment of precursor selection and form hydrogen bonds with the wrong nucleoside triphosphate.

Proofreading

The great fidelity of DNA replication in bacteria is partially due to a *proofreading function* of the cell that recognizes mismatched base pairs and then corrects errors of incorporating the wrong nucleotide. This proofreading presumably occurs while the replication fork is still in the vicinity, during replication. Proofreading of replica-

tion errors is mediated by the *3′-to-5′ exonuclease activity* found associated with all three DNA polymerases. The polymerase detects the lack of base pairing at the end of the newly synthesized strand and moves backward in the 3′-to-5′ direction, removing the erroneous nucleotide. The polymerase can then move forward again and lay down the correct sequence along the template (Figure 11–11).

One way to demonstrate the importance of this mechanism is to use an analogue of a precursor deoxynucleoside triphosphate that contains a sulfur atom on the α-phosphate instead of an oxygen atom. This analogue can be incorporated by DNA polymerase I at the same rate as the normal precursor but, once incorporated, the analogue is not a substrate for the 3′-5′ exonuclease activity. The survival of errors due to incorporating an incorrect analogue of this type is increased by 20-fold. Proofreading provides a second, and independent, chance to achieve fidelity in DNA replication. When two independent processes are used to remove errors, the overall probability of an error surviving both processes is the product of the probabilities of survival for each process. For example, if both processes had frequencies of surviving errors of 10^{-3}, the overall frequency of surviving errors would be 10^{-6}.

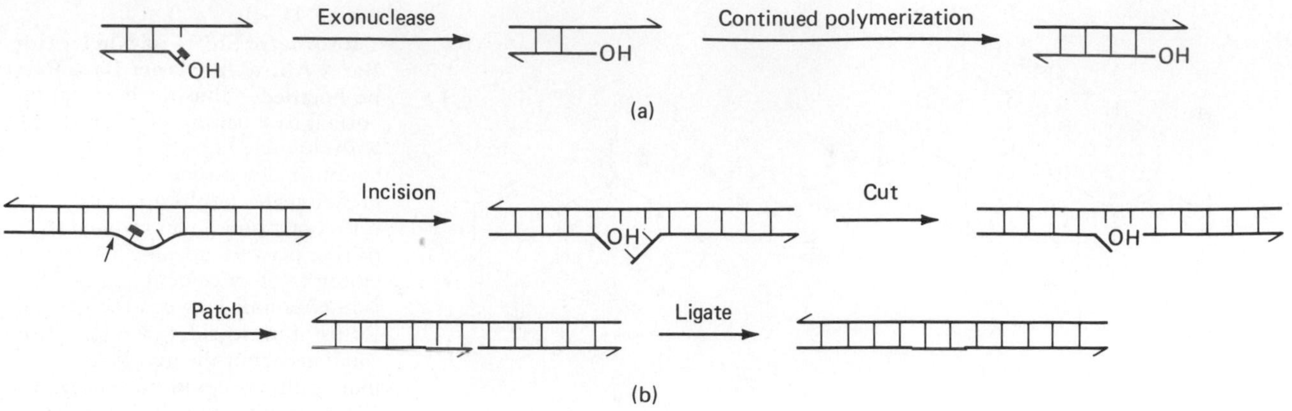

(a)

(b)

FIGURE **11–11**

Proofreading.   (a) Exonuclease repair. The upper and lower lines represent a
DNA double helix replicating; the lower color line represents the newly
synthesized strand; the arrows indicate the 5'-to-3' polarity. The vertical lines
represent complementary base pairing between the bases on the template and
newly synthesized strand. We represent a replication error in the third base pair
from the left by a tilted solid square, indicating a base that is not complementary
to its partner on the template strand. Exonuclease proofreading will remove this
incorrectly paired nucleotide, then polymerization continues with the correct
nucleotides. (b) Endonuclease excision repair. If exonuclease proofreading, the first
line of defense, does not occur, the newly synthesized double helix contains an
incorrect base pair, represented schematically at left by an unpaired bulge in the
double helix, where the incorrect base is symbolized again by a filled square. This
distortion in the double helix is recognized by a repair endonuclease, which cuts
the improperly paired strand and degrades it for some distance beyond. (The figure
indicates degradation for only two bases, but in fact the repair system cuts out a
larger stretch of the improperly paired strand.) The excised nucleotides are replaced
by a repair polymerase, following the dictates of the complementary strand,
indicated by the color arrow in the lower part of the figure labeled "Patch."
Finally, the 3' butt end is sealed by DNA ligase.

Excision Repair

Another activity used to remove damaged DNA in
both eucaryotic and procaryotic cells is *excision repair*.
Repair of damage to DNA can occur at times other than
S phase and leads to the observation of "unscheduled"
DNA synthesis. Among the defects found in DNA are
loss of bases, which can occur spontaneously as well as
being promoted by x-rays, ultraviolet (UV) light, and
alkylations; creation of mistakes by deaminating cytosine
to form uracil, guanine to form xanthine, and adenine to
form hypoxanthine; inappropriate additions to bases by
methylation enzymes; and major alterations to bases,
such as the formation of dimers between adjacent
pyrimidines, stimulated by UV light. (Genetic defects in
the repair of this last damage lead, in humans, to
xeroderma pigmentosum, a condition associated with a high
incidence of sunlight-induced cancers.) Errors of incor-

poration that escape the proofreading function are also
candidates for excision repairs.

Figure 11–11b illustrates a scheme for which substantial evidence exists and which leads to the repair of
defects in DNA. The damage is believed to be detectable
because of a distortion of the usual DNA structure. (The
choice of the daughter strand for repair, in the case of
mismatched bases, is guided by the *asymmetry of methylation* of DNA strands during replication; the daughter
strand will not have had time to be modified if the
damage occurs during or shortly after synthesis.) A
specific endonuclease activity hydrolyzes a phosphodiester bond in the DNA chain near the distortion, a
process known as *incision*. A phosphatase activity removes the 3'-phosphate group at the nick. The distorted
sequence is removed by a 5'-exonuclease activity which
chews along the chain a nucleotide at a time to an
undistorted region. The resulting gap is filled by a DNA

polymerase adding nucleotides to the 3′-hydroxyl end of the gap. Finally, DNA ligase completes the repair by joining the newly synthesized sequence to the rest of the strand.

Both *excision* of the distorted sequence and *resynthesis* of DNA can be performed by many enzymes in *E. coli.* DNA polymerase I is an attractive candidate for the latter role because there is so much of it and since mutations in *polA,* the structural gene for polymerase I, make *E. coli* sensitive to killing by UV light.

Polymerase I contains both a polymerizing activity, which catalyzes the growth of DNA chains in a 5′-3′ direction, and exonuclease activities, which degrade DNA in both the 5′-3′ and the 3′-5′ directions. The last activity, which degrades DNA in the direction opposite to that of synthesis, probably represents a "proofreading" reaction that eliminates mispaired nucleotides immediately after their incorporation into a growing chain. Mutants of polymerase I that lack this exonuclease activity show an enhanced error rate. The 5′-3′ exonuclease activity is also associated with proofreading; the operation of both exonucleases would result in removing small stretches of nucleotides on both sides of a mispaired base.

There seem to be two distinguishable pathways for excision repair, a major one leading to short patches of resynthesized DNA (13–30 nucleotides per patch) and a minor one leading to long patches (hundreds of nucleotides). The repair systems of mammalian cells are qualitatively similar except that the resynthesized patches are often very short (3–4 nucleotides per patch) and there are apparently no long patches. By using this excision repair system, cells minimize the potentially lethal consequences of various environmental poisons.

The Molecular Mechanism of Recombination

In Chapter 9, we briefly noted the curious phenomenon of gene conversion and its relationship to recombination. We are now in a position to consider both of these phenomena at the molecular level. Our starting point is that the genome is a DNA molecule. How does the structure of DNA account for the processes of recombination and gene conversion? Figure 11–12 presents a molecular model of both processes.

The first step is a specific physical interaction between the two homologous DNA molecules (panel 1) destined to recombine with one another. The complementary structure of DNA provides an obvious mechanism for this interaction. If single-strand breaks occur in the two homologous DNA molecules, then each single-strand "tail" will be free to interact with the complementary strand of the *other* DNA molecule, forming a

hydrogen-bonded junction between the two "parental" molecules (panel 2).

This four-armed intermediate in recombination (panel 3) is called a *chi form.* One of its key features is that a strand from one parental molecule is paired with the complementary strand of the other parental molecule. If an allelic difference between the two parental molecules is included in this *heteroduplex region,* then there will be a *mismatch* involving at least one base pair. The excision repair system we mentioned in the last section can then come into action, removing a stretch of nucleotides in one strand of the mismatched region. DNA polymerase (perhaps a special polymerase involved in repair synthesis) can then restore the excised region, following the hydrogen bonding rules for complementarity to the remaining strand. When the excision repair pathway is used to correct undamaged but inappropriately paired regions of DNA, the process is called *mismatch repair.*

Now consider the situation in which one of the parental molecules is wildtype for a gene in the heteroduplex region, and the other carries a mutant allele. If the repair system restores the wildtype nucleotide sequence in both of the heteroduplex regions, the result is *conversion* of the mutant information to its wildtype counterpart. (Of course, conversion to the mutant genotype could occur instead; this is usually much more difficult to detect.) The occurrence of mismatch repair thus explains the phenomenon of gene conversion in the stretch of DNA close to a crossover point.

The next steps are the resolution of the tangled, four-armed intermediate to produce two normal, continuous DNA double helices. These steps, which involve cutting and splicing, are diagrammed in Figure 11–12a (panels 4 and 5). The outcome, as shown in panel 6, is a pair of DNA molecules which are reciprocally recombinant for parental information flanking the point of exchange.

The general features of this model have been confirmed by a wealth of experimental evidence obtained over the past twenty years. One early observation relates to the fact that mismatch repair is not 100% efficient, and some mismatched heteroduplex molecules may escape repair. Such an unrepaired heteroduplex is *internally* heteroduplex at the mismatch site, and will, on replication, give rise to daughter DNA molecules of both parental genotypes. Studies of bacterial viruses and molds have confirmed the occurrence of such internally heterozygous chromosomes, in association with crossing-over.

Mismatch repair has also been detected by examining the fate of mismatched heteroduplex DNA constructed in vitro. M. Meselson and his colleagues have used bacteriophage lambda chromosomes to construct such heteroduplex DNA molecules, carrying mutant

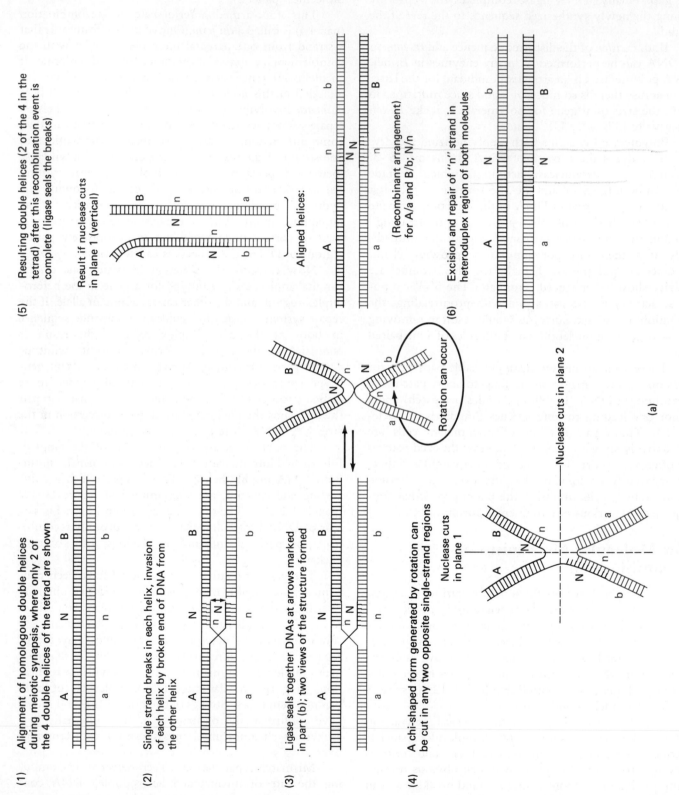

(1) Alignment of homologous double helices during meiotic synapsis, where only 2 of the 4 double helices of the tetrad are shown

(2) Single strand breaks in each helix, invasion of each helix by broken end of DNA from the other helix

(3) Ligase seals together DNAs at arrows marked in part (b); two views of the structure formed

(4) A chi-shaped form generated by rotation can be cut in any two opposite single-strand regions

Rotation can occur

Nuclease cuts in plane 1

Nuclease cuts in plane 2

(a)

(5) Resulting double helices (2 of the 4 in the tetrad) after this recombination event is complete (ligase seals the breaks)

Result if nuclease cuts in plane 1 (vertical)

Aligned helices:

(Recombinant arrangement) for A/a and B/b; N/n

(6) Excision and repair of "n" strand in heteroduplex region of both molecules

FIGURE 11–12

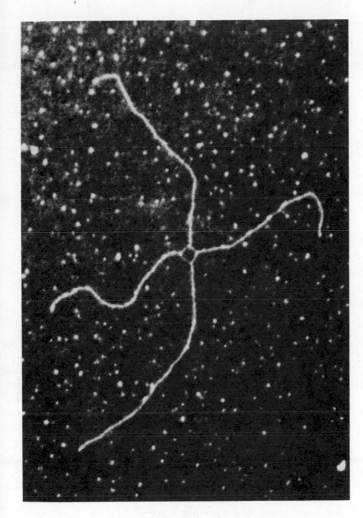

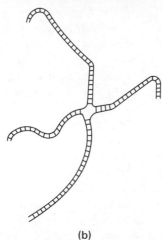

(b)

FIGURE **11-12**

A Molecular Model of Recombination and Gene Conversion. (a) Diagram of the model. Panel (1) shows two parental, homologous DNA molecules paired during the first division of meiosis; each parent is shaded differently, and marked genetically by different alleles at three genes, A/a, N/n, and B/b. The short vertical lines represent hydrogen bonding between complementary base pairs of each DNA double helix. Breaks at homologous positions on one strand of each DNA molecule (arrows) permit one strand of each to associate with the complementary strand of the other (2). The breaks are sealed by ligase (3), resulting in the four-armed chi-form intermediate shown in two geometrically equivalent views in (3) and in (4). Next, the chi form is cut and religated to generate two complete, disentangled DNA double helices. When the cutting and religating proceeds along plane 1, the outcome is two DNA molecules which are reciprocally recombinant for regions flanking the crossover point, as indicated by the disposition of genes A and B in (5). This model accounts for recombination. Note that gene N, close to the crossover point, is *internally* heterozygous with a mismatch of noncomplementary bases at the point of the allelic difference between N and n. The mismatch at gene N triggers a repair system to excise a stretch of one strand in the mismatched region, and resynthesize it according to strict complementarity to the remaining strand, thus accounting for gene conversion. Panel (6) shows the outcome where n is converted to N in both DNA molecules, with the converted strand shown in dashed lines. Recall that the other two DNA molecules in the tetrad were not involved in crossing over, and one retains genotype N while the other retains genotype n. The result is a tetrad with a 3:1 ratio of N to n. (Of course, mismatch repair in the other direction will give a 1:3 ratio of N to n.) (b) Electron micrograph of chi-form recombination intermediate in *E. coli*. The recombinant molecules are those of a bacterial plasmid. Note the open form and the corresponding interpretative drawing, which exactly resemble the open form in part (4) of (a) above. (Micrograph courtesy of D. Potter and D. Dressler.)

genetic information on one strand and wildtype on the other, and then introduced them into bacterial cells by transformation. Under appropriate conditions, the internally heterozygous DNA is converted to either one genotype or the other through mismatch repair.

The most striking confirmation of the model comes from cytological studies of DNA with the electron microscope. Figure 11–12 presents a direct visualization of the chi-form intermediate in recombination. The association of this four-armed DNA molecule with the process of recombination is proved by mutant analysis (see Chapter 9). Mutants have been isolated which are defective in genetic recombination; in these mutants the chi form cannot be found.

Jumping Genes

The occurrence of recombination demonstrates that DNA segments are not inexorably bound at one chromosome location, but can be moved from one homologue to another. Moreover, the occurrence of chromosomal rearrangements such as inversions, duplications, and translocations (see Chapter 9) demonstrates that DNA segments can be moved to different sites within a chromosome or on another nonhomologous chromosome. A special class of genetic elements termed *transposable elements* or *transposons* move about with particularly high frequency.

In the 1940s, B. McClintock described such elements in maize. As can readily be imagined, their analysis by conventional genetic methods demanded intricate and very laborious experiments. As a result, McClintock's pioneering work (for which she was awarded the Nobel Prize in 1983) was not followed up until the methods of molecular genetics provided a more direct approach. Transposons have now been identified in all organisms studied, and are perhaps best characterized in *E. coli*.

This bacterium's genome contains about a dozen different transposons, each one present in several copies. The location of copies of a given transposon often differs from one strain of *E. coli* to another, presumably due to differing transposition events. This process of movement from place to place has been followed in detail, particularly when the transposition is from the chromosome to a distinguishable plasmid (see Chapter 10), or vice versa, or from one plasmid to another.

A remarkable feature of the process is that it is generally coupled to replication of the element, so that one copy remains at the old location while another copy is inserted at the new location. This limited replication of a DNA sequence within the chromosome is thus somewhat analogous to the limited replication associated with DNA repair or with recombination. Indeed, the recombination system of *E. coli* is itself involved in the movement of transposons.

The size of the *E. coli* transposons ranges from several hundred to several thousand base pairs. All of them are bordered by *repeated sequences* (generally inverted, but in some cases direct), which are essential for transposition (Figure 11–F). These repeated sequences presumably target the ends of the element for the transposition machinery. This machinery, an enzyme termed a *transposase,* is in most cases encoded within the element itself, between the repeats.

One of the viruses that parasitize *E. coli,* phage Mu, is itself a transposon; or rather, its genome is a transposon. After the virus has injected its DNA into the bacterial host, the DNA inserts into the host chromosome and then replicates exclusively by the transposition process, distributing copies of itself to more and more sites in the bacterial chromosome. The host cell, its own genome now riddled with copies of Mu, dies eventually; the Mu segments are clipped out, wrapped up in viral structures, and released as a new crop of progeny viruses. There are indications that other viruses multiply in their animal host cells by a similar mechanism.

Transposons have been referred to as *selfish DNA* because of this insidious property of spreading throughout the genome. It must be, however, a self-limiting process, illustrated by the fact that the phage Mu kills its host. Most transposons do not share Mu's independent existence as a virus, but are simply DNA sequences. Their survival thus depends upon survival of the host genome that contains them. Their evolutionary persistence, therefore, undoubtedly reflects a balance between transposition (which, in principle, could go on indefinitely) and the selective disadvantage to the host cell of containing too many copies of a particular selfish DNA sequence.

In higher organisms, many families of transposable elements have been identified, cloned, and sequenced. Some of them have been given whimsical names, such as the *mariner* and *gypsy* elements of fruit flies. Each element has a characteristic sequence, always with repeats (sometimes not quite exact) at each end. Multiple copies of each element are found in the genome, presumably reflecting past and continuing replicative transposition. For example, there are about 30 copies of a group of related sequences, collectively termed *copia,*

Chromosome Duplication in Eucaryotes

As we discussed in Chapter 2, DNA synthesis occurs in eucaryotic cells during only part of the cell division cycle, the S period. After DNA synthesis is complete there is a period of time, G_2 of the cell cycle, whose length varies with cell type, before the beginning of mitosis. During prophase of mitosis the chromatin condenses to form the coiling thread-like structures we identify in the light microscope as chromosomes, doubled structures consisting of two paired chromatids. These are the two units that will be distributed to the

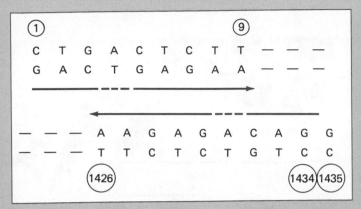

FIGURE **11–F**

A Typical Inverted Repeat at the Ends of a Transposable Element. The terminal DNA sequence of element IS50-L of *E. coli* is shown. The element consists of 1435 base pairs; the numbers of some of the base pairs are shown in circles; the dashes represent schematically the whole interior sequence from number 10 to number 1425. Note the inverted repeat at each end, indicated by the arrows: the sequence of the lower strand, reading left to right from base 1 to base 9, is *almost* identical to the sequence of the upper strand, reading right to left from base 1434 back to base 1426. One base of the nine, indicated by a dotted line in the arrow, is not identical. Most terminal repeats in transposons are, like this one, almost but not quite identical. The sequence also illustrates the fact that the inverted repeats are, in some cases, not quite terminal, as is the case here for the sequence at the right end, where the repeat starts one base back from the terminus.

in the genome of *Drosophila melanogaster*. The human genome is riddled with several hundred thousand copies of *alu* sequences, a particularly prolific and restless family of related transposable elements.

Different laboratory stocks of *Drosophila* have their *copia* sequences at different positions in the genome, indicating movement of the sequences while the stocks were maintained separately. In fact, spontaneous mutations are most often due to movement of a transposable element: most of the classical *Drosophila* mutants (such as those in the white eye gene) are not single base pair changes, but rather insertions of a transposable element into the gene at one position

or another. Direct evidence of transposition has been obtained in the case of the P element of *Drosophila*. When a fly harboring this element is crossed with one lacking it, the progeny of the cross exhibit a high frequency of mutants (and various chromosome aberrations as well) due to insertion of the P element here and there in the genome. In subsequent generations the germ line of such flies continues to show a high mutation frequency.

In fact, P element transposition has become a very widely used technique for introducing modified DNA sequences into *Drosophila*. Cloned or experimentally modified genes are inserted into P element DNA, using convenient restriction sites; the re-

constructed P element DNA is then micro-injected into *Drosophila* embryos at a very early stage of development. Transposition results in the incorporation of the modified P element, with its inserted gene, into one or another site in the embryo's genome. A remarkably high proportion of the flies that develop from these embryos contain the element at one or another site in the germ line. These *transgenic* flies are essentially *engineered transposition mutants*. They have provided valuable information on gene expression during embryogenesis.

Movement of eucaryotic transposable elements is not confined to the germ line; it can occur in somatic cells as well. In *Drosophila mauritiana*, certain eye color mutants are due to insertion of the *mariner* element into the white eye gene. Reversion of the gene to wildtype, by loss of the *mariner* element, sometimes occurs during embryonic development of the eye, producing flies whose eyes are a mosaic of mutant and wildtype patches. The most striking case of such somatic instability of transposable elements is the Ds (for "dissociator") element of maize plants. When this element wanders into and inactivates a gene for kernel pigmentation, the mutant strain does not simply have colorless kernels. Instead, the kernels are speckled with patches of pigmentation. During the course of somatic development of each kernel, the Ds element may leave the gene, restoring its normal activity. It is this bizarre phenotype which led McClintock, 25 years before the age of molecular genetics, to discover transposable elements.

daughter cells during anaphase. What is the structure of chromatid DNA? Although there is some conflicting evidence it now seems clear that each chromatid is composed of a *single DNA molecule* of prodigious size, up to 4×10^{10} Da in a *Drosophila* chromosome. The molecular weight of this monster has been measured

using a sophisticated physical chemical assay, and it is in good agreement with cytophotometric estimates of the DNA content of a chromosome.

Autoradiography experiments, first performed by J. H. Taylor, allow a look at what is happening during DNA replication in a eucaryote nucleus. Rapidly dividing

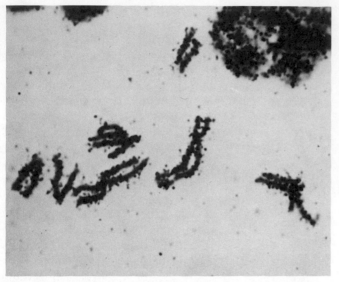
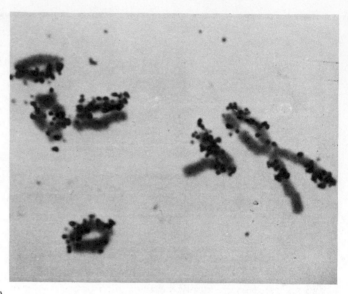

(a)

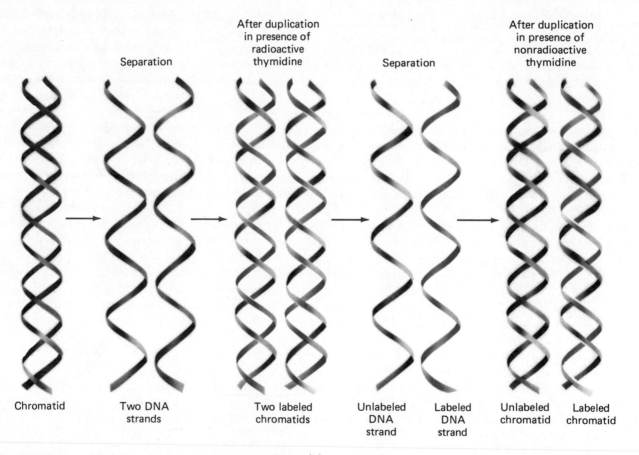

| Chromatid | Separation | After duplication in presence of radioactive thymidine | Separation | After duplication in presence of nonradioactive thymidine |

Chromatid → Two DNA strands → Two labeled chromatids → Unlabeled DNA strand | Labeled DNA strand → Unlabeled chromatid | Labeled chromatid

(b)

FIGURE **11–13**

Chromosome Duplication.   (a) Autoradiograph studies. Root cells of *Vicia faba* were incubated in the presence of tritium-labeled thymidine and examined by autoradiographic means after one (photo a-1) (×2200) and two (photo a-2) (×3200) cell divisions. (Courtesy of J. H. Taylor.) (b) Diagram of autoradiographic studies on chromosome duplication. (After J. H. Taylor.)

cells were given tritiated thymidine and at various times afterward were fixed for microscopic observation. Cells in S phase took up the labeled thymidine into their DNA and some of those which were in mitosis at the time of fixation had radioactive chromosomes. These were visualized by placing a photographic emulsion over the fixed and stained cells. When the emulsion was exposed for some time to the β particles emitted during tritium decay and then developed, an image of the radioactive chromosomes could be seen in the microscope (Figure 11–13). In those cells in which label was taken up, all the chromosomes were found to be labeled and the radioactivity was equally distributed between the two chromatids of each chromosome (photo a-1).

Using the pulse-chase protocol, Taylor placed cultured cells previously exposed to tritiated thymidine into a medium containing excess unlabeled thymidine and allowed them to go through another division cycle before fixing them for autoradiography. This time one chromatid was labeled and one was not (photo a-2). The simplest way of interpreting these observations is to consider each chromatid to contain a single DNA molecule that behaves during replication like the *E. coli* genome in the experiment of Meselson and Stahl. The chromosome in a cell just after division is composed of two strands which separate during S phase and which are joined by two newly synthesized (and, therefore, radiolabeled) strands. The four strands are distributed during S to the two chromatids so that each is labeled. A daughter cell receiving such a labeled chromatid and replicating its DNA during the chase in unlabeled thymidine will conserve both the labeled DNA strand and the unlabeled one, combining each with a newly synthesized, unlabeled strand. Thus, at the second mitosis only one chromatid will be labeled (Figure 11–13b).

Multiple Origins of Replication

The genome of *E. coli* is about 1 mm long and can be replicated in about 40 minutes. In some plants a chromosomal DNA may be 1 m in length, about a thousand times longer, and take 8 hours to be replicated, only about 16 times longer. This could be due to a more rapid DNA chain elongation rate, but in fact the opposite is true: measurements indicate that in eucaryotes 2 μm/min is the rate of movement of a replication fork while *E. coli* has a rate of 15–20 μm/min. To complete the replication of a eucaryote chromosome in the required time at the measured chain elongation rate would therefore require many DNA polymerases working simultaneously on the same DNA, and this is what has been found.

The evidence comes from elegant experiments in which eucaryote cells were labeled with radioactive precursors of DNA for short periods of time. After the cells were broken open the chromosomes were spread out into fibers before being coated with autoradiographic emulsion. The results were clear: there were many patches of label along the fibers separated by unlabeled regions, indicating many initiation points for DNA synthesis. Using the pulse-chase protocol (following the labeling period with a period in the presence of excess unlabeled precursors), it was found that the grain density on the autoradiographic images tapered off at both ends of the labeled region (Figure 11–14a). This suggested a shift from labeled to unlabeled precursor incorporation after the chase was begun and shows that the *replication is bidirectional*. These experiments have been repeated many times on many tissues and the general result is that eucaryote chromosomes have *multiple origins* for replication of their DNA, and the replication proceeds in both directions from each origin with eventual fusion of the replicated strands, presumably by a DNA ligase activity (Figure 11–14b).

The distance between origins varies but generally lies in the range of 10–100 μm. A cell may have thousands of growing sites simultaneously active at any time in the S phase. The origins do not all become active at the same time; some replicate early in S, some late. Thus, there is not only spatial but also temporal control of the activity of the multiple origins.

Chromosome Structure

The chromosomes of eucaryotic cells are highly folded arrangements of chromatin, composed roughly half and half of DNA and protein. The details of the architecture are not known but the folding is absolutely essential to compress the informational content into a package that can fit within the nucleus. The 3-cm length of the fully stretched human chromosome DNA is folded to 10^{-4} of its size in the chromosome. Electron microscopy suggests the DNA that forms the backbone of the chromosome structure is thrown into coils to form, with the accompanying proteins, fibers of between 8 and 30 nm diameter, depending on the source and means of visualization.

The Histones

Two-dimensional gel electrophoresis by the O'Farrell technique (see Chapter 4) reveals the presence in isolated chromatin of hundreds of different kinds of protein. Among them are actin, myosin, tubulin, nuclear membrane, and pore proteins, and a great many whose enzyme activities are detectable while still associated with chromatin.

Chief among the proteins are the histones, which come in these five varieties: H1 (MW 21,000), H2A

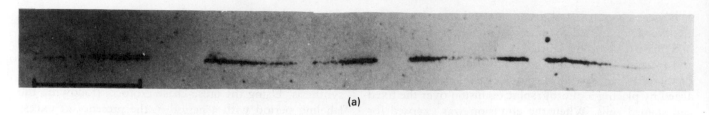

(a)

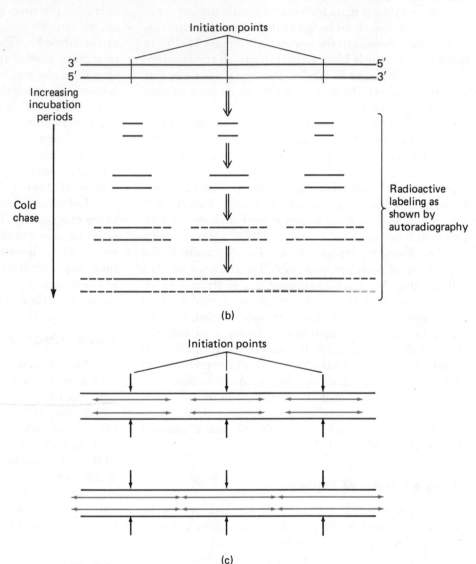

FIGURE **11–14**

Bidirectional Replication of Chromosomal DNA from Multiple Initiation Points. (a) Autoradiographic evidence. Cultured mammalian cells were labeled with radioactive thymidine for 30 minutes, followed by a 45-minute chase with cold thymidine. DNA was then prepared from the cells and submitted to autoradiography. The bar line indicates about 150,000 base pairs (150 kbp) of DNA. The dark, heavily labeled regions reflect DNA synthesized during the period when only radioactive thymidine was present. They are flanked at each end by more lightly labeled regions, reflecting DNA synthesized after the specific activity was reduced by the chase with cold thymidine. (b) An interpretation of the autoradiogram. The color lines represent heavily labeled DNA synthesized in the presence of radioactive thymidine *before* the cold chase. The dotted lines, flanking heavily labeled regions, represent lightly labeled DNA synthesized *after* the addition of cold thymidine. (c) An interpretation of the directions of DNA synthesis at the strand level. The color lines represent the conserved parental DNA strands. The arrows indicate the directions of newly synthesized radioactively labeled strand growth.

(MW 14,000), H2B (MW 13,800), H3 (MW 15,300), and H4 (MW 11,300). A striking characteristic of the histones is that they contain many basic (positively charged) amino acids, which allows them to bind tightly to the DNA through electrostatic interactions. Of the basic amino acids, histone class H1 contains mostly lysine, while the H3 and H4 histones contain more arginine.

The histones are found associated with DNA in all eucaryotes, even viral DNA when it is in a eucaryotic cell. Since DNA associated with histones is a poor template for transcription into RNA, it was thought at one time that histones might be the chief regulatory molecule for gene expression. On reflection, the five classes of histones are too few in number to explain the variety of patterns of genetic expression seen even in differentiated tissues, much less during adaptive responses to environmental perturbations. Nonetheless, the binding of histones to DNA, perhaps controlled by chemical modifications of histone protein, may be involved in controlling the regional pattern of DNA transcription, and undoubtedly plays a role in maintaining or altering the architecture of the chromosome.

There is amino acid sequence heterogeneity in certain of the histones, notably H1, but also H2A and H2B. Some of this is due to different genes for the histones that are differently expressed in different tissues or in different stages of the life cycle of an organism.

H1, H2A, and H2B also show more sequence differences between organisms than do the other histones. Even so, the sequences of H2A and H2B have many regions that are strongly conserved, although isolated from organisms that are evolutionarily distant from one another. H3 and H4 are very strongly conserved; there are only 2 of 102 amino acids that are different in the sequences of H4 from cow and from pea. The sequences of H1 from various organisms are not so well conserved as the others.

The Nucleosome

The details of the architecture of chromosomes are largely unknown, except for an apparently universal substructure. When isolated chromatin is digested briefly with an endonuclease from micrococcal bacteria, only especially susceptible phosphodiester bonds are hydrolyzed. These are apparently found periodically in the chromatin and the fragments produced; therefore, they have sizes that can be arranged as multiples of a basic repeating unit, called the *nu body* or *nucleosome*. Electron microscopy reveals the units to be spherical bodies of about 10 nm in diameter spaced at regular intervals along the DNA helix (Figure 11–15). In mammals these units contain about 200 base pairs of DNA complexed to the

five histones; the exact length of DNA varies with the source of the chromatin. On more complete digestion, a relatively well-defined fragment, the *core particle,* results. It is composed of approximately 140 base pairs of DNA associated with two molecules each of histones H2A, H2B, H3, and H4. The DNA digested away is known as the *linker* and it appears preferentially to be associated with one or two molecules of histone H1. It is the linker whose length varies with the species from which the chromatin is isolated. The core is a reproducible structure in that it can be reassembled from the purified H2A, H2B, H3, and H4 histones, and DNA. All the components are required; antiserum raised against H2B, for example, will block the reassembly. When the three-dimensional structure of the core has been analyzed by x-ray crystallographic techniques, we should know a great deal more about the details of the geometry.

Crystals of cores can be visualized in the electron microscope, and image reconstruction studies have revealed an 11-nm-diameter sphere, in agreement with hydrodynamic measurements of size. The DNA of the core is accessible to nucleases like DNAse I, which yields fragments with lengths that are multiples of 10 base pairs. This observation suggests the DNA is wrapped around the outside of the protein and is still in the usual "B" structure (Figure 10–9) of 10 base pairs per helical turn. Apparently, the DNA is accessible to the nuclease only once per turn when wrapped around the core.

Cross-linking studies have been performed to learn about the topography of the nucleosome, how the components are positioned with respect to one another. Cross-links are formed by a "molecular staple," a bifunctional alkylating reagent that makes covalent bonds

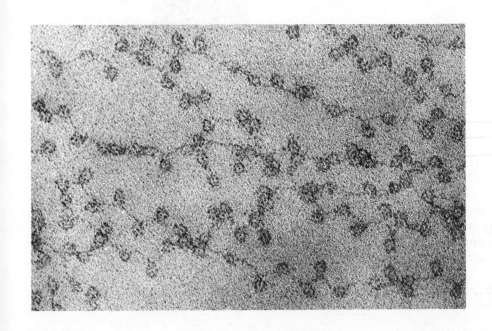

FIGURE **11–15**
Chromatin Fibers Streaming out of a Chicken Erythrocyte Nucleus.
The spherical particles are ~7 nm in diameter and the connecting strand is ~14 nm in length. (Negatively stained with uranyl acetate.) ×570,600. (Courtesy of D. E. Olins and A. L. Olins.)

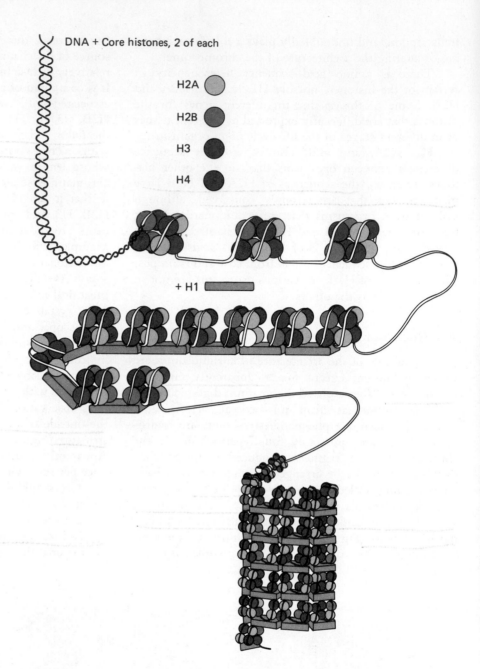

DNA + Core histones, 2 of each

H2A

H2B

H3

H4

+ H1

FIGURE **11–16**

FIGURE **11–16**
**Diagram of Nucleosome Structure
and Postulated Higher Order
Structures in Chromosomes.**
(From original drawing by M.
Cotten.)

with susceptible atoms (see the discussion of the method in Chapter 13). Extensive cross-linking shows that H1 is closer to another H1 than to any of the other four classes of histones and they are all nearer to each other than to H1. A model, quite schematic, that agrees with current structural data is shown in Figure 11–16. Neutron-scattering and x-ray diffraction data indicate that the nucleosomes are arranged in chromatin as a coil 30 nm thick with a pitch of 10 nm. Thus, the proteins of adjacent cores may be in contact with one another, with the DNA forming a more or less continuous coiled coil taking two turns per core.

Similar chromatin structures have been observed in the much smaller "chromosomes" of animal viruses.

These structures are condensed to 1/7 of the length that the viral DNA would have if it were stretched out. Folding the DNA helix around core histones or similar proteins must introduce bends into the DNA helix. These bends might be gradual—a small angle of bend in each successive base pair—or they might involve sharp bends or "kinks" at particular widely spaced base pairs. Anyone who has put together a model railway will appreciate the distinction between gradual bends and sharp kinks, and the problems raised by each. Theoreticians who analyze models of DNA have advanced various speculations as to how bends and/or kinks of different kinds could be introduced with the least distortion of the basic DNA double helix.

In fact, recent high-resolution studies of nucleosome structure indicate sharp, right-angle kinks at intervals of 20 base pairs or so. One functional consequence of kinks is that the sides of the base pairs are exposed in the major groove of the DNA for a few base pairs on either side of the kink. This could allow the sequence of the exposed bases to be detected by binding to proteins. The unperturbed DNA double helix has its bases so well concealed that it is difficult to see how proteins could interact with them easily. And yet, as we shall see in Chapter 14, there is very good evidence that proteins do interact specifically with DNA, and thereby regulate gene activity. Kinky DNA may well be the sites of such interactions. Kinks could also be involved in the condensation of DNA that occurs in procaryotic cells and in the packaging of viral genomes.

Topoisomerases

The condensation of the physical length of the genome that results from forming a nucleosome is only from 46 nm (= 14 turns of B form DNA = 140 base pairs) to an 11-nm core plus a linker, a factor of less than five. What additional mechanisms exist for the necessary further reduction of length? Although this problem is attracting the attention of much research, the answers are not yet in hand. Whatever additional coiling or kinking may be involved, however, must result in distortions of the DNA structure in the form of supercoiling (see Chapter 10 for a definition of supercoiling). Because of the resulting strain, further packing would be impossible. A solution to this difficulty, at least, seems available: the discovery of a class of enzymes, the topoisomerases, which are capable of adding or substracting superhelical turns in DNA and altering the topology of covalently jointed DNA circles.

One of these has been mentioned previously, the DNA gyrase that can add superhelical twists to the circular replicative form of ΦX174 bacteriophage (see Figure 10–21 and the related discussion). Supercoiling is important generally because enzymes that have DNA as a substrate often are unable to catalyze their reactions unless the DNA has a supercoiled structure. We expect that other topoisomerases can catalyze the alteration of supercoiling in response to the requirements of packaging DNA into the chromosome.

But the topoisomerases can do more than add or subtract supercoils. They can also interlock circular rings of DNA and even tie knots in them. This can happen when a length of DNA is passed through a transient double-strand break in another segment of DNA. The two ends produced by the break are thought to remain bound to the topoisomerase and are rejoined after the other DNA length has passed through the break (Figure 11–17). If the length of DNA passed through the break is from the same segment of DNA, the result is the addition to or subtraction from the segment of two superhelical turns. If DNA from a second ring is passed through a break in a DNA ring, the result is two interlocked rings. If DNA from the same ring, but containing an appropriately placed loop, is passed through, the result is a knotted ring. Reactions in vitro with bacterial plasmid circles have produced large interlocking networks of DNA. The ability of topoisomerases to alter the topology of rings is presently easier to measure than their influence on chromatin packaging. But the exploration of such curious enzymatic activities will doubtless aid our search for the structural bases of how the DNA of eucaryotic cells can be compressed to 1×10^{-4} of its molecular length in forming the chromosome.

Nucleosome Replication

What happens to the histones during replication of the eucaryotic chromosome? In synchronized cells (see Chapter 14), histones are synthesized during the S phase and the newly synthesized DNA probably becomes involved in nucleosome structures very quickly. It has also been shown by labeling studies that old histones are not broken down during replication. There still remains the question of how the old and new histones are distributed to the daughter cells. This was addressed by an isotope transfer experiment analogous to the experiment on bacterial DNA replication done by Meselson and Stahl. After briefly labeling histones with ^{3}H-lysine, cells were transferred to a growth medium containing dense isotopes of nitrogen and carbon, and ^{14}C-lysine as a radiolabel. Incubation in the dense isotope medium was carried out while approximately one-eighth of the histones were synthesized, then the chromatin was isolated.

The histones were cross-linked together before being removed, by strong salt solutions, from the DNA. Cross-linking conditions were chosen to maintain the nucleosome structure of eight histone molecules but not to make an insoluble mess of the chromatin. To assess whether there was mixing of the old (light, ^{3}H) and newly synthesized (heavy, ^{14}C) histones, the cross-linked octamers were sedimented to equilibrium in a density gradient of a cesium salt. There were only two octamer bands in the gradient, one all heavy and one all light. Mixed species could have been detected as, for example, bands of intermediate density or ^{14}C label in the light octamer band, but none was observed. These results were interpreted as consistent with conservation of the nucleosome octamer during replication.

Variations on this experimental design gave results suggesting that most of the old histones remained associated with only one of the daughter DNA mole-

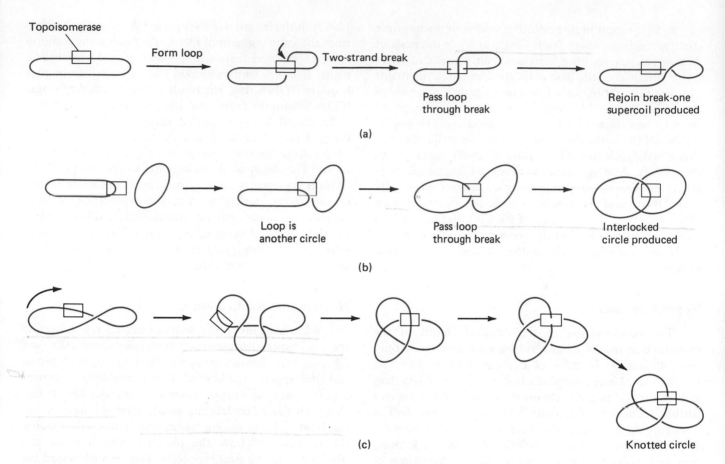

FIGURE 11-17

Action of Topisomerase on Circular DNA. (a) A loop passing through a double-strand break puts a supercoil into the DNA. (b) The "loop" is another circle, producing two interlocked rings. (c) A loop is folded over before being passed through the break, producing a knot.

cules. In the presence of cycloheximide, an inhibitor of new histone synthesis, the nucleosomes appeared to reside on only one daughter molecule; long stretches of naked DNA—the other daughter molecule—could be seen in electron microscope images. Furthermore, in the replication of the chromosome of SV40 virus, it was apparently the daughter molecule with a newly synthesized leading strand (the one capable of being synthesized 5'-to-3' continuously) that mostly retained the old histones. Whether this asymmetrical segregation of old histones will prove to be a general feature of eucaryotic chromosome replication, only time spent in new investigations can tell. How this asymmetry is produced is also an interesting problem to resolve.

Gene Organization

How various genes are ordered along a chromosome and why particular sequences lie on particular chromosomes is the proper concern of geneticists. We shall have

something to say about this subject in Chapter 14, where the regulation of genes with related functional activities is discussed. There are two aspects of gene organization that we address here in connection with DNA replication.

Repetitive Sequences of DNA

Bacterial DNA is composed of a linear series of sequences that are *unique;* that is, found only once in the genome. (For this purpose we regard a sequence to be at least twenty base pairs long. In a random sequence of base pairs the size of the *E. coli* genome, the probability of finding two identical sequences of 20 base pairs would be less than 10^{-5}.) Eucaryotic DNA, on the other hand, has many copies of some DNA sequences. It appears that the larger physical size of the eucaryotic genome may be in part determined by the presence of these *repetitive sequences.*

The observation of repetitive DNA sequences was first made by R. Britten while investigating the rate at

DNA Reassociation Kinetics

It is helpful to express the association of DNA sequences in terms of their concentration C, so the behavior of repetitive sequences can be distinguished from that of unique sequences. Since the association reaction reduces the concentration of single-strand DNA, the rate of association may be expressed as $-\Delta C/\Delta t$, and the rate equation then becomes

$$-\frac{\Delta C}{\Delta t} = kC^2.$$

Separating the variables gives

$$-\frac{\Delta C}{C^2} = k \, \Delta t. \qquad (11\text{--}1)$$

The calculus can be used to show that $-\Delta C/C^2$ is the same as $\Delta(1/C)$, for very small changes in C. With this approximation, the second-order rate equation (11–1) can then be written as

$$\frac{1}{C} - \frac{1}{C_0} = k \, (t - t_0),$$

where C_0 is the concentration of single-strand complementary sequences at the beginning of the association reaction when time $t = t_0$. If we start the measurement of time at the beginning of the experiment ($t_0 = 0$),

$$\frac{1}{C} - \frac{1}{C_0} = k \, t.$$

This equation can be algebraically rearranged to the convenient form

$$\frac{C}{C_0} = \frac{1}{1 + k \, C_0 \, t}.$$

which the separated strands of a DNA molecule can reassociate to form double-strand structures. In a heterogeneous population of single strands, derived from cellular DNA that was fragmented to pieces a few hundred base pairs in length and denatured by heating, he observed that one fraction reassociated to form double-strand DNA much more rapidly than the remainder. The energy of interaction between complementary strands would be expected to be very similar for sequences of a few hundred bases, because in a sequence that long it is unlikely to find a base composition very different from the average. The only plausible explanation for the rapid reassociation of some of the separated strands, therefore, is that it reflected a *higher concentration* of some sequences; that is, *repetitive DNA.*

The reassociation of DNA strands is a bimolecular reaction and its rate depends, therefore, on the concentrations of the two complementary strands. If they are prepared as described above, the concentrations of the complementary strands for a given DNA sequence must be equal. The rate of association will therefore be proportional to the square of the concentration (C) of that particular sequence:

$$\text{rate of association} = k \, C^2,$$

where C is the concentration of a particular sequence of single-strand DNA at a given moment in time and k is the second-order rate constant.

The solution of the rate equation most commonly used (see box on DNA reassociation kinetics) describes the fraction of single-strand DNA (C/C_0) remaining at time t, relative to the concentration of single-strand DNA (C_0) at the beginning of the association reaction ($t = 0$):

$$\frac{C}{C_0} = \frac{1}{1 + k \, C_0 \, t}. \qquad (11\text{--}2)$$

Thus, the fraction of single-strand DNA remaining during the course of reassociation is a function of C_0t (pronounced "cot"), the product of the initial concentration and the elapsed time. Figure 11–18 shows the form of this mathematical function on a linear graph and in the more usual semilogarithmic plot.

When bacterial DNA is sheared into small fragments of a few hundred thousand daltons in mass, the appropriate plot of the fraction of single-strand DNA versus time agrees exactly with Figure 11–18. Such a curve is really the superposition of many similar curves, one for each of the sequences of DNA present in the mixture of fragments. Because the molecular species cannot be followed individually, the concentration plotted is that of a property common to all DNA sequences: the absorbance of light at 260 nm or the phosphorous concentration, etc. Using this convention, a given amount of cellular DNA will have the same molar concentration of all the *unique* sequences, one per mole of complete DNA genome.

When the reassociation experiment is performed on eucaryotic DNA, a more complicated curve is observed.

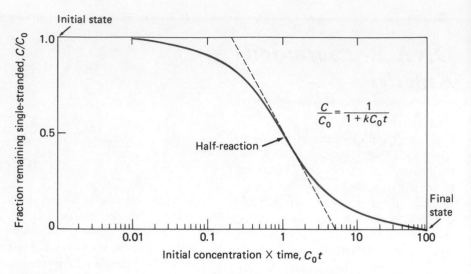

FIGURE 11–18

Renaturation Kinetics (C_0t Plot). The figure shows the fraction of single-strand DNA remaining during the course of renaturation, assuming a theoretical second order reassociation reaction. The fraction of single-strand DNA remaining is assayed by its sensitivity to a specific nuclease or by its chromatographic behavior. The time axis is multiplied by the initial concentration of DNA, C_0, in order to summarize the reaction kinetics with different initial concentrations. This representation gives the plot its acronymic name "Cot." C_0t is plotted on the x-axis along a logarithmic scale which has no theoretical significance but is done in order to display a wide range of data.

A certain fraction of the DNA reassociates rapidly, and complete reassociation occurs over a much wider range of C_0t. It is the repeated sequences of DNA that reassociate rapidly. If a sequence is repeated 10^4 times in the genome, then a given number of grams of fragmented DNA dissolved in a liter of buffer will have a concentration of the repeated sequence some 10^4 times higher than the concentration of a unique sequence. By Equation 11–2, we expect the reassociation to be much faster for the repeated sequences because their C_0 is much higher, and a shorter time is required to achieve the same reduction in concentration of single strands.

Figure 11–19 shows the theoretical curves to be expected from performing the reassociation experiment in *equal weight concentrations* of unique sequences, represented once per genome (a); repetitive sequences, represented 10^4 times per genome (b); and a mixture of equal parts of the two (c). The value of C_0t at which half the single strands have been reassociated is used to characterize the DNA. In curve a, this half-C_0t value ($C_0t_{1/2}$) is 10^3 while it is 10^{-1} for curve b, reflecting the 10^4 larger molar concentration of the repeated sequences in a solution containing equal weights of DNA. Curve c shows that the two populations do not interfere with one another's reassociation: the half-C_0t for the repetitive sequences (the first 50% to reassociate) is the same as for curve b and the half-C_0t of the unique sequences (the second 50% to reassociate) is the same as for curve a.

The reassociation of DNA strands from higher organisms is even more complex than curve c; it is resolved by convention into three roughly distinct classes of DNA sequences: *unique*, *moderately repetitive*, and *highly repetitive*. Table 11–2 summarizes our present information

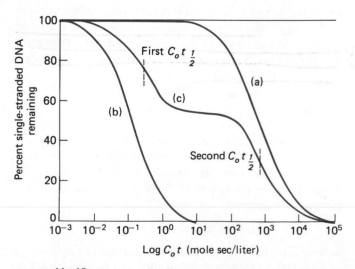

FIGURE 11–19

C_0t **Plot for Unique Sequences (a), for Repeated Sequences of About 10,000 Copies per Genome (b), and for a Half-and-Half Mixture of the Two (c).**

TABLE 11-2

Three Frequency Classes of DNA Sequences Found in Eucaryotic Cells

Frequency Class of DNA	Percentage of Genome	Number of Copies per Genome	Examples
Unique	10–80%	1–10	Structural genes (for hemoglobin, ovalbumin, silk fibroin)
Moderately repetitive	10–40%	10^1–10^5	Genes for rRNA, tRNA, and histones
Highly repetitive	0–50%	$>10^5$	Satellite DNA sequences of 5–300 nucleotides

on the distribution of unique and repeated sequences in the DNA of eucaryotic organisms. It seems clear that many genes specifying protein enzymes are represented only once in the genome. Other gene products, such as ribosomal RNA, are moderately repetitive. However, much of the moderately repetitive DNA seems to have no role in specifying amino acid sequences. The cellular function of the highly repetitive DNA sequences is a mystery for the present.

Reassociation experiments in which the length of the DNA fragments is varied have shown for many species that a common arrangement in DNA is short (100–400 base pairs), moderately repeated sequences interspersed with much longer (10^3–10^4 base pairs), single-copy sequences. An exception is the arrangement in the order of two-winged flies (Diptera), like *Drosophila*, where the moderately repeated sequences are over 5000 base pairs in length and flank unique sequences of at least 10^4 base pairs. The use of genetic engineering techniques allows the production of large amounts of specific sequences of each class. These can be used as probes to locate the positions of complementary sequences in longer fragments of the genome. In the case of sea urchins, for example, it has been found that in a fragment of some 20,000 base pairs, the unique sequences have different moderately repetitive sequences adjacent. We shall return to models that attempt to rationalize this complex informational architecture in Chapter 14.

In Situ Hybridization

The localization of genes, the units of heredity, along the chromosome has been discussed from a genetic point of view in Chapter 9. It is also possible to identify the locations of certain genetically unexpressed structures by using the technique called *in situ hybridization*. Consider, for example, the centromere region of the chromosome, that point where the two daughter chromatids seemingly attach to one another during mitosis. It has been known that this region is heterochromatic in staining (see Chapter 2). We now know that this region contains

repetitive sequences of DNA called *satellite DNA,* and we know this from the following experiment. Satellite DNA can be separated from the bulk of the DNA by a variation of the density gradient method. Using a bacterial RNA polymerase and radioactive precursors, M. L. Pardue and J. G. Gall were able to transcribe this DNA in vitro into a highly radioactive RNA. This RNA was hybridized to a previously denatured chromosome preparation, excess unhybridized RNA was removed by added RNAse, and an autoradiograph was made. Figure 11–20 shows the

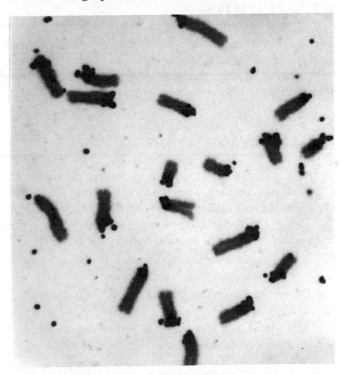

FIGURE 11-20

Chromosomal Localization of Mouse Satellite DNA. Mouse chromosomes where hybridized with complementary radioactive RNA as described in the text. The darkened silver grains of the autoradiogram, lying over the stained chromosomes, show where hybridization has occurred. The mouse satellite DNA is highly repetitive and shows up on the autoradiogram as dark spots. ×3,500. (Courtesy of M. L. Pardue and J. G. Gall.)

result. The radioactive RNA is found only in the heterochromatin centromere regions of all the chromosomes, indicating that this is the location of the satellite sequences. Observing satellite DNA in multiple locations is consistent with its also being among the highly repetitive DNAs revealed by reassociation kinetics (C_0t analysis).

The technique has also been used to locate the moderately repetitive genes for transfer RNAs and ribosomal RNAs on the chromosomes of various eucaryote cells. Genetic engineering techniques have allowed the application of hybridization in situ to locate unique (single-copy) genes that specify proteins. Cloned DNA that is complementary to the protein's messenger RNA can be highly radiolabeled and, after hybridization, located on the eucaryote chromosome by autoradiography (see Chapter 10).

There is much research underway on various problems concerning the replication of DNA and how DNA sequences are organized on the chromosome. If the information in DNA were likened to the books in a library, these processes are analogous to making new books for libraries and understanding how the books are arranged on the shelves. In the next chapter we introduce the idea of circulating the information by beginning the story of how the genes are expressed.

SUMMARY

In this chapter, we have reviewed the transmission of the genetic material from one generation to the next. First, we looked into the process of *mitosis,* in which duplicates of each chromosome are transmitted to each daughter cell at cell division. Next, we described the replication of DNA and its maintenance, repair, and recombination. Finally, we examined the molecular architecture of the chromosome and the large-scale organization of genetic material.

KEY WORDS

Base pair, chromatid, chromatin, chromosome, deoxyribonucleic acid, endonuclease, enzyme, equilibrium density gradient centrifugation, eucaryote, exonuclease, genome, genotype, helix, heterozygous, hybrid, hybridization, information, mitosis, nuclease, polymer, primer, procaryote, recombinant, recombination, replication, ribonucleic acid, template, ultraviolet radiation, virus, wildtype.

PROBLEMS

1. When eucaryotic DNA fragments are denatured to separate complementary strands and then allowed to renature, one fraction reassociates to form double-strand DNA far faster than the rest. What is the special property of this fast-renaturing fraction?

2. What is characteristically found at the ends of a transposon?

3. Define: chi form; mismatch repair; hybridization in situ.

4. All of the DNA structures shown below contain one or more errors except
   A. 5′ . . . ATTGGGGCAT . . . 3′
       5′ . . . TAACCCCGTA . . . 3′
   B. 3′ . . . AUUGGGGCAU . . . 5′
       5′ . . . UAACCCCGUA . . . 3′
   C. 3′ . . . ATTGGGGCAT . . . 3′
       5′ . . . TAACCCCGTA . . . 5′
   D. 3′ . . . ATTGGGGCAT . . . 5′
       5′ . . . TAACCCCGTA . . . 3′
   E. 3′ . . . ATTGGGGCAT . . . 5′
       5′ . . . TATCCCCGTA . . . 3′

5. The enzyme required for the synthesis of DNA from RNA is:
   A. DNA polymerase
   B. RNA polymerase
   C. <u>Eco</u> R1
   D. reverse transcriptase
   E. lambda integrase

6. In order to multiply within hosts that manufacture <u>Eco</u> R1, bacteriophages must be able to:
   A. synthesize a protein coat
   B. synthesize reverse transcriptase
   C. methylate their DNA
   D. synthesize restriction enzymes
   E. synthesize mRNA from their own nucleotides

7. Restriction enzymes create DNA sequences with "sticky" ends by:

A. removing short segments of the nucleotides from one strand
B. cleaving DNA strands at sites several bases apart
C. adding a series of protruding nucleotides
D. catalyzing theta replication
E. Responses A, B, and C are required for a complete explanation

8. The enzymes that catalyze the formation of bonds between a 3′ end and a 5′ end of a DNA segment are called:
A. DNA ligases
B. DNA polymerases
C. restriction enzymes
D. lysogenic enzymes
E. Eco R1 enzymes

9. *E. coli* is grown in ^{15}N (heavy nitrogen) until all the normal light ^{14}N is fully substituted. DNA is extracted and run in a CsCl gradient along with DNA from a ^{14}N-fed culture. The banding pattern obtained is shown below in tube (0). The cells from the ^{15}N culture are washed and transferred to ^{14}N medium, and at three successive cell divisions DNA is extracted and run in a CsCl gradient.
(a) Indicate in the tubes below (1–3) the banding patterns that are expected on the basis of semi-conservative replication after 1, 2, and 3 cell divisions.

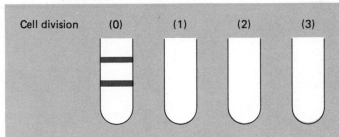

(b) Assuming *conservative* DNA replication, indicate in tube 1′ the banding pattern expected following one round of replication in ^{14}N of the original ^{15}N-substituted culture.

10. In the DNA replication preceding meiosis a cytosine (C) changes to a rare imino form and pairs with adenine (A). Assume that this C=A pair remains intact until the next DNA replication when the C and A bases will pair with their normal partners. If the gamete containing the C=A pair fuses with a normal gamete containing a C≡G pair at this same site on the homologous chromosome, what will the base pair composition of the resulting individual be at this site?

11. The recognition sequences for some restriction enzymes are listed below. Bacteriophage λ has 48,000 base pairs. Which enzyme should cut bacteriophage λ the *least* number of times?
A. GAATTC
   CTTAAG
B. GGCC
   CCGG
C. CGGACCG
   GCCTGGC
D. GGATCC
   CCTAGG
E. AAATTT
   TTTAAA

12. Two newly discovered eucaryotes have the same DNA content per nucleus, but show different time courses of renaturing their DNA. The renaturing experiment was done by shearing the DNA to fragments approximately 350 base pairs long, heating and fast cooling to denature, incubating in standard salt concentrations at an appropriate temperature, and assaying the extent of renaturation by the retention of double-strand molecules on hydroxylapatite columns.

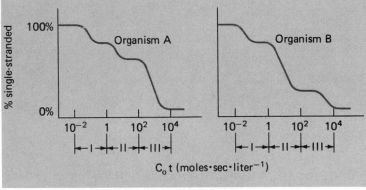

(a) Why are there three distinct phases (I, II, III) in the progress of the renaturation?
(b) How do the two eucaryotes differ in their genetic organization?
(c) Assume one organism is known to be free-living and the other is known to be an intracellular parasite, but the labels designating them have been lost. Predict which is which on the basis of your analysis of the renaturation kinetics. Defend your prediction.

SELECTED READINGS

Britten, R., and D. Kohne. (1981) Repeated segments in DNA. In *Genetics: Readings from Scientific American*. San Francisco: W. H. Freeman.

Crick, F. H. C. (1981) The structure of the hereditary material. In *Genetics: Readings from Scientific American*. San Francisco: W. H. Freeman.

Freifelder, D. (1985) *Essentials of molecular biology*. Boston: Jones and Bartlett.

Grobstein, C. (1981) The recombinant DNA debate. *Genetics: Readings from Scientific American*. San Francisco: W. H. Freeman.

Hanawalt, P. C. (1985) Molecules to living cells. San Francisco: W. H. Freeman.

Howard–Flanders, P. (1981) Inducible repair of DNA. *Scientific American* 245:72–80.

Kornberg, A. (1984) DNA replication. *Trends in Biochemical Science* 9:122–24.

Rosenfield, I., E. Ziff, and B. Van Loon. (1983) DNA for beginners. New York: Writers and Readers.

Wang, J. C. (1982) DNA topoisomerases. *Scientific American* 247:94–109.

The Structure and Synthesis of RNA

12

Genetic information is encoded, as we have discussed in the last two chapters, in the linear sequence of nucleotide bases in DNA. This is also the case for the other nucleic acid, RNA. In fact, the coding scheme is essentially the same. There are four different bases in RNA and they are identical to the four bases found in DNA with one exception—the use of uracil in RNA in place of thymine (5-methyl uracil) in DNA. RNA can also be a repository for genetic information, as is the case for certain RNA viruses. But another function is more striking—the sequence of bases in most RNA molecules is determined by a sequence of bases in DNA. RNA can thus act as an information-carrying intermediate. In particular, the way genetic information is expressed as proteins is always through an RNA intermediate. As we shall see in Chapter 13, the sequence of amino acids in a protein is determined by the sequence of nucleotide bases in a messenger RNA. The sequences of other RNA components in the protein synthetic apparatus are also determined by DNA, but this information does not become incorporated into protein.

The Structure of RNA

Unlike DNA, RNA is not found in the form of a regular double helix of extended length except in certain viruses. This was known for some time because it had been observed that for most cellular RNA molecules the ratios A/T and G/C are not one (Table 12–1) and because RNA when heated does not generally denature into two separate polynucleotide molecules. It was therefore concluded that RNA does not have a regular structure. Even though melting curves of RNA showed that some base stacking does occur, the melting curves observed with most RNA molecules are not as abrupt as those typical of DNA (compare Figure 12–1 with Figure 10–16).

Accounting for these data required a model containing partial base stacking in a single strand of nucleic acid, and it was therefore reasoned that parts of the molecule formed *antiparallel hairpin loops* (Figure 12–2). This prediction of stacked bases in locally double-strand regions has been

TABLE 12–1

Base Composition of Various RNAs

Species Organ (DNA composition in parentheses)	Type	A	G	C	U	ψU★
Rat liver (42% GC)	Nuclear	20.2	25.7	29.5	24.6	
	Nucleolar	16.3	33.0	29.9	20.7	
	Mitochondrial	17.8	31.8	28.4	20.9	
	Ribosomal	20.0	30.5	31.6	20.2	
	Soluble (largely transfer)★	20.9	30.9	28.5	20.4	3.8
	"Messenger"	23.8	28.3	27.4	20.5	
		26.5	20.1	24.0	29.4	
Rat brain	Ribosomal	19.3	32.2	30.4	18.0	
	Soluble	21.3	28.5	27.5	22.5	
Sheep reticulocytes (43% GC)	Ribosomal	18.2	32.7	30.9	18.2	
Yeast (36% GC)	Ribosomal (average of two sets)	25.9	27.7	19.4	26.7	
	Soluble★	18.5	29.2	28.4	20.0	
	"Messenger"	27.5	25.1	20.3	27.1	
Aspergillus niger	Ribosomal	24.3	30.1	22.9	22.7	
	Soluble★	19.9	30.7	27.8	21.6	
Bacillus cereus (35% GC)	Bulk ribosomes	25.2	31.7	21.9	21.2	
	Soluble★	20.5	31.1	28.0	18.8	1.6
Escherichia coli (50% GC)	30S ribosomes	24.6	31.6	22.8	21.0	
	50S ribosomes	25.6	31.4	20.9	22.1	
	70S ribosomes	25.0	31.5	22.1	21.4	
	Soluble (average of three sets)	19.3	32.0	28.3	16.0	5.9
	"Messenger"	24.1	27.7	24.7	23.5	
Micrococcus lysodeikticus (72% GC)	"Large" ribosomes	23.0	33.3	20.8	22.7	
	"Small" ribosomes	22.2	31.5	22.1	24.1	
	Soluble★	19.6	31.7	25.1	23.4	
Tobacco mosaic	Plant virus	29	26	19	27	
Turnip yellow	Plant virus	23	17	38	22	
Wound tumor	Plant virus	31.1	18.6	19.1	31.3	
Polio	Animal virus	29	24	22	25	
Reovirus	Animal virus	28.0	22.3	22.0	27.9	
f2 (MS2)	Bacterial virus	22	26	27	25	

★ψU, when not shown, has been included in the analysis for U. Soluble RNA also contains the methylated bases as well as ψU in much higher concentrations than any of the other types. In soluble RNA from rat liver, 25% of U is replaced by ψU, 10% by 5-methylcytosine, 8.1% by 6-methylaminopurine, and approximately 3% by 1-methylguanine.

confirmed by studies using optical rotation (which is influenced by the helical nature of the structure) and x-ray diffraction (which directly detects the base pairing). The importance of loops was not taken very seriously, however, until it became clear that, at least in some cases, RNA has a definite and precise *secondary* and *tertiary* structure which has important biological consequences. It is useful to know that, because base pairing is *intramo-*

lecular, the melting of RNA (Figure 12–1) is readily reversed. Although the base-paired structure has been disrupted at the higher temperature, the single-strand molecule is still held together by the covalent phosphodiester bonds. In the case of DNA the annealing process is slower because complementary strands need time to find each other and come into proper juxtaposition.

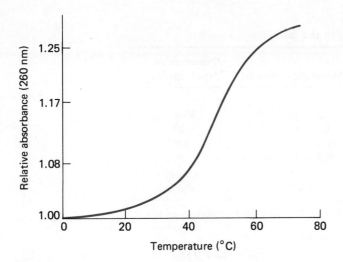

FIGURE **12–1**

Melting Curve of Tobacco Mosaic Virus (TMV) RNA.
The transition is less abrupt and the change in relative
absorbance less extensive than in the case of DNA shown in
Figure 10–16. This is because there is less stacking of bases in
RNA than in DNA. (From P. Doty, et al.)

The Different Kinds of RNA

DNA can be considered a single-minded kind of
molecule, in almost all cases storing information and
making it available for transcription into RNA. Not so
with RNA, which can be divided into several classes on
the basis of both functions and physical properties (Table
12–2). The student should bear in mind that although we
now assign different functions to these categories of
RNA, evidence for their existence was first obtained
from differences in their structural properties. The
principal types of RNA are:

1. viral RNA—normally wrapped in a protein coat;
2. ribosomal RNA (rRNA)—sediments at high speed in
   the ultracentrifuge, by virtue of being part of the
   ribosome particle;
3. transfer RNA (tRNA)—remains in the supernatant
   after ribosomes are sedimented;
4. messenger RNA (mRNA)—heterogeneous in structure and metabolically very active; can be identified
   because it is labeled by radioactive precursors more

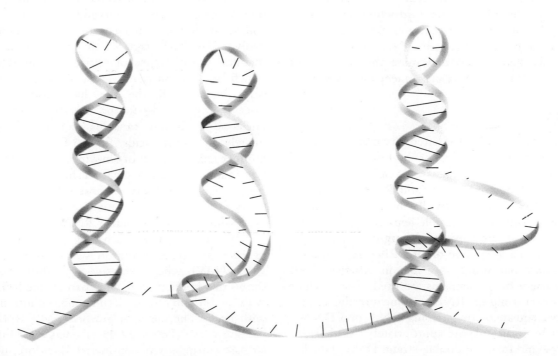

FIGURE **12–2**

**General Model of Base Pairing That Might Occur in a Single Strand of
RNA.** The model consists of unpaired (nonhelical) regions and hairpin loops
forming incompletely paired double helices. The unpaired regions occur where
bases on opposite strands are not complementary.

TABLE 12–2

Relative Abundance of RNA Molecules in the Bacterium _E. coli_

Kind of RNA	Percentage	Molecular Weight	Number of Nucleotides
Ribosomal			
large subunit (50S)	80	1.2×10^6 (23S)	3600
		3.6×10^4 (5S)	120
small subunit (30S)		0.6×10^6 (16S)	1800
Transfer	15	2.5×10^4 (4S)	75
			(approx)
Messenger	5	variable	variable

Note: The S values given in brackets denote the rate at which these molecules sediment in a gravitational field.

rapidly—owing to its more rapid turnover—than ribosomal and transfer RNA.

5. small, nuclear RNA (snRNA)—small (100–300 bases), metabolically stable RNA molecules found in the nucleus, predominantly in ribonucleoprotein particles.

The RNA found in the cytoplasm of eucaryotic cells is synthesized in the nucleus. Much of this RNA product, however, never leaves the nucleus and thus never participates directly in protein synthesis. We shall return to this problem when we discuss the process of transcription later in this chapter. Although the secondary and tertiary structures of RNA molecules are important for their function, these higher-order structural features are determined by the sequence of monomers, the primary structure, as is the case in proteins. Numerous methods have been worked out to determine the sequence of nucleotides in RNA (see box on sequencing RNA).

Viral RNA

Single-strand RNA is the hereditary material of most plant viruses and a number of bacterial and animal viruses. Two groups of animal viruses called _reoviruses_ are known which contain double-strand RNA. It was first thought that RNA was structurally incapable of forming a double helix, but not so; double helix formation seems only to require the presence of the right enzymes. We shall not go into a discussion of viral RNA replication in any detail here but simply say that in different RNA viruses a variety of mechanisms are used. They include various ways of using the RNA genome directly, and the unusual mechanism of transcribing RNA into DNA— with the help of an enzyme appropriately called _reverse transcriptase_—and forming double-strand DNA. This last observation, which created a sensation when it was first discovered, modified the "central dogma" of molecular biology, depicting the major pathways of information flow between macromolecules, from

$$DNA \longrightarrow RNA \longrightarrow protein, to$$

$$DNA \overset{\longrightarrow}{\underset{\longleftarrow}{}} RNA \longrightarrow protein.$$

The RNA to DNA arrow is shown dashed because reverse information flow from RNA into DNA does not occur very frequently. Reverse transcription is now known to occur in uninfected eucaryotic cells, but its normal function remains elusive.

The use of RNA, in RNA viruses, for the storage of genetic information may represent an earlier evolutionary stage of information transfer. It seems plausible that an earlier system of protein synthesis might have involved a self-duplicating form of RNA and that the permanently double-strand DNA, with its added advantage of stability and separation in a nucleus from the rest of the cell, evolved at a later stage. In this connection, we should mention the existence of self-replicating RNA viruses which lack protein coats. These _viroids_ (Figure 12–3) are surprisingly small; the one that causes spindle tuber disease of potatoes—among the largest—has only 359 nucleotides. They are capable of coding only for approximately 100 amino acids, far too few for the minimum number and weight of enzymes necessary for RNA replication but not too few for a short peptide or two that could alter the activity of host cell enzymes. Alternatively, the RNA itself might affect the synthesis or processing of normal cellular RNAs.

We do not know much in detail about the structure of RNA in viruses, except in the case of tobacco mosaic virus (TMV), where extensive x-ray diffraction research shows that in the mature virus particle the RNA interacts with the protein coat molecules and not intramolecularly with itself. The fact that purified TMV RNA shows a melting curve (Figure 12–1)—evidence for intramolecular base pairing—was considered, therefore, not to be of any significance to its structure in the virion. We now know that the formation of intramolecular RNA loops is an important part of the assembly of TMV from the RNA and protein subunits (see Chapter 15).

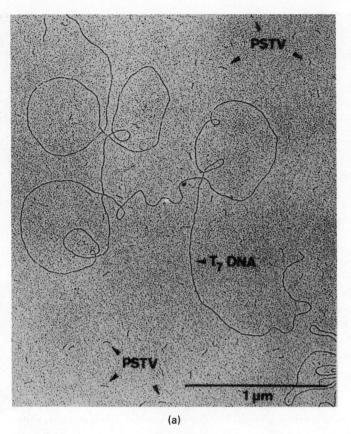

(a)

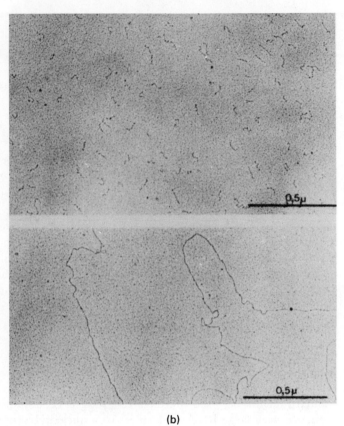

(b)

FIGURE 12–3

Viroids (a) An electron micrograph of the RNA molecule of PSTV, a plant RNA virus. For comparison, DNA molecules of phage T7 (about 40,000 base pairs) were included on the same EM grid. (b) An electron micrograph of the RNA of TobRSV virus, another plant viroid, with DNA of a bacterial virus shown in the lower panel for comparison. (Bars = 0.5 μm; courtesy of T. O. Diener.)

It has been possible for some time to prepare TMV free of other tobacco leaf components, isolate the RNA, and demonstrate that each virus particle (of molecular weight 4×10^7) contains a single strand of RNA of molecular weight 2.2×10^6. TMV RNA has therefore become one of the most intensely studied and best-understood molecules of RNA, having been used not only for physical studies but also for the study of virus assembly, protein synthesis, the mechanism of mutagenesis, and infectivity.

The molecular weight of the RNA enclosed in a virus can vary greatly. Generally, the larger the virus, the larger the RNA molecule it contains, although the precise geometry and architecture do play a role. Thus, as shown in Table 12–3, p. 550, TMV is unusually large for its RNA content because of its cylindrical structure, and human influenza Type A is gigantic in relation to the amount of RNA it contains because it has two protein coats. Most RNA viruses contain a single molecule of RNA—usually single-strand, but sometimes double-strand, as in the reoviruses.

Messenger RNA

In the late 1950s when it became clear that ribosomes are involved in protein synthesis, the question arose if ribosomes are informational—that is, whether each specific type of protein is synthesized by a special ribosome carrying the information required for the synthesis of that protein. There were even then a number of reasons why this seemed unlikely, most having to do with the remarkably rapid response of the protein synthesizing system of the bacterium *E. coli*. For instance, it was observed that within one minute after infection by a bacteriophage, an *E. coli* cell ceases to synthesize bacterial proteins and begins to make those required for viral

Reading the Transcript In Vitro—Sequencing the RNA

For a number of years after F. Sanger (1952) succeeded in determining the order of amino acids in the two polypeptide chains of insulin, the message written into DNA and RNA stood hidden from view. Then in 1965 R. Holley determined the sequence of a transfer RNA by means that were logically similar to those employed by Sanger on proteins. Soon after Holley's achievement, Sanger and his group reported their own ingenious method for sequencing RNA, which has since become a standard technique, used routinely to sequence very large RNA molecules. The method involves the following steps:

1. Organisms from which the RNA is to be isolated are grown in the presence of the radioactive isotope of phosphorus (^{32}P). Since ^{32}P is available commercially at very high specific radioactivity, this permits the use of very small quantities of nucleic acid as a starting material.

2. A variety of RNAses are used to split the polynucleotide specifically at different points in the chain. For instance, ribonuclease T1 splits RNA on the 3′ side of guanylic nucleotides, producing fragments that terminate with Gp-3′.

3. The mixture of oligonucleotides obtained by hydrolysis with a given enzyme is subjected to an ingenious two-dimensional fingerprint method, developed by Sanger and colleagues, that combines electrophoresis and chromatography (Figure 12–A). The nucleotides, after being placed on one end of a strip of cellulose acetate, migrate in one direction in response to an electric field (ionophoresis). The strip of cellulose acetate is then placed on a sheet of DEAE-cellulose paper, and the nucleotides migrate during chromatography at right angles to their ionophoretic migration, at rates determined by their relative affinities for the positive charges of the DEAE groups on the paper. The important result of applying this method is that small fragments, up to five or six nucleotides, will move to particular positions relative to each other on the sheet of paper. Each position is characteristic for a particular base composition. Figure 12–A1 shows a grid obtained by connecting these various positions, and a little thought will enable the student to puzzle out how this grid is read. Thus, this method gives the base composition of the fragments directly by a single experiment.

4. Since the position of each radiolabeled fragment can be determined simply by placing the sheet against a photographic film (autoradiography), it is possible to cut out the spot and to recover (by elution) the fragment for further study. One can, for example, determine the sequence of nucleotides by digestion under limiting conditions with exonucleases. These remove the nucleotides one by one and make it possible to obtain in one experiment all the products of a stepwise digestion. Analysis of the nucleotides removed as a function of the time of digestion gives the sequence.

5. The order, with respect to each other, of the fragments obtained from the first enzyme digestion can be determined by repeating the whole procedure, but this time digesting with a different enzyme. This gives an entirely different set of fragments and when these are matched with the first set they provide overlapping data which allow one to arrange the fragments into a unique sequence.

Using the above procedure, Sanger and G. Brownlee sequenced the 5S ribosomal RNA from *E. coli,* which has a length of 120 nucleotides (for discussion of "S" values, see Chapter

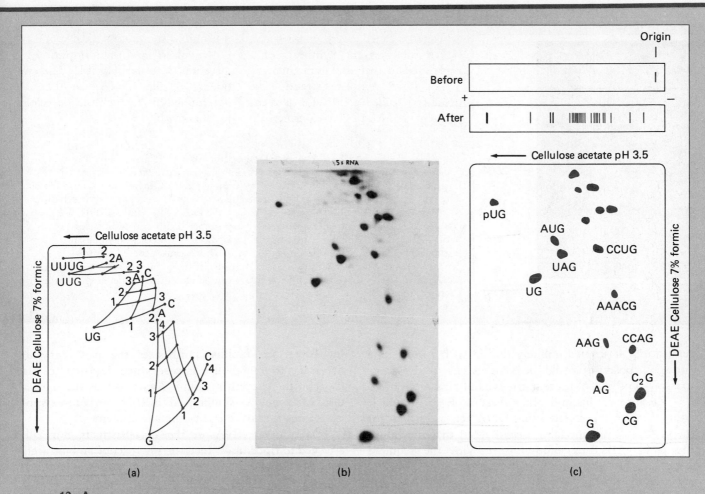

FIGURE **12–A**

Two-dimensional (Ionophoresis/Chromatography) Fingerprint of T1 Ribonuclease Digest of 5S RNA. (a) Final position of various oligonucleotides when fingerprinted by themselves. These positions lie along a grid. One can thus determine the composition of an oligonucleotide from its position on the grid. (b) Autoradiograph of a T1 ribonuclease digest of 5S RNA which has been labeled by growing *E. coli* cells in a ^{32}P medium. Students might try to identify the nucleotide composition of the oligonucleotides by using the standard grid in part (a) and then comparing their results with part (c). (c) The autoradiograph shown in (b); the position of each radioactive oligonucleotide found in the digest is labeled. (From G. Brownlee and F. Sanger.)

4). In the intervening years much larger molecules have been sequenced, and the complete sequence of 16S and 23S ribosomal RNAs from *E. coli* (~1800 and ~2900 nucleotides, respectively) have been reported. With larger RNA molecules an intermediate step must usually be introduced which involves splitting the molecules into a few large fragments. This can be achieved by using two properties of the RNA molecule:

1. The secondary and tertiary structure of some RNA molecules (see below), especially at low temperatures, makes a few bonds more susceptible to ribonuclease digestion than others; digestion under limiting conditions can thus give a few fragments split at specific points.
2. Specific interactions with proteins protect some RNAs from being split by ribonucleases and ensure

that specific bonds are more susceptible to digestion.

These fragments are then separated from each other by ion exchange chromatography or, more frequently nowadays, on large preparative polyacrylamide slab gels (Figure 12–B).

Another method exists for indirectly determining RNA sequences. This method exploits the existence of

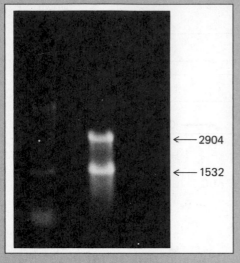

← 2904

← 1532

an enzyme called *reverse transcriptase,* which catalyzes the synthesis of a DNA strand that is complementary to an RNA template strand. The product is called a *cDNA* (for complementary). It can be converted into a double-strand DNA, by DNA polymerase I, and then amplified by insertion into a plasmid. The sequence of the cDNA can then be determined by the methods described in Chapter 10.

FIGURE 12–B
The Use of Polyacrylamide Gel Electrophoresis (PAGE) to Separate RNA Fragments According to their Sizes. The lane at left contains a set of RNA fragments of defined lengths, prepared by transcription of linearized plasmids *in vitro.* The lane at right contains total high-molecular-weight RNA from *E. coli* cells; the overwhelmingly predominant types, yielding the two well-defined bands, are the two ribosomal RNA species of lengths 2904 bases (upper band) and 1532 bases (lower band). After electrophoresis, the gel was stained with ethidium bromide and photographed with ultraviolet light (260 nm), which causes the ethidium bromide–RNA complexes to fluoresce. Portions of the gel containing any desired RNA band can be sliced out, and the RNA can be eluted for sequencing or other chemical manipulations. (Courtesy of R. Libby.)

replication. It seemed unlikely that in such a short time enough ribosomes could be synthesized to allow the massive synthesis of protein involved in viral replication.

Because of this and other reasons, F. Jacob and J. Monod in 1961 postulated that DNA acts as *template* for the synthesis of a special *messenger RNA* that attaches to the ribosome, thus endowing it with the required specificity to synthesize a particular protein. The first critical experiment confirming this postulate was performed that very same year by S. Brenner, Jacob, and M. Meselson, who demonstrated by using radioactive and density-labeling techniques that after phage infection no new ribosomes are synthesized. Instead, a very short-lived form of RNA is synthesized, which becomes attached to the pre-existing ribosomes. This was in fact the postulated messenger RNA.

The same phenomenon was soon demonstrated in uninfected *E. coli* cells through *enzyme induction,* in which

the bacterium rapidly synthesizes the new enzymes necessary to metabolize lactose when exposed to that sugar in the growth medium. Again, it was the translation of a newly synthesized mRNA on pre-existing ribosomes that gave rise to the new enzymes.

Messenger RNA in the cytoplasm is normally attached to ribosomes. Electron microscope and ultracentrifuge studies have shown that ribosomes which are actively engaged in protein synthesis are strung loosely together like beads on an mRNA string (Figures 13–18 and 13–19). One of the problems encountered in purifying mRNA is that it is unstable in most bacterial cells. It now appears that in many eucaryotic cells mRNA is much more stable. There are also structural differences between the mRNAs of procaryotic and eucaryotic cells, and we therefore discuss them separately.

Procaryotic mRNA varies considerably in size for a number of reasons. First, since the polypeptide chains of

TABLE 12–3
Molecular Properties of Some RNA Viruses

RNA Virus	Particle Weight $\times 10^6$	Molecular Weight of RNA $\times 10^6$	Percent of RNA	Shape and Dimension (in nm)
Tomato ring spot	1.5	0.66	44	polyhedral
Tomato bush stunt	10.6	1.65	15	polyhedral, d = 28
Poliomyelitis	16.7	4.8	22–30	polyhydral, d = 30
Tobacco mosaic	40	2.2	5–6	cylindrical. d = 15 × 300
Human influenza type A	280	22.2	0.8	polyhedral, d = 75

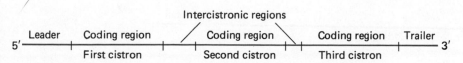

FIGURE 12–4

The General Structure of a Typical Procaryotic mRNA. The transcript of an operon containing three genes (or cistrons) is illustrated. There are three coding regions that specify the three polypeptide gene products, separated by two intercistronic regions; the intercistronic regions are of variable length, and in some cases, where one coding region immediately follows another, there is no intercistronic region. The coding regions are preceded by a 5' leader sequence and followed by a 3' trailer sequence.

proteins vary greatly in size ($5 \times 10^3 - 5 \times 10^5$ Da), it follows that the length of mRNA coding for them must vary accordingly, the ratio of their respective masses being approximately 10 Da of RNA for every dalton of protein. Second, sequencing work on a number of known messengers shows that there is a "leader" at the 5' end of the molecule (translation proceeds from the 5' to the 3' end of mRNA) and this leader varies in length and sequence from one kind of mRNA to another. Third, there are many cases in which one mRNA molecule codes simultaneously for more than one kind of polypeptide chain. The messages are arranged sequentially, with untranslated regions between (Figure 12–4). Frequently, the proteins specified by a single mRNA molecule have related metabolic functions. For instance, one mRNA molecule of 4×10^6 Da in mass codes for ten enzymes involved in histidine metabolism.

Although the mRNA in eucaryotic cells is sufficiently long-lived to allow its isolation, it is still much less stable than other RNAs. However, in metabolically inactive cells like unfertilized eggs, or in cells specialized to synthesize proteins at a constant rate, such as liver cells (which synthesize the proteins of blood plasma), or in immature red cells (which synthesize large amounts of hemoglobin), mRNA molecules can be stable for days.

As in procaryotic cells, the size of cytoplasmic mRNA varies with the size of the protein to be synthesized. In eucaryotic cells, however, it is rare to find an mRNA that codes for more than one protein. There are also untranslated *leaders* at the 5' end and *trailers* at the 3' end. Almost all cytoplasmic mRNAs also contain at the 3' end a string of approximately 100–300 adenylic acid residues, the poly A tail. The function of this poly A sequence, synthesized in the nucleus, is not yet known, although a specific inhibitor of the poly A addition inhibits transport of mRNA from the nucleus to the cytoplasm and the stability of some mRNAs when injected into cells is influenced by the length of the poly A tail. Among the messenger RNAs that lack poly A are those for the histone proteins.

Finally, the 5' end of the eucaryotic mRNA molecule is fitted with a guanine nucleotide linked not with the usual 5'-to-3' phosphodiester bond, but with a 5'-to-5' condensation of the nucleotide triphosphate (Figure 12–5). Methyl groups are added to the guanine, and often to two of the adjacent nucleotides as well. The role of this unusual modification of the 5' end of the messenger, which is called the cap, is not yet completely understood. It is believed that when the messenger has achieved its final primary structure, it is also complexed with protein, a 22,000-Da protein binding to the 5' end and a 52,000-Da protein probably attached to the 3' end. It is in this complexed nucleoprotein state (*mRNP* = messenger ribonucleoprotein particle) that it appears in the cytoplasm of the eucaryotic cell.

The structure of mRNAs in chloroplasts resembles that of procaryotic mRNAs: there is no cap nor is there a poly A tail. The poly A tail is found on mRNAs isolated from mitochondria but, unlike the cytoplasmic mRNAs, there is no 5' cap structure.

Ribosomal RNA

Ribosomes are the protein-synthesizing machines of the cell. We shall discuss their structure and function in great detail in Chapter 13. We need only mention here that they are made of two subunits, unequal in size. In procaryotes—for example, *E. coli*—the larger subunit contains one very large (1.2×10^6 Da, 23S)[1] RNA chain and one small (4×10^4 Da, 5S) RNA chain (Table 12–2), in addition to 34 different protein molecules; the smaller subunit contains a 0.6×10^6 Da (16S) RNA chain and 21 different protein molecules. All three RNA chains are single-strand.

Ribosomes of procaryotes are somewhat smaller than those of eucaryotes (Table 13–5). The ribosomes found inside mitochondria and chloroplasts are more like those of procaryotes than those found in the cytoplasm of

---

[1]The S value measures the speed of sedimentation (see Chapter 4).

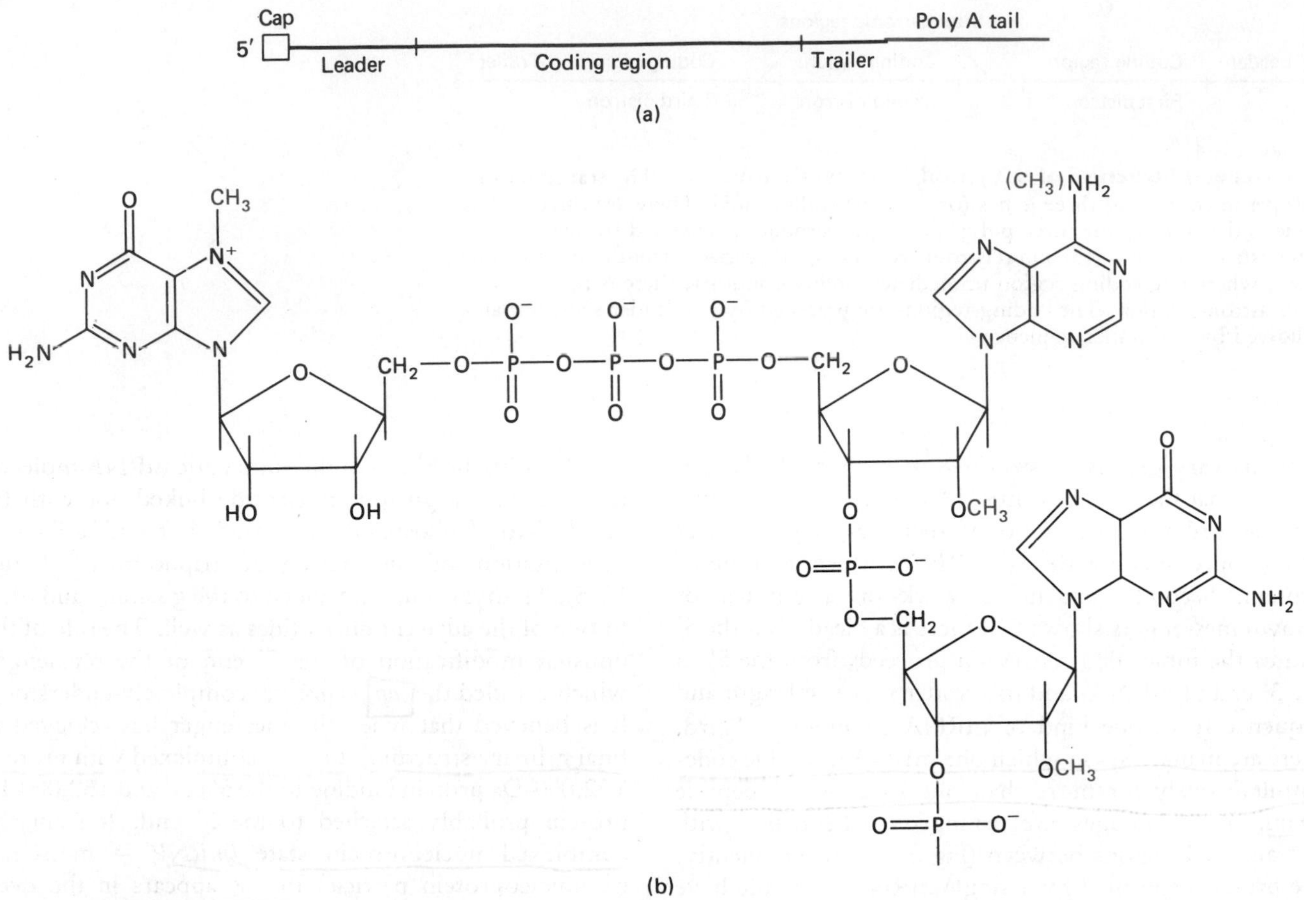

FIGURE 12-5

(a) The General Structure of A Typical Eucaryotic mRNA.   Unlike procaryotic mRNA (Figure 12–B) there is only one coding region (i.e., the typical eucaryotic mRNA is monocistronic). Downstream of the 3′ trailer sequence, there is usually a long poly A tail. (b) The detailed structure of the 5′ cap. Note the methyl modifications (indicated in color) and the 5′-to-5′ linkage between the ribose portions of the first two guanylate residues.

eucaryotic cells—another piece of evidence consistent with the endosymbiont theory of the origin of eucaryotic organelles. The sizes of their respective rRNAs vary similarly.

The sequence of the small 5S RNA molecule in *E. coli* was determined by G. Brownlee and F. Sanger in 1966. Since then a number of 5S RNA molecules from a variety of other sources have been sequenced and from such comparative data it has been possible to identify some general features of this molecule (Figure 12–6).

1. In procaryotes there are four helical regions. They are believed to be universal because they form base pairs even if the precise primary sequence in these regions is not conserved. In other words, it does not matter that position 1 be U; what matters is that if it is U, then 119 must be A, or if it is C, that 119 must be G. Thus,

what is conserved in these cases is the helix, not the primary structure forming the helix. In eucaryotes, helices 82–86 and 90–94 are missing.

2. In all procaryotes studied so far, positions 43–47 have the sequence CGAAC; these nucleotides might interact with the common complementary sequence GTΨCG in loop III of tRNA (see Figure 12–11). This could therefore represent a site of interaction between the ribosome and the transfer RNA molecule.

3. Also in procaryotes there is a strongly conserved region (positions 67–80) protected from enzymatic hydrolysis in vitro by three 50S ribosomal proteins. This region probably represents a site of interaction with other 50S ribosome components.

The 0.6×10^6 Da (16S) RNA molecule from the small (30S) subunit of *E. coli* ribosomes has also been

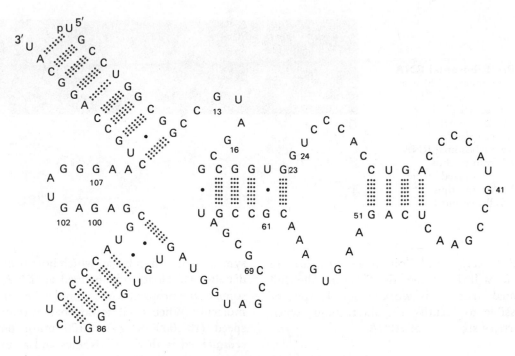

FIGURE **12–6**

Sequence and Primary Structure of Procaryotic 5S Ribosomal RNA.
Potential intramolecular Watson-Crick type base pairing (shown by color H-bonds between the complementary bases) and "wobble" pairing between U and G (shown by black dots) could generate four double-helical regions.

completely sequenced. Comparative studies of the sequence of 16S RNA molecules from several different procaryotic organisms show, like 5S RNA, a number of conserved regions which likely will help us unravel important aspects of ribosome structure and function. An interesting example is the sequence at the 3' end of the 16S RNA molecule, largely conserved in all procaryotes. J. Shine and L. Delgarno noted that this region can base pair with a viral mRNA sequence known as the *initiator region*. This implied that the initiator region could be recognized by *E. coli* ribosomes during the beginning steps of protein synthesis. It has since been shown that similar base pairing occurs with many different mRNA molecules, strongly suggesting that the 3' end of the 16S rRNA molecule is involved in the initiation of protein synthesis.

5′ ······ A U C A C C U C C U U A—OH 3′ End of 16S rRNA
⫽ ⫽ ⫽⫽ ⫽ ⫽⫽ ⫽⫽ ⫽⫽ of *E. coli*
3′ ····· U G G A G G A ····· 5′
Initiator region of cistron for
"A" protein of R17 virus

The sequence of the 1.2×10^6 Da (23S) RNA molecule from the large (50S) subunit of *E. coli* ribosomes has been determined, both by RNA-sequencing methods

and DNA methods applied to the rRNA gene in a cloned form. No doubt we can eventually analyze its sequence in terms of certain aspects of ribosome function.

What do we know about the secondary structure of rRNA? (By *secondary structure* of an RNA molecule we mean the formation of hairpin loops between nucleotides which are fairly close to each other on the nucleotide sequence *primary structure*.) It has been possible to spread out, in the presence of formamide and urea, a precursor to the eucaryotic rRNA molecule and observe in the electron microscope a secondary structure (Figure 12–7).

Since ribosomes are compact, almost spherical particles, it follows that the RNA must be significantly folded; that is, it must also have a *tertiary structure*, which we think of as the folding together of the hairpin-looped molecule (Figure 12–7) into a compact ball. There is mounting evidence suggesting a tertiary structure for rRNA. For instance, the reagent *kethoxal* reacts with guanine but, when ribosomes are exposed to high concentrations of this reagent, less than 10% of the guanines are free to react. Furthermore, evidence indicates that portions of 16s rRNA molecules quite far apart in the primary structure nevertheless interact with the same protein subunit on the ribosome.

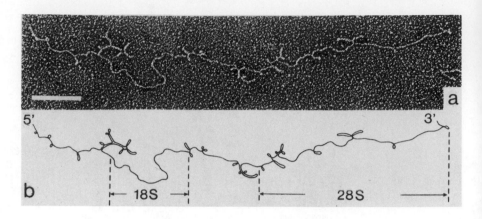

FIGURE 12–7
A Eucaryotic Pre-Ribosomal RNA.
(a) is an electron micrograph; (b) is an artist's rendering of the RNA. The secondary structure of hairpin loops are so characteristic of the precursor molecule that they can also be observed in the two ribosomal RNA molecules derived from it, thus allowing them to be localized in the pre-ribosomal RNA transcript. (Courtesy of P. Wellauer and I. David.)

Though these early results on rRNA structure are encouraging, it will be some time before the full three-dimensional structure is worked out. In spite of this, it is possible to identify the interactions which stabilize the tertiary structure of rRNA:

1. Base pair formation between distantly placed bases as seen, for instance, between the two ends of the 5S RNA molecule in Figure 12–6.
2. Stacking interactions between paired nucleotides, as we have seen in DNA and shall encounter in tRNA.
3. Interaction with specific proteins. As we shall see in Chapter 13, studies of the assembly of ribosomes show that proteins play a specific and necessary role in ribosome structure.

Transfer RNA

Once it was clear that mRNA is the template providing the information for the synthesis of proteins, the question arose of how the nucleotides in the mRNA "select" or interact specifically with the proper amino acids. F. Crick, on purely theoretical grounds, argued it was unlikely that the nucleotides would interact directly and in a specific manner with the amino acids. There was nothing, for instance, in the structure of the four bases to distinguish between two similar hydrophobic amino acids, such as leucine and isoleucine, or glycine and alanine. So Crick predicted that there must be a relatively small *adaptor molecule,* which could recognize both the proper amino acid and the "code" for that amino acid. He argued that the ideal material for such an adaptor molecule would be nucleic acid (presumably RNA), which could utilize base pairing to achieve the specificity required. This brilliant hypothesis, now abundantly confirmed, solved the problem about the nature of interaction between adaptor and messenger (mRNA).

Soon after Crick's suggestion—and following up earlier hints about the involvement of RNA with protein synthesis—a search began which bore fruit almost immediately. M. Hoagland obtained an RNA fraction with some of the properties one would expect from an adaptor molecule. When a cell homogenate is centrifuged at high speed (100,000 × g for 120 min), bound RNA is sedimented with the ribosomes and a supernatant containing unbound RNA, which used to be called soluble RNA, remains. Most, possibly all, of the RNA in this supernatant fraction has a "transfer" function—it combines specifically with amino acids or with peptides and interacts specifically with the codon on the messenger. We distinguish tRNAs on the basis of which of the 20 common amino acids can be covalently attached; that is, tyrosine tRNA can accept tyrosine. Some tRNAs that have different sequences can nevertheless accept the same amino acid; these are distinguished by subscript numbers ($tRNA_2^{Tyr}$). The structure of tRNA is critically important for understanding its function.

In 1965, Robert Holley and colleagues reported the nucleotide sequence of $tRNA_I^{Ala}$ from yeast. This was the first complete sequence of a nucleic acid to be established, an exciting event in molecular biology (earning the Nobel Prize in 1968), and with it, tRNA emerged as a convenient material for precise molecular study of nucleic acid structure.

One earlier observation, which later proved of great value in the sequence analysis, was that tRNA contains a number of unusual bases (Figure 12–8). Many of these differ from A, G, C, and U in having one or more methyl groups substituted at various positions in their ring structures. There are over 50 of these bizarre modified bases present in one tRNA or another. The modified bases themselves have, in some cases, functional roles as plant hormones (known as *cytokinins*).

The method that Holley and colleagues utilized was similar in principle to that employed by Sanger to determine the structure of the protein insulin. Purified yeast alanine tRNA was hydrolyzed under a variety of conditions, catalyzed by a number of enzymes (RNA

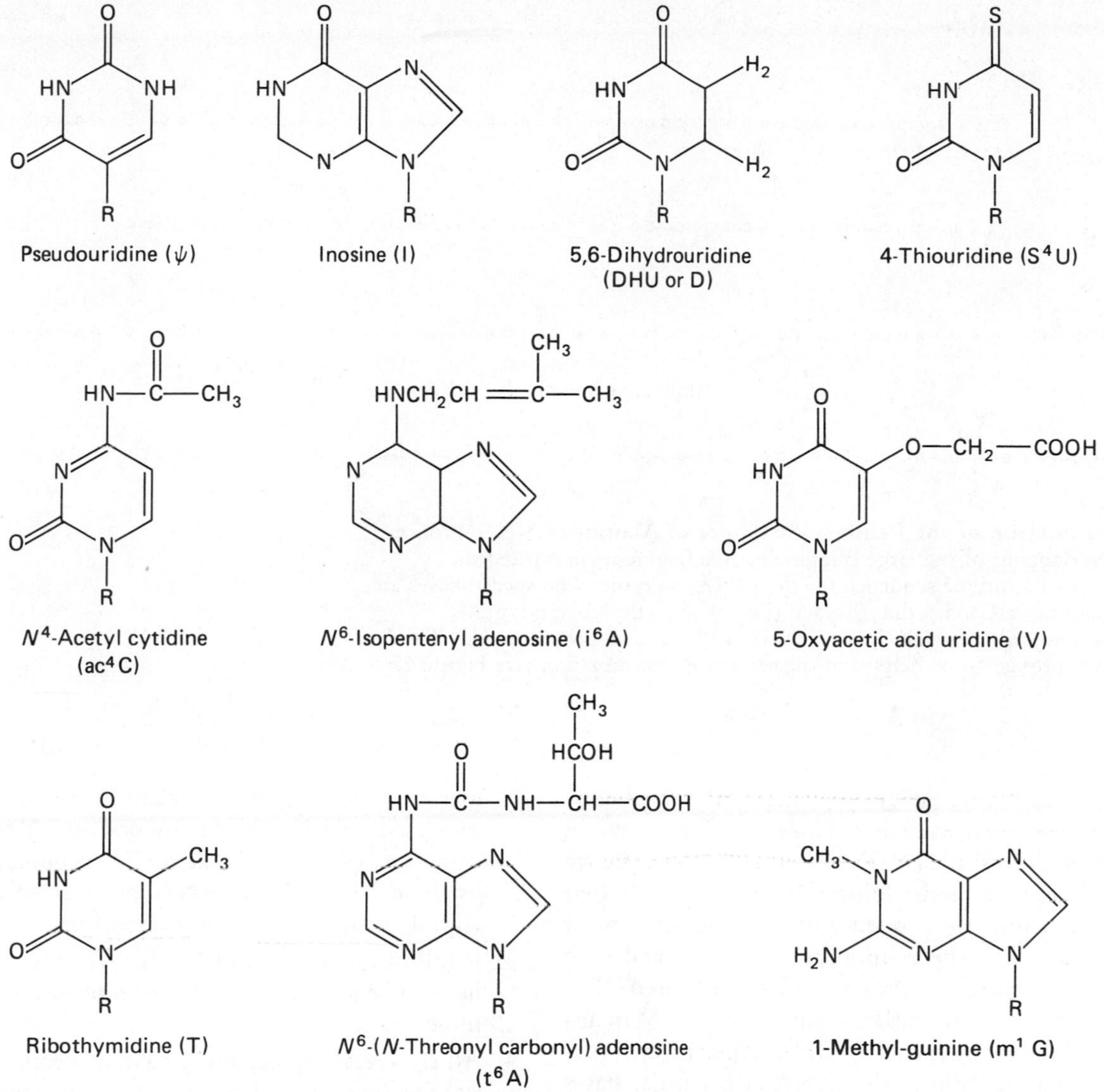

FIGURE 12–8
Modified Bases Found in Transfer RNA. These are only a few of the more than 50 modifications known. Following each name in parentheses is a commonly used symbol for the modified base.

nucleases) of varying specificity. Enzyme digestion at a low temperature yielded only a few large fragments, whereas at a higher temperature larger numbers of smaller fragments were found. By utilizing very long ion exchange columns and a solvent containing urea (which interferes with hydrogen bonding and thus decreases interactions between fragments), they purified the fragments from each other (Figure 12–9). Each purified fragment was again subjected to enzyme or alkaline hydrolysis, the resulting fragments again purified, and so on. By this procedure the sequence of the smaller oligonucleotides was determined, and since the hydrolysis yielded overlapping sequences, it was possible to arrange the order of the small oligonucleotides within the larger oligonucleotides. This was considerably aided by the presence of the unusual bases. Finally, some overlap among the larger oligonucleotides revealed their order (Figure 12–9) so that the complete sequence of yeast alanine tRNA could be determined.

Not long after this important achievement several

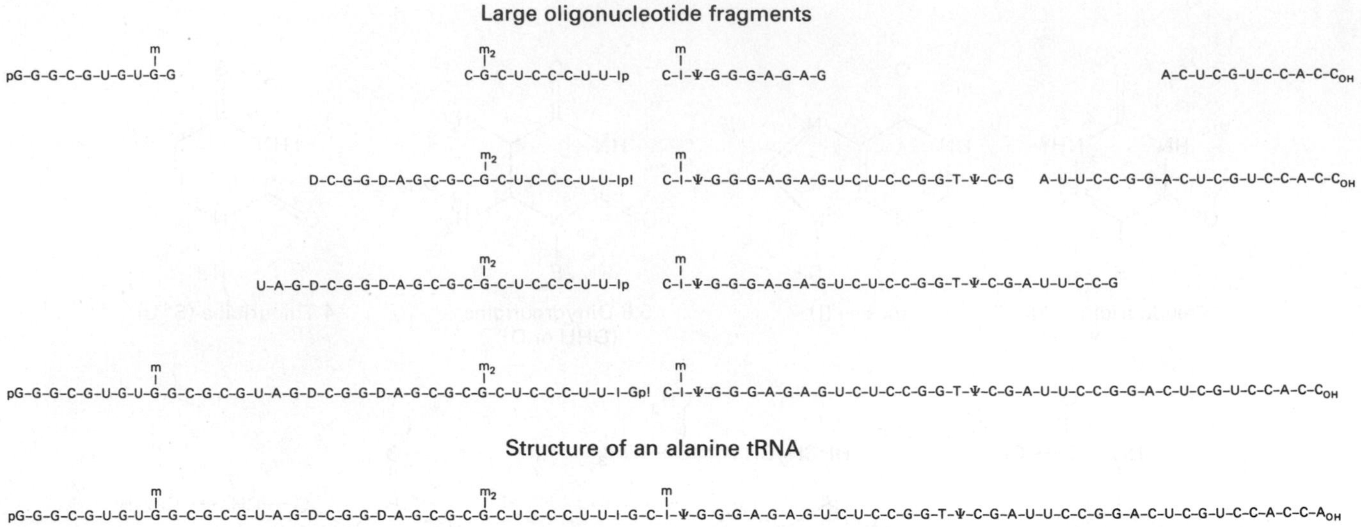

Large oligonucleotide fragments

pG–G–G–C–G–U–G–U–G̅–G   C–G–C–U–C–C–C–U–U–Ip   C–I–Ψ–G–G–G–A–G–A–G   A–C–U–C–G–U–C–C–A–C–C$_{OH}$

D–C–G–G–D–A–G–C–G–C̅–G–C–U–C–C–C–U–U–Ip!   C–I–Ψ–G–G–G–A–G–A–G–U–C–U–C–C–G–G–T–Ψ–C–G   A–U–U–C–C–G–G–A–C–U–C–G–U–C–C–A–C–C$_{OH}$

U–A–G–D–C–G–G–D–A–G–C–G–C̅–G–C–U–C–C–C–U–U–Ip   C–I–Ψ–G–G–G–A–G–A–G–U–C–U–C–C–G–G–T–Ψ–C–G–A–U–U–C–C–G

pG–G–G–C–G–U–G–U–G̅–G–C–G–C–G–U–A–G–D–C–G–G–D–A–G–C–G–C̅–G–C–U–C–C–C–U–U–I–Gp! C–I–Ψ–G–G–G–A–G–A–G–U–C–U–C–C–G–G–T–Ψ–C–G–A–U–U–C–C–G–G–A–C–U–C–G–U–C–C–A–C–C$_{OH}$

Structure of an alanine tRNA

pG–G–G–C–G–U–G–U–G̅–G–C–G–C–G–U–A–G–D–C–G–G–D–A–G–C–G–C̅–G–C–U–C–C–C–U–U–I–G–C–I–Ψ–G–G–G–A–G–A–G–U–C–U–C–C–G–G–T–Ψ–C–G–A–U–U–C–C–G–G–A–C–U–C–G–U–C–C–A–C–C–A$_{OH}$

FIGURE 12–9

The Determination of the Primary Structure of Alanine tRNA. Notice how the overlapping of the large oligonucleotide fragments permitted the establishment of a unique sequence for the tRNA molecule. The special bases are: 1-methylguanine (mG), 5,6-dihyrouracil (D), N^2-dimethylguanine (m2G), 1-methylinosine (mI), inosine (I), pseudouracil (Ψ), and ribothymine (T). These specific bases proved to be helpful in the sequence determination (see Figure 12–8).

more tRNA molecules were sequenced, and now hundreds of sequences are available. This explosive growth in the determination of tRNA sequences is the consequence of radical changes in methodology. It now takes only two weeks to determine the sequence of a minute amount of purified material, where months of effort and one hundred times more material were once required. The current techniques are rather similar to the Maxam-Gilbert method for sequencing DNA. Specific enzymes are used to partially digest the tRNA at particular bases and, after end-labeling with radioactivity, the mixtures of digestion products are separated by electrophoresis on slab gels of polyacrylamide. These "ladder" gels allow the sequence to be read off with great ease.

Before we discuss the important consequences of studying the structure of tRNA, let us list the properties of the tRNA molecule we hope to illuminate. Studies of protein synthesis show that tRNA must have the following properties:

1. It must have a *ribosome binding site* which is universal, since all tRNA molecules must fit the same sites on the ribosome.
2. It must have an *mRNA binding site* which is specific for each kind of tRNA molecule. By the time the structural studies began, it was already known that the protein information on the mRNA was written in groups of three adjacent nucleotide bases; each of these groups is called a **triplet** or **codon.** A triplet is the code for a given amino acid. To recognize the codon by base pairing, the tRNA must have a triplet of complementary bases, or **anticodon.**

3. It follows from (1) and (2) that all anticodons would have to be situated at a common position in all tRNA molecules.

4. All tRNA molecules must have a specific attachment site for their respective amino acids and the site would have to be situated at the same position in all tRNA molecules.

5. All tRNA molecules must have a binding site for the enzyme which catalyzes the attachment of the amino acids. These binding sites must differ in structure so that they can interact specifically with their respective enzymes.

Let us see whether these predictions have been borne out by the structural studies.

In their original article, Holley and his colleagues examined a number of intramolecular base-pairing possibilities raised by the primary structure of yeast alanine tRNA, concluding that a "cloverleaf" structure would maximize the base-pairing possibilities (Figures 12–10 and 12–11). The elucidation of additional primary structures of other cytoplasmic tRNA molecules confirmed

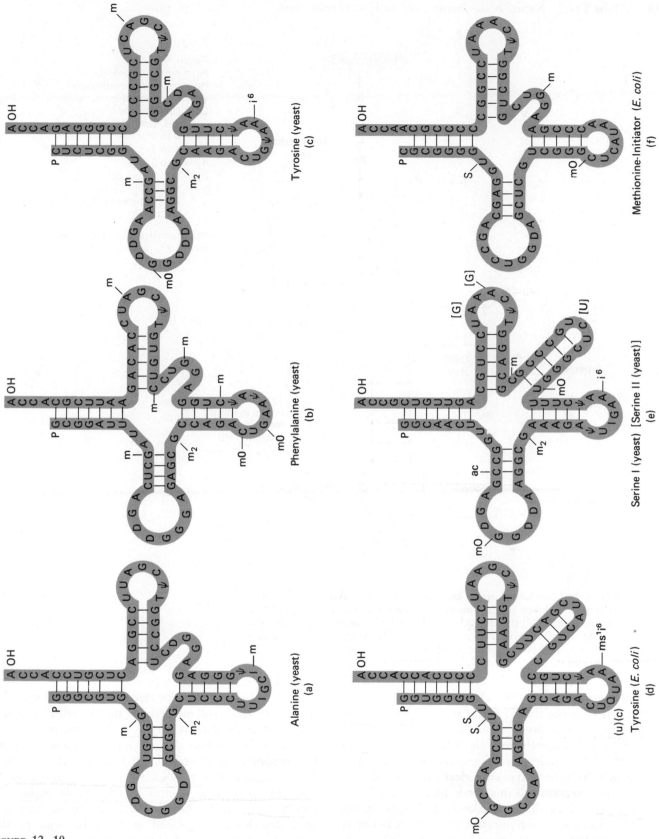

FIGURE 12–10
Cloverleaf Pattern of Six Different tRNA Molecules.
The same secondary structure can be generated from widely
differing base sequences. The differences between two yeast
serine tRNAs are shown in brackets adjacent to their position
in the sequence. The *E. coli* tyrosine tRNA shown, with Q in
the anticodon, recognizes normal tyrosine codons UAU and
UAC (Q is a modified G). Mutants of this tRNA have been

isolated with C at that anticodon position, which allows
recognition of codon UAG, or with U at that position,
which allows recognition of UAA. The usual symbols for the
four standard bases are used: the structures of several
modified bases are shown in Figure 12–8. Other symbols
include: Om, methyl group attached to ribose; m², dimethyl;
ms¹i⁶ and Y are complex modifications.

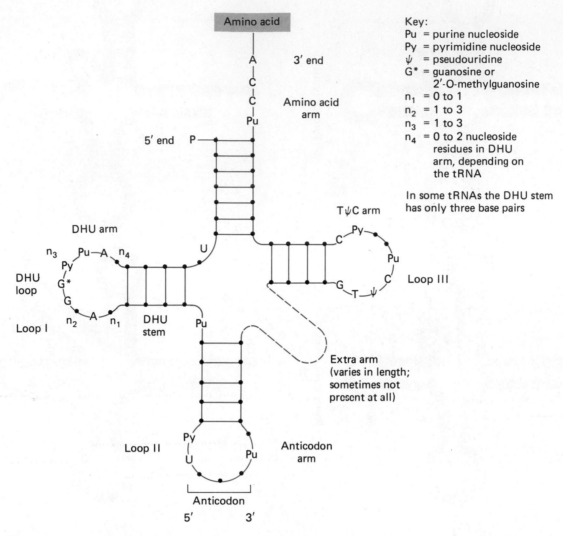

FIGURE **12–11**

General Cloverleaf Structure and Common Features of All Cytoplasmic tRNA Molecules.   Notice (1) the amino acid attachment site at the end of one arm (the acceptor stem); (2) the anticodon at the end of another arm (always the same distance from the amino acid attachment site); (3) the extra arm of variable length; and (4) the other two loops of highly conserved composition.

this hunch beautifully, for in each and every subsequently determined sequence the cloverleaf pattern turned out to be the structure that gave the maximum number of base pairs. Even more convincing, the sequence of bases in each of the tRNA molecules differed widely, but the base pairing which formed the cloverleaf structure didn't. There is one exception to this generalization and that is the *extra arm* which differs markedly in length among tRNAs; in some it is barely present at all, while in others it is composed of twenty or more bases (Figure 12–11). In fact, this turned out to be the exception which proves the rule, for cytoplasmic tRNA molecules vary in size from less than 75 to more than 90 nucleotides, and it is important for them to have a uniform overall length if

they are to bind to the same ribosomal site. The way this is achieved while still maintaining the cloverleaf structure is to contain the variable length in the extra arm.

Another question that Holley considered when he first proposed the cloverleaf structure was where to look for the anticodon. The —CCA$_{OH}$—3′ end of the molecule had already been identified as the site of amino acid attachment, so it seemed logical to guess that the anticodon must be located at the other end of the cloverleaf. It was already known at the time that four different mRNA triplets coded for alanine: GCU, GCC, GCA, and GCG (see Chapter 13 for a discussion of the **genetic code,** the rules that relate codons to amino acids). Therefore, the anticodon would have to be the complementary sequence 3′—CGX—5′, and indeed their alanine tRNA has the sequence CGI at the farthest portion of the loop opposite the amino acid attachment end. Since I (inosine) is an analogue of G, we would expect it to pair with C and we therefore can predict that a proper code in the messenger RNA for this particular alanine tRNA is GCC.

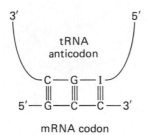

When more tRNA molecules were sequenced, the first thing one looked at, after arranging the nucleotides into the cloverleaf structure, was the three bases in what became known as the *anticodon loop*. Each time they turned out to be *complementary to one of the codons* expected for that tRNA's amino acid, truly a dramatic confirmation of the entire fabric of the biology of information transfer woven over a period of 30 years.

Let us stop for a moment and elaborate on this statement. In the work we have cited there have been two broad underlying avenues of approach. One of these has to do with assigning to each tRNA molecule a particular amino acid transfer function. The other has to do with assigning to each amino acid a given triplet code. Structural work on tRNA succeeded in connecting these two avenues. It is remarkable that all that was needed was to determine the primary structure of tRNA; the rest literally fell into place, because the primary structure revealed a common secondary structure in which the anticodon was always in the correct place.

Molecules of tRNA have a number of other common features, which emerged as more molecules were sequenced. Figure 12–11 shows what they are: the invariant amino acid attachment site, the variable extra arm, and the DHU and TΨC loops with their highly conserved base composition. The relative invariance of these latter two loops suggested at the time that they played an important role in the folding of the molecule into a precise tertiary structure, common to all cytoplasmic tRNA molecules. The observation that tRNA molecules with different nucleotide sequences nevertheless crystallize with each other in the same crystal lattice confirmed the notion of a highly conserved tertiary structure. The presence of the unusual bases in tRNA led to the suggestion that they might be involved in conferring on tRNA the protein-like property of forming a precise three-dimensional structure.

Finally, in 1974, there appeared an article in *Science* by S. H. Kim, A. Rich, and co-workers giving the three-dimensional structure of yeast phenylalanine tRNA at 0.3-nm resolution. Another article based on independent work by A. Klug and co-workers with essentially similar results soon followed in *Nature*. What did the three-dimensional structure of tRNA reveal?

1. The most striking feature of the three-dimensional structure of tRNA is that it is a compact L-shaped molecule with the amino accepting site located at one end and the anticodon at the other (Figure 12–12).

2. The DHU loop and the TΨC loop are indeed folded, interacting closely with each other and stabilizing the L-shaped structure by forming the corner of the molecule.

3. The tertiary structure is stabilized by sets of hydrogen bonds that are unusual—that is, hydrogen bonding between some bases occurs in a way different from that found in DNA. These tertiary structural bonds involve many of the conserved bases as well as some of the variable ones (Figure 12–13). Since most of the conserved bases are modified, this confirmed the notion that modifications play a role in folding the tRNA into a three-dimensional structure.

4. Another important interaction for stabilizing the tertiary structure is the stacking interactions between the planar purines and pyrimidines (Figure 12–13). This stacking is an example of the hydrophobic bonding discussed in Chapter 3.

5. Since most of the tertiary interactions in the crystal structure involve nucleotides which are highly conserved in cytoplasmic tRNA molecules, we can hypothesize that the same structure is relevant to all of them.

6. The anticodon stem is held by only a few tertiary hydrogen bonds to the rest of the molecule, which suggests that there might be some movement of the anticodon during protein synthesis.

As more and more structural analyses of cytoplasmic tRNAs have accumulated, the basic picture outlined here has become more refined but has not changed in its essential features. The same picture is also appropriate for the tRNAs found in many chloroplasts. Mitochondria from mammalian cells, on the other hand, can contain some exceptionally small tRNAs—one example is missing the DHU loop and has a sedimentation rate of only 3S. The invariant —CCA amino acid acceptor is still present and there is also an appropriate anticodon. Exactly how such minimal tRNAs bind to ribosomes and participate in protein synthesis will be the subject of future research.

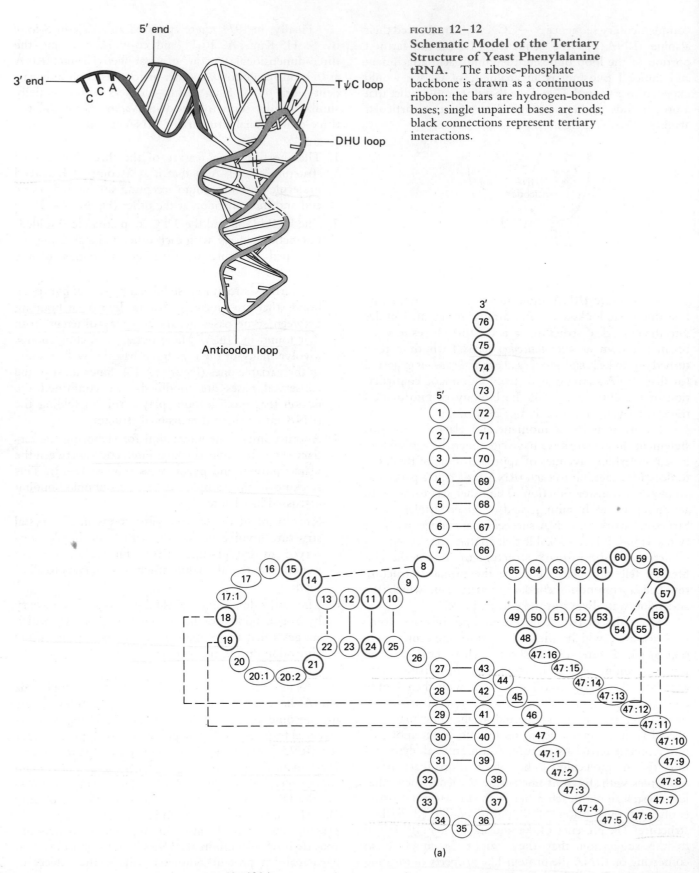

FIGURE **12–12**
Schematic Model of the Tertiary Structure of Yeast Phenylalanine tRNA.   The ribose-phosphate backbone is drawn as a continuous ribbon: the bars are hydrogen-bonded bases; single unpaired bases are rods; black connections represent tertiary interactions.

(a)

FIGURE **12–13(a)**

Synthesis of RNA

Early work on RNA synthesis in mammalian cells and tissues used cell fractionation and autoradiography to follow the incorporation of radioactive precursors into RNA molecules in different parts of the cell. These studies demonstrated that most precursors are first incorporated into RNA molecules in the nucleus, after which most of the newly synthesized RNA is transported into the cytoplasm, although a fraction remains in the nucleus. Localizing RNA synthesis to the nucleus provided an early hint that DNA in the nucleus was the template for RNA synthesis.

This notion was strengthened by the work of L. Astrachan and L. Volkin, who took advantage of a special situation in which a new species of DNA is introduced into a cell. This occurs when a bacterial virus injects its DNA into a host bacterial cell. Volkin and Astrachan found that the RNA labeled by radioactive precursors after virus infection has a base composition similar to that of the invading viral DNA, rather than that of the host bacterial DNA.

The template role of DNA in RNA synthesis was first shown decisively by S. Weiss and J. Hurwitz, who identified an enzymatic activity in cell-free extracts of bacteria and mammalian cell nuclei which synthesized

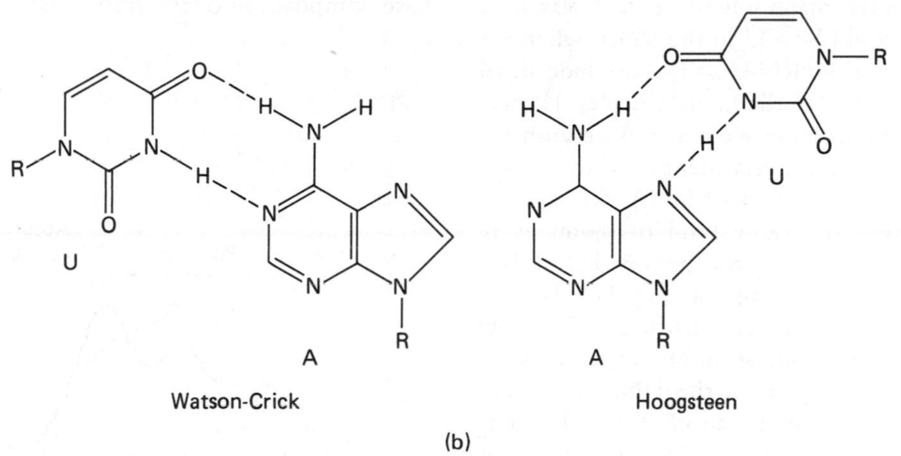

Watson-Crick   Hoogsteen

(b)

FIGURE 12–13

The General Structure of tRNA Molecules, Based on Yeast tRNA^phe
(a) Bases common to all tRNAs at a given position of the cloverleaf secondary structure are circled in boldface. Many tRNAs have one or more additional bases between certain conserved positions; these are indicated by the base they follow (in yeast tRNA^phe) and a colon. For example, the "variable loop" in some tRNAs consists of up to 16 bases between the positions of base 47 and base 48 of yeast tRNA^phe. These are numbered 47:1, 47:2, and so on. Intramolecular hydrogen bonding between complementary base pairs is indicated by a solid line between the circles. The dotted line between bases 22 and 13 indicates that an unusual pairing (such as "wobble" pairing between G and U) is involved in some tRNAs. All tRNAs exhibit Watson-Crick complementary base pairing between the positions connected by solid lines. The dashed lines indicate hydrogen-bonding interactions thought to stabilize the molecule's three-dimensional tertiary structure. Several other pairing interactions between bases in the loops are thought to occur, but these have not been indicated in order to avoid clutter. The outcome of these base-pairing interactions is a standard cloverleaf structure, shown here in two dimension; the three-dimensional L-shaped structure is dependent on the tertiary interactions (dashed lines). A picture of the three-dimensional L-shaped structure is shown in Figure 12–12. (b) Hoogsteen pairing between A and U, compared with Watson-Crick pairing. This hydrogen bonding is one of the types that stabilize the tertiary structure of tRNA.

RNA in vitro, in a reaction absolutely dependent on the presence of added DNA. The enzymatic reaction utilizes the nucleoside triphosphates, ATP, CTP, GTP, and UTP; inorganic pyrophosphate is split from each of these, concomitant with the formation of the phosphodiester bond that links the resultant nucleoside monophosphate to the RNA chain. The enzyme, called *RNA polymerase,* is specific for the ribonucleoside triphosphates; the deoxyribonucleoside triphosphates are not suitable as substrates. Moreover, as in the case of DNA synthesis, the nature of the RNA product formed is dependent on the DNA added.

The base ratios of the RNA synthesized were discovered to be the same as the base ratios of the added DNA, as in the case of the phage-infected bacteria. If one added a different DNA, having different ratios of its nucleotide bases, the resultant RNA contained the same base ratio as the added DNA. The enzyme was therefore making an RNA molecule with a sequence influenced by that of the DNA molecule. It was suspected that the DNA was acting as a template: wherever there was an A in the DNA, there would be a U in the RNA; wherever a G in the DNA, a C in the RNA; and so on. Indeed, all the types of experiments outlined in Chapter 11 performed with DNA polymerase were also done with the RNA polymerase. All the criteria mentioned to test the base-pairing rules in the synthesis of DNA were applied to RNA synthesis and the same kind of results were obtained. (It turned out that this was because both DNA strands were able to act as a template for the relatively crude RNA polymerase preparations used. We now know it is normal for only one strand to act as template for transcription in one region of the DNA; the other strand may be the template in another region.) The product of the reaction was called "informational" or "messenger" RNA to denote that it could carry a message from the nuclear DNA to the cytoplasmic protein-synthesizing ribosomes. The message would control the synthesis of a specific protein.

Histologic and autoradiographic evidence suggest that DNA duplication and RNA synthesis in the nucleus generally take place at different times during the nuclear division cycle. At one time during the S phase a given sequence of DNA acts as template for the enzymatic synthesis of a duplicate DNA molecule; at another time (or times), usually not in S, it acts as a template for the enzymatic synthesis of an RNA molecule.

Does the DNA-mediated RNA polymerase function in the cell as it does in the test tube? When a bacterial culture is given a short pulse of a radioactive RNA precursor, one can easily observe an RNA fraction that is rapidly labeled and, if the pulse is followed with a chase of unlabeled precursor, that rapidly loses label. When the macromolecules of the bacterial cell are separated from

one another by zone sedimentation through a sucrose gradient, ribosomes are found mostly in aggregates called *polysomes.* Single ribosomes sediment more slowly at 70S, as shown in Figure 12–14. Despite the fact that ribosomes contain much RNA, the pulse-labeled radioactivity does not parallel the concentration of ribosomes in the gradient, measured by the absorbance at 260 nm, but sediments instead over that part of the gradient where the polysomes are found. The rapidly labeled RNA is also not transfer RNA, which is much smaller and sediments at the top of the sucrose gradient.

The original and most likely interpretation is that the rapidly labeled RNA is almost all messenger RNA; after synthesis it associates with the ribosomes, thus forming the polysomes, and acts as a template specifying protein synthesis. The same sort of profile as in Figure 12–14 is found if one uses bacteriophage-infected cells, as Astrachan and Volkin did. In that case the rapidly synthesized RNA has a base composition that reflects the infecting bacteriophage DNA and can be quite different from the base composition of the ribosomal RNA.

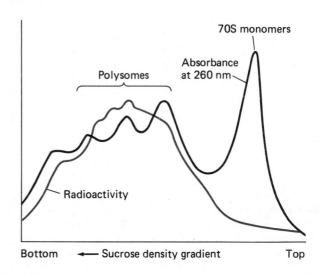

FIGURE **12–14**
Evidence for Messenger RNA.   Bacterial cells were labeled briefly with a radioactive precursor of RNA, then broken open and subjected to zone sedimentation through a sucrose gradient. Rapidly synthesized RNA, indicated by the radioactivity, tends to be in zones toward the bottom of the tube. This RNA cannot, therefore, include much tRNA (of size 4S) since it would sediment very near the top under these conditions. The size distribution of ribosome aggregates (called polysomes) is indicated by the absorbance at 260 nm which detects the rRNA. The rapidly synthesized RNA does not include ribosomal RNA since the 70S single ribosome zone is not labeled. Instead, the radioactivity is associated with the polysomes, which are also found to be most active in protein synthesis. (Pulse-labeled eucaryote cells give a similar pattern except that the monomers sediment at 80S.)

A messenger RNA should have sequences of bases along its length that are complementary to sequences of bases in its template DNA. Thus, it should be possible to demonstrate hydrogen bonding between RNA and its template DNA, very similar to the hydrogen bonding that holds the two complementary strands of DNA together. This was done by S. Spiegelman and B. Hall. They used *hybridization,* in which DNA and RNA are heated above their melting temperatures, then slowly cooled to allow formation of hydrogen bonds. The DNA and RNA will anneal only if there are long stretches of complementary sequences, so the demonstration of a DNA/RNA hybrid molecule represents a far more stringent test of relatedness than base composition similarity.

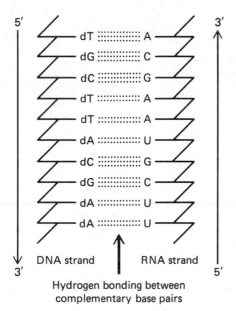

DNA strand   RNA strand

Hydrogen bonding between
complementary base pairs

Hall and Spiegelman used T2 bacteriophage-infected cells labeled for a short time with ³²P as the experimental RNA. They isolated labeled bacteriophage DNA from T2 grown in *E. coli* in the presence of ³H-thymidine, a precursor of DNA. The two labeled materials were mixed at room temperature and loaded into a centrifuge tube with CsCl. The mixture was sedimented to equilibrium, at which time the denser RNA formed a band toward the bottom of the tube and the lighter DNA banded toward the center in the density gradient (Figure 12–15). If the DNA and RNA preparations were mixed, then heated and slowly cooled before sedimentation, the pattern shown in Figure 12–15 was observed. A substantial fraction of the pulse-labeled RNA was then associated with DNA and banded at a density characteristic of double-strand hybrid molecules. After treatment with a nuclease to digest single-strand RNA, essentially all of the ³²P label was found in the hybrid band. As controls, Hall and Spiegelman tried

mixing and heating ³H-labeled bacterial DNA, or DNA from unrelated viruses, with the pulse-labeled RNA from T2-infected cells; no formation of hybrids was observed. They concluded that there was a close relation between the sequences, and therefore the information, of the viral DNA and the RNA from the phage-infected cell.

The technique of hybridizing RNA to DNA as a test of sequence relatedness has been improved and expanded. Single-strand DNA can be immobilized on nitrocellulose filters and hybridization of RNA tested simply by pouring radioactive RNA onto the filter. Unhybridized RNA passes through but hybridized RNA is retained by virtue of hydrogen bonding to the immobilized DNA. The fraction of radioactivity retained by the filter is then a measure of the relative amount of RNA sequences complementary to the DNA. This easy test is made even more sensitive by exposing the filter after the hybridization reaction to ribonuclease. Hybridized RNA is protected from digestion but any RNA that is merely adsorbed to the filter will be hydrolyzed and is easily removed during the subsequent washing. This refinement lowers the background radioactivity of the filter and allows for the detection of smaller amounts of hybrid.

The specificity of the hybridization assay can be exploited to count the number of genes for a particular RNA. The total amount of an RNA hybridized in conditions that force the reaction to its maximum *(saturation hybridization)* allows one to calculate the number of DNA sequences for that RNA. Using this strategy it was possible for Spiegelman to show that ribosomal RNA could be hybridized with 0.3% of the DNA of *E. coli*. Given the size of rRNAs, this result indicates that there are approximately seven copies of the rRNA genes in the *E. coli* chromosome.

One early example of using genetics to elucidate cell function is the work of F. Ritossa and S. Spiegelman on the relation between nucleoli and ribosomal RNA. They exploited a complex mutation in the fruit fly *Drosophila melanogaster* that allowed them to breed flies with one, two, three, or four *nucleolar organizer regions*—the region on one of the chromosomes associated with reestablishing the nucleolus after cell division. They took the DNA from cells of the four kinds of fly and performed saturation hybridization experiments, using purified and radiolabeled ribosomal RNA; that is, they annealed samples of DNA immobilized on filters to increasing concentrations of ribosomal RNA until no more DNA-RNA hybrids could form. Such an experiment gives a measure of the fraction of the DNA sequences that are complementary to the test RNA. Their results clearly showed that the fraction of hybridized DNA from flies with four nucleolar organizers was four times greater

FIGURE **12–15**

Formation of Complementary Helices Between DNA and Messenger RNA. Hybridization shows that the sequences of mRNA and DNA are related. *E. coli* were infected with bacteriophage T2 and labeled briefly with ^{32}P-phosphate. The cells were broken open and the extract was combined with purified T2 viral DNA labeled with ^{3}H-thymidine, and with unlabeled DNA and RNA in sufficiently large amounts that they could be located by their absorbance at 260 nm. (a) The mixture was sedimented to equilibrium in a CsCl density gradient. The T2 DNA forms a band at the position of the DNA marker but the RNA from the infected cell is much more dense. (b) The mixture was heated and slowly cooled before being sedimented as in (a). The ^{32}P-labeled RNA now bands partly at a density consistent with its being hybridized to T2 DNA. (From B. Hall and S. Spiegelman.)

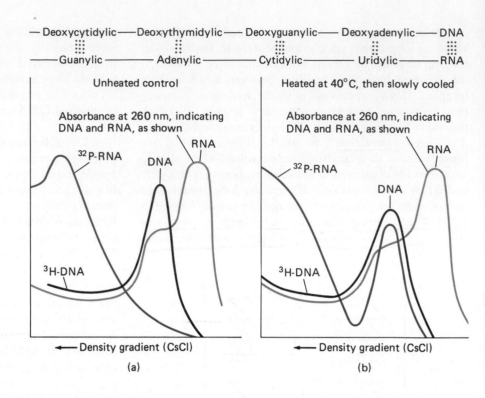

than the fraction of hybridized DNA from flies with only one nucleolar organizer; DNA from flies with two and three nucleolar organizers hybridized in the appropriate ratios (Figure 12–16). Ritossa and Spiegelman concluded that the nucleolar organizer DNA must specify ribosomal RNA and thus solidified the cytological evidence that nucleoli are the cellular loci of ribosomal RNA synthesis.

The method can also be used to test relatedness of two RNA preparations. If one is radioactive and can be hybridized to a given test DNA, adding a related unlabeled RNA will compete with the labeled species and reduce the amount of radioactivity retained as hybrids on the nitrocellulose filter. This *competition hybridization* has found wide use and in fact constituted the first evidence that the two large ribosomal RNAs are not related, since neither would compete with the other.

Hybridization of RNA to DNA immobilized on nitrocellulose has been refined to exploit genetic engineering techniques. DNA can be digested with a restriction endonuclease and the set of fragments produced can be separated by electrophoresis on an agarose gel. When a sheet of nitrocellulose is laid against the gel, a portion of the fragments is transferred to the sheet, the process called Southern blotting (see Chapter 10). After immobilizing the DNA on the nitrocellulose, radiolabeled RNA is hybridized to it, as described above. DNA fragments containing sequences complementary to the test RNA are then detected by autoradiography; they are

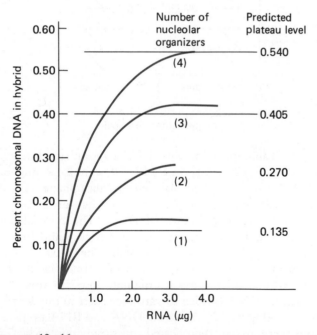

FIGURE **12–16**

Saturation Hybridization Used to Estimate the Number of rRNA Genes in Different Strains of *Drosophila*. By increasing the amount of rRNA in the assay it is possible to involve all the complementary DNA sequences in the formation of hybrids. The different strains were constructed by selective breeding to contain different doses of a standard amount of rRNA genes (shown at right). The saturation hybridization data confirmed the genetics.

said to "light up." Direct RNA hybridization to restriction fragments is only used for RNAs that can be easily purified: tRNAs, rRNAs, and very abundant mRNAs.

Another variation on RNA-DNA hybridization is used to detect the transcription in cells of specific DNA sequences. The cellular RNA is separated by electrophoresis on gels and then blotted onto paper. This procedure has come to be called, whimsically, *Northern blotting,* by analogy to the Southern blotting of DNA on nitrocellulose paper. The DNA sequence of interest is then hybridized to the RNA on the paper. Such *DNA probes* are usually amplified by cloning and radiolabeled. Different RNA bands will light up on autoradiography if the DNA sequence was being transcribed by the cells. Using this technique it is possible to follow the processing of precursor RNAs that become reduced sequentially to a final mature size (see later discussion in this chapter). Northern blotting is also used to detect the response of cellular transcription to perturbations—for example, exposing the cell to a hormone (see Chapter 14).

The last elaboration of the hybridization assay we wish to mention is the kinetic analysis of the formation of hybrids. This is analogous to the renaturation kinetics discussed earlier for DNA. A similar experimental arrangement, although technically more difficult, is used and there is a similar mathematical analysis of the results. These experiments have been used to show, for example, that there are more kinds of single-copy DNA sequences represented in nuclear RNA than in the messenger RNA that is ultimately involved in protein synthesis. We shall return to this puzzle later on.

RNA Polymerases

A great deal is known about the protein in *E. coli* that catalyzes the synthesis of RNA using DNA as a template, the *RNA polymerase.* The *holoenzyme* is composed of five noncovalently bonded polypeptides: two α subunits with MW 40,000; β with MW 150,000; β′ with MW 160,000; and σ with MW 90,000. The *core polymerase,* capable of synthesizing the phosphodiester bonds, does not contain the σ subunit and has the composition $\alpha_2\beta\beta'$. Reconstitution of the active core from purified subunits has been achieved. The σ (sigma) subunit is required for proper initiation of transcription and another protein, called ρ (rho), for proper termination.

In eucaryotic cells three RNA polymerases are typically found, each of great subunit complexity. They can be distinguished from one another by their cellular location and by their sensitivity to the fungal toxin α-amanitin. RNA polymerase 1 (also known as A) is found in the nucleolus, is resistant to α-amanitin, and transcribes the large ribosomal RNAs. RNA polymerase 2 (B) is found in the nucleoplasm, is sensitive to low concentrations (10^{-9}–10^{-8} M) of α-amanitin, and is the main enzyme responsible for synthesizing the heterogeneous nuclear RNA that later becomes processed into messenger RNA. RNA polymerase 3 (C) is also in the nucleoplasm, is sensitive only to high concentrations (10^{-5} – 10^{-4} M) of α-amanitin, and synthesizes tRNAs and 5S rRNA. There are also RNA polymerases found in mitochondria and chloroplasts. Two other enzymes exist that catalyze the synthesis of RNA polymers, but neither uses DNA as a template.

RNA Virus Replicases

The small icosahedral RNA bacteriophages such as Qβ, R17, MS2, etc., have as their genomes a single linear strand of RNA. This RNA is synthesized by a large *RNA replicase* enzyme (MW 215,000) composed partly of viral and partly of host proteins. Curiously, the specific

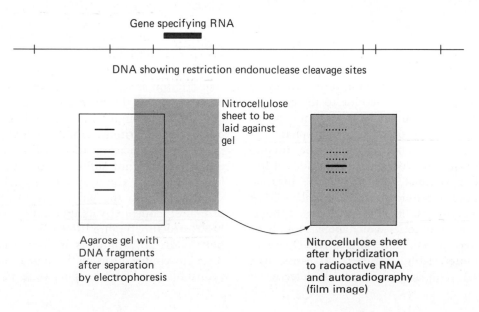

Gene specifying RNA

DNA showing restriction endonuclease cleavage sites

Nitrocellulose sheet to be laid against gel

Agarose gel with DNA fragments after separation by electrophoresis

Nitrocellulose sheet after hybridization to radioactive RNA and autoradiography (film image)

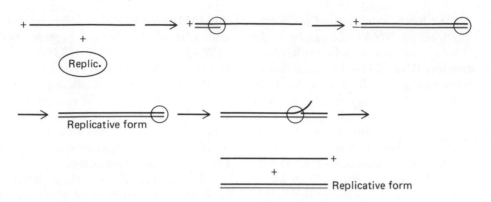

Action of RNA replicase

proteins of *E. coli* used in the RNA bacteriophage replicase are all involved in protein synthesis: elongation factors *Tu* and *Ts,* and ribosomal protein *S1* (see Chapter 13). The fourth protein is one of the four gene products specified by the virus. A single-strand RNA-binding protein known as *host factor* is also required for replication. Synthesis of the virus "plus" strands, the form injected into the host, proceeds first through the formation of a linear *replicative form* containing both a plus and complementary minus strand. This minus strand then serves as a template for the synthesis of more plus strands. The same replicase catalyzes the synthesis of both strands.

RNA Transcription

We turn now to the actual chemistry of polymerizing RNA, which in large part is similar to the polymerization of DNA. The precursors are the four ribonucleoside triphosphates (NTPs), and they embody both the mass and energy needed for synthesis. The information source, again like DNA synthesis, is a strand of DNA that acts as a template.

Elongation

Chain growth occurs at a free 3′ hydroxyl group, the unshared electrons of which attack the α–phosphorous atom of the incoming NTP, displacing inorganic pyrophosphate. The addition of a nucleotide to the growing RNA chain would be at equilibrium were it not pulled forward by the hydrolysis of the pyrophosphate byproduct. This reaction is catalyzed by the ubiquitous pyrophosphatase and proceeds with a standard free energy change of approximately −30 kJ/mole. Information for the synthesis is provided by the template DNA and by the RNA polymerase enzyme, as discussed above.

There are several antibiotics which interfere with the synthesis of RNA. Some, like *actinomycin-D,* block transcription by intercalating between successive base pairs in the DNA template. Actinomycin intercalates specifically between the base pairs formed by antiparallel -G-C- sequences, as shown by H. Sobell (Figure 12–17). Other poisons with this mode of action include aflatoxin-B, a carcinogen associated with a peanut fungus, and ethidium bromide, which at low concentrations shows specificity for binding to mitochondrial DNA. Among the antibiotics which interact with the RNA polymerase are α-amanitin, mentioned above, and streptolydigen.

The spatial relationships between the RNA polymerase and the template DNA are not known in any detail. The holoenzyme, when bound to DNA, can protect about 65 base pairs from nuclease digestion. RNA polymerase stabilizes the melting of DNA into a structure with partially separated strands. The fitting of this enzyme to the template and product must be capable of allowing the rate of 30–50 nucleotides polymerized per second observed in *E. coli.*

Initiation

We shall discuss the control of transcription initiation in Chapter 14. Here we will consider some of the mechanics. Initiation of RNA synthesis is a complicated process that involves at least three steps: *recognition, melting-in,* and *catalysis* of the formation of the first phosphodiester bond.

The DNA sequence that signals the initiation of transcription is called the **promoter.** The polymerase must recognize the promoter as a double-strand DNA structure. Although DNA can undergo local melting—it can exchange tritium from tritiated water with base-paired hydrogen atoms (detected by radiolabeling of the DNA)—this is not sufficient to expose the number of bases necessary for a specific signal to initiate RNA synthesis. Thus, the polymerase probably assesses the promoter sequence by its appearance in the grooves of the DNA structure (Figure 10–9). Since the rate of binding of polymerase to promoters is faster than would be predicted by simple diffusion through a solution, the enzyme probably spends most of its time in contact with the

Polynucleotide Phosphorylase

Another RNA-polymerizing enzyme, *polynucleotide phosphorylase,* is also discussed in the next chapter. It uses ribonucleoside diphosphates as precursors to synthesize RNA and does not require a template. As a result the sequence of bases in the product RNA is random and the base composition is determined primarily by the relative concentrations of the precursors. This enzyme's metabolic role is not yet fully agreed upon; it may represent a scavenging pathway for the degradation of RNA. The synthesis and degradation are essentially at equilibrium because nucleoside diphosphates are the precursors for synthesis and the products of degradation.

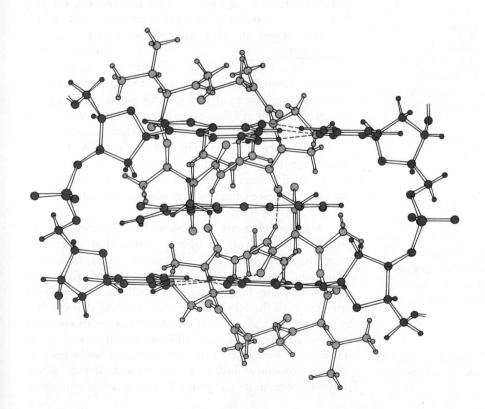

FIGURE 12–17

Intercalation of Actinomycin-D Into DNA. Two GC base pairs in DNA are shown, with hydrogen bonding between them indicated by dotted lines. The actinomycin-D molecule (in light gray) binds along the minor groove of the DNA helix in such a way that the planar phenoxazone ring (color) intercalates between the two G≡C base pairs in a plane parallel to that of the GC base pairs.

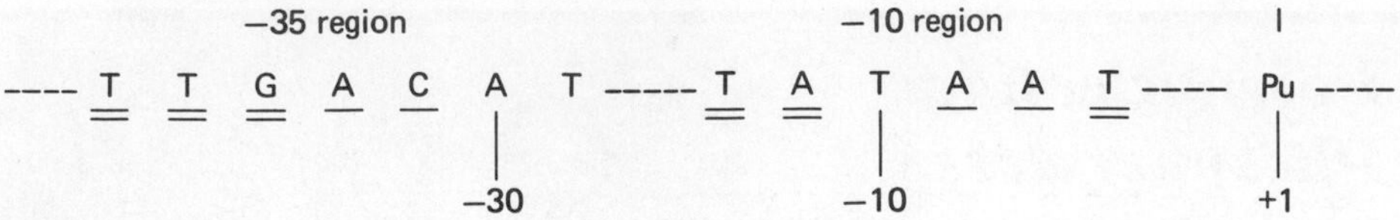

FIGURE 12–18
An Ideal Procaryotic Promoter Sequence. Three regions are denoted: I, the position of the first nucleotide triphosphate (position +1); the "−10 region" (counting down from +1), recognized by the core enzyme; and the "−35 region," recognized by the sigma factor. The sequence TATAAT in the −10 region is often called the "Pribnow box." This is a "consensus sequence," a generalization from over 100 promoters whose sequences have been determined. The bases have been underlined to show how often they appear in promoter sequences: a double underline implies in more than 75%; a single underline in more than 50%; others in more than 40%. (From W. R. McClure.)

DNA, wandering along the structure looking for promoters. The σ subunit of the polymerase in *E. coli* plays a decisive role in the recognition step, by interacting noncovalently with DNA over approximately 12 base pairs.[2] (In the absence of the σ subunit, initiation of RNA synthesis still occurs, but at numerous and aberrant locations in the DNA. The σ subunit confers specificity on the initiation process.) The sequence of the ideal promoter, first postulated by D. Pribnow, is shown in Figure 12–18.

After the polymerase has found the correct sequence, by virtue of noncovalent bonding in the grooves of the DNA structure between bases and the enzyme, the strands must be separated to allow matching of precursor to template. Melting-in of the RNA polymerase involves the formation of ionic bonds between basic amino acid residues of the enzyme and acidic phosphate groups of the DNA. Using the energy released by forming the ionic bonds, the strands are actually pulled apart. This has been demonstrated directly in a small circular DNA by observing the formation of supercoils, due to the "overwinding" caused by polymerase binding and the resultant strand separation. The data suggest that approximately seven base pairs are melted by the polymerase, near the site of template-precursor interaction.

The position of the polymerase on the promoter is such that the first phosphodiester bond is usually formed between a purine nucleoside triphosphate and the next NTP, both hydrogen bonded to the template. The nascent RNA chain thus has a 5′ terminal triphosphate; no primer is required. As elongation continues the σ subunit dissociates and further synthesis is performed by the core polymerase.

Termination

To continue RNA synthesis, the core polymerase and nascent RNA chain must maintain satisfactory contacts with the DNA sequence. When a stretch of DNA nucleotides with lower affinity (a *terminator* sequence) is met, there is a tendency for the dissociation of the polymerase and RNA, and termination of transcription. Two elements play important roles in this process. First is the secondary structure of the nascent RNA chain. As can be seen from Figure 12–19, most bacterial DNA termination sequences have an *inverted repeat* that allows the product RNA to fold into a hairpin-loop structure. This probably acts as a *brake* on the polymerase, causing it to slow or pause in its progress along the DNA template. When the nucleotide sequence that is synthesized downstream from the brake is a run of Us, the weaker base-pairing interactions between product and template favor dissociation and this termination of transcription. The evidence for this model includes the observation that if ribonuclease is present in a cell-free transcription system, synthesis (and degradation) do not stop at the normal places. The lack of the RNA structure means the brake has failed and termination does not occur. Furthermore, mutations have been isolated that alter the sequence of the stem of the hairpin loop and that prevent normal transcription termination.

Another element in bacterial cells, the rho (ρ) protein, plays a role in aiding termination when the terminator DNA sequence is not by itself sufficient. Thus, *rho-dependent termination* is expected for *weak terminators*. The exact way in which the rho protein interacts with the RNA polymerase core enzyme has yet to be elucidated. At present we know that the rho protein has an RNA-stimulated ATPase activity, and it is tempting to believe that the free energy of hydrolysis of ATP is somehow linked to the dissociation of RNA polymerase from its template. The rho protein represents

[2]A specific sequence of this size would be expected to occur by chance only once in 4^{12} (over 10^7) base pairs in a completely random DNA.

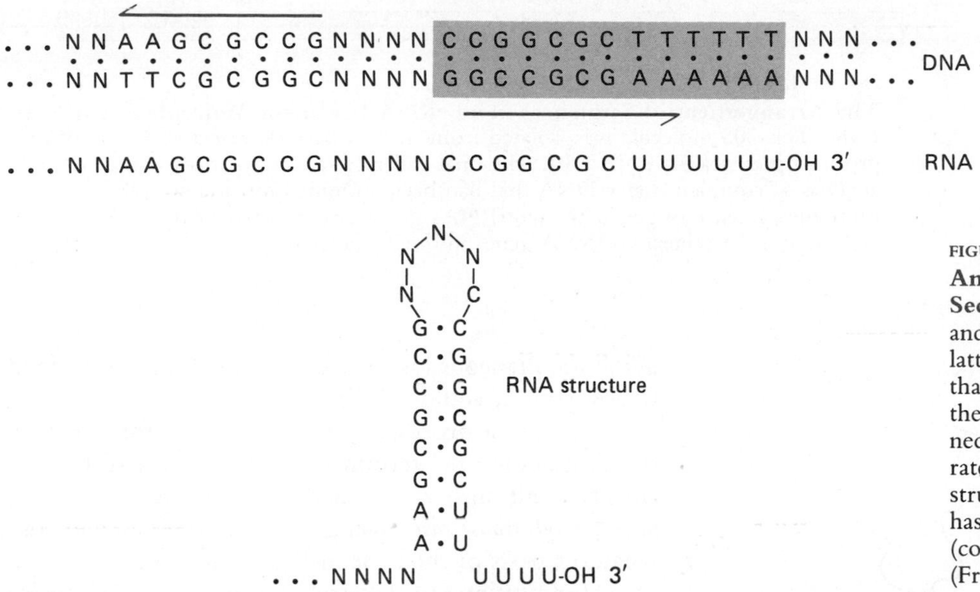

```
...N N A A G C G C C G N N N N C C G G C G C T T T T T T N N N...
   • • • • • • • • • •         • • • • • • •             • • •
...N N T T C G C G G C N N N N G G C C G C G A A A A A A N N N...   DNA
```

```
...N N A A G C G C C G N N N N C C G G C G C U U U U U U U-OH 3'   RNA
```

```
              N
           N     N
           N     C
            G • C
            C • G
            C • G        RNA structure
            G • C
            C • G
            G • C
            A • U
            A • U
  ...N N N N      U U U U-OH 3'
```

FIGURE 12–19
An Ideal Strong Terminator Sequence. Both the DNA sequence and the RNA product are shown. The latter illustrates the hairpin structure that results from base pairing within the RNA. This hairpin is thought necessary for RNA "braking" of the rate of polymerization. The hairpin structure is formed because the DNA has an inverted, repeated structure (color arrows). N = any nucleotide. (From D. Pribnow.)

another means through which regulation of transcription could occur, a topic to which we shall return in Chapter 14.

The Processing of RNA Transcripts

Recent work on RNA synthesis in procaryotic and eucaryotic organisms has revealed that examples can be found among all three sorts of RNA (tRNA, rRNA, and mRNA) in which the mature species is transcribed as part of a larger precursor. In fact, almost every case which has been carefully investigated revealed the existence of just such precursors. In Chapter 14, we will consider the possible reasons for this situation in terms of cell regulation. However, no matter what the function of these extra sequences, their existence forces us to add an additional step to describe the transcription of information from DNA. The first step involves the creation of the *primary transcript,* or precursor. The second involves specific cleavage of the precursor to generate the exact sequence of the mature species. The specific enzymes that catalyze these conversion steps (so far isolated principally from E. coli) cleave RNA with exquisite specificity at very precise locations. One of the most studied pathways of such RNA metabolism in both procaryotic and eucaryotic cells is that of ribosomal RNA.

Because a large proportion of the RNAs in bacteria are ribosomal RNAs, it has been relatively easy to isolate them as newly synthesized molecules. In E. coli there are seven separate genes for the rRNAs and the DNA sequences for several are known. In each gene are found sequences for all three rRNAs in the order 16S, 23S, 5S.

This order was first established by hybridization studies when it was found that short fragments of DNA able to hybridize with 16S rRNA could also hybridize with 23S but not with 5S, while longer DNA fragments could hybridize with all three.

A schematic map of the primary transcript of one of the rRNA genes is shown in Figure 12–20. This RNA has a sedimentation constant of 30S, is some 6000 bases (6 kilobases, or kb) in length, and 2.1×10^6 Da in mass. There are considerable amounts of extra sequences in the precursor that are removed during maturation of the final rRNAs. The processing of E. coli rRNA usually occurs in conjunction with synthesis, and intermediates in the maturation pathway are therefore difficult to observe. In mutant bacterial strains with a defective *RNAase III* nuclease, however, the entire 30S primary transcript accumulates so it can be isolated and its later processing studied in cell-free extracts. It has been found, for example, that active purified RNAse III will convert the 30S RNA to three molecules, precursors to 16S, 23S, and 5S rRNAs, and some fragments. The cutting out of extra sequences is a very precise process, and the structural signals that the processing enzymes recognize are presently under intensive investigation. In many cases the enzymes appear to recognize secondary structure rather than nucleotide sequence. The data suggest that the 16S precursor is cut out of the 30S molecule by RNAse III because the enzyme recognizes a long stretch of paired bases that are due to complementary sequences on either side of the 16S rRNA sequence proper. This would require that a hairpin structure form with the loop containing the approximately 1600 bases of 16S rRNA!

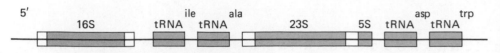

FIGURE 12–20

The Arrangement of Sequences in an rRNA Precursor Molecule From *E. Coli* This 30S molecule was isolated from mutant bacteria defective in one of the processing nucleases, RNAse III. The final order of sequences was worked out by analysis of complementary DNA that had been amplified by cloning techniques. Note the presence of genes for two tRNAs in the spaces between 16S and 23S rRNA, and the other two tRNA genes in the 3′ trailer.

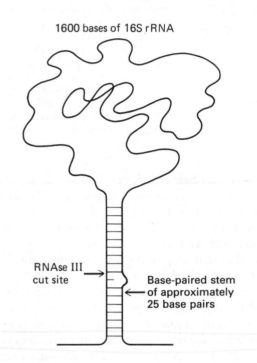

1600 bases of 16S rRNA

RNAse III cut site →

← Base-paired stem of approximately 25 base pairs

Many other enzymes are involved in completing the maturation of the rRNA sequences, among them *RNAse P* (for *p*recursor). This enzyme, which includes in its structure an RNA component essential for its function, plays an important role in processing the precursors of tRNAs. It cleaves a precursor RNA to generate the 5′ terminus of the mature tRNA. The 3′ terminus is matured through the action of *RNAse D,* an exonuclease.

One interesting precursor for tRNAs is also the precursor for 30s rRNA. As can be seen in Figure 12–20, there is embedded within the sequences separating 16S from 23S rRNA sequences the information for two tRNAs. Another pair of tRNAs is found in the 3′ trailer of this 30S precursor RNA. The presence in pre-rRNAs of sequences that specify one or more tRNAs provides a mechanism for maintaining stably the several sets of rRNA sequences found in *E. coli*. If a mutation or an aberrant recombination event resulted only in the deletion of one of the seven rRNA genes, it could probably be tolerated. But if the loss of the rRNA gene resulted also

in the simultaneous loss of the only copy of an embedded tRNA gene, it could well be lethal.

Among the processing steps required to form mature tRNA is adding the terminal -CCA sequence if it is not already built into the precursor. The enzyme *tRNA nuycleotidyl transferase,* using CTP and ATP, can add whatever parts of the 3′ sequence are missing.

Modifications to the structure of the normal nucleotides in RNA occur during processing of the primary transcript:

• Specific *methylase* enzymes add methyl groups to the bases and sugars in rRNA and tRNA.

• Another enzyme catalyzes the isomerization that yields *pseudouridine* in tRNA.

• Still other enzymes cause the more bizarre modifications found in tRNA (see Figure 12–8).

These modifications can begin before the completion of synthesis of the precursor RNA. Some of the modified bases and sugars may represent signals to other processing enzymes.

Eucaryotic cells also synthesize their rRNAs as large precursor molecules. In fact, the maturation pathway for rRNA was first worked out in that favorite of the cell biologist, the HeLa cell.[3] A 45S precursor molecule is cleaved to a 41S intermediate, then cut successively to finally yield the 28s, the 18S, and the 5.8S rRNA molecules (Figure 12–21); the 5S rRNA found in eucaryote ribosomes is derived from a different precursor. The maturation cleavage removes preferentially the sequences without methyl modifications. Whether the sequences removed during this processing in the nucleus contain tRNAs or other RNAs of functional significance is not yet known. The 45S precursor associates rapidly with ribosomal proteins in the nucleus and is processed, therefore, as a ribonucleoprotein complex.

[3]This cell line was derived originally from a cervical cancer affecting a patient named *H*enrietta *L*acks. It has been cultured ever since 1952 and is a model system commonly used for investigating the metabolic properties of human cells.

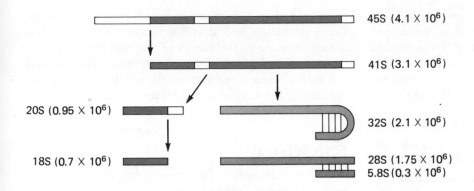

45S (4.1 × 10⁶)

41S (3.1 × 10⁶)

20S (0.95 × 10⁶)

32S (2.1 × 10⁶)

18S (0.7 × 10⁶)

28S (1.75 × 10⁶)
5.8S (0.3 × 10⁶)

FIGURE **12–21**

Processing of Eucaryotic Precursor Ribosomal RNA Molecule. ■ = mature ribosomal RNA sequences, ending up in 18S and 28S RNAs; □ = discarded ribosomal RNA precursor fragments. Arrows indicate cleavage points. Numbers denote sedimentation constants (in Svedberg units, indicated by S) and molecular weights in parentheses.

Multiplicity of Templates for Ribosomal RNA

The requirement for a large number of ribosomes to synthesize protein means that an equally large number of rRNA molecules must be synthesized. Since there is only room on a DNA sequence specifying rRNA (called rDNA) for so many RNA polymerase molecules, and since they can move only so fast along the template, there are limits to how rapidly rRNA can be produced by transcribing a single gene. The limits are well below the numbers required even by bacterial cells, and the universal solution to the problem is multiplication of the number of rRNA genes.

We mentioned earlier that hybridization of rRNA to the *E. coli* genome gave an estimate of seven copies of the complementary sequences in the DNA. In animal cells the estimate ranges from several hundred to over a thousand and in some plants it may well reach to over ten thousand copies of the rRNA genes. The extra copies are associated with special regions on one or more chromosomes, the *nucleolar organizing regions*. These in turn are associated during RNA synthesis with the *nucleolus* of the cell. In each nucleolar organizer, clusters of genes for rRNA are arranged in *tandem repeating units* with nontranscribed regions of variable length, called *spacers,* between.

Not all the copies of the rRNA genes are located on chromosomes. In some specialized cases there is amplification of the number of genes—synthesis of extrachromosomal copies, perhaps by the rolling circle mechanism. During oogenesis in amphibians the amplification is carried to monumental extremes. The developing egg contains ribosomes sufficient for all the protein synthesis during early embryonic development; no new ribosomes are made until later. During the maturation of the amphibian oocyte there is a prodigious synthesis of rRNA from an estimated 10⁶ copies of the rRNA genes. This activity was exploited by O. Miller in making a now-famous electron micrograph that dramatically illustrates the process of transcription (Figure 12–22). It shows long central fibers (3-nm diameter) along which

FIGURE **12–22**

Electron Micrograph of rRNA Genes Undergoing Transcription in Amphibian Oocyte Nucleolus. The long fibers are made of DNA. The fibrils running at right angles to the fibers are the growing chains of rRNA precursors. At the base of each fibril is the RNA polymerase. The initiation point can be identified near the position along the fiber where the shortest fibrils are found. The termination point is near where the longest fibrils lie. Between a termination and the next initiation are untranscribed spacer sequences of DNA. × 21,500. (Courtesy of O. L. Miller, Jr.)

smaller fibrils emerge at right angles and increase in length from one point to another to form an arrowhead shape. The fibrils have been identified as RNA associated with protein because they are sensitive both to RNAse and to proteases; also, radioactive RNA precursors label the fibrils. The long fibers are sensitive to DNAse and are presumably DNA.

The interpretation of the images in Figure 12–22 is that each fibril represents a nascent RNA chain, attached at its base by RNA polymerase to the long DNA template fiber. About 100 polymerase molecules can transcribe a single template simultaneously. The polymerization of RNA begins at fixed points (promoters) along the DNA and as the polymerase traverses the template the RNA chain becomes longer, giving rise to the arrowhead appearance. There are also fixed terminator regions at which transcription stops. Between a terminator and the following promoter is a spacer, of constant length, where transcription was not taking place at the time of fixing the oocytes for electron microscopy. The functional significance of these spacer regions remains to be seen.

The 5S rRNA of eucaryotes is not synthesized in the nucleolus. As many copies are required as of the other rRNAs, however, and the multiple copies of the rDNA are apparently scattered about the genome. If DNA is fragmented and then hybridized to 5S rRNA, the hybrid molecules can be separated from the others. This DNA can then be partially denatured and photographed with an electron microscope (Figure 12–23). A regular repeated pattern of denatured regions—"bubbles"—can be seen with undenatured regions between. The length from the beginning of one bubble to the next is about six times the size of a 5S rRNA molecule. Thus, the DNA of each repeat has 1/6 5S rDNA and 5/6 devoted to other

sequences, the exact nature of which remains unknown. From saturation hybridization measurements it is estimated there are 24,000 copies of this repeating sequence. The genes are located on various chromosomes, depending on the species, and not usually on the chromosomes containing the nucleolar organizers.

RNA Splicing

In our discussion of processing thus far very little has been said about mRNA. In fact, very little was known about the processing of mRNA until the past few years. The subject is now one of the most exciting in cell biology because the initial observations were almost entirely unexpected, partly because of the influence of work on procaryotes. As it happens, processing mRNAs in procaryotes and their viruses appears to be relatively rare. Such is not the case in eucaryotes.

The primary transcripts in the eucaryotic nucleus destined to become mature mRNA molecules are larger and have a wide distribution of sizes. This *heterogeneous nuclear RNA (hnRNA)* mystified cell biologists for years, not least because it represents almost 10 times the mass of RNA that ultimately becomes mature mRNA. If carefully isolated, hnRNA is found in complex associations with proteins, forming *heterogeneous ribonuclear proteins (hnRNPs)*. The normal association of hnRNA with proteins has been demonstrated by cross-linking the two species in vivo using ultraviolet light. At least 25 different proteins have been identified in hnRNP but their function is unknown. Protein remains associated with this RNA through processing in the nucleus, through transport to the cytoplasm, and possibly even during the initial stages of translation.

FIGURE 12–23

5S DNA Molecule From Amphibian Erythrocyte Nucleus. Radioactive 5S RNA was isolated, purified, and hybridized with DNA; the DNA fragments binding to the radioactive RNA were then separated from the bulk of the DNA. This DNA double helix was partially denatured and treated for electron microscopy. The repeat units can easily be seen. × 80,000. (Courtesy of D. D. Brown.)

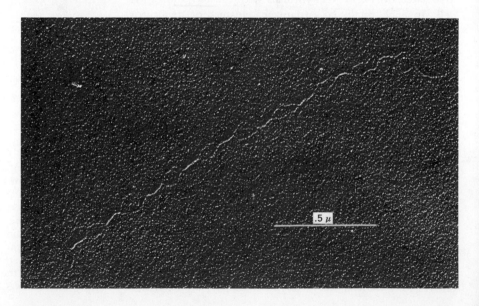

.5 μ

As well as removing extra sequences from the ends of eucaryote hnRNAs, processing results in *removing sequences from inside*. Let us see how this works, first by examining a relatively simple case—the messenger RNA for the β chain of mouse globin, the protein part of hemoglobin (Figure 12–24). The DNA of the gene has been cloned and its sequence determined by P. Leder and his colleagues. The mRNA coding for β globin can be isolated from mouse reticulocyte cells as a 9S RNA and hybridized with the genomic DNA. When this is done, loops of single-strand nucleic acid bulge out of the double-strand structure seen in the electron microscope (Figure 12–24). These *R loops* are due to *intervening sequences (IVS, introns)* in the DNA that are not found in the mature mRNA. One of these, some 116 base pairs long, lies between the 30th and 31st amino acid codons in the mRNA. The other, 646 base pairs long, lies between the 104th and 105th amino acid codons. The DNA sequences that are found in the mature mRNA are called *exons*. These details were discovered by a comparison of the sequence of the β globin DNA with that of the protein.

Although precursors for mRNA were the first noticed to have intervening sequences, they are found as well in the genes for eucaryotic cellular tRNAs and rRNAs. An early concern was how the IVS in the DNA fails to appear in the mature RNA. Is it transcribed and then removed, or merely ignored by the RNA polymerase? Progress in understanding the mechanism of this process came from analysis of a small intervening sequence in the gene for the yeast tRNA that accepts tyrosine. There is a 14-base intervening sequence next to the anticodon. The existence of a mutant yeast strain that is defective in processing allowed the isolation and characterization of yeast tRNA precursor molecules. The important observation was that the 14-base sequence is present in the precursors and, therefore, it must have been synthesized during transcription of the tRNA gene. Work on the isolated precursors has shown that the intervening sequence can be removed and the ends covalently joined in the sequence of reactions we now call *splicing*. The enzymes involved in tRNA splicing have been purified and the process requires ATP.

Splice Site Selection

A question central to the splicing concept is how to recognize the proper place to cut and rejoin the RNA sequence. This is handled in at least three different ways, depending on the RNA to be spliced. In precursors for eucaryotic cellular tRNAs the information is apparently supplied by the cloverleaf secondary structure of the mature tRNA. This is believed because the intervening sequence is always found in the same place, in the

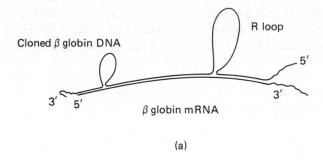

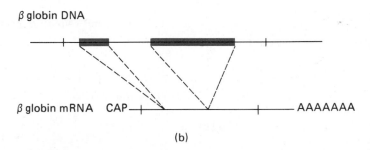

FIGURE 12–24
R Loop Mapping in the Electron Microscope.
(a) Schematic representation. Hybrid molecules are formed by annealing mature mouse β globin mRNA and cloned mouse β globin DNA. (b) Map of mouse β globin gene and mRNA showing the positions and sizes of intervening sequences (color). (From J. Abelson.)

anticodon loop on the 3′ side of the anticodon sequence (see Figure 12–11). The secondary and tertiary structures of the tRNA are apparently not altered much by the presence of the intron; the specific shape of the tRNA can therefore be used as a guide to locate the proper place to cleave the phosphodiester chain and to rejoin (ligate) the ends to form the mature tRNA. In support of this idea, there was an experiment in which the intervening sequence was altered but the processing was unaffected: by genetic engineering techniques, a 21-nucleotide sequence was added to the intron of a tRNA precursor gene; the larger precursor RNA was nevertheless spliced to yield the usual mature tRNA.

In the case of mRNA precursors (pre-mRNAs), the information for splicing selection is probably in the primary sequence of the splice site junctions. The DNA sequences of many genes with intervening sequences have now been determined by genetic engineering techniques. Gradually we are assembling a library of sequences around the junctions between introns and exons. Inspection of the library leads to the conclusion that there exists a **consensus sequence** for the intron-exon junction (Figure 12–25a).

A consensus sequence is a prototype base sequence which summarizes the bases most often found at each position amongst members of a large library of individual sequences. The concept is illustrated by the base sequences shown at the 5′ and 3′ splice sites in Figure 12–25a. The numbers in parentheses below each base refer to the percent of total cases with which the indicated base is found at that position. For example, the G and T immediately following the 5′ splice site are invariant,

found in 100% of the eucaryotic gene sequences tabulated. The base to the left of the 5′ splice site is G in 73% of all cases, so G is the *consensus* for this position, although a considerable number of sequences have a different base here. Similarly, the base on the right (or 3′) side of the invariant G–T dinucleotide is A in most (62%) but not all cases. In other words, members of the library

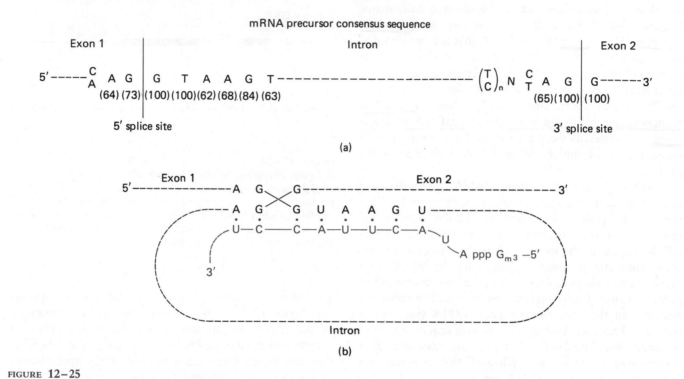

(a)

(b)

FIGURE **12–25**
Splicing of Eucaryotic Nuclear Pre-mRNAs. (a) A consensus sequence of the junctions between an intron and adjacent exons. A typical mRNA precursor, or primary gene transcript before splicing, is shown reading 5′ to 3′ from left to right. (The data come from DNA sequencing of genes, and therefore T is conventionally used in place of U.) Vertical lines indicate the junctions between the left and right exons to be spliced together, and the intron between them that is to be eliminated. The numbers in parentheses below the bases of the consensus sequence specify the frequency with which the indicated base is found at that position, in percent of total cases amongst a large library of eucaryotic gene transcripts. (b) Possible structure of a splicing complex involving a small nuclear RNA. The abundant snRNA called U1 (color) has a sequence at its 5′ end which is complementary at most positions to the sequences typically found at the right and left ends of introns found in many eucaryotic nuclear mRNAs. As a result, the 5′ end of U1 could pair with the right and left ends of the intron as shown, where dots indicate base pairings. This structure could maintain the right and left ends of the intron in close proximity to one another, to facilitate precise excision of the intron and precise splicing of the exon ends to one another. (c) Intermediate steps in splicing. In the first step, the exon-intron boundary is broken on its left (5′) side, and the 5′ G nucleotide of the intron is joined to a specific A residue *within* the intron by an unusual intra-chain 5′-2′ phosphodiester link. In this intermediate, the exon on the right side of the intron remains linked to the intron, which is now in a "lariat." In the next step, the left and right exons are linked up, releasing the intron lariat, which is subsequently degraded.

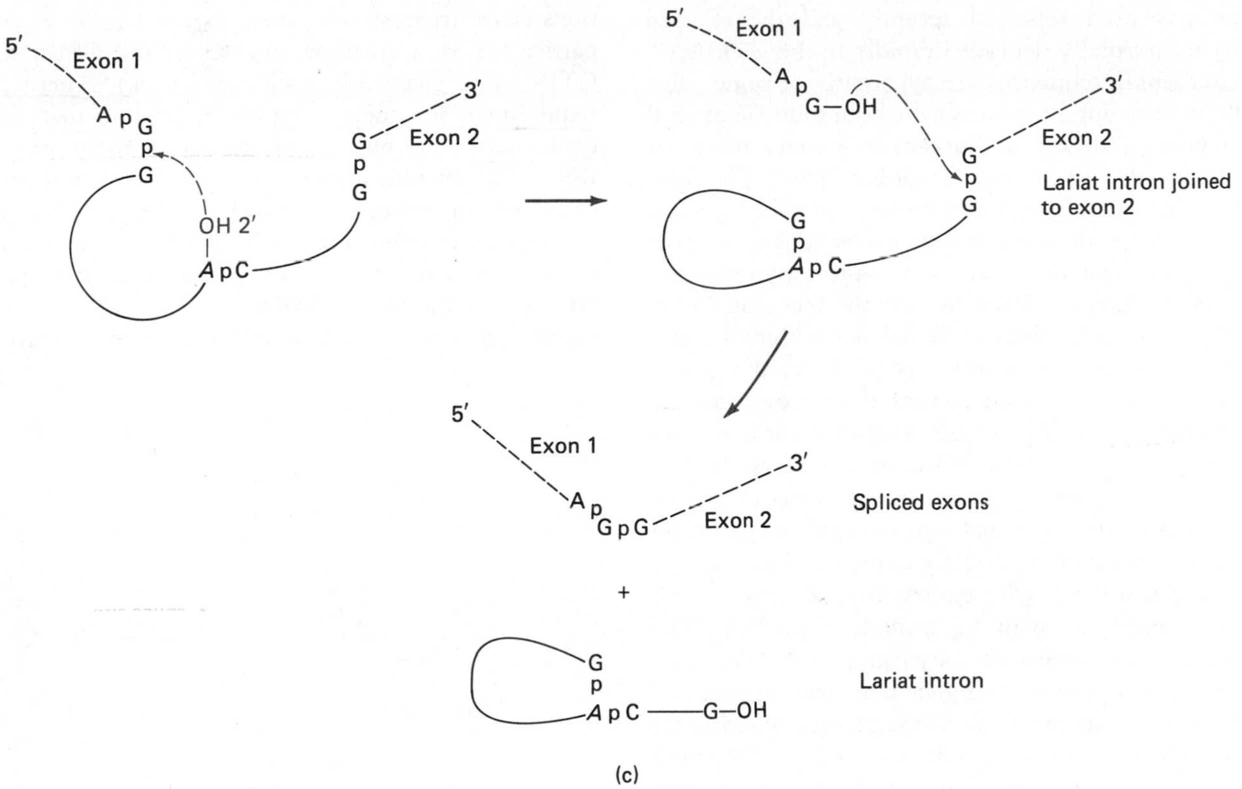

(c)

FIGURE **12—25** *(continued)*

of gene sequences all agree with the consensus at invariant positions, and agree most often but not in every case with the consensus at the other positions.

There have been several suggestions that a second RNA molecule could, by forming complementary hydrogen bonds with a precursor, hold its splice-site sequences together. A specific candidate for this role of splicing jig is one of the *small nuclear RNAs (snRNAs)* found in eucaryotic cell nuclei. These range in size from 100 to 300 nucleotides and many species have been well-characterized biochemically, some even sequenced. They are found in essentially all eucaryotic cells and often in combination with proteins, forming complexes known as *small nuclear ribonucleoproteins (snRNPs,* pronounced "snurps"). Several lines of evidence suggest these molecules are not merely the partial breakdown products of other RNAs but have special functions in the cell, albeit not directly in protein synthesis.

It was pointed out by J. Steitz and her colleagues, and by J. Rogers and R. Wall, that the consensus sequence for the intron-exon junction is complementary to one of the more abundant snRNAs, known as U1, whose sequence is also known. This suggested a scheme for hydrogen bonding the splice junction in the precursor RNA with U1 snRNA (Figure 12–25b). The intron forms a loop whose crossing point is so held in register by

the U1 snRNA that it is easy to imagine a processing enzyme being able to cut out the intron and rejoin the pieces. Subsequent experiments showed that the U1 snRNA is indeed found in RNA-protein complexes associated with the splicing process and that U1 snRNA binds to the 5′ splice site of pre-mRNAs. There is little evidence that U1 snRNA binds to 3′ splice sites, however. Nevertheless, we believe that the U1 snRNA is important for splicing of pre-mRNAs since deleting eight nucleotides from its end abolishes splicing activity and since antibodies raised against U1 snRNA also inhibit splicing. The importance of the consensus sequence to the splicing process is demonstrated by finding that mutations which alter the sequence interfere with normal splicing. (One such mutation leads to the blood disease in humans known as β-thalassemia, which affects the synthesis of β globin protein.) Deletions in the middle of a pre-mRNA intron have little effect on splicing.

Further investigation of splicing turned up another surprise, the formation of a novel structure for the intron during the splicing process (Figure 12–25c). This RNA, containing a branch point where the 5′ end is joined to the middle of the intron, has become known as the *lariat*. After the lariat forms, the intron is split out and the two exons are joined. There is a further breakdown pathway for the lariat-containing intron. A debranching enzyme activity has been reported recently and the released introns are normally degraded rapidly in the nucleus.

Intervening sequences are apparently of some value to cells; intervening sequences have been found in other β globin genes (rabbit and human) in exactly the same position as in the mouse β globin gene. The base sequences of the introns from mouse and rabbit are not similar, except at the ends, but their positions in the gene are identical. The observation of divergent sequences within the introns is consistent with the accumulation of mutations during evolution. When one observes structural features—like the introns' positions in the gene—that are conserved in two species during evolution, it seems likely that the loss of the features must have been selected against, especially if the evolutionary distance between the species is great. In the case of certain mitochondrial genes (see below), the importance of the intervening sequences is directly demonstrable.

One can still ask why eucaryotic transcription uses such a roundabout path to a mature mRNA. This question could be approached by using genetic engineering techniques to construct a gene that lacks an intervening sequence but contains all the sequences found in the mRNA. This experiment was done for a gene of the virus SV40; it was found that deletion of the intervening sequence prevented expression of the gene. The malfunction occurred after synthesis of the primary transcript of the gene.

Self-splicing

RNA processing is full of surprises. The simple beads-on-a-string organization of chromosomes led biologists to expect spacer DNA between genes but not intervening sequences. The linear chemistry of the usual nucleic acids did not prepare biochemists for branched structures like the lariat. Perhaps the biggest challenge to traditional beliefs, however, was the discovery that some splicing requires no protein enzymes to catalyze the process—that is, *the RNA can splice itself!*

This phenomenon was discovered by T. R. Cech (Nobel Prize, 1989) in the processing of the ribosomal RNAs of the ciliate protozoan genus *Tetrahymena*. The information for the selection of the splice site in these precursors lies in the sequence of nucleotides in the intron. There are several relatively short (10–12 nucleotides) sequences which can interact to form a secondary structure believed to aid in bringing together the exons to be spliced. Similar sequences are found in the rRNA introns of the cellular slime mold *Physarum* and in the introns of mitochondrially coded genes for mRNAs and rRNAs in yeast and fungi. These so-called "Group I" introns may all be self-splicing.

The pathway that leads to excision of the intron and ligation of the two exons involves a series of reactions called transesterifications (Figure 12–26). A critical participant is a guanine nucleotide (5′ GMP, GDP, GTP—even guanosine itself will work). There is no requirement for energy, however, beyond that already contained in the phosphodiester chain of the precursor RNA. The guanine nucleotide is transferred to the 5′ end of the intron, either after cleavage at the 5′ splice site or as a concerted action with the cleavage. Another guanine nucleotide in the 5′ position of the next exon is then attacked by the free 3′-hydroxyl of the previous exon, ligating the two exons and splitting out the intron with

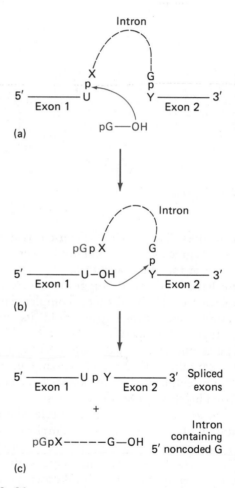

FIGURE 12–26
Self-Splicing Reaction Pathway. Conserved sequences in the intron maintain a complex secondary and tertiary structure of the precursor RNA. A guanine nucleotide (color) is transferred to the 5′ end of the intron, a reaction closely associated with cleaving the intron from exon 1 (a). The two reaction products (b) remain closely associated because of structural features of the precursor RNA or because of proteins (maturases). The 3′-OH of exon 1 is then transferred to the phosphate associated with the 3′ G of the intron, displacing it and ligating the exons (c). The noncoded guanine remains as the 5′ end of the released intron.

The Evolution of Introns

The contrast between the continuous structure of procaryotic genes and the mosaic structure of eucaryotic genes has provoked speculation on the question: "Which came first?" One possibility is that introns are a relatively recent evolutionary acquisition, reflecting the movement about the genome of nonessential DNA sequences, perhaps related to the demonstrable mobility of transposable elements.

Another possibility is that the eucaryotic jumble of exons and introns is actually the earlier evolutionary arrangement. In the prehistory of organic evolution, DNA sequences coding for useful functional *domains* of proteins might well have evolved in separated stretches of DNA. For example, one ancestor of the globin genes might have evolved the sequence coding for effective binding of the heme cofactor, and another sequence, located elsewhere in the genome, might have evolved the sequence effective for association of alpha and beta chains. Given the transposability of DNA sequences, natural selection would have favored any rearrangements that brought these sequences close to one another. But it would take many generations during which selection could operate on random rearrangements for one to appear with an exact registration of the sequences to make a continuous gene.

Eucaryotic nuclear mRNA splicing and processing might thus represent the evolutionary descendant of a mechanism for constructing a message, for a useful protein, from bits of transcripts of separated DNA sequences. Moreover, the continuous structure of procaryotic genes might represent a more recent evolutionary stage in which the sequence domains landed next to one another and the cumbersome machinery for splicing and processing of separated coding sequences had at last become unnecessary. On this view, it is procaryotes, often thought to be ancient life forms, which have the more modern streamlined system of genetic structure and expression. Perhaps this is not so surprising; after all, the procaryotes, ancient though their origins may be, have generation times measured in minutes rather than days or years and have therefore gone through many more generations than eucaryotic organisms under the shaping force of natural selection.

If the streamlined organization of procaryotic genes is the more recent evolutionary development, one might expect to find an occasional laggardly procaryotic gene, one which had not yet rid itself of introns. This is indeed the case: introns have recently been found in two or three genes of the bacterial virus T4. Their intron-exon junctions show a striking similarity to the group I eucaryotic junctions, and the isolated T4 mRNA can splice itself in vitro.

the extra guanine on its 5′ terminus. In support of this reaction scheme, the intermediate containing the intron plus next exon has been isolated and the uncoded guanine has been detected, both from reactions using purified components and from nuclei.

The specificity of the splicing process and the requirement for a guanine nucleotide argued for an enzyme-catalyzed reaction sequence. But the activity of splicing requires only the purified precursor RNA, a buffer system, the guanine nucleotide, and a divalent metal ion (like Mg^{2+}). The activity survives extractions with the strong detergent sodium dodecyl sulfate and phenol, boiling in SDS and mercaptoethanol (both of which usually denature proteins), and treatment with proteases. No evidence of proteins has been discovered so far and it seems inescapable that this splicing is a property of the intron itself. In support of this, when the *Tetrahymena* intron, and bits of the adjacent exons, are inserted into a gene of the bacterium *E. coli,* the intron seems to be cleaved out in the usual way.

The notion that all Group I introns can self-splice is supported by finding at least one example in the mitochondrial precursor RNA for cytochrome *b* which can remove its intron without protein being present. Moreover, mutations have been found in the conserved sequences of the introns that affect splicing only of the gene in which they occur (cis-acting mutations). These can be explained as disrupting the complex RNA structure necessary for the exons to be brought together. But there are inhibitory trans-acting mutations found in genes coded in the nucleus, and these probably encode the structure of proteins. There are also trans-acting mutations that map *in the introns* of the mitochondrial genes and these are also likely to inactivate proteins. Since the precursor RNA can be shown to self-splice in extracts from which proteins have been removed, these proteins (called maturases) may not be involved in catalyzing the bond breaking and making of the splicing. Rather, it is thought they could play a role in forming the complex structure that is necessary for self-splicing to occur. The finding that some maturase proteins are coded by information in the intron provides another reason why these intervening sequences have survived during evolutionary time. Other functions may be coded by other introns; the results of further research on this aspect of gene organization will be very interesting indeed.

Before leaving splicing we should emphasize that <u>not all eucaryotic genes contain introns</u>. In yeast some tRNA genes have none and the histone genes are another set that

lacks these extra internal sequences. On the other hand, at least one gene in the T4 bacteriophage that infects procaryotic *E. coli* cells has been shown to contain an intron. Introns come in variable sizes and numbers. The five introns in the gene for one mouse enzyme contain over 3×10^4 base pairs; the gene for chicken pro-α_2 collagen has 49 introns. One intron in a gene affecting development is 10^5 bases long, and takes over an hour to transcribe. Clearly, much additional work will need to be done before we understand how cells manage to splice accurately over such long distances and with so many potential sites of joining RNA. We shall return to the possible functional significance of transcribed but untranslated RNA sequences in Chapter 14.

KEY WORDS

Anticodon, base pair, chloroplast, codon, consensus sequence, deoxyribonucleic acid, enzyme, genome, genetic code, hybridization, information, mitochondrion, nuclease, nucleoprotein, primary structure, processing, promoter, ribonucleic acid, ribosome, RNA transcription, secondary structure, sequencing, template, tertiary structure, triplet, virus.

PROBLEMS

1. Specify the different classes of RNA found in eucaryotic cells, and what distinguishes them.

2. What was the initial evidence that ribosomes do not themselves carry the information for the structure of the proteins they make?

3. Eucaryotic messenger RNA has distinctive structural features at its 5′ and 3′ ends. Describe them. Do procaryotic messenger RNA molecules have any comparable structural features?

4. Until quite recently, most biochemists believed that proteins were the only biological macromolecules capable of catalyzing chemical reactions. They were wrong. What discoveries revealed that at least one other type of biological macromolecule has catalytic activity?

5. The DNA strands for a particular gene are arranged as shown below:

If the gene is transcribed from right to left, then the template strand for transcription is

A. the lower strand, and the promoter is located at X.

B. the upper strand, and the promoter is located at Y.

C. the lower strand, and the promoter is located at Y.

D. the upper strand, and the promoter is located at X.

E. the lower strand, and the promoter can be located at either X or Y.

The drawing below shows three RNA polymerases (denoted as Pol) transcribing a gene. The answers to questions 6–8 can be found as the letters (a through e) on the drawing.

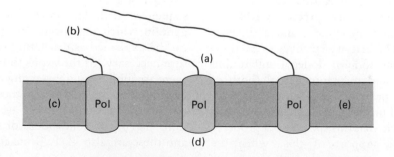

6. The 5′ end of a nascent RNA chain is found at _____.

7. The promoter for this gene is located at _____.

8. The 3′ end of a nascent RNA chain is found at _____.

9. Listed below are properties which a class of cellular RNA molecules might have. Indicate for each property whether it is a characteristic of messenger RNA (m), transfer RNA (t), and/or ribosomal RNA (r). Circle as many as apply, or none if appropriate.

   (a) By total mass, the major RNA in the bacterial cell.   m t r
   (b) The greatest number of distinguishable kinds, in terms of sequence.   m t r
   (c) The smallest (lowest molecular weight).   m t r
   (d) Carries information to specify proteins.   m t r
   (e) Is double-stranded over all its length.   m t r
   (f) Contains modified bases in large numbers.   m t r
   (g) Synthesis inhibited by α-amanitin.   m t r
   (h) Covalently bonded to amino acid.   m t r
   (i) Hybridizes fastest to nuclear DNA.   m t r
   (j) Found in mitochondria but not in chloroplasts.   m t r
   (k) The "adaptor" predicted by F. H. C. Crick.   m t r
   (l) Contains codons.   m t r
   (m) Contains anticodons.   m t r
   (n) Contains a poly A "tail."   m t r
   (o) Capable of forming stem and loop structures.   m t r
   (p) Complete base sequence complementary to contiguous DNA sequence in eucaryote nucleus.   m t r
   (q) Synthesis inhibited by actinomycin D.   m t r
   (r) In procaryotes, contains the Shine-Dalgarno sequence.   m t r
   (s) Template for reverse transcriptase.   m t r
   (t) Attacked by kethoxal.   m t r

10. More nuclear RNA hybridizes to nuclear DNA than messenger RNA, even after saturating the DNA with transfer and ribosomal RNAs. Explain why.

SELECTED READINGS

Freifelder, D. (1985) *Essentials of Molecular Biology.* Boston: Jones and Bartlett.

Hanawalt, P. C., and R. H. Haynes, eds. (1973) *The Chemical Basis of Life.* San Francisco: W. H. Freeman.

Miller, O. L., Jr. (1973) The visualization of genes in action. *Scientific American* 228:34–42.

Spiegelman, S. (1964) Hybrid nucleic acids. *Scientific American* 210:48–56.

The Ribosome and the Translation of Information

13

Genetic information in the form of nucleic acids is similar to information stored in a library or on blueprint copies: it is relatively static and inactive. The final expression of genetic information occurs during protein synthesis, the subject of this chapter. Some of the information coded by the sequence of nucleotides in a messenger RNA determines the sequence of amino acids in a protein that, therefore, also embodies genetic information. Furthermore, protein structures are a means by which genetic information creates cellular architecture and, through enzymes, controls metabolic processes.

The requirements for making a protein, as for other biosynthetic processes, are mass, energy, and information. The *mass* of a protein is contained in its constituent *amino acids;* the *energy* required to join them into a polypeptide comes from *ATP* and *GTP;* and the *information* that determines the sequence of the amino acids is contained in the *sequence of nucleotides in a messenger RNA.* Initiation, elongation, and termination of protein synthesis all take place on a complex organelle, the *ribosome,* whose structure is the object of intensive current research.

After the amino acids have been linked together, or during this process, the polypeptide can be modified or processed. This includes directing the protein to its final destination: in the cytoplasm, as part of a membrane, in an organelle, or—via secretion—to the extracellular space. We begin our discussion by describing early experiments on the cellular localization of protein synthesis.

The Cytological Basis of Protein Synthesis

Early in the history of biochemistry it was discovered that soluble proteins are precipitated in hot acid, while their constituent amino acids are not. This observaton became the basis for a crude analysis of proteins: trichloroacetic acid (TCA, final concentration 5%) is added to a soluble extract; this is heated to approximately 90°C for a few minutes; and the precipitated protein is either sedimented by centrifugation or filtered off (Figure 13–1). After radioactive amino acids became available, some 40 years ago, it was possible to use their incorporation into hot acid-precipitable form as an assay for the cellular localization of protein synthesis.

Mammalian liver cells were shown to be highly active in the incorporation of amino acids into protein, largely the blood proteins such as albumin. A homogenate of liver can be fractionated by differential centrifugation into the morphologically identifiable components of the liver cell (see Chapter 2). At low speed the pellet contains mitochondria and nuclei, plus undisrupted cells. If one spins the low-speed supernatant at high speed, the pellet is called the *microsome* (small particle) fraction. It contains most of the cytoplasmic RNA of the cell, as well as substantial amounts of phospholipid. If the low-speed supernatant is brought to pH 5, then sedimented, the resultant pellet (pH 5 microsomes) contains all the cytoplasmic RNA of the cell.

Most protein synthesis takes place in the microsome fraction. After incubating a whole liver homogenate for various times with radioactive amino acids, plus other necessary cofactors (see below), the protein in the different subcellular fractions contains different amounts of radioactivity. It was shown in very early experiments that the protein in the microsome fraction has the highest specific radioactivity (Figure 13–2). Similar results were obtained when a whole animal was labeled with radioactive amino acids.

The term "microsome" describes the results of a fractionation method: at high speed, spin the supernatant from the low-speed mitochondrial fraction. If a cell has cytoplasmic membranes with bound particles, then the microsome fraction is found to consist of fragments of these membranes, with particles still on them (Figure 13–3). If the cytoplasm of the cell has only particles, then the microsome fraction consists only of particles; if it has only membranes, then the microsome fraction is a membranous fraction. The microsomal particles, composed of RNA and protein, are named *ribosomes;* they are

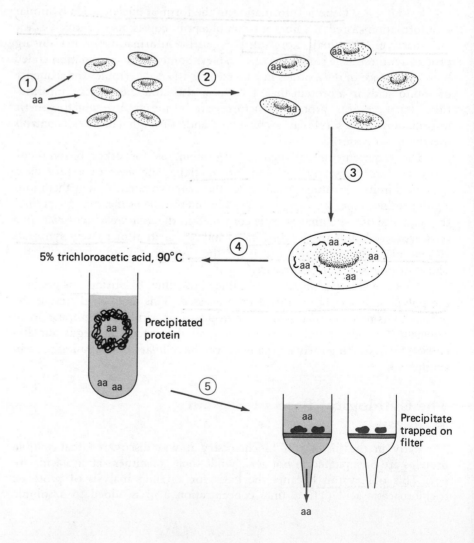

FIGURE **13–1**
Radioactive Amino Acids (aa) Incorporated Into Protein in Vitro.
Radioactive amino acids (1) are incubated with cells (as shown) or with whole cell homogenate, taken up by the cells (2), and incorporated into protein (3). The cells or homogenate is then brought to a final concentration of 5% trichloroacetic acid (TCA), heated at 90°C, and the precipitated proteins (4) are either centrifuged or filtered. The free amino acids remain in the supernatant after centrifugation or are eluted in the filtrate (5). The radioactive protein pellet or radioactive precipitate trapped in the filter paper is then counted for radioactivity.

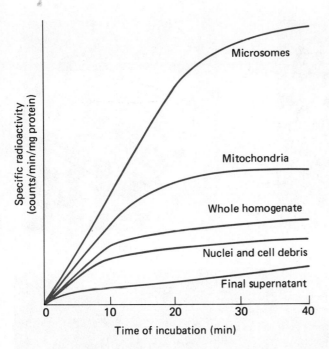

FIGURE 13–2

Incorporation of Radioactive Amino Acids into Proteins in Vitro by Various Subcellular Fractions. The liver of a rat was homogenized and then incubated for varying times with radiolabeled amino acids. Subcellular fractions were then separated by differential sedimentation (see Chapter 2). Each fraction was assayed for hot TCA-precipitable radioactivity (newly synthesized protein) and for total protein. The ratio of these two, the specific radioactivity of the newly synthesized protein, is graphed against the time of incubation of the whole cell homogenate. (Data after P. Siekevitz.)

the structures involved in the synthesis of proteins. The membranes are fragments of the *endoplasmic reticulum* (*ER,* see Chapter 2).

The ribosomes contain most of the cellular RNA. It seems strange today that only 50 years ago no biochemist had a really good idea as to the function of RNA. It was known that certain cells can be stained with basic dyes (such cells are said to be *basophilic* because they contain large concentrated amounts of acidic macromolecular compounds that combine with basic dyes). It was also known that basophilic cells are precisely those that are active in protein synthesis. This observation, by T. Caspersson in 1941 and by J. Brachet in 1942, was one of the first links in the chain of evidence suggesting that RNA is involved in protein synthesis. It is in fact the microsomal particles in the cell that are responsible for the basophilia, for it is the acidic RNA of the ribosomes that combines with the basic dyes.

Cell-Free Protein Synthesis

After RNA had become implicated in cellular protein synthesis, it was found that the incorporation of radioactive amino acids into protein could take place in a test tube. A crude homogenate of an active tissue such as liver, when combined with the appropriate factors, can catalyze the incorporation of added amino acids into liver proteins. In 1952, P. Siekevitz showed that certain isolated subcellular fractions can perform this reaction (Table 13–1). (There is actually a net breakdown of

TABLE 13–1

Relative Ability of Various Isolated Liver Subcellular Fractions to Oxidize Substrate, Form ATP, and Incorporate Radioactive Amino Acids into Their Proteins

	Oxygen Consumption (μ atoms)	ATP Formed (μ moles)	Specific Radioactivity (counts/min/ mg protein)
Homogenate	37.4	4.0	10.8
Mitochondria★	8.2	4.2	1.3
pH 5 Microsomes	0.4	0.0	1.1
pH 5 Supernatant	0.8	0.0	0.4
Mitochondria plus pH 5 microsomes	14.0	4.1	10.2
Mitochondria plus pH 5 supernatant	9.7	4.1	1.5
Mitochondria plus pH 5 microsomes plus pH 5 supernatant	18.8	3.8	4.3
Mitochondria plus boiled pH 5 microsomes	9.7	4.4	1.2

★(Mitochondria were added together with oxidizable substrates and cofactors.)
Microsomes have the protein-synthesizing machinery but cannot make ATP, while mitochondria (plus oxidizable substrates and cofactors) show little incorporation of amino acids and make a lot of ATP. It is only by adding them together that maximum incorporation is obtained. (When the pH 5 supernatant is added to this mixture, inactive protein is also added, thus reducing the value—by increasing the denominator—of the specific radioactivity. (Data from P. Siekevitz.)

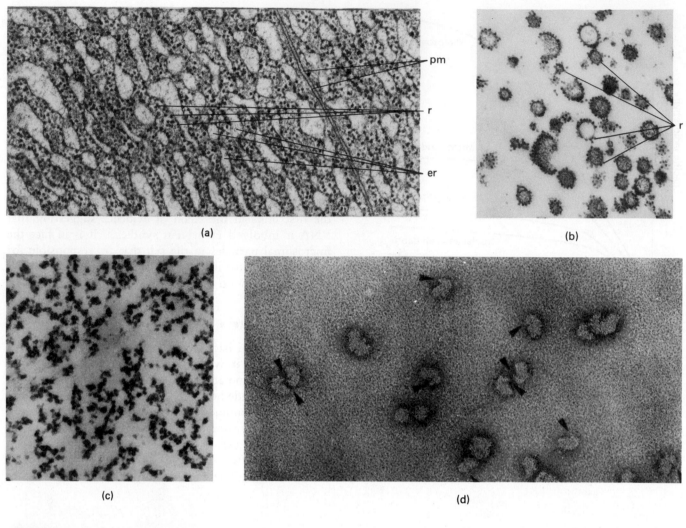

(a)

(b)

(c)

(d)

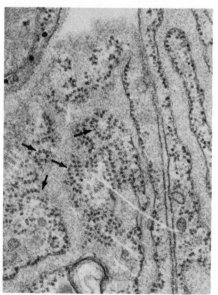

(e)

FIGURE 13–3

Subcellular Fractions Separated by Differential Sedimentation, As Seen in the Electron Microscope.   (a) The endoplasmic reticulum in a pancreas acinar cell. In these cells the cytoplasm consists of membrane-bound vesicles and spaces. Most of the membranes have particles on their surfaces but some are bare, as shown in Figure 2–17. The microscome fraction from pancreas is composed of fragments of these membrane-bound vesicles, some with and some without particles on their surfaces. r = ribosomes; er = endoplasmic reticulum; pm = plasma membrane. ×46,000. (b) Microsomes isolated from a pancreatic acinar cell. The microsomes are in the form of membrane-bound vesicles, with ribosomes on the outer surface. r = ribosomes. ×39,000. (c) Ribosomes isolated from pancreatic acinar cell microsomes. ×52,000. (a, b, c courtesy of G. E. Palade.) (d) Ribosomes isolated from *E. coli* and negatively stained by phosphotungstate. The dark groove (arrows) between the two subunits can be clearly seen. ×215,000. (Courtesy of J. A. Lake.) (e) Endoplasmic reticulum in a plasma cell showing polysome clusters on membranes, as strings or semicircles. ×100,000. (Courtesy of G. E. Palade.)

protein in such a crude reaction mixture, but the presence of radioactive amino acids makes it possible to detect the small amount of new protein synthesized.) The components that stimulated the reaction most were found to be an oxidizable substrate and cofactors for the oxidation—in other words, components required for oxidative phosphorylation, the synthesis of ATP.

This finding was important because it contradicted the idea, held by many biochemists, that the mechanisms for protein synthesis and protein degradation were one and the same, that is, protein synthesis was just the reversal of protein degradation. For years, the main reason that this was thought to be the case had to do with the need for the cell to synthesize specific proteins, rather than just any protein. We know of many proteoloytic enzymes that are very specific in hydrolyzing peptide bonds only between certain amino acids. These proteolytic enzymes, like all enzymes, catalyze a reversible reaction; thus, in principle, they could not only degrade specific proteins to amino acids but also synthesize proteins from amino acids. However, the stimulation of cell-free protein synthesis by conditions that enhance oxidative phosphorylation suggested that energy is required. At that time this implied that protein synthesis could not be simply the reverse of protein breakdown, since the latter was believed to be a spontaneous process.

Many refinements have been made to the crude system used originally to demonstrate the cell-free synthesis of proteins. Instead of using a mitochondrial fraction to make the large amounts of ATP required, ADP, phosphoenolpyruvate, and purified pyruvate kinase can be added. Instead of whole microsomes, a pure preparation of ribosomal particles obtained from the fractionation of microsomes can be used (Figure 13–3). Ribosomes reasonably free of membranes can be obtained merely by homogenizing bacteria or reticulocytes (which have no internal membranes) and differentially centrifuging the resulting suspension. As methods improved it became possible to detect in reticulocyte cell-free synthetic reactions a single radioactive protein, hemoglobin, rather than having to obtain amino acid radioactivity in a mixture of proteins. If a purified mRNA (which contains the information transcribed from a specific gene) is added to suitably prepared reticulocyte extracts nowadays, the specific protein can be obtained with an ease that must amaze the early workers in this field.

The history of the study of protein synthesis illustrates beautifully the confluence of previously separate disciplines—cytology, cell biology, genetics, and biochemistry—into the exuberant intellectual activity we now call molecular cell biology. Before continuing our discussion of the mechanics of protein synthesis, however, we turn to the basic problem of information transfer.

The Genetic Code

Genetic information was found to be encoded by DNA as the result of the research of O. Avery, C. MacCleod, and M. McCarty and of A. Hershey and M. Chase (see Chapter 10). This work and the later one gene–one enzyme hypothesis of G. Beadle and E. Tatum (see Chapter 6) focused the attention of geneticists, biochemists, and cell biologists on the coding problem: how is the information in a sequence of nucleotides in DNA translated into a sequence of amino acids in protein?

This issue was addressed by mathematicians and physicists, as well as by biologists, because the formal problem of coding does not require much specialized knowledge of the mechanics of encoding or decoding. There are in DNA only four nucleotides of importance; modifications are so infrequent that it was felt they could be safely ignored. The amino acid content of proteins, on the other hand, was not nearly so clear-cut. For years biochemists had vied for attention with their reports of ever more baroque amino acids. It was Francis Crick, the preeminent theoretical biologist of our era, who in 1956 first recognized the importance of the *20 amino acids* we have already shown in Figure 4–14. Since four different nucleotides taken one at a time could encode only four different amino acids, it was clear that larger combinations were required. Taken two at a time, there are only $4^2 = 16$ combinations, still too few to encode 20 amino acids. Taken three at a time there are $4^3 = 64$ distinguishable ordered combinations, which is more than sufficient.

The excess combinations worried the decoders, and there were many ingenious schemes invented to reduce the meaningful combinations from 64 to 20. Among these was the *overlapping code*. (Though probably never more than a formal possibility, it is instructive to consider this hypothesis as an example of the style of thinking used by the early molecular biologists.) In this code a long sequence of nucleotide "letters" would be divided into "words" by overlapping one word with the next, as shown in Figure 13–4. A little thought will show that choosing one word for one amino acid restricts the possible options for the following word. This coding hypothesis could be tested, therefore, by enumerating the number of different amino acids that can follow a particular amino acid in the sequences of known proteins. This test was performed in 1957 by S. Brenner, using the rather limited set of peptide sequences then known for insulin, glucagon, cytochrome *c,* trypsinogen, and hemo-

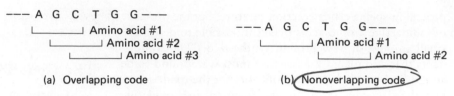

(a) Overlapping code   (b) Nonoverlapping code

FIGURE 13–4
Genetic Coding Schemes.   (a) The overlapping code. The DNA sequence is
divided into three-nucleotide words, each symbolizing an amino acid. Adjacent
amino acid words overlap, as shown by the brackets. Only certain amino acids can
follow others in this coding scheme; for example, only amino acids whose code
words begin with G can follow amino acid number 1 and begin amino acid
number 2. (b) Nonoverlapping code, shown for comparison.

globin, among others. The results showed that the
hypothesis could not explain the observed variety of
amino acid pairs and that a nonoverlapping code is the
correct one. The overlapping code was consigned to the
burial ground that receives all "beautiful hypotheses slain
by ugly facts."[1]

Deciphering the genetic code could have been
accomplished by any of a number of approaches. As it
turned out, all the possibilities were used simultaneously.
The interaction of these various experimental styles
produced one of the most exciting periods in the history
of biological research.

The Adaptor Hypothesis

One of the approaches, as we have discussed above,
was theoretical. When Crick proposed restricting attention to only 20 amino acids, he also confronted directly
the mechanical problem of recognizing at a molecular
level sequences both of nucleotides and of amino acids.
Because base pairing in DNA was a known mechanism
for recognizing nucleotide sequences (see Chapter 10), it
seemed likely that there must be some sort of base pairing
involved in protein synthesis, rather than a direct molecular interaction between amino acids and nucleotides. To
accomplish this Crick proposed in 1958 the existence of
an *adaptor* molecule made of RNA. Part of the adaptor
would recognize nucleic acid sequences, and the rest
would be large and specific enough that only one of the
20 amino acids could be associated with it. Experimental
confirmation of this idea was not long in coming. In the
same year M. Hoagland reported the discovery that
amino acids could become covalently attached to a small
(4S) RNA, then known as *soluble RNA (sRNA)*. We now
call this molecule *transfer RNA (tRNA)*; when the amino
acid is attached it is called *aminoacyl-tRNA (aa-tRNA)*. As
described in Chapter 12, there is at least one kind of

tRNA for each amino acid and each tRNA can decode
one or more of the code "words" in the messenger RNA.

The Message

Since ribosomes were known to be associated with
protein synthesis, it was natural to think that the
information for the synthesis of a particular protein was
in a particular ribosome. Unfortunately, ribosomes appeared to be very similar to one another when isolated
from cells in different states, even though the proteins
being synthesized in those states were very different. A
dramatic example of this was the bacterium *E. coli* when
infected with bacterial virus (bacteriophage) T4. During
the later stages of the infection only proteins associated
with the bacteriophage are synthesized, but the purified
ribosomes are indistinguishable from those in uninfected
E. coli. This was demonstrated decisively by Brenner, M.
Meselson, and F. Jacob, who labeled the ribosomes with
denser ^{15}N before infection and showed that no new
ribosomes were synthesized after infection. When they
added the T4 bacteriophage, they also performed an
isotopic transfer of the bacteria to growth medium containing the lighter ^{14}N. Sedimenting the purified ribosomes
to equilibrium in a CsCl density gradient revealed that
after infection no new ribosomes of lighter density had
been synthesized.

These results were combined with earlier observations concerning the base composition of RNA that was
synthesized during T4 infection. This RNA could be
labeled by adding radioactive precursors at the time of
infection. The base composition was shown in 1965 by E.
Volkin and L. Astrachan to be similar to the base
composition of T4 DNA and different from the base
composition of the RNA from *E. coli* ribosomes. This
rapidly labeled RNA was later named *messenger RNA
(mRNA)*. This is the immediate source of nucleic acid
sequences used in protein synthesis, and it is this RNA
that is recognized by the adaptor tRNA.

---

[1]Paraphrased from T. H. Huxley, *Collected Essays,* viii, "Biogenesis
and Abiogenesis."

The Code Is Triplet

Knowing there is a molecule that contains the code that relates nucleotide sequences to amino acid sequences is not the same as knowing the details of the code. Determining which of the mathematically possible schemes for coding is correct was the result of one of the most conceptually brilliant genetic experiments of this century. This work was carried out by Crick, Brenner, L. Barnett, and R. Watts-Tobin in 1961 on T4-infected *E. coli* and involved an analysis of mutations induced by the dye *acridine orange*. The mutations studied all lay in the bacteriophage r$_{II}$B gene, a gene of vital significance to the infection only in certain strains of *E. coli*. In other *coli* strains the action of this gene is irrelevant, and so mutant viruses can be grown for further study.

It turned out that acridine orange produces mutations as small additions or deletions by intercalating between adjacent base pairs of DNA. During subsequent replication this structural aberration would result in too many or too few bases in the DNA sequence. Of course Crick, Brenner, and their colleagues did not know the molecular mechanism of the mutagenesis; they had to infer the nature of the mutations from the genetic consequences. The mutations were divided arbitrarily into two classes: *plus* and *minus*. These classes were distinguishable genetically because a double mutant with two *pluses* or two *minuses* still showed the mutant phenotype, but a double mutant composed of a *plus* and a *minus* showed almost a wild phenotype. Most importantly, some triple mutants of either all *pluses* or all *minuses* showed a wild-like phenotype, while all triple mutants containing both *plus* and *minus,* in any combination, were still mutant.

These complicated observations were brilliantly interpreted as the results to be expected if the *plus* mutants had a single base addition, the *minus* mutants had a single base deletion, and the genetic code was read off in contiguous *groups of three*. (The interpretation would have been just as satisfactory if the roles of *plus* and *minus* were reversed. Fortuitously, the original nomenclature was correct: the *plus* mutations did result from the addition of a single base and the *minus* mutations from the deletion of

a single base.) Figure 13–5 shows what was thought to be happening. A *plus* mutant would shift the *reading frame* (represented by brackets) in one direction *(phase),* while a *minus* mutant would shift the reading frame in the other direction. A *plus* and a *minus* could compensate one another; the reading frame would be restored to the correct phase after the second mutation (see Figure 13–5). Between the two mutations there would be an *out-of-phase* (and thus altered) message, but the r$_{II}$B gene product is especially tolerant of even large regions of such mistakes, as long as the C-terminal sequence is appropriate. One can also see how a triple *plus* or triple *minus* mutant would restore the reading frame to the correct phase after the third mutation (though the message between the first and third mutation would again be garbled). Subsequent work showed that quadruple and quintuple all *plus* or all *minus* mutants were inactive but the sextuple mutant had the wild phenotype.

This work, entirely genetic, showed the message was most probably written in a *triplet code* without special signals to show the beginning or end of each word (a *comma-free code*). The message would be read in groups of three from a fixed starting point, the *initiator signal.* The biochemical work we next describe was therefore interpreted in terms of these groups of three, which were named *codons* (a legacy of the physicists-turned-biologists, who felt that fundamental units should be named with words ending in "-on," like proton and electron).

Deciphering the Code In Vitro

At the International Congress of Biochemistry in 1963 a startling paper was read by the American biochemist M. Nirenberg. He had been studying the incorporation of amino acids into protein by cell-free extracts and trying to show a stimulation by purified cellular RNAs. This would, it was hoped, allow the molecular identification of the newly discovered messenger RNAs. As a control, Nirenberg and his collaborator H. Matthai added to the extracts what they expected to be an inactive man-made RNA, polyuridylic acid. This polymer contains only one nucleotide repeated many times: . . .UUUUUUU. . . . The control RNA

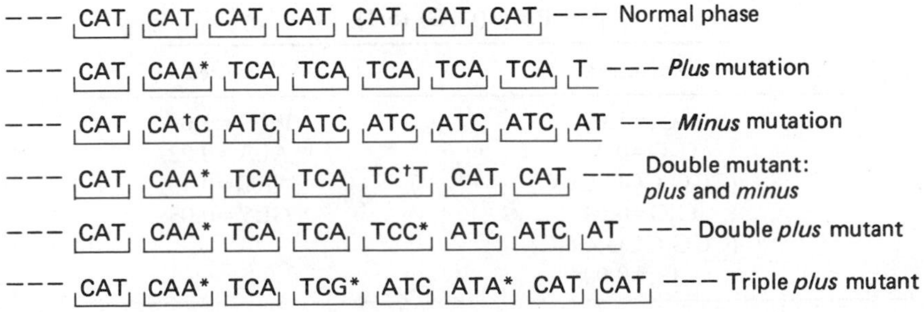

––– [CAT] [CAT] [CAT] [CAT] [CAT] [CAT] [CAT] ––– Normal phase

––– [CAT] [CAA*] [TCA] [TCA] [TCA] [TCA] [TCA] [T] ––– *Plus* mutation

––– [CAT] [CA†C] [ATC] [ATC] [ATC] [ATC] [ATC] [AT] ––– *Minus* mutation

––– [CAT] [CAA*] [TCA] [TCA] [TC†T] [CAT] [CAT] ––– Double mutant: *plus* and *minus*

––– [CAT] [CAA*] [TCA] [TCA] [TCC*] [ATC] [ATC] [AT] ––– Double *plus* mutant

––– [CAT] [CAA*] [TCA] [TCG*] [ATC] [ATA*] [CAT] [CAT] ––– Triple *plus* mutant

FIGURE **13–5**
The Effect of Frame Shift Mutations on the Reading of a DNA Sequence. A simple sequence of nucleotides in DNA is used to illustrate the effects. The location of a single base addition is marked with a *, a single base deletion with a †. See text for additional discussion.

turned out to be the most interesting part of the experiment, however, since it *did* stimulate incorporation, and of only one amino acid, phenylalanine! Thus, the first triplet code word could be assigned: UUU = Phe. The cell-free extracts used in these experiments were straightforward to make, and soon several laboratories were in the race to crack the code.

On the biochemical side, the next experiments were done with man-made mRNAs containing only A nucleotides, which stimulated the incorporation of lysine; of C nucleotides, which stimulated the incorporation of proline; and of random mixtures of nucleotides like poly (U,C), which stimulated the incorporation of leucine and serine as well as phenylalanine and proline. These mRNAs were made by incubating the nucleotide *diphos*phates in the presence of an enzyme discovered in 1955 by S. Ochoa and M. Grunberg-Manago, *polynucleotide phosphorylase*. Presented with UDP only, the enzyme obligingly makes poly U; presented with a 5:1 mixture of UDP and GDP, the enzyme makes a poly (U,G) polymer with approximately five times as many U as G residues, distributed randomly. Similarly, the enzyme makes other polymers of random sequence whose composition depends mostly on the ratio of nucleotide diphosphates used as substrates.

With the aid of this enzyme, a library of synthetic messages could be made and then tested for coding specificity in vitro. Table 13–2 shows the results of a couple of these experiments, summarizing their basic design. The ratio of two nucleotides in a random copolymer determines the relative frequencies of triplets of different composition. For example, in the random poly (U:G = 5:1) polymer, there are one-fifth as many G residues as U residues, so any triplet containing two Us and one G (U_2,G) should be 0.2 times as frequent as the U_3 (UUU) triplet. Since we know that the U_3 triplet specifies the incorporation of phenylalanine, we expect each of the amino acids specified by one of the three different (U_2,G) triplets to be incorporated into protein at 0.2 times the rate of phenylalanine. Similarly, a (U,G_2) triplet should be present at $0.2 \times 0.2 = 0.04$ times the

frequency of U_3 triplets, and each amino acid specified by one of the three (U,G_2) triplets should be incorporated 0.04 times as often as phenylalanine.

Table 13–2 shows that the relative incorporation of certain amino acids agrees very nicely with these quantitative expectations. The incorporation of other amino acids was not detected. Thus, in the case of poly (U,G) (5:1), permutations representing the codons for valine and cysteine (U_2,G) and glycine (U,G_2) were detected. Similarly, in the case of poly (C,G) (5:1), the codons for alanine and arginine (C_2,G) and glycine (C,G_2) were detected. The data of Table 13–2 also reveal an important characteristic of the codon dictionary. Note that glycine incorporation responds to one or more permutations of two chemically different sets of codons, (U,G_2) and (C,G_2). This pattern, in which more than one codon specifies the same amino acid, is termed *coding degeneracy*. Since there are 64 possible triplets for only 20 amino acids, a degenerate code structure was to be expected.

The results of experiments using random copolymer mRNAs identify the nucleotide *composition*, but not the sequence, of triplets coding for individual amino acids. The next and most decisive step in the biochemical analysis of the code came from a technique devised by Nirenberg and P. Leder. They found that synthetic trinucleotides—in effect, individual codons—could be used directly to stimulate the binding of specific aminoacyl-tRNA molecules to ribosomes. Aminoacyl-tRNA molecules can be labeled specifically with an amino acid residue that is radioactive; the cell contains enzymes to catalyze the attachment of amino acids only to their correct adaptor tRNA. The association of aa-tRNA radioactivity with ribosomes, in the presence of one trinucleotide or another, is taken as evidence that the tRNA was stimulated to bind to the ribosome by the presence of the codon. The amino acid attached to the tRNA is, therefore, coded by the trinucleotide. Successful binding is detected by filtering the incubation mixture through nitrocellulose: the ribosome · trinucleotide · aminoacyl-tRNA complex is retained on the filter (presumably because of electrostatic attraction between the

TABLE 13–2

Data on Amino Acid Incorporation Programmed by Random Copolymers Poly (U,G) (5:1) and Poly (C,G) (5:1)

Poly (U,G) (5:1)		Poly (C,G) (5:1)	
Predicted	Detected	Predicted	Detected
U_3=1.0	PHE=1.0	C_3=1.0	PRO=1.0
U_2G=0.2	CYS=0.2	C_2G=0.2	ALA=0.22
(UUG,UGU,GUU)	VAL=0.2	(CCG,CGC,GCC)	ARG=0.19
UG_2=0.04	GLY=0.04	C_2G=0.04	GLY=0.05
(UGG,GUG,GGU)		(CGG,GCG,GGC)	
G_3=0.008		G_3=0.008	

positive charges on the ribosomal proteins and the negative charges on the filter), while unbound aminoacyl-tRNA passes through.

This neat, simple assay provided a way to test the coding capacity of each individual trinucleotide triplet. For example, radiolabeled valine-tRNA became bound to ribosomes in the presence of the triplet GUG but not in the presence of the isomeric triplets UGG or GGU (reading from 5′ to 3′). Hence, it can be concluded that GUG is a codon for valine. In similar fashion, much of the dictionary of the genetic code was worked out (Table 13–3).

The code derived from these experiments, largely done with ribosomes and tRNAs from *E. coli,* is extensively degenerate. Some amino acids (leucine, serine, and arginine) enjoy as many as six different, synonymous codons. The rest of the amino acids all have two to four codons except for methionine, which usually has only AUG, and tryptophan, which usually has only UGG.

Perhaps the most remarkable feature of the genetic code is its incredibly broad distribution. Almost every living thing uses the same dictionary of coding symbols, as judged by tRNA binding experiments with material prepared from the cytoplasm of bacterial, fungal, plant, and animal cells. The code must have developed during the most primordial stage of organic evolution that preceded the emergence of distinctive cell types. Exceptions to this coding dictionary are found in certain mitochondria (see Chapter 15) and in some protozoa.

It is not unreasonable to speculate that the code evolved in step with the evolution of metabolism. This is suggested by the fact that structurally similar amino acids generally have similar coding symbols. For example, the three aromatic amino acids phenylalanine, tyrosine, and tryptophan share a similar structure and their codons are all found in the top horizontal row of the dictionary, for their codons all start with U. Leucine, isoleucine, and valine also share a similar aliphatic side chain structure, and their symbols are all found in the first vertical column, for they all contain U in the second position. One could imagine, for example, that the earliest aromatic amino acid had a codon containing little more than U. As metabolism evolved and other aromatic amino acids were synthesized, additional codons evolved from the first and they all began with U. This hypothesis is strengthened by the fact that the aromatic amino acids share not only a similar structure but also a common biosynthetic pathway (see Chapter 6).

Codon and Anticodon

During the translation of a normal messenger, successive nucleotide triplets presumably act like Nirenberg and Leder's synthetic triplets and dictate the binding of their corresponding aminoacyl-tRNA molecules. The specificity of this process is based on the fact that each species of tRNA contains an *anticodon* triplet of nucleotides that is complementary to its specific codon. The nucleotide sequence of many different tRNA molecules has been worked out, and each has been shown to contain

TABLE 13–3
The Genetic Code

1st ↓ 2nd →	U	C	A	G	↓ 3rd
U	PHE	SER	TYR	CYS	U
	PHE	SER	TYR	CYS	C
	LEU	SER	Ochre★	Opal★	A
	LEU	SER	Amber★	TRP	G
C	LEU	PRO	HIS	ARG	U
	LEU	PRO	HIS	ARG	C
	LEU	PRO	GLN	ARG	A
	LEU	PRO	GLN	ARG	G
A	ILE	THR	ASN	SER	U
	ILE	THR	ASN	SER	C
	ILE	THR	LYS	ARG	A
	MET	THR	LYS	ARG	G
G	VAL	ALA	ASP	GLY	U
	VAL	ALA	ASP	GLY	C
	VAL	ALA	GLU	GLY	A
	VAL	ALA	GLU	GLY	G

★These are the names given to the chain termination triplets by geneticists.

the appropriate anticodon sequence at a particular point in the cloverleaf structure (see Chapter 12, Figures 12–12, 12–13, and 12–14).

The codon : anticodon interaction is not perfectly accurate; no information transfer process can be. A low frequency of misreading can be detected both in vivo and in vitro. For example in whole cells, genes in which a mutation has created an internal stop signal (called a *nonsense* mutation) may nevertheless be completely translated, though at low frequency—about 10^{-4}. Presumably, this represents natural misreading of the stop signal, as if it were a related amino acid codon. A few measurements of missense misreading, based on the detection of trace quantities of an incorrect amino acid at a position normally occupied by another, have been reported; the estimated frequency of missense misreading depends on the type of error and ranges from about 10^{-4} up to 3×10^{-3}.

In cell-free systems, much higher frequencies of misreading often occur. In fact, the classic studies of poly U translation actually showed significant incorporation of leucine and isoleucine in addition to phenylalanine, although at much lower efficiencies. Presumably, these results reflect occasional misreading of a single base in the phenylalanine triplet UUU, so that leucine (CUU or UUA or UUG) or isoleucine (AUU) is accidentally incorporated into protein. Misreading of this type is greatly increased by environmental conditions that lead to a reduction in the specificity of the ribosome, such as excessive magnesium ion concentrations, or the presence of alcohols or certain drugs in the reaction mixture. The antibiotic *streptomycin* has a particularly marked effect on the decoding specificity of procaryotic ribosomes: when streptomycin binds to the ribosome, the frequency of certain one-base reading errors is increased to very high levels. In whole cells, it has been demonstrated that the presence of streptomycin gives rise to the synthesis of aberrant proteins. In short, the specificity of decoding, like other examples of biological specificity, resides in several levels of organization. The primary and most decisive level is the interaction between the nucleotide sequences of codon and anticodon. A secondary level of organization, inherent in the structure of the ribosome, governs the exactness of this primary interaction.

Deciphering the Code In Vivo

It was perhaps a foregone conclusion that the genetic code, deciphered through studies in vitro with isolated components, would in fact correspond to the way information is transferred in living cells. Nonetheless, investigators have gone to great lengths to substantiate this point. Three independent lines of evidence confirm that the coding dictionary of Table 13–3 is the real thing.

First, the amino acid substitutions dictated by specific mutations have been analyzed in a great variety of proteins. *Missense mutations,* which replace one amino acid by another, occur fairly frequently. Thus we would expect them to represent the most common change in the DNA, a single base pair substituted for another. The codons for the wildtype and mutant amino acids should therefore differ at only one position. This has been confirmed in a great many cases. More importantly, exceptions where a mutational substitution of one amino acid for another demands more than one base change to be compatible with the code have been very rare. Table 13–4 presents the results of analyses of this kind in three different proteins from three very different organisms: (1) the enzyme tryptophan synthetase of *E. coli;* (2) the coat protein of tobacco mosaic virus, a plant virus whose genome is RNA rather than DNA; and (3) the hemoglobin of human beings. (The last two proteins are discussed in Chapter 4.)

The studies of tryptophan synthetase are especially convincing because they include many independent mutational events affecting the amino acid at a single position in the protein; this family of mutational substitutions, with the number of independent mutations of each kind observed, is shown in Table 13–4, part 1. It can be seen that all but one of 128 separate mutational events involved an amino acid replacement compatible with single base changes in the relevant codons listed in Table 13–3. Similarly, among 20 amino acid replacements in TMV coat protein, 19 reflect single base changes in appropriate codons; also, among 11 mutant human hemoglobins, all 11 reflect single base changes. This nearly perfect agreement between the pattern of amino acid replacements and the code table cannot be an accident. It provides strong evidence that the code dictionary is the one actually used in living cells and by organisms as different as humans, microbes, and viruses.

A further test is provided by the frame shift mutations described earlier. In the region between two of these mutations, the entire amino acid sequence should be altered. If the code dictionary is used in real life, then the normal and altered amino acid sequences should relate to a unique series of code symbols, as changed by the insertion of a base at one end and the deletion of a base at the other. Figure 13–6 shows the amino acid sequence of a region in the lysozyme protein of bacterial virus T4 and the altered amino acid sequence, which was determined by G. Streisinger and his colleagues, of the same region in the gene when enclosed by two frame shift mutations of opposite sign. The two amino acid sequences agree exactly with a particular coding sequence of bases shifted out of phase by a deletion of a base at the left side and the insertion of a base at the right side. This result confirms, at one stroke, both the pertinence of the code dictionary and the interpretation of frame shift mutations proposed originally by Crick and Brenner.

A still more decisive test comes from the direct comparison of the amino acid sequence of a protein with the nucleotide base sequence of its gene or messenger RNA. In a few cases, this gigantic analytical project has been carried out. One is the bacterial virus MS2, an RNA virus in which, as in TMV, the genes serve as their own messengers. Using methods worked out by F. Sanger (see Chapter 12), W. Fiers and his associates have sequenced the entire RNA molecule of MS2, including the three major genes, and also one of their three protein products. Figure 13–7 shows the sequence of one gene and its protein. As you can see, each one of the 129 amino

TABLE 13–4

Amino Acid Replacements at Many Different Positions amongst Mutants in Three Different Proteins

1) Tryptophan Synthetase A Protein

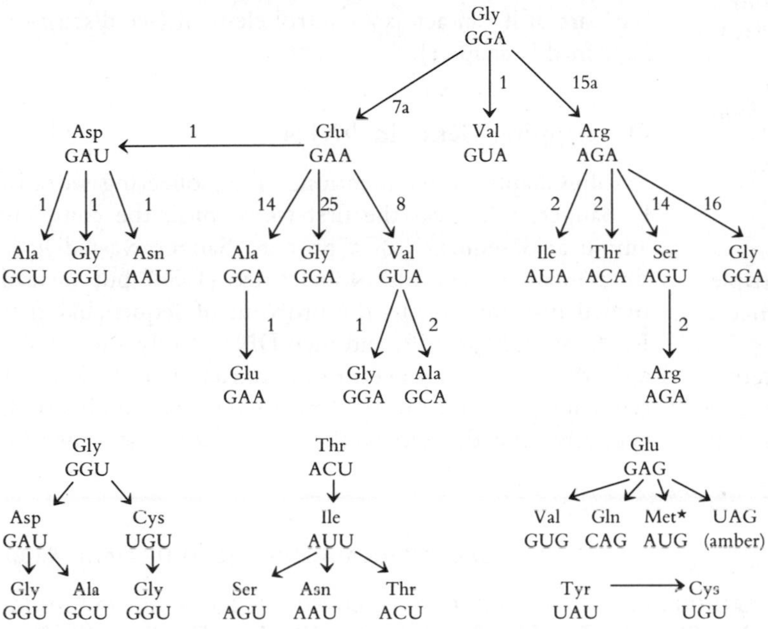

2) TMV Coat Protein

Nitrous acid		*Spontaneous*	
Asp → Ala GAU GCU	Pro → Leu CCU CUU	Thr → Ser ACU UCU	Ile → Thr AUU ACU
Asn → Ser AAU AGU	Ser → Leu UCA UUA	Asp → Asn GAU AAU	Arg → Lys AGA AAA
Ile → Met AUA AUG	Thr → Ile ACU AUU	Asn → Lys AAU AAA	Arg → Gly AGA GGA
Ile → Val AUU GUU	Thr → Met ACG AUG	Asn → Arg* AAU AGA	
Glu → Gly GAA GGA	Ser → Phe UCU UUU	Tyr → Phe UAU UUU	
Asp → Gly GAU GGU	Leu → Phe CUU UUU		

3) Hemoglobin Mutants

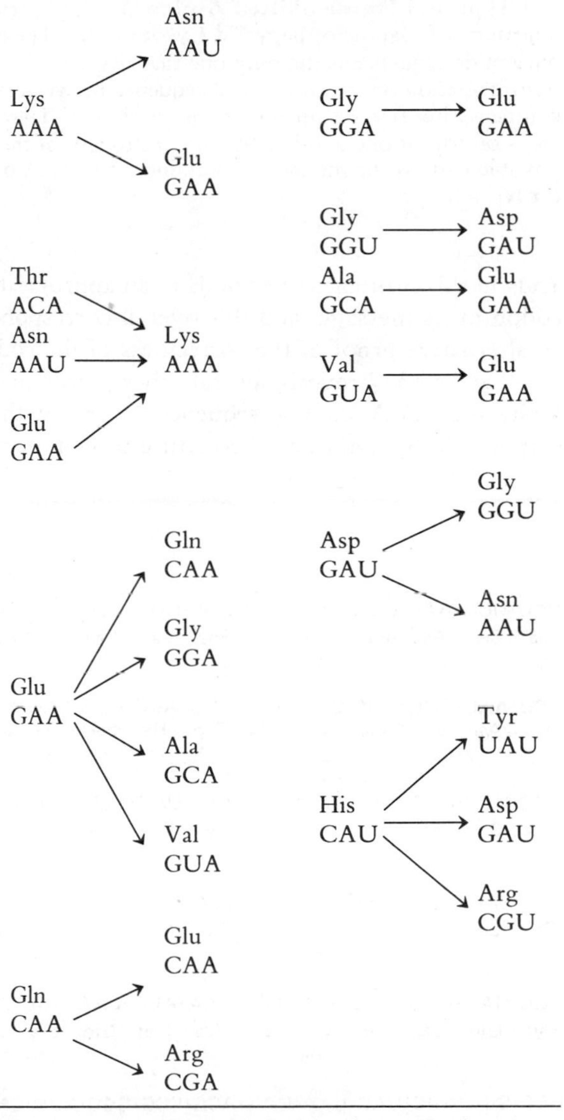

Numbers along arrows refer to the number of independent cases in which the replacement was observed. The "family trees" of positions in tryptophan synthetase represent several generations of mutational change. For example, at one position in the protein Gly is present in the wildtype, indicated by the Gly (GGA) at the top of part (1). Independent mutations to Glu, Val, and Arg were detected the number of times indicated. The mutant with Glu at the position under study in turn sported second-generation mutants with Ala, Gly, or Val at that position, with the number of independent occurrences again shown. Some third-generation mutational changes are also tabulated. Mutational changes of spontaneous origin and those induced by nitrous acid in TMV coat protein are shown in part (2). Spontaneous mutational changes affecting human hemoglobin alpha and beta are tabulated in part (3). Asterisks mark two cases in which the amino acid replacement is not consistent with single base changes. In the cases of the human hemoglobin mutants, only one of two possible synonymous codons is shown in each case. Since the mutations in hemoglobin were detected as changes in its electrophoretic mobility, all the tabulated replacements are between amino acids with different electrostatically charged side chains.

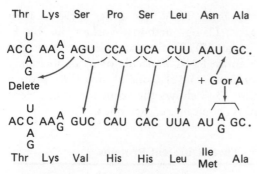

FIGURE **13-6**

Wildtype and Frame Shifted Amino Acid Sequence of a Segment of Bacteriophage T4 Lysozyme. The inferred nucleotide sequence is the only one that fits the transformation of the amino acid sequence between two compensating frame shift mutations, as shown. Two or more bases on top of one another (AA_G^A) refer to any of the possible codons for the indicated amino acid (AAA or AAG, for lysine).

acids in this protein corresponds to an appropriate triplet codon in its message, and this overall correspondence is an alternative proof of the correctness of the code.

Figure 13–7 reveals several other points of interest. First, the RNA coding sequence begins with a start symbol, AUG, although the mature protein begins with the next amino acid in line, alanine. Presumably the initiating formyl-methionine (see discussion later in this chapter) is cleaved off during or after translation, as has been demonstrated to occur for other proteins. Second, the gene ends with two stop codons in tandem: UAA-UAG. This combination has been found at the ends of certain other genes as well, perhaps providing a fail-safe signal to ensure termination at the right place. Finally, notice that there is a sizeable sequence of nucleotides before the next AUG symbol for starting the translated portion of the next gene. The entire significance of the nucleotide sequence between two genes is not known, but part of it can act as a control element (see discussion later in this chapter).

Overlapping Genes in ΦX174

In Chapter 4, we mentioned the pioneering work of F. Sanger, who was the first to establish the complete amino acid sequence of a protein. Sanger wasted little time musing over his Nobel prize (1958) but instead turned his attention to the problem of sequencing first RNA (see Chapter 12) and then DNA. Only a few years ago, this seemed an even more audacious task than the sequencing of a protein. But Sanger was undeterred, typically, and devised methods of analysis so powerful

```
. . . AUA·GAG·CCC·UCA·ACC·GGA·GUU·UGA·AGC· AUG·
```

```
GCU·UCU·AAC·UUU·ACU·CAG·UUC·GUU·CUC·GUC·GAC·AAU·GGC·GGA·ACU·GGC·GAC·GUG·ACU·GUC·GCC·CCA·AGC·AAC·UUC·
Ala Ser Asn Phe Thr Gln Phe Val Leu Val Asp Asn Gly Gly Thr Gly Asp Val Thr Val Ala Pro Ser Asn Phe
1          5                    10                     15                     20                25
```

```
GCU·AAC·GGG·GUC·GCU·GAA·UGG·AUC·AGC·UCU·AAC·UCG·CGU·UCA·CAG·GCU·UAC·AAA·GUA·ACC·UGU·AGC·GUU·CGU·CAG·
Ala Asn Gly Val Ala Glu Trp Ile Ser Ser Asn Ser Arg Ser Gln Ala Tyr Lys Val Thr Cys Ser Val Arg Gln
          30                    35                     40                     45                50
```

```
AGC·UCU·GCG·CAG·AAU·CGC·AAA·UAC·ACC·AUC·AAA·GUC·GAG·GUG·CCU·AAA·GUG·GCA·ACC·CAG·ACU·GUU·GGU·GGU·GUA·
Ser Ser Ala Gln Asn Arg Lys Tyr Thr Ile Lys Val Glu Val Pro Lys Val Ala Thr Gln Thr Val Gly Gly Val
          55                    60                     65                     70                75
```

```
GAG·CUU·CCU·GUA·GCC·GCA·UGG·CGU·UCG·UAC·UUA·AAU·AUG·GAA·CUA·ACC·AUU·CCA·AUU·UUC·GCU·ACG·AAU·UCC·GAC·
Glu Leu Pro Val Ala Ala Trp Arg Ser Tyr Leu Asn Met Glu Leu Thr Ile Pro Ile Phe Ala Thr Asn Ser Asp
          80                    85                     90                     95                100
```

```
UGC·GAG·CUU·AUU·GUU·AAG·GCA·AUG·CAA·GGU·CUC·CUA·AAA·GAU·GGA·AAC·CCG·AUU·CCC·UCA·GCA·AUC·GCA·GCA·AAC·
Cys Glu Leu Ile Val Lys Ala Met Gln Gly Leu Leu Lys Asp Gly Asn Pro Ile Pro Ser Ala Ile Ala Ala Asn
          105                   110                    115                    120                125
```

```
UCC·GGC·AUC·UAC·UAA·UAG·ACG·CCG·GCC·AUU·CAA·ACA·UGA·GGA·UUA·CCC·AUG·UCG·AAG·ACA·ACA·AAG·AAG . . .
Ser Gly Ile Tyr                                                           Ser Lys Thr Thr Lys Lys
          129                                                              1              5
```

FIGURE **13-7**

Nucleotide Sequence of the Coat Protein Gene of Virus MS2 and Amino Acid Sequence of the Coat Protein. AUG signifies the signal for start; UAA,UAG signifies a combined ochre and amber stop signal. The numbers designate the amino acids in the mature protein, starting with alanine (1) and ending with tyrosine (129).

and ingenious (see Chapter 10) that it is now actually simpler to establish the nucleotide sequence of a gene, which determines the amino acid sequence of its encoded protein, than to use the classical methods of amino acid sequence analysis. In 1977, Sanger and his colleagues established the complete sequence of the DNA of bacterial virus ΦX174, comprising no fewer than nine genes. In two cases, the complete amino acid sequences of the protein products were also established. The amino acid sequences of these proteins agree, triplet by triplet, with the nucleotide sequence of their genes via the genetic code dictionary.

This feat of biochemical analysis, which in part led to the award of a second Nobel prize to Sanger (1980), has also revealed a truly extraordinary feature of the decoding process. The DNA of virus ΦX174 consists of about 5400 nucleotides. But the aggregate length of the nine protein products comes to about 2000 amino acids, which should require 6000 nucleotides to contain the requisite information via a triplet code. How does the nucleotide sequence of ΦX174 DNA encode a larger number of amino acids than seems possible?

The secret is that the informational content of a nucleotide sequence in DNA depends on the _phase_ in which it is read. In two cases, Sanger and his colleagues found that a single nucleotide sequence is read in two different reading frames, to encode two different proteins. To put it another way, the _two genes overlap_. Figure 13–8 presents one example, taken from a portion of the ΦX174 virus DNA molecule that encodes both the D-protein and the E-protein and, therefore, corresponds to two different genes in the same place. The amino acid sequences of the two proteins, read off the same nucleotide sequence in two different reading frames, are shown above and below the nucleotide sequence.

This example of the versatility of the genetic code is probably exceptional. If it weren't, Beadle and Tatum could never have proposed the one gene–one enzyme axiom (see Chapter 6). Also, genetic mapping studies have shown that different genes generally occupy topologically distinctive positions (see Chapter 9). Possibly the constraints of very small size have favored the maximum use of the nucleotides available, since this same trick has also been demonstrated in the RNA bacterial virus R17 and in the genome of human mitochondria.

Open Reading Frames

The analysis of cellular DNA sequences through the use of genetic engineering techniques also suggests that overlapping genes are rare. Often, when the sequence of a DNA fragment is examined, two of the three reading frames contain frequent stop signals that would interrupt translation. Usually, only one of the reading frames is "open"—free of stop signals over a long distance—and can code for a protein. Finding such an _open reading frame (ORF)_, extending from an AUG start codon to a stop codon, permits one to deduce the amino acid sequence of a protein without ever isolating it. Indeed, the bulk of the amino acid sequence data now known was deduced from DNA sequence data in this way, rather than from the direct but more laborious methods of protein sequence analysis (see Chapter 4). Computer programs in wide use will specify the six different amino acid sequences predicted by a given DNA sequence (three reading frames encoded on each strand), scan for long ORFs, and even predict some of the chemical properties of the protein encoded by a given ORF.

There were several ORFs found in the DNA of human mitochondria when the sequence was worked out

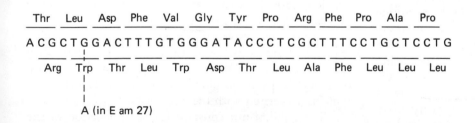

FIGURE 13–8
Gene and Two-gene Products in Virus ΦX174. The nucleotide sequence of a small portion of the "DE region" (all of which has been sequenced) is shown in large letters. Above it is the corresponding amino acid sequence of the D-protein, with the nucleotides appropriately grouped into triplet codons. Below is the predicted amino acid sequence of the E-protein. The location of the E-protein within the D gene, and the phase in which it is read, is proved by the existence of amber mutations of gene E that lie within the nucleotide sequence of gene D. One example is indicated by a dotted line: an E amber mutant (DNA code = TAG) which affects the sixth nucleotide in the sequence above.

by Sanger and B. Barrell. Some of them corresponded in their predicted amino acid sequence to known mitochondrial proteins. The products, if any, coded by the other ORFs were mysterious for a long time. Recently, most of them have been identified as proteins in the bovine mitochondrion. The methods used are an instructive application of the power of new immunological techniques: the DNA sequence was used to predict the amino acid sequence of a protein. Then a peptide was synthesized, using chemical techniques, with a sequence identical to a run of amino acids in the predicted protein. Antibodies were raised against this peptide so that a protein containing the same sequence of amino acids might be detected with the antibodies (see Chapter 17). Finally, the antibodies were used as probes to test whether proteins isolated from one or another mitochondrial respiratory complex could be identified as the product of the ORF. In this way, six ORFs were tentatively identified as genes for proteins associated with the NADH dehydrogenase complex (complex I, see Chapter 8).

Another example of ORFs is found in some of the lengthy introns of certain mitochondrial genes. Mutations that map in some of these ORFs are associated with defects in the splicing of mitochondrial RNAs; the proteins, not yet isolated, coded by such ORFs are perhaps among the RNA maturases discussed in Chapter 12.

Recently, T. Caskey and his colleagues cloned and sequenced the gene for a protein involved in protein chain termination, and they were astonished to discover that the DNA sequence contained no ORF sufficiently long to code for the complete protein! Comparison of the DNA sequence with partial amino acid sequence data for the isolated protein revealed that the protein must be synthesized by means of an obligatory ribosome frameshift at a UGA terminator codon early in the gene's coding sequence. This finding implies that certain "shifty" base sequences in mRNA may actually program the ribosome to change its reading frame, thus permitting a seemingly closed reading frame to code for a protein. Frameshifts are also required for proper gene expression during infection by retroviruses, including the AIDS virus. There are even examples of ribosome "hopping" from one codon on a message to another, but many bases farther along. In this case, studied by R. Weiss and his colleagues, the ORF has more codons than the protein has amino acids.

The Mechanism of Protein Synthesis

We now turn to the mechanics of protein synthesis, a subject that developed in parallel with the solving of the genetic code and that remains an intensely active area of research today. Most of the details of the description are from studies of *E. coli,* but the general principles are apparently common to the mechanics in eucaryotic cytoplasm and organelles.

Since the hydrolysis of proteins to give amino acids is a spontaneous process, amino acids must be activated before they can be coupled together to form a polypeptide. This process occurs in two stages, the final product of which is an aminoacyl-tRNA (aa-tRNA). The correct aminoacyl-tRNA is chosen by its ability to base pair its anticodon to the codon of the mRNA that is associated with the ribosome. Transfer of a peptide (from a peptidyl-tRNA) to the aminoacyl residue of the aa-tRNA elongates the peptide. Thus the aminoacyl-tRNA represents the confluence of the flows of mass (amino acids), energy (activation), and information (anticodon:codon binding) in the synthesis of a protein.

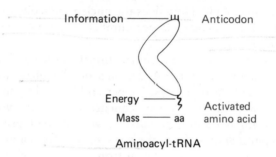

Aminoacyl-tRNA

The Activation of Amino Acids

F. Lipmann originally proposed the idea that, in preparation for its role as precursor in protein synthesis, an amino acid must be brought into a "high energy" state. It was hypothesized that this might be done through the mediation of ATP. Looking for just such a reaction, Hoagland discovered the *amino acid activating enzymes (aminoacyl-tRNA ligases).* Each amino acid has a different specific activating enzyme. ATP and the specific amino acid are both bound to the enzyme, where they react to form an *aminoacyl-adenylate (AMP-aa),* splitting off inorganic pyrophosphate (Figure 13–9). The AMP-aa remains associated with the enzyme.

The equilibrium constant for the formation of the aminoacyl-adenylate and inorganic pyrophosphate is nearly equal to one. Activation of the amino acid proceeds, however, by the energetically favorable hydrolysis of pyrophosphate (after its release from the ligase) to inorganic phosphate, catalyzed by cellular pyrophosphatase:

$$P\!-\!P + H_2O \rightleftharpoons 2\,P_i \qquad \Delta G' = -30 \text{ kJ/mole}$$
$$(-7 \text{ kcal/mole})$$

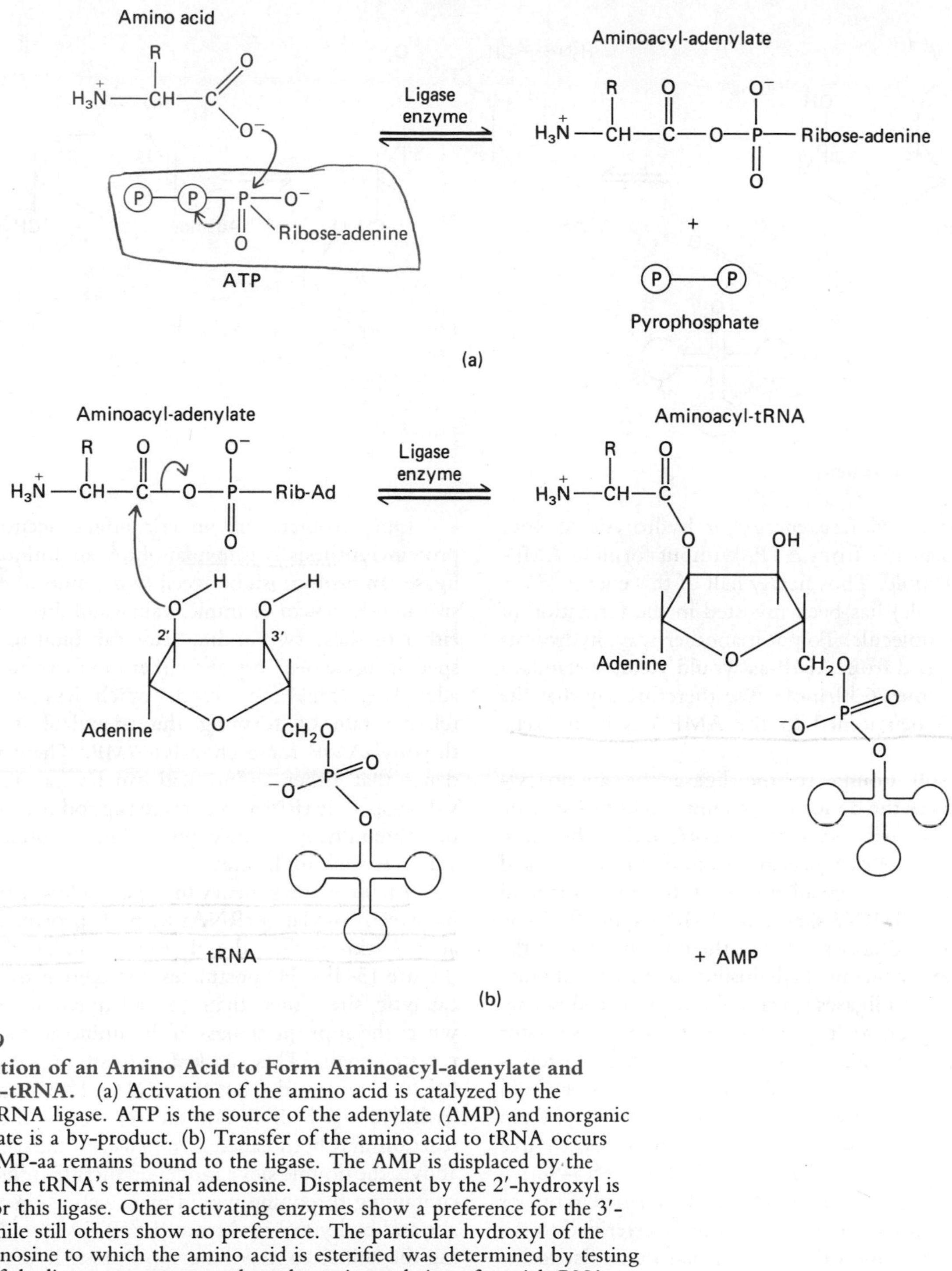

(a)

(b)

FIGURE 13–9

The Activation of an Amino Acid to Form Aminoacyl-adenylate and Aminoacyl-tRNA. (a) Activation of the amino acid is catalyzed by the aminoacyl-tRNA ligase. ATP is the source of the adenylate (AMP) and inorganic pyrophosphate is a by-product. (b) Transfer of the amino acid to tRNA occurs while the AMP-aa remains bound to the ligase. The AMP is displaced by the hydroxyl of the tRNA's terminal adenosine. Displacement by the 2'-hydroxyl is illustrated for this ligase. Other activating enzymes show a preference for the 3'-hydroxyl while still others show no preference. The particular hydroxyl of the terminal adenosine to which the amino acid is esterified was determined by testing the ability of the ligase enzyme to catalyze the aminoacylation of special tRNAs. One such tRNA contained a chemically synthesized terminal adenosine with the 3'-oxygen omitted (tRNA-3'-deoxy A), the other with the 2'-oxygen omitted (tRNA-2'-deoxy A). (c) Isomerization of the amino acid from the 2'- to the 3'-hydroxyl of the tRNA's terminal adenosine. Spontaneous migration of an amino acid on the terminal adenosine of tRNA can occur if the aminoacyl-tRNA is free in solution. This reaction is restricted when the aa-tRNA is bound to a protein.

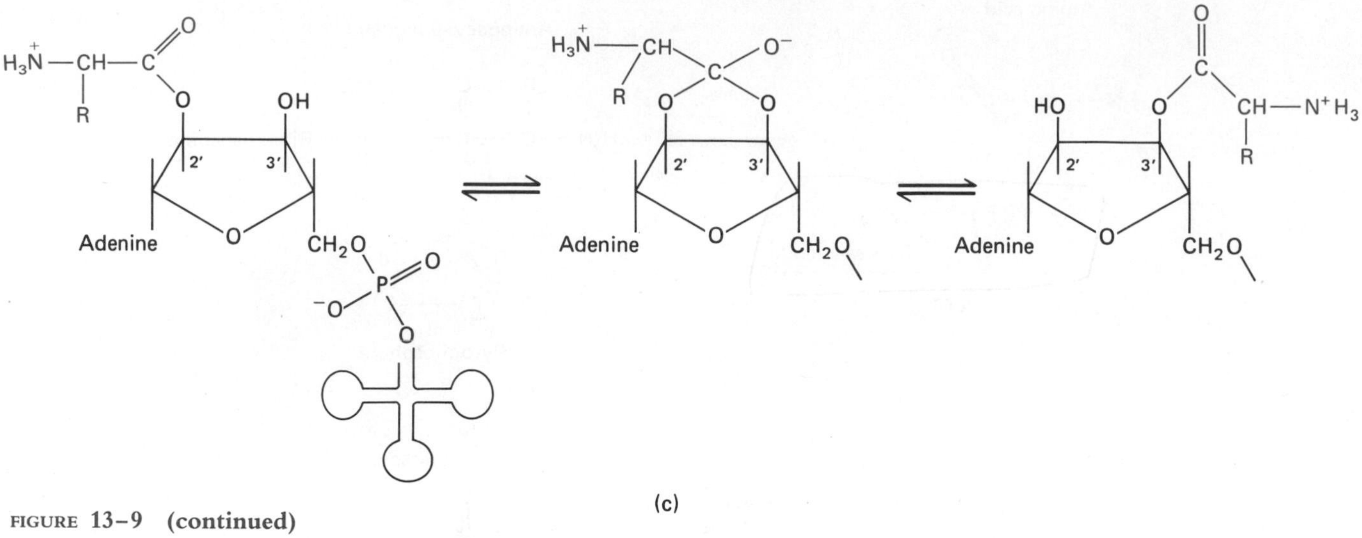

(c)

FIGURE 13-9 (continued)

Now, the standard free energy of hydrolysis of both terminal phosphates from ATP, without forming AMP-aa, is -56 kJ/mole. Thus nearly half of this energy ($56 - 30 = 26$ kJ/mole) has been invested in the formation of the AMP-aa molecule. To put it another way, hydrolysis of the amino acid from AMP-aa would yield, at standard conditions, some 26 kJ/mole. We therefore say that the amino acid when joined to the AMP has been "activated."

While still bound to the ligase the aminoacyl-adenylate acts as the donor of the amino acid to a specific tRNA, forming *aminoacyl-tRNA (aa-tRNA)* as shown in Figure 13-9b. The carboxylate group of the amino acid becomes esterified to the adenosine at the —CCA (found at the 3′ end of all tRNAs) and the AMP is split off. Some aminoacyl-tRNA ligases catalyze the esterification of the *2′-hydroxyl* of the terminal adenosine, as shown in Figure 13-9b. The *E. coli* ligases specific for arginine and valine, for instance, catalyze this reaction. Other ligases—for example, those for alanine and serine—catalyze esterification at the *3′-hydroxyl* while still others—cysteine and tyrosine—seem to show no preference. After release from the ligase, these positions of esterification can, in free solution, be exchanged spontaneously (Figure 13-9c).

The observation that amino acids esterified to one or the other site in *E. coli* are preferentially esterified to the same site in yeast and calf liver cells has given rise to the question of why such specificity might occur. This question has been related to another one concerning the accuracy of protein synthesis. The general notion is that the two sites could be used for different enzymatic purposes. This is in fact the case for later stages of protein synthesis, as we shall see below. But even during the activation of an amino acid, the site not used for transfer to tRNA might represent structurally another function that could help to achieve accuracy.

One problem in understanding accuracy during protein synthesis is to explain how an aminoacyl-tRNA ligase can distinguish between two amino acids of similar size and shape, for example, valine and threonine. In fact, either of these two amino acids can bind to the valine-specific ligase of *E. coli* and either can form an aminoacyl-adenylate. Threonine binds much less well and the relative rate of forming the adenylate is 4×10^{-3} threonyl-AMP for each valyl-AMP. There is also evidence that either amino acid can be transferred to the valine-specific tRNA. Yet there is good reason to believe that threonine is not incorporated into protein in place of valine at such high rates.

An interesting theory to explain a lower frequency of incorrectly acylated tRNAs than of incorrectly activated amino acids is the "double sieve" model of A. Ferscht (Figure 13-10). He postulates that there is on the ligase a catalytic site other than the adenylation/transfer site, where the appropriateness of the amino acid to the tRNA is tested again. This is a *hydrolytic* site: if the amino acid fits, it is removed from the tRNA. The ligase-catalyzed hydrolysis (which is not the reverse of the activation reaction) has been called *verification*. In the case discussed above, the double sieve model predicts that the hydroxyl-containing threonine would bind well to the verification site and be hydrolyzed from the tRNA^{Val}.[2] Valine, lacking the ability to make the appropriate hydrogen bonds, would not fit the verification site well and the correctly acylated tRNA would therefore survive. Ferscht found that the thr-tRNA^{Val} was 40 times more likely to be hydrolyzed than the correctly acylated val-tRNA^{Val}.

---

[2]The superscript in tRNA^{Val} denotes a tRNA that normally accepts Valine (see Chapter 12).

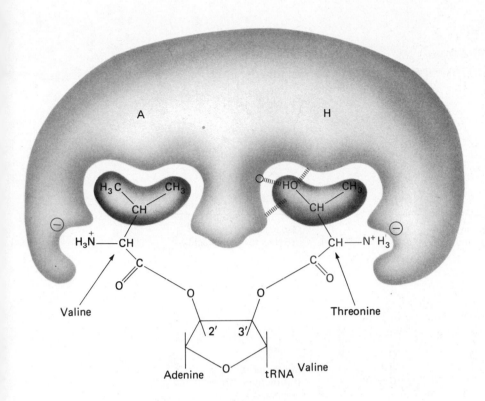

FIGURE 13-10
The Double Sieve Model. Two active sites are shown for the ligase (color), one (A) for activating the amino acid, in this case valine, and transferring it to the tRNA^{Val}, and the other (H) for hydrolyzing the amino acid, in this case threonine, from the tRNA. The 3' terminal ribose of the tRNA is shown with two amino acids esterified, to indicate the positions occupied by a single amino acid whose side chain could fit in both sites. Actually, only one amino acid is bound at a time, but it might be able to migrate between the two sites if its side chain could fit both.

Instead of hydrogen bonding, steric hindrance can be used as the basis for the second sieve. For example, valine is activated by the ile-tRNA ligase but the presence of tRNA^{Ile} causes hydrolysis and the release of valine. The smaller valine can probably fit into the hydrolytic site but isoleucine cannot. It is possible that before being tested in the hydrolytic site an amino acid must migrate to the hydroxyl of the tRNA that is not used for transfer from the adenylate (Figure 13-10). If this migration were also required before dissociation of the aa-tRNA from the ligase could occur, testing in the hydrolytic site would be more likely.

Trying to distinguish between valine and threonine thus implies using more than one ATP per correctly acylated tRNA; after incorrectly acylated tRNA is hydrolyzed, additional ATP must be used to activate another amino acid. Spending energy (in this case, as ATP) to achieve higher accuracy is one of the essential elements of the *kinetic proofreading* model proposed by J. J. Hopfield. Whatever the specific mechanisms, which are still under active investigation, the frequency of surviving inappropriate (misacylated) aminoacyl-tRNA combinations is quite low, much less than 10^{-4}.

Although each amino acid has only a single ligase in *E. coli,* there is usually more than one kind of tRNA that can accept the same amino acid. Clearly, members of such an *isoaccepting family* of tRNAs must share some structural elements that can be recognized by the appropriate ligase and rejected by the wrong ones. The

placement of these elements in the tRNA structure determined by x-ray crystallographic methods is still controversial, though the region extending from the anticodon loop to the accepting stem along the inside of the L-shape has been suggested by A. Rich as the most likely spot (see Figs. 12-12 to 12-14). This is the region most often joined to the ligase by *molecular staples (bifunctional cross-linking reagents).* (See illustrations, p. 598)

If the wrong amino acid has been joined to the tRNA and the aminoacyl-tRNA is released from the ligase, the remainder of the synthesizing machinery seems unable to stop the incorporation of the mistake into protein. This was demonstrated most convincingly by F. Chapeville and others, working in the laboratory of Lipmann. They isolated ^{14}C-labeled cys-tRNA^{Cys} by mixing radioactive cysteine, ATP, and tRNA with a crude preparation of ligases (Figure 13-11a). The cysteinyl-tRNA ligase present in the preparation performed the specific labeling required, illustrating one of the advantages of doing *biological* chemistry. Next they chemically reduced the cysteine residue, removing the —SH group. This formed ^{14}C-labeled *alanyl*- tRNA^{Cys} (Figure 13-11b). This chemically misacylated tRNA was then used as a source of amino acid in cell-free protein synthesis programmed by a chemically synthesized mRNA, random sequence poly (U,G). Such a message usually stimulates the incorporation of cysteine into hot acid-precipitable peptides, as we discussed earlier. But in this case the peptides contained radioactive alanine, whose codons all require C.

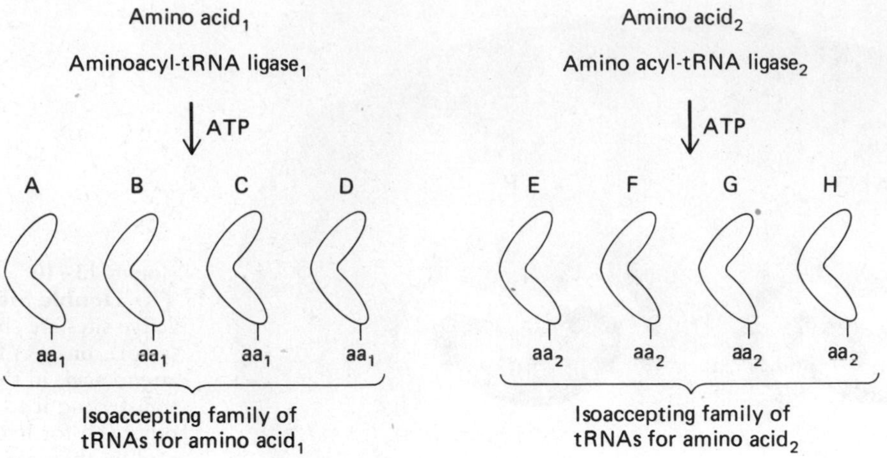
Amino acid$_1$
Aminoacyl-tRNA ligase$_1$
ATP
A
B
C
D
aa$_1$
aa$_1$
aa$_1$
aa$_1$
Isoaccepting family of tRNAs for amino acid$_1$
Amino acid$_2$
Amino acyl-tRNA ligase$_2$
ATP
E
F
G
H
aa$_2$
aa$_2$
aa$_2$
aa$_2$
Isoaccepting family of tRNAs for amino acid$_2$

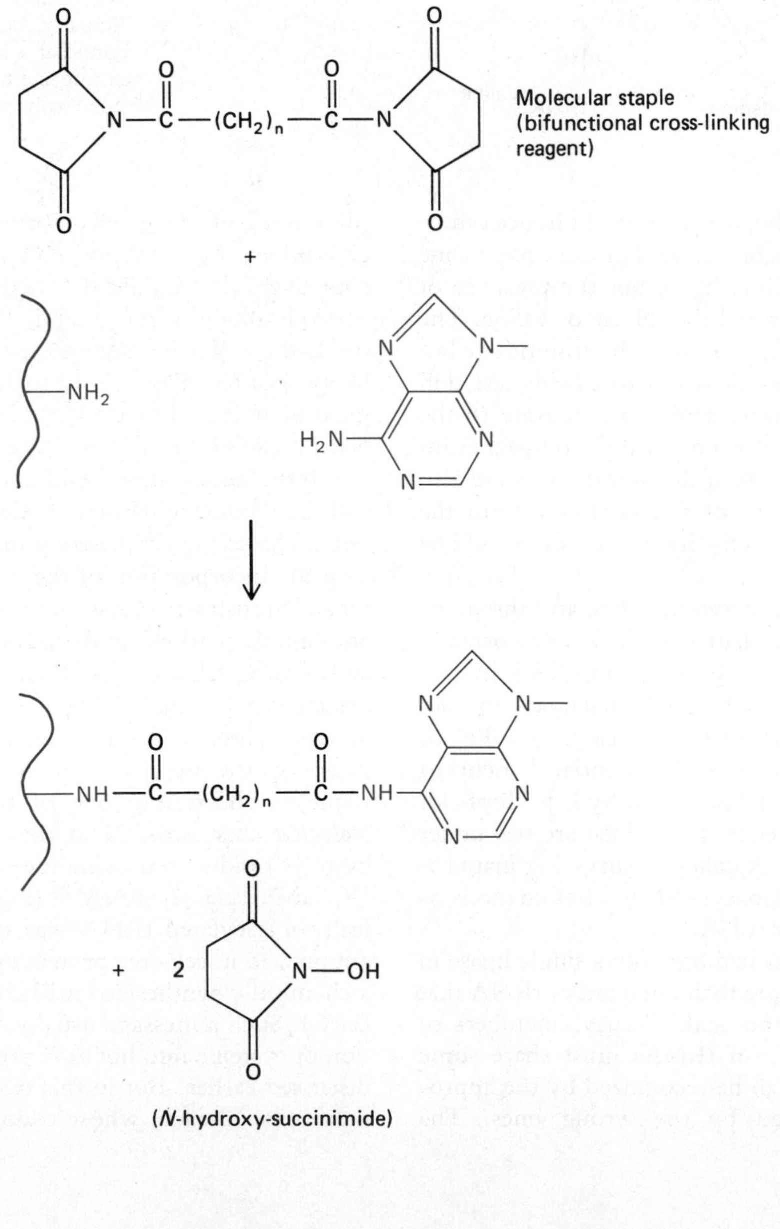
Molecular staple (bifunctional cross-linking reagent)
N—C—(CH$_2$)$_n$—C—N
+
NH$_2$
H$_2$N
NH—C—(CH$_2$)$_n$—C—NH
+ 2
N—OH
(*N*-hydroxy-succinimide)

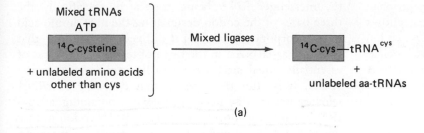

(a)

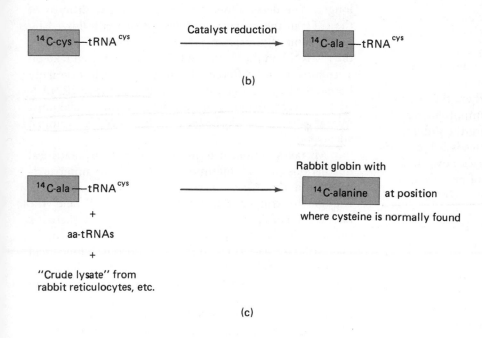

(b)

(c)

FIGURE 13–11

Chapeville's Experiment Showing That aa-tRNA Is Sufficient As a Source of Amino Acids for the Decoding Step of Protein Synthesis. (a) Labeling of cysteine-acceptor tRNA with [^{14}C]-cysteine. (b) Reduction of the cysteinyl side chain so that the sulfur is removed. The product is a radiolabeled alanyl-tRNA that normally accepts cysteine. (c) Use of the [^{14}C]alanyl-tRNA^{cys} in cell-free protein synthesis. Radiolabeled globin, programmed by the mRNA found in the reticulocyte extract, was fingerprinted. Peptides that normally contain cysteine were found to be radiolabeled and to contain alanine instead.

Later, G. von Ehrenstein and others extended this work to an even more dramatic demonstration. They used the [^{14}C]alanyl-tRNA^{Cys} as a source of amino acids in the cell-free synthesis of rabbit globin, catalyzed by a crude lysate of immature red blood cells (reticulocytes) from a rabbit (Figure 13–11c). To the lysate were also added various salts and energy sources ATP and GTP (see below). The result was a ^{14}C-labeled protein that gave, after digestion by trypsin, a series of peptides almost identical to those from authentic rabbit globin, as judged by their two-dimensional fingerprint (see Chapter 4). The differences lay only in those tryptic peptides that usually contain cysteine; after this kind of labeling they were shown to contain radioactive alanine instead. This observation neatly confirmed the sufficiency of the adaptor tRNA to carry an amino acid into protein, unverified by any further part of the synthetic machinery. It also showed that the information for the incorporation of a specific amino acid is decoded on the basis of the tRNA structure and not that of the amino acid itself.

The information that specifies the sequence of nucleotides in a tRNA molecule lies in a DNA sequence, that is to say, in a *tRNA gene*. Mutations in tRNA genes have been used to establish other aspects of tRNA function. For example, the decoding process certainly involves the bases of the anticodon (see Chapter 12), but the genetic evidence strongly suggests that other parts of the tRNA structure also play a role in decoding. When the anticodon suffers a mutation, the altered tRNA can often still accept its usual amino acid, but installs it into protein in response to a different codon. Such a mutant tRNA should make a mistake in protein synthesis every time it is used, and indeed, strains of *E. coli* having such mutated tRNAs often grow more slowly than wildtype strains. But the anticodon is apparently not the only structure that can affect decoding. A mutation in the dihyrouridine loop (see Figure 12–10) of a tryptophan tRNA, isolated and characterized by D. Hirsch, allows the altered trp-tRNA to recognize the UGA codon (which ordinarily means "terminate synthesis") as well as its normal UGG. The complicated folding of a tRNA presumably permits bases far from the anticodon to influence the decoding function.

The Elongation Epicycle

We turn now to the mechanics of elongating a growing protein. (We shall discuss initiation of protein synthesis, a more complicated process, after introducing the participants in the cyclic process of elongation.) Our knowledge of this subject is derived largely from an intensive study of protein synthesis by cell-free extracts of the bacterium *E. coli*. The major features of the mechanism are probably the same in all cells, and in mitochondria and chloroplasts, too. Where the details of eucaryotic protein synthesis differ in important respects from the *E. coli* story, we shall indicate the differences. However, protein synthesis in eucaryotes is being investigated by a large number of laboratories now, and our models will likely change in the light of new data.

During elongation, the aminoacyl-tRNA binds to the ribosome at a special place. Chain growth occurs by transfer of the peptide from a peptidyl-tRNA, bound to the ribosome in another place, to the amino acid of the aminoacyl-tRNA. This forms a new peptidyl-tRNA, longer by one amino acid and bound where the aminoacyl-tRNA was. This peptidyl-tRNA then moves to the appropriate site for donating its peptide to the next aminoacyl-tRNA, and the cycle repeats itself.

Binding aa-tRNA to the Ribosome

Before binding to the ribosome, an aminoacyl-tRNA first forms a complex with a protein elongation factor (Figure 13–12). This factor has a molecular weight of 45,000, comprises a large fraction of the cellular protein in *E. coli* (2–5%), and is called *EF-Tu* (Elongation Factor-*T*emperature *u*nstable). When ready to combine with aa-tRNA, EF-Tu is already in a *binary complex,* with GTP. Since there is such a large amount of EF-Tu and since the EF-Tu · GTP has such a high affinity for aa-tRNA (the equilibrium dissociation constant $K_D = 10^{-7}$), almost no free aa-tRNA is found in *E. coli* cells. (Thus any misacylated tRNAs that escape the surface of their ligase will be trapped in erroneous aa-tRNA · EF-Tu · GTP *ternary complexes,* and the error will be incorporated into protein with a high probability.)

The ternary complex is the form that binds to the ribosome at a specific location, the *A-site*. (This site is named for the later function of the aminoacyl-tRNA in accepting the growing peptide.) The A-site also contains

the messenger RNA being translated, specifically the three bases of the codon designating the next amino acid to be incorporated. Binding is diffusion-controlled—that is, all the aa-tRNAs, in the form of ternary complexes of essentially equal size, have equally rapid access to the A-site. It is after the binding that decoding or *tRNA selection* occurs. Ternary complexes containing an aa-tRNA that is inappropriate to the mRNA codon diffuse out of the A-site rapidly. Ternary complexes containing an aa-tRNA that is appropriate to the codon diffuse out of the A-site slowly. Put another way, ternary complexes containing the appropriate aa-tRNA remain in the A-site longer. The delay allows, or triggers, the hydrolysis of the GTP in the ternary complex, which is followed by dissociation of phosphate and EF-Tu · GDP. This leaves the aa-tRNA in the A-site, which now may well take on a conformation different from that which originally bound the larger ternary complex. Before the aa-tRNA is ready to accept the growing peptide, the amino acid must migrate to the *3'-hydroxyl* of the tRNA's terminal adenosine.

Mistakes can occur in protein synthesis by statistical fluctuations in the lifetimes of erroneous aminoacyl-tRNAs in the A-site. Occasionally one remains long enough to trigger the hydrolysis of GTP and the dissociation of P_i and EF-Tu · GDP. Afterward, there is a certain risk that the peptide will be transferred to the inappropriate aa-tRNA and an erroneous amino acid incorporated. Before this happens, however, the inappropriate aminoacyl-tRNA may dissociate and diffuse away from the ribosome. (The need to move the amino acid to the 3'-hydroxyl of the tRNA's terminal adenosine could provide a delay to make this dissociation more likely.) The dissociation of an aa-tRNA from the ribosome is kinetically irreversible, since it would lead almost immediately to the formation of another complex with EF-Tu · GTP. To reenter the A-site and become ready again to accept a peptide, a ternary complex must be formed and another GTP must be hydrolyzed. It has been found that inappropriate aminoacyl-tRNAs dissociate preferentially from the ribosome, a process called *proofreading*. Thus there are two tRNA selection steps at the A-site: the initial binding of ternary complexes and the proofreading of uncomplexed aa-tRNA. This latter is another example of kinetic proofreading; on the average, more than one GTP is hydrolyzed during attempts to incorporate an erroneous aa-tRNA, and cellular energy is thus spent to increase the accuracy of protein synthesis.

Let us pause to consider how some of these properties were established by experiment. The elongation factor Tu can be bound and retained on a nitrocellulose filter as the binary EF-Tu · GTP complex. The ternary complex with aa-tRNA is not retained on the filter. This property and the ability of EF-Tu to stimulate the

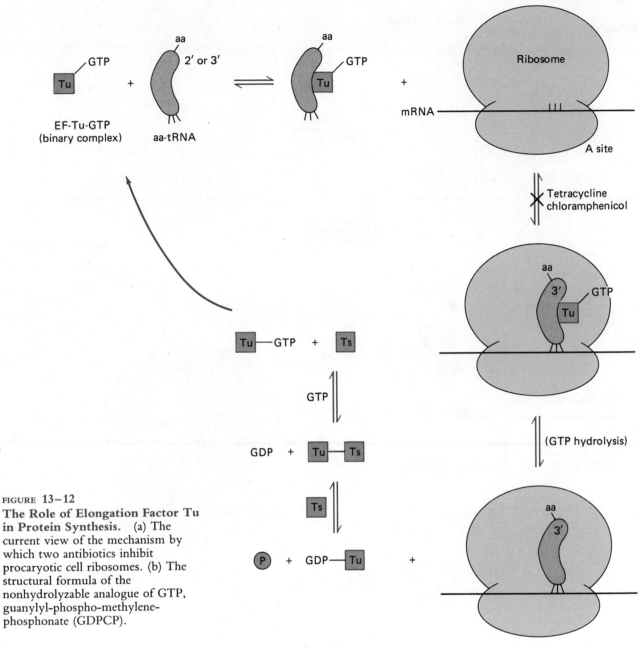

FIGURE 13–12
The Role of Elongation Factor Tu in Protein Synthesis. (a) The current view of the mechanism by which two antibiotics inhibit procaryotic cell ribosomes. (b) The structural formula of the nonhydrolyzable analogue of GTP, guanylyl-phospho-methylene-phosphonate (GDPCP).

(a)

GDPCP

(b)

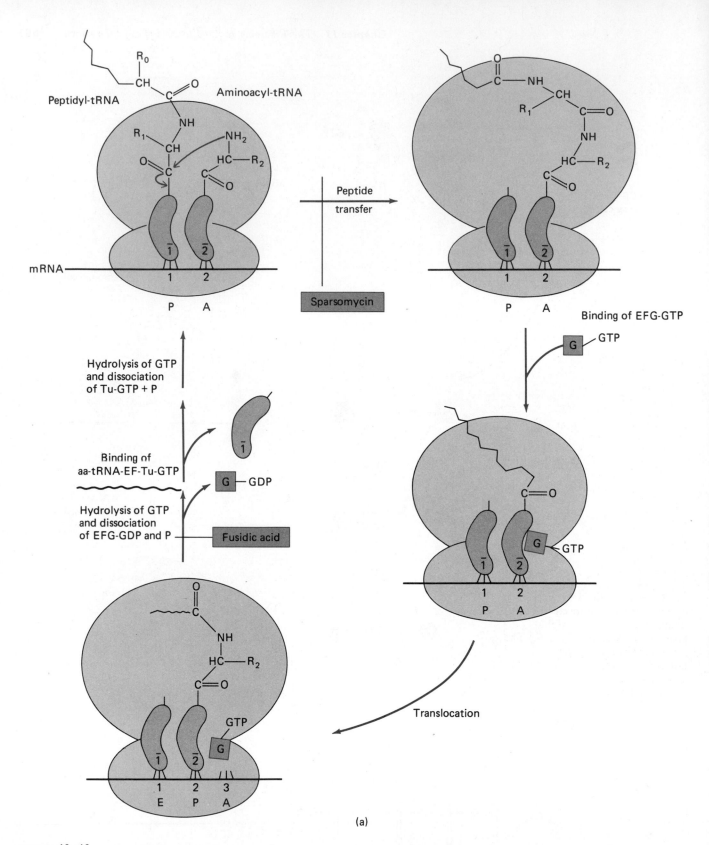

(a)

FIGURE 13–13

Peptide Elongation and Translocation of Peptidyl-tRNA on Bacterial Ribosomes. (a) The P- and A-sites of the ribosome are indicated. The three codons of the mRNA (1,2,3) are located in the small ribosomal subunit. The L-shaped tRNAs have anticodons designated $\bar{1}$, for example, for the tRNA appropriate for binding to codon 1. The exit site (E-site) is proposed by K. H. Nierhaus to be occupied by the deacylated (former peptidyl-) tRNA after translocation. This site loses its tRNA when aa-tRNA binds to the A-site. This step marks the beginning of a new elongation cycle (beyond the wavy black line). (b) The chemical structure of puromycin. Compare with the 3′ end of tyrosyl-tRNA.

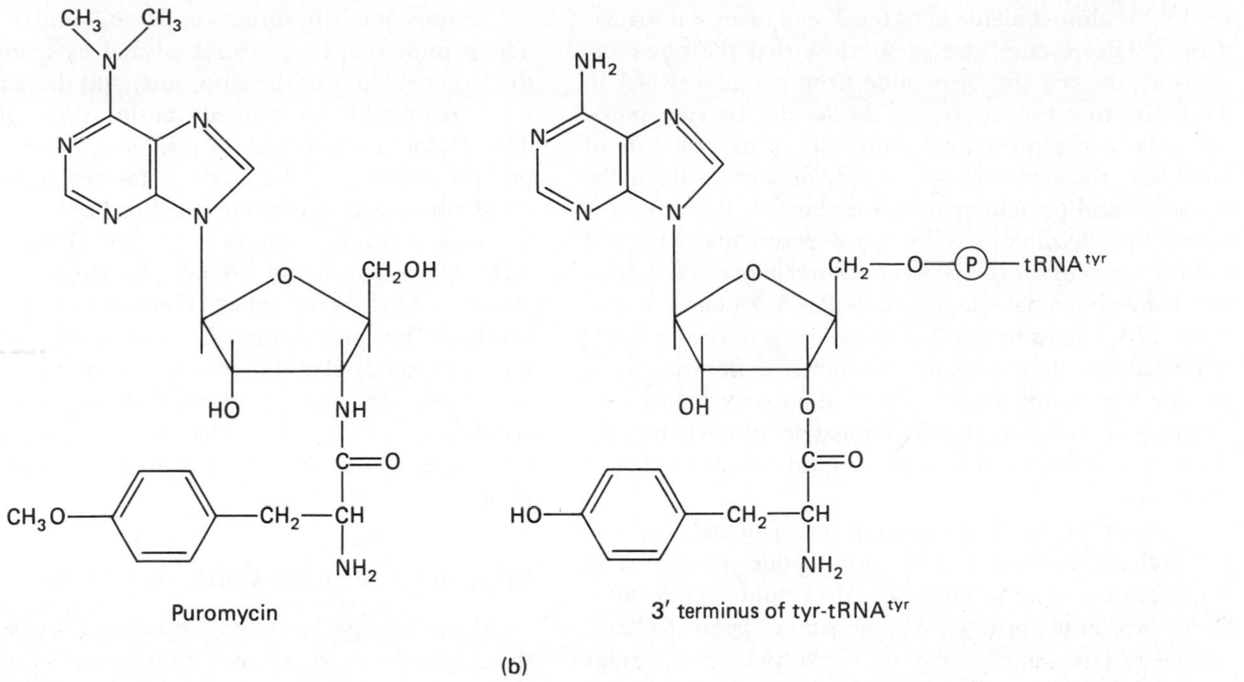

Puromycin   3' terminus of tyr-tRNA^{tyr}

(b)

FIGURE 13–13   (continued)

cell-free synthesis of protein provided assays for the purification of the elongation factor. Once purified and radiolabeled, it was possible to establish the kinetics and specificity of binding to GTP, to various other factors, and to aa-tRNAs. Thus we know that all aminoacyl-tRNAs from *E. coli* form complexes with EF-Tu · GTP, even when it is derived from other bacterial species. Hydrolysis of the terminal phosphate must precede dissociation of EF-Tu · GDP from the ribosome. This was shown by using a *nonhydrolyzable analogue* of GTP (*GDPCP,* guanylyl-phospho-methylene-phosphonate), which can form the ternary complex and bind to the ribosome but which cannot dissociate as a binary complex.

To restore the EF-Tu · GTP complex in *E. coli* requires two steps: (1) displacement of the GDP from EF-Tu by another protein elongation factor (EF-Ts, which is *T*emperature *s*table) to form EF-TU · EF-Ts; this is followed by (2) displacement of the EF-Ts by GTP (to form EF-Tu · GTP, see Figure 13–12).

In eucaryotes, there seems to be only one factor involved at this stage of elongation, called eEF-1 (eucaryotic *E*longation *F*actor). This factor is composed of more than one protein and probably combines the functions of the procaryotic EF-Tu and EF-Ts.

Peptide Transfer

The ribosome has two functionally distinguishable sites, each of which contains both mRNA and tRNA: the A-site discussed above and the P-site, in which lies a *peptidyl-tRNA* (a tRNA covalently bound to the C-

terminus of a peptide—Figure 13–13a). This latter site is sometimes described as the "donor site" since peptides can be donated only from peptidyl-tRNA in the P-site to aminoacyl-tRNA in the A-site. Growth of the protein chain occurs by such *peptide transfer,* a chemical reaction catalyzed by the *peptide transferase center* of the large subunit of the ribosome. The result is a peptidyl-tRNA, longer by one amino acid residue, in the A-site of the ribosome. No additional energy is required for the peptide transfer; the peptide is activated already by virtue of being part of a peptidyl-tRNA. Since the peptide is transferred to the aminoacyl-tRNA, the direction of protein chain growth is from N-terminal to C-terminal. By careful consideration of Figure 13–13a one can see that peptide transfer to aminoacyl-tRNA is the reaction where the flows of mass, energy, and information merge in the synthesis of a protein.

We have several times before pointed out the usefulness of specific inhibitors in unraveling the complexities of metabolic pathways. The most helpful of these in the study of protein synthesis have been certain of the compounds known as antibiotics. The sites of action of two of these, chloramphenicol and tetracycline, are shown in Figure 13–12a, and we now wish to discuss two others, *puromycin* and *sparsomycin*. Puromycin and sparsomycin affect both procaryotic and eucaryotic ribosomes, so they can be used to investigate common features of the mechanism of protein synthesis. Sparsomycin blocks the peptide transfer step and freezes ribosomes with a peptidyl-tRNA in the P-site.

The chemical structure of puromycin (Figure

13–13b) is almost identical to the 3′ end of an aminoacyl-tRNA. The resemblance is so close that the ribosome catalyzes the transfer of peptide from peptidyl-tRNA in the P-site to a puromycin in the A-site. Because puromycin lacks the substantial remainder of the structure of aa-tRNA, the peptidyl-puromycin dissociates from the ribosome and protein synthesis is aborted. Puromycin is useful to cell biologists in the experimental study of cell-free protein synthesis: the puromycin reaction defines to which ribosomal site peptidyl-tRNA is bound. Peptidyl-tRNA bound to the A-site, where it normally exists immediately after peptide transfer, cannot donate a peptide to puromycin. The ability of puromycin to accept a peptide is, therefore, the functional definition (currently the only simple definition) of a peptidyl-tRNA being in the P-site.

Until recently it was thought that if an inappropriate peptidyl-tRNA were formed, by peptide transfer to an inappropriate aminoacyl-tRNA, the resulting error must be retained in the protein. The ribosome appears to have, however, a second chance to ensure accuracy at this stage of protein synthesis. This follows from the observation that peptidyl-tRNA can dissociate from the ribosome during synthesis. A dissociated peptidyl-tRNA is rapidly split into a peptide and an unacylated tRNA (a reaction catalyzed by *peptidyl-tRNA hydrolase*), so the dissociation is sufficient to abort protein synthesis. If the dissociation occurs preferentially to inappropriate peptidyl-tRNAs, those whose structures do not correctly complement the mRNA in the A- or P-sites, fewer errors will be found in completed proteins. This process is called *ribosome editing*.

Translocation of the Peptidyl-tRNA

The restoration of the ribosome to a state capable of continuing peptide transfer requires the participation of another elongation factor, EF-G (for GTP). This common protein (3–6% of *E. coli* soluble protein) binds to the ribosome as a binary complex with GTP, EF-G · GTP. In a way not presently understood, this allows or catalyzes the movement, the *translocation,* of the new peptidyl-tRNA from the A-site to the P-site and the deacylated tRNA to an exit site (E-site) (Figure 13–13a). Simultaneously, the mRNA moves so as to maintain decoding interactions with the peptidyl-tRNA and deacylated tRNA and to bring the next codon into the A-site. Hydrolysis of the GTP then occurs, followed by the dissociation of an EF-G · GDP complex (a process blocked by another antibiotic, *fusidic acid*). Again, the need for hydrolysis of GTP before dissociation of the elongation factor was demonstrated by observing that undissociated EF-G accumulated on ribosomes when nonhydrolyzable GDPCP was substituted for GTP in cell-free reactions. The deacylated tRNA dissociates from the E-site when a new aa-tRNA binds to the A-site.

Eucaryotic ribosomes undergo similar processes. The peptide transfer is also catalyzed by components of the larger subunit of the ribosome, and there is a protein factor required for continued translocation, called eEF-2. This factor is the target of *diphtheria toxin.* The active protein subunit of the toxin catalyzes transfer of the ADP-ribose part of enzyme cofactor NAD^+ (see Chapter 6) to a modified amino acid on eEF-2. With the ADP-ribose covalently bound, the elongation factor is inactive. Diphtheria toxin, therefore, poisons protein synthesis by preventing the action of eEF-2 in the translocation of peptidyl-tRNA. Eucaryotic cytoplasmic ribosomes can also be blocked during elongation by *cycloheximide,* an antibiotic that does not affect procaryote ribosomes or the ribosomes of mitochondria or chloroplasts.

Initiation of Protein Synthesis

Now that we know how proteins grow and some of the mechanics involved, it is appropriate to ask how this process begins. In fact the process of initiation is quite complicated and its final description is not yet in hand. From what we have discussed already there are two questions to be answered: (a) What is transferred to the aminoacyl-tRNA in the ribosome A-site during the first elongation step? and (b) How is the correct beginning point of the mRNA chosen?

The Initiating tRNA

Let us take these in turn. The experiments that first shed light on the initiating mechanics were done by K. Marcker and Sanger in 1963. They showed that *E. coli* cells contain a special kind of methionine-accepting tRNA, one acylated with *N-formyl-methionine* (Figure 13–14). The formyl-methionine group resembles a minimal kind of peptide, and it was immediately suggested that N-formyl-met-tRNA might act to initiate protein synthesis. That turned out to be the case: initially, all polypeptides in procaryotes, and probably in mitochondria and chloroplasts too, begin with N-formyl-met. The N-formyl-met-tRNA is positioned at the ribosome P-site at the beginning of the first elongation epicycle, and formyl-methionine is transferred to the first aminoacyl-tRNA in the A-site.

The tRNA that participates in protein chain initiation is specific for the purpose. It is denoted $tRNA^{FMet}$ to distinguish it from $tRNA^{Met}$, which carries methionine to interior positions of the protein chain. The sequences of these two tRNAs are different in all the cells where they have been compared. In *E. coli* a single ligase acylates both with methionine but only met-$tRNA^{FMet}$ can be formylated, by a special transformylase. Also, met-$tRNA^{FMet}$ will not form a complex with EF-Tu · GTP. Thus the cell keeps the two tRNAs separated in their

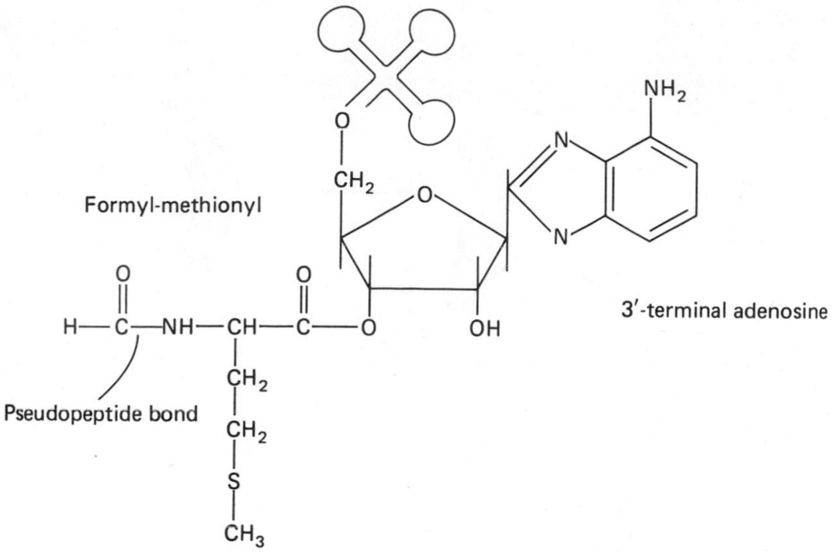

Formyl-methionyl

Pseudopeptide bond

$3'$-terminal adenosine

FIGURE **13-14**
**The $3'$ Terminus of
Formyl-Methionine-tRNA^{FMet}.**
This structure shows the resemblance
to a peptidyl-tRNA.

functional activities. Another difference in the two methionine-specific tRNAs is their decoding specificity. Met-tRNA^{Met} is quite specific at decoding AUG in the mRNA. Met-tRNA^{FMet}, on the other hand, is less specific and will recognize GUG and even UUG and AUA in initiation signals, despite having an anticodon sequence identical to that of tRNA^{Met}. This again illustrates the point that the tRNA structure has features other than the anticodon that control codon recognition.

Although EF-Tu · GTP does not form a complex with the procaryotic initiator tRNA, there is another protein factor that does; it is the *Initiation Factor 2* (IF-2). GTP may be a part of the complex: formyl-met-tRNA^{FMet} · IF-2 · GTP (Figure 13–15).

Eucaryotic cells also have a special initiator tRNA that accepts methionine, but there seems to be no transformylase activity in their cytoplasm and so protein synthesis begins with met-tRNA$_F^{Met}$ in the P-site. (The subscript "F" differentiates the initiator tRNA from the "M" molecule that holds methionine destined for the interior of the protein, and it also reflects the fact that some eucaryotic met-tRNA$_F^{Met}$ molecules can actually be formylated by the *E. coli* transformylase.) The eucaryotic initiator tRNA is very specific for the triplet AUG. There also are protein initiation factors analogous to the procaryotic IF-2. The details of initiation in eucaryotic cells are still being worked out.

The Initiation Signal

The selection of the mRNA region at which to begin protein synthesis is a complicated process, involving the small (30S) subunit of the ribosome, protein factors, and possibly the Fmet-tRNA · IF-2 complex. The diagram in Figure 13–15 illustrates one of several possible schemes.

The 30S ribosomal subunit, an initiation factor (IF-3), and the mRNA all participate in the choice of the correct beginning point. There are at least two ribosomal proteins that are involved, S1 and S12 ("S" means Small subunit; see the discussion of ribosome structure below). These proteins are believed to act in part by *stabilizing base pairing* between the mRNA and a short sequence of bases near the $3'$ end of the 16S ribosomal RNA (Figure 13–16). This mechanism of *messenger selection* was first suggested by J. Shine and L. Dalgarno in 1975. Since then, sequences have been found in many mRNAs that are both near the AUG (or GUG) codon that signals the beginning of translation, with formyl-met-tRNA, and can make several base pairs with the special sequence near the end of 16S rRNA (the *Shine-Dalgarno sequence*). Some of these initiator sequences in the mRNA were identified by J. Steitz and her collaborators after isolating that part of the mRNA in an initiation complex that is protected by the ribosome from nuclease digestion. Although there is much evidence to corroborate the importance of the 16S rRNA in messenger selection, the precise roles of the ribosomal proteins and initiation factors are still being sorted out. The model seen in Figure 13–15 must therefore be regarded as tentative.

The large 50S ribosomal subunit is joined to the *30S initiation complex* (30S · IF3 · mRNA · Fmet-tRNA · IF2 · GTP) to form a *70S initiation complex,* a process aided by the participation of another protein initiation factor (IF-1). Upon the formation of the 70S initiation complex, hydrolysis of GTP occurs and the initiation factors leave the ribosome. The result of this sophisticated molecular choreography is a ribosome with a "peptidyl"-tRNA in the P-site that is poised to donate its Fmet to the first aminoacyl-tRNA, thus beginning elongation of the protein specified by the mRNA (see Figure 13–15).

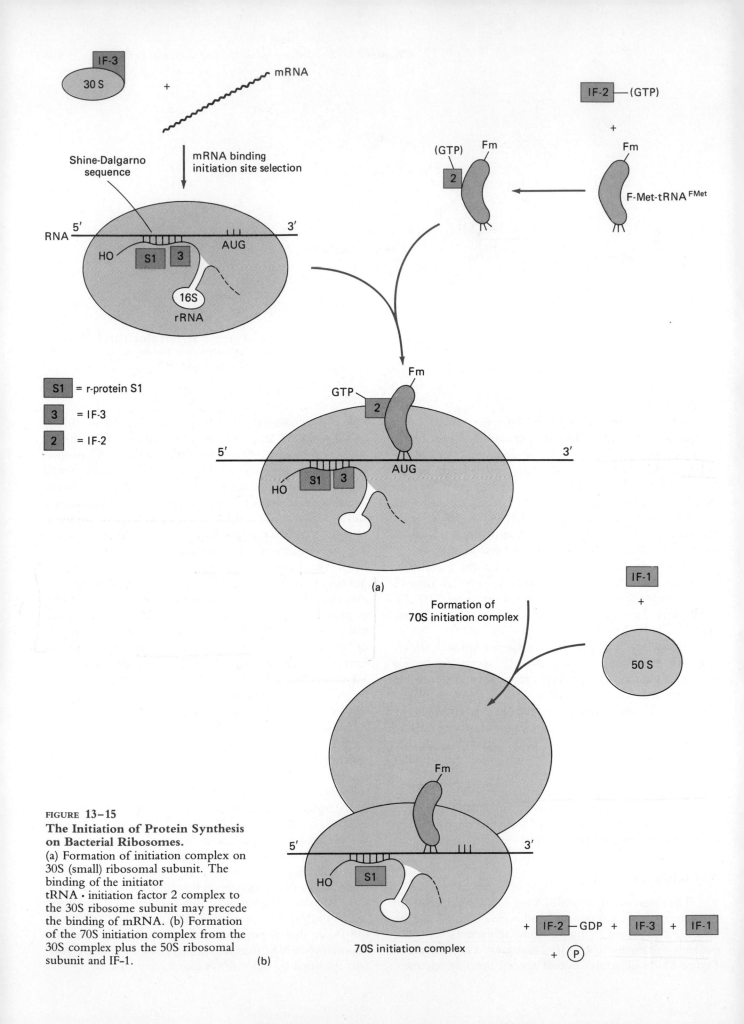

FIGURE 13–15
The Initiation of Protein Synthesis on Bacterial Ribosomes.
(a) Formation of initiation complex on 30S (small) ribosomal subunit. The binding of the initiator
tRNA · initiation factor 2 complex to the 30S ribosome subunit may precede the binding of mRNA. (b) Formation of the 70S initiation complex from the 30S complex plus the 50S ribosomal subunit and IF–1.

S1 = r-protein S1

3 = IF-3

2 = IF-2

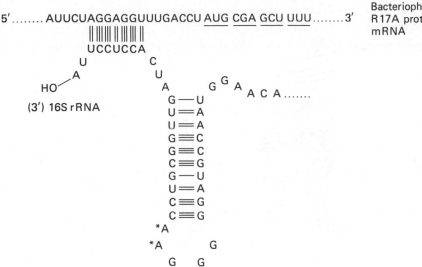

FIGURE 13-16
Postulated Base Pairing Between the Sequence at the 3' Terminus of the 16S rRNA (the Shine-Dalgarno Sequence) and an Initiator Region of a Bacteriophage mRNA. The first four amino acid codons are underlined. ★A is a modified A residue. (Redrawn with permission from J. A. Steitz and K. Jakes.)

You may be wondering whether all proteins isolated from *E. coli* and other procaryotes begin with formyl-methionine. That is not the case. The formyl group, and usually the entire formyl-methionine, is removed from proteins, often before their synthesis is complete. Thus the evidence for a universal procaryotic initiator tRNA was concealed from the investigators who examined the N-terminal sequences of proteins. This *posttranslational processing* also occurs in eucaryotes, where the N-terminal methionine is almost always removed. We shall return to other kinds of processing a little farther on in this chapter.

In eucaryotic cells there is no convincing evidence for an initiation signal in the mRNA that recognizes an analogue to the Shine-Dalgarno sequence. Instead, a preinitiation complex that includes the 40S ribosome subunit binds to the 5' end of the mRNA, aided by the presence of the cap structure (see Chapter 12). The complex then moves down ("scans") the mRNA until it reaches the first AUG triplet, assembles the remainder of the initiation complex, and initiates. The initiation process in eucaryotic cells is much more complicated than in procaryotic cells—it involves many more initiation factors—and is being intensively studied.

Termination of Protein Synthesis

Once begun, elongation of the growing protein chain continues until a special signal is reached. The *termination signal* is composed of one of the termination codons, UAG (amber), UAA (ochre), or UGA (opal), and perhaps additional bases. (At the ends of the message there are in principle no requirements of phase, so initiation and termination signals may be longer than three bases.)

The whimsical color designations for the termination codons were assigned by bacteriophage geneticists before the chemical nature of the code was understood.

The names were chosen so as not to prejudice thinking about the likely mechanism underlying the phenotypic effects of certain mutations. These mutations, which were extraordinarily severe in their effects, were also called *nonsense,* both to distinguish them from missense mutations involving the substitution of one amino acid for another and to denote their special ability to be suppressed by mutations in other genes. To summarize briefly a good deal of hard work by S. Benzer, Brenner, A. Garen, and many others, the suppressor mutations turned out to be in genes specifying the structure of tRNAs. In most cases the mutation changed the anti-codon of the tRNA so that it could recognize one of the termination codons. Thus when a nonsense mutation created a termination signal in the middle of a protein message, the aminoacylated *suppressing tRNA* could, with a certain probability, bind in the ribosome A-site and accept the transfer of the growing peptide. The completed peptide would usually have the wrong amino acid substituted at the point of suppression, but a slightly erroneous enzyme usually has far more activity than does the fragment produced when the termination codon is correctly translated.

The nature of the nonsense-suppressing molecule, a tRNA, was known before the mechanism of protein chain termination was worked out. It was at first thought there might be a special chain-terminating tRNA, with which the suppressor tRNAs would compete for the ribosome A-site during suppression. As it turned out, only protein *Release Factors* (RF-1 and RF-2 in *E. coli*) are associated with codon-specific termination events. The two factors in bacteria incompletely share termination codon specificity: RF-1 is associated with UAG- and UAA-specific termination, and RF-2 is associated with UGA- and UAA-specific termination.

It was natural to assume that the release factors are themselves responsible for recognizing the codons.

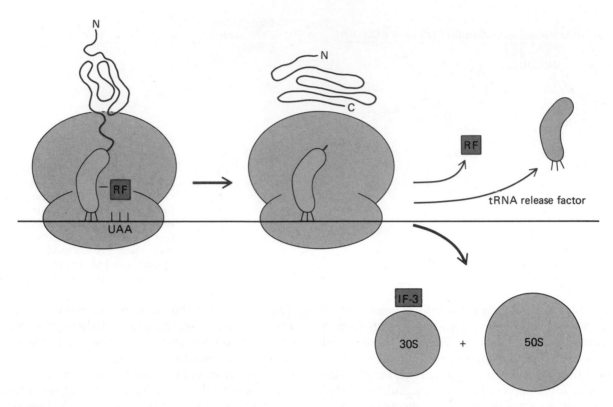

FIGURE **13–17**
The Termination of Protein Synthesis. The termination codon is recognized by a protein release factor (RF), leading to hydrolysis of peptidyl-tRNA and the release of the completed protein. Later the deacylated tRNA dissociates (perhaps aided by another factor), the ribosome subunits separate (perhaps aided by IF-1, and IF-3 binds to the free 30S subunit to stabilize the separation prior to initiating synthesis of the next protein.

Recent studies by E. J. Murgola and A. E. Dahlberg make that seem unlikely. Instead, there is strong reason to believe that the 16S ribosomal RNA recognizes the chain termination triplets. By making base pairs with complementary sequences, the rRNA probably changes its tertiary structure, providing a binding site for RF-1 or RF-2. This conformation, doubtless different from the one resulting from aminoacyl-tRNA binding in the A-site, promotes the chain termination chemistry.

In eucaryotes there is apparently only one release factor (eRF-1) associated with all three termination codons. Also in eucaryotes, the termination step involves the hydrolysis of GTP, a requirement not yet demonstrated in procaryotes.

There is no evidence that release factors themselves perform protein chain termination; instead, they somehow trigger the ribosome's peptide transferase to hydrolyze the peptidyl-tRNA ester linkage. (This is formally equivalent to transferring the peptide from tRNA to —OH, Figure 13–17). The deacylated tRNA, with the help of a tRNA release factor protein, then dissociates from the ribosome.

The Ribosome Cycle

After the termination of protein synthesis, the ribosome subunits separate from one another. In procaryotes the 30S subunit binds to IF-3, which stabilizes the separation. This helps provide free 30S subunits to participate in the initiation phase of protein synthesis. There is free mixing of the large and small subunits as they separate and recombine between rounds of protein synthesis, a fact established by R. Kaempfer with density-labeled ribosome subunits. After shifting the cells growing in medium containing the dense isotope of nitrogen (^{15}N) to medium containing the usual isotope (^{14}N), Kaempfer showed that *hybrid ribosomes* of the intermediate density were quickly formed as the old (heavy) and new (light) subunits recombined.

Our diagrams of protein synthesis may have given the impression that single ribosomes are the usual form found in cells synthesizing protein. The protein-forming apparatus actually consists of clusters or chains of ribosomes, as can be shown by the following simple experiment. A short pulse of a radioactive amino acid is provided either to cells or to an in vitro system, and an extract is then zone centrifuged through a sucrose density gradient to determine the size of the structures with which the newly formed protein is associated. As Figure 13–18 shows, most of the radioactivity is not associated with single ribosomes, but rather with heavier structures that sediment faster. These larger structures, called *polysomes*, consist of trains of ribosomes engaged in the translation of individual messenger RNA molecules. There are essentially two lines of evidence for this interpretation. First, the size of the polysomes has been shown to reflect the size of the messenger RNA being translated. In the case of the bacterial enzyme β-galactosidase, with 1100 amino acids, the polysomes normally contain about 50 individual ribosomes. However, when the length of the messenger RNA is reduced because of a deletion in the gene (see Chapter 9), then the size of the polysomes, and the number of ribosomes they contain, is reduced in proportion.

More decisive evidence for this model comes from some elegant work in electron microscopy. Figure 13–19a shows a highly magnified picture of liver ribosomes, indicating the shapes and arrangements of the large and small subunits and the groove between them, while Figure 13–19b shows what is thought to be the polysome structure, with ribosomes linked together on a messenger RNA strand. Figure 13–20 shows the simultaneous synthesis on a specific portion of the *E. coli* genome of

mRNA by RNA polymerase (transcription) and reading of this mRNA by ribosomes to produce proteins (translation). It is evident that the polysomes increase in size as the length of the mRNA increases (see also Figure 12–22).

The synthesis of proteins in eucaryotic cells also occurs on polysomes, although there is not the linkage of transcription and translation shown in Figure 13–20; the mRNA of eucaryotic cells must leave the nucleus before it can be translated. Indeed, the nuclear membrane may act to delay translation until the complex processing of mRNA is completed, as has been suggested by J. Cairns. Just how the mRNA penetrates the nuclear envelope is not as yet understood. There is some evidence that mature mRNA associates with proteins—perhaps ribosomal proteins—before being translocated through the nuclear membranes. The antibiotic *cordycepin* reduces the appearance of new mRNAs in the cytoplasm and also blocks the formation of the poly A sequence on the 3′ end of eucaryote mRNAs (see Chapter 12). For this reason it has been postulated that the function of the poly A sequence is to aid in transport of mRNA out of the nucleus. Against this idea is the fact that the mRNAs for the histones (the primary protein constitutent of chromosomes) do not have poly A sequences and yet seem to be translated normally after being translocated into the cytoplasm. Also, mammalian mitochondrial mRNAs have poly A tails but need go nowhere else to be translated.

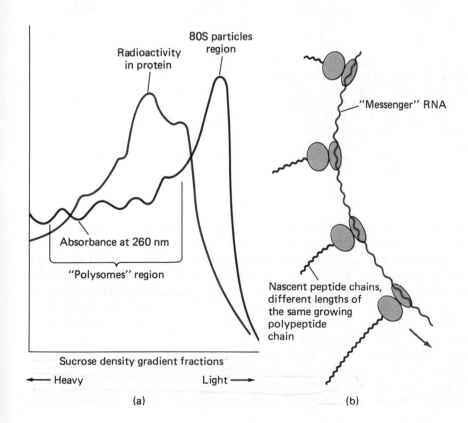

Radioactivity in protein

80S particles region

"Messenger" RNA

Absorbance at 260 nm

"Polysomes" region

Nascent peptide chains, different lengths of the same growing polypeptide chain

Sucrose density gradient fractions

← Heavy   Light →

(a)   (b)

FIGURE 13–18
Evidence for Protein Synthesis by Polysomes. (a) Liver cells were incubated for a short time in the presence of radioactive amino acids, then broken open, and the contents sedimented as zones through a sucrose density gradient (see Chapter 2). The bulk of the radioactivity sediments faster than 80S ribosomes, detected by their absorbance at 260 nm. (Data from P. Siekevitz). (b) The polysome model, in which several ribosomes translate simultaneously a single mRNA.

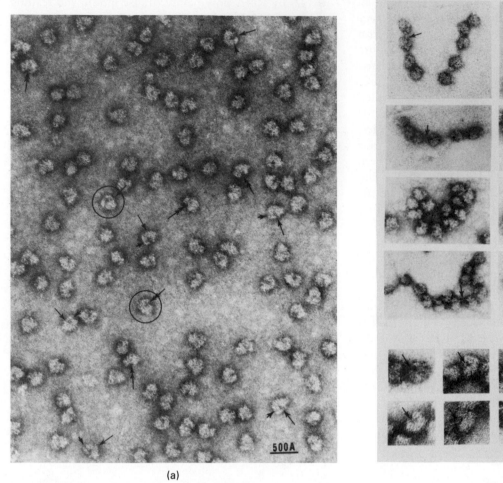

(a)

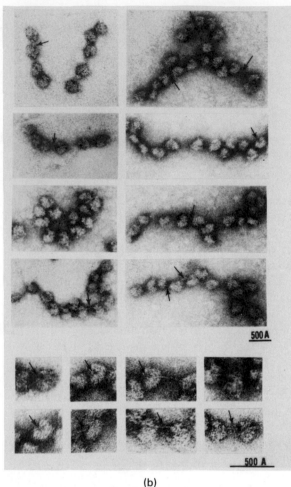

(b)

FIGURE **13–19**
High-Resolution Electron Micrograph Images of Liver Ribosomes.   (a) A
field of monomeric ribosomes, negatively stained, illustrating the separation (dark
groove) between large and small subunits (long arrow) and the groove in the small
subunit (short arrow). The circle with single arrow encloses a right lateral image
and the circle with double arrow encloses a left lateral image. (b) A field of
polysomes, with arrows indicating the strand (presumably messenger RNA)
between the ribosomes. All the samples are negatively stained with
phosphotungstic acid, so that the lighter ribosomes stand out against a darker
background. (Courtesy of Y. Nonomura, G. Blobel, and D. Sabatini.)

Another fundamental difference between procaryotic and eucaryotic cell mRNAs is their information content. Transcription in *E. coli* often produces an mRNA containing information for more than one protein, a *polycistronic message* (see Figure 12–4). This happens rarely, if ever, in eucaryotes.

Finally, the 5′ end of eucaryotic mRNAs has an unusual methylated G structure (see Chapter 12). The presence of this "cap" is thought by some to control the lifetime of the mRNA; when the cap is removed it may expose the rest of the message to exonucleases. The cap also aids in the translation of messages, probably at the initiation stage. The presence of an analog of the cap structure, methylated GDP, has been reported to inhibit the formation of initiation complexes on 40S ribosomal subunits. A *cap binding protein* (CBP) has been detected in the cytoplasm of eucaryotic cells. Preparation of antibodies that inhibit the function of CBP should allow a finer analysis of the role played by the cap in protein synthesis.

Taken together, the poly A "trailer" and the methylated G cap with its associated 5′ "leader" represent noncoding structures in the sequence of the eucaryotic mRNAs. In procaryotic messages there are also noncod-

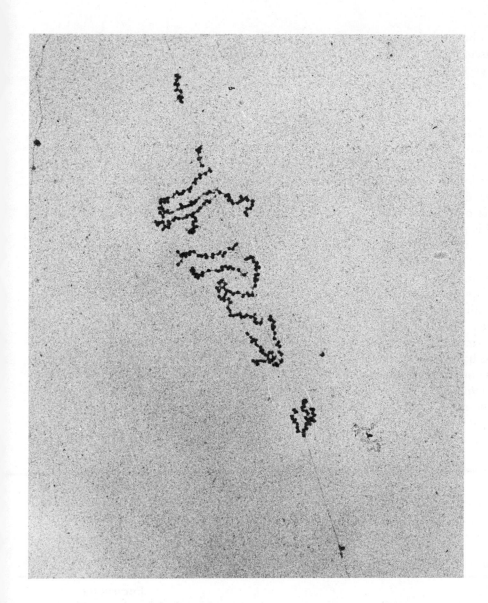

FIGURE 13—20
**Electron Micrograph of an *E. coli*
DNA Fiber.** This micrograph shows
increasing numbers of ribosomes on
increasing lengths of messenger RNAs.
×200,000. (Courtesy of O. Miller.)

ing sequences: leaders, trailers, and the nucleotides between the coding sequences of a polycistronic mRNA. With the advent of easier techniques for determining nucleic acid sequences will come a greater knowledge of the kinds of noncoding sequences. The ones at the 5′ end are surely involved at least in protein chain initiation and the ones at the 3′ end perhaps in protein chain termination. But sequence knowledge will have to be combined with a more sophisticated understanding of the mechanics of protein synthesis before we can claim to have penetrated all the mysteries of the mRNA.

The Structure of the Ribosome

Historically, the ribosome was treated as a passive structure on which the processes of protein elongation occur, the peptidyl-tRNA and aminoacyl-tRNA taking the principal roles, with the various protein factors and GTP being ancillary. In fact the ribosome undoubtedly is an active participant in all the important functions of protein biosynthesis. We have already discussed the role of the 16S rRNA and some ribosomal proteins in the selection of the region on the mRNA at which to begin translation and the catalytic activity responsible for the transfer of the growing peptide to aminoacyl-tRNA. We wish now to present our current view of the structure of this ubiquitous cellular organelle. This will be done with a description of the methods used to reach the present model because the experimental methods developed to describe the ribosome's structure are sure to be used in exploring even more complicated parts of the cell.

Ribosomes are found in the cytoplasm of all cells that synthesize protein and also in the mitochondria and chloroplasts of eucaryotic cells (see Chapter 15). As we mentioned before, they are sometimes associated with membranous structures and sometimes lie freely in the cytoplasm. The functional differences between free and membrane-bound ribosomes are discussed in a later section. Because of their relatively larger size and greater density, ribosomes are easily purified from other components of the postmitochondrial supernatant by differential sedimentation. Treatment of membranes with mild detergents frees the membrane-bound ribosomes.

All ribosomes are composed of two ribonucleoprotein *subunits* that can become dissociated from one another in solutions of low magnesium ion concentration (see Figures 13–3 and 13–19; Table 13–5). The subunits can then be separated from one another by virtue of their different sizes; the usual method used is zone sedimentation through sucrose gradients. We have already indicated in a pictorial fashion the different functional roles played by the large and small subunits of the ribosomes. This knowledge was obtained by attempting to carry out the various subreactions of protein synthesis with one or the other purified ribosomal subunits. This is another example of the general biochemical strategy of reconstruction experimentation. Once one has an assay, the system can be divided in various ways until the functional abilities of the subdivisions are known. Then further subdivisions can be made to locate the sites of subreactions, and so on. Carrying this approach to its extreme is the ultimate goal of the study of ribosome structure.

Composition

The ribosomal subunits are composed of RNAs and proteins *(r-proteins),* which are held together by nonco-valent bonds. The rRNAs were obtained first but now techniques are available for the separation and purification of all the components of the subunits of procaryotic ribosomes, and the work on eucaryotic subunits is going ahead rapidly. This work progressed in stages, from cruder separations to the more finely detailed resolution of all the ribosomal proteins by a single technique. First came the removal of the rRNA from the protein by *phenol extraction,* the vigorous shaking of the purified subunits in the presence of a two-phase system composed of liquid phenol and an aqueous buffer; the two phases are separated by centrifugation. This is a traditional way of isolating nucleic acids, which dissolve in the aqueous phase, from proteins, which denature and tend to lie at the interface between the two phases. A gentler method of separating the two macromolecular components is to suspend the ribosomal subunit in a concentrated solution of a salt such as LiCl (Figure 13–21). Certain LiCl concentrations split away some of the proteins *(split proteins),* leaving a *core* of proteins still associated with the rRNA. Thus was a substructure of the subunit inferred. Those proteins that remained with the RNA were hypothesized to lie next to the rRNA, where they would make more contacts with it and be harder to split away by a given salt concentration.

Purifying the individual members of a small population of proteins is easier than dealing with all the r-proteins, and the usual ion exchange chromatography methods produced enough material to attempt reconstruction of the ribosomal subunits from purified split proteins and the cores. In some cases, omitting one of the

TABLE 13–5

Composition and Properties of Ribosomes, Their Subunits, and Their RNA and Protein

	Procaryotes		Eucaryotes (In Cytoplasmic Matrix)	
	Mol Wt	Sedimentation Value	Mol Wt	Sedimentation Value
Complete ribosome	2.5×10^6	70S	4.8×10^6	80S
Large subunit	1.6×10^6	50S	3.2×10^6	60S
Small subunit	0.9×10^6	30S	1.6×10^6	40S
RNA (1/2 to 2/3 Mass)				
Large subunit	1.2×10^6 2904 nucleotides *(E. coli)*	23S	1.6×10^6	28S
	0.03×10^6 120 nucleotides *(E. coli)*	5S	0.03×10^6	5S
Small subunit	0.6×10^6 1542 nucleotides *(E. coli)*	16S	0.9×10^6	18S
Proteins (1/3 to 1/2 Mass)				
Large subunit	34 (L-proteins)		?	
Small subunit	21 (S-proteins)		?	
Different kinds in complete ribosome	52 *(E. coli)*		75–90	
Dimensions (Approximate)				
Large subunit	23-nm diameter		23-nm diameter	
Small subunit	23×11 nm		9-nm diameter	

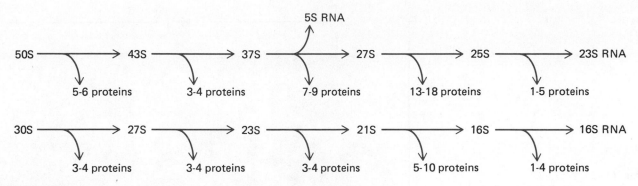

FIGURE 13−21
Progressive Degradation of Bacterial 50S and 30S Subribosomal Particles.
By using EDTA, urea, and increasing concentrations of monovalent salts, usually
CsCl or LiCl, progressively smaller particles are produced that have lost some of
the ribosomal proteins. The pathway followed during this degradation is indicated
by the S values of the isolated subparticles. At the end, only the naked 23S and
16S RNAs remain. (After A. S. Spirin.)

proteins prevented the reassociation of others. The
pattern of interdependencies that ultimately emerged was
summarized in an *assembly map* (Figure 13–22) by M.
Nomura, whose group did most of the experiments. The
map indicates those r-proteins that must be present before
other r-proteins can be added to form a larger structure.
The assembly is shown as a hierarchy: some of the
proteins interact first with the rRNA and others assemble
in the structure afterward. It is tempting to believe that
the map is telling us a great deal about how the ribosome
is assembled in vivo. The reconstruction process turns
out not to require any macromolecular components other

than those found in the final structure; it is, therefore, a
self-assembly process. We shall return to this subject in
Chapter 15.

The assembly map of the 16S subunit of *E. coli* refers
to the various proteins by an internationally agreed
nomenclature based on the positions of the proteins after
a two-dimensional separation (Figure 13–23). The first
dimension is an electrophoresis done in a 4% polyacrylamide gel that has been cast in a glass tube. After the
proteins are separated into groups according to their
mobility at pH 5.5, the cylindrical gel is removed from
the tube and laid along the top of a slab of polyacrylamide

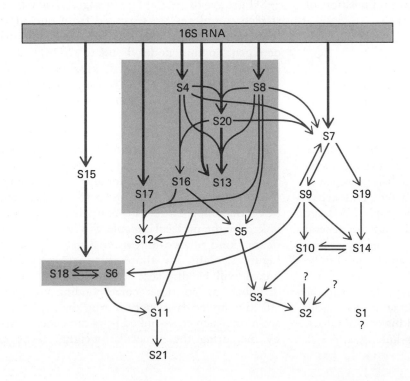

FIGURE 13−22
**Assembly Map of Proteins in the
Small (30S) Ribosomal Subunit of
E. Coli.** The terms S7, etc., refer to
*s*mall subunit proteins. Seven proteins,
S15, S17, S16, S20, S8, S13, and S7,
bind directly (thick arrows) to the 16S
RNA. After this primary binding step,
other secondary proteins—for example,
S16 or S9—begin to bind. Some
proteins, such as S3 and S10, bind to
secondary proteins. Thin arrows
between proteins refer to the
facilitation of the binding of one
protein by another; thus, S8 aids in
binding S20, S13, S12, S5, and S7.
The binding properties of S1 and S2
are unknown. (After M. Nomura and
H.-G. Wittman.)

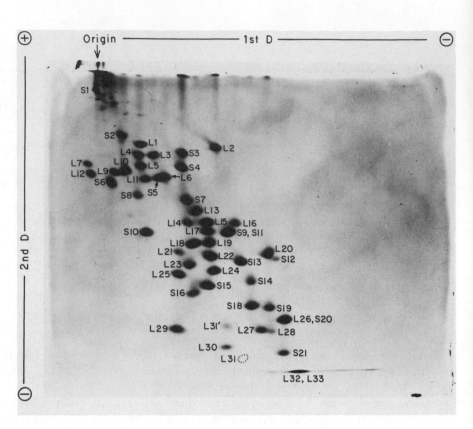

FIGURE **13-23**

The Use of Two-Dimensional Polyacrylamide Gel Electrophoresis for the Analysis of a Complex Mixture of Ribosomal Proteins. In this experiment the entire protein content of 70S ribosomes from *E. coli* was analyzed. Protein was extracted from 70S ribosomes with 66% acetic acid, dialyzed, and freeze-dried. For the first electrophoresis, the acrylamide concentration was 4% and the pH was 5.5. For the second electrophoresis, the acrylamide concentration was 18% and the pH was 4.6. Both gels contained 6 M urea to enhance the separation of small proteins. (Courtesy of R. Traut.)

(18%) that has been cast between glass plates in a pH 4.6 buffer. An electric field is applied at right angles to the first gel, and the proteins migrate into the slab gel at rates determined by their mobilities at the new pH. This two-dimensional gel electrophoresis method is now almost universally used to analyze the protein composition of ribosomal subunits or fragments thereof.

Topography

Many methods were applied to the problem of identifying which r-proteins lie at the surface of the subunits. Most of these depend on the reactivity of amino acid side chains toward chemical probes. After forming covalent bonds between the probe and the side chains, the proteins are separated on the two-dimensional gel system. Those that reacted are identified by the presence of some feature of the attached probe, usually a radioactive label that can be detected by autoradiography. A large number of such probes has been used, and the general picture that has emerged is that those proteins occupying a peripheral position in the assembly map also tend to be more easily labeled by the probes.

Nearest neighbors among the r-proteins can be identified by cross-linking probes. One of the more clever ones, exploited by R. Traut and his collaborators, is used to form a bifunctional cross-linking agent (a

molecular staple). This is iminothiolane, which contains both a reactive *imidate group* (see Figure 7–10) and an —*SH group* (Figure 13–24). Two such molecules, having reacted with two lysine residues on near neighbor proteins, can be "stapled" together by oxidizing the —SH groups to —S——S— linkages. The r-proteins and stapled complexes are separated from one another on a tube gel of polyacrylamide made up in the denaturing detergent sodium dodecyl sulfate *(SDS)*. The proteins held together by the staple will travel at a rate appropriate to a molecular weight that is the sum of the molecular weights of the two cross-linked r-proteins. After separation, the proteins are unstapled while still in the tube gel by diffusing in an —SH reagent like β-mercaptoethanol (HO—CH_2—CH_2—SH), which reduces the —S——S— bonds to form separate —SH groups again. The gel is laid across a slab of polyacrylamide and a second dimension of electrophoretic separation is performed, again in the SDS-containing buffer. Proteins that were never cross-linked by the molecular staple will lie along the diagonal of this kind of two-dimensional gel (a "diagonal gel," Figure 13–24b), but those that were previously cross-linked will be displaced to the left of and below the diagonal, at positions corresponding to their higher effective molecular weight during the first electrophoresis. Many such cross-linked pairs can be identified simply by measuring the molecular weight of the pairs and seeing which of the single r-proteins are involved; others were identified, after isolation from the gel, by the standard two-dimensional electrophoresis for r-proteins. These experiments reveal the *topography* of the ribosome: how the various parts connect with one another.

Another way of determining topography is to dissect the ribosome by very gentle treatment with RNAase. By suitably choosing the digestion conditions it is possible to make a few specific cleavages of the rRNA and thus to produce defined fragments of the subunit. After separating these from one another, by zone sedimentation for

example, their composition can be determined in the standard two-dimensional r-protein polyacrylamide gel system. Moreover, the RNA fragment can be isolated and its sequence determined. In this way *domains of structure,* proteins associated with specific regions of the rRNA and with one another, have been discovered.

Immunological Analysis

One of the most fruitful methods of elucidating the ribosomal structure has been immunological. Antibodies

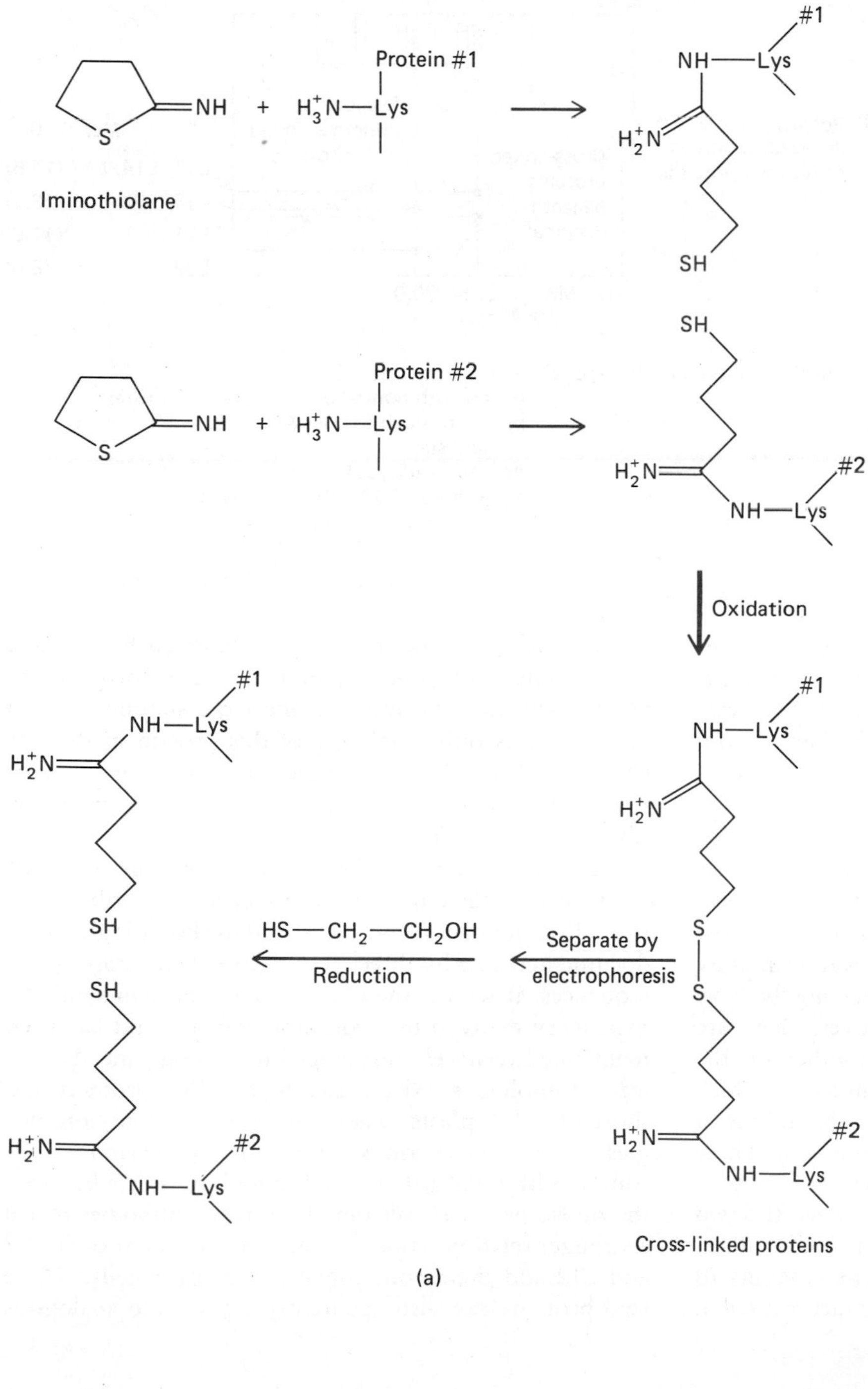

FIGURE 13-24
Diagonal Polyacrylamide/Dodecyl Sulfate (SDS) Gel Electrophoresis of Cross-Linked Ribosomal Proteins. (a) The cross-linking of neighboring proteins. Iminothiolane reacts with lysine residues, freeing its —SH group. Proteins within approximately 2 nm can form a covalent cross-link via an —S—S— bond. After separation, stapled proteins can be separated by reduction of the cross-link back to —SH groups. (b) Diagonal gel electrophoresis. Proteins that are not cross-linked move at equal rates during electrophoresis in both directions. These proteins ultimately lie along the diagonal. Proteins that were cross-linked during the first electrophoresis move more slowly than during the second electrophoresis, when they are not cross-linked. As a result they lie below the diagonal. The molecular weight scales were established with both standard proteins and purified ribosomal proteins. The "L" numbers at the lower right refer to specific proteins of the large 50S ribosomal subunit.

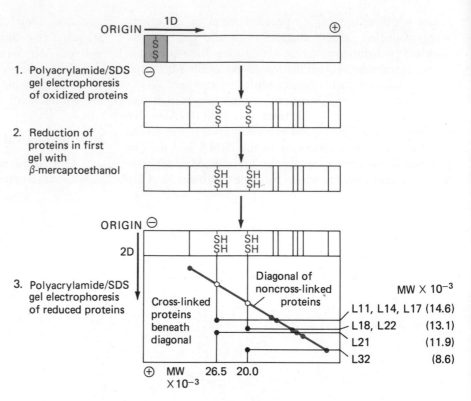

1. Polyacrylamide/SDS gel electrophoresis of oxidized proteins

2. Reduction of proteins in first gel with β-mercaptoethanol

3. Polyacrylamide/SDS gel electrophoresis of reduced proteins

Radioactive spots beneath diagonal: eluted with 1% SDS
mixed with nonradioactive total 50S protein
passed through Dowex AG1-X8 to remove SDS
    and stain
dialyzed and lyophilized
analyzed by 2D electrophoresis in
    polyacrylamide/urea gels

FIGURE **13–24**   (continued)   (b)

against purified r-proteins make extremely useful probes for determining topography, for identifying the participation of proteins in particular functions, and for assessing the relatedness of components of ribosomes from different cell sources. The massive undertaking of purifying all 52 different proteins from the *E. coli* ribosome was undertaken by H.-G. Wittman and his group. They have also worked out the primary structure of all these proteins. We can now determine which part of an r-protein can be cross-linked to another or to the rRNA. There is a very little, if any, immunological homology among the different r-proteins, and they mostly have very different primary structures. However, there are two r-proteins that separate from one another on the two-dimensional gel system, named L7 and L12, which have identical amino acid sequences, except that L7 has an acetyl group at the N-terminus while L12 does not. There are four copies of L7/12 per 70S ribosome.

An r-protein lying at the interface between the two ribosomal subunits could separate with one or the other, depending on the method of inducing the subunits to dissociate. This potential ambiguity was in fact realized in one case where there are two sites assigned for a single protein: one in the small subunit, where the protein was named S20, and the other in the large subunit, named L26. There is only one copy of this protein in the 70S ribosome. Aside from two copies of S6 and four copies of L7/12, there is only one copy of each r-protein in the ribosome of *E. coli*.

Antibodies against the purified r-proteins from *E. coli* revealed many homologous r-proteins in ribosomes from different procaryotic cells, and the homologies were documented also by direct comparison of the amino acid sequences. It seems, therefore, that the protein synthetic machinery evolved in a common ancestor and has been maintained relatively unchanged for a long time. Significant homologies exist between the *coli* r-proteins and those of chloroplasts. There are also extensive homologies among the various eucaryotic cytoplasmic ribosomes. Although little general homology exists between the ribosomes of *E. coli* and of rat liver, antibodies reveal a stronger relation among the individual *coli* r-proteins L7 and L12 and their homologues from other cells. These two proteins can also apparently replace the analogous proteins in some functions, as judged by reconstitution experiments.

The most visually dramatic use to which the antibodies against r-proteins have been put is in the technique of *immuno-electron microscopy*. The general shape of the ribosomal subunit can be seen in negatively stained electron micrographs (Figure 13–25). When antibodies with two antigen-combining sites each are added to purified subunits, two of them become joined by an antibody molecule at the same position on each subunit. The location of the antigenic determinant can thus be visualized relative to some general feature of the subunit. Gradually this technique has produced a picture of the 30S ribosomal subunit with the positions of various proteins demarcated. In combination with the other topographic techniques we are on the verge of having a complete geographical knowledge of the ribosome of *E. coli.*

But this structural information is only half the story needed. Topography must be related to function before we can produce an illustration like Figure 13–26, p. 620, with the roles of all the r-proteins and rRNAs depicted in detail. This we are already beginning to do, using the same techniques discussed previously. For example, it is possible to reassemble a ribosomal subunit from a mixture that omits one of the proteins. The functional defects of such a structurally defective particle can be assayed using the cell-free reactions we have already described. In this way it was shown that r-proteins S2, S3, and S14 are involved in the binding of aminoacyl-

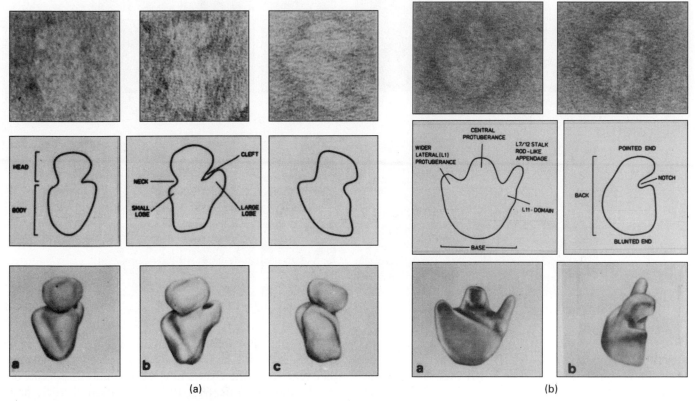

(a)   (b)

FIGURE **13–25**
Immuno-Electron Microscopy of Ribosomes. (a) Electron micrographs, interpretive drawings, and photographs of a three-dimensional model of a small ribosome subunit of *E. coli.* Three different views are shown. The 30S subunits in the electron micrographs were negatively stained with uranyl acetate. Various features of the model are identified in the drawings. (b) Similar illustrations of large 50S ribosomal subunit. (c) Ribosome subunits cross-linked by antibodies to three different 30S proteins: S10, S6, and S15. In the electron micrographs on the left, the cross-linking antibodies are indicated by arrows. Enlarged views of cross-linked complexes are shown in the center panels. The right panels show an interpretive drawing of the cross-linked complex and the region on the three-dimensional model where the cross-linking occurs. The S6 protein cross-links to two regions of the 30S subunit. (d) Similar illustrations of the 50S subunit, cross-linked by antibodies to ribosomal proteins L1, L18, and L23. (Horizontal bar = 50 nm; courtesy of G. Stöffler and M. Stöffler-Meilicke.)

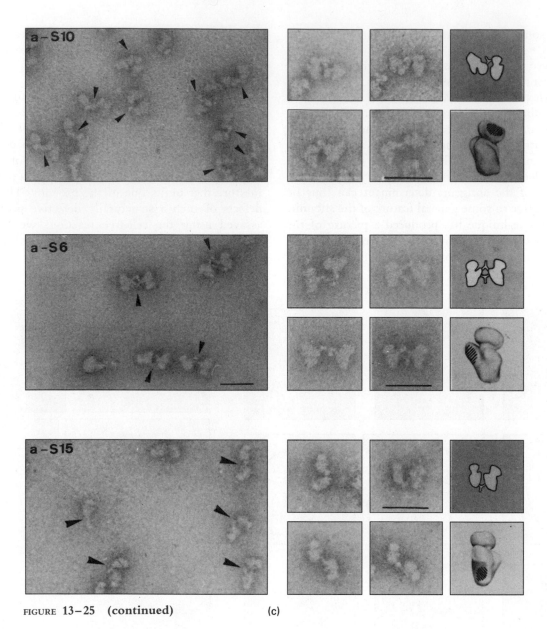

FIGURE 13–25 (continued)   (c)

tRNA. Or, one may use affinity-labeling techniques to attach covalently a peptidyl-tRNA, for example, to nearby structures. Thus it was shown that the ribosome P-site is composed in part of proteins L2 and L27, and the 23S rRNA. Or, one may add a fragment with a single antigen-combining site of an antibody directed against a specific r-protein to see what functional deficit results. This method was used to demonstrate that L7 and L12 are involved in the GTPase activities of the ribosome.

Ribosomal RNA Structure and Function

It should not be thought that only the r-proteins are important in ribosomal function; the rRNA also plays a major role. For example, ribosomes lacking certain r-proteins are still active in protein synthesis, and particular regions of the rRNA have been found important to the binding of mRNA, tRNA, initiation factors, and elongation factors. Biologists are shifting more and more to the belief that ribosomal RNAs are structures with significant functions during protein synthesis. There is, for example, fairly good evidence that most if not all the antibiotics that interfere with protein synthesis in bacteria have binding sites on rRNA. Moreover, mutations that confer resistance to the antibiotics often occur in genes that specify the rRNA sequence.

It is now clear that there is a well-defined structure for the 16S rRNA. The sequences of over 500 different genes specifying this molecule have been determined, and all of the predicted rRNA products can be folded into the

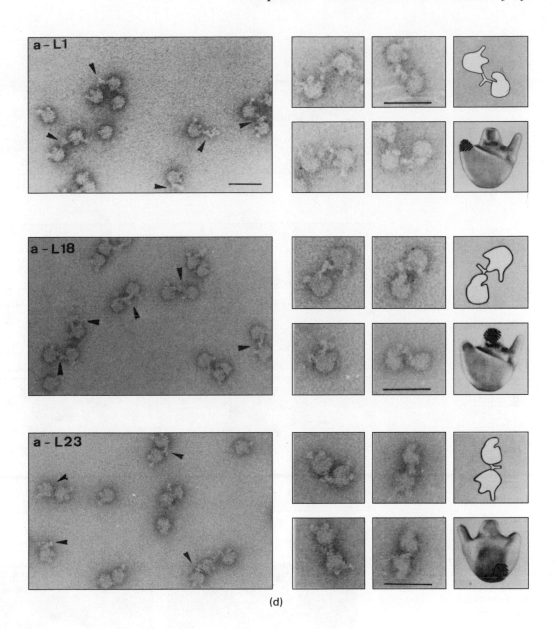

(d)

same secondary structure, composed of numerous stems and loops (Figure 13–27a). Although the primary sequences may be different, the secondary structure of 16S rRNA is highly conserved. It is interesting from an evolutionary point of view that 12S mitochondrial rRNA fits a very similar structure but with some features missing, and that the 18S rRNA of eucaryotic cytoplasm can be represented by the 16S procaryotic rRNA structure with some extra features added at the edges. Among the striking features of the analysis of 16S rRNAs are highly conserved sequences; some regions have exactly the same sequence in all the examples studied.

The positions of the r-proteins have been located relative to the 16S rRNA sequence by various techniques. When a particular protein is bound, the site of binding on the rRNA tends to be protected from digestion by

nucleases. The protected sequence(s) can be determined by methods discussed in Chapter 12 and fit into the overall model for the 16S rRNA. Alternatively, a cross-linking reagent can attach the protein to a site (or sites) in the rRNA and the bases involved in the cross-link can be determined by sequence analysis.

Using these methods, R. Brimacomb, H. F. Noller, H.-G. Wittman, their colleagues, and others have shown that tRNA binds to a particular set of structural regions when it is in the functionally defined A-site and to another set when it is in the functionally defined P-site. Both these sets are among the highly conserved sequences in the 16S rRNA. By applying these techniques further, the E-site has also been located in a relative location appropriate to the functional model (P-site lying between E-site and A-site, see Figure 13–13).

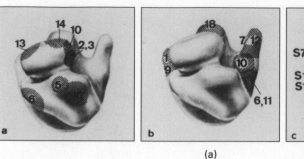
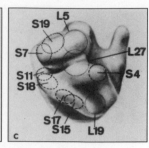
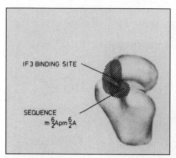

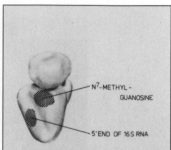

(a)

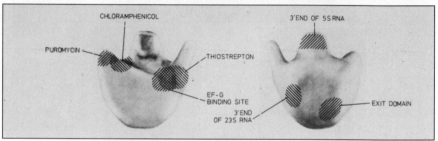

(b)

FIGURE 13–26

Three-Dimensional Model of *E. Coli* Ribosome. (a) Three views of the 70S ribosome, showing the locations of various r-proteins, as determined by immuno-electron microscopy. (b) Models of the 30S (upper panels) and 50S (lower panels) subunits, showing the location of various functional and structural features—for example, the binding sites for antibiotics puromycin, chloramphenicol, and thiostrepton. The "exit domain" is the region where the growing polypeptide emerges from the ribosome during protein synthesis. (c) Model of the 70S ribosome, showing the regions important in protein synthesis: the factor-binding site, the decoding site, and the peptidyl-transferase center. The center panel indicates the relative orientation of the tRNA to the various domains. One possible pathway for the mRNA is shown in the lower right panel. (Courtesy of G. Stöffler and M. Stöffler-Meilicke.)

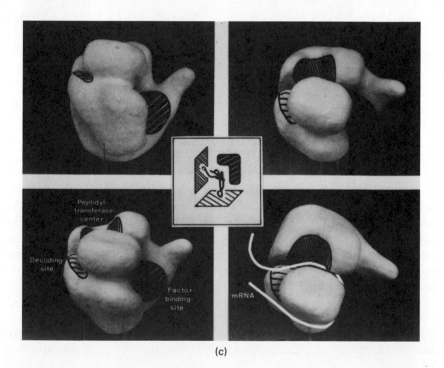

(c)

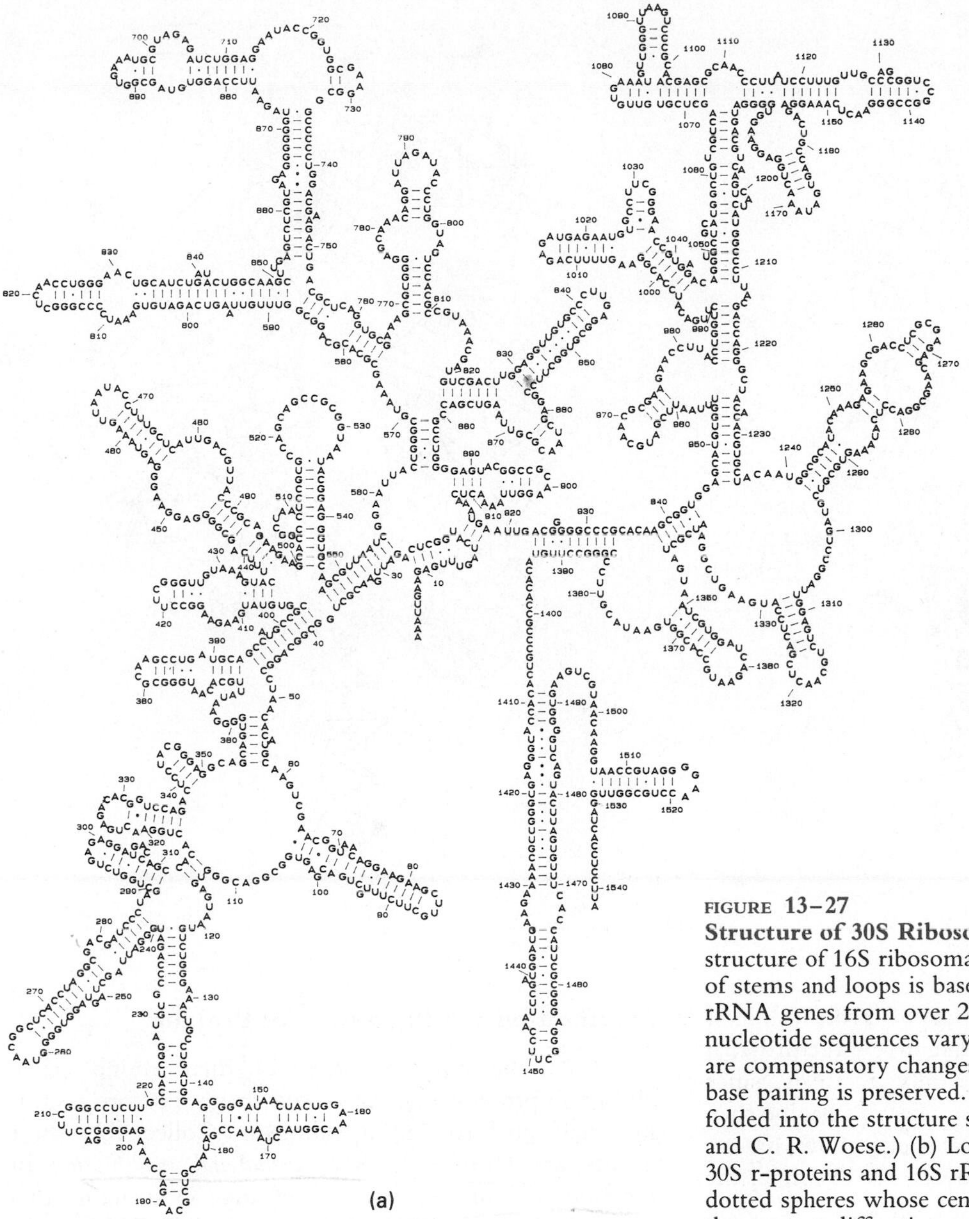

(a)

FIGURE **13–27**
Structure of 30S Ribosome Subunit. (a) Secondary
structure of 16S ribosomal RNA. This complex arrangement
of stems and loops is based on sequence analyses of the
rRNA genes from over 200 organisms. Although the primary
nucleotide sequences vary from organism to organism, there
are compensatory changes in the helical stem regions so that
base pairing is preserved. Each sequence determined can be
folded into the structure shown. (Courtesy of H. F. Noller
and C. R. Woese.) (b) Location in three-dimensional space of
30S r-proteins and 16S rRNA. The r-proteins are shown as
dotted spheres whose centers are located in space according to
the neutron diffraction analyses by P. Moore and his
colleagues. The 16S rRNA is folded to bring binding regions
near their respective r-proteins and to be consistent with
cross-linking and protection experiments. At the top is the 3'
domain, at the lower left is the "central" domain, and at the
right is the 5' domain. (Courtesy of H. F. Noller.)

The stems and loops of the secondary structure
model can be located in three dimensions using data
generated from neutron diffraction by the r-proteins. P.
Moore and his colleagues have done a systematic study
with this sophisticated technique and have located all the
16S r-proteins in space, relative to one another. Since the
rRNA sequences to which the proteins bind are known,
the nucleic acid structural model can also be made three-
dimensional (Figure 13–27b).

These methods, the techniques discussed earlier, and
sophisticated electron microscopic analyses are currently
being applied to the structure of the 50S subunit, where
the peptide transferase activity, among others, is found.
There is growing speculation that 23S rRNA, and not
large subunit r-proteins, may be the active agent in
catalyzing the transfer of peptides from peptidyl-tRNA
to aminoacyl-tRNA. If true, this would eliminate the
chicken-egg problem of explaining how protein synthesis

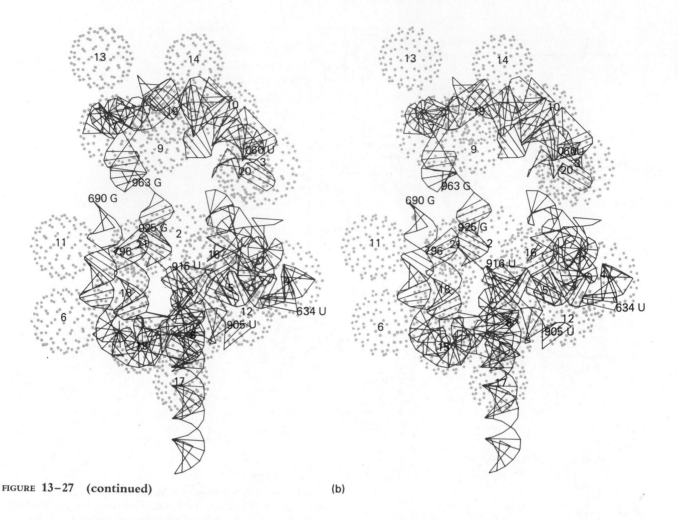

FIGURE 13–27 (continued)   (b)

Post-Translational Fate of Proteins

could be catalyzed in the absence of proteins. It is plausible that during evolution the first structure that acted like a ribosome was made only of RNA, since adding proteins to an RNA core is easier to imagine than the reverse (this idea is even more attractive now that we know some RNAs can act as catalysts, for example, the self-splicing RNAs discussed in Chapter 12).

As the dissection of the ribosome is accomplished at a finer and finer level, we shall eventually reach an understanding of the structural mechanics of protein synthesis that rivals our present knowledge of the active sites of enzymes. The student is referred to the Suggested Readings at the end of the chapter for more detailed discussions of ribosome structure.

Post-Translational Fate of Proteins

Proteins differ from one another in what happens to them after synthesis on the ribosome. Some of the differences result in changes in the structure of the protein. Other differences have to do with where in the cell the protein ultimately expresses its function.

Modification and Processing of Proteins

After the polymerization of their amino acids, almost all proteins undergo some changes before assuming their final role in metabolism. Collectively these events are known as *post-translational modification* or *processing* (in this context, any change in structure that occurs after the amino acids involved are linked into the protein is regarded as post-translational; such changes may occur before the last amino acid is added). We have discussed already the removal of the initiator amino acid and, in procaryotes, the formyl group from the N-terminal methionine. There may be additional amino acids removed from the ends of proteins synthesized on free ribosomes. (The N-terminal proteolysis for the case of proteins synthesized on membrane-bound ribosomes is discussed below.) The enzymes responsible for these steps—deformylases, aminopeptidases and carboxypeptidases, and endopeptidases—have only begun to be studied in detail and we must wait a bit longer for the full explanation of how their processing specificity is controlled.

Amino acids may be modified after being added to a protein. This may occur during or after termination of

Archaebacterial Peculiarities

Archaebacteria synthesize proteins using the same general scheme that we have described for eucaryotic cells and the usual bacteria (eubacteria). It has proved useful to compare the structures involved to see how closely related they are to those of other cells. Some structures share properties in common with the eubacteria, while others seem more like those of the eucaryotic cytoplasmic protein synthesizing molecules. For example, methionine is used as the initiating amino acid by archaebacteria, like the eucaryotic cytoplasm. The ribosomes of archaebacteria, however, are small, like those of other procaryotes. The elongation factor from archaebacteria interacts with diphtheria toxin, like eEF-2 of eucaryotic cytoplasm. The sensitivity of archaebacterial ribosomes to various antibiotics is different from both eucaryotes and procaryotes: some species (the thermophilic *Thermococcus celer,* for example) are not sensitive to sparsomycin; some are not sensitive to fusidic acid *(Thermoplasma acidophilum);* several are sensitive to the eucaryotic-specific antibiotic α-sarcin (a polypeptide from plants that catalyzes the cleavage of a conserved rRNA sequence in the large ribosome subunit); and several are sensitive to the procaryotic-specific thiostrepton. There is a Shine-Dalgarno sequence in the 16S rRNA of archaebacteria, as there is in the eubacteria.

On the other hand, the ribosomal RNA sequences from archaebacteria do not seem to be related to those of either the eubacteria or the eucaryotes. Also, there is in the transfer RNAs of archaebacteria almost no thymine (derived by modification from uracil) in the TΨC loop, while this is almost always found in a common position in the tRNAs of both eucaryotes and eubacteria (see Chapter 12). Taken together, these data argue for assigning the archaebacteria to a kingdom separate from both the eucaryotes and the eubacteria. There is continuing hot debate on the question of whether the archaebacteria gave rise to any other kind of cell. Further fuel for the flames can be expected to come from more complete analyses of the protein synthetic structures in these unusual cells.

protein synthesis or after subsequent processing. The formation of the disulfide linkages that stabilize tertiary folding of the proteins is an example of modification of cysteinyl residues. Another example is hydroxylation of proline, to form hydroxyproline, in the processing of collagen, the main protein constituent of connective tissues. There are many proteins whose N-terminal amino group is modified by an acetyl or other acyl ligand or whose C-terminal residue becomes modified to an amide. Finally there are some amino acids that accept methyl groups (e.g., lysine) or phosphoryl groups (e.g., serine) during the lifetime of the protein of which they are a part. There are some 40 different modifications of lysine in one protein or another. The addition of phosphate to specific parts of a protein and its removal are used to control enzyme activity (see Chapter 14).

Proline   Hydroxyproline   Serine   Serine phosphate   Lysine   ε-Acetylated lysine

Removal of whole peptides from the middle of a protein's sequence of amino acids is another type of post-translational processing. This can be minor, such as the breakage of a single peptide bond or the removal of one or two amino acids, or it can involve the complete separation of peptides with different functions from a large precursor protein. An example of this is the synthesis by anterior pituitary gland cells of *proopiomelanocortin,* the precursor for several peptide hormones. Depending on the particular cell, this large polypeptide is cleaved into one or more hormones (e.g., adrenocorticotropin) or peptides active in the central nervous system (e.g., β-endorphin, Figure 13–28). The discovery of such endoproteolysis also explained how some multichain proteins like insulin are synthesized. The A and B chains of insulin are joined by disulfide linkages that do not re-form specifically once broken, as those of RNAse do (see Chapter 4). The correct pairings are normally guaranteed by the rest of the sequence of a precursor protein, *proinsulin.* After the —S—S— bonds are formed in proinsulin, proteolytic enzymes remove the excess amino acids (Figure 13–28) and the matured insulin can then adopt a new, more active, conformation. Processing also occurs outside the cell, where it has been extensively studied in the case of the precursors of digestive enzymes (e.g., chymotrypsinogen, see Chapter 4). Such "activation" of extracellular proteins protects the cell's interior from the deleterious effects of, for example, digestive enzymes. Activation of hormones like insulin and melanocyte-stimulating hormone, on the other hand, seems to occur mostly intracellularly, probably in the Golgi apparatus (see below).

The Compartmentalization and Secretion of Proteins

We have several times alluded to the activities of extracellular and organellar proteins without explaining how, after they are synthesized, they arrive at their sites of action. For example, there are at least three distinct locations within the mitochondrion where proteins are found, proteins synthesized by cytoplasmic ribosomes (see Chapter 15). Unraveling the mechanism of this protein "sorting" is a fascinating story that still remains to be completed.

The first experiments to try to answer the question were done on the secreted proteins of liver, pancreas, and plasma cells. First came the electron microscopic observation that in secretory cells the ribosomes are associated with membranes of the endoplasmic reticulum (ER). Next came the demonstration in the late 1950s and early 1960s by Siekevitz and G. E. Palade that secreted proteins are in fact synthesized by such membrane-bound ribosomes and not by ribosomes that are free of membranes. These two classes of ribosomes were isolated separately from the pancreas of a guinea pig that had been injected with a short (1–2 minute) pulse of radioactive leucine. From each of the classes of ribosomes (bound and free), the newly formed secretory protein α-chymotrypsinogen was extracted, purified, and assayed for radioactivity. The specific radioactivity of the α-chymotrypsinogen obtained from the bound ribosome fraction was some eight times higher than that of the protein obtained from the free ribosome fraction, so it was inferred that this secreted protein was synthesized on bound ribosomes. This in turn led to experiments in 1966 showing that secreted proteins cross the ER membrane into the cisternal space of the ER after their synthesis (Figure 13–29).

The plot thickened when C. Milstein and his collaborators found in 1972 that extracts from myeloma tumor cells synthesized an immunoglobulin protein as a much larger molecule than is ordinarily found after secretion. In the presence of membranes, proteins synthesized in vitro were found to be smaller, the size of the usual immunoglobulin. The extra amino acids were located at the N-terminal end of the protein, and later analysis revealed them to have an unusually large fraction of hydrophobic side chains. These investigators referred to these extra amino acids as a "signal" because they seemed to indicate that the protein was to be cleaved, starting the secretion process. They thus hypothesized that a part of the newly synthesized molecule itself specifies its final destination.

The first formal model that was developed to interpret these observations, due mainly to G. Blobel and D. Sabatini, is shown in Figure 13–30. The N-terminal, partly hydrophobic region acts as a *signal sequence* that marks the ribosome for binding to the ER membrane and allows penetration and transport *(vectorial discharge)* of the growing protein through the membrane. When this sequence has emerged from the bulk of the ribosome, after ~80 amino acids have been synthesized, a *signal recognition particle (SRP)* binds to the ribosome. The SRP is a complex of six proteins and a 7S-RNA; it interacts with the ribosome and the emerging peptide. The binding of the SRP *inhibits further translation* until the ribosome · SRP complex binds to a 72,000-Da *docking protein* in the ER membrane. There are also special structures in the membrane that form a channel for the transport through the membrane of the rest of the protein. Another protein, *ribophorin,* binds to ribosomes in a one-to-one complex during these processes, but its exact role is not yet agreed. The functional roles of the RNA and the various proteins in the SRP are also

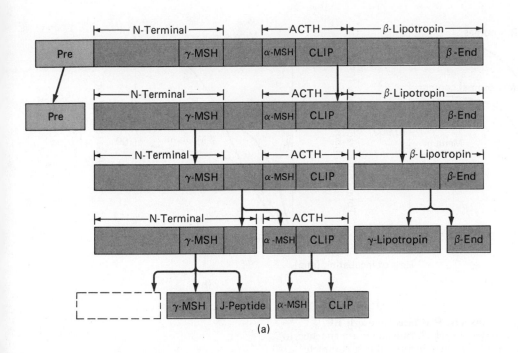

(a)

FIGURE 13-28
Processing by Proteolysis. (a) The topology of proopiomelanocortin. The structure of the polyprotein precursor (approximately 250 amino acids long) is shown with some of the biologically active peptides that result from processing. The dark bars represent basic amino acids (usually in pairs) that are the sites of attack by trypsin-like enzymes. Different types of pituitary cells process the same precursor differently. Other modifications often occur before the peptide is completely mature and active—for example, glysosylation, phosphorylation, methylation. ACTH is adrenocorticotropin; γ-MSH and β-MSH are the melanocyte-stimulating hormones; CLIP is corticotropin-like peptide. (b) The topology of proinsulin. After synthesis and formation of the three disulfide bonds, the 30-amino acid C-peptide (color) is removed to form the active hormone insulin, consisting of the A and B chains. The processing requires breaking peptide bonds (arrows) in the middle of proinsulin (endoproteolysis, like the breaks catalyzed by trypsin) and further removal of amino acids from the exposed ends (exoproteolysis, like the breaks catalyzed by carboxypeptidase B). Also see Figures 4–23, 4–24, 4–70.

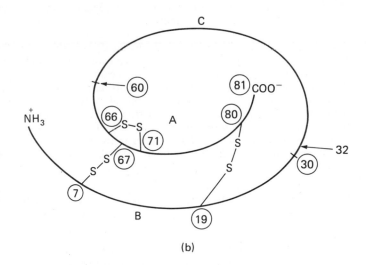

(b)

unknown, but are a subject of intense current research. After docking, the SRP dissociates from the complex and translation resumes.

As the N-terminal signal peptide arrives in the lumen of the endoplasmic reticulum, a membrane-bound proteolytic enzyme (the *signal peptidase*) removes it and thus generates an N-terminal sequence closer to that seen in the secreted protein. Additional proteolytic processing can occur later. In some cases, removal of the signal sequence occurs after additional amino acids have entered

the ER space but almost certainly before synthesis of the protein is complete; for this reason this process is referred to as *cotranslational* processing. The protein moves through the ER space and then into the Golgi complex.

Many of the experiments that confirmed this model were done in cell-free extracts capable of synthesizing proteins that are normally secreted. It was found that if rough ER membranes (in the form of empty vesicles) are included in the extracts, the growing protein is resistant

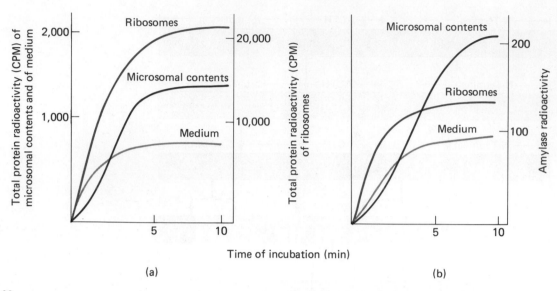

Time of incubation (min)

(a)   (b)

FIGURE **13-29**

Evidence for Vectorial Discharge of Nascent Proteins through ER Membranes.   Pigeon pancreas microsomes, in which ribosomes are attached to membrane-bounded vesicles (see Figure 13–3), were incubated in a complete protein synthesis medium including radioactive leucine. After various times of incubation, the mixture was centrifuged to yield the microsomes, as a pellet, and the medium. The microsomes were washed free of contaminating medium by another centrifugation and then treated with deoxycholate, which solubilized the vesicle membranes and the contents of the ER vesicles. The ribosomes were then separated from the contents by another centrifugation. Total proteins were assayed for each fraction: the medium, ribosomes, and vesicular contents. The secretory protein amylase was also purified from each fraction, and the radioactivity of total proteins (a) and of amylase (b) was determined. In both cases radioactivity first appeared in the ribosome fraction. The graph for amylase indicates that as radioactive protein left the ribosomes, it appeared firstly not in the medium but in the contents of the microsomal vesicles. This demonstrates vectorial movement from the ribosome through the membrane and into the vesicle contents. The radioactive amylase in the medium fraction probably comes from broken microsomal vesicles: the specific radioactivity of the amylase (amylase radioactivity divided by total protein) is always higher in the microsomal contents fraction than in the medium. (Data after C. Redman, P. Siekevitz, and G. E. Palade.)

to gentle proteolytic digestion. The resistance is not seen if the membranes are omitted, and it is reduced if the SRP complex is absent. If, in the absence of the SRP, the membranes are added after synthesis has gone part way to completion, there is no protection from added proteases. These results were interpreted as showing that the signal peptide folds on itself if the membrane receptors are not available for binding and thus cannot lead the way into the vesicular space.

The techniques of genetic engineering have also been applied to the subject of protein secretion in bacteria, by creating fused genes that give rise to fusion proteins: an N-terminal sequence like that of a secreted protein and a C-terminal sequence with the essential features of a cytoplasmic enzyme. Provided enough of the N-terminal sequence is included—more than the signal peptide has sometimes been found necessary—the cytoplasmic por-

tion is secreted and its enzyme activity can be detected outside the bacterial cell.

Many other experiments have been done in bacteria, where secreted proteins are extruded outside the cell during synthesis on ribosomes that are bound to the cell membrane. Removal of the protruding peptide with proteolytic enzymes does not lead to dissociation of the ribosomes from the membrane. Since treatment with puromycin does lead to dissociation of the ribosomes, they are probably held on the membrane largely by association with the peptidyl-tRNA. Apparently the transmembrane channel structure is tightly associated with the growing protein so it cannot slip back. There is evidence in bacteria that a transmembrane energy gradient drives transport so that movement is one-way.

Several signal sequences have been determined, but there is no simple consensus sequence among them

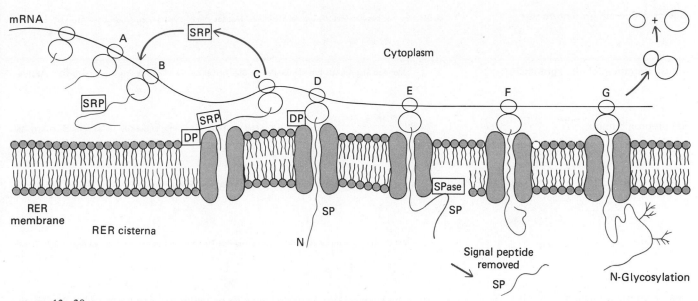

FIGURE 13-30
The Signal Peptide Hypothesis. The ribosome, free from membranes, initiates protein synthesis on an mRNA specifying a protein to be secreted. The signal peptide emerges from the bulk of the ribosome (A) and is recognized (B, C) by the signal recognition particle (SRP). Translation stops (C) until the ribosome · SRP complex binds to the rough endoplasmic reticulum (RER) membrane via docking protein (DP). The signal recognition particle dissociates and translation resumes. The signal peptide (SP) travels through a channel in the RER membrane (D), emerging in the RER cisterna where eventually it is attacked by a proteolytic enzyme, the membrane-bound signal peptidase (SPase, E). Protein synthesis continues, the growing peptide folds up inside the RER (F) and can also begin to be glycosylated (G). (Further details are in the text.)

(Figure 13–31). They are approximately 16–26 amino acids long and in the middle contain at least nine hydrophobic amino acids; methionine, leucine, and phenylalanine are common. But there are also charged amino acids (e.g., lysine) and polar ones (e.g., serine). Clearly the structure-function relations between the signal peptides and the channel proteins will be interesting, if complex, to determine. Further, the sites at which proteolysis removes the signal peptide vary greatly in structure.

Although the signal hypothesis is now firmly established, it has been refined in the light of more data: (1) In many cases, proteins are synthesized completely on free ribosomes and only afterward are inserted into a membrane; this is called *post-translational* processing. These proteins could be destined for an organelle, to be localized either in one of its membranes or in a compartment of soluble proteins, or for the plasma membrane. Post-translationally processed proteins have different kinds of signal sequences (see discussion of organelle proteins in Chapter 15). (2) Not all proteins that are cotranslationally transported into the ER have N-terminal sequences that are removed. The signal sequences of these proteins interact with the SRP, but not with the signal peptidase.

(3) Recombinant DNA techniques allow test proteins to be constructed with sequences changed or omitted. Experiments using this approach show that in some cases the N-terminal sequence, although necessary, is not itself sufficient to direct a protein into the ER space. The interior signal sequences implied by these results have not yet been completely identified.

Some of the proteins that enter the ER are processed further. The most striking change is the addition of sugar groups to amino acid side chains: asparagine, *N-glycosylation*, and serine (or threonine), *O-glycosylation*. Additional modification by sugars occurs later in the secretion pathway, in the Golgi apparatus. After glycosylation, proteins destined for secretion would be more easily retained in the ER space—the sugars would prevent any return through the ER membrane—but nonglycosylated proteins, notably the serum albumins, are apparently secreted in the same way. That way leads from the rough ER to the smooth ER and then to the Golgi system (Figure 13–32). The pathway was established in cell fractionation experiments by the kinetics of passage of radioactive proteins from one compartment to another and by autoradiography of radiolabeled proteins at the electron microscopic level of resolution (Figure 13–33a).

Bovine preparathyroid hormone:
Met-Met-Ser-Ala-Lys-Asp-Met-Val-Lys-Val-Met-Ile-Val-Met-Leu-Ala-Ile-Cys-Phe-Leu-Ala-Arg-Ser-Asp-Gly-Lys-
↑

Mouse immunoglobulin light chain:
Met-Asp-Met-Arg-Ala-Pro-Ala-Gln-Ile-Phe-Gly-Phe-Leu-Leu-Leu-Leu-Phe-Pro-Gly-Thr-Arg-Cys-Asp-Ile-
↑

Rat preproalbumin:
Met-Lys-Trp-Val-Thr-Phe-Leu-Leu-Leu-Leu-Phe-Ile-Ser-Gly-Ser-Ala-Phe-Ser-Arg-Gly-
↑

Rat preprolactin:
Met-Asn-Ser-Gln-Val-Ser-Ala-Arg-Lys-Ala-Gly-Thr-Leu-Leu-Leu-Leu-Met-Met-Ser-Asn-Leu-Leu-Phe-Cys-Gln-Asn-Val-Gln-Thr-Leu-Pro-
↑

Rat relaxin:
Met-Ser-Ser-Arg-Leu-Leu-Leu-Gln-Leu-Leu-Gly-Phe-Trp-Leu-Phe-Leu-Ser-Gln-Pro-Cys-Arg-Ala-Arg-Val-
↑

Sea raven preproinsulin:
Met-Ala-Ala-Leu-Trp-Leu-Gln-Ser-Phe-Ser-Leu-Leu-Val-Leu-Leu-Val-Val-Ser-Trp-Pro-Gly-Ser-Gln-Ala-Val-Ala-
↑

Chicken lysozyme:
Met-Arg-Ser-Leu-Leu-Ile-Leu-Val-Leu-Cys-Phe-Leu-Pro-Leu-Ala-Ala-Leu-Gly-Lys-Val
↑

E. coli outer membrane lipoprotein:
Met-Lys-Ala-Thr-Lys-Leu-Val-Leu-Gly-Ala-Val-Ile-Leu-Gly-Ser-Thr-Leu-Leu-Ala-Gly-Cys-
↑

FIGURE **13–31**

Complete and Partial Amino Acid Sequences of NH$_2$-Terminal Extensions of Some Preproteins. Precursor sequences are approximately 20–30 amino acids in length and consist predominantly of hydrophobic amino acids. Arrows denote sites of proteolytic cleavages resulting in the formation of either mature proteins or the proprotein intermediates.

FIGURE **13–32**

Scheme Showing the Passage of a Newly Synthesized Protein to the Outside of a Pancreatic Acinar Cell. The protein is synthesized on a membrane-bound ribosome and, during synthesis, passes across the membrane of the rough endoplasmic reticulum. Once inside this ER space it moves through the cisternal spaces to the smooth ER, via a connection (possibly a transition vesicle) to the membrane-bound spaces of the *cis*-side of the Golgi complex, through a passageway within the Golgi to its distal *(trans)* face, which then buds off to form the zymogen granules. By means of a fusion of the zymogen granule membrane with the cell membrane, it is finally discharged into the extracellular space, in this case the pancreatic duct (see Figure 7–44a).

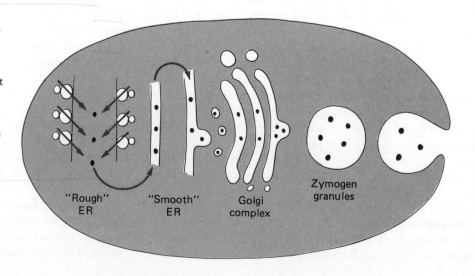

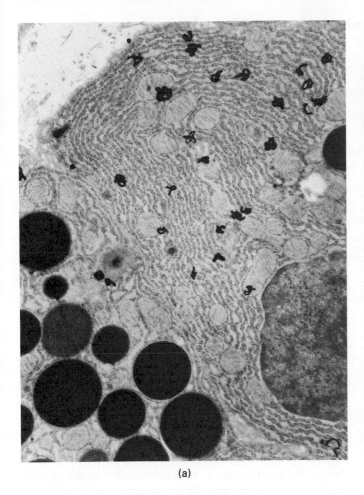

(a)

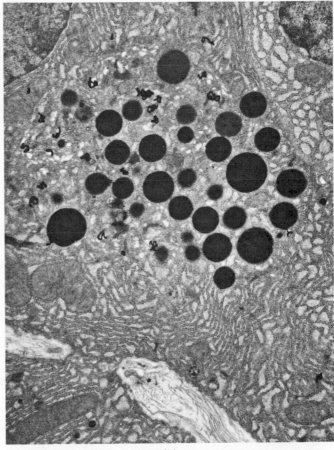

(b)

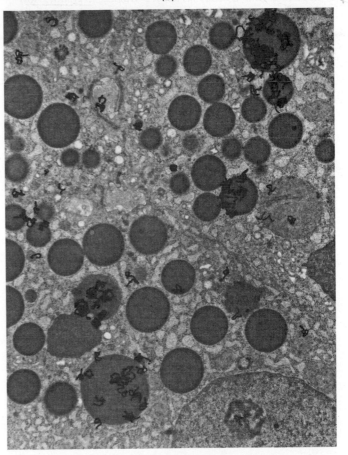

(c)

FIGURE 13–33

Intracellular Passage of Secretory Proteins, as Visualized by Autoradiography of Pancreatic Acinar Cells. The experiment was performed as follows: guinea pig pancreas slices were incubated with [³H]-leucine to label their proteins and then processed for autoradiography at various times thereafter. (a) The processing was performed at 3 minutes after the introduction of radioactive leucine. The majority of autoradiographic grains (very dark splotches), denoting newly synthesized secretory protein, are over the rough endoplasmic reticulum membranes with ribosomes attached (cf. Figure 13–3). (b) The processing was performed after 3 minutes of radioactive incubation plus 10 minutes of incubation with nonradioactive leucine (a pulse–chase experiment). The majority of autoradiographic grains are over the smooth membranous elements of the Golgi complex, indicating migration of labeled secretory protein from the RER to the Golgi. (c) The processing was performed after 3 minutes of radioactive incubation plus 60 minutes of incubation with nonradioactive leucine. The majority of the autoradiographic grains are concentrated over the contents of mature zymogen granules, denoting movement of the newly synthesized proteins from the Golgi to the zymogen granules. The zymogen granules are the large, dark, circular elements in all the figures. They move towards that part of the membrane of the pancreatic acinar cell that faces the pancreatic duct; a fusion of the cell membrane and the zymogen granules membrane occurs, and the contents of the granules are discharged into the duct. ×16,000. (Courtesy of J. D. Jamieson.) See also Figure 7–44.

Movement along the secretion pathway has been inhibited in some cells by poisons of both oxidative phosphorylation and microtubule action, so both energy and mechanochemical coupling are probably involved. The movement of the proteins may be direct, through interconnecting spaces, or it may be mediated by the budding off of *transition vesicles* from the smooth ER and later fusion with Golgi vesicles on the *cis* (convex) side of that organelle.

After further modification, proteins are transported to the *trans* (concave) side of the Golgi complex where they are packaged directly into *secretory granules* for export or concentrated into *condensing vacuoles* for intracellular storage, later to be exported via secretory granules.

The various membrane-enclosed spaces of the Golgi seem to be in constant and complex motion. The result of all this activity varies with the fate of the protein. We have discussed until now only proteins to be secreted, but other proteins are synthesized on the rough ER, namely those destined for the membranes of the cell, the ER, the Golgi, the lysosomes, and the peroxisomes. The Golgi is where the routes to these various destinations diverge. Proteins destined for inclusion in membranes stay associated with the rough ER membrane during synthesis (or become associated with it after synthesis), folding into something like a final structure during translocation—anchoring sequences stop the transfer of the peptide through the membrane and self-inserting domains penetrate the bilayer (see discussion of membrane architecture in Chapter 7). The route to the Golgi would be essentially the same as for a secreted protein except that the membrane proteins would travel attached to the membrane, rather than in the interior, of the transition vesicle.

Proteins destined for exile leave the Golgi by budding off into secretory vacuoles or granules; proteins destined for the plasma membrane are also carried by vesicles. These membrane-bound structures travel across the cytoplasm, possibly along microtubules, and fuse at the frontier with the plasma membrane in a reversal of endocytosis (Figure 7–44). Proteins to be secreted are then deported into the extracellular space, sometimes the duct of a gland, while membrane proteins are left on the border. The extra plasma membrane inevitably produced by this process is taken in again, by endocytosis. Various cell types may temporarily store secretory granules, others pass them promptly along. Fusion of lysosomes with secretory vesicles and digestion of the contents may occur as a method of controlling the release of secreted protein. The many variations of the basic scheme are discussed in more advanced texts. The student is referred to the discussion of neurosecretion in Chapter 17 for more information on the control of these processes, since the general mechanisms are likely to be similar.

The general problem of how various proteins are sorted into organelles is not yet solved. A popular idea is that there are *address markers* on the protein that specify its ultimate location. One marker that controls the localization of enzymes destined for the lysosome involves a sugar residue, *mannose-6-phosphate* (M6P). Lysosomal hydrolases are synthesized on membrane-bound ribosomes and processed by the ER and the Golgi complex. Among various sugar residues that are added is M6P. There are receptors in the Golgi membrane for proteins that contain this relatively unusual M6P modification. The M6P-proteins apparently collect at the region of the Golgi membrane where the receptors are, and the vesicles that bud off this region, the primary lysosomes, preferentially contain proteins with the M6P marker. This model leaves open the question of how the M6P receptors are clustered in special regions of the Golgi.

An interesting feature of the sorting process is that the Golgi complex (GC) maintains the characteristic appearance and lipid composition of its membranes even though ER membranes in the form of transitional vesicles fuse with it and secretory vesicles analogous in appearance to the plasma membrane leave it. The GC thus organizes its own membrane, which is in keeping with a popular view that the GC is involved in the production of the plasma membrane. J. E. Rothman hypothesizes that the distinct *cis* and *trans* compartments of the GC act sequentially to remove ER membrane proteins. This process starts at the *cis* site; when membrane reaches the *trans* site only those proteins remain that are destined for secretion, for specific intracellular sites, or for the plasma membrane. The ER proteins removed at the *cis* Golgi site would then be recycled somehow back to the ER. Maintaining the individuality of the GC membrane, despite its being placed in continuity with other membranes and despite potential mixing of components due to membrane fluidity, is a fascinating phenomenon not yet understood, but deserving our special attention.

Collagen

A specific example of many of the above steps in the processing of a protein is illustrated by what is known about the synthesis and secretion of collagen (Figure 13–34). Collagen is a secretory product of many types of cells, such as fibroblasts, bone, tendon, cartilage, and the various epithelia associated with basement membranes. It is the most abundant protein in mammals. The structure of collagen is based on a helix of three chains, and the molecule ends up as highly insoluble protein. In cartilage, the collagen chains are identical; in most other tissues they are not. Other features of collagen are the very high content of hydroxylysine and hydroxyproline (far higher than in noncollagen proteins), its content of carbohydrate

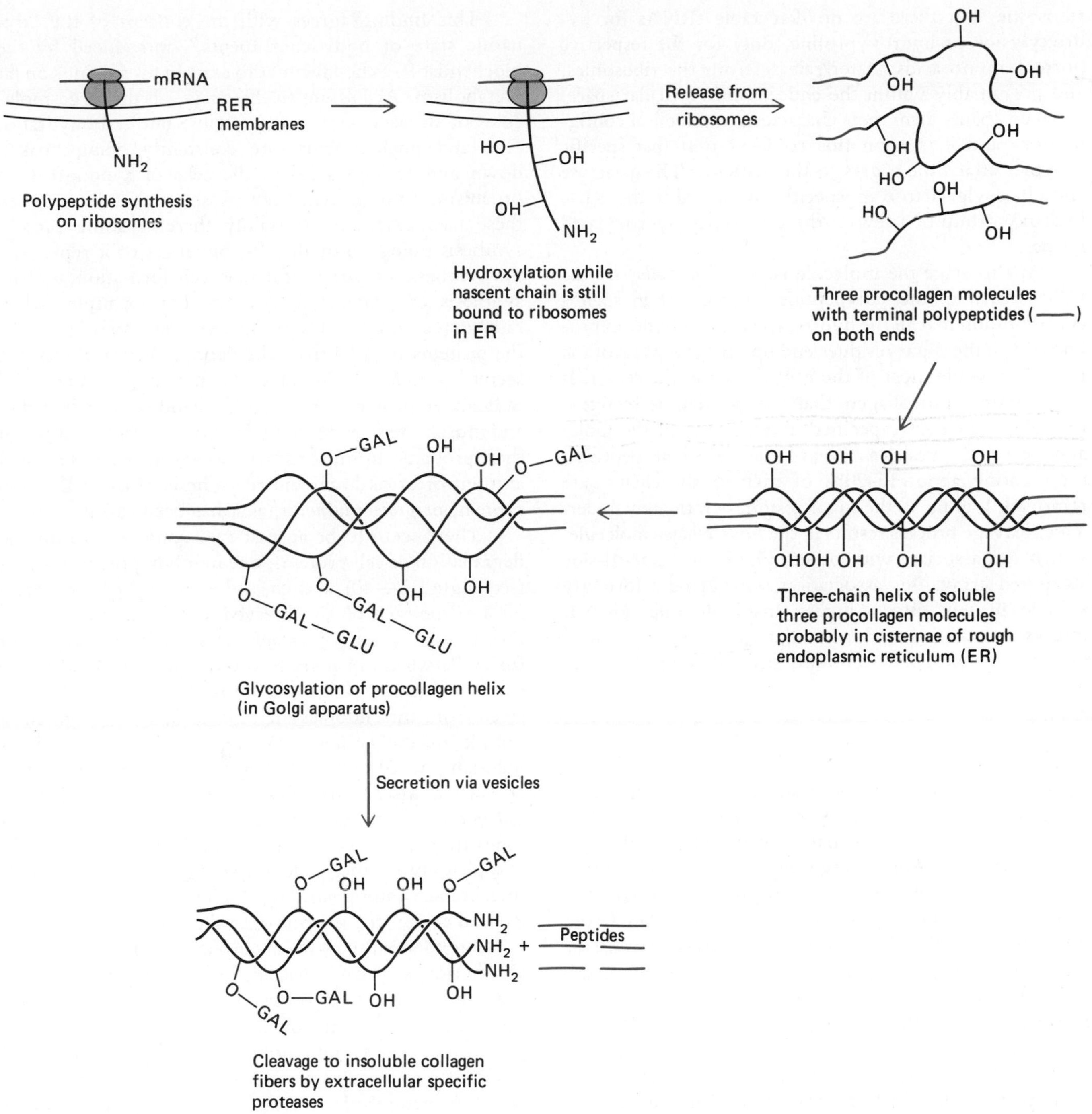

FIGURE 13-34
Steps in the Synthesis of Cartilage Collagen.

(mainly galactose and the disaccharide glucosyl-galactose), the regularity of the amino acid sequence (nearly every third residue is glycine), and the very large size of the polypeptide chain (over 1000 amino acids). How does such a peculiar protein get synthesized and how does such an insoluble protein get secreted?

Collagen is synthesized as a precursor protein with extra polypeptides at both ends of the chain. The hydroxylation of the lysine and proline residues occurs, via specific enzymes, after these amino acids are in the polypeptide chain: not only has this hydroxylation been found to occur while the growing chain is on the ribosome, but there are no detectable tRNAs for hydroxylysine or hydroxyproline, only for the respective normal amino acids. After release from the ribosomes, and presumably within the endoplasmic reticular space, collagen chains form their characteristic trihelical configuration, and it is upon this configuration that specific enzymes attach the sugars to the proteins. The galactose and glucosylgalactose are specifically bound to the delta-hydroxy group of a few of the previously hydroxylated lysines.

At this stage the molecule is *soluble procollagen*—the helical form of the polypeptide complex is in such a configuration that all the hydroxylysine and hydroxyproline and all the sugar residues end up on the outside of the molecule, while most of the glycines lie in the center. It is as soluble procollagen that the protein is secreted, following the route specified above through the Golgi and secretory vesicles. Next, the terminal peptides, representing about 15–20% of each of the chains, are removed, leaving all the sugar residues on the remainder. This cleavage process results in the *tropocollagen* molecule, which can associate with others like it in a side-to-side staggered array; this association transforms a formerly soluble protein into a highly insoluble one. Finally, intermolecular and intramolecular *cross-links* are formed between certain lysine residues, strengthening the collagen fiber.

Protein Turnover

About 40 years ago biochemists began using radioactive amino acids to study the processes involved in protein synthesis. Up to that time it had been thought there was very little protein synthesis and breakdown in the cells of adult, nongrowing organisms. It was assumed that once the cellular proteins were formed, they lasted the lifetime of the cell. There is, however, a constant nutritional requirement for protein or amino acids and a constant excretion of ammonia and urea in the urine of organisms. Since these nitrogen-containing compounds could have come only from the catabolism of proteins, it was assumed that the steady excretion represented that small proportion of cells that were breaking down, their proteins being hydrolyzed, and the resultant amino acids deaminated to form urea and ammonia. This exception to the relative metabolic stability of proteins was thought to be a consequence of "wear and tear"—that is, the only proteins being metabolized were those resulting from the breakup of dead and dying cells. It was thus a distinct surprise to find that if radioactive amino acids are injected into an animal and then the proteins isolated from the various tissues, these proteins are found to be radioactive. Since the radioactive amino acids become incorporated into protein molecules, there must be synthesis of proteins in the cells of the adult organism.

This finding agrees with the concept of the "dynamic state of body constituents," introduced by the biochemist R. Schoenheimer to explain his findings on fat metabolism. According to this view, all the large molecules of the cell—not only proteins, but carbohydrates, fats, and nucleic acids—are constantly being broken down and resynthesized in the cells of a nongrowing organism. Our present view is somewhat in between these two extremes. Certainly there is some protein synthesis going on in all cells, but most of it represents the synthesis of protein for new cell formation and the synthesis of proteins for export. For example, when radioactive amino acids are injected into an animal and the proteins isolated from the various tissues, those that secrete protein are found to contain high amounts of radioactive protein. On the other hand, tissues like skin and muscle are found to contain very little radioactivity in their proteins. In most tissues, however, there is a small amount of breakdown and resynthesis of the cell's own protein, or *protein turnover,* as it has been called.

There seem to be at least two major routes for the degradation of cell proteins, one involving the lysosomes (see Figure 7—40) and the other not. Approximately 30% of the proteolysis observed in cultured liver cells is due to the action of *autophagolysosomes,* the result of fusion between a primary lysosome and an *isolation vesicle* or *vacuole* (a membrane-surrounded space, possibly derived from the endoplasmic reticulum, which contains a sample of the contents of the cytoplasm, sometimes including whole organelles like peroxisomes or mitochondria). Inside, the cell's protein can be attacked by *cathepsins* and other degradative enzymes found in lysosomes that are active in the low-pH environment of the autophagolysosome. These enzymes do not seem to discriminate among various proteins; any control must be exerted at the step of entry into the isolation vesicle.

Another pathway occurs inside the cell cytoplasm or at interior surfaces adjacent to the plasma or other membranes. Here discrimination occurs among various normal proteins on the one hand, and among abnormal proteins, peptides, and fragments, which are all degraded faster than their normal counterparts. This degradation occurs at neutral pH and can occur in supernatants that are free from lysosomes.

We now believe that one of the roles played by protein turnover is adaptation: since the cell operates its metabolism by providing a set of protein catalysts, an alteration in the metabolic activity of nondividing cells is aided by an alteration in the set of enzymes. Rapidly dividing procaryotic cells can dilute out the unwanted enzymes by growth, but a resting eucaryotic cell can only rid itself of an irrelevant or counterproductive protein by degrading it.

A secondary purpose of such proteolysis is the removal of abnormal proteins or peptides, as we mentioned above and will discuss in more detail below. Moreover, under duress a starving cell adapts by proteolyzing its contents for food during the emergency (rates have been measured as high as 4.5% of total liver cell protein degraded per hour, mostly via lysosomes). Finally, proteins that enter a cell from its environment are also degraded. (The cleavage of specific peptide bonds during post-translational processing was discussed earlier.)

Current research on intracellular proteolysis is concerned with the questions of which degradative enzymes are involved, in which cellular compartment they are located, and how the rates of degrading different proteins can be controlled.

The primary experimental approach to the study of proteolysis is to label a protein with a radioactive tag and to follow its progress through various stages of decay into amino acids (see Figure 7–13). This can easily be done by giving a cell a pulse of a radioactive amino acid and locating the protein of interest by purifying it or reacting it with specific antibodies. Several variations of this approach have been developed in order to avoid the perturbing effects of the later addition of excess unlabeled amino acid during the chase. These kinds of experiments have been widely applied to whole organisms, tissue slices, cells in culture, and cell-free extracts. The measured rates of degradation vary over a wide range, from lactate dehydrogenase with a half-life of six days to thymidine kinase with a half-life of 0.1 day (see Chapter 7 and Figure 7–14 for half-lives of membrane proteins).

Much work on proteolysis has been done in cultured mammalian cells and bacteria. This is because cell-free extracts, which have proved to be simpler in so many other metabolic studies, typically do not respond to inhibitors and enhancers of proteolysis in the way that whole cells do.

One of the striking features of intracellular proteolysis is the requirement for metabolic energy. Inhibitors of ATP production also inhibit protein degradation. In itself this observation suggests an important metabolic role for turnover since there is the implication that the pathways used are subject to control. Another set of inhibitors that affect proteolysis are antibiotics that block protein synthesis on ribosomes. This suggests that the degradative enzymes may themselves turn over and, therefore, need to be resynthesized.

One of the ways ATP is used in proteolysis is as part of a protein-marking process involving *ubiquitin*. This 74-amino acid polypeptide is widely distributed in both procaryotic and eucaryotic cells, hence its name. Ubiquitin is activated by attachment of AMP, donated by an ATP. The activated polypeptide is conjugated to proteins destined for rapid degradation; there is often more than one ubiquitin-AMP attached to a single target protein. The digestion of ubiquitin-labeled proteins is probably also ATP-dependent. Although the details remain to be worked out, it is easy to imagine that proteases recognize the ubiquitin label and preferentially digest the target protein. Still mysterious are the structural features that are recognized by the ubiquitin-conjugating enzyme(s).

Proteins containing errors are preferentially degraded by the ATP-dependent pathways. (An error in this context is one of the 19 alternative amino acids found in the place of the correct one.) This has been established by labeling with $[^3H]$-leucine, for example, and measuring the half-life of hot acid-precipitable label—that is, of labeled proteins. Under conditions that lead to the synthesis of erroneous proteins, the observed half-life is much shorter than when protein synthesis is normal. Among the errors that are caught by this process in bacteria are those committed during protein synthesis in the presence of the antibiotic streptomycin. In procaryotic or eucaryotic cells, when the amino acid analogue *canavanine*—which resembles arginine—is incorporated into proteins, they are also degraded preferentially by the ATP-dependent pathway(s). Although preferential degradation will reduce the number of erroneous proteins, the extent of the contribution this process makes to the overall accuracy of protein synthesis is not yet known.

SUMMARY

We have reached the end of our accounting of the expression of genetic information in the form of proteins. There is much more to be said about all these subjects, of course, and interested readers should consult the Selected Readings to extend their learning. Our description thus far of the mechanics of protein synthesis leaves a lot of questions unanswered. Proteins, as such, are not always the end product of genetic expression; they can combine with one another to form larger aggregates or assemblies. How this happens is the subject of Chapter 15. The problem of controlling the mechanics of protein synthesis is treated in Chapter 14.

KEY WORDS

Anticodon, base pair, chloroplast, codon, cross-link, deoxyribonucleic acid, electrophoresis, endoplasmic reticulum, energy, enzyme, eucaryote, genetic code, genetic map, information, initiator, lysosome, mitochondrion,

N-terminal, organelle, plasma membrane, polymer, polysome, procaryote, processing, ribonucleic acid, ribosome, RNA transcription, translation, triplet, virus.

PROBLEMS

1. A region of DNA corresponding to a particular procaryotic gene has the structure shown below:

    3'... TTTCCTAGGGCAGGTACATG-
    GACCTTAGTAGGGGGCCTTA ... 5'
    5'... AAAGGATCCCGTCCATGTAC-
    CTGGAATCATCCCCCGGAAT ... 3'

    The gene is transcribed from *left* to *right* with the first C in the template strand directing the synthesis of the *first* nucleotide in the mRNA chain. Assuming there is no postsynthetic processing of the polypeptide chain, the second amino acid in the protein encoded in this gene is
    A. Asp.
    B. Pro.
    C. Leu.
    D. Ile.
    E. Tyr.

2. To begin translation, the ribosome binds near the
    A. 5' end of the promoter.
    B. 3' end of the promoter.
    C. 5' end of the mRNA.
    D. 3' end of the mRNA.
    E. 3' end of the operator.

3. If two adjacent genes are transcribed in the same direction and the two genes overlap, then
    A. the two genes must be encoded in the same reading frame.
    B. the two genes could be encoded in either the same or different reading frames.
    C. the two genes must be encoded in different reading frames.
    D. the two genes must be transcribed into a polycistronic mRNA.
    E. the promoters for the two genes must overlap.

4. The first codon in an mRNA codes for the amino acid found at the _____ of the completed polypeptide chain.
    A. 5' end
    B. carboxyl terminus
    C. ribosome
    D. 3' end
    E. amino terminus

5. What is the anticodon found on the tRNA that carries the amino acid tryptophan (Trp)?
    A. 5'-UGG-3'
    B. 5'-GGU-3'

C. 5'-CAA-3'
D. 5'-CCA-3'
E. 5'-ACC-3'

6. Compare and contrast the mass, energy, and information (including control) requirements for the cellular biosynthesis of an enzyme with those for the cellular biosynthesis of an amino acid.

7. Respond to the following statements with TRUE or FALSE; if you respond FALSE, explain why the statement is not true; if TRUE, give a reason why the statement is true.
    (a) Signal sequences allow the association of ribosomes with endoplasmic reticulum membranes.
    (b) In procaryotic cells an mRNA molecule cannot begin to direct protein synthesis until its synthesis is completed.
    (c) The amino acid sequence of a protein is determined by an intimate complementary interaction between the amino acid and the codon sequences in the mRNA.
    (d) Judging by their response to inhibitors of protein synthesis, procaryotic ribosomes are similar to the ribosomes of eucaryotic mitochondria but different from those of eucaryotic cytoplasm.
    (e) Addition to a cell-free protein synthesizing system of compound GDPCP, a nonhydrolyzable analogue of GTP, will result in all peptidyl-tRNAs becoming stuck in the ribosome A-site.
    (f) All proteins must terminate with the amino acid proline, since all mRNAs terminate with a CCA sequence at their 3' end.
    (g) The first indication that the genetic code cannot be universal was the absence in the dictionary derived from experiments in *E. coli* of a codon for hydroxyproline, a major constituent of vertebrate collagen.
    (h) GTP hydrolysis is necessary for the binding of aminoacyl-tRNA to the ribosome.
    (i) After the action of diphtheria toxin, a eucaryotic ribosome would have two tRNAs bound, one peptidyl-tRNA and one aminoacyl-tRNA.
    (j) The formyl group on the methionine of a protein destined for secretion by a eucaryotic cell is removed by the "signal peptidase."

8. Treatment with the mutagen "acridine gold" during replication causes, we shall assume, the deletion (−) or insertion (+) of one base pair in viral DNA. Two mutants caused by acridine gold treatment have been

combined by genetic crossing. The resulting "pseudorevertant" double mutant has a phenotype almost, but not quite, like that of the wildtype parent and thus must have a + − or − + genotype. The protein product of the mutated gene has been sequenced in the wildtype and double mutant forms:

wildtype: N—Lys-Cys-Ala-Asp-Pro- Lys—C
double mutant: N—Lys-Val- Leu-Ile- Arg- Lys—C

(Residues elsewhere in the protein are identical.)

(a) What is your best estimate of the mRNA sequence of the wildtype gene in this region?

(b) What is the order of the mutations in the double mutant?

(c) What would the protein product be if the order (but not the signs) of the two mutations were reversed?

9. H. G. Khorana (Nobel laureate) could have synthesized by chemical means the following single strand of DNA:

5′− dG dA dG dT dC dG dA dG-
dT dC dG dA dG dT dC−3′

This can be used as a template for DNA polymerase I, which in the presence of the four deoxyribonucleotide triphosphates would synthesize a two-strand product. Slippage of the template allows the enzyme to produce a product that is much longer than the 15-mer shown. The double-strand DNA can be used as a template with RNA polymerase to produce RNA. In suitable conditions, this kind of RNA will act as a message whether or not there is a proper ribosome binding site and initiation codon.

(a) What will be the nucleotide sequence of the RNA product(s)?

(b) What will be the amino acid sequence of the peptide product(s)?

10. When a cell-free extract from *E. coli* is programmed with the mRNA A-U-G-U-U-U-U-U-U-U-U-U-U-U-U-U, it synthesizes fmet-phe-phe-phe-phe. After addition of the drug frampomycin, a newly discovered antibiotic from the Grundy Center Drug and Live Bait Company, the same message directs the synthesis only of fmet-phe.

(a) What step(s) in polypeptide synthesis does frampomycin inhibit?

(b) Will either the product oligopeptide (uninhibited case) or the product dipeptide (inhibited case) be found attached to tRNA at the end of the reaction?

(c) What known inhibitor of protein synthesis appears to act at the same step as frampomycin?

11. Bromoacetic acid is very reactive, forming covalent bonds with many organic compounds. Bromoacetic acid can be attached by its carboxyl group to lysyl-tRNA at two positions of the lysine: the α amino group or the ϵ amino group (see drawing). It was found that when a radiolabeled α-bromoacetyl derivative was added to ribosomes in the presence of polyadenylic acid as an mRNA, most of the radioactivity could be donated to the antibiotic puromycin; under similar conditions, the ϵ-bromoacetyl derivative was not donated to puromycin. In other experiments the α- and ϵ-derivatives were bound to ribosomes in the presence of poly adenylic acid but without any elongation factor G. After sufficient incubation, radiolabeled lysine from the α-derivative was covalently bound to protein L2 of the large ribosomal subunit, while the ϵ-derivative was covalently bound to protein L27 of the large subunit

and to the 23S rRNA. Interpret these experimental results in the light of your knowledge of the structure and function of the ribosome.

12. Compare the rate of loss of a particular enzyme in a eucaryotic cell that degrades the protein at a rate of 30% per hour with a procaryotic cell that does not degrade the protein at all, but which has stopped its synthesis while continuing to grow with a doubling time of 100 minutes.

SELECTED READINGS

Crick, F. H. C. (1981) The genetic code. In *Genetics. A collection of readings from Scientific American.* San Francisco: W. H. Freeman.

Crick, F. H. C. (1981) The genetic code III. In *Genetics. A collection of readings from Scientific American.* San Francisco: W. H. Freeman.

Freifelder, D. (1985) *Essentials of molecular biology.* Boston: Jones and Bartlett.

Hanawalt, P. C., and Haynes, R. H., eds. (1973) *The chemical basis of life.* San Francisco: W. H. Freeman.

Hardesty, B., and Kramer, G., eds. (1986) *Structure, function, and genetics of ribosomes.* New York: Springer-Verlag.

Hopfield, J. J. (1974) Proceedings of the National Academy of Sciences, USA. 71: 4135–4139.

Kirkwood, T. B. L., Rosenberger, R. F., and Galas, D. J., eds. (1986) *Accuracy in molecular processes: Its control and relevance to living systems.* London: Chapman and Hall.

Yanofsky, C. (1981) Gene structure and protein structure. In *Genetics. A collection of readings from Scientific American.* San Francisco: W. H. Freeman.

Regulation of the Flows of Mass, Energy, and Information

14

In the last few chapters we have concentrated our attention on the functioning of individual parts of the metabolic machinery. These might be thought of as analogous to the individual operations involved in the manufacture of an industrial product such as an automobile. More impressive still is the marvelous ability of the cell to coordinate its many functions in the face of environmental fluctuations so as to maintain itself or, even more spectacular, to grow and divide. The automobile industry could learn much by studying how a cell balances the controls of inventory, subcontractors, raw materials, and energy resources, not only to make extracellular products but also to build another factory in its own image.

We now know some general principles of cellular control at the molecular level. In this chapter we shall discuss (sometimes with specific examples) the current state of our understanding of regulation, draw attention to some avenues of approach, and suggest the direction in which future work is likely to proceed.

Control of Enzyme Activity

By regulation of enzymatic activity we mean both those control mechanisms that determine the amount of an enzyme present in a cell and those that control the reaction rate of existing enzymes. The former operate at the level of genetic expression (see Chapters 12 and 13), while the latter operate on the enzyme molecule itself (see Chapter 5) and therefore affect metabolic flow rates more rapidly. These two types of control are often coordinated with each other.

As an example, consider an *E. coli* bacterial cell growing in an environment that lacks tryptophan. The synthesis of this essential amino acid requires the activity of several enzymes: the ones catalyzing the metabolic pathway from chorismate to tryptophan (see Chapter 6). To grow under these conditions the cell must transcribe several genes into mRNAs, translate the messages into proteins, and ensure that the activity of these enzymes is sufficient to supply all the needed tryptophan. Now imagine that the

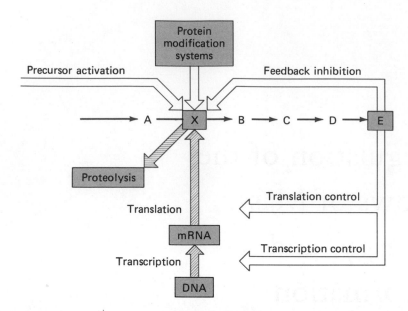

FIGURE **14–1**

Sites of Control of Cellular Metabolism. The X is an enzyme that catalyzes the transformation of metabolite A into B, the first step in a biosynthetic pathway leading to end product E. Information flow is shown schematically by large cross-hatched arrows. Control of the activity of enzyme X is shown in the top part of the figure, control of its synthesis at the level of transcription and translation in the lower part. Information flow in the control circuits is represented by large arrows.

environment changes and tryptophan becomes available (the mammal with the *E. coli* in its intestine has a protein-rich meal). The *E. coli* responds to this change by reducing the activity of the pathway that makes tryptophan and by inhibiting the synthesis of enzymes that catalyze the pathway. This control allows the cell to use the metabolic resources that were devoted to maintaining the tryptophan biosynthetic pathway for other purposes: more rapid growth and division.

The existence of coordination among metabolic pathways allows an unperturbed cell to maintain constant concentrations of metabolites, despite their synthesis and utilization by several pathways (see Chapter 6). Moreover, the cell exhibits *homeostasis;* if the environment changes there is a small transitory change in metabolite concentrations followed by a relatively rapid return to typical levels. Damages can often be put right. Even mutations that inactivate a whole biosynthetic pathway sometimes can be compensated by channeling material through an alternative route to produce normal amounts of a vital substance. It is possible that this is the significance of seemingly redundant pathways for certain metabolites (see discussion in Chapter 6).

When discussing control, the concern is primarily with information flows, within the cell and between the cell and its environment. The flow of information that controls enzyme concentrations originates in DNA, passes into RNA via transcription, and into protein via translation (Figure 14–1). Control is exerted at each of these stages and perhaps also in the degradation of proteins by proteolysis.

Enzyme-Metabolite Interactions

Of the small molecules that control catalytic activity, some are specific to the enzyme—substrate, product, cofactors, inhibitors, activators, and so on—and some are general—pH or ion concentrations. The substrates and products of an enzymatic reaction have great influence on the rate of the reaction. In cells, an enzyme typically is *not* saturated by its substrate, so as substrate concentration is increased, the rate is augmented. As the reaction proceeds, the concentration of product increases, the net forward reaction tends to slow, and eventually the reaction will come to equilibrium. In many cases the product of an enzymatically catalyzed reaction also inhibits the enzyme by competing with the substrate for the active site. The degree of this inhibition depends on the concentrations of both substrate and product and on their relative affinities for the enzyme (see Chapter 5).

Feedback Inhibition by End Products

Most enzymes are involved in a linear sequence of chemical reactions in which the product of one reaction is a substrate for the next enzyme in the sequence. Considering a short unbranched segment of metabolic pathway,

$$A \rightarrow B \rightarrow C \rightarrow D \rightarrow E,$$

it has often been found that the product of the reaction D → E inhibits the enzyme catalyzing the reaction A → B. This is an example of *negative feedback control* (*negative* because A → B is inhibited, and *feedback* because the result of a later event controls the rate of an earlier one). Thus

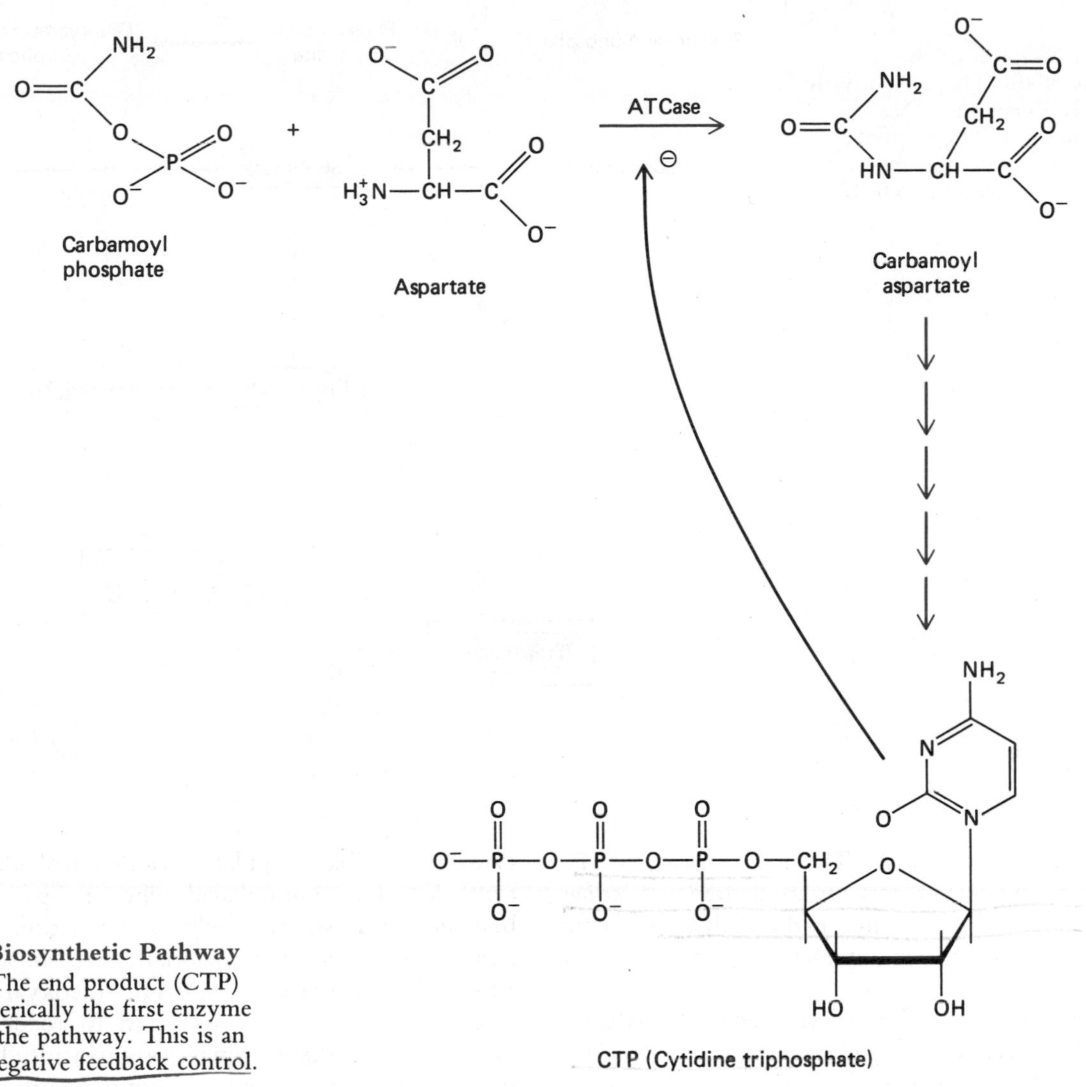

FIGURE **14–2**
Simplified Biosynthetic Pathway for CTP. The end product (CTP) inhibits <u>allosterically</u> the first enzyme (ATCase) in the pathway. This is an example of <u>negative feedback control</u>.

CTP (Cytidine triphosphate)

when the concentration of E rises to a certain level, the flow of mass through the whole pathway segment is inhibited, by virtue of the diminished flow of A → B, and the further formation of E is consequently slowed. As the concentration of E falls, usually by its being used in one or another subsequent chemical reaction, the inhibition of A → B is relaxed and faster synthesis of E can occur.

A small molecule that affects the function of an enzyme is called an **effector** of that enzyme. The intriguing problem is how a molecule as different from A as E may be (if it is separated by a large number of intermediate chemical transformations, it can be very different indeed) nevertheless influences the enzyme catalyzing A → B. <u>The answer does not lie in E competing with A for an active site</u> (where E cannot bind, since it has the wrong shape), <u>but rather in the existence of two different enzyme-combining sites, one for the substrate and one for the feedback inhibitor</u>. This

state of affairs has now been demonstrated for a large number of enzymes.

The theory of this mode of changing enzyme activity was first formally stated by J. Monod, J. P. Changeaux, and F. Jacob; we have already discussed it in Chapter 5 both in relation to the function of enzymes generally and for the particular case of aspartyl transcarbamoyl synthetase (ATCase). Proteins like ATCase whose actions can be modified by small molecules that do not especially resemble the substrate, and thus have an *other shape (allosteric)*, are called <u>*allosteric proteins*</u>. If we once again examine the catalytic pathway of which ATCase is a part (Figure 14–2), we notice that this enzyme is the first in a linear sequence of reactions that synthesize the RNA precursor CTP. The pathway begins with the condensation of carbamoyl phosphate and aspartate, an amino acid precursor of protein synthesis, and then goes through five more steps, each catalyzed by a different enzyme, before

FIGURE 14–3

A Schematic Version of the Biosynthesis of the Three Aromatic Amino Acids Tyrosine, Phenylalanine, and Tryptophan. Compare the information in this illustration with that in Figure 6–12.

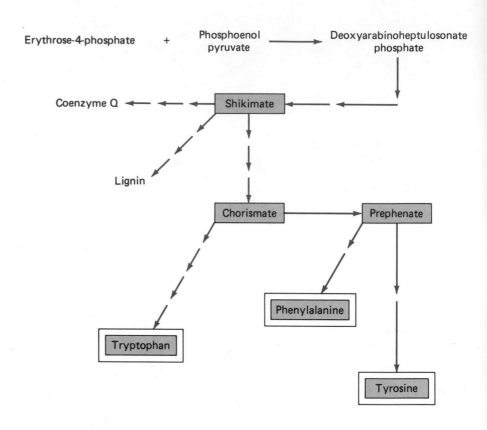

CTP is finally synthesized. The *end product, CTP,* regulates its own biosynthesis by an *allosteric interaction* with ATCase, the first enzyme in the pathway specific for CTP. (Details of this interaction were given in Chapter 5.)

The net result of the negative feedback mediated through ATCase is that the cellular concentration of CTP is closely regulated. If the concentration rises, the activity of ATCase is further inhibited and the synthesis of CTP is diminished. A slowing of CTP synthesis leads to a reduction in CTP concentration since CTP is used constantly for the synthesis of RNA. Conversely, a fall in CTP concentration reduces CTP binding to the allosteric inhibition site, ATCase activity rises, the rate of CTP synthesis increases, and so does CTP concentration. Thus, homeostasis is assured.

More Complicated Pathways

In Chapter 6 we discussed the biosynthetic pathways for the aromatic amino acids (Figure 6–12). A more schematic version of these pathways in bacterium *E. coli* is illustrated in Figure 14–3. Even this simple version is a more complicated scheme than the unbranched linear sequences of reactions in CTP synthesis. The complexities arise because several of the reaction steps are shared among the biosynthetic pathways leading to the three aromatic amino acids: phenylalanine, tryptophan, and

tyrosine. Feedback regulation can control adequately the synthesis of an amino acid when its pathway is unbranched. But simple feedback regulation by a single amino acid of the first enzymatic step shown in Figure 14–3, the synthesis of DAHP (deoxyarabinoheptulosonate phosphate), would lead to reduced rates of synthesis for all three aromatic amino acids. The problem the cell faces, then, is how to control both the levels of intermediates in the common pathway and the levels of all three amino acids, in the face of differing demands.

The first seven reactions of the pathway to aromatic amino acids are shared (Figure 14–4). In *E. coli* cells the first reaction is catalyzed by three different enzymes, which are coded by three different genes. Each enzyme uses the same precursors and produces the same products. Enzymes catalyzing the same reaction are called *isozymes,* and their occurrence in cells is widespread. In the case of aromatic amino acid biosynthesis by *E. coli,* each of the three DAHP synthetase isozymes responds to a different allosteric regulator: one is inhibited by phenylalanine, one by tryptophan, and one by tyrosine (indicated by feedback arrows crossing the three reactions in Figure 14–4). When the concentration of phenylalanine rises, for example, only one of the DAHP synthetase isozymes is inhibited and material can still flow rapidly through the common pathway for the synthesis of tryptophan and tyrosine.

Since the common pathway in these conditions will produce chorismate, a branch point in the scheme in

Erythrose-4-phosphate + Phosphoenol pyruvate

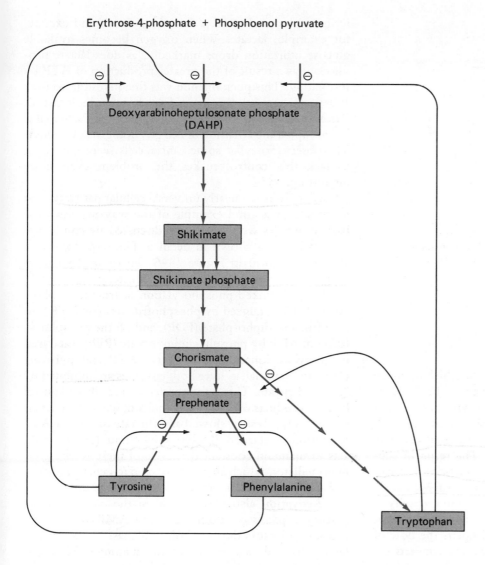

FIGURE 14–4
**Synthesis of Tyrosine,
Phenylalanine, and Tryptophan
from Common Precursors.** Black
arrows and (−) signs indicate points of
feedback inhibition. (Adapted from
P. H. Clarke.)

Figure 14–4, additional control is necessary to prevent the synthesis of phenylalanine when it is plentiful. This regulation occurs by allosteric interaction of phenylalanine with one of the branch enzyme pathways leading from chorismate. In *E. coli* the chorismate mutase and prephenate dehydratase activities are present on one *bifunctional enzyme,* and it is subject to feedback inhibition by phenylalanine. The path to tyrosine is similar in that a bifunctional enzyme combines chorismate mutase and prephenate dehydrogenase activities and is regulated by the tyrosine concentration. The tryptophan branch is controlled by an allosteric interaction of tryptophan with the enzyme catalyzing the reaction from chorismate to anthranilate (Figure 14–4).

The problem of controlling the flow of mass through the common and branch pathways has not been solved in the same way by all cells. In some other procaryotes (e.g., *B. subtilis*), there is only one DAHP synthetase and it responds not to the end-product amino acids but to

intermediates of the common pathway, chorismate and prephenate. In *Pseudomonas* species a single DAHP synthetase responds to tryptophan, tyrosine, and phenylpyruvate (a substrate for the transaminase that finally forms phenylalanine).

Precursor Activation and Feedforward Control

In Chapter 5 we discussed the activation of enzymes. This often occurs in response to binding a precursor in the pathway containing the enzyme. Especially simple examples are found among the first enzymes in degradative pathways where the activation is by the substrate. In the general pathway

$$S \rightarrow T \rightarrow U \rightarrow V \rightarrow I,$$

which degrades substrate S to tricarboxylic acid cycle intermediate I, for example, it will often be found that the activity of the first enzyme does not follow Michaelis-

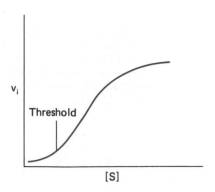

FIGURE 14-5
The Positive Homotropic Kinetics of a Typical Pacemaker Enzyme in a Degradative Pathway. Note the relatively low initial velocity at low substrate concentrations and the rapid rise in velocity above a threshold substrate concentration.

Menten kinetics. If it did, the substrate S, which might be an amino acid, would be funneled away to the degradation path from other uses (e.g., protein synthesis), even if S were present at relatively low concentrations. Instead, the concentration of S *feeds forward* to the first enzyme, which experiences *substrate activation*. The result of substrate activation is a *sigmoid* (S-shaped) kinetic response to increasing substrate concentration, rather than Michaelis-Menten kinetics. That is, the first enzyme's activity is kept low until an excess of substrate S is present. At that concentration, the threshold, the cell opens the flow of mass through the degradative pathway and converts the extra substrate into other useful molecules (Figure 14-5).

The activation need not be by the immediate substrate of the responding enzyme. In the glycolytic pathway of mammalian liver cells, pyruvate kinase is activated allosterically by fructose diphosphate. There are several intervening enzymatic conversions between the formation of fructose diphosphate and the phosphorylation of ADP by pyruvate kinase, so this feedforward control is known as *precursor activation*. As it turns out, mammalian liver pyruvate kinase is also substrate-activated by phosphoenol pyruvate. Again, different cells may have different control mechanisms: the pyruvate kinase of mammalian muscle cells does not respond to fructose diphosphate.

The Pasteur Effect and the Regulation of Glycolysis

Cells that both can grow without oxygen and can use it as the terminal electron acceptor of respiration are called *facultative anaerobes*. In the absence of oxygen, they ferment glucose by the glycolytic pathway and excrete, for example, lactate. When oxygen becomes available glucose utilization drops markedly, as does lactate production, as a result of the greater production of ATP by respiration. This phenomenon was first investigated over 100 years ago by the great French microbiologist L. Pasteur, and the mechanism of control of the *Pasteur effect* has occupied the attention of cell physiologists ever since. The observation that some cancer cells seem somehow to lack this control makes the problem even more interesting.

Glycolysis is a nearly universal cellular pathway, and its control is a good example of the way in which the basic principles we have already discussed are combined. The pathway is shown in detail in Figure 6-15 and in abbreviated form in Figure 14-6. There are two main steps where control is exerted by allosteric interactions: (1) the ATP-linked phosphorylation of fructose-6-phosphate (F6P), catalyzed by phosphofructokinase (PFK), to give fructose diphosphate (FDP); and (2) the phosphorylation of ADP by phosphoenol pyruvate (PEP), catalyzed by pyruvate kinase (PK), to give ATP and pyruvate (Pyr). In mammalian liver both enzymes are inhibited by ATP and citrate. This is interpreted as a feedback control by later products of energy metabolism; when ATP levels are high glycolysis is slowed and the rate of further ATP formation is retarded. Inhibition by citrate (or isocitrate) acts to maintain constant the levels of these Krebs cycle intermediates, which are not only used to generate energy but also act as precursors for biosynthetic pathways.

Activation also plays a role in the control of the glycolytic pathway. Both ADP and AMP are positive allosteric effectors of phosphofructokinase. This can also be understood as a means of maintaining ATP levels constant during varying demands by the many metabolic transformations requiring energy. As ATP is used, ADP and AMP are both produced and when their concentrations rise the glycolytic pathway is stimulated so as to generate more ATP. We have discussed above the feedforward activation of pyruvate kinase by FDP and PEP.

Some cancer cells have very high rates of glucose uptake despite having a functioning respiratory pathway. (Their defect in control offers a possible avenue of intervention in therapy, as do most observed deviations from normal in tumor metabolism.) The differences could lie in a reduced response by glycolytic enzymes to ATP levels or it could be, as has been suggested by E. Racker, that massive utilization of ATP keeps its levels below those necessary to control glycolysis.

Glycolysis is not the only pathway to respond to levels of ATP, ADP, or AMP. The energy status of the cell can influence the flow of mass through many reactions. One way of evaluating the level of cellular

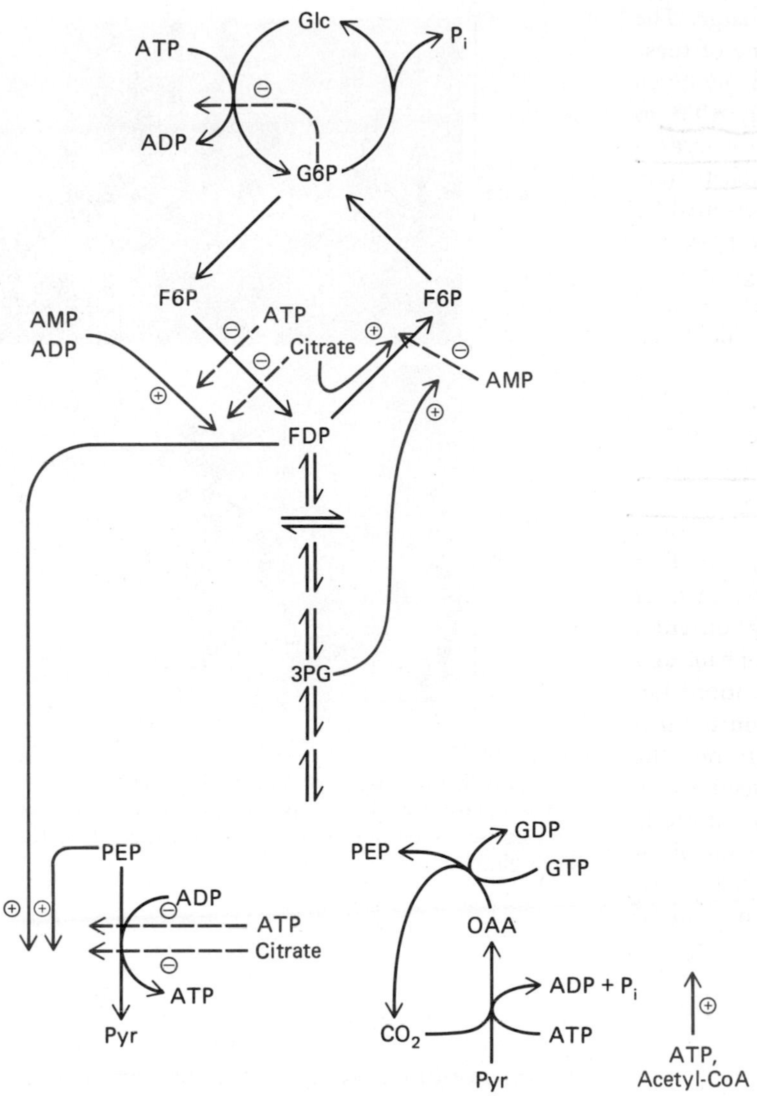

FIGURE **14–6**
An Abbreviated Representation of the Glycolysis and Gluconeogenesis Pathways in Mammalian Liver Cells. The main points of control are shown: Glc = glucose; G6P = glucose-6-phosphate; F6P = fructose-6-phosphate; FDP = fructose diphosphate; 3PG = 3-phosphoglycerate; PEP = phosphoenol pyruvate; Pyr = pyruvate; OAA = oxaloacetate. Feedback inhibition by ATP, G6P, and citrate is shown (color lines, negative sign), as well as activation by ADP and AMP (color arrows parallel to reaction arrows, positive sign) and feedforward activation by FDP (positive sign). Inhibition of gluconeogenesis by AMP is shown, as well as activation by ATP, citrate, acetyl-CoA, and 3-phosphoglycerate.

energy is the *energy charge,* a concept introduced by D. Atkinson:

$$\text{energy charge} = \frac{\frac{1}{2}[\text{ADP}] + [\text{ATP}]}{[\text{AMP}] + [\text{ADP}] + [\text{ATP}]}.$$

The denominator is the total available concentration of adenine nucleotides while the numerator is a measure of high-energy phosphates available for metabolic reactions. (The factor ½ comes from considering the reaction catalyzed by adenylate kinase: $2\,\text{ADP} \rightleftharpoons \text{ATP} + \text{AMP}$.) If all the adenine nucleotides are in the form of ATP, the energy charge is unity; if all are AMP, the energy charge is zero. It seems as though many enzymes respond, by whatever detailed mechanism, to the energy charge itself: when it is close to one, an enzyme like phosphofructokinase is inhibited but when the energy charge falls

toward zero, the enzyme is stimulated. In some cells an energy charge of 0.85 seems to be where metabolism is poised. Below, ATP-generating pathways are stimulated, while an energy charge above 0.85 stimulates utilization of ATP.

Among the enzymes that appear to be stimulated by a high energy charge are the ones specific to gluconeogenesis (see Figure 14–6). The synthesis of glucose from pyruvate shares many, but not all, of the enzymes of glycolysis. Two reactions of gluconeogenesis that are not shared are the hydrolysis of fructose diphosphate, catalyzed by fructose diphosphatase, and the formation of phosphoenol pyruvate from pyruvate via pyruvate carboxylase, an oxaloacetate/malate shuttle across the mitochondrial membrane (see Figure 8–23), and phosphoenol pyruvate carboxykinase. Fructose diphosphatase is inhibited by AMP and stimulated by citrate and 3-phosphoglycerate. It thus has a response generally opposite to that of phosphofructokinase in respect of energy charge. The control has evolved to the point that only one of these two enzymes normally shows much activity at any given time, thus avoiding the *futile cycle* of using ATP to synthesize fructose diphosphate only to have it hydrolyzed again immediately to fructose-6-phosphate (see Figure 6–2). A PEP/pyruvate futile cycle is prevented by the requirement of pyruvate carboxylase for acetyl-Co A as a positive effector. Unless the energy charge is high, acetyl-Co A will be used rapidly in the tricarboxylic acid cycle and thus be unavailable for activating the initial step of gluconeogenesis.

It is easy to imagine ever more complicated regulatory schemes based on the principles we have discussed so far. Enzymes occupying critical points in the metabolism of a cell may be subject to modulation by many allosteric effectors. The current champion of complexity for a single enzyme is the glutamine synthetase of *E. coli*. This protein is composed of 12 identical subunits, each of 50,000-Da mass, and responds to at least eight different metabolites including the AMP, CTP, tryptophan, and carbamoyl phosphate that we have already mentioned in connection with other enzymes. The evolution of such baroque control appears to be a specialty of *E. coli;* the glutamine synthetases of mammalian cells tend to be much simpler and respond to only a few allosteric effectors. It can be supposed that careful control of glutamine synthetase is of great value to *E. coli,* and that these advantages are gained in other ways by mammals.

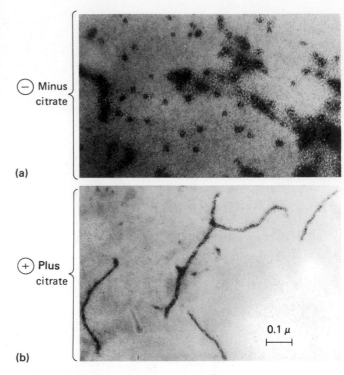

FIGURE 14–7

Electron Micrographs of Purified Acetyl-CoA Carboxylase.   (a) In the absence of citrate. ×160,000. (b) In the presence of citrate. ×240,000. (Courtesy of A. K. Kleinschmidt.)

We close our discussion of allosteric control with an example of conformational change whose consequences are detectable by the electron microscope (Figure 14–7): the case of acetyl-Co A carboxylase. This enzyme catalyzes the carboxylation of acetyl-Co A by CO_2 to form malonyl-Co A, the first and key step in fatty acid synthesis. A high citrate concentration enhances fatty acid synthesis and acetyl-Co A carboxylase is the site of this activation. Electron microscopic investigation has revealed an exciting visualization of this allosteric control. Figure 14–7a shows the purified enzyme as it appears in

the absence of citrate; it has the form of small particles, dimensions from 10 to 30 nm, having a sedimentation constant of 20S. In the presence of the appropriate citrate concentration, one that activates the enzyme, a remarkable change takes place (Figure 14–7b): the individual enzyme particles line up to form filaments from 7 to 10 nm in width and up to 400 nm long, having a sedimentation constant of 46S. The change in the structural relations between particles is easy to interpret as a change in the structure of the enzyme, induced by the binding of allosteric effector citrate.

Covalent Modification of Enzymes

The activity of enzymes in one part of the metabolic network can be altered by information from another part. This information is always communicated in the form of a material substance. The binding and dissociation of ligands, such as end products, precursors, or other allosteric effectors, are rapid events and the target enzymes respond to the equilibrium concentrations of the metabolites. In many cases, however, coordination of the cell's activities requires an enzyme to respond to some process that does not make a freely diffusible metabolite. This situation arises, for example, when a cell must respond to the presence in its environment of a hormone that cannot penetrate the cell membrane. The pathway of information flow under these conditions is accomplished by the attachment or detachment of a ligand, by making or breaking a covalent bond. The control of an enzyme by this mechanism results from the steady-state balance of activity of two other enzymes, one that attaches the ligand and the other that detaches it. The source of the ligand is always an activated intermediate, so this kind of control dissipates free energy.

A typical ligand for this purpose is a phosphate group, although in various contexts acetyl, methyl, and other more complicated ligands have been observed in analogous roles. A specific example is the attachment of a phosphate group to the L-isozyme of pyruvate kinase; this reduces catalytic activity by altering the enzyme's affinity for one of its substrates, phosphoenol pyruvate. The enzymes that catalyze the attachment of a phosphate are called *protein kinases;* the phosphate donor is usually ATP:

$$\text{protein} + \text{ATP} \xrightarrow{\text{protein kinase}} \text{protein-}\boxed{P} + \text{ADP}.$$

Some proteins can accept more than one phosphate group. Phosphate is usually esterified to a serine residue, although threonine and even histidine, lysine, and tyrosine have been observed as phosphate acceptors. Small peptides containing the same sequence of amino acids that surrounds the acceptor residue are often acceptors also. This suggests that secondary or tertiary folding of the protein may not always be required for activity as a phosphate acceptor.

Detachment of the phosphate group is catalyzed by enzymes known as *phosphoprotein phosphatases:*

$$\text{protein-}\boxed{P} + \text{H}_2\text{O} \xrightarrow{\substack{\text{phosphoprotein}\\\text{phosphatase}}} \text{protein} + \text{P}_i.$$

While the protein kinases show significant specificity toward their substrate proteins, the phosphatases show little. The same phosphatase can remove multiple phosphates from a single phosphoprotein and single phosphates from a large variety of phosphoproteins.

Protein kinases and phosphoprotein phosphatases have been found in both procaryotic and eucaryotic cells. The attachment and detachment of ligands by covalent bonds thus seems to be a common mode of regulating the activity of enzymes. It should be realized, however, that the presence of the attached ligand does not overwhelm the responsiveness of an enzyme to reversible allosteric

effectors. In the case of pyruvate kinase from mammalian liver, inhibition by the phosphate group can be overcome by increases in the concentrations of positive effectors such as fructose diphosphate or the substrate phosphoenol pyruvate. In this case the presence or absence of the phosphate sets a general level of enzyme activity but does not eliminate rapid control by reversibly bound ligands. In fact, the concentrations of these allosteric effectors affect the rate of phosphorylation by the protein kinase and also may alter rates of dephosphorylation by the phosphatase. This brings us to the important question of what regulates these two activities.

Regulation of Protein Kinases

The main known regulators of protein kinases are ligands that bind reversibly. First and foremost among these is *3',5'-cyclic AMP* (*cAMP,* Figure 14–8). A protein kinase that responds to this effector is known as *cAMP-dependent* (sometimes as a *protein kinase A*). In the presence of appropriate concentrations of cAMP the (negative) regulatory protein subunits (R) bind cAMP and undergo a conformational change; $R \cdot (cAMP)_2$ then dissociates from the catalytic (C) subunits.

$$R_2C_2 + 4\,cAMP \rightleftharpoons R_2(cAMP)_4 + 2\,C.$$
(inactive holoenzyme)   (active catalytic subunit)

(An alternative mechanism is that R first dissociates from R_2C_2, spontaneously undergoes a conformational change, and then binds cAMP, which stabilizes the dissociated state.) Without the R subunit, there is a change in the conformation of C, which can then express its kinase activity (see Figure 14-10). As cAMP concentrations fall, the equilibrium is displaced to the left and kinase activity becomes inhibited.

Regulation of Cyclic AMP

The reader will already have noticed that regulation of a metabolic enzyme's activity by phosphorylation-dephosphorylation is rather less direct than other modes of regulation we have discussed. To follow the path of information flow back to its source, we must now consider the synthesis and degradation of cAMP. The mass and energy for cAMP synthesis is supplied by its precursor ATP and the reaction is catalyzed by enzymes known as *adenyl cyclases* (Figure 14–8). Degradation of cAMP to 5'-AMP is catalyzed by *phosphodiesterases*. The concentration of cAMP inside a cell, and thus the relative activity of the cAMP-dependent protein kinases, is determined by the competing activities of adenyl cyclases and phosphodiesterases.

The importance of cAMP as a member of the chain of regulation of enzyme activity was first appreciated by

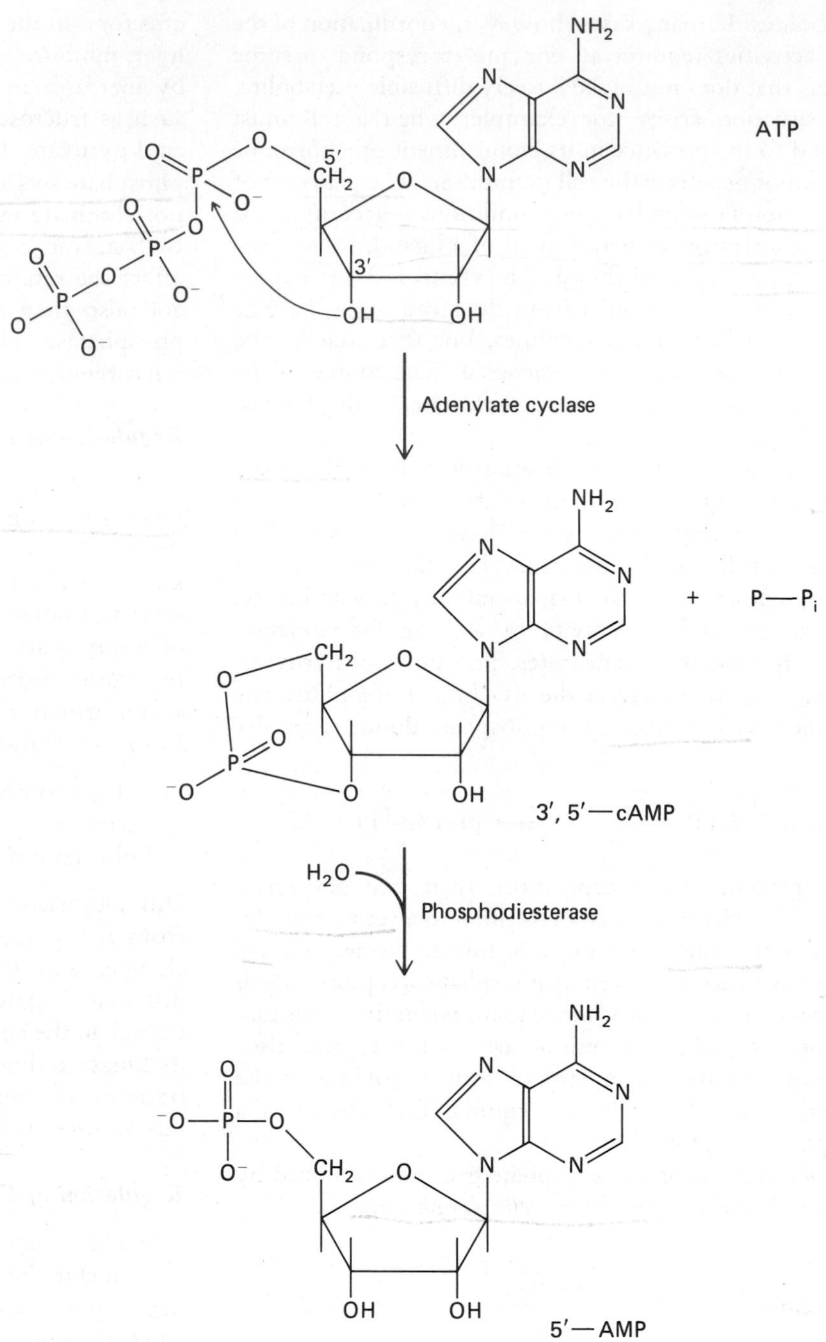

FIGURE 14–8
Synthesis and Degradation of cAMP. ATP is cyclized to form 3′,5′-cyclic AMP, splitting off inorganic pyrophosphate. Hydrolysis of cAMP by phosphodiesterase yields 5′-AMP.

E. Sutherland, who was investigating the stimulatory effects of the peptide hormone glucagon on the conversion of glycogen to glucose in mammalian liver cells. Working with T. Rall and W. Butcher, Sutherland discovered that glucagon increases the concentration of cAMP in liver cells by activating an adenyl cyclase associated with the liver cell membrane. These researchers also discovered that in liver cell homogenates, free of membranes, glucagon had no direct effects but cAMP did. Furthermore, an inhibitor of phosphodiesterase, caffeine, potentiates the effects of the hormone in whole liver cells, presumably by allowing higher steady-state concentrations of cAMP to be maintained. Sutherland called cAMP a *second messenger*, meaning an intermediate between the first hormone messenger (in this case, glucagon) and the cell's metabolism. From the discussion above we now know there are at least third messengers in the form of protein kinases (and one-and-one-half messengers in the form of adenyl cyclase!).

It turns out that many peptides, acting as hormones, growth factors, and neurotransmitters, affect target cells by binding to specific *receptor proteins* on the outer surface

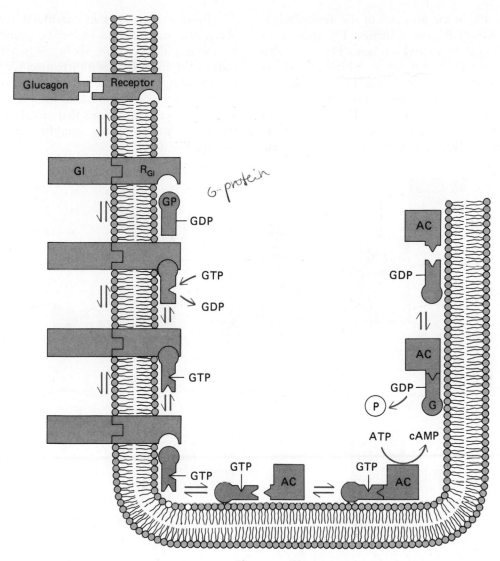

FIGURE **14–9**
Transduction of Information Across the Cell Membrane. Glucagon, like other peptide hormones and like epinephrine, binds reversibly to its specific receptor, an integral membrane protein. The binding induces a conformational change in the receptor, which allows it to interact with a membrane-bound G_s-protein (GP). In turn this binding weakens the association of GP with GDP, allowing GTP to bind. This activates the G_s-protein which can combine with membrane-bound adenyl cyclase (AC), stimulating it. Adenyl cyclase catalyzes the synthesis of cAMP which acts as an intracellular second messenger. The stimulated synthesis of cAMP continues until a GTPase activity on the G_s-protein catalyzes the hydrolysis of GTP to GDP, which brings adenyl cyclase back to unstimulated levels of activity. The G_s-protein must then diffuse through the membrane and become reactivated by binding to a hormone-receptor complex. Inhibitory hormones work similarly. Their receptors activate inhibitory G_i-proteins which, when GTP is bound, inhibit adenyl cyclases.

of the cell membranes (see Chapter 7). After the receptor binds its hormone (factor or transmitter) there is stimulation of an adenyl cyclase associated with the inner surface of the membrane. Coupling of the function of the receptor to the adenyl cyclase is accomplished by another protein that is bound to the inner surface of the membrane. This protein, known as the *G-protein* (or *N-protein*), strongly binds the *n*ucleotide GTP when the receptor has bound hormone. With GTP bound, the G-protein (actually, its α subunit) interacts with and stimulates the activity of adenyl cyclase.

The G-protein also has a GTPase activity. Since adenyl cyclase is stimulated when a G_s-protein (N_s-protein) binds GTP, but not GDP, the GTPase activity tends to stop the stimulation of adenyl cyclase (Figure 14–9). One experiment that established this behavior was the observation that in the presence of the nonhydrolyzable analogue GDPCP (see Chapter 13) there is a persistent activation of adenyl cyclase. There are also G_i-proteins (N_i-proteins) that act to inhibit adenyl cyclase. It appears that one hormone · receptor complex can activate several G-proteins to bind GTP and thus stimulate or inhibit several adenyl cyclases. This arrangement would also allow more than one kind of receptor to influence adenyl cyclase, by using a common G-protein coupling system.

Protein hormones in the blood carry information about the state of the endocrine glands secreting the hormone and, therefore, about the general physiological state of the animal. This information is communicated to a responding cell via its membrane and the synthesis of cAMP. As the cAMP concentration rises it can affect the activity of protein kinases that covalently modify other enzymes, thus producing amplified physiological effects (Figure 14–10).

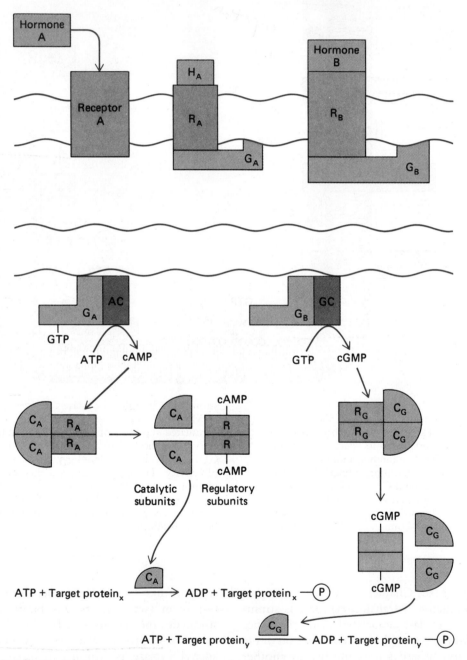

FIGURE 14–10
The Binding of a Hormone to Its Specific Membrane Receptor Leads to Phosphorylation of Target Proteins.   Both cyclic AMP and cyclic GMP-activated protein kinases are shown.

Cholera Toxin and the G-Protein

Cholera is a serious infectious disease in many parts of the world. The cause of damage to the patient is high intracellular cAMP. A stimulating G_s-protein is poisoned by the active subunit of *cholera toxin,* which is excreted by the bacteria that cause the disease. Once inside the cell, this subunit catalyzes the transfer of ADP-ribose from NAD^+ to an arginine residue on the G-protein (recall our discussion in Chapter 13 of a similar transfer to eEF2, catalyzed by the active subunit of diphtheria toxin). The covalently attached ADP-ribose inactivates the GTPase activity of the G-protein so it continually activates adenyl cyclase and the cell experiences very high concentrations of cAMP.

In the epithelial cells of the intestine, these high concentrations of cAMP activate the transport of chloride into the intestine. Sodium ions follow, to maintain electrostatic neutrality, and the cells lose NaCl into the intestinal fluid. The ions are replaced from the extracellular fluid and, therefore, from the rest of the body. The flow of Na^+ and Cl^- leads to a loss of water into the intestine, to maintain osmotic equilibrium. The afflicted patient experiences a massive diarrhea which, left untreated, is lethal due to dehydration.

Treatment of the diarrhea exploits the cotransport process we discussed in Chapter 7. The patient is given antibiotics to control the bacteria that produce the toxin, and an isotonic NaCl solution containing glucose. In the intestine, the glucose participates in cotransport with sodium ions; both enter the cells. Once inside, the sodium pump sends Na^+ into the body; chloride follows to maintain electrostatic neutrality and water follows to maintain osmotic equilibrium. This reduces the dehydration and spares the patient until the infection is brought under control.

Protein Kinase C and Calcium Signaling

Not all peptide signaling agents act by influencing intracellular cyclic AMP. There are over two dozen membrane receptors that, when activated by their specific hormones, neurotransmitters, or growth factors, lead to the production of second messengers via a degradative pathway. For example, the activated receptor stimulates a membrane-associated phospholipase to hydrolyze a membrane lipid: phosphatidylinositol (Figure 14–11). The products of the hydrolysis are both second messengers: diacylglycerol, which remains associated with the membrane, and *phosphoinositol-bis-phosphate* (IP$_3$), which diffuses into the cytoplasm. The presence of IP$_3$ in the cell leads to a rapid release of Ca^{2+} ions from an intracellular store (probably the endoplasmic reticulum). Later there is probably also influx of Ca^{2+} from the extracellular fluid. Diacylglycerol activates, in the presence of the elevated intracellular $[Ca^{2+}]$, a membrane-associated protein kinase (*protein kinase C*). The physiological effect of binding the hormone, etc., to its receptor is finally expressed by the phosphorylation of cellular enzymes, catalyzed by protein kinase C.

The intracellular concentration of calcium ion is normally quite low, $<10^{-7}$ M, due to the presence of active calcium pumps in the cell membrane and in organelle membranes. A rapid rise in cytoplasmic $[Ca^{2+}]$, to $\sim10^{-5}$ M, is a typical cellular response to many stimuli. Usually this stimulus is mediated through the cell membrane, by binding a small effector molecule to a specific receptor, for example. The conformational change elicited in the receptor by binding its ligand can lead to opening of calcium channels in the cell membrane, as well as the pathway involving IP$_3$ that was described above. The Ca^{2+} ion may bind directly to a target protein and cause an activation or inhibition.

More commonly, however, calcium acts indirectly, after binding to the ubiquitous protein *calmodulin*. This small acidic protein has four binding sites for calcium ions. It has survived in a wide variety of organisms, from protozoans and coelenterates to flowering plants and humans, with remarkably little variation in its sequence of approximately 150 amino acids. When intracellular $[Ca^{2+}]$ rises to $\sim10^{-6}$ M, it binds to calmodulin and the complex undergoes a conformational change. The activated complex can bind to various target proteins, induce a conformational change, and stimulate (typically) their activity. Among the enzymes that are activated by the calmodulin $\cdot$ Ca^{2+} complex are protein kinases. Another example is cyclic nucleotide phosphodiesterase—the enzyme that catalyses the cAMP $\rightarrow$ AMP reaction; thus, this cellular response interacts with the response mediated via cAMP.

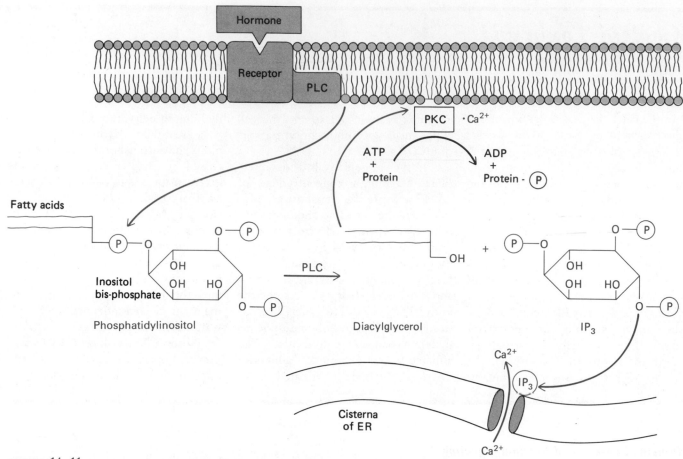

FIGURE **14-11**

Formation of Second Messengers Diacylglycerol and IP$_3$, and the Activation of Protein Kinase C.   When hormone (or neurotransmitter or growth factor) binds to a specific receptor, it activates a membrane-associated phospholipase C (PLC). This enzyme catalyzes the hydrolysis of phosphatidylinositol, a relatively rare phospholipid in the cell membrane (color). The products, diacylglycerol and phosphoinositol bis-phosphate (IP$_3$), act as second messengers. IP$_3$ stimulates the release of calcium ions from intracellular stores, most likely the endoplasmic reticulum. The increased intracellular [Ca^{2+}] and diacylglycerol activate the membrane-associated protein kinase C, which catalyzes the phosphorylation of specific target proteins.

Control of Phosphoprotein Phosphatases

The regulation of phosphoprotein phosphatase activity seems much less complicated, at least in our present state of knowledge. There is competition among the various phosphoprotein substrates that would, in conditions of limiting phosphatase, provide some specificity of action; those phosphoproteins with the highest affinity for the phosphatase would be dephosphorylated fastest. There also appear to be allosteric effects on the susceptibility of a phosphoprotein to dephosphorylation. Allosteric effectors, by binding to the substrate protein, may also affect rates of phosphorylation. Finally, there have been reports of heat-stable proteins whose presence can inhibit the phosphatase activity.

Activation of Glycogen Phosphorylase

As a final example of the complicated path of regulation induced by the binding of glucagon (or epinephrine) to a mammalian liver cell, consider Figure 14–12, which shows how activation of adenylate cyclase stimulates the breakdown of glycogen. The increased concentration of cAMP first activates a protein kinase, called *phosphorylase kinase kinase* because it phosphorylates

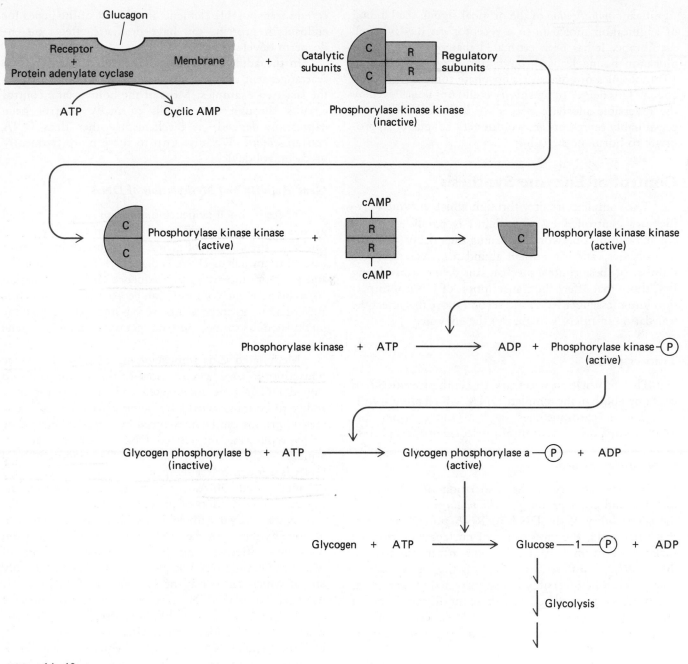

FIGURE **14–12**
Action of Glucagon on the Liver Cell to Regulate the Breakdown of Glycogen. Control is exerted through the intermediary cyclic AMP and an enzyme cascade. To begin the cascade, cAMP binds to the regulatory subunits of phosphorylase kinase kinase.

a second protein kinase: *phosphorylase kinase*. It is the second kinase, which is activated by the addition of the phosphate group (and by calmodulin · Ca^{2+}), that actually phosphorylates the glycogen-splitting enzyme, known as *glycogen phosphorylase*. After phosphorylation

of glycogen phosphorylase to the active *a* (phosphorylated) form, it acts to split glucose from glycogen, using inorganic phosphate (which is why the enzyme is a phosphorylase), to form glucose-1-phosphate.

The successive enzymatically catalyzed steps allow a

significant *amplification* of the original signal, the binding of a glucagon molecule to a receptor on the liver cell membrane. It has been estimated that one molecule of glucagon bound leads through this *enzymatic cascade* to 10^8 molecules of glucose-1-phosphate formed from glycogen. This degree of sensitivity could not be achieved by the reversible allosteric effects we discussed earlier and presumably represents an evolutionary adaptation appropriate to hormone signaling.

Control of Enzyme Synthesis

The complex circuitry through which enzyme activities are controlled by metabolites is paralleled by an equally complex network governing the rates of synthesis of enzymes, and thus their abundance. We discuss a number of these control mechanisms below, considering first those that affect the transcription of DNA information into messenger RNA and second those that affect the translation of mRNA to specify the enzyme.

Transcription

The information necessary to synthesize mRNA is contained both in the template DNA, which in eucaryotic cells lies in chromosomes, and in the enzymes that catalyze the polymerization of nucleoside triphosphates.

Chromosome Structure

In eucaryotic cells, the association of DNA with histones and other proteins in *chromatin* probably affects the accessibility of the DNA to RNA polymerases. On the other hand, the uniformity of nucleosome structure makes it seem unlikely that histones affect transcription differently in different regions of the chromosome. Histones can be reversibly phosphorylated or acetylated, and it has been suggested that these modifications might affect gene expression. There is little experimental evidence, however, for a causal relationship between histone modification and gene transcription.

Another way in which regulation of expression could occur is through physical rearrangement of the DNA itself, for example, inversion or translocation events that link a gene with a stronger or weaker promoter. Inversions of this kind have been shown to regulate the expression of the gene for the flagellar protein of the bacterium *Salmonella typhimuriam*. An example of translocation of a structural gene leading to its transcriptional activation is the joining of the DNA sequences for the variable and constant regions of immunoglobulins (see Chapter 17). Some kinds of tumors result from translocations of DNA fragments that lead to overexpression of normal genes (see discussion later in this chapter). In maize plants, movement of certain transposable elements in and out of the genes for endosperm proteins can have dramatic effects on endosperm development.

In the adult organism the control effects of DNA rearrangement are often abnormal or pathological, as in the last two examples. Most of the homeostatic control circuits through which cells normally control gene expression depend on mechanisms other than DNA rearrangement. We now turn to these more frequently used control devices.

Gene Activity and Methylation of DNA

There is much evidence that the presence of methyl groups on cytosine bases in the DNA of eucaryotic cells can affect the activity of genes. The primary methylated base in mammalian DNA is 5-methyl-cytosine (5-mC), and it occurs mostly in the sequence CpG (see Chapter 10). The frequency of 5-mC can be very high, as much as 50% of all Cs in some plants, so it is likely that not all the methylated cytosines in the genome influence gene expression.

Methylated C is important in the control of gene transcription. One way to show this is to compare two cloned DNAs, one methylated and one not, for their ability to be transcribed after being placed in a cell. The methyl groups can be introduced by a specific methylase or by replicating the cloned DNA in the presence of 5-methyl-dCTP as a precursor. The nonmethylated DNA is a much better template for transcription than is the methylated DNA. Another test is to compare the degree of methylation of genes from cells in which the gene is transcribed with cells in which it is not. This can be done by digesting the DNA from the cells in question with two different restriction endonucleases, both of which can cut the DNA sequence C · C G · G, but only one of which can cut at the sequence C · 5-mC · G · G. After separating the DNA fragments by electrophoresis it can be determined by the Southern blot technique whether the gene was cut (see Chapter 10). By probing with the cloned DNA for the gene in question, a different pattern of fragments would be seen if the gene were methylated in one or the other cell (Figure 14–13). Cells in which the gene is not transcribed have DNA sequences that are methylated while cells in which the gene is active do not.

Additional evidence that methylation might inactivate genes comes from treating cells with the cytidine analogue 5-azacytidine (5-azaC), which has a nitrogen substituted for the carbon at position 5 of the pyrimidine ring. After incorporation into cellular DNA, the 5-azaC, lacking the carbon at position 5, cannot be methylated. Moreover, the analogue also reduces the general level of methylation in the treated cell, probably by inhibiting the methylating enzyme(s). An example of a gross regulation

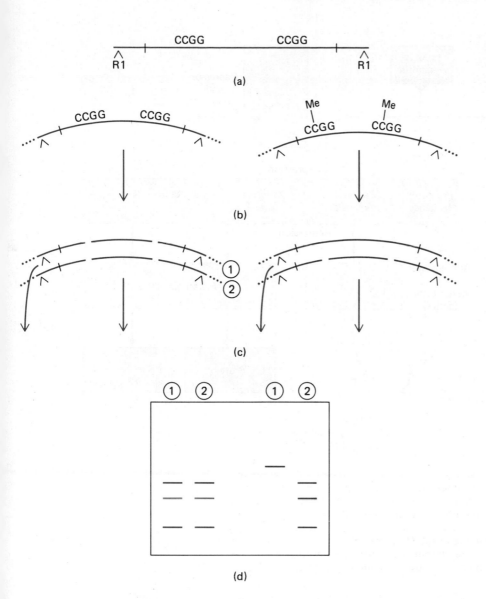

FIGURE 14–13
Assay for the Extent of Methylation in a Gene Being Transcribed Differently in Different Cells. (a) A cloned fragment of DNA showing the restriction sites (R1) used to cut the fragment from bulk DNA and the positions of two $C \cdot C \cdot G \cdot G$ sequences in one strand of the DNA (for clarity, the other strand is not shown). (b) The DNA from two different cells, in one of which the gene is methylated and the other not. (c) The same DNA after digestion with two different restriction endonucleases, one of which cuts at $C \cdot C \cdot G \cdot G$ or $C \cdot mC \cdot G \cdot G$ (2) and the other only at $C \cdot C \cdot G \cdot G$ (1). The unmethylated sequence is cut by both enzymes within the gene but the methylated sequence can only be cut by enzyme (2). (d) After digestion the resulting DNA fragments are separated by electrophoresis, then transferred to a nitrocellulose filter and hybridized with a radioactively labeled probe of the cloned DNA. The radioactive probe will bind to three fragments of cellular DNA if the gene sequence was cut by enzymes 1 or 2, but to only one fragment if it was not.

of gene expression that is influenced by the incorporation of 5-azaC is the *Barr body*, found in each cell of female mammals. This is the structure that results from inactivation of an entire X chromosome. Genes in the inactive X chromosome can be turned on after treatment with 5-azaC. Even more dramatically, the presence of 5-azaC causes curious morphological changes to occur in cultured cells, for example the conversion of fibroblasts to muscle cells!

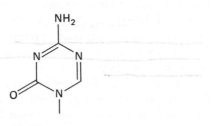

5-Aza cytosine

Methyl groups are added to newly replicated DNA shortly after it has been synthesized. The parent strand is used as a guide to show which sequences are the appropriate substrates—opposite a completely methylated $5\text{-mC} \cdot G$ in the parental DNA is an antiparallel $G \cdot 5\text{-mC}$. Newly replicated DNA, which contains only one parental strand, is said to be *hemimethylated* before methylation of the daughter strand. The enzymes that perform this function are called *maintenance methylases*, since they maintain the previous methylation pattern. The pattern of methylated sequences is thus transmitted to successive daughter cells—that is, it is inherited.

Repression and Induction — The Operon Model

In the face of a naturally fluctuating environment, bacterial cells regulate not only the activity of their enzymes but also the amounts. One of the ways this

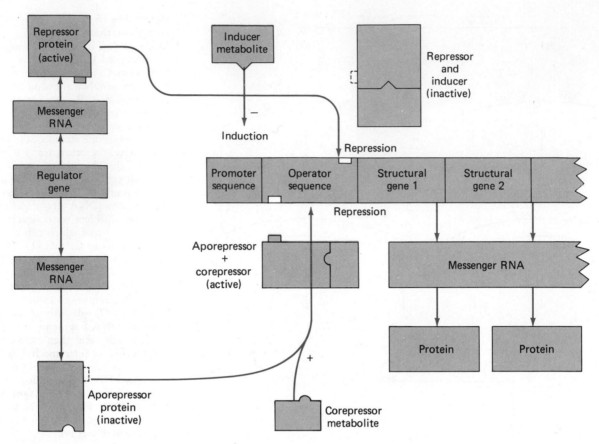

FIGURE 14–14

The Operon, a General Model for the Regulation of Gene Expression.   A group of contiguous structural genes (SG_1, SG_2, etc.) is transcribed by RNA polymerase into a single messenger RNA molecule in procaryotes. RNA polymerase begins transcription at a site termed the promoter. The regulatory protein (or repressor) binds to the operator site of the DNA and blocks transcription. This activity of the repressor may be either inhibited (−) by the binding of a small molecule inducer or activated (+) by the binding of a small molecule corepressor.

happens is by control of the rate of transcription of the structural genes specifying the enzymes. For example, the five enzymes necessary to convert chorismate to tryptophan (see Figure 14–3) are present in only very small amounts *(basal levels)* if tryptophan is present in the growth medium of *E. coli.* When exogenous tryptophan becomes limiting for growth, synthesis of the enzymes is observed, preceded by synthesis of mRNA containing the information of the appropriate genes. The levels of enzymes in the presence of tryptophan are said to be *repressed*, while the higher levels seen in the absence of tryptophan are called *derepressed*. The differences in these rates of synthesis can be as much as several hundredfold.

By maintaining only repressed levels of the tryptophan biosynthetic enzymes when tryptophan is plentiful, the cell conserves cellular resources: mass and energy. Virtually all enzymes concerned with the biosynthesis of

essential metabolites in bacterial cells are subject to *repression* control of this type.

Enzymes that are catabolic, engaged in breaking down food molecules to obtain mass and energy for other cellular needs, are controlled slightly differently. The archetypal example is the β-galactosidase of *E. coli,* an enzyme catalyzing the hydrolysis of lactose into glucose and galactose. When lactose is absent from the growth medium and *E. coli* are growing on glucose, only basal or repressed levels of β-galactosidase are found. When glucose is absent and lactose present, the rate of synthesis of β-galactosidase is *induced* and increases to a thousand times the basal rate.

In each of the above examples the concentration of a small molecule is the signal, either to repress or to induce the synthesis of an enzyme. The small molecule is specific to the regulatory response. The mechanistic question,

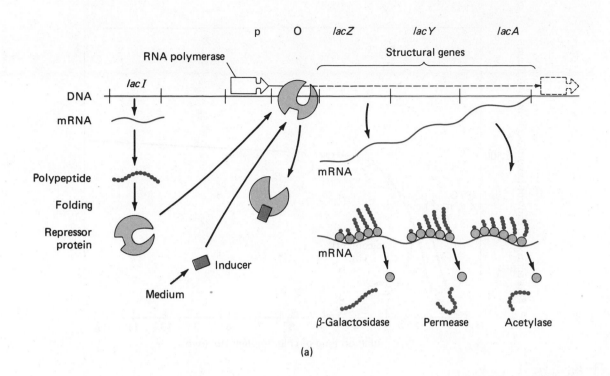

(a)

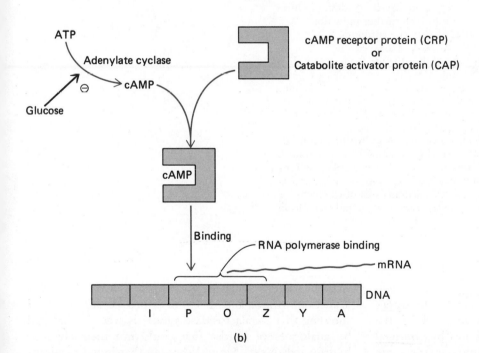

(b)

FIGURE **14–15**
Regulation of the Lactose Operon.
(a) The *lacI* gene (which is adjacent to the lactose operon, but not part of it) produces the *lac* repressor at a constant rate. The repressor binds to the operator (O) region and blocks transcription. The small-molecule inducer binds to the repressor and alters its conformation in such a way that it binds less well to O. Transcription then proceeds much more frequently through the operator and into the structural genes (Z, Y, and A) for the three enzymes of lactose catabolism. (b) Positive control is exerted by the CRP (*cyclic AMP receptor protein*). When it binds cAMP, binding of CRP to the *lac* promoter is enhanced and this binding activates the RNA polymerase. In the presence of excess cellular glucose the concentration of cAMP is low and little transcription of the *lac* operon gene occurs, even in the presence of inducers.

therefore, is how the small molecule can intervene in the expression of a gene.

The study of microbial metabolism was initiated almost a century and a half ago by Louis Pasteur. It is particularly fitting, therefore, that the Pasteur Institute in Paris was the scene of the most illuminating experimental work on induction and repression. This was done in the

1950s by Monod and Jacob, who in a brilliant collaborative effort dissected the physiological and genetic basis of these control mechanisms. The result was the general operator-repressor model illustrated in Figure 14–14 and the model for control of the lactose catabolism system in Figure 14–15a. In brief, the model proposes that regulation of transcription begins when RNA polymerase binds

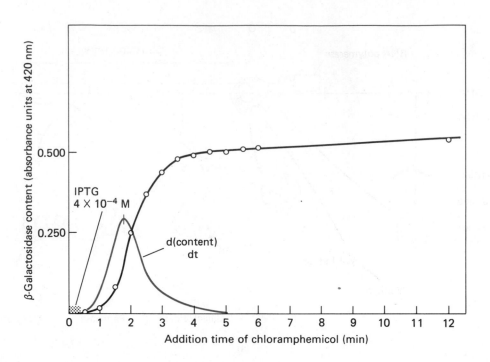

FIGURE 14–16
Induction of β-galactosidase Synthesis. The graph shows the kinetics of appearance of *lacZ* mRNA by measuring its ability to program the synthesis of β-galactosidase. A culture of bacteria was induced at time zero by adding inducer (IPTG); after 20 seconds, inducer was diluted out to stop further induction. At subsequent points, aliquots were removed, treated with the antibiotic chloramphenicol to prevent further enzyme synthesis, and assayed for β-galactosidase activity to quantitate the amount of the enzyme made. Note that enzyme synthesis begins about 1 minute after induction; the 1-minute lag reflects the time needed to transcribe and translate the β-galactosidase gene, *lacZ*. In the absence of inducer (which was diluted out at 20 seconds), enzyme synthesis proceeds for about a minute, then declines rapidly to a rate of zero. The color curve shows the derivative of the black curve versus time, that is, the instantaneous rate of enzyme production. This curve rises as *lac* mRNA induced during the first 20 seconds accumulates, and then falls as the mRNA is degraded. Direct measurements of *lac* mRNA by hybridization to *lac* DNA exhibit the same kinetics. The burst of enzyme synthesis following brief exposure to an inducer is termed an *elementary wave* of enzyme induction. Detailed analysis of elementary wave kinetics reveals the time necessary to complete transcription and translation of the gene and the stability of the mRNA. (After A. Kepes.)

to a special DNA sequence, the *promoter*. Transcription must then proceed past an adjacent sequence, the **operator.** A protein **repressor,** when active, binds to the operator and blocks transcription of nearby *structural genes,* those specifying enzymes. Some repressors are activated when they bind a small metabolite, others are inhibited. The set of structural genes together with their promoter and operator is called an **operon.** Let us now discuss a specific example of the model, the lactose operon, together with the experimental results that led Monod and Jacob to propose their scheme.

A plausible early idea was that lactose enhanced β-galactosidase synthesis by virtue of its role as substrate;

enzyme catalysis itself might be the signal that causes induction. Monod and M. Cohn disproved this theory by showing that β-galactosidase synthesis was also induced by analogues of lactose that could not themselves be hydrolyzed. Next, Monod showed that inducers stimulate the synthesis of two other proteins in addition to β-galactosidase: lactose permease, which facilitates the transport of lactose and other galactosides into the cell, and a galactoside transacetylase, whose metabolic function is still obscure. Genetic mapping of the three genes specifying these proteins showed that they lie next to one another in the order *lacZ* (β-galactosidase), *lacY* (lactose permease), *lacA* (transacetylase). The induction and repression of these three proteins proceeds *coordinately,* because the three genes are transcribed onto a single messenger RNA. (In bacterial cells, genes whose products comprise a metabolic unit, such as a biosynthetic pathway, are often located in a contiguous cluster and transcribed on a single polycistronic mRNA—see Chapter 13. Such is not the case in eucaryotic cells.)

If *E. coli* cells are exposed to an appropriate concentration of inducer for a few seconds and it is then removed, an *elementary wave* of β-galactosidase synthesis begins about a minute later and continues for some five minutes (Figure 14–16). A. Kepes showed that if an inhibitor of RNA chain initiation is present during the pulse of inducer, there is no wave of enzyme synthesis; if the RNA inhibitor is added after the inducer is removed, a normal wave of enzyme synthesis ensues. These experiments suggest that the inducer stimulates the initiation of synthesis of an mRNA coding for the enzyme. The one-minute lag before enzyme activity is seen (Figure 14–16) corresponds to the time required to complete the transcription of the *lacZ* gene into an mRNA molecule and to translate the mRNA to form β-galactosidase. The brief duration of the elementary wave (less than 5 minutes in Figure 14–16) reflects the fact that the mRNA is unstable. When the transacetylase is measured in a similar experiment, the lag preceding its elementary wave is slightly longer, suggesting that the *lacA* gene is farther from the point of initiation of RNA synthesis, and is transcribed slightly later.

The most important conclusion that follows from experiments of this type is that induction has its effect at the level of RNA synthesis. We note with some amazement that this important conclusion was based on biological, not chemical, data. The interpretation placed on his experiments by Kepes has since been confirmed by DNA:RNA hybridization studies (see Chapter 12). Viruses can be isolated with the bacterial *lac* genes as part of the viral genome. DNA is conveniently isolated from these viruses and used as a probe to detect mRNA complementary to this *lac* DNA. Brief exposure of *E. coli* to inducer elicits the synthesis of RNA that hybridizes with the *lac* DNA.

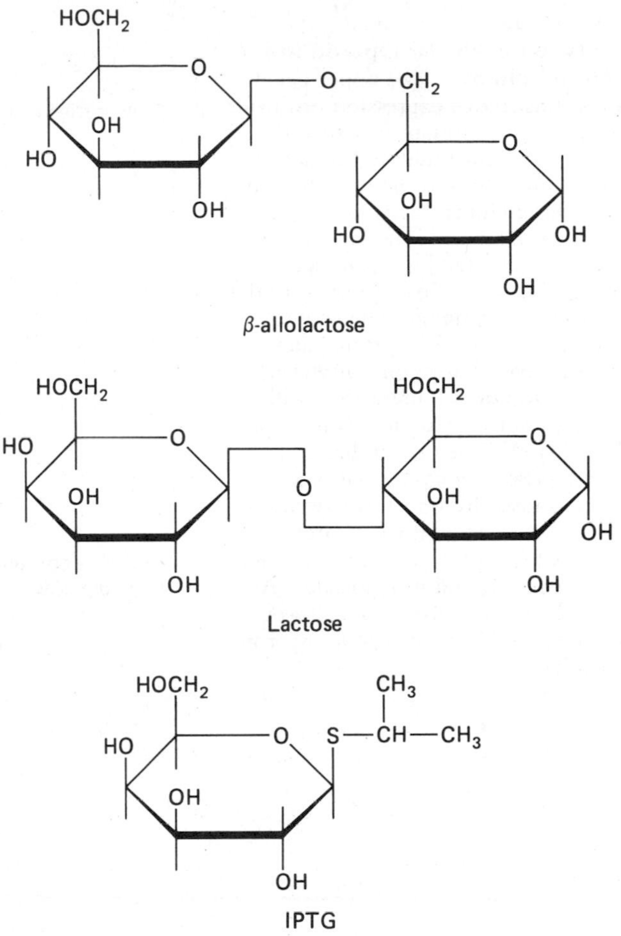

Inducers and Corepressors

Exactly how does a small molecule, like an inducer (Figure 14–17), affect the coordinated transcription of a contiguous group of genes? In finding the answer, Monod and Jacob focused on mutant *E. coli* in which the regulation of the operon was defective and the expression of the *lac* genes was *constitutive:* that is, the three proteins were produced at a high rate whether inducer was present or not. The first group of constitutive mutants studied all contained a lesion in a single gene, designated *lacI* (for

β-allolactose

Lactose

IPTG

FIGURE **14–17**

Inducers of the *lac* Operon. Allolactose is the "natural" inducer; it is made from lactose by the action of β-galactosidase present in the cells at basal levels (see text). *Isopropylthio-*D-*g*alactoside (IPTG) is often used in experiments because it cannot be broken down and, therefore, its intracellular concentration can be determined more easily.

inducible). The important conclusion from the isolation of such mutants is that there exist *regulatory genes,* the sole function of which is to control the expression of other genes. The distinction between a structural gene for an enzyme and a regulatory gene for a control element has since become the basis for all investigations of gene expression. (In the special case of an enzyme itself serving as a control element, the gene is both structural and regulatory and the control is described as **autogenous,** discussed later in this chapter.)

The way in which the state of the *lacI* gene influences expression of the *lac* structural genes was elucidated by further genetic experiments. Monod and Jacob constructed strains of *E. coli* with two sets of *lac* genes; one set contained a mutation in *lacI* and the other did not. The

FIGURE **14–18**

Regulation of the *lac* Operon in i⁻/ i⁺ Merodiploids. The upper panel shows constitutive expression of the *lac* operon in an i⁻ mutant in which the inactive repressor (indicated by a half circle) cannot bind to the operator. In i⁻/i⁺ merodiploids (lower panel), the i⁺ gene encodes a normal, active repressor, indicated by a complete circle with a small slot representing the site of inducer binding. The active repressor binds to the operator and shuts off gene expression. However, it is subject to normal interaction with the small-molecule inducer (small filled square). When the inducer binds to the repressor, a conformational change (shown by transformation of the repressor circle into a square) abolishes the repressor's ability to bind to the operator, and it dissociates. As a result, the i⁻/i⁺ merodiploid shows normal inducible control of *lac* operon expression.

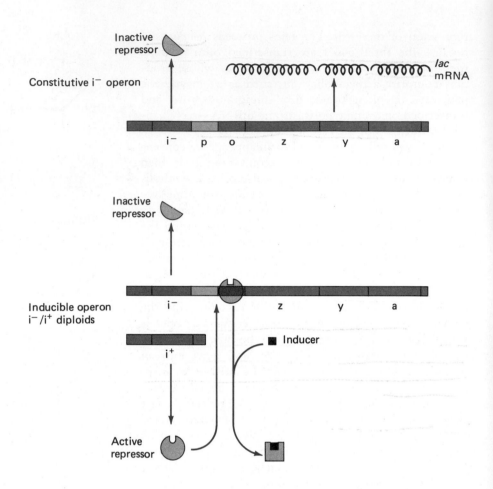

question was, whether the presence of the *lacI* constitutive mutation would make the cell containing two *lac* gene sets synthesize β-galactosidase constitutively. The results were quite clear-cut; only basal levels of the enzyme were synthesized unless an inducer was present (Figure 14–18). In other words the wildtype *lacI* gene was dominant and the *lacI* mutation was recessive in this "heterozygous" strain of *E. coli*. From this result it could be concluded that the product of the *lacI* gene could diffuse throughout the cell and inhibit (repress) the synthesis of β-galactosidase from both *lacZ* genes. Thus, the protein specified by the *lacI* gene was a repressor of *lac* RNA synthesis (see Figure 14–16a). Since the small molecule inducers had the same effect on the expression of β-galactosidase as did the constitutive mutations in the *lacI* gene, it was concluded that the inducer inactivated the repressor protein and prevented it from blocking transcription (Figure 14–18).

Monod and Jacob pointed out that other repressor molecules, specific for other genes, could explain repression of biosynthetic enzyme syntheses. Within a short time this idea was corroborated by the discovery of recessive constitutive mutants affecting biosynthetic en-

zymes. For example, mutations in the gene designated *trpR* have the effect of allowing maximally derepressed rates of synthesis of the enzymes of tryptophan biosynthesis, even in the presence of excess tryptophan (Figure 14–19). These mutations are recessive and evidently do not produce an active repressor. In this case, the normal protein product of the *trpR* gene must be inactive as a repressor unless it is combined with tryptophan, which acts as a *corepressor*. (The protein part of the repressor is called the *aporepressor*.) The difference between induction and repression control thus lies in the effect of the small molecule on the repressor protein. An inducer inactivates its repressor and allows transcription to proceed, while a corepressor activates its repressor and thus inhibits transcription. In both cases the active repressor blocks the synthesis of the mRNA specifying the appropriate structural genes (see Figure 14–14).

Operators

Since different repressors act on different genes, there must be recognition elements that determine the specificity of the interaction between the repressor and its

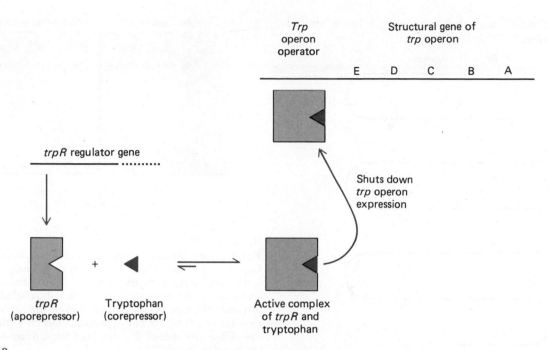

FIGURE **14–19**
Regulation of the Tryptophan (trp) Operon. The product of the *trpR* gene, the *aporepressor* is inactive as a repressor by itself. It has a site (shown as a triangular slot) for specific interaction with tryptophan (shown as a dark triangle). The complex of aporepressor with tryptophan binds to the *trp* operator sequence at the 5′ end of the operon and shuts off the expression of the operon.

target gene(s). One such element is a specific surface region in the repressor protein. The elements in the target gene(s) were also defined by genetic experiments, through the discovery of another class of constitutive mutants. This class did not have lesions within the *lacI* gene; instead the mutations were located in a small region of the genetic map immediately adjacent to the *lacZ* gene specifying β-galactosidase. Monod and Jacob postulated that this region, which they designated *lacO* (for *o*perator), was the site or target that the *lac* repressor protein recognized and at which it bound to block transcription (see Figure 14–15a). By again using "heterozygous" strains with two sets of *lac* genes, Monod and Jacob showed that a constitutive mutation in *lacO* was dominant, but only in respect of the structural genes in its own set. That is, a constitutive *lacO* gene could activate transcription of the genes immediately adjacent to it but not those on the other set if their *lacO* gene was normal.

This property, called *cis-dominance*, demonstrates two things. First, the constitutive mutation in the operator gene does not produce its effect through being translated into a protein, since that protein could diffuse throughout the cell and affect the other gene set. Second, since the expression of all the structural genes in a set is affected by a constitutive mutation in the operator, they must comprise a functional control unit. This unit contains the operator and the immediately adjacent structural genes *(lacZ, lacY,* and *lacA)* whose mRNAs are synthesized (coordinately) together. In the mutant *lacO* gene the altered DNA sequence no longer binds the repressor protein as well. To the extent the operator DNA is not occupied by repressor, transcription occurs.

Numerous other operons have been identified in bacterial cells. The tryptophan operon includes the genes specifying the five proteins that catalyze the transformation of chorismate to tryptophan and an adjacent operator to which the tryptophan repressor binds (see Figure 14–19). A variation on the theme is illustrated by the enzymes for arginine biosynthesis. The genes specifying the nine proteins necessary to synthesize arginine are located in seven groups, rather than one. Each group is preceded by an operator that can interact with a single arginine aporepressor protein. The control of the transcription of the seven gene groups is "coincident" rather than "coordinate" and the arrangement has a special name, **regulon.**[1]

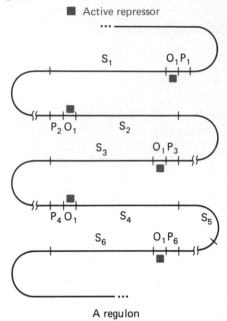

A regulon

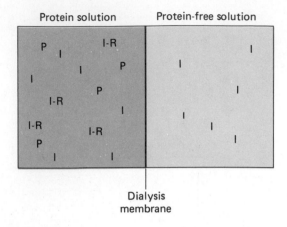

FIGURE 14–20

Equilibrium Dialysis Detects Binding of Radioactive Inducer by Repressor.   A solution to be analyzed, containing proteins (P) and repressor (R), is dialyzed against an equal volume of inducer (I) that is radiolabeled and can penetrate the membrane. At equilibrium the concentration of radioactivity is higher in the protein solution to the extent that it contains species that can bind the inducer—that is, repressor molecules.

The control of expression of separated genes is also illustrated by the tryptophan repressor that can, in the presence of tryptophan, block transcription of the gene *(aroH)* specifying the tryptophan-specific DAHP synthetase at the beginning of the common aromatic amino acid biosynthesis pathway. The tyrosine repressor has even more versatility since it can, in the presence of tyrosine, block transcription of the gene *(aroF)* specifying the tyrosine-specific DAHP synthetase and, in the presence of phenylalanine, block transcription of *aroG* specifying the phenylalanine-specific DAHP synthetase (see Figure 14–4).

The Molecular Nature of Repressor and Operator

The scheme of gene regulation that Monod, Jacob, and their collaborators inferred from genetic studies (see Figure 14–14 and 14–15a) and presented in 1961 could not be subjected to biochemical tests until the repressors were identified and isolated. This was first accomplished in 1966 when W. Gilbert and B. Mueller-Hill isolated the *lacI* gene product. To do so, they took advantage of the fact that the *lac* repressor is an allosteric protein and must interact directly with inducer (see Figure 14–17). They therefore isolated and purified a protein from crude cell

extracts that could bind radioactively labeled inducer, using the technique of *equilibrium dialysis*. In this method the protein fraction is dialyzed against a solution of labeled inducer. After equilibrium is reached, an increased concentration of radioactivity inside the dialysis bag implies binding of the inducer to protein that is retained inside the bag (Figure 14–20). Proof that this protein was the *lacI* gene product came from comparing the alterations in inducer-binding properties of the repressors that were isolated from various *lacI* mutant strains. The alterations were in accord with the physiology of the mutant strains: alterations that led to progressively weaker binding of inducer were from mutant strains with progressively higher uninduced rates of β-galactosidase synthesis.

With the isolated repressor in hand, Gilbert and Mueller-Hill and others were able to confirm the basic features of the Monod-Jacob model. Radioactive repressor does bind with very high affinity to *lac* DNA containing a normal operator (Figure 14–21a) and with much less affinity to *lac* DNA containing a constitutive mutation in the operator (Figure 14–21b) or to nonoperator DNA. Inducers, when combined with a repressor, cause it to undergo a conformational change and thus lower the affinity for operator DNA by a thousandfold (Figure 14–21c). The extraordinarily tight binding between the *lac* repressor and the *lacO* gene means that only very few molecules of repressor need be synthesized. The slow initiation of RNA synthesis that determines the rate of synthesis of the *lacI* protein allows a concentration of

[1]Coordinate regulation occurs when the genes are all adjacent and are transcribed on a single mRNA; coincident regulation occurs when the gene groups all share a common repressor and operator sequence.

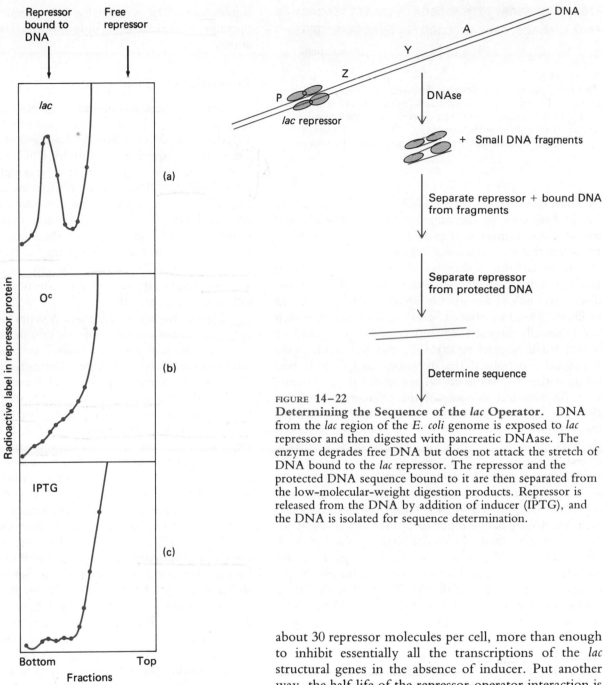

FIGURE 14–21
Binding of Radioactive Repressor Protein to *lac* DNA.
(a) A mixture of radioactive repressor and normal *lac* DNA is
sedimented through a glycerol gradient; some of the
radioactive repressor binds to the DNA and sediments with
it. The arrows at the top indicate the positions to which *lac*
DNA and free repressor sediment. (b) When the same
experiment is performed with *lac* DNA from an operator
constitutive (O^c) mutant, no repressor binds to the DNA. (c)
When the same experiment is performed with normal *lac*
DNA but in the presence of the inducer IPTG, then again
repressor does not bind to the DNA. (From Gilbert and
Mueller-Hill.)

FIGURE 14–22
Determining the Sequence of the *lac* Operator.   DNA
from the *lac* region of the *E. coli* genome is exposed to *lac*
repressor and then digested with pancreatic DNAase. The
enzyme degrades free DNA but does not attack the stretch of
DNA bound to the *lac* repressor. The repressor and the
protected DNA sequence bound to it are then separated from
the low-molecular-weight digestion products. Repressor is
released from the DNA by addition of inducer (IPTG), and
the DNA is isolated for sequence determination.

about 30 repressor molecules per cell, more than enough
to inhibit essentially all the transcriptions of the *lac*
structural genes in the absence of inducer. Put another
way, the half-life of the repressor-operator interaction is
about 5000 s in the absence of inducer but drops to 5 s in
its presence.

Gilbert and his colleagues made use of operator-
repressor binding to obtain samples of operator DNA.
After mixing the repressor with *lac* DNA, the unpro-
tected sequences were digested away with deoxyribonuclease (Figure 14–22). After removing the debris and the
nuclease, the operator DNA was freed from repressor by
adding inducer. This set of experiments is an elegant
example of the use of biological specificity to aid in its
own analysis. The sequence of base pairs in the operator

5′—TGTTGTGTGGAATTGTGAGCGGATAACAATTTCACACA—3′
3′—ACAACACACCTTAACACTCGCCTATTGTTAAAGTGTGT—5′

↑ ↑↑↑↑↑ ↑ ↑ O^c mutations

FIGURE 14–23

The *lac* Operator Nucleotide Sequence. Explanation is given in the text. The shaded bases indicate tracts of identical nucleotides running in opposite directions on the opposite strands (i.e., palindromic sequences). The sites of known operator constitutive (O^c) mutations are shown by vertical arrows.

has been determined. Figure 14–23 shows a sequence of the 21 base pairs in the protected region. The classical operator-constitutive mutations have been shown to be base-pair changes within this region.

A remarkable feature of this structure is that it contains a tract of identical bases running in opposite directions on opposite strands (an *inverted repeat,* marked in Figure 14–23 by shaded bases). The chance that such symmetrically disposed sequences could occur at random is very small, suggesting that they may play a role in the biological function of the operator. Originally it was thought that the symmetrical sequences might interact with the identical subunits of the tetrameric repressor. But some operator-constitutive mutations do not involve the symmetric sequences, and the repressor has been shown to make several nonsymmetrical contacts with the operator (Figure 14–23). Chemical modification reagents have been used to identify operator base pairs, whose access to such probes is altered significantly by binding to repressor. Base pairs have been removed from the 21 that are protected by bound repressor, to see which might not be necessary for the high-affinity binding. Various DNA sequences have been synthesized with single base-pair replacements at specific locations, to see if the affinity for repressor is significantly altered. Taken together, these studies suggest the importance of the central 17 base pairs for the specific binding of the *lac* repressors (Figure 14–23).

The *lac* repressor is found inside the cell as a tetramer of identical protein subunits, each with 38 kd molecular mass. The complete monomer sequence of 347 amino acids has been determined, and a number of studies have been made of fragments of the repressor to dissect its functional anatomy. The inducer binds to a single site on each subunit, and binding is associated with a change in conformation of the subunit that reduces the affinity of the tetrameric repressor for operator DNA.

The first 58 amino acid residues of the subunit are the ones involved in binding to operator DNA. The details of the interaction await the determination by x-ray crystallographic methods of the three-dimensional structure of the repressor-operator complex. In the meantime the significance of the symmetrical DNA sequence of the *lac* operator is approachable only by theoretical considerations.

Promoters and Their Control

Given the linear continuity of DNA, genes must be demarcated by sequence information that tells the transcription machinery where to start and where to stop. The "start" sequences that direct RNA polymerase to bind and to initiate transcription at a defined site are termed *promoters*. Promoters were first identified through the isolation of mutant bacteria that have an increased or decreased efficiency for transcribing particular genes. Genetic analysis showed that these promoter mutations were *cis*-dominant—that is, they affected the expression only of contiguous genes on the same DNA segment—and mapped just upstream (on the 5′ side) of the gene they controlled.

The techniques of molecular genetics have permitted definition of the sequence information in promoters that RNA polymerase reads as "start." In procaryotes, the information consists of three elements: two separate sequences and a particular spacing between them (Figure 14–24). About 10 bases upstream of the start of transcription (a position referred to as −10), bacterial promoters contain the characteristic sequence: T A T A A T. (This sequence is also known as the *Pribnow box,* named for D. Pribnow, who first described it.) Further upstream, at about position −35, most bacterial promoters contain another characteristic sequence: T T G A C A.

We call these sequences "characteristic" because they are *not* absolutely uniform in all promoters. The sequences specified above represent the prototype or **consensus sequence** of a family of related but not quite identical sequences. For example, all bacterial promoters so far examined have the final T residue of the −10 sequence; most (but not all) have T in the first position; most (but not all) have A in the second position; only about half have T in the third position; and so forth. In other words, all bacterial promoters have a sequence at about −10 that follows the consensus at most of the six positions, but often departs from it at one or two positions. There is a somewhat wider distribution around the consensus sequence at −35. Finally, the positions of the two consensus sequences follow the same rule of variation within a small allowed range: they occur at *about* −35 and −10 (referring to the middle of each), varying by a couple of bases or so from one promoter to another.

We suppose that these minor departures from the prototype are responsible for the wide range of transcription frequencies among different genes. In bacteria, "weak" promoters dictate transcription only about once per hour, whereas "strong" promoters are transcribed a

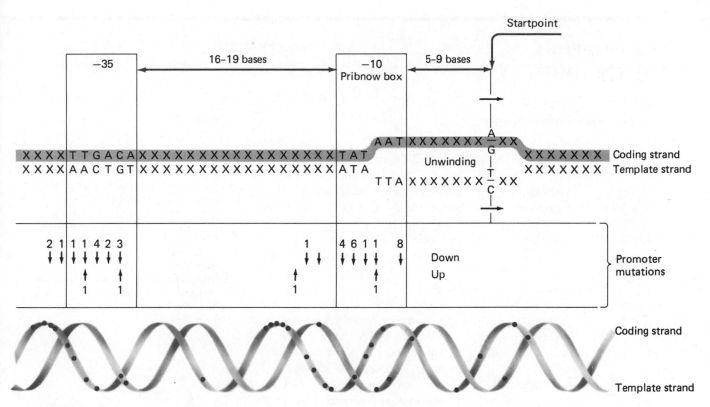

FIGURE 14–24
The DNA Sequence of a Typical Bacterial Promoter. The consensus sequences of the −35 region and the −10, or Pribnow box, region are specified; X indicates that any base pair will do. The positions of promoter mutations are shown below the base sequence, with an arrow pointing down, indicating mutations that reduce promoter efficiency, and an arrow pointing up, indicating mutations that increase promoter efficiency (these summarize mutations in several different promoters; the numbers indicate the relative frequencies of the mutations). The bottom double helix, viewed from one side, shows contact points with RNA polymerase as dots on one strand or the other of the double helix. Note that most, but not all, contact points lie on the coding strand, which is not transcribed.

thousand times more frequently. Deliberate modification of the promoters of cloned genes has confirmed that transcription efficiency depends upon the sequences in the two consensus regions and their spacing.

In bacteria, the sigma (σ) subunit of the RNA polymerase confers specificity for the promoter (see Chapter 12). In the case of spore formation by *B. subtilis,* an example of a procaryotic developmental pathway studied thoroughly by R. Losick and his colleagues, there are several σ factors involved that turn on the expression of several promoters. Variation in the activities of these proteins, via proteolysis, is part of the mechanism for regular progress through the developmental pathway. Since different genes are expressed in the mother cell and the forespore, this system is a model for more complicated development—for example, of tissues in a multicellular organism.

The promoter signals in eucaryotic genes are more complicated still; unifying principles, even of the consensus type, are still unclear. The sequence T A T A, or one agreeing with it at three of four positions, is found at about −20 in most eucaryotic genes that have been examined so far. This short sequence cannot supply a unique start signal, as it occurs here or there in many other stretches of DNA. Other specificities must be involved, but they remain to be discovered, despite having sequenced many eucaryotic genes. Even the experimental modification of cloned eucaryotic genes has yielded a confusing list of results (see discussion later in this chapter). Of course there must be general rules of specificity—the three different RNA polymerases of eucaryotic cells must recognize some particular signals. In principle, deliberate modification of cloned eucaryotic genes should provide the answers.

The Footprints of DNA Control Regions

Control sites, within or near genes, are DNA sequences that bind to proteins involved in gene expression—that is, repressors, activators, and RNA polymerases. The precise nucleotides or base pairs that bind to such proteins can be identified by a method known as "footprinting." The method flows from an ingenious extension of the end-labeling and ladder gel techniques employed in DNA sequencing (see Chapter 10). Footprinting works as follows.

A restriction fragment (actually a large number of identical fragments) that contains the gene of interest is isolated and end-labeled on one strand, as in DNA sequence analysis. The end-labeled fragment is then incubated with a regulatory protein that binds to it—for example, a repressor protein. Next, the mixture is submitted to chemical treatment that will cut the DNA (for example, endonuclease degradation or certain simple chemical hydrolytic reactions) about once per molecule at all points that are *not*

protected by association with protein. Finally, the resultant products are separated on a sequencing gel, identified in the usual way by autoradiography (due to the radioactive label at the end of the original restriction fragment), and separated by size, just as in DNA sequence analysis (Figure 14–A).

The sizes of these degradation products directly identify those stretches of DNA that are protected by interaction with the regulatory protein. Suppose, for example, that the protein protects the DNA from degradation at positions between 9 to 14 and 21 to 28. In that case, the labeled products revealed on the gel will run only up to position 9, and then there will be no other labeled products up to position 15; products from 15 to 20 will be found, but none from positions 21 to 28 (Figure 14–A). The method is identical to DNA sequencing but, applied in this way, can reveal the points of contact between a regulatory protein and a DNA sequence.

FIGURE **14–A**
Footprinting. The example is RNA polymerase complexed to a promoter sequence. The arrows indicate randomly located single breaks produced by partial digestion with DNAse. No breaks, however, are produced in the region protected by the RNA polymerase. Therefore, denatured fragments running from the labeled end to positions in this region are *not* seen on a sequencing gel (bottom left side).

Positive Control

The binding of active repressor to an operator leads to inhibition of transcription of the genes in the operon. Inhibition can be classified as negative control. There is another class of regulatory systems that is involved in the *positive activation* of gene expression, and an example is found again in the control of the *lac* operon. Induction of β-galactosidase synthesis works very poorly in cells growing on glucose, but it becomes normal when the glucose is removed. This "glucose effect," discovered over forty years ago, also diminishes the response to induction of a variety of other catabolic operons; it can be described as a control system that allows *E. coli* prefer-

entially to exploit glucose as a food despite the presence of other resources. In their original model, Monod and Jacob did not address the glucose effect and its mechanism remained a mystery for several years. Then in 1965, R. Makman and Sutherland reported a seemingly unrelated phenomenon: glucose also inhibited the synthesis in bacteria of cyclic AMP. The very presence of this nucleotide in bacteria was something of a surprise; it had previously been implicated only in the very different control mechanism that affects the activity of enzymes in eucaryotic cells (see discussion earlier in this chapter).

Shortly after cAMP was discovered in bacteria it was found that normal induction of the *lac* operon occurs

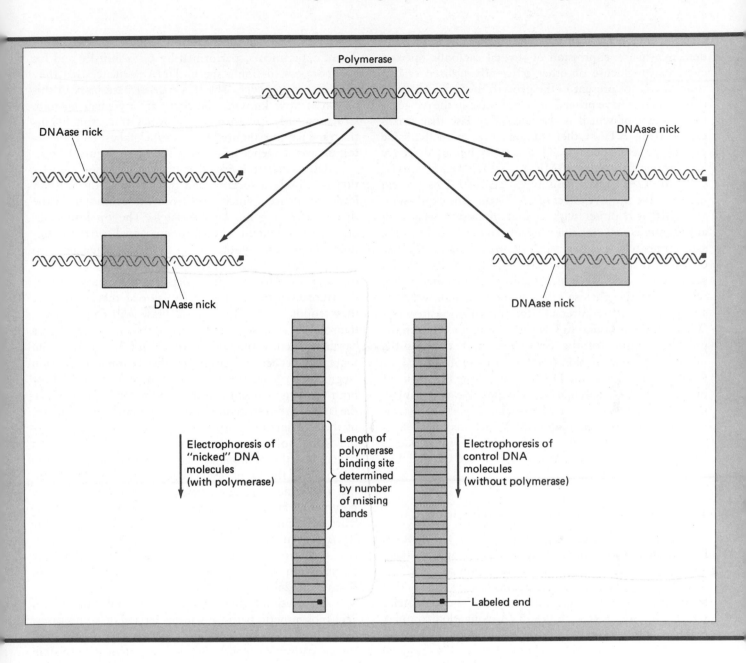

even if the cells are growing in glucose, on addition of inducer in the presence of cAMP. The same is true of the inducibility of other catabolic operons. The response can be even more strikingly demonstrated in cell-free extracts, in which the addition of cAMP stimulates dramatically the synthesis of mRNA corresponding to these operons. The physiological glucose effect thus turned out to be mediated through the requirement of cAMP for transcription of the affected operons. Cells growing in glucose have very low intracellular concentrations of cAMP. It is an intermediate, so far unidentified, of glucose metabolism that leads to the inhibition of cAMP synthesis. Other carbon sources do not lead to such low

cAMP concentrations, and induction of the *lac* operon, as well as others, can still occur. Even these carbon sources can lead to very low cAMP concentrations, however, if the growth rate of the cell is restricted by other means—the availability of an only slowly utilizable nitrogen source, for example. If the cell is slowed generally, the intermediates of glucose metabolism increase sufficiently to reduce the concentration of cAMP. There is thus a complicated control over the utilization of food resources that is only now beginning to be understood.

Again the unraveling of the mechanistic details of this kind of positive control began with genetic studies.

Certain mutants of *E. coli* were discovered that are defective in the expression of several catabolic operons even when glucose or other efficiently utilized carbon sources are not present in the growth medium. Some of these mutants have proved to be defective in the *cya* gene, the product of which is the adenyl cyclase that makes cAMP from ATP. Other mutants *(crp)* with the same phenotype retain adenyl cyclase activity but are defective in a protein that binds cAMP, the *cAMP receptor protein, CRP* (or *CAP, catabolite activator protein*). This protein mediates the positive effect of cAMP on gene expression.

CRP is a dimer with a total molecular weight of approximately 50,000; it is now available for study by the same approaches we have already mentioned for the case of the *lac* repressor. Mutants with an altered DNA sequence near the RNA polymerase binding site are less sensitive to the glucose effect in vivo and show a less stringent cAMP requirement for transcription in vitro. When cAMP is bound to CRP, the complex binds in the promoter region of the *lac* operon and activates the initiation of transcription (see Figure 14–15b). Since the CRP binding site on the DNA is very close to the site of binding of RNA polymerase, it is possible that a direct interaction between the two proteins effects the activation. Alternatively, binding cAMP-CRP may help separate the strands of DNA and thus aid in the formation of an open complex by RNA polymerase (see Chapter 12).

Transcription Termination and Attenuation

Despite the persuasive logic of efficiency in controlling transcription at the initiation stage, it appears there are advantages to controlling termination as well. We discussed in Chapter 13 the possibility of modulating transcription termination mediated by the ρ-protein, which itself interacts with the RNA elongation complex (nascent RNA chain, DNA template, and RNA polymerase). An example of this kind of control is provided by the *trp attenuator*.

As mentioned earlier, the operon for the enzymes synthesizing tryptophan from chorismate is subject to negative repressor control. There is also another control element, located between the promoter/operator and the structural genes. During starvation for tryptophan a large fraction of *trp* RNA transcripts are completed and synthesis of the structural gene products is therefore high. When tryptophan is present in the cell, not only are there fewer transcripts initiated—the *trp* repressor protein (aporepressor) is activated by binding corepressor tryptophan—but also about 85% of the *trp* RNA transcripts that are made terminate before they include structural gene information. By genetic manipulation it is possible to remove some of the DNA sequence in this region, after which termination is significantly reduced. These experiments, performed by C. Yanofsky and his collaborators, define a special DNA sequence that they named the *attenuator*. The sequence of base pairs in this region is now known, and there are two hairpin loop structures that can form in the RNA transcript by intrastrand pairing (Figure 14–25a). One of these hairpins is a terminator sequence of the sort shown in Figure 12–21.

Attenuation requires that transcription and translation be closely associated and that some of the mRNA leader sequence, which runs from the promoter to the first structural gene, be translated. The *trp* leader sequence does contain an appropriate signal for the initiation of protein synthesis. The leader sequence also contains two adjacent tryptophan codons. When tryptophan (or tRNA^Trp) is in sufficiently low concentration, the ribosome will be forced to *pause* at the location of these codons. The RNA polymerase will forge ahead, transcribing the sequence needed to form the longer hairpin structure first (segments 2 and 3) and then the sequence needed to complete the terminator hairpin (segment 4). Since the bases in segment 3 are needed for both the large hairpin and the terminator hairpin, when the large hairpin structure forms it prevents the formation of the termination hairpin (Figure 14–25b). In this way the RNA polymerase can continue beyond the region containing the terminator sequence and complete the synthesis of the *trp* mRNA. When tryptophan is present the ribosome continues translating behind the RNA polymerase and, by steric hindrance, prevents the larger hairpin (segments 2, 3) from forming. Presumably the RNA polymerase is sufficiently ahead of the ribosome that the terminator region can fold up and, with the participation of the ρ-protein, terminate transcription (Figure 14–25b).

Attenuation has been found for several other amino acid biosynthetic pathways in bacteria. For some cases, attenuation is the only way that availability of enzymes for the pathway is controlled. In each attenuation system there is a leader sequence that can be translated and that contains codons specifying the critical amino acid. Since coupled transcription and translation of information does not occur in eucaryotes, we do not expect to find translationally coupled attenuation as a control mechanism in those cells. Nevertheless, there may be other transcriptional termination control systems that at present are completely unexplored.

Antitermination

Given that transcription termination can be controlled, the possibility exists that blocking normal termination would allow the expression of the genes beyond.

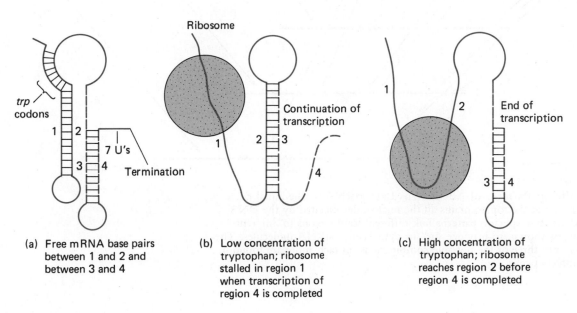

| | Leader polypeptide | | trpE protein |

Met Lys Ala Ile Phe Val Leu Lys Gly Trp Trp Arg Thr Ser — Stop Met Gln Thr Gln
pppAAG (23) AUG AAA GCA AUU UUC GUA CUG AAA GGU UGG UGG CGC ACU UCC UGA (91) AUG CAA ACA CAA

(a) **Free mRNA base pairs between 1 and 2 and between 3 and 4**

(b) **Low concentration of tryptophan; ribosome stalled in region 1 when transcription of region 4 is completed**

(c) **High concentration of tryptophan; ribosome reaches region 2 before region 4 is completed**

FIGURE **14–25**

Attenuation of *trp* Operon Transcription. The upper panel shows the DNA sequence of the leader polypeptide, with the two critical tryptophan (UGG) codons boxed. The lower panel shows attenuation control in schematic form. (a) Base pairing between complementary regions of the mRNA produces helical stems involving region 1 paired with 2, and 3 paired with 4. The 3–4 stem preceding a stretch of U residues is a termination signal, so transcription stops there under normal conditions. (b) During tryptophan starvation, ribosomes stall at the "hungry" tryptophan codons of the leader sequence, at the beginning of region 1. These stalled ribosomes prevent pairing of 1 with 2. This permits 2 to pair with 3, as shown. As a result, 3 cannot pair with 4, abolishing the termination signal, and transcription proceeds into the structural genes of the operon. (c) Restoration of tryptophan permits ribosomes to translate onward to region 2. This blocks pairing of 2 with 3. As a result, 3 is once again free to pair with 4, producing the termination signal that halts transcription.

In fact such phenomena occur in the life cycle of some bacterial viruses. The product of the *N* gene of bacteriophage lambda is an *antiterminator protein*. It blocks the effect of the ρ-protein on the RNA polymerase elongation complex. Transcription thus continues past the ρ-dependent terminator(s) until encountering a strong ρ-independent terminator, where transcription does stop.

The effect of the *N* gene protein is specific for certain sequences of DNA that lie between the promoter and the first ρ-dependent terminator. The *N* gene protein must bind to the sequence before the polymerase has reached it. A consensus sequence has been found for these *nut* (*N*

utilization) sequences in *E. coli:* C G C T C T T (T)A. The action of the lambda *N* gene protein also requires the participation of bacterial proteins, at least one of which influences termination and antitermination in bacterial transcripts. Since *N* does not prevent termination as such, but only ρ-assisted termination, a reasonable guess was that the *N* gene protein blocks the access of ρ to the RNA polymerase. Since mutations in the ρ-protein interfere with normal *N* gene protein function, this guess is probably wrong. Instead, the binding of *N* gene protein to the DNA must interfere with its secondary structure or with its relation to the polymerase, including ρ.

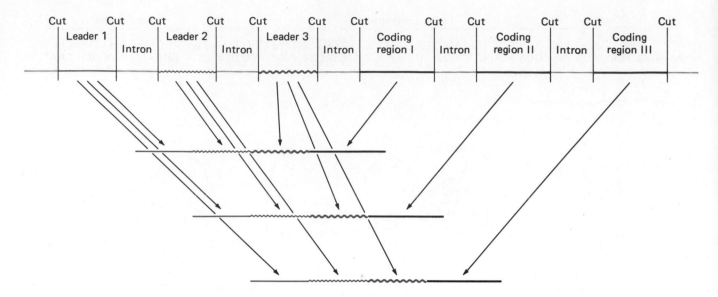

FIGURE **14-26**

Multiple Splicing Patterns of Adenovirus Late mRNA. The initial nuclear transcript, shown at the top, contains all the nucleotides dictated by the DNA sequence. Alternative splicing patterns link different leader exons to different coding exons, as shown, to produce various cytoplasmic mRNA molecules. The figure shows only three of the possible combinations; in fact, more are found in adenovirus-infected cells.

Transcription in Eucaryotes

Our knowledge of the control of transcription in cells other than bacteria is increasing explosively. Until the debris has settled, however, the view of eucaryotic transcription is still difficult to appreciate in detail. The most obvious difference in transcription between the two classes of cells is the necessity in eucaryotes of transporting RNAs across the nuclear membrane. This barrier provides the potential for delay while more subtle preparations are made prior to the use of RNA in translation. We have already described in Chapter 12 the more complex processing of mRNA in eucaryotes, in particular the removal of intervening sequences. It is now agreed that this splicing mechanism can be regulated.

The evidence for this comes from a study of gene expression in SV40 and adenovirus 2, viruses that infect mammalian cells. During infection by either virus, there appear mRNAs composed of nucleotide sequences transcribed from noncontiguous regions of the viral genome. Thus, the immature mRNA must be processed by splicing (Figure 14–26). The novel feature is that of over a dozen mRNAs found late in the infection by adenovirus 2, all have the same 150-nucleotide leader sequence at the 5′ end. This leader is coded by DNA some distance away from each of the structural genes whose information is in the mRNAs. The splicing is variable; different intervening sequences are spliced out of the large precursor for different mRNAs.

Although numerous genes in several organisms are now known to be processed by differential splicing, it is still too soon to know to what signal(s) this complicated splicing system is responding, so our description of the regulation is incomplete. Another interesting question is what functional role is played by the common leader of viral mRNAs. An intriguing possibility is that it is somehow involved with transport through the nuclear membrane. A common leader sequence binding to a transport structure could allow the coordinated entry to the cytoplasm of those mRNAs whose translation must be closely synchronized. It will be recalled that most if not all mRNAs in eucaryotes code for only one protein, so the convenient procaryotic coordination mechanism of a polycistronic message is not available.

If one radioactively labels the mRNA made in different cell types of higher eucaryotes and then detects by hybridization the complementarity to various cloned DNA probes, one finds that the pattern of transcription is in many cases specific to cell type—that is, a given cell type will transcribe gene A and not gene B, while another cell type will do the reverse. Thus, there is no doubt that control of transcription analogous to that observed in procaryotic cells occurs in eucaryotic cells as well.

So far, however, regulatory proteins—activators or

repressors—have been identified in only a few cases. A particularly well-documented example is the transcription of genes for 5S ribosomal RNA in frog oöcytes, where a specific activating protein has been found and purified. In another example, steroid hormones turn on the gene for tyrosine-amino-transferase (TAT) in mammalian liver cells. The steroid hormones, with their high lipid solubility, penetrate easily through the cell membrane, then bind to an *intracellular receptor* protein in the cytoplasm. The hormone · receptor complex then moves into the cell nucleus and activates transcription of the TAT gene. The steroid · receptor complex is thus functionally analogous to the cAMP-CRP complex of *E. coli* that we described earlier.

Thanks to the explosive growth of recombinant DNA technology over the last decade, cell biologists have now identified many sequences that control transcription of eucaryotic genes on the same DNA (*cis-acting*). The placement of these control elements differs quite remarkably from the placement of analogous promoter and operator sequences in bacteria. Recall that bacterial operator sequences are located near the start of transcription, 20 base pairs or so upstream of the coding sequence, and the promoter sequences that dictate the efficiency of transcription are located about 10 and 35 base pairs upstream of the transcription start. This placement makes functional sense in that a single RNA polymerase molecule (which is large enough to cover about 60 base pairs of DNA) could simultaneously contact the control sites and the site at which it initiates transcription. Indeed, footprinting analysis (see preceding box) demonstrates that bacterial RNA polymerase binds to precisely this short promoter-operator stretch of DNA, just upstream of the gene.

In contrast, *cis*-acting control sequences of eucaryotic genes have been found at a bewildering variety of locations. The site of activator protein binding in the above-mentioned case of frog 5S rRNA genes is located *within* the coding sequence, *downstream* (on the 3′ side) of the transcription start site. Control sequences in most other cases are on the 5′ side of the transcription start, but are often hundreds or even *thousands* of base pairs away. A sequence that is 2500 base pairs upstream controls the activation of the TAT gene by the steroid · receptor complex. Sequences that affect the response of other genes to this activator have been found several hundred base pairs upstream of the gene in question. Finally, the genomes of certain animal viruses contain *enhancer* sequences that activate the transcription of genes located on either their 5′ or 3′ sides, and at distances of up to 2000 base pairs away. Enhancers, also found in cellular genomes, can be present in either orientation and still exert their stimulating effects (Figure 14–27). Finally, there are complex sequences that act as *silencers*, turning

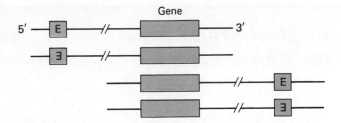

Gene

FIGURE 14–27

Enhancer Sequences Can Activate Genes Regardless of Their Mutual Orientation. An enhancer (E or Ǝ) may lie on the 5′ or the 3′ side of a gene, and may have either orientation. This was established by genetic engineering experiments in which the various orientations were compared for their ability to stimulate expression of the gene. The intervening sequence of nucleotides can be hundreds, even thousands, of base pairs long. Enhancers sometimes interact with activating proteins; in some cases, sufficient activating protein can stimulate gene expression even in the absence of the enhancer.

off the nearby gene, on either side, by acting as sites for binding proteins.

The mechanism by which eucaryotic control sequences affect the transcription of genes far removed from them remains a mystery to be explored. One possibility is that the control sites are not that far separated in three-dimensional space from the structural genes—that is, the DNA-protein complex is so arranged that the two sites are much closer than would be surmised from a linear arrangement of the intervening DNA. Alternatively, the control sites perturb the local structure of chromatin, or they may bind accessory proteins (like those mentioned earlier) that in turn direct the transcription machinery to move along the DNA in search of transcriptional start sites.

The final level of regulation of mRNAs is their differential degradation. In procaryotes most messages have half-lives of only a few minutes and continual resynthesis of mRNA is necessary to synthesize continually the appropriate protein(s). The survival of mRNA in eucaryotes is considerably more variable. Not only do different messages in the same cell have widely differing half-lives but there are wide variations from cell to cell. Illustrative examples can be found in cells specialized for the synthesis of single proteins. The mRNA for fibroin, the major protein of silk, has an unusually long half-life in the silk gland of the silkworm *Bombyx mori*. The most dramatic example is the large amount of RNA that is synthesized in developing egg cells. In many organisms there is little if any RNA synthesis during the first cleavage divisions after fertilization. Proteins are synthesized using mRNA that was stored in an inactive form during oögenesis. The details of the molecular mechanism of this storage remain to be discovered although protein-RNA complexes are probably involved.

Surgical Operations on Cloned Genes

Genes contain two sorts of information. One, of course, is the nucleotide sequence that encodes the amino acid sequences of the gene's protein product. Another is the "control" region of the gene that dictates the conditions under which the gene will be expressed and the rate of its expression. Experimental modification of a cloned gene at defined sites

enables the molecular geneticist to identify the control region.

Mutagenesis can be targeted to a specific gene, or a portion of a gene, by treating a selected restriction fragment from that gene with a chemical mutagen in vitro. The mutagenized restriction fragments are then cloned back into an expression vector, and mutational inactivation of the gene is

detected by its phenotypic effect. The precise site of mutational change must then be established by clipping out and sequencing the mutagenized restriction fragment. This procedure resembles conventional genetics, except that mutation alteration is confined to a chosen stretch of DNA.

The choice can be much more precise in the procedure sometimes

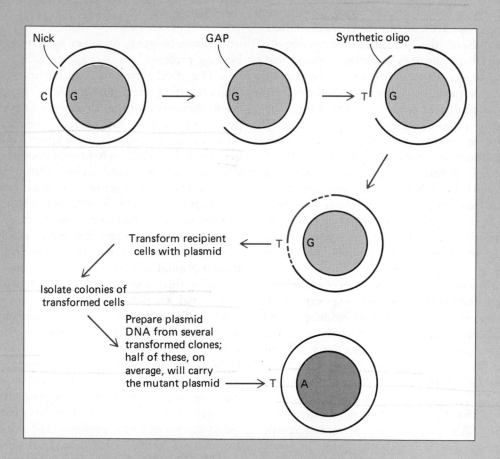

FIGURE 14-B

An Example of Site-Directed Mutagenesis. A plasmid carrying the gene of interest is first nicked on one strand near the site to be mutagenized, the indicated G-C base pair in this example. Part of the nicked strand is then degraded by exonuclease treatment to create a gap. A synthetic oligodeoxyribonucleotide (color line) containing a T rather than a C at the chosen site is then hybridized to the plasmid, the rest of the gap filled in by DNA polymerase (dotted line), and the ends sealed by DNA ligase. When this reconstituted, heteroduplex plasmid replicates in host cells, half the descendent plasmids will have a T-A mutant base pair at the site. During growth the mutant and wildtype plasmids will tend to segregate into separate cells, and pure clones with the mutant plasmid can be isolated.

called "reverse genetics," or *site-directed mutagenesis*. This term covers a variety of procedures that are used to create a specific mutation at a chosen site. Here is a representative protocol. A plasmid carrying the gene of interest is first nicked by a restriction enzyme near the site to be altered, under conditions in which the enzyme generally cuts only one strand of the DNA rather than both. Then treatment with an *exonuclease,* which degrades DNA from one end, removes DNA nucleotides from the nicked strand for some distance into the region where a mutation is to be created. The result is a "gapped" plasmid, in which only one strand is intact in the region of interest. One may then simply insert a *near*-complementary DNA strand into the gap.

Technical advances in nucleotide chemistry have made possible the organic synthesis of short single-strand DNA sequences. For example, microgram quantities of a synthetic polymer of 20 or 30 bases (an *oligo*deoxyribonucleotide) of defined sequence can be manufactured in a few hours at a cost of about $100. In our example, the "oligo" will carry the normal sequence at all positions but the one to be mutated. It will therefore be sufficiently complementary to the opposite, intact single strand to bind to it. DNA polymerase is then used to fill in the rest of the gap (primed by the synthetic oligo), and DNA ligase is used to reseal the circle of the strand that has undergone the surgical operation.

The result (Figure 14–B) is a plasmid with a single incorrect base in one strand, a deliberately created heteroduplex mismatch. When the reconstructed plasmid replicates in a host bacterium, half its descendants will carry a base-pair mutation at the directed site.

Another approach to deliberate DNA sequence modification relies on a different use of exonuclease en-

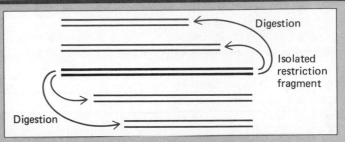

FIGURE 14–C
Construction of Deletions in DNA by Exonuclease Trimming.   A restriction fragment containing the gene of interest is represented by the central heavy double lines, each line symbolizing one strand of the DNA fragment. Exonuclease digestion cuts in from the right end (above) or the left end (below) to produce a collection of shortened fragments represented by the color double lines. The fragments shortened at either end are first trimmed, then ligated into a cloning vehicle, and then the expression of the gene is assessed. The extent of exonuclease digestion at one end or the other (which cannot be precisely controlled) is assessed by DNA sequence analysis after recloning the shortened fragments.

zymes. Incubation of a restriction fragment with an exonuclease for varying lengths of time yields fragments with more and more trimmed off one end or the other. The result is formally identical to the *deletion mutations* of conventional genetics (Figure 14–C). The trimmed fragments can then be isolated by electrophoresis, cloned back into a convenient vehicle, and the expression of the gene assayed. Those deletions that reach into the sequence essential for control typically reduce gene expression and thus can be used to define the limits of the control region. In most bacterial genes, the major control region defined by such experiments comprises a few score of nucleotides on the 5′ side of the coding region, as we have described above. In eucaryotic genes (and some bacterial genes), the control region is more complex; it involves regions both to the 5′ side and within the coding sequence. The definition of these latter domains of gene control information would have been difficult through conventional genetics but is now easily done by molecular surgery and cloning.

Conversely, it can be asked whether adding (rather than subtracting) nucleotides to a gene control sequence affects its behavior. Enzymes have been isolated that can add one or more nucleotide residues to one strand of a DNA molecule. Thus, a restriction fragment can be extended at one end by, for example, a single deoxyadenylate (dA) residue, or by two, three, or many, and the effect on gene expression can be determined. DNA polymerase will fill in the other strand, making a "blunt end."

DNA ligase will join up any pair of blunt-ended DNA molecules that happen to collide in its proximity, as well as fragments with the sticky ends generated by cutting with restriction enzymes. At very high DNA concentrations, the ligase will join up virtually anything with anything else, albeit at low efficiency. By use of these methods a fragment with modified ends can be spliced back into a plasmid vehicle for further investigation.

These methods for linking DNA segments together permit the construction of *hybrid genes* (or *chimeric genes*) with the control region of one linked to the coding region of another. The expression of the constructed hybrid gene can then be studied by cloning it into a convenient plasmid vehicle and assaying gene expression in cells transformed by the engineered plasmid.

This approach may sound rather complex at first, but in fact it represents an efficient way to deploy scientific information in order to gain further information. Three examples illustrate what we mean. One of the fundamental questions of cell physiology is the manner in which genes for cell surface proteins are controlled and how their products are exported into or through the cell membrane. Unfortunately, most cell surface proteins have no enzymatic activity and are therefore laborious to detect. On the other hand, an easy, reliable, and very sensitive assay for the enzyme β-galactosidase was developed many years ago. It is tactically efficient, therefore, to hook the coding sequence of the β-galactosidase gene, *lacZ*, to the control sequences of cell surface genes. A chimeric gene of this type is called an *operon fusion.* The β-galactosidase *activity* of the fused gene product then provides a convenient assay of the *expression* of the cell surface gene's control region, and the processing and transport of the hybrid product. In a sense, this exercise in genetic engineering is an end run around conventional biochemistry, for it circumvents the need to work out biochemical assay methods for cell surface proteins.

In fact, genetic engineering can be used to circumvent the need to work out a purification procedure for a given protein, if its gene can be isolated instead and fused to the *lacZ* gene. The resulting chimeric gene leads to a *protein fusion,* in which the hybrid gene product has the amino acid sequence of one protein up to the junction site and the amino acid sequence of the second protein from there on. The first 30 amino acids of β-galactosidase are irrelevant to the enzyme's activity, and so protein fusions that join practically any amino acid sequence to β-galactosidase within this region retain its enzymatic activity. Beta-galactosidase is one of the largest proteins in the *E. coli* cell, and for this reason can be isolated in a single step by gel electrophoresis. Hybrid proteins consisting of most of β-galactosidase hooked to something else—the product of a gene fusion—are larger still, and equally easy to isolate. They can also be purified by using antiserum against β-galactosidase to precipitate the desired product.

Conversely, the control region of *lacZ* responds to well-understood environmental signals, through which its expression can be manipulated easily over a thousandfold range of rates. One interesting question in cell physiology is the effect of high levels of ribosomal proteins on cell behavior. To examine this question, the control region of *lacZ* can be spliced onto the coding region of a ribosomal protein gene of interest, the fused gene cloned into a plasmid, and then

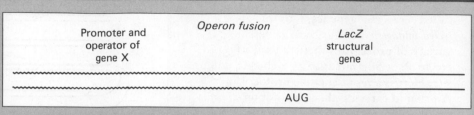

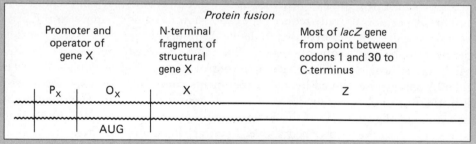

The Control of Protein Synthesis

Although in principle it should be feasible to regulate the availability of enzymes by modulating the process of translation, it seems that most cells exercise control at the RNA level. This is economically defensible, of course, since any regulation that stopped the production of an enzyme by blocking the synthesis of its mRNA would spare all the energy required for that synthesis. Nevertheless there are special circumstances in which such frugality is apparently not so important, and there we do indeed find examples of regulation at the level of protein synthesis.

A major mechanism for preventing the translation of a messenger RNA is steric hindrance, by a protein bound to the initiation region. One of the first cases to be described is that of the regulation of viral proteins synthesized during infection with the small icosahedral RNA bacteriophages, like R17 or Qβ. The mRNA for these proteins is the viral genomic RNA itself, and since it must be included in the new virus particles, shutting

its expression turned up and down at will by manipulating the response of the *lacZ* control region. It is as if one could graft the control of a familiar device—let us say the Pac-Man video game—onto any car, truck, boat, airplane, or spaceship and control them all by the same set of commands. Only one set of manipulations would have to be learned in order to run each device on land or sea or in the sky or outer space.

This strategem is also useful for the isolation of gene products that are normally produced in small quantities, making them difficult to purify. Once under *lac* control, and carried on a multicopy plasmid, a gene that is normally expressed weakly can be throttled up so that its protein product comprises as much as 10–20% of total cell protein, and it can then be purified by very simple procedures.

Such a trick has enormous practical implications for the pharmaceutical industry. In the past, a medically important mammalian protein such as insulin could only be produced by the laborious and expensive procedure of dissecting out pancreas organs after slaughtering great numbers of pigs and purifying insulin from the tissue by multiple steps. Genetic engineering has made it possible to clone the insulin genes into a bacterial plasmid under control of a bacterial regulatory system, turn the genes up to full throttle, and then isolate the

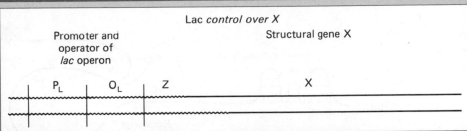

insulin produced, in great abundance, in a few steps.

Finally, operon and protein fusions have been used to analyze bacterial regulatory circuits by making β-galactosidase the signal of the activity of other operons. The work of B. Wanner and colleagues illustrates this approach. They wished to analyze the several genes whose expression is repressed by inorganic phosphate. To do so, they used molecular genetic techniques to fuse the β-galactosidase coding sequence to a *random collection* of sites throughout the genome, and then isolated many independent clones of this collection of *lacZ* fusions. They then tested for increased *lacZ* expression on low phosphate medium, taking advantage of a histochemical stain called XG, which is converted to a blue dye by the action of β-galactosidase. It is easy to screen thousands of clones, spotting 60 or so per petri plate, for the few that turn blue only on low-phosphate medium. In this way they identified 54 clones in which *lacZ* had been fused to genes that are subject to phosphate repression. The *lacZ* activity of these fu-

sions could then be used in genetic crosses as a *phenotypic marker* to map the sites of these genes, through their linkage to other genes at known map positions. Note that this procedure permits mapping of genes that are subject to a particular regulatory regime without knowledge of the genes' normal functions, a striking departure from the procedures of conventional genetics.

This departure is precisely what makes chimeric genetic methods so powerful. The physiology and enzymology of *lacZ* are well understood and provide a battery of standard techniques: control of gene expression by known inducers of the *lac* operon, simple and very sensitive assays for β-galactosidase, and histochemical stains for detecting the enzyme's activity on a petri plate. By grafting either the control sequence or the protein coding sequence of *lacZ* onto other genes, the behavior and location of these genes can be analyzed by means of experimental tools already well worked out for *lacZ*.

down RNA synthesis is not practicable. The coat protein is needed in large amounts (approximately 180 copies per virion) but the viral replicase protein subunit acts catalytically (see Chapter 12), so only a few molecules need be made. A third protein, the viral A protein, although included in the virion, is also needed in much smaller amounts than the coat protein. How can the differential expression of these RNA genes be achieved?

After the infecting viral RNA enters the host bacterial cell, small amounts of coat, replicase, and A

protein are made. The replicase protein organizes four bacterial proteins that are required for the complete replicase enzyme complex (see discussion in Chapter 12). As larger amounts of this complex become available it binds to the viral RNA preferentially in the region of the initiation sequence of the coat protein. The effect of binding the replicase to the viral RNA is to inhibit translation of the coat protein gene and to leave the RNA relatively unlittered with ribosomes so that replication can proceed (Figure 14–28).

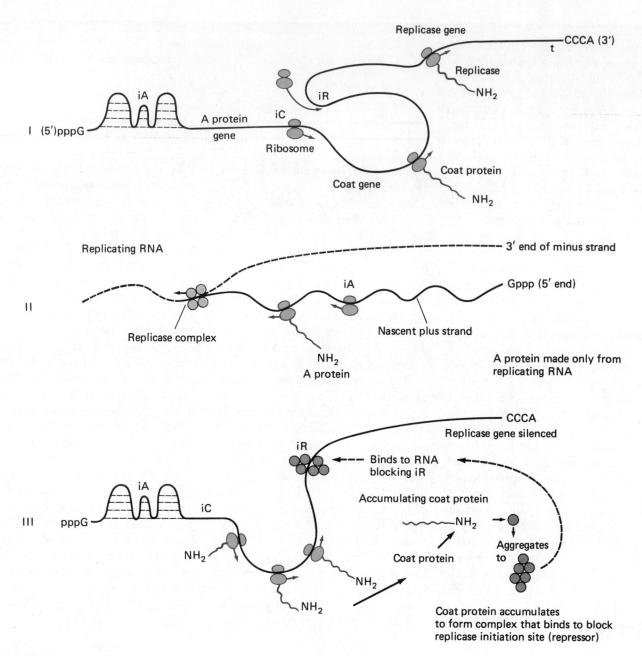

FIGURE **14–28**
Control of RNA Phage Translation. The RNA phage is shown from its 5′ end (beginning with pppG) to its 3′ end. The three genes, reading from 5′ to 3′ (left to right), code for the A protein, the coat (C), and the replicase (R). Panel I shows the translation early in the phage growth cycle. The initiator sequence of the A-protein gene (iA) is occluded by intramolecular base-paired stem structures, shown schematically with dotted lines representing the base pairs; as a result, A-protein is not made. Ribosomes merrily initiate translation at the initiation sequences of the coat gene (iC) and replicase gene (iR) to produce these proteins. After sufficient replicase has been made, it carries out synthesis of the complementary RNA strand, shown schematically in panel II. In this replicating structure, the A gene initiator sequence (iA) is opened up, permitting translation of the A gene. After sufficient coat protein has accumulated, aggregates of coat protein bind near the replicase initiator sequence (iR) and block translation of the replicase gene, shown in panel III.

The inhibition of coat protein synthesis is not complete and later in infection, as coat protein inevitably increases, it binds preferentially to the initiator sequence in front of the replicase gene, as seen in Figure 14–28. The coat protein recognizes a hairpin, a base-paired helical region that forms spontaneously in isolated viral RNA. The bound coat protein inhibits further initiation of the replicase protein synthesis. Since viral RNA is still being replicated, the inhibition of coat protein synthesis is lessened. Thus the coat protein can be synthesized in the large amounts required for assembly of virus particles.

The reason why the viral A-protein is synthesized slowly is probably to be found in a peculiarity of the total viral RNA structure. Despite the fact that the initiator region of the A gene (its Shine-Dalgarno sequence) can make very good base pairs with the 16S rRNA, ribosomes attach to the initiator region at only one-twentieth the rate of attaching to the coat protein initiator region. This is probably because the A-protein initiator sequence is involved in structural interactions with other parts of the viral RNA molecule and is, therefore, not as accessible to the ribosomes. Data that support this conclusion include the fact that fragments of viral RNA are better initiators of A-protein synthesis than is the intact viral RNA. The fragments would not contain the distant regions of the RNA sequence that normally interact with the initiator sequence. Also, H. Lodish showed that mild treatment with formaldehyde, which unfolds base-paired regions, increases the relative rate of A-protein synthesis.

Even in the absence of interfacing secondary and tertiary structures, different initiator sequences in mRNA may differ in the strength of the contacts made with the various proteins and rRNA sequences involved in the initiation complex (see Chapter 13). When ribosomes are limiting, the initiators with the greatest affinity will be preferential sites of binding and it is their genes that will be translated. This is another way in which differential gene expression can be achieved at the translational level and one that could involve the monocistronic messengers of eucaryotes as well as the polycistronic messages of bacteria and viruses.

Finally, it is possible to modify one or more of the various components involved in initiation of protein synthesis. A notable example of this is the phosphorylation of the initiation factor eIF2 in reticulocytes, immature blood cells that synthesize large amounts of hemoglobin. After modification, the formation of the GTP · eIF2 · met-tRNA^F complex is inhibited and so is the initiation of globin synthesis. This is certainly a regulatory response since it occurs preferentially if *hemin,* the precursor of the prosthetic group of hemoglobin, is in limiting concentration. A falling hemin concentration activates a protein kinase, sometimes known as HRI (hemin-regulated *i*nhibitor). This enzyme catalyzes the ATP-linked phosphorylation of the α subunit of eIF2. There is also in reticulocytes a phosphoprotein phosphatase that removes phosphate groups; the fraction of phosphorylated eIF2 is the result of the difference in the rates of phosphorylation and dephosphorylation. The HRI can phosphorylate itself, a reaction that is inhibited by hemin, and the phosphorylated enzyme is more active in phosphorylating eIF2. In addition, perturbations in energy metabolism of the reticulocyte can result in the phosphorylation of eIF2. Altogether this seems a rather complex control system at the translational level. Of course this too is a special situation: most globin synthesis occurs after the reticulocyte has lost its nucleus so it can no longer regulate gene expression at the transcriptional level, and globin is the major product of protein synthesis in the reticulocyte, so inhibiting translation initiation generally affects mostly globin.

A model for the regulation of eIF-2 phosphorylation in reticulocytes

Autogenous Regulation

In discussing the operon model we chose examples in which the synthesis of repressor or activator proteins is constitutive. That is, no control is exerted by small effector molecules on the production of those regulatory proteins. But not all repressors are produced constitutively; some also regulate their own synthesis. These are examples of *autogenous regulation*.

The best-studied case is the regulation of the enzymes used in *histidine utilization*, the *hut* system of *Salmonella typhimurium* bacteria. There are four enzymes involved, coded by two transcriptional units: one contains two structural genes and the other contains the other two structural genes and the gene for the repressor of the system (Figure 14–29). The transcription of both these operons is inhibited by the repressor until the natural inducer urocanate binds to and inactivates it, thus inducing the operons. Work by B. Magasanik and his collaborators has shown that the expression levels of the two transcriptional units are coincidentally regulated. Binding to the two operators is independent, as can be shown by the differential effects of mutations on the repressor's binding affinity.

Autogenous regulation can occur at the translational level as well. Throughout infection of *E. coli* by bacteriophage T4, there is synthesized a protein specified by viral gene 32. The gene-32 protein is needed for viral

DNA synthesis and for genetic recombination. It has the ability to bind preferentially to single-strand regions of DNA and thus can act as an *unwinding protein*. In cell-free extracts programmed by gene-32 mRNA, synthesis of protein is inhibited by added gene-32 protein. The presumption is that the protein binds to its own mRNA and thus represses its own synthesis. The hypothesis can be tested by seeing what levels of gene-32 protein are made in cells infected with virus containing mutations in gene 32. In most cases, the mutant gene-32 protein is found in abnormally large amounts, consistent with the hypothesis that it is not able efficiently to repress its synthesis.

Regulation of Ribosomal Protein Synthesis

The ribosomes in *E. coli* bacteria contain 53 different proteins and three RNAs. The syntheses of all these macromolecules are closely coordinated regardless of the growth conditions in which the cell finds itself, growth conditions in which the number of ribosomes per cell varies over a fortyfold range. The coordination is maintained despite the fact there is only one gene for each of the proteins and seven genes for the rRNAs. The problem of how to manage this balancing act attracted the attention of many cell physiologists. On theoretical grounds it seemed unlikely that a single repressor,

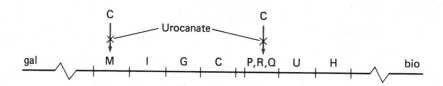

FIGURE 14–29

Histidine and the *hut* Operons of *Salmonella Typhimurium*. Histidine is degraded in four enzymatic steps to glutamate, ammonia, and formamide (HCONH₂). *HutMIGC* constitutes the left-hand operon and codes for the third enzyme (I), the fourth enzyme (G), and the repressor protein, the C-gene product. *Hut(P, R, Q)UH* constitutes the right-hand operon and codes for the second enzyme, urocase (U), and the first enzyme, histidase (H). The promoter of the left-hand operon is M; P, R, and Q make up the promoter-operator region for the right-hand operon. Urocanate, the first decomposition product of histidine, is the inducer and, presumably, inactivates the repressor. (From D. C. Hagen and B. Magasanik.)

analogous to the lactose repressor, could bind to numerous operators and control adequately all the r-protein syntheses. Even if this were possible the lifetimes of the r-protein mRNAs would also have to be held closely together. Moreover, the control of the rRNAs, which are used stoichiometrically in ribosome assembly, would have to be carefully adjusted relative to the r-protein mRNAs, which are used more than once; each mRNA can direct the synthesis of numerous r-protein products.

Once methods were available for determining the sequences of the promoter regions of r-protein mRNAs, it was possible to compare them to see whether they have a common repressor binding site. This work showed there were not sufficient similarities to explain the observed coordinated control by a classical operator-repressor mechanism. Another test involved placing into the *E. coli* cell extra copies of the genes for certain r-proteins. When the extra gene copies were present the specific mRNAs were found to be increased in the expected amounts, but their encoded r-proteins were not significantly increased over those coded by genes present in normal amounts. With these results in hand, attention then shifted toward the possibilities of controlling r-protein synthesis at the translational level.

The simplest model for translation control would be for a ribosomal protein to bind to its own mRNA and thereby block further translation. Since these proteins bind to rRNA during assembly of the ribosome, it could even be that similar nucleotide sequences were involved. This turned out to be the case. The most direct proof was to add a purified ribosomal protein to a suitable cell-free extract and observe a reduction in the rate of synthesis for that protein. The added r-protein acted like a repressor but at the translational level, a *translational repressor*. Not all ribosomal proteins act this way. It is still possible to control all the syntheses of r-proteins, however, because they are coded by polycistronic mRNAs (Figure 14–30). When one mRNA codes for several r-proteins, only one of them need bind to the *ribosome entry site* to shut off all the translation. This protein could be any of those coded by the mRNA, not necessarily the first.

There are several variations on this theme. Some polycistronic mRNAs encode more than one r-protein that acts as a translational repressor. These messages contain more than one *regulatory unit* and each such unit has its own translational repressor. Some of the regulatory units encode other proteins that are used in translation but not as parts of the ribosome, for example, elongation factor Tu (see Chapter 13).

For this system to be successful at coordinating the syntheses of r-proteins, there must be a competition for the translational repressor between its mRNA and the rRNA. The competition must favor the assembly of

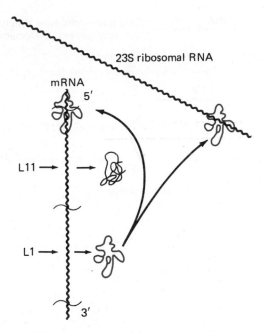

FIGURE 14–30

Translational Repression of a Polycistronic mRNA that Encodes Ribosomal Proteins. The illustration depicts translation of the single mRNA that encodes ribosome proteins L1 and L11. The proteins fold into their final conformation as they are synthesized. Protein L1 (in color) can bind to 23S ribosomal RNA in an early step of ribosome assembly. It can also bind, with a lower affinity, to the 5′ end of its own mRNA. The presence of L1 bound to the 5′ end (the ribosome loading site) of the mRNA blocks entry of new ribosomes and thus blocks the translation of both cistrons of this operon. Most ribosome protein messengers encode several different ribosome proteins, one of which is capable of blocking the ribosome loading site of its own messenger in this way. Thus, an excess of these regulatory ribosome proteins, when not associated with ribosomal RNA in completed or assembling ribosomes, inhibits feedback for continued translation of ribosome protein messengers.

ribosomes; only when rRNAs are not available should further translation of r-proteins be stopped. In some cases this competition has a simple structural basis. There is a binding site for r-protein S8 on the 16S rRNA that is similar in its nucleotide sequence to one found in the mRNA encoding S8. (See figure on next page.)

Once the possibility is admitted of controllable synthesis of regulatory proteins, more complex control schemes may be expected. Many have in fact been found, including cases of repressors with multiple promoters, and operators with multiple binding sites for proteins and with both repression and activation responses to binding, depending on the protein bound and its concentration. Complex control over the many aspects of the life cycle of the small bacteriophage lambda, for example, mimics

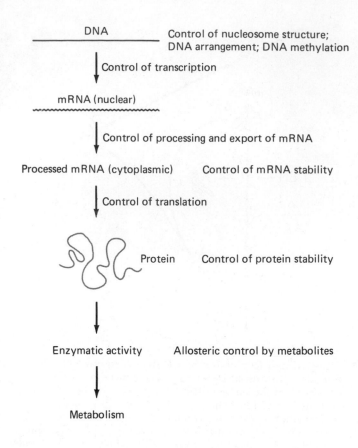

DNA   Control of nucleosome structure;
DNA arrangement; DNA methylation

Control of transcription

mRNA (nuclear)

Control of processing and export of mRNA

Processed mRNA (cytoplasmic)   Control of mRNA stability

Control of translation

Protein   Control of protein stability

Enzymatic activity   Allosteric control by metabolites

Metabolism

the differentiation into distinct cell types that one associates with development in multicellular organisms.

Integrated Control

By discussing them separately we may have given the impression that control over gene transcription, messenger translation, and enzyme activity are mutually exclusive, in the sense that a given regulatory response would involve only one of these three levels. While this is true in some cases, control circuits in biology often include simultaneously more than one different level. Consider, for example, the regulatory effects of the end product of a biosynthetic pathway, such as an amino acid. At one level the end product governs the activity of the pathway by feedback inhibition of the first enzyme, thus controlling the flow of mass into the amino acid. At another level, the end product may also activate a repressor protein, allowing it to bind to the operator controlling the genes for the synthesis of the amino acid, thus inhibiting their transcription.

Stringent Control

In bacterial cells, an interesting regulatory circuit called *stringent control* offers an instructive example of integrated control at all three levels. When one or more

of the 20 amino acids is in short supply, a larger fraction of the tRNAs that accept that amino acid will be unacylated. This leads to a profound adjustment of many of the cell's metabolic activities. Various pathways are inhibited, such as those of nucleotide and lipid biosynthesis; the overall rate of protein synthesis is reduced, even of proteins that do not contain the amino acid in short supply; the expression of genes for tRNA, rRNA, and ribosomal proteins, among others, is severely restricted, while expression of other genes is enhanced, for example those of amino acid biosynthetic operons. The total effect of stringent control is to transform cellular activity from a mode of active proliferation to a more conservative pattern of turnover without mass increase. Bacterial cells show a very similar response when a source of metabolic energy, such as glucose, is suddenly withdrawn.

Mutations that prevent the stringent control response have been discovered, and some lie in a gene called *relA* (for "relaxed"). The product of the *relA* gene turns out to be a ribosome-associated enzyme that responds to the binding of unacylated tRNA to the ribosome A-site by catalyzing the formation of an unusual nucleotide, guanosine-5'-diphosphate-3'-diphosphate (ppGpp, also known as "magic spot," Figure 14–31). Magic spot, or a product or products derived from it, mediates most of the diverse aspects of stringent control.

At the metabolic level, ppGpp simply inhibits a variety of enzymes subject to stringent control. At the translational level, ppGpp inhibits certain translation factors, very probably because of its resemblance to the GTP that normally interacts with them. At the gene transcription level, there is evidence that ppGpp affects the ability of RNA polymerase to recognize different promoters. The response of bacterial cells to a metabolic energy crisis also involves a reduction of various activities that is mediated through the accumulation of ppGpp; this shows that natural selection is a more persuasive advocate of energy policy than are scientists.

The production of ppGpp is the signal for a number of integrated responses concerned with bringing back into balance the metabolism of the cell. The first integration center is the ribosome, where signals concerning the status of 20 amino acid pools are received and processed. Another integrating center is the RNA polymerase, which can respond not only to ppGpp, or similar nucleotides, but also to other molecules concerned with protein synthesis (Figure 14–32). RNA polymerase from *E. coli*, for example, can bind formyl-methionyl-tRNA[Fmet] and protein initiation factor IF-2. If RNA polymerase binds the initiator tRNA (because it is present in excess over the initiation factor), transcription of *lac* operon genes is favored over ribosomal RNA genes; if it binds the initiation factor, the opposite is true.

FIGURE 14–31
Synthesis of ppGpp from GDP and ATP. The enzyme catalyzing the reaction, sometimes known as "stringent factor" and coded by the *relA* gene, can also catalyze a similar reaction between GTP and ATP, producing pppGpp. Both these products are degraded by a specific hydrolysis whose enzyme has been found by mapping techniques to be specified by gene *spoT*.

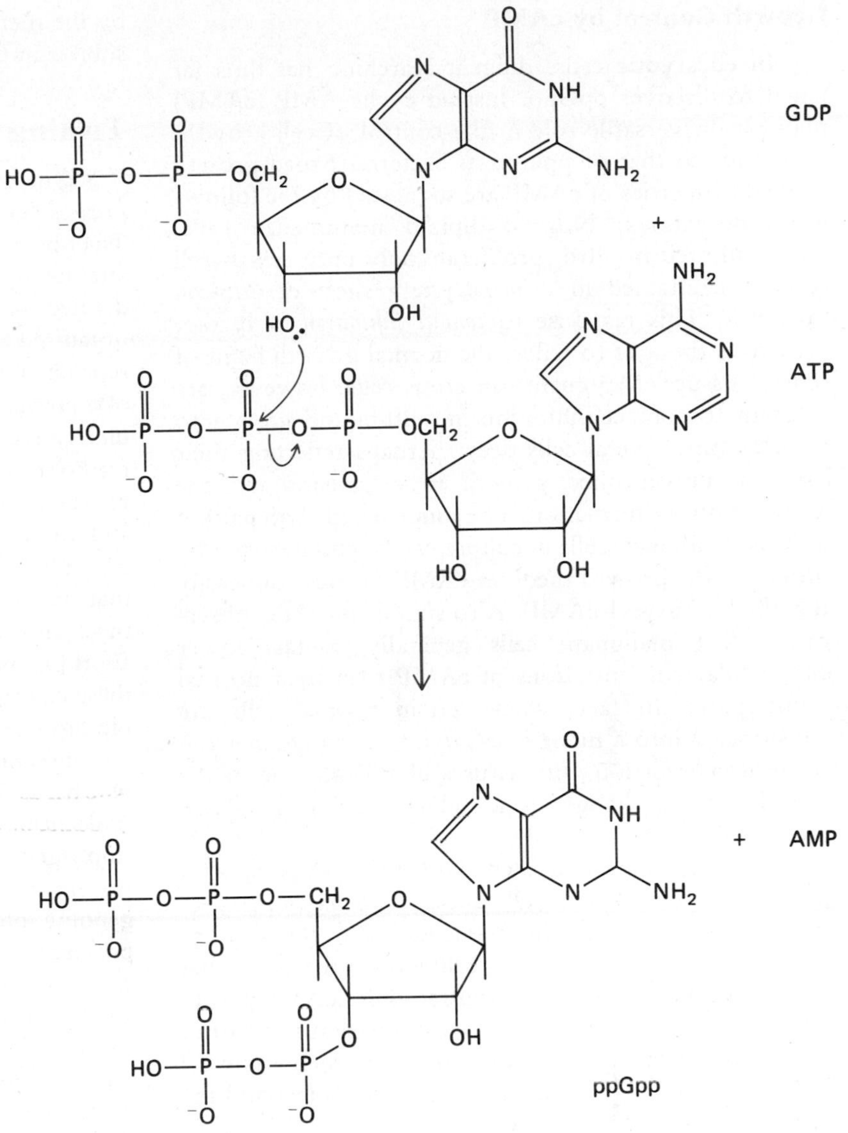

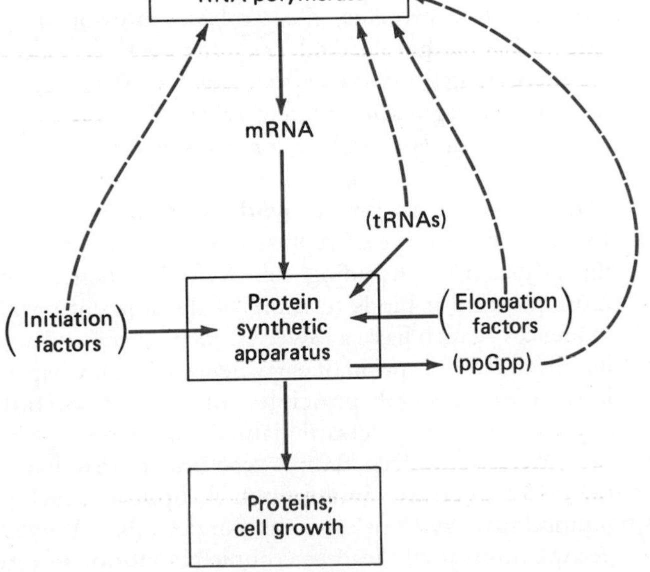

GDP

+

ATP

ppGpp   + AMP

FIGURE 14–32
A General Model, Partly Hypothetical, of Controlling Relationships Between Components of the Protein Synthetic Apparatus and the Activity or Specificity of RNA Polymerase. The dotted lines with arrows converging upon RNA polymerase represent feedback controls by protein synthesis factors upon RNA polymerase.

Growth Control by cAMP

In eucaryotic cells, diligent searching has thus far failed to discover ppGpp. Instead cyclic AMP (cAMP) may play a versatile role in the control of cell growth, analogous to that of ppGpp in bacteria. Broad growth control properties of cAMP are suggested by the following observations. Normal diploid mammalian cells, when cultured in vitro, proliferate only until cell-to-cell contact is attained in a *monolayer,* a sheet of one-cell thickness. This response to *contact inhibition* of further growth is thought to reflect the normal growth habit of cells in tissues. Malignant cancerous cells, however, are resistant to contact inhibition in culture and grow into thick clumps several cells deep, perhaps reflecting their habit of uncontrolled growth in organisms and the resulting tumor formation. The abnormal growth pattern of some malignant cells in culture can be counteracted by adding to the growth medium cAMP or the more easily absorbed dibutyryl-cAMP. Also significant is the observation that malignant cells generally contain lower intracellular concentrations of cAMP than their normal counterparts. In fact, when certain normal cells are transformed into a malignant variant, a change that can occur after infection with viruses like SV40, one of the first detectable changes is a sudden decline in cAMP concentration.

These facts are consistent with the suggestion that contact between normal cells causes an elevated intracellular concentration of cAMP—analogous to the result of binding peptide hormones to cell surface receptors that we discussed earlier—and that intracellular cAMP participates in an integrated growth control mechanism. Part of what makes a cancer cell malignant might then involve a defect in the linkage between cell-to-cell contact and the increase in intracellular cAMP.

The Influence of Cell Structure on Control

Before leaving the subject of inventory control, the regulation of metabolites, it is well to point out that the cell is not adequately modeled by the chemist's beaker. Eucaryotic cells have a complex internal architecture, and this may complicate the analysis of control mechanisms. We need to remember this all the more because most of the examples of regulation that have been worked out in detail are from procaryotes, which have a simpler overall structure and fewer compartments. What might be thought to be compounds freely accessible to enzymes, or operators freely accessible to regulatory proteins, may not be so within the architectural framework of the eucaryotic cell. The fact that individual enzymes and even whole metabolic cycles can be compartmentalized should warn us that cell structure may play a role in regulation

by the mere fact of its existence. We shall return to this subject in Chapter 16.

Timing Control

Our general experience with living systems impresses one with the *reproducibility of biological processes.* The observable events from fertilization of an egg to the attainment of sexual maturity follow a reasonably well-defined schedule in trial after trial. These events are organized in both time and space. It is fundamentally the reproducibility of what we observe that evokes the idea of a *program* to explain these events. Most cell biologists understand the program to be a set of instructions that are read out in a defined order and at a controlled rate. The instructions must include the specifications for reading and timing the program and the instructions are undoubtedly related to the information contained in genes; but that is as far as our current understanding of the mechanics of programs goes: instructions and a timer. Working out the material and energy composition of these concepts is possibly the most important task of cell biology for the next decade.

It seems likely that the fundamental timing processes will be complex and, therefore, difficult to dissect at the molecular level. An example of timing control in bacteriophage lambda-infected *E. coli* serves to illustrate what we are up against. Shortly after the entry of the viral genome into the host cell (or after activation of a latent provirus already there) a group of "early" genes is expressed. Later on in the infection cycle the expression of early genes is typically reduced. What controls this rising and falling transcription of certain DNA sequences? According to M. Ptashne and his colleagues, bacteriophage lambda can control these events through the differential affinity of adjacent operator sequences for two regulator proteins, the lambda repressor and one specified by a gene called *cro*. After activation of the provirus, which occurs via proteolysis of the repressor protein, the regulatory protein specified by the *cro* gene starts to accumulate. At low concentrations of *cro* protein, only one of the set of adjacent operators is occupied, which turns off any further synthesis of repressor (Figure 14–33). The absence of repressor allows transcription of the early genes, including *cro*. As *cro* protein becomes more plentiful it binds to more of the adjacent operator sequences, which have a lower affinity for *cro,* and acts to inhibit the transcription of early genes. We may expect to learn more general principles of control as further experiments on the versatile lambda virus are reported.

There have been many experiments that have attempted to perturb timing control, both in developing multicellular systems and in single cells. Very few perturbations if any lead to completely chaotic progress.

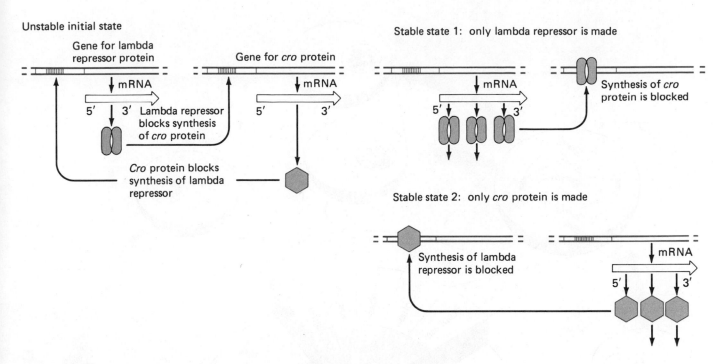

FIGURE 14-33

Reciprocal Control of Repressor and *cro* Genes of Virus Lambda. The repressor blocks transcription of the *cro* gene (left-hand panel and upper right-hand panel), whereas the *cro* gene represses transcription of the repressor gene (left-hand panel and lower right-hand panel). As a result, two stable states are possible, depending on whether the repressor or the *cro* protein first reaches a threshold concentration. If the repressor wins the race, so to speak, then it shuts off the *cro* gene and repressor synthesis continues (upper right-hand panel); if *cro* wins, it shuts off the repressor gene and *cro* synthesis continues (lower right-hand panel).

It may be that the doses necessary to perturb seriously the linkage in timing control kill the cell or organism. Although the experiments on developing embryos are of great interest, we leave that task of description to others and approach here the most obviously timed series of events shared by all cells, the *division cycle*.

The Cell Division Cycle

The history of our description of the cell cycle is an instructive example of how concepts in biology are limited by available experimental technique. At first the cycle was described in terms of only two stages—cell division and interphase—and later by three stages—mitosis (including prophase, metaphase, etc.), cell division, and interphase—because these were the only events observable with microscopes. After radioisotopes became available for use in biological investigations, it was discovered that DNA synthesis, detected by ^{32}P incorpo-

ration, occurred during a discrete part of interphase. This finding allowed the identification of five stages of the division cycle: the DNA *s*ynthesis (S) phase, separated from the *m*itosis (M) and *d*ivision (D) phases by two *g*ap phases, G_1 and G_2. The use of genetic analysis has for some cells allowed the dissection of further substages. (See illustration on next page.)

The most detailed genetic and electron microscopic description has been made by L. Hartwell, B. Byers, and their colleagues on the budding yeast *Saccharomyces cerevisiae* (Figure 14-34). They made an exhaustive search for *temperature-sensitive* mutants that could not progress properly through the division cycle. Named *cdc* mutants (*cell division cycle* mutants), these yeast cells grow normally at the *permissive temperature* of 23°C. When shifted to *nonpermissive* 36°C, cultures of the mutants stop dividing and eventually all the cells appear to be blocked at a particular stage of the cycle, the *termination point*. This

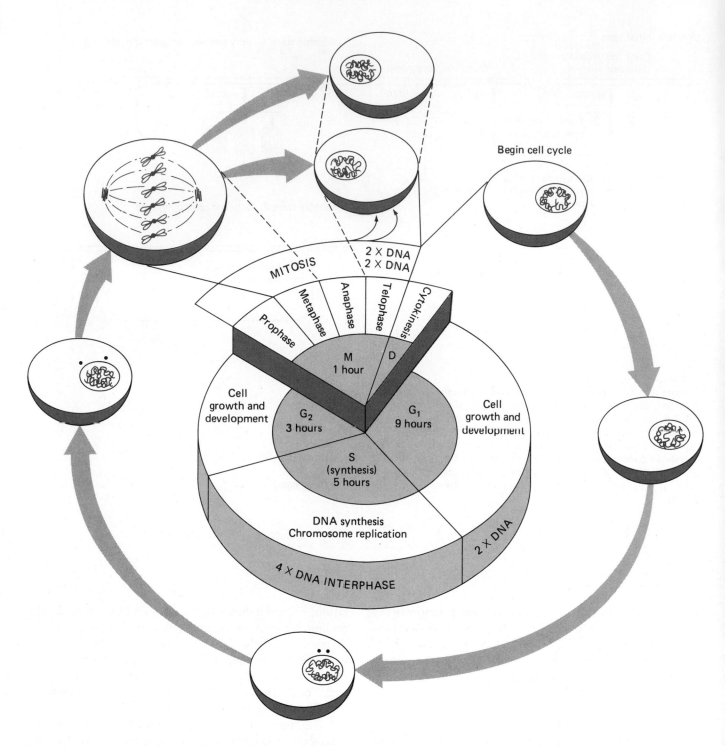

stage can be identified partly through a characteristic morphology seen in electron microscopic images (Figure 14–35). Simple microscopic observation of a culture shifted from 23°C to 36°C shows that some cells stop immediately while others finish dividing but do not complete a second division. This behavior is to be expected of a group of cells that are distributed over all the stages of the division cycle when the temperature shift occurs. At the nonpermissive temperature, cells which no

longer need the function specified by the mutated *cdc* gene can complete their division cycles. Cells that have not yet reached the stage in which the *cdc* gene's function is necessary must stop dividing when that stage is reached. Thus, for each mutant *cdc* gene, an *execution point* in the cycle can be defined, beyond which the function of the gene is no longer needed for completion of the division cycle. (Other investigators have used other terms for the same concept: *transition point, block point,* and execution

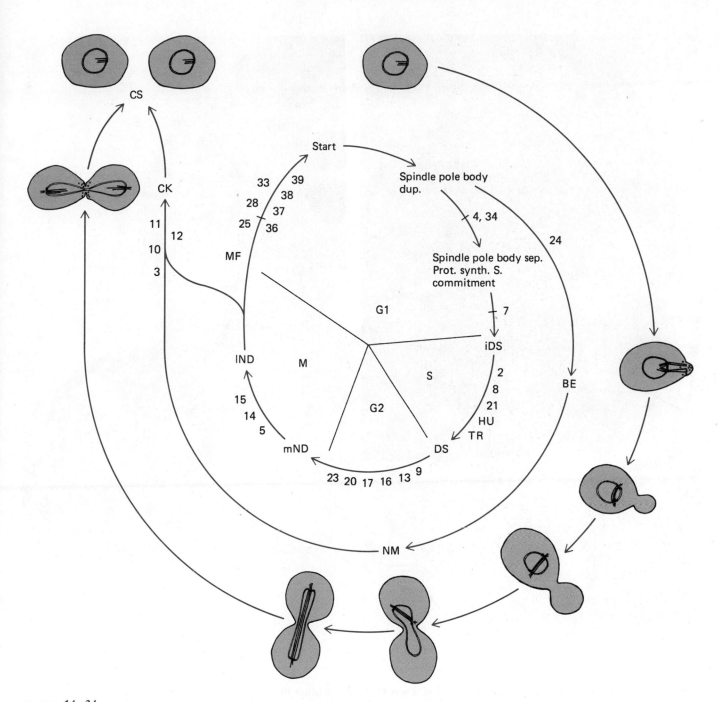

FIGURE **14—34**
The Cell Division Cycle of *Saccharomyces Cerevisiae*. Abbreviations include:
BE = bud emergence; NM = nuclear migration; CK = cytokinesis; CS = cell
separation; spin. pole dup. and sep. = spindle-pole duplication and separation;
prot. synth. S = protein synthesis required for DNA synthesis; iDS = initiation
of DNA synthesis; mND = medial nuclear division; lND = late nuclear division;
HU and TR = hydroxyurea and trenimon (inhibitors of DNA synthesis); and
MF = mating factor. The numbers refer to the execution points of various *cdc*
genes whose proper expression is required for progress from stage to stage.
Further details are in the text.

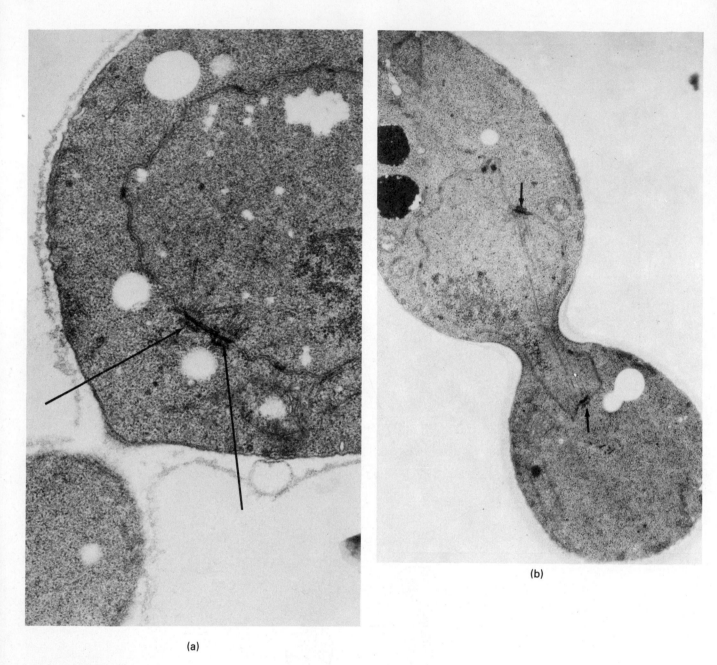

(a)

(b)

FIGURE **14–35**

Electron Micrographs of Yeast Cells Arrested at Two Different Points in the Cell Division Cycle. The cell divisions are arrested because of temperature-sensitive mutations in two different *cdc* genes. Each panel shows a parental cell and a smaller bud in which cell division has arrested. (a) The block is in *cdc* 4. The bud (lower left) forms, but the two spindle-pole bodies remain in the parent cell (arrows); in the normal division cycle, one of the spindle-pole bodies migrates to the bud cell. (b) The block is in *cdc* 7, whose execution point follows that of *cdc* 4. In this case, one of the spindle-pole bodies *has* migrated to the bud cell (arrows indicate the two spindle-pole bodies, and the bud cell is at lower right), but the next step is blocked: mitosis does not occur, and the bud cell does not receive a set of chromosomes. (Courtesy of B. Byers.)

point all refer to the same kind of experimental observation, namely the stage of the cell cycle beyond which a given perturbation no longer prevents completion of division.)

The definition of these execution points requires a series of assays for identifying stages in the cell cycle. Several assays are now available, including morphology, as viewed in both the light and the electron microscopes, and inhibitions by drugs or by physiological molecules. Before describing the method for ordering execution points in time, however, we must pause to consider a problem associated with the chemical study of the cell cycle. Detecting chemical changes in a single living cell is a technical feat that requires highly specialized equipment. Thus, most cell biologists are faced with the necessity of studying cultures of many millions, even billions, of cells in order to amplify sufficiently the chemical processes of interest. The cells of such cultures, however, are generally not all in the same stage of the cell cycle. This is not very worrisome if the primary interest is general metabolism, but if the subject of the research is to identify differences that occur during the cell cycle, it is a disaster.

The way around this difficulty is to *synchronize* the cells so they travel around the cycle together. Even a synchronized culture must be studied over a relatively brief interval, however. This is because the timing of the cell cycle, while reasonably reproducible, varies slightly from cell to cell. Even progeny from a single cell, therefore, will not long remain synchronized with one another, and after a few generations, a formerly synchronized culture has large numbers of cells in every stage of the division cycle.

Synchronizing Cell Cycles

There are two general approaches to the synchronization of cell cycles: *induction synchrony* and *selection synchrony*. An example of the first, developed by E. Zeuthen and co-workers, is the use of heat shocks to delay the division of the ciliate protozoan *Tetrahymena*. This single-celled organism grows normally at 29°C. The cells cannot divide at 34°C, so division is delayed while the cells are at that temperature. In addition, *excess delay* occurs after returning the cells from 34°C to 29°C. The amount of excess delay depends on the stage of the cell cycle that the cells had reached before they were placed at 34°C: the excess delay is short just after division but increases with progress through the cycle. Shortly before division, there is a transition point (execution point) and the higher temperature no longer causes excess delay.

The effect on a culture of *Tetrahymena* of a sequence of alternations of the temperature, between 29°C and 34°C, is to redistribute the cells into a similar stage of the cycle—that is, to induce synchrony—because cells originally at an earlier stage, having suffered less delay, catch up to those that were originally farther along and so were delayed longer. The result of this induction synchrony on a culture of *Tetrahymena* is shown in Figure 14–36.

The advantage of the induction strategy is that essentially all of the culture is synchronized and population-average chemical measurements give information about a single stage of the cycle of an individual cell. The disadvantage is that the induction method itself has an effect on the physiology of the cell. Thus it can never be certain that any changes observed are not the result of the

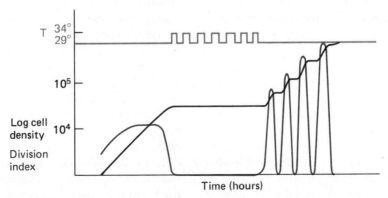

FIGURE **14–36**
Synchrony of *Tetrahymena Pyriformis* Induced by Heat Shocks. Regular exponential growth (doubling time, 2.5 hours) is seen before temperature alterations. Afterwards, there is a stepwise increase in cell number, doubling every 1.5 hours at first, but gradually returning to the normal 2.5-hour division time. The division index (fraction of cells in mitosis, shown in color) also shows the effects of synchrony. (Redrawn from E. Zeuthen and G. H. Scherbaum.)

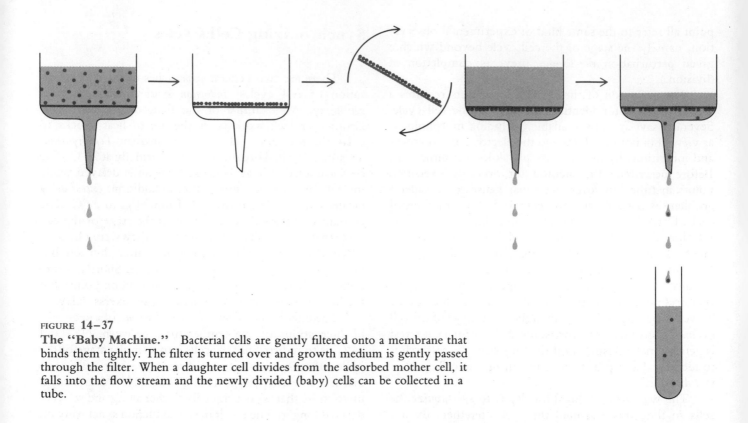

FIGURE **14–37**
The "Baby Machine." Bacterial cells are gently filtered onto a membrane that
binds them tightly. The filter is turned over and growth medium is gently passed
through the filter. When a daughter cell divides from the adsorbed mother cell, it
falls into the flow stream and the newly divided (baby) cells can be collected in a
tube.

induction, rather than of progress through the cycle. This
is certainly true for the example of inducing synchrony in
Tetrahymena by heat shocks. After the end of the series of
temperature alternations, the cells are abnormally large
and the time between the successive synchronized divisions is shorter than in an unperturbed culture.

Other induction methods include the use of agents
that block at certain stages of the cycle, inhibitors of
DNA synthesis for example. If a mammalian cell is in
the S phase, then addition of high concentrations
(2 mM) of thymidine will stop DNA synthesis and block
further progress through the cycle. If the cell is outside
the S phase, it will continue until the end of the G_1 phase
but then be blocked. When thymidine is removed many,
but not all, of the cells are at the beginning of the S phase
and are therefore synchronized. By waiting until all the
cells are finished with the S phase and again applying the
thymidine block, all the cells can be brought to the same
point, at the end of the G_1 phase. The *double thymidine
block* is often used to synchronize mammalian cells in
culture but its effect on their physiology is, like all
induction methods, a worry.

Selection synchrony attempts to meet this difficulty
by separating newly divided cells from the rest of the
culture. The gentler the separation method, the less
perturbation of the cells' physiology will occur. One of
the most benign methods developed thus far is the "baby

machine," designed by C. Helmstetter and D. Cum-
mings for the study of the division cycle in bacteria
(Figure 14–37). This apparatus relies on the adsorption of
bacteria to a thin filter. After bacteria are attached, the
flow of culture medium through the filter is reversed.
Thus one of the new (baby) daughter cells produced at
each division has a good chance of falling into the stream
of culture fluid and being collected in a vessel for study.
Another example of this tactic exploits the behavior
during mitosis of certain eucaryotic cells in culture.
Normally these cells adopt a flattened morphology with
many points of attachment to the substratum. As the cells
enter mitosis they "round up" and become less well
attached. They can thus be selectively detached from the
wall of the culture vessel by gentle mechanical stress,
"shaken off," and recovered from the culture fluid. The
obvious disadvantage of these methods is that the yield of
cells that can be studied is low. Filtration of small young
cells or distribution of cells by size during velocity
sedimentation through a sucrose gradient have also been
used, among other techniques, for selection synchrony.

Both methods produce populations of cells with less
than perfect synchrony. After even the most successful
techniques, the synchronization breaks down after three
to four divisions. This is due less to a failure of technique
than to the natural variation in the period required for
division to occur.

Analysis of the Control of the Cell Cycle

With synchronized populations it is possible to study the effect of various perturbations on the timing control. For example, in *Tetrahymena*, cycloheximide can be added and the question asked: at what time after the end of the series of heat shocks does this inhibitor of cytoplasmic protein synthesis no longer prevent the expected cell division? This defines a transition point (execution point) at a certain time relative to the time of the next division. The transition point is interpreted as the time when the last protein required for division has been synthesized; the experiment can thus be used to define "division proteins." It has also been found that actinomycin D, an inhibitor of RNA synthesis, has an earlier transition point. From this it can be concluded that the sequence RNA → protein → division must occur, at least in *Tetrahymena* synchronized by heat shocks. Of course there is no proven precursor-product relation between the critical RNA(s) and protein(s) in this sequence.

The assayable property of the timed process that fails first is known as the *initial defect*. Initial defects other than division can be used—for example, DNA synthesis or some morphology that can be assessed by microscopy. The temperature-sensitive mutants of *S. cerevisiae* can be ordered with respect to several initial defects by raising the temperature and seeing which chemical or morphological process fails first. With assays that can be done on individual cells, such as autoradiography after the incorporation of tritiated thymidine, it is not even necessary to synchronize the cultures.

To establish the order of two execution points not separated by an assayable event, it is necessary to apply the two perturbations sequentially (Table 14–1). For example, in the scheme shown in Figure 14–34, the execution point of the *cdc* 8 gene precedes that of hydroxyurea (HU, an inhibitor of DNA synthesis). This was established by first applying hydroxyurea to the *cdc* 8 cell culture, then raising the temperature to nonpermissive 36°C and diluting out the hydroxyurea, and then observing that division occurred. It could occur because the cell had already passed the *cdc* 8 execution point when it became blocked by hydroxyurea. Once the HU was removed, division could occur even at the high temperature. Students should convince themselves that (a) if the heating and hydroxyurea treatments were reversed (the *reciprocal shift*) the cells would not divide, and (b) if the original order of treatments were maintained but *cdc* 9 mutants used, the cells also would not divide.

The analysis of relative execution points of a large number of *cdc* mutations and other chemical and morphological events led to the cell cycle scheme shown in Figure 14–34. It is not appropriate to review all those experiments, but it is necessary to point out the general features of the scheme and how they were established. First, there are *two dependent sequences of events,* one relating to DNA synthesis and nuclear division, the other to bud emergence and nuclear migration. The justification for proposing two independent pathways lies in the existence of mutations that block DNA synthesis but do not affect bud emergence, for example *cdc* 9; there are also mutations that do not block DNA synthesis but prevent bud emergence at nonpermissive temperatures, such as *cdc* 24. Since cytokinesis and cell separation do not occur in cells blocked by *cdc* mutations in the bud emergence path, the two paths must converge, as shown in the

TABLE 14–1

Sequencing by Reciprocal Shifts

Relation			Completion of Developmental Program	
		1st Incubation: 2nd Incubation:	Restrict A, Permit B Permit A, Restrict B	Restrict B, Permit A Permit B, Restrict A
(1) Dependent	A B → →		−	+
(2) Dependent	B A → →		+	−
(3) Independent	A → B →		+	+
(4) Interdependent	A,B →		−	−

scheme. The separate path to cell separation is included because of mutants like *cdc* 10, which cannot perform cytokinesis at high temperature but which nevertheless undergo successive cycles of nuclear division and bud emergence.

The cell cycle scheme for *S. cerevisiae* is the most thoroughly analyzed, but its general features are not very different from those of other cells. Nuclear division cycles that are independent of cell division cycles are normally seen in multinucleate cells. The same state can be induced by drugs like cytochalasin. In most regularly growing cells, the cell division and nuclear division pathways converge before completion of the division. One result of not doing this would be anucleate cells; these do occur, but not normally.

The advantage of inhibitors, like hydroxyurea, in perturbing progress through the cell cycle is that their effects are, or can be, relatively well understood. The disadvantage is that there are too few available for a fine dissection of the pathways. There are, on the other hand, potentially as many mutants that can be isolated as there are gene products necessary for continuing along the dependent pathways. Unfortunately, the nature of the block for any particular mutant can take a long time to work out. Thus we expect that both inhibitors and mutations will be used to unravel the control of the division cycle.

We should mention that syntheses of DNA and division proteins are not the only changes that can be accurately timed. The activity of many enzymes has been measured in synchronized cell cultures and has been found to change regularly with progress through the cycle. Some enzyme activities, like that of α-glucosidase in yeast, appear to increase in a stepwise manner in synchronized cultures. Other enzymes, like those associated with DNA synthesis, show a peak of activity during a small fraction of the cycle. These have been referred to as *step enzymes* and *peak enzymes,* respectively. It had once been assumed that the increase in activity reflected an increased synthesis of the enzyme; the falling phase of a peak activity was supposedly the result of an unstable enzyme that turned over rapidly. Direct examination of labeled proteins, however, suggests that the bulk of proteins synthesized in synchronized cultures of yeast increase in a smooth manner, doubling during one cell cycle. The "step" and "peak" enzymes observed previously were apparently reflections of stage-specific control of the enzyme activity and not changes in enzyme amounts.

Some proteins, however, are synthesized at special times during the cell cycle—for example tubulin, the monomer for the microtubules used during mitosis (see Chapter 15) and "cylin" (see problem 4 and Figure 14–38). Of course, well-timed changes in activity can be used as markers against which to measure the execution points of *cdc* mutants and thus to dissect more finely the pathway of control.

Models of Control of the Cell Cycle

With description of the cycle becoming finer and finer, several models have been proposed for its control. One influential effort by J. M. Mitchison is described here because it is both simple and yet explains rather disparate experimental results. A fuller account can be seen in the article cited in the Selected Readings.

Basically the model concerns the last member of a dependent sequence of timers and is concerned with the actual trigger for cell division (Figure 14–38). During interphase there is a steady accumulation of some impor-

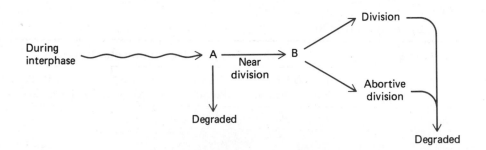

FIGURE **14–38**
Model of a Mitotic Timer. Substance A (perhaps a protein) accumulates during interphase. It can be degraded by various treatments, for example, heat shocks. Normally, however, it reaches a critical concentration near the end of the cycle. This triggers its conversion into a structure B that is essential for division. After division, the structure is degraded. If there is an insufficient quantity of substance A to allow complete construction of structure B, division is abortive and does not occur. (After J. M. Mitchison.) A recent model for the mitotic timer identifies the compound B-like trigger as a complex between a cyclin and a protein kinase coded by a *cdc* gine (*cdc* 28 in *S. cerevesiae; cdc* 2 in *Schizosaccharomyces pombe*). (See Selected Readings.)

tant substance, perhaps a division protein. The accumulation occurs because synthesis of the substance A is faster than its degradation. When a critical concentration is reached, near mitosis, A is converted into B. It is best to consider B as a structure or assembly, whose composition is dictated by geometrical constraints. After completing the assembly of B, division begins; at the completion of division, B is degraded so as to allow resetting of the timing for the succeeding cycle.

This model explains the results of fusing cells at different stages of the division cycle. In the multinucleate plasmodial slime mold *Physarum,* the nuclei normally divide in unison; there is a natural synchrony of division. When two plasmodia are fused, the nuclei from both still divide synchronously, but at a time between that expected for either plasmodium before fusion; some nuclei are delayed while others are accelerated. This behavior suggests that the concentration of a freely diffusible substance (A) must reach a critical level before division can occur. After fusion the concentrations of all substances will become the volume-averaged mean of the concentrations in the plasmodia before fusion. Similar fusions have been performed on mammalian cells, with roughly similar results.

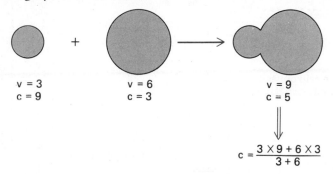

$$c = \frac{3 \times 9 + 6 \times 3}{3 + 6}$$

The experiments we discussed earlier, those on the delay of division in *Tetrahymena* after heat shocks, are consistent with the degradation of substance A during time at the high temperature. This delay also occurs if inhibitors of protein synthesis are added. The transition point found for heating and for cycloheximide would correspond in the model to the conversion of A to B. The requirement in the model for a structure of defined composition (B) is based on the results of amputating *Amoebae.* If a large part of the cytoplasm of *Amoebae* is removed by microsurgery, there will be a delay of the next cell division. If the surgery is done daily, division can be delayed for months, during which time unperturbed *Amoebae* will divide tens of times. Removing part of a well-stirred cytoplasm cannot reduce the *concentration* of substance A. Thus, it can be hypothesized that this cell division requires the conversion of A into a structure requiring a defined *amount* of A. If the conversion cannot be completed because of insufficient quantities, the

structure and its components are assumed to be degraded ("abortive division," Figure 14–38).

Aging

The phenomena of aging and death are among the most important unsolved problems of biology today. Are there manifestations at the cellular level of these properties of multicellular organizations? The answer is yes, certainly. It is, however, difficult to choose what to describe among the many possible theories and experiments that attempt to get at these problems. We shall therefore discuss work on only one cell type from the many used, the diploid human fibroblast.

In 1961, L. Hayflick and P. S. Moorhead reported that lung fibroblasts, isolated from a human embryo, would not grow indefinitely in tissue culture (see following box). At the time this was a surprising result, since many investigators had grown vertebrate cells for long periods of time with little or no loss of vigor. In fact, A. Carrel, who was one of the pioneers of cell-culture techniques, reported he had serially cultivated chicken cells for over 25 years, much longer than the normal life span of the organism. But after many repetitions by laboratories throughout the world, it is now agreed that normal diploid fibroblast cells have a *spanned life.* (Carrel's early results are now thought to have been the result of inadvertent reseeding of the cultures with fresh cells that were present in the chicken serum used to supplement the growth medium.)

The growth of eucaryotic cells in culture has become an almost routine laboratory exercise. It is easiest when the growth medium contains *serum*—the fluid part of blood left after clotting. Cultured cells deprived of serum often stop growing and remain *quiescent,* alive and able to be stimulated to grow again when serum is added to their medium. Since serum is expensive and not very consistent in its behavior, a lot of effort has gone into identifying those components of serum that are essential for cell growth. By trial and error several *growth factors* have been identified. These are compounds, often proteins, that either enhance the division of cells in culture or inhibit it. Among those that promote growth are peptide hormones, like insulin and glucagon, and steroid hormones, like hydrocortisone. A factor that inhibits growth is the adrenocorticotropic hormone, ACTH, which stimulates the adrenal glands in vertebrates.

Certain of the growth factors were identified for the first time as *mitogens;* they can stimulate quiescent cells into mitosis. These include the proteins epidermal growth factor *(EGF),* platelet-derived growth factor *(PDGF),* fibroblast growth factor *(FGF),* and nerve growth factor *(NGF).* The growth factors bind to specific receptors on the surface of cells and exert much of

Culturing Mammalian Cells

We describe here some of the technical details of growing in culture cells whose normal habitat is a multicellular organism. To grow cells that behave most nearly like those in the intact organism, they must be allowed to attach to a suitable surface, a *substratum*. Cells that need the substratum, and do not grow when free in solution, are known as *anchorage-dependent*. (Exceptions include blood cells.) Thus, cells are typically grown on a specially treated plastic surface, in a convenient vessel, bathed with a liquid medium chosen by trial and error to give as vigorous and physiological a growth as possible. Connections between the substratum and the cell are thought to involve ions, both Mg^{2+} and Ca^{2+}, and proteins, for example *fibronectin* (see Chapter 17). Fibroblasts grown in this way lie flattened on the surface, with long processes thrust into the surrounding area. As the cells divide, the daughters tend to wander away from one another until prevented from doing so by the increasing cell density. Growth slows considerably when cells come in contact with one another, as indeed they must when the vessel's available surface is covered with a *monolayer* of cells. This phenomenon is known as *contact inhibition* or *density-dependent regulation of growth*.

At this point the investigator usually *splits* the culture. This involves dislodging the cells from the surface—usually with the aid of EDTA, which binds strongly the Mg^{2+} and Ca^{2+} ions that help hold the cells to the substratum, or trypsin, which hydrolyzes peptide bonds in extracellular proteins like fibronectin, or both. After bringing the freed cells to a known volume, one-half the cells are seeded into another culture vessel like the first. Growth to a monolayer after successive such "one-to-two" splits is often described as successive *population doublings,* but the prudent observer will also make a direct estimate of the cells introduced to and derived from a culture flask. Splits may also be made as one-to-four, one-to-eight, and so forth.

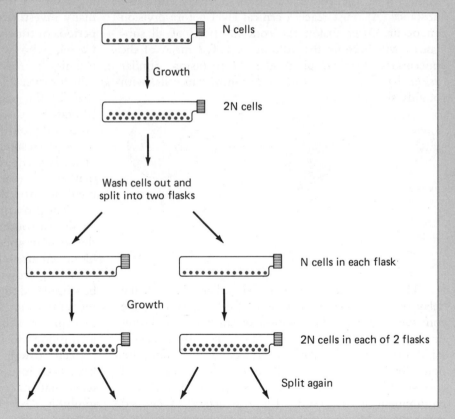

their effect while outside. In this respect they behave very much like the peptide hormones we discussed earlier in this chapter. One of the early responses to binding EGF is the activation of an intracellular protein kinase that catalyzes the transfer of phosphate from ATP to tyrosine residues on target proteins, including itself. We shall have more to say about this enzyme activity in the next section of this chapter. The critical question of how protein kinase activity is linked to mitogenic activity, however, remains unanswered.

What Hayflick and Moorhead found was that embryonic fibroblasts cultured in this way will grow up to a new monolayer only for some 60 population doublings. Until the end of this growth the cells retain their characteristic morphology and, more importantly, their usual number and kind of chromosomes. Other investigators have repeated these experiments with fibroblasts derived from donors of different ages. A rather extensive series of this type was done by G. M. Martin. He found a significant correlation between the age of the human donor and the number of population doublings that the fibroblasts would undergo before ceasing to grow. The correlation was negative: fibroblasts from older donors were able to sustain fewer population doublings in culture.

Moreover, there is a strong correlation between the life span of different organisms and the division potential of their somatic cells when cultured in vitro. The correlation extends even to genetic variants within a species, including our own. Humans who are homozygous for certain recessive single-gene defects are afflicted with a tragically premature onset of many of the characteristics of senescence, and they die young; fibroblasts from the victims of these inherited *progeroid* diseases also show a markedly reduced division potential in vitro.

These results have given rise to the use of cultured fibroblasts as a model system for the study of aging. It can be hoped that an understanding of why fibroblasts do not grow indefinitely in culture, and why their division potential is correlated with the life expectancy of the donor, will illuminate the murky question of why the donors themselves age. Experimental work in this area has focused on a search for biochemical differences between young and old cells and, more importantly, attempts to relate these differences to division potential. Two general theories have been proposed to explain how such differences might arise: the *program theory* and the *error theory*.

The program theory postulates that senescence (and ultimately death) is written in the genes; it reflects the unfolding of a genetic program, like that which determines the course of embryonic and early adult development (Figure 14–39). The theory is consistent with the genetic determination of life span and cellular division potential in vitro, but it is difficult to test because the nature of such a senescence program remains entirely mysterious.

The error theory, in contrast, holds that senescence is not programmed but accidental. Under this view, the genetic program for adult existence comprises only a repeating loop of instructions for maintaining the organism; it says, in effect, "live forever." Unfortunately, errors inevitably intrude on this interesting program, and the system for maintaining the organism or its cells inevitably breaks down. Many types of errors, in gene replication, gene expression, and homeostasis, have been

FIGURE 14–39

The Program Theory of Aging. The idea of a program is elicited by the reproducibility of observed events that occur from the fertilization of an egg, through infancy, sexual maturity, and senescence, to death. It is as if there were a set of instructions (here depicted as a strip of cinema film frames) that the organism follows in a determined sequence and at a determined rate. (Courtesy E. L. Menninger.)

proposed as causes of the deterioration. The error theory seems very plausible, at first thought, but it is weakened by the observation that breakdown is not absolutely inevitable. For example, most single-celled microorganisms do not show mortality, but go on dividing indefinitely. The fundamental difference between mortal and normally immortal cell lines remains a tantalizing problem.

Cancer

Not all cultured eucaryotic cells have limited life spans. Generally, however, immortal cells that grow indefinitely in culture are not normal; in particular, they have abnormal complements of chromosomes. Often these cells can cause malignant tumors when introduced into an appropriate animal. Conversely, when cells isolated from a tumor can be grown in culture they can often be cultured indefinitely. While the study of cancer began at the level of tumors in whole animals, cell biologists have found ways to model many of the pathological phenomena with cultured cells. One of the fascinating questions of cell biology today is exactly how normal cells differ from those that resemble a cancer.

Cells from a *malignantly transformed cell line* can be propagated indefinitely in culture and they cause tumors when placed (at a dose of less than 10^6 cells) into an animal that would accept normal cells of the same type. The process that converts a normal cell into a tumor-producing immortal cell is called *transformation*.[2] There are other patterns of cell behavior that manifest only some of the properties of tumor cell lines: *immortal* cells like the mouse-derived 3T3 line that are density- and contact-inhibited but do not produce tumors, and *partially transformed* cells that produce tumors but senesce in culture. Usually a cell line that is able to be grown in soft agar—that is, independent of a substratum—will produce a tumor when introduced into a suitable host. When such cells are grown on a substratum in culture they do not stop when a monolayer is reached; instead they "pile up" into disordered aggregates. Not only are they anchorage-independent, they also are no longer contact-inhibited. Both these functional properties are associated with special structures at the tumor cell surface. These can sometimes be detected by certain proteins derived from plants—lectins—which bind to the sugar residues of surface glycoproteins. Transformed cells often *agglutinate,* stick together, when treated with lectins, while normal cells rarely do. Transformation-associated surface structures can also be detected by antibodies raised against the abnormal cells.

The Pathology of Cancer

There are two fundamental properties of malignant cells that allow them to produce tumors. *Uncontrolled growth* is one that we discussed above. In an animal with a functional immune system, tumor growth means being able to escape surveillance by that immune system. It turns out that cultured cells are often rejected by an animal's immune system, even if the same type of animal was the original source of the cultured cells. To avoid this kind of rejection, specially treated animals with defective

immune responses are often used to test cultured cells for their ability to form tumors.

The second property that makes cells malignant is the ability to form *metastases,* other tumors that are products of the original one and are separate from it. This is related to the lack of both contact inhibition and cell cohesiveness that is associated with cancer cells. The malignant cells can *invade* other tissues, migrating like embryonic cells do during development. The most invasive types of malignant tumor cells generally exhibit an unstable karyotype and are called *aneuploid*.

A cancer is a malignant type of *neoplasm,* a new growth of tissue that is independent relative to the tissue of origin. Benign neoplasms are named for the tissue of origin—if the benign tumor arises from a gland, for example, it is called an adenoma (the suffix "oma" means tumor). Malignant invasive tumors are named both for the embryonic origins of the tissue and for the tissue itself. If the cancer arises from the embryonic mesoderm, it is called a *sarcoma;* if it arises from the ectoderm or the endoderm, it is called a *carcinoma*. A malignant tumor that arises in a gland is thus called an adenocarcinoma.

Carcinogenesis Is a Multistep Process

Carcinogenesis, the process that leads to the formation of a malignant neoplasm, involves more than one step. Carcinogenesis is the term used to describe the process elicited by the application to an organism of some external agent, the *carcinogen*. The term "transformation" is used to describe what are probably the same events at the cellular level. The first step is *initiation;* it occurs as a rapid response to the carcinogen, perhaps requiring its presence during only one cell division cycle. The state that results from initiation is long-lived, lasting many months, but by itself is usually insufficient to produce a tumor. Another step, *promotion,* is required for that. Changes in internal physiology elicit promotion in the organism but the material basis for these changes—the progression factors—have not been identified. In the cultured cell model, however, agents called *promoters* have been found that influence progression to full transformation; promoters, while present, can also elicit malignant behavior from cultured cells.

Examples of initiating agents include x-rays and polycyclic hydrocarbons like 3-methyl cholanthrene (Figure 14–40a). Examples of promoters include carbon tetrachloride and phorbol esters—for example TPA, the cocarcinogenic principal in croton oil (Figure 14–40b). In the organism, carbon tetrachloride damages liver cells, eliciting a proliferative response from normal liver and numerous tumors if an initiator was previously applied. TPA and other phorbol esters are able to substitute for the diacyl glycerol that is necessary to activate protein

---

[2]The term "transformation" is also used by molecular geneticists to refer to a recipient cell taking up a DNA molecule and being altered genetically as a consequence. That process in the context of cancer research is called "transfection."

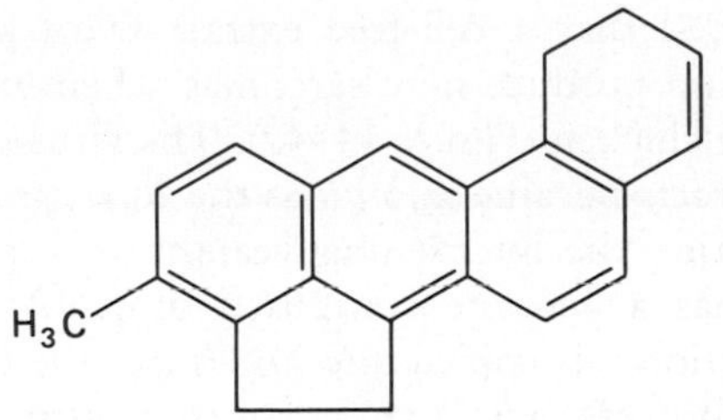

3-methyl cholanthrene

(This must be oxidized to an active carcinogen)

(a)

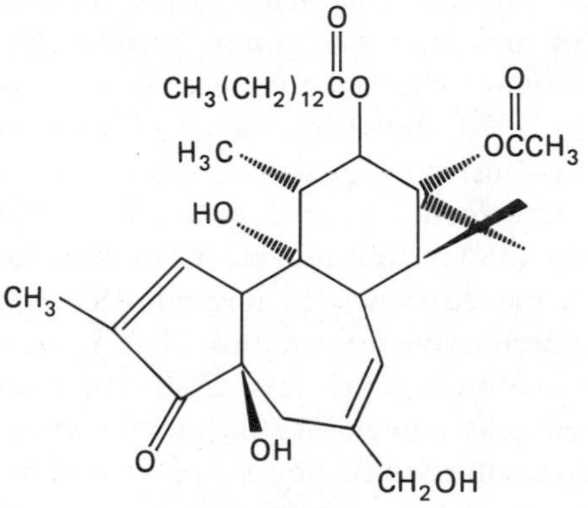

TPA

(12-O-tetradecanoyl phorbol-13-acetate)

(b)

FIGURE 14–40
Cancer-Initiating Chemicals.   (a) The structural formula for the polycyclic hydrocarbon 3-methyl-cholanthrene, an initiating agent. This must be oxidized to an active carcinogen. (b) This is the active principal of the oil derived from seed of *Croton tiglium*. The high incidence of esophageal cancer on the island of Curaçao may be related to the widespread use of these seeds.

kinase C (see earlier discussion in this chapter). Initiation and promotion are most likely both composed of distinguishable substeps, but these have not yet been identified. Finally, as we shall see below, the continued activity of certain genes is necessary for some tumors to persist.

The Genetic Basis of Cancer

There is much evidence to suggest that cancers begin by the transformation of a single cell. The tumor may eventually have cells of varying appearance but they are all a clone of the single *founder cell*. Since the cancer property is inherited, the fundamental difference that arises during transformation is a change in the content or arrangement of genetic information in the founder cell.

This could arise by an *epigenetic change,*[3] one involving, for example, a loss of methyl groups in the control region of a gene that affects cell growth (see the earlier discussion in this chapter). One way to initiate the malignant transformation of cells is to expose them to x-radiation. Such treatment damages the DNA, producing many single-strand breaks in the molecule. These breaks may resemble the appearance of DNA replication forks (which also have single-strand breaks due to discontinuous synthesis of the DNA), so that methylase enzymes bind to the free ends of the breaks and thus become less available for the normal maintenance methylation function. With normal methylation inhibited by this competition, loss of methyl groups and consequent heritable epigenetic defects would occur during replication. This model, suggested by R. Holliday, is consistent with the finding that x-rays are relatively less effective at producing single mutations in mammalian cells than at producing a predisposition for malignant transformation.

Oncogenes

The more common view of transformation is that it arises from *genetic changes,* those that alter the sequence of nucleotides in the DNA of the transformed cell. For example, many carcinogens also can cause mutations. Sometimes the external agent must be metabolized by the cell before becoming a mutagen (Figure 14–41). Other genetic alterations associated with cancer are chromosome rearrangements—movement of one part of a chromosome to another location, either within the same chromosome (inversion) or on another chromosome

[3]Epigenetic changes alter the phenotype but not the genotype—for example, stable changes in gene expression during development.

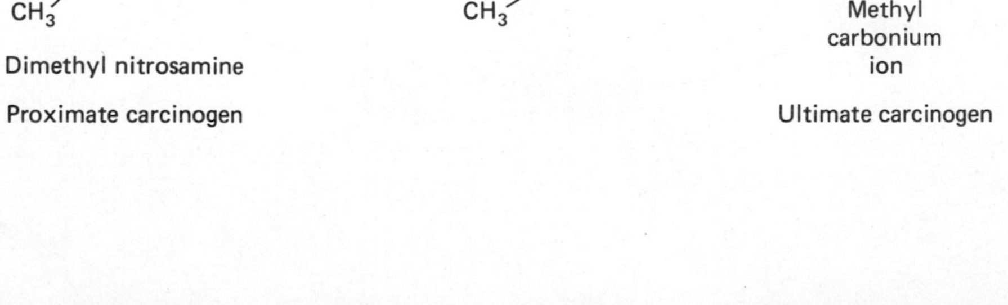

Dimethyl nitrosamine

Proximate carcinogen

Methyl carbonium ion

Ultimate carcinogen

FIGURE 14–41
The Carcinogen Dimethyl Nitrosamine. This carcinogen is itself inactive, but is oxidized in cells to a methyl carbonium ion that damages DNA.

(translocation)—and amplifications of parts of chromosomes to yield numerous copies of the DNA involved. Another long-standing observation is that in many animals, including humans, there is a genetic basis for either a predisposition to cancer or for the disease itself. An example of the former is the much higher incidence of skin cancers in humans with the inherited disease *xeroderma pigmentosum,* a condition in which the ability to repair DNA damaged by the ultraviolet radiation in sunlight is seriously impaired. The association of heritability with many aspects of malignant tumors gave rise to the concept of cancer genes.

Viral Oncogenes

In the early part of this century it was suspected that cancer is caused by an infectious agent. Several parasites were nominated as possible carcinogens, but it was the viruses that provided the clearest evidence for being associated with tumors. Although viruses have been shown to be linked to many tumors in other animals, the current view is that only a few human cancers are caused by infectious agents (a significant exception is the involvement of hepatitis B virus with liver carcinoma, the most common malignancy worldwide). In 1911, P.

Rous reported that a cell-free extract from a chicken sarcoma could produce new sarcomas when introduced into another chicken (Figure 14–42). The virus contained in these extracts became known as the *Rous sarcoma virus (RSV)*. During the intervening years it was established that RSV has a genome composed of RNA and that during infection it is copied into DNA (see the following box). This process is known as *reverse transcription;* it is catalyzed by an enzyme coded by the virus and known, with less than perfect elegance, as *reverse transcriptase* (or RNA-dependent DNA polymerase). The group of viruses that have this property are known as *retroviruses.*

It was possible for many years to believe that retroviruses are *oncogenic,* tumor causing, because the infection irritates the cell into a state of uncontrolled growth. In 1970, however, G. S. Martin isolated a temperature-sensitive mutant of RSV that affects its ability to transform cultured cells. At the permissive temperature (35°C) this mutant virus transforms cells with the same efficiency as normal RSV. But at the higher nonpermissive temperature (41°C), the efficiency of transformation is much less. Even more impressive, transformed cells containing the mutant virus regain a normal appearance within hours after raising the temperature to the nonpermissive level (Figure 14–43). These

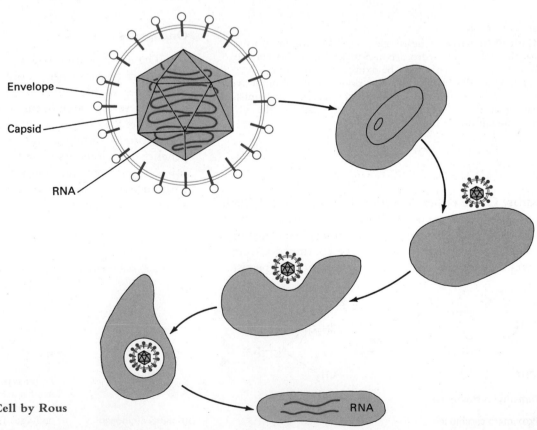

FIGURE **14–42**
Infection of a Chicken Cell by Rous Sarcoma Virus

The Life-style of Rous Sarcoma Virus

Like all viruses, the extracellular form of a retrovirus—the *virion*—is composed of a genome made of nucleic acid and surrounded by a protein coat, the *capsid*. Retroviruses have an additional covering, the *envelope* (see Figure 14–42). During infection, the virion binds to the surface of a cell, aided by specific proteins in the envelope, the particle enters the cell, probably by receptor-mediated endocytosis, and the envelope and capsid are removed. The virion contains two RNA molecules, usually identical, that act as messages for the virus-encoded reverse transcriptase and as templates for the synthesis of DNA by reverse transcription. Eventually a double-strand DNA is made; it forms a covalently closed circle, using host cell enzymes, and in that form *integrates* into the nuclear DNA, using a virus-encoded enzyme. The integration is aided by a special repeated sequence in the viral nucleic acid that resembles those found in transposons (see Chapter 11), and the complex mechanism of integration is similar to that used by certain

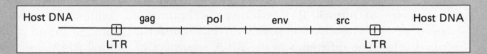

FIGURE 14–D
The Genome of Rous Sarcoma Virus in the Provirus State.   The four viral genes are flanked by host DNA lying just outside the two long terminal repeats (LTR), directly repeated sequences that aid integration and that also affect the expression of nearby host genes.

("retro") transposons. Once integrated, in which state it is called a *provirus*, the viral genes are flanked by the repeated sequence, which is a few hundred base pairs in length and known as the *long terminal repeat* (Figure 14–D). This repeated sequence acts as a regulator for transcription of the viral genes and, consequently, for the viral RNA itself.

Some of the RNA transcribed from the proviral DNA is used as a message to make viral proteins that will be incorporated into the virion: the *gag* gene codes for a precursor protein that is cleaved to form the capsid components; the *env* gene codes for a glycoprotein that becomes associated with the cell membrane. In

a productive infection there is a high rate of synthesis of viral RNA and the *gag* and *env* proteins, the capsid proteins assemble around the two RNAs, and this particle *buds out* through the cell membrane, becoming coated with the membrane-derived envelope in the process. The virus-specific glycoproteins (the products of the *env* gene) are concentrated in the region of the cell membrane through which the progeny viruses bud and thus become a prominent feature of the envelope. The virion does not include another important protein, the *src* gene product. Instead it becomes associated with the inner surface of the cell membrane and transforms the cell into a cancerous state.

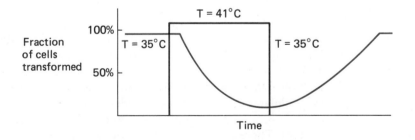

FIGURE 14–43
The Behavior of Rous Sarcoma Virus.   A temperature-sensitive mutation of the viral *src* gene produces temperature-sensitive transformation of infected cells. At the permissive temperature (35°C) most cells in the culture are transformed. After shifting to a high, nonpermissive temperature (41°C), the fraction of transformed cells falls rapidly as shown. The change in transformation is reversible if the temperature is again lowered, as shown at the right.

experiments suggested that the protein product of a single gene is sufficient to cause and maintain a tumor. The discovery of this gene, which we now call a *viral oncogene* (or *v-onc*), led to an explosion of research into the genetic basis of cancer.

The oncogene in RSV is called *src* (for *sarcoma*) and its molecular structure was established first by comparing the sequence of normal RSV with strains bearing a deletion of *src*. The deletion strains do not transform cultured cells. The difference in the RNA sequences of the two viruses showed which set of nucleotides was most likely the *src* gene. Later on, when the techniques of gene cloning were available, DNA copies of the RSV genome were mapped to identify *src* more precisely (see Figure 14–D). Using these methods it was possible to isolate and clone a fragment of DNA that when introduced into a chicken cell would transform it. The induction of malignant transformation, or any expressed phenotype, by isolated DNA is known as *transfection*.

Cellular Oncogenes

The activation of oncogenes present in normal cells was suggested by R. J. Huebner and G. J. Todaro as a way carcinogens might work. Their original conception was that the oncogenes were left behind by viral infections during evolution. A search was begun in 1972, by J. M. Bishop, H. Varmus, and their colleagues, to see whether a gene like *src* could be identified in normal chicken cells. They used as a probe a highly radioactive DNA sequence prepared by reverse transcription of normal RSV. Fragments of this DNA were hybridized with RNA from a deletion mutant of RSV lacking *src;* radioactive *src* DNA fragments could then be isolated because they did not hybridize to the RNA. After further purification, the *src* DNA was added to a digest of cellular DNA and hybridization was attempted. Not only were hybridizable sequences found in chickens and other birds but also in mammals (including humans) and fish. In fact, a cellular analog of the *src* gene—called *c-src* to distinguish it from the viral *src* gene or *v-src*—has been found in all vertebrates tested, and in the fruit fly *Drosophila* as well (cellular oncogenes are also known as *protooncogenes*). Bishop and Varmus were awarded the Nobel Prize in 1989 for their work on retroviruses and cellular oncogenes.

This has turned out to be a general pattern. There are more than three dozen retroviral oncogenes known, and all but one of them has been shown to have a widely distributed *c-onc* equivalent. But the original idea that the cellular oncogenes were left behind by an ancient retroviral infection is probably wrong. This seems likely because cellular oncogenes have introns in their DNA sequences (see Chapter 12) while the viral oncogenes do not. This difference is easier to explain if *c-oncs* were the source of *v-oncs*. One scenario that explains the origins of *v-oncs* involves a translocation of the *c-onc* into a viral genome in the provirus state—that is, integrated into the host DNA. During expression of the altered provirus RNA, processing of the *c-onc* RNA sequences would remove the introns before wrapping the RNA into its protein capsid. Subsequent reverse transcription of the viral RNA could not reconstitute the lost introns. This origin for oncogenes is also consistent with their impressively broad distribution in different organisms.

There is a more direct way to search for cellular oncogenes. It involves isolating the DNA from cancer cells and attempting to transfect cultured cells with it. (If the DNA comes from normal cells, transformation to malignancy almost never occurs.) Experiments of this kind, many by R. A. Weinberg and his colleagues, have been repeated with DNA from several different kinds of tumor cells. Molecular cloning techniques have allowed significant refinements to the transfection assay (Figure 14–44). The cancer cell DNA can be digested into fragments by a restriction endonuclease, the fragments cloned into a suitable plasmid or other vector, and the cloned fragments isolated and amplified and then used to test for transfection activity. By successive applications of these techniques it has been possible to show that quite small DNA fragments isolated from tumor cells are capable of transfecting susceptible cultured cells.

The hypothesis that active transfecting DNA might be novel genetic information, because it had been isolated from a tumor, was disproven when it was found that the nucleotide sequence of some of these transfecting fragments matched the sequence of one or another of the *c-onc* genes already identified by hybridization to retroviral oncogenes. Of course, once a cloned DNA fragment is known to have transfection activity, that fragment can be used as a probe (after suitable radiolabeling) to search in normal cells for an analogous sequence. In this way numerous new oncogenes have been identified. DNA fragments from these normal cell oncogenes have very low transfecting activity. A comparison between the sequences of the strongly transfecting DNA from the tumor and its normal counterpart has allowed the critical differences to be identified. In the case of an oncogene from a bladder carcinoma known as EJ, there is but a single nucleotide change, in a 350-base-pair segment, that makes the difference between forming a tumor and not. This change has been identified at the protein level as well: a G-to-T nucleotide change leads to a glycine-to-valine change in the active oncogene product. Sequence analysis has revealed that this transfecting DNA fragment is from the cellular gene that gave rise to the retroviral oncogene known as *ras* (for *rat* *sarcoma*).

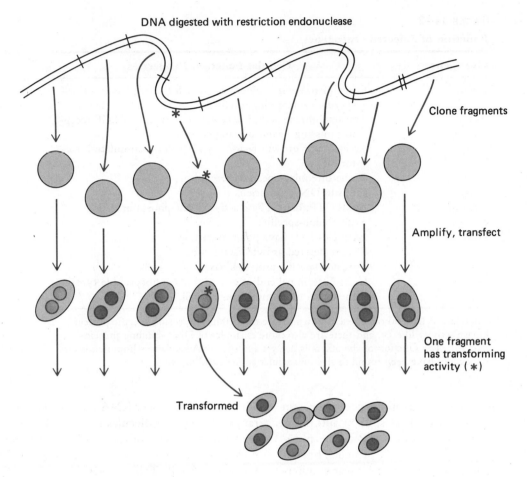

DNA digested with restriction endonuclease

Clone fragments

Amplify, transfect

One fragment
has transforming
activity (*)

Transformed

FIGURE **14–44**
Cloning of Oncogene Fragments. Total genomic DNA from a tumor cell
line is cut with restriction enzyme, and the fragments are cloned in bacterial
plasmids to produce a gene "library" of the tumor cell line. DNA is then isolated
from each of a large number of these clones and used to transfect normal
mammalian cells. The rare plasmid that contains an inserted oncogene will confer
the transformed phenotype (detected by growth properties) on the mammalian
cells it transfects.

A Normal Function for Cellular Oncogenes

The important questions that emerge from finding
so many oncogenes in so many different cells are (1) what
is the normal function of these genes and (2) how is the
behavior of the gene changed to produce a tumor. There
has been some progress in answering these questions but
the complete explanation of the normal function of
oncogenes and their role in carcinogenesis has yet to be
found.

The first oncogene to yield up its function was the
v-*src* from Rous sarcoma virus. An early finding was
that *src* coded for a *phosphoprotein* of *60,000* molecular
weight: *pp60v-src*. Again, a critical technique was to
compare normal RSV with a mutant virus containing a
deletion of the *src* gene. Soon after the discovery of

pp60v-*src*, it was shown to be associated with an enzyme
activity; it is a *protein kinase* that catalyzes the transfer of
phosphate from ATP to a *tyrosine* residue on the substrate
protein. Among the known v-*onc* genes, at least 20 have
tyrosine-specific protein kinase activity. By using probes
made of v-*src* DNA, it is possible to search in normal cells
for RNA transcribed from c-*src*. Such RNA sequences
have been found, so we know the c-*src* gene is at least
expressed by transcription. In addition, antibodies raised
against pp60v-*src* have been found to precipitate a protein
from normal cells. This protein, known as pp60c-*src*, also
has tyrosine-specific protein kinase activity.

A pair of other viral oncogenes, v-Ha-*ras* and
v-Ki-*ras* (from the *Harvey* and *Kirsten* strains of a rat
sarcoma virus), code for 21-kd proteins that have a high
affinity for binding GTP and a GTPase activity. The

TABLE 14–2

Function of Selected Oncogenes

Oncogene	Cellular Function/Properties
c-*erbB*	Receptor for EGF (epidermal growth factor); has tyrosine-specific protein kinase activity
v-*erbB*	Transmembrane and protein kinase domains of EGF receptor; missing ligand-binding domain
c-*fms*	Receptor for growth factor CSF-1 (colony-stimulating factor-1)
c-*mos*	Serine/threonine-specific protein kinase
c-*myc*	Binds to DNA in nucleus
c-Ha-*ras*	Binds GTP; has GTPase activity; self-phosphorylating
c-Ki-*ras*	(threonine-specific)
v-*ras*	Examples (>1) have point mutations
c-*sis*	Platelet-derived growth factor subunit
c-*src*	Tyrosine-specific protein kinase
v-*src*	Lacks C-terminal end of c-*src* and phosphorylatable tyrosine there; has other mutations as well

There are at least 50 oncogenes known. More than 20 have tyrosine-specific protein kinase activity like *src,* and there are also more examples of DNA-binding proteins like *myc.* The function of the others is not yet agreed. DNA sequences homologous to oncogenes have been found in all multicellular animals and in yeasts.

p21v-*ras* proteins also have a threonine-specific protein kinase activity for themselves—that is, they autophosphorylate a threonine residue, using GTP as the phosphate donor. There may be other target proteins, but for the present these remain a mystery. The cellular counterparts of the viral oncogenes are transcribed into RNA and translated into protein in normal cells, but the c-*ras* proteins have a higher GTPase activity and protein kinase activity has not yet been detected. DNA sequence comparisons suggest that the c-*ras* proteins are related to the G-proteins that couple peptide hormone receptors to adenyl cyclase (see discussion earlier in this chapter).

DNA sequences resembling *ras* genes have also been detected in *Drosophila* and yeast, again suggesting that these genes have some basic cellular function that has been highly conserved throughout evolution. There are two such genes in *Saccharomyces cerevisiae,* each of which is significantly larger than the mammalian c-*ras.* If both are inactivated, the yeast cells die, but either v-Ha-*ras* or v-Ki-*ras* can complement the lethal defect. Using the sophisticated genetics available to yeast research, suppressors of the lethal defect have been isolated and shown to be components of the cAMP-dependent protein kinase pathway. The evidence favors the hypothesis that the yeast c-*ras product* activates adenyl cyclase. The c-*ras* product has also been shown to accept fatty acid residues that are both necessary for its activity and help to anchor it in the cell membrane.

Almost all the known c-*onc*s are transcribed in normal cells, in many tissues, and in many organisms.

The amounts of c-*onc* RNA are usually very small, in the range of 1 to 10 molecules per cell. Although RNAs and proteins for several other c-*onc* genes have been found in normal cells, only a few of their enzymatic functions are known in detail (Table 14–2).

The normal function for the cellular oncogenes with protein kinase activity is presumably to phosphorylate target proteins of critical importance to growth control. This expectation is met for those tyrosine-specific protein kinases. An early response of cells to some of the protein hormones known as growth factors—for example, epidermal growth factor (EGF) and platelet-derived growth factor (PDGF)—is the phosphorylation of tyrosine residues on certain cellular proteins, including the membrane receptors for the growth factors. There is good evidence that such receptors themselves possess tyrosine-specific protein kinase activity. At least a few of the proteins that become phosphorylated in response to EGF and PDGF are also targets of the oncogene protein kinases. But the connection between growth factors and oncogenes is even closer: the oncogene known as c-*erbB* codes for the EGF receptor and another oncogene, c-*sis,* codes for PDGF itself.

Both the protein products of v-*src* and c-*src* have been localized by microscopy to the plasma membrane (both are covalently linked to a membrane-bound fatty acid), the site of the growth factor receptors, and especially to *adhesion plaques*—specialized regions of the membrane that form contacts between the cell and its substratum (Figure 14–45). It turns out that one of the components

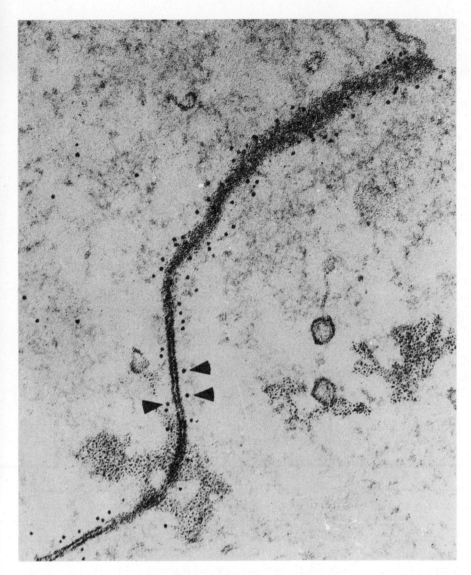

FIGURE **14-45**
Localization of v-*src* Protein of an Avian Sarcoma Virus at the Plasma Membrane. The image shows a gap junction between two virus-transformed cells in tissue culture. V-*src* (arrows) concentrates on the inner surface of the plasma membrane, at the gap junction. The protein was visualized by electron microscope immunocytochemistry, using ferritin as a marker. (Courtesy of M. Willingham.)

of adhesion plaques—*vinculin,* a cytoskeletal protein (see Chapter 16)—is also phosphorylated by pp60v-*src*. A plausible, but neither unique nor proven, scenario for the normal function of c-*src* would be as a growth factor receptor with protein kinase activity. On receipt of its (unknown) normal stimulus, c-*src* phosphorylates itself, other proteins, and vinculin. These alterations in membrane structure in turn prepare the cell to divide.

The existence of so many cellular oncogenes suggests that they must have distinguishing features of functional significance. In this connection it may be important that different retroviruses show some specificity for the tissues in which tumors are induced. Tissue specificity is also a common property of cancer: c-*oncs* identified by transfection (except *ras* and *myc*) also show it, as well as the heritable predisposition to certain cancers seen in humans. Taken together these observations implicate differentiation as an important feature of oncogene function.

If c-*oncs* are involved in differentiation, they should vary in their expression from one differentiated cell type to another and from one time to another during development. There is some evidence supporting these expectations. Active v-*src* suppresses the expression of genes found in the differentiated host cell and induces the expression of a gene normally active only in the embryo. Two oncogenes that are expressed in an immature blood cell cancer can be turned off rapidly by the application of chemicals that induce differentiation of the cells to a more mature form. During the development of a mouse embryo, the expression of several c-*oncs* varies over time.

Carcinogenesis by Pathological Oncogenes

What goes wrong with a cell when it becomes cancerous? There are two models that try to account for carcinogenesis by changes in the cellular oncogenes: the *dosage* (or *quantitative*) model, which states that increases in the activity of the c-*onc* lead to transformation, and the *qualitative* model, which states that a change in the behavior of the c-*onc*—perhaps a change in the specificity of its catalytic activity—leads to transformation. Some clues that lead to a choice between these two models can be obtained by comparing the behavior of viral oncogenes with their normal counterparts.

Many of these results are consistent with the dosage model. It has been found, for instance, that the activity of v-*src* in infected cells is many times higher than the activity of c-*src* in normal cells. Moreover, when cells containing the temperature-sensitive mutant v-*src* are placed at the nonpermissive temperature, the enzymatic activity due to pp60v-*src* drops and the cells lose the transformed phenotype. In some tumors there is found an increased number of copies of the DNA for c-*onc*s: there has been an *amplification* of the genes and, presumably, increased expression of their information.

The cancer known as Burkitt's lymphoma is associated with a translocation of part of one chromosome—number 8, which is known to contain the cellular oncogene c-*myc*—to one of a group of other chromosomes, each of which includes an active gene for the synthesis of an immune protein. The B-type lymphocytes that give rise to this tumor normally synthesize large amounts of antibodies. The c-*myc* gene thus comes under the influence of the same enhancer sequence that activates the heavy chain genes (see Chapter 17). An alternative rearrangement, seen in other Burkitt lymphomas, has the immune protein gene (with its associated enhancer) translocated to chromosome 8, near the normal location of c-*myc*.

Consistent with the quantitative model, data suggest that retroviruses with v-*onc*s cause tumors by introducing an activated oncogene into the cell. But there are some retroviruses—for example, *avian leukosis viruses* (ALVs)—that do not code for any known oncogenes, yet nevertheless cause tumors. When DNA from various tumors of this kind is examined, it is found that the ALV DNA is always integrated into the same location in the host cell's genome. The location is in fact in the control region of an already known cellular oncogene, c-*myc*. Apparently the c-*myc* gene is activated by the presence there of the viral DNA. This mechanism of carcinogenesis, called *insertional mutagenesis,* occurs in other virus-induced tumors as well. It is likely that viral insertion leads to cancer when the integrated viral DNA brings a cellular oncogene under the control of a viral promoter. (This would be analogous to the construction of an operon fusion, although the engineer in this case is the virus rather than a cell biologist.)

There is also some evidence favoring the qualitative model for carcinogenesis. We have already mentioned that most known carcinogens also cause mutations. The previously described EJ bladder cancer is associated with a mutation in the coding sequence of c-Ha-*ras* at amino acid position 12. A human lung carcinoma was found to be associated with an altered c-Ha-*ras* gene bearing information for an amino replacement at position 61 of the p21 protein.

The behavior of oncogenes can be also fitted into the multistep process observed in natural carcinogenesis. There is evidence that a change in more than one oncogene enhances the probability of transformation. Activation of the c-*myc* gene by itself can in certain cases *immortalize* a normal cell—allow it to grow indefinitely—but not confer the property of producing tumors. By itself, an activated c-*ras* gene will lead a normal cell to anchorage-independent growth (in soft agar), but the cells die out before too many more generations. When an activated c-*myc* is combined with an active c-*ras*, however, a fully transformed cell results, one that can cause tumors in an immune-deficient animal.

Other properties of tumors in whole animals are also mimicked by cultured cells. Admittedly, there is a considerable simplification represented by the cell culture model. Despite these dangers, the advances we have summarized here suggest optimism for applying cell methods to the study of cancer. Additional good news is that many experiments designed to probe the mechanisms of cancer may turn out to aid our understanding of the normal control of information flow in cells.

SUMMARY

The chemical reactions that define cell activity are closely controlled in rate and coordinated with one another. These controls operate at virtually every level of cell function. At the level of intermediary metabolism, many metabolites bind to enzymes in the pathways of their synthesis or utilization to decrease or increase the enzymes' catalytic activities. In some cases, metabolites exercise an indirect control by triggering covalent modification of groups of enzymes, for example by protein phosphorylation and dephosophorylation.

Cells also control the rate at which enzymes are produced at the level of gene expression. Some genetic controls operate through methylation of cytosine residues at specific positions on DNA that affect the rate of transcription. In many cases, regulatory proteins bind to specific control sites in or near genes to activate or repress their transcription. Other control mechanisms govern the probability that transcription will either terminate before reaching a given gene or will proceed into the gene and produce a complete mRNA transcript. Once an mRNA is produced, the rate at which it associates with ribosomes to be translated can also be subject to regulation.

The coordinate regulation of the expression of large numbers of genes underlies complex cell behaviors such as orderly progress through the cell division cycle. We are beginning to understand the circuitry of this complex regulation system and the genetic and molecular causes of its occasional failure in the pathology of cancer.

KEY WORDS

Allosteric, clone, consensus sequence, deoxyribonucleic acid, effector, energy, enzyme, eucaryote, exonuclease, feedback, genetic code, information, kinase, ligand, mass, mitosis, oncogene, operator, operon, polymer, procaryote, promoter, repressor, ribonucleic acid, ribosome, RNA transcription, substrate, translation, virus.

PROBLEMS

1. Define the following terms: constitutive synthesis; repressor; autogenous regulation; site-directed mutagenesis; promoter; footprinting; attenuation.

2. Describe the experimental method through which Gilbert and Mueller-Hill identified and eventually purified the *lac* repressor.

3. The table below gives the genotype at the *lacI*, *lacO*, and *lacZ* loci of various haploid and partially diploid strains of *E. coli*. Fill in the columns labeled "beta-galactosidase synthesis," indicating a (+) for the cases where enzyme is made or a (−) if it is not made.

	Beta-Galactosidase Synthesis	
Genotype	No Inducer Added	Inducer Added
$I^+O^+Z^+$		
$I^+O^+Z^-$		
$I^-O^+Z^+/I^+O^+Z^-$		
$I^+O^+Z^+/I^-O^+Z^-$		
$I^+O^cZ^+$		
$I^+O^cZ^+/I^+O^+Z^+$		
$I^+O^cZ^+/I^+O^+Z^-$		
$I^+O^cZ^-/I^+O^+Z^+$		

4. It has been reported that protein synthesis is required for mitosis to occur in embryonic cells of the surf clam. This was somewhat unexpected since these embryos divide without increasing in size and have considerable stockpiles of most of the components that are normally duplicated during the cell cycle. By examining the synthesis of various proteins during the embryonic cell divisions, one protein was found whose concentration fluctuates during the cell division cycle (see figure below); the protein was named "cyclin." One possible hypothesis to explain the role

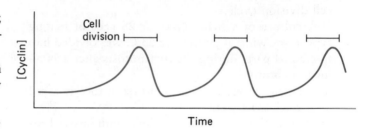

of this protein is that regular changes in cyclin concentration represent the cell "clock" that defines the length of the cell cycle.

(a) Explain why you accept, reject, or conditionally accept this hypothesis.

(b) Propose at least one experiment that could be used to test the hypothesis. State clearly the possible results of the experiment, assuming the hypothesis is true and assuming the hypothesis is false.

(c) It has also been observed that the mRNA for cyclin is already present in the surf clam egg before fertilization. Propose a mechanism to explain how expression of the information in this mRNA could be controlled.

5. Analysis of the control of the cell division cycle in the budding yeast *Saccharomyces cerevisiae* has been aided by the isolation of special temperature-sensitive mutants, the *cdc* (cell division cycle) mutants.

(a) What phenotype distinguishes a *cdc* mutant from other temperature-sensitive mutants—for example, one that stops all protein synthesis when placed at a high temperature?

When a haploid yeast *cdc* 28 mutant is placed at nonpermissive 36°C, it seems unable to progress

Culture	Change in Culture Conditions	Time under New Conditions	Change in Culture Conditions	% Dividing at 4 Hours
A	No change		No change	100
B	Raise T to 36°C	4.0 hrs		< 5
C	Add mating factor α	4.0		< 5
D	Raise T to 36°C	0.8	Add mating factor α; lower T to 23°C	85
E	Add mating factor α	0.8	Raise T to 36°C; remove mating factor	5

through stage G_1 of the cell cycle. A similar behavior is elicited when haploid yeast of mating type *a* are exposed to the peptide mating factor released by cells of mating type α (alpha). The following data are the results of a "reciprocal shift" experiment designed to see which function, the one associated with *cdc* 28 or the one associated with absence of mating factor α, is executed first during normal progress through the cell division cycle.

A culture of synchronized *cdc* 28 cells of mating type *a,* growing at permissive 23°C was divided into five equal parts during the first mitosis after achieving synchrony.

(b) Which function is executed first in the yeast cell cycle? Briefly explain.

(c) Could this experiment be done with unsynchronized cells? Briefly explain.

6. Respond to the following statements with TRUE or FALSE. If you answer TRUE, explain why you think the statement is true; if you answer FALSE, explain why the statement is not true.

(a) Carbamoyl phosphate binds to the effector site of the enzyme aspartyl transcarbamoyl synthetase (equals aspartyl transcarbamylase).

(b) DAHP synthetase activity in the pathway for synthesizing aromatic amino acids is likely to be controlled by the concentration of shikimic acid-5-phosphate.

(c) The activity of adenyl cyclase is modified by the action of the $G_s \cdot GTP$ complex, which acts by catalyzing the transfer of ADP-ribose from NAD^+ to the adenyl cyclase.

(d) Part of the information needed to control cell metabolism is manifested as the cellular concentration of small molecules.

(e) A covalent modification of a G_s protein that stimulates its GTPase activity would tend to increase cellular [cAMP].

(f) A normal cell becomes transformed into a tumor cell after infection by *E. coli.*

(g) Normal animal cells can grow forever, or rather their descendants can, provided they are kept in culture rather than in the animal.

(h) The racemization over time of L-amino acid residues in proteins to D-amino acid residues is a kind of "error" that could explain aging.

7. Control of enzyme activity requires communicating information from one location to another. In multicellular organisms, the two locations may be in different cells and the communication may involve hormones, growth factors, neurotransmitters, and so forth. These molecules often need not enter a cell to elicit a response; the information is carried along the signaling pathway by a second messenger.

(a) List three different forms of "second messengers."

(b) Identify what molecule receives the "information" carried by the second messenger.

(c) Briefly explain how one molecular form of a second messenger can elicit different responses in two different cells.

8. Allosteric activator A binds reversibly to enzyme E: $A + E \rightleftharpoons E \cdot A$. The equilibrium *dissociation* constant (K_D) has the value 2×10^{-6} molar. In the presence of a physiological concentration of substrate S, the enzyme has the following properties:

State of Enzyme	Relative Rate of Reaction Catalyzed by E
Activator bound	100
A not bound	10

If [A] is 1×10^{-6} molar, what is the relative rate of the reaction catalyzed by E at the same substrate concentration?

9. In a particular procaryotic cell the biosynthetic pathway for amino acid E is

$$\text{Pyruvate} \xrightarrow{1} B \xrightarrow{2} C \xrightarrow{3} D \xrightarrow{4} E$$

The degradative pathway for amino acid E is

$$E \xrightarrow{5} F \xrightarrow{6} G \xrightarrow{7} H \xrightarrow{8} \text{Krebs cycle intermediate.}$$

The flow of mass through both pathways is regulated

by [E]. Using the numerals above the arrows to designate enzymes:

(a) Predict which enzymes' activities will be modified by [E] and how they will be modified.

(b) Predict which enzymes' concentrations will be modified by [E] and how they will be modified.

(c) Assume there is an analogue of amino acid E (call it E'), which resembles E structurally but which cannot be metabolized once it enters the cell. Adding E' inhibits the growth of cells using either glucose or E as a carbon source. Cells grow normally in the presence of E' if both glucose and E are also present. Interpret these results.

10. Purified mRNA (e.g., rabbit globin mRNA) can be microinjected into a cell (e.g., *Xenopus* oöcyte) that responds by synthesizing the appropriate protein (e.g., rabbit globin). When a fibroblast in the G_1 phase of the division cycle is fused with a fibroblast in the S phase, the G_1 nucleus immediately begins DNA synthesis. Using these two observations, design an experimental project to identify a soluble cellular factor that triggers entry into the S phase of the cell division cycle. Specify the assays you would use and tell how you would narrow the choices among possible molecular forms of the postulated factor.

SELECTED READINGS

Atkinson, D. E. (1977) *Cellular energy metabolism and its regulation.* New York: Academic Press.

Berridge, M. J. (1985) The molecular basis of communication within the cell. *Scientific American* 253 (10):142–152.

Bishop, J. M. (1982) Oncogenes. *Scientific American* 246(3):80–92.

Croce, C. M., and Klein, G. (1985) Chromosome translocations and human cancer. *Scientific American* 252(3):54–60.

Draetta, G. (1990) Cell cycle control in eukaryotes: Molecular mechanisms of *cdc* 2 activation. *Trends in Biochemical Sciences* 15:378–383.

Freifelder, D. (1985) *Essentials of molecular biology.* Boston: Jones and Bartlett.

Hanawalt, P. C., ed. (1980) *Molecules to living cells.* San Francisco: W. H. Freeman.

Hanawalt, P. C., and Haynes, R. H., eds. (1973) *The chemical basis of life.* San Francisco: W. H. Freeman.

Hunter, T. (1984) The proteins of oncogenes. *Scientific American* 251(8):70–79.

Land, H., Parada, L. F., and Weinberg, R. A. (1983) Cellular oncogenes and multistep carcinogenesis. *Science* 222:771–778.

Maniatis, T., and Ptashne, M. (1981) A DNA operator-repressor system. In *Genetics. A collection of readings from Scientific American.* San Francisco: W. H. Freeman.

Mitchison, J. M. (1974) Sequences, pathways and timers in the cell cycle. In *Cell cycle controls.* B. M. Padilla, I. L. Cameron, and A. Zimmerman, eds. New York: Academic Press, pp. 125–142.

Murray, A. W., and Kirschner, M. W. (1989) Dominoes and clocks: Two views of the cell cycle. *Science* 246:614–621.

Pitot, H. C. (1978) *Fundamentals of oncology.* New York: Marcel Dekker.

Weinberg, R. A. (1983) A molecular basis of cancer. *Scientific American* 249(11):126–142.

Yanofsky, C. (1981) Attenuation in the control of expression of bacterial operons. *Nature* 289:751–758.

Gerbil fibroma cell (ATCC CCL146) in tissue culture, with stress fibers stained by the indirect immunofluorescence method: mouse monoclonal antibody against tropomyosin plus fluorescein isothiocyanate-conjugated goat antimouse IgG. The stained stress fiber structures show characteristic interruptions along their length, participate in peripheral polygonal networks, and surround the nucleus. ×2800 (Courtesy of J. J. C. Lin.)

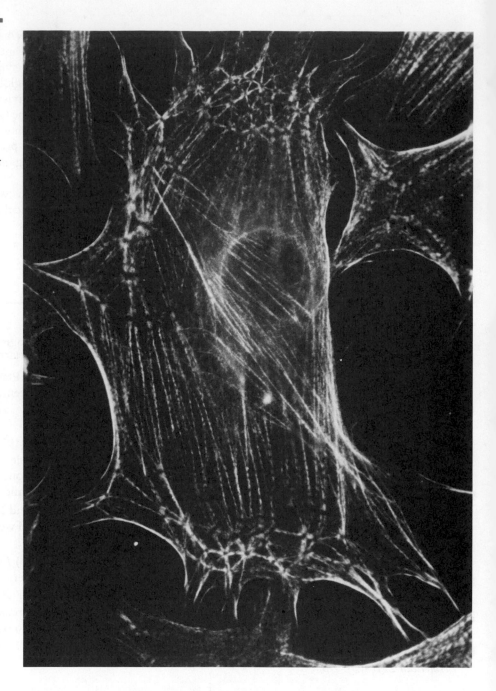

We show how, by using the available energy and information, the cell engages in a number of integrated processes such as assembly of complex structures, growth, movement, and interactions between cells and their environment.

Our understanding of the molecular bases of integrated processes—for example, development and the operation of the nervous system—is woefully incomplete, and it is these areas that most need the attention of future cell biologists.

PART

4

THE INTEGRATION OF CELL BEHAVIOR— THE USE OF ENERGY AND INFORMATION

From Molecules to
Biological Structures

15

We have discussed in previous chapters how protein molecules differ in their structural complexity, consisting in the simplest cases of a single polypeptide chain and in more complicated cases of one or more subunits, each composed of one or more polypeptide chains (see Chapter 4 for definitions). The highly specific structure of proteins is determined by their linear sequence of amino acids. After denaturation, some proteins fold back spontaneously to their native structure, because it has the lowest free energy. For these proteins, the amino acid sequence determines the native structure. There remains the likely possibility for other proteins that the pathway of folding during synthesis will "kinetically trap" a native structure that is not the lowest free energy configuration. The native structure in these cases would be determined by the sequence of amino acids added at the beginning of the protein's synthesis. An energy barrier in the completed protein structure could make attaining the lowest free-energy configuration so slow that it would not be reached during the molecule's expected life span in the cell. (See figure on p. 708.)

Even though we cannot yet specify all the rules for determining the structure of a given protein, it is nevertheless appropriate to ask similar questions about the assembly of more complicated structures, such as enzyme complexes, ribosomes, viruses, cytoplasmic fibrils, and beyond these in complexity, organelles and the cortical regions of cells. Our knowledge of these structures is relatively primitive but current techniques and approaches hold great promise, and it is important to indicate both the present level of understanding and the direction this work will take in the future.

Let us begin by drawing formal distinctions among the processes of *self-assembly, aided assembly,* and *directed assembly*. (See figure on p. 709.)

In *self-assembly* the only information needed for the formation of a structure is contained in the macromolecules that themselves form the assembled product. In the purest form of self-assembly, no other macromolecular component is necessary for the assembly process. This does not exclude all contributions by the cell; it maintains an appropriate environment of micrometabolites, pH, and salts, which are almost certainly necessary for a self-assembly process.

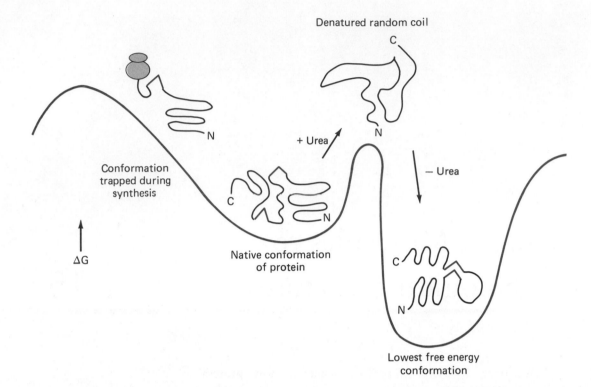

In an *aided assembly* process, macromolecules other than those in the final structure are needed for the assembly process. In certain cases, enzymes may be required specifically to modify the macromolecules during the formation of a structure. This should be distinguished from the general post-translational processing discussed in Chapter 13; the action of such an enzyme in aided assembly is directed toward components after they have interacted to form a partially assembled structure.

In *directed assembly* there is a requirement for a *pre-existing structure* to organize components so that more of the structure can be made. There are two different versions of directed assembly. One which we have discussed already is the template-directed process of DNA replication in which detailed structural information is duplicated. The other version is more analogous to the seeding of a crystal into a saturated solution; the presence of a given structure acts as an *initiator* of an otherwise prohibitively slow assembly process.

But why have complex structures made of multiple components evolved? Why does not the cell synthesize a single molecule with all the desired characteristics? It is possible to recognize several advantages to the assembly of larger structures from smaller subunits. First of all it is easier to make accurate copies of a small part than of a large structure. Consider two simple cases of making a large structure containing 10^4 amino acids, (1) by making one polypeptide chain 10^4 amino acids long, or (2) by assembling 10^2 identical polypeptide chains of one hundred amino acids each. If the average probability of

making an error in protein synthesis is 10^{-3} per amino acid, then the probability of making perfect copies of the subunit is $(0.999)^{100} = 90\%$. This should be contrasted with the similarly calculated probability of making a perfect copy of the large structure: less than 0.01%. If errors in the small chains reduce their likelihood of combining with others, then there is a good chance of assembling a hundred error-free chains into a perfect large structure. There will also be much less DNA required to specify the sequence of amino acids in a smaller chain than in a larger one.

Certain disadvantages must also attend the use of a construction strategy using subunits. A certain part of each subunit must be devoted to combining with neighbors and is therefore unavailable to evolve toward more activity in its catalysis or more versatility in its regulation. Also the rules of geometry limit the number of ways identical subunits can combine to form more complicated structures. These ways may or may not be optimal for some biological functions. Thus, one expects a balance in cells between the use of large complicated structures composed of identical subunits and large single chains.

Self-Assembly

Self-assembly processes may involve the coming together of protein molecules only or they may involve the interaction of protein with nucleic acids or lipids. Let us discuss each of these in turn.

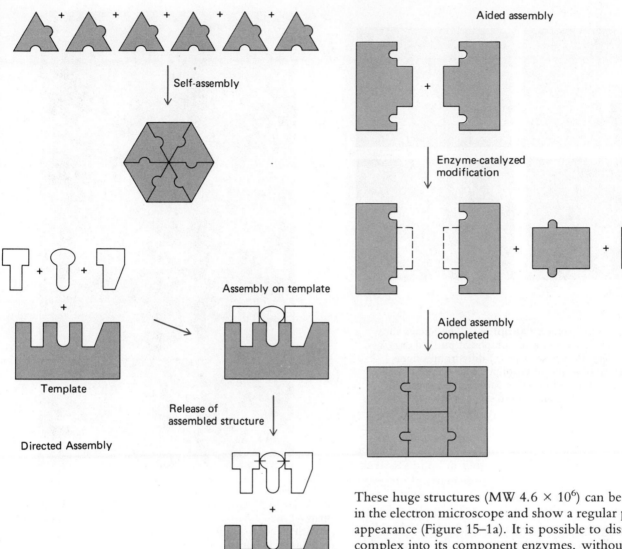

Self-assembly

Template

Directed Assembly

Assembly on template

Release of
assembled structure

Aided assembly

Enzyme-catalyzed
modification

Aided assembly
completed

Multienzyme Complexes

The biochemists of the 1930s often treated the cell as if it were a "bag of enzymes in solution." We have known for some time, however, that many enzymes are not merely independent molecules in solution but are associated with structures—for example, membranes or filaments—or with other enzymes in *multienzyme complexes*. In fact, this association might be more common than is sometimes believed. When enzymes are studied in the laboratory, the concentrations used are usually much lower than are found inside cells. Consequently, weak associations of enzymes with each other or with other cell components could be missed under laboratory conditions.

One example that has been studied in considerable detail is the pyruvate dehydrogenase complex in *E. coli*.

These huge structures (MW 4.6×10^6) can be visualized in the electron microscope and show a regular polyhedral appearance (Figure 15–1a). It is possible to dissociate the complex into its component enzymes, without breaking any covalent bonds, and Table 15–1 shows a current view of the stoichiometry of the complex. After extensive purification the separated components can be reassembled under the appropriate conditions of temperature, pH, and salts to form a practically fully active structure with strong morphological similarity to the original complex (Figure 15–1b). It thus appears that the case of the pyruvate dehydrogenase complex is a straightforward example of self-assembly in which only the molecules forming the structure are involved in its assembly.

The biological importance of segregating enzymes that are involved in metabolically related processes has already been mentioned when we considered the functions of organelles such as chloroplasts, mitochrondria, and ribosomes. Another example of multienzyme complexes is found in the biosynthetic pathway of aromatic amino acids in *E. coli*, discussed in Chapters 6 and 14. If the enzymes of a pathway are clustered in close proximity, metabolic intermediates can react more rapidly than if they were equilibrated throughout the cell volume.

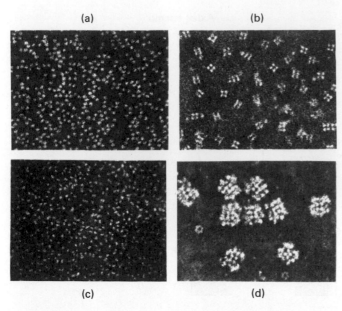

(a)   (b)

(c)   (d)

FIGURE 15–1
Pyruvate Dehydrogenase Complex.   The figures show
negatively stained electron micrographs of purified enzymes.
(a) Pyruvate dehydrogenase (E_1); (b) dihydrolipoamide
acetyltransferase (E_2); (c) dihydrolipoamide dehydrogenase
(E_3); (d) pyruvate dehydrogenase complex, composed of
(a), (b), and (c). ×200,000. (Courtesy of R. M. Oliver and
L. J. Reed.)

There are several levels of understanding that may be
achieved when studying the assembly of complex structures. After the composition has been analyzed, in terms
of various kinds of structural subunits, something needs
to be known about their relationships, the bonds that
hold them together, and the pathway(s) taken in building
up the final shape. Our next example of self-assembly
reveals progress that has been made on all these issues.

Tobacco Mosaic Virus

The study of viruses has proved to be a fertile area
for research on the mechanism of assembling biological
structures. The reasons for this are clear: viruses can be
grown in large quantities, their infectivity is a convenient
biological assay to evaluate the success of the assembly,

FIGURE 15–2
A Tobacco Mosaic Virus Particle.   The shadow-casting
method used here permits the calculation of the height of the
particle. The small piece is a fragment of the virus.
×200,000. (Courtesy of R. C. Williams.)

they are generally large enough to be seen clearly under
the electron microscope, and they provide a broad range
of structural complexity—from the relatively simple
tobacco mosaic virus (TMV) to the complex "T-even"
coliphages such as T2 and T4. Viruses also differ in their
assembly processes, ranging from TMV, a classic case of
self-assembly, to the aided and directed assembly of the
more complex bacterial viruses.

TABLE 15–1
The Composition of the Pyruvate Dehydrogenase and Other Enzymatic Complexes of *Escherichia coli*

	Number of Polypeptide Chains	Molecular Weight of Each Chain	Subunit Structure	Molecular Weight of Complex
Dihydrolipoamide acetyltransferase	24	36,000	Trimer	2.6×10^6
Pyruvate dehydrogenase	24	96,000	Dimer	4.6×10^6
Dihydrolipoamide dehydrogenase (flavoprotein)	12	56,000	Dimer	1.3×10^6

A subunit is a separable, observable component of the native oligomeric complex.

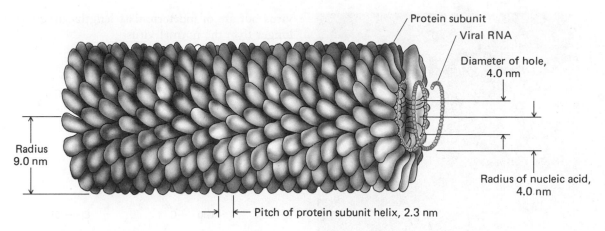

FIGURE 15–3
Model of Structure of Tobacco Mosaic Virus.   The complete particle contains 2130 identical protein subunits, forming a cylindrical structure. The RNA strand is embedded in the protein, in the form of a gigantic helix. (Redrawn from A. Klug.)

Tobacco mosaic virus, the first virus to be crystallized (by W. Stanley in 1935), is a cylindrical structure with a molecular weight of 40 million and dimensions of 16 × 300 nm (Figure 15–2). Extensive structural studies of the virus—which used techniques such as x-ray diffraction, electron microscopy at various stages of dissociation, solution properties (sedimentation, diffusion, light scattering), as well as studies of the separated protein and RNA components—have yielded a remarkably detailed and precise model of the virus particle.

The protein coat of the virus consists of 2130 identical protein monomers, each with a molecular weight of 18,000. This monomer is a single chain of 158 amino acids of known sequence; the folding of the chain produces a blimp-like ellipsoid. The RNA of the virus is a single strand of 6400 nucleotides with a molecular weight of 2.4 million. The RNA strand forms a helix with a radius of 4.0 nm that lies embedded in a larger helix (radius 8.5 nm) of identical pitch (2.3 nm) formed by the protein subunits (Figure 15–3). Down the middle of the molecule runs a cylindrical hole with a radius of 2.0 nm that can be seen in the electron microscope if a special staining technique is used.

In the case of pyruvate dehydrogenase, the complex is disassembled, the various components are isolated and extensively purified, and they are then allowed to reassemble into the final multienzyme complex. In the case of TMV, an alternative strategy is instructive: the virus is dissociated and then it is allowed to reassemble only partially. Early in reassembly, under the appropriate conditions, there appear small "doughnut-shaped" structures, showing holes in the center if viewed end on (Figure 15–4). Using hot detergent, TMV can be disso-

FIGURE 15–4
Demonstration of Cylindrical Structure of TMV, With Hole at Center.   TMV can be dissociated into protein and RNA. The protein can be reassembled in the absence of RNA. After partial reassembly, small doughnut-shaped particles can be observed which, if viewed end on, show clearly the circular cross section of the particle and the hole in the center. ×150,000. (Courtesy of R. C. Williams.)

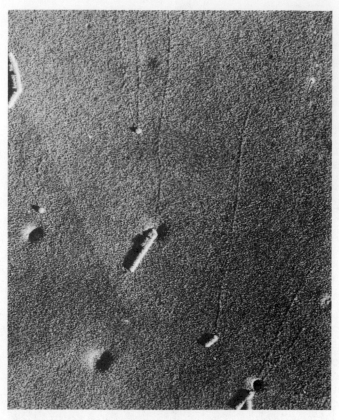

FIGURE **15–5**
Stepwise Dissociation of TMV, Showing Strands of RNA Extending From One End of the Particle. The virus was partially dissociated with a solution of hot detergent. From the position and the length of the RNA strand, it must be concluded that the strand is wound in a helical conformation inside the virus particle, a conclusion that was first suggested by R. Franklin on the basis of her x-ray studies. ×150,000. (From R. Hart.)

virus but are of indeterminate lengths, frequently much longer than the normal virus.

$$Glu-C \overset{O \ \text{⦀} \ H-O}{\underset{O-H \ \text{⦀} \ O}{}} C-Glu$$

$$+2H^+ \underset{-2H^+}{\Updownarrow}$$

$$Glu-C \overset{O}{\underset{O^-}{}} \qquad \overset{O^-}{\underset{O}{}} C-Glu$$

The effects of differences in pH on the assembly of the protein subunits alone suggest that the process involves two sets of hydrogen-bonded carboxyl groups, probably of glutamic acid residues. The hydrogen bonding appears to raise the pK of these groups to 7.5, approximately. It is thought that these residues remain bound to one another, despite being in an aqueous medium, because they are constrained to be close to each other by other interactions in the assembled structure. Since the pK of these groups is 7.5, they will dissociate their protons in neutral or slightly alkaline solutions—thus becoming negatively charged—and the assembly of the protein coat alone will not occur because of electrostatic charge repulsion. If, however, the virus RNA is mixed with the protein, then assembly will occur in neutral or slightly alkaline solutions, and, furthermore, the assembly process will terminate when the entire length of RNA is used up. The infectivity of such reconstituted particles is almost as high as that of the normal virus. These and other results suggest that the RNA interacts with the protein and has some guiding influence in the assembly process. Specificity for the sequence of bases in TMV RNA was shown by the finding that nonviral RNAs (either synthetic or natural) would not assemble with the viral protein.

A troubling feature of the early reassembly experiments was the lengthy time, some hours, required to regain maximum infectivity. This seemed too long for the natural process of infection and replication inside the tobacco cell where the RNA would be exposed to the risk of nuclease digestion. The problem was resolved by the elegant work of A. Klug and his collaborators. The key to the mystery is the spontaneous formation of protein monomers into a *subunit (disk) assembly* containing two layers of 17 monomers each (Figure 15–6). The disks contain practically the same number of proteins as is found in two turns of the helical virus structure. Disks are the dominant form of the protein under physiological

ciated in stages and EM photographs taken of the process. Figure 15–5 shows TMV particles, with the strand of RNA extending from them, at various stages of dissociation. These images constitute a dramatic visual confirmation of the early ideas regarding the structure of TMV and also give clues to a possible mechanism of assembly.

Fractionation procedures for TMV have been worked out whereby it is possible to prepare native protein free of RNA and, conversely, undegraded RNA free of protein. The protein-free RNA can infect tobacco leaves (although with an efficiency of only 1 to 2% that of the normal virus), showing that the protein is not necessary for virus replication but that it is useful for the infection process. At acid pH and low ionic strengtn, conditions that differ from those found naturally in the tobacco leaf cell, the protein subunits alone can aggregate; they form rods that are identical in diameter to the TMV

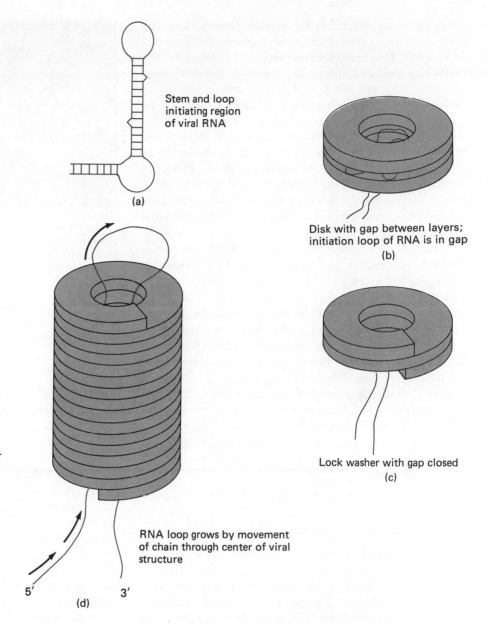

Stem and loop
initiating region
of viral RNA

(a)

Disk with gap between layers;
initiation loop of RNA is in gap

(b)

Lock washer with gap closed

(c)

RNA loop grows by movement
of chain through center of viral
structure

5′   3′

(d)

FIGURE 15–6

Spontaneous Self-Assembly of TMV RNA with Protein Subunits.
A plausible scheme for the assembly of TMV (after P. J. G. Butler and A. Klug). A hairpin loop of the viral RNA (a) inserts into the central hole of a protein disk. This loop intercalates between the two layers of protein subunits, binding around the disk in a turn (b). This interaction results in the formation of a helical "lock-washer" (c), which traps the viral RNA between the protein subunits. New protein disks are added to elongate the virus structure (d). The result is the rod seen in Figure 15–2.

conditions. Of course, if the disks are an intermediate in the pathway of assembly and are readily available, the whole process could go much faster. This is because the coming together of preformed subassemblies is more likely than the coming together of 34 individual monomers. When conditions were chosen to favor the presence of disks, the time required for reassembly was speeded up to minutes instead of hours.

The disk structure also contributes to the initiation of virus assembly. When TMV protein and RNA are mixed in a ratio sufficient to provide only one disk per RNA, the disks protect a fragment of RNA from the attack of added nucleases. (This "protection" strategy is often used to analyze specific nucleic acid sequences that interact with proteins—see the discussion of the ribosome binding during initiation of protein synthesis in Chapter 13.) The

protected fragment is 65 nucleotides in length, just about the length of one turn of RNA in the helical virus structure (see Figure 15–6). The sequence of this fragment has been determined, and it can form a stem and loop structure by base pairing. The loop at the end of the long base-paired stem is believed to interact with the protein disk to initiate the assembly of the virus. Disks are then added to the initiation subassembly, consistent with the finding that RNA sequences are protected during assembly progressively in groups, in this case of 100 nucleotides each.

The details of the interaction of the RNA with the disk structure are beginning to become clear. The loop of RNA probably enters the disk through the hole in the center (see Figure 15–4). Then the bases of the RNA are bound between the two layers of the disk, a reaction that opens the stem and binds those bases as well. The disk then undergoes a conformational change to the form of a *lock washer*. This provides two structural features that aid further assembly: it brings the layers of the disk closer together, trapping the bound RNA, and it produces a definite place on the otherwise indefinite circular disk on which further disks can assemble. The lock-washer structure can also be produced by the disk alone if it is placed at low pH. The disk structure under those conditions is believed to be stabilized by the hydrogen bonds between the two carboxylic acid residues mentioned earlier. As the RNA is bound, there is sufficient energy of attraction to form the lock-washer conformation at neutral pH.

An unexpected feature of the initiating sequence is that it is not near an end of the RNA. This means that subsequent disks are added to the structure by binding to a loop of RNA that arises from threading one length of the RNA through the hole in the center of the growing virus structure (see Figure 15–6). Electron micrographs show clearly that there are two lengths of RNA protruding from partly assembled viruses, in agreement with the model proposed by Klug. Among details to be worked out are how the virus structure can assemble to cover the other end of the RNA chain and the specific geometry of the RNA-protein interactions of the disk. There are not sufficient data to decide, but we should not be surprised to learn that other helical filamentous viruses assemble by similar pathways.

Spherical Viruses

There is another group of viruses that from the point of view of their geometry appear to have a simple structure. They have *icosahedral symmetry,* the same as is shown by the 20-sided, triangle-faced, regular icosahedron. (This figure has vertices where five triangles come together and thus has fivefold rotational symmetry, see Figure 15–8b. Pentamers that form a vertex in the viral structure may be an important assembly intermediate.)

D. L. D. Caspar and Klug formulated an ingenious theory which states that the icosahedral symmetry "represents the optimum design for a state of minimum energy of a closed shell built of regularly bonded, identical subunits." Their theory predicted a number of icosahedral arrangements that were subsequently discovered in a variety of viruses. An ingenious demonstration of the icosahedral symmetry of Tripula iridescent virus was made by P. Horne. He used the technique of double-shadow casting in the electron microscope and compared the shape of the shadows with those cast by an icosahedral model (Figure 15–7).

FIGURE **15–7**

Demonstration of the Icosahedral Structure of Tripula Iridescent Virus.   (a) EM photograph of virus particles, shadowed from two different directions. ×200,000.
(b) Cardboard model of equivalent icosahedron, shadowed by two light sources. The two types of shadows, one blunt and the other pointed, are characteristic of the icosahedral geometry. The same shapes appear in the EM photograph of the virus particle. (Courtesy of P. Horne.)

Icosahedral viruses are composed of a shell of apparently identical morphological subunits (*capsomeres*); the shell encloses a single, often circular, molecule of RNA or DNA. They vary in size and complexity from the tiny virus associated with, and therefore called *s*atellite of, *t*obacco *n*ecrosis *v*irus, STNV (18-nm diameter, 2×10^6 MW, 60 identical proteins arranged to give 12 capsomeres), to the bacterial virus ΦX174 (25-nm diameter, 6.3×10^6 MW, 32 capsomeres of more complex structure), to structures as large as the adenovirus (70-nm diameter, 2×10^8 MW, 252 capsomeres of very complex structure). Figure 15–8 shows electron micrographs of ΦX174 and adenovirus as well as the respective icosahedral structures to which these viruses seem to conform.

Early attempts to reproduce in vitro the self-assembly of viruses with icosahedral symmetry had proved fruitless. It was only when the conditions of reassembly were carefully controlled that a number of these viruses were demonstrated capable of self-assembly. One of the first successes was the case of *c*owpea *c*hlorotic *m*ottle *v*irus (CCMV), an icosahedral plant virus of 4.5×10^6 MW. It contains a single molecule of RNA (1×10^6 MW) surrounded by a protein shell composed of 180 subunits, each containing 183 amino acid residues (19,600 MW). These 180 subunits form an icosahedral structure composed of 32 groups, 12 of which contain 5 subunits and 20 of which contain 6 subunits.

The key to the successful demonstration of self-assembly of CCMV turned out to be proper disassembly.

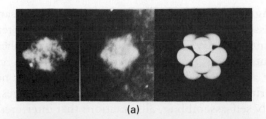

(a)

(b)

FIGURE 15–8
Two Icosahedral Viruses and Their Respective Structural Models. Visualization of the individual capsomeres is facilitated by the use of negative staining with phosphotungstate. This technique allows the electron-dense stain to surround structures, leaving them less electron dense. (Courtesy of P. Horne.) (a) Φ X 174 and its Ping-Pong ball model, composed of 12 spheres (× 400,000). (b) Adenovirus and its model, composed of 252 spheres (×85,000).

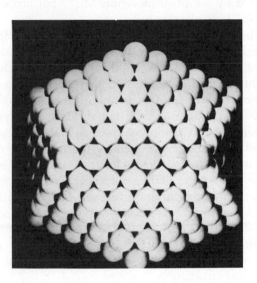

The details of the method illustrate how laboratory procedures can be interpreted in terms of molecular structure. The normal virus particle (sedimentation rate of 88S) is stable at pH 5 in 0.1 M salt. When the pH is raised to 7 the virus swells (manifested by a slower sedimentation, 78S), possibly because the increase in pH creates uncharged amino and charged carboxyl groups, instead of paired ionic bonds, on adjacent subunits. If dialyzed overnight at pH 7 against a strong salt solution (1.0 M NaCl), the swollen virus will disassemble into its protein and RNA components, which can then be separated by ultracentrifugation. This implies that the bonds holding the RNA and protein together are primarily ionic. The protein and RNA can be purified to remove all macromolecules other than those found in the final virus structure.

These highly purified preparations can be reassembled into viruses by mixing the RNA and protein in a strong salt solution at pH 7.4 and 5 mM Mg^{2+}, then gradually lowering the salt concentration by dialysis. It can be shown that at this point the nucleic acids and protein fractions sediment as nucleoprotein particles at rates slightly slower than normal virus, implying a slightly larger than normal structure. However, if the Mg^{2+} is now removed and the pH lowered to 5 (by dialysis), then particles are obtained that, from the point of view of sedimentation rate, appearance under the electron microscope, and infectivity, are indistinguishable from native CCMV.

Even more convincing evidence of self-assembly is that the infectivity of the reassembled virus is not affected by a powerful ribonuclease, snake venom phosphodiesterase. Nucleases will destroy the infectivity of incompletely assembled particles in which some of the RNA is left uncovered and susceptible to enzyme action. For example, if Mg^{2+} is left out of the neutral reassembly solution, then a more slowly sedimenting heterogeneous aggregate is formed in which infectivity is slight to begin with and totally destroyed by the phosphodiesterase. Just as in the case of TMV, the RNA plays a role in stabilizing the virus structure because the various protein aggregates formed by lowering the pH of the purified protein preparation can be disassembled by raising the pH to 7. It would therefore seem that it is the RNA that holds together the "swollen virus," presumably by electrostatic interactions between the RNA and protein which are stabilized by the divalent cation Mg^{2+} but broken by a high NaCl concentration.

These results demonstrate that it is possible for a simple icosahedral virus to be formed by a self-assembly process, but we also learn that in this case the self-assembly process must follow a particular ionic pathway if it is to be successful. Presumably, it is important for the various bonds to form in a certain order. It is not yet clear whether the pathways in vivo bear any relationship to the steps of assembly determined by the laboratory experiment. It is possible that the cellular environment, with its aqueous and hydrophobic interfaces, may expose the virus to conditions entirely different from those devised in the laboratory. It is nevertheless significant for us to recognize that the protein and nucleic acid components of even the icosahedral viruses have the intrinsic ability, given proper conditions, to form a specific functional (infectious) structure.

Ribosomes

Because of the complexity of many cellular structures, the study of their assembly processes has lagged behind that of simpler forms such as some multienzyme aggregates and the smaller viruses. There are, however, a number of simple cell structures that are clearly capable of self-assembly. In Chapter 13 we discussed architectural details of the *E. coli* ribosome and experiments done by M. Nomura and his colleagues in which the 30S subunit reassembled from purified proteins and the 16S rRNA. Their assembly map shows the pattern of requirements for the inclusion of a particular ribosomal protein in the final structure. As with most of the reassemblies done by the research scientist, however, it can never be certain whether the requirements demonstrated in vitro are the same as those inside the cell.

One way to resolve this uncertainty is to apply genetic analysis to assembly processes occurring in vivo. There are in *E. coli* certain mutations that interfere with the proper assembly of the ribosome. Nomura and his

colleagues found that a heating step is necessary for self-assembly of ribosomes in vitro (see Chapter 13). Mutants defective in ribosome assembly were therefore sought among *cold-sensitive cells*—those which cannot grow at a reduced temperature that still allows normal bacteria to grow, albeit slowly. If the ribosomes were strictly self-assembling, all mutations that interfered with assembly would have to involve the components of the assembly, and some altered components would be expected to have other detectable changes in function. For example, certain cells resistant to the antibiotic spectinomycin have an alteration in r-protein S5, are cold sensitive, and have defects in ribosome assembly at low temperatures.

Analysis of the composition of the incomplete structures that accumulate at low temperature (20°C) in cold-sensitive mutants should also give information about the assembly pathway in the cell. The intermediate structure that accumulates at low temperature in certain cold-sensitive mutants has the same composition of r-proteins as one of the intermediates (the "RI" complex) of the reconstitution pathway in vitro. This finding is consistent with the idea that the assembly proceeds along similar pathways in the test tube and in the cell.

Other perturbations that affect ribosome assembly, however, suggest the importance of proteins that are not found in the final assembly. Cells that require the amino acid *meth*ionine and that are instead given *eth*ionine during growth accumulate inactive 30S and 50S ribosomal subunits. The absence of methionine prevents post-transcriptional modifying enzymes from methylating the rRNAs. Without these modifications the rRNA apparently cannot participate correctly in assembly. Furthermore, a different spectinomycin-resistant mutant accumulates a precursor to 16S rRNA that is longer by some 90 nucleotides and has fewer methyl groups than the mature 16S rRNA isolated from ribosomes. From this it is possible to infer that the final processing cleavages of rRNA, as well as some of the methylations, occur during actual assembly of the ribosome.

Methionine   Ethionine

If post-transcriptional processing of rRNA is coupled to ribosome assembly, the processing enzymes can be considered to *aid* the assembly in vivo. Proteins (including, presumably, ribosomal proteins) are generally found associated with the rRNA precursors, both in procaryotes and in eucaryotes (where the proteins must enter the nucleus after having been synthesized in the cytoplasm). Assembly of the ribosome during processing of its components seems likely to be the rule. The ability of the isolated and purified components to be reconstituted may thus disguise the complexity of the cell's assembly pathway, as revealed by the properties of various assembly-defective mutations.

Membranes

In Chapter 7 we gave the experimental evidence for the current model of membrane structure. There is not yet a clear-cut detailed mechanism for how this structure is assembled, but all the available experimental data lead to the conclusion that membranes are assembled from their components, without the intercession of other enzymes. Certain membranes can be assembled de novo from soluble components. In other cases, the presence of a pre-existing membrane structure may be necessary for the assembly of additional components. For the present, we regard membrane assembly as a self-assembly, rather than as a directed assembly, process. Experiments on membrane assembly are of two types: (1) those having to do with membrane growth and differentiation, and (2) those attempting to reconstruct a functioning membrane from its constituent molecules.

Membrane Growth

As an example of the first type, the results of experiments using two suitable systems are shown in Figures 15–9 and 15–10. From Figure 15–9 it is clear that as the chloroplast membrane increases in mass, assayed as the amount of chlorophyll per culture, the ratio of its constituent cytochromes to its constituent chlorophyll molecules is constantly changing. This change is reflected in variations in the photosynthetic efficiency (activity/chlorophyll) of the membranes during the cell cycle (cf. Chapter 8). From Figure 15–10 it is apparent that somewhat the same thing happens during the growth and development of the rat liver endoplasmic reticulum membranes. The ratios of different protein components are constantly changing with time. If we assume that, even when synchronized, *Chlamydomonas* cells still have normal chloroplast growth, the results shown in Figure 15–9 are consistent with the notion that a fully functional (differentiated) membrane does not form as a single discrete event. Rather, the membrane seems to be assembled from its specific components in a series of

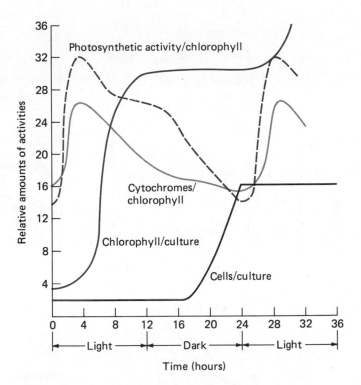

FIGURE **15–9**

Growth and Differentiation of Cholorplast Membranes in Synchronous Culture of an Alga, *Chlamydomonas Reinhardtii.* Cells can be synchronized by alternating cycles of 12 hours light, followed by 12 hours dark. Such cultures contain a large fraction of the cells at the same stage of the division cycle, but at the risk of having perturbed the physiology of the cells by the synchronization method (see Chapter 14). These cells increased in size in the light and divided three times in the dark, resulting in an eightfold increase in cells. In the light, the chloroplast increased in mass per cell, and the number of chloroplast membranes per cell (as measured by chlorophyll per culture) also increased. The number of cytochromes in the chloroplast membranes and their photosynthetic activity also rose, but not at the same rate as did the chlorophyll. Thus, the ratios of cytochromes to chlorophyll and of photosynthetic activity to chlorophyll (a measure of differentiation) were constantly changing during the growth of the membranes in the light. (From S. Schor, P. Siekevitz, and G. E. Palade.)

steps—that is, there may be a preferred assembly pathway for membranes in somewhat the same manner as described earlier for ribosomes.

The above experiments do not indicate whether, during the increase in membrane mass, the new components are inserted only in new membrane material, or into pre-existing membrane as well. This question can be approached by labeling newly synthesized material and then identifying by various assays the location in the membrane of the new components. Since membrane structures can move laterally within the bilayer (see Chapter 7), results from this approach can only be

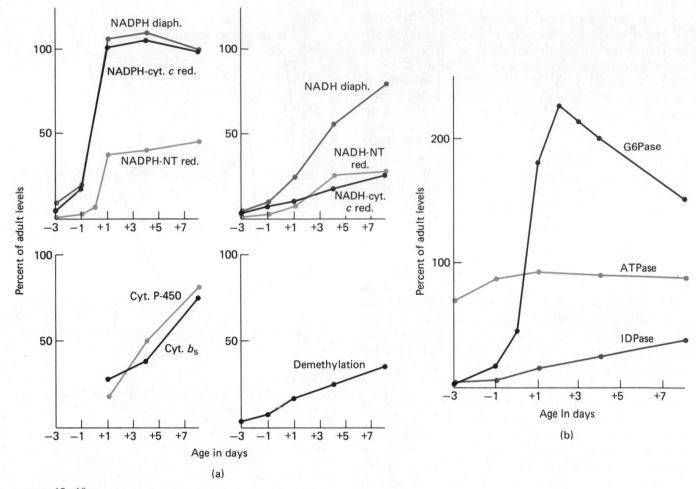

FIGURE 15–10

Differentiation of Endoplasmic Reticulum (Microsome) Membranes in Liver of Perinatal Rats. Membrane electron transport enzymes (a) and phosphatases (b). Age refers to days before (−) and after (+) birth. Adult level refers to the specific activity of the enzyme in question in the microsomes of a full-grown animal. All the enzymes are constituents of the microsomal membranes. During prenatal development, the amount of endoplasmic reticulum membranes per cell increases. Before birth, only ribosome-studded rough endoplasmic reticulum membranes are present (cf. Chapter 13); after birth, smooth endoplasmic reticulum membranes both appear for the first time and increase in amount. The results show that the enzyme content of the differentiating membranes is constantly changing, probably as a result of differences in expressing the genes of these proteins. G6Pase = glucose-6-phosphatase; ATPase = adenine triphosphatase; IDPase = nucleoside diphosphatase; Cyt. = cytochrome; NADPH diaph. = NADPH diaphorase activity; NADPH-cyt. red. = NADPH-cytochrome *c* reductase; NADPH-NT red. = NADPH-neotetrazolium reductase; NADH diaph., NADH cyt. *c* red., NADH-NT red. = same as for NADPH activity; Demethylation = oxidative demethylation of aminopyrine. (Courtesy of G. Dallner, P. Siekevitz, and G. E. Palade.)

interpreted when the time required to localize components is much shorter than would allow them to move significantly from their original sites of insertion in the membrane.

When the location of membrane-associated glucose-6-phosphatase was determined (Figure 15–11), it was found that the deposition of lead phosphate, marking the

site of the enzyme, was not localized at discrete sites along the membrane. Such sites might have been expected if they functioned as growing regions. Instead, the enzyme was present over the entire endoplasmic reticulum membrane system (Figure 15–11; compare with Figure 2–K). As closely as the grains of lead phosphate can be resolved, there is no evidence for "old" or "new"

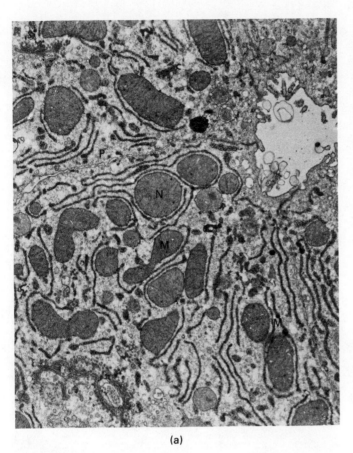

(a)

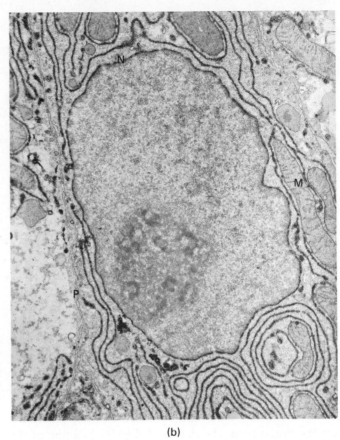

(b)

FIGURE 15-11

Histochemical Localization of Glucose-6-Phosphatase in Differentiating Rat Liver Endoplasmic Reticulum Membranes. (a) Sample removed trom livers approximately one day before birth. (b) Sample removed from liver just after birth. During this time, there is a large increase in the enzymatic activity (Figure 15–10b). The enzyme activity is localized by the deposition of insoluble, electron-dense lead phosphate (after incubating glucose-6-phosphate + lead → glucose + lead phosphate). Phosphatase activity is found all over the endoplasmic reticulum membranes, in contrast to restricted regions which could thus be identified as "new" membrane. There are no lead deposits over the mitochondrial membranes (M), nor over the plasma membranes (P); the outer membrane of the nuclear envelope (N) does show phosphatase activity. These histochemical observations agree with the biochemical localizations (see Chapter 13). ×15,000. (Courtesy of A. Leskes, P. Siekevitz, and G. E. Palade.)

regions of the growing membrane. Either the newly synthesized enzyme is inserted at many sites in the growing membrane system, as opposed to a few sites that would define "new" membrane, or the enzyme distributes itself uniformly in the membrane before the fixation for electron microscopy can reveal a preferential growing region.

A final question has to do with the incorporation into a growing membrane of the two major classes of membrane components, protein and lipids. Are these two incorporated together or can their incorporation into a membrane be separated in time? There are two types of

indirect information that lead us to suspect a relatively loose coupling between protein and lipid incorporation into membranes. The first comes from the observations, particularly in the case of bacteria, that the lipid-to-protein composition of membranes can vary with the growth conditions. The second comes from the data of Figure 7–14, which show that the turnover of the protein moieties and of the lipid moieties in a membrane are not coordinated with each other. More direct evidence was obtained by L. Mindich from experiments using a mutant of *B. subtilis* that requires glycerol for growth. When glycerol was removed from the medium, this mutant could still increase the amount of protein (and consequently increase specifically the activity of succinate dehydrogenase) in its membranes, in the absence of lipid synthesis. These cells could only divide once, slowly, and lost viability after a few hours. During the first few hours in the absence of glycerol, however, net protein synthesis continued, giving membranes that had 88% protein and 12% lipid in them, as compared to 80% protein and 20% lipid in cells supplemented with glycerol.

Membrane Reconstitution

The second type of experiment examining the problem of membrane assembly is reconstitution in vitro. This process involves solubilizing the membrane components, usually by a detergent, and then attempting to reassemble a functioning membrane from the solubilized components. As an illustration of the type of experiment that can be performed, Figure 15–12 and Table 15–2 give the results of an early attempt with rat liver mitochondria. The detergent, in this case deoxycholate, completely clarifies the mitochondrial suspension. Upon dilution of the detergent, a reassembly takes place and the reassembled membranes can be centrifuged down from the diluted suspension, as shown in Figure 15–12.

That these membranes are only partially functional can be noted from the data in Table 15–2. Some of the mitochondrial enzymes (cf. Chapter 8) are localized in the soluble matrix and are easily lost, as is the case for adenylate kinase that is believed to be a soluble enzyme localized between the outer and inner membranes. Succinate dehydrogenase and cytochrome oxidase activities, due to integral membrane enzymes, are found at high levels in the reconstituted membrane fraction. The NADH-cytochrome c reductase activity, which requires several intermediary proteins (see Chapter 8), is lost from the membrane fraction by detergent treatments. One or more of the intermediary proteins are probably removed by the detergent. It is clear that the isolated "membrane" fraction is truly membranous; that can be demonstrated from the electron micrographs (Figure 15–12b) and from the concentration of phospholipid (Table 15–2). These membranes, however, have completely lost the capacity to couple phosphorylation to oxidation.

It is significant that in the presence of the deoxycholate concentration (0.3%) used to solubilize the membrane, practically all the enzyme activities are inhibited; upon removal of the detergent, by dilution and centrifugation, some of the activities are regained. The important point is that a demonstration of even partial reassembly indicates the capacity for partial self-assembly; some of the proteins and lipids can combine to give a partially functioning membrane with some of the attributes of the original intact mitochondrion.

A persistent question about reconstitution experi-

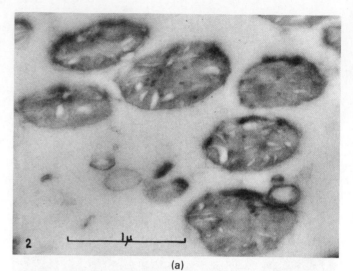

(a)

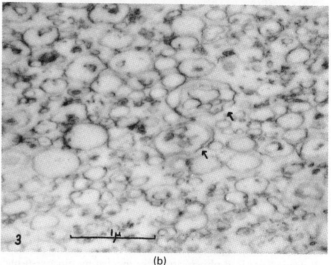

(b)

FIGURE 15–12
Reconstitution of Mitochondrial Membranes.
(a) Electron micrograph, from an isolated rat liver mitochondrial fraction, shows outer and cristae membranes, a highly electron-dense matrix, and some mitochondrial granules. ×43,000. (b) Electron micrograph, showing a membrane fraction isolated as a pellet after solubilizing the mitochondria in 0.3% deoxycholate, then diluting the detergent and centrifuging at 105,000 × g for one hour. The membranes are probably a mixture of outer and inner membranes. ×30,000. (Courtesy of M. L. Watson and P. Siekevitz.)

ments of this type is the degree of solubilization that has occurred during the detergent treatment. Are the individual proteins and lipids separated from each other or are there lipoprotein or protein-protein complexes still present in the detergent-treated suspension? In other words, is there a basic core structure, composed of some proteins or some lipoproteins, which is necessary in order for reassembly to occur, perhaps by a process of specific binding of other proteins and lipids onto this core? The answer to this question seems to be that the disassembly

TABLE 15–2

Comparison of Some Biochemical Properties of Isolated Mitochondria with Those of a Membrane Fraction Derived from It

	mg Mitochondria Protein/ g Tissue	mg Mitochondria Phospholipid/g Tissue	Phospholipid/ Protein	Succinate Dehydrogenase (Activity/ mg Protein)	Other Krebs Cycle Substrates* (Activity/ mg Protein)	Cytochrome Oxidase (Activity/ mg Protein)	Adenylate Kinase (Activity/ mg Protein)	NADH-cytochrome c Reductase (Activity/ mg Protein)
Mitochondria	9.5	2.1	0.22	870	28,40,30 72,14,22	150	59	340
Membrane fraction	0.56	0.33	0.59	5600	41,28,12 0, 0,23	880	0	0

Note: The mitochondria used were those shown in Figure 15–12 (upper), while the membrane fraction is that shown in Figure 15–12 (lower).

*Refers to the oxidation activity, in relative units, for substrates citrate, alpha-ketoglutarate, fumarate, pyruvate, malate, and beta-hydroxybutyrate, respectively.

Data taken from P. Siekeritz and M. L. Watson.

and reassembly processes vary with the type of membrane studied.

Perhaps the best studied case is that of a unicellular organism, *Mycoplasma,* used in experiments by S. Razin and his colleagues. The plasma membrane of this organism can be completely solubilized by the detergent deoxycholate into separate protein and lipid components. They can be reconstituted in the presence of a divalent cation such as Mg^{2+}, after diluting away the deoxycholate, into a membrane that has the same density, gross chemical composition, and electron microscopic structural appearance as do the native membranes and that still has some enzyme activity. By physical measurements, however, it has been found that the assembly is not perfect; the ways the proteins interact with each other and with the lipids are not quite the same as in the native structure. A divalent cation is believed to be needed in the reassembly process to neutralize the negatively charged groups on lipids and proteins; such charges would otherwise interfere with the assembly process by electrostatic repulsion. Indeed, if the assembly is allowed to occur at a lower pH, little or no Mg^{2+} is found to be necessary. Most membrane proteins bear a net positive charge at a lower pH, while the phospholipids are still negatively charged; under these conditions there can be electrostatic attraction between the membrane components. The stability of the reconstituted membranes at lower pH is nevertheless enhanced by the presence of Mg^{2+}, indicating electrostatic bonding via Mg^{2+} of the proteins and lipids of the membranes (see figure on top of p. 722).

In more complicated membranes—those of the mitochondria, the thylakoids of chloroplasts, chromatophores of photosynthetic bacteria (see Chapter 8), or the sarcoplasmic reticulum membranes of muscle—there is evidence for a basic lipoprotein core that remains intact after the usual solubilization procedures and upon which the solubilized proteins and lipids are complexed to reconstitute a membrane. The usual solubilization of sarcoplasmic reticulum membranes yields lipoprotein particles that still retain the Ca^{2+}-activated ATPase of the muscle membrane. It is hypothesized that these particles exist as such in the native membrane and are a core upon which the entire membrane is assembled in vivo, although more work is needed to prove or disprove this idea.

Native membranes can be deleted of some of their constituents, and these constituents can be added back to the residual protein-lipid core. As mentioned in Chapters 7 and 8, phospholipid can be extracted from membranes by organic solvents, or the fatty acids can be removed by phospholipase action. In some cases phospholipids, in the form of micelles, can be added back into the membrane. The criterion of successful assembly derives from the fact that some membrane enzymes are only active in the presence of lipid. If the lipid is removed by organic solvents or hydrolyzed by phospholipases, enzymatic activity is lost. When these lipids are added back, the activity is restored, as if the lipid has been reassembled into an appropriate relation with the enzyme.

One of the more interesting examples of reconstituting membranes is the work by L. Rothfield, M. J. Osborn, and their colleagues on the outer membrane of gram-negative bacteria. This membrane contains, in addition to lipid and protein, a lipopolysaccharide that is characteristic for each species of bacteria. The components of the outer membrane can be extracted and purified to yield three main fractions: enzymes that transfer sugars (glucose, galactose, manose), a complicated lipopolysaccharide "core," and the main phospholipid of the membrane, phosphatidylethanolamine. These three separate components can be reassembled to form a

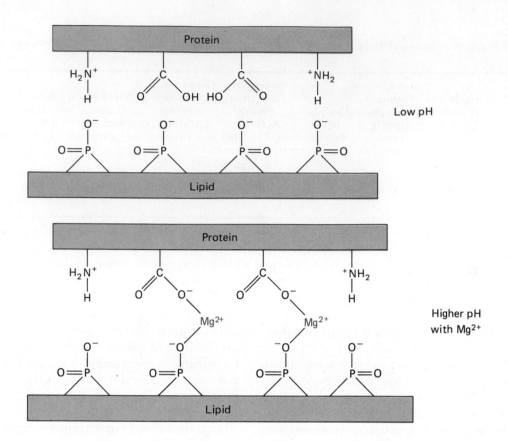

membrane fragment that is fully active with regard to its enzymatic properties.

The reassembly of the membrane-associated sugar transfer activity takes place in steps and in a certain order: the lipopolysaccharide and the phosphatidylethanolamine must first combine under certain conditions to form a binary complex; this can be separated as such by zone sedimentation through a density gradient. The purified sugar-transfer enzyme fraction can be added to this complex and a tertiary complex is then formed, again separable by zone sedimentation. All this is shown in Figure 15–13. These researchers also used the neat trick of forming a phosphatidylethanolamine monolayer film, to which was added in succession the lipopolysaccharide and

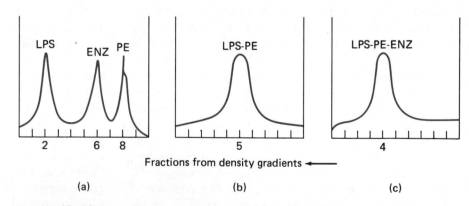

FIGURE 15–13

Assembly of a Membrane Fragment. The outer membrane components of a gram-negative bacterium, *Salmonella typhimurium,* were extracted and purified, giving lipopolysaccharides (LPS), sugar-transferring enzymes (ENZ), and phosphatidylethanolamine (PE). These can be separated by zone sedimentation through a sucrose density gradient (a). When the LPS and PE are mixed, a complex is formed that sediments differently than its components (b). When ENZ is added to this complex, a tertiary complex is formed, sedimenting yet differently from any other component or mixture (c). (Data after L. Rothfield.)

the enzyme fraction. This allows an active complex to reassemble, this time directly in the form of a membrane film. Again, the components came together in a self-directed manner. This experiment is a more powerful demonstration of membrane assembly since even the basic lipid topology was altered, yet an active function could be demonstrated.

Aided Assembly: Fibrin

So far we have discussed examples of molecular assembly in which we lack clear evidence for involvement by external agents. In the formation of fibrin we have the best understood and perhaps the simplest example of an aided assembly process that is controlled, in part, by external factors not incorporated into the final structure. Fibrin formation is the molecular basis for the clotting of blood. A complex series of events, resulting from the interaction of a number of factors, leads to the conversion of the precursor *prothrombin* to the active proteolytic enzyme *thrombin*. This enzyme then acts on the protein *fibrinogen* to convert it to *fibrin,* a massive branched polymer responsible for the structural rigidity of the blood clot (Figure 15–14).

Fibrinogen is a protein of 330,000 molecular weight composed of six polypeptide chains (two Aα, two Bβ, and two γ) arranged in a three-beaded structure (Figures 15–15a, b). Recent work in electron microscopy (Figure 15–15c), x-ray diffraction, and calorimetry has resulted in a revised structure of fibrinogen. The outside beads are now seen as composed of two domains, each consisting of folded Bβ and γ chains. The central bead is composed of two loosely interacting domains, one of which is formed by the folded N-terminal ends of all six chains and the other by the folded C-terminal domains of the Aα chains (Figure 15–15d). Thus, the fibrinogen molecule has a twofold symmetrical structure, composed of two subunits of three chains. Each molecule is held together by numerous interchain and intrachain disulfide bonds (Figure 15–16).

Thrombin is a proteolytic enzyme with a trypsin-like specificity—that is, it can hydrolyze peptide bonds in which arginyl or lysyl residues donate the carboxyl group. Interestingly, of all such bonds in fibrin, only four arginyl-glycyl peptide bonds are cleaved by thrombin, releasing four peptides—two A peptides and two B peptides from the two Aα chains and the two Bβ chains, respectively (Figure 15–17). The release of the peptides changes the properties of fibrinogen in such a way that these molecules (now called *fibrin monomers*) aggregate in an overlapping manner to produce the fibrin polymer (Figure 15–18).

The forces involved in this aggregation are probably both hydrophobic and hydrophilic since the fibrin polymer is soluble in detergents (which will break up

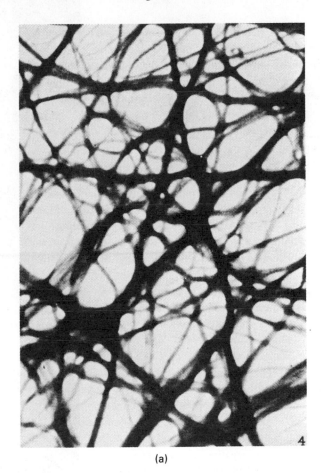

(a)

FIGURE 15–14
Structure of the Fibrin Clot.
(a) Low-resolution picture showing the type of branching that occurs (×2500). (b) High-resolution picture showing precise banding and fine structure. As will be seen in Figure 15–18, the 23-nm spacing of the dark bands can be explained by the half-staggered arrangement proposed for fibrin polymerization (×15,000). (Bar = 0.1 μm; courtesy of C. Hall and H. Slayter.)

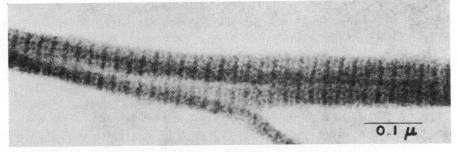

(b)

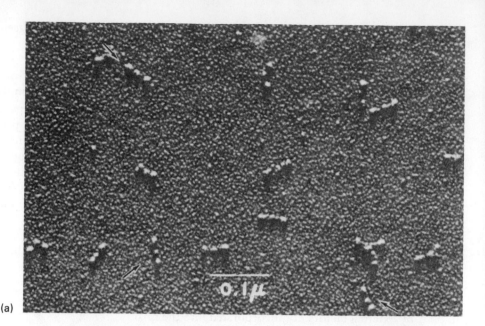

(a)

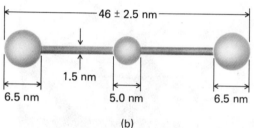

(b)

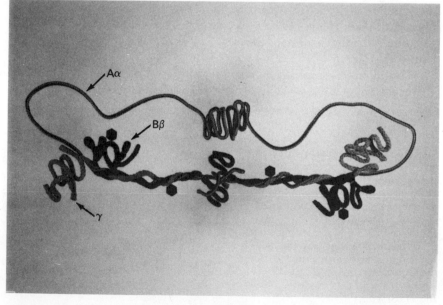

(c)

FIGURE 15–15

Structure of the Fibrinogen Molecule. (a) Shadow-cast electron micrograph of fibrinogen showing structure composed of three "beads" linked to each other by thin fibrils. The molecular weight of the molecule is 340,000; it is composed of three pairs of chains—two Aα, two Bβ, and two γ chains—cross-linked by disulfide bonds. (Bar = 0.1 μm; courtesy of C. Hall and H. Slayter.) (b) Classical model of fibrinogen based on early electron microscopy, showing dimensions of the molecule. (c) Recent electron micrographs of fibrinogen, visualized by negative staining. The subdivision of the outer beads into two distinct domains and of the central beads into two loosely associated domains can be clearly seen. (Courtesy of H. P. Erikson and W. E. Fowler.) (d) New model of fibrinogen showing (1) terminal beads, each consisting of two domains of the folded C-terminal ends of the Bβ and γ chains, and (2) the central bead, the two loosely associated domains of which consist of the folded N-terminal ends of all six chains and folded C-trminal ends of the Aα chains. The hexagonal structures are carbohydrate moieties associated with the Bβ and γ chains. (Courtesy of M. Carson.)

(d)

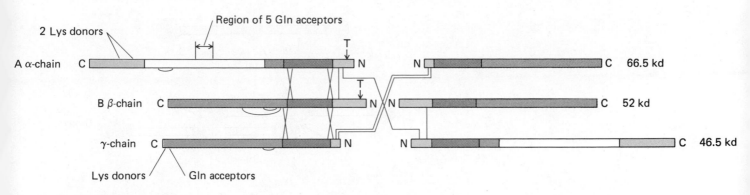

FIGURE 15–16
The Polypeptide Chain Structure of Human Fibrinogen. The molecule has two fold symmetry; the N-terminal regions of the six chains are located in the folded domain at the center of the molecule. Three disulfide bonds (color lines) hold the two halves of the molecule together. Seven disulfide bonds cross-link the chains and six disulfide bonds form intrachain cross-links in each half molecule. The 29 disulfide bonds account for all 58 cysteinyl residues in the molecule. The three color regions in each half molecule are arranged as coiled coils of α-helices; they form the two filaments that connect the three spherical lobes of the fibrinogen structure. On either side of the coiled-coil filaments there are three disulfide bonds that hold the three chains together near the ends of the spherical domains. Four stippled regions, three N-terminal and one C-terminal, form the two domains of the central bead. The three cross-hatched regions of the Aα, Bβ, and γ chains form the globular domains of the two terminal beads. Note where the thrombin cleaves to allow the dissociation of the four fibrinopeptides and the location of the lysine donors and glutamine acceptors which form the isopeptide bonds (see Figure 15–19) that convert soluble into insoluble fibrin. (Modified from R. R. Hantgan, C. W. Francis, H. A. Scheraga, and V. J. Marder.)

hydrophobic interactions), urea, and even in buffers of low or high pH (all of which interfere with intramolecular hydrogen bonding). Clots formed from blood plasma, however, are not soluble unless the solvents contain agents that break covalent peptide or disulfide bonds. The conversion of soluble to insoluble fibrin is brought about by another enzyme *(plasma transgluta-* *minase)* which, in the presence of Ca^{2+}, forms C—N cross-links between fibrin monomers. The reaction involves the ε-amino groups of lysyl residues and the γ-carboxylamide groups of glutaminyl residues; ammonia is released in the process of forming this *isopeptide* bond (Figure 15–19). Thrombin is used to convert the inactive form of plasma transglutaminase (clotting Factor

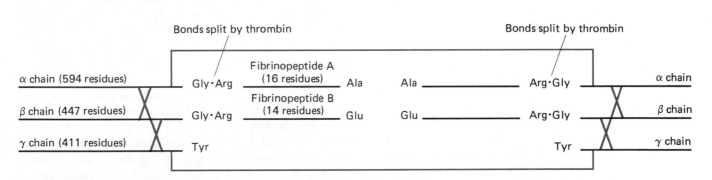

FIGURE 15–17
The Effect of Thrombin on Fibrinogen. Thrombin converts the fibrinogen molecule into the fibrin monomer by specifically splitting four arginyl-glycyl peptide bonds in the Aα and Bβ chains of fibrinogen, thereby converting them into the α and β chains of fibrin. The solid color lines are disulfide bonds holding the two halves of the molecule together. The broken color lines are disulfide bonds holding the α, β, and γ chains together. The monomers thus formed associate with each other to form the fibrin polymer.

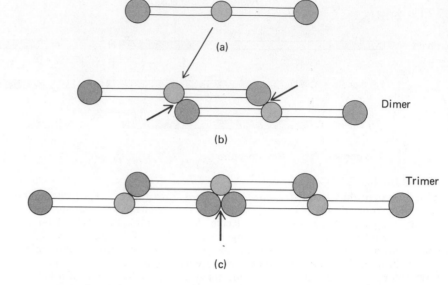

FIGURE 15–18
**The Geometry of Fibrin
Polymerization.** (a) Schematic
drawing of fibrin monomers in which
the domains composing each lobe are
simplified and shown as single beads.
(b) Dimer formed by the staggered
association of two monomers. Dark
arrows indicate the so-called "stag
contacts" between a central lobe and a
terminal lobe. (c) Trimer making stag
contacts and "long contacts" (light
color arrow) between terminal lobes. It
is these long contacts that become
cross-linked by isopeptide bonds
between the γ chains (see Figure 15–19
for explanation of isopeptide bond
formation). (d) Formation of
protofibril by addition of monomer to
both ends of a trimer. (e) Formation of
branched fibrils by random association
of protofibrils (see Figure 15–14a).

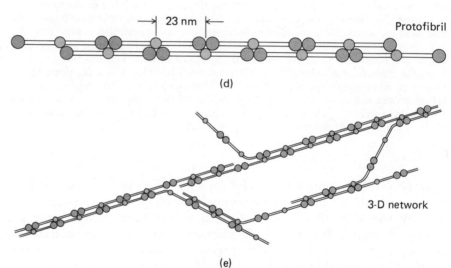

XIII) to the active enzyme, as well as to form soluble
fibrin.

Thus, we have:

$$\text{prothrombin} \xrightarrow[\text{factors} + \text{Ca}^{2+}]{\text{a number of}} \text{thrombin}$$

$$\text{fibrinogen} \xrightarrow{\text{thrombin}} \text{soluble fibrin}$$

$$\text{Factor XIII} \xrightarrow{\text{thrombin}} \text{plasma transglutaminase}$$

$$\text{soluble fibrin} \xrightarrow[\text{transglutaminase} + \text{Ca}^{2+}]{\text{plasma}} \text{insoluble fibrin}$$

The presence of cross-links between the polymerizing molecules has enabled us to obtain information

regarding the geometrical relations between the subunits
of the fibrin polymer. By breaking disulfide bonds with
a sulfhydryl compound (R—SH) and dissolving the clot
in a detergent at various stages of polymerization, it is
possible to show that two of the three chains of the fibrin
subunit are involved in the covalent cross-links (Figure
15–20).

The γ chains form γ—γ dimers in which each chain
supplies both a glutaminyl donor and a lysyl acceptor
group to the isopeptide bond. R. Doolittle and M. Chen
succeeded both in isolating the cross-linked segments of
the γ—γ chains and in sequencing them; they demonstrated that the double cross-link occurs between two γ
chains (presumably in neighboring fibrin subunits), oriented to each other in antiparallel manner (Figure 15–21).
The reciprocal cross-linking occurs close to the C-
terminal end of the two γ chains. Furthermore, by
building models, they were able to demonstrate that two

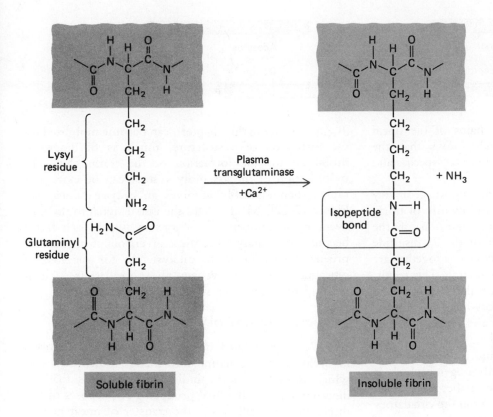

FIGURE **15–19**

Molecular Mechanism for Transforming Soluble Fibrin Into Insoluble Fibrin. The enzyme plasma transglutaminase catalyzes the formation of *isopeptide bonds* from γ-carboxyl groups of glutaminyl residues and ε-amino groups of lysyl residues, releasing ammonia. It is likely that similar mechanisms of cross-linking are found inside cells (from A. G. Loewy).

α-helices allow perfect contact betwee the appropriate donors and acceptors of the γ chains.

The other chains in the fibrin subunit that are found to be involved in the cross-linking process are the α chains. But here, instead of a dimer, a polymer is formed. One or more bonding sites are involved in this polymerization and the alignment is staggered in such a way that the donor of one chain joins with an acceptor in a different chain. Such a staggered arrangement fits both observations on the early stages of polymerization of fibrin molecules made by J. Ferry as long ago as 1952 and subsequent electron microscopic studies. To this day, a staggered arrangement of fibrin molecules provides the best explanation for the 22.5-nm repeat unit observed in

FIGURE **15–20**

The Covalent Cross-Linking of Polypeptide Chains During Insoluble Fibrin Formation. Shown are the polypeptide chains of fibrinogen, soluble fibrin, and several stages of insoluble fibrin formation. After initiating clotting, samples were dissolved at intervals by adding a mixture of urea, 2-mercaptoethanol, and sodium dodecyl sulfate (SDS); they were then analyzed by polyacrylamide slab gel electrophoresis in SDS. Samples run in the various lanes are as follows: (1) Molecular mass markers, in kilodaltons. (2) Fibrinogen. Note the Aα, Bβ, and γ chains. (3) Soluble fibrin 30 minutes after adding thrombin and ethylenediamine tetraacetate (EDTA), a reagent that chelates Ca^{2+}. (4–7) Insoluble fibrin, obtained by adding thrombin and Ca^{2+}, and dissolved after 1.5 minutes (4), 5.0 minutes (5), 10 minutes (6), and 30 minutes (7). (From a class experiment at Haverford College.)

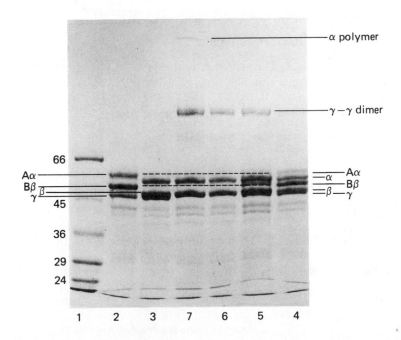

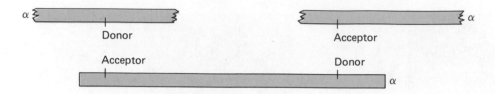

fibrin (see Figure 15–16). The β chains of the fibrin molecule do not cross-link, which shows that the cross-linking process is not random but very specific and orderly.

What does all this tell us about the specific geometry of the fibrinogen molecule and the mechanism of fibrin assembly? Studies on the enzyme digestion of the fibrinogen molecule show that half of the 28 disulfide bonds of the molecule holding all six chains together are located very close to their N-terminal ends. This result, when taken together with the antiparallel nature of the polymerization and of the γ chain cross-linking, confirms our view of the geometry of fibrin and fibrin polymerization.

The blood clotting system is thus a precisely regulated assembly process whereby fibrinogen, which is constantly present in circulating blood, becomes transformed into a structure that plugs leaks in the circulatory system. The fibrinogen secreted into the blood does not have all the structural properties required to form insoluble fibrin. The sequential action of two enzymes (thrombin and plasma transglutaminase) is needed before the final insoluble product, fibrin, can be made; this permits the process to be regulated in a highly precise manner. Clotting is an example of how post-translational modification—the covalent modification of proteins after the translation process—is used to regulate assembly processes (Figure 15–22).

It is likely that the eucaryotic cell uses such mechanisms extensively. As we saw in Chapter 14, the metabolism of procaryotic cells is often regulated in response to environmental conditions. This response involves the expression of genes being turned on or off, which leads to the rapid addition of a protein to the metabolism of the cell. Eucaryotic cells, on the other hand, are likely to be more sluggish. It is possible that some temporarily stored proteins, such as tubulin (see

discussion later in this chapter), can become mobilized for the formation of a structure, much as fibrinogen is mobilized for the formation of the fibrin clot. The mobilization of the assembly system can be extremely rapid when assembly involves an *enzyme cascade* (see Figure 15–22), which relates the initial signal to the final event by a number of reactions in which each factor becomes the catalyst for the conversion of an inactive proenzyme to an active catalyst. As for the case of glucagon's effect on glycogen phosphorylase (see Chapter 14), this results in a substantial amplification.

Directed Assembly

We have defined directed assembly as a process involving some structural information other than that which is built into the individual molecules that are incorporated into the final product. For purposes of this discussion we shall ignore the transfer of linear genetic information into linear protein sequences, which was described in Chapters 12 and 13. Transfer of structural information for directed assembly is more nearly analogous to the seeding of a crystal, involving a given three-dimensional arrangement of molecules—the *initiator structure*—which is necessary to initiate or increase the rate of assembly of a particular structure. The initiator structure often is derived from a source separate from that which provides the bulk of the final structure and may not be included in the final structure.[1]

Bacterial Flagella

The structure and assembly of bacterial flagella has been studied very thoroughly, and it is clear that an

---

[1]We do not regard membrane assembly in vitro as directed, because the phospholipids can spontaneously form a closed vesicle that incorporates proteins. In the cell, however, newly synthesized membrane components are added to pre-existing membranes.

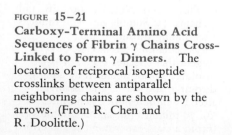
FIGURE 15–21
**Carboxy-Terminal Amino Acid
Sequences of Fibrin γ Chains Cross-
Linked to Form γ Dimers.**   The
locations of reciprocal isopeptide
crosslinks between antiparallel
neighboring chains are shown by the
arrows. (From R. Chen and
R. Doolittle.)

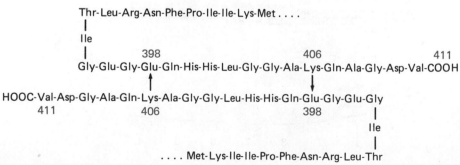

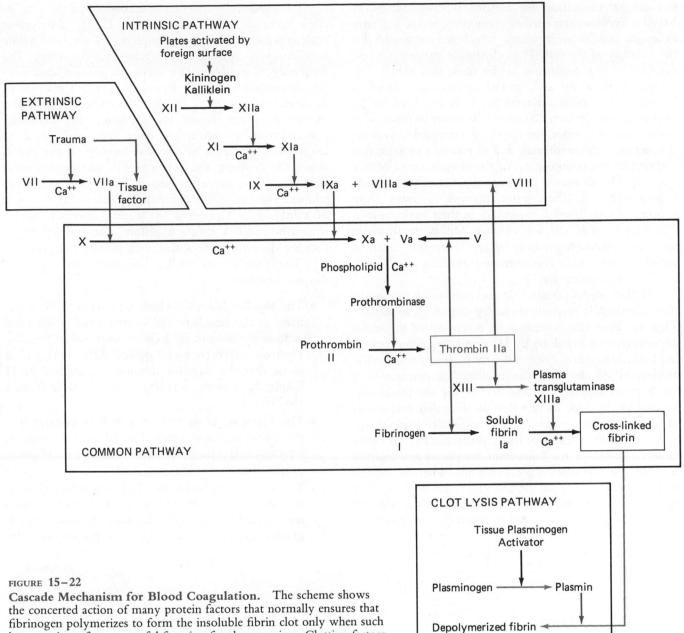

FIGURE 15–22

Cascade Mechanism for Blood Coagulation.   The scheme shows
the concerted action of many protein factors that normally ensures that
fibrinogen polymerizes to form the insoluble fibrin clot only when such
hemostasis performs a useful function for the organism. Clotting factors
are normally found as inactive precursors (color Roman numerals)
circulating in the blood and are converted (horizontal black arrows) to
active enzymes (black Roman numerals, followed by the letter "a") at
the site of a clot. The term "cascade" expresses the relationship whereby
an enzyme acts on a precursor (vertical color arrows), converting it to
an enzyme that then acts on another precursor, converting it to an active
enzyme, and so on. A cascade brings about considerable amplification of
the rate of a given process and of the quantity of materials used in it.
There are two pathways for clotting, an *intrinsic pathway,* in which a
foreign surface and factors from platelets activate the clotting
mechanism, and an *extrinsic pathway,* in which factors are released from
tissue as a result of trauma. These pathways converge on a common
pathway in which Factor X is converted to Factor Xa. Many reactions
require Ca^{2+} as a cofactor. Thrombin acts on three other inactive
precursors as well as converting fibrinogen to fibrin. Another cascade,
in which the enzyme *tissue plasminogen activator* converts the inactive
precursor *plasminogen* to the fibrin-specific proteolytic enzyme *plasmin,*
brings about the breakdown of the insoluble fibrin clot.

initiator or nucleation mechanism is involved. Many bacteria have on their surface wavy projections, 5–10 μm in length and 20 nm in width, which are responsible for the motility of the cell. Though simple structures, these flagella are true organelles in the sense that they are an integral part of the cell, and their structure, number, distribution, and behavior are precisely regulated (in the case of *E. coli,* by some 20 genes). In order to function in a coordinated manner, the flagella are coupled to systems that energize the membrane and to receptor proteins that respond to the composition of the environment. Motile bacteria, like all motile cells, are capable of a *chemotactic response*—that is, they move toward or away from sources of certain small molecules in their environment. The bacterium *E. coli,* for instance, will normally swim *up* a concentration gradient of an "attractant" such as aspartate and *down* a concentration gradient of a "repellent" such as isoleucine.

It now seems certain that the swimming motion of the bacterium is brought about by *rotation of the flagella.* This has been demonstrated by a variety of ingenious experiments (many done by H. Berg), including tethering bacteria to microscope slides by means of antiflagellar antibodies and observing the resulting counterrotation of the bacteria themselves! The motor for the rotation is located at the base of the flagella. Flagellar rotation is normally counterclockwise; it occasionally stops or reverses for short periods of time. When the flagella rotate counterclockwise, the bacterium moves in practically a straight line; this motion is called a *run*. When the flagella reverse, they tend to tangle and an erratic motion called *tumbling* (or *twiddling*) occurs. The following run will thus be at an angle random to the first run (Figure 15–23).

Chemotaxis works via modulation of the rate of these reversals of rotation (Figure 15–23). Attractants such as aspartate reduce the frequency of reversals, while repellents such as isoleucine increase their frequency. The frequency of tumbling also seems to be correlated with the chemotactic response. Consider a cell in a gradient of aspartate concentration. If, after a twiddle, the direction of the next run is up the gradient, the increasing concentration of attractant over time will tend to lengthen the interval to the next twiddle. If the run is down the gradient, the next twiddle, and therefore the next chance to move toward the gradient, will occur sooner. The conclusion that flagellar reversal is the cause of tumbling is supported by studies showing that in organisms with a single flagellum, the reversal of its rotation does correlate with the chemotactic response.

The flagellar organelles are composed of three distinct structures:

- The *flagellar filament,* which represents 95% of the mass of the structure and is composed of identical protein monomers, 40 kDa in mass, called *flagellin.* Electron microscope and optical diffraction studies show that the flagellin subunits are aligned in 11 longitudinal rows, forming a hollow tube (Figure 15–24a).

- The filament, at its proximal end, terminates in a short *hook* region composed of subunits differing in structure and arrangement from those of the filament proper (Figure 15–24b).

- The hook is attached to the *basal body,* which consists of a short rod or shaft on which two rings or disks are situated in gram-positive bacteria, four rings in gram-negative bacteria like *E. coli* (Figure 15–24b).

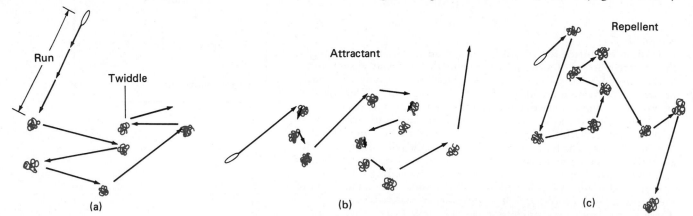

FIGURE 15–23
Movement of *E. Coli* Bacteria by Flagellar Rotation.
(a) When the bundle of flagella rotate counterclockwise the bacterium moves several lengths in almost a straight line: a "run." When the flagella reverse their rotation sense, they tangle with one another and there is an erratic motion of the bacterium: the "twiddle." The next run is at an unpredictable angle to the previous run. (b) When a bacterium is moving toward an attractant, the frequency of twiddling is reduced; when moving away, the twiddles occur more often. The result is an erratic progress toward the attractant. (c) When moving away from a repellent, the frequency of twiddling is reduced and the bacterium tends to increase its distance from the repellent.

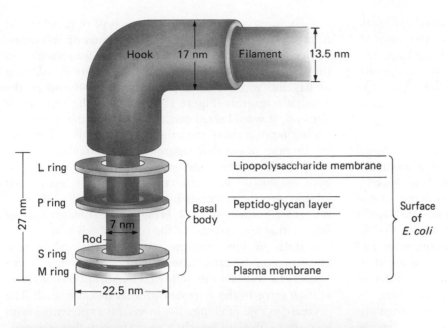

FIGURE 15–24
Structure of Bacterial Flagellar Hook and Basal Body. (a) Diagram of flagellar components, showing the attachments of the basal body components to the various outer wall–membrane layers of *E. coli*.
(b) Electron micrographs of the basal body and hook parts of the flagellum, connecting to the flagellar filaments at the arrow. The L, P, S, and M rings of the basal body can be clearly seen. The site of attachment to the cytoplasmic membrane is the lower M ring, from which the filament extends. (Scale bar = 25 nm; from M. L. de Pamphlis.)

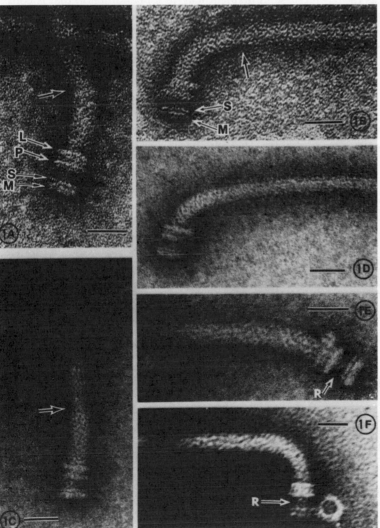

(b)

The upper one or two rings are bound to the cell wall of the bacterium and the lower one or two rings to the plasma membrane. The hook and basal structure are composed of at least six different protein chains, distinguishable by acrylamide gel electrophoresis in sodium dodecyl sulfate. A protein chain with mass of 42 kDa constitutes the major component.

The assembly of the filament portion of bacterial flagella is clearly of the directed variety. When purified flagella are heated they dissociate into the flagellin monomers. S. Asakura and his co-workers demonstrated that flagellin molecules will not aggregate into flagella unless some flagellar fragments are added (Figure 15–25). They also showed that polymerization occurs at the end of the flagellar fragments, which therefore act as initiators for the polymerization reaction. Flagellins of different strains of *Salmonella* polymerize at different rates. F. Oosawa and Asakura showed that the different flagellins determine mainly the rate of polymerization, whereas the fragments determine whether or not polymerization will occur. Some strains of *Salmonella* have flagella with a tighter wave than others and are denoted as "curly." Fragments from curly flagella tend to impose the curly pattern on flagellins from normal flagella, but fragments from normal flagella yield curly flagella if the monomers are of the curly type. This suggests that both the initiator and the monomer units are important in determining the morphology of the final product.

By using antibodies against one type of flagellin and reconstituting filaments with this monomer on fragments composed of another type of flagellin, it was possible to show that flagellar reconstitution only occurs via addition of flagellin molecules to the concave *distal* end of the flagellar fragments (Figure 15–26). (Since the motor is at one end, it would seem easier to lengthen the flagella by adding monomers to the other end.) No growth occurs by joining oligomeric aggregates of flagellin to one another. Also, when the normal growth of flagella has been inhibited with chloramphenicol (an agent that inhibits protein synthesis in bacteria), flagellin added to the growth medium will bring about the growth of flagella that are attached to the bacterial cell body.

Both in vivo and in vitro, flagella exhibit the phenomenon of growth termination. Growth in vitro occurs at a constant rate followed by abrupt termination, as if an error in the assembly process had occurred. The decrease in rate of elongation in vivo is exponential with increase in filament length, a phenomenon that is consistent with the hypothesis that the flagellin necessary for growth must diffuse through the flagellar canal: the longer the path for the supply of monomers, the slower the growth. Final termination of growth in vivo may, however, be caused by an error in assembly or possibly by a termination factor produced by the cell.

To explain how the initiator-induced assembly of bacterial flagella is induced by the initiator and why assembly can only occur by the addition of flagellin

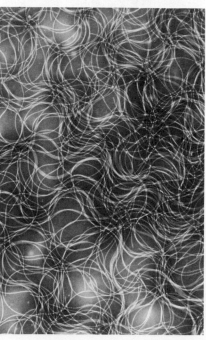

FIGURE 15–25
Flagella Reconstituted From the Purified Protein Flagellin Obtained From the Bacterium *Salmonella*.   In both the normal (a) and the curly (b) strains, flagellar fragments are necessary for the assembly of flagella from the flagellin molecules. ×24,000. (From S. Asakura et al.)

(a)   (b)

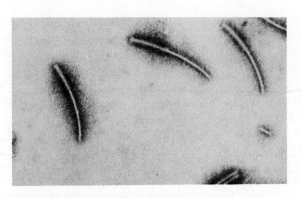

FIGURE 15–26

Flagellar Growth Occurs at its Distal End. Two variants of flagellin were used; the 1.2-type does not cross-react with antibodies prepared against the i-type. Filaments were reconstituted from 1.2-type flagellin, but with i-type initiator fragments, and labeled by antibody specific for i-antigen. This preparation was negatively stained with 1% phosphotungstic acid. The antibody can be seen as a fuzzy outline on the initiating fragments. Flagellar growth only occurs at one end of the fragment. ×30,000. (Courtesy of T. Iino.)

subunits to the distal end of the fragments, the following hypothesis can be made: attachment of flagellin requires a specific binding site that is present at only one end of an initiator, and after a flagellin molecule binds at that surface, a conformational change is induced in the molecule that provides the proper surface for binding the next molecule. This would explain why no binding of monomers occurs at the proximal end, and more critically, why growth cannot occur by flagellar fragments joining together or by self-assembly of flagellin monomers.

The initiator-induced assembly of flagellin into flagellar filaments can also be characterized as _self-replicative_. The information in the flagellar initiator is stored in the form of a three-dimensional structure, much the way structural information is stored in a seed crystal before the process of crystallization. This can be contrasted with the storage of genetic information, which can be considered one-dimensional (the linear sequence of nucleotide bases in a DNA strand) and which controls the formation of a complementary, rather than an identical, structure.

It seems likely that the cell requires the presence of some sort of three-dimensional structure, probably in the basal body or the hook, to initiate flagellar synthesis. Thus, the DNA specifies the information required for making flagellin molecules, but the latter do not contain sufficient information to polymerize rapidly into flagella. Apparently some more information must come from structures other than the unpolymerized flagellin molecule. This perhaps is as it should be, since otherwise

flagellin might polymerize into flagella inside the bacterial cell itself. The requirement of an initiator mechanism may well be necessary for the orderly polymerization of bacterial flagella. If this is so, then it might be expected that the initiated or directed assembly of structures is a widely occurring phenomenon in the growth and development of the cell. At least one other intracellular example of directed assembly is known, that of microtubules.

Microtubules

Microtubules are unbranched hollow cylinders (24-nm outside diameter, 14-nm central hole) found in many different kinds of eucaryotic cells (see Chapter 2). They are major constituents of mitotic spindles, cilia, and flagella (the last have a structure very different from procaryotic flagella). Microtubules are also found throughout the cytoplasm, especially in slender protrusions or in portions of cells that have elongate shapes.

Both assembly and disassembly of microtubules— for example, that associated with the spindle of the mitotic apparatus—occur normally and often in the cell. Also, hydrostatic pressure, low temperatures, and certain plant alkaloids, such as colchicine, cause microtubules to depolymerize both in vivo and in vitro. A variety of experiments have demonstrated that microtubules can be formed by cells for some time after protein synthesis has been inhibited, which suggests that microtubules are formed from a pool of precursors that have already been synthesized by the cell. Microtubules assemble at specific locations in the cell—for instance, near the centriole during mitosis. Thus, if we are eventually to understand microtubule assembly, we have to explain not only the *time* of assembly but also its *location*.

Microtubule structure is remarkably uniform. The cylinder is usually composed of 13 *protofilaments*, lying parallel to the central axis. The protofilaments are polymers of a dimer composed of two different subunits (a _heterodimer_). Each dimer is slightly out of register with the subunit on the adjacent protofilament (Figure 15–27).

The microtubule monomers, called α and β *tubulin* (55 kDa), were originally purified by using as a marker radioactively labeled colchicine, which binds to tubulin specifically. The amino acid sequences of α and β tubulin bear sufficient resemblance to one another that they probably evolved from a common ancestor protein. Comparison of the amino acid sequences of tubulins from distantly related species shows that tubulin, like actin, has also conserved much of its primary structure over evolutionary time. Microtubules purified by centrifugation, when depolymerized, yield the same tubulin as those isolated by using their colchicine-binding properties. In vertebrates there are genes for half a dozen closely

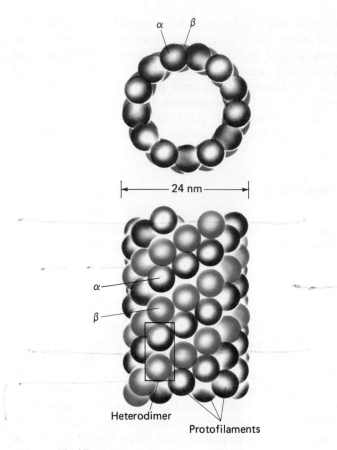

FIGURE **15–27**
Microtubule Substructures. The longitudinal view shows the protofilaments, made of αβ heterodimers, running parallel to the axis of the filament. Morphological units in adjacent protofilaments are displaced with respect to one another. The transverse view shows the 13 protofilaments of the microtubule wall. Adjacent rows of subunits are slightly rotated with respect to the row above.

related α and β tubulins; these differ slightly in amino acid sequence but functional differences have yet to be observed.

Tubulin monomers have a sedimentation coefficient of 6S. A faster sedimenting fraction of material can also be shown to consist of tubulin. Under the electron microscope this heavier fraction turns out to be tubulin molecules organized into curved or ring-shaped filaments. Assembly of the dispersed 6S tubulin into microtubules is rather slow. But when the 6S fraction is mixed with the heavier fraction, microtubule assembly proceeds rapidly. Assembly requires GTP, one molecule of which binds to one tubulin subunit, becomes hydrolyzed to GDP during polymerization, and is then released. The precise pathway of polymerization is still subject to debate but it seems likely that tubulin molecules assemble on pre-existing protofilament spirals or rings, or on other

special nucleation structures—microtubule *o*ganizing *ce*nters (MOTCs)—like the centrosome. The rate of assembly is affected by other proteins that are typically found associated with even highly purified microtubules, the microtubule *a*ssociated *p*roteins (MAPs). Another model for assembly is that protofilaments first aggregate laterally to form sheets that, when 13 filaments wide, roll up into microtubules (Figure 15–28). The addition of αβ dimers at one end (the "plus" end) elongates the microtubule. Assembly of microtubules in vivo is thus a precisely regulated form of directed assembly; a structure is first laid down that controls not only the time and location but also the direction of growth of the microtubules.

The Many-Level Assembly Processes of Some Complex Structures

We have discussed, so far, three levels of assembly, illustrating each level with examples chosen either for the clarity with which they illustrate the principles involved or for the biological importance of the structures being assembled. The second half of this chapter is devoted to an examination of more complex assembly processes that utilize a combination of principles or levels of assembly. Here, again, we have chosen some structures (T-even bacteriophages) because they clearly illustrate the principles involved and other structures (mitochondria and chloroplasts) because they are organelles of considerable biological importance. We shall conclude the chapter with an analysis of the assembly process of cortical structures in ciliates, because they provide us with the clearest available evidence for the storage of hereditary information, in structures *other* than DNA, being utilized for the control of assembly.

The Morphogenetic Pathway of the T-Even Bacteriophages

The bacteriophages are among the most intensely examined assembly systems because they are ideally suited for the study. They are large and therefore easily observed in the electron microscope; they are readily susceptible to manipulation by genetic techniques; and there is a large body of data regarding the biochemistry of their maturation processes. In short, their complexity makes them interesting and yet their mechanism of assembly is eminently accessible to experimental analysis.

Among the most complex bacteriophages are T2 and T4 (Figure 15–29). Each virus consists of a single DNA molecule of 166 kb, carrying some 180–200 genes and lying tightly condensed inside a polyhedral head. Protruding from one end of the head is a complex tail assembly consisting of a collar with whiskers, tail tube,

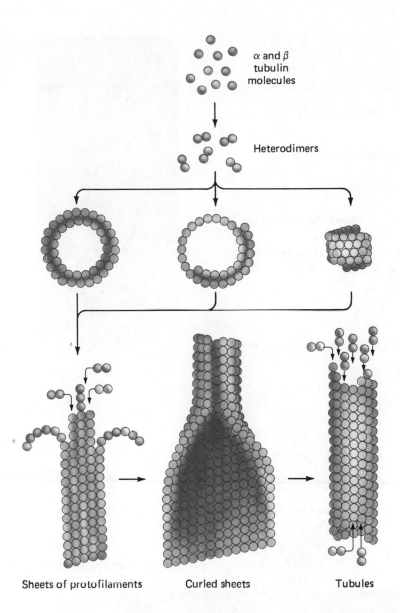

α and β
tubulin
molecules

Heterodimers

Sheets of protofilaments   Curled sheets   Tubules

FIGURE **15–28**
**Model for the Assembly of
Microtubules.**   Tubulin subunits
form αβ heterodimers, which form
rings or short helices. After these
open, the resulting sheets of
protofilaments grow by elongation at
one end and by adding more
protofilaments. When 13
protofilaments are in the sheet, it
closes to form the microtubule.
Further elongation occurs at one end
by the addition of αβ heterodimers.

sheath, baseplate, and tail fibers. The viruses attach
themselves to the surface of an *E. coli* bacterium by the
tail fibers and the spikes of the end plate, after which
the sheath contracts and drives the tail tube through the
bacterial cell wall and plasma membrane, allowing
the DNA to pass into the cell (Figure 15–29d).

The T4 bacteriophage structure is made of some 40
different proteins, and several additional proteins seem to
participate in its construction. Genetic studies show that
about 50 different genes are involved in the construction
of the virus (Figure 15–30). These numbers are likely to
become larger as further research results become available.

When a mutation occurs in a gene specifying a
certain virus protein, the protein will be either absent or
nonfunctional. Since assembly of the virus occurs in a
number of steps, the absence of a particular protein will

cause the assembly process to halt at the point at which
the protein is used in the pathway of assembly. If such a
mutant virus cannot be recovered, it is not useful for
experimental purposes. On occasion, a lethal mutation
that does not kill the cell under an alternate set of
conditions, such as a different temperature, is found.
These *conditional lethals* can be grown under *permissive*
conditions and then studied under the *restrictive* conditions, in which their lethal character is expressed. By
performing the appropriate genetic crosses under permissive conditions, R. Edgar and R. Epstein mapped many
genes affecting T4 assembly (Figure 15–30).

Under restrictive conditions, it is possible to determine biochemically as well as morphologically at what
point in the assembly process a particular mutation is
expressed and, therefore, how a particular gene product
functions. This was done by W. Wood and Edgar, and by

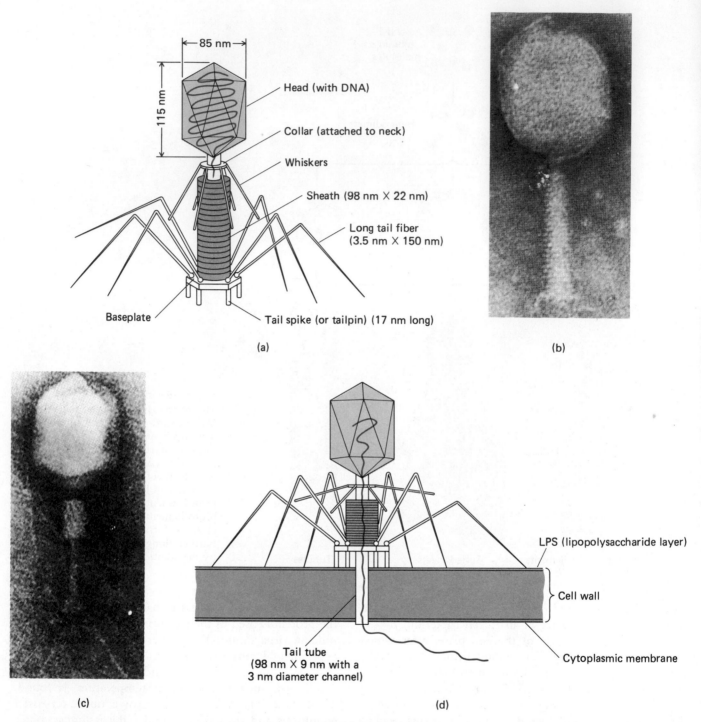

FIGURE **15-29**

The T4 Bacteriophage and its Infection of *E. Coli*.   (a) Diagram of this complex bacterial virus which consists of 166 kb of DNA, coding for ~200 proteins of which ~40 are found in the T4 virion (viral particle). During infection, the tips of the long tail fibers bind to specific receptors in the outer lipopolysaccharide layer of the bacterial host cell. Contact with the receptor causes a change in the kink angle of the tail fibers and the rest of the virion is lowered to the cell surface (like a spacecraft landing). When the tail spikes touch the cell surface, the base plate expands, freeing the first row of sheath subunits and forming a hole. The base plate contains some 16 different proteins, including one with an enzymatic activity like lysozyme. As the sheath contracts, this activity digests a hole in the rigid peptidoglycan layer of the cell wall and the tail tube goes through. After the tail tube contacts the cell membrane, viral DNA is injected into the cytoplasm. (b) Normal intact form (×600,000). (c) Contracted (triggered) form (×680,000). (d) Diagram illustrating structural changes during early stages of infection.

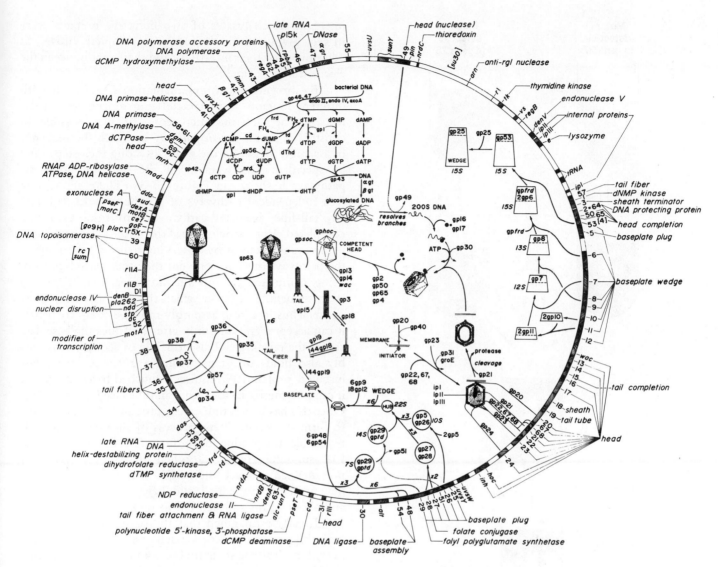

FIGURE 15–30

Genetic and Assembly Maps of T4 Bacteriophage. The 166-kb genome is shown as a circle. A coordinate system (not shown) starts at the junction between genes r$_{II}$A and r$_{II}$B, at approximately nine o'clock and proceeds clockwise around the circle. The functions of many gene products are indicated next to the letter or number symbols designating the genes. Simplified pathways for nucleotide and DNA synthesis, and for viral assembly, are shown within the circle. Below the center of the map, protein gp20 (gene 20 product) forms the initiator structure on the cell membrane, assisted by gp40. This structure combines with gp23 and gps22, 67, and 68 to form an intermediate on the membrane that is cleaved with gp21 protease. This combines with the viral DNA, combined with gps 16 and 17 in an ATP-dependent process that packs the head with DNA. Starting on the circular map at about eight o'clock is the tail assembly pathway, starting at about six o'clock is the base plate assembly pathway. These all converge towards the center in the final pathway toward the complete virus. (Courtesy of B. S. Guttman and E. M. Kutter.)

E. Kellenberger, not only by following the accumulation of different virus components during defective assembly but also by *complementation experiments* in vitro. In this latter test a culture of bacteria is infected with a defective

(mutant) phage; then the bacteria are broken open (lysed) and the contents (lysates) of two such infections are mixed, to see whether normal phages are produced. Figure 15–31 illustrates a typical complementation exper-

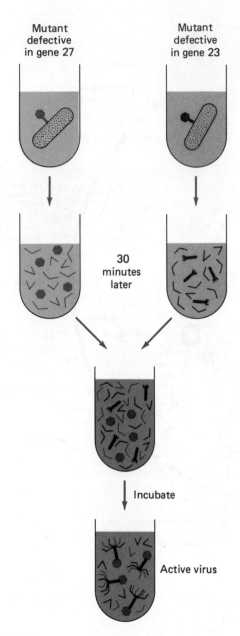

FIGURE 15–31
Complementation Experiment in Vitro With Mutants Defective in Gene 27 and Gene 23.   Two assembly reactions occur in this experiment: union of heads and tails and attachment of the tail fibers. One virus (color), with a defective tail gene, produces heads and tail fibers. Another (black), with a mutation in a head gene, produces tails and tail fibers. When the two extracts are mixed and incubated, the parts assemble to produce infectious virus. The reaction requires two steps: the head must be attached to the tail before the tail fibers because the tail requires the presence of the head for alignment. (Courtesy of W. Wood and R. Edgar.)

iment in which lysates of two defective mutants were mixed, giving within a short time complete viruses, as determined by their infectivity and their appearance in the electron microscope.

By carrying out complementation experiments with different mutants and characterizing morphological defects in the electron microscope, it was possible to construct a *morphogenetic pathway* in which the sequence of assembly steps and the genes affecting each step could be specified. One version of the T4 morphogenetic pathways is shown in Figure 15–30. There are at least four independent pathways of assembly (head, tail, and two tail-fiber assembly pathways) converging to form a completed virus particle. Proof that these are independent pathways is obtained by observing that a block in one of the pathways does not prevent the other pathways from being completed. It appears, however, that there is a strict sequential order in which the final assembly must occur. Thus, the head assembly must be complete before it can accept a tail, and only after this does the baseplate accept the tail fibers. This principle of sequential assembly also holds for steps within a given pathway. For instance, protein P19 requires a finished baseplate before it can polymerize to form the tail tube (see Figure 15–30). Once this has begun to happen, protein P18 can assemble to form the sheath. This process of sheath formation is reversible, the product being in equilibrium with the free P18 subunit. However, when the sheath has fully polymerized to reach the length of the tail tube, then a new protein P15 is added, stabilizing the whole structure and allowing the finished tail to join with the head.

It can be imagined that the sequential ordering of assembly ensures the correct self-assembly of the phage, even though all the structural components are synthesized simultaneously. As the assembly proceeds, new structural arrangements are produced that allow further components to be added. The kinds of chemical linkages (hydrophobic bonds, hydrogen bonds, salt linkages, etc.) are the same ones we have already encountered in the formation of macromolecular structures. The molecular surfaces allow stable specific binding, and the bound forms have new surfaces that can bind other components. Thus, precisely timed control of synthesis of components is not needed to achieve the assembly of most viruses.

The stringency of the sequential order is a *necessary condition* for an accurate assembly, but it does not appear to be a *sufficient condition* since a number of post-translational control mechanisms are interposed at various points in the assembly process of the bacteriophage particles. Rather than give a systematic account of the T4 assembly, we shall list the various post-translational mechanisms that have been identified in different bacteriophage assemblies and indicate where they occur. We shall organize our account by using the categories of aided and directed assembly.

Cleavage—A Case of Aided Assembly

Cleavage of both DNA and protein has been observed in the assembly of many viruses. It has been demonstrated in a number of bacteriophages that the DNA synthesized during infection is several phage genomes in length. This large molecule is composed of the usual genome joined end-to-end many times (a *concatamer*). The DNA is cut into the lengths found in the mature phage only if the appropriate protein precursors of the phage head (proheads) can encapsulate the DNA. If the proheads are defective or absent, then the DNA is not cut. One can hypothesize from these results that the DNA of some viruses is cut only after it is fitted into the prohead, the latter being a kind of measuring vessel for determining the appropriate amount of DNA.

Protein cleavage occurs at a number of stages in virus assembly, probably performs a variety of different functions, and seems to occur in all viruses, with the possible exception of plant viruses. (In the case of polio virus, the entire genome is translated into a single gigantic polypeptide chain, which is subsequently cut into the different protein chains utilized for virus assembly.) In bacteriophage assembly, protein cleavage acts on a variety of virus precursor particles, rendering them capable of engaging in the next step of assembly. One of the more dramatic events in the formation of the DNA-filled head of T4 is the disappearance of protein gp22 (*gene 22 product*), the "core protein" that is assembled inside the head prior to the packaging of the DNA (see Figure 15–30). The gp22 seems to be completely hydrolyzed during head assembly into small trichloroacetic acid-soluble peptides. Another equally striking event is the conversion by hydrolysis of protein gp23, the major protein of the outer shell of the head (capsid protein), into protein gp23★, which is 10.6 kDa lighter.

There are a number of other steps in the assembly of T4 in which it can be demonstrated that a particular gene product acts catalytically, although no direct proof exists as yet that protein cleavage is involved. For instance, in the assembly and attachment of tail fibers, eight gene products are involved, four of which are structural components. Of the remaining four, one (gp63) is responsible for the attachment of the tail fibers and appears to act catalytically.

The function of cleavage reactions may be manifold. Cleavage may remove certain proteins necessary for a previous assembly step in order to allow a subsequent assembly step to occur (as with gp22). Cleavage may also modify a protein in such a way as to prepare it for a given assembly step after the protein has already been incorporated into a structure. Thus cleavage offers possibilities of control which a self-assembly mechanism could not possibly achieve.

There is much yet to learn about the mechanism of cleavage. As is the case in the specific conversion by thrombin of fibrinogen to fibrin (but unlike general proteolysis by nonspecific proteases such as trypsin), some of the cleavages in virus assembly pathways are catalyzed by proteolytic enzymes specific for a substrate with a very particular three-dimensional structure. In other instances the cleaving enzyme can act on a range of substrates, although even here some degree of specificity must be involved. It is likely that as assembly systems in the eucaryotic cell become more familiar, large numbers of proteolytic enzymes will be found to have high degrees of specificity for particular protein substrates.

Fusion—Another Case of Aided Assembly

In bacteriophage λ, genes C and E are necessary for head formation. The protein products of these genes, pC and pE, can be recovered from lysates, but pC is absent in mature heads. Two proteins, X1 and X2, can be identified in mature heads, and tryptic maps (fingerprints, Figure 15–32a) suggest that X2 is a cleavage product of X1. Furthermore, comparison of tryptic maps of X1 and X2 with those of pC and pE shows that X1 and X2 have homologies with *both* pC and pE. The simplest explanation of this is that X1 and X2 are *fusion products* of a piece of pC and a piece of pE (Figure 15–32b). It appears that the fusion involves formation of covalent linkages other than disulfide bonds because agents such as 2% sodium dodecyl sulfate or 6 M guanidine · HCl (which break noncovalent bonds) and 10% mercaptoethanol (which breaks disulfide bonds) do not dissociate the X1 and X2 proteins. What kinds of covalent bonds are formed in the fusion reaction are not yet known, but it is possible that they include peptide bonds between N-terminal and C-terminal groups, or perhaps covalent bonds between amino acid side chains.

Fusion reactions have been reported in other assembly systems. The first clearly documented fusion reaction, discussed earlier in this chapter, was observed in the formation of insoluble fibrin. Covalent bonds are also known to form between subunits of collagen fibers (see Chapters 13 and 17).

It may well turn out that the fusion of polypeptide chains is a significant post-translational mechanism for the control of assembly processes. In eucaryotic cells some 1–5% of the intracellular protein is insoluble, in that it will not dissolve by boiling in detergent or guanidine · HCl in the presence of a disulfide-splitting reagent like mercaptoethanol. This insoluble material,

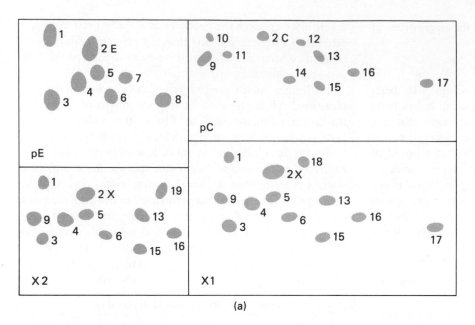

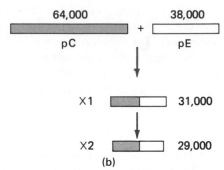

(a)   (b)

FIGURE **15–32**

Assembly of Bacteriophage λ.   (a) Drawing of peptide maps of bacteriophage λ gene products pE and pC, recovered from bacteriophage lysates, and of peptides X1 and X2, identified in mature bacteriophage heads. Corresponding numbers indicate spots of identical mobilities. An explanation consistent with the homologies observed is that X1 and X2 are fusion products of pE and pC. (b) Schematic representation of the fusion between λ pC and λ pE, as suggested by the observed homologies. The lengths of the bars are proportional to the apparent molecular weight of the proteins. The X1 and X2 peptides are held together by covalent bonds, since they cannot be separated by a detergent and a disulfide reducing reagent; this is consistent with their being fusion products. (From R. Hendrix and S. Casjens.)

found in membranes as well as the cytoplasm, must have been assembled by enzymes which knitted the proteins together with covalent bonds.

Nucleation—Forms of Directed Assembly

We have already discussed nucleation in the self-assembly of TMV disks, and as a device of directed assembly when we described the formation of bacterial flagella and of eucaryotic microtubules. The phenomenon of nucleation can be subdivided into two types: (1) nucleation not necessarily involving changes in conformation, and (2) nucleation in which a conformational change is induced. Both types occur in virus assembly.

The first type of nucleation, in which no change of conformation need occur, is very widespread in virus assembly and explains the strict sequence of assembly pathways. In T4 morphogenesis, finished baseplates nucleate assembly of the tail tube, which in turn nucleates sheath assembly, and so on. The structure of the subassemblies does not appear to change before and after nucleation. Furthermore, membranes of the host cell seem to participate in *heterologous* nucleation processes, as

in head and tail fiber assembly (see Figure 15–30). On the subject of host participation, it should be pointed out that there are many examples besides that of nucleation. It may well turn out that each of the viral assembly mechanisms has its host-controlled counterpart.

The second type of nucleation, in which conformational changes of proteins might occur, is more difficult to demonstrate experimentally. In the life cycle of T4 there are two large-scale structural changes that appear to involve conformational alterations of the subunits, the contraction of the tail sheath during infection and the expansion of the capsid during formation of the mature head. For bacteriophage λ, R. Williams has proposed an ingenious crystallographic rearrangement to account for the increase in head size observed prior to DNA packaging. The new arrangements of protein subunits do not prove that conformational changes have occurred. But changes in other properties (stability to detergents, pH, etc.), combined with the apparent bonding changes between subunits, strongly suggest that the individual molecules have changed shape and, therefore, the surfaces on which they interact. Other examples of conformational change to ensure, for instance, that growth occurs only at one end of a rod or that the distal end of one rod joins to the proximal end of another are much more difficult to demonstrate. There is little doubt, however, that conformational change coupled with nucleation is likely to be an important mechanism in the assembly of structures.

Transient Structures—Another Feature of Directed Assembly

The analysis of the assembly of bacteriophages turned up a mechanism that had not previously been encountered in any other system. This is the formation of *transient structures,* which are necessary for an initial stage in the formation of a structure but are later removed and thus do not form part of the final structure.

Transient structures were first actually observed by Kellenberger and his colleagues, who described them as shape-determining (morphopoietic) cores in T4 bacteriophage head precursors. M. Showe and L. Black, as well as others, showed that the product of T4 gene 22 and the internal proteins form a transient core around which the permanent proteins of the head are assembled. Protein gp22 is removed by proteolysis after formation of the head. Certain gene 22 mutants form aberrant structures—

for example, *polyheads*—which seem to be assembled by variation of the angle between subunits in the head (Figure 15–33). Thus, an aberrant assembly results from the lack of the transient core component.

The presence of transient cores has been reported in the assembly of several other bacterial viruses and may play a role also in animal virus assembly. J. King and his collaborators have shown that gene products p8 and p5 of *Salmonella* bacteriophage P22 assemble together into a precursor shell (prohead). Protein p8 is located on the inside and is thought to perform the function of a *scaffolding protein,* while p5 forms the outside shell. When DNA condenses inside the head, all the p8 proteins are displaced and are then recycled to help direct the assembly of other heads. As mentioned above, the membrane of the host cell may also perform certain transient functions in the assembly process. The use of structures that transiently supply information for given steps in an assembly process may well be a widely occurring phenomenon.

Assembly of Organelles: Chloroplasts and Mitochondria

Both the chloroplast and the mitochondrion are, like the nucleus, surrounded by a double-membrane envelope. In mitochondria and chloroplasts the main metabolic functions are localized in an inner membrane system, the cristae of the mitochondrion and the thylakoid of the chloroplast, which lie adjacent to matrices containing soluble enzymes auxiliary to this main function. This whole is enclosed by an outer membrane. For years biologists have been very curious as to how these organelles are formed and how they increase in number during the growth of the cell.

There have been numerous cinematographic observations that both chloroplasts and mitochrondria divide; division furrows appear in the outer membrane and

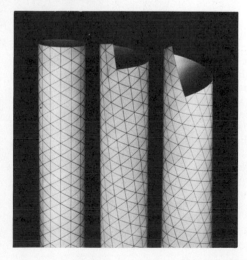

FIGURE **15–33**
Effect of Mutation on T4 Head Assembly. Certain T4 bacteriophage mutants defective in gene 22 form aberrant structures such as tubular polyheads in which the angular pitches of the gp23 subunits are different. Shown are the normal T4 head, the normal packing arrangement of head subunits, a section of tubular polyhead, and some abnormal packing arrangements that could produce tubular polyheads. (Courtesy of E. Kellenberger, F. Eiserling, and E. Boy de la Tour.)

divisions occurs by constriction and separation. D. Luck has shown that the membrane components of new "daughter" mitochondria are derived as a whole from the "mother" mitochondrion during division. An increase in the number of mitochondria does not occur by synthesis of new, separate mitochondrial components which are then complexed to form a mitochondrial structure. Instead, the completed membranes of daughter mitochondria are derived in part from the membranes of the

pre-existing mitochondrion and in part by the synthesis of new components; these latter are inserted into the pre-existing membrane, which increases in size until the mature mitochondrion is formed.

This conclusion was based on experiments with a mutant of the fungus *Neurospora crassa* that requires added choline for growth. If the cells are grown on a low-choline medium, the isolated mitochrondria can be shown to have a certain density by sedimenting them in

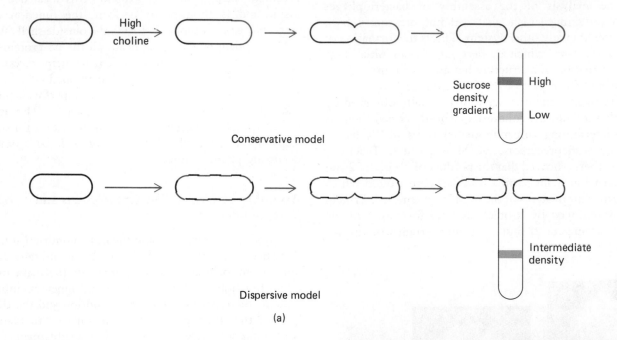

(a)

FIGURE **15–34**

Duplication of Mitochondria.   (a) Models to explain mode of adding new material to growing mitochondria. New mitochondrial structures are synthesized after a shift from low-choline to high-choline medium. In the conservative model the new material is added at a small number of locations (here only one) in the mitochondrial structure and segregated from old (low-choline) material during division. Mitochondria grown in low and high choline are distinguishable by their differing densities during sedimentation in a sucrose gradient. The conservative mode of duplication results in two populations of mitochondria. The dispersive mode of duplication, in which new material is added at many growing points, results in a single population of mitochondria that has a density between those associated with growth in low- and high-choline media. (b) Buoyant density of mitochondria (white bands) isolated from choline-requiring *N. crassa* cells. The tubes contain mitochondria from equal numbers of cells grown in low-choline medium, and after transfer to a high-choline medium. (1) In low-choline medium, (2) 20′ after transfer to high-choline medium, (3) 40′ after transfer, (4) 60′ after transfer, (5) 90′ after transfer. There is apparently only one population of mitochondria, decreasing in density with time after the shift to high-choline medium. If the lateral transfer in the mitochondrial membranes of new material is slow, these data support the dispersive model. (Courtesy of D. L. Luck.)

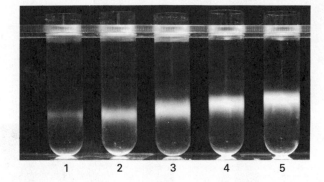

sucrose density gradients (Figure 15–34). If they are grown on a high-choline medium, the extra choline allows a large increase in the phosphatidylcholine of the lipids of the cells, including those of the mitochondria, and the mitochondria isolated from these cells have a lower buoyant density because of the higher content of lipid. What happens to their mitochrondria when cells on a low-choline medium are shifted and allowed to grow on a high-choline medium? If lateral diffusion of the phosphatidylcholine is not extensive and if soluble precursors only are used to form new mitochondria, their density should reflect only the high-choline medium and, therefore, be low. A population of mitochondria extracted from the culture should show a mixture of heavy mitochrondria (from the low-choline medium) and light mitochondria (from the high-choline medium). This model for growth of mitochondria can be characterized as *conservative* (Figure 15–34a). On the other hand, the new mitochondria may be formed by random addition of new material (including the added choline) to pre-existing mitochondria, which then divide. In that case, only one population should be visible, a population that decreases steadily in density as the cells continue to grow in the high-choline medium. This kind of growth would be consistent with a *dispersive* model. The experimental data are more consistent with the dispersive model, as can be seen in Figure 15–34b.

Mitochondrial growth and division appear to resemble the self-directed multiplication mechanism that occurs in bacteria. As a result, many investigators began to look for a source of internal control for the process—a DNA that might contain all the information necessary for a self-replicating organelle. It was not surprising, although no less dramatic, when it was found that mitochondria and chloroplasts do indeed contain their own DNA, different in many respects from the nuclear DNA of the cell.

From this beginning the study of just how autonomous the mitochrondria and chloroplasts really are has continued to the present day. We shall review this work here, even though some of the results have already been mentioned in previous chapters. First of all, the DNA in many organelles is so different from nuclear DNA in base composition that the two kinds of DNA can be separated by means of equilibrium density gradient centrifugation. This can be seen in Figure 15–35 for mitochondrial DNA (compare Figure 10–10a) and in Figure 15–36 for chloroplast DNA.

The gross differences in the chemical and physical properties of these DNAs lent strong support to the idea that the DNA found in all isolated mitochondria and chloroplasts is really a constituent of these organelles and is not a contaminant from the cell nucleus. Organelle DNAs have several other peculiar properties: they are

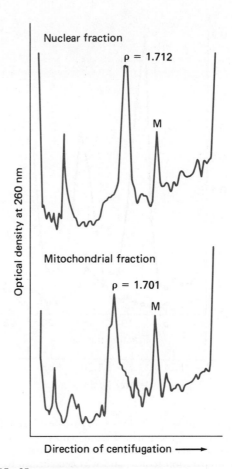

FIGURE 15–35
Optical Absorbance After Sedimenting *Neurospora* DNA Fractions to Equilibrium in a Cesium Chloride Density Gradient. The density (ρ) of *Neurospora* nuclear DNA (1.712 g/cm^3) and mitochondrial DNA (1.701 g/cm^3) is indicated. (M denotes a bacteriophage DNA put in as a marker.) (Data from D. L. Luck and E. Reich.)

small, with molecular weights from 0.3 to 2.6×10^7 in animal mitochondria, up to 5×10^7 for plant, fungal, and protozoan mitochondria, and from 10^8 to 10^9 for the chloroplast DNAs of algae and plants. While this DNA is double stranded, it has been isolated in the form of *closed circles* from animal mitochondria (Figure 15–37) and from algal and plant chloroplasts. There are about six of these circles per mitochondrion, perhaps up to 12 per chloroplast, and all are apparently identical to one another. The significance of the circular structure to the DNA's function is not clear at all.

The relative amount of organelle DNA per total cellular DNA can vary greatly, from only 1–2% in animal mitochondria (cf. Figure 15–35), up to 10–20% in mitochondria of certain yeasts, and about 10% in algal chloroplasts (cf. Figure 15–36). Hybridization experiments (see Chapter 11) have shown there is very little similarity in base sequences between either mitochondrial

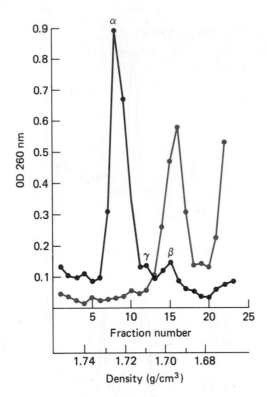

FIGURE 15–36
Separation of Chloroplast DNA From Nuclear DNA.
DNA was extracted and purified from whole cells of the alga
Chlamydomonas reinhardtii (black) and from a purified
chloroplast preparation (color) and was centrifuged to
equilibrium on a CsCl density gradient. The β band (density,
1.693 g/cm^3) in the whole cell DNA preparation is
chloroplast DNA; the position of the β band in the gradient
coincides with the major band in the chloroplast preparation,
and it is quite different from the nuclear α band at density
1.727 g/cm^3. (Courtesy of B. de Petrocellis, P. Siekevitz, and
G. E. Palade.)

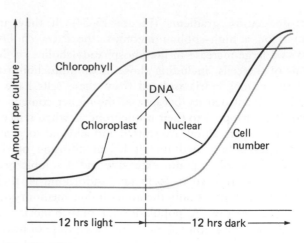

FIGURE 15–38
**Time Course of Properties of a Culture of the Alga
C. Reinhardtii That has Been Synchronized by Light-
Dark Cycles.** During the light cycle, the cell and the
chloroplast increase in size, as measured by the increase in
chlorophyll content and as can be seen in the microscope. In
the middle of the light period, chloroplast DNA increases.
This can be demonstrated by isolating total DNA and
separating the chloroplast DNA from the nuclear DNA (see
Figure 15–36). Only in the dark does nuclear DNA increase.
This is usually followed by two to three cell divisions,
resulting in four to eight daughter cells. (After S. Schor, P.
Siekevitz, and G. E. Palade.)

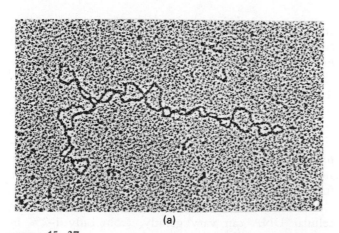

(a)

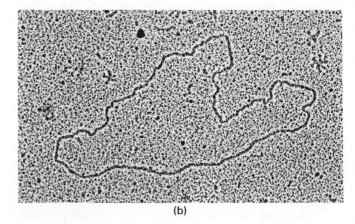

(b)

FIGURE 15–37
Circular Mitochondrial DNA. A preparation of DNA from rat liver
mitochondria, prepared by the aqueous protein monolayer technique, and rotary
shadowed with platinum-palladium is shown. (a) Supercoiled circular molecule.
(b) Open circular molecule. (Magnification ×44,000; courtesy of D. R.
Wolstenholme.)

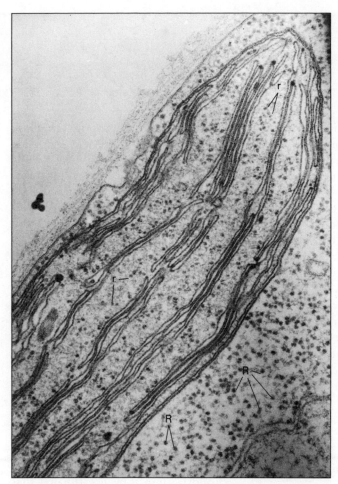

FIGURE 15–39
Electron Micrograph of Part of the Alga *Chlamydomonas*
Reinhardtii. Note the 70S chloroplast ribosomes (r) and the
80S cytoplasmic ribosomes (R) (×82,000). (Courtesy of G. E.
Palade.)

or chloroplast DNA and nuclear DNA. Thus, the DNA
of the organelles is not simply a large piece of nuclear
DNA that has been replicated and has accidentally found
its way into the organelle. It is an entirely different
molecule that has its own coding sequences. This has
been amply confirmed by sequence analysis of the
organelle DNAs.

What is the function of mitochondrial and chloroplast DNA? First of all, it is certain that it is replicated
within the organelle. In the mitochondrion, the DNA
polymerase involved is different from the nuclear enzyme, but replication still occurs by the same semiconservative mechanism (see Chapter 12). This replication
can be assayed by the incorporation of radioactive
nucleotides into the DNA of these organelles. In those
situations in which cell division can be synchronized, the
replication of organelle DNA has sometimes been observed to be out of phase with that of nuclear DNA. This

has been demonstrated for yeast mitochondria and, as
shown in Figure 15–38, for algal chloroplasts. The
question of how mitochondrial or chloroplast DNA
replication is controlled, so as to keep pace with nuclear
DNA replication and with cell division, is not yet
answered.

Insight into the functional importance of mitochondrial DNA first came from the observation of "petite"
mutants, both of yeast and of the fungus *Neurospora
crassa*. Petites, named for their smaller size under the usual
growth conditions, have a defective mitochondrial respiratory system, involving losses of cytochrome *b* and
cytochrome oxidase. The defect is inherited in these
mutants in a non-Mendelian extranuclear fashion. Since
the first observation of petite mutations, the evidence has
become overwhelming that they involve a deletion of
mitochondrial DNA. Indeed, chemical compounds such
as ethidium bromide can induce petite mutants of yeast
that are almost completely missing their mitochondrial
DNA. (Since petite yeast can be grown anaerobically in
the presence of appropriate substrates, the absence of
mitochondrial DNA can be tolerated under those conditions. As we shall see, the necessary mitochondrial
function of supplying Krebs cycle intermediates does not
depend on the mitochondrial DNA.) These results also
provide evidence that mitochrondrial DNA has the
information for the synthesis of at least some of the
proteins required for cytochrome *b* and cytochrome
oxidase activities. The observation that mitochondrial
DNA seems to code for the synthesis of certain proteins
led investigators to look for a protein-synthesizing
system in mitochondria and chloroplasts.

Protein Synthesis in Mitochondria and Chloroplasts

Early electron micrographs of chloroplasts showed
numerous particles resembling ribosomes, and these were
later isolated and shown in fact to be ribosomes. In the
case of mitochondria it was much more difficult, for the
morphological evidence was mostly negative, but eventually it was possible to isolate ribosomes from animal,
plant, algal, and yeast mitochondria. The latter three
mitochondrial ribosomes resemble in size those from
bacteria, being about 70S in size and containing 23S +
16S RNAs in the two ribosomal subunits (see Table
13–5). In animal mitochondria the ribosomes are much
smaller, sedimenting at 55S. In all cases, the ribosomes of
mitochondria are smaller than those in the eucaryotic cell
cytoplasm. Plant and algal chloroplasts also have ribosomes smaller in size than their cytoplasmic counterparts,
as is shown in Figure 15–39.

These ribosomes are functional; purified mitochondria and chloroplasts can incorporate radioactive amino
acids into some of their proteins under appropriate

Genetic Code of Mitochondria

The simple notion of a genetic code shared in common between the cytoplasm and the mitochondrial matrix was lost in 1981. F. Sanger, B. Barrell, and their colleagues determined the complete sequence of the circular human mitochondrial DNA and found there a number of nucleotide sequences apparently specifying proteins but using a variation of the usual coding dictionary (see Table 13–3). The codon UGA, for example, is used to specify amino acid tryptophan and not as a signal for the termination of protein synthesis (Table 15–A). On the other hand, two of the six codons that specify arginine in cytoplasmic protein synthesis, AGA and AGG, are used instead to signal protein chain termination. Finally, codon AUA is not used to specify isoleucine in human mitochondrial genes, but rather to specify methionine. Since there are other isoleucine, arginine, and stop codons, this variation does not present the mitochondria with any difficulties in its protein synthesis. Another variation from the usual coding is that AUA (and AUU) can also be used as part of the signal to initiate protein synthesis. Mitochondria from yeast have other peculiarities in their coding dictionary.

There are less than two dozen different tRNAs in human mitochondria. This means the number of anticodons is too few to recognize all the codons possible in the way cytoplasmic tRNAs do, through the "wobble" discussed in Chapter 12. Instead, a single tRNA may recognize all four codons sharing the same first two nucleotides ("two-out-of-three" decoding), for example the codon CUX, where X is any of the four ribonucleotides. These coding differences have also been found in other mitochondria: bovine, yeast, and *Neurospora*.

The significance of these striking differences between mitochondrial and cytoplasmic protein synthesis is far from clear. Doubtless it is important that much genetic information must be included in a small DNA molecule. This interpretation also helps to explain the absence of introns in RNAs synthesized in animal mitochondria. It has been estimated that the rate of evolution in the sequences of mitochondrial DNA is as much as ten times faster than for nuclear DNA. The details of how faster evolution and selective pressure for small size led to the curious state of mitochondrial decoding will not be understood until much more research has been done.

TABLE 15–A

Exceptions in Mitochondria to Universal Cytoplasmic Genetic Code

Codon	Cytoplasmic Code	Mammalian Mitochondria (Human)	Yeast Mitochondria (*Saccharomyces cerevisiae*)
UGA	Terminate (opal)	Trp	Trp
AGA	Arg	Terminate	Arg
AGG	Arg	Terminate	Arg
AUA	Ile	Met (initiate)	Met
AUU	Ile	Ile (initiate)	Ile
CUA	Leu	Leu	Thr

conditions. Also, ribosomes isolated from mitochondria and from chloroplasts have been shown to incorporate radioactive amino acids into protein in vitro when appropriately supplemented with cofactors (cf. Chapter 13). The organelles have within them all the necessary macromolecules for protein synthesis, including the individual tRNAs and the corresponding aminoacyl-tRNA ligases. The tRNA molecules, however, are different from the corresponding tRNAs found in the cytoplasm. The organelles contain a DNA-dependent RNA polymerase and can synthesize ribosomal and transfer RNAs from their DNA template, as was shown by work with isolated organelles. Generally, the products of organelle transcription and translation remain associated with the organelle and its membranes.

The rRNAs of mammalian mitochondria are significantly smaller than those of procaryotes (954 nucleotides in the small subunit rRNA of human mitochondria versus 1542 nucleotides in the 16S rRNA of *E. coli*). In the case of yeast and human mitochondria, the rRNAs are all coded by the organelle DNA. The tRNAs found in organelles and used for protein synthesis there are generally the products of transcribing organelle DNA. (In the mitochondria of the ciliate *Tetrahymena,* however, there have been reports of *imported tRNAs,* transcribed in the nucleus.)

The sequences of many mitochondrial tRNAs have been determined and have produced at least two surprises. The highly conserved pair of guanine residues in the D-loop of the usual tRNA sequence (see Figures 12–13 and 12–15) is not present in at least one mitochondrial tRNA species from *Neurospora.* There is also a case where two different mitochondrial tRNAs accept glutamic acid at their 3′ end but one later accepts the transfer of an amine to form glutaminyl-tRNA. There is apparently no gln-tRNA ligase activity in those mitochondria. The biggest departure from the common structure of cytoplasmic tRNAs (see Figure 12–12) is a very much smaller serine-accepting tRNA from human mitochondria in which the entire D-loop and stem are missing!

The proteins of yeast mitochondrial ribosomes are probably all coded by the nucleus, as are the aminoacyl-tRNA ligases and various factors. The ribosomes of organelles are quite different from those in the cytoplasm, however, being generally smaller. A striking difference is their sensitivity to certain antibiotics. Protein synthesis by mitochondrial and chloroplast ribosomes is inhibited by chloramphenicol and not by cycloheximide, while for the cytoplasmic ribosomes the reverse is true. This property has often been exploited to differentiate the protein synthesis by cytoplasmic ribosomes from that of mitochondrial or chloroplast ribosomes. Thus in the presence of cycloheximide, the products of mitochondrial or chloroplast ribosomes will be synthesized preferentially. Finally, the initiation of protein synthesis in eucaryotic organelles involves formyl-methionyl-tRNA.

The initiator tRNA, the size of the ribosomes, and the spectrum of antibiotics that inhibit protein synthesis are all similar to those found in most procaryotic (eubacterial, not archaebacterial) cells and not to the structures and functions found in eucaryotic cytoplasm. Moreover, comparison of the sequences of ribosomal RNAs from chloroplasts and blue-green algae (cyanobacteria) shows such a strong similarity that the two must have an evolutionary relationship. (There is also a significant relation between the rRNA sequences of eubacteria and plant mitochondria.) These similarities have been used as corroborative evidence for the endosymbiote theory of the evolution of chloroplasts and mitochondria (see Chapter 1).

Given the presence in chloroplasts and mitochondria of enzyme systems capable of expressing genetic information, it was not unreasonable to suppose that these organelles might be able independently to duplicate themselves, analogous to parasitic bacteria residing within what might be regarded as a host cell. A simple calculation, however, shows that the DNA within either the mitochondrion or the chloroplast does not contain enough information to code for all the proteins found in the organelle. If we assume, for example, a molecular weight of 10^7 for the mitochondrial DNA (length approximately 17,000 base pairs), we can calculate, after subtracting that amount of DNA necessary for the coding of the rRNAs and tRNAs, that there is left only enough to code for about 15 proteins of molecular weight 50,000—a very small number considering the many functional enzymes we know mitochondria to possess. In animal mitochondria, in fact, there are only about 13 proteins coded by the mitochondrial DNA. The solution to this problem is that the information for synthesizing the remaining organelle proteins resides in the nuclear DNA. This had been inferred early on from genetic findings: certain mutants of yeast and of *Neurospora* exhibit alterations in mitochondrial function that are inherited in a strictly Mendelian fashion, thus implicating the nuclear chromosomes as the site of the affected genes.

In 1890, R. Altmann described mitochondria as autonomous organelles, capable of self-replication within the cell. Modern knowledge requires modification of this statement, to say that the replication of mitochondria and of chloroplasts occurs by a *cooperation* of genetic material within and without these organelles, via a cooperation of the protein-synthesizing machinery within and without these organelles. Thus, we can describe these organelles as *semi*autonomous; they replicate their entire functional structure, but only with the aid of components outside themselves.

Contributions by Cytoplasm and Organelle

Many of the details of the contributions made to organelle structure by cytoplasm and by organelle protein synthesis were discovered through exploiting the differential sensitivity of the respective ribosomes to antibiotics, as described above. The data from antibiotic sensitivity largely agree with the mode of inheritance of mutations affecting the proteins: Mendelian, implying a nuclear gene and a cytoplasmic site of synthesis, or non-Mendelian, implying an organelle gene and synthesis site. Experiments measuring the sensitivity to antibiotics have shown that the respiratory enzymes of the mitochondria are synthesized partly on cytoplasmic ribosomes *(cytoribosomes)* and partly on mitochondrial ribosomes *(mitoribosomes).* For example, the single polypeptide chain of cytochrome *b* is synthesized by mitoribosomes, while the single polypeptide chain of cytochrome *c* is synthesized by cytoribosomes. Cells grown in the presence of a radioactive amino acid can normally incorporate radioactivity into all the protein subunits of the cytochrome oxidase. If cells are also grown in the presence of cycloheximide, however, or of chloramphenicol or erythromycin, only some of the subunits become labeled. Thus, G. Schatz found that cycloheximide prevents the synthesis of the low-molecular-weight polypeptides of cytochrome oxidase, indicating that these are synthesized by the cytoribosomes, while erythromycin inhibits the synthesis of the higher-molecular-weight polypeptides, indicating that these are synthesized by the mitoribosomes (Figure 15–40).

A similar situation exists for the mitochondrial ATPase complex (the stalk and knobs that can be seen in the electron microscopic images of mitochondria (cf. Figures 8–3 and 8–18). This huge structure can be broken down into three large fragments:

- A soluble catalytic unit that has cold-labile ATPase activity (in contrast to the activity of the whole complex, which is cold-stable) and insensitive to the

inhibitor oligomycin (in contrast to the sensitivity of the complex when it is part of the mitochondrion—cf. Figure 8–7);
- An oligomycin-sensitivity-conferring protein (OSCP) that, when added together with the membrane fraction to the soluble ATPase, confers oligomycin sensitivity on the ATPase activity;
- A membrane fraction.

The OSCP is a single polypeptide chain, the soluble ATPase has five subunits, and the membrane part of the complex contains two or three protein subunits as well as lipids. When all three fractions are mixed, a reconstituted complex can be formed, having all the properties of the activity in the whole mitochondrion. By using specific inhibitors of protein synthesis such as those mentioned above, A. Tzagaloff found that all five subunits of the

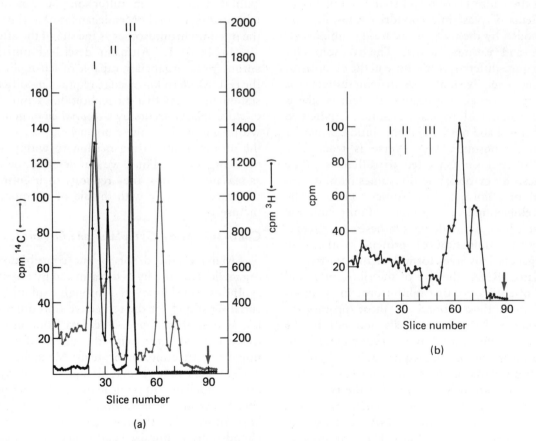

(a)

(b)

FIGURE 15–40

Cytochrome Oxidase Polypeptides Synthesized by Yeast Cells. (a) One batch of wildtype cells was grown in the presence of ^{14}C-leucine and in the absence of inhibitors (color), while another batch was grown in the presence of both ^{3}H-leucine and cycloheximide (black). (b) Wildtype cells were grown in the presence of ^{3}H-leucine and erythromycin. Mitochondria from ^{14}C-labeled cells and from ^{3}H-labeled cells were isolated and mixed for the separate cases (a) and (b), and the cytochrome oxidase peptides were isolated and analyzed as in Figure 8–10. Roman numerals I, II, and III represent the higher-molecular-weight polypeptides synthesized in the presence of cycloheximide (a), but not made in the presence of erythromycin. (Courtesy of G. Schatz.)

soluble ATPase as well as the OSCP are made in the cytoplasm, while probably all the subunits of the membrane fraction are made by the mitochondria.

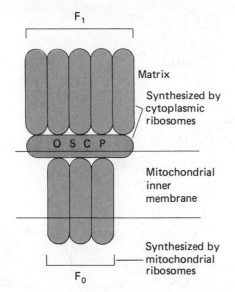

The story of chloroplasts is the same: ferredoxin and many of the Calvin-Benson C3 enzymes are synthesized in the cytoplasm while the *b* and *f* cytochromes are synthesized in the organelle. Like some of the mitochondrial enzymes mentioned above, the important chloroplast enzyme ribulose bisphosphate carboxylase has differentially synthesized subunits. The large subunit is synthesized by the chloroplast, but the small subunit by the 80S cytoplasmic ribosomes. Isolated pea chloroplasts are capable of incorporating radioactive amino acids into the large subunit, but not into the small subunit, of this enzyme.

Accounting for the source of each organelle protein will not resolve all the complexities, however. Synthesis of proteins by organelles can be influenced by cytoplasmic products. Just how protein synthesis in the two compartments is controlled to produce the coordinated development and metabolism of the eucaryotic cell is one of the more exciting puzzles to be solved in the future.

Importing Cytoplasmic Proteins into Organelles

The mechanisms by which proteins synthesized in the cytoplasm are transferred into organelles are necessarily complex, since there are so many different potential destinations. (In mitochondria there are four: the inner and outer membranes and the intermembrane and matrix spaces.) Proteins synthesized in the cytoplasm and destined for use in chloroplasts or mitochondria seem mostly to be synthesized on free cytoplasmic ribosomes. The transport into the organelle does not begin until the protein is completed and is, therefore, termed posttranslational, to distinguish it from the cotranslational

transport of proteins into the ER. The structure (or a part of the structure, a *targeting peptide*) of a protein destined for the organelle can apparently recognize some transport system normally part of the organelle membranes. An example of an imported protein is the small subunit of the chloroplast enzyme, ribulose-bisphosphate carboxylase (see Chapter 8). This protein is synthesized as a large precursor in the cytoplasm. The mechanisms by which the precursor recognizes the transport system in the chloroplast membrane and is cleaved to the finished small subunit is not known. The movement of proteins from the cytoplasm into organelles is substantial; approximately 10% of human body proteins flow along this pathway into mitochondria.

Work in the laboratories of Schatz and R. Butow has given us more information about the import of proteins into yeast mitochondria. In the mitochondria of *Saccharomyces cerevisiae,* the cytochrome oxidase complex is composed of nine polypeptide subunits, six of which are coded in the nucleus and translated in the cytoplasm; the other three are coded and synthesized in the mitochondrion itself. At least three of the nuclear-coded peptides are synthesized as longer precursors. After its synthesis in the cytoplasm on free ribosomes, subunit V, for example, has a 20-amino acid extra sequence (a "presequence") at the N-terminal end. This precursor binds, via its N-terminal sequence, to a specific receptor that seems to be present in regions where the outer mitochondrial membrane is juxtaposed to the inner membrane. From there it is translocated into the organelle, a process that requires the presence of a matrix–negative potential across the inner membrane. Once it has arrived at the matrix, the precursor is cleaved between amino acid residues 20 (phe) and 21 (ala), probably by a protease located in the matrix of the mitochondrion (Figure 15–41).

One way to explain specific translocation to the mitochondrial matrix is to propose the existence of a special sequence of amino acids that acts as a *matrix-targeting domain*. By analogy with the signal sequence of secreted proteins (see Chapter 13), this might be expected at the N-terminus of matrix protein precursors. In fact, such precursors do have similar N-terminal sequences, although they are not sufficiently alike to be described as a "consensus sequence." There are usually several basic residues—lysines and arginines with positively charged side chains—and several hydroxylated residues—serines and threonines; there are no negatively charged (acidic) residues. The positively charged residues are believed to help in the movement of the N-terminal sequence into the matrix, which under conditions of active respiration has a relatively negative electrostatic charge.

Proof that a matrix-targeting domain exists came from gene fusion experiments. By genetic engineering techniques, the N-terminal sequence of a matrix protein

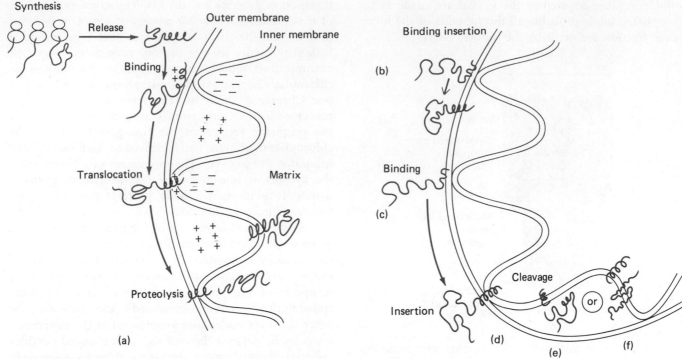

FIGURE **15–41**

Importing Proteins Into Mitochondria.   Proteins encoded in the nucleus and destined for mitochondria are synthesized in the cytoplasm on free ribosomes, usually with extra amino acids at the N-terminus, the "presequence." This contains several basic (positively charged) and hydroxylated amino acid residues. The protein with a matrix-targeting presequence binds to a receptor region on the outer mitochondrial membrane where it is juxtaposed to the inner membrane. (a) Import into the mitochondrial matrix. The bound protein is translocated through the membranes by an unknown mechanism, a process that requires the electrical potential component of the protonmotive force (see Chapter 8). After the protein is inside the matrix the presequence is removed by proteolytic enzymes there. The targeting presequence may remain associated with the membrane; its ultimate fate is not known. (b) Importation into the outer membrane. After binding to the receptor region, the protein inserts into the outer membrane and remains there. This is probably the result of a stop-transfer sequence of amino acids (see Chapter 7), a stretch of uncharged residues followed by charged residues. The inserted protein moves laterally away from the receptor region. (c, d) Importation into the intermembrane space or the inner membrane. After binding to the receptor region, the protein is transferred toward the matrix but a stop-transfer sequence results in its insertion into the inner membrane. If there is cleavage of the peptide chain the protein may end up in the intermembrane space (e). If not, it stays associated with the inner membrane (f). (After E. C. Hurt and P. G. M. van Loon.)

was joined onto another protein whose location can be easily identified—for example, β-galactosidase from *E. coli*. The bacterial enzyme ended up in the mitochondrial matrix. Conversely, when the gene for the matrix protein was engineered so the N-terminal sequence was missing, the protein did not enter the mitochondrial matrix. These experiments provide strong evidence for the N-terminal sequences being both necessary and sufficient to target a protein to the matrix. Similar experiments suggest that inner and outer membrane proteins, as well as intermembrane-space proteins, have targeting sequences.

Recombinant DNA techniques have been applied to this problem to identify many more matrix-targeting domains. Comparing the amino acid sequences that work reveals that 12 N-terminal residues or fewer suffice for the importing function. The domains have few negative charges and a distinct bias in favor of hydrophobic residues distributed so that the presequence can form an amphiphilic helix. In fact, N-terminal artificial peptides using only four different kinds of amino acids, provided they form amphiphilic helices in the membrane environment, cause an attached protein to be imported. Thus, the matrix-targeting domains do not seem to need a specific sequence of amino acids to have a specific function.

There are variations on these themes. Some mitochondrial proteins can spontaneously insert into lipid vesicles, thus apparently not requiring a specific receptor for beginning the translocation process. Removal of an N-terminal extension peptide is not always required for import since, after synthesis in the cytoplasm, the hemeless precursor to cytochrome *c*—apocytochrome *c*—is imported into the mitochondrion (to the intermembrane space) but undergoes no proteolytic processing. ATP hydrolysis has been proposed to play a separate role in the energetics of import: to help unfold proteins in the cytoplasm before they are translocated through the membranes. ATP hydrolysis has also been proposed to aid in detaching the imported protein from a complex with "chaperonins," intramatrix proteins that assist the translocation process. Further research should resolve the several additional mysteries of organelle targeting.

Influence of Light on Chloroplast Development

The mechanism controlling the duplication of mitochondria and chloroplasts is a mystery still to be solved. In the latter case there is a handle to the problem, namely the response of chloroplast growth and development to light. Figure 15–42 shows the development of the lamellar system of the chloroplast upon illumination. There is now a consensus concerning the interpretation of images seen during the formation of the fully differentiated chloroplast in higher plants. Based on observations at the light and electron microscope levels, Figure 15–43 shows schematically the formation of the chloroplast internal membrane system from a progenitor called the *proplastid* (or *etioplast*) that contains no chlorophyll. The internal membranes of the chloroplast invaginate in the dark as sheet-like porous lamellae from the inner membrane of the proplastid envelope. These sheets contract into a regular ordering of tubules, the whole structure being called the *prolamellar body,* which in many cases organizes itself into an almost crystalline array of membranes (Figure 15–42). This body can stay at this stage almost indefinitely within the proplastid if the plant is kept in the dark. Once light shines on the leaves, however, chlorophyll begins to be formed from protochlorophyll, and at the same time the prolamellar body breaks down to an irregularly arranged structure such as seen in the 5-day dark cell. Longer periods in the light result in the streaming out of long membranous loops from the prolamellar body, which form concentric circles around it. As the body disappears, the membranes lengthen to form flattened sacs and later many fuse to form the grana.

The biochemical correlates of this structural differentiation in response to light include the formation of chlorophyll from stored protochlorophyll and a continuous synthesis of protochlorophyll. At the same time, an increase in protein synthesis occurs; if chloroplast enzyme amounts or activities are measured, rather large increases of ferredoxin, chloroplast cytochromes, and other enzymes are found (representative data are iven in Table 15–3). This effect of light is mediated by a rather specific protein, phytochrome P_{730} (so-called because its pros-

TABLE 15–3

Components in Proplastid and in Chloroplast

Component	Proplastid	Chloroplast
Membrane Area*		
Thylakoid	21	137
Prolamellar body	22	0
Total internal	43	137
Molecules		
Protochlorophyll	200×10^4	0
Chlorophyll	0	$20{,}000 \times 10^4$
Ferredoxin	13×10^4	110×10^4
Cytochrome *f*	4.2×10^4	45×10^4
Cytochrome b_{559}	5.5×10^4	180×10^4
Cytochrome b_{563}	4.6×10^4	110×10^4
Ribulose bisphosphate carboxylase	35×10^4	54×10^4

Note: The beans were grown in the dark for 14 days before proplastids were obtained. Some of the beans were then illuminated for 45 hours and chloroplasts were then obtained. The values are listed per plastid. Both proplastids and chloroplasts contain DNA and ribosomes. (After J. W. Bradbeer.)

*Relative units

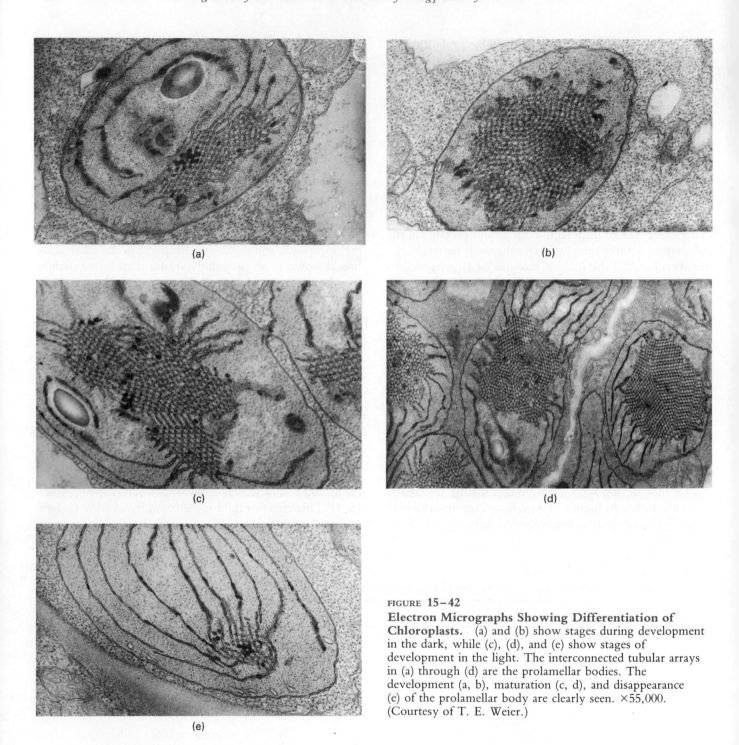

(a)

(b)

(c)

(d)

(e)

FIGURE 15–42

Electron Micrographs Showing Differentiation of Chloroplasts. (a) and (b) show stages during development in the dark, while (c), (d), and (e) show stages of development in the light. The interconnected tubular arrays in (a) through (d) are the prolamellar bodies. The development (a, b), maturation (c, d), and disappearance (e) of the prolamellar body are clearly seen. ×55,000. (Courtesy of T. E. Weier.)

thetic group absorbs light at 730 nm, in the far red). How this pigment mediates between the light and the responsive protein-synthesizing system is not known. In general, one can state that no thylakoid membranes can be found without chlorophyll formation, which requires light and subsequent protein synthesis; conversely, protein synthesis is necessary not only for membrane formation but also for continuous chlorophyll synthesis.

Self-Replicative Properties of the Cortical Structures of Ciliates

In our discussions of directed assembly and of self-replicating organelles, we have not considered whether there is information for control that might reside in molecules other than DNA. In a number of cases of directed assembly of viruses, it is quite clear that the

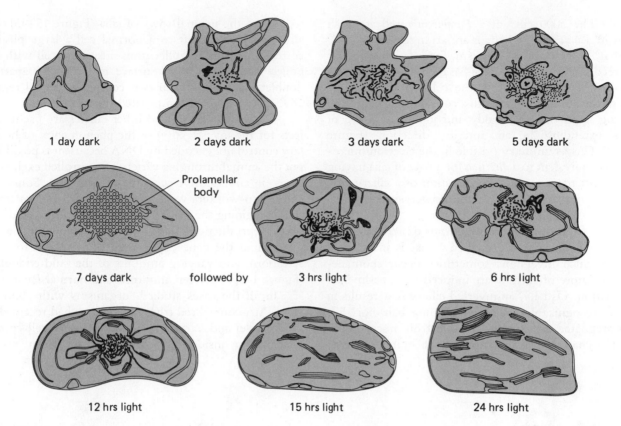

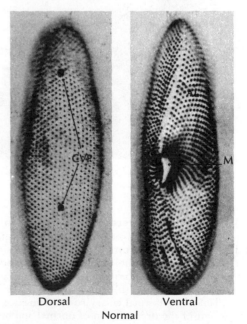

FIGURE 15–43
Diagrammatic Representation of Chloroplast Development. A safflower seedling is grown in the dark and then the light. (After A. H. P. Engelbrecht and T. E. Weier.)

information responsible for the synthesis of the directing molecules is encoded by the DNA of the virus or host. In self-replicating organelles such as chloroplasts and mitochondria, the complexity of the contributions by the nuclear and organellar genomes makes it difficult to determine whether some of the information necessary for development might not be encoded by DNA.

The most convincing evidence regarding the existence of inherited control that is not directly due to DNA comes from the extensive studies by T. Sonneborn and his co-workers on the development of cortical structures in ciliates such as *Paramecium aurelia. Paramecium,* like other protozoa, has a well-defined *cortical layer* of the cytoplasm, located beneath the plasma membrane. Unlike the fluid cytoplasmic interior, the cortical layer is stiff, or gelated (having the consistency of a gel).

The cortical layer forms the matrix or supporting structure for a number of morphological features of ciliates, such as the contractile vacuole pores, the opening (vestibule) to the gullet (Figure 15–44), and the very precisely marked *kinety fields,* visible in silvered preparations as rows of dots where the cilia and the basal bodies (known in ciliates as *kinetosomes*) are joined (Figure

FIGURE 15–44
Organization of Cortex of *Paramecium Aurelia* Visualized by Silver Impregnation. The rows of attachment sites for cilia are illustrated by the black dots. CVP = contractile vacuole pore; M = mouth.

15–45a). The 5000 cilia of a *Paramecium* cell are each located in a *unit territory;* these are arranged in some 70 rows that, in the normal organism, form a characteristic, easily recognizable pattern. The basal bodies of a row are positioned asymmetrically in the cell, to the right of center of the unit territory, with a connecting strand both emerging from the basal body and joining a group of strands extending in an anterior direction (Figure 15–45b). This asymmetry (visible in the electron micrographs) is equivalent to a *direction* for a row of cilia having the same orientation; thus the direction of a ciliary row determines the orientation of the power stroke of the ciliary beat (see Chapter 16).

It is possible to isolate a *Paramecium* or a *Tetrahymena* (a related ciliate) with a ciliary row that is inverted in direction. Such inversions sometimes occur spontaneously, but more often they are induced by experimental manipulation. Usually, an inverted ciliary row results in an easily recognizable, erratic swimming behavior. Silver-impregnated specimens and electron micrographs show the change in orientation of the unit territories,

relative to the normal rows of cilia (Figure 15–45). It is also possible to graft to a normal cell a large piece of cortex containing a gullet, thus creating a cell with two gullets and two ventral surface patterns, 90° apart. A doublet *Paramecium* with two complete cortical regions 180° apart has also been isolated (Figure 15–46).

Paramecium aurelia and other ciliates are useful subjects for the investigation of the phenomenon of hereditary control not encoded by DNA because it is possible to put the animal through a variety of controlled exchanges. Thus, in an extensive series of experiments, Sonneborn and his co-workers were able to exchange the micronuclei (containing the gene copies that are combined during sexual reproduction), most of the macronuclear material (containing the gene copies used as templates for transcription), and varying amounts of the fluid endoplasm (Figure 1–5a) between abnormal and normal *Paramecia*.

In all the cases studied, organisms with abnormal cortex structures bred true, even if they had received the micronuclei and varying amounts of other cell components lying inside the cortical region from normal

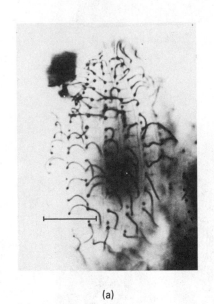

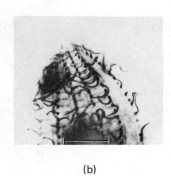

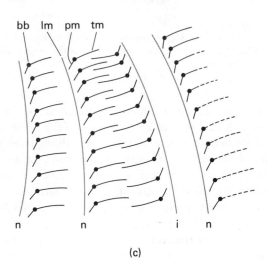

(a)   (b)   (c)

FIGURE **15–45**

Normally Oriented and Inverted Ciliary Rows in *Tetrahymena.*
(a) Photograph of a protein silver preparation of normal cell. The black dots are basal bodies, most of which bear cilia (dark, curved structures). The more lightly stained structures are microtubule bands that are embedded in the cortex. The feeding structure of the cell, known as the oral apparatus, is out of focus in the upper left of the photograph. (b) Anterior end of a cell that includes one inverted ciliary row, flanked by normal ciliary rows. The scale bars in (a) and (b) are 10 μm. (c) Diagram of the organization of normal and inverted ciliary rows. The three normal (n) and one inverted (i) row are depicted in (b). In the inverted rows, the longitudinal microtubule bands (lm) are on the "wrong" side of the basal bodies (bb). Notice especially that the microtubule bands directly associated with the basal bodies, the transverse (tm) and postciliary (pm) microtubules, are rotated 180° in the inverted rows, relative to the nearby normal rows. (Courtesy of S. Ng and J. Frankel.)

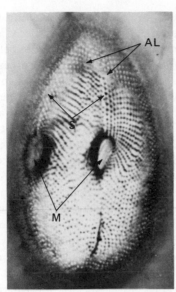

Doublet

FIGURE 15–46
Doublet Form of *Paramecium Aurelia.* Observe the
location of the kinety fields of the normal animal (Figure
15–44) and how some of the kinety fields are duplicated in
the doublet animal. Notice that the doublet animal has two
mouths (M), two preoral sutures (S), and two anterior left
kinety fields (AL). (Courtesy of T. Sonneborn and R.
Dipple.)

Paramecia. Furthermore, what bred true was not only the
gross morphology of the cortex but also the precise
location and orientation of basal bodies and other structures in the unit field. The behavior of cilia was also
inherited in the structure of the cortex: the cilia of
inverted rows always beat in a direction opposite to that
of normally oriented cilia during forward and backward
swimming.

A truly striking characteristic of these bizarre structures is their persistence through cell divisions. Sonneborn propagated one clone of *Paramecium* with some
inverted ciliary rows for 100 generations. As he so
dramatically put it, "If all the descendants were laid
end-to-end, they would reach from here to the sun; and
back again; over a thousand, million, million times!"
This is not by itself very strong proof of stability, since
mutation frequencies are low (10^{-5} per generation, say)
and not all the 10^{30} progeny will really have been tested
for maintenance of the phenotype. Nevertheless, appropriate genetic experiments were done to show that
neither ciliary row reversals nor doublet cells were
mutations that could be inherited in the usual way.
Moreover, sensitive chemical tests for the presence of
DNA have failed to detect it in cortical structures.

George Bernard Shaw said, "If all the economists of

the world were laid end-to-end, they still would not
reach a conclusion." Can we do better with our understanding of the behavior of the cortical structure of
Paramecium? The inference one can draw from these
experiments is that the assembly of a number of structures in the cortical region, as well as some of its
behavior, are controlled directly by the cortex rather than
by anything in the cell interior.

What, then, in the cortex determines the structure,
pattern, and behavior of cortical structures? There is not
yet a definitive answer to this question in molecular
terms, but it is possible to outline what the general
properties of the *cortical determiners* are likely to be:

- The determiners (there can of course be more than
  one kind) are probably not DNA because all efforts
  to demonstrate the presence of DNA in the cortex
  have failed.

- The determiner cannot be the basal body itself. This
  has been demonstrated in the ciliate *Oxytricha fallax.*
  Occasionally in this organism there is a partial failure
  of cytokinesis; the resulting *doublet cell* goes on to
  grow by binary fission, yielding two doublet daughter cells at each division. *Oxytricha* forms cysts under
  conditions of nutritional deprivation. Electron microscopic studies of the cysts have failed to detect any
  kinetosomes (basal bodies) or other gross cortical
  structures. Yet when a doublet cell encysts, it
  emerges from the cyst still as a doublet, and when a
  singlet cell encysts, it emerges as a singlet. The
  controlling factor is unlikely to be the size of the
  cyst, because some doublet cysts have been found
  which are smaller than some singlet cysts.

- There is a localized cortical differentiation, or micro-
  geography, in which the determiners are arranged.
  Studies with doublet forms of *Oxytricha* show that
  longitudinal microsurgical cuts that separate the
  doublet into two similar halves always yield singlets,
  whether they are allowed to regenerate or pass
  through encystment. When doublets are sliced trans-
  versely, however, then both fragments produce
  doublets.

- The determiner is probably associated with developing basal bodies. In *Paramecium,* if a portion of the
  cortex is removed, leaving an area free of cilia and
  kinetosomes, this area becomes repopulated with
  these structures, but only during preparation for cell
  division and only immediately anterior to an existing
  kinetosome. Basal bodies do not arise in the middle
  of cortical "bald spots." It would seem that the
  region immediately anterior to a basal body is the
  only region of the cortex provided with the
  means—at a molecular level—of nucleating basal
  body assembly.

The discussion above forces us to conclude that there are molecular determiners that influence the structure, orientation, and behavior of a number of morphological features of the cortex. Each unit territory contains one or more determiners that exert their specific influence locally and independently of the agencies, nuclear or otherwise, in the cell's interior. This is not to imply that the materials, probably protein, of which the determiner is made do not derive their structure from information encoded in DNA; if the determiner is made of a number of proteins, we can be practically certain that the information for the synthesis of these proteins is encoded in the DNA of the cell. Indeed, we know of a number of gene mutations that affect the structure of the cortex. What is suggested, however, is that the specific *organization* of the determiner is self-replicative and it is this self-replicative organization that specifies the self-replicative properties of the cortex that we have discussed.

Figure 15–47 illustrates a hypothetical scheme that utilizes familiar assembly principles to account for the self-replicative properties of the cortex. We have assumed a determiner composed of three proteins: A, B, and C. The information for the synthesis of each of these proteins derives from the DNA. Proteins A, B, and C cannot assemble spontaneously to form the determiner but require an assembled determiner as a template or nucleation device to provide the information for the specific assembly. This is a form of directed assembly with which we are already familiar. It is analogous to the bacterial flagella system where some polymerized flagellin is required for the polymerization of additional flagellin. Once the determiner has been assembled, it must first be removed from the template before it can act as a directing agent in the assembly of a basal body. We suggest that determiner acts as an agent in the directed assembly of two structures: (1) of another determiner and (2) of the product of the determiner, say a kinetosome (basal body). (This is entirely analogous to the use of nucleic acids as genetic determiners: they act as an agent in the directed assembly of themselves and of a product, a primary transcript in the case of DNA genes or a protein in the case of RNA genes.)

This kind of thinking creates a picture of heritable control of cell structure that goes beyond current dogma, that is, morphogenetic regulation through the control of protein synthesis. We suggest that some forms of *molecular organization,* which reside in the spatial arrangement of macromolecules, carry specific information from cell to cell. Of course this scheme explains only transmission of the cytoplasmic determiners from generation to generation. The origin of the organization is not obvious from the scheme, even as the origin of DNA genetics is not obvious from its current state. It can be speculated that the origin of some forms, including possibly the kinety fields of ciliates, was the result of random events. The cytoplasmic determiners that survive today represent the result of natural selection operating on changes in the original form. Thus, the inheritance of structural determinants and of genes are seen not to be conceptually so far apart as might have first been supposed. They are both subject to change and selection and, through time, evolve.

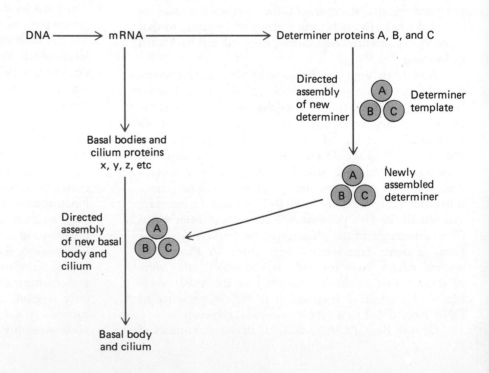

FIGURE **15–47**
Hypothetical Assembly Scheme.
This pathway illustrates how a determiner acts both as a nucleation device or template for the assembly of another determiner and then acts as a directing structure for the assembly of a basal body.

It remains to be seen whether self-replicating cytoplasmic determiners, the organization of which is not directly dependent on the DNA of the cell, are widespread in eucaryotic cells. Widespread or not, the molecular mechanisms of phenomena such as these represent a great challenge to cell biology in the immediate future, and we can expect that new approaches to the mechanism of cellular assembly processes will emerge from our present efforts.

SUMMARY

The purpose of this chapter has been to prepare the student to think about approaches to the study of assembly pathways of supramolecular structures. In the 1930s and 1940s students were trained to think about how to study metabolic pathways, and in the 50s, 60s, and 70s, they worked on the numerous approaches to studying macromolecular biosynthesis. Today cell biologists focus on the mechanisms by which complex macromolecular structures are assembled and on how cell-cell interactions form the multitude of organismic morphologies. We are a long way from being able to account in molecular terms for the facial resemblance between parents and their children, but there cannot be any doubt that such resemblances are common and reliable biological phenomena that must eventually be susceptible to molecular analysis.

KEY WORDS

Assembly, chloroplast, codon, conformation, cortex, cross-link, cytochrome, filament, genetic code, information, initiator, microtubule, mitochondrion, monomer, organelle, phospholipid, polymer, replication, ribosome, template, translation, virus.

PROBLEMS

1. What makes it appropriate to describe ribosome assembly as a form of aided assembly?

2. Why does one classify microtubule assembly as a form of directed assembly?

3. How would you demonstrate that fibrinogen is composed of six chains: 2 Aα, 2 Bβ, and 2 γ chains?

4. How would you use insoluble fibrin as a substrate to assay for an enzyme capable of breaking glutamyl-lysine isopeptide bonds?

5. What are the criteria that show a virus reassembly procedure has been successful?

6. What is the advantage to the cell in building a structure from numerous identical subunits, rather than from fewer and larger subunits?

7. What are the lines of evidence that the RNA of tobacco mosaic virus plays a role in the assembly of the protein coat?

8. What is meant by describing the replication of mitochondria as "semiautonomous."

9. What aspect of the cortex structure in *Paramecium* is self-replicative?

10. Polyacrylamide gel electrophoresis, cross-linking, and immunoelectron microscopy are techniques that can be applied to the analysis of complex cellular structures. Consider the following hypothetical analyses on an object that appears under the electron microscope as a sort of "duffel bag" (floppy cylinder).

    (a) A purified preparation of the structure was obtained, using electron microscopy as an assay. This preparation was denatured in SDS and analyzed by polyacrylamide gel electrophoresis in SDS.

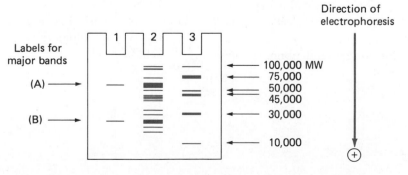

Lane 1: small amount of preparation; lane 2: large amount of preparation; lane 3: marker proteins of known molecular weight. Interpret the bands detected on this gel, which was stained with Coomasie Blue.

(b) Molecular cross-linking agent iminothiolane (see Chapter 13) was used to analyze the structure. First the iminothiolane was added to the preparation in low concentrations, then the resulting —SH groups were oxidized to —S—S— covalent bonds and the reaction mixture was electrophoresed in an SDS tube gel in oxidizing conditions. Next the tube gel was turned through 90° and reelectrophoresed in reducing conditions, so the —S—S— bonds were broken again. The results of the second (slab) gel are shown below (the expected positions after the second electrophoresis of components A and B are indicated beside the gel image). Interpret these results.

(c) Finally, immunoelectronmicroscopy was used with antibodies raised against the primary constituents found in the preparation. Shown below are sketches of the electron microscopic images found at lower concentrations of antibodies, when pairs of structures were detected. Interpret these results and give a plausible model for the structure.

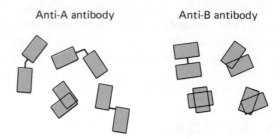

Anti-A antibody   Anti-B antibody

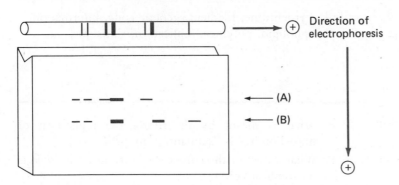

Direction of electrophoresis

(A)
(B)

SELECTED READINGS

Bretscher, M. S. (1985) The molecules of the cell membrane. *Scientific American* 253(4):100–108.

Butler, P. J. G., and Klug, A. (1978) The assembly of a virus. *Scientific American* 239(5):62–69.

Caspar, D. L. D., and Klug, A. (1962) Physical principles in the construction of regular viruses. Cold Spring Harbor Symposium on Quantitative Biology: Cold Spring Harbor, NY, 27:1–24.

Cleveland, D. W. (1987) The multitubulin hypothesis revisited: What have we learned? *Journal of Cell Biology* 104:381–383.

Doolittle, R. F. (1981) Fibrinogen and fibrin. *Scientific American* 245(6):126–135.

Dustin, P. (1980) Microtubules. *Scientific American* 243(2):67–76.

Kirschner, M., and Mitchison, T. (1986) Beyond self-assembly: From microtubules to morphogenesis. *Cell* 45:329–342.

Mosig, G., and Eiserling, F. (1988) "Phage T4 Structure and Metabolism." In *The Bacteriophages,* Volume 2. Calendar, R., ed. New York: Plenum Publishing Corp.

Unwin, N., and Henderson, R. (1984) The structure of proteins in biological membranes. *Scientific American* 250(2):78–94.

The Cytoplasmic Matrix and the Conversion of Chemical Energy Into Work

16

Performance of mechanical work is a universal property of living things. Such diverse phenomena as injection by phage T2 of DNA into a bacterium, rotation of the bacterial flagella of *Salmonella*, vibrations of the blue-green alga *Oscillatoria*, beat of the cilia of *Paramecium*, shuttle flow of the cytoplasm in the slime mold *Physarum*, circular flow of the cytoplasm in the water plant *Elodea*, transport of materials up and down the nerve axon, movement of the chromosomes during cell duplication, constriction of the animal cell during cell division, and contraction of smooth or striated muscle are but a few examples of the great variety of biological *mechanochemical transductions*—processes that transform chemical energy into mechanical motion. These processes are an important property of the cytoplasmic matrix, or *cytomatrix*—a structural complex that may be viewed as a special organ of the cell. Other properties of the cytomatrix include supporting the cellular organelles and defining the cell shape.

The cytomatrix is composed of a number of fibrous structures, the *cytoskeleton*, that interact with one another and with the cell membranes and that confer on the cytomatrix many of its interesting properties. It is also becoming clear that the cytomatrix *regulates the localization and interaction of many macromolecules and assemblies* that previously were believed to be simply dissolved and randomly distributed in a largely aqueous cytosol. This function too is believed to be due to the cytoskeleton fibers.

We shall begin our discussion of the cytomatrix by describing the filaments involved in mechanochemical transductions. These filaments include the actin-myosin system and the tubulin-dynein system (the structure of filaments involved in the rotational motion of bacterial flagella has already been discussed in Chapter 15; the details of the motor that drives the rotation are still being investigated). We shall then discuss the *intermediate filaments* and *superthin filaments* that with actin filaments and microtubules form the cytoskeleton. Finally, we close the chapter with a description of the interaction of all these filaments with one another and with membranes, and the integration of cellular function that emerges from such interactions.

The Actin–Myosin System in Striated Muscle

Although actin and myosin are found widely distributed in various eucaryotic cell types, the actin–myosin mechanochemical system was first discovered and has been most actively studied in vertebrate striated muscle cells. This is because in striated muscle these proteins are present in their highest concentrations and in their greatest degree of organization. It is appropriate, therefore, to begin discussing mechanochemical phenomena with a study of striated muscle, and afterwards to discuss actin- and myosin-based cell motility at a more generalized level.

Overall Structure of the Muscle Fibril

Striated muscle, the voluntary muscle that moves the bones of animals, is composed of bundles of numerous elongated multinucleate cells. Motor nerve axons are attached to the surface of the muscle cells and transmit to them the electrical impulse that initiates contraction. Under the microscope, especially in phase-contrast, these muscle cells show a series of highly regular and distinct bands or *striations* (Figure 16–1); these can be resolved with even greater clarity under the electron microscope.

Figure 16–2 shows various aspects of the structure of striated muscle fibers. One can discern three major features, each playing an important role in muscle contraction.

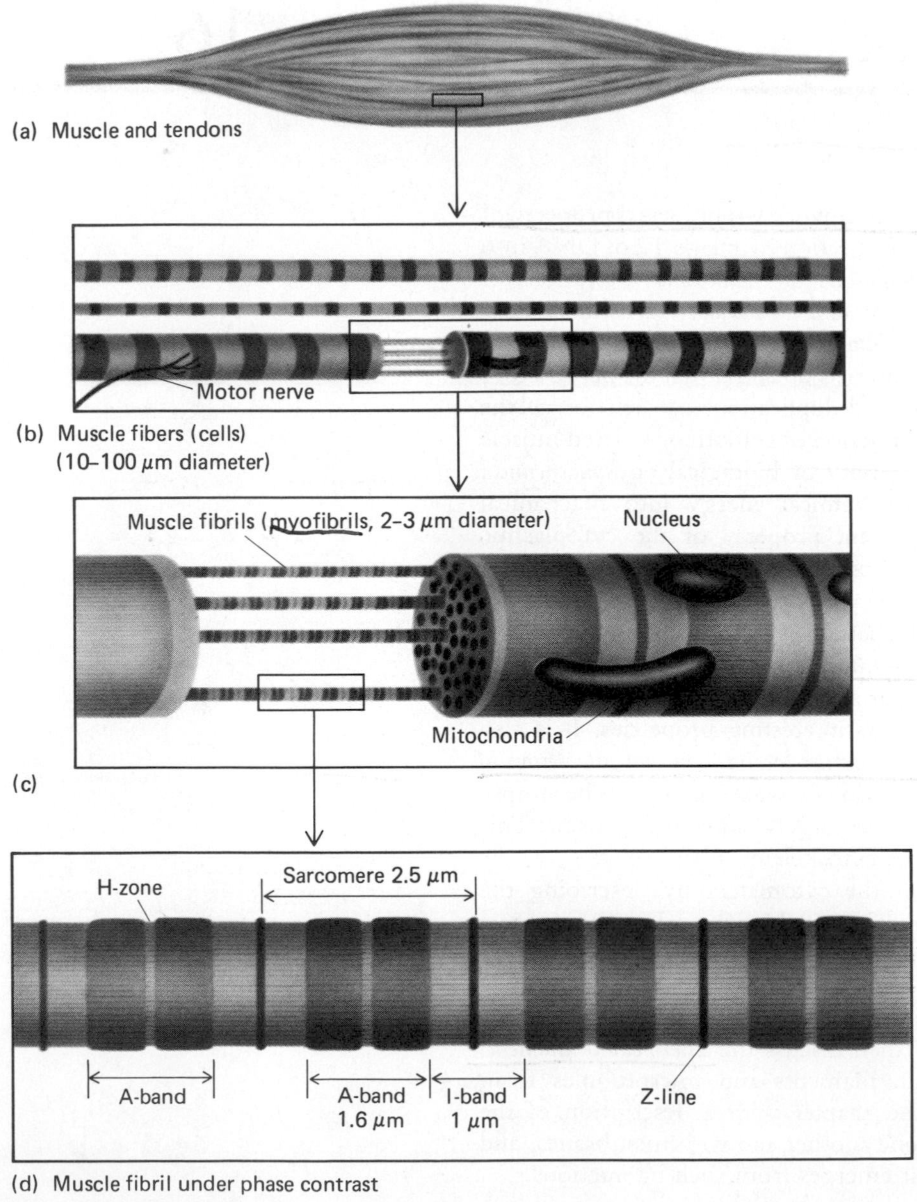

(a) Muscle and tendons

(b) Muscle fibers (cells)
(10–100 μm diameter)

Motor nerve

Muscle fibrils (myofibrils, 2–3 μm diameter)

Nucleus

Mitochondria

(c)

H-zone

Sarcomere 2.5 μm

A-band

A-band 1.6 μm

I-band 1 μm

Z-line

(d) Muscle fibril under phase contrast

FIGURE 16–1
The Organization of Striated Muscle, As Seen in the Light Microscope. The muscle (a) is an organ composed of numerous elongated cells or fibers (b). The fiber in turn contains numerous contractile elements or fibrils (c), which under the phase-contrast microscope (d) can be seen to have a striated structure of repeating units (sarcomeres). The sarcomeres are composed of two types of bands—the A-band and the I-band—the latter being divided by a structure called the Z-line. (The names of the bands are descriptive of their appearance under the polarizing microscope: through crossed polarizers, the *a*nisotropic band (A-band) is darker and the *i*sotropic band (I-band) is brighter.)

FIGURE 16–2 (Opposite)
The Fine Structure of Muscle
(a) Muscle fibrils composed of muscle filaments. The fine structure of the bands is clearly visible (see Figs. 16–3 and 16–5). (Courtesy of H. E. Huxley.) (b) Fibrils of cardiac muscle and their relation to mitochondria. These latter organelles are the primary source of ATP and provide a continuous energy input to power the work output of the cardiac muscle cell. (c) Muscle fibrils and their relation to the muscle cell membrane system (sarcoplasmic reticulum). Note both the longitudinal and the transverse elements of the sarcoplasmic reticulum, and the precise matching of their periodicity in relation to that of the muscle fibrils. (b and c, Courtesy of D. Fawcett.) (d) Schematic representation of the structural elements of striated muscle. (Courtesy of D. Fawcett and G. S. Bloom.)

X 30,000

×7,500

(a)

(b)

×7,500

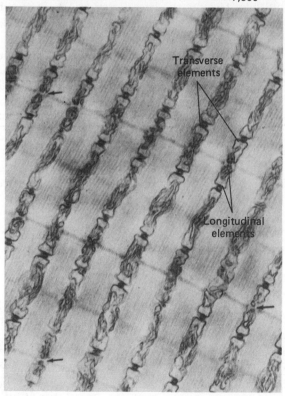

Transverse
elements

Longitudinal
elements

(c)

Myofibrils

Sarcolemma

Z line

Triad of the
reticulum

Transverse
tubule

Sarcoplasmic
reticulum

A band

Mitochondrion

Longitudinal
reticulum

I band

Transverse
tubule

Terminal
cisternae

Sacrotubules

(d)

- A system of elongated elements called *myofibrils,* arranged longitudinally in the muscle cell (Figures 16–2a and 16–2d). The myofibrils are the contractile machinery of muscle.

- Between the myofibrils, mitochondria that supply the myofibrils with ATP (Figures 16–2b and 16–2d): the more active the muscle, the more numerous the mitochondria.

- Also lying between the myofibrils, two sets of vesicles that do not appear to connect with each other (Figures 16–2c and 16–2d). These are the thin *transverse tubules* (*T-tubules*), which are invaginations of the plasma membrane, and the *sarcoplasmic reticulum,* with its interconnected *longitudinal vesicles* and *terminal cisternae.*

The interconnection of the walls of the transverse tubules with the plasma membrane and the lack of connection between the transverse tubules and the sarcoplasmic reticulum was demonstrated by H. E. Huxley; he immersed living muscle fibrils in a solution of ferritin, an iron-containing protein that is readily seen in the electron microscope as a small dense granule. After a short period, he sectioned the muscle fibril and found the ferritin throughout the muscle in the interior of the transverse tubule, but not in the sarcoplasmic reticulum. This demonstrated that the transverse tubule space is continuous with the exterior of the muscle cell but does not have connections with the sarcoplasmic reticulum that are large enough to pass the ferritin (MW = 650,000).

The *muscle cell* or *muscle fiber* is the smallest unit of muscle that can give a normal physiological response when activated by a nerve cell. Figure 16–3 shows diagrams and micrographs of the bands, zones, and lines of the myofibril at two levels of resolution. It can be seen that the various structures visible under the light microscope can be interpreted in terms of two sets of filaments (thin and thick) that overlap partially with each other.

The *sarcomere,* the smallest unit in the contractile structure, lies between two *Z-lines* (*Z-disks,* see Figure 16–3). Connected on either side of the Z-line are the *thin I-filaments.* Overlapping with two sets of thin filaments are the *thick A-filaments,* the centers of which are demarcated by the *M-line.* The thick filaments carry projections or *cross-bridges* on their surface; these are absent near the center of the filaments, giving rise to the *pseudo-H-zone.* The proper *H-zone* is a region at the center of the sarcomere where there is no overlap of thick and thin filaments. Thus, the A-band is the region represented by the thick filaments, and the I-band is the region where the thin filaments of two adjacent sarcomeres do not overlap with thick filaments.

The precise interdigitation of thin and thick filaments can be appreciated even better by examining their arrangement in cross section. Figure 16–3 also shows this view. Every thick filament is surrounded by six thin filaments. The figure also shows why two thin filaments between every pair of thick filaments can be seen in longitudinal section.

Muscle Contraction Is Not Contraction at the Molecular Level

On shortening, adjacent Z-lines move closer together; this is the basic structural change during striated muscle contraction. Early theories proposed that shortening of the muscle cell was caused by shortening of the molecules forming the structural framework of the cell (Figure 16–4a). Since no convincing evidence for such theories ever emerged, H. E. Huxley and J. Hanson proposed an ingenious alternative model: shortening of the muscle fiber is caused by a *sliding movement of interdigitating filaments,* bringing about an increase in overlap between them (Figures 16–4b and 16–5). Mechanical linkage between the two filaments occurs via multiple, temporary, lateral *cross-bridges* Figure 16–4c). Neither filament contracts along its length; the Z-lines are brought together by a relative motion of the filaments, driven by small, cyclically repeated conformational changes of the cross-bridges (Figure 16–4d). The sliding filament theory thus accounts for the amount of contraction observed as depending on *numerous repetitive small conformational changes,* rather than a single large one.

The sliding filament theory can best be understood after first introducing the mechanochemical proteins of the muscle fiber. Then it will be possible to provide a unifying explanation for the structural basis of contraction.

Mechanochemical Proteins of Muscle

Historical Survey

Since the muscle cell is such a highly efficient chemical-mechanical work transducer, it seems reasonable to assume that a considerable proportion of its proteins are concerned with this role. As early as 1864 the great physiologist W. Kühne suggested a method of extracting muscle, utilizing 10% NaCl, which yielded a concentrated protein extract. Unfortunately, Kühne considered muscle contraction to be analogous to blood clotting, with uniform contraction of the participating components. This analogy, which also dominated the thinking of other muscle physiologists, seriously inter-

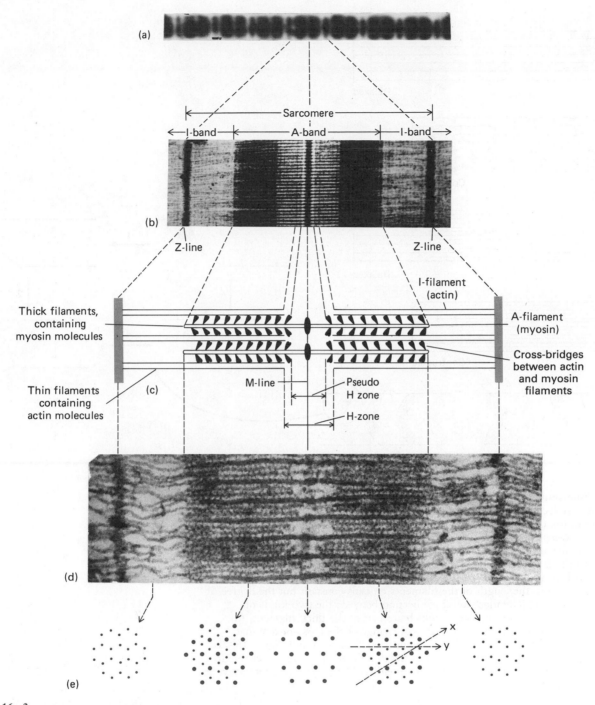

FIGURE 16-3

Relationship Between the Structure of the Myofibril as Seen in the Phase-Contrast Microscope and the High-Resolution Structure as Seen in, and Deduced from, Electron Microscopy. A single sarcomere of the myofibril is illustrated: (a) as seen in phase-contrast microscopy; (b) under the electron microscope at low magnification; (c) diagram that interprets the longitudinal views; (d) high-resolution electron microscopy; and (e) diagram that interprets the structure in cross section. X and Y are cross sections through two different planes of the myofibril; they show why it is possible to see either one or two thin filaments between every thick filament.

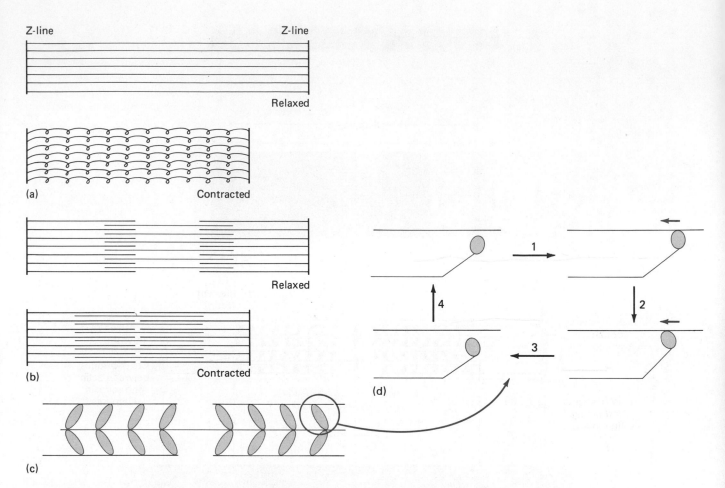

FIGURE 16–4
Sliding Filament Theory of Striated Muscle Contraction. (a) Older theory presumed a molecule connected to the Z-disks and running the whole length of the sarcomere to the opposite Z-disk. The contraction of this molecule was supposed to bring the Z-disks together. (In longitudinal views of the sarcomere, the Z-disk appears as a line and is called the Z-line.) (b) The thin filaments are connected to the Z-disk but do not run the whole length of the sarcomere. There are thick filaments (color) in the middle of the sarcomere that overlap the thin filaments. During contraction, the length of the filaments remains constant but the degree of overlap increases. (c) The mechanical connection between the two kinds of filaments is through "cross-bridges," which are part of the thick filament. (d) Contraction is produced by a conformational change of the cross-bridge that occurs while it is attached to the thin filament (step 2, see text and Figure 16–21 for details). Repeated cyclically, this causes the thin filament to move, relative to the thick filament. Since the thin filaments on either side of the sarcomere move in opposite directions, the sarcomere shortens.

fered with further progress in the study of muscle proteins.

A. Danilevskii (1881–1888) made important contributions by distinguishing among various classes of proteins that could be extracted from muscle using water, dilute alkali and acid, and 6–12% NH_4Cl solutions. He anticipated modern work by extracting muscle fibrils with various solutions and noting the disappearance of the A-bands (see Figures 2–32 and 16–1) under the microscope.

At the beginning of this century a number of workers did much to increase our understanding of a protein component of muscle that was extracted most conveniently in strong salt solutions at alkaline pH. O. von Fürth called this material "myosin." It precipitated at low ionic strength[1] and could thus be purified

[1]Ionic strength: $\Gamma/2 = \frac{1}{2} \Sigma c_i z_i^2$, where the sum is taken of the products of the molar concentration c_i and the square of the electrostatic charge z_i^2 of all ions in the solution.

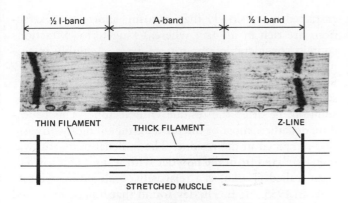

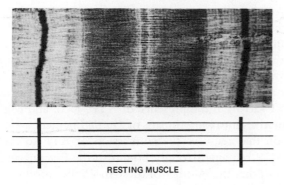

FIGURE **16–5**

The Effect of Stretching on the Width of the I-band and the H-zone. The sliding filament theory explains how the sarcomere at various stages of contraction can show differences in the size of the I-band and the H-zone, while the width of the A-band remains constant. The electron micrographs illustrate the experimental observations and the diagrams interpret the images in terms of the sliding filament model. (Courtesy of H. E. Huxley.)

from other globulin proteins, which do not precipitate as readily under those conditions. Physical studies soon showed that such "myosin" is not homogeneous. In 1942, F. Straub discovered another protein in the mixture. The crude preparation known as "myosin" was renamed *actomyosin* since it is composed of two proteins, which Straub named *actin* and *myosin*. These two proteins have very different physical properties and can associate with each other strongly.

In 1948, K. Bailey isolated from muscle another major protein component, which he called *tropomyosin;* in the 1960s, S. Ebashi and his co-workers added *troponin* and α-*actinin* to the list of proteins involved in the contraction of striated muscle.

The discovery of myosin and actin, though important, was overshadowed by exciting discoveries concerning the interaction of these proteins with each other and with the high-energy compound ATP. In 1939, V. A. Englehardt and M. N. Lyubimova made the dramatic discovery that what we now know as actomyosin has

ATPase activity. After the discovery of actin, I. Banga purified myosin and showed that it contains the enzymatic activity. Because the activity of purified myosin is 100 times slower than that of other ATPase enzymes, a number of workers considered the possibility that the activity might be due to an impurity. Efforts to separate the activity from myosin proved unsuccessful, however, and there is no longer any doubt that myosin is capable of catalyzing the hydrolysis of ATP;

$$ATP + H_2O \xrightarrow{\text{myosin}} ADP + P_i.$$

The importance of the discovery of Engelhardt and Lyubimova cannot be overemphasized; it focused the attention of muscle physiologists on the molecular properties of myosin and actin and thus helped transform muscle physiology into a molecular science.

Another important discovery regarding the interaction of these proteins with ATP was made by Albert Szent-Györgyi and his co-workers in the 1940s. They showed first that mixing actin with myosin in concentrated salt solutions produces a great increase in viscosity, which suggests that these molecules interact to form large polymers. The effect is reversed by adding ATP, which causes the viscosity to drop back to the original level—that is, the viscosity found in separate actin and myosin solutions (Figure 16–6). These results can be summarized in the following way:

$$\left.\begin{array}{l} \text{actin} + \text{myosin} \rightarrow \text{actomyosin} \\ \text{actomyosin} + ATP \rightarrow \text{actin} + \text{myosin} \end{array}\right\}$$

Of course, since myosin hydrolyzes ATP, the drop in viscosity is not permanent but slowly reverses as the ATP disappears from the preparation. These results showed that ATP has an effect on the physical relationship between actin and myosin.

Since, myosin and actin are not found in solution in the muscle cell, it would seem appropriate to study the myosin-actin-ATP interaction in the precipitated or solid state. H. H. Weber (1934) had shown that if a solution of actomyosin is extruded from a capillary tube into a solution of low ionic strength, it forms a thread-like precipitate. A number of workers then studied the effect of ATP on these threads, but it was not until Albert Szent-Györgyi (1941) precipitated actomyosin preparations that were richer in actin that something truly exciting happened—the threads, on addition of ATP, contracted! To be sure, these early attempts yielded threads that contracted in all directions rather than just shortening. Later, however, threads prepared from higher actomyosin concentrations and extruded at sufficiently high rates so as to orient the actomyosin molecules turned out to be much better models of the muscle fibril; they truly shortened, and were even capable of performing work.

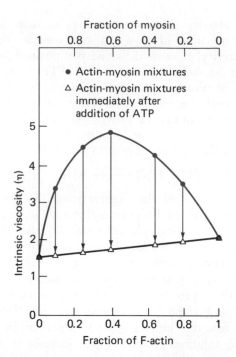

FIGURE 16–6
The Effect of ATP on the Viscosity of Actomyosin.
The graph is a summary of experiments in which ATP was added to different mixtures (actomyosin) of F-actin and myosin, after which the decrease in viscosity was measured. The results show that solutions of pure actin or pure myosin do not decrease in viscosity upon addition of ATP, but mixtures containing varying amounts of both proteins do. The observation that the values for mixtures, after the addition of ATP, lie on a straight line connecting the pure actin and pure myosin values is consistent with the interpretation that ATP dissociates an actomyosin complex into actin and myosin. Other physical measurements have confirmed this interpretation. (From B. Kerekjarto.)

The analogy between the actomyosin thread and the muscle tissue fiber was strengthened by the preparation of *glycerinated fibers* (by Albert Szent-Györgyi in 1949). When a muscle fiber is placed into 60% glycerol at −10°C, the cell membranes break down, and over a 10-day period the soluble proteins are slowly extracted. What remains behind is the contractile machinery of muscle, composed mainly of actin, myosin, tropomyosin, and troponin. Glycerinated muscle fibers were found to have some properties analogous to the synthetic actomyosin threads and some analogous to muscle fibers. This provided further support for a molecular approach to the problem—that muscle contraction can be studied by purifying the protein components and examining their interactions.

Actin

Actin can be extracted from muscle with solutions of high ionic strength and slightly alkaline pH. Since actin is extracted more slowly than myosin, short-term extractions are rich in myosin whereas long-term extractions contain considerable amounts of actin. A highly imaginative procedure for preparing actin was developed by Straub. He extracted the muscle fibrils first to remove some of the myosin; then, after treating with an alkaline solution, he dried the residue with acetone. This procedure denatures most of the proteins including the myosin that remained in the residue. The undenatured actin was then extracted from the powder with water. This form of actin, called *G-actin,* is a relatively small *g*lobular molecule. In 1955, H. E. Huxley found that muscle extracted with KI solutions loses the thin filaments seen in electron micrographs. The extract contains primarily actin.

Actin represents about 15% of the muscle cell protein. The complete 41.7-kilodalton (kDa), 374-amino acid sequence of the polypeptide chain of G-actin is known. There is one residue of a rare modified amino acid, 3-methyl-histidine; since it is also found in myosin, it may be important in the contraction mechanism. G-actin, in the presence of ATP, KCl, and Mg^{2+}, polymerizes into long threads of *F-actin* (filament actin), a reaction that requires the conversion of ATP to ADP. The ADP remains strongly bound to the F-actin, and it turns out that the number of these molecules produced is equal to the number of G-actin units in the F-actin polymer:

$$n\ \text{G-actin} + n\ \text{ATP} \xrightarrow{Mg^{2+}} (\text{F-actin-ADP})_n + n\ P_i$$

$$n\ (\text{F-actin-ADP}) + n\ \text{myosin} \rightarrow n\ \text{actomyosin} + n\ \text{ADP}$$

The ADP can be released from F-actin by its reacting with myosin. The fact that energy is necessary to polymerize actin is likely to be of importance to cell function.

Examination of F-actin under the microscope shows a rigid filament of helical structure (Figure 16–7). X-ray analysis of F-actin prepared in vitro, and also as isolated from the muscle cell, conveys a picture of the filament as a polymer of two strands wrapped around one another. Figure 16–7 shows an electron micrograph and a model that gives a plausible interpretation of what can be seen in the micrograph. The model does not show the polarity of the filament, which is discussed later.

Myosin

Myosin can be purified by short-term extraction at pH 6.8 in KCl solutions of high ionic strength. This preparation contains about 35% of the muscle cell protein and comprises most of the A-band material and the thick filaments, which are seen in electron micrographs to have disappeared after the extraction. Further purification, by precipitation at low ionic strength, gel permeation chromatography, and ultracentrifugation, have been used to

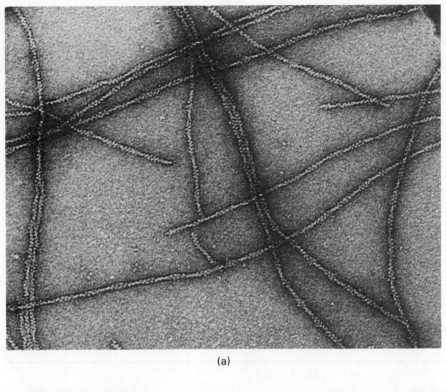

(a)

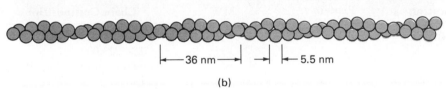

(b)

FIGURE **16–7**
The Structure of F-actin. (a) An electron micrograph of F-actin reconstituted from purified rabbit muscle G-actin. If one interprets this picture to be produced by two filaments of globular units twisted around each other, then one can count the number of subunits per turn of the helix as well as identify the places where the two strands cross over one another. (b) Model of F-actin filaments showing approximately 14 subunits per turn. (Courtesy of J. A. Spudich.)

remove the much larger actomyosin complexes that contaminate such myosin preparations. The molecular mass of rabbit skeletal muscle myosin is approximately 530 kDa; the molecule is composed of two identical heavy chains of 223 kDa each and four light chains of 16–25 kDa (Figure 16–8, Table 16–1).

The two 18-kDa light chains are identical and will dissociate from the myosin molecule if it is treated with the reagent DTNB—5,5′-dithiobis-(2-nitrobenzoate)—a sulfhydryl-containing reagent capable of breaking disulfide bonds. The removal of the two "DTNB chains" (also called LC2 or "regulatory MLCs") does not have any as yet discernible effects on the properties of myosin, and their particular role, if any, in muscle contraction remains to be discovered. The other two light chains can be removed from the myosin molecule by exposing it to alkali (pH 11). Removal of these two chains (called A1 and A2 or LC1 and LC3 or alkali MLCs) causes myosin to lose its ATPase and actin-binding activities. That this effect is due to the removal, and not to the high pH, can be demonstrated by adding these light chains back to

myosin and regenerating its enzymatic and actin-binding properties.

The myosin molecule has a unique structure (Figure 16–9a). It consists of a long *tail* composed of the two heavy chains—α-helices that twist around each other in the form of a coiled coil—and two globular *heads* at one end of the molecule.

An important property of the skeletal muscle myosin molecule is that, at the salt concentration and pH of living muscle cells (ionic strength 0.15, pH 7.0), it associates with itself to form a filament. An interesting aspect of this association is that the myosin molecules *pack in opposite directions* to form a tapering filament, with projections caused by the heads of each molecule and a bare zone in the middle of the filament (see Figure 16–14). The structure of these myosin filaments is described further when we discuss the fine structure of the muscle thick filament.

By digesting myosin with trypsin or papain, it has been possible to develop an understanding of the topography of the myosin molecule. Brief trypsin digestion

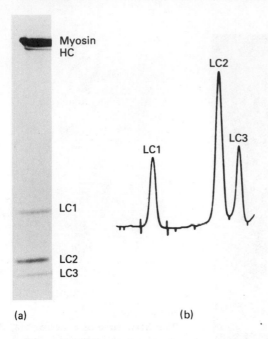

(a)   (b)

FIGURE 16–8

The Chain Composition of Chicken Breast Muscle Myosin.   (a) Polyacrylamide gel showing the heavy chain (HC) and the light chains (LC) of myosin. (b) Densitometer tracings showing the relative amounts of light chains. LC2, the "regulatory" chain, is found as a single copy in each of the myosin heads. LC1 and LC3, the "essential" light chains, have the same amino acid sequence except that LC3 is missing a 41-amino acid sequence form the N-terminal end. Each myosin head contains a single copy of either LC1 or LC3. (Courtesy of S. Lowey.)

TABLE 16–1

The Proteins of the Myofibril

Protein	Function	Location	% of Myofibril Protein	MW	Chain Composition
Myosin	ATPase, structural, conformational, mechanical	Thick filament	43–45	521,000	2 heavy chains 2 × 223,000 4 light chains (see Fig. 16–8) 1 LC 1 1 × 25,000 2 LC 2 2 × 20,000 1 LC 3 1 × 16,000
Actin	Myosin ATPase activation, structural, conformational	Thin filament	21–23	41,900	1 × 41,900
Trypomyosin	Regulational, actin binding	In grooves of thin filament	5	70,000	2 × 35,000
Troponin	Tropomyosin binding, regulational, Ca^{2+} binding	Thin filament at 400-Å intervals	5	70,500	1 TN-T 1 × 30,500 1 TN-I 1 × 22,000 1 TN-C 1 × 18,000
C-protein	Unknown, possibly in the assembly of thick filament	Thick filament; eleven 430-Å intervals on both halves of thick filament	1–2	140,000	1 × 41,900
α-Actinin	Structure of Z-disk	Z-disk	1	180,000	2 × 90,000
M-line protein	Structure of M-line and attachments of thin filaments	M-line	2	165,000	1 × 165,000
Titin	Third longitudinal filament of sarcomere	Probably mostly in I-band	8–10	2,500,000–3,000,000	2,500,000–3,000,000
Nebulin	?	In I-band	3–5	700,000	700,000

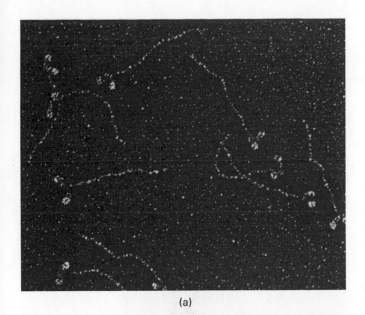

(a)

FIGURE 16-9
The Myosin Molecule. (a) Electron micrograph of myosin molecules from smooth muscle. Myosin samples were sprayed on freshly cleaved mica and rotary shadowed with platinum. (Courtesy of S. Lowey.) (b) Drawing of the quaternary structure of myosin and fragments produced by various treatments.

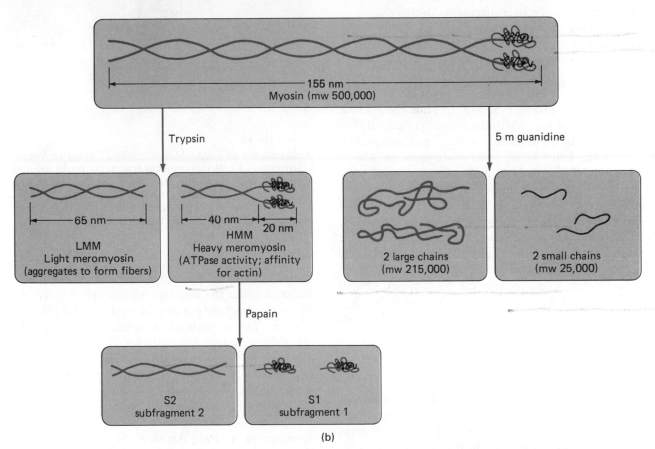

(b)

results in the hydrolysis of a number of peptide bonds, lying close to each other in a region about halfway up the tail of the myosin molecule. This produces a relatively specific break in the myosin tail, yielding two fractions (Figure 16–9b)—*light meromyosin (LMM)* and *heavy mero-* *myosin (HMM)*. Each of these two fractions, first described by Andrew Szent-Györgyi, retains some of the properties of the myosin molecule.

The LMM (126 kDa) is almost 100% α-helical and retains the low solubility properties of the myosin molecule at low ionic strength, aggregating with ease to form long smooth fibers (Figure 16–10). The HMM includes some of the tail region and the two globular heads; it is soluble at low ionic strength and retains both the ATPase property and the actin-combining capacity of the native myosin molecule.

The myosin molecule can also be broken with papain, which splits peptide bonds preferentially near the two heads of the molecule, thereby liberating first one and then the second head (see Figure 16–9b). These heads, called *subfragment 1* (*SF1* or *S1*), retain the ATPase and actin-binding properties of myosin and have been utilized extensively for a variety of enzymatic and physical experiments. The study of S1 is especially helpful under the conditions of low ionic strength at which the native myosin molecule normally aggregates. If the myosin rod obtained by papain digestion is treated with trypsin, LMM and *subfragment 2* (*SF2* or *S2*) are obtained. Even though it is very α-helical, S2 does not aggregate at low ionic strength (see later discussion in this chapter).

The myosin ATPase activity is inhibited by Mg^{2+} and activated by Ca^{2+}. Under ionic conditions similar to those of resting muscle (0.15 M KCl, 5 mM Mg^{2+}, 5 mM ATP, pH 7.0), myosin is a very sluggish enzyme. The ATPase activity of myosin can be raised by adding actin, even in the presence of Mg^{2+}. Thus the ATPase properties of myosin and actomyosin are radically different, the former being inhibited by Mg^{2+} and activated by Ca^{2+}, and the latter being greatly activated by Mg^{2+}. Actomyosin ATPase activity differs from myosin ATPase activity also in a number of other respects: pH optimum, response to high ATP concentrations, and response to the antibiotic oligomycin. Most importantly, ATP has only minor effects on the structure of the myosin molecule but very major effects on the physical properties of the actomyosin complex. These results led to the view that actomyosin is the fundamental unit of mechanochemical function and that it must be studied directly to understand the conversion of chemical energy into mechanical work.

Actomyosin Structure

When F-actin and myosin are mixed, the viscosity of the mixture rises, an effect that can be reversed by adding ATP. This behavior can be studied directly, using actin and myosin at high ionic strength (0.60; at low ionic strength actomyosin precipitates) or using actin and HMM at low ionic strength (0.15). It is possible to learn the general structural features of these reactions by direct observation with the electron microscope. When F-actin and myosin are mixed, the globular heads of the myosin molecule attach themselves to—*decorate*—the actin filament. The heads attach at an angle of less than 90°,

FIGURE **16–10**
Aggregation of Light Meromyosin (LMM) to Form Filaments.   Light meromyosin fragments retain the ability of muscle myosin to aggregate and form filaments at low salt concentrations. The surface of these filaments is smooth and lacks the projections observed with filaments aggregated from myosin. (Courtesy of H. E. Huxley.)

forming an *arrowhead* appearance that can be easily detected under the electron microscope (Figure 16–11). The same structure can be observed better if HMM or S1 is used. The arrowhead structure, which can also be observed in intact glycerol-extracted muscle, suggests that the actin filament is *polar*—that is, its properties differ with direction along the filament. Upon the addition of ATP and Mg^{2+} the myosin dissociates from the actin and this too can be clearly seen in the electron microscope.

Actomyosin ATPase Activity

The other important functional property of actomyosin that can be measured, of course, is the ATPase activity, and it has been studied intensively for a number of years. Measurements of ATPase activity have often been performed with HMM or S1 because it is possible to work at physiological ion concentrations without having the enzyme aggregate.

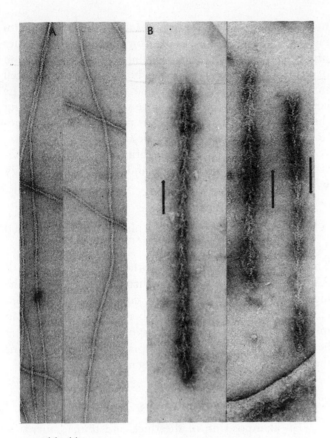

FIGURE **16–11**
Decoration of Actin Filaments with Heavy Meromyosin. (a) Actin filament, showing coiled structure of globular units. (b) Electron micrograph of a thin filament forming an aggregate with heavy meromyosin (HMM). The HMM attaches at an angle, forming "arrowheads," showing that the actin has an inherent polarity. (Courtesy of H. E. Huxley.)

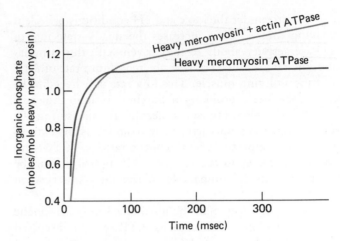

FIGURE **16–12**
The "Early Burst" of Phosphate Release, Characteristic of the ATPase Activity Observed in Heavy Meromyosin and Heavy Meromyosin Plus Actin. After combining the proteins and ATP for varying times and precipitating the proteins with trichloroacetic acid, inorganic phosphate was measured in the supernatant. The curves demonstrate the stimulating effect of actin on heavy meromyosin ATPase. After the early burst, the slope of the heavy meromyosin ATPase curve appears low because the rate of phosphate release is being measured over millisecond intervals. (Courtesy of E. W. Taylor and the American Chemical Society.)

The production of inorganic phosphate (P_i) that is catalyzed by myosin can be measured either by coupling the ATPase to another reaction that requires phosphate or by precipitating the protein with trichloroacetic acid at various intervals and assaying for P_i. Using the former method, a steady increase in the production of P_i can be observed. Using the latter method, an *early rapid burst* of P_i product, followed by a much slower steady-state rate (Figure 16–12), is observed. These results are interpreted as showing that the early burst arises from the formation of *bound phosphate*, which can be released by precipitating the protein with trichloroacetic acid. (Bound phosphate would not be available as substrate for a subsequent reaction, so the first method does not detect the early burst.) Since no evidence has so far been obtained for a covalent attachment, it is generally assumed that the phosphate is held on the myosin by secondary forces. By using gel permeation chromatography and ³H-labeled ATP, it is also possible to show that [³H]ADP also

remains bound to myosin during the early burst of P_i production.

On the basis of the above as well as other data, E. W. Taylor and his associates have proposed a scheme to explain the apparent slowness of the myosin ATPase activity; it is not the ATP hydrolysis step but the spontaneous dissociation of products (ADP + P_i) that is slow. In its simplest form this scheme is as follows:

$$M + ATP \xrightarrow{\text{fast}} M\ ATP \xrightarrow{\text{fast}} M\ ADP\ P_i \xrightarrow{\text{slow}}$$
$$M + ADP + P_i.$$

By the use of a variety of techniques such as ultraviolet spectroscopy, fluorescence, and ¹⁸O exchange, D. Trentham and his colleagues have detected a number of additional steps involved in the reaction. There is an early burst of proton release, which can be shown to precede the early burst of P_i release. Also the binding of ATP to myosin is very tight; it involves a larger free energy change than the subsequent hydrolysis of ATP. This binding step can be thought of as providing the energy for the simultaneous dissociation of actin from myosin. Phosphate release is the slow step that governs the slow, steady-state rate of the overall reaction. It is this step that would be expected to be accelerated by the presence of actin.

In vitro under the ionic and pH conditions prevalent in living muscle, actin increases the steady-state rate of ATPase activity about 20-fold. Even with this stimulation, the rate is much less than the maximum rate of ATP splitting in living muscle. This low rate is believed to be due to the steric problem of making an actin molecule available to every myosin molecule. By measuring the ATPase rate at various actin-to-myosin ratios and extrapolating to saturating actin concentrations, a 200- to 500-fold increase in the rate of ATP hydrolysis can be predicted; this is comparable to the rates observed in intact muscle.

Now, how can actin be introduced into the kinetic scheme developed for myosin ATPase? As discussed before, the addition of ATP to actomyosin causes the actin and myosin to be dissociated. Measurement of the rate of this process, by following the decrease in turbidity of an actomyosin solution, shows that this is a very fast reaction. The simplest hypothesis is that the ATP hydrolysis occurs after the dissociation of actin and is a reaction of the myosin rather than actomyosin. On the other hand, the product-release step by myosin alone is very slow and so it makes sense to suggest that this is the step that is accelerated by the formation of actomyosin. The following scheme is the simplest explanation of these results; it was first proposed by R. W. Lymn and Taylor:

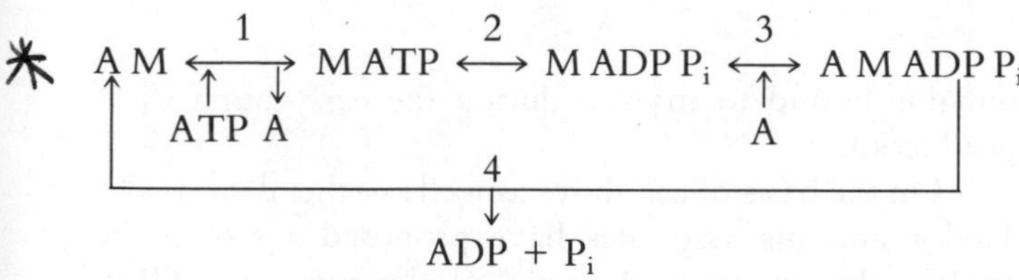

$$A\,M \underset{\substack{\uparrow\\ \text{ATP}\quad A}}{\overset{1}{\longleftrightarrow}} M\,ATP \overset{2}{\longleftrightarrow} M\,ADP\,P_i \underset{\substack{\uparrow\\ A}}{\overset{3}{\longleftrightarrow}} A\,M\,ADP\,P_i$$

$$\overset{4}{\longrightarrow} ADP + P_i$$

Experimental studies have supported this scheme and data are accumulating that, as in the scheme for myosin alone, introduce a number of conformational intermediates into an expanded sequence of events. There are still a number of uncertainties: The two heads of the myosin molecule may influence one another, either positively or negatively, in their reactions with actin and ATP. More importantly, there is not yet available a sufficiently detailed picture of the conformational states of the various intermediates that can be detected.

ATP is the primary energy source for a number of transductions occurring in the cell but ATP also exerts a regulatory role on a number of metabolic processes. This means that the exact concentration of ATP is of critical importance to muscle contraction and numerous other processes. Muscles perform not only steady work but, on occasion, must also operate in bursts. The muscle cell must therefore make energy from ATP available to the contractile machinery while at the same time maintaining the concentration required for its many regulatory functions. The *creatine phosphate shuttle* achieves this by storing ATP energy in a "pool" of creatine phosphate, synthesized from ATP and creatine, and releasing the energy when needed.

$$\text{ATP} + \text{creatine} \underset{\text{phosphokinase}}{\overset{\text{creatine}}{\rightleftharpoons}} \text{phosphocreatine} + \text{ADP}$$

This reaction helps to maintain the steady-state level of ATP despite its being rapidly utilized (Figure 16–13). The high demand for ATP during contraction is reflected structurally by the large number of mitochondria in muscle (see Figure 16–2b) to supply that ATP, another beautiful example of the regulatory system that governs the relations between cellular organelles.

Composition of the Filaments

The composition of the thick and thin filaments is particularly important to understand if we are to learn how the force that brings about the sliding movement is generated. It is here that we begin to bridge the gap between microscopy and biochemistry, a development that is so characteristic of recent trends in molecular cell biology.

Hanson and Huxley demonstrated very convincingly that after treatment of muscle with solutions that extract myosin, A-bands will disappear (as observed in the phase-contrast microscope; an experiment which Danilevskii had anticipated in 1888) and thick filaments will vanish (as observed in electron micrographs). This led Hanson and Huxley to the conclusion that the thick filaments are composed mostly of myosin.

If the remaining material is extracted with 0.6 M KI, the thin filaments disappear. Quantitative analysis showed that actin did not account for all the material in the thin filaments. From a variety of independent lines of evidence, it is clear that tropomyosin and troponin are also part of the thin filament.

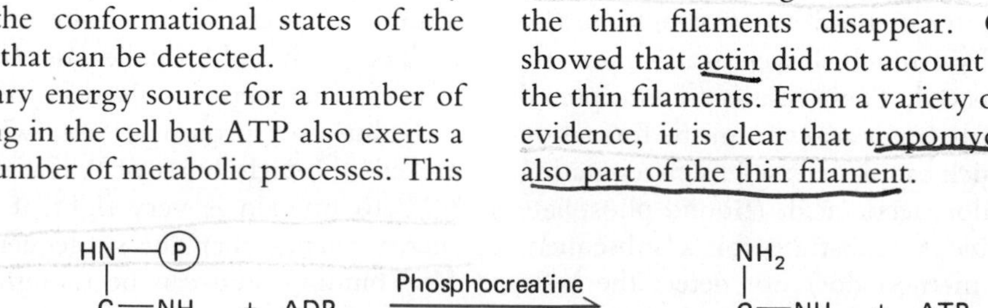

Phosphocreatine + ADP ⇌ (Phosphocreatine kinase) Creatine + ATP

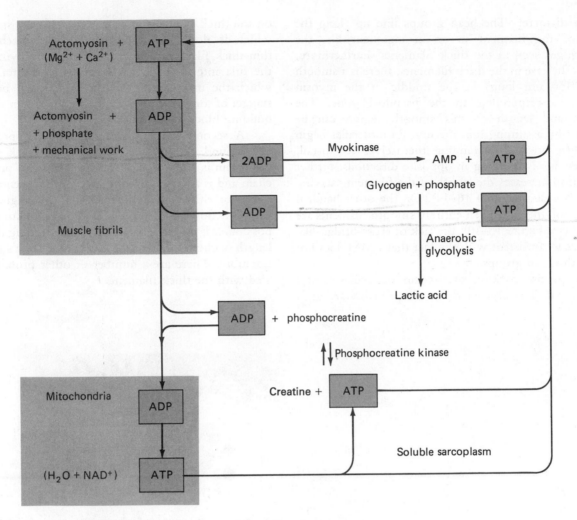

FIGURE 16–13
Energy Flow in Muscle. The muscle is seen here as a mechanochemical transducer, using energy from ATP to produce movement. The ATP is generated locally by four separate systems: (1) aerobically by oxidative phosphorylation in the mitochondria, (2) from phosphocreatine, (3) from ADP via myokinase, and (4) from glycogen-to-lactic acid anaerobic fermentation. (Redrawn from P. Seikevitz.)

Another prediction of the sliding filament theory is that the sarcomere must be *bipolar.* This could be due to an opposite polarity in the two halves of the thick filament, since the thin filaments slide over it in opposite directions; or opposite polarities could reside in the two sets of thin filaments of a sarcomere; or it might be that bipolarity is found in both the thick and the thin filaments. As it turns out, both filaments demonstrate the required bipolarity.

It can be demonstrated that actin filaments exhibit polarity. In order to account for the sliding mechanism of muscle contraction, however, there must be polarity of the thick filament as well as the polarity of the two sets of thin filaments in a sarcomere.

Structure of the Thick Filament

The myosin molecule has two small bulbous heads and a long tail. Heavy meromyosin—which contains the heads—has the ATPase and the actin-combining activity. Under the electron microscope, the thick filaments have easily detectable *cross-bridges* that interact with the actin-containing thin filaments (see Figure 16–3). It is therefore tempting to conclude that the cross-bridges represent the bulbous heads of the myosin molecule. In a dramatic series of experiments, H. E. Huxley was able to demonstrate that this is so. He showed that when myosin is allowed to aggregate at low ionic strength, it forms elongated filaments of variable length but reasonably uniform diameter. The head groups line up along the length of the filament in an arrangement like the cross-bridges seen in the thick filaments. Furthermore, just as is the case in the thick filaments, there is a smooth region (150 nm long) in the middle of the myosin filament, corresponding to the pseudo-H-zone. The presence and length of this smooth region can be explained by assuming that the myosin molecules begin their aggregation by arranging themselves tail-to-tail, with their heads pointing in opposite directions; further aggregation increases the length of the filament carrying the cross-bridges (Figure 16–14). On the other hand, if LMM is polymerized, then entirely smooth filaments are obtained (see Figure 16–10); the lack of cross-bridge-like structures is consistent with the fact that LMM does not contain the head groups.

The precise packing of myosin molecules is still controversial. The myosin heads are probably arranged on the thick filament in a helical array, the structure of which is doubtless important for the mechanism of thin-thick filament interaction. During polymerization, the first intermediate that forms is a myosin dimer in which the molecules point the same way, but with a stagger of some 43 nm. Presumably this dimer is the basic building block of the thick filament.

A second "inner" protein, named C-protein, has been found to be associated with the thick filaments. C-protein is composed of a large (140-kDa) polypeptide chain and is found at 43-nm intervals in seven transverse stripes on either side of the thick filament (Figure 16–15). The function of C-protein is as yet unknown. One possibility is that it plays a role in determining the precise length of the thick filament during the latter's process of assembly. There are a number of other proteins associated with the thick filament.

FIGURE **16–14**

Relationship Between the Structure of the Myosin Molecule and the Morphology of the A-band Filaments. (a) Aggregation of purified myosin molecules in vitro. Note the smooth area at the center, explained by assuming tail-to-tail aggregation as illustrated in the diagram. (b) The thick filament (myosin) has a smooth region (pseudo-H-zone) at the center, similar to the one seen in the aggregate above. The diagram, based on many such micrographs, illustrates the thick-filament morphology. (Courtesy of H. E. Huxley.)

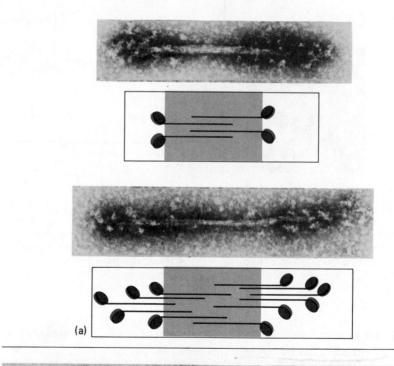

(a)

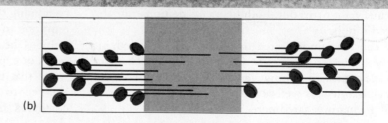

(b)

FIGURE **16-15**
Localization of C-protein on the Thick Filaments. The stripes are produced by antibodies that combine specifically with certain proteins to produce regions of high electron density, because of localized increase in mass, in the EM image. The seven stripes indicated have been identified as being specific for C-protein. The eighth and ninth stripes are each specific for a different protein. (Courtesy of F. A. Pepe and B. Drucker.)

Structure of the Thin Filament

The aggregation experiments by H. E. Huxley, described above, indicate the geometric basis for opposite polarity in the thick filaments, but what about the thin filaments? To answer this question, Huxley again performed an ingenious experiment. The assay was to mix actin filaments with heavy meromyosin and detect with the electron microscope the attachment of HMM to actin, to form "arrowheads." The direction in which the arrowheads point demonstrates the polarity of the actin filament. It can now be asked whether the thin filaments on opposite sides of the Z-line have opposite polarity. When the test was made on thin filaments that were extracted from muscle while still attached to their Z-lines, HMM formed arrowheads that pointed in opposite directions on opposite sides of the Z-line (Figure 16–16). Thus, Huxley was able to show that both types of filaments have the requisite polarity to account for the sliding theory.

The Z-line appears to have the function of attaching and properly orienting the thin filaments. It is composed mainly of the protein α-actinin, which consists of two 90-kDa chains. The thick filaments, on the other hand, appear to be held in register by the M-line, which contains a protein composed of two 44-kDa chains.

X-ray analysis definitely shows that the helical arrangement of F-actin (see Figure 16–11) is the structural basis for the thin filament. But the thin filaments do not only contain actin, they also contain the regulatory proteins tropomyosin and troponin.

Regulatory Proteins of Muscle

When myosin is in a solution of ionic content of pH similar to that of muscle, it has very low ATPase activity; when actin is added, a considerable increase in the rate of ATP hydrolysis is obtained. This is only true for pure actin and myosin preparations. If the actomyosin preparation is not very pure, then the high ATPase rate is not observed, unless Ca^{2+} is added to the solution. This effect of Ca^{2+} in bringing out "actin activation" can also be observed in the phenomenon of tension production. Threads prepared from pure actomyosin will exert tension when ATP and Mg^{2+} are added. However, threads of impure actomyosin will not exert tension unless Ca^{2+} is also included in the ATP and Mg^{2+} solution.

Perhaps the most reliable system for demonstrating the effect of Ca^{2+} on tension production is the glycerinated muscle. When ATP and Mg^{2+} are added to glycerinated myofibrils, no tension is developed. When

FIGURE 16–16
Opposite Polarity of Thin Filaments on Opposite Sides of the Z-disk. Thin filaments are decorated by heavy meromyosin. When thin filaments are still attached to the Z-disk, it can be shown that they have opposite polarity, because arrowheads of HMM point away from the Z-disk on either side of it. This opposite polarity is consistent with the sliding filament model, which demands that during contraction the thick filaments in adjacent sarcomeres converge towards the Z-disk. (Courtesy of H. E. Huxley.)

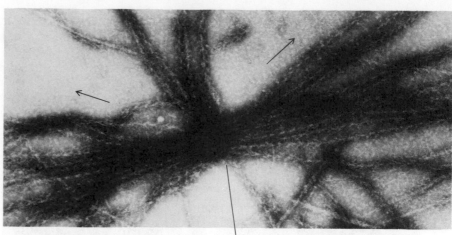

Z-line attachments of thin filaments

Ca^{2+} is added as well as ATP and Mg^{2+}, then as much tension can be developed as in live muscle.

Ebashi and his co-workers demonstrated that the influence of Ca^{2+} on ATPase activity and on tension production in (impure) actomyosin preparations depends on the presence of two proteins: *tropomyosin* and *troponin*. Tropomyosin and troponin together inhibit actin activation of ATPase activity and also tension production, an effect which Ca^{2+} reverses. This is why muscle is at rest even though ATP and Mg^{2+} are present in the muscle cell. Contraction is brought about when the muscle cell is stimulated by a nerve to release stored Ca^{2+} into the cell interior. The increase in $[Ca^{2+}]$ reverses the inhibition of actomyosin ATPase by tropomyosin and troponin. This phenomenon is referred to as the "regulation of muscle contraction." The properties of these two proteins allow them to be the agents of Ca^{2+}-induced actin-based regulation of contraction.

Tropomyosin

Tropomyosin was the first mechanochemical protein of muscle to be crystallized, and, in fact, it was this property that led to its discovery by Bailey. Tropomyosin is a 70-kDa fibrous protein (40×2 nm) composed of two 35-kDa polypeptide chains; it comprises about 8% of muscle cell protein. Like LMM, the two chains of tropomyosin form α-helices, which are wrapped around each other in a coiled coil. Most of the amino acid sequence of tropomyosin is known, and it substantiates a prediction by F. Crick that a coiled coil is likely to be stabilized by the presence of hydrophobic residues placed at 3–4 residue intervals. Tropomyosin shows a tendency to form strands through end-to-end association. Tropomyosin also binds to actin. Both these properties will be discussed later. It appears that tropomyosin has two Ca^{2+} binding sites of high affinity (association constant, K_A, about 2×10^6 M^{-1}) and two of low affinity (K_A about 4×10^4 M^{-1}).

Troponin

Troponin is an 80-kDa protein composed of three different polypeptide chains. Troponin has three important properties: (1) it binds to tropomyosin, (2) it leads to inhibition of actomyosin ATPase in the absence of Ca^{2+}, and (3) it is the only mechanochemical protein that binds Ca^{2+} in sufficient amounts and with sufficient tightness to account for most of the Ca^{2+} binding by the cell interior during active contraction. Interestingly, each of the three chains of troponin (TN) is responsible for one of the above properties:

- TN-T (MW 37,000) binds to *tropomyosin*,
- TN-I (MW 23,000) is involved in the *i*nhibition of actomyosin ATPase, and
- TN-C (MW 18,000) binds Ca^{2+}, $K_A = 10^6$ M^{-1}.

Tropomyosin, TN-I, and TN-T combine to inhibit the actinomyosin ATPase, with or without the presence of Ca^{2+}. The TN-C, when present, binds Ca^{2+} and relieves the inhibition.

Regulatory Proteins in the Thin Filament

Tropomyosin binds to actin directly; x-ray diffraction evidence suggests that the long (2×40 nm), fibrous tropomyosin molecules fit into the two grooves of the actin filament (Figure 16–17a). Each tropomyosin molecule appears to be in contact with seven actin monomers. Troponin shows a strong affinity for tropomyosin, and the TN-T chain of the troponin molecule is mainly responsible for the binding. Electron microscopic studies of tropomyosin crystals in the presence and absence of troponin suggest that the troponin molecule binds about a third of the way along the tropomyosin molecule

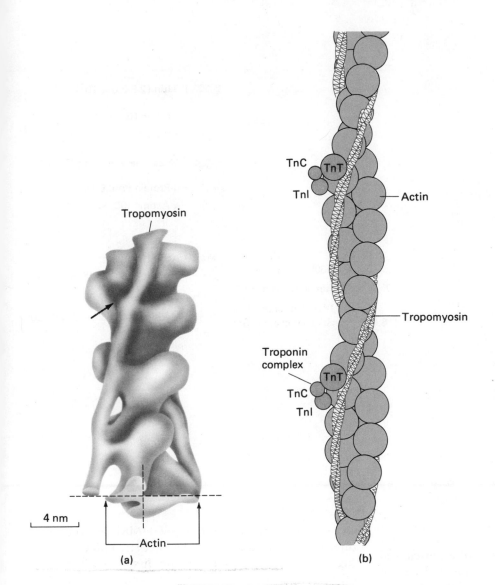

FIGURE 16–17

The Organization and Composition of the Thin Filament of Striated Muscle.   (a) Three-dimensional reconstruction of an optical diffraction pattern obtained from electron micrographs of a negatively stained paracrystal: actin-binding tropomyosin. The actin monomer is slightly elongated, with a larger and a smaller domain. The arrow marks the smaller domain of the actin monomer; the tropomyosin is attached there. (Courtesy of E. J. O'Brien, J. Couch, G. R. P. Johnsen, and E. P. Morris.) (b) Model summarizing a number of structural studies, showing the arrangement of actin, tropomyosin, and the troponin complex: TnC, TnT, and TnI.

(Figure 16–17b). The structure of the thin filament that emerges from these various studies agrees well with stoichiometry of the proteins present in the thin filament, which is actins:tropomyosin:troponin = 7:1:1.

Other Proteins in the Filaments

Figures 16–3 and 16–5 show the remarkable uniformity in length of the thick and thin filaments in the striated muscle cell and the precision of their alignment. When actin and myosin filaments are aggregated in vitro, they form structures of indeterminate length. Clearly there must be some additional structures, probably proteins, that allow the assembly of the filaments in muscle to proceed correctly.

Table 16–1 summarizes the information regarding the myofibrillar proteins discussed above. The polyacry-lamide gel of myofibrillar protein chains shown in Figure 16–18 suggests that there may be some additional minor components, the location and function of which must yet be determined. Two very high molecular weight proteins in the myofibril have recently been described by K. Wang. One of these is *titin*, which usually appears on polyacrylamide gels as a doublet with an apparent molecular weight of $2.5–3.0 \times 10^6$. The smaller member of the pair is probably a partially degraded form of this huge protein chain. Titin appears to be located mainly in the I-band region; it may anchor the thick filaments by linking them to the Z-disks, thus providing structural continuity in the sarcomere. Titin may be a component of the so-called third filament that has been observed in the myofibril by some workers. The other large protein, *nebulin,* is less well understood.

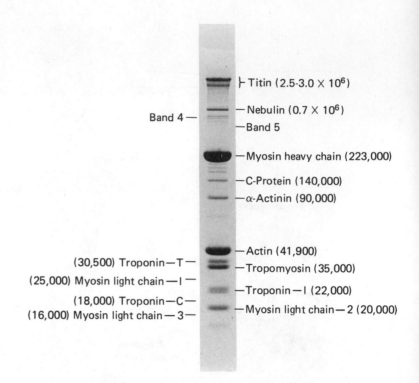

FIGURE **16-18**
Polyacrylamide Gel of the Major Proteins of the Skeletal Myofibril.
Numbers in brackets represent molecular weights of the protein chains. (Courtesy of K. Wang.)

Band 4 —

├ Titin (2.5-3.0 × 10⁶)
— Nebulin (0.7 × 10⁶)
— Band 5

— Myosin heavy chain (223,000)

— C-Protein (140,000)
— α-Actinin (90,000)

— Actin (41,900)

(30,500) Troponin—T —
(25,000) Myosin light chain —I —

(18,000) Troponin—C —
(16,000) Myosin light chain—3 —

— Tropomyosin (35,000)

— Troponin—I (22,000)
— Myosin light chain—2 (20,000)

Mechanism of Muscle Contraction

Having described the molecules of the muscle cell and the structures in which they are found and having given some preliminary details about the interactions of the molecules and about the sliding filament mechanism, we can now combine all this information to give an overview of current ideas on the mechanism of muscle contraction.

Resting State

Muscle tissue in the resting or relaxed state is soft and pliable; it can be *stretched* to almost twice the length of the contracted state. The intracellular fluid surrounding the myofibrils has an ionic concentration of 0.15 M, a pH of approximately 7.2, and an Mg^{2+} and ATP concentration of a few mM. Muscle cytosol Ca^{2+} concentrations are very low (10^{-8} to 10^{-7} M), but Ca^{2+} is stored at higher concentrations inside the sarcoplasmic reticulum membrane compartment. Studies with purified myosin, actin, tropomyosin, and troponin interactions in solutions of high ionic strength (where myosin is soluble),

and studies of the interactions of HMM and the other three proteins in solutions at the ionic strength of muscle, suggest that myosin and actin are dissociated from each other under resting conditions. Low-angle x-ray diffraction studies confirm that this is also the situation in resting muscle; that is, there does not seem to be any interaction between the thin and thick filaments. This explains why relaxed muscle is capable of being stretched. Furthermore, the ATPase activity of a relaxed muscle is very low.

When a muscle cell dies or when it is extracted with glycerol, and consequently loses its ATP, it becomes stiff; it is described as in a state of *rigor*. Studies of extracted actomyosin in solution show that the myosin heads are attached to the actin thin filament. High-resolution electron microscope pictures of muscle in rigor reveal cross-bridges sticking out from the myosin (thick) filaments in the neighborhood of (perhaps attaching to) the thin filaments (Figure 16-19). Low-angle x-ray diffraction studies of whole muscle in the rigor state confirm the above, showing that a certain proportion of the mass of the thick filament is then closer to the thin filament and probably attached to it.

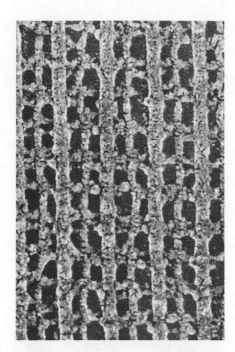

Muscle in Rigor.   The electron micrograph shows cross-bridges sticking out perpendicular to the myosin fiber. (Courtesy of J. E. Heuser.)

Initiation of Contraction

When a muscle cell is stimulated by a single nerve action potential, it responds within 1–3 milliseconds with a single *twitch*, lasting about 20–200 milliseconds. A. V. Hill pointed out that the response interval is so short there is not enough time for a substance to travel from the surface of a muscle fiber to its interior by diffusion. The connection of the transverse tubules to the plasma membrane, as well as the physical proximity of the transverse tubules to the sarcoplasmic reticulum, can explain the rapidity of the response of the muscle fibril to stimulation. The wave of excitation traveling down the nerve leads to the release of acetylcholine at the neuromuscular junction. This generates another excitation wave that travels rapidly along the membrane of the muscle cell and down into the transverse tubules; the excitation triggers the release of Ca^{2+}, causing contraction.

The Ca^{2+} appears to be released mostly from the terminal cisternae of the sarcoplasmic reticulum into the cytoplasm surrounding the myofibrils. This release is not understood in detail, but it can probably be described as due to the opening of a voltage-dependent permeability pathway that allows the Ca^{2+} to flow down its electrochemical gradient. The concentration of Ca^{2+} in the sarcoplasm rises from resting levels, between 10^{-8} and 10^{-7} M, to between 10^{-6} and 10^{-5} M; this allows the

actomyosin ATPase activity to be expressed and tension development to proceed. After the wave of excitation has passed, a pump in the sarcoplasmic reticulum membrane reduces the sarcoplasm Ca^{2+} concentration to the resting levels. The Ca^{2+} is believed to enter the longitudinal vesicles and thence to return to the terminal cisternae. One of the interesting features of Ca^{2+}-release is that it appears to be regulated by the load on the muscle; the heavier the load, the more Ca^{2+} is released.

Phosphocreatine Shuttle

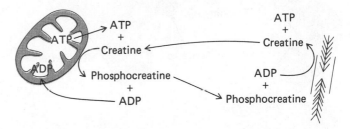

To maintain the muscle in a contracted state (tetanus), it must be stimulated repeatedly. A variety of lines of experimental evidence have demonstrated that contraction is initiated by the release of Ca^{2+} from the sarcoplasmic reticulum into the interior of the cell. There are therefore three distinct states in which muscle can find itself:

1. relaxed, Mg^{2+}, ATP;
2. rigor, no ATP;
3. contracting, Mg^{2+}, ATP, Ca^{2+}.

When muscle contracts, after the release of Ca^{2+} into the sarcoplasm, two things happen: ATPase activity goes up as much as 2000-fold and contraction occurs, which can be measured by the generation of tension.

How does Ca^{2+} exert this effect? It can be shown that Ca^{2+} binds mainly to troponin. X-ray diffraction analysis shows that relaxed and contracted muscle differ in the position of the tropomyosin molecule on the actin filament (Figure 16–20). In contracting muscle, the tropomyosin lies close to the groove formed by the double helical structure of the actin polymer (see Figure 16–17), but in relaxed muscle, the tropomyosin has moved away from the groove. The interpretation illustrated in Figure 16–20 suggests that when tropomyosin lies close to the groove, it allows the myosin heads to bind to actin, and, conversely, when tropomyosin lies away from the groove, it prevents the myosin heads from binding to actin.

The question now is, how does the binding of Ca^{2+} to troponin affect the movement of the tropomyosin molecule? It might be predicted that binding Ca^{2+} must produce a conformational change in the troponin

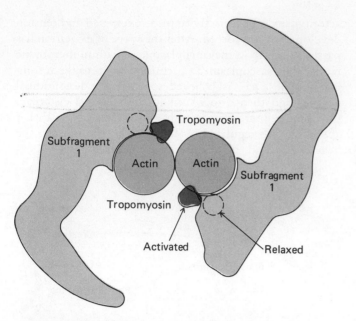

FIGURE **16-20**
**End-on View of the Relationship Between Actin,
Tropomyosin, and the S1 Heads of Myosin.** The
diagram shows how the position of tropomyosin shifts from
the activated to the relaxed state, thereby blocking the
attachment of the myosin heads. Based on a number of x-ray
diffraction studies. (Courtesy of H. E. Huxley.)

molecule, and extensive studies confirm this prediction.
Much must yet be done, however, to understand the
precise mechanics of the conformational change in troponin and the effect this has on the movement of
tropomyosin on the thin filament.

Movement of the Cross-bridges

The myosin heads (cross-bridges) can either be
attached to or detached from the actin filament. As the
muscle cell contracts, thin and thick filaments slide with
respect to each other, thereby increasing their degree of
overlap. The question is: how does the attachment and
detachment of cross-bridges relate to the sliding motion?
This question was answered by including two steps
involving *movement of the cross-bridges* in the sequence of
reactions that lead to sliding of the filaments, giving the
following four steps (Figure 16–21):

- Step 1. Attachment of myosin cross-bridges to actin.
- Step 2. Change in conformation of the attached
  cross-bridges to produce sliding movement, or if the
  muscle is held at constant length, to produce tension
  (the *working* or *power stroke*.)
- Step 3. Detachment of the cross-bridges.

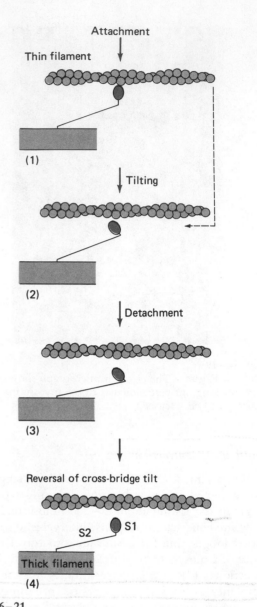

FIGURE **16-21**
Model Proposed by H. E. Huxley. This model suggests
how the sliding movement of filaments can be brought about
by a change in tilt of the cross-bridges. (Courtesy of H. E.
Huxley.)

- Step 4. Reversal of the cross-bridge conformational
  change (*recovery stroke*).

What evidence is there for this model of muscle
contraction? Let us take each step in turn.

1. *Attachment of the cross-bridges.* X-ray diffraction studies
   indicate that the myosin cross-bridges are unattached
   in the relaxed state and attached to the thin filaments in
   the rigor state. Though it is harder to obtain unequiv-
   ocal data on the contracting state, loss of order in the
   x-ray diffraction pattern associated with the cross-

bridges (consistent with their being in motion) and some apparent transfer of mass from the thick filament to the thin filament can be observed. This transfer of mass would account for one fifth of the cross-bridges being attached at any one time; this is not an unreasonable proportion, since in a cycle of attachment and detachment only a fraction of cross-bridges would be attached at a given time.

2. *Movement of the attached cross-bridges.* The cross-bridges in vertebrate muscle are not easily resolved in the electron microscope. In insect muscle, however, they can be seen much more clearly. In this material it is possible to observe that the cross-bridges in the relaxed state stick out at right angles and that in the rigor state they are tilted at an angle of 45°. This angle is very similar to the one observed in HMM-actin complexes (see Figure 16–11), and it therefore is reasonable to conclude that *the myosin head changes its angle of tilt during the power stroke*. Admittedly, the evidence for such a conclusion is limited and the change in the angle of tilt of the cross-bridges may not be the only conformational change occurring during the power stroke—for example, some conformational changes may occur in the actin of the thin filament and others may occur in the main body of the thick filament. Figure 16–21 illustrates Huxley's view of how a change in tilt of the cross-bridges might produce sliding motion of thin and thick filaments with respect to each other. This model has several features:

(a) It interprets the observation that a small region in the myosin tail is unusually susceptible to trypsin digestion by suggesting that this is the place where the subfragment 2 portion of the tail can bend away from the main body of the thick filament.

(b) Since subfragment 2, unlike LMM, does not associate strongly with itself, it is reasonable to expect that it can lift itself off the thick filament, like a hinge.

(c) Subfragment 2 is a thin structure that may not have sufficient rigidity to act as an "oar," but a change in tilt of the cross-bridge will exert a tension along the axis of the subfragment and it is not unreasonable to expect that the subfragment can withstand such tension.

(d) It is unlikely that the cross-bridges tilt in unison and it is therefore expected that when one cross-bridge tilts and the neighboring ones do not, some of the energy will go into deforming adjacent structures, such as stretching of subfragment 2, bending of the myosin head, and so forth. Such deformation would store energy that could be utilized for sliding of the filaments when a nearby cross-bridge detaches.

(e) When a myofibril contracts, it does so at constant volume, which means that the filaments move further apart during contraction. The hinge aspect of SF2 in the model allows the myosin heads to tilt over the same angle, even if the separation of the filaments increases during contraction.

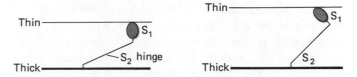

3. *Detachment of the cross-bridges.* In the presence of Mg^{2+} and ATP, detachment of cross-bridges occurs in solution, in glycerinated fibers, and in the living muscle cell. Without ATP the cross-bridges remain attached and the muscle cell is fixed in length; it is in rigor.

4. *Reversal of cross-bridge movement.* Electron spin resonance spectroscopic evidence suggests that the binding and hydrolysis of ATP are followed by some change in conformation of the cross-bridges. There are other data based on the fluorescence of tryptophan residues in HMM that suggest the same general conclusion. Furthermore, addition of the analogue of ATP known as AMPCPP (adenosine triphosphate with a CH_2 group replacing the oxygen between the α and β phosphorus atoms) allows dissociation of myosin from actin and a loss of rigor, but the dissociated cross-bridges are still tilted! This can presumably be observed because AMPCPP is hydrolyzed only 10^{-3} times as fast as ATP and the hydrolysis must occur before the conformational change from tilted to perpendicular. It is the perpendicular conformation that is seen, in the presence of ATP, when the chelator EGTA has been added to reduce $[Ca^{2+}]$ so actin cannot reattach.

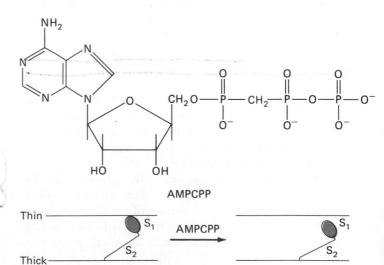

AMPCPP

Biophysics of the Striated Muscle Cell

The striated muscle cell is a highly evolved, extremely effective piece of machinery for the conversion of some of the energy of ATP into mechanical work. For instance, the flight muscle that powers the wing of a bee has a continuous energy output equivalent to that of a piston aircraft engine—it hydrolyzes about one half of its weight of ATP per minute (2.8 kJ/kg/s = 2400 kcal/kg/hr). It contracts at a speed equivalent to 10 times its length per second, reaching its maximum power output in milliseconds. A muscle 1 cm in cross section can exert a tension equivalent to 3 kg. In optimal conditions, the overall mechanical efficiency of muscle reaches 25%. Muscle is not only a very efficient piece of transducing machinery but also a delicately regulated mechanism in which the molecular events must be controlled with the greatest of precision.

Muscle has some unique physical properties, which were studied in the earlier part of this century:

- Maximum tension is exerted by muscle when it is held at a constant length. Although in the formal sense, mechanical work is equal to force times distance, muscle maintaining a force at constant length gives off heat and is said to do "internal work."

- If stimulated muscle is allowed to shorten, the tension it can exert is less than that exerted at constant length. The faster the rate of shortening, the lower the tension exerted.

- Muscle that is allowed to shorten liberates more heat than muscle held at constant length. The difference, called the "shortening heat," is proportional not to the rate of shortening but to the distance of shortening.

Since, for a constant distance of contraction, the shortening heat remains constant and since a muscle lifting a heavy weight does more work than when it is lifting a light weight, the total energy (heat plus work) that a muscle must expend will vary with the weight it lifts. Thus, the machinery determining the energy release in muscle is controlled not only by the distance of contraction but also by the tension the muscle experiences during contraction. This is a beautiful example of machinery that is regulated so as to adjust energy expenditure to the work that must be performed.

Many of these biophysical parameters can be explained in terms of the sliding filament theory. First, assume that the directional motion of the filaments is due to a cycle of reactions between the cross-bridges and the actin filaments: attachment, conformational change producing a directional force, detachment, and restoration of the conformational state that is present just before attachment. Thus, the more cross-bridges attached at any one time, the greater the force. Since there is a certain amount of time required to complete the cross-bridge cycle, the greatest number of attachments at a given sarcomere length occurs when there is no change in length; as the muscle shortens, some of the cross-bridge cycles will be in progress and the number of attachments that can take place at any one time will be smaller. This effect will be greater with increasing rate of shortening, and the force that is generated will be correspondingly lower. This is then how one might explain why the force exerted by contracting muscle decreases with increasing rate of shortening.

According to the above model, it would seem reasonable that energy is liberated only when a cycle of attachment-detachment occurs. Thus, only when a cross-bridge can contribute to tension is work being done, and there is a necessary connection between the work done by the muscle and the energy released.

Finally, the sliding mechanism can explain very nicely the shape of the length-tension curve observed in muscle (Figure 16–22). The amount of the isometric tension a muscle can exert depends on the degree to which it is stretched. Maximum tension is obtained when there is maximum overlap between thin and thick filaments. If muscle is stretched beyond this length, then the tension exerted decreases with increase in degree of stretching. This can be explained by the decreasing amount of overlap between the thick and thin filaments. Beyond a certain degree of stretch the tension reduces to zero, and this can be interpreted to represent the degree of stretch when the filaments cease to overlap. At the other extreme, when muscle has been allowed to shorten considerably, then tension also decreases. This can be shown to be due to the overlap of thin filaments coming from opposite directions. Tension falls to zero when the thick filaments come into contact with the Z-line and resist further shortening.

The evidence for the moving cross-bridge, sliding filament model of muscle contraction is steadily accumulating, but it would be misleading to claim that our understanding is complete. For instance, the possibility that conformational changes occur in parts of the system other than the cross-bridges has not been eliminated. Another mystery is whether the activity of one myosin head affects the activity of the other head on the myosin molecule. In other words, is there any cooperativity in the relationship between the two cross-bridges? Another issue that is not finally settled is whether the various intermediate states involving actin, myosin, ATP, and its hydrolysis products might include some covalent inter-

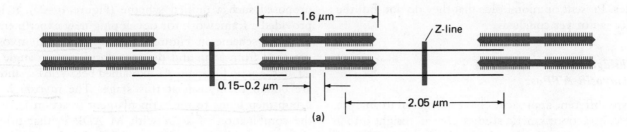

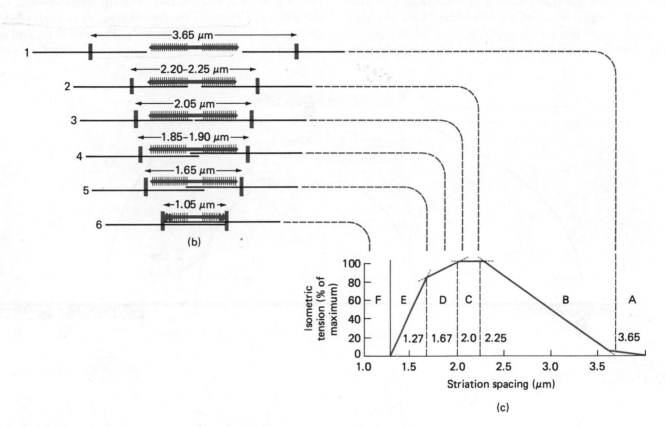

FIGURE 16–22

Length-tension Diagram of Striated Rabbit Muscle, in Relation to Degree of Filament Overlap.   (a) Precise dimensions of relevant filament structures. (b) Six progressively shorter sarcomere lengths that can explain the shape of the length-tension diagram: (1) no overlap between thin and thick filaments; (2) maximum overlap between thin filaments and region of thick filaments with cross-bridges; (3) maximum overlap between thin and thick filaments; (4) overlap between thin filaments equivalent to width of bridgeless region of thick filaments; (5) length associated with contact of thick filaments with Z-lines; and (6) length associated with crumpling of thick filaments. (c) Length-tension diagram. The tension associated with the stimulated muscle is graphed against the equivalent length of the sarcomere. The shape of the curve is related to degree of overlap between the thin filaments and the region of thick filaments that contain the cross-bridges. (Redrawn from A. Gordon, H. E. Huxley, and F. Julian.)

mediates. Present opinion holds that they do not, but the evidence is not yet conclusive.

Cross-bridge Cycle and Kinetics of Actomyosin ATPase

Two different approaches have been used to analyze muscle contraction: kinetic studies give an insight into a number of steps involved in the breakdown of ATP, and optical studies provide evidence for the sliding of filaments and movement of cross-bridges. Can the data obtained from these two separate lines of inquiry be unified into one consistent mechanism? Taylor and Lymn

proposed such a unified scheme (Figure 16–23); it has provided a framework for developing new experiments.

The scheme in Figure 16–23 begins with myosin detached from actin and the cross-bridge at an angle of 90° to the actin filament. As explained below, ADP and P_i are bound to myosin at this stage. The myosin head group then binds to the actin filament (reaction 1). It is the combination of actin with M ADP P_i that brings about the rapid release of P_i and ADP, thereby raising the rate of myosin ATPase action. During dissociation of P_i, there occurs a tilting of the cross-bridge to an angle of 45° (reaction 2, the "working stroke"). The state at the end of reaction 2 is analogous to a muscle in rigor. Upon

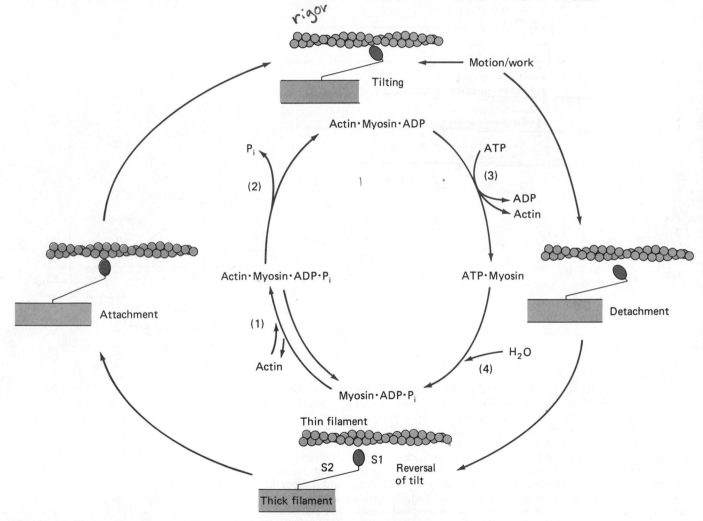

FIGURE 16–23
Scheme Similar to That Proposed by E. W. Taylor and R. W. Lymn That Attempts to Unify the Structural and Biochemical Information on Muscle Contraction.   Myosin, with bound ADP and P_i, attaches reversibly to actin (1). Upon dissociation of P_i from actin·myosin·ADP, a change in the angle of tilt of the cross-bridge occurs; this brings about relative motion of the actin filament with respect to the myosin filament (the power stroke, 2). Myosin (and ADP) dissociates from actin upon the addition of ATP (3). During hydrolysis of the ATP to ADP and P_i, the cross-bridge becomes perpendicular to the actin filament (the recovery stroke, 4). Myosin then attaches to actin again (1). Other schemes differ in allowing myosin·ATP and other forms of myosin also to bind to actin, but with different affinities.

addition of ATP, a rapid dissociation step occurs (reaction 3) leading to the detachment from actin of M ATP. This is followed by the hydrolysis of ATP (reaction 4) and the formation of M ADP P$_i$. At the same time the "recovery stroke" occurs with the myosin head group assuming a 90° angle to the actin filament.

It is worth pointing out that in the scheme shown in Figure 16–23, the hydrolysis of ATP and the actual working, or power, stroke do not occur simultaneously. This may be a general characteristic of mechanochemical coupling involving high-energy phosphates (recall the discussion in Chapter 13 of the translocation of mRNA during protein synthesis). While elaborate, this model of the cross-bridge cycle still leaves many unanswered questions. For example, the details of changes in the conformation of myosin that give rise to mechanical work are not yet understood.

The Actin-Myosin System in Smooth Muscles and in Nonmuscle Cells

So far we have discussed the striated muscles of vertebrates such as the rabbit, chicken, and frog. These animals have two other basic types of muscle; they differ from striated muscle not only in their fine structure but also in the structure and function of their mechanochemical proteins. These are the ceaselessly functioning muscles of the heart and the intermittently acting smooth muscles of the blood vessels, stomach, intestines, and uterus. The contraction of smooth muscles resembles in a number of ways the mechanochemical mechanisms of nonmuscle cells.

Contraction in Smooth Muscle Cells

Smooth muscle differs from striated muscle in that the entire cell, rather than the sarcomere, is the unit of contraction. Also, the fine structure of smooth muscle is much less regular than that of striated muscle. The myosin-containing thick filaments of smooth muscle are frequently oriented in parallel arrays, but they are in much lower concentration (actin:myosin=15:1) than those of striated muscle. Transverse sections of smooth muscle (Figure 16–24) show thick filaments (15-nm diameter) surrounded by an irregular array of actin-containing thin filaments (7–8-nm diameter). Smooth muscle also contains *intermediate filaments* (10-nm diameter). Both the thin and the intermediate filaments are very long; they traverse large portions of the cells and eventually attach to dense bodies (plaques) found in the cytoplasmic matrix or adjacent to the plasma membrane (Figure 16–25). The most prominent intermediate filaments in smooth muscle, as demonstrated by immunofluorescent techniques, contain the protein *desmin*

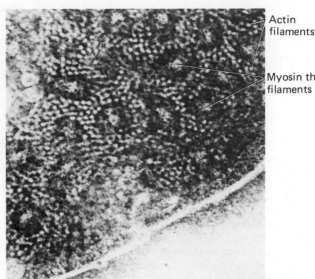

Actin filaments

Myosin thick filaments

FIGURE 16–24
Myosin and Actin Filaments in Smooth Muscle.
Transverse section of smooth muscle showing myosin thick filaments surrounded by actin thin filaments. (Magnification ×100,000; courtesy of A. V. Somlyo and A. P. Somlyo.)

(50 kDa), a protein also found at the Z-lines of striated muscle.

It is not known whether the intermediate filaments are actively involved in the contraction process; they might provide a framework for the attachment of the force-generating filaments, and/or they might be involved in the control of changes in cell shape. Another protein, *filamin,* has also been identified in smooth muscle. Filamin is a very large molecule composed of two protein chains, each of mass 250 kDa. It associates with actin and is found in a constant ratio to actin in various smooth muscles. In view of the low amount of myosin in smooth muscle, it is possible that filamin also participates in the force-generating mechanism.

Myosin-Based Regulation of Contraction by Ca^{2+}

Just as in striated muscle, the calcium ion is the initiator of the contraction process. In smooth muscle, however, there is *myosin-based* control by calcium. When the concentration of intracellular Ca^{2+} rises, the 20-kDa light chain of myosin becomes phosphorylated, by ATP, in a reaction catalyzed by a *myosin light chain*-specific protein kinase (MLC kinase). Phosphorylated smooth muscle myosin exhibits an actin-stimulated ATPase activity and is active in contraction. Relaxation of the muscle occurs via a phosphatase activity that removes the phosphate from the light chain (Figure 16–26). The contraction activity of a smooth muscle cell thus depends

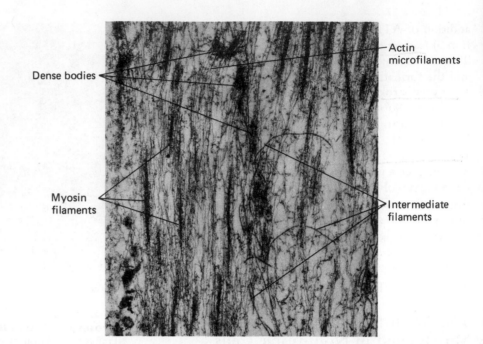

FIGURE **16–25**
Interaction of Actin with Dense Bodies and Intermediate Filaments.
Longitudinal section of smooth muscle showing relationship between actin microfilaments, dense bodies, and intermediate filaments. The actin filaments insert on both sides of the dense bodies and run to the myosin filaments. The intermediate filaments interconnect the dense bodies and probably hold them in register with respect to each other. (Courtesy of A. V. Somlyo and A. P. Somlyo.)

on the balance of activities of the kinase and the phosphatase.

Calcium control is exerted by its effect on the activity of the kinase. When the intracellular Ca^{2+} concentration rises to about 10^{-5} M, a complex is formed between calcium and the small calcium-binding protein known as *calmodulin* (see also Chapter 14). The $Ca^{2+} \cdot$ calmodulin complex, in association with the MLC kinase catalytic protein chain, forms the active enzyme. When intracellular $[Ca^{2+}]$ falls to the usual resting levels of 10^{-7} M, the kinase is inactive, probably because the calmodulin dissociates from the enzyme. Similar myosin-based mechanisms of actomyosin control occur in non-

muscle cells such as platelets, macrophages, HeLa cells, and possibly many types of cells in lower organisms.

Another protein found in smooth muscle cells, known as *caldesmon* (140 kDa), may play a role in actin-based contraction control. Caldesmon binds to actin and tropomyosin. This combination does not stimulate myosin ATPase. In the presence of Ca^{2+} concentrations sufficient to bind calmodulin, the caldesmon tends to dissociate from the actin and the myosin ATPase can be activated. Caldesmon is also found in nonmuscle cells where it presumably plays a similar role. Antibodies against caldesmon are available and will allow further dissection of its function.

FIGURE **16–26**
Scheme Proposing How Ca^{2+} Regulates Contraction in Smooth Muscle Through the Phosphorylation of Myosin.   The phosphorylation of myosin presumably promotes contraction, since actin-activated myosin ATPase can be detected in the phosphorylated system. (After R. S. Adelstein and D. J. Hartshorne.)

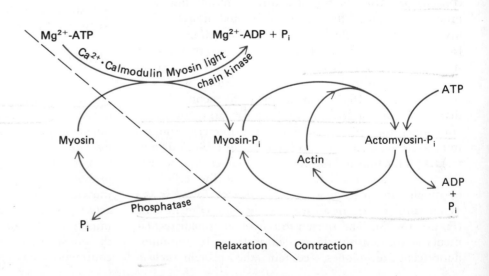

Primitive Motility in Nonmuscle Cells

The contraction of striated muscle is just one of many examples of mechanochemical transduction found in biological organisms. In recent years cell biologists have begun to study a large variety of phenomena involving the conversion of chemical energy into work, utilizing some of the techniques and insights gained from the study of muscle. In doing so, they have not only extended greatly the applicability of the actin-myosin system to a large number of cell biological phenomena but they are also showing that a comparative study of these systems aids the basic understanding of the mechanism by which chemical energy is changed into mechanical work.

One of the earliest observations in this broader study of mechanochemical phenomena was the extraction in 1952 of an actomyosin-like protein from the plasmodium of the slime mold *Physarum polycephalum*. This organism has the appearance of a yellow amoeba of gigantic size—it can weigh several grams—and is capable of very rapid *protoplasmic streaming;* rates as high as 1 mm per second can be observed. The actomyosin-like extract from this organism exhibits a very rapid drop in viscosity upon addition of ATP, followed by a slower rise in viscosity (Figure 16–27). The rise in viscosity parallels a release of inorganic phosphate from the ATP, suggesting that the ATPase activity of the extract accounts for the viscosity increase. S. Nakajima purified the protein complex responsible for the ATP-induced viscosity change and showed that the very same complex also carried the ATPase activity.

In 1966, S. Hatano and F. Oosawa succeeded in purifying an actin and a myosin from the slime mold; thus the modern era of general, or "primitive," cell motility began in earnest. Amoeboid cells like the slime mold are believed to have undergone very little change during a long span of evolution. The presence of actin and myosin in an amoeboid organism is, therefore, taken as a dramatic example of the principle that recent evolutionary developments in higher organisms have often made use of molecular systems that had developed in generalized cells at very much earlier periods in the history of life. The question therefore arises as to the similarity between the "primitive" and modern counterparts of actin and myosin, and the distribution and activity of these proteins in cells of varying lineage and function.

Actin and Its Distribution in the World of Life

Hatano and Oosawa purified slime mold actin by exploiting some of its biological properties. They added myosin from a rabbit muscle to actin from the slime

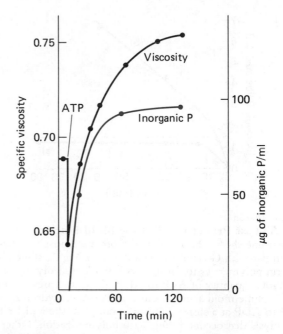

FIGURE 16–27

Evidence for the Presence of an Actomyosin-like System in a Protein Extract of the Slime Mold *Physarum Polycephalum*. Upon addition of ATP, a very rapid drop in viscosity is obtained, which is followed by a slow rise. ATP hydrolysis, as indicated by the appearance of inorganic phosphate, parallels the rise in viscosity. (From A. G. Loewy.)

mold; the myosin reacted with the actin to form "decorated" thin filaments, such as the ones shown in Figure 16–11. These filaments could then be purified by ultracentrifugation or by precipitation at low ionic strength; finally, the actomyosin complex could be dissociated with Mg^{2+} plus ATP and the actin then separated from myosin by ultracentrifugation.

Slime mold actin prepared in this manner, though low in yield, has startlingly similar properties to the actin of striated muscle. It polymerizes from the G- to the F-form, converting 1 mole of ATP to ADP for every mole of G-actin:

$$n \text{ G-actin} + n \text{ ATP} \xrightarrow{0.3 \text{ M KCl}} (\text{F-actin ADP})_n + n \text{ P}_i.$$

Furthermore, slime mold F-actin has the same molecular weight and forms a thin filament with the same helical appearance as muscle actin. Studies of the amino acid sequence of slime mold actin show that it is very similar to the actin of rabbit skeletal muscle.

There was, however, one surprise that came out of the study of slime mold actin. When its G-actin polymerizes in the presence of Mg^{2+}, the actin itself becomes an ATPase, continuing the breakdown of ATP at a steady rate (Figure 16–28).

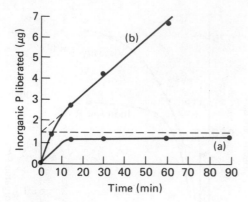

FIGURE **16–28**
The ATPase Properties of Slime Mold F-actin. (a) The broken line shows the amount of inorganic phosphate that is equimolar with G-actin. In the absence of Mg^{2+}, slime mold G-actin polymerizes to form F-actin while hydrolyzing an equimolar quantity of ATP to ADP. (b) In the presence of Mg^{2+}, slime mold actin becomes an ATPase, hydrolyzing ATP to ADP at a steady but lower rate than the initial rate of hydrolysis that occurs during actin polymerization. Other results confirm that actin polymerization does indeed go to completion in the first (15-minute) interval of rapid ATP hydrolysis. (Courtesy of F. Oosawa.)

$$ATP + H_2O \xrightarrow[\text{2 mM MgCl}_2]{\text{actin, 0.1 M KCl}} ADP + P_i.$$

Hatano discovered that this ability of actin to break down ATP is brought about by the presence of another protein that was present in the actin preparations and that he named *plasmodium actinin*. The "Mg-polymer" formed by actin in the presence of plasmodium actinin has a much more flexible structure than the F-actin thin filaments of skeletal muscle. This flexibility appears to decrease, however, upon addition of ATP. As discussed later, actin can play mechanochemical roles in the absence of myosin and it could well be that the Mg^{2+} polymer is of importance in this connection.

Since this important work by Hatano and Oosawa, actin has been identified as one of the most widely distributed proteins of the eucaryotic cell and one that is being implicated in a large variety of important biological functions. Three technical developments have helped establish the ubiquitous occurrence of actin and the numerous roles it plays:

- The use of the actin-myosin interaction to aid the fractionation of actin led to an interesting modification of this technique in the hands of H. Ishikawa, R. Bischoff, and H. Holtzer. They treated thin-sectioned biological material with HMM or with subfragment 1 of myosin, and were thereby able to identify actin filaments in the electron microscope by

FIGURE **16–29**
The "Arrowhead" Test for the Presence of Actin Thin Filaments in the Cytoplasm of Nonmuscle Cells. An actin filament from the plasmodium of the slime mold *Physarum polycephalum* was decorated with S1 heads of rabbit muscle myosin. Arrows indicate periodicity. ×200,000. (Courtesy of V. T. Nachmias, H. E. Huxley, and D. Kessler.)

virtue of their "decoration" by HMM or SF1. This "arrowhead test" for actin was the first frequently utilized, reliable assay for the presence of F-actin filaments in various parts of the cell (Figure 16–29).

- Acrylamide gel electrophoresis of most eucaryotic cells reveals a major band in the 42-kDa region (Figure 16–30). In a number of cases, independent evidence based on direct fractionation of cellular proteins or reactivity with antibodies raised against muscle actin has confirmed that this band is actin. In the slime mold, for instance, 15–20% of the total protein is actin and high amounts (5–10%) are present in such different cell types as guinea pig granulocytes, human platelets, and in a variety of nerve tissues, such as bovine brain.

- For many years, it was practically impossible to prepare antibodies of high titer against actin, presumably because it is so universally distributed and

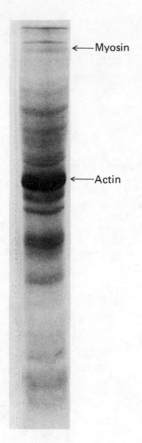

← Myosin

← Actin

FIGURE 16-30
The Presence of Actin in a Nonmuscle Cell.
Polyacrylamide gel of an extract of the slime mold *Physarum polycephalum.* Approximately 20% of the total protein of the multinucleate plasmodium of this slime mold is actin and only 0.75% is myosin. (Courtesy of D. Kessler, V. T. Nachmias, and A. G. Loewy.)

so highly conserved in all organisms. It was therefore a great surprise when K. Weber and E. Lazarides succeeded in obtaining antibodies against actin that had been denatured by detergent. By using *indirect immunofluorescence* techniques (see Chapter 7), it is possible to visualize numerous bundles of actin filaments in cells and to relate these to details in cell structure and function.

The above experimental approaches have been used to study the distribution and behaviors of actin in nonmuscle cells at various stages of their life history. What emerges from this work is a striking difference between muscle and nonmuscle actin, especially in the latter's dynamic behavior. The thin filaments of skeletal muscle have very low turnover in either structural appearance, or chemical composition. But the actin-containing structures of a nonmuscle cell are continually forming and breaking down, under exquisite control.

During interphase, for example, a cultured cell such as a fibroblast can be in a resting (nonmotile) state. In this condition the cell is spread out and lies close to the surface of the culture dish, making numerous points of contact with it. When such a cell is stained with the indirect antiactin immunofluorescent technique, its actin filaments are found to be organized into linear bundles, or *stress-fibers,* which often extend the entire length of the cell (Figure 16–31a).

When a nonmotile fibroblast changes into a moving cell, the leading edge flattens into sheet-like extensions (*lamellipodia*), which exhibit a wave-like motion resembling the ruffle of a dress (Figure 16–32). These lamellipodia are also able to extend finger-like projections (*filopodia*) which make contact with the substratum and shorten, thereby pulling the main cell body forward. When a moving cell is examined with the antiactin indirect immunofluorescence technique, one can see, in addition to some stress-fibers in the main body of the cell, a diffuse mesh of much thinner actin filaments, especially in the region of the lamellipodia (see Figure 16–31b). These cells do not appear to contain myosin filaments in their cytoplasm.

Equally dramatic differences in actin filament structure can be seen in other cells. In cultured kidney cells (Pt-K2), stress-fibers are located in the nonmotile central region of the cell; whereas, in the ruffles or pseudopods, which are the motile portions of the cell, the actin is present in a diffuse mesh of much thinner filaments (Figure 16–33). The role played by actin in these variously organized cytoskeletal structures is not clear, but it is nevertheless obvious that actin can be arranged into a variety of structures during interphase, depending on whether the cells are motile or stationary, isolated or confluent.

As the cell rounds up before mitosis, the stress-fibers disappear and some of the actin shows up in the pole-to-chromosome fibers of the mitotic spindle (Figure 16–34a). After mitotic division has been completed, a contractile ring forms which can be shown to contain actin by using indirect immunofluorescence (Figure 16–34b). Finally, when cell division is complete and the daughter cells flatten against the substratum, actin-containing stress-fibers reappear. Clearly, actin must be playing a variety of roles in the structure and motion of cells and cell components at various stages of their life cycle. Furthermore, the situation must be even more complex than described above because it is likely that at any one time the cell may be engaged in more than one type of force- or structure-generating function in which actin is involved. The molecular mechanisms of these structural gymnastics are one of the most fascinating areas of future research in cell biology.

(text continued on page 792)

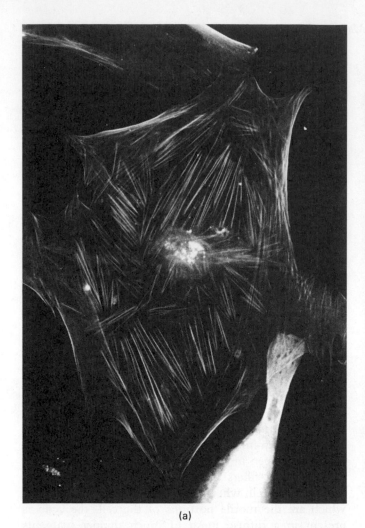

(a)

(b)

FIGURE **16–31**
Cells Stained with the Indirect Antiactin Immunofluorescent Technique.
Cells are first exposed to rabbit antiactin antibody; then the antibody-coated actin
fibers are stained with goat antirabbit antibody (directed against rabbit antibody)
that is coupled to a fluorescent dye. (a) A stationary cell showing actin filaments
extending to the edge of the cell. (b) A motile cell showing actin filaments
extending to the edge of the cell. (×6000; courtesy of E. Lazarides.)

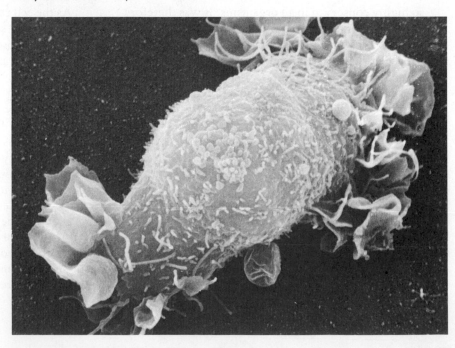

FIGURE **16–32**
**Scanning Electron Micrograph of a
Rat Sarcoma Cell.** The cell is
forming ruffles (lamellipodia) and
projections (filopodia). (Magnification
×4,000; courtesy of K. R. Porter.)

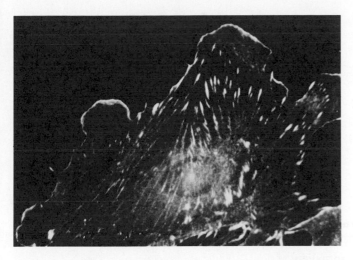

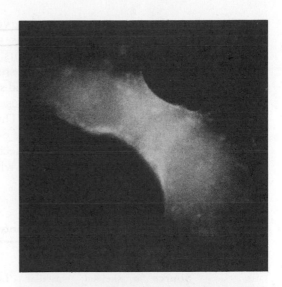

FIGURE **16–33**
The Effect of Cell Motility on the Organization of Actin Filaments. This Pt-K2 kidney cell was reacted with an antibody directed against α-actinin, which stains the dense bodies to which the stress-fibers are connected. The stress-fibers are composed of actin. The attachment plaques form a punctate pattern along the stress-fiber. Note that the stress-fibers are located in the central nonmotile portion of the cell. The ruffles or pseudopods are the motile portion of this cell, and there the α-actinin is attached to a diffuse mesh of much thinner actin filaments. (×4000, courtesy of J. W. Sanger and J. M. Sanger.)

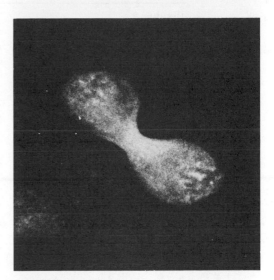

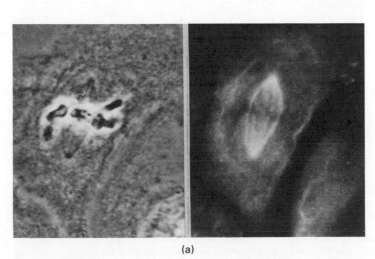

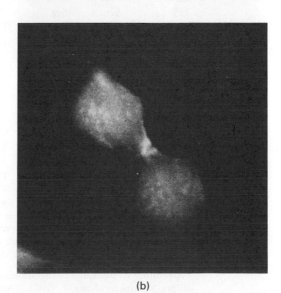

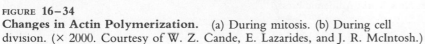

(a)   (b)

FIGURE **16–34**
Changes in Actin Polymerization. (a) During mitosis. (b) During cell division. (× 2000. Courtesy of W. Z. Cande, E. Lazarides, and J. R. McIntosh.)

The universal presence of actin in eucaryotic cells is one of the most striking qualities of this remarkable protein. Another is the invariability of its molecular weight and amino acid content. Indeed, the studies of M. Elzinga and his co-workers on the amino acid sequence of various actins reveal it to be one of the most conserved protein molecules in nature (Table 16–2). This may not be overly surprising if the constraints on the evolution of a protein that plays so many roles in the mechanochemical process are considered: actin interacts with itself to form a helical polymer; it binds ATP and converts it to ADP; it forms a complex with tropomyosin and troponin; it associates with myosin and modifies its ATPase activity; and it interacts with a number of other proteins. On the other hand, how can the highly conserved amino acid sequence of actin be reconciled with the many functions it has both in the same cell and in different cell types? Two possibilities come to mind, both of which are at present under intensive investigation: (1) a variety of slightly different kinds of actin molecules exist, each with

TABLE 16–2

Variability of Amino Acid Content of Different Actins. *

Actin Source	Mammalian		Lamprey	Mammalian β-nonmuscle	Sea Urchin	Scallop
	skeletal	vascular				
Residue #						
1	Asp	Glu	Asp	–	–	–
2	Glu	Glu	Asp	Asp	Asp	Asp
3	Asp	Glu	Glu	Asp	Asp	Asp
4	Glu	Asp	Glu	Asp	Glu	Glu
5	Thr	Ser	Thr	Ile	Val	Val
6	Thr	Thr	Thr	Ala	Ala	Ala
10	Cys	Cys	Cys	Val	Val	Val
16	Leu	Leu	Leu	Met	Met	Met
17	Val	Cys	Cys	Cys	Cys	Cys
76	Ile	Ile	Ile	Val	Val	Val
89	Thr	Ser	Thr	Thr	Thr	Thr
103	Thr	Thr	Thr	Val	Val	Val
129	Val	Val	Val	Thr	Ala	Ala
153	Leu	Leu	Leu	Met	Leu	Leu
162	Asn	Asn	Asn	Thr	Thr	Thr
176	Met	Met	Met	Leu	Leu	Leu
201	Val	Val	Val	Thr	Thr	Thr
225	Asn	Asn	Asn	Gln	Gln	Asn
235	Ser	Ser	Thr	Ser	(Ala)†	Ser
260	Thr	Thr	Thr	Ala	Thr	Ser
265	Ser	Ser	Ser	Ser	Ala	Ser
267	Ile	Ile	Ile	Leu	Ile	Leu
272	Ala	Ala	Ala	Cys	Ala	Ala
279	Tyr	Tyr	Tyr	Phe	Tyr	Tyr
287	Ile	Ile	Ile	Val	Ile	Val
297	Asn	Asn	Asn	Thr	(Thr)	Thr
299	Met	Leu	Leu	Leu	(Leu)	Leu
303	Thr	Thr	Thr	Thr	(Ser)	Thr
306	Tyr	Tyr	Tyr	Tyr	Tyr	Phe
324	Thr	Thr	Thr	Thr	Ser	Thr
358	Thr	Ser	Thr	Ser	Ser	Ser
365	Ala	Ala	Ala	Ser	Ser	Ser

Residues between parentheses mean residues whose position has to be confirmed by direct sequence analysis. The cytoplasmic actins have a deletion at position 1. Residues formerly designated 234a–374 (Lu and Elzinga, 1977) are now numbered here as 235–375.

*A number of positions in the amino acid sequence, showing degree of conservation in various actins. (From Vandekerckhove, J. (1983) "Actin—structure and function." In *Muscle and non-muscle cells*, G. G. dos Remedios and J. A. Barden, eds. New York: Academic Press.)

its own specific functions, or (2) actin can interact under different conditions with different types of accessory proteins (*actin modulator* or *binding proteins*), thereby increasing its range and variety of activities. We shall discuss the various actin-binding proteins later.

Different Actins

The separation of actin from other cellular proteins by acrylamide gel electrophoresis in an anionic detergent like sodium dodecyl sulfate (SDS) revealed not only that actin is a major component of the cell's proteins but also that its molecular weight is remarkably constant throughout the plant and animal kingdoms. A different separation technique, however, one that is sensitive to differences in the isoelectric points of proteins (isoelectric focusing), has revealed that there are at least three different isoelectric forms of actin in vertebrate cells. Skeletal muscle contains the acidic α-actin, while brain, liver, and platelets contain the more basic β- and γ-actins. Analysis of nonvertebrate cells has revealed other isoelectric species of actin—for example, the cellular slime mold *Dictostelium discoidum* has an actin form that is even more acidic than α-actin.

Careful comparisons of the small differences in amino acid sequence among various actins reveal differences that may be related to differences in function. Although the majority of amino acid substitutions are structurally conserved changes—that is, they are changes between amino acids of similar physical properties—there are a few substitutions that could be functionally significant. Consistent with this hypothesis is the finding that differences in amino acid sequence among actins obtained from different tissues in the same organism are greater than sequence differences in actins from the same tissue but from different species.

These subtle changes in actin amino acid sequence may eventually provide a clue to the difference in behavior between skeletal muscle actin, smooth muscle actin, platelet actin, and brain actin. But how can the presence of a variety of actin-based processes within the same cell be explained? The approach to this problem was hampered for some time by the experience that purification of nonmuscle actin yielded only a small fraction of the amount revealed by acrylamide gel electrophoresis. Since the isolation methods depended on interaction with muscle myosin fractions, it was possible that nonmuscle actins with different combining properties might be missed. More general isolation techniques have now been developed and more general assays for actin can be used. One of the latter involves the curious finding that actins specifically inactivate pancreatic deoxyribonuclease I (DNAse I). Whatever functional relevance this property might have in vivo, it provides a way of assaying the actin content of a cell fraction that is independent of

interactions with muscle-derived proteins. Using these general techniques it has been possible to increase the yields of extracted actin to over 30%, and the protein biochemists are now more reassured that they are studying a representative sampling of the important nonmuscle actins.

These increased yields have made it possible to compare the different forms of actin prepared from nonmuscle cells, but so far no functional differences have been found among them; they all are decorated with HMM to the same extent, for example. Why there should be multiple forms thus remains a mystery for the moment. It is suspected that critical differences may reside in the interactions with a variety of nonmuscle proteins, involving functions other than the ones presently studied (see Figure 16–51).

Myosin and Its Distribution in the World of Life

While actin has been shown to be a ubiquitous component of eucaryotic cells, it has been much more difficult to demonstrate the presence of myosin in various nonmuscle cells. The reason for this is twofold: (1) Myosin is present in much lower concentrations than actin in nonmuscle cells (0.3–1.5% of total protein) and consequently is very difficult to observe by electron microscopic techniques. (2) Myosin and the regulatory proteins controlling its activity are much more variable in their properties than is actin. Nevertheless, it has been possible to isolate myosins from various cell types and they all share at least three properties: the ability to bind with muscle actin filaments, the possession of an actin-activated ATPase activity, and a composition of both heavy and light chains.

It is too early to provide a thorough catalogue of the range of properties observed in myosins from different cell types: Molecular weights can vary, from 140,000 (*Acanthamoeba* Myosin I) to 460,000 (*Physarum polycephalum*); the ability to form bipolar thick filaments can vary, from skeletal muscle, which forms very large ones, to *Physarum*, which forms very small ones, to *Acanthamoeba* Myosin I, which apparently does not polymerize at all; the specific activity of myosin ATPase, its actin activation, and its response to ionic strength and pH vary greatly in myosins from different origins.

Regulation of the Actin-Myosin Interaction in Nonmuscle Cells

We have already discussed the actin-based regulation (by Ca^{2+}) of the actin-myosin interaction in striated muscle and the myosin-based regulation (also by Ca^{2+}) in smooth muscle. It has become increasingly clear that there is a great deal of variation in the regulatory machinery of the muscles of lower forms and in non-muscle cells. Some of these mechanisms involve Ca^{2+}-binding proteins, others appear to be related to phosphorylation mechanisms.

Ca^{2+}-Dependent Regulation

Many mechanochemical processes in the cell appear to be influenced by calcium. Evidence for the role of calcium includes the requirement of calcium in the growth medium for fibroblast motility; the effects of calcium on the division of glycerol-extracted cells; and the requirement in the presence of ATP of 10^{-6} M calcium for the contraction of demembranated microvilli isolated from the intestinal epithelium.

To implicate calcium more directly in the regulation of actomyosin, three results are relevant:

- Tropomyosin or tropomyosin-like proteins have been identified in nonmuscle cells—for example, platelets, fibroblasts, and brain cells (Figure 16–35).
- A major Ca^{2+}-binding protein, calmodulin, has been found in a number of cells derived from chick embryos, including brain and other neural tissues.
- There is experimental evidence that Ca^{2+}-binding proteins are involved in activating the actin-myosin ATPase activity but not the myosin ATPase activity.

Phosphorylation-Dependent Regulation

The regulation of the actin-myosin interaction in smooth muscle is mediated by a kinase that phosphorylates one of the myosin light chains and a phosphatase that dephosphorylates it. The kinase requires Ca^{2+} plus calmodulin to stimulate its activity. Similar mechanisms of control have been found to operate in nonmuscle cells. For instance, R. S. Adelstein has done a detailed study of the platelet system and was able to show that, in the presence of dephosphorylated myosin, activation of the myosin ATPase by actin could not be observed.

These results on phosphorylation have attracted considerable attention, and much recent work is concerned with demonstrating how many of the various proteins of the actin-myosin mechanochemical system are capable of being phosphorylated by specific kinases or otherwise modified covalently by regulatory enzymes. (The covalent modifying enzymes themselves are also capable of being regulated, as discussed in Chapter 14.)

In summary, research on the multiplicity of mechanisms involved in nonmuscle mechanochemical activities is still in its infancy. Enough has already been achieved, however, to conclude that actin and myosin play an important role in a number of these cell processes.

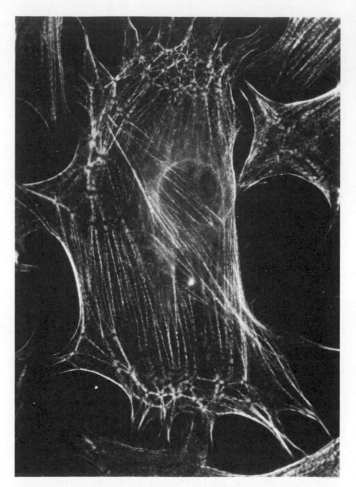

FIGURE 16–35
Location in Nonmuscle Cell of Tropomyosin. Gerbil fibroblast cells were fixed and then stained with a monoclonal antibody against muscle tropomyosin. The location of the antitropomyosin antibodies was determined in the ultraviolet microscope after staining again with antibodies that react with the structure of the first antibodies and to which fluorescent dye was attached, the technique known as indirect immunofluorescence. The banded fluorescence on bundles of microfilaments is characteristic of tropomyosin. ($\times 6000$; courtesy of J.J.C. Lin.)

Motility and the Tubulin-Dynein System

The structure of cilia and flagella was discussed briefly in Chapter 2 and the assembly of microtubules in Chapter 15. This section briefly describes some of the experiments that demonstrate that the *bending of cilia and flagella* is brought about by the *sliding of microtubules against each other*.

In spite of the much earlier origins of cilia and flagella, as compared with that of muscle, the phenomenon of ciliary motion is much more complex than that of muscle. The motion of muscle is essentially a contraction and relaxation. Cilia, in contrast, can move in planar, helical, and three-dimensional space; they can pull the cell forward or push it backward; the waves of the ciliary beat can be propagated from tip to base or from base to tip; furthermore, the ciliary response can be abruptly turned on or off, as well as modulated by a large variety of phototactic or chemotactic signals. It is no wonder, therefore, that as the studies of the structure of cilia and its relevant proteins have proceeded in the last decade, we have witnessed the unfolding of a bewildering complexity of insights and problems connected with this ancient and fascinating phenomenon. To illustrate how complex the motion-generating machinery must be, consider the studies of D. Luck and his associates. They have investigated the ^{35}S-labeled axoneme proteins of *Chlamydomonas reinhardtii* flagella by two-dimensional gel electrophoresis and have resolved approximately 200 different polypeptide chains.

The "9 + 2" arrangement of microtubules seen in cross section of a cilium or flagellum (Figure 2–41) is one of the most conserved organelle structures in eucaryotic cells. Figure 16–36 summarizes the present state of our ideas regarding this complex structure. It should be noted that, although at first glance radially symmetrical, the axoneme of the cilium is, in reality, bilaterally symmetrical: its inner sheath, C1 fibers, and, in certain cases, its doublets 5 and 6 are connected by permanent bridges. This bilateral symmetry may be related to the variability of ciliary motion.

As is frequently found, however, some variants of even the most "invariant" structures are eventually discovered. Thus, a 9 + 1 structure has been observed in flatworms, 9 + 0 in the mayfly and some fish, even 9 + 9 + 2 in the sperm of some insects, and 6 + 0 in the male gamete of a parasitic marine protozoan. The significance of these variations in structure to the mechanism of the flagellar or ciliary beat are currently under active investigation.

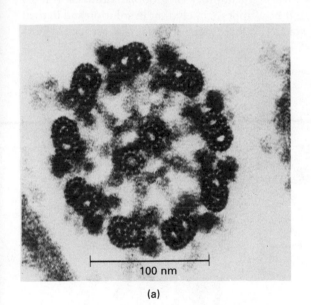

(a)

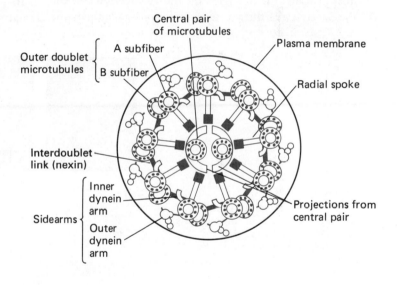

FIGURE 16–36
The Fine Structure of the Axoneme. (a) Transmission electron micrograph of a ciliary axoneme. (Courtesy of L. Tilney.) (b) Schematic representation of the complex detail of the axoneme structure, as deduced from a variety of microscopic studies. Represented are two central singlet microtubules consisting of 13 protofilaments (C1 and C2) and nine outer doublet microtubules consisting of 10 (or 11) + 13 protofilaments arranged in the form of a complete subfiber A to which a partial subfiber B is attached. Each doublet carries two arms. The remaining structures include the inner sheath, spokes, spokeheads, C1 fibers, and bridges between C1 and C2. The dynein arms all point clockwise when viewed from the axoneme base. There is also regularity in the vertical direction: the outer dynein arms are attached to every third tubulin dimer, the inner arms to every fourth tubulin dimer. The outer doublet microtubules are connected by a highly elastic filament that contains the protein nexin. (Modified from J. Darnell, H. Lodish, and D. Baltimore.)

Mechanism of Ciliary and Flagellar Motion

It has been known for some time that eucaryotic cilia transduce energy all along their structure because, for instance, both demembranated flagella and severed flagella can resume their beat upon addition of ATP. Even more convincing, when ATP is applied to partially demembranated flagella, bending occurs only at points where the membrane has been removed.

The molecular approach to the study of ciliary motion was begun by the work of I. Gibbons. He was able to remove the membrane from the contractile apparatus of the cilium (the *axoneme*) by using the detergent digitonin. When he treated suspensions of cilia with EDTA (a chelator of divalent cations like Ca^{2+} and Mg^{2+}) and recovered the cilia by sedimentation, the axoneme had lost its ATPase activity while the supernatant had gained it. Electron microscope observation of the EDTA-treated axonemes revealed that they had lost the *arms* projecting from the A subfiber. Mixing the supernatant with the axoneme brought about both the reappearance of the arms on the A-subfiber and the ATPase activity of the axoneme. Gibbons gave the name

dynein to the protein fraction responsible for the ATPase activity and the A-subfiber arms. Dynein has been purified and shown by SDS-polyacrylamide gel electrophoresis to contain in *Chlamydomonas* three distinct very large polypeptide chains (>300 kDa), as well as several intermediate and light chains. In other species, such as sea urchin sperm, only two heavy chains have been observed. For a while these discrepancies were interpreted as related to the possibility that the outer arm dynein might have three heavy chains and the inner arm two. Although this explanation appears to be correct for *Chlamydomonas,* there appears to be species variation in the number of heavy chains found in the outer and inner dynein arms.

Recent electron microcopy studies by U. W. Goodenough, J. E. Heuser, and others, using quick-freeze, deep-etch, and rotary shadowing techniques, have given us a beautifully precise picture of the dynein molecule. In *Chlamydomonas,* spreading the outer arm on a mica substrate reveals a complex structure of three major strands carrying a number of globular domains (Figure 16–37a). It now appears likely that in solution and in their native state, the three dynein heads are associated with

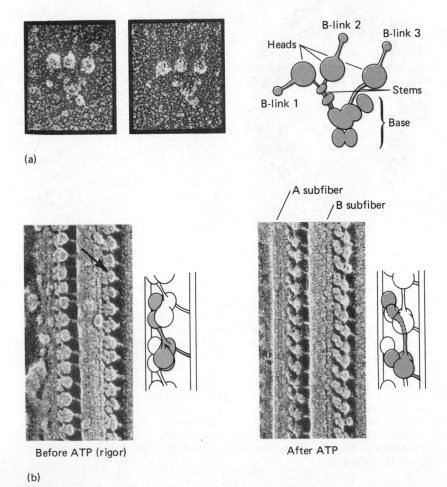

FIGURE **16–37**
Structure of the Dynein Outer Arm.   (a) Two electron micrographs of the dynein outer arms as they appear when spread out on mica substrates. The drawing summarizes the results of studying many EM images. Each head and B-link is associated with a given heavy polypeptide chain. In their native state, dynein arms show their heads and B-links in close association. Dynein inner arms are similar except that, depending on the organism, they may contain only two heads and B-links. (b) The effect of ATP on the orientation of dynein outer arms. Two quick-freeze, deep-etch EM images of outer dynein arms connecting A subfibers to B subfibers before (rigor) and after addition of ATP are shown. The drawings summarize many EM images and clarify the structures of the dynein arms. (Courtesy of U. W. Goodenough and J. E. Heuser.)

each other (Figure 16–37b). Interestingly, each of the three heavy polypeptide chains present in their respective heads and β-links have a unique ATPase activity.

The dynein outer arm extends diagonally from the A subfiber of one doublet to the B subfiber of its neighboring doublet. In doing so, its beads lie on the feet of its neighboring outer arm (Figure 16–37b). Goodenough and Heuser have reported dramatic changes in the orientation of the arms when axonemes are exposed to ATP. As Figure 16–37b shows, ATP brings about a significant change in the orientation of the three heads with respect to the A and B subfibers. Unfortunately, in spite of the precision with which we can delineate the change in conformation brought about by ATP, we do not have as yet an understanding of the mechanism whereby this change brings about ciliary motion. Nevertheless, a number of significant observations have begun to give us an insight into the overall mechanism of ciliary motion.

Firstly, how can the ciliary beat be related to the structure of the cilium? Two mechanisms can be imagined: differential changes in the lengths of the microtubules and sliding motion of the microtubules with respect to each other. Detailed EM evidence by P. Satir and by F. Warner gave strong evidence that the sliding theory is correct (Figure 16–38). By examining cross sections of the tips of cilia, they were able to determine how far the tips of outer doublets reached. It turned out that in a straight cilium, all the outer doublets extend the same distance; in a bent cilium, the doublets on the inside of the bend extend farther than those on the outside. Thus, the pattern of movement of the subfibers observed in the tips of cilia is consistent with the hypothesis that microtubules maintain their lengths but slide with respect to each other during bending.

Since the outer doublets are connected to each other by nexin-containing filaments, the sliding motion is limited by their elasticity. The experiment by Satir and

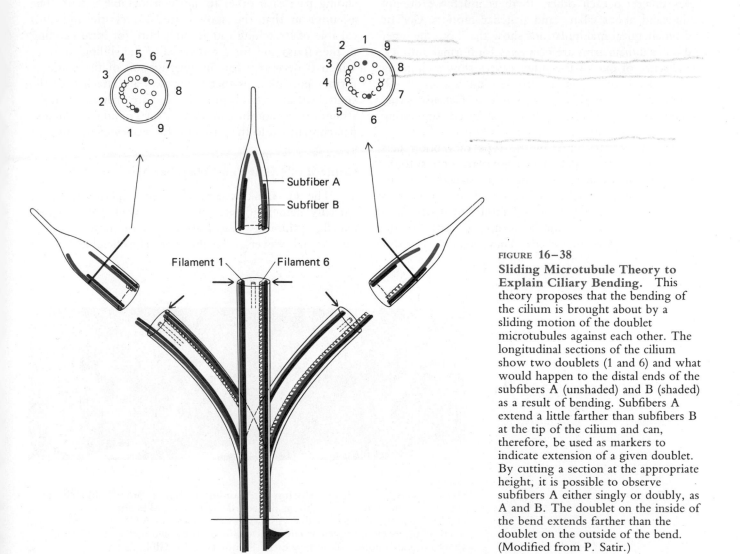

FIGURE **16–38**
Sliding Microtubule Theory to Explain Ciliary Bending. This theory proposes that the bending of the cilium is brought about by a sliding motion of the doublet microtubules against each other. The longitudinal sections of the cilium show two doublets (1 and 6) and what would happen to the distal ends of the subfibers A (unshaded) and B (shaded) as a result of bending. Subfibers A extend a little farther than subfibers B at the tip of the cilium and can, therefore, be used as markers to indicate extension of a given doublet. By cutting a section at the appropriate height, it is possible to observe subfibers A either singly or doubly, as A and B. The doublet on the inside of the bend extends farther than the doublet on the outside of the bend. (Modified from P. Satir.)

Warner, although demonstrating that sliding occurs, did not prove whether sliding is the cause or the result of bending.

Secondly, a beautiful experiment demonstrating that the sliding motion of flagellar microtubules is an active, rather than a passive, result of bending was performed by K. Summers and Gibbons. They treated axonemes with trypsin, under conditions that allowed only partial digestion, and showed in the EM that this destroyed the radial spokes. The treated cilia could no longer bend, proving that radial spokes are required for bending. When these axonemes were exposed to ATP, entire axonemal microtubules were extruded. Presumably one microtubule doublet would "walk" along its neighboring doublet, which in turn would walk along its own neighbors (Figure 16–39), thereby producing the sequential extrusion of entire subfibers.

Although we know that the dynein arms power the limited sliding movement of outer doublet microtubules with respect to each other, there is much we do not understand about ciliary and flagellar motion. Genetic and biochemical manipulations show that both the inner and outer dynein arms are necessary for normal motility; the absence of one or the other causes slow swimming even if bend formation or propagation is normal. D. Luck and his group have shown that *Chlamydomonas* mutants lacking radial spokes are paralyzed, suggesting that radial spokes are necessary for motility.

The great variability in the type of motion performed by cilia and flagella are in apparent contrast to the uniformity and conservaton of their structure in the animal and plant kingdoms. Satir and his co-workers have advanced a "switch-point" model of dynein arm activity that may provide a common basis for the different behaviors observed. They hypothesize that during the wave motion of a cilium, there is an effective stroke followed by a recovery stroke. Based on observations on the lateral cilia of the mussel gill, they concluded that during each effective stroke, arms on doublets 1–4 become active and arms on doublets 6–8 are turned off, whereas during each recovery stroke, arms on doublets 6–8 become active and arms on doublets 1–4 are turned off. They also suggested that spoke-central sheath attachment is influenced by the cyclic arm activity. By modulating these two types of attachment activities (dynein and spoke), it is conceivable that a variety of complex movements could be generated.

Sliding movement alone does not account for *local bending,* which is necessary for the propagation of the wave motion observed in ciliary and flagellar motion. To explain local bending it is necessary to invoke the radial spokes of the axoneme; the cilia of certain nonmotile mutants of the alga *Chlamydomonas* lack these structures. These spokes are believed to prevent the subfibers from sliding past each other to any considerable extent. The assumption that the spokes are also elastic and thus capable of stretching can account both for local bending of the cilium and for the reversal of the sliding motion, which is necessary for the propagation of the bend. A further possibility is that the spokes are attached to the central sheath in the bent region but are unattached in the straight region of the axoneme. Cycles of attachment and detachment could then account for waves of bending.

Control of Ciliary and Flagellar Motion

The C. Brokaw and Gibbons experiments with partially demembranated flagella showed that cilia and flagellae exhibit a great deal of local autonomy in their motion. However, cells do exert considerable control

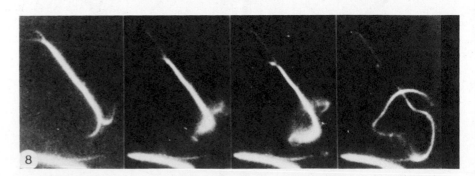

FIGURE 16–39
Demonstration of Sliding Motion of Axonemal Microtubules.   In a classic experiment, Summers and Gibbons destroyed the radial spokes and the nexin-containing filaments with trypsin. On the addition of ATP, axonemal microtubules were extruded at rates that conform to the sliding rate that brings about normal bending. (Courtesy of K. Summers and I. Gibbons.)

over the wave form, the direction of the effective beat, the direction of bend propagation, and even over the initiation and termination of motion. Our detailed understanding of cellular control of ciliary and flagellar motion is still minimal, but it is already clear that the cytosol [Mg^{2+}] and the effect of membrane depolarization on [Ca^{2+}] are some of the factors affecting the control.

The Cytoplasmic Matrix: An Integrating Organelle

Early students of the transparent "ground substance" of the cell divided themselves into two groups. On the one hand there were the "fixers and stainers": they discovered all kinds of filamentous structures in the cytoplasm but these were soon shown to be nothing more than fixation and staining artifacts. On the other hand there were the "pokers and squeezers": during the first 30 years of this century they argued that the ground substance was an elastic gel. When strands of cytoplasm were pulled out of the main body of a cell, they snapped back; when iron particles were introduced into the cytoplasm, they moved erratically (rather than smoothly) under the influence of magnetic fields. Since cytoplasm is generally thought to be a dilute protein solution, the pokers and squeezers concluded that the molecules responsible for its gel-like properties are elongated filaments, a "brush heap" as G. Scarth used to call them. But not only was the cytoplasmic gel *elastic* it was also capable of *flowing* at the same time. This ability to show properties of both a solid and a liquid simultaneously (*thixotropy*) was much emphasized by early students of cell structure, such as A. Frey-Wyssling, because it suggested that the bonds holding the filaments together were capable of rapid breakdown and re-formation (cross-link modulation).

The early light microscopists never succeeded in resolving the filamentous structure of the cytomatrix. It is only recently that immunologically enhanced images of microtubules and bundles of microfilaments and intermediate filaments have been detected. In fact, even the early years of electron microscopy revealed little regarding the presence of the cytoskeletal filaments. They were first discovered in structures in which their regularity of organization revealed them more easily. Thus, striated muscle was the material in which microfilaments were first observed and cilia were the structures in which microtubules first became evident. These early studies were followed by improvements in procedures, after which rapid progress was made in the study of cytoskeletal filaments in less ordered samples of cytoplasm. The recently developed use of immunological staining at the electron microscope level combines high-resolution optics with the specificity of immunological interactions.

High-voltage electron microscopy allows the greater depth of focus required to study an interconnected cytoskeleton, and the use of quick freezing, deep etching, and rotary shadowing is reducing artifacts of preparation and increasing resolution (see Chapter 2). These developments have elicited a renaissance in the study of the cytomatrix.

The vision of the cytomatrix that emerges from this study is that of an integrated cell organ: one that controls the maintenance and changes of its shape and generates its motion, as well as that of the organelles and cell membranes. The cytomatrix also controls the specific localization of enzymes and, hence, exerts an important influence on cell metabolism. There is even evidence that messenger RNAs for specific proteins—for example, actin—are localized to special places in the cell by being bound to the cytomatrix. Finally, the cytomatrix may mediate signals coming from the cell's exterior to interior regions of the cell, such as the nucleus.

In the following sections we shall examine several aspects of the cytomatrix: the filaments found in the cytoplasm, the proteins associated with these filaments, the interactions between them, and the interactions with the cytosol. Finally, we shall discuss a number of integrated cell processes in which the cytomatrix is an important participant.

The Intermediate Filaments

For a number of years, electron microscopic and immunofluorescence studies of the cytoplasmic matrix have revealed the presence of unbranched filaments approximately 10 nm (8–11 nm) in diameter. These *intermediate filaments* (IFs) fall into five distinct classes that resemble each other in a number of ways:

- Each class is encoded by a family of genes.
- Each gene family is expressed preferentially in a particular cell type, although interesting exceptions to this rule have been observed.
- IF proteins bear considerable structural resemblance to each other.
- IFs are insoluble in all but the most potent protein unfolding reagents, such as sodium dodecyl sulfate (SDS).

In the electron microscope, the IFs are long, wavy structures that are dispersed singly or aggregated in bundles in the cell. Generally they form a dense cage-like network that surrounds the nucleus and extends to the cell surface (Figure 16–40). At the cell surface, some IFs aggregate at desmosomal junctions (Figure 16–41); they also appear to form close contacts with the nuclear pore complexes.

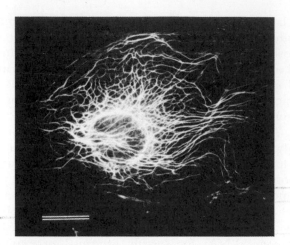

FIGURE **16–40**
Indirect Immunofluorescence Image of IFs of a Tissue-cultured Pt-K2 Cell Using an Antikeratin Polyclonal Antibody. The intermediate filaments form a dense cage-like structure around the nucleus. (Bar = 10 μm; courtesy of P. M. Steinert.)

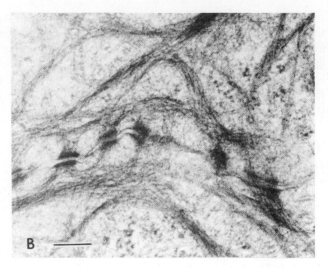

FIGURE **16–41**
Ultrastructure of Keratin IFs in Cultured Mouse Epidermal Cells. Filaments often appear to extend from cell to cell. In fact, high-resolution studies show that the IF bundles of each cell terminate abruptly at intercellular (desmosomal) junctions. (Bar = 0.2 μm; courtesy of P. M. Steinert.)

The Five Classes of Intermediate Filaments

Intermediate filaments are classified according to the types of vertebrate cells in which they are found or according to their chain structure (see Figure 16–43). These classifications also relate to the specific properties of the IF proteins. There is available a "universal" antibody that binds to all the classes of IFs. Proteins found in plants, yeasts, and other simple eucaryotes also bind this antibody.

Epithelial IFs (*keratins,* class I and class II) are found in epithelial cells such as epidermal cells, pancreatic acinar cells, and hepatocytes, as well in the products of these cells such as skin, hair, and nail. The keratin IF proteins can be divided according to their electrofocusing properties into acidic (class I) and neutral-basic (class II) proteins; these in turn fall into numerous molecular weight classes, ranging from 40 kDa to 70 kDa, that are differentially expressed in various stages of embryonic development and in various epithelia. At least one acidic and one neutral-basic keratin are found in each epithelial cell; the keratin IFs are heteropolymers of the two types of chain.

Mesenchymal IFs (class III), composed of the protein *vimentin* (55 kDa), have been found distributed in most tissues of mesenchymal origin and in cells cultured from these tissues. In some cell types, vimentin has been observed to coexist with keratin; in other cell types vimentin coexists with desmin, the intermediate filament protein of muscle. A characteristic property of vimentin IFs is that they seem to be in close association with microtubules and respond to microtubule inhibitors, such as colchicine, by withdrawing from the cytoplasm to form a perinuclear cap. This effect is reversible.

Muscle IFs (class III), composed of the protein *desmin* (53 kDa), are found in skeletal, cardiac, and smooth muscle. In striated muscle these IFs are located in the region of the Z-disk and M-line and are believed to interconnect the myofibrils, keeping them in register in the muscle fibril (Figure 16–42). In muscle cells, the IFs are less soluble in high salt concentrations than actin (thin) and myosin (thick) filaments; the IFs can be visualized by dissolving the latter filaments in KI.

Glial IFs (class III), found in the supporting cells of the brain, the spinal cord, and the peripheral nervous system, are composed of an acidic protein of 51 kDa.

Neuronal IFs (class IV) are found in very large amounts in neurons and dendrites. The IFs are found in close alignment with microtubules and they carry small projections which appear to connect with each other and with microtubules. Neuronal IFs are composed of three protein chains, the "neurofilament triplet," of about 200, 150, and 68 kDa.

Lamins (class V), found in the nucleus of eucaryotic cells, are a newly recognized type of intermediate filament. The lamins form the *nuclear lamina complex*—sometimes called the *karyoskeleton*—a meshwork of fibers on the inner surface of the nuclear membrane. The lamins range from 60–75 kDa and there are at least four different kinds of lamins found in vertebrate cells.

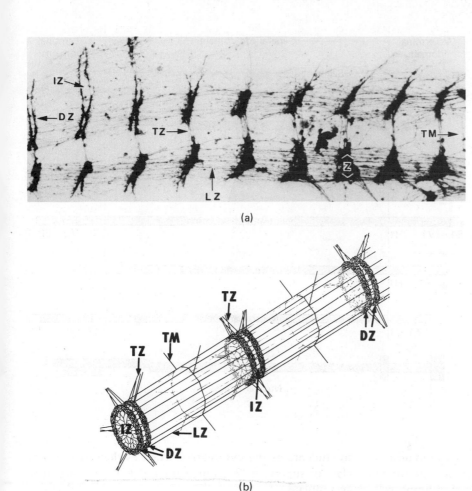

(a)

(b)

FIGURE **16–42**

The Role of Intermediate Filaments in Maintaining Striated Muscle Cell Structure. (a) KI-extracted muscle fibrils. A number of filaments are present: TZ, transverse filaments connecting M-lines to each other; LZ, longitudinal filaments connecting Z-disks within a sarcomere to each other. DZ show the doublet nature of the Z-disks. IZ are presumptive filaments holding the Z-disk doublet together. (b) Schematic diagram of sarcomere-associated cytoskeletal network of intermediate filaments. (Courtesy of K. Wang and R. Ramirez-Mitchell.)

The Structure of Intermediate Filament Proteins

The various classes of IF proteins can be distinguished from each other by immunological methods. Since, however, they form morphologically similar filaments, it is possible that they have some common molecular properties. This, indeed, was found to be the case. X-ray analysis shows that all IFs contain coiled-coil alpha-helices, lying parallel to the axis of the filament. The coiled-coil region forms a rod-like central domain of invariant length, interrupted by short nonalpha-helical stretches (Figure 16–43). At either end of the alpha-helical regions, terminal domains of varying structure are found. Optical studies suggest that the rod-like alpha-helical stretches of the IF proteins are involved in the assembly of the filament and that the terminal portions extend from the filament into the cytoplasm. Differences in the size and amino acid sequences of these terminal domains probably account for the observed variations in apparent diameter and antigenic properties of IFs.

What Is the Function of Intermediate Filaments?

The function to IFs is not yet clear. Surprisingly, when antibodies to IFs have been injected into a cell, they cause marked changes in the cell's appearance without causing marked changes in its motility or capacity to divide. With the exception of the lamins, IFs do not undergo major structural changes during cell division. This suggests that the role of IFs must be different and more subtle than the major functions attributed to the other filamentous components of the cytomatrix. Normally, depolymerization of IFs seems not to occur; to remove an intermediate filament may require its hydrolysis.

At present the only agreed functional roles of the IFs are (1) that of providing the cell and the tissue with mechanical strength, suggested by the involvement of some IFs such as keratins with the desmosomes, and (2) that of positioning the nucleus in the cell, suggested by the cage-like concentration of some IFs around the

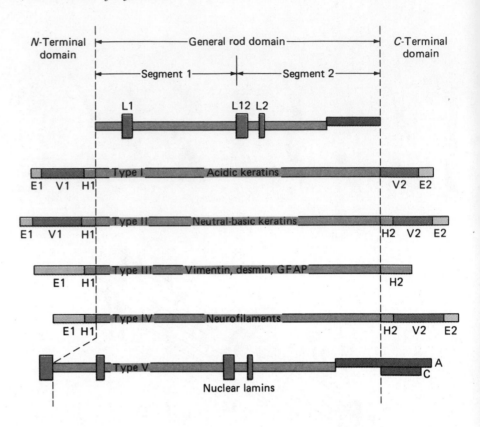

FIGURE **16–43**
Structure of Intermediate Filament Proteins.   All intermediate filament proteins possess a common central domain that consists of an alpha-helical, coiled-coil structure interrupted by three linkers (L1, L12, L2). The lamins have a central domain that is 42 amino acid residues longer than the other IFs. The central domain is involved in polymerizing IF proteins into the 10-nm filaments. At both ends of the central rod domain there are "terminal" domains; these protrude from the axis of the filament structure. The terminal domains can be classified into subdomains of homologous sequences (H), variable sequences (V), and highly charged N-terminal (N) and C-terminal (C) sequences. (Courtesy of P. M. Steinert.)

nucleus. The almost universal presence of IFs in eucaryotic cells, their variability, and their tissue specificity would argue for their involvement in significant cell functions. IFs may also play a role in cell-cell interactions.

One useful application that has emerged from the study of IFs is that of distinguishing between tumor groups in medical diagnostic procedures. Thus, by using immunofluorescence methods based on monoclonal antibodies against specific proteins in IFs, it is possible to distinguish between tumors of epithelial origin containing keratins, tumors of glial origin containing one of the glial filament proteins, tumors of muscle origin containing desmin, tumors of lymphomal origin containing vimentin, and also tumors of muscle origin containing vimentin. These techniques have been especially useful for the diagnosis of tumors that are difficult to distinguish by the more conventional morphological methods.

The Superthin Filaments

Until very recently it was generally accepted that only three major filaments comprised the basic structural framework of the cytoskeleton: the microtubules, the microfilaments, and the intermediate filaments. To be sure, the thick myosin filaments might have been added to the list, but in most tissues other than muscle they are either present as very short bipolar structures or not in evidence at all. It therefore came as a considerable surprise

to students of the cytomatrix when evidence for a new class of superthin (2–3 nm diameter) filaments began to accumulate.

Early evidence for superthin filaments emerged when M. Schliwa utilized the nonionic detergent Triton X-100 to extract a number of cell types. This treatment removes a considerable amount of protein from the cell matrix and reveals not only beautifully clear images of microtubules, intermediate filaments, and microfilaments but also varying lengths of 2–3 nm filaments connecting the larger filaments to each other (Figure 16–44). Schliwa was able to demonstrate that the superthin filaments do not contain actin, by showing that they are not decorated by S1 myosin heads.

Could it be that the "connector" filaments seen by Schliwa are only bare portions of a continuous network? This possibility is raised by another approach to the study of superthin filaments, one derived from the use of strong protein unfolding reagents. D. Gassner and his co-workers removed droplets from the moving portion of the plasmodium of the slime mold *Physarum polycephalum* and extracted them with high concentrations of urea and sodium dodecyl sulfate (SDS). Examination of the extracted plasmodial samples revealed a network of filaments, the thinnest of which measured 2–3 nm in cross section, as well as some globular domains (Figure 16–45). The insolubility of this network in urea and SDS suggests that the superthin filaments are covalently

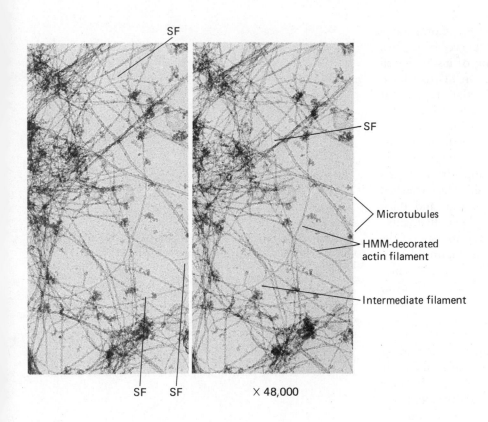

SF

SF

Microtubules

HMM-decorated
actin filament

Intermediate filament

SF   SF   × 48,000

FIGURE **16–44**
Stereo View of Superthin (2–3 nm, Denoted SF) "Connector" Filaments.   African green monkey kidney cells were extracted with Triton X-100, fixed, critical point-dried, and viewed in a high-voltage electron microscope. A three-dimensional network of microtubules (20-nm diameter), intermediate filaments (9–11-nm diameter), and actin microfilaments (6–8-nm diameter, decorated with heavy meromyosin) can be seen. Interconnecting these filaments are a number of 3-nm superthin filaments. ×48,000. (Courtesy of M. Schliwa.)

0.25µm

FIGURE **16–45**
The Presence of a Covalently Cross-linked Cytoskeletal Matrix of Superthin (2–3 nm) Filaments in the Plasmodium of the Slime Mold *Physarum Polycephalum.* Samples of streaming endoplasm in the plasmodium were rapidly removed and frozen in liquid N_2, then extracted in 7 M urea and 4% SDS. The plasmodial "ghosts" were further purified by electrophoretic extraction, then prepared for microscopy by critical point-drying, dry-cleaving, and rotary shadowing. Note the interconnected network interrupted by irregularly spaced nodules. (Courtesy of D. Gassner, Z. Shraideh, and K. E. Wohlfarth-Bottermann.)

cross-linked. An exciting aspect of this work is that the covalently cross-linked matrix of superthin filaments was derived from an actively streaming portion of the slime mold plasmodium, indicating that the cross-links are capable of rapid breakdown and re-formation in vivo.

The use of strong protein unfolding reagents has revealed in a number of tissues the presence of a covalently cross-linked meshwork that retains the shape of the original cell, a "cell ghost." Extraction of striated muscle with high concentrations of guanidine·HCl and Triton X-100 results in a "tissue ghost" that, when treated with high-purity collagenase, produces "muscle fiber ghosts." Although consisting of less than 0.5% of the original protein, the ghost still retains the general shape of the muscle fiber (Figure 16–46). Comparable studies on smooth muscle cells and nerve axons also reveal the presence of ghosts consisting of insoluble cell matrices. Muscle fiber ghosts disperse in SDS to form large subunits of many millions molecular weight that cannot enter a dilute (2.5%) acrylamide gel but can diffuse through 1% agarose (Figure 16–47).

A third superthin filamentous material in striated muscle has been proposed by a number of workers. The clearest evidence for such filaments comes from studies by R. Locker and also by A. Magid, who stretched muscle to the point of separating the thin and thick filaments, thereby bringing superthin "gap filaments" into view. The gap filaments are believed to connect the tip of the thick filament with the Z-disks (Figure 16–48). These gap filaments may in fact form a covalently cross-linked matrix that extends from Z-disk to Z-disk.

Motion via Villipodia

The above studies were performed using various extracting reagents that removed a part or all of the more soluble proteins of the cytomatrix. As is the case frequently in biological research, the great diversity in the world of life aided experimental progress. In this in-

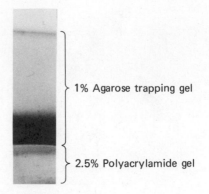

FIGURE 16–47
A Specially Constructed Electrophoretic Gel Consisting of a Layer of 1% Agarose and Another Layer of Very Dilute (2.5%) Polyacrylamide. The agarose layer acts as a "trapping gel," allowing particles too large to enter a 2.5% polyacrylamide gel to be preserved. In the absence of such a trapping gel, very high molecular weight particles are lost during the staining and washing processes. The dark band consists of a preparation of muscle fiber ghosts dispersed in 2% sodium dodecyl sulfate and 5% 2-mercaptoethanol. The fact that the dispersed particles cannot enter an acrylamide gel suggests that their particle weight is at least several millions.

stance, it provided investigators with an experimental system lending itself especially well to the optical study of superthin filaments.

The sperm of nematodes do not possess a flagellum but, when activated, form a pseudopod from which projections called "villipodia" emerge (Figure 16–49(a)). These attach to the substratum and move backwards from the tip of the pseudopod toward the cell body, causing the sperm to "walk" on multiple "legs" (Figure 16–49(b)). Biochemical analyses and electron microscopic observations in the laboratories of H. Ris and T. M. Roberts demonstrated the absence of microtubules, IFs, and microfilaments, as well as their respective constituent proteins. Instead they reported the presence of an extensive branching network of superthin filaments that are 2 nm thick in *Caenorhabditis elegans*. In *Ascaris suum,* these filaments appear to be as thick as 5–10 nm, although the possibility remains that these structures represent side-to-side associations of superthin 2-nm filaments.

In the villipodia, the filaments are attached to the membrane system and organized into complexes that extend back into the pseudopod (Figure 16–49(c)). At times, these complexes are radially organized like test tube brushes, while at other times they lie in parallel arrays (Figure 16–49(d)). The filaments consist of a family of 14-kDa polypeptide chains that are expressed only in male gametes. Motion appears to involve the assembly of materials to form membrane and superthin

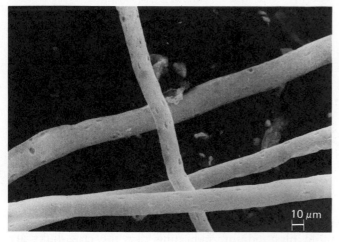

FIGURE 16–46
Muscle Fiber "Ghosts." These remnants, representing less than 0.5% of the total protein, were prepared by extracting chicken pectoral muscle with 6 M guanidine·HCl and 5% 2-mercaptoethanol. The resulting "tissue ghost" is treated with high-purity collagenase to leave cell ghosts. These resemble the original muscle fibers in a number of morphological characteristics, including the indentations left by nuclei. Under phase-contrast microscopy, striations are visible, indicating that the covalently cross-linked matrix of the fiber ghosts continues to be organized in repeating sarcomeres. (From A. G. Loewy and H. S. Kaufman.)

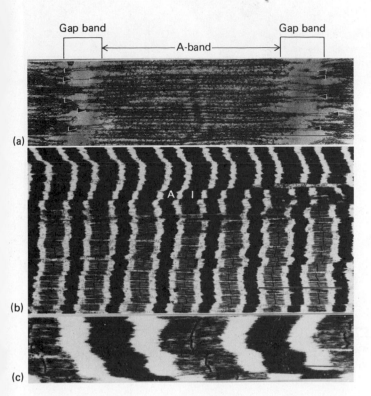

Gap band — A-band — Gap band

(a)

(b)

(c)

FIGURE **16–48**

Evidence in Stretched Frog Muscle for a Third Filament that Connects Thick Filaments with the Z-disk. The muscle was stretched until the thin and thick filaments no longer overlapped. (a) In the gap thus created, it is possible to see at very high magnification very fine filaments extending from the thick filaments. Pairs of arrowheads indicate examples that can be followed from the edge of the I-band, across the "gap band," A-band, and the other gap band, to the other edge of the I-band. (b) The muscle fiber was briefly soaked in myosin S1 to label actin filaments. A gradient of labeled I-band density was established that can be seen to extend downward from the top of the micrograph. Since no label appears in the gap band region, it can be concluded that the thin filaments crossing the gap band are not actin. (c) Z-lines and adjacent A-bands remain parallel despite following a curved path. This suggests there is a connecting link of constant length between these two structures. Scale: (a) 0.5 μm; (b) 10 μm; (c) 2.5 μm. (Courtesy of A. Magid.)

filament complexes at the leading edge of the pseudopod and the disassembly of these structures in the cell body (Figure 16–49(b)). Whether these superthin filaments are capable of carrying out mechanochemical transductions is not known, but the anatomy of the villipodia strongly suggests that the "motor" for the movement of these structures is located close to the site of movement rather than in the cell body.

There is mounting evidence for the existence of superthin filaments in the cytomatrix of a variety of cell types. These filaments may participate in making and breaking covalent cross-links with other filamentous

structures. Such properties may well prove to be useful in explaining forms of motion not readily accounted for by the involvement of microtubules or microfilaments.

Proteins Associated with the Filaments of the Cytomatrix

It is the various proteins associated with the cytoplasmic filaments that allow the cytomatrix to engage in its numerous activities. Although not yet supported by conclusive experimental evidence, this hypothesis can explain how, for instance, the very same actin molecule is involved in a number of very different cell processes at the same time.

Microfilament-Associated Proteins (MFAPs)

Actin, the bulk of the protein in the microfilament, interacts reversibly in vitro with over 60 other proteins that have various effects on its structure and function. These are called *microfilament-associated proteins (MFAPs).*

In the cytoplasm of most cells, actin (taken as the monomer) occurs in very high concentration and would be expected, therefore, to polymerize and form a gel. In fact, only part of the actin is filamentous because many of the proteins that combine with actin regulate the polymerization and organize the filaments into more complex structures. Some of these structures are remarkably ordered, such as the cross-linked bundles in microvilli; others are relatively disordered, such as those found in lamellipodia and in the cytoplasm, where much of the actin is in the form of a randomly cross-linked network of filaments.

Years ago a number of laboratories reported that crude extracts of actin obtained from a variety of cells can form *gels* capable of supporting their own weight. Using this reaction as an assay, a number of proteins (generically named *gelactins*) have been isolated that, in the presence of Mg^{2+} or Ca^{2+}, are able to cross-link actin and form loose gels. The gelation can be measured by an increase in viscosity of the actin solution and the concentrations of gelactins that cause this increase (or even a precipitation) can be evaluated (ratios of gelactins:actin = 1:50 to 1:500). Gel formation depends on three variables: the concentration of the actin polymer, the average length of the polymer, and the concentration of functional cross-linking gelactin proteins. The last variable is also in many cases affected by $[Ca^{2+}]$, which influences the affinity of certain gelactins for actin. Some of the cross-linking proteins themselves form polymers—for example, filamin and spectrin—and this also affects their ability to bind actin.

The binding of all these proteins to actin changes the properties and functions of actin, as noted in Table 16–3

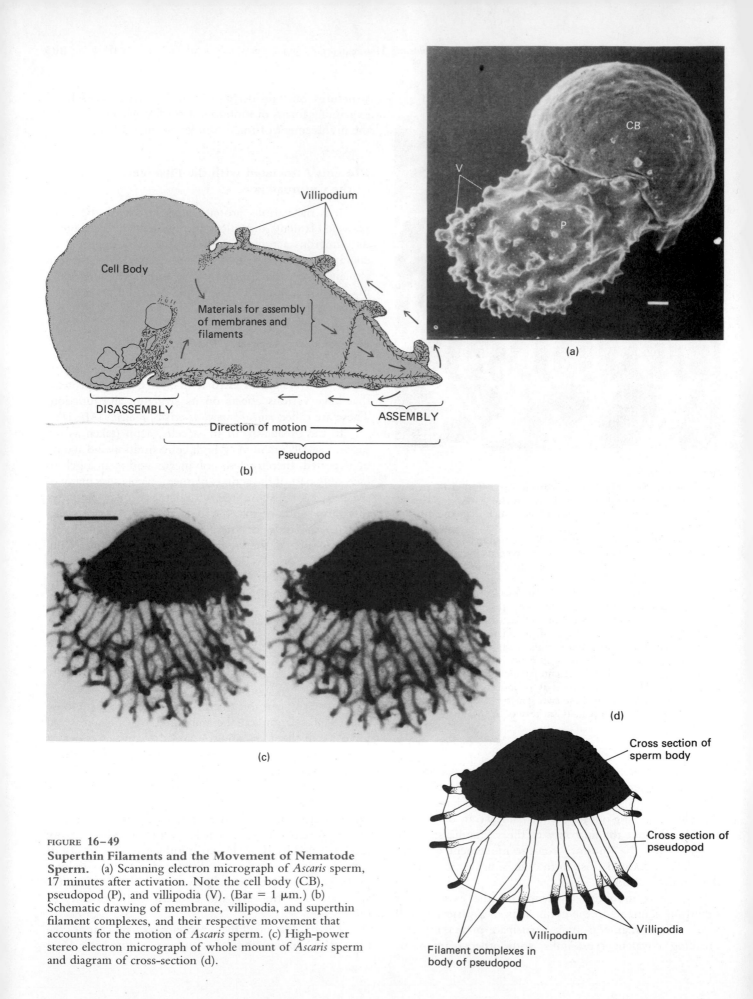

Cell Body

Villipodium

Materials for assembly of membranes and filaments

DISASSEMBLY

ASSEMBLY

Direction of motion

Pseudopod

(b)

CB

V

P

(a)

(c)

(d)

Cross section of sperm body

Cross section of pseudopod

Villipodia

Villipodium

Filament complexes in body of pseudopod

FIGURE 16–49
Superthin Filaments and the Movement of Nematode Sperm. (a) Scanning electron micrograph of *Ascaris* sperm, 17 minutes after activation. Note the cell body (CB), pseudopod (P), and villipodia (V). (Bar = 1 μm.) (b) Schematic drawing of membrane, villipodia, and superthin filament complexes, and their respective movement that accounts for the motion of *Ascaris* sperm. (c) High-power stereo electron micrograph of whole mount of *Ascaris* sperm and diagram of cross-section (d).

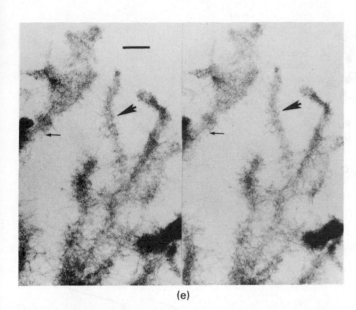

(e)

FIGURE **16–49 continued**

Note how numerous filament complexes branch within the pseudopod and terminate in the villipodia. The pseudopod can be seen, preferably with a stereo viewer, as lightly grained material. The filament complexes in the villipodia are darkly stained but continue into the body of the pseudopod as more lightly stained structures. (Bar = 5 μm.) (e) Stereo electron micrographs of sperm whole mounts. The organization of superthin filament complexes is shown. Normally, filaments are organized radially, like a test tube brush (large arrow). In aging cells, filaments can collapse close to the core of the complex, forming parallel bundles. (Bar = 0.5 μm.) (Courtesy of S. Sepsenwol, H. Ris, and T. M. Roberts.)

and visualized in Figure 16–50. These effects include "capping," which prevents lengthening of the actin polymerization; "severing," which chops up the filamentous actin; "stabilizing," which prevents changes; "bundling," which binds actin filaments rigidly in parallel; "gelating," which forms amorphous gels of actin; "spacing," which forms loose, parallel arrays; "contraction" or sliding, which brings about movement; and side-on or end-on membrane anchorage, which are self-explanatory.

The protein *profilin,* found in spleen, thymus, brain, lymphocytes, and platelets, binds reversibly with monomeric actin (G-actin), thereby inhibiting its polymerization. Its function could be to maintain a pool of unpolymerized actin, so as to be available for rapid changes in the mechanochemical activities of cells. The profilin-actin complex (*profilactin*) has been crystallized and it is likely that this will allow the solution of its structure in the near future.

Vinculin acts to collect the actin filaments into foci at adhesion sites in the plasma membrane. Farther from the membrane, the rod-like protein α-*actinin* binds to actin to "splay out" the actin filaments. Other rod-like proteins, members of the *spectrin* family, are found at the interior face of the plasma membrane of many cell types. Spectrin is made of two subunits; one, the α-subunit, is the same in all cells and binds to actin and to the Ca^{2+}-binding protein calmodulin. Another protein, one without actin-binding activity, binds to spectrin; in red blood cells this protein is known as *ankyrin*. The three proteins—ankyrin, spectrin, and actin—form a gel-like structure

TABLE **16–3**

Some Characteristics of Some of the Major Actin-Binding Proteins

Proteins	Where Found (So Far)	Apparent Molecular Weight	Subunits	Ca^{2+}-sensitivity
Gelation proteins				
Filamin	Smooth muscle, fibroblasts	250,000–270,000	2	No
Bundling proteins				
α-Actinin	Muscle, HeLa cells	100,000–105,000	2	Yes (HeLa); No (muscle)
Fascin	Sea urchin eggs	58,000	1	No
Fimbrin	Microvilli	68,000	1	No
Severing proteins				
Gelsolin	Macrophage, platelets	90,000	1	Yes
Villin	Microvilli	95,000	1	Yes
β-Actinin	Skeletal muscle	34,000, 37,000	1 each	No
Fragmin	Slime mold	43,000	1	Yes
Sequestering depolymerizing proteins				
Profilin	Lymphocytes, amoeba	12,000–15,000	1	No
Membrane-protein binding				
Vinculin	Muscle, HeLa cells	130,000	1	Yes (HeLa); No (muscle)
Spectrin family	Red blood cells, microvilli, neurons	220,000–260,000	1 each	Yes
Capping protein	Amoeba, slime mold	29,000, 31,000	1 each	No

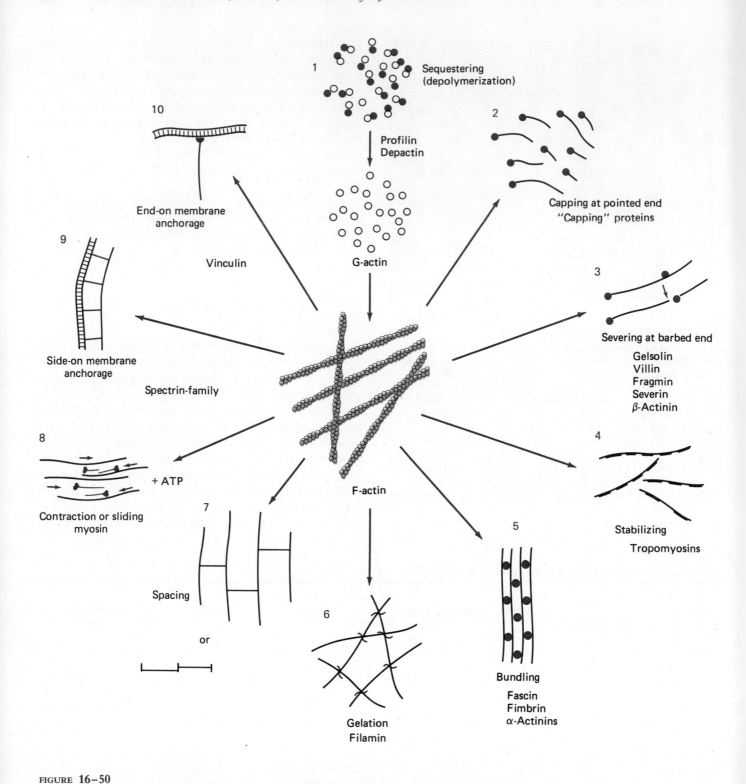

FIGURE 16–50

The Roles Played by a Number of Actin-binding Proteins on the Behavior and State of Aggregation of Actin Monomers and Filaments.   "Pointed" and "barbed" ends are named according to the "arrow" that results from decorating actin filaments with subfragment 1 of myosin. Assembly of actin monomers into polymeric filaments occurs at the barbed end; disassembly occurs at the pointed end. (Modified from K. Weber.)

that strengthens the plasma membrane and may help to preserve the shape of the cell.

An example of bundling is found in the microvillus of the intestinal epithelial brush border (see Figure 16–60). The microvillus contains villin-actin-fimbrin bundles. *Villin,* in the presence of Ca^{2+}, seems to limit the length of the actin filaments by capping of the "barbed" end, while *fimbrin* binding to actin accelerates the formation of parallel fibers. The actin-containing bundles of the microvillus extend into the terminal web at the plasma membrane and at that site interact with another protein, spectrin. The bundles formed by these proteins in vitro are very similar to the structures seen with the electron microscope in the tissue. All the capping proteins act by binding at the "barbed" end of actin filaments, blocking the addition of monomers and halting filament growth. Since depolymerization can occur from the other end, injecting capping proteins into cells leads to the rapid disassembly of actin bundles. This result is also seen if antivinculin antibodies are injected.

It is already clear that many of the effects of the actin-modulating proteins seen in vitro have analogies in vivo that can be visualized in the electron microscope. For example, in most nonmuscle cells a higher proportion of the actin is found in the unpolymerized G-form. At the advancing edge of a motile cell, actin is frequently found as single filaments in close juxtaposition to the membrane, both perpendicular and parallel to it. Further back from the leading edge, actin is frequently found as large bundles—the "stress fibers"—that end at points where the cell is attached to the substratum.

A number of the changes of actin structure that can be observed in vitro have been hypothesized to explain certain mechanochemical changes observed in vivo. Thus in certain cases, contraction is believed to be brought about by a bipolar myosin filament, containing two or more myosin molecules, that cause a sliding motion of actin filaments, much the same as in striated muscle.

Movement of the ends of actin filaments might also be brought about by *treadmilling,* in which G-actin monomers are added to one end of a filament (the "barbed" end) and removed from the other (the "pointed" end). Treadmilling will lead to the appearance of movement of the whole filament, but structures attached to the middle of such a filament will not move.

Microtubule-Associated Proteins (MAPs)

Alpha and beta tubulins, the fundamental proteins of the microtubule, hydrolyze the nucleotide GTP during their assembly and form a polar linear structure, characteristics that resemble actin. As is the case with microfilaments, a number of proteins (the MAPs) interact specifically with microtubules in vitro, affecting their polymerization and their interaction with other cytoskeletal elements and causing them to treadmill. High-resolution optical studies have identified MAPs as slender projections sticking out at intervals from the walls of microtubules (Figure 16–51).

Dynein is the most prominent MAP of the ciliary microtubules. The proteins in the cell interior that resemble dynein most closely are called *MAP1* and *MAP2*. These are assemblies of high-molecular-weight protein chains (≈ 300 kDa) that contain also some low-molecular-weight components. Recent reports by R. Vallee and his co-workers make a convincing case that MAP2 is restricted to the dendritic processes and the cell bodies, where it is found in association with IF cables. They were able to demonstrate this latter interaction by depolymerizing the microtubules with vinblastine and showing by indirect immunofluorescence that the MAP2 remained attached to the IFs. The mitotic spindle is another structure in which important progress has been made regarding the identification of MAPs. Monoclonal antibodies prepared against these proteins stain the mitotic spindle (Figure 16–52).

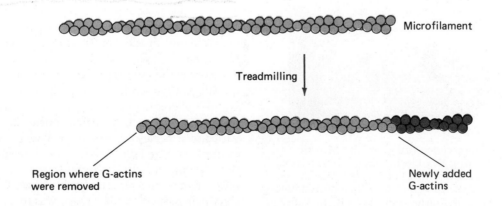

Microfilament

Treadmilling

Region where G-actins were removed

Newly added G-actins

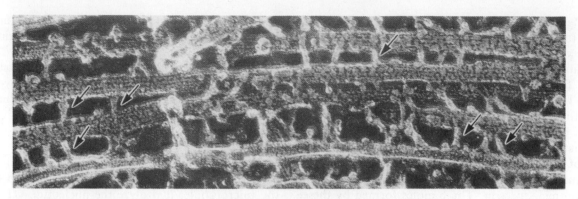

FIGURE 16-51

Electron Micrograph of MAPs Extending as Projections from Microtubules.   Quick-frozen, deep-etched suspensions of microtubules saturated with tau proteins. Note the frequent cross-bridges (arrows) between the microtubules. (Bar = 100 nm; courtesy of N. Hirokawa.)

Intermediate Filament-Associated Proteins (IFAPs)

Just as with the other classes of filaments, IFs have a number of associated proteins (the IFAPs), some of which are generally distributed and others that are tissue-specific. In hair, for instance, there are relatively low molecular weight (10–45,000) proteins that act to aggregate keratins laterally via disulfide and noncoavlent bonds. Two relatively high molecular weight proteins

that cross-link IFs into networks are *paranemin* (280 kDa) and *synemin* (230 kDa), found associated with desmin and vimentin in the muscle cells of birds. Some IFAPs bind neurofilaments to microtubules. Other IFAPs, for example *ankyrin* (230 kDa), act to "cap" IFs, anchoring the ends to various structures at the cell surface.

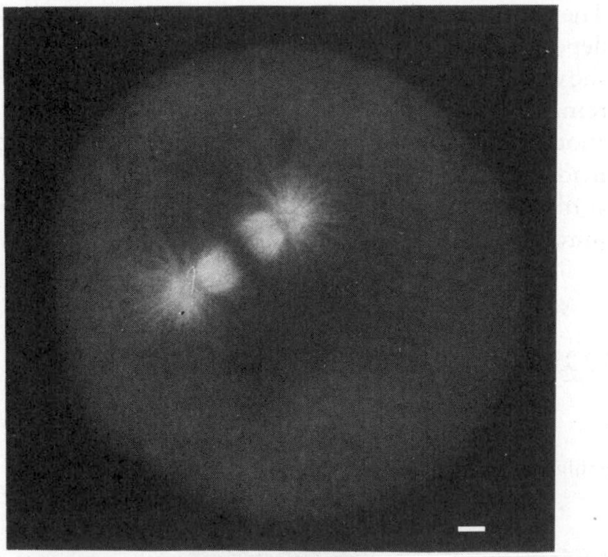

Filament-Filament and Filament-Membrane Interactions: Is There a Microtrabecular Lattice?

The concept of the *microtrabecular lattice (MTL)*, first proposed by K. Porter, was based on observations of cells grown in culture and spread out in a thin layer on a surface. The cells were fixed with gluteraldehyde, stained, again fixed in osmium tetroxide, dehydrated with alcohol and acetone, transferred to liquid Co_2, dried by the critical point method, and examined with the high-voltage (1000 kV) electron microscope. The images that emerged from this method showed a highly interconnected lattice extending throughout the cell, the *microtrabeculae* (see Figure 2–38). A thin layer of material lay adjacent to membranes and coated the filaments of the cytoskeleton. In addition, a large number of thin connections, ranging from 2–3 nm to 10 nm, appeared to interconnect all the structural elements; polysomes frequently lay at the junctions of these connections.

To overcome early skepticism regarding the reality of the microtrabecular lattice, Porter and his colleagues varied their preparative procedures over a wide range—for example, using rapid freezing and deep-etching rather than fixing. All these methods revealed an MTL that retained the characteristics listed above. Perhaps the most convincing aspect of their work was to expose the cells to different conditions before fixation—a lower temperature

FIGURE 16-52

Indirect Immunofluorescence Image of the Mitotic Apparatus of a Dividing Sea Urchin Egg.   The sectioned and fixed cells were stained with a fluorescently labeled monoclonal antibody directed against a 37-kDa microtubule-associated protein. (Bar = 10 μm; courtesy of R. B. Vallee and G. S. Bloom.)

or reagents known to affect the structure of the cytoskeleton—that alter reversibly the appearance of the MTL, without eliminating it.

Skepticism regarding the existence of the MTL is based on two separate issues:

- A number of workers doubt that filaments other than microtubules, IFs, and microfilaments exist in the cell. The difficulty here may lie in definitions. There does not seem to be any disagreement regarding the existence of connector filaments such as dynein, MAPs, and the biochemically undefined connector filaments, seen by Schliwa after Triton X-100 extraction (Figure 16–45). Numerous other workers have observed filaments as small as 2–3 nm in diameter that vary in length and interconnect the elements of the cytoskeleton with each other and with membranes. However, these filaments have been viewed as short, unbranched, and smooth "connectors," rather than the more extensive, branched microtrabeculae of variable width. Recent work with the quick-freeze method is beginning to reveal much more extensive branching and merging filaments, varying in diameter and granular surface structure (Figures 16–53 and 16–54). The similarities between these structures and Porter's microtrabeculae are striking and probably will not be wished away by definitional choices.

- Another objection to the MTL is that no major protein fraction can be identified that would correspond to what must amount to a major structural component of the cytoskeleton. Two considerations are relevant here. First, it is unlikely that the microtrabeculae are composed of a single protein. In fact, considerable variation in their chemistry is likely, depending on what they interconnect and what region of the cell they happen to occupy. If the MTL plays a role in the development of the heterogeneous local structures in the cytomatrix, a great deal of molecular heterogeneity in the MTL may be expected. Second, it is likely that a portion of the MTL is covalently cross-linked and therefore does not appear in the major analytical system of contemporary biochemical research, which is the polyacrylamide gel.

Figures 16–55(a) and 16–55(b), p. 816, are electron micrographs of a peripheral nerve trunk that was extracted in urea and SDS, then washed in buffer and treated with collagenase. The remaining nerve ghost was treated with guanidine·HCl and fixed. This preparation of the highly insoluble remainder of the cell shows vesicular, fibrillar, and filamentous cross-linked structures, topologically similar to the ones shown in Figures 16–53

and 16–54. Figure 16–55(c) shows a polyacrylamide gel of these insoluble structures after boiling them in SDS and mercaptoethanol. A great deal of the material is still not capable of being broken down to molecular weights less than 10^7. It is not easy to escape the conclusion that a large portion of the structures seen in Figures 16–55(a) and 16–55(b) are in fact covalently cross-linked.

In summary, the filamentous elements of the cytoskeleton cross-link with each other and with membranes to form the cytoplasmic matrix. The structure is in constant movement and change, and certainly varies in composition in different parts of the cell and at different stages of development. Whether or not the cytomatrix is described as a microtrabecular lattice depends on whether a relationship is seen between the generally accepted experimental observations and the structure originally described by Porter.

The Cytomatrix and the Cytosol

The early experimental cytologists, who broke open cells and centrifuged down the various organelles, called the aqueous supernatant portion of the centrifuged homogenate the *cytosol*. This material is rich in proteins, many of them enzymes, transfer RNA, and the low-molecular-weight metabolites of the cell.

Observations in recent years have provoked a reevaluation of the relation of the cytosol fraction, as defined above, to the aqueous phase of the cytoplasm. Before commenting on this issue, we must first define carefully the question. The cytomatrix can be considered an interconnected system (or gel) in which proteins and other molecules are held in some state of aggregation. The aggregates are probably not permanent and the molecules in the cytomatrix are likely to be in equilibrium with free molecules in the solution around them. One question regards the nature of this equilibrium. If a given molecular species forms part of the matrix for most of its life in the cell, it must be judged to be part of the cytomatrix; if it is present in a freely diffusible state for most of its life, then it must be judged to be part of the aqueous cytoplasm. It seems likely that most small metabolites will not generally be sequestered in the cytomatrix. The main question we must ask, therefore, is whether the aqueous cytoplasm of the living cell is as rich in protein and nucleic acid as the cytosol, the layer of organelle-free liquid obtained after homogenization and centrifugation.

There are many lines of evidence showing that the aqueous phase of the intracellular cytoplasm contains relatively few macromolecules and that many proteins in the cytosol are bound by secondary interactions to the filaments of the cytomatrix for a large portion of their lives:

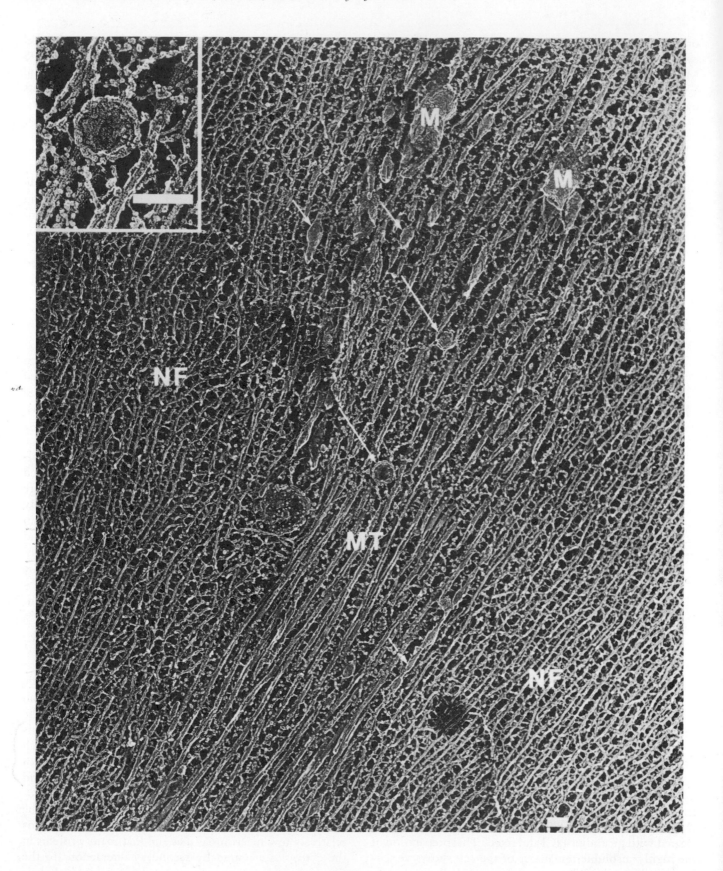

FIGURE **16-53**
Connector Filaments Between Microtubules, Neurofilaments, and Membrane-bound Vesicles and Organelles, Revealed by Quick-freeze, Deep-etch Microscopy of Axons. After treatment with the neurotoxic agent IDPN (β,β'-iminodipropionitrile), bundles of microtubules (MT) remain in the central region of axons, and neurofilaments (NF) are found in the periphery. This makes it possible to examine the microtubule connector filaments separately from those of neurofilaments. The respective connector filaments differ in their morphology; those of the neurofilaments are short and unbranched while those of the microtubule are much longer and branched. Mitochondria (M), membrane-bound organelles (long arrows, inset), and smooth endoplasmic reticulum (short arrows) are associated with the microtubules. (Bar = 0.1 μm; courtesy of N. Hirokawa, G. S. Bloom, and R. B. Vallee.)

- Studies of glycolytic enzymes over the last 10 years by C. Masters and others show that a number of these enzymes bind to the thin filaments of skeletal muscle as well as to the microfilaments of other cell types. The degree of binding in muscle cells is profoundly affected by conditions—for example, continuous stimulation to contract. Enzyme binding can also show considerable specificity: aldolase, for example, though normally considered a "soluble" enzyme, binds strongly (the association constant = 4.1×10^5 M^{-1}) to actin filaments, once every 10–12

heptameric repeat units. Interestingly, binding to microfilaments fundamentally affects a number of enzyme kinetic parameters; the degree of binding also varies with the developmental state of the tissue.

- Recently a series of thorough and very ingenious studies on the diffusion of proteins in intact cells has been reported by P. Paine. He injected a small drop of gelatin into the large oöcyte of the frog *Xenopus laevis* (Figure 16–56) and, after a period of equilibration, analyzed the total protein in the gelatin drop as well as in comparable portions of cytoplasm and nucleus. Paine measured the relative concentrations of 90 polypeptides, by two-dimensional polyacrylamide gel electrophoresis, and found that over 80% of the oöcyte polypeptides assayed exist, at least in part, in a nondiffusing form. He also made the interesting observation that diffusible actin in the cytoplasm has an unusually high concentration, higher than the critical concentration for polymerization. This was interpreted to mean that factors inhibiting polymerization must be present in the oöcyte. Paine also utilized his ingenious method to demonstrate that dramatic changes occur during meiosis in the abilities of proteins to diffuse.

- The ability of a radioactive DNA probe to hybridize to complementary RNA has allowed cell biologists to test the hypothesis that special mRNAs might be localized to certain regions of the cell. This kind of

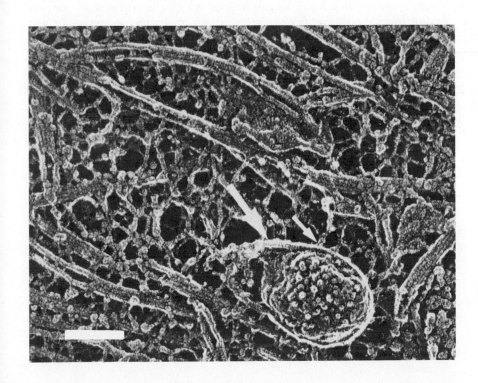

FIGURE **16-54**
Connector Filaments Between Microtubules of a Saponin-treated Axon from an IDPN-treated Rat. The quick-freeze deep-etch method reveals the length of the connector filaments and the extensiveness of their branching. The network of these structural elements is strikingly similar to Porter's microtrabecular lattice. Notice that cross-bridges between a membrane organelle (thick arrow) and microtubules are clearly preserved (short arrow). (Bar = 0.1 μm; courtesy of N. Hirokawa, G. S. Bloom, and R. B. Vallee.)

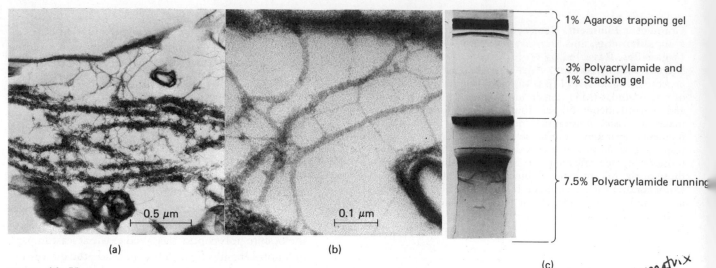

(a) (b) (c)

1% Agarose trapping gel

3% Polyacrylamide and 1% Stacking gel

7.5% Polyacrylamide running

FIGURE 16–55

EM Images of a Covalently Cross-linked Cytomatrix from a Frog Sciatic Nerve. (a) A "nerve trunk ghost" was prepared by extracting the nerve with 6 M urea, 5% SDS, and 5% 2-mercaptoethanol. The ghosts were then washed in buffer, treated with collagenase and washed in 6 M guanidine·HCl, and fixed in gluteraldehyde. Three highly insoluble structural elements, in the form of vesicles, fibers, and thin filaments, survive this treatment and can be seen to interconnect. The arrows points to 2–3 nm filaments. (b) Higher magnification of the same EM image, showing that the thinnest filaments are well below 6 nm in diameter. (From A. G. Loewy and P. S. Klainer.) (c) When nerve trunk ghosts are boiled in 5% SDS and 5% 2-mercaptoethanol and then electrophoretically fractionated, they can be shown to contain a number of very high molecular weight protein chains, one of which cannot even enter a stacking gel consisting of 3% polyacrylamide and 1% agarose. (From A. G. Loewy and K. Wolfe.)

cytoplasm = cytosol + cytomatrix

experiment has revealed that mRNAs for actin are found preferentially in the periphery of nonmuscle cells and that frog oöcytes have different mRNAs in different regions. Other experiments have shown that ribosomes tend to be bound to the cytomatrix proteins instead of diffusing freely in the aqueous cytoplasm.

The studies cited above and others that we do not have space to discuss show that the cytomatrix contains an extensive gel phase in which the components do not diffuse freely. What about the aqueous phase of the cytomatrix? Does it consist of a dilute solution of "bulk" water with properties similar to the water encountered in test tube solutions? Or, does cell water mostly bind to the surface of one or another protein and thus possess properties different from the dilute aqueous solutions with which we are familiar?

This is a complicated problem which is only beginning to be faced. In this connection it is important to

FIGURE 16–56

Experimental System for Measuring the Diffusion of Proteins Through the Cytoplasmic Matrix. (a) A reference phase, consisting of a droplet of gelatin, is injected into the cytoplasm of a frog oöcyte. (b) After allowing time for diffusion, the oöcyte is frozen and the reference phase, the nucleus, and an equivalent piece of ctyoplasm are removed. The proteins from these three samples are then analyzed by two-dimensional gel electrophoresis. (Redrawn from P. L. Paine.)

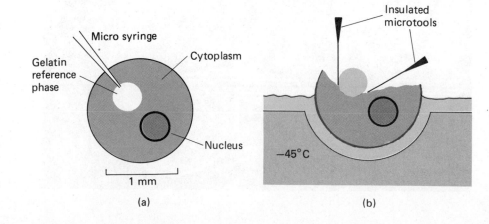

Micro syringe
Gelatin reference phase
Cytoplasm
Nucleus
1 mm
(a)

Insulated microtools
−45°C
(b)

realize that the cytomatrix is a structure with a tremendously large surface area. One estimate by J. Clegg is that a cell 16 μm in diameter would have a cytoskeletal surface area of 50,000 to 100,000 μm^2 (this should be contrasted with the surface of a sphere of the same diameter: ≈800 μm^2). The question is how far from its surface does the protein gel phase influence the solvent properties of the water. Estimates of the fraction of cytoplasmic water with altered solvent properties have ranged from 2% to 100%. So far no unambiguous answers have been obtained, but Clegg's measurements with the cysts of brine shrimp *Artemia* suggest that most of their cytoplasmic water exhibits rotational and translational motions that differ appreciably from those of the pure liquid.

In summary, the cytomatrix is a highly structured, finely divided gel, containing and surrounded by water with special properties. Apart from controlling a number of integrated processes (discussed in the next section), the cytomatrix supplements the role of the cell organelles and membrane-bound vesicles by adding another dimension to the segregation of cellular materials. As membrane-bound structures separate their contents into a compartment by restricting diffusion through membranes, the cytomatrix brings about a form of compartmentalization by selective binding of molecules. This binding may vary from one position of the cell to another and from one developmental or physiological state to another.

Movement in Nonmuscle Cells

A number of integrated cell processes that result in movement are associated with the cytomatrix; they involve the cooperative interaction of many molecules. The literature describing these events is enormous; progress in achieving a molecular understanding of these processes, however, has been relatively slow. In the following discussion the various kinds of nonmuscle cell movements are divided into three broad categories; further understanding might well change the nature of this classification. We shall begin with a discussion of movement seemingly brought about by a contraction of a solid matrix, a process that appears to resemble the mechanism already discussed for striated muscle. We shall then discuss instances in which movement is brought about by an interaction between molecules bound to cellular organelles and actin filaments. Finally, we shall discuss a newly discovered motor system that is causing cell biologists to rethink the role of microtubules in nonmuscle cell movements.

Solid-State Contraction of the Cytomatrix

In muscle and in ciliary motion there occurs a form of contraction in which the filaments involved are fixed with respect to each other in two dimensions but engage in a limited sliding motion in the third dimension. Similar forms of "contraction" through sliding in one dimension can be observed in the cytomatrix of other cells.

Cell Cleavage

During mitosis, nuclear division in animal cells is normally followed by rapid cell cleavage, resulting in the formation of two daughter cells. Just before cleavage, a constriction appears in the center of the dividing cell, the *cleavage furrow*. Examination by electron microscopy shows that during formation of the cleavage furrow a *ring of microfilaments* appears and then rapidly disappears after cleavage is completed. These microfilaments are oriented parallel to the plane of cleavage and lie close to the cell membrane (Figure 16–57a). The presence of actin in these microfilaments has been demonstrated by injecting cells with fluorescently labeled myosin light chains (Figure 16–57b) and fluorescently labeled actin (Figure 16–57c).

A plausible theory to explain cleavage is that it results from the contraction of the ring of microfilaments in the cleavage furrow. This was first suggested by D. Marsland in the 1950s, long before the presence of microfilaments had been observed. It is likely that the mechanism of contraction of the cleavage ring is based on an actin-myosin interaction; the injection of antibodies raised against myosin inhibits cleavage. Exposure to cytochalasin B brings about the depolymerization of actin and has been used in a number of studies to demonstrate the necessity of actin in a given process; cytochalasin B also inhibits cleavage. The observation that colchicine does not inhibit cleavage suggests that microtubules are not actively involved in cell cleavage.

Structural formula of cytochalasin B

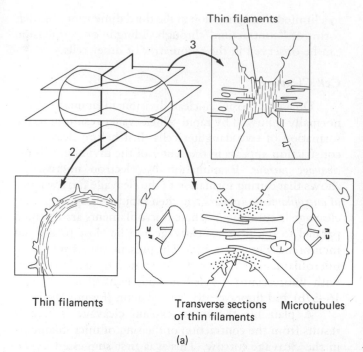

Thin filaments

Thin filaments

Transverse sections of thin filaments

Microtubules

(a)

FIGURE 16–57

Division in Animal Cells by the Formation of a Cleavage Furrow. (a) Drawing showing orientation of actin microfilaments. Panel at upper left shows cell in the process of cleaving and three possible planes of sectioning: (1) longitudinal plane through the center of the dividing cell, showing transverse sections of the actin filaments; (2) transverse plane through the cleavage furrow, showing the actin filaments longitudinally; (3) longitudinal section in the plane of the cleavage ring, showing the actin filaments arranged longitudinally. (Courtesy of T. Schroeder.) (b) Dividing Pt-K2 cell previously injected during interphase with (iodoacetamido tetramethylrhodamine) fluorescently labeled myosin light chains. (a) A-band of myosin spans the cell from one side to the other before any obvious signs of cleavage. (2) Cell begins to constrict (arrow). Note the greater concentration of myosin at the furrow. (3 to 6) As upper furrow deepens, lower furrow begins and continues to develop. (6) Lower furrow also shows high concentration of myosin (see arrowhead in (4)). (Bar = 10 μm; courtesy of J. M. Sanger, B. Mittal, J. S. Dome, and J. W. Sanger.)

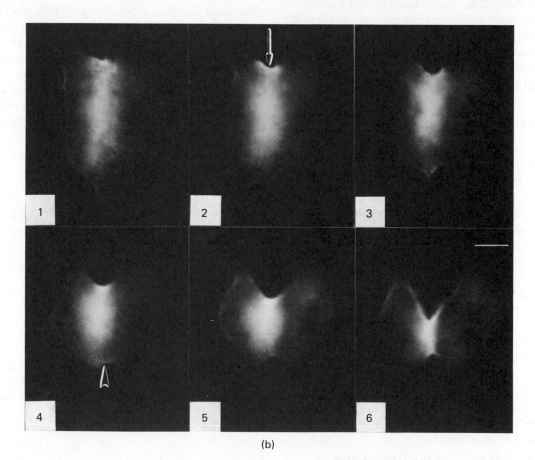

(b)

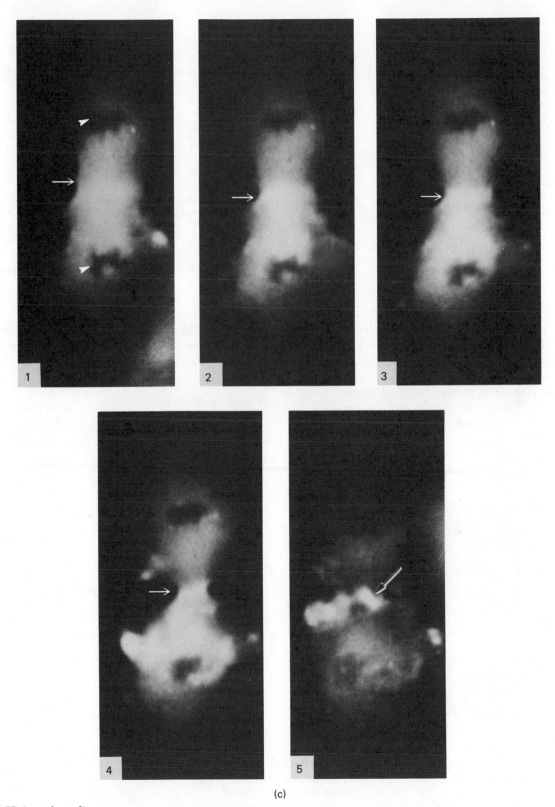

(c)

FIGURE **16–57 (continued)**
(c) Dividing Pt-K2 cell injected during interphase with (lissamine
rhodaminesulphonyl chloride) fluorescently labeled actin. (1) At the end of
anaphase, a bright band of actin is visible (arrow) midway between the dark
regions in which separating chromosomes are located (arrowheads). (2 to 5) Actin-
containing cleavage ring contracts, showing increasing concentrations of actin.
(×4000; courtesy of J. M. Sanger, B. Mittal, J. S. Dome, and J. W. Sanger.)

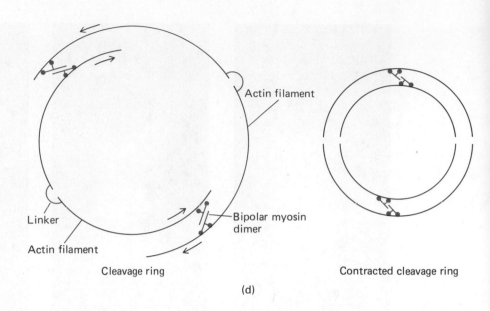

FIGURE 16–57 (continued)
(d) Hypothetical scheme showing how contraction of the cleavage ring could be brought about by the action of a bipolar myosin dimer "walking" in opposite directions on pairs of actin filaments in antiparallel orientation. The actual contraction of the cleavage ring can only occur if the actin filaments are anchored to each other as proposed by the colored linkers. However, the same effect could be brought about if the filaments were embedded in a coherent matrix or gel.

The myosin in the cleavage ring has not been detected in the form of thick filaments. In that respect the cleavage ring resembles smooth, rather than striated, muscle. In the absence of direct evidence, it must be assumed that myosin is present in small bipolar assemblies that bring about the contraction of the ring by causing the microfilaments to slide relative to each other (Figure 16–57d). Since the contractile ring remains constant in width throughout its contraction, depolymer-

ization of actin must occur during the process, resulting finally in the total disappearance of the ring once cleavage has been completed.

Amoeboid Motion

The motion of *Amoeba* (Figure 16–58), and of other shape-changing animal cells such as fibroblasts, has been studied for many years; a large body of evidence has

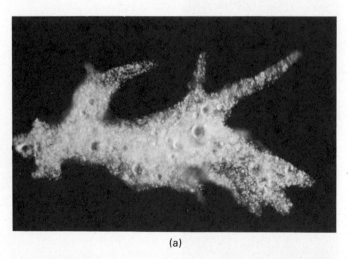

(a)

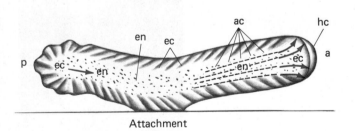

(b)

FIGURE 16–58
Amoeboid Motion. (a) A phase-contrast micrograph of *Amoeba proteus.* (b) Generally accepted observations regarding amoeboid motion. At the anterior end (a) there is a hyalin or transparent cap (hc); the posterior end (p) is crenated, showing the effect of loss of endoplasm (en) bringing about a shrinkage in the gelated ectoplasm (ec). At the posterior end there is conversion of ectoplasm to endoplasm (ec → en); at the anterior end there is a conversion of endoplasm to ectoplasm. Actin microfilaments (ac) are present in the ectoplasm and are shown to begin polymerization in the endoplasm of the anterior portion of the cell. (Modified from D. L. Taylor, J. A. Rhodes, and S. A. Hammond.)

accumulated concluding that some form of contractility of the cytoplasm is involved. Even though there are significant differences between the bulk movements of cytoplasm seen in *Amoeba* and the movements of other so-called amoeboid cells, there are sufficient similarities to discuss them together. All theories of amoeboid motion have to take account of four observations: (1) the cells attach at given points to the substratum (the adhesion plaques); (2) there is an *ectoplasmic layer*—gel-like in behavior—just below the cell surface, in which actin microfilament bundles are present; (3) there is an *endoplasmic core,* a less viscous region, in which the actin filaments are fewer and not organized; (4) during movement, there is a conversion of gel-like ectoplasm into less viscous endoplasm at the retreating portion of the cell and a conversion of endoplasm into ectoplasm at the advancing portion of the cell (Figure 16–58b).

There is at present no agreement as to the precise location of any contractions nor whether the motive force is generated in the ectoplasm or endoplasm. An early view regarding this controversy is the *posterior contraction theory* (Figure 16–59a), advanced in the 1920s by C. Pantin and S. O. Mast. This theory proposes that

amoeboid motion is caused by a contraction of the ectoplasm at the rear of the cell, producing an increase in pressure in the fluid endoplasm, relief of the pressure by expansion at the advancing portion of the cell, and bulk movement of the endoplasm towards the front. This model is consistent with the observation that ectoplasm is converted to endoplasm in the rear while the reverse process occurs in front, thus presumably accounting for the mass flow of endoplasm from rear to front. Also consistent with the posterior contraction theory is the observation that when ATP is injected into the posterior region of an *Amoeba* forward motion is enhanced, while injection of ATP into the anterior end reverses the streaming of the endoplasm. The posterior contraction theory does not by itself explain why suction applied to a cell does not result in retraction of protrusions. There must be some stabilization of the evaginated structure, perhaps by cytoskeletal elements, shortly after protrusion (see later discussion of fibroblast movement).

A more recent *frontal contraction theory,* proposed by R. D. Allen, suggests that contraction occurs in the leading portion of the cell during the conversion of endoplasm to ectoplasm (Figure 16–59b). The moving

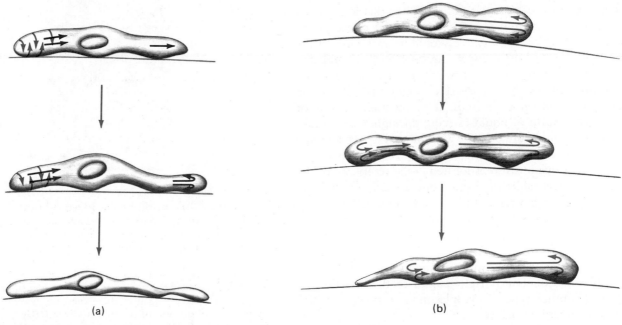

(a)   (b)

FIGURE **16–59**
Contraction Theories of Amoeboid Motion.   (a) Posterior contraction theory. At the rear of the cell, toward the center, gel-like ectoplasm is converted into more fluid endoplasm. The ectoplasm in the outer cortex contracts, increasing the pressure and forcing the fluid endoplasm forward. At the front of the cell, the endoplasm is converted into ectoplasm, which migrates backward. (b) Anterior contraction theory. The endoplasm is believed sufficiently stable to withstand tension. As the endoplasm arrives at the front of the cell and bends sharply, it is converted into ectoplasm and exerts tension, which pulls the remaining endoplasm forward.

(streaming) endoplasm, while less gel-like than the ectoplasm, is believed to be sufficiently stable to propagate a tension from the contraction and thus to be pulled forward. In spite of masses of data assembled on amoeboid motion, no experiment so far has been able to eliminate decisively either one of these theories.

Some of the properties of this motion can be demonstrated outside the cell. The work of J. Condeelis, P. Moore, and D. Taylor showed that at appropriate concentrations of Ca^{2+}, isolated cytoplasm from *Amoeba* would stream outward and form a pseudopod-like appearance. Similar to the behavior of the intact cell, the cytoplasm thickens after bending at the forward end and then contracts in a backward direction. The cytoplasm of *Amoeba* contains microfilaments made of actin (7 nm in diameter, decoratable by heavy meromyosin) and thick filaments made of myosin. At the appropriate calcium concentration, filaments form and contraction (streaming) occurs.

Cytoplasmic Streaming

The very active cytoplasmic streaming in the plasmodium of the slime mold *Physarum polycephalum* (Figure 16–60) has been studied for many years by N. Kamiya. The most generally accepted theory explaining this dramatic form of cytoplasmic motion is that it is caused by *waves of contraction* traveling along veins of ectoplasm that surround the flow (these would be analogous to the peristaltic waves that propel digesting food through the intestine). Very ingenious work by Kamiya and K. Kuroda supports this theory. They demonstrated that the distribution of velocities in the stream is similar to that of a passive stream of liquid moving through a tube under hydrostatic pressure. K. E. Wolhlfarth-Bottermann appears to have demonstrated that the active contraction generating this pressure is located in the ectoplasm of the vein. He was able to dissect out a vein, replace the endoplasm with a salt solution, and continue measuring by tensiometric means the cycle of contraction and relaxation observable in the veins before the endoplasm had been removed. Electron microscope analysis shows the presence of actin microfilaments in the veins of ectoplasm, oriented at right angles to the direction of flow. (Another type of cytoplasmic streaming, in plant cells, is described below.)

Movement of Microvilli

The epithelial cells of the intestine, which are responsible for the transfer of nutrients and electrolytes from the contents of the gut to the blood supply of the organism, perform this task with great efficiency by virtue of possessing an enormously expanded surface area (the *brush border*) that faces the intestinal contents. This expansion is brought about by the formation of about

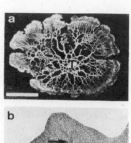

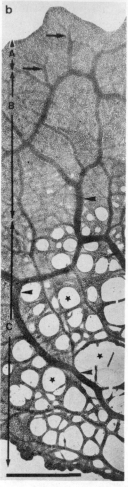

FIGURE **16–60**

Cytoplasmic Streaming in the Plasmodium of the Slime Mold *Physarum Polycephalum*. This organism adopts the form of a large multinucleate plasmodium that spreads into a thin sheet of cytoplasm; within the plasmodium there is an active "shuttle" streaming motion, reversing its direction approximately every minute. Forward motion of the plasmodial mass is brought about by changing the relative lengths of the streaming periods in the two directions. (a) Micrograph of portion of the plasmodium. (Bar = 2 cm; courtesy of K. E. Wohlfarth-Bottermann.) (b) Detail of the morphological organization of a plasmodium as shown in (a): A, Anterior region consisting of a homogeneous protoplasmic sheet. B, Intermediate region containing endoplasmic pathways in the ectoplasmic sheet (arrows). C, Posterior region consisting mainly of single protoplasmic strands (arrowheads), because the former homogeneous protoplasmic sheet is no longer continuous (free areas marked by stars). (Bar = 3mm; courtesy of W. Naib-Majani, W. Stockeror, and K. E. Wohlfarth-Bottermann.)

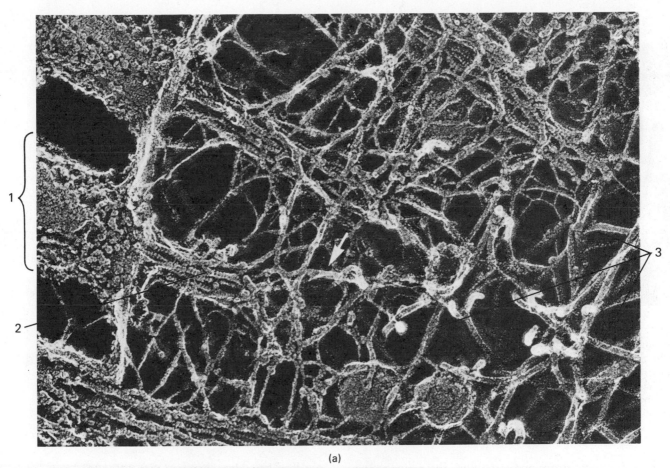

(a)

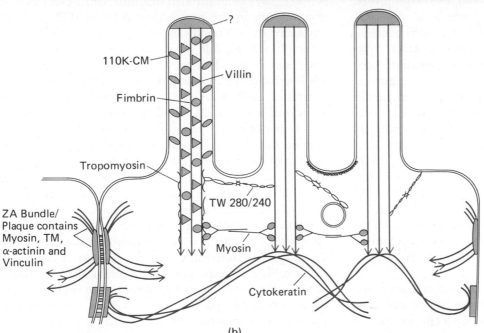

(b)

FIGURE 16-61
The Intestinal Epithelial Brush Border. (a) Quick-freeze, deep-etch electron micrograph showing the cytoskeletal assemblies present in this remarkable cell. (1) Microvillus, (2) extensions of bundles of actin filaments from the core of the microvilli, (3) intermediate filaments. (Bar = 0.1 μm; courtesy of N. Hirokawa and J. E. Henser.) (b) Diagram of intestinal epithelial brush border identifying some of the proteins associated with the cytoskeleton. (Courtesy of M. Mooseker.)

1000 slender finger-like projections (*microvilli*) per cell (Figure 16–61a). These structures are known to move; they extend and retract as well as fan out. Presumably such movements help to stir the solution near the surface of the gut, thereby aiding diffusion.

The cytoskeletal assemblies responsible for the structure and motion of the intestinal brush border have been studied intensively over the last decade, and remarkable progress has been made in the identification and characterization of the component proteins (Figure 16–61b).

(a)

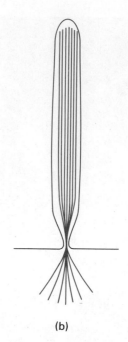

(b)

FIGURE 16–62

The Stereocilia on the Surface of the Hair Cells Found in the Cochlea and Vestibule of the Inner Ear.   (a) Scanning electron micrograph showing the surface of the hair cells and their "pipe organ" arrangement of the stereocilia. (Courtesy of L. Tilney.) (b) Diagram of a stereocilium in longitudinal section showing extension of actin filaments from stereocilium into the body of hair cell.

Among the most prominent are villin, the major protein found in brush borders; actin, which is the main component of a bundle or core of 20–30 microfilaments (the bundle of filaments extends into the apical cytoplasm, forming a structure known as the terminal web); fimbrin, the protein mainly responsible for the association of the actin filaments in their ordered and parallel array; and myosin and tropomyosin, which are associated with the portion of the actin bundle that penetrates the terminal web (much of the myosin is present in the form of bipolar dimers). There are also proteins that attach the actin bundle to the membrane and to the underlying cytoskeleton (Figure 16–61). Between the brush border cells is a specialized region of the cell surface known as a *junctional complex* (*z*onula *a*dherens, ZA) that contains the contractile proteins actin, myosin, and tropomyosin, as well as α-actinin and vinculin (see further discussion of junctions in Chapter 17).

The role of actin and associated proteins in maintaining the structure of the microvilli can be deduced from direct observation of their components. The mechanism by which movement of the microvilli is brought about is not well understood. It is difficult, for instance, to imagine how the interaction between the myosin bipolar dimers and the basal end of the actin bundles can bring about movement. Indeed M. Mooseker and his co-

workers have demonstrated that differential extraction of that myosin in the presence of ATP at 0°C (without extracting the ZA bundle myosin) produces a filamentous assembly still capable of contracting in the presence of Ca^{2+} and ATP. The observation that the ZA bundle is involved with the contraction process of the brush border correlates well with the observation that the actin associated with the ZA bundle is in an antiparallel array.

Other hair-like protrusions of the cell surface that contain bundles of actin are the *stereocilia* found in the cochlea of the inner ear (Figure 16–62a). These structures also contain a core of actin filaments that extends into the body of the cell (Figure 16–62b). The stereocilia are not involved in spontaneous motion; instead, sound waves cause them to move, a motion that is converted into an electrical signal that is then transmitted to the brain. Like organ pipes, stereocilia come in a variety of sizes, presumably each responding characteristically to different sound wave frequencies. The stereocilium is built so that its actin filaments pass through a narrow constriction at the base of the structure (Figure 16–62b); the constriction renders the stereocilium exquisitely sensitive to perturbation by sound waves. The mechanism whereby the motion of the stereocilium causes the production of an action potential is as yet unknown.

Motion of the Cell Surface

Cells engage in numerous forms of movement in which the cell surface folds inwards (invagination), outwards (evagination), or both (crenation). One of the most thoroughly investigated cell surfaces is that of the mammalian red blood cell that, in many species, normally assumes the shape of a biconcave disk. This shape is considerably deformed when the cell passes through tiny blood capillaries. Also, under a variety of conditions, the surface can assume a number of different shapes (Figure 16–63). Although considerable progress has been made regarding the identity and disposition of the proteins anchored to the cytoplasmic face of the erythrocyte membrane, the precise molecular mechanism of these cell surface changes is not yet understood.

Movement by Fibroblasts

Brush border microvilli and stereocilia of the inner ear are examples of permanent extensions of the cell surface. There are also many examples of temporary extensions—the small *filopodia* and the larger *lamellipodia* (see Figure 16–32)—in which bundles of actin filaments, or at other times microtubules, are found in parallel array. At times an array of filopodia will represent a large fraction of the total surface area. Both these structures represent evaginations of the cell surface and both are capable of movement.

The amoeboid motion of fibroblasts (and of other cells growing in tissue culture or during embryonic migration) is based on these temporary extensions of the cell surface. Unlike the motion of *Amoeba,* however, there is in this "amoeboid" motion very little major movement of cytoplasm. The fibroblast cell is attached tightly to the substratum by many points at the rear and front. Forward progress is accomplished by motion of the leading edge of the cell. There is protrusion of one or more lamellipodia, followed by contact with the substratum and attachment to it. The initial attachments are of limited extent (*focal contacts* or *adhesion plaques*) but they are strong (see Figure 16–32). There are also larger and looser attachments (*close contacts*). As the cell moves forward, the close contacts are detached but the focal contacts often remain, after the cell has pulled away, as the sites of attachment of *retraction fibers* (Figure 16–64).

High-voltage transmission electron microscopy has revealed that a bundle of filaments terminates at each adhesion plaque. These bundles (stress fibers) lie in the cortex of the flattened cell and run from the focal contact towards the nucleus. The microfilaments in the bundles have been shown to contain actin by their ability to bind heavy meromyosin. Fluorescent antibodies against several other muscle proteins have detected immunologically similar proteins in the bundles: myosin, tropomyosin, α-actinin, and filamin. The α-actinin protein has also been detected in the adhesion plaque itself, perhaps playing a role analogous to its function in the Z-disk of

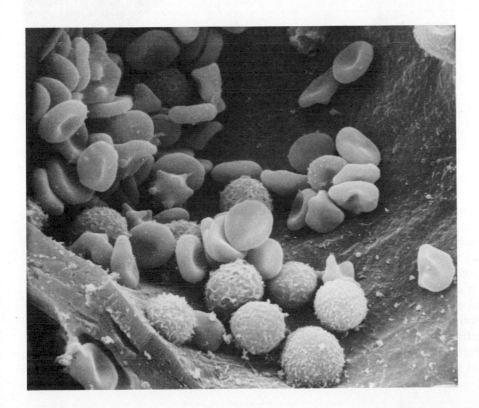

FIGURE 16–63
Scanning Electron Micrograph of Red Blood Cells on the Inner Surface of a Capillary Wall.
Under the influence of a variety of reagents, red cells can change their shape, from biconcave to round, and their surfaces, from smooth to crenated. Magnification ×3500. (Courtesy of R. G. Kessel.)

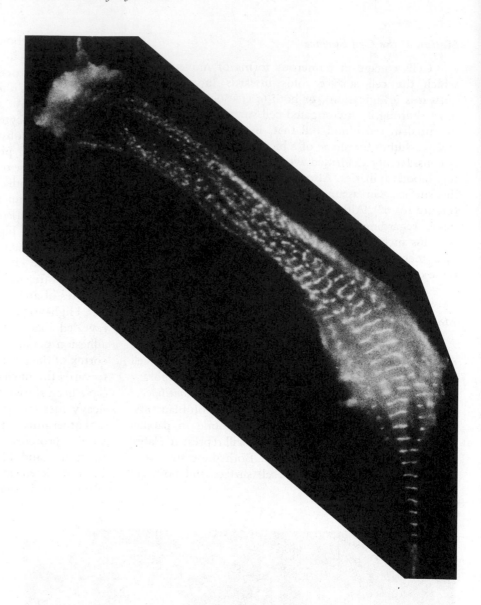

FIGURE **16-64**

The Retraction Fiber. As a cell moves, certain areas (focal contacts) remain attached to the substratum, which causes a portion of the cell to be pulled out into a retraction fiber. This chick embryonic cardiac myocyte was stained with an antibody directed against α-actinin, a protein associated with the Z-disks of myofibrils. Note that the distance between Z-bands in the terminal portion is larger than that in the proximal portion of the retraction fiber (RF), suggesting that it has been stretched by the forward movement of the cell. Note also the ruffle at the leading edge of the cell and its diffusely distributed α-actinin, which is probably attached to diffusely distributed thin filaments of actin. (Courtesy of J. W. Sanger and J. M. Sanger.)

striated muscle. Finally, a fiber bundle has been isolated by microlaser surgery from a glycerinated fibroblast and shown to contract after the addition of ATP.

These findings suggest a model in which an actin-myosin-like interaction causes a contraction that brings the cell body towards the focal attachments at the leading edge. The force developed by the actin–myosin is propagated as tension in the filament bundles; the attachment of the bundles to the cytoskeleton allows this tension to be expressed as movement of the bulk of the cell. It seems likely that contraction occurs mostly at the front of the cell, although tension is developed over the entire length of the cell. This is because the rear of the cell retracts very rapidly when the tension is relieved, by disruption of the rearmost contacts (the trailing edge of a cell seems always to rupture its contacts, rather than loosening them). This movement is too rapid to be the result of contraction alone.

An actin-myosin mechanism for contraction is not the only possible explanation for fibroblast motility. There are microtubules in the lamellipodia and they too are joined to the rest of the cytomatrix. A contractile protein like dynein could be arranged on microtubules to produce the required contraction. The real puzzle about movement of nonmuscle cells, however, is to explain the protrusive activity of the lamellipodium. So far, all the mechanisms we have discussed explain only contraction and not protrusion.

There are several kinds of protrusive actions that have been observed in cells. The simplest is *blebbing,* the formation of a very small protrusion (bleb) that extends and retracts over a time scale of a few seconds (the small evaginations of the cell surface that result from blebbing have also been called microspikes). Blebs can be made to retract by adding sorbitol to the surrounding medium. The fibroblast membrane is relatively impermeable to sorbitol, so its addition makes the surrounding medium hypertonic and water leaves the cell, reducing the pressure bearing on the cell membrane. This leads to the notion that blebbing occurs because of intracellular hydrostatic pressure "blowing out" the cell membrane at regions that are weakened. How the cortical layer is weakened in a small area to allow the blebbing is not known. On the other hand, the retraction of blebs might be due to the contraction of filaments attached to the plasma membrane of the bleb.

Another force that could explain protrusion is the polymerization of actin filaments. The model for this is the formation of the acrosomal process of the sperm of the marine invertebrate *Thyone,* studied by L. Tilney and his associates. The extension of this protrusion is very rapid (10 μm/sec) and involves only the polymerization of G-actin to F-actin. A similar process could produce the protrusions necessary to move the lamellipodium forward.

Addition of new membrane at the leading edge of the cell may also play a role in protrusion. This suggestion follows from observations by M. Abercrombie that particles move backward on the dorsal surface of a cell that is moving forward. The flow of membrane that is thus revealed presumably arises from sources that add membrane material near the cell's leading edge and take it up at sinks of membrane material in the vicinity of the nucleus (since particles stop moving there). The flow on the ventral surface appears to go only to the rear of the lamellipodium. The addition of membrane at the leading edge may be due to the active exocytosis that occurs there.

A final possibility for explaining protrusion is that a slug of gel-like cytomatrix, attached to myosin, may be able to move relative to actin filaments in the cortex, thrusting forward (Figure 16–65). This mechanism seems more plausible in the light of data showing that myosin-coated objects can be moved along actin filaments (see discussion below of cyclosis). The mechanical stability of the gel would be capable of providing the protrusion.

Even though there are several possible forces to explain protrusion and contraction in nonmuscle cells, the exact mechanism for their movement is still not understood.

Movement of Cellular Structures Along Filaments

Large structures that are produced in one region of a cell cannot be transported efficiently to another part of the cell by diffusion; the rates are too slow. Recent research has revealed that many such structures—vesicles, organelles, and so forth—are moved along a filament that leads from one place to another within the cell. Some of these filaments are made of actin, some are microtu-

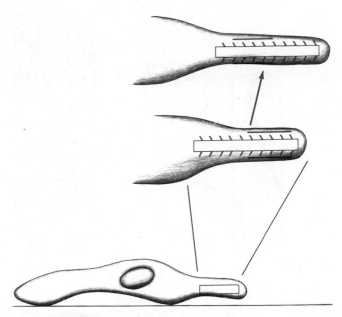

FIGURE 16–65
Protrusion by a Cross-linked Gel Slug. A small volume of the cytomatrix is hypothesized to be sufficiently cross-linked to be able to withstand compression. To the outer surface are attached myosin molecules. The slug "walks" along the actin filament bundles lying in the cortical layer of the cell. As the slug contacts the membrane, protrusion results. To continue this movement, actin filaments must be laid down at the forward margin.

bules. Attached to the structure to be moved are *motor proteins;* these also bind to the filaments, and depending on the polarity of the filament, motion occurs in one direction or the other, driven by energy from the hyrolysis of ATP.

Cyclosis in Plants

The cytoplasm of most mature plant cells is shaped like a barrel, lying in a thin layer between a large central vacuole and the rigid cell wall, and exhibits a circular motion called *cyclosis.* This movement is especially dramatic in the giant, multinucleated cells of green algae such as *Nitella* and *Chara.* The cytoplasm of these cells (which can reach a length of several centimeters!) exhibits a steady motion; in the presence of sunlight, the velocity of some cell components is as high as 0.1 mm/sec. The direction of the stream is at an angle to the axis of the cylindrical cell, causing the streaming motion to follow a gently helical path along the periphery of the cell (Figure 16–66a). A stationary cortex (or plasma gel) lies directly against the cell wall; the rapidly moving endoplasm is more toward the interior. Since the moving layer of endoplasm is barrel shaped, movement in one direction must be compensated for by an equivalent reverse

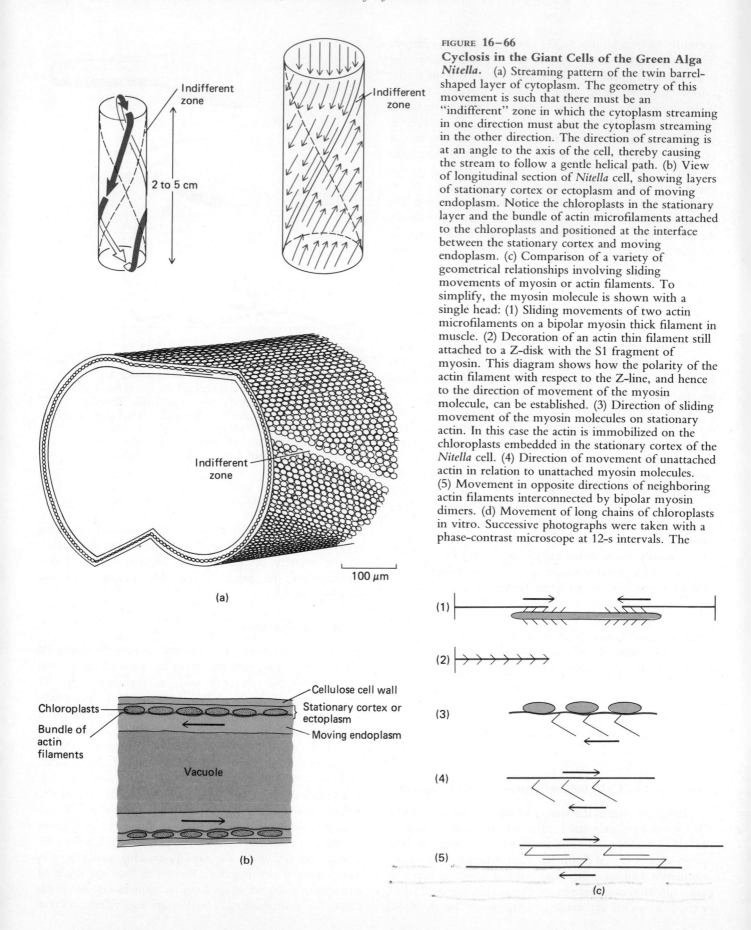

FIGURE 16-66
Cyclosis in the Giant Cells of the Green Alga *Nitella.* (a) Streaming pattern of the twin barrel-shaped layer of cytoplasm. The geometry of this movement is such that there must be an "indifferent" zone in which the cytoplasm streaming in one direction must abut the cytoplasm streaming in the other direction. The direction of streaming is at an angle to the axis of the cell, thereby causing the stream to follow a gentle helical path. (b) View of longitudinal section of *Nitella* cell, showing layers of stationary cortex or ectoplasm and of moving endoplasm. Notice the chloroplasts in the stationary layer and the bundle of actin microfilaments attached to the chloroplasts and positioned at the interface between the stationary cortex and moving endoplasm. (c) Comparison of a variety of geometrical relationships involving sliding movements of myosin or actin filaments. To simplify, the myosin molecule is shown with a single head: (1) Sliding movements of two actin microfilaments on a bipolar myosin thick filament in muscle. (2) Decoration of an actin thin filament still attached to a Z-disk with the S1 fragment of myosin. This diagram shows how the polarity of the actin filament with respect to the Z-line, and hence to the direction of movement of the myosin molecule, can be established. (3) Direction of sliding movement of the myosin molecules on stationary actin. In this case the actin is immobilized on the chloroplasts embedded in the stationary cortex of the *Nitella* cell. (4) Direction of movement of unattached actin in relation to unattached myosin molecules. (5) Movement in opposite directions of neighboring actin filaments interconnected by bipolar myosin dimers. (d) Movement of long chains of chloroplasts in vitro. Successive photographs were taken with a phase-contrast microscope at 12-s intervals. The

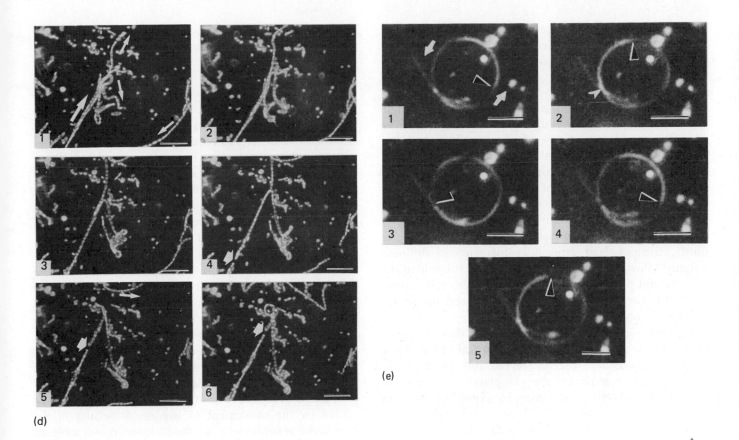

(d)

(e)

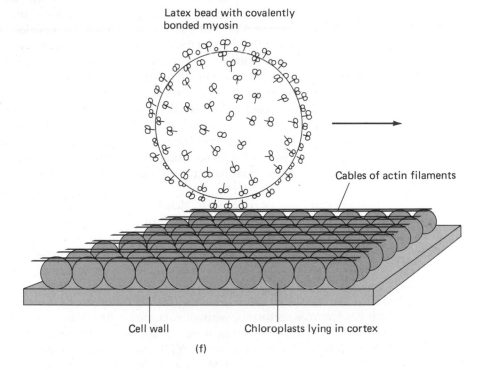

FIGURE **16–66 (continued)**

three long chloroplast chains were moving at ∼10 mm/s. The direction of movement is indicated by long arrows in (1) and (5). The movement of a gap in the chain is indicated by short arrows (4–6). (Bars = 50 mm; courtesy of S. Higashi-Fujime.) (e) Rotation of rings of chloroplast chains in vitro. Successive photographs were taken with a dark-field microscope at 1/8-s intervals. The speed of rotation was ∼160 rpm. The direction of movement is indicated by a white arrow. A black and white arrow indicates a particular point on the rotating ring. (Bars = 5 mm; courtesy of S. Higashi-Fujime.) (f) Latex bead experiment by M. P. Sheetz and J. A. Spudich. The illustration shows a spread-out section of the cell wall and cortex of the alga *Nitella*. Lying on top of the cell wall is the layer of chloroplasts and on top of them are ribbons (cables) of actin filaments (color). Muscle myosin was covalently bonded to latex beads. Even though the orientation of the myosin molecules on the bead would be expected to be random, when the beads were allowed to come into contact with the actin ribbons, the beads moved, without rotation, along the ribbons.

Latex bead with covalently bonded myosin

Cables of actin filaments

Cell wall

Chloroplasts lying in cortex

(f)

motion; this requires the presence of two "indifferent zones," each representing the boundary along which the streaming cytoplasm can move in opposite directions.

In the cortical layers directly inside the plasma membrane there is a single layer of microtubules, running parallel to cellulose filaments in the cell wall. The orientation of these microtubules differs from the direction of the streaming motion, leading to the conclusion that they influence the synthesis of the cell wall rather than the motion of the cytoplasm. Below the microtubules there is a single layer of tightly packed chloroplasts, lying in linear arrays and oriented in parallel to the direction of the stream (Figure 16–66b). The chloroplasts of a given row are tied to each other by 0.2-μm-diameter fibrils, composed of actin microfilaments of uniform polarity. Furthermore, the direction of microfilament polarity and the direction of streaming motion is consistent with the direction of sliding observed in muscle (Figure 16–66c).

Kamiya and Kuroda demonstrated that the actin filaments are part of the stationary cortical phase. They cut open one end of the *Nitella* cell and, by applying gentle pressure, extruded the cytoplasm. Under these conditions long ribbons of chloroplasts could be observed in the extruded cytoplasm, held together by the bundles of actin filaments. Kamiya, Kuroda, and S. Higashi-Fujime were able to remove the chloroplasts and showed that the fibrils were capable of autonomous motion along a microscope slide in the presence of ATP and Ca²⁺. This motion was translational if the fibrils maintained their linear shape (Figure 16–66d), or rotational if they formed circles (Figure 16–66e). The rate of movement of the fibrils was enhanced by the addition of ATP and heavy meromyosin, the rate being comparable to that of cytoplasmic streaming in vivo. *Nitella* therefore provides the opportunity of obtaining bundles of actin filaments of similar polarity that can be manipulated to study the mechanism of movement brought about by actin-myosin interaction.

An experiment that moves one step closer to explaining the motion of the endoplasm with respect to a stationary cortex was performed by M. Sheetz and J. Spudich. They slit open a *Nitella* cell, flattened it inside out against a surface, and washed off the endoplasm. Latex beads to which muscle myosin had been covalently attached were then added to the exposed inner surface of the cortex. Such beads move uniformly without rotation along the ribbons of exposed *Nitella* cortex at over 1 μm/second! Movement depends on the presence of ATP and is inhibited by N-ethyl maleimide (NEM), a myosin ATPase inhibitor. If a bead near the indifferent zone wanders off its actin bundle and, by Brownian motion, comes into contact with an actin bundle of opposite polarity, the direction of motion of the bead also reverses.

This bead motion can be used as an assay for critical structural features of the myosin molecule: fragments of myosin can be attached to the beads and tested for their ability to propagate motion; antibodies against specific regions of myosin can be evaluated for their ability to inhibit the motion. Myosins from different sources can also be evaluated. For example, when myosin from *Acanthamoeba* is used to coat the beads, movement only occurs if the myosin has been phosphorylated. Since *Acanthamoeba* myosin cannot form thick filaments, the myosin head group (without the hinge region) is apparently sufficient for the motion. It is perhaps ironical that this plant cell may turn out to play a critical role in gaining an understanding of mechanochemical transduction by skeletal muscle.

How does myosin by "walking" along actin filaments bring about the motion of an entire layer of endoplasm? As yet there is no clear answer. Kamiya and Kuroda measured the flow rate of the endoplasm relative to the distance from the stationary cortex and found that the entire endoplasm, from cortex to vacuole, moves at the same rate. This suggests that the force causing the motion is generated at the interface between cortex and endoplasm. The question of whether the viscous drag caused by the movement of myosin molecules in the endoplasm is sufficient to set the endoplasm into uniform motion is still unanswered. Kuroda suggests that the myosin molecules may be attached to large structures such as vesicles, thereby increasing considerably the magnitude of viscous drag. Another possibility is that the myosin molecules are attached to an interconnected lattice of fine filaments in the endoplasm. It has recently been demonstrated, in carrot cells grown in suspension culture, that plant cells also contain a network of fine filaments in their cytoplasm. If the myosin were bound to particles so large that they could not be removed from the cytomatrix, their movement could drag along the entire endoplasm.

The Movement of Pigment Granules

Many lower vertebrates such as fish are able to change their external color by moving pigment granules within specialized epidermal cells, called *chromatophores*. An evenly distributed state, which is colored, can change to a state in which the pigment granules cluster in the center of the cell, a state which is less colored (Figure 16–67a). The movement of pigment granules has previously been attributed to microtubules; however, recent studies point to an additional role for another cytomatrix component.

Two types of pigment granule movement can be distinguished: *aggregation,* in which pigment granules move smoothly and rapidly (5–10 μm/s) toward the cell center, and *dispersion,* which is both slower (2–5 μm/s)

FIGURE 16–67

The Movement of Pigment Granules in the Cultured Erythrophores of the Squirrel Fish.
(a) Dispersed (left) and aggregated (right) distribution of pigment granules in erythrophores, critical-point dried and viewed by 10^6-electron-volt microscopy. The pigment granules are organized into long radial files along microtubules (bar = 5 μm). (b) Cells quick-frozen during aggregation, with the water substituted by an osmium tetroxide-methoxyethanol mixture at $-90°C$. This treatment fixes cell components and extracts the carotenoid-containing pigment granules. Note the presence of two matrices: (1) a β-cytomatrix at the periphery, also containing microtubules (Mt) and smooth endoplasmic reticulum (SER), and an α-cytomatrix (MTL) forming a more compact structure around the cell center (PM). Nucleus (N) and mitochondria (M) can also be seen clearly (bar = 2.5 μm). (c) High-resolution image of α-cytomatrix showing thickened and somewhat beaded structure of filaments. Magnification ×150,000. (All courtesy of K. Porter.)

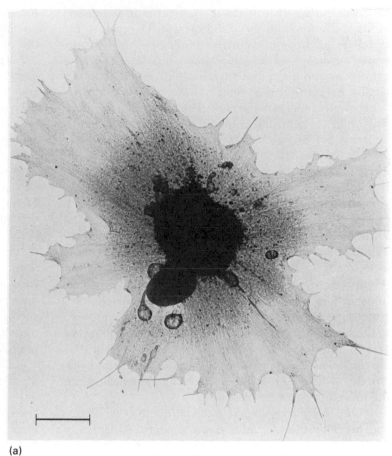

(a)

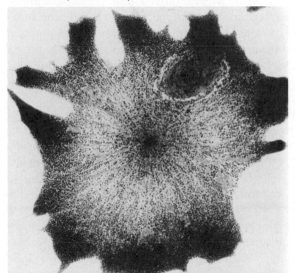

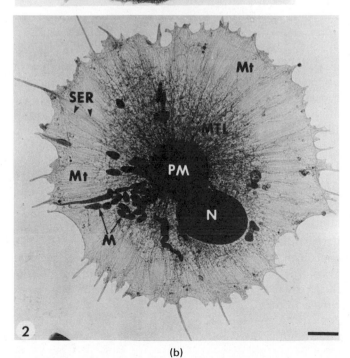

(b)

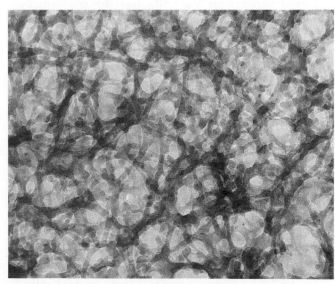

(c)

and "saltatory"—that is, the movement is accomplished in fits and starts both away and toward the center of the cell. This saltatory movement continues throughout the period of dispersion but stops upon the onset of aggregation. In cultured pigment cells, pigment motion can readily be controlled by epinephrine, which stimulates aggregation, and caffeine, which supports contraction. This ability to control chromatophore movement, as well as a number of other properties of the chromatophore system discussed below, have made it a useful experimental material for investigating the role of the cytomatrix in the directed movement of organelles.

Structurally, the fish chromatophore consists of a central region, containing a pair of centrioles enclosed in some dense material, from which thousands of microtubules radiate. These microtubules clearly play some role in orienting the pigment granules in linear columns (Figure 16–67a). Interestingly, few or no intermediate filaments are found in various kinds of chromatophores, and only small amounts of filamentous actin are present, located at the periphery of the cell. Inhibitors of actomyosin—for example, antibodies to myosin, N-ethylmaleimide-modified heavy meromyosin, and cytochalasin B—have no effect on pigment movement or on the morphology of the lattice.

Under the high-voltage electron microscope, Porter and his colleagues found a three-dimensional lattice of fine filaments (2–6 nm in diameter) in which the pigment granules seem to be encased. Cells in the dispersed and aggregated states exhibit two different lattices. These are visualized after preparing cells by quick freezing followed by substituting the water with an osmium tetroxide-methyoxyethanol mixture at −90°C (this mixture both acts as a fixative and extracts the pigment in the granules). As can be seen in Figure 16–67b, the aggregated states show two matrices: the β-cytomatrix, which remains dispersed throughout the cell, and the α-cytomatrix, which has contracted to the central region. High-resolution analysis of the α-cytomatrix shows that its filaments are thin in its dispersed form and become thickened and beaded in its aggregated form (Figure 16–67c).

Porter made careful studies of the movement of individual granules through several cycles of aggregation and dispersal. He found that upon dispersal the granules always return exactly to the same position in the cell from which they had aggregated. This suggests that the α-cytomatrix in chromatophores is a continuous structure and that pigment granules, although capable of movement, remain attached to it at specific points.

What about the role of microtubules in the motion of the pigment granules? If chromatophore microtubules are depolymerized, in a variety of ways, pigment granules aggregate, but in clumps instead of uniformly into the central region. These clumps cannot be induced to redisperse. From these and other experiments, Porter and his co-workers have concluded that the α-matrix supplies the motive force for aggregation and that the microtubules play a role only in "organizing" the movement.

An interesting question that requires clarification concerns the polarity of the microtubules. A technique developed by R. McIntosh allows the identification of microtubule polarity; at high ionic strength, added tubulin copolymerizes with existing microtubules to form curved sheets that appear as hooks in cross section. Viewed from the plus end of a pre-existing microtubule, the hooks show a clockwise curvature. A study of pigment cells revealed that the microtubules all point with their plus ends (the fast-growing ends) away from the central mass. Surgical procedures in which the arms of pigment cells were removed and replaced showed that when the polarity of some of the microtubules reversed, so did the direction of the associated pigment dispersal movement. How this radial polarity is set up and maintained is as interesting a question as how the motion of the pigment granules is guided along the microtubules.

Finally, what is known about the energetics of pigment granule movement? It turns out that Ca^{2+} is necessary for aggregation; the removal of Ca^{2+} causes a reversible halt of aggregation. When ATP is depleted with poisons of energy metabolism in colored pigment cells or chromatophores, aggregation occurs and dispersal is prevented. In another cell type carrying black pigment (melanophores), however, detergent-permeabilized cells required the addition of ATP for aggregation. It may well be that both dispersal and aggregation require energy.

This specialized form of motion of intracellular organelles reveals with some clarity the role played by a fine-filament (microtrabecular) matrix. It seems highly unlikely that the role of this matrix in the motion of organelles is restricted to this particular system. Another process likely to involve such a matrix is the movement of chromosomes during mitosis. The future may bring additional insights gained from an understanding of the structure and function of the cytomatrix of fine filaments.

Chromosome Movement

Perhaps one of the most important forms of movement in biology, one which must have been present at the very beginning of the evolution of eukaryotic cells, is the coordinated motion of chromosomes during mitosis. Two types of chromosome motion occur during mitosis: the irregular (saltatory) and relatively slow movement of chromosomes from various positions of the prophase nucleus to the metaphase equator, and a much more rapid, smooth, and synchronized progression of chromosomes away from the metaphase plate to the opposite poles. In spite of many decades of study and effort devoted to explaining chromosome motion during mitosis, the proteins providing the motive force and even the nature of the filaments directly involved in the movement are still shrouded in mystery. Nevertheless, considerable progress has been made in recent years in characterizing the spindle structure through the use of detergent-permeabilized and extracted cells.

The spindle is a collection of microtubules (spindle fibers) that converge at opposite poles toward asters; these consist of a pair of centrioles from which microtubules radiate (Figure 16–68a). (Plant cells do not form asters, but since we are not concerned here with the

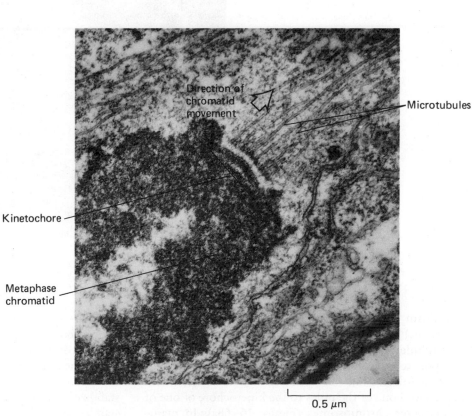

(a)

(b)

FIGURE 16–68
The Mitotic Spindle and Chromosome Motion. (a) Cross section of spindle at metaphase, showing the two types of microtubules: the polar microtubules running from pole to pole (even though individual microtubules may be interrupted) and chromosomal microtubules running from pole to kinetochore. (b) Association of the metaphase chromatid between microtubules and the kinetochore in mammals. The kinetochore is a three-layered disk, the outer electron-dense layer being the structure into which the microtubules are apparently inserted. (Courtesy of J. D. Picket-Heaps.)

Spindle pole

Centriole pair

Chromosomal microtubules

Polar microtubules

Kinetochore

Metaphase chromatids before separation in anaphase

Direction of chromatid movement

Microtubules

Kinetochore

Metaphase chromatid

0.5 μm

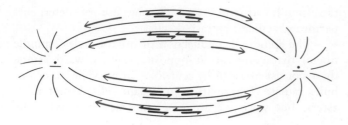

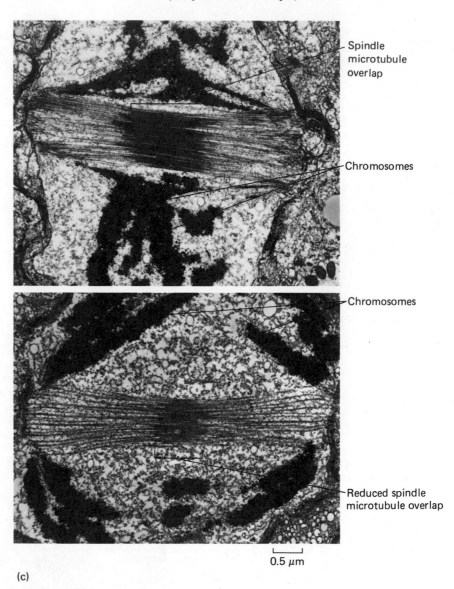

FIGURE **16–68 (continued)**
(c) Proposed mechanism for spindle elongation, bringing about a separation of daughter chromosomes during anaphase: (1) sketch suggesting how spindle might elongate by sliding motion, brought about by bipolar myosin dimers causing movement of spindle fibers of opposite polarity. (2) Electron micrograph of two spindles at different stages of spindle elongation, showing decrease of spindle overlap. (Courtesy of J. D. Picket-Heaps.)

Spindle microtubule overlap

Chromosomes

Chromosomes

Reduced spindle microtubule overlap

0.5 μm

(c)

mechanism of spindle assembly this difference is not important.) The spindle as such contains two kinds of microtubules: *polar (interzonal) microtubules,* which extend from pole to pole (even though individually they may not cover the full length), and *chromosomal microtubules,* which extend from one of the poles to the kinetochore of one of the daughter chromosomes (Figure 16–68a). In mammals, the kinetochore is a three-layered disk, the outer

electron-dense layer being the structure into which the microtubules are apparently inserted (Figure 16–68b).

Chromosomal microtubules initiate with their "minus" ends at the poles. During elongation, certain microtubules will contact a kinetochore, which tends to stabilize the microtubule against depolymerization. This stabilization is important, because microtubules normally polymerize and depolymerize relatively rapidly, a property termed *dynamic instability*. The chromosomal microtubules are stabilized at both ends: the minus end is bound to the centrosome and the plus end to the kinetochore. The property of dynamic instability explains how elongating microtubules could follow relatively random paths and still participate in chromosome movement; microtubules that fail to contact a kinetochore or some other stabilizing structure will tend to depolymerize and thus allow new microtubules to nucleate at the pole. To form the metaphase plate, the kinetochore must contact and attach to microtubules from both poles.

In forming the metaphase plate, chromosomes move away from the equatorial plane as well as towards it, a saltatory motion. There seems to be little question that chromosomal microtubules direct the motion of chromosomes during prometaphase. Experimentally severing chromosomal microtubules connected to a given chromosome causes it to stop moving. Alignment of chromosomes at the metaphase plate is the result of a balance of tensions, propagated along the microtubules. If one of the kinetochores of a pair of metaphase chromatids is destroyed with a very fine laser beam, then both chromatids move rapidly to the opposite pole. If ATP is depleted during prometaphase, by the addition of metabolic inhibitors, chromosomes move back to the poles, an effect that is reversible. Furthermore, movement towards the equatorial plane is accompanied by elongation of the chromosomal microtubules. Tubulin labeled so it can be visualized in the EM with a gold-conjugated antibody is found to be incorporated into chromosomal microtubules during prometaphase.

The balance that characterizes chromosomes lining up at metaphase could be due to tension exerted by chromosomal microtubules. At the equatorial plane, the tensions would be equal and net motion would stop. This is consistent with observations on the spermatocytes of a grasshopper where some chromosomes have three kinetochore regions, two oriented towards one pole and one towards the other. Such chromosomes do not lie in the equatorial plane at metaphase; instead they are found nearer the pole attached to two kinetochores. Moreover, the sum of the lengths of the two chromosomal microtubules connected to one pole equals the length of the microtubule connected to the other pole. These and other data suggest that chromosomal microtubules act as tension-producing structures with the magnitude of the tension proportional to the length of the microtubule.

The ATP-dependence of prometaphase motion probably reflects the involvement of a motor protein that is part of the kinetochore structure. It has been found in model systems that plus-end motion by chromosomes is inhibited by vanadate ion. This presents a paradox, since vanadate inhibits dynein but that motor protein moves structures towards the minus end of a microtubule.

Another unanswered question is how a motor protein located in the kinetochore can sense the length of the microtubules to which it is attached.

There appears to be a similarity between pigment dispersion in chromatophores and prometaphase movement of chromosomes to the metaphase plate. Both are slower, saltatory processes and both are reversed by depletion of ATP. The movement of chromosomes to opposite poles during anaphase could also be analogous to the movement of pigment granules during aggregation; both movements are relatively fast, smooth, and independent of ATP depletion by metabolic poisons. Thus, it it may be useful to think of the asters during cell division as containing the equivalent of two chromatophore cytomatrices; on the other hand, the chromatophore could be viewed as a single aster system. Additional evidence to support the analogy is the fact that microtubules are oriented with their minus, slower-growing ends toward the center of the pigment aggregation, as they are toward the spindle poles.

Can any additional insight be gained from the analogy of chromatophore movement to chromosome movement? The process of anaphase begins when the kinetochores of two sister chromatids split apart. This process is not related to microtubule activity because it occurs even if the microtubules have been depolymerized with colchicine. There are two ways the separation of daughter chromosomes is accomplished during anaphase; the relative importance of these differs from one type of cell to another.

One mechanism of daughter chromosome separation involves the elongation of the spindle, thereby moving the asters away from each other. This movement is correlated with an equivalent separation of the daughter chromosomes and, at its simplest level, is explainable as the movement of the asters pulling the chromosomes passively by their kinetochore microtubules. The movement of the asters, on the other hand, appears to be associated with a sliding motion of the polar microtubules. Microscopic analysis of elongating spindles shows that the degree of overlap between polar microtubules is decreased during the period of spindle elongation (Figure 16–68c). There is also microscopic evidence of fine interconnections between the overlapping microtubules, suggesting a dynein-like mechanism for the sliding of microtubules past each other.

There are differences, however, between chromosome separation and the mechanism of microtubule interaction in ciliary axonemes. One difference is that axonemal dynein as such is not involved. D. Luck has found a mutant in *Chlamydomonas* that lacks dynein, yet can engage in the elongation of the spindle in anaphase. Another difference is that antibodies to dynein do not inhibit spindle elongation in permeabilized cells. Yet another difference is that in spindle elongation, unlike the ciliary axonemal system, the overlapping microtubules are of opposite polarity. If a sliding mechanism is involved in the process of spindle elongation, it must be unique to antiparallel microtubules.

The second process by which daughter chromosomes move away from each other is correlated with shortening of the chromosomal microtubules. S. Inoue has assembled a great deal of experimental evidence demonstrating that such shortening does occur, being associated with depolymerization of the microtubules at their kinetochore attachment ends. One convincing experiment involves incubating a detergent-permeabilized dividing cell in excess tubulin, a treatment that was intended to inhibit depolymerization of the chromosomal microtubules but not the sliding of the polar microtubules. It turns out that under these conditions, poleward movement *is* inhibited more than spindle elongation. Furthermore, careful experiments indicate that agents affecting the rate of depolymerization also affect the rate of chromosome movement.

The problem with the chromosomal microtubule depolymerization hypothesis is that we cannot be sure which is cause and which is effect. Do depolymerizing microtubules actively pull the chromosomes or are the chromosomes pulled by something else?

Recently, two different groups reported an important finding based on the observation by J. R. McIntosh and M. E. Porter that, in addition to the dynein associated with the axoneme of cilia and flagella, there is another dynein molecule found in the cytoplasm. Monoclonal antibodies raised against subunits of cytoplasmic dynein localize this protein to the mitotic spindle and, particularly, to the kinetochore of chromosomes. Although much work must yet be done, it is entirely possible that cytoplasmic dynein is the motor that translocates chromosomes toward their minus end at the centromere during anaphase.

As in the case of pigment granule motion, microtubules play an important structural role. We emerge with a picture of chromosome movement in which energy is expended in moving chromosomes to the metaphase plate by performing work against tension exerted by an elastic matrix or via a motor protein associated with the kinetochore. At the metaphase plate, this tension is exerted in both directions and represents a state of dynamic equilibrium that comes to an end when the kinetochores of daughter chromatids separate. From then on, it is likely that cytoplasmic dynein acts as a motor and moves the chromosomes toward the poles. The rate and direction of this movement may well be controlled by the depolymerization of the microtubules.

Axonal Transport

Nerve cells have a unique geometry (Figure 16–69a). They consist of very long and finely divided structures (dendrites, axon, and terminal branches), the proteins for which are manufactured in a cell body containing the nucleus, the ribosomes, and the rough endoplasmic reticulum. In large mammals such as the whale or the giraffe, some neurons may be 10 meters long while other neurons are so finely divided that they can make contact with 100,000 other cells (Figure 16–69b). Neurons have a highly specialized cytomatrix, capable of developing and maintaining their complex structures while simultaneously transporting the proteins synthesized in the cell body to the tips of the terminal branches.

R. Lasek and his co-workers have devised an ingenious method for studying axonal transport in situ. They injected radiolabeled precursors of protein, such as the amino acid [^{35}S]methionine, into the vicinity of the cell body and then measured the rate of movement of proteins that incorporated the label as they migrated down the axon: they removed the nerve after given time intervals, cut it into small segments, extracted the proteins, and identified the individual labeled chains by two-dimensional gel electrophoresis. The great length of neurons allows small differences in the rates at which different proteins and organelles move along the axon to be measured more easily. The results are stunning in the complexity of the transport systems they reveal (Figure 16–69c–e).

At least five systems of transport operate, independently, along the axon. Each of these systems can be characterized by protein chains moving at characteristic rates. The transport rates fall into two broad categories: fast (50–400 mm/day) and slow (0.2–8 mm/day).

The proteins carried by the *fast* component (FC) of axonal transport are all associated with membrane-bound organelles: small vesicles, secretory granules, dense bodies, multivesicular bodies, and mitochondria. Measures of the movement of these vesicles, whether studied by light microscopy or by the isotope method, all show rates of 200–400 mm/day (≈ 4 μm/s). Different types of vesicles, however, move at different rates as well as in different directions. As might be expected, vesicles that supply the synaptic terminal—secretory granules—move in the *anterograde (orthograde) direction* (from the cell body toward the terminal), while endocytotic vesicles originating in the axon terminus move in the *retrograde direction* (from the terminal towards the cell body).

The proteins carried by the *slow* component (SC) of axonal transport fall into two distinct classes. Proteins in the SCa class—for example, tubulin and the subunit

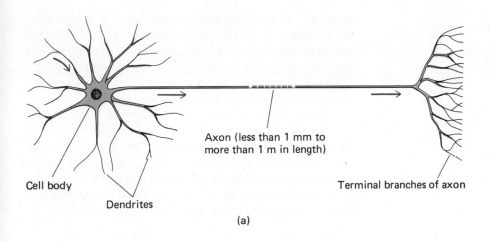

(a)

FIGURE 16–69

The Phenomenon of Axonal Transport.   (a) Diagram of a typical neuron showing dendrites, cell body with nucleus, axon, and terminal branches. A signal is transmitted from the dendrite to the terminal branches. (b) Complexity of the dendrite system of a Purkinje neuron from the human cerebellum. This cell is connected to 100,000 other neurons from which it can receive signals. The system of neuronal transport must receive adequate supplies of proteins and lipids manufactured in the cell body. (Courtesy of S. Ramon y Cajal.) (c) Sketch of neuron showing orthograde and retrograde vesicle movement.

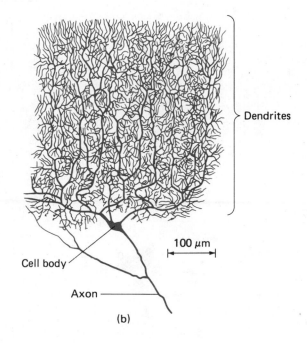

(b)

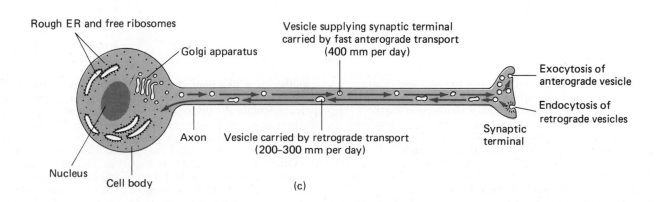

(c)

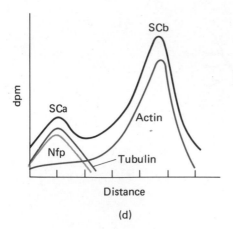

(d)

proteins of neurofilaments—move at 0.2–1 mm/day, while proteins in the SCb class—over 100 different polypeptides, including actin, spectrin, calmodulin, and myosin-like proteins—move at 2–8 mm/day (Figure 16–69d). Figure 16–69e illustrates the very large number of proteins that move in the SCb system.

What is arresting about these results is the strong correlation in movement rates by the various proteins in each of the two slow components. The simplest hypothesis to explain the correlation is that there is some primary mover molecule in each component to which the various proteins are bound. A good guess for the primary mover is one filament or another of the cytomatrix.

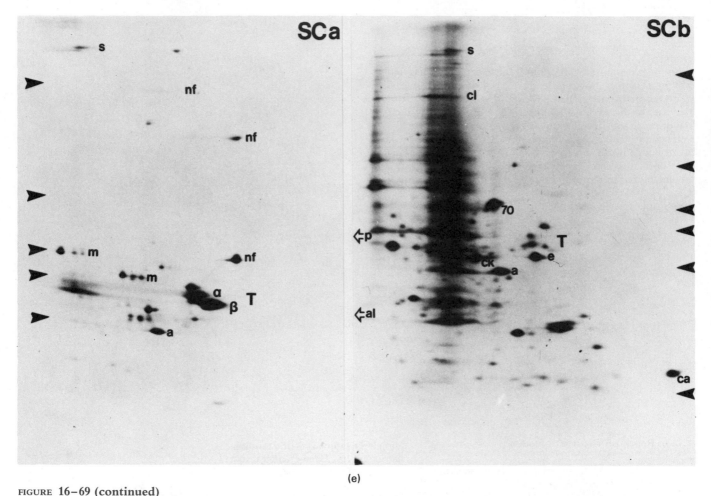

(e)

FIGURE 16–69 (continued)
(d) Profiles of the rates of transport of the SCa and SCb components in retinal ganglion cells of the guinea pig. The black curve is the distribution of total transported radioactivity. The curves of tubulin and neurofibular protein radioactivity parallel perfectly the SCa component. In the case of actin, the peak of radioactivity corresponds well with that of the total radioactivity, especially the orthograde (anterograde) portion of the curve. However, the retrograde portion of the curve trails well into the SCa component, a fact which can be interpreted to mean that the polymerized and moving portion of the actin in the SCb component is in

equilibrium with a nonmoving, presumably unpolymerized, form of actin. (Courtesy of R. Lasek.) (e) Two-dimensional polyacrylamide gel electrophoresis of proteins characterizing the two slow components of axonal transport. The SCa group contains relatively few protein chains but includes the three neurofilament (nf) and two tubulin chains (T, α, and β). Other chains are MAPs (m), spectrin (s), and a small amount of actin (a). The SCb group consists of a large variety of protein chains, particularly actin (a), calmodulin (ca), creatine kinase (ck), clathrin (cl), and spectrin (s). (Courtesy of R. Lasek.)

The kinetic studies thus offer the opportunity to measure the degree of association of various cytoplasmic proteins with different primary mover molecules. If the association is tight, then the migration profile of a given protein would be expected to parallel closely the distribution of the transport component. If the association is loose, then labeled proteins that start out with the primary mover will tend to dissociate and gradually be left behind. Figure 16–69e shows how the profiles of the microfilament and tubulin proteins resemble that of the SCa proteins collectively. Actin, on the other hand, has a broader distribution than the collection of SCb proteins, leaving a long "trail" as it moves forward. Since transported actin parallels the SCb distribution at the leading edge, the broadening is probably due to actin in solution not being swept along by the SCb system.

How likely is it that part of the cytomatrix is moving at the rates observed for the components of axon transport? The slowest component of axon transport, the SCa system, transports the subunit proteins of neurofilaments and microtubules. High-resolution (quick-freeze, deep-etch) electron microscopy shows that these two types of filaments themselves are interconnected, by numerous superthin filaments. The question is whether the labeled filament proteins are polymerized into the cytomatrix or whether they are merely bound to the primary mover molecule.

Treating neurons with neurotoxin IDPN (β,β'- iminodiproprionitrile) causes the neurofilaments to segregate towards the perimeter of the axon, leaving the microtubules at the center (see Figure 16–54). It can be shown that IDPN also selectively blocks the axonal transport of the neurofilament subunit proteins, without similarly affecting the slow movement of microtubule proteins or the fast movement of vesicle-related proteins. This result can be explained if SCa motion is powered by microtubules and neurofilament proteins are passive participants. The IDPN must uncouple the movement of labeled neurofilament proteins from the primary mover. In agreement with the hypothesis that microtubules are the primary mover of the SCa component is the fact that agents disrupting microtubules also inhibit the movement of the SCa system. The faster-moving SCb system would seem to be powered by actin and myosin-like or other motor proteins: agents disrupting actin microfilaments appear to disrupt the SCb system, and also the movement of vesicles.

Recently, Sheetz and his colleagues have reported the isolation from squid axon of a new motor protein— called *kinesin*—that acts in conjunction with microtubules to produce motion (Figure 16–70). Their original method of isolation relied on a physiological response to a nonhydrolyzable analogue of ATP, AMPPNP. When added to a preparation showing fast movement of

FIGURE 16–70

Movement of Latex Beads Along Polar Microtubules, Driven by Motor Proteins. Microtubules with known polarity were prepared by nucleation from centrosomes; the microtubule plus ends were oriented away from the centrosome. At the right is the elapsed time, in seconds. (a) Kinesin-induced movement of latex beads along centrosomal microtubules. All beads marked with letters move away from the centrosome (C) towards the plus end of the microtubule and would correspond to anterograde transport in the intact axon. Beads *f* and *g* move to the right, out of the field of view, at between 0 and 3 seconds. ×10,000. (b) Bidirectional movement of beads along a centrosomal microtubule, induced by a crude fraction of squid axoplasm. A bead at 0 seconds moves rapidly to the left (arrowhead) towards the centrosome (C), pauses, and then moves more slowly in the opposite direction. Another bead marked with an asterisk moves in a discontinuous fashion towards the centrosome and then stops. The bead marked with the arrowhead passes the bead marked with the asterisk between 1 and 3 seconds. ×10,000. (Courtesy of R. D. Vale, B. J. Schnaff, T. Mitchison, E. Stener, T. S. Reese, and M. P. Sheetz.)

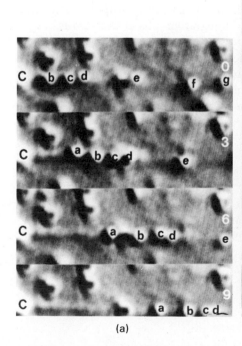

(a)

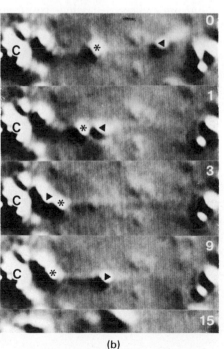

(b)

organelles along microtubules, the analogue halted the motion and led to decoration of the microtubules by the organelles. After purifying the decorated microtubules, ATP was used to release the components of kinesin. Progress was assayed at various stages of the purification by the ability of latex beads coated with the kinesin preparation to move along purified microtubules in the presence of ATP (Figure 16–70).

Kinesin is a 600-kDa complex of six polypeptide chains, four of 110 kDa and one each of 70 kDa and 65 kDa. This complex is not sensitive to N-ethyl maleimide nor to 20-μM vanadate, inhibitors of ciliary dynein, and is much smaller than any dynein yet reported. Moreover, the motion associated with the protein is in a direction (anterograde) opposite to that associated with dyenin (retrograde). An analogous protein has been isolated from mammalian brain, and it appears likely that the kinesin motor for microtubules will prove to be as ubiquitous as the myosin motor for microfilaments. Antibodies raised against squid axon kinesin have been used to detect proteins with similar structures in other cells. An exciting result is that kinesin-like proteins are found in the spindle of dividing cells and that a mutation affecting chromosome movement during meiosis in *Drosophila* lies in a gene specifying a kinesin homologue.

Another use to which the antibodies have been put is in an affinity column, to retain kinesin protein from a supernatant fraction from squid axoplasm. This fraction will ordinarily support bidirectional movement of organelles on microtubules (Figure 16–70b). After being passed through the antikinesin column, however, only retrograde movement is supported. This movement is sensitive to 20-μM vanadate and to N-ethyl maleimide; the size of the protein complex involved has been estimated as 1500 kDa. The simplest hypothesis to explain these results is that squid axons contain in their cytoplasm a cellular dynein-like protein. By having two motor proteins, a single polar microtubule can support motion in both directions.

We can be confident from these results that the molecular basis for axon transport will be unraveled soon. Among the problems that remain to be solved are how the direction of movement is chosen, which subunits or domains of subunits are responsible for the various functions (binding to the organelle, binding to the microtubule, force generation), whether these motors play a role in the slower components of axon transport, and how many other cellular motions are driven by these motors.

SUMMARY

In this chapter we have discussed the properties of the cytoplasmic matrix, the phase that supports both the plasma membrane and the membrane-bound organelles of the cell. Since organization and control of mechano-chemical processes is a major and dramatic property of the cytomatrix, they have been the unifying themes of the discussion. It should be recognized, however, that the cytomatrix plays numerous other roles in the life of the cell. Some of these, such as the compartmentalization of molecular composition, have only begun to be studied; others, such as the transmission of signals from cell surface to nucleus, are barely suspected. Cell biology is passing from an analytical phase that identified the properties of the major protein components of the cytomatrix to an integrative phase that promises to explain major processes of cell function in terms of the interactions of their component macromolecules. This chapter is indicative of this transition; it tells much of what has been found while it holds out the hope of very much more yet to be discovered.

KEY WORDS

Alpha helix, assembly, chloroplast, chromatid, chromosome, conformation, cortex, cross-link, electrophoresis, energy, enzyme, fibril, filament, helix, intermediate filament, kinase, microfilament, microtubule, mitosis, monomer, organelle, plasma membrane, polymer, sarcoplasmic reticulum, transduction, tubule, vesicle, x-ray diffraction.

PROBLEMS

1. Respond to the following statements with either TRUE or FALSE. If you respond TRUE, justify your response with a specific example. If you respond FALSE, explain why you think the statement is not true.
   (a) Microtubules are smaller than microfilaments.
   (b) The flagella of bacteria contain dynein and ATP.
   (c) Actin/myosin contraction mechanisms are necessary to explain nonmuscle cell motility.
   (d) α-actinin is a protein found only in the Z-disk of striated muscle.
   (e) The movement of cross-bridges during contraction in a striated muscle fiber must be closely synchronized, because muscle tension drops when the cross-bridges all detach at the same time.
   (f) The Z-disks of a sarcomere are brought closer together through shortening of the thin filaments.
   (g) Colchicine interferes with the process of mitosis by preventing the assembly of microfilaments.
   (h) The contraction-specific ATPase sites on skeletal muscle actin are blocked until the sarcoplasmic calcium concentration rises to $\approx 10^{-5}$ M.
   (i) Removal of ATP from bacterial cells will result in a rigor-like state with stiff bacterial flagella.
   (j) When a vertebrate striated muscle fiber is stretched too far, stimulating it produces no tension because the thin filaments are pulled out of the Z-disk attachment sites.

2. After a contraction, the passive stretching of a myofibril is accompanied by transfer of heat from the fibril to the environment. As a first approximation, assume this stretching is a quasi-equilibrium process —that is, $\Delta G = 0$.
   (a) What is the sign of the fibril's enthalpy change during the stretching?
   (b) Interpret from your knowledge of thermodynamics and cell structure the entropy change in the fibril during the stretching.

3. The actual hydrolysis step catalyzed by vertebrate actomyosin, $ATP + H_2O \rightarrow ADP + P_i$, does not proceed with a large negative free energy change. Yet we say that this hydrolysis provides the energy to drive muscle contraction. Explain why.

4. In a rapidly working muscle cell, the total number of contraction cycles (myosin cross-bridge binding/dissociation cycles), from initial activation until exhaustion, is much larger than the number of ATP molecules in the cell. Give as complete an account as you can for this apparent paradox in bookkeeping.

5. Explain why the latex beads coated with myosin head groups by Sheetz and Spudich did not rotate when moving along cables of actin filaments.

6. How do you distinguish between the cytomatrix and the cytoskeleton?

7. Describe how you can use morphological investigation—for example, electron microscopy—to determine the interconnectedness of membrane-bound spaces.

8. (a) Upon addition of ATP, actomyosin undergoes a dramatic drop in viscosity. Draw pictures of the two states in which actomyosin can be found that account for the large drop in viscosity.
   (b) After the actomyosin stands awhile, why does the viscosity gradually increase again?

9. What is the evidence that the head groups of myosin are responsible for its ATPase activity and that the tails are responsible for the ability of myosin to aggregate into a bipolar thick filament?

10. In the resting muscle, sarcoplasmic Ca^{2+} is sequestered in the sarcoplasmic reticulum while ATP and Mg^{2+} are present in the sarcoplasm. Most commercially available Mg^{2+} salts contain traces of Ca^{2+}, generally in amounts sufficient to activate actomyosin. In studying the control of muscle contraction in vitro, how can the role of the sarcoplasmic reticulum in removing Ca^{2+} from the sarcoplasm be duplicated?

11. What is meant by saying that muscle contraction is regulated with great precision?

12. What is the evidence that energy transduction in eucaryotic flagella occurs along the entire length of the flagellum?

13. What is the role of the radial spokes of the eucaryotic flagellum?

14. What assay could be used to test whether a certain protein is an actin-capping protein?

15. What is the evidence that contraction of the cleavage furrow during the division of animal cells is based on the interaction between actin microfilaments and myosin?

16. List the basic techniques used for studying various mechanisms of cell movement.

17. What forms of movement must have arisen early in the evolution of eucaryotic cells of
   (a) both plants and animals
   (b) animal cells only.

SELECTED READINGS

Allen, R. D. (1981) Cell motility. *Journal of Cell Biology* 91:148s–154s.

Allen, R. D. (1987) The microtubule as an intracellular engine. *Scientific American* 256(2):42–49.

Bretscher, M. S. (1987) How animal cells move. *Scientific American* 257(6):72–90.

Cohen, C. (1975) The protein switch of muscle contraction. *Scientific American* 223(5):36–45.

Dembo, M., Harlow, F. H., and Alt, W. (1984) The biophysics of cell surface motility. In *Cell surface dynamics: Concepts and models*. A. S. Perelson, C. DeLisi, and F. W. Wiegel, eds. New York: Marcel Dekker, Inc.

Franzini-Armstrong, C., and Peachey, L. D. (1981) Striated muscle—Contractile and control mechanisms. *Journal of Cell Biology* 91:166s–186s.

Gibbons, I. R. (1981) Cilia and flagella of eukaryotes. *Journal of Cell Biology* 91:107s–124s.

Haima, L. T., and Rosenbaum, J. D. (1981) Cilia, flagella, and microtubules. *Journal of Cell Biology* 91:125s–130s.

Hoyle, G. (1970) How is muscle turned on and off? *Scientific American* 222(4):84–93.

Huxley, H. E. (1965) The mechanism of muscle contraction. *Scientific American* 213(6):18–27.

Hyams, J. S., and Brinkley, B. R. (1989) Mitosis—Molecules and mechanism. London: Academic Press, Ltd.

Ishikawa, H., Hatano S., and Sato, H. (1986) Cell motility: Mechanism and regulation. New York: Alan R. Liss, Inc.

Lazarides, E. (1982) Intermediate filaments: A chemically heterogeneous, developmentally regulated class of proteins. *Annual Review of Biochemistry* 51:219–250.

Lazarides, E., and Revel, J.-P. (1979) The molecular basis of cell movement. *Scientific American* 240(5):100–113.

Murray, J. M., and Weber, A. (1974) The cooperative action of muscle proteins. *Scientific American* 230(2):58–71.

Pollard, T. D. (1981) Cytoplasmic contractile proteins. *Journal of Cell Biology* 91:156s–165s.

Preston, T. M., King, C. A., and Hyams, J. S. (1990) The cytoskeleton and motility. New York: Chapman and Hall.

Satir, P. (1974) How cilia move. *Scientific American* 231(4):44–63.

Schliwa, M. (1986) The cytoskeleton—An introductory survey. New York: Springer Verlag.

Weber, K., and Osborn, M. (1985) The molecules of the cell matrix. *Scientific American* 235(4): 110–121.

Wessels, N. K. (1971) How living cells change shape. *Scientific American* 225(4):76–82.

Beyond the Cell

Life cannot be maintained in a system at chemical equilibrium. Energy and mass constantly flow into a living cell from its environment and out of the cell to its environment. In a multicellular organism, the environment is dominated by the presence of other cells and their products. To understand a cell in such surroundings, therefore, the relations among cells and how they affect one another must be examined.

Biological information, as well as mass and energy, is exchanged among cells. As we have stated before, this information is always carried by a material substance. Some of the information travels long distances—for example, a hormone carries through the circulation information about the state of the cell that secreted it. But information is also exchanged over much smaller distances—for example, between adjacent cells. One example of this is the excitation communicated by a nerve cell to a muscle cell; this occurs via a small secreted neurotransmitter molecule that diffuses rapidly between the two cells (see description later in this chapter). Some nondiffusing (or very slowly diffusing) information lies in structures resting on or near the cell surface. These structures are used to form and maintain the associations of cells that characterize tissues and organs in multicellular organisms.

In this chapter we shall first describe the immediate environment of the cell; the connections that the cell makes with its environment; and the noncellular matrix and adjacent cells, including the problem of how cells can recognize one another during the formation of tissues by exchanging information. Then we treat specific structures that join together adjacent cells. Finally, we discuss interactions among the partly fixed, partly wandering cells of the immune system.

Connections Between the Cell and Its Environment

It is not a simple task to identify where the cell ends and its environment begins. We discussed in Chapter 7 the limiting membrane of the cell, the

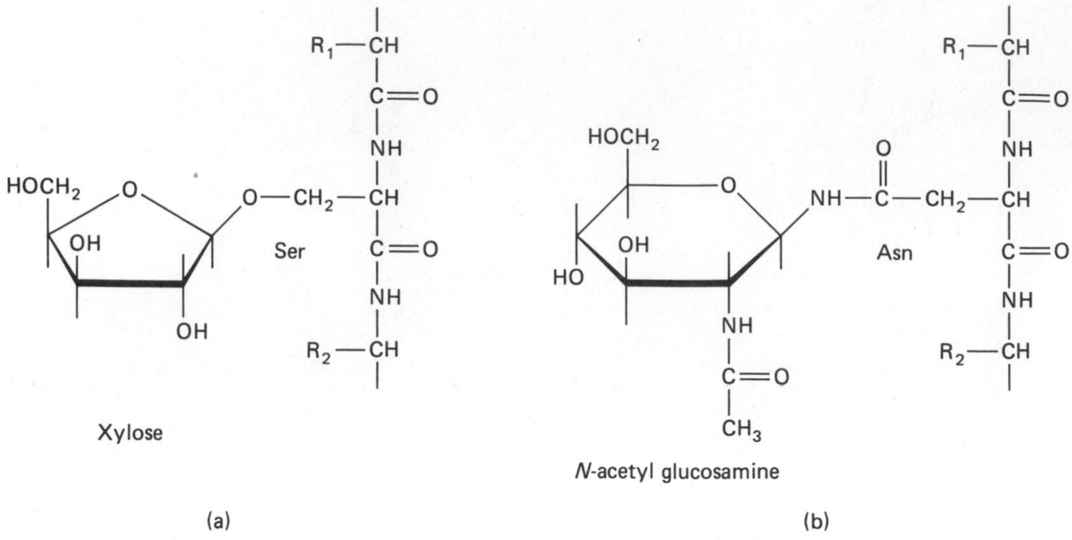

(a)   (b)

FIGURE 17–1

Carbohydrate-Protein Links in Glycoproteins.   (a) O-link to serine. The sugar xylose is joined at its 1-carbon in glycosidic linkage to the hydroxyl oxygen of a serine side chain. O-linked sugars are also linked to threonine side chains. (b) N-link to asparagine. The sugar N-acetyl glucosamine is joined at its 1-carbon to the amide nitrogen of an asparagine side chain. N-linked carbohydrates are usually synthesized as a core structure of several sugars before being transferred as a block to the protein (see Chapter 13). If the first O-linked sugar is a xylose, the glycoprotein is called a *glycosaminoglycan (GAG, or proteoglycan);* if the first sugar is not xylose, the glycoprotein is sometimes called a *mucin.* Collagens are glycoproteins in which the first sugar is a glucose or galactose, O-linked to a hydroxylysine.

plasmalemma. But cells contain structures that extend beyond the surface of the lipid bilayer. We shall try to develop a picture of the cell that distinguishes these structures from a noncellular layer of variable thickness that surrounds most cells in an organism—the *extracellular matrix*.

The fluid-mosaic model for membrane structure (see Chapter 7) distinguishes between integral proteins—those embedded in the bilayer—and peripheral proteins—those that bind to the head groups of the phospholipids and/or to integral proteins. For a cell in a tissue, it is difficult, both in principle and by experiment, to distinguish between a peripheral protein and a protein that belongs to the extracellular matrix. For our present purposes, an integral protein—one that remains with the cell unless it is treated with a detergent (to solubilize the bilayer lipids)—may be defined as belonging to the cell. A nonintegral protein is regarded as peripheral if it is associated with the phospholipid head groups or with integral membrane proteins; otherwise it is regarded as belonging to the matrix.

The Glycocalyx

A common feature of the outer surface of most animal cells is not primarily protein in nature; it is the layer of carbohydrate that surrounds the cell—the *glycocalyx*.

Integral Glycoproteins

Membrane proteins and lipids that have covalently bonded carbohydrate are always arranged so the hydrophilic sugars face outward, away from the cytoplasm. This is also true for integral glycoproteins in the membranes of organelles. The outer surface membrane of a cell may contain 10^1–10^3 different integral glycoproteins, many of which are present in only a few copies per cell. These differ from one another in their amino acid sequences and are coded by different genes. The carbohydrate parts of these glycoproteins are also diverse. In fact, there is evidence that the processes that add carbohydrate to proteins are not especially accurate. The result is a distribution of similar but not identical carbohydrate groups on proteins with the same amino acid sequence. This variation is called *microheterogeneity*.

There are two general ways that carbohydrates are attached to proteins: through O-glycosidic links to serine or threonine side chains or through N-glycosidic links to asparagine side chains (Figure 17–1). The structures of some integral glycoproteins are known.

One of the most completely studied is *glycophorin* (glycophorin A) from the human red blood cell. Both O-linked and N-linked oligosaccharides are present on this protein, which spans the bilayer; the sugars are all on the outside surface (Figure 17–2). The red blood cell surface contains about 4×10^5 molecules of glycophorin, so it has been relatively easy to isolate and to study. The carbohydrate has some interesting functions: *sialic acid* residues on the oligosaccharides give the cell a negative electrostatic charge that tends to prevent them from sticking together. Glycophorin carbohydrate also contains information that determines the ABO and the MN blood group antigens and acts as the binding site for certain viruses (e.g., influenza). The essential functions of glycophorin can apparently be performed by other surface structures (perhaps by proteins known as glycophorins B and C); there are individuals with an inherited condition in which glycophorin A is completely absent from the red blood cells, which nevertheless behave normally.

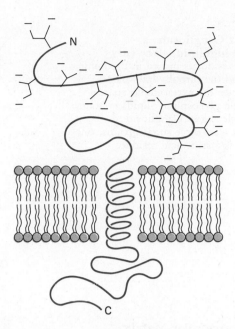

FIGURE 17–2

Glycophorin, the Major Glycoprotein of the Red Blood Cell Membrane. The N-terminal region lies outside the cell and contains the carbohydrate. There are over a dozen carbohydrate groups attached, mostly O-linked, and they all contain negatively charged sialic acid. There are 23 residues that pass through the lipid bilayer, probably in the form of an α-helix and about three dozen residues lying inside the cell.

Sialic acid
*N-acetyl*neuraminic acid
= NANA)

Glycolipids

Another contribution of carbohydrate to the glycocalyx is made by glycolipids in the outer leaflet of the membrane bilayer. The most prominent group of these are the *glycosphingolipids* (GSL, Figure 17–3). The red blood cell is again the model from which much of our knowledge of glycolipids was obtained; there are at least

four dozen different types found in the erythrocyte membrane. Glycolipids are now known to be a typical feature of the animal cell plasmalemma; there are many fewer glycolipids in the internal membranes of the cell.

When glycolipids with small carbohydrate groups are not accessible to surface labeling, they are said to be *cryptic*. The fraction of cryptic carbohydrate can change markedly with changes in cell behavior, becoming much less in malignantly transformed cells. It seems likely that the inaccessibility of some glycolipids is due to an intimate association with integral membrane proteins, which is consistent with the finding that adding glycosphingolipids can stimulate membrane-associated enzyme activities—for example, the sodium–potassium pump. It is also possible to demonstrate an association of glycolipids with proteins by attaching them together with bifunctional covalent cross-linking reagents: "molecular staples."

Although there is not universal agreement about the functions of glycolipids, the GSLs are suspected to play several roles: certain blood group antigens (for example, Lewis and H-D) are associated with GSLs and a ganglioside is a receptor for cholera toxin in some cells.

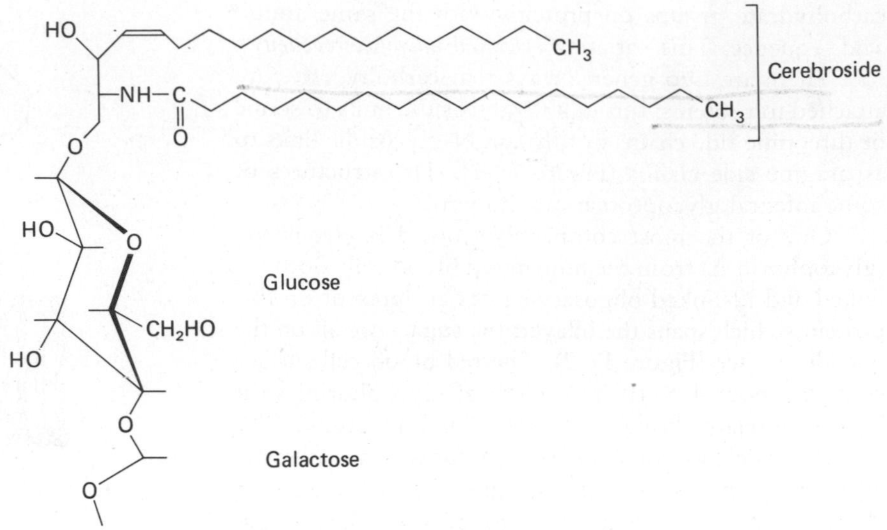

FIGURE **17–3**
The Gm1 Ganglioside, a Glycosphingolipid. The fatty ceramide residue is buried in the lipid bilayer and the carbohydrate part projects into the extracellular space. The simple glycosphingolipids contain but a single sugar. Gangliosides contain at least three sugar groups, one of which is a sialic acid residue.

Cerebroside

Glucose

Galactose

The Extracellular Matrix

The environment of the cell, beyond the lipid bilayer and integral proteins, contains both protein and carbohydrate components, all of which are products of nearby cells. The water of the extracellular matrix is definitely not in the form of a dilute aqueous solution; instead, water structure and function are strongly influenced by the macromolecules there. This is because water molecules in the extracellular matrix are more likely to be near protein and carbohydrate than near other water molecules. As a consequence, the resistance to movement by diffusion of small molecules is greater in matrix water than in bulk water.

Collagen

The collagens are probably the best understood proteins in the extracellular matrix because in certain tissues, like vertebrate tendon, they represent most of the protein present and are therefore easy to isolate and study. For example gelatin, obtained from the connective tissue of animals, is almost pure denatured collagen. Collagen-like proteins are found in all eucaryotic organisms, except protozoa. Collagen is a long (300 nm), thin (1.5 nm) fibrous molecule, the center of which (95% of the length) is in the form of a characteristic triple helix (see Chapter 4). At both the N- and C-terminal ends of each polypeptide chain are short (~20 amino acids) nonhelical peptides, called "telopeptides." The telopeptides and two locations within the triple helix are sites of intermolecular cross-links, involving lysine and hydroxylysine side chains, which stabilize the fibers formed from several collagen molecules (Figure 17–4). Collagen fibers are capable of withstanding great tensions and lend strength to the extracellular matrix.

Twelve types of collagen have already been recognized. Of these, there are seven main types, distinguished originally on the basis of the tissues in which they are most abundant (Table 17–1). Skin and tendon contain mostly collagen of type I. Cartilage contains type II collagen, a second skin collagen is called type III, and type IV and VII collagen are found in basement membranes (see discussion below). Cell surfaces contain other collagens.

Each collagen molecule is made up of three polypeptide chains, of which there are over a dozen kinds presently known (Table 17–1). Actually, there is good reason to believe that many more kinds of collagen chains exist. We expect these to have very similar sequences, but since each chain has over 1000 amino acids, it may be some time before the rare collagen chains are catalogued. Collagens isolated from vertebrates are glycoproteins; they all contain carbohydrate, although the amounts and kinds are variable (Figure 17–5).

The synthesis, secretion, and assembly of collagen fibers was discussed in Chapter 13. The genes that specify collagens are notable for the large number (~50) of introns they contain; the gene sequence is about ten times longer than necessary to code for the protein. The exons almost all have lengths that are a multiple of nine bases, suggesting the gene (at least in the triple helical region) arose from an ancestral DNA sequence that was duplicated many times.

Elastin

The extracellular matrix has the property of elastic recoil—after deforming under stress, a tissue usually regains its original shape. In vertebrates, this behavior is due to the presence of elastic fibers or sheets that are made mostly of the protein elastin. Only a few tissues contain

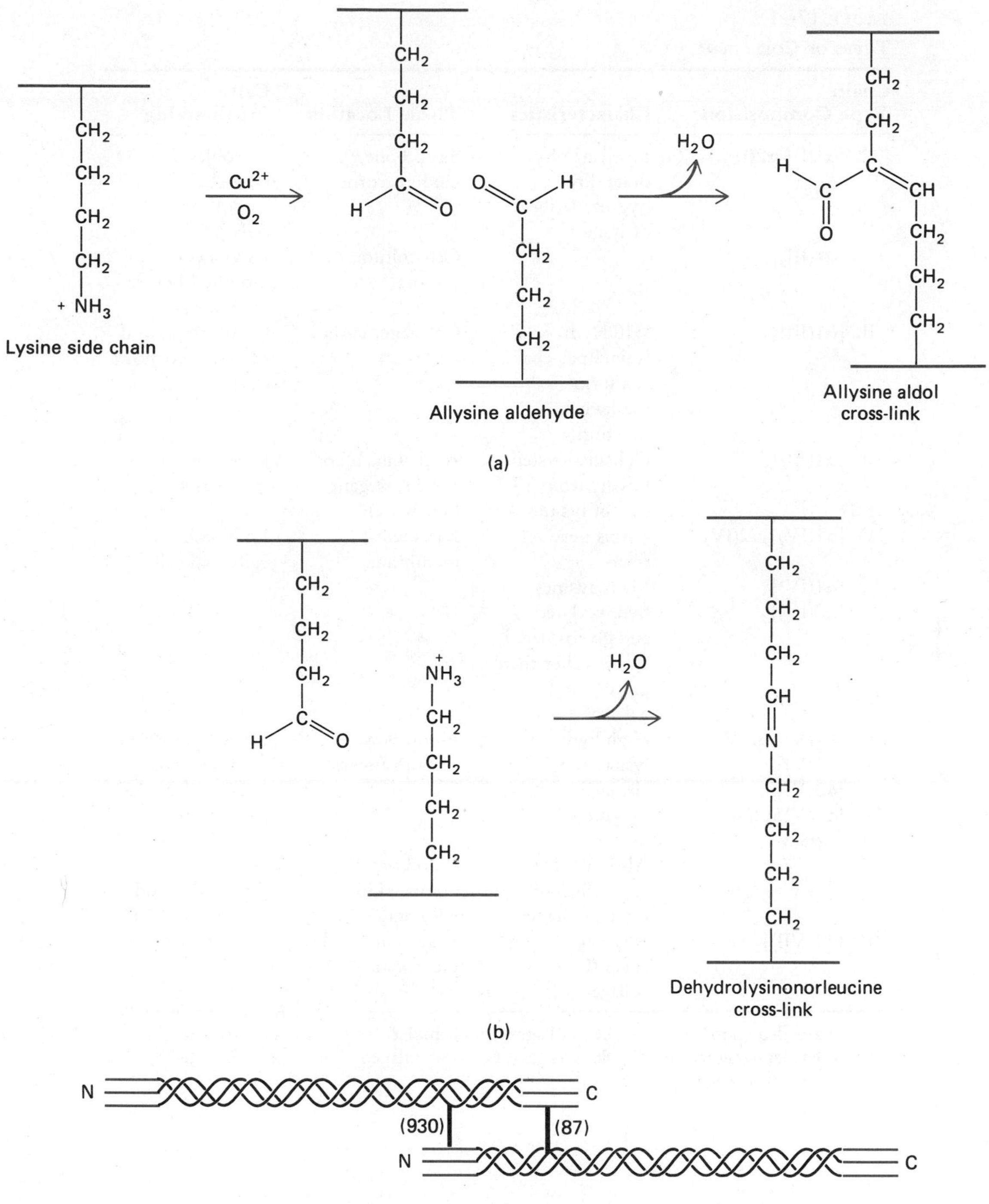

FIGURE 17-4

One of Several Kinds of Cross-Links That Occur Between the Terminal Regions of Collagen Molecules. (a) The enzyme peptidyl lysyl oxidase first catalyzes the formation of allysine aldehyde. Two such aldehydes can react spontaneously to form the aldol cross-link. (b) The active aldehyde group can condense with lysine or with hydroxylysine, as well as with another aldehyde. (c) General location of interhelix cross-links. The site on the N-terminal telopeptide is (color) shown cross-linking near the C-terminus (residue 930 in $\alpha1(I)$ collagen); the C-terminal telopeptide cross-links with a site near the N-terminus (residue 87).

TABLE 17–1

TYPES OF COLLAGEN*

Chain Type Composition	Characteristics	Tissue Location	Cells Synthesizing
I. $[\alpha 1(I)]_2 \alpha 2(I)$	Low carbohydrate, low hydroxylation of lysine	Skin, bone, tendon, cornea	Fibroblasts, epithelia, smooth muscle cells
$[\alpha 1(I)]_3$		Cell culture, tumors	Extensively modified before secretion
II. $[\alpha 1(II)]_3$	>10 Hydroxylysines per chain; can form meshwork of fine fibrils	Cartilage, cornea	Cartilage, neural retina, notochord cells
III. $[\alpha 1(III)]_3$	Contains cysteine, low hydroxylation of lysines	Fetal skin, blood vessels, organs (not bone)	Fibroblasts, myoblasts
IV. $[\alpha 1(IV)]_2 \alpha 2(IV)$	Forms network of fibers	Basement membrane	Endothelial and epithelial cells
$[\alpha 1(IV)]_3$ $[\alpha 2(IV)]_3$	Many lysines hydroxylated and glycosylated, sugars other than glucose and galactose		
V. $[\alpha 1(V)]_2 \alpha 2(V)$ $[\alpha 1(V)]_3$ $[\alpha 3(V)]_3$ $[\alpha 1(V)]\alpha 2(V)$ $\alpha 3(V)$	High hydroxylysine, low alanine, 3-hydroxyproline	Blood vessels, smooth muscle	Smooth muscle cells; cartilage cells, under certain conditions
VI.	Also called short chain (SC) or intima collagen	Blood vessels, uterus, skin, placenta	Forms 10–20 nm fibrils of blood vessels
VII. $[\alpha 1(VII)]_3$	Also called long chain (LC) collagen	Basement membranes	Human amnion

*There are also endothelial cell (EC) collagens, high-molecular-weight (HMW) and low-molecular-weight (LMW) collagens, and G chain collagen. These are present in small amounts in a variety of tissues.

large amounts of elastin: the large blood vessels and the lungs (Figure 17–6).

Elastin and collagen are the least soluble proteins in the extracellular matrix. Elastin also lacks the amino acid methionine, so treatment with cyanogen bromide breaks the polypeptide chains (at methionine residues, see Chapter 4) of collagen and other proteins, yielding small soluble peptides that are easily removed from the insoluble elastin.

Of course, such a highly insoluble protein is not synthesized in that form. The primary translation product is likely to be a molecule called *tropoelastin,* an approximately 72-kDa protein that is rich in glycine, alanine, proline, valine, and other hydrophobic amino acids. Peptide maps of tropoelastin are similar to those of elastin. Some characteristic peptides include ala-ala-ala-lys and ala-ala-lys, repeated six times per tropoelastin.

The cross-links that give the elastin its characteristic

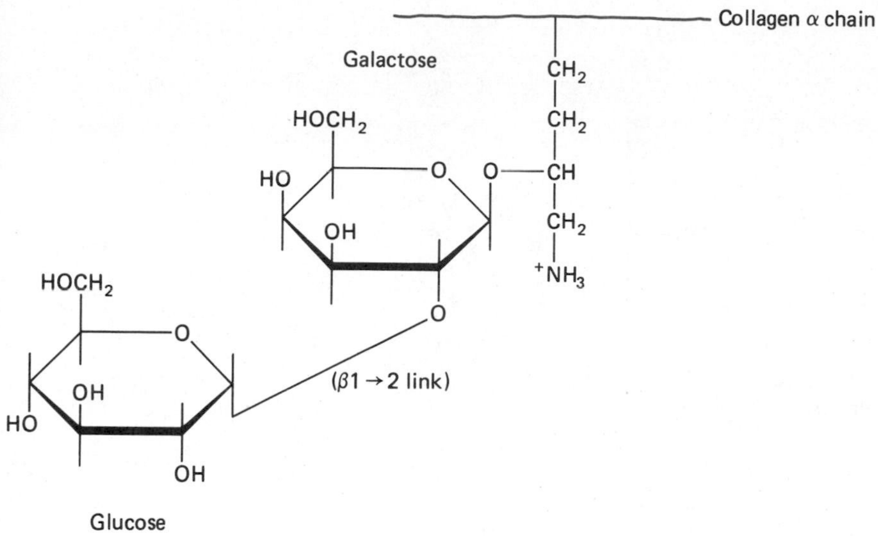

Collagen α chain

Galactose

(β1 → 2 link)

Glucose

FIGURE **17–5**
A Collagen-Specific Carbohydrate.
Only two kinds of carbohydrate are
found attached to the hydroxylysines
of collagen; one is shown here. The
other does not have the glucose
residue. In some collagen chains a very
large fraction of hydroxylysines carry
sugars, in others only a small fraction.

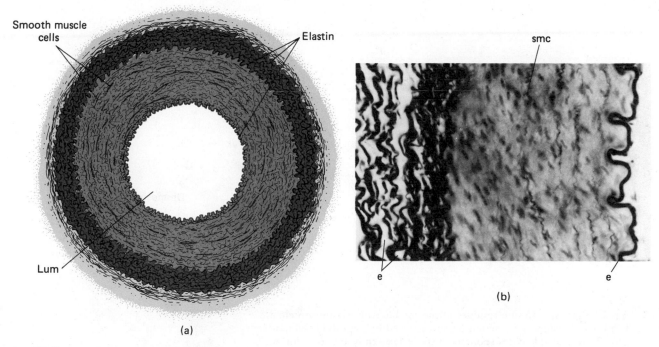

(a)

(b)

FIGURE **17–6**
Elastin in the Aorta. (a) The lumen (Lum) of the blood vessel is surrounded by
a thick elastic wall. Curving bands of elastin (e), alternate with smooth muscle
cells (SMC). × 235. (b) × 1400. (Courtesy of M. H. Ross.)

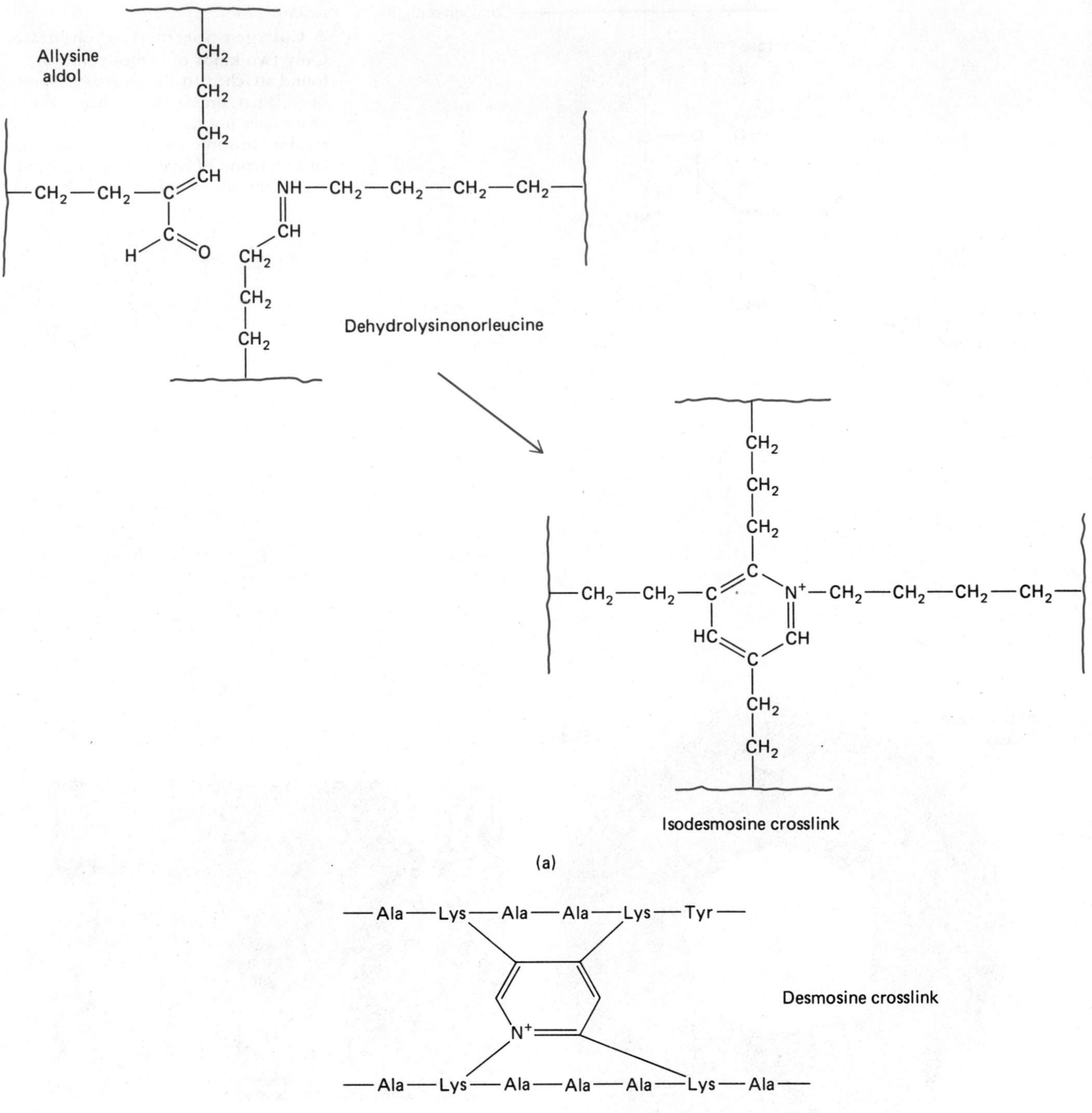

Isodesmosine crosslink

(a)

Desmosine crosslink

(b)

FIGURE 17–7

Crosslinks in Elastin. Lysine residues in the soluble tropoelastin protein are oxidized to form allysine, the reaction being catalyzed by peptidyl lysyl oxidase (see Figure 17-4). The aldehydes spontaneously react with lysine, or with one another, to form *isodesmosine,* or *desmosine,* characteristic cross-link structures. (a) Reaction between allysine aldol and dehydrolysinonorleucine to form isodesmosine cross-link. (b) Typical amino acid sequence of two tropoelastin chains in region of desmosine cross-link.

elasticity are made after the protein has been secreted (Figure 17–7). The mechanism is similar to the one described earlier for collagen (except that there is no hydroxylysine in elastin). The enzyme peptidyl lysyl oxidase catalyzes the formation of aldehydes from lysine side chains (see Figure 17–4). Inhibition of this enzyme, which occurs in animals eating a diet that is deficient in the element copper, leads to reduced cross-linking, reduced elasticity, and easily damaged blood vessels (aortic aneurisms). Such animals were the first sources of uncross-linked tropoelastin. The cross-linking reactions reduce the free lysine residues from 38 in tropoelastin to less than 10 in elastin.

The cross-linked structure of elastin creates a network of peptides that cannot be moved apart from one another. The other property that gives this network elasticity is the conformation of the peptides. There is almost no regular secondary or tertiary structure; instead, the peptides are in a *random coil*. Exactly what leads to this highly atypical conformation is not known, but physical measurements confirm that elastin has very little order in its structure. When structures containing random coils are deformed, peptides tend to become *more ordered*. It is the increase in entropy that accompanies the rerandomizing of the structure that gives the protein its characteristic elastic recoil.

Fibronectin

In the developing embryo, cells are organized either into sheets, known as *epithelia,* or into clusters, known as *mesenchymes.* Epithelial cells produce a tight layer of extracellular matrix, the *basement membrane* or *basal lamina,* while mesenchymes are embedded in an interstitial extracellular matrix.

Fibronectins (fibre = fiber; nectere = to bind) are found on the surfaces of mesenchyme cells ("cellular fibronectin"), in the interstitial extracellular matrix, and in the circulating blood ("plasma fibronectin"). Fibronectins are usually formed as a disulfide-lined dimer (or polymer) of 220–250 kDa peptide subunits. There are binding sites on fibronectins for a cell surface receptor, for collagen, for the clotting protein fibrin, for heparin (see below), and for several other molecules of the extracellular matrix. The cellular fibronectins tend to be insoluble, and the soluble plasma forms, therefore, have been studied more thoroughly. There are, on the other hand, many structural and functional similarities between the two forms.

Antibodies to fibronectin can be used to localize the protein by the direct immunofluorescence method (Figure 17–8). In vitro, fibronectin is found as aggregates and fibers, beneath and between cells while they are growing

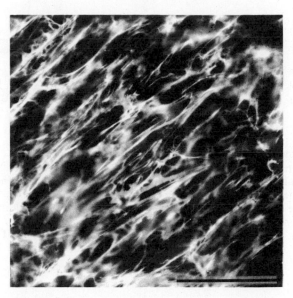

FIGURE 17–8

Fibronectin in the Extracellular Matrix. A confluent culture of chick embryo fibroblasts was stained with antifibronectin antibodies conjugated to fluoresceine. Fibronectin was detected as fibrils on the cell surface and in the extracellular matrix. (Bar = 50 μm.) (Courtesy of K. M. Yamada.)

and completely surrounding cells when they become confluent. One of the locations of fibronectin in vivo is in the basement membrane found below sheets of cells that form an epithelium. Fibronectin is also associated with collagen fibrils.

Plasma fibronectin is composed of an A and a B subunit, cross-linked by disulfide bonds at their C-terminal ends (Figure 17–9). Both chains have sites that are hypersensitive to proteases, and gentle treatment with trypsin, chymotrypsin, or thermolysin produces large fragments that can be separated from one another and then tested for their functions. The results from such experiments have given rise to the concept of domains of structure and function, as illustrated in Figure 17–9. This domain arrangement is also seen in the nucleotide sequence of the gene that specifies chicken cellular fibronectin. There are more than four dozen small exons of about 150 base pairs each, a size that correlates nicely with that expected to code for a 45-amino acid repeating sequence found at least one dozen times in bovine plasma fibronectin. Other repeating sequences of amino acids are also found. It appears that fibronectin may have evolved by duplication of small nucleotide sequences.

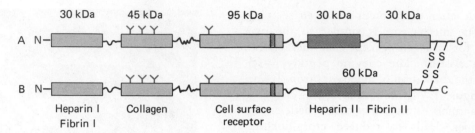

FIGURE 17-9

Domain Structure of Plasma Fibronectin.   The two chains are cross-linked by disulfide bonds near their C-terminal ends. Domains are indicated by blocks, joined by parts of the polypeptide chain that are especially sensitive to proteolysis. The approximate size of each domain is given in kilodaltons (kDa). Carbohydrate attachment sites are indicated as gray Y-shaped structures. The various binding functions of the domains are indicated below the blocks. The site where fibronectin chains can be cross-linked by transglutaminase is near the N-terminus.

The various binding activities of fibronectin are separated into the structural domains. The cell surface-binding regions are in the middle of each protein chain. By further proteolysis the cell binding activity has been localized to an 11.5-kDa fragment. Peptides with amino acid sequences found in the fragment have been synthesized in the laboratory and tested for their ability to bind to cells; this allowed the identification of a 30-amino acid sequence as highly active in the binding function. Further dissection of the fragment has narrowed the number of critical amino acids to just three: Arg-Gly-Asp (RGD, using the single letter code for amino acids; actually, the smallest peptide with cell-binding activity is RGDS, but amino acids other than serine—the one found in fibronectin—still permit binding activity).

A site on the cell surface of fibroblasts that interacts with fibronectin has also been identified. There are at least two integral membrane proteins involved, each with a mobility on SDS polyacrylamide gels appropriate to 140 kDa. When these proteins are added to artificial lipid bilayers, in the form of spherical liposomes, fibronectin binds to the surface of the liposome. In addition, this receptor binds peptides and other proteins that contain the RGD sequence—fibrinogen and fibrin, for example. By the same token, there are other RGD receptors on the fibroblast cell surface, as well as on other cells. The RGD receptor on platelets can bind fibronectin and fibrinogen, as well as at least two other proteins. There is also evidence for membrane ganglioside glycolipids acting as binding sites for fibronectin. The various protein receptors that bind fibronectin comprise a family known as the *integrins*.

The collagen-binding domain of fibronectin can interact with most types of collagen, although denatured collagen is more tightly bound. In fact, binding to affinity columns made of gelatin is one of the methods for purifying fibronectin. Because fibronectin can bind both to the cell surface and to collagen, it can mediate the attachment of cells to collagen layers (Figure 17-10). The ability of fibronectin to mediate binding between cells and fibrin, which is a major component of blood clots, suggests an important role for fibronectin in the healing of wounds.

Fibronectin stimulates cells to attach (adhere) to a variety of surfaces, consistent with the numerous binding sites on fibronectin. The amount of fibronectin associated with the surface of normal cells varies with the phase of the cell cycle, being smallest during mitosis. When added to malignantly transformed cells, which typically are associated with relatively little fibronectin, the cells spread out and attach to the substratum of a culture flask and regain their normal shape. This adherence response can also be elicited by the cell-binding domain of fibronectin; it is blocked by antibodies to that domain or by soluble peptides containing the RGD sequence (insoluble peptides containing RGD mediate cell attachment, similar to intact fibronectin). RGD peptides can also interfere with cell migration.

Two other phenomena that are believed to be influenced by fibronectin are cell polarity and cell differentiation. An example of the latter is the observation that smooth muscle cells lose their contractile phenotype when plated on a fibronectin layer. Fibronectin and its cell surface receptors, or analogous proteins, may also be involved during bacterial and viral infections. Some strains of the pathogenic bacterium *Staphylococcus aureus* bind to fibronectin, an activity that could either make an infection worse by keeping the bacteria near a susceptible

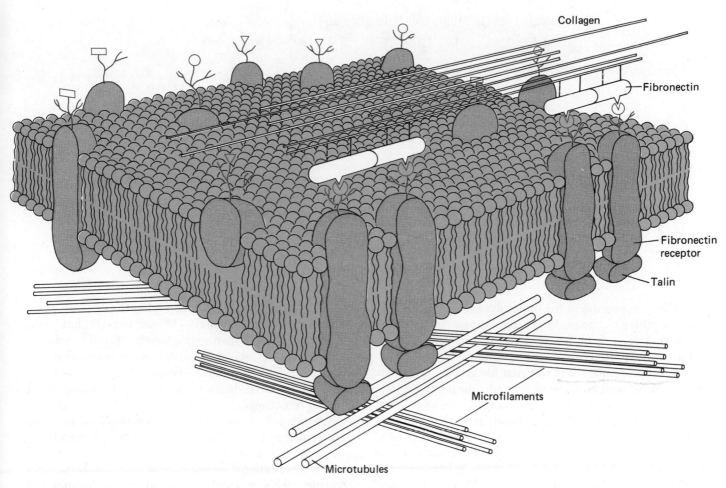

FIGURE **17–10**
Fibronectin Can Bind Cells to Collagen Layers.   This illustration shows the
binding of fibronectin to receptors, shown as integral membrane proteins that span
the bilayer, and to collagen fibrils in the extracellular matrix. The receptor proteins
interact with actin microfilaments in the layer of cytoplasm lying adjacent to the
cell membrane. Talin is associated with end-on termination of microfilament
bundles (See Figure 17–22).

cell or help trap the bacteria for disposal by cells of the
immune system.

Laminin

Lying beneath the layer of cells called the epithelium,
a thin sheet of special structure known as the *basal lamina*
can be detected in the electron microscope. In the light
microscope this structure is seen as a *basement membrane*
and includes the underlying collagenous layer. There is a
two-layered appearance of the basal lamina, a thin (20

nm) *lamina rara* (or lamina lucida) immediately adjacent to
the epithelial cells and a thicker (20–50 nm) *lamina densa*
that runs parallel to the overlying cell sheet or tube. Basal
laminae are also found underlying the endothelial cell
layer of capillaries, and around fat cells, muscle cells, and
so forth—wherever a layer of cells abuts on connective
tissue. There are two cases where a basal lamina separates
two epithelial cell layers: the kidney glomerulus, where a
specialized membrane separates the glomerular epithelium from the capillary endothelium, and the lung
alveolus, where the alveolar epithelium faces capillaries.
In those cases there are two lamina rara.

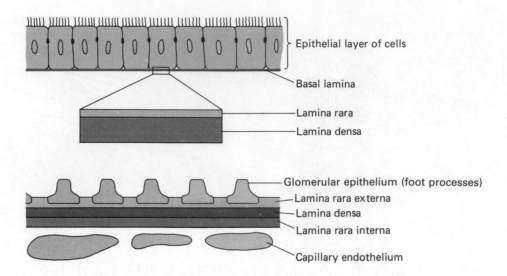

Epithelial layer of cells

Basal lamina

Lamina rara

Lamina densa

Glomerular epithelium (foot processes)

Lamina rara externa

Lamina densa

Lamina rara interna

Capillary endothelium

Characterization of the molecules in the basal lamina was simplified by the study of certain tumor cell lines that produce large quantities of basal lamina that can be released by appropriate treatments into the culture fluid. The glycoprotein *laminin* was first identified by such an approach, as well as a special type of collagen (type IV). Laminin, found only in the lamina rara, can be split by reagents that reduce disulfide bonds into two kinds of peptide chains, an A (or α) chain of 200 kDa and a B (or β) chain of 400 kDa. The assembled molecule is composed of three A chains and one B chain, held together at their ends by disulfide linkages (Figure 17–11a). By using the rotary shadowing technique, it is possible to visualize laminin in the electron microscope. It has a cross-like appearance, consistent with its chain composition (Figure 17–11b).

Laminin contains 12–15% carbohydrate, including 4–6% sialic acid, mostly N-linked to asparagine side chains. Antibodies have been prepared against the protein; using immunofluorescence and immune electron microscopy, it has been shown that laminin is present in the lamina rara of a variety of tissues, immediately adjacent to the membranes of the overlying cells. Despite antigenic similarities, there is reason to believe that laminins from different tissues, or from different organisms or from the same organism during development, have slightly different amino acid sequences.

Using the partial proteolysis approach, described earlier for the case of fibronectin, structural and functional domains have been identified in laminin (see Figure 17–11a). The main function proposed for laminin is to mediate the binding of cells to type IV collagen. Certain epithelial cell lines (e.g., PAM-212) adhere preferentially to type IV collagen-coated surfaces; the attachment of these cells is significantly enhanced by laminin, but not by fibronectin. The effect is specific for type IV

collagen—attachment to types, I-, II-, III-, and V-coated surfaces are not enhanced. While laminin has been demonstrated to enhance adherence of other cells to collagen surfaces, it is difficult to obtain clear evidence of specificity. This is because fibronectin is also a component of the basal lamina and can also participate in binding to collagen.

Another kind of molecule containing protein and carbohydrate—the proteoglycan known as heparan sulfate (see below)—is also found in the lamina rara, and there is a heparan sulfate-binding domain on the laminin structure. This is another way in which laminin can mediate the binding of cells to the basal lamina.

Some of the proteins we have described above exist in more than one form. We have discussed the most common form of the molecules associated with the extracellular matrix; an alternate form might actually be an integral membrane protein. There is a steadily increasing list of other proteins in the extracellular matrix that either confer special structural properties to the matrix or act to bind cells to the matrix, or both. Among these are the chondronectins, cell-spreading factors that are found in serum, entactins, vitronectins, and so forth. Using the examples discussed above and the general principles they illustrate, it should be possible to place into the context of cell-environment interactions a large number of as yet incompletely described glycoproteins.

Proteoglycans

The glycoproteins contain a few oligosaccharides, bound in covalent linkage to the protein at serine, threonine, and asparagine side chains. *Proteoglycans* also contain such oligosaccharides, but in addition, they contain at least one large polysaccharide chain, called a *glycosaminoglycan (GAG)*. Carbohydrate often dominates

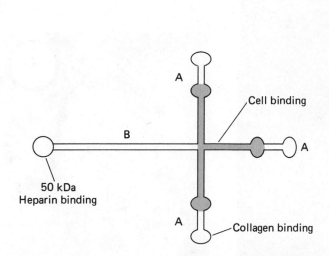

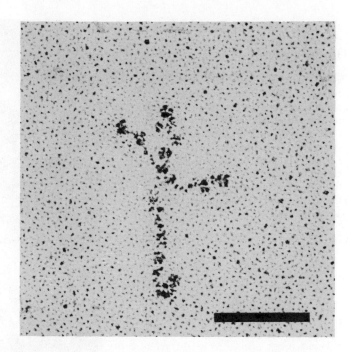

FIGURE **17–11**
Structure of Laminin.   (a) Structural and functional domains of laminin. Laminin consists of A and B chains linked by disulfide bonds and terminating in globular domains. Likely sites of ligand-binding activity are indicated. The central region (color) and some of the globular domains can be obtained as protease-resistant fragments. (b) Rotary shadowing electron microscopy of laminin. The cross-like structure of the molecule can be seen, as well as the more densely staining globular domains. (Bar=35 nm; courtesy of J. A. Madri.)

the structure of the proteoglycan, which can be up to millions of daltons in mass. The proteoglycans from cartilage have been studied in great detail, since they comprise a large fraction of the substantial extracellular matrix in that tissue. Protein is a minor part of cartilage proteoglycans and is therefore called a *core protein;* attached to it are up to 100 very large glycosaminoglycan chains (Figure 17–12).

The glycosaminoglycans were studied first by extracting them from tissues using proteolytic enzymes. This had the effect of removing most of the core protein, so its presence and its covalent attachment to the polysaccharides was missed for a long time. The structure of the core proteins is in fact still incompletely described. Another approach is to translate mRNAs from cartilage tissue in cell-free extracts. The product proteins that interact with antibodies raised against chondroitin sulfate proteoglycan from cartilage have an apparent molecular mass of 340 kDa. This may be an overestimate of the core protein size, since post-translational removal of peptides seems likely to occur before the core protein is finally modified into the proteoglycan. The mature core protein

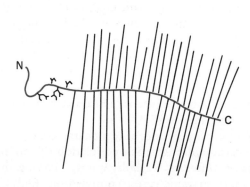

FIGURE **17–12**
Structure of Cartilage Proteoglycan.   The core protein (color) is dominated by many O-linked carbohydrate chains known as glycosaminoglycans (black). These have negative electrostatic charges and repel one another. Near the N-terminus there are usually found smaller N-linked oligosaccharides.

FIGURE 17–13

Synthesis and Secretion of Proteoglycans. (a) Golgi area of cartilage-forming cell. The Golgi apparatus (G) is shown, with the rough endoplasmic reticulum (RER) on one side and secretory vesicles (SV) on the other. These secretory vesicles seem to contain a filamentous material. ×50,000. (b) Exocytosis into the extracellular matrix (ECM). The contents of the secretory vacuole have the appearance of dense granules on filaments. ×50,000. (Courtesy of V. C. Hascall, Jr.)

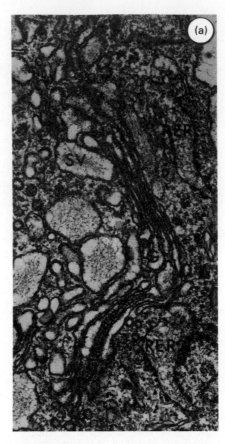

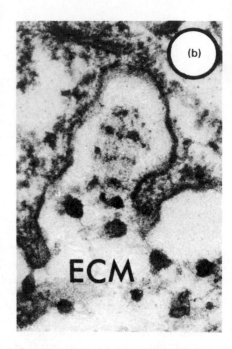

is thought to be only 200 kDa. Core proteins are probably synthesized on membrane-bound ribosomes. Some oligosaccharide groups are installed in the rough endoplasmic reticulum, as occurs for other glycoproteins, but most are probably added in the Golgi apparatus. After packaging into vesicles, the proteoglycan is secreted into the extracellular space by exocytosis (Figure 17–13).

The repeating unit of the polysaccharide polymer is a disaccharide; one part is an *amino sugar* that gives the structure its name, glycosaminoglycan (glycose = sugar, + amino). The amino sugar is usually joined in ester linkage to a sulfate, which has a strong negative charge. The other part of the disaccharide is a *uronic acid,* whose carboxyl group also has a negative charge. The high density of negative charges confers many unusual properties on proteoglycans. Different GAGs have different sugars (Figure 17–14).

Glycosaminoglycans have a characteristic variability in their carbohydrates. Each population of core proteins has many different carbohydrates, varying in size, charge, and, in some cases, composition. Only average values can be given for such molecules. Moreover, variation is also possible in the post-translational proteolytic processing of the core proteins themselves. Some of the variation in the proteoglycan structures apparently can occur after secretion.

Synthesis of the GAGs, except hyaluronic acid, occurs as a post-translational modification of the core

FIGURE 17–14

Structures of Different Glycosaminoglycans (GAGs). (a) Chondroitin-6-sulfate and chondroitin-4-sulfate. These two GAGs differ only in the carbon of the galactosamine to which the sulfate (about 0.8 groups per disaccharide) is esterified. One GAG may have both forms in a single polymer. The polysaccharide is attached to a serine (or threonine) side chain in the core protein by a linker sequence: glucuronate-galactose-galactose-xylose-serine. (b) Dermatan sulfate. This GAG differs from chondroitin sulfates only in that some disaccharides contain the L-isomer iduronate instead of the D-isomer glucuronate. The ratio of the two isomers can vary over a wide range. There is about one sulfate residue per disaccharide. (c) Heparan sulfate and heparin. The carbohydrates of these two GAGs are the same, both disaccharides shown are found in the same polymer, but they are found on different core proteins. Heparin generally has more sulfates and more iduronates than heparan sulfate. Both these GAGs, unlike all the others, can have sulfation of the amino group. (d) Keratan sulfate. The polysaccharide can be O-linked or N-linked, depending on the core protein. The O-galactosamine linking sugar is branched, being bonded to both a sialic acid-galactose disaccharide and, via a galactose-galactosamine, to the polymer. (e) Hyaluronic acid (HA). This GAG contains no sulfate esters and is the longest polymer; the disaccharides of HA may be repeated as many as 10^4 times. HA is not linked covalently to a core protein; instead there are extensive secondary bonds involving a 50-kDa "link protein." Tissues containing hyaluronic acid are swollen from taking up water. This property results in the formation of spaces between cells and tends to cushion them from mechanical shocks.

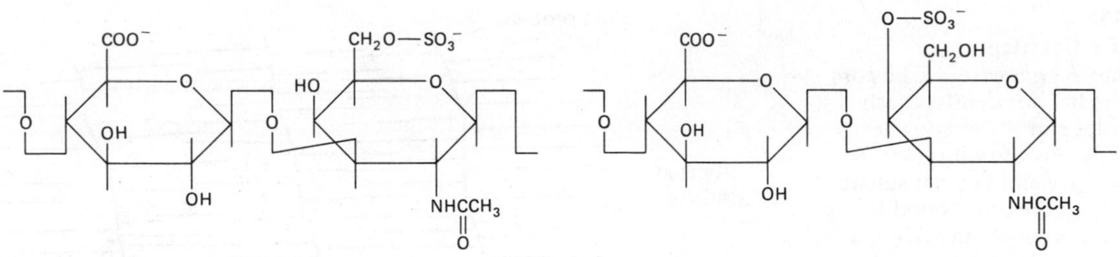

Glucuronate
β1→3

N-Acetyl
galactosamine β1→4
6-sulfate

Glucuronate

N-Acetyl
galactosamine β1→4
4-sulfate

Chondroitin 6-sulfate
Typical repeat length
20–60

Chondroitin-4-sulfate
Typical repeat length
20–60

(a)

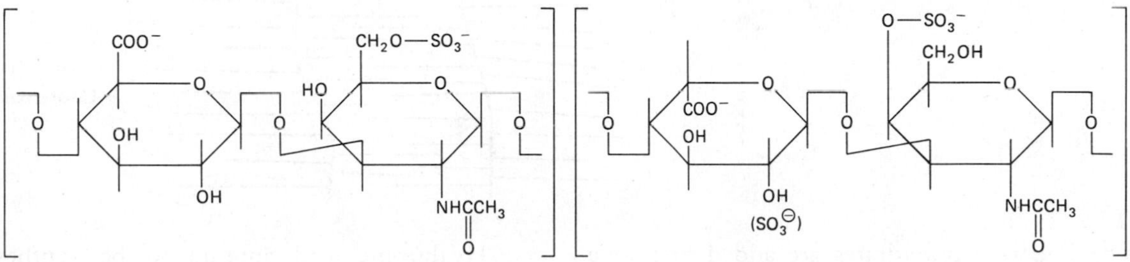

Glucuronate

N-Acetyl
galactosamine

α-L-iduronate

N-Acetyl
galactosamine

Dermatan sulfate
Typical repeat length
30-80

(b)

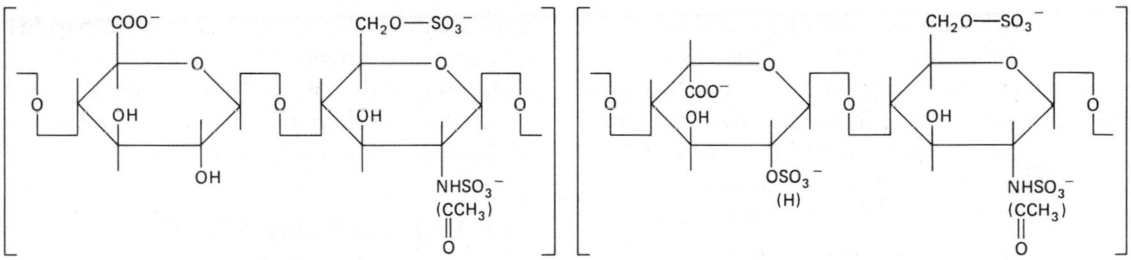

Glucuronate

N-Acetyl
α-D-Glucosamine

Iduronate

N-Acetyl
glucosamine

Heparan sulfate (heparin)
Typical repeat lengths
10–60

(c)

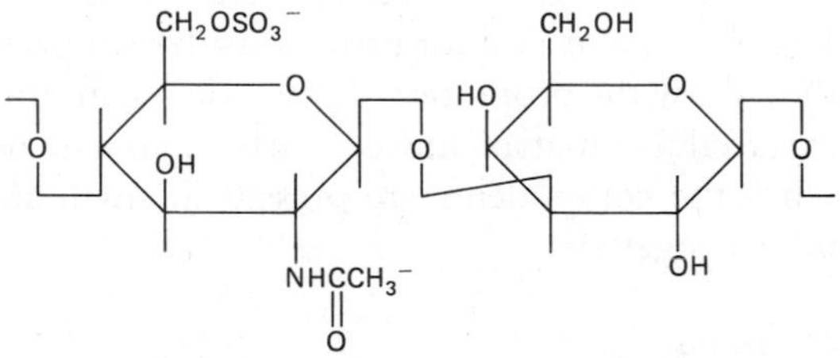

N-Acetyl
glucosamine

Galactose

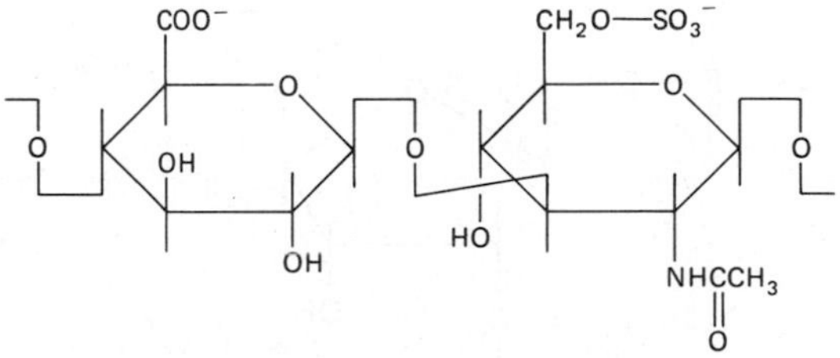

Glucuronate

N-Acetyl
β-Glucosamine

Keratan sulfate
Typical repeat length
5–40

Hyaluronic acid
Typical repeat length
50–10000

(d)

(e)

FIGURE **17–15**

Diagram of a Cartilage Proteoglycan Aggregate.   The core protein (color) has covalently attached oligosaccharides and glycosaminoglycans of two types: chondroitin sulfate and keratan sulfate. The core proteins are also bound by noncovalent interactions (involving a linker protein) to a hyaluronic acid polymer, only a small length of which is illustrated.

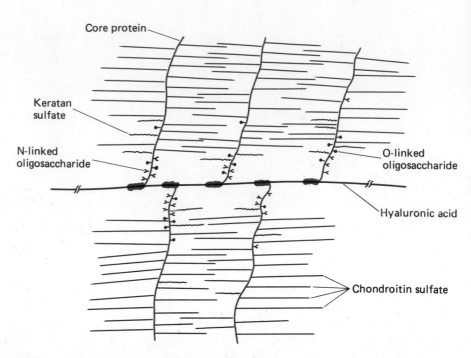

protein. The linker carbohydrates are added first, one sugar at a time, then the polymer is built on, again one sugar at a time. Chondroitin sulfate provides a specific example. The first sugar is activated by forming uridine-diphosphate-xylose (UDP-xylose); all the monomers are activated before incorporation into the polymer by the formation of a UDP-sugar. This is added to a serine or threonine side chain. Then two galactoses are added, then the glucuronate, and finally the N-acetyl galactosamine is added. Successive additions of glucuronate and N-acetyl galactosamine make up the polymer. Sulfations occur after the sugars are added; the activated form of sulfate is phosphoadenyl phosphosulfate. Epimerizations to L-iduronate also occur after the glucuronate is incorporated (this is not very common on chondroitin sulfate but occurs often in other GAGs).

Phosphoadenine phosphosulfate (PAPS)

Hyaluronic acid appears to be synthesized and secreted separately from the proteoglycans. The precursor UDP-glucuronate and UDP-N-acetyl glucosamine are added alternately to a growing UDP-hyaluronic acid in the cytoplasm near the plasma membrane. The growing chain is translocated through the cell membrane, perhaps via a lipid-based channel. Such a lipid is involved in the synthesis of the cell wall polysaccharide in bacteria, a process that shows many similarities to the synthesis of hyaluronic acid. The polysaccharide emerges from the membrane and there interacts with specific receptor sites on the core proteins of proteoglycans.

The Extracellular Matrix of Connective Tissue

Although all cells in a multicellular organism have an extracellular matrix, those of connective tissues were studied first because the extracellular matrix is such a prominent feature of the tissue. Moreover, some connective tissues are dominated by certain of the components. It is therefore useful for us to consider connective tissues; they show the properties of individual components of the extracellular matrix more clearly than can be observed when the components are present in small amounts and mixed together.

Cartilage

Cartilage has large aggregates of proteoglycans, formed on hyaluronic acid (Figure 17–15). These aggregates are very large: as many as 100 proteoglycans can bind to a single hyaluronic acid, giving a length of almost

10 μm and a diameter of 500–600 nm. The large volume occupied by the GAGs is due to the extended nature of the negatively charged polysaccharides; electrostatic repulsion keeps the chains extended and separated from one another. Another consequence of the extended structure is the effect on surrounding water molecules, which tend to be more immobilized than in bulk water. This is due primarily to the electrical charges on the glycosaminoglycans. The negative charges tend to organize water in multiple layers around the GAGs. Such proteoglycans are capable of being compressed; the water can be squeezed out under pressure, but this brings the negative charges closer together. When the pressure is relieved, the structure springs back, by regaining water, toward its original shape so as to separate the negative charges farther from one another (Figure 17–16).

The advantages of this structural arrangement are more marked in the cartilage tissue from the load-bearing joints found between bones. There the cartilage has larger proteoglycan aggregates than most other extracellular matrices; moreover, the tissue is smaller than the volume the aggregates would occupy in solution (a network of collagen fibers keeps the tissue compressed) and the negative charges are held near one another (Figure 17–16). This means the cartilage acts like a partially compressed spring; it is able to withstand loads with less compression of the structure (since the negative charges would then be brought even closer to one another). When there is a deficit in the sulfation of the cartilage proteoglycans, as happens to the mutant mice known as "brachymorphic," the limbs are shorter because of a reduction in the width of the cartilage growth plates at the end of the long bones. Certain forms of arthritis are caused by abnormalities in the collagen network of joint cartilages.

Tendon

Collagen fibers dominate the structure of the tendons that connect muscles to bones. The fibers lie parallel to one another and are cross-linked extensively into bundles (Figure 17–17). The aggregation of collagen

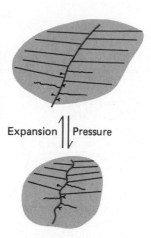

FIGURE 17–16

Proteoglycans from Cartilage Show Reversible Compressibility. The glycosaminoglycans (black) have almost one negative charge per monomer and, as a result, are extended and lie as far from one another as possible. Water fills the space between adjacent polymers. When pressure is applied, the structures can deform by squeezing out water, but this brings the negative charges closer together. When the pressure is removed, electrostatic repulsion drives the GAG chains farther apart, water reenters the structure, and its volume increases.

FIGURE 17–17

Appearance of Tendon in the Polarizing Microscope. The type I collagen fibrils in tendon (bottom panel) are oriented in parallel bundles. There is a regular crimp in the fibers, observed as an alternation in intensity of the image. This is believed to allow tendons to stretch a little, under load, without breaking the covalent bonds of the collagen fibers. Compare this appearance with that of skin (top panel) and intervertebral disk (middle panel). The former is irregular in structure; the latter shows regularity of structure, but is more appropriate to resist compression, rather than tension, forces. (Bar = 25 μm.) (Courtesy of R. L. Trelstad.)

FIGURE **17–18**
Secretion of Collagen Fibers.
(a) Electron micrograph of embryonic
chick tendon cell. The densely staining
condensation vacuole contains at least
two sets of collagen molecules since it
is longer than the length of a fully
processed collagen molecule. (Bar
between arrows = 300 nm.) (b)
Diagram illustrating model of collagen
secretion. After synthesis in the
endoplasmic reticulum (er),
procollagen is transported to the Golgi
apparatus (ga); there the procollagen
begins to aggregate. Condensation
vacuoles (cv) carry the aggregates to
the cell surface and fusion with the
plasmalemma discharges them into the
extracellular space. The aggregates are
secreted into deep recesses in the cell
surface; this may allow control of the
growing end of the fibril. (c) Electron
micrograph showing the deep recesses
and the secretion of collagen fibrils.
(Bar = 300 nm.) (Courtesy of R. L.
Trelstad.)

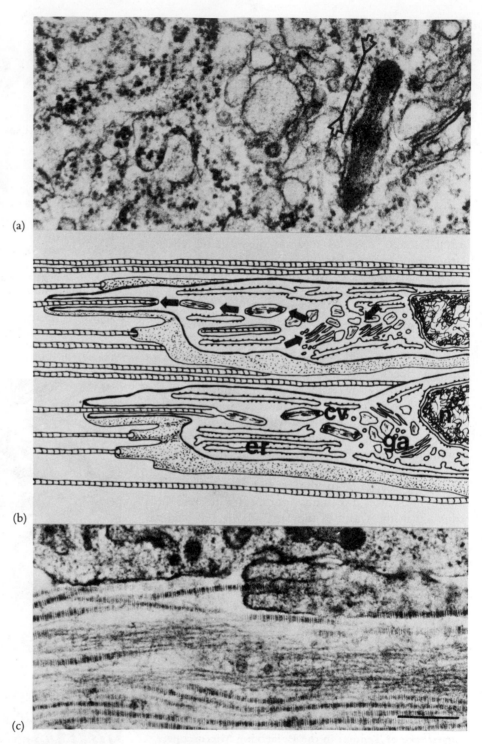

(a)

(b)

(c)

molecules into fibers (see Chapter 15) may begin before
secretion (Figure 17–18). Moreover, the release of collagen into the extracellular space, which requires energy
and an active cytoskeleton, is not a random process. In
the developing chick embryo, tendon cells secrete collagen fibers in a parallel array at characteristic infoldings of
the cell membrane that are restricted to a small area of the
cell surface (Figure 17–18). It seems likely that the

pre-existing collagen fibers also help to direct the assembly of the tendon bundles, but the exact way this happens
is not known. The parallel, cross-linked nature of
collagen fibers in tendons gives them great stability under
tension stress; there is very little stretching in a tendon
that holds a muscle to a bone, for example, and the
cross-links prevent the collagen fibers from slipping past
one another.

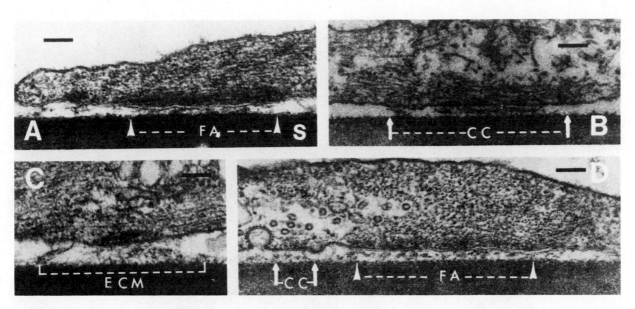

FIGURE 17–19

Regions Where the Surface of a Fibroblast Approaches a Cross-Linked Gelatin Substratum. The vertical sections show sites of focal adhesion (FA) and close contact (CC), classifications based on interference microscopy (see Figure 16–64). ECM is a different contact structure, where the cell makes a close approach to aggregates of the extracellular matrix. Panel D also shows a coated pit forming in the cell surface. (Bars = 100 nm.) (Courtesy of W.-T. Chen.)

Adhesion of Cells to the Substratum

For many cells growing in vitro, attachment to a substratum is required for normal metabolism and for growth. This is believed also to be true for cells in multicellular organisms. In this section we shall discuss the methods used to investigate the adhesion of cells to surfaces and the current picture that cell biologists have of this phenomenon.

Adhering to a substratum is a multistep process that involves not only binding between surface structures but also reorganization of the cytoskeletal components within the cell. On a culture dish, two kinds of connections between cells and their substrates are observed when examined with the interference reflecting microscope, *close contacts* and *focal contacts* (see Figure 16–60). Close contacts are regions of the cell surface that approach within 30–50 nm of the substratum surface; focal contacts are more well defined sites of closer (less than 20 nm) approach of the cell to the substratum. Each of these can also be observed in the electron microscope (Figure 17–19). (The structure known as the *adhesion plaque,* seen in the electron microscope, is believed to be equivalent to the focal contact identified by interference reflection microscopy.) In addition to these contacts with the

substratum, there are *extracellular matrix contacts,* where the cell surface joins with aggregates of extracellular matrix other than those on the surface of the culture dish.

Numerous microfilaments lie inside the cell, terminating opposite the areas of close approach to the substratum or to the extracellular matrix. End-on terminations of bundles of microfilaments, perhaps attached to the protein known as *talin,* lie opposite focal contacts. When immunofluorescence microscopy is used in conjunction with interference reflection microscopy, it can be demonstrated that microfilament-associated proteins, specifically α-actinin and vinculin, lie in the regions of substratum contact (Figure 17–20). The protein product of the cellular oncogene *c-src* is also found in the region of the focal contact (adhesion plaque, see discussion in Chapter 14). Vinculin is found in only some of the extracellular matrix contacts.

The arrangement of extracellular fibronectin and of intracellular actin is closely correlated (Figure 17–21). The association of microfilament bundles and fibronectin fibers with regions of contact is not fortuitous: the organization of the intracellular components can be altered by disturbing the extracellular components, and vice versa. For example, treatment of cells with cytochalasin B, which disrupts microfilaments, leads to rapid

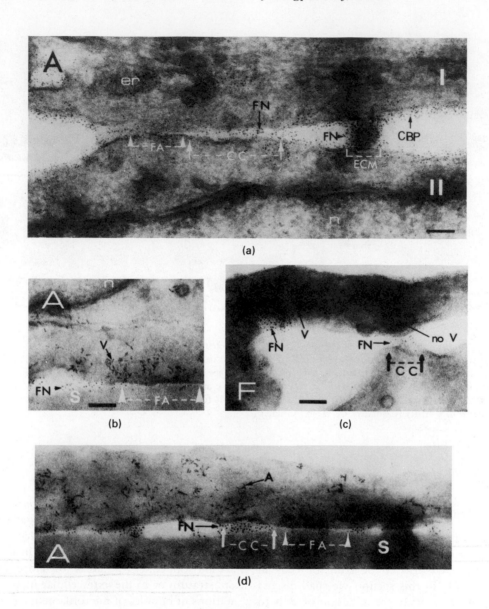

(a)

(b)   (c)

(d)

FIGURE **17-20**

Immunoelectron Microscopy of Cell Adhesion Sites.   Indirect
immunostaining was used to examine proteins found outside the cell—fibronectin
(FN) and concanavalin A-binding proteins (CBP)—and proteins found inside the
cell—vinculin (V) and α-actinin (A). (a) Cell-cell adhesion sites. The two fibroblast
cells (I, II) contact at focal adhesion (FA), close contact (CC), and extracellular
matrix (ECM) sites. Fibronectin was detected by secondary antibodies conjugated
with Imposil; fibronectin by secondary antibodies conjugated with ferritin.
Concanavalin A-binding proteins, but not fibronectin, are found in the focal
adhesion site; both extracellular proteins are found in the other two sites. (er =
endoplasmic reticulum; n = nucleus.) (b) Cell-substratum contact. The focal
adhesion contact with the substrate (S) has significant amounts of intracellular
vinculin (Imposil-antibody), but little extracellular fibronectin (ferritin-antibody).
This is also true for cell-cell focal adhesion contacts. (c) Cell-cell adhesion site. The
close contact region stains for fibronectin, but not vinculin. (d) Cell-substratum
contact. The close contact site is strongly labeled for fibronectin (ferritin-antibody)
and α-actinin (Imposil-antibody), while the focal adhesion site has little label for
either. This labeling pattern is also seen for cell-cell contacts. (Bars = 100 nm.)
(Courtesy of W.-T. Chen.)

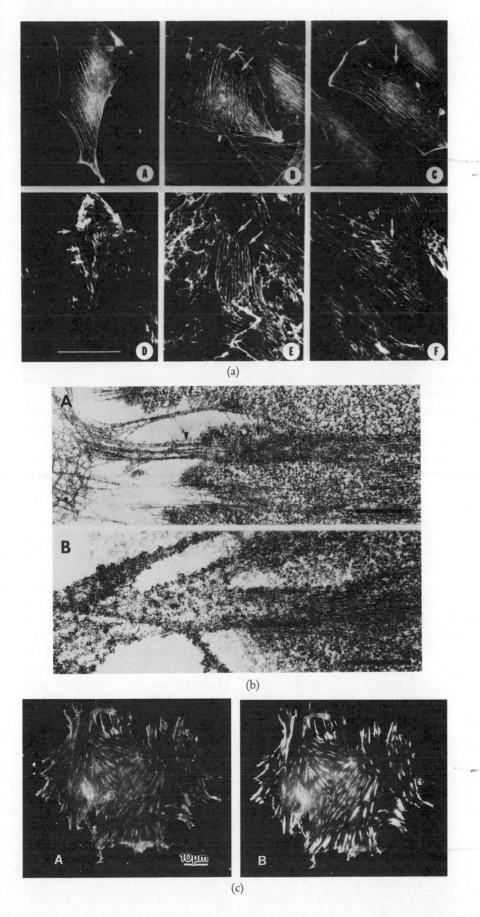

(a)

(b)

(c)

FIGURE **17–21**
Correlation Between Fibronectin and Actin. (a) Double-labeled immunofluorescence. The upper row of micrographs shows the distribution of actin and the lower row the distribution of fibronectin in NIL8 hamster cells, spread on a substratum. There is a good correspondence between antiactin and antifibronectin staining. For example, the rightmost pair of panels shows actin bundles (arrows) terminating before the edge of the cell and fibronectin fibrils (arrows) continuing beyond the cell boundary. (Bar = 50 μm.) (Courtesy of R. O. Hynes.) (b) Electron micrographs of NIL8 hamster cells. (top, A) Grazing section from the bottom of the cell. The extracellular space is to the left. Fibrils (arrowhead) approach the membrane and appear colinear with microfilament bundles inside the cell (arrow). (bottom, B) Fibronectin localized by indirect immunoferritin-staining technique. Ferritin particles can easily be seen outside the cell, staining fibrils. (Bar = 0.5 μm.) (Courtesy of I. I. Singer.) (c) Correlation between the CSAT antibody and antitalin. A chick embryo fibroblast is stained (A) with the CSAT monoclonal antibody, which localizes the family of cell surface glycoproteins known as $\beta 1$ integrins, and (B) with antibody against talin, an intracellular protein that interacts with actin. Secondary antibodies conjugated with fluorescein (anti–CSAT) and rhodamine (anti–antitalin) allow both the integrins and talin to be localized in the ultraviolet microscope (×2000). (Courtesy of C. H. Damsky.)

release of the fibronectin from the cell surface. Alternatively, treatment of cells with antibodies specific for fibronectin leads to disorganization of the microfilament bundles. The influence of added fibronectin on the shape and adherence of transformed cells is accompanied by an increase of the microfilament bundles at the cell margins. Fibronectin is also found at the site of those extracellular matrix contacts where vinculin is missing.

Cells have the ability to recognize and bind to more than one kind of extracellular molecule. Fibroblasts can adhere to a collagen-coated substratum in the absence of added fibronectin, even in the presence of antifibronectin antibodies. This observation suggests that fibroblasts have on their surface a structure that can recognize and bind to collagen without the mediation of fibronectin (this structure has recently been identified as a 31-kDa integral glycoprotein called "anchorin"). Fibroblasts also adhere to laminin-coated substrata. Liver cells will spread onto laminin-coated substrata in the presence of anti-fibronectin antibodies, or onto fibronectin-coated substrata in the presence of antilaminin antibodies. Also, liver cells will attach directly to collagen-coated surfaces. These attachment processes will occur even in the presence of cycloheximide, suggesting that new protein synthesis is not necessary to produce the appropriate cell surface receptors.

That there is more than one kind of cell-substratum binding system was also indicated by the use of mutant cell lines. One such system is found in the material left behind when a cell monolayer is detached from its substrate with the Ca^{2+}-chelator EGTA. This *substratum-attached material (SAM)* will support the binding of normal cells in the presence of antifibronectin antibodies. A mutant line has been isolated from Chinese *hamster ovary (CHO)* cells that does not adhere to fibronectin-coated surfaces, but that does adhere to SAM-coated surfaces. This implies that there are at least two separate cell-surface binding molecules, one for fibronectin and one for SAM. It has also been found that mild trypsinization inactivates the SAM-specific adherence faster than the fibronectin-specific adherence. By examining membrane proteins from the trypsinized cells on SDS polyacrylamide gels, it should be possible to identify an altered protein that is associated with the loss of SAM-specific binding.

The attachment of cells to a substratum in vitro proceeds in three steps. First there is *binding* of cell surface integral proteins (receptors) to one or another component of the extracellular matrix. Then there is *recruitment* of other receptors to the region of binding, with the formation of additional interactions and the stabilization of the attachment. Finally, there is a *reorganization* of the cytoskeleton, specifically the microfilament bundles, and

the formation of close and focal contacts (adhesion plaques). Similar events probably occur as a cell migrates, during embryogenesis, onto the substratum where it comes to rest.

There is much still to be explained about these processes. One of the remaining significant questions concerns cell surface receptors for the various components of the extracellular matrix. These are expected to be integral membrane proteins, perhaps glycoproteins. There must also be some linkage by which information flows across the cell membrane, to explain how the cytoskeleton reacts to or influences the adhesion process. This presumably requires a transmembrane molecule(s). The means by which receptors cluster into the structures called close and focal contacts is also a mystery.

One method that can be used to identify molecules involved in adhesion is to raise antibodies against the cell surface and see which ones interfere with adhesion. The antibodies of interest could block the attachment of cells to a surface or they might cause adherent cells in culture to round up and, perhaps, to detach from the substratum. More specific reagents are *monoclonal antibodies,* which bind to a single structural feature on a surface antigen (see later discussion in this chapter).

One example of a monoclonal antibody that interferes with adhesion is "CSAT" (*Cell Substratum Attachment*), isolated by N. T. Neff and co-workers (there are several other antibodies known with similar properties). The CSAT antibody inhibits the attachment of embryonic chick skeletal muscle cells (*myoblasts*) to collagen-coated surfaces and causes the detachment of these cells from a substratum. It is now known that CSAT blocks the cell surface interactions with fibronectin and laminin.

Besides its effects on myoblasts, CSAT antibody is also able to induce some kinds of attached chick embryonic fibroblasts to round up; these cells, however, do not detach from the substratum, even at 50-fold higher concentrations than are sufficient to detach myoblasts. In addition to fibronectin and/or laminin, there is apparently a different group of molecules involved in the attachment of chick fibroblasts. Using immunofluorescence techniques, the CSAT antibody was found to stain cells around the adhesion plaques, defined as the sites where antivinculin antibodies also stained the cells.

The CSAT antibody can be used to purify cell surface proteins that interact with it, by affinity chromatography. Such proteins can be isolated from a nonionic detergent extract of chick fibroblasts. This kind of extraction results in the formation of a cytoplast (see Chapter 16), a fibrous network having the same shape as the cell and its extracellular matrix. Proteins from these fibroblast extracts will block the detaching effect of CSAT antibodies on myoblasts. There are three glycoproteins, isolated as a complex by CSAT antibody affinity chromatography, that may be involved in the attachment process; they have mobilities on SDS polyacrylamide gels in the 140-kDa region and seem to differ in their protein components. A structural model that might explain the influence of the extracellular matrix on the cytoskeleton has these proteins forming an integral membrane receptor for fibronectin, on the outer surface, and for talin, on the inner surface of the membrane (Figure 17–22).

Cell-Cell Recognition

During the formation of a multicellular organism the numerous descendants of the fertilized egg must interact with one another. These interactions are specific, as can be demonstrated by the formation of the correct connections between various neurons, even when the cells start out their growth at some distance apart. Specificity must be manifested at the cell surface by interactions between molecules, the material substances by which recognition information is exchanged between cells. In this section we discuss the kinds of molecules that show this specificity and point out some of the experiments that identify such molecules. Then we point out ways in which cell-derived molecules or structures may become long-lived signals, capable of guiding the movement of a migrating cell during embryogenesis. Finally, we address the question of how the modification of surface molecular signals might lead to the formation of tissues, organs, and organisms. The correct adhesion of cells to one another is necessary to form tissues; the lack of correct adhesion by cancer cells is responsible for tumor metastases. We discuss first this critical aspect of cell-cell interactions.

Cell-Cell Adhesion

There are two general methods that have been used to identify structures that are important in cell-cell adhesion. The first is to isolate various molecules from the cell surface, by proteolysis for example, and to test whether they will perturb adhesive interactions. The second is to raise antibodies to cell surface components and to test them similarly. Both approaches have been fruitful. Before describing either, however, we need to know how such adhesion measurements are performed.

The feature that distinguishes measures of adhesion to cells from adhesion to other surfaces, like collagen-coated glass, is the presence on the test surface of cells. The degree of adhesion can be estimated by the force necessary to dissociate cells that have adhered to a test cell layer or by the retention of cells to a surface containing one type of test cell layer relative to the same surface containing a layer of different cells (Figure 17–23).

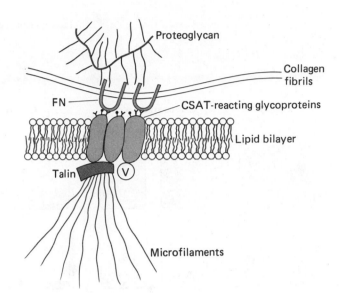

FIGURE 17–22

A Model to Explain How the Extracellular Matrix Can Influence the Cytoskeleton at Adhesion Plaques. The information must be communicated via a structure which is believed to include the three glycoproteins that bind the CSAT antibody. These glycoproteins normally bind to fibronectin (FN) in the extracellular matrix, and to talin (a 217-kDa protein) and vinculin (V) at the cytoplasmic surface. Talin also interacts with the microfilament bundles of the cytoskeleton. Talin and vinculin can be isolated together, so their association is reasonably strong.

Biologically important cell-cell adhesion should show some degree of specificity. Another typical assay thus involves the proper reaggregation of dissociated cells from mixed populations. These experiments were first done by H. V. Wilson early in this century. He used the loosely associated cells of the sponge, which were dissociated by forcing them through a silk mesh. Cells prepared this way will reaggreagate to form new sponges, with formerly outer cells ending up outside. To re-establish the organism, both movement and sorting have to occur. When cells from two differently colored species of sponge were mixed, the reaggregated sponges were never of mixed types; the cells sorted themselves out into sponges of the appropriate color while reaggregating (Figure 17–24). In general, cells from adult organisms are difficult to dissociate; for this reason, embryos have been the objects of study for cell-cell adhesion phenomena. Early experiments on the importance of cell adhesion for embryo development were done by J. Holtfreter on the cells of the three germ layers of the early amphibian gastrula. He dissociated and thereby mixed the cells of the embryo by elevating the pH of the medium. Neutralizing the medium allowed the reassociation of the cells into the original germ layers and allowed

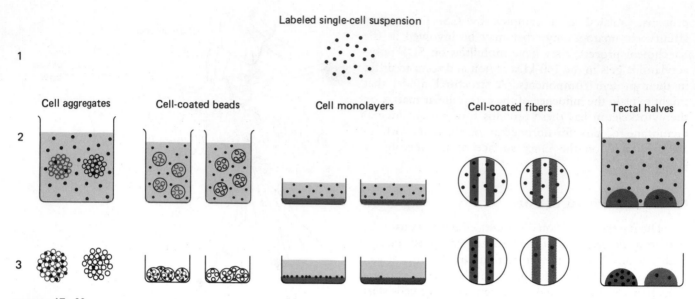

FIGURE **17–23**

Methods of Studying Specific Cell-Cell Interactions.   In step 1, labeled tissues are dissociated into single cells. In step 2, these cells are added to a surface on which other cells are present. In step 3, the surface is washed and the labeled cells that adhere are counted. Specificity is measured by comparing the adhesion of different-labeled cells on identical surfaces or of the same-labeled cells on the same-shaped surfaces, bearing different cells. In each of the different cases in the drawing, the cell suspension specifically adheres to the color surface.

the germ layers thus formed to take up correct positions relative to other germ layers. The work by Wilson and by Holtfreter was quite important because it demonstrated that the information necessary for sorting cells into correct associations is part of the cells themselves.

One of the first examples of a cell-derived molecule with the ability to perturb adhesion specifically is a 50-kDa glycoprotein studied by A. Moscona and his colleagues. This molecule is released into the culture medium when cells from the developing eye of the chick (chick embryo neural retina cells) are placed in a medium lacking serum. The 50-kDa glycoprotein (also known as a "cognin") is able to stimulate the adhesion of neural retina cells to one another, but it has no effect on cells from other organs. Another glycoprotein, derived from chick neural retina cells by treating them with Ca^{2+}, inhibits their aggregation; this 10-kDa protein is called "ligatin." Numerous other cell adhesion molecules have

been identified in the chick neural retina system; we shall discuss some of them later.

Disrupting Cell-Cell Adhesion

Another way to get information about the kinds of molecules involved in cell-cell recognition is to treat cells with enzymes and then to test the cells for the ability to adhere. The earliest application of this method was the use by Moscona of crude preparations of trypsin to dissociate cells; the normal adhesive interactions were disrupted. This result has given rise to the notion that proteins are important in cell-cell adhesion processes. More modern methods of dissociating cells, using brief treatment with purified trypsin, allow adhesive abilities to be expressed after a brief recovery period. Some dissociated cells, when treated with the enzyme neuraminidase (which removes sialic acid residues), re-aggregate more rapidly. Galactosidases are also able to

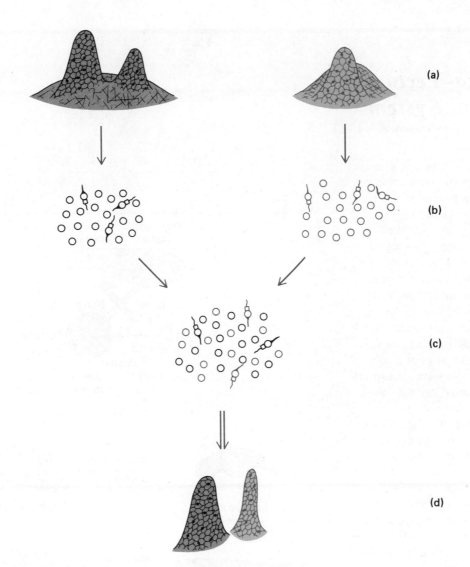

FIGURE **17–24**
Specific Aggregation of Sponge Cells. Loosely associated cells of adult
sponges from two species with different colors (a) were dissociated into single cells
by forcing them through a silk mesh (b). After combining the disaggregated cells
(c), they sorted themselves specifically into organisms (d). Sponge cells have the
ability to recognize and associate with other cells of the same species.

perturb reaggregation, suggesting that galactose sugars
are important.

Processes leading to adhesion and their response to
disruption by trypsin can be distinguished on the basis of
whether they require calcium ion or not. When Ca²⁺ is
required for the aggregation of cells, the adhesion is called
calcium-dependent; the ability of the cells to participate
in calcium-dependent adhesion is usually protected in

the presence of Ca²⁺ from attack by trypsin (0.01% =
100 μg/ml), but is lost if Ca²⁺ is absent. If aggregation
can go on in the absence of Ca²⁺, the adhesion is called
calcium-independent. The ability of cells to show calcium-
independent aggregation is usually insensitive to attack
by low trypsin concentrations (0.0001% = 1 μg/ml),
whether Ca²⁺ is present or not; but this concentration of
trypsin destroys the ability of the cell to participate in

The Retina-Tectum Recognition System

One of the more dramatic examples of specificity in the association of cells during development is in the development of the visual system in vertebrates. The retina of the eye forms from an outgrowth of the neural tube. Nerve cells leaving the retina send their axons into the brain. These axons terminate in lower vertebrates in a region known as the optic tectum. There the cells from various regions of the retina, which will "see" specific areas of the visual field, make connections (projections) with specific regions of the optic tectum (Figure 17–A).

The experiments that established the specificity of the retina-tectum connections were mainly done in amphibian embryos. The strategy was to remove the retina, rotate it, and replace it during the development of the eye, but before the axons from the retina had made their connections. If this is done early during development, there is no effect on the response of the mature organism to visual stimuli. If done beyond a critical stage, however, the response of the organism is abnormal. If the retina of a salamander, for example, is rotated 180° after the critical stage, the adult animal will attack a food source inappropriately, by 180°. The same kind of experiment can be done on adult amphibians, which can extensively regenerate neural connections. In this situation, axons regrow from the replaced retina into the brain and find the appropriate target neurons in the tectum. These experiments illustrate the determination of the retina-tectum connections, as well as their exquisite specificity.

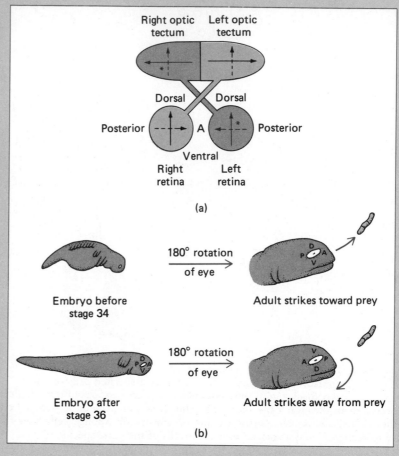

FIGURE **17–A**

Retina-Tectum Cell Connections.   (a) In the adult amphibian, various retinal neurons make position-specific connections to a group of cells known as the optic tectum. The cells of the right retina make connections in the left optic tectum and vice versa. The color arrows show the way in which cells in the retina "project" onto the surface of the tectum. (b) Position specificity within the neural retina becomes fixed at a particular stage of development. If the embryonic eye is rotated by 180° before stage 34, the connections to the retina are normal and the adult amphibian can strike prey accurately. If the rotation occurs after stage 36, however, the adult strikes away from the prey. The connections of the neural retina cells are apparently correct but project images from an inappropriate part of the visual field. It can be supposed that the information determining anterior, posterior, dorsal, and ventral in the neural retina becomes fixed during stages 34 and 35. To explain the retina-tectum specificity, both the cell-cell recognition factors and their time of fixation must be accounted for.

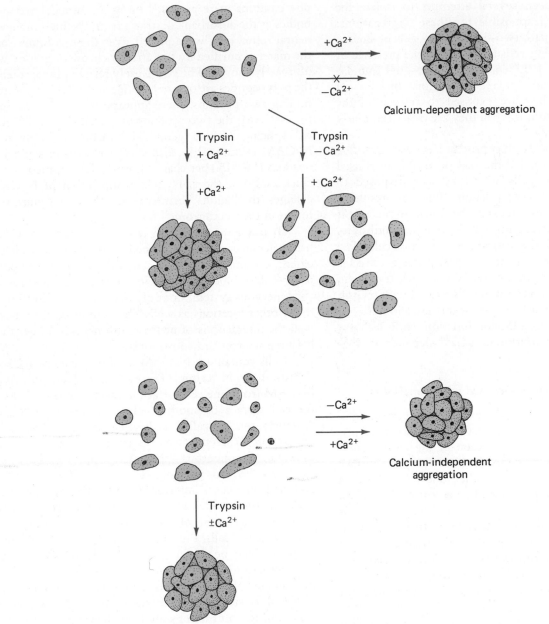

Calcium-dependent aggregation

Calcium-independent aggregation

calcium–dependent aggregation. Calcium–dependent adhesions are regarded as likely candidates for important processes during the development of embryos.

A cell that exhibits calcium–dependent aggregation can be incubated in the presence and in the absence of trypsin, and then the proteins of the surface membrane can be compared. By this means it should be possible to identify those proteins that survive the trypsin in the presence of Ca^{2+} and are, therefore, candidates for participating in calcium–dependent cell–cell aggregation processes. Once isolated, such a protein should compete for binding with structures on the cell surface and, therefore, block cell–cell aggregation. Alternatively, antibodies raised against such a protein should bind to it on the cell surface and might also block cell–cell aggregation.

Finally, when added back to cells treated with trypsin in the presence of calcium, the protein should require the presence of calcium to stimulate cell–cell aggregation. Although calcium–dependent aggregation has been demonstrated in several cells, no surface protein has yet been identified that meets all these tests.

A specific example is the chick neural retina cells that, having been exposed to trypsin in the presence of Ca^{2+}, show calcium–dependent aggregation. After exposure to trypsin in the absence of Ca^{2+}, the aggregation of chick neural retina cells is calcium–independent. The ability to express calcium–dependent aggregation changes during development of the chick embryo, decreasing from day 7 to day 18. In contrast, calcium–independent aggregation does not show this developmental regulation. There have been several attempts to isolate the molecules that are responsible for these aggregations. One way is to compare two-dimensional gels of surface proteins after exposing cells to trypsin in the presence or absence of calcium. J. Lilien and his colleagues found a 130-kDa glycoprotein that is retained on chick neural retina cells when they are exposed to trypsin in the presence of Ca^{2+}, but not when they are treated similarly in the absence of Ca^{2+}.

This same 130-kDa glycoprotein has been identified as critical to adhesion by the use of antibodies raised against neural retina cells after they were exposed to trypsin in the presence of calcium. Since the resulting mixture of antibodies blocked the calcium-dependent aggregation, the mixture must have included antibodies that could react with the critical adhesion molecule. The antibodies could also bind to and precipitate a 130-kDa glycoprotein in cells previously treated with trypsin in the presence of Ca^{2+}, but not in cells treated with trypsin alone. It turned out, however, that this protein does not depend on Ca^{2+} for its adhesion function, so it does not meet the complete definition of a Ca^{2+}-dependent adhesion protein.

Cell Adhesion Molecules Detected with Antibodies

If chick neural retina cells are injected into rabbits, they will raise antibodies against cell surface proteins. These antibodies block aggregation of the chick neural retina cells during embryogenesis. Identifying the surface molecule(s) that are critical to the adhesion process is usually difficult since they need be present in only small amounts. One characteristic of such a molecule is that it reacts with the antibodies and reduces their ability to block aggregation. This property can be used as an assay to follow purification of the molecule. Alternatively, the molecule can be detected by precipitation with the antibodies. In the mixture of antibodies that results from such experiments, it is never clear if the antibodies that precipitate a molecule are the same ones that inhibit aggregation, especially since proteolysis, which produces fragments that also react with the antibodies, is common. Nevertheless, G. M. Edelman and his colleagues used this method to isolate and partially purify a *neural cell adhesion molecule (N-CAM)*. Other such molecules have been isolated from liver (L-CAM), cultured epithelial cells (Cell-CAM 120/80), and so forth.

Monoclonal antibodies (see discussion below) have proved to be extremely useful in such studies because they are directed against a single antigenic determinant. Using the partially purified N-CAM as antigen, monoclonal antibodies were obtained that react with the protein mixture and that block aggregation of neural retina cells. This constituted evidence that the same protein is responsible for both activities. Chromatogra-

phy columns were prepared using the monoclonal antibodies as the absorbant. Extracts from the membranes of neural retina cells were passed over these columns, and the material purified by this affinity chromatography was ultimately shown to be two closely related glycoproteins; the polypeptides are of size 160 kDa and 130 kDa. Despite finding two proteins, there appears to be only one gene for N-CAM; the two proteins may result from differential splicing of this gene. The carbohydrate part of N-CAM is unusual in that it contains many sialic acid residues (130–150 per chain). During development of the chick embryo, the sialic acid composition of N-CAM changes; the ''adult'' form contains about one third that found in the ''embryonic'' form.

By using appropriately labeled antibodies, N-CAM has been detected on various neural tissues and, to a lesser extent, on muscle and liver cells. Monoclonal antibodies to N-CAM block the development of the embryonic chick nervous system, not only in the neural retina but in many other locations as well. The inhibition in each case is in the interactions of neuron with neuron, rather than between neuron and substratum.

Cells containing N-CAM bind to purified N-CAM and also to an N-terminal fragment of the protein. Thus, N-CAM from one cell surface interacts with N-CAM at the cell surface of another cell, holding the two cells together; this is called a *homophilic* (same molecule) *interaction*. When the two cells are of the same type—for example, two neurons—the interaction is called *homotypic*. By limited proteolysis a domain structure for N-CAM has been inferred (Figure 17–25). Evidence favoring this model includes the observation that an N-terminal fragment of N-CAM blocks the aggregation of chick neural cells. Purified N-CAM, or the N-terminal fragment, can be reconstituted into lipid vesicles, and these vesicles also bind to embryonic cells that are known to contain N-CAM. This supports the notion that N-CAM is sufficient to explain the observed cell-cell adhesion. Recent findings implicate heparan sulfate binding to N-CAM as also being important in homophilic interactions between neurons, perhaps by inducing a conformational change in the N-CAM molecule(s).

Surface Enzymes as Adhesion Molecules

To explain the events that occur during development, the strength and specificity of cell-cell adhesion must vary over time and space. For example, the formation of tissues from migrating cells might be expected to require de-adhesion at one stage or another. One way in which this could occur is if the adhesion process involves surface enzymes. In fact, *glycosyl transferase* activities have been found on the surfaces of embryonic cells. One enzyme activity studied extensively in the chick neural retina cells is GalNAc transferase,

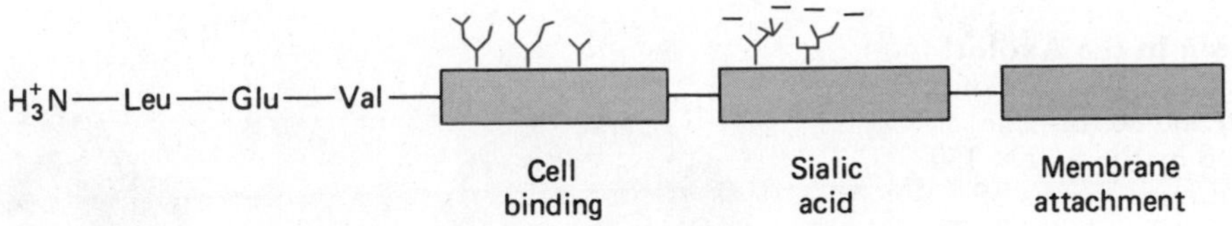

Cell
binding

Sialic
acid

Membrane
attachment

FIGURE **17–25**
Domain Structure of N-CAM. The N-terminal region is involved in
homophilic binding. There are three N-linked carbohydrates attached. The
C-terminal domain attaches N-CAM to the cell membrane. Polysialic acid is
attached to the central domain. Experiments involving the binding to cells by
vesicles containing N-CAM suggest that less sialic acid improves the binding,
presumably because there is less repulsion by the negative charges. N-CAM from
which sialic acid has been completely removed still binds well, however.

which transfers N-acetyl galactosamine to a cell surface
acceptor. This acceptor, which is likely a glycoprotein, is
a substrate of the enzyme and would be expected to bind
to it. If the acceptor and enzyme lie on adjacent cells,
respectively, and if the kinetic binding and catalytic
constants have the appropriate values, enzyme-substrate
binding could be part of the adhesion process. It has also
been observed that transfer of GalNAc is sometimes
followed by release of the newly glycosylated acceptor
from the cell; this could be a mechanism for de-adhesion.

Evidence for the involvement of GalNAc transferase
in adhesion is still somewhat circumstantial, but it seems
likely that this enzyme activity will be shown to be
important in cell-cell interactions. It has been shown, for
example, that treating chick neural retina cells with
trypsin, in the absence of Ca^{2+}, reduces the apparent
activity of the transferase; if calcium ion is present during
the trypsin treatment, the transferase activity can still be
demonstrated and acceptors added afterward can still be

glycosylated. Also, a covalent inhibitor of the enzyme
(UDP-galactose, oxidized to a dialdehyde) will still bind
specifically to trypsin-treated cells. It appears, therefore,
that trypsin removes the other substrate, the acceptor
normally found on the cell surface, and protein synthesis
is required before the cells regain a new acceptor.

The acceptor can be labeled if the UDP-galactose
used in the transferase reaction is radioactive. By this
means the acceptor has been identified as a glycoprotein
of approximately 140 kDa. This is intriguingly similar in
size to the 130-kDa glycoprotein already implicated in
cell-cell adhesion (see discussion above). The two proteins are both removed from the cell surface by digesting
with trypsin in the absence of Ca^{2+}. Both the acceptor
and the adhesion activity return, with similar time
courses, if protein synthesis can occur. In the presence of
Ca^{2+}, both proteins are protected from trypsin digestion.
If these two proteins are one and the same, it should be
possible to demonstrate the reaction of each with the

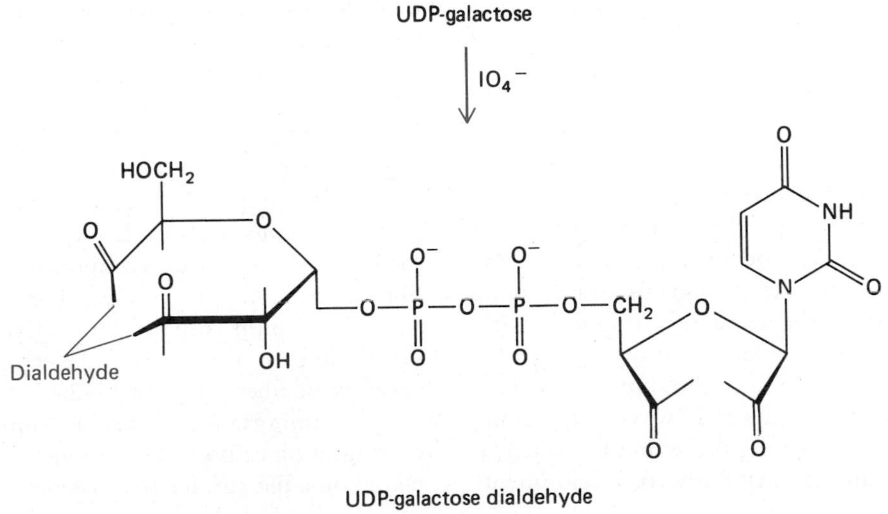

UDP-galactose

IO_4^-

UDP-galactose dialdehyde

FIGURE 17–26

Neural Crest Migration in the Axolotl Embryo.   (a, b) External appearance of the embryo at stages 25 (a) and 30 (b). The trunk region is indicated by the bar. × 150. (c, d) Scanning electron micrographs of trunk region at stages 25 (c) and 30 (d). The neural tube (NT) lies on top of the somites, numbered 4 through 8 in the trunk region. On top of the neural tube lie the neural crest cells (NC). By stage 30, migration of the neural crest cells has begun; they can be seen moving over the neural tube and onto the somites. × 150. (Courtesy of J. Löfberg.)

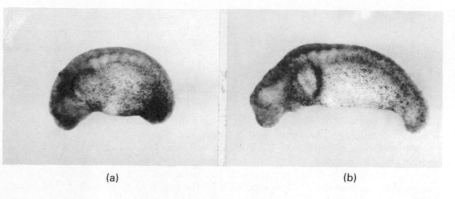

(a)   (b)

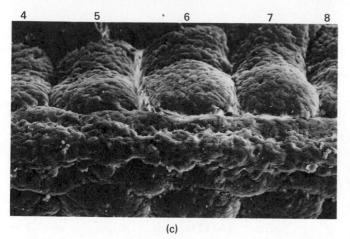

(c)

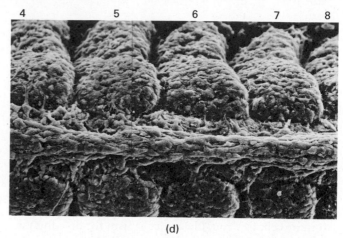

(d)

same monoclonal antibodies; they should also, of course, have the same amino acid sequence. To explain all the specific cell-cell adhesions necessary to account for development of a multicellular organism will of course require several more adhesion systems, probably acting in combination.

Cell Migration During Embryogenesis

During the development of a multicellular organism, cell-cell adhesion forces do not remain constant. If they did the developing embryo could only have the form of a spherical ball of cells. De-adhesion, the disruption of intercellular junctions, and migration of cells must occur to account for the final shape of the adult organism. It is generally believed that cell migration is influenced decisively by the extracellular matrix (ECM).

The area of the developing chick embryo known as the *neural crest* provides a good example of migrating cells. The neural crest is formed by a group of cells lying along the dorsal surface of the neural tube (Figure 17–26). The neural crest cells migrate away from the neural tube, eventually to form pigment cells of the skin or groups of nerve cells. The migration occurs as a "wave," beginning near the head of the embryo and proceeding rearward, at a characteristic time during chick embryo development.

At an earlier stage of development, neural crest cells adhere to one another and to the neural tube. These associations must be weakened before migration can begin. Gap junctions (see discussion in the next section) are disrupted, and electrical coupling between the cells can no longer be demonstrated. The basal lamina over the neural tube also breaks down, either because of mechanical stress exerted by cell proliferation or because of digestion by locally acting enzymes. The amount of cell adhesion molecules, including N-CAM, diminishes. All these processes contribute to a loosening of the cells in the neural crest.

The migration occurs in two directions, between the ectoderm and the somites and between the neural tube and the somites (Figure 17–27). The region between the ectoderm and the somites contains numerous extracellular matrix components, laid down mainly by the ectoderm cells, and is also relatively available to the experimenter. Among the components in the ECM are fibronectin, collagens of types I and III, hyaluronic acid, and chondroitin sulfate proteoglycan. The matrix is visualizable in the scanning electron microscope as a network of fibers (Figure 17–28).

The importance of various components of the ECM for migrating cells can be demonstrated in vitro. Cells are placed on a flat surface that has previously been prepared

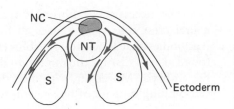

FIGURE **17–27**

Routes of Migration of Neural Crest Cells. The illustration shows a cross section through the dorsal region of the axolotl embryo. Shown are the somites (S), the neural tube (NT), and the neural crest (NC, color). One route of neural crest cell migration proceeds laterally, between the epidermis and the somites; these cells form the pigment cells in the skin. The other route is ventral, between the neural tube and the somites; these cells form groups of cells (ganglia) in the peripheral nervous system.

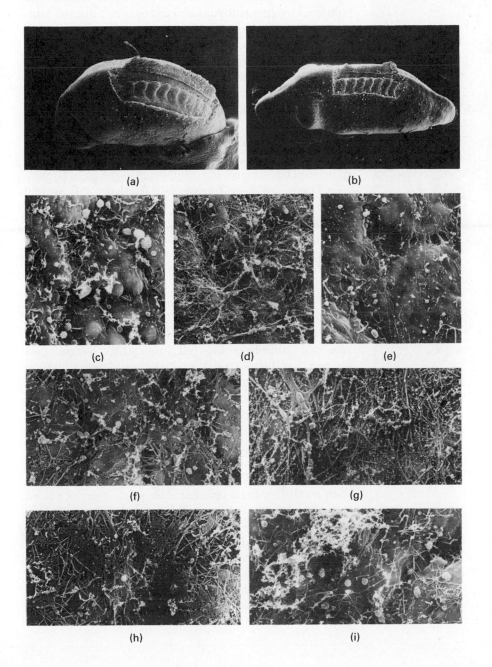

FIGURE **17–28**

Extracellular Matrix Lying Between the Epidermis and the Somites of the Axolotl Embryo. (a, b) Scanning electron micrographs of stage-25 (a) and stage-30 (b) embryos, after removing the overlying epidermis covering the trunks. The positions of areas shown at higher magnification start at the top (c, e) and run about halfway down (h, i) the exposed somites. × 25. (c–i) Representative areas from the surface of the axolotl embryo in the region of the neural tube and fifth somite. There is less extracellular matrix lying under the ectoderm in the ventral region (e). A general increase in extracellular matrix fibers is seen between stage 25 and stage 30. (c)–(e) × 1500; (f)–(i) × 3000. (Courtesy of J. Löfbert.)

FN

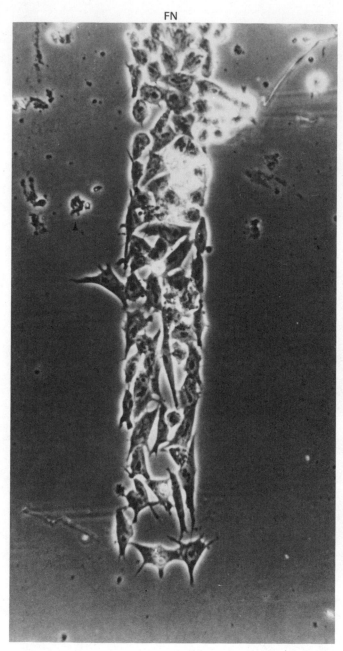

FIGURE 17-29
Neural Crest Cells Migrate on Adherent Surfaces.   The cells can be seen located on a narrow region coated with fibronectin (FN). Cells that wander away from that region round up and cease migrating (arrowhead). (Courtesy of K. M. Yamada.)

with streaks of various ECM components—for example, a glass surface on which a streak of fibronectin has been placed (Figure 17–29). Motile cells tend to collect on the streak of fibronectin and migrate along it. Streaks of collagens, hyaluronic acid, chondroitin sulfate, and laminin are less effective.

The necessity of fibronectin for the migration of neural crest cells can be assessed by specific inhibitors. Antibodies that block fibronectin-cell binding have been found also to disrupt the migration of neural crest cells in the embryo. Moreover, a peptide that competes for the cell-binding domain of fibronectin reduces the migration as well.

Recently, evidence has been reported by B. D. Shur and his colleagues suggesting that surface galactosyl transferases may be important in cell migration. Neural crest cells from quail were allowed to migrate randomly on artificial extracellular matrices in vitro, and the migration was perturbed by adding inhibitors of or substrates for galactosyl transferase activity. The migration on matrices that resemble the basal lamina was inhibited by agents that inhibit the enzyme, but the same agents did not significantly affect migration on fibronectin matrices. The addition of the substrate UDP-galactose stimulated migration, but UDP-glucose had no effect. Finally, these investigators demonstrated that basal lamina-like matrices and the glycoprotein laminin could accept galactose via a similar enzyme activity. It now seems likely that at least two cell-matrix interactions affect migration of neural crest cells, one involving binding to fibronectin and one involving recognition by galactosyl transferases of components in the basal lamina.

There are two further characteristics of normal embryonic migration that need explanation: the stimulus that begins the process and the information that guides the migrating cells to their final destination. Complete descriptions of both these processes are not yet available, but it seems likely that the ECM plays an important role in each.

The timing control that must be operating to trigger neural crest cell migration has at least two components: one that initiates and controls the loosening process and one that stimulates the cell movement. Recent data suggest that the loosening is largely complete well before the cell movement begins. The experiments that demonstrate this involve transplanting pieces of tissue from older embryos, where migration is already underway, to earlier embryos, where migration has not yet begun. The result is to stimulate a precocious migration (Figure 17–30).

The experiment can be done even more convincingly with a cell-free piece of plastic. A small sheet is placed overnight between the ectoderm and the somites of an older embryo. The sheet is then removed and checked by microscopy to ensure there are no adherent cells. When such a sheet is placed into the same location in an earlier-stage embryo, it is fully capable of inducing precocious migration of the neural crest cells, even though the only component on the sheet is some extracellular matrix (Figure 17–31).

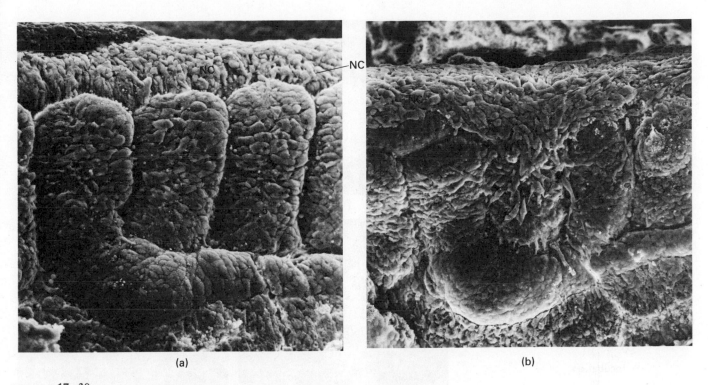

(a)   (b)

FIGURE **17–30**
Precocious Migration of Neural Crest Cells, Stimulated by Epidermis from Older Embryo.   (a) Scanning electron micrograph, lateral view, of a stage-25 embryo that received a graft of epidermis formerly lying over somite number 4 of another embryo at the same stage of development. The neural crest cells (NC) have just begun to migrate; most still lie over the neural tube. × 130. (b) Same view of a stage-25 embryo that received a similar graft of epidermis, but from an embryo at stage 30 of development. The neural crest cells are seen to be migrating along the surface of the fourth somite, behavior appropriate to the older embryo. ×130. (Courtesy of J. Löfberg.)

Guidance of Migrating Cells by the Extracellular Matrix

The presence of ECM components in the path on which a migration occurs and the stimulating influence of the ECM on migration have led to the notion that the matrix may direct the neural crest cells to their target locations. In favor of this hypothesis is the increase in ECM components that occurs before migration begins and the oriented nature of the fibers (compare Figures 17–21 and 17–28). It has been shown in vitro that cells will remain on and move over regions of the substratum that are adhesive (Figure 17–29). If a gradient of adhesiveness exists in embryos, then moving cells would be expected to travel up the gradient until the adhesiveness is too great for further movement to occur (Figure 17–32). We imagine that such gradients could be set up by cells present in the vicinity.

Different degrees of adhesiveness can arise through the interaction of different adhesion molecules on a cell's surface with the molecules of other cells. There are demonstrable changes in the kinds of cell adhesion molecules associated with neural crest cells during their development.

A similar kind of specificity in adhesion can explain how developing nerve cells make the proper connections with their target cells. In this case, the cell does not migrate as a whole; instead, a growing tip (the *nerve growth cone*) migrates while the rest of the cell extends behind. In the developing nervous system of insects, connections are made between appropriate cells despite the presence of a large number of other neurons. C. Goodman and his collaborators have used antibodies to identify specific cell surface structures. In one example, a monoclonal antibody known as *Mes*-2 can recognize a structure present on the surface of only four cells in a population of 1000 neurons. This structure is not present all the time; it is expressed only for a short interval during embryonic development. However the timing may be controlled, having a specific adhesion system present

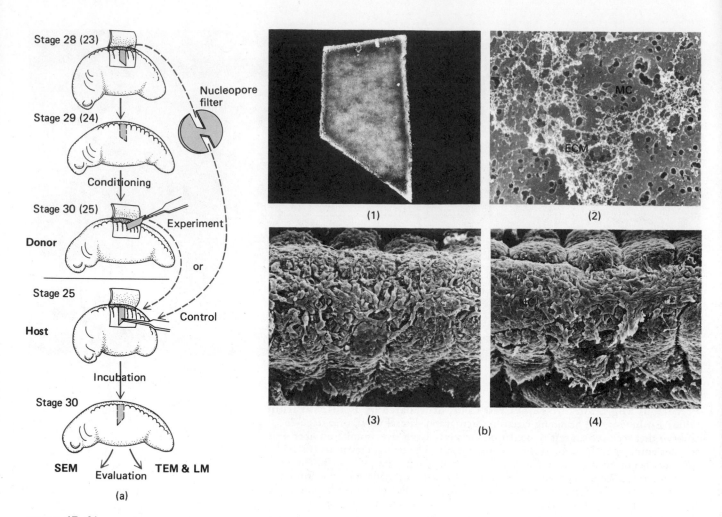

FIGURE 17-31

Precocious Migration of Neural Crest Cells, Stimulated by Extracellular Matrix from Older Embryo.   (a) Digram of experiment. A small microcarrier (made of a piece of special plastic) was placed under a flap of epidermis in an embryo at stage 28 of development and allowed to become "conditioned" by remaining there overnight. The conditioned microcarrier was removed, verified to be free of cells, and surgically implanted into a stage-25 embryo. After further incubation, the embryo was prepared for microscopic examination. (b) Scanning electron micrographs of experiment. (1) The microcarrier, untreated. × 150. (2) Extracellular matrix (ECM), normally lying under the epidermis, attached to a microcarrier (MC) (compare with Figure 17–28, which shows the ECM from the same region). × 5000. (3) Control experiment. The microcarrier, conditioned in a young embryo, was implanted at the position of the asterisk and allowed to remain there overnight. The migration of neural crest cells is the same, on the side of the neural tube where the microcarrier was implanted and on the opposite side of the neural tube. × 130. (4) Experiment. The microcarrier was conditioned in an older embryo, then implanted at the position shown by the asterisk and allowed to remain overnight. The migration of neural crest cells on the side where the microcarrier was implanted is more extensive than the migration on the opposite side of the neural tube. × 130. (Courtesy of J. Löfberg.)

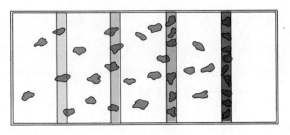

FIGURE 17–32

Cells Migrating/Resting Preferentially on Regions of Increasing Adhesiveness. Cells placed on a surface begin migrating. As they encounter regions of varying adhesiveness, the cells should preferentially remain in the region of maximum adhesiveness.

during development does explain one way in which the proper nerve cell connections can be established.

Finally, it is important to realize there are possibilities for information transfer in an ECM that are different from expectations based on the familiar aqueous solution. Information is a property of molecular structure, as we have pointed out many times previously. But in the ECM, molecular structure can be altered even when chemical structure is unchanged. This can arise when a large molecule is placed under *mechanical stress*, and can occur if it is attached to a cell which is contracting or moving. Since the ECM molecules are connected, a force generated in one region will create a structural change (deformation) that is propagated some distance before being dissipated by elastic properties of the matrix. This is equivalent to a gradient: the "concentration" of the deformed structure decreases more or less smoothly over some distance. Examples of mechanical stress eliciting a biological process include the stretch receptors studied by neurobiologists. More recently, A. Harris has studied the guidance of migrating cells on a matrix deformed by a mechanical force.

Cell-Cell Connections

Cells communicate with one another indirectly by affecting their noncellular environment—either locally, as we have discussed above, or at some distance, via hormones (for example) that circulate through the body in the blood (discussed in Chapter 7). Both these modes of communication require the receiving cell to possess specific receptors for the signal that is emitted by the sending cell(s). Direct communication between cells is also possible; this requires the cells to be very close together or to be physically connected in some manner. Receptors are again necessary for information transfer to occur.

Some cell-cell connections can also be understood from another point of view. By anchoring two cells together so they can remain attached along a substantial area, the movement of materials between cells by passive diffusion processes can be greatly slowed. Under such conditions the cells as a group can control the material movements, to the benefit of the organism. All connections between cells also act to hold the participating cells in an ensemble. This structural cohesion is one of the bases for defining groups of cells as an organ. There are other functional advantages to this togetherness, but they are the subject of courses and texts on general physiology.

Cell-Cell Junctions

Some of the best examples of cell-to-cell connections occur in *epithelia*, which are sheets of cells that cover or line organs, like the intestine in animals. Figure 17–33 is a drawing of a set of typical epithelial cells, as seen under the electron microscope. Notice that the cells are not symmetrical, but they have a *polarity* or sidedness in the

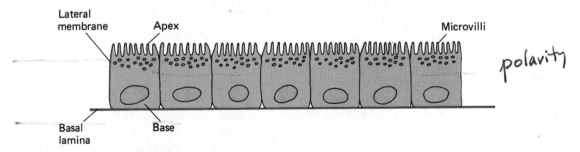

polarity

FIGURE 17–33

Typical Epithelial Cells, Showing Polarity. The cells of an epithelium have an asymmetric shape; this structural polarity is also expressed as different morphologies and different functions at each end of the cell. The edge of the cell that faces the "outside" is called the apex and the opposite edge the base, which is firmly attached to the underlying basal lamina. Epithelial cells are joined to one another at their lateral membranes by junctions. The surface of the apical membrane is often folded extensively into microvilli.

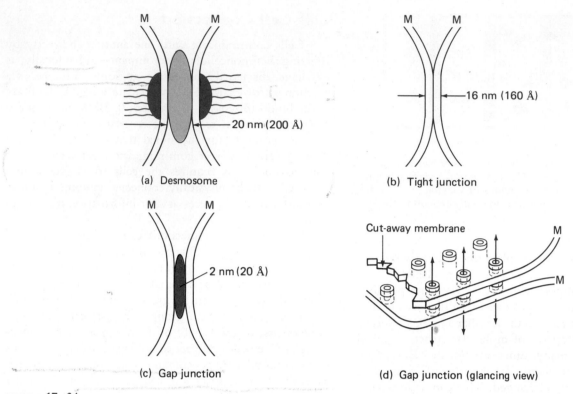

(a) Desmosome

—20 nm (200 Å)

(b) Tight junction

—16 nm (160 Å)

(c) Gap junction

2 nm (20 Å)

Cut-away membrane

(d) Gap junction (glancing view)

FIGURE **17–34**
Junctional Complexes.   Illustration showing the various junctional complexes
between epithelial cells.

appearance of the membranes as well as in the orientation of organelles, such as mitochondria. At the end known as the *apex* (the free surface facing the region from which material is absorbed or to which secretions are delivered), the presence of a fuzzy coat of microvilli and numerous vesicles are the structural correlates of rapid absorption or secretion of solute molecules. At the opposite end, known as the *base* (which is attached to the organ covered or lined by the epithelium), materials move into or out of the organ through the *basal lamina*. The *lateral membranes* are the ones that carry the various connecting or junctional structures. A *junctional complex* consisting of two types of structures, the *tight junction* and the *desmosome,* is found at the apical end (Figure 17–34).

Tight Junction

The *tight junction* (*zonula occludens*) consists of regions in which the adjacent plasma membranes appear fused together (Figure 17–34). Freeze-fractured membranes in this region show one face of the participating membrane displaying a network of ridges and the other face a

network of corresponding grooves (Figure 17–35). These are probably outward-facing integral membrane proteins; equivalent regions in two cells abut to form the tight junction. The function of the tight junction is to prevent the diffusion of water and solutes along an *inter*cellular route. An example is found in the abdominal skin of a frog where epithelial cells are involved in active transport of Na^+ ions; the barrier represented by the set of tight junctions prevents the back-leakage of ions through the space between cells. Other examples occur largely in those situations where a sharp separation between two organ-level compartments is essential: the intestinal epithelium, the bladder, the blood-brain barrier, and in the anterior chamber of the eye. The tight junction also acts as a barrier to the mixing of membrane proteins between the apical and basal regions of the cell.

Desmosome

The *desmosome* is a structure in which the membranes of the two cells diverge some 20 nm from each other (see Figure 17–34). Both membranes appear thickened and a

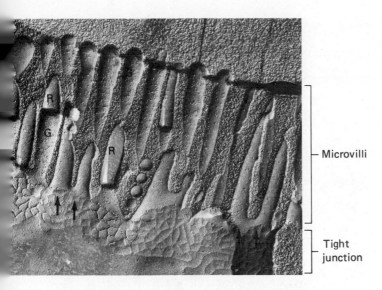

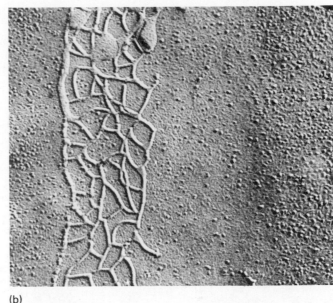

(b)

FIGURE **17–35**

Tight Junction (Zonula Occludens).   (a) Freeze-fracture preparation showing meshwork arrangement of ridges (R) on inner fracture face and grooves (G) on outer fracture face. The position where adjacent ridges contact one another is the position of the permeability barrier to diffusion by large macromolecules (see also Figures 7–6 and 7–11). × 70,000. (Courtesy of D. S. Friend and N. B. Gilula.) (b) Tight junctions of intestinal epithelium visualized by electron micrograph of a replica of a frozen-fracture surface. The rope-like strands are on the surface of the cleaved membrane. (Courtesy of J. P. Revel.)

denser structure appears between them (Figure 17–36). *Spot desmosomes* occupy a small area of the cell surface; when the desmosome forms a complete ring of attachment running around the cell (see Figure 17–36), it is known as a *belt desmosome* (*zonula adherens*). High-resolution electron micrographs suggest that very thin filaments run through the space between the two membranes, while on the cytoplasmic side a dense mat of fibers and long cytoplasmic filaments can be observed. The intracellular fibers and filaments have been demonstrated to contain actin (belt desmosome) and cytokeratins (spot desmosome). The function of the desmosome would seem to be purely a mechanical one, holding cells together under mechanical stress. Calcium ion is probably necessary for this attachment. The importance of the desmosome for welding cells together can be demonstrated by shrinking the cells osmotically and showing that they remain attached in the region of the desmosome.

Gap Junction

Farther toward the base of epithelia cells, beyond the junctional complex, *gap junctions* are usually found (see Figures 17–34 and 17–37). In surface view, these are irregular flat areas of variable size. In cross section the intercellular space is reduced to 2 nm and numerous particles bridge the gap. These particles can best be seen in freeze-cleaved sections (Figure 17–37c). Gap junctions actually connect the two cells together, as can be demonstrated by detecting the flow of ions (as an electric current) and larger molecules (fluorescent dyes) from one cell to another. These connections permit coordinated cellular function by allowing signaling molecules to pass directly between adjacent cells, even through a series of cells. X-ray diffraction and electron microscopic analyses have yielded a more detailed structure for the gap junction (Figure 17–37d). Two structures known as *connexons*, one associated with each bilayer, form the

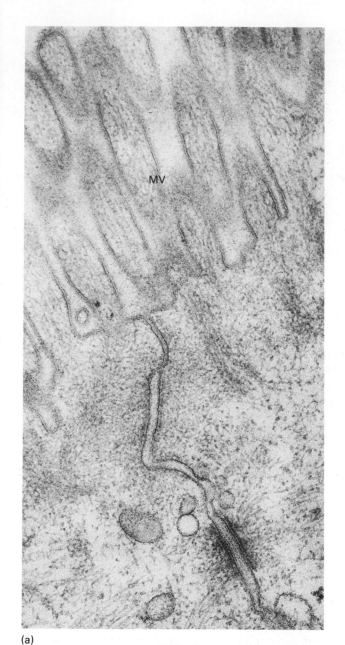

(a)

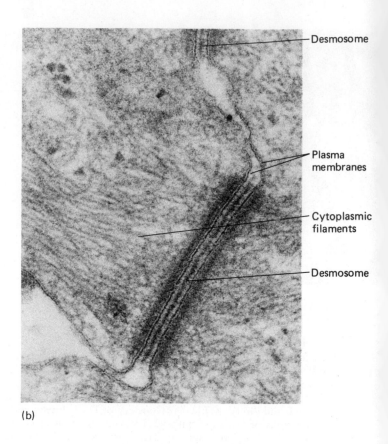

(b)

FIGURE **17–36**

Structure of Desmosome. (a) Microvilli (MW) at surface of two cells in intestinal epithelium, separated by plasma membranes that form part of a desmosome. × 87,000. (b) A magnified image of a desmosome in amphibian skin. × 270,000. (Courtesy of G. E. Palade.)

FIGURE **17–37**

Gap Junctions in Liver Cells. (a) The membranes of two adjacent cells come together and reduce the space between them to 2 nm; in this space are protein aggregates that extend to both membranes and enclose a hole or pore. Ions and molecules as large as dyes can penetrate the junction from one cell to the next. × 37,500. (b) In this higher magnification view the gap is clearly visible. On each side of the junction are cisternae of the endoplasmic reticulum. × 105,000. (Courtesy of G. Pappas.) (c) Replica of the frozen-fractured particle-lined face of a gap junction. The particles are the "connexons" that form the pore structure that allows molecules to pass from one cell to its neighbor. There is a complementary array of depressions or pits on the other membrane face. (Courtesy of D. W. Fawcett.) (d) Schematic drawing of gap junction structure derived from x-ray diffraction and electron microscopic analyses. Two connexons are formed from six gap-junction proteins, each of which is composed of cytoplasmic, transmembrane, and gating domains. (Illustration copyright I. Geis.) (e) Schematic drawing of the channel formed by two connexons. (right) The gate structure is open and molecules can flow through the channel. (left) The gate structure is closed and the channel is blocked. (Illustration copyright I. Geis.)

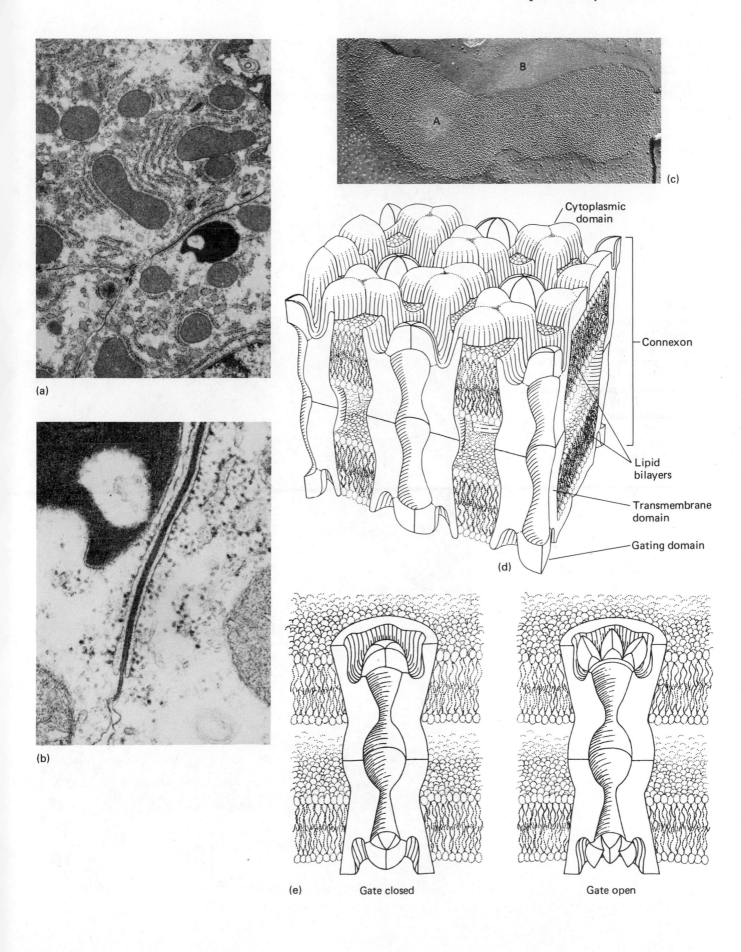

(a)

(b)

(c)

Cytoplasmic
domain

Connexon

Lipid
bilayers

Transmembrane
domain

Gating domain

(d)

(e)　　Gate closed　　　　　　　　　　Gate open

FIGURE 17–38
Isolated Gap Junctions from Rat Liver Cells. The photograph shows a preparation of gap junctions, which are readily seen among the other membrane components because of the dark gap material between the two membranes: the 28-kDa connexon protein. × 30,7000. (Coutesy of N. B. Gilula.)

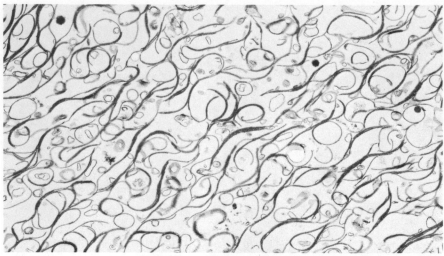

(a)

FIGURE 17–39
Electrotonic (Gap) Junction Between Two Nerve Cells. These cells are from a nucleus in the central nervous system of a toad. (a) The small vesicles in the upper cell are not clustered near the junction (compare with Figure 17–41). × 80,000. (b) At higher magnification, the approximately 2-nm cleft between the two membranes can be seen. × 220,000. (Courtesy of G. D. Pappas and J. Keeter.)

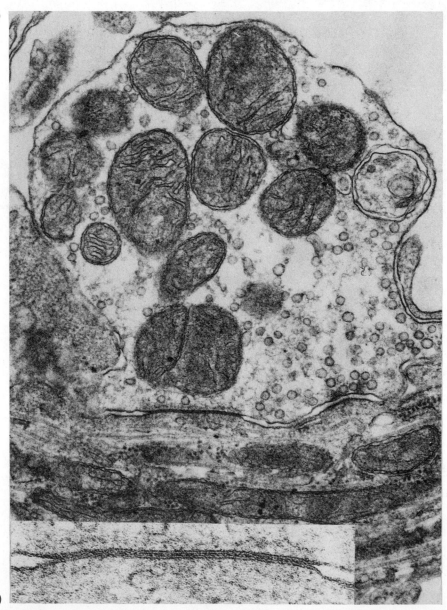

(b)

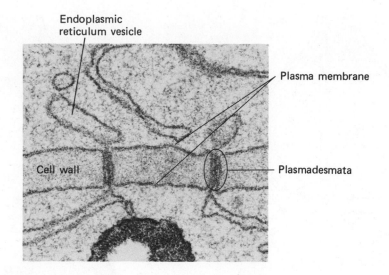

FIGURE **17–40**
Plasmadesmata in the Root Tip Cells of Maize.   These large interconnections between plant cells that traverse the cell wall and the plasma membranes of the two cells are continuous so that large molecules can pass from cell to cell. Channels connecting the cytoplasm of plant cells are usually associated with endoplasmic reticulum vesicles. × 50,000. (Courtesy of H. H. Mollenhauer.)

channel. Physiological analyses give reason to believe that the channels are gated, opening and closing in response to changes in the environment of the connexons (Figure 17–37e).

The gap junctions are sufficiently stable that they can be isolated intact (Figure 17–38). Gap junctions are found in smooth muscle and heart muscle, where coordinated contraction is apparently mediated by ions traversing the connections. They are also found in nerve-nerve connections, where they are called *electrotonic junctions* (Figure 17–39). In these circumstances the gap junction communicates membrane depolarization signals directly from one cell to the next.

In plants, communication between cells can be even more direct. Adjacent cells are often interconnected by membrane-enclosed strands of cytoplasm passing through 50–100 nm holes in the cellulose cell wall (Figure 17–40). Experiments indicate that molecules, but not organelles, are able to pass from one cell to another through these strands.

The Synapse

The typical communication connection between a transmitting nerve cell and a receiving cell is not direct. Instead, the signal is carried by a special chemical called a *neurotransmitter*. The signal transmission and reception is associated with a special structure, the *synapse* (or *synaptic junction*), found in nerve-muscle and nerve-nerve interactions (Fig. 17–41 and 17–42). This is a highly specialized

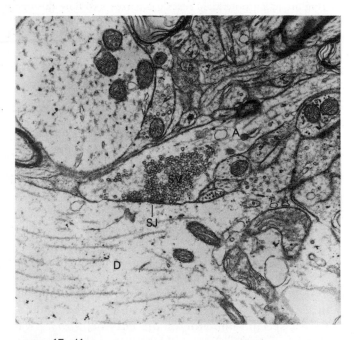

FIGURE **17–41**
A Chemical Synapse.   The electron micrograph shows the connection between a dendrite (D) and an axon (A) in the central nervous system. At the area where the two cells are in contact, the apposing membranes appear thickened (SJ). Within the axon are numerous small synaptic vesicles, clustered near the presynaptic membrane. × 36,000. (Courtesy of G. D. Pappas and S. G. Waxman.)

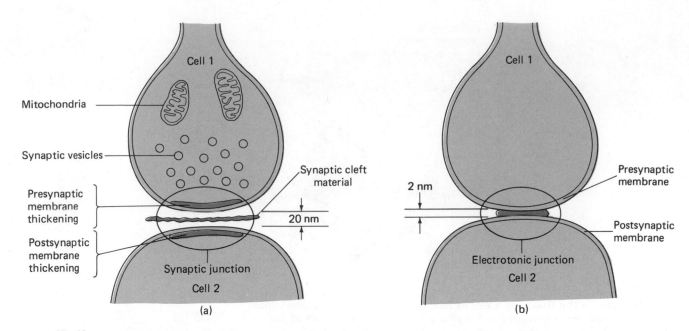

FIGURE **17-42**

Junctions Between Nerve Cells or Between a Nerve and a Muscle Cell.
(a) The chemical junction or synapse. Neurotransmitter is released from synaptic
vesicles—after they have fused with the nerve cell membrane—into the synaptic
cleft and is bound to receptors in the postsynaptic membrane. This allows ions to
flow through the receptor and causes the depolarization of the postsynaptic cell
membrane. (b) The electrotonic junction, a type of gap junction. No
neurotransmitter is involved. Depolarization of the presynaptic cell membrane
during an action potential causes ion currents to flow through the gap junction (see
Figures 17–34 and 17–37), which causes depolarization of the postsynaptic cell
membrane. The space between the cells is approximately 20 nm in the chemical
junction, 2–3 nm in the electrotonic junction.

form of cell junction; it has certain features, such as the
cleft material and the presynaptic and postsynaptic thick-
enings, which are found nowhere else and whose structure and function undoubtedly have to do with the transmission of the electrical signal via chemical transmitters.

The chemical nature of the transmission of most
signals from nerve to nerve or nerve to muscle was first
shown by Sir Henry Dale and his collaborators. They
observed, in the case of nerve-muscle interaction, that
acetylcholine is liberated upon stimulation of the motor
nerve. (They also found that the acetylcholine is rapidly
hydrolyzed by an enzyme, acetylcholinesterase; indeed,
unless a specific inhibitor of the enzyme, *eserine*, was
added, no acetylcholine could be detected upon nerve
stimulation.)

It was observed that stimulation of a *denervated*
muscle, one from which the nerves had been removed,
caused no appearance of acetylcholine; this showed that
the acetylcholine was released by the nerve, and not by
the muscle, upon stimulation. Even if the nerve is
stimulated in the presence of *curare*, a poison preventing
the contraction of the muscle, acetylcholine still appears.
Other chemical transmitters, such as norepinephrine,
behave the same way in nerve-nerve interactions. The
transmitter is released by the sending (presynaptic) cell
and causes the receiving (postsynaptic) cell to change its
physiological state. The transmitter is destroyed or
removed to stop the action.

Synaptic Vesicles

In the 1950s, B. Katz and R. Miledi found that
release of neurotransmitter at the neuromuscular junction
occurred in quantum steps. When it was found that the
presynaptic nerve endings in neuromuscular junctions or
in neuron-neuron junctions had within them tiny vesicles, it was postulated that the neurotransmitters are
contained in the vesicles. Fusion of an individual vesicle
with the synaptic membrane could account for the
quantal nature of the release. The vesicles were subse-

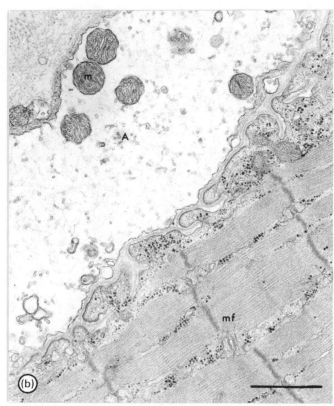

FIGURE 17–43

Morphological Evidence for the Role of Synaptic Vesicles During Signal Conduction. Both electron microscopic images show a neuromuscular junction in the cutaneous pectoris muscle of the frog. There are numerous infoldings of the axonal membrane into the muscle cell. (a) Control specimen showing numerous synaptic vesicles (v) at the terminus of the axon (A). Also shown are mitochondria (m) and myofibrils (mf). × 20,000. (b) The same region after electrical stimulation of the nerve for eight hours. There are many fewer vesicles in the axon. × 32,000. (Courtesy of B. Ceccarelli, W. P. Hurlbut, and A. Mauro.)

quently named *synaptic vesicles.* Since the time between fusion of the vesicle and response by the postsynaptic nerve is short, too short for long-range diffusion, transmitter release must occur at or very near that point in the synaptic junction where the receptor for the transmitter lies.

The synaptic vesicles can be seen in an axon-dendrite synapse in the central nervous system shown in Figure 17–41 and in a neuromuscular junction synapse shown in Figure 17–43. They always lie within the nerve terminal and near that area of the membrane where it makes close contact with the membrane of an adjacent cell. The idea that the neurotransmitter is contained in the synaptic vesicles is based on two types of experiments. First, synaptic vesicles have been isolated and have been found to contain the neurotransmitter—in the case of the

neuromuscular junction, acetylcholine. Second, when the nerve is stimulated continuously until there is no further electrical response from the exhausted muscle, there is also observed a decreased number of synaptic vesicles in the presynaptic cell (compare Figure 17–43b with 17–43a).

An interesting electrical property of the isolated nerve-muscle preparation was found in 1950 by Katz and P. Fatt. They observed that in an unstimulated ("resting") muscle there occurred a series of small spontaneous depolarizations, with a frequency of about one per second and with an amplitude of 0.4–0.5 mV. Because they occurred at the *end-plate,* the synaptic connection between nerve terminal and muscle cell, and because of their small amplitude, they were called *miniature end-plate potentials* (*mepps*). A recording of them is shown in Figure 17–44).

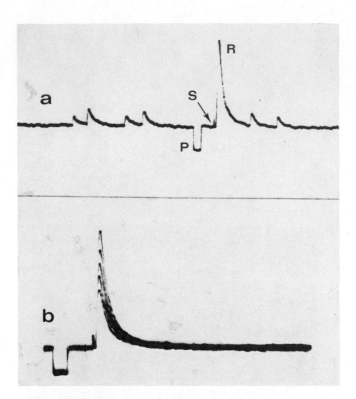

FIGURE **17—44**

Quantal Release of Neurotransmitter. Shown are intervals during an intracellular recording of membrane potential in the region of the motor end-plate of a frog muscle cell. Both spontaneous and evoked activity of the neuromuscular junction are shown. The downward square pulse P was the result of a calibration signal, 1 mV in amplitude and 10 ms in duration. (a) At S, a stimulus was delivered to the motor nerve. This was followed within 1–2 ms by a response R, the evoked end-plate potential. In the solution used to bathe the preparation (0.5 mM Ca^{2+}, 4 mM Mg^{2+} Ringer's), the end-plate potentials were too small to elicit a muscle cell action potential. (b) A superimposed record of five successive end-plate potentials recorded several minutes later from the same preparation. Note the fluctuation in amplitude; the average of the fluctuations roughly equals the average amplitude of a miniature end-plate potential. (Data courtesy of B. Ceccarelli and W. P. Hurlbut.)

These mepps can only be recorded at the neuromuscular junctions and not elsewhere in the muscle cell. We know they arise from the activity of the nerves because when the nerves to the muscle are cut, the muscle mepps disappear in a few days, concomitant with the disintegration of the nerve endings attached to the muscles. Also, the frequency of the discharge is controlled by the membrane potential of the nerves and not by the potential of the muscle fiber. Finally, mepps can be correlated with discharge of the neurotransmitter substance from the nerve to the muscle; when release of the neurotransmitter acetylcholine is blocked by botulinus toxin, it also abolishes the mepps of the muscle.

These findings led Katz, J. del Castillo, and their colleagues to the idea of a *quantum release* of the chemical transmitter. They hypothesized that the regular discrete discharges of the mepps are the result of the release of a packet containing thousands of acetylcholine molecules. Thus, it was hypothesized that a mepp is due to spontaneous fusion with the presynaptic membrane of a single synaptic vesicle and consequent release of its neurotransmitter content into the synaptic cleft.

When nerve stimulation occurs, there is no change in the size or amplitude of the mepps, only in their frequency of occurrence. This frequency rises by a factor of several hundred. A nerve impulse causes the mepp activity to be greatly increased in a short space of time and this clustering then results in a large end-plate potential made up of an integral multiple of unit occurrences (Figure 17–44). Evidence for this idea was strengthened when it was discovered that reducing extracellular $[Ca^{2+}]$ and increasing $[Mg^{2+}]$ (which inhibits vesicle fusion with the synaptic membrane) causes reduction of the end-plate potential, but in *discrete steps*. Conversely, raising the K^+ concentration (which depolarizes the presynaptic cell) causes a graded increase in the discharge rate. In summary, all the evidence seems to indicate that at the junction between many excitable cells (nerve-nerve or nerve-muscle) there is transmission of a signal in the form of a chemical transmitter that is contained in vesicular structural elements. It is this occurrence in vesicles and the subsequent release from these vesicles that accounts for the quantal nature of the response.

If, during discharge of the transmitter, the synaptic vesicle membrane fuses with the cell membrane, what happens to the "extra" cell membrane thus formed? It apparently is recycled again into the cell, in the process called endocytosis, to form intracellular vesicles. Perhaps the best piece of evidence for this is illustrated in Figure 17–45, where it is shown that the extracellular marker peroxidase—which cannot penetrate membranes—is nevertheless brought into the cell, within vesicles, after transmitter discharge.

Signal Transmission

Acetylcholine is not the only chemical substance that acts as a neurotransmitter. There are also the *monoamines*, 5-hydroxytryptamine (serotonin) and histamine; the *catecholamines*, norepinephrine and dopamine; and some *amino acids*, such as glutamic and aspartic acids, glycine, and gamma-aminobutyric acid (Figure 17–46). These do not occur universally; norepinephrine acts as transmitter in smooth muscle but acetylcholine plays that role in striated muscle. While most of these chemicals are excitatory, stimulating a response in the recipient cell,

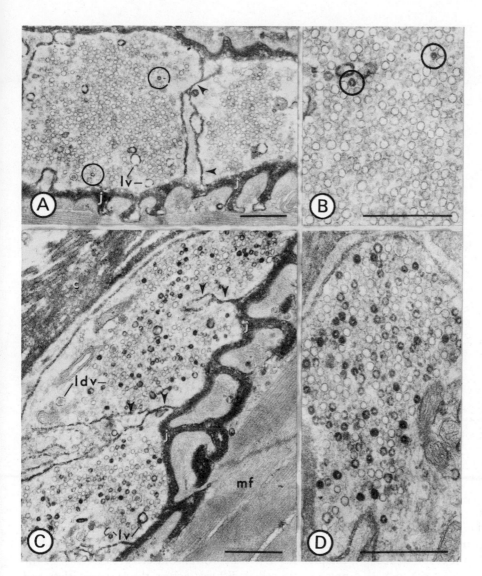

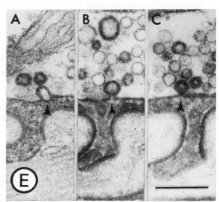

FIGURE **17–45**

Recycling of Synaptic Vesicle Membranes.   A nerve–muscle preparation was incubated in the presence of horseradish peroxidase, an enzyme that cannot cross the plasma membrane and whose location can be visualized by means of a histochemical stain; it is most easily seen as the dark material within the clefts (j) between the nerve and muscle cells. (a, b) In the unstimulated state only a small amount of the peroxidase is found in the synaptic vesicles (encircled). (c, d) After stimulating the nerve and allowing the preparation to rest for a while, numerous vesicles contain the peroxidase. The muscle myofibrils (mf) can be seen in (c). (e) Images suggesting the endocytotic infolding of the presynaptic cell membrane that would account for the entry of the peroxidase. Magnifications: (a) × 29,000; (b) × 59,000; (c) × 35,000; (d) × 58,000; (e) × 140,000. (Courtesy of B. Ceccarelli, W. P. Hurlbut, and A. Mauro.)

FIGURE 17–46
Neurotransmitters. The molecular structures of several classes are shown.

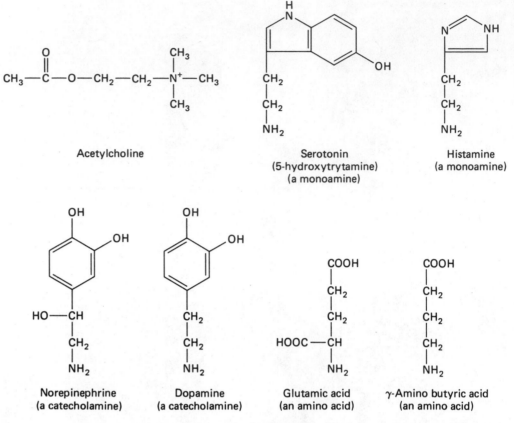

Acetylcholine

Serotonin
(5-hydroxytrytamine)
(a monoamine)

Histamine
(a monoamine)

Norepinephrine
(a catecholamine)

Dopamine
(a catecholamine)

Glutamic acid
(an amino acid)

γ-Amino butyric acid
(an amino acid)

glycine and gamma-aminobutyric acid are inhibitory. It is thought that all of these chemical transmitters are stored in specific synaptic vesicles in cells and that a single cell can have only one type of synaptic vesicle with one species of transmitter, although other chemical substances with functions other than neurotransmission may be released from a presynaptic cell when it is stimulated. What is the biochemical mechanism of the signal transmission? Excitation of the nearby axon causes an influx of Ca^{2+} from the exterior solution into the nerve terminal, and it is this influx that triggers the morphological-biochemical event of secretion. Neurotransmitter release is thus very similar to the secretion of hormones, digestive enzymes, and the like that we discussed in Chapter 13. When neurotransmitter is released from the presynaptic cell it diffuses across the synaptic cleft and

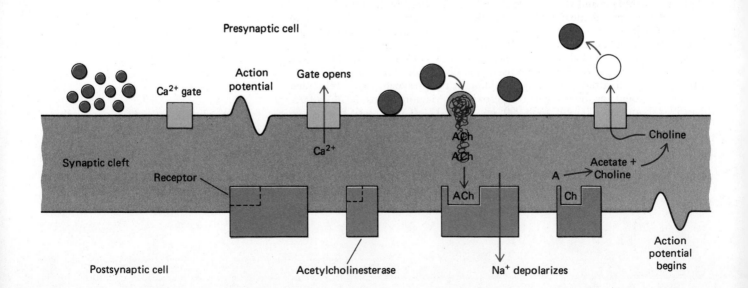

binds specifically to a particular protein, the receptor, in the postsynaptic membrane.

The effect of transmitters that stimulate the postsynaptic cell is to depolarize its membrane. Transmitters act by changing ion conductances: excitatory transmitters increase sodium conductance and thus depolarize the postsynaptic cell, while inhibitory transmitters may increase potassium conductance and lead to hyperpolarization. Alternatively, an inhibitory impulse might increase the conductance of another ion (e.g., chloride) and thus tend to maintain the membrane potential close to the equilibrium potential of that ion. Conductance changes caused by neurotransmitters are not due to the voltage-dependent channels whose opening and closing cause the action potential, since reagents that poison those channels do not affect the postsynaptic potentials.

An excitatory neurotransmitter, when it is bound to its receptor on the postsynaptic membrane, causes a conformational change in the receptor and thus allows a large inward flux of Na^+ into the cell. Unlike the Na^+ channel of the squid axon, there is no inactivation and the Na^+ current will flow as long as transmitter is bound to the receptor. In the case of acetylcholine, stopping the chemical signal is the function of the plasma membrane-bound acetylcholinesterase, which rapidly hydrolyzes the acetylcholine. The choline perhaps diffuses off the postsynaptic membrane, across the synaptic cleft, and back into the neuron through the presynaptic membrane, where it is reformed again in the cytoplasm into acetylcholine by means of the enzyme choline acetylase. In the case of the amine transmitters, the same general sequence is presumed to occur, but with the mitochondrial enzyme monoamine-oxidase hydrolyzing the transmitter amine after it has been taken up by the postsynaptic cell. There is evidence that specific mechanisms transport amino acid transmitters back into pre-synaptic cells, thus stopping the chemical signal.

The Acetylcholine Receptor

The only receptor for a neurotransmitter that has been purified and well characterized, thanks to M. Raftery and his colleagues, is the acetylcholine receptor (AChR). There are two reasons for this success. One is the availability of large amounts of the receptor in the electric organ of *Torpedo*, an electric ray. This organ is really a large array of neuromuscular junctions arranged in series like a battery. The other reason is the use of *α-bungarotoxin*, a specific receptor antagonist that has been purified from a snake venom. This protein molecule binds with high affinity specifically to the receptor and it can be labeled to high specific radioactivity with iodine. Used as a probe, α-bungarotoxin has been used to monitor the purification of the receptor. It has also been

used to localize the receptor at the postsynaptic membrane of the neuromuscular junction.

The purified AChR consists of five subunits (two α, and one each β, γ, δ) surrounding a central channel (Figure 17–47). The subunits all have similar molecular masses, from 40 kDa to 57 kDa, and they all have several identical amino acid sequences interspersed along the protein chains. They are all transmembrane proteins; each has five α-helices going through the membrane (Figure 17–47a). Electron micrographs and x-ray diffraction studies indicate that the complete AChR has a dumbbell shape, with the narrow part spanning the membrane (Figure 17–47c). The binding site for acetylcholine is probably on the α-subunit, since only this subunit was found to bind the α-bungarotoxin.

Examination of the five α-helical regions of each of the subunits showed there are regions of uncharged hydrophobic amino acids, such as valine, interspersed with regions of charged amino acids, such as glutamate. Molecular models have been built that suggest the hydrophobic regions can be on the outer surface of the receptor, next to the membrane lipids, and the hydrophilic regions can line the central channel. This arrangement makes sense because the AChR acts as a Na^+ channel. The hydrophilic amino acids would thus line the surface that interacts with the Na^+ ions as they move into the cell. The channel would normally have to be closed, but open when acetylcholine binds to the receptor and elicits a conformational change in the proteins.

By means of electrophysiological measurements of current flow the number of Na^+ ions transversing the membrane during synaptic transmission can be estimated. In the case of the electric organ of the eel, the estimate is that there are 10^6 Na^+ ions translocated per second per receptor site. This high a flow is consistent with a structure acting as a Na^+ channel. Since the snake venom binds very tightly to the receptor, the estimation of how many receptor sites there are can be made by counting the number of radiolabeled α-bungarotoxin molecules. A density of $3 \times 10^4/\mu m^2$ has been found in a frog neuromuscular junction.

By labeling α-bungarotoxin with a fluorescent label it is possible to localize the AChR to a particular site on a postsynaptic cell membrane by visualizing the fluorescence in a microscope illuminated with ultraviolet light. In Figure 17–48, a muscle cell has been treated with two fluorescent probes: (1) an antibody to the heparan sulfate proteoglycan, conjugated with green fluorescing fluorescein, and (2) α-bungarotoxin, conjugated with red fluorescing tetramethyl-rhodamine. By using filters that pass only certain colors, it is possible to visualize only one probe even though both are present. The images in Figure 17–48 show that the acetylcholine receptor is located on the muscle cell membrane in the same place as the heparan sulfate.

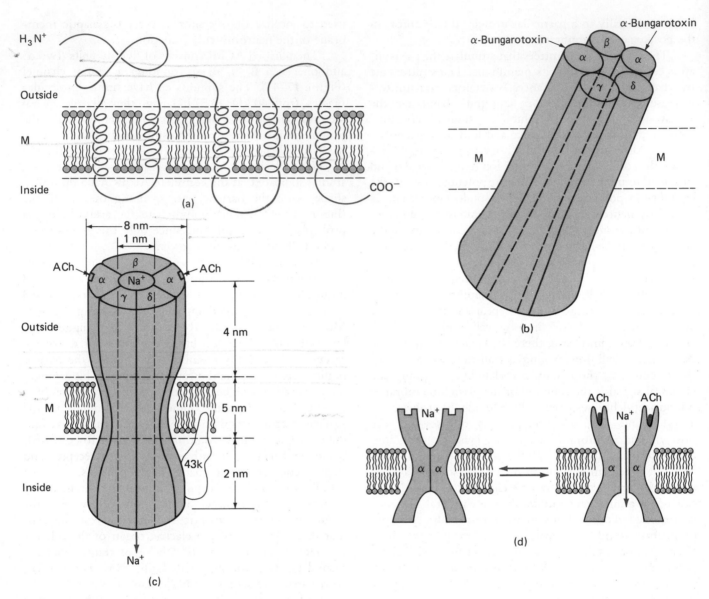

FIGURE **17–47**

Diagrams of the Acetylcholine Receptor Structure. (a) The protein subunits are arranged so as to span the membrane (M). The N-terminal regions are on the outside of the cell, the C-regions inside. The membrane-associated regions are in the form of α-helices. (b) The five subunits surround a central cavity, the functional Na^+ channel. Also shown is the binding site on the α-subunit for the acetylcholine antagonist, α-bungarotoxin. (c) The overall dumbbell shape and dimensions of the receptor, the binding site for acetylcholine (ACh), and the suggested site of binding of a 43-kDa protein that anchors the receptor in the membrane. (d) A possible mechanism for how the binding of ACh to the α-subunit causes a conformational change, opening the Na^+ channel.

Regeneration of Synapses

The synapse between a nerve and a muscle, the neuromuscular junction, can regenerate after disruption. Experiments that study this process can give clues to the puzzle of how these synapses form during development. Much research of this type has been done by U. J. McMahan and his colleagues on frog neuromuscular junctions (Figure 17–49). They found that the basal lamina plays a critical role in specifying the areas of the muscle cell membrane to which nerves form synapses.

Acetylcholinesterase is attached to the basal lamina of the junctions; little of this enzyme activity can be found on the basal lamina away from synapses. Acetylcholinesterase activity persists in its fixed location for weeks after removing the cellular parts of the junction. Thus, using histochemical staining techniques, it is possible to localize the original positions of junctions in a muscle that is regenerating its neural connections. The positions of new neuromuscular junctions can be visualized by staining their acetylcholine receptors with, for example, labeled α-bungarotoxin. The importance of the basal lamina in making specific connections during regeneration was demonstrated by removing the cellular parts of the synapse and observing that neuromuscular junctions would form in their original locations. The simplest of the many experiments done was to damage muscles by brief freezing. After this treatment, all the cellular components of the junction—the muscle fiber, nerve terminal, and the overlying Schwann cell (see Figure 17–49)—disintegrate and are removed by phagocytosis. Only the basal lamina remains. When the muscle fibers regenerate, the acetylcholine receptors cluster at the sites where the acetylcholinesterase activity marks the original locations (Figure 17–50).

The aggregation of acetylcholine receptors can be used as an assay to purify the components of the basal lamina responsible for the long-lived signal that guides regenerating nerve axons to the sites of former synapses. McMahan and his group tested extracts of the electric ray *Torpedo* and found in the insoluble fraction a trypsin-sensitive and heat-sensitive subfraction that has much higher AChR aggregation activity than the original extract. By gel permeation chromatography, the active principal(s) has a molecular mass of between 50 and 100 kDa. It is estimated that this molecule must be active at a concentration of approximately 10^{-10} M, which means it has the effectiveness of a hormone or growth factor. Antibodies raised against the partially purified subfraction block AChR aggregation. Labeled antibodies stain specifically the frog neuromuscular junction synapse.

By expanding and refining these approaches, cell biologists should soon be able to identify the structures that give rise to the function of making specific connections during development. Understanding how communication between nerve cells works at the level of the synapse is a first step to understanding how networks of nerves produce the phenomena of behavior, consciousness, memory, and inspiration. However fascinating that search is, and will be, we must leave its description to others. **oh.boy!**

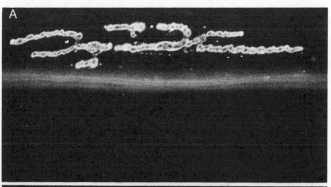

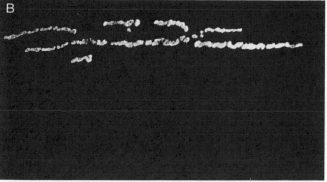

FIGURE **17–48**

Localization of the Acetylcholine Receptor at the Neuromuscular Junction.   (a) A fluorescent-labeled antibody to heparan sulfate (see text) shows the pattern of nerve "fingers" in the basal lamina covering the surface of the muscle cell. (b) Fluorescent-labeled α-bungarotoxin binds to acetylcholine receptors in the membrane; they are intimately linked to the muscle cell surface where the nerve "fingers" lie. The banded appearance is due to the accumulation of receptors at the tips of the folds of the muscle membrane (lighter regions); the darker regions are due to the invaginations of the muscle surface (see Figure 17–45). × 800. (Courtesy of M. J. Anderson and D. M. Fambrough.)

Cell-Cell Interactions in the Immune System

A multicellular organism provides a good environment for the growth of cells. Unfortunately, the carbohydrates, proteins, lipids, and small molecules that nourish the organism's own cells can also satisfy the mass, energy, and information requirements of other cells, even other organisms, that invade from the outside. When other cells start to grow in a vertebrate, reducing the resources available for the organism, certain of its cells respond by secreting protein *antibodies* that bind specifically to the surface of the foreign cells. Marked in this way, the foreign cells become susceptible to phagocytosis, to lysis, and to cell death. The series of cellular events that detects the presence of foreign cells or cell parts and leads to their destruction is called the *immune response*. (Since antibodies are proteins of the class called

(a)

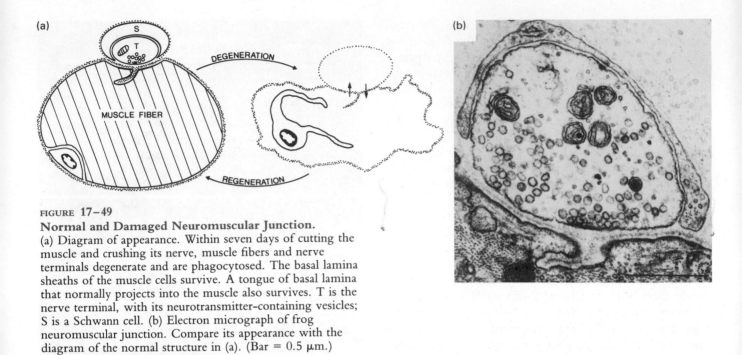

(b)

FIGURE **17–49**

Normal and Damaged Neuromuscular Junction.
(a) Diagram of appearance. Within seven days of cutting the muscle and crushing its nerve, muscle fibers and nerve terminals degenerate and are phagocytosed. The basal lamina sheaths of the muscle cells survive. A tongue of basal lamina that normally projects into the muscle also survives. T is the nerve terminal, with its neurotransmitter-containing vesicles; S is a Schwann cell. (b) Electron micrograph of frog neuromuscular junction. Compare its appearance with the diagram of the normal structure in (a). (Bar = 0.5 μm.)
(Courtesy of U. J. McMahan.)

globulins, they are also called *immunoglobulins*.) A foreign structure—a macromolecule (a toxin, for example) or a cell—that elicits the formation of antibodies is called an *antigen*. The antibody combines with a small part of the surface of an antigen, the **antigenic determinant** or **epitope**.

We usually think of the *immune system*, the collection of cells and extracellular molecules that participate in the immune response, as helping an organism to withstand, for example, infections by foreign bacteria or viruses. (Virus-infected cells, even though derived from an organism's own cells, are treated as foreign; when the infected cell dies its ability to propagate new viruses dies also.) Even in the absence of infection, one of an organism's own cells may begin to behave in an extraordinary way and, as a result, become a target for the immune system. This could happen when a cell becomes cancerous: new surface structures that give the tumor cell its malignant properties might also be recognized by the immune system and the cancer neutralized at an early stage. Should such *immune surveillance* fail, the cancer would continue to grow, to the detriment of the organism.

The immune system does not, however, always work to the organism's advantage. Immune surveillance also makes it difficult for a body to accept a transplanted

organ, even when that organ is necessary to replace a malfunctioning original. Furthermore, the immune system itself can go wrong, mistaking the organism's own cells as foreign, and leading to conditions known as *autoimmune diseases*. Much current research and medical practice is devoted to controlling the immune system of humans: stimulating its activity to reject cancer or to recover from AIDS, and inhibiting its activity to control autoimmune diseases or to allow organ transplants to function.

The immune response is notable for its display of cell-cell interactions. There are two major classes of cells involved, the B cells and the T cells. These are different types of the small unpigmented cells found in the blood (white blood cells), known as *lymphocytes*. Of the 10^{14} cells in an adult human, 1% are lymphocytes. Like other blood cells they differentiate from stem cells; these are located in the bone marrow of adults and in various tissues during embryonic development. Differentiation of the immune system cells is a complex process. Final maturation occurs in two general locations: in mammals the *bone* marrow[1], giving rise to the B cells, and the *thymus* gland, giving rise to the T cells. The two types of cells can be distinguished because they have different

[1]In birds, the *bursa* of Fabricius.

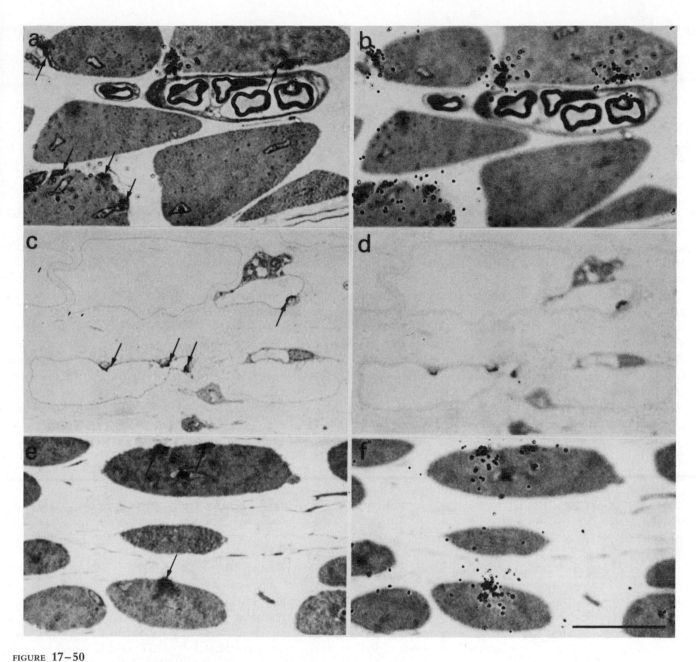

FIGURE **17–50**
Regeneration of Neuromuscular Junctions.   Sections of muscle tissue treated
with ^{125}I-labeled α–bungarotoxin to mark the site of acetylcholine receptors in the
neuromuscular junction (autoradiographic grains, right panels) and stained to locate
cholinesterase (arrows, left panels). (a, b) Normal muscle. (c, d) Basal lamina
sheaths, left after damaging cells with brief freezing. (e, f) Regenerating
neuromuscular junctions, 30 days after damage. The sites of cholinesterase activity,
which mark the locations of former neuromuscular junctions, are the sites where
new acetylcholine receptors form. (Courtesy of U. J. McMahan.)

Recombinant DNA Technology Applied to the Acetylcholine Receptor

The use of recombinant DNA techniques is revolutionizing the study of cellular physiology, as well as the study of regulation of gene expression. This approach has been used by the laboratories of S. Numa and B. Sakmann to elucidate the function of the AChR subunits. These investigators constructed recombinant DNA plasmids that contained the protein-coding sequences of each of the four subunits of the *Torpedo* AChR (appropriate DNA sequences were derived from mature mRNAs and did not, therefore, contain introns). After transfecting the plasmids into monkey cells in culture, poly A-containing mRNAs were isolated and injected into frog (*Xenopus*) oöcytes, cells that do not normally contain AChR. Translation of the AChR mRNAs by the oöcytes was verified by two methods. Labeling the oöcytes with ^{35}S-methionine and precipitating the protein products with antibodies specific for the AChR subunits revealed four radioactive proteins with apparent molecular weights like those of the subunits. More convincing was the demonstration that oöcytes injected with the mRNAs had binding sites for α-bungarotoxin and responded to acetylcholine with an inward electrical current.

This method of incorporating the acetylcholine receptor into a naive cell provides an assay with which to measure the requirements for successful reconstitution. Numa and Sakmann exploited this assay to test whether all four subunits were necessary for installing the proteins in the membrane. When only three mRNAs were injected, the antibodies detected only lower amounts of the subunit proteins in the cell membrane. The absence of any one of the subunits apparently inhibited the normal assembly, or resistance to degradation, of the remaining ones. When only three mRNAs were injected there were also marked changes in the oöcyte membrane potential, again indicating that all four subunits are necessary for normal activity of the receptor/channel.

Plasmids were also made that contain the coding sequence of one or another AChR subunit from calf muscle. These have been used in conjunction with the *Torpedo* AChR mRNAs to test the activity of *chimeric receptors,* containing some subunits from calf muscle and some from the electric ray. It was found that any subunit of *Torpedo* AChR can be replaced by the corresponding one from calf muscle and the chimeric receptor will have roughly the same steady-state ion conductance properties. This is consistent with the high degree of homology in the amino acid sequences of their respective subunits. Differences were detected, however, in the chimeric receptor's gating properties: channel opening, channel closing, and acetylcholine dissociation. The average duration that a channel remains open in a calf muscle AChR is about ten times as long as in a *Torpedo* AChR. The equivalent duration of a channel formed by a chimeric AChR containing only the δ-subunit from calf muscle is like that of the all-calf AChR, while an AChR containing only the α-subunit from calf muscle has a duration lying between the all-calf and all-*Torpedo* values. These results suggest the δ-subunit is mostly responsible for the gating behavior of the AChR.

By using the appropriate restriction enzymes, it is possible to construct a *chimeric gene*, containing some sequences from one source and some from another. This approach was used to form a chimeric δ-subunit coding sequence, partly from calf muscle and partly from *Torpedo*. By injecting the mRNA from this sequence into oöcytes, along with mRNAs for the other subunits from *Torpedo*, it was possible to pinpoint specific regions in the δ-subunit that are important in various gating properties.

Recombinant DNA techniques can also be used to introduce single base changes at specific locations in protein coding sequences. If such a change were found to have occurred naturally, it would be regarded as a mutation, so the technique is called *site-directed mutagenesis*. This approach allows an even finer resolution than the construction of chimeric genes for working out the structure-function relations of proteins. Applied to the α-subunit, it has revealed critical sites for binding acetylcholine. These methods have been applied to many problems of cell biology and the results are being published monthly. Taken together, recombinant DNA techniques provide important tools for research on numerous aspects of cell structure and function.

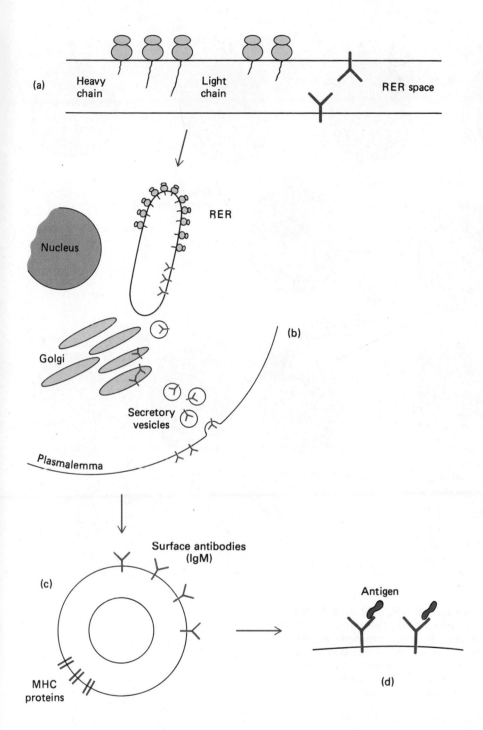

FIGURE 17–51
Antibodies Lying in the Membrane of B Cells. (a) The two types of antibody protein chains are synthesized on ribosomes that are associated with the membranes of the rough endoplasmic reticulum (RER). Stop transfer sequences in the heavy chain protein keep it from entering entirely into the RER space. (b) Assembled antibody travels to the Golgi apparatus and is packaged into vesicles, which fuse with the cell membrane. (c) The Y-shaped antibody protein (of the IgM class) projects its combining sites into the extracellular space. The MHC proteins on the cell surface (see Figure 17–53) are also indicated. (d) When part of the structure of an antigen (Ag) combines with the antibody, the B cell can become activated for growth and form a clone.

surface structures. They also participate in the immune response in different ways.

B Cells and T Cells

The B cells give rise to the cells that produce extracellular antibodies. Although lumped together by a single name, antibodies are actually a population of protein molecules with different primary structures. This diversity derives from a variety of different B cell types, each of which synthesizes a single type of antibody that is determined during the maturation of the cell. This protein synthesis occurs on ribosomes bound to the endoplasmic reticulum, and the antibody in immature B cells remains associated with the cell surface membrane (Figure 17–51). Since the specific binding regions of the surface antibody face the extracellular space, it can combine with antigen in the cell's environment.

Further development of the B cell requires antigen binding. Growth and division then occur so as to form a *clone* of cells, all arising from a single ancestral B cell and all synthesizing the same type of antibody. The antigen

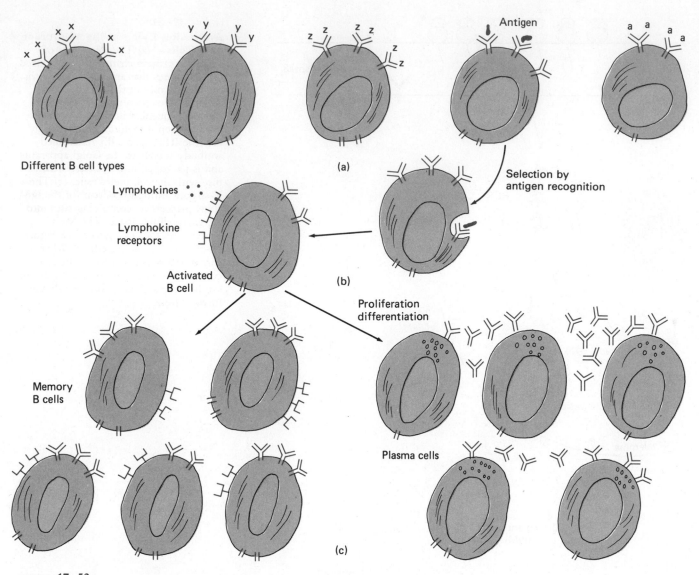

FIGURE **17–52**
Clonal Selection.   (a) Different types of immature B cells exist, each with a specific type of antibody on its surface. There are relatively few immature B cells with any particular surface antibody. (b) Selection by antigen recognition. After exposure to the antigen, which will contain several epitopes, some B cells become activated. Activation involves synthesis and expression of receptors for various protein factors (lymphokines) that stimulate growth and differentiation. (c) Two types of cells result: memory B cells and plasma cells. The former bear surface antibody with the same specificity as the parental B cell; the latter secrete large quantities of antibody with the same specificity as the parental B cell.

thus *selects* a pre-existing B cell and helps to stimulate the production of antibodies. This process is called *clonal selection* and was first suggested as the mechanism for the large diversity of antibodies by M. Burnet and N. K. Jerne (Figure 17–52). The clonal selection mechanism requires there to be at least one B cell for all the antigens an organism is likely to encounter throughout its life. A foreign cell or macromolecule will contain numerous antigenic determinants on its surface, and under normal conditions, numerous clones of B cells will be stimulated

to form antibodies. The ensemble of resulting antibodies is therefore called *polyclonal antibody* (also *polyclonal antibodies*).

The members of a single clone of stimulated B cells have different fates (Figure 17–52). Some differentiate into *memory B cells*, capable of being stimulated again by binding antigen. Since there are more appropriate memory B cells than there were immature B cells when antigen was first encountered, the immune response to a second challenge with antigen is faster and stronger. The

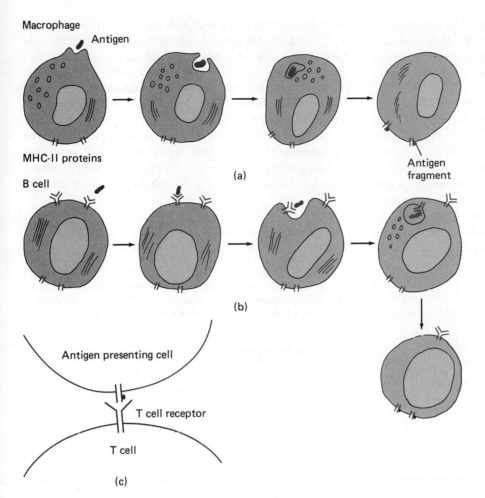

Macrophage
Antigen
MHC-II proteins
(a)
Antigen fragment

B cell
(b)

Antigen presenting cell
T cell receptor
T cell
(c)

FIGURE 17–53

Activation of T Helper Cells by Antigen. (a) Macrophages engulf antigen and process it. A fragment of the antigen is displayed on the cell surface in combination with class II MHC protein. (b) After antigen binds to surface antibody on B cells, the combination is brought inside by endocytosis and processing of the antigen occurs. The partially digested antigen becomes associated with an MHC protein. (c) T cells recognize the combination of antigen fragment and MHC protein by virtue of an antigen receptor that is specific for both. Binding the combination to the receptor stimulates T helper cells to synthesize growth factors (lymphokines) that are released into the circulation and also to synthesize receptors for lymphokines. The presence of lymphokines stimulates growth and maturation of both T cells and B cells with the appropriate receptors.

organism then shows *immunity* to the antigen because the more intense response is usually sufficient to block the activity of the foreign structure before the organism shows any symptoms. The memory B cells differ from their naive counterparts both quantitatively—a higher proportion respond to the antigen—and qualitatively. Creating memory B cells is the goal of vaccination.

Some B cells in a stimulated clone differentiate into *plasma cells.* These large cells synthesize antibodies with the same antigen-combining site as the ancestral B cell. These antibodies are secreted into the extracellular space, where they can circulate throughout the organism and combine with the antigen wherever it may be found (there are about 10^{20} antibody molecules in an adult human). Massive synthesis and secretion of antibodies is the primary activity of plasma cells in the immune response. The plasma cells live for only a few days, so continued synthesis of antibodies requires continued presence of antigen to give continued proliferation and differentiation of the appropriate B cells. Since antibodies are proteins and are, therefore, present in cell-free extracts, the *humoral* (noncellular) immune response is mediated by the B cells and the plasma cells.

The critical event in the immune response by B cells

is *activation.* This involves two processes, recognition of antigen and interaction with a *T helper cell.*

Antigen recognition occurs in two parts, first by direct binding of the antigen to the membrane-bound antibody on the surface of the B cell. There is also recognition of antigen fragments by specific antigen receptors on the surface of certain T cells. The fragments are present on the surface of *antigen presenting cells*—B cells or macrophages, for example.

Phagocytic macrophages take up large antigens and partially digest (process) them. One or more digestion products become associated with the macrophage cell surface; they are said to be "displayed" there (Figure 17–53a). The exact structure on which the fragment of antigen is displayed is not agreed, but it is in close proximity, or bound, to another cell surface molecule known as a MHC protein. The *MHC* (*m*ajor *h*istocompatibility *c*omplex, see further discussion later in this chapter) proteins are specific to the organism and are one of the important ways a cell can be recognized as "self," rather than as "foreign." The T helper cell has antigen receptors that can recognize the combination of the MHC protein and the foreign antigen fragment (Figure 17–53c). Binding at these receptors activates the T helper cell and the macrophage. A T helper cell can also be activated by binding to a B cell surface structure that contains processed antigen in association with MHC protein (processing of antigen by B cells occurs after endocytosis of antigen–antibody complex).

Recognition of antigen by B cells results in the formation of cell surface receptors for factors that stimulate growth and differentiation: *B cell growth factors* (*BCGF*) and *B cell differentiation factors* (*BCDF*), collectively known as *lymphokines*. In the presence of lymphokines, which are synthesized by activated macrophages and T cells, the B cells become activated. Proliferation results in a 10^3-fold increase in the population of responding B cells. Differentiation results in the synthesis and secretion of various classes of antibody, all able to combine with the same antigenic determinant.

Recognition by T cells of antigen presented by a macrophage activates both cell types. The macrophage responds by secreting the lymphokine known as *interleu-*

kin-1 (IL-1). The T cell responds by secreting *interleukin-2* (IL-2) and by expressing cell surface receptors for this lymphokine. These and other lymphokines are the factors that complete the activation of B cells. T cells also respond to these factors but their continued expression of at least some lymphokine receptors requires continued recognition of antigen.[2]

Information is exchanged between various cells of the immune system in two ways, direct binding of one surface structure to a surface structure in another cell and secretion by one cell of a lymphokine that binds to a

[2]Activated T cells also produce a lymphokine known as GM-CSF (granulocyte-macrophage colony stimulating factor). GM-CSF stimulates the production of HIV, the AIDS virus, from a lymphocyte cell line that is chronically infected with HIV. Another lymphokine, gamma interferon (IFN-γ) inhibits the HIV production. Observations of this kind are useful guides to develop strategies for the treatment, perhaps the prevention, of AIDS.

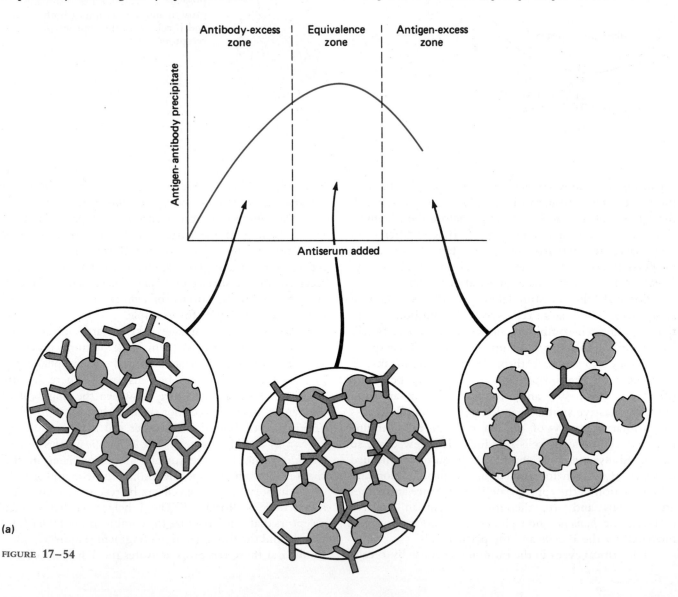

(a)

FIGURE **17–54**

specific receptor in another cell. The specificity of these interactions is determined by the detailed molecular architecture of the surface structures. We have already described the receptors for hormones, neurotransmitters, and growth factors, and how they behave (see Chapter 14 and earlier discussion in this chapter). Our description of cell-cell interactions in the immune system must also include some information about the molecules that provide specificity for antigen recognition.

Immunoglobulins

The three major types of molecule involved in antigen recognition, immunoglobulins, T cell receptors,

and the MHC proteins, share several structural properties in common. Their underlying genetic basis is complex and unusual. We shall discuss the immunoglobulins in some detail since they were the first to be studied and are the most well understood.

A few weeks after a foreign protein or polysaccharide is injected into a vertebrate, its blood will be found to contain IgG (*immunoglobulin G*) antibodies, the simplest and the most prevalent of the circulating antibodies. These antibodies bind specifically to and precipitate the foreign molecule. The formation of a precipitate requires that the antibodies and antigen be mixed in the proper proportions. This is called the *precipitin reaction* (Figure 17–54; in the laboratory today, other assays are used to

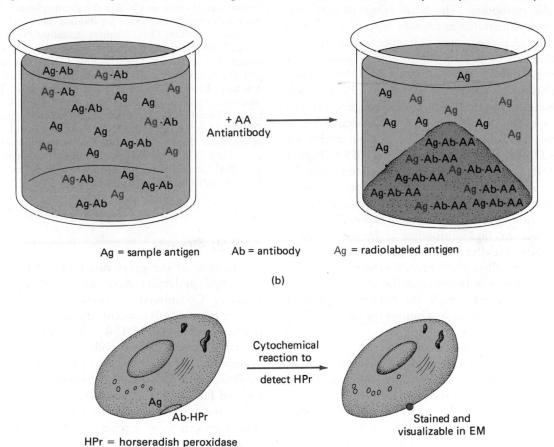

FIGURE **17–54**
Reactions Between Antigen and Antibody.   (a) Precipitin reaction. In this example the antigen has three different antigenic determinants, and three different antibodies have been elicited during immunization. Each antibody has two identical combining sites. Maximum precipitation is obtained in the "equivalence zone" because a more regular lattice is obtained in which every antibody reacts with two antigen molecules and every antigenic determinant has reacted. In the zone of "antibody excess," some antibody sites have not reacted with antigen, and in the zone of "antigen excess," some antigenic determinants have not reacted with antibody. (b) Radioimmunoassay (RIA). A known amount of radioactive antigen (color) is mixed with an unknown amount

of sample antigen and a constant amount of specific antibody. The more antigen in the sample, the less radioactive antigen will combine with the constant amount of antibody. The amount of radioactive complex is determined after removing the antibody from the mixture, by combining it with a specific anti-antibody or with protein A (from *Staphylococcus aureus),* which forms tight nonspecific complexes with antibodies. (c) Other assays. Enzymes can be covalently bound to antibodies and the location of a fixed antigen can be identified by an enzyme-catalyzed cytochemical reaction. Antigen-antibody complexes activate an enzyme cascade in a process known as complement fixation. When activated, complement lyses red blood cells that then release their hemoglobin, an easily detected endpoint.

detect antibodies or antigens). The precipitate consists of a *lattice* containing antigen and antibody molecules. The stoichiometry of the reaction, as well as other evidence, indicates that the IgG antibody has two similar combining sites, whereas the antigen can have two or more, usually nonidentical, combining sites.

Antibody Diversity

Since there are several million species of organisms living today, since each species contains some 10^3–10^5 different proteins, and since each protein from a different species is likely to elicit an immune response, a vertebrate has the capacity to synthesize many millions of different kinds of IgG antibodies. The huge range of antibody diversity can be illustrated in another way. It is possible to take a purified protein such as bovine serum albumin (BSA) and attach to it a small, foreign organic compound, for example a dinitrophenyl group (DNP, Figure 17–55a). (This can be done conveniently by reacting BSA with the Sanger reagent, fluorodinitrobenzene, as described in Chapter 4.) If the BSA-DNP derivative is injected into an animal, it makes specific antibodies against the DNP group as well as against BSA (Figure 17–55b). DNP alone will not elicit antibodies; it must be attached to a large molecule such as a protein or a polysaccharide before it stimulates an immune response. A small molecule so attached is called a *hapten*. The most covenient place to find antibodies is in the animal's serum, the liquid left after blood has been allowed to clot; this preparation is called, therefore, an *antiserum*. Proof of specificity for DNP is obtained from the observation that the antibodies will react with the hapten when it is attached to another protein, ovalbumin for example, but not to ovalbumin alone.

The study of the immune response to chemically well-defined molecules like haptens gave a clearer picture of the specificity of antibodies. That the hapten-antibody reaction is specific can be demonstrated by showing that it fails to occur in the presence of an excess of free DNP and that precipitates of BSA-DNP IgG can be dissociated by free DNP (Figure 17–55b). The versatility of the immune response, however, is even greater than indicated so far. If the dissociation constant of the BSA-DNP IgG immune complexes is measured, a distribution of values ($K_D = 10^{-6}$ M to 10^{-10} M) is found, showing that the animal had synthesized a large number of different kinds of anti-DNP antibodies.

$$BSA\text{-}DNP \cdot IgG \rightleftharpoons BSA\text{-}DNP + IgG$$

$$K_D = \frac{[BSA\text{-}DNP]\,[IgG]}{[BSA\text{-}DNP \cdot IgG]}$$

Dissociation of immune complexes

Dinitrophenol   Dinitrophenol-BSA

(a)

FIGURE 17–55
Immunological Response to a Hapten. (a) Dinitrophenol (DNP). This structure will not by itself elicit the formation of antibodies. (b) After coupling DNP to a macromolecule, like bovine serum albumin (BSA), the dinitrophenol-BSA does cause an immune response. Antibodies can be obtained by injecting a substance into a rabbit, usually twice with two weeks between, then after a few weeks isolating the rabbit's serum. The antiserum raised against DNP-BSA can be partially purified by reacting it with BSA alone; a precipitate will form between the BSA and antibodies to it. The supernatant still contains antibodies specific for DNP; this can be demonstrated by the ability of the serum to precipitate DNP-BSA or ovalbumin-DNP. Suitable control experiments demonstrate that the antiserum has no specificity to proteins alone (neither BSA nor ovalbumin) and that excess DNP will disrupt the precipitates. (c) Alone, DNP does not elicit the formation of effective antibodies.

Antibody Structure

In spite of the great variability of their combining sites, IgG molecules have fundamental similarities in structure. Considered as proteins, they all precipitate in a narrow range of salt concentration, they all have the same molecular mass (150 kDa), and they all have a fairly similar electrophoretic mobility—they used to be called γ-globulins because they move the slowest during electrophoresis of the serum globulins. These initially discovered biochemical properties suggested that IgG molecules must have a uniform general shape. Since IgG antibodies can form a precipitin lattice, they also must have at least two similar combining sites that bind the antigen for which they are specific. Intensive studies during the last 30 years have revealed an IgG structure with precisely these properties.

Two of the most important discoveries were made in 1954 and paved the way for the later structural work. R. Porter found that by treating a collection of IgGs with the proteolytic enzyme papain, all the molecules were cleaved into three fragments, two of which (Fab) retained *a*ntigen *b*inding ability. The Fab fragments were not, however, able to precipitate the antigen and Porter therefore correctly concluded that each fragment carried only one combining site. The third fragment (Fc)

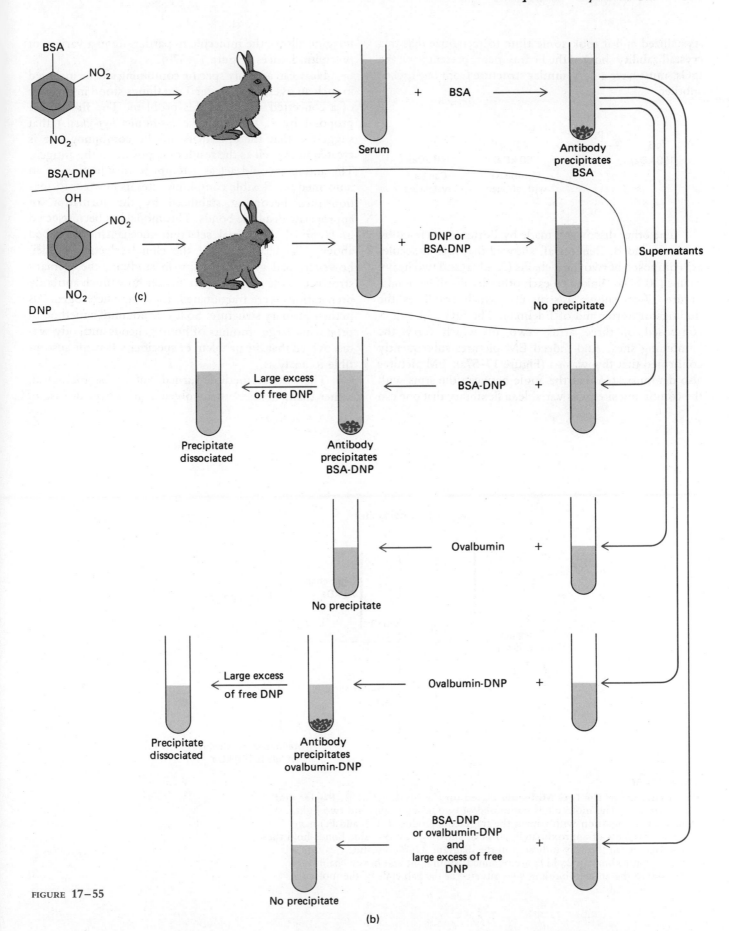

FIGURE 17–55

crystallized and it took some time to recognize that this crystallizability showed the Fc fragment, present in all the IgGs, must have a very similar structure from one IgG to another.

$$IgG \xrightarrow{\text{Papain}} 2 F_{AB} + F_C$$

IgG	2 F$_{AB}$	F$_C$
(150 kDa)	(50 kDa)	(50 kDa)
	Combines with antigen	Can be crystallized

The other discovery, made by Porter and refined by Edelman and B. Benacerraf, showed that IgG molecules are composed of two light chains (25 kDa) and two heavy chains (50 kDa), linked to each other by disulfide bonds. Figure 17–56 shows a structure which combines the findings of Porter and of Edelman. The IgG molecule is composed of three arms, two of which carry the combining sites. And indeed EM pictures subsequently confirmed that this was so (Figure 17–57a). EM pictures also demonstrated that the angle between the arms with the combining sites was variable, a flexibility that one can imagine allows the molecule to participate in a variety of precipitin lattices (Figure 17–57b).

How can a highly specific combining site be included in such an overall structure? Two things stood in the way of a concerted attack on this problem. The first was a proposal by L. Pauling (the *instructive hypothesis*) that suggested that the specificity of the combining site is created in the cell as the result of exposure to the antigen. The antigen would act as a template to which is fit an unformed but flexible combining site, its new conformation then becoming stabilized by the formation of appropriate disulfide bonds. This model has been rejected in favor of the clonal selection mechanism described above. The champions of the clonal selection model, however, had great difficulty in studying the primary structure of the molecule. No matter how much antibody preparations were fractionated, they were heterogeneous in their primary structure. So it was not until a method of producing large amounts of homogeneous antibody was discovered that the problem of specificity became susceptible to analysis.

The critical method turned out to be biological, rather than chemical. It involved a malignant disease of

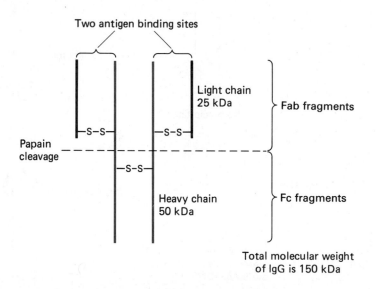

FIGURE 17–56

The Structure of the IgG Molecule Based on the Findings of R. Porter and G. Edelman.   The molecule is composed of two heavy chains and two light chains. Upon digestion with papain the fragments produced (Fab and Fc) have molecular weights of approximately 50 kDa, showing that disulfide bonds hold the chains together in the way indicated in the diagram. Upon disulfide cleavage, chains of approximately 25 kDa are obtained. Both light and heavy chains contribute to the antigen binding sites situated at the Fab ends of the molecule.

antibody-forming cells called *multiple myeloma*, in which a single antibody-synthesizing plasma cell proliferates to form a large population of cells. Typically in this disease, specific immunoglobulin light (L) chains are produced in such excess that they are excreted in the urine, where they had been described as early as 1847 by the physician H. Bence Jones. It was eventually discovered that while each patient excretes homogeneous L chains in urine (as a *Bence Jones protein*), no two patients produce the same L chain. In other words, a given plasma cell tumor makes only one kind of antibody.

The systematic investigation of this phenomenon began by determining the amino acid sequences of a large number of Bence Jones proteins. Then the same was done with heavy (H) chains fractionated from the plasma cells in the blood of myeloma patients. A number of fascinating results emerged from these studies. The sequences fell into groups of approximately 110 amino acids, the members of each group bearing a certain degree of homology to one another. The light chains are composed of two such units, while the heavy chains are composed of four (Figure 17–58). Edelman proposed that these homology units arose from the duplication of an ancestral gene and then became modified during the course of evolution to satisfy a number of functions. Indeed, evidence has accumulated to suggest this actually happened. It turned out that the light (L) chains consist of two parts: a *variable region* of about 110 residues (starting from the N-terminus of the chain), which is different for each Bence Jones protein, and a *constant region* consisting of the remaining 105 (approximately) residues of the chain, which is essentially the same. H chains also have a variable region, of about the same size, but the constant region is of course much longer (Figure 17–58). After working out the exact location of the disulfide bonds, an arrangement of the chains was obtained which fitted beautifully the three-armed structure hypothesized by Porter and later seen in the EM.

We would expect the combining site to be a region of great variability in sequence, but we would also expect that, to bring about the necessary conformation to support the combining site in its proper position, certain specific residues would be needed that would therefore be *highly conserved* from one molecule to another. And indeed this is what is observed in comparing the amino acid sequences of Bence Jones proteins (Figure 17–59). The variable regions contain certain positions where residues are highly conserved (framework segments), other regions where residues vary somewhat, and other *hypervariable regions* (*complementarity determining regions, CDR*) in which residues vary greatly (Figure 17–60).

That the amino acid residues located in the hypervariable regions are located in combining sites was

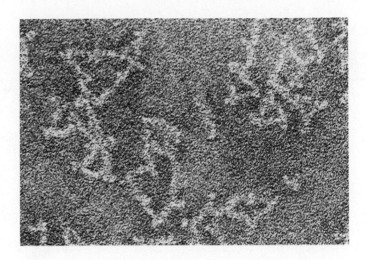

(a)

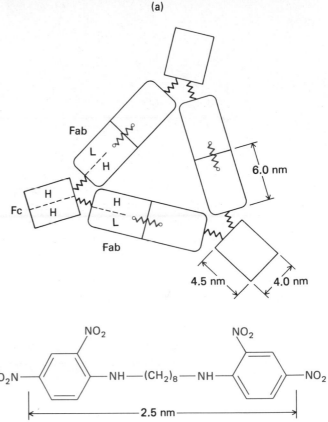

(b)

FIGURE **17–57**
Electron Microscope Evidence for the Y-Shaped Structure of the IgG Molecule. (a) Electron micrograph of dimers, trimers, and tetramers of IgG molecules binding to antigen. This demonstrates that the ends of the Fab fragment contain the binding sites. It also shows the variable angle of the arms, hence the existence of a "hinge." (b) Diagram of the trimer structure, showing the Fc and Fab domains, the latter held together by the antigen (color). Also shown is the structural formula of the antigen, two DNP-like residues connected by an eight-carbon liner. (Courtesy of N. M. Green.)

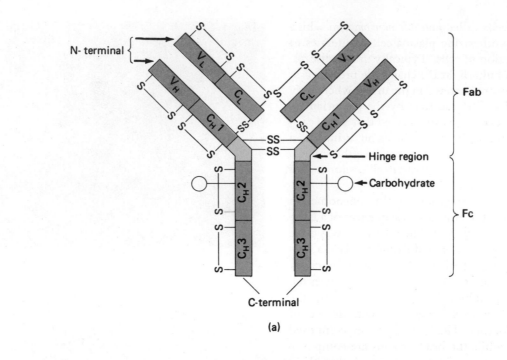

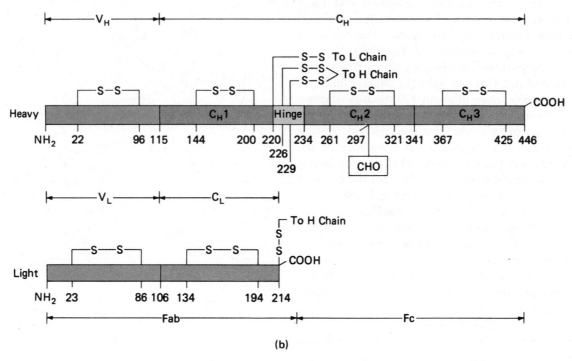

(b)

FIGURE 17–58

Structure of the IgG Molecule.   (a) Schematic summary, as derived from amino acid sequencing data. There are twelve domains; each is stabilized by a single intrachain disulfide bond. At the ends of the arms of the Y-shaped molecule are the variable (V) regions; this is where an antigenic determinant is bound. The light chain is attached to the heavy chain at the hinge region, again by a disulfide bond. Two interchain disulfide bonds in the hinge regions hold the two halves of the molecule together. A carbohydrate oligosaccharide is linked to each of the Asn297 residues in the C_H2 domains. (b) Schematic summary of the precise location of each cysteine residue and the carbohydrate moiety in the human IgG_1 molecule. Different "classes" of immunoglobulins, each with a different immunological function, result from combinations of the same heavy chain V region with C regions of various types (μ, γ, ϵ, α), forming immunoglobulins of various classes (IgM, IgG, IgE, IgA).

```
                          120                                    130
C_H1 (Residues 119–220)   Ser Thr Lys Gly Pro Ser Val Phe Pro Leu Ala Pro Ser Ser Lys Ser  -   -  Thr Ser Gly Gly Thr
C_H2 (Residues 234–341)   Leu Leu Gly Gly Pro Ser Val Phe Leu Phe Pro Pro Lys Pro Lys Asp Thr Leu Met Ile Ser Arg Thr
C_H3 (Residues 342–446)   Gln Pro Arg Glu Pro Gln Val Tyr Thr Leu Pro Pro Ser Arg Glu Glu  -   -  Met Thr Lys Asn Gln

140                                    150                                    160
Ala Ala Leu Gly Cys Leu Val Lys Asp Tyr Phe Pro Glu Pro Val Thr Val  -   -  Ser Trp Asn Ser  -  Gly Ala Leu Thr Ser
Pro Glu Val Thr Cys Val Val Val Asp Val Ser His Glu Asp Pro Gln Val Lys Phe Asn Trp Tyr Val Asp Gly  -  Val Gln Val
Val Ser Leu Thr Cys Leu Val Lys Gly Phe Tyr Pro Ser Asp Ile Ala Val  -   -  Glu Trp Glu Ser Asn Asp  -  Gly Glu Pro

                170                                    180                                    190
Gly  -  Val His Thr Phe Pro Ala Val Leu Gln Ser  -  Ser Gly Leu Tyr Ser Leu Ser Ser Val Val Thr Val Pro Ser Ser Ser
His Asn Ala Lys Thr Lys Pro Arg Glu Gln Gln Tyr  -  Asp Ser Thr Tyr Arg Val Val Ser Val Leu Thr Val Leu His Gln Asn
Glu Asn Tyr Lys Thr Thr Pro Pro Val Leu Asp Ser  -  Asp Gly Ser Phe Phe Leu Tyr Ser Lys Leu Thr Val Asp Lys Ser Arg

                          200                                    210
Leu Gly Thr Gln  -  Thr Tyr Ile Cys Asn Val Asn His Lys Pro Ser Asn Thr Lys Val  -  Asp Lys Arg Val  -   -  Glu Pro
Trp Leu Asp Gly Lys Glu Tyr Lys Cys Lys Val Ser Asn Lys Ala Leu Pro Ala Pro Ile  -  Glu Lys Thr Ile Ser Lys Ala Lys
Trp Gln Glu Gly Asn Val Phe Ser Cys Ser Val Met His Glu Ala Leu His Asn His Tyr Thr Gln Lys Ser Leu Ser Leu Ser Pro

          220
Lys Ser Cys
Gly
Gly
```

FIGURE 17–59

Amino Acid Sequences of the Constant Regions of Three Heavy Chains Obtained From the Plasma Cells of Human Myeloma Patients.   The chains are aligned in such a way as to maximize the homologies. Amino acids that are identical in two or more chains are underlined. Even in the constant regions a few differences in sequence occur. (Courtesy of G. M. Edelman.)

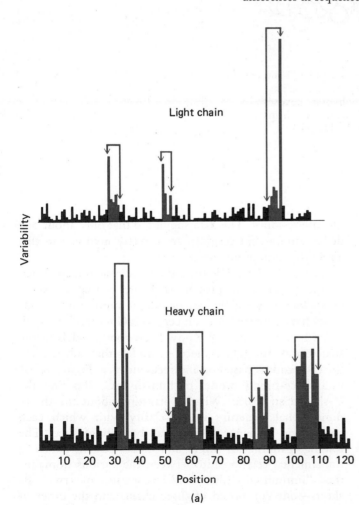

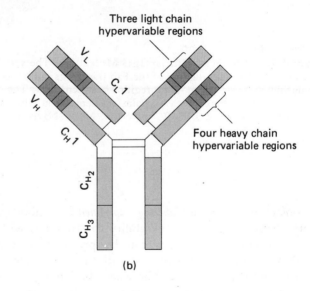

FIGURE 17–60

Hypervariable Regions in the Variable Domains of Light and Heavy Chains.   (a) Amino acid residue variability of different positions in the V regions of light and heavy chains. A large value in the histogram indicates a residue position where many different amino acids have been found, a low value where only a few different amino acids have been found. Arrows indicate where affinity labels have been localized. These positions tend to correspond to the regions where hypervariability occurs. (Courtesy of A. L. Wasserman.) (b) Schematic diagram of the IgG molecule. The positions of the four hypervariable regions in the heavy chain and of the three hypervariable regions in the light chain.

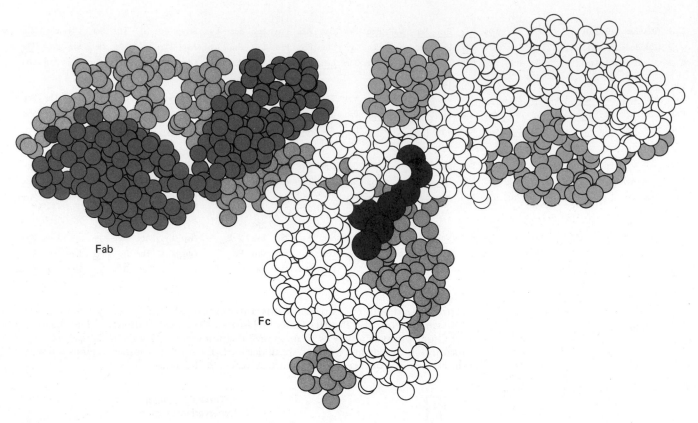

FIGURE **17–61**
Space-filling Model of Human IgG Molecule.   The variable (V_H and V_L) and
constant (C_H and C_L) regions of the Fab fragments can be seen clearly, as well as
the unfolded chains of the hinge regions between them. The carbohydrate hexoses
are seen as dark spheres. Heavy chains are depicted in white and gray; light chains
are colored. (Courtesy of E. W. Silverton, M. A. Navia, and D. R. Davies.)

demonstrated by *affinity-labeling experiments*. Antibodies
against haptens that have the ability to cross-link with
certain amino acid residues can be prepared. If the
antibody thus produced is reacted with the free hapten, a
covalent link can be formed between the hapten and a
susceptible amino acid residue in the combining site. It
turns out that amino acids in or very near the hypervariable regions are the ones that become covalently linked to
the affinity-labeling haptens (Figure 17–60).

Careful studies of the binding by specific haptens
demonstrated that the interaction between hapten and
binding site is noncovalent and is brought about by the
same polar and nonpolar interactions that are responsible
for the tertiary and quaternary structure of proteins.
Using antigenic polysaccharides of various lengths, E.
Kabat showed that one binding site is a relatively small,
probably shallow, cleft about 2.5 nm long which can
accommodate a polysaccharide molecule consisting of six

glucose residues. The data suggested that only about one
dozen amino acid residues are directly involved in this
antibody-combining site.

As we have seen before, the most exacting approach
to the problem of protein structure is that of x-ray
crystallography. In the last decade, a number of laboratories have reported data on certain IgG molecules which
could be crystallized. We now have good models of the
structure of the IgG molecule, models that substantiate
the earlier ideas based on indirect evidence. Figure 17–61
is a space-filling model of human IgG, showing the
Y-shaped structure. What is arresting about the three-
dimensional structure is the clarity with which each
amino acid sequence domain can be distinguished. The
folding patterns of light and heavy chains show a
fundamental relationship between the various domains;
the "immunoglobulin fold" consists of two β-
sheets—one composed of three chains and the other of

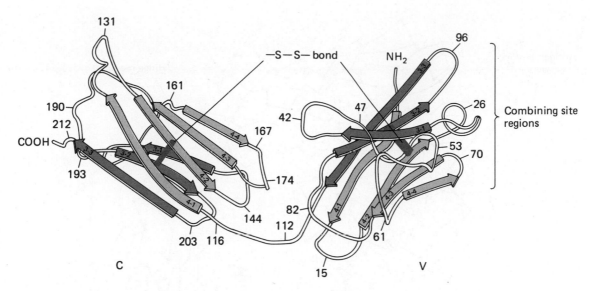

FIGURE **17–62**
Folding Pattern of the V and C Regions of the Light Chain.   Two "immunoglobulin folds" can be seen, composed of two layers of antiparallel β-sheets stabilized by an S—S bond and lying over one another. One layer contains three chains and the other contains four. A similar fold in the C_H2 region contacts the C region of the light chain, forming a tight domain with a hydrophobic interior between the two sets of four-chain β-sheets. The V_L region contacts the V_H region via the three-chain β-sheet of each. In this case there is a fairly loose association, producing hydrophilic clefts that form the antigen binding site. Notice the extra loop in the vicinity of residue 26 of the V region, forming part of the combining site; residues 53 and 96, both hypervariable, also form part of the combining site. (Courtesy of M. T. Schiffers; copyright American Chemical Society.)

four—lying on top of and linked to each other by a disulfide bond (Figure 17–62). Figure 17–63 shows how the entire combining site is formed by the association of a light and heavy chain, and how the hypervariable regions of both variable regions are located near the antigen-combining site.

The interpretation of the binding-site structure has been significantly aided by the lucky discovery of hapten-like molecules that bind strongly to myeloma proteins and by crystallization of protein hapten complexes. In one such study, antibodies against the hapten phosphorylcholine were analyzed by x-ray crystallography and the precise location of the combining site in relation to the hapten was delineated. Figure 17–64 shows how the phosphorylcholine fits into a cleft formed by several amino acids that do not lie adjacent to one another

in the primary structure: Tyr (33H), Glu (35H), Arg (52H), Lys (54H), Gln (56H), Asp (90H), and Trp (104H). In this instance, there appears to be only one light chain amino acid contribution to the binding site (in position 96L). In other cases, the hypervariable regions of the light chain also form important parts of the binding site. Two points emerge from these studies: (1) The combining sites are made of loops of the polypeptide chain, containing many (though not all) of the hypervariable amino acids, fixed to a relatively rigid structure generated by the conserved amino acids in the variable region; (2) The combining sites can vary greatly in both size and shape; wedge-shaped cavities have been observed as well as the shallow groove mentioned above.

An epitope or antigenic determinant is the small site on a large antigen to which an antibody is specific. A

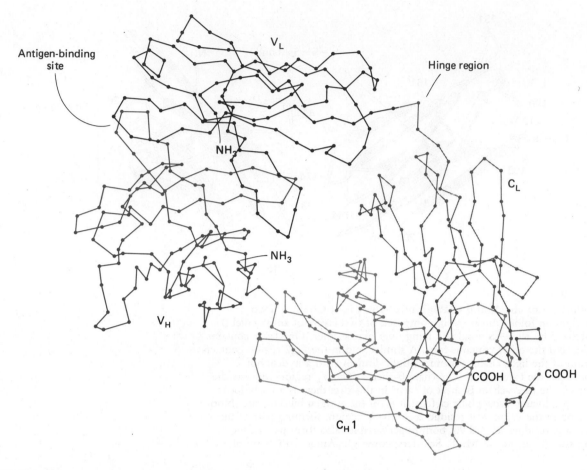

FIGURE **17–63**
Carbon Skeleton Representation of an Fab Fragment. The combining site contains the hypervariable amino acid residues (closed circles). Both the light and heavy chains contribute to the structure of the combining site. (Used with permission from J. D. Capra and A. Edmundson, based on original data from D. R. Davies and E. A. Padlan.)

foreign macromolecule will contain many such epitopes and antibodies can in principle be raised against each. Further, a single epitope can select different antibody-producing cells, and each type of antibody will combine with the epitope with a different specificity (strength).

One of the foremost technical advances of this century is the method for preparing large quantities of antibodies of a single type. The method does not exploit the techniques of chemical engineering to synthesize proteins of a single amino acid sequence. Rather, the antibodies are secreted by members of a single clone of cells resembling plasma cells; these immunoglobulins are called, therefore, *monoclonal antibodies.* C. Milstein and

G. J. F. Köhler invented this technique, for which they received the Nobel Prize in 1984.

The method relies fundamentally on the ability of plasma cells to synthesize only one type of antibody. The problems were to select out the proper plasma cell and to amplify its activity. Both problems were solved by the use of myeloma cells, exploiting their unrestricted growth (a property of transformed cells) and their ability to stimulate the synthesis of a single immunoglobulin. Milstein and Köhler found that if a culture of homogeneous myeloma cells is mixed in the presence of polyethylene glycol with a heterogeneous mixture of antibody-producing cells (spleen cells from an immunized mouse do nicely), fusion of the two cell types will occur (see Chapter 7 for a discussion of this phenomenon). The fusion products can be separated and each *hybridoma* cell, as they are called, will grow to form a clone of identical

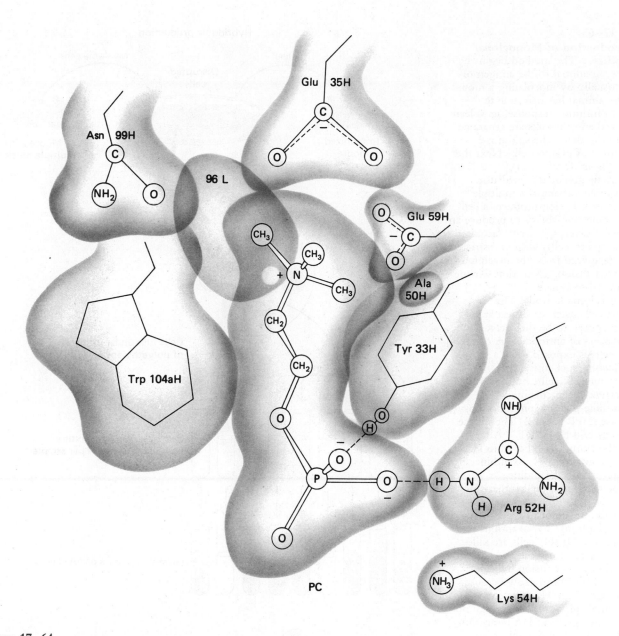

FIGURE **17–64**
Model of the Binding Site of an IgG Molecule and Its Hapten, Phosphoryl choline (PC). In this instance the contribution to the combining site comes almost exclusively from hypervariable amino acid residues of the heavy chain. In other cases the light chain makes important contributions. (Used with permission from J. D. Capra and A. Edmunson; based on original data of D. R. Davies and E. A. Padlan.)

cells (Figure 17–65). The amazing and useful property of these hybridomas is that they secrete only one kind of antibody, in addition to that formed by the parent myeloma cell, the antibody that was formerly the product of the spleen cell from the immunized mouse.

The method is made more efficient because antigen-activated B cells are more likely to fuse than B cells that are not actively dividing. By using myeloma variants that do not produce their own IgG, the output of the hybridoma can be restricted to the antibody of interest.

FIGURE 17–65

The Production of Monoclonal Antibodies.   The method begins by exposing an animal to the antigen of choice, usually by inoculating a mouse. After the animal has had time to mount an immune response, its spleen is removed and the cells are separated from one another. The spleen is a good source of plasma cells. Next the spleen cells are fused, using polyethylene glycol as a mediator, with a special variant of a multiple myeloma cell. Fusion produces a cell that combines the ability to produce an antibody of given specificity (acquired from the spleen cells) with unrestricted growth (acquired from the myeloma cells). After fusion, a selection scheme is often used to isolate mouse-myeloma hybrids. The myeloma cells used for fusion are carefully chosen to be themselves nonproducers of antibodies. A culture of such cells is exposed to 6-thioguanine, a chemical that generally will kill cells, but not if they are defective in the enzyme hypoxanthine-guanine phosphoribosyl transferase (HPRT$^-$ cells). The mouse spleen cells will be HPRT$^+$. After fusion, the culture is exposed to HAT medium, containing *h*ypoxanthine, *a*minopterin, and *t*hymidine. Aminopterin blocks the synthesis of purines and pyrimidines, so cells in HAT medium cannot grow unless they can use hypoxanthine to make purines. HPRT$^-$ cells are killed because they can't use the hypoxanthine. The fusion hybrids (HPRT$^-$/HPRT$^+$) can grow because they contain active HPRT enzyme. The parental spleen cells, although not killed, do not grow in HAT medium (or, if they do, they die out because they are mortal), so after a time the only things growing rapidly are the hybridoma cells. By testing for antibody production with the appropriate antigen, the clone producing the desired antibody can be identified. Each clone produces a single type of antibody directed against a specific epitope or antigenic determinant.

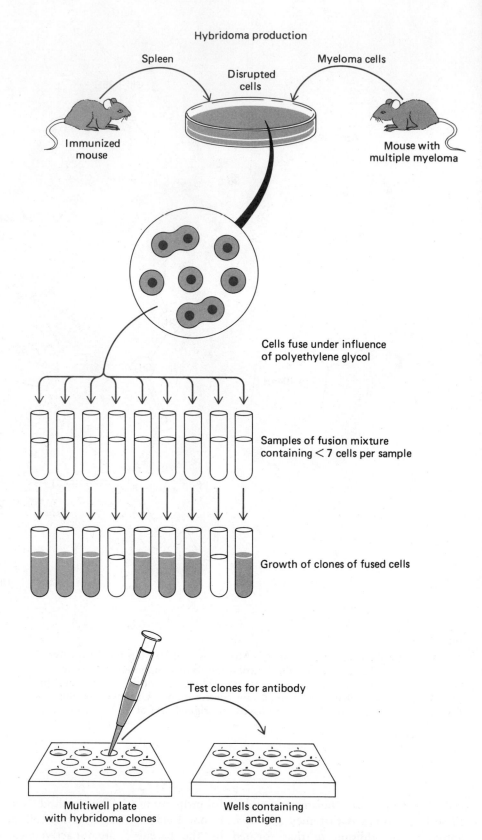

Hybridoma production

Spleen

Myeloma cells

Disrupted cells

Immunized mouse

Mouse with multiple myeloma

Cells fuse under influence of polyethylene glycol

Samples of fusion mixture containing < 7 cells per sample

Growth of clones of fused cells

Test clones for antibody

Multiwell plate with hybridoma clones

Wells containing antigen

$$\text{PRPP} + \text{Hypoxanthine} \xrightarrow{\text{HPRT}} \text{IMP}$$

$$\text{PRPP} + \text{Guanine} \xrightarrow{\text{HPRT}} \text{GMP}$$

PRPP
Phosphoribosyl
pyrophosphate

Guanine

Hypoxanthine

IMP

6-Thioguanine

Although clonal selection has already occurred, the cell biologist is faced with the need to distinguish which clone is the one secreting antibodies against the antigen of interest. This is done by brute force, as they say: after a few weeks, each hybridoma clone is tested with antigen and the ones that can react are identified (Figure 17–65). This testing may not take very long; if the original antigen preparation was pure, many spleen cells will have been producing antibodies to that antigen. One of the beauties of this method, however, is that the antigen need not be pure; since each hybridoma secretes antibody to only one specific epitope, the hybridoma will do the purification itself! If the investigator is willing to test lots of hybridomas, eventually one will be found that is specific to the antigen of interest. After cloning the desired hybridoma, some of the cells can be frozen in liquid nitrogen to be used in the future.

A hybridoma is like a factory for the production of antibody. A simple way to exploit these cells is to inoculate some into the abdomen of a mouse (Figure 17–66). There they will grow, secreting their antibodies into the peritoneal cavity. The mouse becomes bloated in

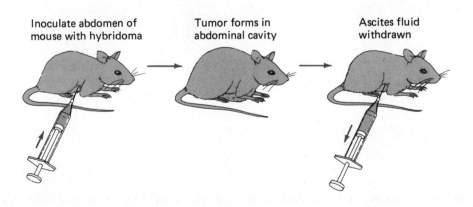

Inoculate abdomen of
mouse with hybridoma

Tumor forms in
abdominal cavity

Ascites fluid
withdrawn

FIGURE 17–66
Production of Monoclonal Antibodies in Mouse Ascites Fluid. Hybridoma cells are inoculated into the abdomen of a mouse. As the tumor grows, the cells secrete antibody into the peritoneal cavity. The increased protein concentration leads to an increased volume of fluid, to maintain osmotic equilibrium. When removed, this ascites fluid is a rich source of relatively pure monoclonal antibodies.

FIGURE 17–67
Affinity Chromatography in Which a Monoclonal Antibody Is Used to Purify a Given Protein.   First, a monoclonal antibody directed against a given protein is produced. The antibody is then immobilized by covalently attaching it to some inert beads. These beads are poured into a column and then an impure protein solution containing the given protein to be purified is poured through the column. The antigen attaches to the antibody. The column is washed thoroughly to remove all the other proteins in the mixture, after which the antigen is eluted by using a solution (usually containing a higher salt concentration) that dissociates the antigen-antibody complex.

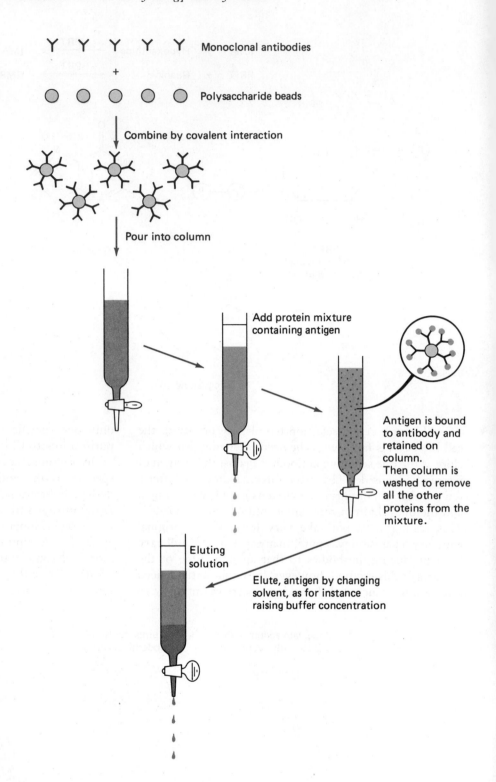

appearance from the accumulation of *ascites fluid*. This fluid can be removed and its protein content—on the order of 0.1 gram—is practically pure monoclonal antibody.

Thousands of clones may have to be tested if the original antigen was not pure. As a reward for the hard work of isolating the proper clone, monoclonal antibodies can be used to purify the antigen, by *immunoaffinity chromatography*. The antibodies are immobilized by crosslinking them to the solid support of a chromatography column, and the impure preparation containing antigen is passed through such a column (Figure 17–67). The antigen will bind to the antibody, while contaminating structures pass through. The antigen can usually be eluted by changing the pH or salt concentration of the buffer flowing through the column, which disrupts the secondary interactions by which antigen and antibody combine. The purification of the antigen can be practically accomplished in a single step!

Monoclonal antibodies are the quintessential example of biological specificity being used to unravel the mysteries of cell structure and function. They are used in biomedical research as labeling agents for various biological structures. We have seen numerous examples of fluorescent-labeled antibodies used to locate, by ultraviolet microscopy, parts of living and fixed cells. Monoclonal antibodies are also coming into commercial use as the active agent in kits for the medical diagnosis of various kinds of infectious agents—bacteria, viruses, fungi, yeasts, and parasites. Monoclonal antibodies can also identify tumor cells and, by bearing appropriate ligands, deliver therapeutic agents to kill those cells. That a reagent derived from a cancer can be used to cure other cancers is perhaps ironic, but it is also illuminating of how basic research on cell biology can be applied in significant ways. Monoclonal antibodies can be elicited against cell surface structures; this is how cell adhesion molecules and markers for various differentiated cells were identified (see discussion earlier in this chapter). In particular, B cells can be distinguished from T cells on the basis of surface structures that are detected by monoclonal antibodies.

The Genetic Basis of Antibody Diversity

An important question was inherent in the discovery of the very large number of antibodies that a vertebrate is capable of synthesizing: how is the information for so many proteins encoded in the genome? One answer was provided by the *germ line theory*, which proposed that each immunoglobulin was encoded by its own particular gene. Each organism would inherit the genes for all its antibodies from its parents. This theory was in complete agreement with the one gene–one polypeptide theory of Beadle and Tatum (see Chapter 9). But if there were one gene for each of the immunoglobulins, another problem became important: there might not be enough DNA in an organism to account for the observed immunological diversity. It was also difficult to explain how the evolution of an animal's immunoglobulin genes could keep up with the evolution of all the potential antigens in other organisms.

Another theory proposed that there were only a few genes for immunoglobulins; they were passed on from one generation to the next, but they were subject during the development of an organism to change (diversification). Each original germ line gene would become

modified into a large number of kinds, enough to meet the need of the organism for different antibodies. The two theories merged in part when W. J. Dreyer and J. C. Bennett proposed that the organism need have only a few genes for the constant regions of immunoglobulins and many genes for the variable regions. The observed antibody diversity could arise from a random assortment of the genes for the V regions with those for the C regions. This would answer the problem of a limited genome. It contradicted the one gene–one polypeptide theory, however, and also raised the difficult issue of how to explain recombination of genes during growth of somatic cells (*somatic recombination*), where such a process had not previously been observed.

All this speculation was resolved when the techniques of recombinant DNA analysis were applied to immunoglobulin genes. The story of deducing our current picture of how genes assort to give antibody diversity is a fascinating one, but we cannot tell it all here. Instead, we shall give the final answer and let the interested reader consult a text on immunology for the experimental details.

Genes for various regions of the immunoglobulin molecule are found in widely separated parts of the genome (Figure 17–68). There are four kinds of genes, one group for the variable regions (V genes), another for the constant regions (C genes), another for the "joining" regions (J genes), and, in the heavy chains, a set for "diversity" regions (D genes). In mice, for the kind of light chain known as kappa, there are approximately 250 V_K genes, four active J_K genes, and one C_K gene. If these various genes could combine randomly, this would imply the ability to synthesize approximately 1000 ($= 250 \times 4 \times 1$) different kappa light chains. For the heavy chain, there are a couple hundred V genes and four J genes, but 15–20 D genes to combine with the appropriate C gene. This would imply approximately 12,000–16,000 different heavy chains. If the kappa light chains assort at random with the heavy chains, there are potentially over ten million different antibodies that can be formed.

The mechanism by which the genes are moved around is not yet understood. The movement is not simple, however. In the case of the light chains, one of the V genes is moved so that it is juxtaposed to one of the J genes (Figure 17–68). This arrangement is transcribed into RNA and the primary transcript is spliced to yield the final mRNA for the antibody.

Recombining various V, J, and D genes is not the only mechanism for generating diversity. The mechanism that performs this function is not completely specific. The point at which two genes join can vary by a few nucleotides, and sometimes there are extra nucleotides added. Both these aberrations will have

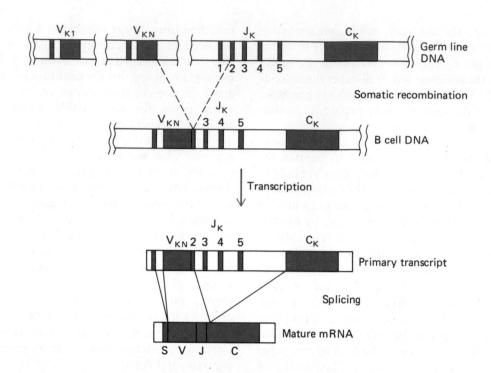

FIGURE 17–68

Organization of Genes for Immunoglobulins and Synthesis of an Antibody Messenger RNA.   The example illustrated is the mouse light chain known as kappa. Numerous genes (between 19 and 300) for the variable region (V_K genes) are located in one part of the DNA of chromosome 6, while the five genes for the joining regions (J_K) and the constant region (C_K) are located in another (only four of the J genes are believed to be active). There are two exons for the V genes; an intron lies between the signal sequence(s) and the rest of the coding information. Somatic recombination during development of the B cell brings together one of the V genes and one of the J genes; the intervening DNA appears to be deleted. Each V gene contains two flanking sequences—each with highly conserved nine-base and seven-base sequences and separated by a 12-base spacer sequence—that are believed to be involved in the recombination. The primary transcript of this DNA contains the two exons of the V gene followed by the J gene and then considerable unneeded information before the C gene. The excess RNA is removed by splicing to form the mature mRNA. Similar genetic reorganization and RNA splicing occurs in the expression of the heavy and the other light chain genes.

consequences for the structure of the final antibody and thus lead to increased diversity. In addition, the genes for the V regions can undergo mutation during development (*somatic mutations*). The rate of mutation has been estimated to be very much higher than for ordinary genes. This supplies additional diversity over that available from the genes inherited from the parents. It is possible that over 10^9 different antibodies can be produced as the result of all these processes.

The T Cell Receptor

T cells are essential for a responsive immune system. Their name derives from the requirement that T cells must spend part of their development in the thymus gland. There are at least three different kinds known, the *T helper* cells, the *T suppressor* cells, and the *cytotoxic (killer) T* cells. All three kinds are specific for a particular antigen and so all three must, like B cells, have some surface structure or receptor that manifests that specificity. Such a T cell receptor has now been identified; its structure is somewhat different from the membrane-bound immunoglobulin that forms the B cell receptor, but both may combine with antigen in similar ways. The T helper and cytotoxic T cells, however, only respond to antigen if it is accompanied by another cell surface structure, a major histocompatibility complex (MHC) protein (Figure 17–69). Activating a cytotoxic T cell

The Immune System as a Paradigm for Development

There is another reason why it is instructive to consider the way in which genetic information is expressed during the maturation of cells in the immune system. There may be clues to help cell biologists decipher the general mechanism by which differentiation and development occur in multicellular organisms. This problem is one of the outstanding mysteries of biology today and solving it is the occupation of numerous laboratories throughout the world. Let us take a few moments to discuss some of the possibilities suggested by studying the cell biology of the immune system.

First of all, we may have been misled by the results of hybridizing DNA from the different cells of an organism. The ability to hybridize such DNA essentially completely has been interpreted as showing that all an organism's cells contain the same DNA sequences. What those hybridization experiments did not show, however, is the position within the chromosomes of particular stretches of DNA; those sequences need *not* be in the same order in all cells to give generally complete hybridization. If the aggregate of cellular DNA sequences is represented by all the books in a library, it may be that there is a normal process for shuffling the books on the shelves. We have learned from a detailed analysis of the restriction maps of B cells that the position of V genes, for example, changes during development. Is this a general mechanism for differentiation? As particular B cells are selected for growth by interacting with antigen, it may be that practically random processes produce numerous possible routes to differentiation, only some of which are selected for encouragement by external stimuli.

Another aspect of the expression of genetic information by lymphocytes is that certain genes do not give rise to messenger RNA until they are moved to another part of the genome. It is as if those books were brought from the archive stacks of the library into the general reading room. Identifying the control elements that allow transcription of active genes in their new environment is essential to an understanding of development. After the primary transcript has been made, there is still the problem of differential splicing to form the final correct mRNA. Both these phenomena are illustrated by cells of the immune system and both are studied intensively in that context.

Finally, we see the influence that cells, especially their surface proteins, have on one another during their maturation. One reading of the genetic information influences later activity by other readers, and a cooperative behavior characterizes the interactions.

The vision of development that lymphocyte behavior provides is both provocative and disturbing. Models based on the linear reading of a developmental program seem too simplified to explain the maturation of B and T cells. Even though the control of gene expression by cells of the immune system is complicated when compared to *E. coli*, it can be hoped that studying them will pave the way for understanding the much more difficult problems of development.

requires that an antigen be displayed on a stimulating cell in combination with a particular class I MHC protein; the T helper cell requires antigen and a class II MHC protein.

Cells from outside the organism will not have the appropriate MHC proteins on their surfaces and cannot, therefore, activate T cells. Cytotoxic T cells are able to recognize an animal's own cells because essentially all the cells of an organism bear class I MHC surface proteins. If any of these cells are also displaying a foreign antigen—a protein associated with an infecting virus or parasite, for example—they will be killed by cytotoxic T cells. Cytotoxic T cells do not engulf their targets like macrophages do, but the exact lethal mechanism is not yet known. Helper T cells are presumably specific for other cells of the immune system because only those cells have the necessary class II MHC proteins on their

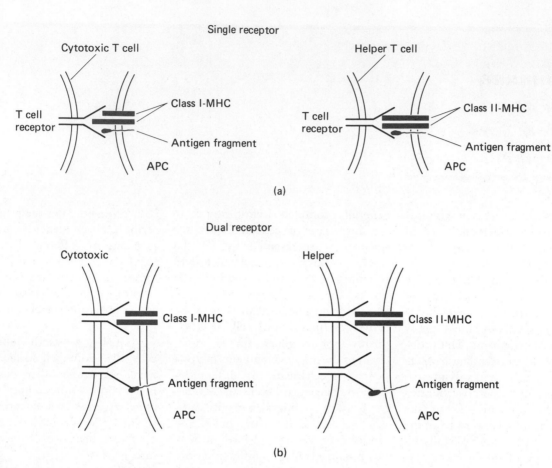

FIGURE 17–69

Two Theories for Antigen Activation of T Cells. The T cell responds to both processed antigen and an MHC protein. (a) The single-receptor model. In this scheme, a single T cell receptor recognizes both antigen fragment and the MHC protein on the antigen-presenting cell (APC) by having binding sites for both on a single receptor molecule. (b) The two-receptor model. Two T cell surface structures comprise the recognition system. One is specific for processed antigen and the other for MHC protein.

surfaces. Although recognition by suppressor T cells probably works in a similar fashion, the details are not yet understood. The action of suppressor T cells is to reduce the activity of B cells. Such a reduction is essential in the case of the numerous antigens that are "self," part of the organism's own repertoire of cell surface structures and macromolecules. The specific reduction of immune responsiveness, in both B cells and T cells, is called *tolerance*. It is widely believed that suppressor T cells play a role in producing tolerance; again, the mechanism is not understood.

Activation by antigen leads to the presence on the T cell surface of receptors for the lymphokine *interleukin 2* (IL-2). The gene for this growth factor has been cloned and sequenced. IL-2 is composed of a polypeptide of 133 amino acids and oligosaccharides of variable structure. There are about 10^4 molecules of the IL-2 receptor on

activated T cells. Although coded by a single gene, receptors manifest both a high affinity ($K_D \approx 10^{-11}$ M) and a low affinity ($K_D \approx 10^{-9}$ M); the reason for two functional behaviors is not understood. There are several lymphokines known already and more expected; we leave description of their specific functions to advanced accounts of immunology.

Structure of the T Cell Antigen Receptor

The molecular nature of the T cell antigen receptor was worked out by applying recombinant DNA techniques to the analysis of recombined genes in T cells, by exploiting tumors of T cells, and by methods, like those pioneered by Milstein and Köhler, which amplify the activity of T cells. Once a large source of identical T cells was available, antibodies against surface structures could be raised (using methods similar to those discussed earlier to distinguish surface structures on adhering cells, etc.). Certain antibodies against T cell surface proteins would block the ability to help a B cell, detected by a reduced secretion of lymphokine growth factors; the surface antigen in this case was the T cell antigen receptor (Figure 17–70).

The structure of the T cell receptor has features that are similar to a membrane-bound antibody, of the sort found on B cells. Some of these features have been determined by direct amino acid sequencing, others by an analysis of the genes that encode the receptor. There are two glycoprotein chains of 40–45 kDa each, called alpha and beta. Antibodies against the receptor coprecipitate other species and there may be five more glycoproteins involved in the complete receptor complex. Other experiments have suggested even more complicated structures with accessory proteins, and so forth.

Each α and β chain contains an N-terminal variable region and a C-terminal constant region. In the C-terminal region, there is a "stop transfer" sequence of hydrophobic amino acids that acts to anchor the receptor in the cell membrane. A significant homology exists between the nucleotide sequences of the T cell receptor and immunoglobulins, about 30%. Genes for variable, joint, and diversity regions are present; they are not contiguous in the DNA of germ line cells but they are found together in mature T cells (Figure 17–71). As in the case of immunoglobulin information, the V genes are probably subject to somatic mutations during development and imprecise recombination to form the DNA sequences of the T cell receptor protein. It has been estimated that over a million different combinations of α and β chains can be generated.

MHC Proteins

When cells from one vertebrate organism are introduced into another, the usual response by the recipient animal is to mount an immune reaction against the

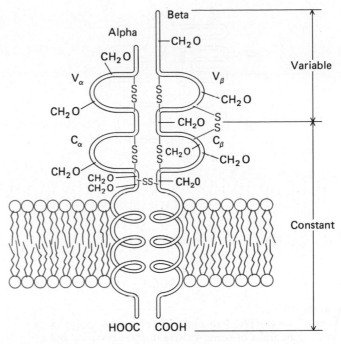

FIGURE **17–70**

Antigen Recognition Chains of the T Cell Receptor.
The receptor is made of two polypeptide chains, α and β, which combine with antigen, and other glycoproteins whose functions are not well understood. Both α and β chains have N-terminal variable regions and C-terminal constant regions that include a 17-residue transmembrane segment and a short (5-residue) cytoplasmic region. Extracellular V and C regions have carbohydrate (CH_2O) attached. The structure was determined by analyzing cloned genes and by studying the receptors of T cell lines that grow indefinitely, either leukemia tumors or hybridomas. Antibodies raised against these cells included some that blocked the immune response of these T cells; these antibodies were specific for the T cell receptor.

FIGURE **17–71**
The Organization of the Genes for the Beta Chain of the T Cell Receptor.
The V genes are on different stretches of DNA from the D, J, and C genes. T cell V genes are flanked by sequences similar to those found on immunoglobulin V genes. The C_β genes have four exons (a similar D, J, C set lies downstream of the first). The V region of the β chain is composed of residues coded by the V_β, D_β, and J_β genes. It is believed that recombination and splicing events occur during expression of these genes, as in the case of the immunoglobulin genes, but the details are not yet clear.

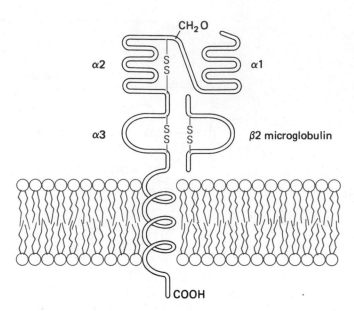

FIGURE 17–72

The Class I MHC Protein. There are two polypeptide chains, an alpha of some 45 kDa in three domains and a smaller chain (11.6 kDa) known as β_2 microglobulin that forms a single domain. Attached carbohydrate is indicated by CH_2O. Class I MHC proteins are found on the surface of all nucleated cells in a vertebrate organism; exceptions are the nonnucleated red blood cells and sperm.

foreign cells and reject them. This observation has been studied intensively, since such immune responses limit the ability to transplant organs, like kidneys and hearts, into human patients. Incompatibility of tissues (histoincompatibility) from different organisms implies a detectable difference between the cells of one organism and another. The most significant difference is that each organism bears on the surface of its nucleated cells proteins that identify them as "self." These proteins are largely coded by the genes of the *major histocompatibility complex* and are known as *MHC proteins*.

In mice, the MHC proteins are known also as H-2 antigens; they were discovered over 50 years ago, by P. A. Gorer, as blood group antigens that are also associated with tissue incompatibility. Study of the genetics of the H-2 proteins revealed a complex organization that includes genes which affect *immune responsiveness* (I genes, some of which code for the cell surface structures called class II MHC proteins) and regulators of the complement system. The human MHC proteins (also known as HLA antigens) are encoded by a similarly complex set of genes.

A class I MHC glycoprotein is composed of two polypeptide chains (Figure 17–72): a *heavy chain* of 45 kDa, which contains the carbohydrate and whose amino acid sequence is different from one organism to another, and a smaller chain (11.6 kDa), whose sequence

is similar from one organism to another and which is also known as β_2 *microglobulin*. Even though the heavy chain is where the differences that define "self" are located, there is a great deal of homology (~80%) among the heavy chains from different mice. There is also significant homology (~70%) between human and mouse class I MHC heavy chains.

Variability occurs in certain regions of the heavy chain structure. These regions have been correlated with the three-dimensional structure of the class I molecule, determined by D. Wiley, J. Strominger, and their colleagues (Figure 17–73). There are several genes that specify heavy chains and there are over 50 alleles known of one of the genes that specifies the heavy chain. Each nucleated human cell bears at least three such proteins on its surface. Counting the diversity of MHC proteins in populations of organisms is still in progress, but there are probably over 10^6 different variants among humans. Not all antigens can be successfully recognized when presented in combination with a single allele of a class I MHC protein (this was shown by comparing the immune responsiveness of mice to a single antigen combined with different alleles of the I genes). A large number of alleles of class I MHC proteins may be required, therefore, to allow an effective immune response to the many different antigens that an organism is likely to encounter.

The class II MHC glycoproteins are also made up of four domains from two polypeptide chains; each contains carbohydrate and each is about the size of an immunoglobulin light chain (Figure 17–74). There are several genes that specify these proteins and numerous alleles of each. The diversity of class II MHC proteins is increased by unequal crossing-over and by a process called "segment transfer" that has results like those of gene conversion (see Chapter 9). Certain regions of the genes have very high rates of recombination.

There are two features of the class II MHC proteins that are important to keep in mind. First, they are expressed only on cells that participate in the immune response: B Cells, T cells, macrophages, certain cells in the skin, and so forth. Second, they are part of an *associative recognition system*—that is, they trigger activation of an immune response only when they are accompanied on the cell surface by a foreign antigen. Helper T cells, for example, must simultaneously detect class II MHC protein and antigen on the surface of a B cell before they induce proliferation and maturation into plasma cells. The helper T cells themselves are activated by coming into contact with a cell that presents the antigen in the company of a class II MHC protein. A foreign class I MHC protein is presumably recognized as such a combination, since T cells are the primary actors in tissue rejection, a cellular immune response.

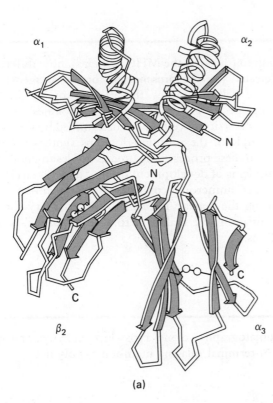

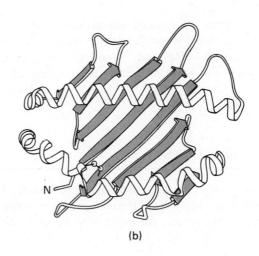

(a)

(b)

FIGURE **17–73**

Three-Dimensional Structure of Class I MHC Protein. (a) Side view,
showing four regions (the membrane-attachment region was removed before x-ray
analysis). The β_2 microglobulin and the α_3 region lie close to the membrane. The
α_1 and α_2 regions form a binding pocket. (b) The antigen-binding pocket, top
view. Eight segments of β-sheet (ribbon arrows) form a flat blottom, and two α-
helices form the sides of a cleft, which can hold a peptide of 12 amino acids if they
are extended, 20 if they are in an α-helix. The cleft is also the site of the variable
region of the heavy chain, expected if the antigen is to bind there. A fragment of
antigen could bind in the pockct and be recognized simultaneously as a complex
with other structural aspects of the MHC protein. (Courtesy of D. Wiley and J.
Strominger.)

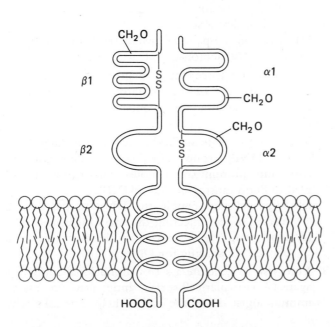

FIGURE **17–74**

The Class II MHC Protein. There are two polypeptide
chains, alpha and beta, each of approximately 30 kDa and
each comprising two domains. Carbohydrate (indicated as
CH_2O) is attached to the extracellular part of both chains.
The variable regions are in the alpha 1 and beta 1 domains.
Note the general resemblance to the structure of the T cell
antigen receptor (Figure 17–70) and the class I MHC protein
(Figure 17–72).

SUMMARY

To attempt to cover in a single chapter all the interesting biological phenomena that exist beyond the outer leaflet of the cell membrane is a daunting prospect. Instead of a comprehensive account, we have described the kind of environment in which cells find themselves in multicellular organisms. Primarily, this is other cells and their products.

The theme of cell-cell interactions is an important one to recognize if we wish to study the differentiation of cells during development. Some of the best studied examples of this lie in the interactions among cells of the immune system. There we see how colleagues are recognized (through the MHC proteins), how differences are detected (by the presence of foreign antigens), and how one cell influences the growth of another (by secreting specific growth factors, lymphokines). We also see how diversity of cell type is determined by genetics in the example of the immunoglobulins synthesized by B cells. Can these processes be used as guides to understand the problems of development or how the nervous system makes its numerous connections? Is it possible that re-sorting the genetic information is necessary for such differentiation? We shall have to wait for answers to these questions.

KEY WORDS

Allele, antibody, antigen, antigenic determinant, C-terminal, chromatography, clone, cross-link, enzyme, epitope, filament, genome, helix, information, microfilament, monomer, N-terminal, plasma membrane, polymer, recombination, solute, substrate, vesicle, x-ray diffraction.

PROBLEMS

1. Give two examples of how biological information can be transferred from cell to cell in a specific way. Be sure to indicate how the signal is received by the recipient cell.

2. (a) What does the chemistry of a vertebrate tendon have in common with that of a gelatin dessert (Jell-O)?
   (b) What does the structure of the protein elastin have in common with a teenager's bedroom?

3. You have added a monoclonal antibody with specificity for fibronectin to a culture dish coated with fibronectin and containing a layer of cells adhering to the fibronectin. You fix the cells at intervals thereafter and stain for immunofluorescence microscopy, using fluorescein-conjugated antiactin antibodies. You have a control slide to which an irrelevant antibody has been added. What would you expect to observe, and why?

4. (a) How might you determine whether a cell surface molecule is important in the adhesion of one cell to its neighbor?
   (b) Distinguish between a homophilic and homotypic interaction among cells.
   (c) Name two components of the extracellular matrix that have been shown to be important in guiding cell migration during embryogenesis. What is the experimental evidence for their being required?

5. (a) Which of the following structures serves to facilitate the flow of signaling molecules or ions through a series of cells?
   (i) tight junction
   (ii) gap junction
   (iii) belt desmosome
   (b) How can the passage of material from one cell to another be demonstrated?

6. What is the evidence that the neurotransmitter responsible for passage of the excitatory signal across the neuromuscular junction is contained in synaptic vesicles?

7. You wish to isolate antibodies specific for the dinitrophenol (DNP) group. You begin by generating an antiserum, immunizing a mouse with a protein coupled to fluorodintrobenzene (FDNB).
   (a) Why did you couple the DNP to the protein instead of just injecting the small molecule itself?
   (b) How would you prove that the antibodies you elicited are specific for DNP?

8. You wish to generate a set of monoclonal antibodies against a particular cell surface antigen and you ask an immunologist down the hall to help. She asks why

you want to generate monoclonals rather than a polyclonal serum against the same antigen and insists that you tell her precisely what you consider to be the relative merits of monoclonals versus polyclonals before she spends time helping. You are desperate for advice so you think hard about your reply. What is it?

SELECTED READING

Ada, G. L., and Nossal, G. (1987) The clonal-selection theory. *Scientific American* 257(2):62–69.

Cohen, I. R. (1988) The self, the world and autoimmunity. *Scientific American* 258(4):52–60.

Cooper, M. D., and Lawton, A. R. (1974) The development of the immune system. *Scientific American* 231(5):58–72.

Frazier, W., and Glaser, L. (1979) Surface compounds and cell recognition. *Annual Review of Biochemistry* 48:491–523.

Gottlieb, D. I. (1988) GABAergic neurons. *Scientific American* 258(2):82–89.

Hay, E. O. (1981) Extracellular matrix. *Journal of Cell Biology* 91:205s–223s.

Herzberg, E. L., Lawrence, T. S., and Gilula, N. B. (1981). Gap junctional communication. *Annual Review of Physiology* 43:479–491.

Hynes, R. O. (1986) Fibronectins. *Scientific American* 254(6):42–51.

Kennedy, R. C., Melnick, J. L., and Dreesman, G. R. (1986) Antiidiotypes and immunity. *Scientific American* 255(1):48–56.

Marrack, P., and Kappler, J. (1986) The T cell and its receptor. *Scientific American* 254(2):36–45.

Milstein, C. (1980) Monoclonal antibodies. *Scientific American* 231(4):66–74.

Raff, M. C. (1976) Cell-surface immunology. *Scientific American* 234(5):30–39.

Rothman, J. E. (1985) The compartmental organization of the Golgi apparatus. *Scientific American* 253(3):74–89.

Tonegawa, S. (1985) The molecules of the immune system. *Scientific American* 253(4):122–131.

Glossary

Numbers in parentheses indicate the chapter in which the term predominantly appears.

active transport A process requiring energy input to move a molecule across a membrane toward an area of higher concentration of that molecule. **(7)**

allele One of two or more alternative, randomly assorted forms of a particular gene. **(9)**

allosteric effector (Other shape.) Refers to a small molecule whose structure does not resemble that of an enzyme substrate but which binds to the enzyme and modifies its activity. **(5)**

alpha helix Regular spiral structure of contiguous amino acids. **(4)**

Angstrom unit (Å) Unit of length, 10^{-10}m, about the length of a C-H bond in methane. **(2)**

anneal To slowly relax to form the native structure; for example, slow cooling to allow the two heat-separated strands of DNA to reform the base-paired helical structure. **(10)**

antibody A protein secreted by specialized cells of a vertebrate organism in response to encountering a structure recognized as foreign. **(17)**

anticodon The three bases in a tRNA molecule that pair with a three-base codon of mRNA to decode the message. **(12)**

antigen A structure recognized as foreign by the immune system of a vertebrate animal. **(17)**

antigenic determinant Small part of antigen structure that elicits formation of antibodies and to which an antibody binds. **(17)**

assembly Product or process of bringing together several macromolecules into a complex structure. **(15)**

autotroph An organism capable of growth and maintenance from a source of energy not derived from life processes, specifically, not from reduced carbon compounds. **(1)**

base pairing Associating a nucleotide base with a complementary one by the formation of specific hydrogen bonds (e.g., A with T or U, G with C). **(10)**

C-terminal Refers to the end of a protein that has a carboxyl group that is not attached to another amino acid. **(4)**

calorie The amount of energy that, in the form of heat, will raise the temperature of one gram of water by one degree (from 14.5°C to 15.5°C). (1 kilocalorie = 1 kcal = 1000 calories.). **(3)**

carrier A lipophilic structure that facilitates movement across a membrane of a molecule by binding to it and diffusing as a complex through the lipid bilayer. **(7)**

cell theory M. J. Schleiden and T. Schwann's 1838 generalization of the theory that all living systems are composed of cells and cell products. **(1)**

chloroplast A clearly defined, membrane-surrounded organelle containing chlorophyll and found in the cytoplasm of eucaryotic cells capable of photosynthesis. **(8)**

chromatid One of the two strands of a duplicated chromosome still held together at the centromere. **(9)**

chromatin The complex structure of protein and nucleic acid in which the nuclear DNA of eucaryotic cells is found. **(9)**

chromatography A technique for separating, on columns of solid resin, molecules by differences in surface charge, surface adsorption, size, etc. **(4)**

chromosome   Densely staining threadlike structures in the nucleus of eucaryotic cells, containing the hereditary material of the cell. **(9)**

clone   Identical cells descended from a common ancestor; to isolate a gene by inserting its DNA into a plasmid, introducing the plasmid into a bacterial cell, and thus amplifying the genetic sequence by growth of a clone of the bacterial cells containing the plasmid. **(10)**

codominance   Genetic expression in which two alleles interact or cooperate to produce the phenotype; thus, the phenotype of the heterozygote is distinct from the phenotype of either the dominant or recessive homozygote. **(9)**

codon   Three contiguous bases in an mRNA that are decoded to specify one amino acid during protein synthesis. **(13)**

coenzyme   A small organic compound that is required, but is not alone sufficient, for enzyme activity. **(5)**

cofactor   An atom or ion (e.g., ultratrace element) that is essential for enzyme activity. It either interacts directly with the protein portion of the enzyme or is part of the coenzyme. **(5)**

conformation   The overall shape of a macromolecule. The macromolecule's function depends on its conformation. **(4)**

consensus sequence   Prototype that summarizes the nucleotide bases most often found at each position of the sequence in members of a large library of sequences having the same function. **(12)**

cortex   Outer layer of a globular object. **(2)**

coupling   Functionally joining; two chemical reactions show coupling if they share a common intermediate. **(6)**

crossing-over   The process by which genetic characteristics on separate members of a homologous pair of chromosomes are rearranged (recombined) onto the same chromosome by the breaking and rejoining of chromosome segments (thus yielding a new combination of characteristics in the offspring). **(9)**

cross-link   To join together by covalent bonds, often after an experimental manipulation. **(2)**

cytochrome   Metal-containing colored protein, found in all cells, that is able to be oxidized and reduced as part of its function. **(8)**

cytogenetics   A subdiscipline of genetics uniting cytological observation of chromosomes and the statistical analysis of genes. **(9)**

dalton (Da)   A unit of molecular mass named after John Dalton, 19th-century English physicist, and equal to one-twelfth the mass of an atom of the most abundant form of carbon (1.66×10^{-24} g). The abbreviation for kilodalton is kDa (also commonly, if inaccurately, kd or kD). **(4)**

denature   To unfold by heat or chemicals the native, physiologically active structure of a macromolecule. **(4)**

density gradient   A solution in which the buoyant density (measured in gram/ml) increases smoothly in one direction. **(4)**

density gradient centrifugation   (*See* equilibrium density gradient centrifugation.)

deoxyribonucleic acid (DNA)   Linear polymer of nucleotides, containing genetic information. Each nucleotide is composed of a nitrogenous base (purine or pyrimidine), a deoxyribose sugar, and a phosphate group. **(10)**

dialysis   Process of differential movement, and thus separation, of smaller molecules through pores of a membrane that restrict the passage of larger molecules. **(4)**

diploid   Having a double set of genes in the form of paired homologous chromosomes. **(9)**

dominant   Describing an allele whose effect is expressed in the phenotype both when present in the homozygous state, as expected, and when present in the heterozygous state, where the effect of the other (recessive) allele is masked. **(9)**

effector   A small molecule that elicits a change in the functional state of the macromolecule to which it binds. **(5)**

electron microscope   A microscope, developed in the 1940s, that uses beams of electrons instead of light for visualizing material at greater magnification than previously possible. **(2)**

electrophoresis   A technique for separating molecules by use of an electrical field, according to charge, size, and shape, which determine at what rate a molecule moves through a given medium (e.g., a polyacrylamide gel or filter paper); sometimes called ionophoresis. **(4)**

endocytosis   Bulk-transport process operating by invagination of the cell membrane and budding off inside of a vesicle, thereby bringing into the cell a sample of the extracellular environment. **(7)**

endonuclease   An enzyme that catalyzes the hydrolysis of a nucleic acid, breaking a covalent phosphodiester bond at an internal position of the polymer. **(10)**

endoplasmic reticulum (ER)   A system of membrane-limited channels distributed throughout most eucaryotic cells, first discovered with the electron microscope by Keith Porter in 1945. Two forms may be present: the rough, or granular, in which ribosomes are associated with the ER, and the smooth, or agranular, which is free of ribosomes. **(2)**

energy   Fundamental property of a system that allows it to do work on another system. The amount of this property is fixed; if work is done on another system, the capacity to do further work is diminished by the amount of work done. Measured in SI unit joules (J) and also in calories. **(1)**

energy transduction The conversion of one form of energy into another: chemical energy into kinetic energy and heat; light energy into chemical energy and heat; etc. **(1)**

enthalpy (H) A term used in thermodynamics to describe the heat content of a system. The change in enthalpy, ΔH, is the heat absorbed ($\Delta H < 0$) or given off ($\Delta H > 0$) in a chemical reaction or process at constant pressure; measured in joules (J) or J/mole ($J \text{ mole}^{-1}$). **(1)**

entropy (S) A term used in thermodynamics to describe the degree of disorder of a system. In any spontaneous process, the entropy of an isolated system without internal fields (e.g., electric, magnetic) always increases or remains the same; measured in joules (J) per degree (Kelvin), or J/K·mole ($J \text{ K}^{-1} \text{ mole}^{-1}$). **(1)**

enzyme A protein that catalyzes a chemical reaction in a highly specific manner. **(5)**

epitope Antigenic determinant. **(17)**

equilibrium State of a chemical reaction or process in which forces are balanced so there is no net tendency to go forward (toward product formation) or backward (toward the formation of reactants); all measurable parameters (e.g., concentration, temperature, pressure) remain constant with no input of mass or energy and regardless of how long one measures. **(1)**

equilibrium density gradient centrifugation A technique for separating macromolecules by distribution in a density gradient created by sedimenting a dense salt (e.g., CsCl) to equilibrium. The macromolecules form a band centered at the buoyant density of the molecule. **(10)**

eucaryote An organism having genetic material surrounded by a nuclear membrane, having other membrane-bound organelles in the cytoplasm, and using a mitotic apparatus to apportion the chromosomes during cell division. **(1)**

evolution A process, first proposed as a theory by Darwin and Wallace, of change by natural selection of randomly occurring mutations. Present-day organisms arose during evolution by successive changes from a common ancestor. **(1)**

exocytosis Bulk-transport process operating by fusion of a vesicle with the cell membrane, thereby releasing the vesicle contents into the extracellular environment. **(13)**

exonuclease An enzyme that catalyzes the hydrolysis of a nucleic acid, breaking a covalent phosphodiester bond at the end position of the polymer. **(10)**

feedback Influence on one step in a chain of events by a product of a later step in that same chain. **(5)**

fibril Long, thin structure of cross section smaller than a fiber. **(2)**

filament Long, thin structure of cross section smaller than a fibril. **(2)**

free energy (G) (Gibbs) The part of the energy available to do work during processes occurring at constant temperature and pressure. In any system (isolated or not) undergoing a spontaneous process in these conditions, the Gibbs free energy always decreases or remains the same. If there is no free energy change in a system capable of change, the system is at equilibrium. **(1)**

gamete The haploid sex cells of an organism; the ova and sperm that, after fertilization, form a diploid cell. **(9)**

genetic code The set of 64 possible nucleotide triplets and the amino acid or stop signal to which each triplet corresponds. **(13)**

genetic map A quantitative description of the linear arrangement of genes on a chromosome, determined by genetic crosses using marker genes whose locations on a chromosome are known or by using the frequency of recombination of two allele pairs in a cross to define the map distance between them. **(9)**

genome The set of genes making up the usual number found in a given cell, haploid or diploid; the DNA molecule(s) of which the genes are composed. **(9)**

genotype The genetic constitution of an organism. **(9)**

haploid Having only a single set of chromosomes (unpaired); the sperm and the egg cells are haploid. **(9)**

helix Regular spiral; the shape spontaneously adopted by two complementary strands of nucleic acid that have been base-paired together. **(10)**

heterotroph An organism that must derive its energy for growth and maintenance from living things or from their products, specifically, from reduced carbon compounds. **(1)**

heterozygous composed of differing kinds of something; in genetics, having different alleles for a given gene on each member of a homologous chromosome pair. **(9)**

high-energy phosphate A phosphate ester whose hydrolysis proceeds with a large negative free energy change. ATP is the most universally important of the high-energy compounds (see p. 111 for others). **(6)**

homologous Having a similar structure and origin; in genetics, chromosomes that are of similar shape and size, and that have alleles for the same genes, are said to be homologous. **(9)**

homozygous Composed of the same kind of something; in genetics, having identical alleles for a given gene on a pair of homologous chromosomes. **(9)**

hybrid Arising from mixed sources, races, or species; in genetics, the progeny of genetically different parents. **(9)**

hybridization The formation of a macromolecule by structural subunits of different origins. **(10)**

hydrogen bond An interaction, weaker than a covalent bond but stronger than the van der Waals force, that occurs between a hydrogen atom covalently bonded to an electronegative atom (the hydrogen donor) and a second electronegative atom (the hydrogen acceptor). **(3)**

information A property of a system that allows a reduction in the number of possible choices. In cells, information is contained in the nucleotide sequence of DNA from where it is both expressed in the control of metabolism and replicated prior to distribution to daughter cells during cell division. **(1)**

initiator The structure or molecule that is necessary to begin the synthesis of a polymer. **(13)**

intermediate filaments Long, thin (10-nm diameter) structures of various composition found inside animal cells, forming part of the cytoskeleton; they are intermediate in diameter between microtubules (20 nm) and microfilaments (7 nm). **(16)**

joule (J) The SI unit of energy, named after the English physicist James Prescott Joule (1818–1889), equal to work done by a force of one newton moving through a distance of one meter; equivalent to 10^7 ergs, one watt-second, and 0.24 calories. **(3)**

kinase An enzyme that catalyzes the addition of a phosphate to an acceptor molecule using, usually, ATP as the donor. **(6)**

Krebs cycle Citric acid cycle or tricarboxylic acid cycle; the circular metabolic pathway found in most cells that successively oxidizes two-carbon fragments (acetyl-CoA) to Co_2 and transfers the electrons to NAD^+ and FAD. **(6)**

ladder gel A gel through which fragments of nucleic acid, whose lengths differ by one or a small number of nucleotides, have been electrophoresed. **(10)**

ligand Small molecule that binds to an enzyme at other than the active site responsible for catalysis, thus bringing about a functional effect by a change of the protein's structure. **(4)**

linkage Describing the phenomenon of genes tending to be transmitted together during meiosis, as though joined, because they are located on the same chromosome. Linkage may be complete (always transmitted together) or partial (transmitted together more often than expected from independent assortment). **(9)**

lysosome A widely distributed membrane-bound organelle of eucaryotic cells with variable structure and containing digestive enzymes capable of breaking down most cellular components. **(2)**

macromolecule A large molecule of molecular weight greater than one thousand. **(1)**

mass Fundamental property of material substances, conserved during chemical and biological reactions, measured as the ratio of a force to the acceleration that force produces. The SI unit is the kilogram. **(1)**

meiosis A process consisting of two successive nuclear divisions during which the genetic material is reduced by half from the diploid state to that of haploid sex cells. **(9)**

microfilaments Cytoplasmic fibers of approximately 7-nm thickness and varying length and containing actin. Microfilaments are found almost universally in eucaryotic cells and are involved in the cytoskeleton and in cell movement. **(16)**

micrometer (μm) 10^{-6}m; 1000 nm; about the length of an *E. coli* cell. **(2)**

microtubules Straight, long, cylindrical structures consisting of polymers of the protein tubulin; found in the cytoplasm of eucaryotic cells and associated with motion (flagella, mitotic spindle) or with providing structural stiffness (avian erythrocyte). **(16)**

mitochondrion A DNA-containing organelle, bounded by two membranes. It is found in the cytoplasm of aerobic eucaryotic cells. Its primary function is the generation of ATP driven, via a proton gradient, by the transfer of reduced electrons to molecular oxygen. **(8)**

mitosis The division of a eucaryotic cell's set of chromosomes into two sets, each identical to the original. **(9)**

mitotic apparatus The structures within a eucaryotic cell (e.g., microtubules, spindles) that organize to cause the division of that cell's set of chromosomes into two, each identical to the original. **(2)**

molecular cell biology An approach to the study of biology in which biological phenomena are analyzed in terms of physical and chemical principles utilizing a variety of complementary approaches: genetic, biochemical, physical, optical, etc. **(1)**

monomer The basic subunit of a larger molecule (polymer) with monomers attached to each other by the repetition of one or a few kinds of covalent bonds; for example, amino acids are the monomers of proteins and are joined by peptide bonds. **(4)**

mutagen A chemical compound or treatment that enhances the frequency of mutations. **(9)**

mutagenesis The process by which the frequency of mutations is enhanced. **(9)**

N-terminal Refers to the end of a protein that has an amino group that is not attached to another amino acid. **(4)**

nanometer (nm) 10^{-9} m; 10^{-3} μm; about the distance between bases in DNA at the points where the bases attach to deoxyribose. **(2)**

natural selection A process by which an organism having genetic traits that enhance the survival of its offspring is more likely to perpetuate those traits into future generations. **(1)**

nuclease An enzyme that catalyzes the hydrolysis of a nucleic acid. **(10)**

nucleoprotein A macromolecular structure containing stably associated protein and nucleic acid. **(10)**

operator A gene controlling the expression of adjacent structural genes; when a repressor protein binds to the DNA sequence of the operator, transcription of the adjacent genes is blocked. **(14)**

operon A cluster of adjacent genes including both structural genes specifying proteins and control genes (e.g., operator, promoter) that affect the expression of the structural genes. **(14)**

organelle compartments of eucaryotic cells bound by one or two membranes, differing in content and structure, and containing enzymes for specialized biological functions: for example, nucleus, chloroplast, mitochondria, Golgi complex, endoplasmic reticulum. **(2)**

osmolar The sum of molar concentrations of all molecules and ions which, in dilute solutions, determines the osmotic pressure. **(7)**

osmosis The tendency for a solvent to move through a semipermeable membrane to a region of lower solvent chemical potential energy (concentration). **(7)**

osmotic balance Osmotic equilibrium. **(7)**

osmotic equilibrium A state where there is no net tendency for water to move across a semipermeable membrane because the chemical potential energy (concentration) of the water is equal on the two sides of the membrane. **(7)**

osmotic pressure The equilibrium hydrostatic pressure that must be applied to prevent the osmotic flow of water; estimated by the van't Hoff equation: $\Pi = RTC_s$. **(7)**

oxidant A chemical compound capable of accepting electrons from a second compound that becomes oxidized as a result. **(8)**

oxidation The removal of electrons (and often of accompanying protons as well) from a molecule, the reductant. **(8)**

phenotype The physical appearance or functional behavior of an organism, as distinguished from its genetic makeup or genotype. **(9)**

phospholipid A primary constituent of biological membranes, composed of a phosphate attached through an intermediate glycerol molecule to two fatty acids. **(3)**

plasma membrane A thin (8–10 nm) semipermeable barrier surrounding a cell's cytoplasm, composed of a lipid bilayer, with embedded and attached proteins. **(7)**

polymer A large molecule composed of similar or identical subunits that are joined together by repetition of one or a few kinds of covalent bonds. **(1)**

polysome Polyribosome; ribosomes arranged in rows or clusters complexed with a single messenger RNA molecule and involved in protein synthesis; may be associated with endoplasmic reticulum, the plasma membrane, or the outer membrane of the nuclear envelope, or may be free in the cytoplasmic matrix. **(13)**

primary structure Of a polymer macromolecule, the linear sequence of monomers (e.g., the amino acid sequence of a protein). **(4)**

primer A pre-existing structure to which additional units are added (e.g., the RNA primer to which deoxymononucleotides are added during DNA replication in bacteria). **(11)**

procaryote Cell having no nuclear membrane surrounding the genetic material and no membrane-bound organelles; bacteria and blue-green algae (cyanobacteria) comprise the procaryotes. **(1)**

processing The various alterations to the structure of a protein or nucleic acid that take place after its polymerization. **(13)**

promoter The sequence of DNA nucleotides to which RNA polymerase binds before beginning synthesis of RNA. **(12)**

quaternary structure The level of structural complexity in a protein that results from bringing together a small number of folded polypeptide chains. **(4)**

recessive Describing an allele whose effect is expressed in the phenotype only when it is present in the homozygous state, and, therefore, not masked by another (dominant) allele. **(9)**

recombinant DNA DNA sequences, from different sources, attached in the laboratory by enzymes; the methodology of creating, propagating, amplifying, and expressing such sequences. **(10)**

recombination The result of the process of crossing-over that occurs during meiosis and can result in offspring having different combinations of genes (i.e., new genotypes) than existed in either of the parents. **(9)**

reductant   A chemical compound capable of donating electrons to a second compound that becomes reduced as a result. **(8)**

reduction   The addition of electrons (and often of protons as well) to a molecule, the oxidant. **(8)**

renature   To reform the native, physiologically active structure after its having been unfolded as the result of, for example, heating. **(4)**

replication   Exact copying of a nucleic acid molecule in which one strand serves as a template for formation of another complementary strand. **(11)**

repressor   Protein that binds to a special sequence of DNA (the operator) and thereby inhibits transcription of nearby genes. **(14)**

resolution   A term used in optics to describe the distinguishing of two points or objects as separate entities. **(2)**

ribonucleic acid (RNA)   Polymer molecule of ribonucleotides, each composed of a nitrogenous base (purine or pyrimidine), the sugar ribose, and a phosphate group. **(12)**

ribosome   Small round particle (~20-nm diameter), consisting of RNA and protein, that is the protein synthesizing structure of both procaryotic and eucaryotic cells. **(13)**

sarcoplasmic reticulum   Network of membranes that occurs in muscle cells and surrounds the myofibrils. **(16)**

secondary structure   Regular arrangement of monomers in a linear polymer, for example, an α-helix of amino acids or base pairing of a polynucleotide. **(4)**

sedimentation coefficient/constant (S value)   The speed, measured in units of 10^{-13} seconds = 1 Svedberg unit = 1 S, with which a macromolecule moves in a centrifugal field; the higher the S, the faster the movement. **(4)**

segregation   One of the genetic principles first stated by Mendel, describing the separation of chromosome pairs, and thus of alleles, in meiosis by which arises the random array of gene combinations found in offspring. **(9)**

sequencing   Determining the order of monomers in a linear polymer, for example, of the nucleotides in a DNA strand. **(4)**

solute   A molecule or ion that is dissolved in a solvent to form a solution. **(7)**

steady state   A condition in a chemical reaction or a process in which some parameters (for example, concentrations) remain constant; distinguished from equilibrium by the requirement in the steady state for continued input of mass or energy, or both, to maintain the constancy. **(1)**

substrate   A molecule that is a reactant in an enzyme-catalyzed reaction. **(5)**

sucrose gradient   A solution in which buoyant density changes with position because the concentration of sucrose changes with position in the solution. **(4)**

surroundings   The environment of a system; the rest of the universe, or that part of it that exchanges mass and energy with a system. **(1)**

system   A collection of components that act together; a part of the universe that one is observing. **(1)**

template   A pre-existing structure whose surface guides the synthesis of another structure, for example, a parental DNA strand. **(11)**

tertiary structure   Folding in an irregular manner so as to associate parts of a polymer separated by many monomers in a complex three-dimensional form. **(4)**

tetrad   One of the four (tetra = four) strands (chromatids) of the structure arising during the first division of meiosis when the duplicated chromosomes form into pairs. **(9)**

thermodynamics, first law of   A physical principle stating that energy is always conserved; that it can be transformed but can be neither created nor destroyed. **(1)**

thermodynamics, second law of   A physical principle stating that during any spontaneous process the degree of disorder of the universe always increases or remains the same. **(1)**

transcription   The transfer of genetic information between nucleic acids of different types, especially from DNA to RNA, by using a strand of one as a template for synthesis of the other. **(12)**

transduce   To change from one form into another, as to transduce light energy into chemical energy; transduction is the change. **(1)**

transition state   The most unstable state in a reaction pathway, when bonds are being broken or made. **(5)**

translation   The transfer of genetic information from RNA to protein, during which the sequence of nucleotides in a messenger RNA specifies the (shorter) sequence of amino acids in a protein. **(13)**

triplet   Three contiguous bases in a nucleic acid, especially mRNA, that specify one amino acid during protein synthesis. **(13)**

tubules   Small hollow cylindrical structures, whether straight or curved; microtubules are a ubiquitous feature of eucaryote cells and are the result of polymerizing a tubulin monomer. **(2)**

ultracentrifuge   A machine that spins samples at high angular velocities, over 30,000 rpm (or 3,000 radians per second). **(4)**

ultraviolet radiation   Light energy with wavelengths less than 350 nm. **(3)**

vacuole A membrane-surrounded, often large, space for storage of food and other compounds in the cytoplasm of cells, most frequent in plant cells where a vacuole can comprise nearly the whole volume of a cell. **(2)**

vesicle A small membrane-surrounded space; the cells inside contain enzymes or secretory products, etc. **(2)**

virus A particle containing nucleic acid and protein that is metabolically inert outside a cell but that, inside a cell, is capable of using the cell's processes to replicate new virus particles. **(1)**

wildtype The form of a gene most commonly found in wild populations; also used to describe a standard gene form against which genetic changes in a trait are measured. **(9)**

x-ray diffraction The scattering of x-radiation by regular arrays of a molecule into regular patterns from which the structure of the molecule can sometimes be inferred. **(1)**

Appendix

The following is a reprint of the original journal article by Watson and Crick announcing their discovery of the structure of DNA. It is reprinted with permission from *Nature* (vol. 171, April 25, 1953, pp. 737–738).

Molecular Structure of Nucleic Acids
A Structure for Deoxyribose Nucleic Acid

We wish to suggest a structure for the salt of deoxyribose nucleic acid (D.N.A.). This structure has novel features which are of considerable biological interest.

A structure for nucleic acid has already been proposed by Pauling and Corey[1]. They kindly made their manuscript available to us in advance of publication. Their model consists of three intertwined chains, with the phosphates near the fibre axis, and the bases on the outside. In our opinion, this structure is unsatisfactory for two reasons: (1) We believe that the material which gives the X-ray diagrams is the salt, not the free acid. Without the acidic hydrogen atoms it is not clear what forces would hold the structure together, especially as the negatively charged phosphates near the axis will repel each other. (2) Some of the van der Waals distances appear to be too small.

Another three-chain structure has also been suggested by Fraser (in the press). In his model the phosphates are on the outside and the bases on the inside, linked together by hydrogen bonds. This structure as described is rather ill-defined, and for this reason we shall not comment on it.

We wish to put forward a radically different structure for the salt of deoxyribose nucleic acid. This structure has two helical chains each coiled round the same axis (see diagram). We have made the usual chemical assumptions, namely, that each chain consists of phosphate di-ester groups joining β-D-deoxyribofuranose residues with 3′,5′ linkages. The two chains (but not their bases) are related by a dyad perpendicular to the fibre axis. Both chains follow right-handed helices, but owing to the dyad the sequences of the atoms in the two chains run in opposite directions. Each chain loosely resembles Furberg's[2] model No. 1; that is, the bases are on the inside of the helix and the phosphates on the outside. The configuration of the sugar and the atoms near it is close to Furberg's 'standard configuration', the sugar being roughly perpendicular to the attached base. There is a residue on each chain every 3–4 A. in the z-direction. We have assumed an angle of 36° between adjacent residues in the same chain, so that the structure repeats after 10 residues on each chain, that is, after 34 A. The distance of a phosphorus atom from the fibre axis is 10 A. As the phosphates are on the outside, cations have easy access to them.

The structure is an open one, and its water content is rather high. At lower water contents we would expect the bases to tilt so that the structure could become more compact.

The novel feature of the structure is the manner in which the two chains are held together by the purine and pyrimidine bases. The planes of the bases are perpendicular to the fibre axis. They are joined together in pairs, a single base from one chain being hydrogen-bonded to a single base from the other chain, so that the two lie side by side with identical z-co-ordinates.

[1]Pauling, L., and Corey, R. B., *Nature,* **171,** 346 (1953); *Proc. U.S. Nat. Acad. Sci.,* **39,** 84 (1953).
[2]Furberg, S., *Acta Chem. Scand.,* **6,** 634 (1952).

One of the pair must be a purine and the other a pyrimidine for bonding to occur. The hydrogen bonds are made as follows: purine position 1 to pyrimidine position 1; purine position 6 to pyrimidine position 6.

If it is assumed that the bases only occur in the structure in the most plausible tautomeric forms (that is, with the keto rather than the enol configurations) it is found that only specific pairs of bases can bond together. These pairs are: adenine (purine) with thymine (pyrimidine), and guanine (purine) with cytosine (pyrimidine).

In other words, if an adenine forms one member of a pair, on either chain, then on these assumptions the other member must be thymine; similarly for guanine and cytosine. The sequence of bases on a single chain does not appear to be restricted in any way. However, if only specific pairs of bases can be formed, it follows that if the sequence of bases on one chain is given, then the sequence on the other chain is automatically determined.

It has been found experimentally[3,4] that the ratio of the amounts of adenine to thymine, and the ratio of guanine to cytosine, are always very close to unity for deoxyribose nucleic acid.

It is probably impossible to build this structure with a ribose sugar in place of the deoxyribose, as the extra oxygen atom would make too close a van der Waals contact.

The previously published X-ray data[5,6] on deoxyribose nucleic acid are insufficient for a rigorous test of our structure. So far as we can tell, it is roughly compatible with the experimental data, but it must be regarded as unproved until it has been checked against more exact results. Some of these are given in the following communications. We were not aware of the details of the results presented there when we devised our structure, which rests mainly though not entirely on published experimental data and stereochemical arguments.

It has not escaped our notice that the specific pairing we have postulated immediately suggests a possible copying mechanism for the genetic material.

Full details of the structure, including the conditions assumed in building it, together with a set of co-ordinates for the atoms, will be published elsewhere.

We are much indebted to Dr. Jerry Donohue for constant advice and criticism, especially on interatomic distances. We have also been stimulated by a knowledge of the general nature of the unpublished experimental results and ideas of Dr. M. H. F. Wilkins, Dr. R. E. Franklin and their co-workers at King's College, London. One of us (J. D. W.) has been aided by a fellowship from the National Foundation for Infantile Paralysis.

J. D. Watson
F. H. C. Crick

Medical Research Council Unit for the
Study of the Molecular Structure of
Biological Systems,
Cavendish Laboratory, Cambridge.
April 2.

[3]Chargaff, E., for references see Zamenhof, S., Brawerman, G., and Chargaff, E., *Biochim. et Biophys. Acta,* **9,** 402 (1952).
[4]Wyatt, G. R., *J. Gen. Physiol.,* **36,** 201 (1952).
[5]Astbury, W. T., Symp. Soc. Exp. Biol. **1,** Nucleic Acid, 66 (Camb. Univ. Press, 1947).
[6]Wilkins, M. H. F., and Randall, J. T., *Biochim. et Biophys. Acta,* **10,** 192 (1953).

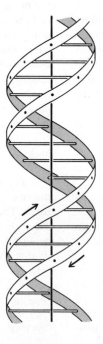

This figure is purely diagrammatic. The two ribbons symbolize the two phosphate—sugar chains, and the horizontal rods the pairs of bases holding the chains together. The vertical line marks the fibre axis.

Answers

Chapter 1

1. (a) The advice to Pat should be, "Don't waste your money; cell line X is surely a fraud." This is because they are claimed to grow in contradiction to the first law of thermodynamics; for 5000 units of food energy only 5000 units can be converted into other energy forms, not the 6000 units claimed as growth plus heat.

   (b) Don't invest; this cell line Y is not behaving according to the second law of thermodynamics. Even though energy is conserved (5000 units in = 5000 units out), every spontaneous process must result in an entropy increase which is mostly heat released to the environment. A claimed energy efficiency of 100% contradicts the second law of thermodynamics.

2. Prions can't be like other cells, which contain their genetic information in the linear structure of nucleic acid DNA.

3. The experiments of Jeons and Jeons (Jeons, K. W. and Jeons, M. S. (1976) Endosymbiosis in amoebae. *J. Cell Physiology*, 89:337–344.) show that it takes a relatively short time to evolve a dependence on symbiont bacteria. This model could explain the origins of mitochondria in a eucaryotic cell, but it does not help to understand how the eucaryotic cell nucleus came into being.

4. The artificial cell would be man-made, by definition, and thus not *directly* evolved. By the criterion of evolutionary history, it would seem not to meet the definition of life. But observation shows that it is alive, and further consideration of its "evolutionary"

history so as to include that of the humans who made it could lead to a different interpretation.

5. (a) The destination states will be CO_2 + 2 H_2O at room temperature and pressure in both cases. In the case of the internal combustion engine there will be a larger final volume, due to the outward movement of the piston. By proper choice of the piston size this increase in volume can be kept very small.

   (b) No, the heat flows are not the same. Since the origin and destination states are roughly the same the change in internal energy must be roughly the same, but in the second case work is done on the surroundings. Since $\Delta U = Q - W$ this means there is less heat flow to the surroundings.

   (c) This is consistent with the thermodynamic laws because neither the heat nor the work flow of energy is a variable of state; they depend on the path between states, as exemplified by these different burnings of methane.

6. (a) Since the system is at constant pressure and ΔH is positive (endothermic process), heat flows into the ice.

   (b) Since $\Delta G = \Delta H - T\Delta S$ and under the stated conditions ice melting is spontaneous, with $\Delta G < 0$, the sign of ΔS must be positive.

   (c) ΔG must be $+6.4 - 6.5 = -0.1$ units/mole.

   (d) Since ΔG is negative, ΔS must be positive and the system must have less organization as the result of melting. This is consistent with our expectation that liquid water has a greater disorganization than crystalline ice.

7. (a) The rubber band is cooler than the lips.

(b) Heat is flowing from the lips (higher temperature) to the rubber band (lower temperature).

(c) For the rubber band, ΔH (the heat flow at constant pressure) is positive (heat flowing into the rubber band).

(d) Since the contraction is spontaneous, ΔG must be negative.

(e) Since ΔG = ΔH − TΔS and since ΔH is positive, ΔS must be positive so −TΔS is sufficiently negative to give a negative ΔG.

8. The earth orbiting the sun is practically at equilibrium. The tendency of the earth to fly away from the sun is balanced by the gravitational attraction between the two bodies. An automobile moving down the highway requires a continuing input of energy to overcome wind resistance and friction between the tires and the road. It is therefore not at equilibrium and would be modeled more accurately by a steady state.

9. Mass requirements include the bicycle and the rider; energy must be supplied by the rider (unless the trip is all downhill) and information includes the ability to balance the bicycle during ordinary maneuvers.

10. The glucose molecule might be changed into another molecule by chemical processes in the receiving cell. In this case the flow would be of mass. Or the glucose molecule might be oxidized by the receiving cell to provide metabolic energy, in which case there was a flow of energy. Or the metabolic state of the synthesizing cell might be sensed as an increase in glucose concentration by the receiving cell, which could respond by performing some appropriate process; in this case there was a flow of information.

Chapter 2

1. Relevant experiments are those using brine shrimp eggs where it was found that cooling to −271°C (2 K) does not decrease their subsequent rate of hatching. Since it is likely that at this low temperature the rate of chemical processes is practically zero, the results show structure is a sufficient condition for ⋅ initiating life processes.

2. Through the microscope, the observer sees a cross section of the object, not a volume. If the observer pays attention to the area, then the eucaryotic cell will appear 100 times larger than the procaryotic cell. If the observer pays attention to the diameter, then the eucaryotic cell will appear only ten times larger.

3. The wavelength of blue light is shorter than that of red light. The relation between resolution and wavelength [1/R = d = (0.61 λ)/N.A.] shows that the shorter the wavelength the greater the resolution—

that is, the shorter the distance between two points that can still be distinguished as separate.

4. The tangent (tan) of the angle θ is X/Y, so X = Y tan θ.

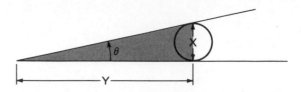

5. Fixation results in different degrees of dehydration of different cell constituents; this enhances differences in refractive index, allowing certain structures—for example, the nucleus—to stand out. By denaturing enzymes, fixation prevents self-digestion (autolysis) of cell contents. Finally, fixation cross-links macromolecules, thereby preserving their spatial relationships and the native cell structure.

6. The higher the voltage the greater the depth of focus that can be used. The greater the depth of focus the easier it is to see how structures are connected to one another along the direction of observation. By taking pairs of electron micrographs at appropriate angles and viewing the photographs separately with each eye, it is possible to see a three-dimensional image.

7. Even a 50-kV microscope has a theoretical resolving power of 0.005 nm. This is 100 times shorter than can be achieved with the quality of existing lenses and present techniques of achieving contrast. The even shorter theoretical resolving power of higher voltage microscopes is therefore of no practical significance.

8. Neutral red accumulates in the vacuole of living plant cells. It enters dead cells but is not accumulated, so that the intensity of the color in dead cells is no greater than that of the background solution. To show that accumulation is correlated with the living state, you can take samples of living onion epidermis stained with neutral red and subject them to heat, ethanol, metabolic poisons, etc. If the accumulated dye equilibrates after treatments thought to kill the cells, that would constitute evidence that the accumulation depends on the cells being alive.

9. A stain for the light microscope is a colored substance or dye which absorbs some portion(s) of the visible spectrum of light. A stain for the electron microscope is a substance containing heavy atoms—lead, osmium, silver, or platinum—which absorb electrons.

10. Plausible answers include the following:
(a) The eucaryote is good at maintaining constant its functions in the face of varying environmental conditions. This requires a more complex, therefore

larger, cellular machinery than necessary for the procaryotic cell, which tends to be found in a more constant environment (niche).

(b) The eucaryotic cell stores much larger amounts of DNA, requiring a special mechanism for replication and segregation into daughter cells. This and other special functions can only be accomplished in a larger cell.

Chapter 3

1. (a) Na^+, K^+; (b) Cl^-; (c) P (as phosphate: PO_4^{3-}) or S (as sulfate: SO_4^{2-}).

2. Carbon dioxide (O=C=O); acetic acid

$$\overset{\displaystyle O}{\overset{\displaystyle \|}{(CH_3C-OH)}}$$ or formaldehyde ($H_2C=O$); methyl chloride (CH_3Cl).

3. (a) There will be a partial positive charge on the carbon atom, because oxygen is electronegative and the bond between carbon and oxygen is, therefore, polar.

(b) There are no partial charges on either oxygen atom in O_2. Even though oxygen is electronegative, the structure is symmetrical and there is no polarity of the bond.

4. Covalent radius is associated with covalent bond distance and is roughly half the length of the bond. Van der Waals radius is associated with the distance between atoms in contact but not bonded together. It is longer than the covalent radius.

5. No; neither of the carbon atoms is bonded to four different groups.

6. H—O—H is 109°; O—H—O is strongest when 180°, but can vary.

7.

	[Na⁺]	[CH₃COOH + CH₃COO⁻]
1.	0.32 M	0.50 M
2.	0.50 M	0.78 M

8. There are four isomers altogether: cis-cis, cis-trans, trans-cis, and trans-trans.

9. According to the data in Table 3–12, and on page 114, the imidazole group (pK = 6–7), the phosphate group ($pK_2 = 6.7$) and the bicarbonate ion (HCO_3^-), with pK = 6.4.

10. Since the pH of the gel buffer is 2 pH units above the pK of the M_3 molecule (pH = $pK_3 + 2.0$), essentially all the M_3 molecules will be in the unprotonated (and therefore uncharged) form and will tend to stay at the origin, as is seen in drawing B. (Another way to look at the dissociation equilibrium is to say that M_3 molecules are in the uncharged form for most of the time and in the charged form for very little of the time.) Since the pH of the gel buffer is 2 pH units below the pK of the M_1 molecule (pH = $pK_1 - 2.0$),

essentially all the M_1 molecules will be in the protonated (and therefore positively charged) form and will tend to move towards the negative pole during the electrophoresis. The scale on drawing B shows they moved from the origin to position 1.0 during the experiment. This means that a fully positively charged molecule will move 1.0 unit, relative to an uncharged molecule. The problem then is to determine the average electrostatic charge on M_2 molecules. The numerical value of the average charge will be the same as the distance moved in the units shown (drawing B).

From the equilibrium relation, the dissociation constant $K_2 = ([M_2][H^+]/[M_2H^+])$, or $[M_2]/[M_2H^+] = K_2/[H^+]$. In the logarithmic form this can be expressed: log ($[M_2H^+]/[M_2] = pK_2 - pH = -0.5$. This can be solved for $[M_2H^+]/[M_2] = 10^{-0.5} = 0.316$. The average electrostatic charge on the M_2 molecules is the ratio of $[M_2H^+]$ to the total concentration of M_2 molecules, protonated or not, $[M_2H^+] + [M_2]$. This can be calculated as $[M_2H^+]/([M_2 H^+] + [M_2H^+]/0.316)$, or 0.24 units.

11. Since the pH of of the gel buffer is 2 pH units above the pK_3 of M_3, essentially all the M_3 molecules will be uncharged and will tend to stay at the origin, as seen in drawing B. Since the pH of the gel buffer is 2 pH units below the pK_1 of M_1, essentially all the M_1 molecules will have a positive charge and will tend to migrate toward the negative pole during electrophoresis, as seen in drawing B. Since M_2 molecules end up at position 0.7, this means they have an average positive charge $[M_2H^+]/([M_2H^+] + [M_2]) = 0.7$, from which it can be calculated that $0.7 [M_2] = 0.3 [M_2H^+]$. From the equilibrium equation, $[M_2]/[M_2H^+] = 0.3/0.7 = K_2/[H^+]$. Since the electrophoresis was carried out at pH 8.0, $[H^+] = 10^{-8}$. This allows one to calculate that $K_2 = 4.3 \times 10^{-9}$.

Chapter 4

1. (a) Asp, Glu, His, Lys, Arg. Residues from Cys and Tyr are protonated at pH 6, but the protonated forms are uncharged. The residue from His is only half protonated.

(b) negative.

(c) negative.

(d) His.

2. Asp (low hydrophobicity, two negative charges, a little more acidic than Glu)

Ser (hydrophilic because of —OH group)

Glu (two negative charges, a little more hydrophobic than Asp)

Ala (more hydrophobic than Ser)

Val (more hydrophobic than Ala)

Lys (two positive charges)
Arg (two positive charges, more basic than Lys)

3. At pH 3 the carboxyl groups of the side chains are uncharged. An α-helix can form by hydrogen bonding between the appropriate residues. At pH 5 these groups have a negative charge and repel one another, thereby opening up the helix into a random coil.

4. Congratulations; you have sequenced the A chain of insulin:
Gly-Ile-Val-Glu-Gln-Cys-Cys-Ala-Ser-Val-Cys-Ser-Leu-Tyr-Gln-Leu-Glu-Asn-Tyr-Cys-Asn.

5. (a) 3; (b) 3; (c) 1; (d) 4; (e) 4; (f) 7.

6. α-helix on left; antiparallel β-pleated sheet on right.

7. Outside: Arg, Asp, Gln, Glu, His, Lys, Ser, Thr, Tyr. Inside: Ile, Met, Val.

8. A dimer requires substitutions of hydrophobic for hydrophilic amino acids at the surface(s) where the two monomers contact one another.

9. It will take only 15 minutes. The molecule spun at the same speed would take twice as long (20S/10S), but spinning at twice the speed applies four times the force.

10. Asp-Arg-Val-Tyr-Ile-His-Pro-Phe-His-Leu-Leu-Val-Tyr-Ser.

11. Ser-Tyr-Ser-Met-Glu-His-Phe-Arg-Trp-Gly-Lys-Pro-Val.

12.

	Human	Dog	Rabbit	Cat
(a)	1	2	1	1
(b)	3	4	1	2
(c)	1	2	4	3
(d)	4	1	2	3
(e)	1	2	1	1
(f)	2	2	2	2

Chapter 5

1. (a) Under mild conditions activation energies are high and reactants do not have sufficient activation energy at room temperature.
(b) By using extremes of temperature, pH, pressure, etc., or using inorganic catalysts.
(c) Extreme conditions would denature cellular macromolecules, in particular proteins and nucleic acids.
(d) Protein enzymes catalyze cellular reactions.
(e) Amino acid side chains form a special surface structure on the enzyme to which only specific reactants can bind and be catalyzed to react.

2. If 10^{-3} moles urea are consumed per *minute* by a solution of 8 μg/ml urease, then 60 times that concentration (480 μg/ml) would consume 10^{-3} moles urea per *second*. For every mole of urea consumed per second there are 10^{-4} moles urease, which implies that 100 ml of 480 μg/ml urease contains 10^{-7} moles of urease, or that 1000 ml would contain 10^{-6} moles of urease. If 480 μg/*ml* is 10^{-6} molar, then 480×10^3 μg/*liter* is 10^{-6} molar, and the molecular weight is 480,000.

3. Let the rate constant for $E + P \rightarrow E \cdot S$ be k_4. The equation is then:
$$\frac{\Delta[ES]}{\Delta t} =$$
$$k_1 \cdot [E] \cdot [S] + k_4 \cdot [E] \cdot [P] - (k_2 + k_3) \cdot [ES]$$

4. The activated complex is a transitional form between substrate and product; it is also known as the transition state. The active site is a location on the enzyme to which substrate binds. The substrate must have the activation energy to attain the activated complex state, a task made easier by the action of the enzyme, since it lowers the energy of activation. The enzyme too may have a transition state structure, one that while difficult to detect is different from that found when substrate, product, or reaction intermediate is bound.

5. Horse liver; since it has the lowest K_M of the three enzymes listed and since 10^{-3} M is approximately $2 \times K_M$.

6. True; at [S] much less than K_M, v_i is proportional to [S] (because $v_i \approx V_{max} [S]/K_M$ when $K_M + [S] \approx K_M$). False; when $K_M \approx [S]$, tripling [S] increases v_i much less than a factor of three.

7. The K_M is the same as before treatment with DFP, 4×10^{-5} M. DFP is an irreversible inhibitor that forms stoichiometrically a covalent compound with the enzyme. Survival of enzyme activity, detected as nonzero V_{max} for the preparation, means that there was insufficient DFP to inactivate all the enzyme. The surviving enzyme, not having reacted with DFP, is completely normal.

8. The ratio of K_M^{app}/K_M is $2.7/1.5 = 1.8 = 1 + [I]/K_I$. By algebra, $K_I = 2.5 \times 10^{-3}$ M.

9. The simplest way to approach this problem is to make a double reciprocal (Lineweaver-Burk) plot. K_M in the absence of inhibitor **U** is 1×10^{-3} M and V_{max} is 100 units. In the presence of **U,** the apparent K_M is 0.5×10^{-3} M and V_{max} is 50 units. Since the V_{max} is different in the presence of **U,** it cannot be a competitive inhibitor. It is in fact an *uncompetitive* inhibitor, one which binds only to the E·S complex.

10. The data can be plotted by the method of Lineweaver and Burk or the value of V_{max} can be read from the table of values of initial velocity: 20,000 min^{-1} for both pH 8.5 and 7.5. The Michaelis constant is the concentration at half-maximal velocity; this will be approximately 5×10^{-5} M for pH 8.5, and approximately 10×10^{-5} M for pH 7.5. The enzyme has lower affinity for substrate (higher Michaelis constant) at the lower pH. One hypothesis is that substrate binding depends on a weak acid group in the active site, one that is neutral or negative at pH 8.5 (the higher affinity condition). This group be-

comes positive or neutral as a proton associates with it at the lower pH (the lower affinity condition). There are two weak acid groups that might have a pK_A in this range of pH: the imidazole of histidine and the terminal amino group of a lysine residue or of the peptide chain. If it is a histidine, it might be modified by an irreversible enzyme inhibitor with a structure like TPCK, a tosyl chloromethyl ketone. A radioactive "T?CK" could be attached to the enzyme and the modified histidine residue detected after fingerprinting. Other modifying reagents could be used to detect the amino group. Each such reagent should be arranged to have a structure like that of the normal substrate, to make more likely the modification of residues in the active site.

Chapter 6

1. (a) FALSE. The hydrolysis of proteins is spontaneous; hydrolysis is not the reversal of synthesis, but a different pathway.

(b) FALSE. The enzyme that catalyzes the hydrolysis of phosphate esters is a phosphatase. Phosphorylase enzyme catalyzes the addition of phosphate to a polymer, to yield a phosphorylated monomer and a shorter polymer.

(c) FALSE. If the storage of energy were 100%, there would be no driving force to make the storage pathway go forward, since it would be at equilibrium.

(d) TRUE. Shikimate is phosphorylated by ATP, before condensing with PEP.

(e) FALSE. Arsenate will allow the formation of 3PG without using ADP, but arsenate does not affect the coupling of PEP → PYR with ADP → ATP.

(f) FALSE. The only reactions cellular molecules undergo in a cell are those catalyzed by enzymes. Metabolic pathways *exist* because some cellular molecules can undergo only one reaction. As another example, there are no enzymes to catalyze the simple hydrolysis of ATP.

2. (a) Yes. The reaction proceeds despite a positive ΔG because the reactants and products are not at standard concentrations. It will proceed to equilibrium, where the ratio of $[G-1-P]/[G-6-P] = K_{eq} = \exp (-\Delta G'/RT)$. Since $RT = 2.5$ kJ/mole, $K_{eq} = 0.08$. By conservation of mass of glucose-phosphates, $[G-6-P] + [G-1-P] = 0.10$ M. Solving these two equations yields $[G-6-P] = 0.092$ M, $[G-1-P] = 0.008$ M.

(b) The product of the phosphoglucomutase reaction will be produced if $[G-1-P]$ is kept low, for example by being used in another reaction.

3. $\Delta G = 0 = \Delta G' + RT \ln (10^{-6} \text{[sucrose]}/10^{-9})$.
$\Delta G' = -7 = -RT \ln (10^3 \text{[sucrose]})$.
$\text{[sucrose]} = 10^{-3} \exp (7/RT) = 16$ mM.

4. (a)

						$\Delta G'$ (kJ/mole)
MU—C—K	+	H_2O	$\rightleftharpoons$	MU—C	+ K	−44
MU—C	+	H_2O	$\rightleftharpoons$	MU	+ C	−5
C	+	K	$\rightleftharpoons$	C—K	+ H_2O	+25
MU—C—K	+	H_2O	$\rightleftharpoons$	MU	+ CK	−24

(b)

					$\Delta G'$ (kJ/mole)
A	+ MU—C—K	$\rightleftharpoons$	A—K + MU—C		−15
A—K	+ B	$\rightleftharpoons$	D + K		−8
E	+ MU—C—K	$\rightleftharpoons$	E—K + MU—C		−3
E—K	+ D	$\rightleftharpoons$	F + K		−5
A + B + E + 2 MU—C—K	$\rightleftharpoons$	F + 2 MU—C + 2 K			−31

5. (a) Combining reactions one and three, $\Delta G' = -14$ kJ/mole. Combining reactions two and three, $\Delta G' = -5$ kJ/mole.

(b) PC + ADP $\rightleftharpoons$ C + ATP; $\Delta G' = -14$ kJ/mole. This reaction will go to equilibrium when $\Delta G = 0$. $\Delta G = \Delta G' + RT \ln ([C][ATP]/[PC][ADP])$, but if $[PC]/[C] = 1$, $\Delta G = \Delta G' + RT \ln ([ATP]/[ADP])$. At equilibrium $[ATP]/[ADP] = \exp (-\Delta G'/RT) = 240$.

6. Remember that $\Delta G = \Delta G' + RT \ln ([B]/[A])$ and that $\Delta G' = -RT \ln K_{eq} = -RT \ln ([B]^{eq}/[A]^{eq})$.

7. (a) Yes, there is an ADP effect (slowing of the pathway on starving for ADP), because ADP is used as a co-substrate in the phosphorylation reactions.

(b) Neither pathway consumes NAD^+; each regenerates the NAD^+ used for oxidation during the later stages of the pathway.

(c) Equal amounts of ATP are generated by each pathway, two per precursor (U or glucose), although the BURP pathway involves two oxidation/reductions.

(d) The ratio of glucose flow to PEP flow is 1/2, since there are two PEP molecules produced from each glucose during lactate fermentation. The ratio of U flow to P-Y flow is 1, since each U gives rise to only one P-Y.

(e) There can be no ^{14}C labeling of NADH by the pathway; only electrons and protons are transferred to NAD^+, not carbon atoms.

8. (a) $\Delta G' = -RT \ln K_{eq} = -RT \ln [G6P]_{eq}/[F6P]_{eq} = -1.7$ kJ/mole (Note the reversal of sign from the

table, caused by the reversal of precursor and product.) $K_{eq} = \exp (+1.7/RT) = \exp (1.7/2.5) = 2.0$

(b) First, a pathway from fructose to glucose must be given:

fructose + $P_i \rightleftharpoons$ F–6–P $\rightleftharpoons$ G–6–P $\rightleftharpoons$ glucose + Pi
$\Delta G' = +15.9 - 1.7 - 13.8 = 0.4$ kJ/mole.

$K_{eq} = [gluc]_{eq}/[fruc]_{eq} = \exp (-0.4/RT) = 1.17$

(c) Yes, the reactions are coupled because sorbitol is a common intermediate, and because NADPH and NADH can be interconverted by an appropriate set of reactions.

(d) The desired pathway is one involving reactions 1' and 2', but with $\Delta G' << 0$ (assuming $\Delta G'$ is a rough approximation to the ΔG under cellular conditions). This might be done using ATP as the energy source, a likely guess since ATP is a participant in the reactions listed in the table:

	$\Delta G'$ (kJ/mole)
ATP + H_2O + $NADP^+$ + NADH $\rightleftharpoons$ NADPH + NAD^+ + ADP_i	−31.3
$\left.\begin{array}{l} \text{glucose + NADPH + } H^+ \rightleftharpoons \\ \text{sorbitol + } NADP^+ \\ \text{sorbitol + } NAD^+ \rightleftharpoons \\ \text{fructose + NADH + } H^+ \end{array}\right\}$	0.4★
ATP + H_2O + glucose $\rightleftharpoons$ ADP + P_i + fructose	−30.9

Note the common intermediates: $NADP^+$, NADH, NADPH, NAD^+, and sorbitol.

★The 0.4 kJ/mole indicated is for the sum of reactions 1' and 2', which has the same sum as the last four reactions listed in the table, run in the appropriate directions.

Chapter 7

1. (a) TRUE. At equilibrium the free energy required to move galactose inside the cell is RT ln 10^5 = 29.2 kJ/mole, which is less than the 33 kJ/mole available from hydrolyzing ATP.

(b) FALSE. Raising the osmotic pressure inside a cell is accomplished by increasing intracellular solutes, which is accompanied by water moving *into* the cell.

(c) TRUE. A solution in which a cell loses volume and remains smaller is one that is hyperosmolar and whose solutes do not penetrate the cell membrane.

(d) TRUE. Since cells generally have a high permeability to K^+ and since the Donnan potential is an equilibrium potential, all the permeable ions are already at equilibrium and increasing K^+ permeability cannot change the Donnan potential.

(e) TRUE. The resting potential of a nerve cell is close to the potassium equilibrium potential. Adding valinomycin and thereby increasing the permeability of the nerve cell membrane to potassium would bring the resting potential closer to the potassium equilibrium potential, a more negative value than the already negative resting potential.

(f) FALSE. Membrane phospholipids can diffuse rapidly within the lipid bilayer, but cannot flip-flop rapidly from one side of the bilayer to the other.

(g) TRUE. Both the outer surface of a vesicle and the inner surface of the plasma membrane face the cytoplasm.

(h) FALSE. Reducing the conductance of sodium ion will tend to bring the resting potential closer to the potassium equilibrium potential, making the resting potential more negative.

(i) FALSE. If glycerol can permeate the cell's membrane, it will enter the cell and water will also enter to maintain osmotic balance. This can occur even if the cell's interior concentration is 0.3 M osmolar.

(j) FALSE. A coated pit is a region of the cell membrane, underlying which is a group of clathrin molecules—which gives it the "coated" appearance—and within which are receptors for some extracellular ligand.

2. (I). Found in total membrane preparation even after washing with salt, so probably an integral membrane protein. Labeled by both probes, but more by permeable probe ethyl acetimidate, so exposed on both sides of the membrane, spanning the bilayer.

(II). Not labeled with the impermeable probe isoethionyl acetimidate, but labeled with the permeable probe, so exposed on the inner surface of the bilayer.

(III). Washed off by salt, so probably an extrinsic membrane protein exposed only on the outer surface of the bilayer, since not labeled more by permeable than by impermeable probe.

(IV). Not washed off by salt, so probably an integral membrane protein facing outside only, since not labeled more by permeable than by impermeable probe.

3. Urea has a very low permeability coefficient and has a relatively low solubility in oil, compared with water. Methyl urea and dimethyl urea have higher oil solubility and have higher permeability coefficients. These data are consistent with a solubility-diffusion mechanism of transport through a membrane whose primary permeability barrier is a lipid layer (bilayer). The L-pentose has a low permeability coefficient and a low solubility in oil, consistent with the same mechanism for permeating the cell membrane. The D-pentose, on the other hand, has a very much

higher permeability coefficient (and, of course, the same partition coefficient as for its L-isomer), which means it must have a different, presumably enzyme-mediated, permeation pathway through the membrane.

4. In the following, S = [S], the substrate concentration; subscripts "in" and "out" refer to the side of the membrane.

$$J_{net} = J_{in} - J_{out} = \frac{J_{max} S_{out}}{K + S_{out}} - \frac{J_{max} S_{in}}{K + S_{in}}$$

Very far from saturation with substrate means $K + S \approx K$. Under these conditions,

$$J_{net} = \frac{J_{max} K[S_{out} - S_{in}]}{K^2} = -\frac{J_{max}}{K}[S_{in} - S_{out}]$$

$$J_{net} = -\frac{J_{max}}{K} \Delta C_S$$

5. By passive diffusion through the lysosomal membrane, cholesterol being soluble in the lipid bilayer.

6. (a) 5.0 mM.

(b) 30 mM. Calculate the result by assuming the cotransport system to be at equilibrium:

$$\Delta G = 0 = RT \ln (S_i Na_i)/(S_o Na_o) + zF\Delta E.$$

Since Na^+ is the only charged species, z = 1. This equation can be reexpressed

$$RT \ln S_i/S_o = +RT \ln Na_o/Na_i - F\Delta E$$

or

$$\ln S_i/S_o = \ln Na_o/Na_i - F/RT\ \Delta E$$

F = 92 kJ/V; RT = 2.54 kJ; ln S_i/S_o = 4.1; S_i/S_o = 60; and S_i = 30 mM.

(c) No. Since the net charge of alanine under physiological conditions is neutral the electrical potential will affect only the Na^+ transport: z = +1.

(d) Yes. Since arginine has a net positive charge under physiological conditions its transport inward is favored by the negative membrane potential; z = +2 and the $F/RT\Delta E$ term will contribute 3.6 kJ/mole. This gives ln S_i/S_o = 5.9, S_i/S_o = 360, and S_i = 180 mM.

(e) 0.02 M. The easiest way to see this from the graph is to determine the x-intercept = $-1/K_M$, which is the same for all values of [S].

(f) The inward cotransport rate appears to be saturating at [S] = 500 mM, so J_{max} (1/1000) for that value of [S] must be the limiting maximum velocity for saturating S and Na^+. The value of [S] for half that value of J_{max} (1/500) is [S] = 10 mM.

7. Peptide #1 lies outside the lipid bilayer with its carbohydrate and at least one tyrosine residue exposed to the extracellular medium; this accounts for the labeling with lactoperoxidase/iodine and digestion with neuraminidase. Peptide #3 lies inside the cell with its tyrosine(s) sufficiently exposed to react with lactoperoxidase, since it can be labeled with radioactive iodine if the membranes are disrupted. Since peptide #2 cannot be labeled, it probably lies with its tyrosine(s) within the lipid bilayer. Since peptide #1 is outside and peptide #3 is inside, the glycoprotein spans the bilayer, peptide #2 probably being the connecting link.

8. (a) Yes. Since $[Cl^-]$ is only 0.06 mM, there must be other negatively charged ions to neutralize the 0.16 mM positive ions. These might be bicarbonate, carbonate, sulfate, etc.

(b) There is 240 mM solute contributed by K^+, Na^+, and Cl^- ions; there must be 60 mM solute to explain how 300 mM sucrose is isotonic. Since there are 130 mM positive ions and only 110 mM negative ions, the protein (or other species) must account for 20 mM net negative charge and 60 mM net solute. The relatively high osmotic pressure inside *Chara* is balanced by an equivalent hydrostatic pressure. The alga can withstand a high hydrostatic pressure inside because it has a mechanically stable cell wall.

(c) K^+ has, to two significant figures, an equilibrium potential of −190 mV. This is calculated by solving $\Delta G = 0 = RT \ln [K^+_{in}]/[K^+_{out}] + F\Delta E^{eq}$. Na^+ has an equilibrium potential of −170 mV. Cl^- has an equilibrium potential of +200 mV (remember that z = −1 for Cl^-).

(d) The equilibrium potential for K^+ is −190 mV, which is very close to the observed resting potential of −180 mV. Since the data are only given to two significant figures, at best, the difference between the resting potential and the equilibrium potential may be due to experimental error. One cannot, therefore, rule out the equilibrium hypothesis on the basis of the K^+ data.

(e) The equilibrium potential for Na^+ is −170 mV, again too close to the resting potential to rule out the hypothesis.

(f) The equilibrium potential for Cl^- of +200 mV is very different from the resting potential; they even have different signs. These data are sufficient to rule out the equilibrium hypothesis to explain the resting potential.

(g) An alternative to equilibrium is a steady state with equal net fluxes inward of positive and negative ions. Since the two major positive ions flow inward with $\Delta G < 0$, there must be a pump to explain their outward movement—their concentrations inside the cell remain constant so net K^+ flow and net Na^+ flow must be zero. A pump is also necessary to explain the inward flow of Cl^-, which would otherwise be nonspontaneous; passive flow outward will keep [Cl$^-$] constant inside the cell. The steady-state condition for the resting potential can be modeled by an equation analogous to the one discussed in the text for the squid axon: $I_{Na} + I_K = I_{Cl}$; this implies no net increase in positive charges inside. The equation analogous to 7-L for Cl$^-$ is $I_{Cl}/g_{Cl} = E_m - E_{Cl}$, because $z_{Cl} = -1$. The final equation for the resting potential, analogous to 7-M, is

$$E_m = \frac{g_{Na}}{g_{Na} + g_K + g_{Cl}} E_{Na} + \frac{g_K}{g_{Na} + g_K + g_{Cl}} E_K + \frac{g_{Cl}}{g_{Na} + g_K + g_{Cl}} E_{Cl}$$

(h) Given the resting potential of -180 mV the conductances for the positive ions, with their negative equilibrium potentials, must be much greater than for Cl$^-$, since its equilibrium potential is positive. If g_{Cl} is very much smaller than g_K and g_{Na}, and if $g_K = g_{Na}$, the resting potential would be midway between E_K and E_{Na}, which is the value observed. If g_{Cl} is not so small, the contribution from E_K (-190 mV) must be larger, to explain a resting potential of -180 mV, and g_K in that case would have to be larger than g_{Na}.

(i) Since Cl$^-$ is far from equilibrium, an action potential could begin with opening a Cl$^-$ channel. This would allow Cl$^-$ to leave and result in a depolarization of E_m, making it more positive. Analogous to the Na$^+$ channel in the squid axon, the Cl$^-$ channel would have to be inactivated and shut off, to allow recovery of the resting potential. This could occur by opening voltage-sensitive channels for either K$^+$ or Na$^+$, or both; positive ions would flow out and tend to repolarize the membrane. An alternative hypothesis to explain membrane depolarization would be activated pumping inward of positive ions. An experiment to test the first hypothesis would be to clamp the membrane to zero volts and measure the current flow in the presence of agents that block the movement of K$^+$ (tetraethylammonium?) and Na$^+$ (? some toxin or positive ion). A large observed outward negative ion flow (positive current inward) would be consistent with Cl$^-$ efflux and not with pumping positive ions inward. If this flux could be blocked by analogs of Cl$^-$, that would be persuasive data favoring the first hypothesis.

Chapter 8

1. (a) FALSE. Each photon absorbed is of equal effect in photosynthesis and both blue and yellow photons can be absorbed by either photosystem in a green alga.

(b) FALSE. Since the symport brings into the matrix both a positive and a negative charge there will be no driving force by ΔE, nor will ΔE be perturbed by the symport.

(c) FALSE. $\Delta G' = z F \Delta \epsilon' = (-2)(96.5)(-0.16) = +31$ kJ/mole, nonspontaneous.

(d) FALSE. Ascorbate/phenazine methosulfate adds electrons after the rotenone block, but oligomycin blocks the F_1 ATPase and this will cause a buildup of the proton gradient, ultimately blocking electron transfer at coupling site III, which lies between the source of the electrons and O_2.

(e) FALSE. Dinitrophenol will uncouple electron flow from generating a proton gradient, but this will speed up electron flow and increase the production of oxygen.

(f) FALSE. $\Delta G = \Delta G' + RT \ln ([Fe^{2+}]/[Fe^{3+}])$, so ΔG will be higher (more positive) if $[Fe^{2+}]/[Fe^{3+}] > 1$.

$$\Delta \epsilon = \Delta G/zF = \Delta \epsilon' + (RT/zF) \ln (Fe^{2+}]/[Fe^{3+}]$$

but for electrons, $z = -1$, so $\Delta \epsilon$ gets less positive.

2. (a) FALSE. The electrons in butyrate (at $+0.25$ V) are sufficiently reduced to flow directly into the ETC feeding PS I, without an additional boost by a second photosystem.

(b) FALSE. If butyrate is the electron donor, there will be no oxidation of H_2O and no production of O_2.

(c) TRUE. Some oxygen from CO_2 ends up in glucose.

(d) TRUE. Some oxygen from CO_2 ends up in water.

(e) FALSE (probably). Electrons transferred from butyrate ($+0.25$ V) to oxidized PS I ($\approx +0.4$ V) during noncyclic electron flow will not have sufficient energy to produce ATP: $\Delta \epsilon' \approx +0.15$ V; for two electrons, $\Delta G' = -29$ kJ/mole, too little for ATP synthesis ($\Delta G' = 31$ kJ/mole). At some concentrations of reactants and products there might be sufficient energy for ATP synthesis, but it is more likely that ATP is generated by cyclic photophosphorylation.

3. (a) From reading the graph, 4000 cpm.

(b) TPP$_m^+$, the *amount* of TPP$^+$ inside the mitochondria.

(c) 10 mM, since mitochondrial respiration is blocked and the proton gradient would be zero.

(d) From the specific radioactivity (2000 cpm/μmole) one can calculate the amount of TPP$^+$ inside as 2 μmoles. From the likely matrix concentration, one can calculate the volume represented by the matrix space of the mitochondria in the suspension: 2 μmoles/V_m = 10 mM (= 10^4 μM); V_m = 2/10^4 liter = 2×10^{-4} liter = 20 μl.

(e) From the graph, 40,000 cpm. The higher amount

of TPP$^+$ is presumably due to its being concentrated in the relatively negative matrix during respiration.
(f) TPP$_m^+$ is 40,000/2000 μmoles = 20 μmoles. The volume of the matrix is assumed to be the same as calculated above. This means [TPP$^+$]$_m$ = (20 × 10^{-6})/(2 × 10^{-4}) = 100 mM, which is 100 times the concentration found in the absence of respiration.
(g) Assuming that [TPP$^+$]$_m$/[TPP$^+$]$_c$ is in equilibrium with ΔE, ΔG = 0 and ΔE = (E$_m$ − E$_c$) = −(RT/F) ln 100 = −120 mV.

4. (a) $\Delta G = \Delta G'_{ATP} + RT \ln ([ADP][P_i]/[ATP]) + RT \ln [GP]_i^2/[GP]_o^2$. Maximum [GP]$_i$ will occur when ΔG = 0: [GP]$_i$/[GP]$_o$ = 1.3 × 10^3; [GP]$_i$ = 13 mM.

(b)

$$NADH + H^+ \rightleftharpoons$$
$$NAD^+ + 2 H^+ + 2 e^- \qquad \epsilon' = 0.32 \text{ V}$$
$$NO_3^- + 2 H^+ + 2 e^- \rightleftharpoons$$
$$NO_2^- + H_2O \qquad \epsilon' = 0.42 \text{ V}$$

$$NADH + H^+ + NO_3^- \rightleftharpoons$$
$$NAD^+ + NO_2^- + H_2O \qquad \Delta\epsilon' = 0.74 \text{ V}$$

$\Delta G' = z F \Delta\epsilon' = (-2)(96.5)(0.74)$ kJ/mole = −140 kJ/mole.
(c) (1) Grows; maximum growth from NADH produced in Krebs cycle and flow of electrons to O$_2$ → 2 ATP per NADH oxidized.
(c) (2) Does not grow; GP cannot be fermented since more NADH is produced from NAD$^+$ by transforming GP to GAP and during glycolysis than can be oxidized back to NAD$^+$ by reducing the pyruvate formed.
(c) (3) grows; electrons can be transferred from NADH to NO$_3^-$; less ATP/NADH, however, since electrons are donated to NO$_3^-$ from cyt *b*, which bypasses the last coupling site.
(c) (4) grows; HCN blocks reduction of cyt *a*/*a*$_3$ downstream from cyt *b*, so anaerobic respiration is resistant to HCN.

5. (a) No, because there is no net transfer of electrical charge when 2 Na$^+$ move out and 2 K$^+$ move in: z = 0.
(b) There is apparently a symport, independent of the ouabain-inhibitable Na/K-ATPase, by which H$^+$ and K$^+$ move inward together.
(c) Apparently there is also an antiport which links the movement of H$^+$ and Na$^+$. The transport is demonstrably reversible: in the presence of high internal and low external [Na$^+$], H$^+$ flows out of the cell. The antiport is independent of both the Na/K-ATPase and respiration.
(d) K$^+$ and H$^+$ can equilibrate via their symport pathway: K$_{out}^+$ + H$_{out}^+$ $\rightleftharpoons$ K$_{in}^+$ + H$_{in}^+$. For this

coupled process, ΔG = R T ln ([K$^+$]$_{in}$[H$^+$]$_{in}$)/ ([K$^+$]$_{out}$ [H$^+$]$_{out}$) = R T ln (80/2.5) (10^{-8}/10$^{-6.5}$). 80/2.5 = 32 and 10$^{-1.5}$ = 1/31.6; their product (1.01) is too close to 1.0 to be distinguished, from the precision of the reported data, so these data cannot rule out an equilibrium between K$^+$ and H$^+$. Na$^+$ and H$^+$ can equilibrate via their antiport pathway: Na$_{in}^+$ + H$_{out}^+$ $\rightleftharpoons$ Na$_{out}^+$ + H$_{in}^+$. For this coupled process, ΔG = R T ln ([Na$^+$]$_{out}$[H$^+$]$_{in}$)/ ([Na$^+$]$_{in}$ [H$^+$]$_{out}$) = R T ln (100/3.2)/(10$^{-1.5}$). 100/ 3.2 = 31.2, which when multiplied by 1/31.6 is too close to 1.0 to be distinguished, with the precision of the reported data; again these data cannot rule out an equilibrium between Na$^+$ and H$^+$.
(e) Respiration will move protons out of the cell. The K$^+$/H$^+$ symport will move K$^+$ into the cell and the Na$^+$/H$^+$ antiport will move Na$^+$ out of the cell. When the distribution of K$^+$ and Na$^+$ is sufficiently asymmetric, the Na/K-ATPase pump will be driven backwards, synthesizing ATP. The energy flow is: respiration causing proton flow, whose reversal causes Na$^+$ and K$^+$ flow, which flow backwards through the ouabain-inhibitable pump, synthesizing ATP.
(f) The pump pathway is

$$2 \text{ Na}_{in}^+ + 2 \text{ K}_{out}^+ + ATP + H_2O \rightleftharpoons 2 \text{ Na}_{out}^+ + 2 \text{ K}_{in}^+ + ADP + P_i.$$

$\Delta G = RT \ln ([Na^+]_{out}^2[K^+]_{in}^2 [ADP] [P_i] / [Na^+]_{out}^2 [K^+]_{in}^2 [ATP])$.
If this transport were at equilibrium, ΔG = 0 and [ATP]/[ADP] = 2.0 × 10^4. (Remember that [P$_i$]/ [P$_i$]' = 0.02.)

6. (a,b,c) NADH → d —**CS**→ c —**R**→ b —**Q**→ a —**CS**→ oxygen. The coupling sites are indicated as **CS**. Both sites couple two ATP per pair of electrons.
(d) R T ln ([ADP][P$_i$]/[ATP]) is, assuming [ADP] = [P$_i$], approximately 11 kJ/mole and ΔG for ATP hydrolysis is therefore approximately −42 kJ/mole. There are two ATPs synthesized per pair of electrons per site, so there must be a $\Delta\epsilon$ sufficient to account for approximately −42 kJ/mole per electron, approximately 0.43 volts.

Chapter 9

1. (a,b)

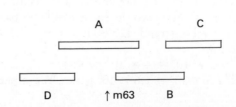

(c) Yes, if it lies to the left of D or to the right of C.

(d) No, since it cannot simultaneously lie within all four deleted regions.

2. R/Y × B/Y → R/Y, R/B, B/Y, Y/Y = orange : purple : green : yellow = 1 : 1 : 1 : 1.

3. The stubby allele is sex-linked dominant.

4. The sequence of the genes is: A C D B.

5. (a) 50%; (b) 25%.

6. (a) The order of genes is A C B.

(b) The back cross will be AaBb × aabb and the progeny:

$$\frac{A\ \ b}{a\ \ b} = 40\%; \quad \frac{a\ \ B}{a\ \ b} = 40\%; \quad \frac{A\ \ B}{a\ \ b} = 10\%;$$

$$\frac{a\ \ b}{a\ \ b} = 10\%$$

7. (a) Purple dominant to green: cut dominant to potato.

(b) (1) $\dfrac{\text{purple}\ \ \text{cut}}{\text{green}\ \ \text{pot}} \times \dfrac{\text{green}\ \ \text{cut}}{\text{green}\ \ \text{pot}}$

(2) $\dfrac{\text{purple}\ \text{cut}}{\text{green}\ \text{pot}} \times \dfrac{\text{purple}\ \ \text{pot}}{\text{green}\ \ \text{pot}}$

8. C.

9. A.

10. E.

11. D.

12. A.

13. E.

14. Yes; 20 centimorgans: (120/600 × 100) = 20.

15. A.

16. A and E.

Chapter 10

1. Fragment 1, because it has a higher GC content.

2. (a) X = 0.5

(b) These values differ with the particular DNA being analyzed and they can therefore be used to characterize the DNA.

3. Treat with a nuclease specific for single-strand DNA, or use density gradient centrifugation to separate the double-strand DNA, which is more dense, from the homologous single-strand DNA.

4. From Table 10–2 you can calculate that the length of DNA is 0.52 μm per million daltons, so the viral DNA molecule of 4×10^6 daltons would be 2.08 μm long, if it were double-strand DNA, approximately twice that long if it were single-strand. Also from Table 10–2, the number of base pairs per million daltons of DNA is about 1500, so that a molecule of 4×10^6 daltons in mass would consist of 6000 base pairs (6 kb) if double-strand, 12,000 base pairs (12 kb) if single-strand.

5. (a) $2 \times 10^9 / 5 \times 10^5 = 4 \times 10^3$ genes.

(b) $8 \times 10^{10} / 5 \times 10^5 = 1.6 \times 10^5$ genes. We expect these numbers to represent upper limits because we know from many different kinds of evidence that only 10% of mammalian DNA codes for genes. The result is that a human genome has only 16,000 genes, not such a very large number, which is surprising when one considers the complexity of some of its products—for example, the human brain with its intricate network of at least 10^9 neurons.

6.

1.1	1.5	3.2	3.9	4.1

7. (a) G C A T T C G T A G C A.

(b) C G T A A G C A T C G T.

(c) The primer remains attached to the nucleotides which have been synthesized.

(d) The primer DNA is not labeled with ^{32}P and therefore does not show up on the photographic film.

8.

4	1.5	5	3
C	A	D	B

9. If A + T = 0.66, then A = T = 0.33 and G = C = 0.17, where the letters A, T, G, C symbolize the fraction of the DNA bases of the designated type and also the probability that a given base will be of the designated type. The probability of finding a series of bases is the product of the probabilities of finding the individual bases in the series.

(a) $(0.17)^4 \cdot (0.33)^2 = 9.09 \times 10^{-5}$

(b) $(0.33)^4 \cdot (0.17)^2 = 3.43 \times 10^{-4}$

(c) $(0.17)^6 = 2.41 \times 10^{-5}$

(d) $(0.17)^4 = 8.35 \times 10^{-4}$

10. A smaller number of colonies should suffice in the case of the six-base cutter, because the cloned fragments will be larger.

11. If $(A+T)/(G+C) = 0.39$, then 2A/2G = 0.39 = A/G; it is also true that A + T + G + C = 1.0 = 2A + 2G. Solving by algebra shows A = T = 0.14 and G = C = 0.36. The length of DNA in which on the average one will find a given base sequence is approximately the reciprocal of the probability of finding the given sequence in a random sequence of bases with the given composition. For Taq I, P = $(0.14)^2 \cdot (0.36)^2 = 2.54 \times 10^{-3}$, so the average length will be approximately 394 base pairs (.39 kb). For Eco RI, P = $(0.14)^4 \cdot (0.36)^2 = 4.97 \times 10^{-5}$, so the average length will be approximately 20121 base pairs (20 kb).

12. The ends of fragments will stick together if the restriction enzyme makes staggered cuts.

13. In the former transgenic mice, the gene was incorporated into the germ line cells.

Chapter 11

1. This fraction contains sequences that appear repetitively in the genome of eucaryotes.

2. Identical or near-identical sequences, in either direct or inverted orientation, are characteristically found at the ends of transposons.

3. Chi form: a four-armed intermediate formed during recombination, in which one strand of each parental DNA molecule has crossed over to base-pair with the other.

   Mismatch repair: the removal of a stretch of DNA around a mismatched base pair, followed by repair synthesis of that strand according to the dictates of the remaining strand.

   Hybridization *in situ*: hybridization of a radioactive DNA probe to a chromosome preparation or a tissue slice.

4. D.

5. D.

6. C.

7. B.

8. A.

9.

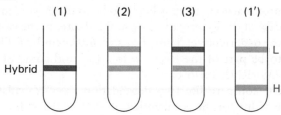

10. 1/2 cells C≡G C≡G; 1/2 cells A═T C≡G.

11. C.

12. (a) The three phases of renaturation occur because there are different concentrations of different sequences of DNA. The sequences that renature early (phase I) are present in higher concentration—that is, they represent highly repetitive sequences. The sequences that renature next (phase II) have intermediate concentration, moderately repetitive. The sequences that renature most slowly (phase III) have the lowest concentration, a single copy per genome.
(b) Both organisms have the same amount of highly repetitive sequences. Organism A has more single-copy DNA, presumably coding for protein sequences, than does organism B. Organism B has more moderately repetitive sequences.
(c) Both organisms must have all the metabolic reactions required for life and consequently all the enzyme activities to catalyze those reactions. The free-living organism must supply the enzymes, and the genetic information to specify them, itself while the intracellular parasite might get some of the enzymatic activities from its host, coded by the host DNA. Since single-copy DNA is probably the class

which includes the coding sequences for protein enzymes, organism A is the more likely to be free-living since it has the larger relative amount of slowly renaturing DNA.

Chapter 12

1. Ribosomal RNA (rRNA), very large molecular weight; mRNA, often "capped" and with poly A tails; tRNA, small molecular weight and having a clover-leaf secondary structure; snRNA, found in protein complexes in the nucleus. Precursors of all the eucaryotic RNAs are synthesized in the nucleus but, except for the snRNAs, most are found in the cytoplasm. There are also rRNAs, tRNAs, and mRNAS in mitochondria and chloroplasts.

2. The experiments by Brenner, Jacob, and Meselson that showed T4-infected *E. coli* cells express virus-specific proteins from pre-existing ribosomes via a short-lived virus-specific mRNA.

3. At the 5′ end there is often a modified guanine nucleotide, the "cap." At the 3′ end there is often a long sequence of repeated adenine nucleotides, the poly A "tail." These features are rarely, if ever, found in procaryotic mRNAs.

4. The discovery by T. R. Cech that ribosomal RNA from *Tetrahyemena* can catalyze, in the absence of protein, its own splicing reactions.

5. A.

6. b.

7. c.

8. a.

9. (a) r (b) m (c) t (d) m (e) none (f) t
   (g) m, t, (5S)r (h) t (i) r (j) none (k) t
   (l) m (m) t (n) m (o) all (p) none
   (q) m, t, r (r) r (s) none (t) all

10. The nuclear RNA contains the primary transcripts, which include the RNA sequences removed during processing, whether from the ends, or from the inside of the primary transcripts.

Chapter 13

1. E. The first amino acid will be the first methionine in the (lower) informational strand.

2. C. The mRNA is translated from 5′ to 3′.

3. B.

4. E.

5. D.

6.

Mass	Enzyme Amino acids, (tRNAs)	Amino acid Krebs cycle or glycolysis inter- mediates, ammo- nia

	Enzyme	Amino Acid
Energy	ATP, GTP	ATP, phospho-enolpyruvate
Information	mRNA, ribosome, initiation, elongation factors	Biosynthetic enzymes, feedback inhibitors, activators

7. (a) TRUE. Although it is the nascent chain that first interacts with the signal recognition particle, the ribosome—SRP complex interacts with the docking protein in the ER membrane.

(b) FALSE. The synthesis of mRNA is from 5′ to 3′ and it is at the 5′ end that translation begins, so the mRNA can bind ribosomes and begin to be translated before its synthesis is complete.

(c) FALSE. The mRNA sequence is decoded by hydrogen bonding with the anticodon of an aminoacyl-tRNA.

(d) TRUE. Antibiotics like chloramphenicol inhibit procaryotic and mitochondrial ribosomes, but not eucaryotic cytoplasmic ribosomes.

(e) FALSE. GTP hydrolysis is not necessary for translocation from the A-site to the P-site.

(f) FALSE. CCA is at the 3′ terminus of tRNA, not mRNA.

(g) FALSE. Hydroxyproline is incorporated during protein synthesis as proline, using proline codons, and hydroxylated afterwards.

(h) FALSE. Aminoacyl-tRNA binds to the ribosome as a ternary complex with elongation factor Tu and GTP; after GTP hydrolysis, the Tu·GDP dissociates, leaving the aminoacyl-tRNA in the A-site.

(i) FALSE. Diphtheria toxin acts by catalyzing the transfer of ADP-ribose from NAD to eucaryotic elongation factor 2, which acts during translocation. Protein synthesis stops with only peptidyl-tRNA bound to the ribosome.

(j) FALSE. The initiating methionine of eucaryotic proteins is not formylated and the signal peptidase removes the entire signal peptide, usually some two dozen amino acids long.

8. (a)—AAA/G-UGU-GCU-GAU-CCG-AAA/G—
   — Lys - Cys - Ala - Asp - Pro - Lys —
(b) Minus (the first U in the Cys codon), plus (A added after the Pro codon).
(c)— Lys - Met - Cys -terminate (opal)
   —AAA/G-AUG-UGC-UGA-UCC-AAA/G—

9. (a) There will be two products, since either strand in the absence of promoters can act as a template:
A: 5′- *GAG   UCG   AGU   CGA   GUC*
      GAG   UCG   AGU   CGA   GUC-3′
B: 5′- GAC   UCG   ACU   CGA   CUC
      GAC   UCG   ACU   CGA   CUC-3′

(b) In principle each RNA can give rise to three peptide products, one for each reading frame. In this case, the sequences of nucleotides repeat and only one repeating peptide product is produced for each RNA:
A:N . . . Glu - Ser - Ser - Arg - Val - Glu - Ser . . . C
B:N . . . Asp - Ser - Thr - Arg - Leu - Asp - Ser . . . C

10. (a) Since a dipeptide is made, peptide bond formation must occur, as well as binding of aminoacyl-tRNA to the ribosome. Translocation does not occur or if it does, the peptide cannot be donated to the next aminoacyl-tRNA.

(b) Yes. In either case the peptidyl-tRNA should remain bound to the ribosome, since there is no chain termination codon in the mRNA.

(c) In procaryotes, fusidic acid inhibits protein synthesis after translocation, by inhibiting dissociation of elongation factor G.

11. The ability to be transferred to puromycin by the α-bromoacetyl derivative of lysyl-tRNA implies it binds to the ribosome P-site. Since the ε-derivative cannot be transferred to puromycin, it must bind in the ribosome A-site. It can be hypothesized that covalent bonding of the bromoacetyl group to a ribosome structure will tend to occur while the derivative is bound, so the attachment site of the bromoacetyl group is likely to be in the A-site, in the case of the ε-derivative, or in the P-site, in the case of the α-derivative. Thus, ribosomal protein L2 is likely to be part of the P-site, while L27 and some of the 23S rRNA form part of the A-site.

12. To compare the two degradation rates, calculate a half-life for the eucaryotic cell: $P/P_0 = 0.7$ = fraction protein surviving after 60 minutes; $P/P_0 = 2^{-60/T}$, where T is the half-life. $T = (-60 \cdot \log 2)/\log (P/P_0)$ = 116 minutes. This means the enzyme will survive longer in the eucaryotic cell than in the procaryotic cell growing with a doubling time of 100 minutes (equivalent to having an enzyme half-life of 100 minutes).

Chapter 14

1. *Constitutive synthesis:* unregulated synthesis of an enzyme, typically in mutants in which the repressor no longer binds to the operator. *Repressor:* the product of a regulatory gene that binds to the operator of a structural gene and represses transcription. *Autogenous regulation:* the control that results when a gene product regulates the transcription of its own gene. *Site-directed mutagenesis:* specific mutational alteration of a chosen site—for example, through the insertion of a synthetic oligonucleotide on one DNA strand which is mismatched to the other strand at a particular site. *Promoter:* The sequence of DNA nucleotides to which RNA polymerase binds before beginning synthesis of RNA.

Footprinting: technique for determining which regions of a DNA molecule are protected from endonuclease digestion by binding a regulatory protein; typically, the DNA is end-labeled, then subjected to random, partial endonuclease digestion in the presence of the regulatory protein, then the resulting fragments are separated by size on a ladder gel. *Attenuation:* a regulatory mechanism that links ribosome movement to the termination of transcription at a site just upstream of a structural gene; the presence of ribosomes in a particular region of the mRNA prevents its forming a base-paired structure with another complementary region, which in turn affects the formation of still another stem-loop structure, which signals transcription termination.

2. Gilbert and Mueller-Hill used radioactive *inducer* to purify a protein which bound the inducer with high affinity. The protein's identity as the *lacI* gene product was confirmed through its alteration in *lacI* mutants.

3.

Genotype	Beta-Galactosidase Synthesis	
	No Inducer Added	Inducer Added
$I^+O^+Z^+$	−	+
$I^+O^+Z^-$	−	−
$I^-O^+Z^+/I^+O^+Z^-$	−	+
$I^+O^+Z^+/I^-O^+Z^-$	−	+
$I^+O^cZ^+$	+	+
$I^+O^cZ^+/I^+O^+Z^+$	+	+
$I^+O^cZ^+/I^+O^+Z^-$	+	+
$I^+O^cZ^-/I^+O^+Z^+$	−	+

4. (a) Conditionally accept. The data are not inconsistent with the hypothesis, so it cannot be rejected. The data do not, however, include any proof that cyclin concentration controls the timing events of the cell cycle, as opposed to cyclin concentration being controlled by the cell cycle. The hypothesis cannot be accepted without further experimentation.
(b) One way to test the hypothesis would be experimentally to alter the concentration of cyclin in the cell at various stages of the cell cycle. A variety of techniques could accomplish this: microinjection, cell fusion, transfection with a DNA vector containing the cloned gene for cyclin under the control of an inducible promoter, etc. If the hypothesis is true, increasing the concentration of cyclin should stimulate a premature cell division. If the hypothesis is false, changing the cyclin concentration should not affect the timing of the cell cycle.
(c) Any mechanism of posttranscriptional control would be sufficient to account for the data: protein binding to mRNA at the translation initiation region,

translational control RNA binding to mRNA, phosphorylation or other covalent modification of a translation initiation factor specific for this mRNA, etc. As it turns out, microinjected mRNA stimulates meiosis in the clam oocyte, so a masked mRNA seems a likely explanation.

5. (a) An asynchronous culture of cells with a temperature-sensitive mutation that interfered with protein synthesis would when placed in nonpermissive conditions stop synthesizing protein, but would stop in various phases of the division cycle. Each cell in a culture of *cdc* mutants would, in contrast, continue through the division cycle relatively normally until it reached the same point, the execution point, and would end up with the same terminal phenotype.
(b) The function that occurs in the absence of mating factor α is executed first. This explains why synchronized *cdc* 28 cells exposed for 0.8 hr to a high temperature (culture D) were not prevented from dividing after 4 hr; they had already executed the critical function. The data from culture E are also consistent with this answer; the α-inhibited function must be executed before the *cdc* 28 function.
(c) Yes, if the time during which the first change in culture conditions is sufficient for all cells to be blocked at the execution point. Then changing the conditions will or will not allow the cells to divide together, depending on whether the new condition is executed before or after the first condition during the normal cycle.

6. (a) FALSE. Carbamoyl phosphate is a substrate and binds to the active site, not to the regulatory effector (allosteric) site of the enzyme.
(b) FALSE. Shikimic acid phosphate is the product of a pacemaker reaction at the beginning of a linear unbranched segment of a pathway and is therefore unlikely to be a regulator of an earlier enzyme in the pathway. A more likely feedback inhibitor is an intermediate at a branch point—for example, chorismate.
(c) FALSE. $G_s \cdot GTP$ does activate adenyl cyclase, but it acts by binding to it reversibly and causing (or stabilizing) a conformation with higher enzyme activity.
(d) TRUE. An example is the concentration of end product, which communicates the status of its biosynthetic pathway by feedback on one of the enzymes in the pathway.
(e) FALSE. Adenyl cyclase is stimulated by $G_s \cdot GTP$. If the G_s GTPase is activated there will be less stimulation of adenyl cyclase, less synthesis of cAMP, and consequently lower [cAMP] inside the cells.
(f) FALSE. Normal cells may become cancerous

after infection by certain viruses, those containing oncogenes.

(g) FALSE. Normal cells with the normal number and kind of chromosomes divide only a defined number of times in culture.

(h) TRUE. Even though slow, racemization would be expected to alter the structure and thus the function of long-lived proteins (such racemization can be detected in the lens of the vertebrate eye).

7. (a) cAMP, cGMP, diacylglycerol, phosphoinositol-bis-phosphate (IP$_3$), Ca^{2+}.

(b) cAMP: protein kinase A; cGMP: protein kinase; diacylglycerol and Ca^{2+}: protein kinase C; Ca^{2+}: calmodulin; IP$_3$: Ca^{2+} channels in the endoplasmic reticulum and cell membranes.

(c) A given protein kinase can phosphorylate only certain target proteins. The population of susceptible target proteins will differ from cell to cell, because different cells synthesize different proteins.

8. [E] · [A] / [E · A] = 2 × 10^{-6} molar: equilibrium condition. When [A] = 1 × 10^{-6}, [E · A]/[E] = 1/2. [E] + [E · A] = constant (required by conservation of mass of enzyme); call the constant 1. Therefore, (1−[E])/[E] = 1/2 or [E] (form when activator not bound) = 2/3 and [E · A] (form when activator bound) = 1/3. Multiplying the relative concentrations of enzyme forms by the relative rates observed for those forms: (100 × 1/3) + (10 × 2/3) = 40 = expected relative rate.

9. (a) Enzyme 1 is likely to be feedback inhibited by E, since it catalyzes the first reaction of the pathway leading to E. Enzyme 5 is likely to be activated by E since this is the first enzyme in the degradative pathway, a pathway that would allow growth of the cells on E as a carbon source.

(b) High concentrations of E are likely to induce the synthesis of enzymes 5–8, the degradative pathway enzymes, and repress the synthesis of enzymes 1–4, the biosynthetic pathway enzymes. Low concentrations of E are likely to induce enzymes 1–4 and repress enzymes 5–8.

(c) E′ acts as a feedback inhibitor of enzyme 1, turning off the biosynthesis of E. In the absence of exogenous E, whether or not glucose is present, the cells will not grow since they are missing a precursor for protein synthesis. Since E′ also inhibits growth of cells using E as a carbon source, E′ must prevent the use of E for protein synthesis, probably by competitive inhibition of the activating enzyme that attaches E to transfer RNA (since if E is present in higher concentrations, when glucose is the carbon source, the cells can grow.

10. The critical experiment is to find conditions in which an extract of soluble factors from S phase cells can, when microinjected into G$_1$ phase cells, elicit premature DNA synthesis. Significant numbers of synchronized cells, all in S phase, will be needed to extract. An appropriate assay for DNA synthesis is incorporation into acid-precipitable form of [^{3}H]thymidine. Extracts can be treated with immobilized enzymes to see whether the soluble factor is inactivated by proteases, for example, in which case it is probably a peptide or protein. Various separation schemes can be tried to purify the factor, etc.

Chapter 15

1. First, a heating step is necessary for self-assembly of ribosomes in vitro. Second, the RNA must be modified by enzymes after transcription—to methylate certain bases, for example.

2. Tubulin subunits do not polymerize at physiological rates by themselves in vitro. In vivo they appear to require a microtubule organizing center.

3. Polyacrylamide gel electrophoresis in the absence of a disulfide reducing reagent reveals the molecular weight of the undenatured molecule is 320,000. When SDS and a disulfide reducing reagent (e.g., mercaptoethanol) are used in the electrophoresis, three chains of MW approximately 50,000 are found in equal amounts. One copy of each chain would account for a MW of only 150,000; two copies of each chain in the molecule would account for 300,000; and three copies would require 450,000.

4. Using denaturing polyacrylamide gels (containing SDS and mercaptoethanol) to separate the various species, one would look for a decrease in the number of γ—γ dimers and a corresponding increase in the number of γ monomers. One can also look for a decrease in the amount of α-chain polymers.

5. The reassembled and the native virus should give similar values for the following assays: molecular weight, image under the electron microscope, infectivity, and susceptibility to nucleases.

6. The probability of synthesizing an error-free subunit is higher for a smaller subunit than for a larger subunit. Rejecting a subunit during assembly because it contains an error (incorporated during synthesis of the subunit) results in less waste for a smaller subunit than for a larger subunit.

7. Protein alone assembles into cylindrical structures of indeterminate length, some longer and some shorter than the native virus. Only when RNA is also present does the length of the reassembled structure resemble that of the native virus. In slightly alkaline pH, assembly of the protein subunits is very slow without RNA. Addition of RNA to an alkaline solution of the viral protein elicits normal assembly.

8. This description reflects the finding that part of the required protein synthesizing machinery is present in the cytoplasm rather than in the organelle itself.

9. The information that specifies the structure of individual proteins of the cortex is, just as with other proteins, found in the DNA. What is self-replicative in the cortex is the relative positioning of the proteins, which brings about highly organized structures such as basal bodies.

10. (a) The structure contains two major protein components, A and B. A has an apparent molecular weight of 55,000 and B an apparent molecular weight of 25,000. The faint bands seen when large amounts of the preparation were electrophoresed are probably contaminants.

(b) The positions of bands after the second electrophoresis (in reducing conditions) give information about the composition of the species separated during the first electrophoresis (in oxidizing conditions). For example, the rightmost band after the first electrophoresis contains only component B, as does the darker stained band to its left. Since the latter has a slower mobility during the first electrophoresis, it has a higher molecular weight. Since both contain B only, the simplest model that accounts for the data is that the rightmost band is B and the band to its left is the dimer B_2. Similarly, it can be argued that the third band from the right after the first electrophoresis is A, the fourth AB, the fifth AB_2, the sixth A_2, and the others heavier combinations of A and B. These results suggest that A is next to B and next to A in the structure, also that B is next to A and next to B.

(c) The anti-A antibodies cross-link the structure at one end. The anti-B antibodies cross-link the structure at the center. A plausible model for the structure has A subunits at the ends and B subunits in the center. Since B_2 was detected, but not B_3, it is likely there are two B subunits in the center. Since A_2 was detected it is likely there are two A subunits at each end. The cross-linked AB and AB_2 species that were observed can also be expected from the model.

Chapter 16

1. (a) FALSE. Microtubules have a diameter of ~24 nm, while microfilaments have a diameter of ~7 nm.
(b) FALSE: The flagella of eucaryotic cells contain dynein and are driven by ATP; the flagella of bacteria are made of flagellin and powered by a proton gradient via a rotating "motor."
(c) FALSE. Actin/myosin contraction may (or may not) be the explanation for how a cell body is brought up towards an extended pseudopod, but the protrusion of the pseudopod in the first place can be explained by completely different mechanisms—for example, actin polymerization or lipid flow, etc.
(d) FALSE. α–Actinin is found in nonmuscle cells—in adhesion placques, for example.
(e) FALSE. The cross-bridges probably act independently of one another, asynchronously. Muscle tension does drop when all the cross-bridges detach, but this is caused by low $[Ca^{2+}]$ and is not observed during the normal contraction cycle.
(f) FALSE. The filaments in the sarcomere do not shorten, the degree of overlap between thick and thin filaments becomes less.
(g) FALSE. Colchicine interferes with the polymerization of micro*tubules*.
(h) FALSE. The ATPase sites lie in the thick filaments, on the cross-bridges.
(i) FALSE. The bacterial flagella will not move—if all the ATP is gone the proton gradient is probably also gone—but since their movement is based on rotation, rather than sliding filaments, the flagella will not be any more or less stiff than usual.
(j) FALSE. Tension depends on overlap between the thick and thin filaments. When the muscle fibril is stretched too far, overlap is lost.

2. (a) Since heat is given off to the environment, $\Delta H < 0$ (positive heat flow is from the environment into the system).
(b) Since $\Delta G = \Delta H - T\,\Delta S$ and since $T > 0$, ΔS must be negative, thus giving a positive value for $-T\Delta S$ and balancing the negative ΔH to give a zero ΔG. The structural basis of the negative ΔS, ordering of the fibril structure, may be due to breaking cross-bridge attachments during stretching and therefore achieving a more consistent orientation of filaments.

3. The hydrolysis of ATP is a necessary step in the cross-bridge movement pathway. Furthermore, there is a later step in the pathway that does proceed with a large negative free energy change. That is why it is appropriate to relate the hydrolysis of ATP to the contraction energy requirement.

4. ATP must be regenerated to account for the amount hydrolyzed in an actively contracting muscle. Regeneration occurs under aerobic conditions mostly in the mitochondria. ADP is phosphorylated using the energy of the proton gradient. The ATP that leaves the mitochondria can be used to charge up creatine to

creatine-phosphate, which in turn acts as a phosphate donor to ADP in the vicinity of the myofribrils. Under anaerobic conditions (which may occur when a very actively contracting vertebrate muscle cuts off its own blood supply), glycolysis supplies the needed ATP.

5. By analogy with vertebrate striated muscle, the conformational change in the myosin head groups must occur effectively in a plane. This requires that the binding site on the actin also lie in that plane. Myosin head groups with a different orientation do not bind to the actin, since the angle between two is inappropriate. Since the actin binding sites lie practically along a line—the cable of actin filaments—the direction of the power stroke of any myosin head group that *can* bind to actin is always along the same line and no rotation occurs.

6. The cytoskeleton is composed of an assembly of filamentous structures interacting with each other and with cellular organelles and membranes. The cytomatrix includes the cytoskeleton, but also includes other proteins that are immobilized—that is, not diffusing freely—presumably because they are bound to the cytoskeleton.

7. Ferritin is a large, electron-dense iron-containing protein that can be detected by electron microscopy. This protein can diffuse from the outside medium into the transverse tubule system, but not anywhere else. This experiment, by H. E. Huxley, showed that the transverse tubule is not interconnected with the sarcoplasmic reticulum of the muscle fiber.

8. (a) In the absence of ATP, myosin is tightly bound to actin (equivalent to the rigor state in intact skeletal muscle) and the resulting aggregate organizes the adjacent water, thus increasing viscosity. In the presence of ATP, myosin dissociates from actin and the viscosity falls.

(b) Myosin has an ATPase activity that is activated by binding to actin. As the hydrolysis of ATP proceeds the concentration of free ATP falls and the myosin binds again to actin, forming large relatively stiff aggregates. This causes the viscosity of the actomyosin solution to rise.

9. Trypsin cleaves the myosin molecule into two fragments. One fragment, light meromyosin, consists of the myosin tail and retains the ability to aggregate into smooth, bipolar thick filaments. The other fragment, heavy meromyosin, consists of the myosin head groups and retains the ATPase activity.

10. The chelating agent EGTA is used to bind free Ca^{2+} while leaving Mg^{2+} in solution.

11. The energy expenditure of vertebrate striated muscle is regulated according to the work it must perform. Energy output in muscle is determined not only by the distance of a contraction but also by the tension developed during contraction.

12. Severed flagella and demembranated flagella resume their beat when ATP is added to the medium. In the case of partially demembranated flagella, bending occurs only at the points where the membrane has been removed.

13. First, the radial spokes of eucaryotic flagella anchor the subfibers to each other and thus prevent them from sliding past one another. Second, they may be involved in the propagation of the bending wave along the flagellum.

14. The addition of monomers during polymerization occurs at the barbed end of the HMM-decorated actin polymer. Adding the presumptive capping protein should prevent polymerization by binding to the barbed end of the microfilament, as could be shown with the electron microscope—by gold immunolabeling, for example.

15. First, the cleavage furrow can be shown in the transmission electron microscope to contain microfilaments, arranged as a ring that decreases in diameter during cleavage. Second, indirect immunofluorescence shows that the microfilament ring cross-reacts with antiactin antibodies. Third, antimyosin antibodies inhibit cleavage. Finally, cytochalasin B depolymerizes the microfilaments and also inhibits cleavage.

16. (a) Identification in the EM of filamentous structures and their positions in relation to the movement.
(b) Purification of proteins suspected to be involved in the mechanism.
(c) Generation of antibodies against the purified proteins.
(d) The use of indirect immunofluorescence or immunostaining at the EM level to prove the presence of these proteins in the observed filamentous structures.
(e) The use of antibodies to inhibit the motion.
(f) The use of specific inhibitors known to act on specific components of various filamentous structures to determine whether they also inhibit motion.
(g) Construction from purified proteins and cofactors of model systems in vitro that mimic the motion in vivo.

17. (a) Chromosome movement during mitosis.

(b) Flagellar motion of sperm, cleavage during cell division, movement of the cell surface for ingesting solid food.

Chapter 17

1. Release of neurotransmitters at a neuromuscular junction, which are then bound by transmitter receptors, and release of hormones (either peptide or steroid), which are recognized by receptors located either on the plasma membrane (peptides) or in the nucleus (steroids). Passage of a signal at a neuromuscular junction is an example of short-range intercellular communication. In the case of hormones, the message may pass through the circulation in what appears to be a diffuse manner, but only those cells that bear the correct receptor molecules will be able to receive the signal.

2. (a) The major constituent of both is a form of the protein collagen, which in Jello is in a denatured form.
(b) The major property that determines the three-dimensional structure of both is entropy. Elastin is made up of a cross-linked peptide network, arranged in a series of random coils. Perturbing such a structure requires energy and results in the conformation becoming more ordered (rather like a compulsively neat parent visiting the bedroom). When the perturbing force is released (the parent leaves the vicinity), the structure rerandomizes.

3. Prior to the addition of antifibronectin antibody, you would expect to see the actin arranged in bundles of microfilaments, terminating at the regions of contact between the cell and its substratum. On binding of antifibronectin antibody, disorganization of the fibronectin at the regions of attachment, leading to disruption of the actin bundles, is expected. This change in the three-dimensional arrangement of the actin can be visualized in the uv microscope.

4. (a) One approach would be to raise antibodies that bind to cell surface proteins and, in doing so, inhibit adhesion. Following solubilization of the membranes, the proteins mediating adhesion can then be purified by immunoaffinity chromatography.
(b) A homophilic interaction is one between two like molecules; a homotypic interaction is one between two like cells.
(c) Fibronectin and surface glycosyl transferases. The involvement of fibronectin in cell migration was shown by streaking a glass surface with fibronectin and demonstrating that motile embryonic cells placed on such a surface tend to collect on the streak and move along it. Streaks of other extracellular matrix components are much less effective. The migration of neural crest cells from quail is perturbed by the addition of either inhibitors or substrates of galactosyl transferase if the substratum is a matrix resembling the basal lamina, but not if migration occurs on a fibronectin matrix.

5. (a) Gap junctions serve to connect adjacent cells together and to allow for the intracellular flow of ions and small molecules.
(b) The flow of ions from one cell to another can be detected by electrical effects—current flow between microelectrodes placed in different cells—and the flow of fluorescent dyes from one cell to another can be detected by ultraviolet microscopy.

6. The first piece of evidence is that isolated synaptic vesicles have been shown to contain the neurotransmitter. Secondly, stimulation of the nerve until the muscle can no longer respond is correlated with a decreased number of synaptic vesicles in the presynaptic cell.

7. (a) In order to be immunogenic, a molecule must be at least a minimum size. DNP is too small. DNP is therefore termed a *hapten*: a substance that cannot by itself elicit an immune response but that will bind to an antiserum generated by immunization with the hapten conjugated to a larger (usually protein) molecule.
(b) Show that the antibodies bind to a second protein when conjugated to DNP, but not to the second protein alone, or show that the interaction between antiserum and DNP-protein conjugate can be inhibited by excess soluble DNP.

8. A monoclonal antibody is directed towards a single determinant (epitope) on an antigen. Its specificity does not alter with time in culture or when grown as an ascitic fluid, and large amounts of a particular monoclonal antibody can readily be generated. Its patterns of cross-reactivity with structurally related antigens can easily be determined and are reproducible, which is particularly important when antibodies are used for purposes of clinical diagnosis and tissue typing. The so-called monoclonals are particularly helpful in purification: an impure antigen can be used to elicit a set of monoclonals, one of which can then be employed to isolate the desired component from the antigen mixture by immunoaffinity chromatography. However, if you want to use the antibodies in assays or procedures that involve precipitation of antigen-antibody complexes, then a monoclonal antibody is less effective than a polyclonal antiserum; being directed at a single determinant, monoclonal antibodies may not cross-link the antigen molecules as efficiently.

Index

A letter following a page number indicates that the reference is to material outside the regular text:
  F indicates a figure;
  B indicates boxed material;
  T indicates a table; and
  R indicates one or more of the Selected Readings.